SOURCES AND EFFECTS OF IONIZING RADIATION

United Nations Scientific Committee on the
Effects of Atomic Radiation

UNSCEAR 2000 Report to the General Assembly,
with Scientific Annexes

VOLUME II: EFFECTS

UNITED NATIONS
New York, 2000

NOTE

The report of the Committee without its annexes appears as Official Records of the General Assembly, Fifty-fifth Session, Supplement No. 46 (A/55/46).

The designation employed and the presentation of material in this publication do not imply the expression of any opinion whatsoever on the part of the Secretariat of the United Nations concerning the legal status of any country, territory, city or area, or of its authorities, or concerning the delimitation of its frontiers or boundaries.

The country names used in this document are, in most cases, those that were in use at the time the data were collected or the text prepared. In other cases, however, the names have been updated, where this was possible and appropriate, to reflect political changes.

UNITED NATIONS PUBLICATION
Sales No. E.00.IX.4
ISBN 92-1-142239-6

CONTENTS

VOLUME I: SOURCES

**Report of the United Nations Scientific Committee on the
Effects of Atomic Radiation to the General Assembly**

Scientific Annexes

Page

VOLUME II: EFFECTS

Scientific Annexes

(Volume II)

ANNEX F

DNA repair and mutagenesis

CONTENTS

INTRODUCTION

1.   Risk estimates for the induction of human disease are obtained primarily from epidemiological studies. These studies can clearly distinguish radiation effects only at relatively high doses and dose rates. To gain information at low doses and low dose rates, which are more relevant to typical human radiation exposures, it is necessary to extrapolate the results of these studies. To be valid, this extrapolation requires a detailed understanding of the mechanisms by which radiation induces cancer and genetic disorders.

2.   Several lines of evidence show that sites of radiation-induced cell lethality, mutation, and malignant change are situated within the nucleus and that DNA is the primary target. When DNA is damaged by radiation, enzymes within the cell nucleus attempt to repair that damage. The efficiency of the enzymatic repair processes determines the outcome: most commonly, the structure of DNA is repaired correctly and cellular functions return to normal. If the repair is unsuccessful, incomplete, or imprecise, the cell may die or may suffer alteration and loss of genetic information (seen as mutation and chromosomal aberration). These information changes determine heritable genetic defects and are thought to be important in the development of radiation-induced cancer. The more complete the knowledge of the ways in which human cells respond to damage and of the mechanisms underlying the formation of mutations and chromosomal aberrations, the more accurate will be the predictions of the oncogenic and hereditary effects of ionizing radiation.

3.   DNA repair is itself controlled by a specific set of genes encoding the enzymes that catalyse cellular response to DNA damage. Loss of repair function, or alteration of the control of repair processes, can have very serious consequences for cells and individuals. It is anticipated that DNA repair plays a critical role in protecting normal individuals from radiation effects, including cancer. Clinical experience has revealed individuals who are both hypersensitive to radiation and cancer-prone; some of these individuals have recently been shown to have defects in genes involved in the response to DNA damage.

4.   In recent years there have been significant advances in the molecular analysis of repair processes and the understanding of the mechanisms that induce genetic changes. Additionally, new methods have been developed to simplify the identification of the genes involved. As the details of damage-repair processes become clearer, it is seen that these processes have considerable overlap with other cellular control functions, such as those regulating the cell cycle and immune defences. In this Annex the Committee continues to review such developments in molecular radiobiology, as it began to do in Annex E, "Mechanisms of radiation carcinogenesis" of the UNSCEAR 1993 Report [U3], in order to improve the understanding of how radiation effects are manifested in cells and organisms.

I. DNA DAMAGE AND REPAIR

A. THE ROLE OF DNA REPAIR GENES IN CELL FUNCTION

5.   The information needed to control cellular functions such as growth, division, and differentiation is carried by the genes. Genes, which are specific sequences of DNA, act mainly through the production of complementary messages (mRNA) that are translated into proteins. Proteins can have a structural role but commonly work as enzymes, each of which catalyses a particular metabolic reaction. Thus specific genes contain the code for (encode) specific cellular functions. The production of proteins can be timed so that they work at specific points in the development of a cell or organism, but protein function can also be controlled by post-translational modifications. These modifications are carried out by other proteins, so that a complex set of interactions is necessary to fine-tune cellular functions. Proteins involved in important aspects of cell metabolism (e.g. DNA replication) may also work in multi-protein complexes [A1]. There is some evidence also that some of the complexes are assembled into larger structures situated in defined regions of the nucleus (e.g. the nuclear matrix) [H2].

6.   Loss or alteration of information in a specific gene may mean that none of that gene product (protein) is formed, or that the protein is less active, or that it is formed in an uncontrolled fashion (e.g. at the wrong time or in the wrong amount). While some minor genetic alterations may not affect protein activity or interactions, others may significantly disrupt cellular function. Since certain proteins work in a number of different processes or complexes, the loss or impairment of one type of protein can affect several different functions of the cell and organism (pleiotropic effect).

7.   A very large number of genes, 60,000-70,000 [F14], are required to control the normal functions of mammalian cells and organisms. However, the genes form only a small part of the genome (the complete DNA sequence of an organism), the remainder of which largely consists of many copies of repetitive DNA sequence. The genes are linked in linear arrays interspersed by non-coding sequences, to form chromosomes located in the cell nucleus. Most genes are present in only two copies, each on a separate homologous chromosome, one inherited from the mother and one from the father. To monitor damage and to maintain the genes without significant alteration is a major concern for the cell. Repair

processes are common to all organisms from bacteria to humans and have evolved to correct errors made in replicating the genes and to restore damaged DNA. This fact has in recent years provided a useful tool for molecular geneticists in the analysis of repair processes; well characterized micro-organisms can serve as model systems to understand the structure and function of repair genes. The information gained in this way can sometimes also be used directly to isolate human genes of related function [L1]. While the structure and function of repair genes appears to be highly conserved from lower to higher organisms, the regulation of their activities may differ in different organisms.

8. The consequences of loss of repair capacity are seen in a number of human syndromes and mutant cell lines. These show hypersensitivity to environmental agents, and the human syndromes often have multiple symptoms, including cancer-proneness, neurological disorder, and immune dysfunction. Good progress has been made over the last few years in mapping and cloning the genes involved.

9. Ionizing radiation damages DNA and causes mutation and chromosomal changes in cells and organisms. Damage by radiation or radiomimetic agents also leads to cell transforma-tion (a stage in cancer development) and cell death. In the light of current research, it is seen that the final response to radiation damage is determined not only by cellular repair processes but also by related cellular functions that optimize the opportunity for recovery from damage. For example, radiation damage may cause an arrest in the cell cycle; this is thought to be a damage-limitation step, allowing time for repair and reducing the consequences of a given dose [L22]. There is now some understanding of the way in which radiation alters cell cycle timing (Section II.B.2), although the roles of a number of enzymatic activities that are induced or repressed shortly after irradiation remain to be clarified (Section III.B).

10. The severity of DNA damage, or the context in which damage occurs (e.g. during DNA replication), will often dictate a repair strategy that places survival first and incurs genetic change. DNA replication may bypass sites of single-stranded DNA damage, inserting an incorrect base opposite the altered or lost base. Additionally, in attempting to repair damage to DNA, enzymes may not be able to restore the structure with fidelity. Thus, mutation and chromosomal rearrangement are not passive responses to damage; rather, they are a consequence of the interaction of cellular processes with damage. The types of genetic change that occur will depend on the types of initial DNA damage, from their potential to miscode at replication and from the probability that specific repair enzymes will act on given types of damage.

B. TYPES OF DAMAGE AND PATHWAYS OF REPAIR

11. DNA is a double-helical macromolecule consisting of four units: the purine bases adenine (A) and guanine (G), and the pyrimidine bases thymine (T) and cytosine (C). The bases

are arranged in two linear arrays (or strands) held together by hydrogen bonds centrally and linked externally by covalent bonds to sugar-phosphate residues (the DNA "backbone"). The adenine base pairs naturally with thymine (A:T base pair), while guanine pairs with cytosine (G:C base pair), so that one DNA strand has the complementary sequence of the other. The sequence of the bases defines the genetic code; each gene has a unique sequence, although certain common sequences exist in control and structural DNA elements. Damage to DNA may affect any one of its components, but it is the loss or alteration of base sequence that has genetic consequences.

12. Ionizing radiation deposits energy in tracks of ionizations from moving charged particles within cells, and radiations of different quality may be arbitrarily divided into sparsely ionizing, or low linear energy transfer (low-LET), and densely ionizing (high-LET). Each track of low-LET radiations, such as x rays or gamma rays, consists of only a relatively small number of ionizations across an average-sized cell nucleus (e.g. a gamma-ray electron track crossing an 8 μm diameter nucleus gives an average of about 70 ionizations, equivalent to about 1 mGy absorbed dose, although individual tracks vary widely about this value because of track stochastics and varying path lengths through the nucleus). Each track of a high-LET radiation may consist of many thousands of ionizations and give a relatively high dose to the cell; for example, a 4 MeV alpha-particle track has, on average, about 23,000 ionizations (370 mGy) in an average-sized cell nucleus [G27, U3]. However, within the nucleus even low-LET radiations will give some small regions of relatively dense ionization over the dimensions of DNA structures, for example, where a low-energy secondary electron comes to rest within a cell.

13. Radiation tracks may deposit energy directly in DNA (direct effect) or may ionize other molecules closely associated with DNA, especially water, to form free radicals that can damage DNA (indirect effect). Within a cell the indirect effect occurs over very short distances, of the order of a few nanometres, because the diffusion distance of radicals is limited by their reactivity. Although it is difficult to measure accurately the different contributions made by the direct and indirect effects to DNA damage caused by low-LET radiation, evidence from radical scavengers introduced into cells suggests that about 35% is exclusively direct and 65% has an indirect (scavengeable) component [R21]. It has been argued that both direct and indirect effects cause similar early damage to DNA; this is because the ion radicals produced by direct ionization of DNA may react further to produce DNA radicals similar to those produced by water-radical attack on DNA [W43].

14. Ionization will frequently disrupt chemical bonding in cellular molecules such as DNA, but where the majority of ionizations occur as single isolated events (low-LET radiations), these disruptions will be readily repaired by cellular enzymes. However, the average density of ionization by high-LET radiations is such that several ionizations are likely to occur as the particle traverses a DNA double helix.

Therefore, much of the damage from high-LET radiations, as well as a minority of the DNA damage from low-LET radiations, will derive from localized clusters of ionizations that can severely disrupt the DNA structure [G27, W44]. While the extent of local clustering of ionizations in DNA from single tracks of low- and high-LET radiations will overlap, high-LET radiation tracks are more efficient at inducing larger clusters, and hence more complex damage. Also, high-LET radiations will induce some very large clusters of ionizations that do not occur with low-LET radiations; the resulting damage may be irrepairable but may also have unique cellular consequences (see paras. 192, 199, and 201) [G28]. Additionally, when a cell is damaged by high-LET radiation, each track will give large numbers of ionizations, so that the cell will receive a relatively high dose and there will be a greater probability of correlated damage within a single DNA molecule (or chromosome) or in separate chromosomes. As a consequence, the irradiation of a population of cells or a tissue with a "low dose" of high-LET radiation results in a few cells being hit with a relatively high dose (one track) rather than in each cell receiving a small dose. In contrast, low-LET radiation is more uniformly distributed over the cell population; at doses of low-LET radiation in excess of about 1 mGy (for an average-size cell nucleus of 8 μm diameter), each cell nucleus is likely to be traversed by more than one sparsely-ionizing track.

15. The interaction of ionizing radiation with DNA produces numerous types of damage; the chemical products of many of these have been identified and classified according to their structure [H4, S3]. These products differ according to which chemical bond is attacked, which base is modified, and the extent of the damage within a given segment of DNA. Table 1 lists some of the main damage products that can be measured following low-LET irradiation of DNA, with a rough estimate of their abundance. Attempts have also been made to predict the frequencies of different damage types from a knowledge of radiation track structure, with certain assumptions about the minimum energy deposition (number of ionizations) required. Interactions can be classified

according to the probability they will cause a single-strand DNA alteration (e.g. a break in the backbone or base alteration) or alterations in both strands in close proximity in one DNA molecule (e.g. a double-strand break), or a more complex type of DNA damage (e.g. a double-strand break with adjacent damage). Good agreement has been obtained between these predictions and direct measurements of single-strand breaks, but there is less good agreement for other categories of damage [C47]. While complex forms of damage are difficult to quantify with current experimental techniques, the use of enzymes that cut DNA at sites of base damage suggests that irradiation of DNA in solution gives complex damage sites consisting mainly of closely-spaced base damage (measured as oxidised bases or abasic sites); double-strand breaks were associated with only 20% of the complex damage sites [S87]. It is expected that the occurrence of more complex types of damage will increase with increasing LET, and that this category of damage will be less repairable than the simpler forms of damage. Theoretical simulations have predicted that about 30% of DNA double-strand breaks from low-LET radiation are complex by virtue of additional breaks [N19] and that this proportion rises to more than 70%, and the degree of complexity increases, for high-LET particles [G29].

16. Some of the DNA damage caused by ionizing radiation is chemically similar to damage that occurs naturally in the cell: this "spontaneous" damage arises from the thermal instability of DNA as well as endogenous oxidative and enzymatic processes [L2, M40]. Several metabolic pathways generate oxidative radicals within the cell, and these radicals can attack DNA to give both DNA base damage and breakage, mostly as isolated events [B46]. The more complex types of damage caused by ionizing radiation may not occur spontaneously, since localized concentrations of endogenous radicals are less likely to be generated in the immediate vicinity of DNA. This theme is taken up in Annex G, *"Biological effects at low radiation doses"*, which considers the cellular responses to low doses of radiation.

Table 1
Estimated yields of DNA damage in mammalian cells caused by low-LET radiation exposure
[L60, P31, W39]

Type of damage	Yield (number of defects per cell Gy^{-1})
Single-strand breaks	1 000
Base damage [a]	500
Double-strand breaks	40
DNA-protein cross-links	150

a   Base excision enzyme-sensitive sites [P31] or antibody detection of thymine glycol [L60].

---

17. Measurement of the endogenous levels of DNA base damage has been difficult because of the artefactual production of damage during the preparation of the DNA for analysis (e.g. by gas chromatography/mass spectrometry) [C55]. This difficulty explains the presence in the literature of considerably inflated (by factors of at least 100) values for background

levels of base damage. Interestingly, the recognition of damage by base excision repair enzymes (paragraph 22) has provided a less discordant method of measurement, although the specificity of the enzymes for different types of base damage is not precisely known. These enzymes cut the DNA at the site of base damage, to give a single-strand break that can be measured accurately by a number of techniques. Using this method, measurement of an important form of oxidative damage, 7,8-dihydro-8-oxoguanine (generally known as 8-oxoguanine)), has given steady-state levels of 500-2000 per cell, depending on cell type [P30]. Using similar measurement methods, the level of 8-oxoguanine induced in cellular DNA by gamma rays is about 250 per cell per Gy [P31]. A newly developed ultrasensitive assay for another type of base damage in human cellular DNA, thymine glycol, couples antibody detection with capillary electrophoresis. This method showed a linear response for yield of thymine glycol with gamma-ray dose down to 0.05 Gy, giving a level of about 500 thymine glycols per cell per Gy against a background of 6 thymine glycols per cell [L60]. The difficulties experienced in measuring base damage accurately in cellular DNA and the relatively low levels now found for the commoner types of damage have also called into question the extent to which some previously identified forms of base damage occur in cells following irradiation.

18.   The measurement of endogenous levels of other types of DNA damage, such as double-strand breaks, has involved similar technical difficulties. Many of the methods used to measure double-strand breaks in mammalian cells introduce this form of damage either inadvertently or deliberately as part of the methodology. This is because the mammalian genome is so large that it had to be reduced in size by random breakage first before useful measurements could be made. This problem has been overcome in part by the introduction of methods based on the gentle release of DNA from cells by their lysis in a gel matrix [C64, O3], but there is commonly still a background level of DNA breakage amounting to a few per cent of the total DNA. However, as documented in Section II.B, it is unlikely that mammalian cells have a high steady-state level of DNA double-strand breakage, since these breaks act as a signal for damage-recognition processes that can block the cell cycle or induce programmed cell death. It is possible that even one unrepaired double-strand break can trigger this cellular response (paragraph 101). It has also been found that one unrepaired double-strand break can cause lethality in irradiated yeast cells (paragraph 108). Thus, tolerance of this form of damage in cells is likely to be very low.

19.   While the precise nature of the damage will influence repairability, it is possible to consider a few general categories of damage in order to describe their consequences. A simplified classification can be based on the ability of enzymes to use the complementary base structure of DNA to facilitate repair of the damage site. Thus, damage to single strands (base modifications, single-strand breaks) can be removed or modified, followed by resynthesis using the undamaged strand as a template. Where the damage affects both strands of a DNA molecule in close proximity (double-strand breaks, cross-links), it is more difficult to repair and requires different enzymatic pathways for its resolution. To resolve successfully more complex types of damage may require enzymes from more than one repair pathway. To illustrate the knowledge of the different repair pathways available to the cell, the following account (to paragraph 34) includes a discussion of the repair of damage caused by various DNA-damaging agents as well as ionizing radiation.

20.   DNA repair enzymes can be characterized as cellular proteins acting directly on damaged DNA in an attempt to restore the correct DNA sequence and structure. These relatively specialized enzymes appear to undertake the initial stages of recognition and repair of specific forms of DNA damage. For example, DNA glycosylases catalyze the cleavage of base-sugar bonds in DNA, acting only on altered or damaged bases [W1]. Further, there are several different types of glycosylase that recognize chemically different forms of base damage. However, enzymes that carry out normal DNA metabolism are also part of the repair process for many different forms of damage. In the latter category there are, for example, enzymes involved in the synthesis of DNA strands (DNA polymerases) and enzymes involved in the joining of the DNA backbone (DNA ligases). Several different types of DNA polymerases and ligases have been identified; it is thought that they have different roles in normal DNA metabolism and that only some are active in DNA repair [L3, P1].

21.   The simplest repair processes directly reverse the damage; for example, many organisms, but not mammals, possess an enzyme that directly photoreactivates the UV-induced dimerization of pyrimidine bases [S1]. Similarly, the enzyme O^6methylguanine-methyltransferase directly removes methyl groups induced in DNA by alkylating carcinogens [P13]. However, most damage types require the concerted action of a number of enzymes, forming a repair pathway. Several apparently discrete repair pathways have been identified, as described below and illustrated in Figure I.

22.   Damage to individual bases in DNA may be corrected simply by removing the base, cleaning up the site, and resynthesis. In this process, termed the base-excision repair pathway, a DNA glycosylase removes the damaged base, a DNA endonuclease cuts the DNA backbone, the sugar-phosphate remnants are removed by a phosphodiesterase, and a polymerase fills in the gap using the opposite base as a template (Figure Ia) [L2, S49]. Even where a single base is damaged, therefore, several different enzymes are required to give correct repair. The latter part of this process may also be used to repair single-strand breaks in DNA. Radiation-induced DNA breaks are generally not rejoined by a simple ligation step, because sugar damage and, often, base loss occur at the site of a break. Base-excision repair is generally localized to the single DNA base and is very rapid [D16, S58]; however, in mammalian cells a minority of repair patches of up to 6 bases have been found, indicating a second "long-patch" pathway (see paragraph 75).

23.   Many DNA glycosylases are specific for the removal of one type of altered base from DNA; for example, uracil-DNA glycosylase removes only uracil and some oxidation products of uracil [F15]. However, there is some overlap in the specificity of some base-excision repair enzymes. An

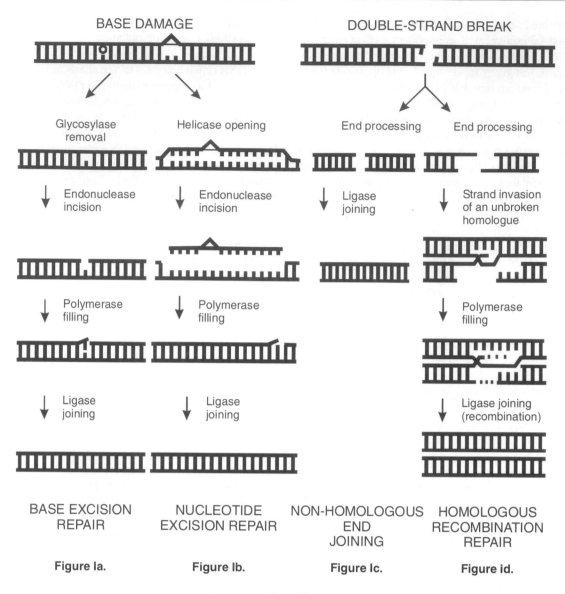

Figure I.
Mechanisms of DNA repair (simplified).

Figure Ia: Base damage is excised by a specific glycosylase: the DNA backbone is cut and the gap filled by a polymerase. The resulting gap is refilled.

Figure Ib: Bulky base damage is removed along with an oligonucleotide (of about 30 bases in human cells). Resynthesis takes place using the opposite strand as a template.

Figure Ic: A double-strand break is rejoined end-to-end.

Figure Id: A double-strand break is repaired with the help of a homologous undamaged molecule (shown in red). Strand invasion allows resynthesis on complementary sequence, followed by a resolution of the strands and rejoining.

example of this is seen in the response to ionizing radiation of bacterial cells defective for one or more nucleases involved in base-excision repair: while cells defective for either nuclease alone were not radiation sensitive, the loss of both nucleases made the cells extremely radiation sensitive [C60, Z10]. Some glycosylases have a very broad specificity for different types of base structure, such as 3-methyladenine-DNA glycosylase II, which acts on a variety of chemically-modified purines and pyrimidines, 5-methyloxidized thymines, and hypoxanthine. It has recently been found that 3-methyladenine-DNA glycosylase II will also remove natural bases from DNA, especially purines, at significant rates. This finding has led to

the novel suggestion that the rate of excision of broad-specificity glycosylases is a function of the chemical bond stability in DNA. On this basis, damaged bases are excised more readily than natural bases because their chemical bonds are less stable [B57].

24.   A specialized form of base-excision repair involves the removal of mismatched DNA bases that occur as errors of DNA replication or from the miscoding properties of damaged bases. For example, 8-oxoguanine is a common product of oxidative damage to guanine bases (paragraph 17); this product is highly mutagenic because of its ability to miscode

(polymerases incorporate adenine instead of cytosine opposite 8-oxoguanine, thus changing the DNA sequence). Cells have evolved three different methods to deal with the formation of 8-oxoguanine: one glycosylase can correct the mismatch by removing 8-oxoguanine from DNA, while a different glycosylase can remove the mismatched adenine after DNA replication [A27, T48]. Excision and resynthesis of the missing base occur as in the general base-excision repair process. Additionally, another enzyme can remove the precursor of 8-oxoguanine (8-hydroxy-GTP) from the nucleotide pool before it is incorporated into DNA [M51]. A number of other mismatches, arising most commonly in DNA replication and recombination, are corrected by a separate "long patch" pathway, known as the MutHLS pathway, that is similarly important in protecting cells from high frequencies of mutation (see paragraph 165) [K38].

25. Where a damaged base is close to another damaged base or to a single-strand break, which is the most simple form of clustered damage, repair may be compromised. Examination of model DNA substrates with base damage on opposite strands, separated by different numbers of bases, has shown that glycosylases do not repair both sites of damage when they are very close to one another (1-3 bases) [C56, C57, H49]. Further, attempted repair of both sites of base damage can result in a DNA double-strand break, because the DNA backbone is cut as part of the repair process.

26. In contrast to base-excision repair, nucleotide-excision repair removes a whole section of single-stranded DNA containing a site of damage, generally a bulky DNA adduct causing distortion of the double helix. These enzymes have to perform such functions as recognition of damage, cutting of the strand at a specified distance either side of the damage, and unwinding and removal of the strand (Figure Ib). As might be expected, at least 11 enzymes have already been identified as components of nucleotide-excision repair [W52], not including polymerase and ligase. The enzymes of nucleotide-excision repair are highly conserved from microbes to humans, and this feature has been used to assist in the isolation of the genes encoding these functions. Several of these genes have been found to be mutated in humans, giving rise to a series of disorders, including xeroderma pigmentosum and Cockayne's syndrome [H3]. Individuals inheriting mutated nucleotide-excision repair genes are generally sensitive to sunlight and chemical agents causing bulky damage in DNA, but a few individuals show cross-sensitivity to ionizing radiation [A2, R15]. Additionally, a small fraction of damage induced by gamma rays is not repaired in cells derived from individuals with xeroderma pigmentosum, which suggests that ionizing radiation induces some bulky damage (e.g. purine dimers) that cannot be removed by the base-excision repair pathway [S56]. Alternatively, a fraction of non-bulky base damage (8-oxoguanine, thymine glycol) may be removed by the nucleotide-excision repair system, particularly in long-lived cells such as neurons that sustain a great deal of endogenous oxidative damage [R22]. These possibilities may explain why severe cases of xeroderma pigmentosum also suffer progressive neurological degeneration.

27. A surprising recent discovery was that some nucleotide-excision repair enzymes are also involved in the normal process of gene expression (transcription). Thus, when genes are actively expressing, they require some of the same functions needed for repair, such as unwinding the DNA helix, and it seems that the same proteins are used. This finding explains the previously puzzling observation of a link between human disorders with sensitivity to sunlight and those with complex defects in gene expression (such as trichothiodystrophy [L70]).

28. Another important discovery about nucleotide-excision repair is that it operates at different rates in different parts of the genome [H1]. Thus, actively expressing genes are repaired much faster than the remainder of the genome. Much of the detail of this process has now been elucidated: it is thought that when damage occurs in a gene that is actively expressing, the proteins involved in this process (the RNA polymerase II transcription complex) stop working, and the stalled complex acts as a signal to the repair proteins to go to the damage site. This two-tier repair system has been found in organisms from bacteria to humans, and the protein mediating the signal to bring repair to the damage site has been identified in bacteria [S2]. The presence in the cell of a fast transcription-coupled repair process has also been found to have genetic consequences: only one of the DNA strands of the duplex is transcribed, and only this strand is repaired rapidly. It has been found in normal cells that most of the mutations induced by DNA damage are in the non-transcribed strand, presumably because lack of fast repair allows the damage to interact with other processes, causing mutations [M2]. In contrast, repair-defective cell lines show a completely altered mutation spectrum, with most mutations recovered in the transcribed strand. Much of this detail has been established using UV-light damage, but the repair of other types of DNA damage, including certain forms of base damage induced by ionizing radiation such as thymine glycols, is influenced by transcription-coupled processes [H1, C41]. Again, certain human sun-sensitive disorders have been found to lack either the fast repair path (Cockayne's syndrome) or the slower overall path of nucleotide-excision repair (xeroderma pigmentosum group C). Loss of the fast or slower pathways may also affect the clinical outcome of sun-sensitive disorders: Cockayne's syndrome does not result in cancer-proneness, while xeroderma pigmentosum patients are highly prone to skin cancers [M2].

29. More severe forms of damage require yet more resources for their correct repair. This is especially true of damage affecting both DNA strands simultaneously, since there is no undamaged strand to act as a template for repair. Severe damage may occur directly by a damaging agent causing complex DNA changes or may arise during the replication of unrepaired single-stranded DNA damage. It is likely that such severe damage will be repaired by recombination enzymes, which rejoin or replace damaged sequences through a variety of mechanisms. In general, there are two main types of recombination repair processes: homologous recombination and illegitimate recombination, although site-specific recombination processes also occur. The principles of recombination repair have been well

established in micro-organisms, and it has recently been found that similar processes occur in human cells.

30. Homologous recombination takes advantage of the sequence identity between certain regions of DNA to repair damage; such regions exist, for example, in the maternal and paternal copies of chromosomes and in the duplicated chromosome (sister chromatids) following DNA replication. The DNA sequence, from which information is derived to repair the damaged copy, must be identical over a considerable length (≥ 200 base pairs). It is known in the budding yeast (*Saccharomyces cerevisiae*), for example, that homologous recombination is the main method for DNA double-strand break repair. Several of the *rad52* group of yeast mutants were isolated on the basis of their extreme sensitivity to ionizing radiation and have been shown to be defective in both DNA double-strand break repair and homologous recombination [G1]. In recombination, the broken 3' end of a DNA strand invades an unbroken double-stranded homologue, and resynthesis on this template re-forms the damaged strand (Figure Id). Separation of the joint product of this reaction requires the activity of enzymes cutting and rejoining the newly-synthesized DNA strands. Depending on which strands are cut and rejoined, this reaction may also result in crossing over (genetic exchange) of DNA strands.

31. Illegitimate recombination (including DNA end-joining processes) is a common mechanism for rejoining broken DNA sequences in mammalian cells (Figure Ic). When foreign DNA molecules are integrated into the genome [R1] or when the genomic breakpoints of deletions and rearrangements are analysed (Section IV.C), it is found that these genomic sites show little sequence homology. There appears to be more than one repair pathway involved, and terms such as non-homologous end joining and direct-repeat end joining are used to describe different pathways in this Annex (see Section II.B.1). It can be argued that illegitimate recombination is a mechanism for rapidly rejoining broken DNA ends without the need for the complex machinery of homologous recombination [R1]. It is also likely that because of the large amount of repetitive DNA sequence in mammalian cells, if the processes of homologous recombination were generally available in cells, there would be an intolerable level of reshuffling of the genome. While homologous recombination is thought to be a mechanism for repairing DNA with little error, illegitimate recombination is likely to cause alteration and/or loss of DNA sequence.

32. It is likely that both homologous and illegitimate recombination processes are able to repair severe damage in the genome. However, a surprising recent discovery in mammalian cells is that the some of the enzymes involved in repairing radiation-induced breakage of DNA also take part in a site-specific recombination process, V(D)J immune-system recombination. This process assembles functional immune genes from separate genomic regions, through somatic gene rearrangement, and is dealt with in more detail in Section II.B.1.

33. There is also evidence that cells have specific surveillance mechanisms for DNA damage and that these mechanisms interface with other aspects of cellular metabolism such as cell-cycle progression [M3]. Thus it is envisaged that when the genome, and perhaps other parts of the cell, sustains damage, a response mechanism is set up to maximize the chance of repairing the damage (or in some cases to commit the cell to a programmed death). The details of these mechanisms, as well as how the overall response is coordinated, are not yet clear.

34. More than 50 genes are already known to affect the repair of DNA damage in lower eukaryotic organisms such as yeasts [F1], but this figure includes genes involved in processes such as cell-cycle checkpoints (Section II.B.2). Additionally, new genes are being found continually, both in searches for homologues of existing repair genes and in genome mapping projects. In view of the numbers already discovered, the multiplicity of types of damage requiring repair, and the recent discoveries of complexity in repair pathways, it would not be surprising if the overall number in humans is a few hundred genes. Therefore, a significant fraction of the genome (paragraph 7) is devoted to maintaining the integrity of DNA. Since the damage to DNA from ionizing radiation is also very diverse, many of these genes will play a role in its repair.

C. SUMMARY

35. Ionizing radiation interacts with DNA to give many different types of damage. Radiation track structure considerations indicate that the complexity of the damage increases with linear energy transfer, and that this complexity may distinguish radiation damage from alterations occurring spontaneously and by other agents. Attempts to measure endogenous levels of damage have suffered from high levels of artefacts, and despite improved methods there are still large margins of error in these estimates. At present, therefore, it is difficult to compare radiation-induced levels of damage with those occurring spontaneously, especially when damage complexity is taken into account. The importance of the relationship between spontaneous and induced levels of damage in the determination of low dose responses is taken up in Annex G, *"Biological effects at low radiation doses"*.

36. A large number of genes have evolved in all organisms to repair DNA damage; the repair gene products operate in a co-ordinated fashion to form repair pathways that control restitution of specific types of damage. Repair pathways are further co-ordinated with other metabolic processes, such as cell cycle control, to optimize the prospects of successful repair.

37. It is likely that the simpler forms of DNA damage (single sites of base damage, single-strand breaks) arising endogenously and from exposure to ionizing radiation will be repaired rapidly and efficiently by base-excision repair processes, so these types of damage are not normally a

serious challenge to biological organisms. However, because of the relatively large amount of base damage and single-strand breaks induced (Table 1), if base-excision repair systems are compromised, the consequences would be very serious for the cell and the individual. DNA damage such as double-strand breaks represents a more difficult problem for cellular repair processes, but more

than one recombination repair pathway has evolved to cope with this damage. Damage caused by large clusters of ionizations in or near DNA, giving more complex DNA alterations, may represent a special case for which separate repair pathways have to come together to effect repair, or where there is a consequent loss or alteration of the DNA sequence as a result of incorrect or inadequate repair.

II. REPAIR PROCESSES AND RADIOSENSITIVITY

A. RADIOSENSITIVITY IN MAMMALIAN CELLS AND HUMANS

1. The identification of radiosensitive cell lines and disorders

38.   Individuals vary in their sensitivity to ionizing radiation. Highly radiosensitive individuals have been detected when they present for cancer therapy; these are seen as rare patients suffering severe normal tissue damage after standard therapy treatments. It has been possible to group some radiosensitive patients into defined disorders, such as ataxia-telangiectasia and the Nijmegen breakage syndrome, but others appear to be asymptomatic (that is, with none of the symptoms of known sensitivity disorders, but also discovered following treatment for cancer). Additionally there are individuals who show less extreme radiosensitivity, some of whom may be variants of known disorders such as ataxia-telangiectasia.

39.   Ataxia-telangiectasia is the best described of radio-sensitive disorders. It has a complex phenotype; cerebellar ataxia, neuromuscular degeneration, dilated ocular blood vessels (telangiectasia), immunodeficiency, chromosomal instability, and a substantially increased incidence of some cancers and neoplasms are common to ataxia-telangiectasia patients [B10]. Ataxia-telangiectasia is inherited primarily as an autosomal recessive trait, although it has been suggested that both radiosensitivity and cancer-proneness behave with some dominance. The disease is progressive, with most affected individuals surviving only to adolescence or early adulthood. Lymphocytic leukaemia and non-Hodgkin's lymphoma appear to be the commonest forms of cancer, but solid tumours in various organs are also associated with ataxia-telangiectasia [H9]. Estimates of the frequency of the disorder vary but suggest an average of about 1 per 100,000 [P8, S12, W10].

40.   The radiosensitive phenotype of ataxia-telangiectasia is also readily demonstrated in cells cultured from patients, using cell survival and chromosome damage assays. For example in a survey comparing the survival of cells cultured from 42 normal individuals with those from 10 ataxia-telangiectasia individuals following x-irradiation, the ataxia-telangiectasia cells were, on average, 2.7 times more sensitive than the normal cells (see Figure II) [C8]. Compared to normal cells, an elevated frequency of chromosomal aberrations is found both spontaneously and after irradiation of ataxia-telangiec-

asia cells. Also, while irradiation of normal cells in the pre-synthesis (G_0) phase of the cell cycle yields only chromosome-type aberrations, both chromatid- and chromosome-type aberrations are found in ataxia-telangiectasia [T1]. A striking feature of ataxia-telangiectasia cells is their resistance to radiation-induced DNA synthesis delay: normal cells show a rapid inhibition of DNA synthesis after irradiation, while ataxia-telangiectasia cells have a delayed and/or much reduced inhibition [P4].

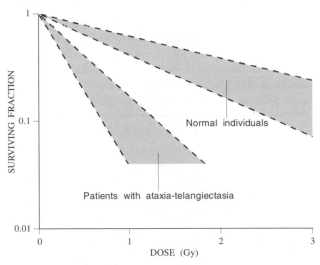

Figure II. Survival of human fibroblast cells after x-irradiation as measured by their colony-forming ability [C8]. *The range of D_0 (the dose required to kill 63% of the cells) is 0.3–0.6 Gy in 10 patients with ataxia-telangiectasia and 1–1.6 Gy in 42 normal individuals.*

41.   Nijmegen breakage syndrome is a clinically separate radiosensitive disorder characterized by variable immune deficiencies, microcephaly, developmental delay, chromo-omal instability, and cancer susceptibility [B4, S7, V7, W4]. Lymphoreticular cancers again seem to characterize this disorder [S8]. Nijmegen breakage syndrome patients show no ataxia or telangiectasia, but their cellular phenotype is very similar to that of ataxia-telangiectasia [A17, J1, N6, T2]. Other patients with similarities to ataxia-telangiectasia and Nijmegen breakage syndrome have been found and in some cases classified separately by genetic analysis ([C12, W5]; see also below). One case has been reported of combined ataxia-telangiectasia and Nijmegen breakage syndrome and called A-T_{FRESNO} [C18]. Also, there are a number of reports of families or individuals who show symptoms that partially overlap

ataxia-telangiectasia and Nijmegen breakage syndrome; an example would be a family with individuals showing either ataxia-telangiectasia or a disorder involving ataxia, microephaly, and congenital cataract [Z3].

42.   To add to this complexity, a number of individuals with a variant form of ataxia-telangiectasia have been described; these patients may show a slower onset of symptoms and have intermediate levels of cellular radiosensitivity [C4, C7, F2, J2, T3, Y1, Z1].

43.   In addition to individuals having multiple symptoms associated with radiosensitivity, some otherwise normal persons have been found to be highly radiosensitive. Woods et al. [W9] described an apparently normal 13-year-old girl who developed multiple complications following radiotherapy for Hodgkin's disease. Cells derived from a skin biopsy showed a highly sensitive radiation survival response, similar to that for ataxia-telangiectasia cells. Plowman et al. [P7] reported similar findings for a 14-year-old boy with acute lymphoblastic leukaemia; again, no ataxia-telangiectasia or Nijmegen breakage syndrome-like symptoms were present, but both whole-body and cellular radiosensitivity were as extreme as for ataxia-telangiectasia. This individual has now provided the first example of radiosensitivity in humans where a defined cellular repair defect (in the ligase IV enzyme) can be readily demonstrated; repair of radiation-induced DNA double-strand breaks and interphase chromosome damage are similarly impaired (see also paragraph 66) [B15].

44.   There is some evidence for ionizing radiation sensitivity in several other cancer-prone disorders, although the published data do not always agree on the degree of sensitivity (see also Section III.A). Bloom's syndrome is a rare autosomal recessive disorder showing severe growth retardation, variable immune deficiencies, and abnormal spermatogenesis [G20]. The age of onset of cancer is considerably earlier than normal; about one third of surviving cancer patients with Bloom's syndrome develop multiple primary tumours with no consistent pattern of cancer type or location. Individuals with Bloom's syndrome develop a distinct facial rash from sunlight sensitivity, and their cells are not only hypersensitive to several different DNA-damaging agents but also show DNA replication abnormalities [L47]. Genetic instability is seen in high levels of spontaneously-occurring chromosomal aberrations and sister chromatid exchanges; chromosomal sensitivity to x rays has been found especially in cells in the G_2 phase of growth [A16, K30]. One case of Bloom's syndrome developed oesophageal stricture following standard radiotherapy treat-ent for lung cancer. This is very rarely seen in such treatment and is suggestive of hypersensitivity to radiation [K31].

45.   Fanconi's anaemia is a cancer-prone disorder, most commonly presenting with acute myeloid leukaemia (15,000-fold increased risk), although solid tumours are also found. Bone marrow failure is a common diagnostic feature of this disorder, although symptoms may include congenital malformations, abnormal skin pigmentation, and skeletal and renal abnormalities [J15]. Fanconi's anaemia cells show high levels of chromosomal aberrations and are hypersensitive to

DNA cross-linking agents (e.g. mitomycin C, diepoxybutane). Additionally, a high proportion of deletions has been reported in certain genes of Fanconi's anaemia cell lines, giving a higher frequency of mutation in assays measuring loss of heterozygosity (see paragraph 172) [S68]. There has been some dispute over the extent to which Fanconi's anaemia cells are sensitive to ionizing radiation; a lack of genetic classification in these experiments may account for some of the variability found (see paragraph 68). However, when they compared published data on the sensitivity of human fibroblasts, Deschavanne et al. [D7] concluded that Fanconi's anaemia was one of the few disorders for which sensitivity to radiation could be distinguished from that of normal cells.

46.   The genes controlling radiosensitivity in humans will not be identified simply by analysing radiosensitive disorders. This is because mutations in many genes are deleterious to the extent that the development of a viable organism is inhibited. This point was illustrated by the creation of a "knockout" mouse for the UV-damage repair gene *ERCC1* (knockout meaning that both copies of the gene are inactivated). No human variant for the *ERCC1* gene has been found, and the knockout mouse dies before weaning, apparently as the result of a massive load of (unrepaired) damage [M7]. Therefore, to examine the full range of genes involved, it has been necessary to derive radiosensitive mutant lines from cells in culture. To this end, more than 50 mutant cell lines sensitive to various genotoxic agents have been identified; many of these show some degree of x-ray sensitivity and are being used to dissect repair pathways in mammalian cells [C16, H11]. These cell lines are especially useful for gene cloning, since this has often proved difficult to achieve using cells derived from human patients. It is possible at the present stage of knowledge to group these mutant cell lines into several categories based on their responses, and recently several of the genes involved have been mapped or cloned (Table 2). As radiosensitive cell lines have been developed in laboratories around the world, almost all have been found to represent defects in different genes. For this reason, and because ionizing radiation produces a diversity of DNA damage, it is anticipated that a large number of the human genes involved in determining radiation resistance remain undiscovered.

47.   The discovery and analysis of the genetic basis of radiation sensitivity is also being pursued through other strategies, including the biochemical analysis of repair reactions and the purification of repair proteins, as well as the identification of human repair genes by homology to their counterparts in lower organisms (Section II.A.3). In many instances, the genes discovered by these routes are found to give rise to a high level of sensitivity when mutated but to affect only a small fraction of the human population or to be inconsistent with life. However, some genes affecting radiation sensitivity will probably have more subtle effects, either because the particular gene mutation only partially reduces gene product activity or because the gene is not vital for cellular response to radiation. Studies exploring the latter types of response, which may affect a much larger fraction of the human population, are described in Section III.A.

Table 2
Classification of radiosensitive disorders and cell lines

Type of defect	Disorder / defective cell line(s)	Human gene designation [a]	Human gene location	Animal model phenotype [b]
Probable DNA break repair defect and loss of cell-cyle control following damage	Ataxia telangiectasia Nijmegen breakage syndrome irs 2/V-series	ATM NBS1 XRCC8	11q23 8q21 ?	AT-like [c] - -
DNA double-strand break repair defective and V(D)J recombination defective	xrs XR-1/M10 V3/scid/SX9 180BR	XRCC5 XRCC4 XRCC7 LIG4	2q35 5q13 8p11-q11 13q33-q34	Immune deficiency [d] Embryonic lethal Immune deficiency [d] Embryonic lethal
Sensitivity to many different agents; some have DNA single-strand break repair defect and/or replication defect	46BR Bloom's syndrome EM9 irs1 irs1SF UV40	LIG1 BLM XRCC1 XRCC2 XRCC3 XRCC9(FANCG)	19q13 15q26 19q13 7q36 14q32 9p13	Viable (acute anaemia) Embryonic lethal Embryonic lethal Embryonic lethal - -
Radiosensitivity inferred from structural homology of genes to those known to be involved in response to radiation damage in lower organisms	- - - - -	ATR hRAD50 hRAD51 hRAD52 hRAD54 hMRE11	3q22-23 5q23-31 15q 12p13.3 1p32 11q21	Embryonic lethal Embryonic lethal Embryonic lethal Viable Viable Embryonic lethal

a XRCC = x-ray cross complementing gene.
b Knockout mice except for *XRCC7;* (-) indicates no model yet available
c Symptoms similar to ataxia telangiectasia (see paragraph 62).
d Severe combined immune deficiency.

2. Mechanisms of enhanced sensitivity in human disorders

48.   In addition to being radiosensitive, ataxia-telangiectasia cell lines show enhanced sensitivity to agents that have in common an ability to damage DNA molecules by producing highly reactive chemical radicals, causing both base damage and sugar damage, leading to breakage of DNA strands. Ataxia-telangiectasia cells have been found to be hypersensitive to a variety of chemicals that cause such DNA damage through radical action (bleomycin, neocarzinostatin, hydrogen peroxide, streptonigrin, phorbol ester [M5]). Also, inhibitors of DNA topoisomerases that can trap these enzymes during DNA-strand passage, leaving open breaks, are more effective at inducing chromosomal damage and cell killing in ataxia-telangiectasia cells than in normal cells [C1, H7, S11]. More recently, ataxia-telangiectasia cell lines have been found to be hypersensitive to restriction endonucleases; these enzymes produce only double-strand breaks in DNA by direct enzymatic cutting [C17, L56]. There is also evidence of modest chromosomal hypersensitivity in some ataxia-telangiectasia lines to agents such as UV light, especially when irradiated in extended G_1 phase, possibly because of the excessive production of breaks when DNA synthesis is attempted [E1, K4].

49.   Experiments varying the time component either during or following irradiation have revealed the general nature of the defect in ataxia-telangiectasia cells. It was shown that normal cells held after irradiation in a non-growing state had some recovery (or sparing) from lethal effects, while ataxia-telangiectasia cell lines showed little or no sparing [C9, U15, W2]. More strikingly, irradiation at low dose rates, where the same dose was given over a period of days instead of minutes (factor of 500 difference in dose rates) showed a very large sparing effect on normal cells and little or no effect on ataxia-telangiectasia cells [C10]. These observations are consistent with an inability of ataxia-telangiectasia cells to recover from radiation damage, and they also show that the defect cannot be abrogated simply by allowing more time for damage restitution.

50.   Lack of a sparing effect appears to be typical of a particular class of radiosensitive cell lines. Thus lines that are known to have a defect in the repair of DNA double-strand breaks, for example the xrs series and XR-1, also lack recovery under irradiation conditions in which their normal counterparts show a large sparing effect [S14, T7]. Similarly, yeast radiosensitive lines that are unable to rejoin double-strand breaks, because of a defect in recombination (*rad50, rad51, rad52*), also lack sparing [R2, R3]. A substantial body of data in yeast supports the contention that the double-strand break is the DNA damage most likely to be lethal to cells, and that its repair is responsible for the recovery seen under sparing conditions [F3].

51.   In contrast to these cellular studies implicating strand breakage as the type of DNA damage involved in the ataxia-telangiectasia defect, it has been difficult to prove that ataxia-telangiectasia cells have a break-repair defect at the molecular

level [M5]. Recently, irradiation at 37°C with low-dose-rate gamma rays has shown a small increase in DNA double-strand breaks following repair ("residual damage") in ataxia-telangiectasia cells relative to normal cells [B9, F10, F16]. However, cytogenetic studies have provided more satisfactory evidence for a break-repair defect: a significantly elevated fraction of unrestituted chromosomal breaks remain in ataxia-telangiectasia cells after irradiation [C11, T1]. Support for these findings has been obtained from the measurement of both DNA double-strand breaks and chromosomal breaks (using prematurely condensed chromosomes to allow rapid analysis) after gamma irradiation of normal and ataxia-telangiectasia lymphoblastoid cells at different phases of the cell cycle [P5]. A consistent decrease in the rapid component of repair was found in ataxia-telangiectasia relative to normal cells; this decrease was small and usually statistically non-significant for DNA double-strand breaks, but larger and significant for chromosomal breaks. Differences in amounts of residual chromosomal damage between normal and ataxia-telangiectasia cells give a close, but not exact, approximation to their relative survival levels after irradiation [C14, P5].

52. Hamster cell lines showing strong similarities to ataxia-telangiectasia have been isolated and their responses characterized (the irs2 line [J5] and the V-series [Z5]). These lines are hypersensitive to agents known to cause DNA breakage, have radioresistant DNA synthesis, and have no measurable biochemical defect in DNA break repair [T8, Z6]. It has also been shown that the radiation sparing effect is absent in irs2, while it is present in other lines that have similar radiosensitivity but do not show ataxia-telangiectasia-like characteristics (such as irs1 and irs3) [T30].

53. Overall, these studies strongly support the view that the increased sensitivity of ataxia-telangiectasia and related cell lines to agents such as ionizing radiation derives from an inability to recover from DNA breakage, leading to a higher level of residual chromosomal damage. However, the molecular mechanisms leading to radiosensitivity in this disorder are still not fully understood, despite considerable recent progress in defining the function of the ataxia-telangiectasia gene product (see Section II.A.3).

54. The functional defects in other cancer-prone disorders have also not been well characterized. Primary cells from Fanconi's anaemia patients have spontaneous delay and arrest in the G_2 phase of growth, as well as an increased frequency of chromosomal aberrations [J15] and recombination [T39]. G_2 delay and aberration frequency increases are corrected by lowering the oxygen tension during growth, leading to the suggestion that reduced detoxification of oxygen radicals may be responsible for the phenotype [C62, J17]. However, immortalized Fanconi's anaemia fibroblasts have lost this oxygen effect, showing that this factor is not a basic (or underlying) defect [S48]. In Bloom's syndrome, the high frequency of sister chromatid exchanges and specific types of chromosome aberrations suggested a defect in DNA repair and/or DNA replication.

3. Analysis of genes determining radiosensitivity

55. The classification of radiosensitive disorders and cell lines into genetic groups, followed by mapping and cloning of the affected genes, has dramatically increased knowledge of the molecular mechanisms of recovery from DNA damage. Once the affected gene has been cloned, its sequence may reveal the nature of the gene product (protein), because of similarities to known genes. Gene sequence data from at-risk groups will also allow deleterious mutations to be identified, and permit analysis of the role of these genes in disorders such as cancer. Manipulation and expression of the gene under defined experimental conditions allow specific functions to be studied in cells and in animals. Animal models of the human disorder can be created by replacing the normal pair of genes with defective copies (a knockout animal; see paragraph 46) and assessing the resulting phenotype.

56. Genetic classification of the ataxia-telangiectasia disorder initially indicated that several different genetic groups might exist [C3, J2], but mapping and cloning of a gene (ATM) found to be mutated in patients has cast doubt on this designation. The genetic mapping data, based on ataxia-telangiectasia family studies, mostly placed the affected gene into one chromosomal region, 11q23.1 [G2, Z2]. Positional cloning procedures in this region led to the identification of the ATM gene, which has homology to a gene family encoding PI-3 kinases [B31, S24, S25]. The PI-3 kinase family contains a number of large proteins involved in cell-cycle checkpoints, the regulation of chromosome-end length, and DNA break repair, including site-specific recombination (see Section II.B.1). It is therefore likely that ATM and other members of this family are involved in the detection of certain types of DNA alterations and may coordinate response by signalling these changes to other regulatory molecules in the cell [J8, K3, S25, T29].

57. Analysis of mutations in the ATM gene of ataxia-telangiectasia patients showed that the majority are compound heterozygotes (i.e. the mutations in the two gene copies derive from independent events) and that these commonly lead to an inactive, truncated protein [G19, M30]. However, individuals from 10 families in the United Kingdom with less severe symptoms (paragraph 42) all have the same mutation in one copy of ATM (a 137-bp insertion caused by a point mutation in a splice site) but differ in the mutation in the other gene copy. The less severe phenotype appears to arise from a low level of production of normal protein from the insertion-containing gene copy. Two more families with this less severe phenotype have mutations leading to the production of an altered but full-length protein, again suggesting that the severity of the symptoms in ataxia-telangiectasia is linked to the genotype of the individual [L41, M30]. However, it is possible that individuals with less severe symptoms have an increased risk of developing specific types of cancer [S78] (see also paragraph 139).

58. A further two families presenting with many of the symptoms of ataxia-telangiectasia (paragraph 39) but

without dilated ocular blood vessels failed to show mutations in the *ATM* gene. The cellular characteristics of family members, such as radiation sensitivity and DNA-synthesis delay (paragraph 40), were mostly intermediate between those of classical ataxia-telangiectasia patients and normal individuals. Examination of other genes implicated in the repair of DNA double-strand breaks revealed that the human homologue of the yeast *MRE11* gene (paragraph 74) was mutated in these families [S82].

59. A rare form of leukaemia, sporadic T-cell prolympho-cytic leukaemia (T-PLL), shows a high frequency of *ATM* mutations; although some of the changes identified in the *ATM* gene were rearrangements, most mutations in T-PLL were single DNA base-pair changes in the kinase region of the *ATM* gene and did not lead to protein truncation [V5, Y8]. This finding has prompted the suggestion that the *ATM* gene acts as a tumour suppressor in cells that may develop T-PLL. There is no evidence for an involvement of the *ATM* gene in T-cell lymphoblastic leukaemias [L58], but in B-cell chronic lymphocytic leukaemia (B-CLL), 34%-40% of tumour samples showed about 50% reduction in levels of ATM protein [S70, S76]. *ATM* mutations were detected in the tumours of 6 out of 32 patients (18%); also 2 of these 6 patients had mutations in both tumour and normal cell DNA, indicating that they were carriers of an *ATM* gene defect (i.e, they inherited a mutation in one copy of the *ATM* gene; see paragraph 137 *et seq.*) [S76]. Similar results were obtained in a separate study [B62], and although they are based at present on small samples, the data suggest that the frequency of *ATM* mutations in the normal cells of patients developing B-CLL may be much higher than in the general population [B62, S76]. Patients with ATM deficiency also had significantly shorter survival times [S70].

60. Study of the ATM protein has shown that it has a nuclear location and is expressed in many different human tissues. Additionally, ATM protein does not increase in amount in response to cell irradiation, consistent with the idea that it is part of a DNA-damage-detection system rather than being regulated in response to DNA damage [L41]. However, the ATM protein does associate with DNA, and this interaction increases when the DNA is irradiated [S77]. The *ATM* gene has a complex structure with multiple transcription start sites, leading to messenger RNAs of different length and predicted secondary structure. This multiplicity of *ATM* transcripts may allow cells to modulate ATM protein levels in response to alterations in environmental signals or cellular metabolism [S40].

61. Initially it was also suggested that the Nijmegen breakage syndrome involves more than one genetic group [J1], but recent evidence shows that only one gene is involved [M41] and that this gene maps to chromosome 8q21 [M41, S59]. These data support clinical findings (paragraph 41) showing that the Nijmegen breakage syndrome is a separate radiosensitive disorder, distinct from ataxia-telangiectasia. The cloning of the gene (*NBS1*) mutated in this syndrome has confirmed this (paragraph 74), with the majority of patients carrying small deletions in this gene [M45, V6].

62. Knockout mice that lack a functional homologue of the ataxia-telangiectasia gene (*Atm*) have recently been bred; they show many of the symptoms of the human disorder, but also give further insights into the action of this gene [B33, E9, X3]. For example, the *Atm* knockout mice are viable but growth-retarded and infertile. They are also very sensitive to acute gamma radiation; at a whole-body dose of 4 Gy, about two thirds of the *Atm* knockout mice died after 5-7 days, while normal and heterozygous mice remained without morbidity after two months [B33]. Primary cells derived from the *Atm* knockout mice also show many of the characteristic features of cells from ataxia-telangiectasia patients [X4]. The cells grow poorly, are hypersensitive to gamma radiation, and fail to undergo arrest of the cell cycle following irradiation (see also paragraph 102). The *Atm* gene product was found to locate to homologous chromosomes as they associate (synapse) at meiosis in germ cells [K27]; loss of fertility in the knockout mice results from failure of meiosis due to abnormal synapsis and subsequent chromosome fragmentation [X3]. Immune defects occur in these mice, and the majority develop thymic lymphomas and die before four months. Based on the understanding to date of the defective response to DNA breakage in ataxia-telangiectasia, the meiotic failure and immune defects in these mice could both relate to an inability to respond to "programmed" DNA double-strand breaks (site-specific breaks that occur in the course of normal cellular processes). Such breaks are thought to be essential in meiosis, to drive the process of meiotic recombination, and they are required for V(D)J recombination in immune system development (see paragraph 82).

63. Unlike in the human disorder, no evidence was found of gross cerebellar degeneration in *Atm* knockout mice aged 1-4 months, but it is possible that the animals are dying too early for this symptom to be revealed [B33, X4]. However, behavioural tests indicated some impairment of cerebellar function [B33], and detailed studies of the brain in these animals showed that *Atm* deficiency can severely affect dopaminergic neurons in the central nervous system [E14]. There is an almost complete absence of radiation-induced apoptotic cell death (Section II.B.3) in the developing central nervous system of *Atm*-deficient mice, while the thymus shows normal levels of apoptosis after irradiation [H48]. Additionally, elevated levels of oxidative damage were recorded in mouse tissues affected by the loss of ATM, especially the cerebellum [B63].

64. In yeast cells, the closest sequence homologue to the *ATM* gene product is the Tel1 protein; mutations in the *TEL1* gene are associated with shortened chromosome ends (telomeres) and genetic instability (three- to fourfold increased levels of mitotic recombination and chromosome loss) [G22]. It has been known for many years that chromosomes in ataxia-telangiectasia cells show a relatively high incidence of telomere fusions, and recent studies have shown that preleukaemic cells from ataxia-telangiectasia patients have an increased rate of telomere loss [M29]. This loss may contribute to chromosomal instability [R40]. However, as yet the relationship between the telomere fusions and tumorigenesis in ataxia-telangiectasia is not clear.

Interestingly, loss of the *TEL1* gene does not lead to x- or gamma-ray sensitivity [G22, M31], but the combined loss of *TEL1* and another gene in the same family (*ESR1/MEC1*; see next paragraph) gives extreme sensitivity to gamma rays, suggesting that these two genes may functionally overlap in protecting the cell against radiation damage [M31]. In fission yeast, following prolonged growth, loss of these two genes led to circular chromosomes lacking telomeric sequences [N26].

65.   A further member of the human PI-3 kinase family, related to *ATM,* has been found through homology searches [B32, C38]. The gene (named *ATR* or *FRP1*) encodes a protein that is most closely similar to the fission yeast *rad3* gene product, which is itself structurally and functionally related to the budding yeast *ESR1/MEC1* and the fruit fly *mei-41* gene products. Mutations in these genes render the yeast or flies sensitive to killing by both ionizing and UV radiation, and are known to play important roles in mitotic and meiotic cell-cycle controls. Expression of an inactive form of *ATR* at a high level in human fibroblasts increased radiation sensitivity by a factor of 2-3 and abrogated radiation-induced G_2 arrest (see paragraph 108). Additionally, cells with defective *ATR* develop abnormal nuclear morphologies, which may indicate further cell-cycle perturbations [C48, W50]. The *ATR* gene product is expressed at a high level in germinal tissue and localizes along unsynapsed meiotic chromosomes [K27], thereby playing a role that is complementary to that of the *ATM* gene product (paragraph 62). Thus in addition to their roles in mitotic cells, *ATM* and *ATR* gene products may be involved in the co-ordination of meiotic chromosome synapsis, perhaps by signalling breaks and monitoring repair synthesis to guard against genetic instability.

66.   Some other genes affecting radiosensitivity in humans have also been mapped and cloned. Following the cloning of the human DNA ligase I gene, it was discovered that this was defective in a unique individual with growth retardation, immunological abnormalities, and cellular sensitivity to a variety of DNA-damaging agents, including radiation [B5]. The genes for other human DNA ligases have also been cloned using homology searches or protein purification [C31, W19], and subsequently another unique radiosensitive individual with impairment of double-strand break rejoining (described in paragraph 43) was found to be defective in DNA ligase IV [R38]. It has been established in yeast and mammalian cells that DNA ligase IV is specifically involved in the repair of DNA breaks by non-homologous end joining (Section II.B.1) [S64, T40, W46]. DNA ligase III has been implicated in the base-excision repair pathway (see paragraph 78).

67.   Bloom's syndrome patients of different ethnic origins (Ashkenasi Jewish, French-Canadian, Mennonite, and Japanese) were found by cell fusion analysis to fall into one genetic group [W36]. The gene defective in Bloom's syndrome (named *BLM*) has been positionally cloned and shown to be similar to the bacterial RecQ helicase, a type of enzyme that opens up the DNA helix and is associated with genetic recombination in bacteria [E6, K39].

Knocking out the gene homologous to *BLM* in mice results in embryonic growth retardation and lethality, apparently because of a wave of increased apoptotic death [C61]. The yeast homologue of the *BLM* gene product interacts with DNA topoisomerase enzymes, known to be required for the resolution of interlocking DNA molecules following replication. In the absence of the *BLM* gene product, it is suggested that replicated DNA (in mitosis or meiosis) is entangled and may give rise to sister-chromatid exchange and chromosomal non-disjunction [W35]. Support for this idea comes from studies with the fission yeast, where the recQ helicase is required for the recovery from cell-cycle arrest during DNA replication following DNA damage; absence of the recQ helicase leads to an increase in the rate of genetic recombination [S60].

68.   Cells derived from Fanconi's anaemia patients have been classified into eight genetic groups [J16, J19, S46], consistent with the heterogeneity of symptoms found (paragraph 45). To date, three of these genes have been mapped, *FANCA* to chromosome 16q24 [G21, P20], *FANCC* to 9q22 [S46], and *FANCD* to 3p22-26 [W37]. The *FANCA* and *FANCC* genes have now been cloned [F9, L43, S47]; these gene sequences predict proteins that are structurally different from each other and from other known proteins. Mutations of the *FANCA* gene were analysed in 97 patients from different ethnic groups with Fanconi's anaemia; the majority of mutations detected were either DNA base-pair alterations or small deletions and insertions scattered throughout the gene, with a smaller number of large deletions [L49]. The *FANCC* gene product is a cytoplasmic protein [Y7], but it interacts with the *FANCA* gene product, and the complex translocates to the nucleus [K40]. While this nuclear localization is consistent with a possible role in DNA repair, the precise function of these gene products remains unclear. *FANCC* also has binding sites for the p53 tumour suppressor protein (paragraph 100), and binding of p53 to the gene can regulate its expression [L46].

69.   One of the more recently identified genetic groups of Fanconi's anaemia, FA-G, has been found to be caused by mutation of a previously-identified gene, *XRCC9* [D18]. This gene was identified by the complementation of a rodent cell line sensitive to a variety of DNA-damaging agents including a twofold enhanced sensitivity to x rays, and is potentially involved in post-replication repair of DNA [L59] (Table 2).

70.   Human genes implicated in the repair of radiation-induced DNA damage have also been cloned by their structural homology to genes involved in specific repair pathways in lower organisms. For example, it is known that the main pathway for the repair of DNA double-strand breaks in lower organisms involves homologous recombination (paragraph 30); in budding yeast, the genes responsible are *RAD50, RAD51, RAD52, RAD53, RAD54, RAD55, RAD57, MRE11,* and *XRS2* (often called the *RAD52* group of genes). Mutations in these genes render the yeast cells defective in mitotic and/or meiotic

recombination and sensitive to ionizing radiation. At least some of the products of these genes act together in a multi-protein complex to effect recombination [H27, J14, U18]. In yeast, functional relationships have also been found between proteins of the PI-3 kinase family and some of the Rad52 group proteins. For example, the activity of the *RAD53* gene product, another protein kinase, is regulated by both *ESR1/MEC1* and *TEL1* gene products, suggesting that the Rad53 protein acts as a signal transducer in the DNA damage response pathway downstream of these ATM-like proteins [S43]. Sequence similarities have led to the cloning of human homologues of some of the yeast *RAD52* gene group: *hRAD50* [D16], *hRAD51* [S10, Y6], *hRAD52* [M10, S44], *hRAD54* [K28], and *hMRE11* [P22]. Preliminary evidence has been obtained from their abilities to partially correct the defects in radiosensitive yeast cell lines that some of these genes, or their mouse homologues, are involved in the repair of radiation damage [K28, M37]. Further, knocking out the *RAD54* gene in both the fruit fly [K41] and mouse [E12] confers radiation sensitivity and a defect in homologous recombination. There is also evidence for regulated expression of these recombination proteins through the cell cycle, with levels of RAD51 and RAD52 increasing during the S phase and peaking in G_2 in human and rodent cells [C5].

71.   The human *RAD51* gene product has been shown to bind DNA and thereby underwind the double helix, an early step in the recombination process [B34]. The *RAD51* gene in yeast is not essential for cell survival, but recently the homologous gene has been knocked out in mice and surprisingly was found to give embryonic lethality [L42, T34]. Knockout embryos arrested early in development but progressed further if there was also a mutation in the gene encoding the tumour suppressor p53 (*TP53*), possibly because of a reduction in programmed cell death (see paragraph 112). This may suggest that *RAD51* in mammals has functions in addition to recombination and repair, and in a separate study the human RAD51 protein has been shown to directly interact with p53 [S45]. Cells from the mouse *Rad51* knockout were hypersensitive to gamma rays, although this response has not been quantified, and show chromosome loss in mitotic cells [L42]. In mitotic cells, the mouse RAD51 protein is found to concentrate in multiple foci within the nucleus at the DNA-synthesis (S) phase of the cell cycle [T37]. In meiotic cells, RAD51 is found on synapsing chromosomes and disappears shortly after this stage [P23], but this pattern of localization of RAD51 is disrupted in *Atm*-deficient mice [B47].

72.   The pattern of cellular localization found for the RAD51 protein has also been found for the product of a gene commonly mutated in familial breast and ovarian cancer patients, *BRCA1*; subsequently, RAD51 and BRCA1 proteins were shown to interact directly [S51]. *Brca1*-deficient mouse cells show a modest increase in sensitivity to ionizing radiation and have a reduced capacity to repair base damage in transcribing DNA (transcription-coupled repair, paragraph 28) [G34]. In human *BRCA1*-deficient cells, increased sensitivity to cell killing by gamma rays and reduced repair of DNA double-

strand breaks can be partially corrected by introduction of the normal gene, while mutant *BRCA1* genes fail to restore these defects [S86]. A product of a second gene associated with familial breast cancer susceptibility, *BRCA2*, has similarly been shown to interact with RAD51 [M42, S61], as well as with p53 [M46]. *BRCA1* and *BRCA2* are highly expressed in rapidly proliferating cells, with expression highest at the start of S phase of the cell cycle [R23]. Analysis of mutations in the *BRCA* genes has shown that families with a high proportion of breast cancer tend to have mutations in different parts of the gene from families having a predisposition to ovarian cancer [G37, G38]. Despite their importance in cancer predisposition, the molecular function of the *BRCA* genes is unknown, and the finding of interaction with RAD51 provided the first clue that their role may be in DNA repair. As noted in paragraph 62, the ATM protein also has a specific pattern of localization to meiotic chromosomes, and loss of ATM can disrupt the localization pattern of RAD51. Thus, through their connections with *RAD51*, these data provide for the first time a mechanistic link between four different genes involved in cancer susceptibility (*ATM*, *TP53*, *BRCA1*, and *BRCA2*).

73.   Mice have been bred with a knockout of the *Brca1* gene or the *Brca2* gene; like *Rad51* knockout mice, they were both found to arrest early in embryonic development [H40, L50, S61]. However, viable mice have been bred with a mutation in *Brca1* or *Brca2* that does not completely inactivate the gene. A conditional deletion of the *Brca1* gene in mice, confining the defect to the mammary glands, gave a low frequency of mammary tumour formation with long latency (see also paragraph 106) [X7]. Embryonic fibroblasts derived from *Brca1*-deficient mice proliferate poorly and show genetic instability [X8] and a reduction in the frequency of homo-logous recombination [M50]. About one third of the *Brca2*-deficient mice survived to adulthood but showed small size, poor tissue differentiation, absence of germ cells, and development of thymic lymphomas [C59]. Cells derived from viable *Brca2*-defective mice proliferate poorly, apparently as a result of spontaneously high levels of p53 and p21, causing cell-cycle arrest (see paragraph 100) [C59, P32]. These cells accumulate high levels of chromosomal aberrations and show enhanced sensitivity to DNA-damaging agents, including x rays [M48, P32]. The evidence linking the *BRCA* genes to DNA repair processes suggests that they may not function as tumour suppressors but are involved in the maintenance of genome integrity.

74.   The yeast *RAD50, MRE11,* and *XRS2* gene products assemble into a multi-protein complex that is implicated in the processing of broken DNA, as well as in a number of other functions, including meiotic recombination and telomere maintenance [H46]. This multiplicity of roles for the complex may arise from its nuclease activities on both single- and double-stranded DNA ends that trim up the DNA in preparation for end-joining and homologous recombination processes. Human gene products with similar functions have been identified (paragraph 70); significantly, the gene mutated in the Nijmegen breakage syndrome (*NBS1*) has been found

to be the functional homologue of *XRS2* (paragraphs 41 and 61) [C54, M45, V6], and individuals with a variant form of ataxia-telangiectasia have mutations in the human homologue of *MRE11* (paragraphs 42 and 58) [S82]. The *NBS1* gene product contains motifs that are commonly present in proteins involved in cell-cycle regulation and DNA-damage response, suggesting that like ATM, this gene has a role in the signalling mechanism following damage to DNA by ionizing radiation. Study of the purified protein shows that DNA unwinding and nuclease activities of RAD50/MRE11 are promoted by the presence of NBS1 [P34]. The human RAD50 and MRE11 proteins co-localize in cell nuclei following ionizing radiation damage (but not after UV irradiation); their sites of localization were distinct from those of the RAD51 protein, consistent with their different roles in DNA repair [D17]. Co-localization of MRE11/RAD50 proteins was much reduced in ataxia-telangiectasia cells, suggesting that *ATM* gene signalling is important for the assembly of the break-rejoin complex [M28]. Additionally, the BRCA1 protein has been shown to colocalize and interact with RAD50, and radiation-induced sites of localization including MRE11, RAD50 and NBS1 proteins were much reduced in breast-cancer cells lacking BRCA1 [Z12].

75.   Human genes encoding enzymes responsible for the repair of the numerous forms of base damage caused by DNA-damaging agents, including ionizing radiation, are also being discovered [L69, S49]. A number of key enzymes (DNA glycosylases, endonucleases) in the base-excision repair pathway have been isolated by biochemical purification, followed by protein sequencing and gene identification, or through their homologies to known enzymes in lower organisms. For example, the hOGG1 glycosylase [R41] removes 8-oxoguanine opposite a cytosine (paragraph 24), while the hNTH1 glycosylase [A28] removes oxidized pyrimidines such as thymine glycol (paragraph 17). The pathway of base-excision repair that results in the incorporation of a single nucleotide ("short-patch repair") has been reconstituted under cell-free conditions, using purified enzymes [K24]. Additionally, a second "long-patch repair" pathway has been identified, in which between two and six nucleotides are replaced following repair of a reduced or oxidized base-less site; this pathway also requires the structure-specific nuclease DNase IV (also known as FEN1) to remove the displaced nucleotides during repair [K42]. The reconstitution of base-excision repair of oxidized pyrimidines in DNA has revealed that hNTH1 glycosylase is strongly stimulated by one of the proteins involved in nucleotide-excision repair, XPG (xeroderma pigmentosum group G protein). The XPG protein binds to non-paired regions of DNA, and acts as a structure-specific nuclease where the unpaired region is greater than five base pairs, as will happen at sites of bulky damage. However, at sites of oxidative base damage it seems that the unpaired region is less than five base pairs, and instead of cutting DNA the XPG protein promotes the activity of the hNTH1 in removing the base damage [K46]. These findings may explain the extreme symptoms of some patients lacking XPG, including growth failure and neurological

dysfunction, and the early death of mice carrying a knockout of the *Xpg* gene [H51] (while mice that are totally defective for nucleotide-excision repair such as those defective in *Xpa* are viable [N27]).

76.   The main endonuclease involved in the repair of base-less sites (Figure Ia) in human cells, known variously as HAP1, APE, APEX, or Ref-1, was cloned in several laboratories following biochemical purification and was found to have a surprising additional function when compared with the bacterial enzyme. It stimulates the DNA-binding activity of transcription factors involved in signal transduction such as Fos, Jun, NFκB, and p53 by reduction/oxidation mechanisms [X5]. Signalling and repair responses to oxidative damage may therefore be coordinated through this one enzyme; this may be especially important during tissue proliferation, since *Ref-1* knockout mice die during embryonic development [X6]. Knockout mice have been produced for several other genes in the base-excision pathways, and a number of these have also proved to be embryonic-lethal, showing their importance for the normal functioning of cells and tissues [W45]. However, a knockout mouse for the broad-specificity glycosylase 3-methyladenine-DNA glycosylase II was recently shown to be viable [E13], possibly because of its mode of action (paragraph 23) and the likelihood that other more specialized glycosylases can substitute for it in removing damaged bases.

77.   The abundant enzyme poly(ADP-ribose) polymerase, also known as PARP, is rapidly recruited to sites of DNA breakage following irradiation, where PARP transiently synthesizes long, branched chains of poly(ADP-ribose) on itself and other cellular proteins. These chains are degraded by another enzyme, poly(ADP-ribose) glycohydrolase, with a half-life of only a few minutes. The role of PARP in response to DNA damage, which has long been in dispute [C49], has been clarified from recent experiments in which the *Parp* gene was knocked out in mice [M9]. These mice are viable and fertile, although adult size is smaller than average and litter sizes are smaller than for normal mice. Following whole-body irradiation (8 Gy gamma rays), the *Parp* knockout mice died much more rapidly than normal mice from acute radiation toxicity to the small intestine; the survival half-time of these irradiated PARP-deficient mice was comparable to that of irradiated *Atm* knockout mice (paragraph 62). PARP-deficient cells or cells in which a mutated PARP is expressed [S63], show increases in chromosomal aberrations, sister-chromatid exchanges, and apoptosis following DNA damage. The average length of telomeres is significantly shorter in PARP-deficient mouse cells, leading to an increased frequency of chromosomal fusions and other aberrations [D21]. Recent biochemical data suggest that PARP has an important function as a molecular sensor of DNA breaks, especially single-strand breaks, and its absence reduces the efficiency of base-excision repair [O4]. In addition, it has been suggested that the synthesis of poly(ADP-ribose) chains causes negative-charge repulsion of damaged DNA strands, preventing accidental recombination between homologous sequences [S62].

78.   To date, studies with radiosensitive mammalian lines have identified more than eight genetic groups [J3, T9, Z13]. The genes responsible have provisionally been named the *XRCC* (for *X-Ray Cross-Complementing*) group, despite the fact that some of the cell lines are not primarily sensitive to x rays. The human gene corresponding to the first group, *XRCC1*, has been cloned and encodes a protein interacting with DNA ligase III [C2]. Further, the XRCC1 protein has been shown to associate with DNA polymerase β [C39, K29], which fills in the gaps created during the repair of damaged bases. It is suggested that XRCC1 may act as a scaffold protein in the final steps of base-excision repair (Figure Ia), supporting the activity of DNA polymerase β and DNA ligase III [K29]. The *XRCC2* gene [C66, L65, T35] and the *XRCC3* gene [L65]have been cloned recently, and both genes have structural homology to the yeast and human *RAD51* genes [T47]; it is likely, therefore that they are involved in repair of damage by homologous recombination. Other human genes recently cloned using mammalian cell lines (*XRCC4, XRCC5, XRCC7*) are involved in the repair of radiation-induced DNA double-strand breaks as well as in immune gene recombination, as detailed below (Section II.B.1). Current mapping data place all of the remaining genes on different chromosomes, and none maps to the location of characterized radiosensitive human syndromes (Table 2).

4. Summary

79.   Several radiosensitive human disorders and examples of individual radiosensitivity have been identified in recent years. The sensitivity in these individuals is characterized by a greatly increased susceptibility to cancer, although this cancer is not necessarily radiation-induced. Many radiosen-

sitive lines of cultured cells have also been established, and these have been useful for identifying the genes involved and for functional studies. In general, it has been found that enhanced sensitivity arises from an inability to recover from DNA damage, because of a reduction in damage detection and/or repair processes. Reduction of repair capacity commonly leads to a lack of low-dose-rate sparing and a higher level of genetic changes.

80.   Considerable progress has recently been made in defining repair gene functions in human and other mammalian cells, and a summary of well-defined repair pathways for damage by ionizing radiation is shown in Figure III. A number of important genes have been cloned, including the gene *ATM*, which determines sensitivity in the ataxia-telangiectasia disorder. Studies with the *ATM* gene product and the production of mice defective for the gene suggest that it participates in the detection of DNA double-strand breaks and passes this information on to other important molecules regulating cellular response processes. Other recently discovered genes act directly in the repair of radiation damage; for example, the *RAD51* gene is vital for repair by homologous recombination. The *RAD51* gene product has also been found to interact with products of the breast-cancer susceptibility genes, *BRCA1* and *BRCA2*, suggesting an unsuspected role for these genes in damage recovery. In accordance with this finding, mice defective for the *BRCA* genes have symptoms very similar to *Rad51*-defective mice, and the *Brca2*-defective mice are radiation-sensitive. Additionally, some repair genes when knocked out in animals give embryonic lethality, showing that these genes have important roles in basic cellular processes influencing tissue development.

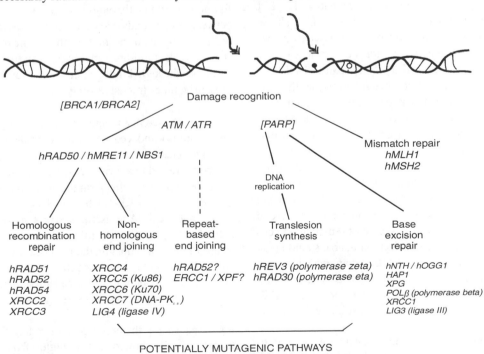

Figure III. Important genes in DNA repair pathways in human cells following damage by ionizing radiation.
Some of the important genes identified in each pathway are shown (the h prefix indicates a human homologue of a yeast gene); names of proteins are shown in brackets if different from the gene names. The BRCA genes and the PARP gene (square brackets) have speculative roles in recombination repair and base-excision repair, respectively. Pathways that can lead to mutation are indicated.

81. The conservation of repair genes from lower to higher organisms and progress in understanding the human genome suggest that further rapid progress will be made in the discovery and analysis of genes influencing radiation sensitivity. These genes will be important tools in understanding the extent of variation in radiation sensitivity in the human population and should help to identify individuals especially at risk.

B. RELATIONSHIP BETWEEN REPAIR AND OTHER CELL REGULATORY PROCESSES

1. Radiosensitivity and defective recombination in the immune system: non-homologous end joining of DNA double-strand breaks

82. It has been noted that radiosensitive disorders are often associated with some degree of immune deficiency. The recent discovery that the *scid* (*severe combined immune-deficient*) mouse strain is radiosensitive has led to a reexamination of this relationship and to rapid progress in understanding at least one mechanism of DNA-break rejoining. In the immune system, functional immunoglobulin and T-cell receptor genes are assembled from separate genetic regions during lymphoid differentiation [A5]. The separate regions (termed "V" for variable; "D" for diversity, and "J" for joining) involved in this type of site-specific recombination are flanked by recognition (signal) sequences at which double-strand DNA breakage occurs prior to the rejoining of the regions. Mice homozygous for the *scid* mutation lack functional T-cells and B-cells because of a defect in V(D)J recombination [H8, L9], and about 15% of the mice develop lymphomas [C50]. It was shown that fibroblasts derived from homozygous *scid* mice are three to four times as sensitive as normal mice to x rays or gamma rays and have a reduced ability to repair DNA double-strand breaks [B8, H10]. Mice heterozygous for the *scid* mutation also show some increased sensitivity to gamma rays, compared to normal mice, and cultured bone marrow cells from *scid* heterozygotes are marginally more radiosensitive than normal cells [K45]. Detailed studies of repair capacity in *scid* cells show that the rate of repair of double-strand breaks is reduced by a factor of about 5, but if sufficient time is allowed (24 h) the final levels of repair are similar to those of normal cells [N20]. Subsequently, similar tests on several radiosensitive mammalian lines derived from cultured cells showed that those with large defects in double-strand break repair are also defective in V(D)J recombination, while radiosensitive lines, including ataxia-telangiectasia cells, with near-normal double-strand break repair are not V(D)J-recombination-deficient (Table 2) [H12, K25, P6, T4, T11].

83. At least four different gene products are common to V(D)J recombination and repair of radiation-induced DNA breaks; three of these products are defined by their respective rodent sensitive lines: xrs, scid/V3, and XR-1/M10 (Table 2). Recently there has been a breakthrough in identifying the genes and gene products complementing the defects in these lines, to define a repair pathway termed non-homologous end joining (paragraph 31). The xrs-complementing gene (also known as *XRCC5*) codes for a subunit of the Ku antigen (p86), a DNA end-binding protein discovered in normal human cells that reacted with sera from patients with certain autoimmune diseases [T5]. The *scid*/V3 protein (*XRCC7* gene product) has been found to be the catalytic subunit of a large DNA-dependent protein kinase, also known as p460, belonging to the PI-3 kinase family and related to the ataxia-telangiectasia gene product (see paragraph 56) [H17]. The Ku antigen consists of two subunits, p70 and p86, which interact to form a dimer; this dimer binds to broken ends of DNA, recruiting the p460 subunit and conferring kinase activity on the complex [G7]. The DNA-dependent protein kinase has been shown to phosphorylate a number of substrates *in vitro*, including the p53 tumour suppressor and RNA polymerase II [A4], but it is not clear whether these constitute important targets *in vivo*. The most recently cloned gene of this series, the human *XRCC4* gene complementing the XR-1 line, encodes a protein unrelated to any other yet described [L30]. Functional studies of the XRCC4 protein show that it is a nuclear phosphoprotein that is a substrate *in vitro* for the DNA-dependent protein kinase and that it associates with the recently discovered DNA ligase IV [C51, G30]. This discovery suggests that XRCC4 acts as a go-between in the assembly of a DNA-break repair complex in which the final step is mediated DNA ligase IV (paragraph 66). Experiments using cell extracts show that the rejoining of breaks by mammalian DNA ligases is stimulated by purified Ku86 protein, especially when the break ends cohere poorly [R24]. This observation suggests that Ku86 may function to stimulate ligation, perhaps through its ability to bridge the gap between broken ends. However, the precise mode of action of these proteins in promoting the repair of DNA double-strand breaks (and presumably in initiating a coordinated response through kinase action on relevant proteins) is presently under intensive study.

84. The different subunits of the DNA-dependent protein kinase are not induced following x irradiation [J22]. However, the gene encoding Ku86 in primates (but not in rodents) can be expressed in a form that has been found to respond to radiation damage. In this form the gene gives rise to a related protein, termed KARP-1, which includes a p53 binding site [M43], and the RNA transcript and protein were increased by a factor of up to 6 at 90 min following x-ray irradiation. Interestingly, this induction did not occur in cells defective in either p53 or ATM, suggesting that at least some of the non-homologous end joining is activated through interactions with p53 and ATM signalling processes [M44].

85. Sequencing the entire gene encoding the p460 kinase from *scid* cells showed that a single DNA base alteration leads to a premature stop codon in the highly conserved terminal region of the gene [A20], as other data had predicted [B37, D11]. This mutation gives a protein truncated by 83 amino acids in *scid* cells, leading to a partial abolition of kinase activity. Mutations in this gene have subsequently been

found in other radiation-sensitive lines of cultured cells; for example, the SX9 line (Table 2) has a mutation giving rise to a single amino-acid substitution at a position before that of the *scid* mutation, and leads to a more severe phenotype than the *scid* mutation [F25]. The partial loss of activity in *scid* [A19, B38, C52, P26] may explain another surprising phenomenon, whereby sublethal irradiation of *scid* mice can transiently rescue V(D)J recombination [D12]. While the mechanism of this rescue is still unknown, it suggests that p460 in *scid* is not completely inactive.

86. The importance of the catalytic subunit of DNA-dependent protein kinase in the development of T-cells and B-cells and in protecting animals against cancer has been illustrated by the chance finding of transgenic mice in which the *Xrcc7* gene was knocked out [J20]. These mice, named *slip*, showed a lack of mature lymphocytes, but most remarkably all animals died within 5-6 months from thymic lymphoblastic lymphomas. No tumours of B-cell or myeloid linkage were found, suggesting that *Xrcc7* acts as a tumour suppressor of the T-cell lineage. Treatment of *scid* mice with gamma rays (1 Gy) at 24-48 h after birth led to 86% developing T-cell lymphoma with very short latency, with no other tumour type observed [G35]. Further, mice having both the *scid* defect and a knockout of the *Parp* gene (paragraph 77) also develop very high levels of T-cell lymphoma, suggesting that the carcinogenic effects of a partial defect in double-strand break rejoining is exacerbated by a reduction in repair capacity for other types of DNA damage [M6].

87. The assessment of radiation sensitivity of different tissues has shown that epithelial cells of the intestine and kidney in *scid* mice show a much greater radiosensitization than bone marrow cells from the same animals, compared to the relative sensitivity of these cells in normal animals [H28]. However, the additional sensitivity of *scid* epithelial cells effectively made these cells similar in overall sensitivity to *scid* marrow cells (e.g. at a dose of 2 Gy), since normal epithelial cells are more radioresistant than normal marrow cells. Irradiation of epithelial cells at low dose rate (16 mGy min^{-1}) altered the survival of *scid* cells much less than it did for normal cells [H28], as expected for this type of radiosensitive cell line (see paragraphs 49–50).

88. In addition to the *scid* mouse as a model of defective DNA double-strand break repair, knockout mice for the gene encoding Ku86 have been created [N21, Z9]. These mice weigh 40%-60% less than normal mice, and further weight loss occurred when newborn mice were gamma-irradiated at whole-body doses in excess of 0.25 Gy. Survival of Ku86-deficient mice irradiated at 2-4 months old with gamma rays was also compromised; doses of 3-4 Gy caused 50% mortality in two weeks. Ku86-deficient mice given 2 Gy gamma rays survived for up to 12 weeks but unlike normal mice showed severe hair loss within one month post-irradiation. Examination of tissues at four days post-irradiation showed severe injury to the GI tract at much lower doses than for the normal mice, along with atrophy of lymphoid organs [N21]. In these mice it was found that T- and B-cell development was arrested at an early stage, as in *scid* mice. The Ku70 subunit has been knocked out in mouse embryonic stem cells, and these cells are sensitive to gamma rays. As might be expected, Ku70-deficient cells are defective in DNA end-binding activity and in V(D)J recombination [G31]. Ku70-deficient mice have similar growth reduction and radiation sensitivity to Ku86-defective mice, but they also develop a high frequency of spontaneous thymic and disseminated T-cell lymphomas [L57].

89. Mice lacking DNA ligase IV do not survive beyond the late embryonic stage of development; ligase IV-deficient embryonic fibroblasts are hypersensitive to gamma rays but not to ultraviolet light and are defective in V(D)J recombination [F20]. Very similar defects have been found in mice and murine fibroblasts lacking the *Xrcc4* gene. Further, lethality in XRCC4- and ligase IV-deficient mouse embryos is associated with severe disruption of the development of the nervous system due to extensive apoptotic cell death [G33]. It is presumed that the enhanced sensitivity of neuronal cells relates to an inability to repair DNA double-strand breaks, although it is not known whether these breaks result from normal metabolism or a specific neuronal process (e.g. a recombination process) required for neuronal function. It is of considerable interest that a single person defective in DNA ligase IV has been identified (paragraphs 43 and 66); this person showed extreme sensitivity to radiation and developed acute lymphoblastic leukaemia at age 14, but was not severely impaired otherwise. Analysis of the ligase IV gene and protein in cells derived from this person showed that ligase function was not completely defective, presumably explaining their relatively normal developmental progress [R38]. This example of repair deficiency is important in revealing that mutations leading to partial activity of the gene product may be permissive for growth and development, but may have undesirable consequences including the possibility of cancer formation.

90. The search for human mutations in the *XRCC4, 5, 6,* and *7* genes among patients known to be compromised in immune functions is beginning to yield some candidates [C28, H26].

91. There is a difference between the "programmed" double-strand breaks generated in the V(D)J recombination process and those caused by radiation damage. The breaks at V(D)J sites have 5'-phosphorylated blunt ends and can be rejoined by a DNA ligase without further processing [S6], while radiation-induced breaks are often not directly ligatable because of extensive damage to the sugars and bases at the break sites (see Section I.B). Since, in the cell, the non-homologous end joining pathway is involved in repairing both types of break, it may be suggested that the exact structure (and possibly complexity) of the breaks does not influence their recognition by the proteins that initiate the rejoining process. However, it seems likely that the context in which the break occurs will influence its repair; in yeast cells, three proteins involved in modulating chromatin structure (Sir2, Sir3, and Sir4) interact with the Ku70 protein and have a role in DNA break rejoining

[T41]. A defect in any of the three *SIR* genes led to increased sensitivity to gamma radiation, providing the homologous recombination pathway was inactive. While the changes in chromatin structure brought about by the Sir proteins are not well understood, it is thought that they may make the broken DNA inaccessible to other DNA-modifying enzymes such as nucleases and thereby protect the damaged DNA from loss. It has also been found that the V(D)J recombination process is restricted to the G_1 phase of the cell cycle [S6]. It is likely that this rejoining process is also cell-cycle-controlled in the repair of radiation-induced double-strand breaks [L51]. Thus, the XR-1 line is highly sensitive and defective in radiation-induced double-strand break repair in G_1 but has nearly normal sensitivity and break repair in late S phase [G5]. One implication of this analysis is that there must be at least two major pathways for the repair of DNA double-strand breaks: the non-homologous end-joining pathway active in G_1 and at least one other pathway operating in other stages of the cell cycle.

92.   As noted in paragraph 70, there is evidence suggesting that the homologous recombination repair pathway is active in late S/G_2 stages of the cell cycle; additionally, chick cells defective in the *RAD54* gene were found to be more sensitive than normal cells in this part of the cycle. Chick cell lines were generated with defects in both the homologous recombination repair pathway and the non-homologous end-joining pathway (RAD54/Ku70-defective), and these showed increased radiation sensitivity and higher levels of chromosomal aberrations than either single defect [T42].

93.   Unlike mammalian cells, in lower eukaryotic organisms such as the budding yeast, repair of DNA double-strand breaks is mainly effected through homologous recombination processes. However, proteins similar to the mammalian Ku70/86 have recently been found in yeast, and these form a dimeric complex binding directly to DNA ends. DNA end-joining in yeast was shown to be impaired when the genes encoding the yeast Ku proteins are mutated, and this process is distinct from the rejoining mediated by homologous recombination. However, it is possible that some activities are shared by both homologous and illegitimate recombination pathways, since it was found that the *RAD50* gene product (paragraph 70) interacts with the yeast Ku proteins [M32]. Mutation of the gene encoding the smaller Ku subunit (*HDF1*) has also been found to lead to shortened telomeres, similarly to *TEL1* mutations (paragraph 64), suggesting that in addition to their end-joining role these proteins may help protect yeast chromosome ends [P21]. Intriguingly, both the Ku and Sir proteins are located at telomeres in undamaged yeast cells, but following the induction of DNA double-strand breaks they relocate to break sites in the genome [M54]. Loss of *HDF1* gene function alone does not, however, lead to radiation sensitivity or a measurable defect in DNA double-strand break repair in yeast, while the combined loss of *HDF1* and homologous-recombination gene functions leads to extreme sensitivity [B35, S50]. Additionally, while the loss of homologous recombination repair in yeast leads to an elevated frequency of chromosomal aberrations, the combined loss of these two pathways gives a reduction in chromosomal aberration frequency [F24]. This observation has been considered as evidence for the involvement of non-homologous end joining mechanisms in chromosomal aberration formation.

94.   While ataxia-telangiectasia cells are not defective in V(D)J recombination, a recurrent feature of the disorder is the appearance of T-cell clones with characteristic chromosomal rearrangements at sites of immunoglobulin and T-cell receptor genes [A7, B2, R4]. Elevated levels of recombination of T-cell receptor genes have been described in lymphocytes from ataxia-telangiectasia patients relative to those from normal individuals [L10]. ATM-deficient mice develop thymic lymphomas and die by 5 months (paragraph 62), but if the gene that normally causes programmed double-strand breaks leading to V(D)J recombination is inactivated, no lymphomas develop [L66]. This striking observation suggests that ATM normally suppresses aberrant recombination events.

95.   The rejoining of DNA double-strand breaks, produced enzymatically, has also been examined using small DNA molecules transfected into mammalian cells or exposed to cell extracts. These studies have shown that almost any sort of break end (with flush ends, or with complementary sequence at the ends, or with mismatched ends) can be rejoined by cellular enzymes [N7, P29, R1]. However, even with complementary ends, a fraction of the breaks are rejoined with a loss of sequence around the break sites. This mis-rejoining process was found to occur by a non-conservative recombination mechanism [T10] that appears to differ from the non-homologous end-joining pathway. The mechanism entails deletion of DNA bases between short (2-6 bp) direct repeat sequences, such that one of the repeats is also lost (Figure IV; see also paragraph 182). Occasionally the mis-rejoining can be more complex, with an insertion of DNA at the deletion site. Using substrates that attempt to model more closely DNA double-strand breaks produced by radiation damage, where the sequences at the break-ends are not complementary, breaks were shown to be rejoined by either this repeat-driven mechanism or by a process of blunting the ends before rejoining [M33, R18]. The repeat-driven process of mis-rejoining has been shown to be independent of the Ku proteins [M33]. Recent data with yeast cells reveal that this mis-rejoining mechanism prevails where both the homologous recombination and non-homologous end-joining pathways are knocked out [B35, B36]. This error-prone mechanism constitutes, therefore, a third pathway for the rejoining of double-strand breaks common to many organisms. Interestingly, in extracts from two ataxia-telangiectasia cell lines, the frequency of mis-rejoining by this mechanism was about 20 times higher than in extracts from normal cell lines [G3, N7]. The genetic basis of this short repeat-driven process is unknown, although in principle it may require similar enzymes to those that recombine adjacent homologous sequences within a chromosome. Adjacent sequences with homology of several hundred base pairs recombine in a process termed single-strand annealing in both mammalian cells and yeast; in budding yeast some of these recombination

events are *RAD52*-dependent, but also require the products of the *RAD1* and *RAD10* genes [O5]. *RAD1* and *RAD10* combine to give a single-stranded endonuclease required to snip off the overhanging DNA strands generated by the recombination of adjacent repeats. In mammalian cells, the homologues of *RAD1* and *RAD10* are the nucleotide-excision repair genes *ERCC1* and *XPF*, respectively [W52], and there is evidence that the *ERCC1* gene is involved in recombination-dependent deletion formation in mammalian cells [S85].

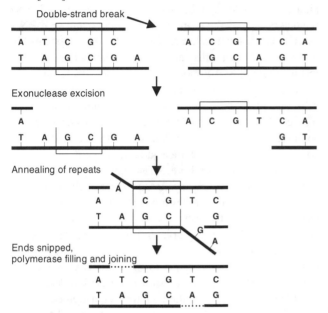

Figure IV. Direct-repeat end joining model.
Direct repeats (shown in boxes) are identified on either side of the initial double-strand break. Exonuclease removes opposite strand beyond repeats. Repeats are annealed, leaving extended tails (snip ends), which are removed. Strands are infilled by polymerase and ligated to give the deleted form with only one copy of the repeat.

96. Ataxia-telangiectasia cell lines were found to have elevated rates (by a factor of 30-200) of intrachromosomal recombination in an introduced DNA construct [M8]. These findings suggest that ataxia-telangiectasia cells have a general disturbance in recombination processes, and that the cell extract studies highlight those events leading to sequence loss and rearrangement (see Section IV.B). It seems likely that an elevated level of non-conservative recombination would also lead to an increase in mutation frequency, provided that the mutation system (i.e. target genomic region) allows detection of relatively large deletions and rearrangements. Evidence of an increased spontaneous frequency of mutations in ataxia-telangiectasia relative to normal persons has been found from direct analysis of their blood cells. Measurements made in ataxia-telangiectasia erythrocytes (loss of heterozygosity in the GPA system, paragraph 213) show loss mutations were elevated, on average, by a factor of 10, while ataxia-telangiectasia lymphocytes (*HPRT* gene mutation) show an average elevation of about 4 [C15]. In each system, however, there was a wide range of frequencies, from nearly normal to more than 100-fold higher than normal. These results are also supported by studies of transfection of small target genes

carrying breaks at specific sites (by endonuclease cutting) on shuttle vectors. The proportion of correctly rejoined target genes was lower in ataxia-telangiectasia than normal cell lines [C13, R5, T6]. It is unclear whether this recombination defect is directly related to radiosensitivity in ataxia-telangiectasia, but disturbance of recombination would tend to give genetic instability and ultimately may lead to cancer-proneness.

2. Radiosensitivity and the cell cycle

97. Mammalian cells vary in their radiosensitivity through the cell cycle, generally showing greatest resistance during the period of DNA synthesis (S phase) and least resistance during G_2/mitosis and late G_1 (where a long G_1 exists) [S37]. To account for this variation in response, different hypotheses were advanced, such as fluctuation in radioprotective substances in the cell and/or duplication of the genetic material in S phase [S38, T27]. However, a number of studies indicate that the fluctuation in response correlates to the ability of cells to repair radiation damage when irradiated in certain cell-cycle phases [I3, W25]. Additionally, it has more recently been found that certain DNA repair-defective cell lines have altered cell-cycle responses (see paragraph 91). These results suggest that the opportunity for repair of damage in relation to DNA synthesis or segregation is an important determinant of response (see paragraph 9). Accordingly, the ability of the cell to optimize repair and to avoid the interaction of damage with other cellular processes by modulating the cell cycle after irradiation may be crucial to recovery and genome stability.

98. It is now well established that some radiosensitive (*rad*) cell lines in yeasts are defective primarily in regulation of the cell cycle. The paradigm for this type of defect is the *rad9* line of budding yeast. Although identified many years ago by sensitivity to x rays and UV light, it was discovered more recently to be defective for a regulatory function (termed a checkpoint) in the G_2 phase of the cell cycle. The loss of this checkpoint allows *rad9* cells carrying DNA damage to proceed into mitosis and die. The normal *RAD9* gene product appears to respond only to damage, since *rad9* mutant cells have normal division kinetics when unirradiated, although they do show spontaneous chromosomal instability [W6, W7]. Irradiated *rad9* cells have been shown to be defective also in extended-G_1 arrest, indicating that the normal *RAD9* gene product operates at more than one transition point in the cell cycle [S9].

99. Numerous other checkpoint genes have now been identified, especially in lower eukaryotes such as yeasts, and homologues of many of these regulatory genes are present in human cells [C42]. It is clear, therefore, that all cells regulate their cycling in relation to a variety of signals to make sure that a proper progression of phases occurs (e.g. that mitosis does not occur until the S phase is finished) but also to ensure that the genome is intact before starting a new phase. Certain gene products, for example, the tumour suppressor p53, appear to be central to this monitoring process in mammalian cells [A21, C43], and the detailed workings of this regulatory network are currently the subject of intense research.

100. It has been known for many years that x irradiation of mammalian cells in extended-G_1 phase (slowly proliferating primary cultures or cells grown until they run out of nutrients ["plateau phase"]) induces a dose-dependent G_1 arrest [L7, N1]. This arrest is evident at doses as low as 100 mGy, but the duration of the arrest is reduced if the G_1 cells are allowed time to recover before restarting the cycle. Kastan et al. [K2] observed increases in the levels of the tumour suppressor protein p53 correlated to a transient G_1 arrest after relatively low doses (500 mGy) of gamma radiation in normal human haematopoietic cells. Cells that were mutant for p53 showed no G_1 arrest but retained the G_2 arrest found typically after irradiation (but see also paragraph 109). The apparent increase in p53 derives from a prolonged half-life of the protein (the protein half-life is normally only about 30 minutes). The link between p53 and G_1 arrest has been generalized for other cell types and for a variety of DNA-damaging agents [F4]. A likely role for p53, given its function as a transcriptional activator, is to activate the genes involved in negative growth control. Radiation-induced G_1 arrest was found to correlate to inhibition of at least two members of the compound family of proteins that controls cell-cycle progression (the cyclin/cdk proteins); this inhibition was dependent on p53 and was mediated by the p53-inducible kinase inhibitor p21 [D4].

101. The signal for p53-dependent G_1 arrest is likely to be DNA breakage; agents causing DNA double-strand breakage (ionizing radiation, bleomycin, DNA topoisomerase inhibitors) are very effective inducers, while UV light appears to cause p53 induction through the processing of base damage into breaks. The introduction of restriction endonucleases into cells has shown that DNA double-strand breakage alone is sufficient for this response [L14, N5]. Microinjection of broken DNA molecules into cell nuclei has suggested that very few double-strand breaks may be required for arrest, and only one or two may be sufficient [H29].

102. Ataxia-telangiectasia cells lack DNA-synthesis arrest (Section II.A.1) and G_1 arrest, and the DNA-damaging agents that strongly induce p53 are also the agents to which ataxia-telangiectasia cells are sensitive [L14, N5]. It has further been shown that p53 induction is reduced and/or delayed in irradiated ataxia-telangiectasia cells and suggested that the ataxia-telangiectasia gene product (ATM) is part of a signalling mechanism that induces p53 [K3, K6, L14]. A similar alteration of p53 response has been found after irradiation of cells derived from patients with Nijmegen breakage syndrome [J21]. ATM has been shown to physically associate with p53; it is responsible for a rapid phosphorylation of a specific residue at the N-terminal end of p53; this phosphorylation occurs within minutes of treatment of cells with ionizing radiation or radiomimetic drugs but not after UV light treatment [B59, C58, K43, S73]. These processes contribute to its increased half-life following irradiation and also lead to the association of p53 with other proteins involved in damage-signalling pathways. In mouse fibroblasts deficient in both ATM and p53 proteins (or ATM and p21), the loss of the G_1/S cell-cycle checkpoint following irradiation is no worse than for either single defect. This again suggests that

ATM and p53 operate in a common pathway of cell-cycle control in response to DNA damage. However, it seems that ATM may regulate only some aspects of p53-dependent responses; for example, the ATM defect leads to premature senescence in mouse fibroblasts, and in this case the combined deficiency of ATM and p53 [W47] or ATM and p21 [W49] alleviates the senescence.

103. The ATM protein has also been shown to interact directly with the proto-oncogene c-Abl [B48, S65], a protein kinase that itself interacts with other important cell-cycle regulators such as the retinoblastoma protein and with RNA polymerase II. When cells are irradiated, c-Abl is activated in normal cells, but this activation is absent in ataxia-telangiectasia cells. Additionally, cyclin-dependent kinases are resistant to inhibition by radiation in ataxia-telangiectasia cells, and this appears to be due to insufficient induction of p21 [B30]. A kinase named CHK1 that links DNA repair with cell-cycle checkpoints in yeast, has recently been found to be conserved in humans and mice and is dependent on ATM for its activity in meiosis. Like ATM (paragraph 62), CHK1 protein localizes along synapsed meiotic chromosomes, probably at the time that they are repairing DNA breaks. It is speculated that CHK1 acts to coordinate signals from both ATM and ATR (paragraph 65) to ensure the correct progression of meiosis [F17]. The activity of a second kinase, CHK2, which is homologous to the yeast RAD53 protein (paragraph 70), has also been shown to be ATM-dependent; this kinase is involved in the negative regulation of the cell cycle [M23]. It seems likely, therefore, that the ATM protein is involved in multiple signal-transduction pathways, by virtue of its interaction with a number of important regulatory proteins.

104. Alterations in the response of cell-cycle regulators in ataxia-telangiectasia cells may be involved in some aspects of the disorder but are unlikely to be responsible for the radiosensitivity of ataxia-telangiectasia. For example, p53-deficient cells have a tendency to be radioresistant rather than radiosensitive [L6], and SV40-transformed ataxia-telangiectasia cell lines in which p53 function is ablated still have the characteristic radiation sensitivity. It is more likely that any compromise of p53 response will affect cancer-proneness; both human patients (Li-Fraumeni syndrome) and $Tp53$-knockout mice show elevated levels of cancer [D3, M4]. Loss of one ($Tp53^{+/-}$) or both ($Tp53^{-/-}$) gene copies in mice also has significant effects on sensitivity to radiation-induced cancers, seen as a much reduced time for tumour onset. When $Tp53^{+/-}$ mice were irradiated with gamma rays to a dose of 4 Gy at 7-12 weeks, the tumour onset time was reduced from 70 to 40 weeks, while $Tp53^{-/-}$ mice showed a decreased latency only if irradiated when newborn (because at later times their frequency of spontaneous tumour development was already extremely high, masking the effect of irradiation). It is of interest that p53-deficient mice show a high incidence of thymic lymphomas, similar to ATM- and BRCA2-deficient mice (paragraphs 62 and 73), although it should be remembered that this result may be influenced by the fact that all of these mice were produced in the same genetic

background. In the $Tp53^{+/-}$ mice, the normal $Tp53$ allele was commonly lost in tumour cells; it was notable that in radiation-induced tumours there was both a high rate of loss of the normal allele and a duplication of the mutant allele [K11].

105. Inactivation of mitotic checkpoint genes has been found in tumour cells derived from both BRCA1-deficient and BRCA2-deficient mice (paragraph 73), suggesting that these defects cooperate in cancer progression. Conditional deletion of the $Brca1$ gene led to a low frequency and long latency for mammary tumours, but $Tp53^{+/-}$ mice carrying this $Brca1$ mutation showed a significant acceleration in mammary tumour formation [X7]. It may be significant that the phosphorylation of BRCA1 in response to DNA damage is mediated by ATM [C67], in view of the potential involvement of ATM in breast cancer (paragraph 137 et seq.). In BRCA2-deficient cells, mutations in $Tp53$ or genes controlling the mitotic checkpoint were implicated in promoting cellular transformation and the development of lymphomas [L64]. Centrosomes, which control the movement of chromosomes during mitosis, are abnormal in both BRCA1- and BRCA-2-deficient cells, leading to unequal chromosome segregation [T46, X8].

106. Mice that lack DNA-dependent kinase (scid phenotype, paragraph 82) accumulate high levels of p53 because of the presence of "natural" DNA double-strand breaks from unrepaired V(D)J recombination intermediates but show a typical G_1 arrest following irradiation. These mice develop thymic lymphoma with low incidence (15%) and long latency. The importance of p53 expression in protecting the animals from cancer was shown in mice that lack both DNA-dependent kinase and p53; these mice show prolonged survival of lymphocyte precursors and onset of lymphoma/leukaemia by 7-12 weeks of age (mice defective in p53 alone develop lymphoma to 50% incidence at 16-20 weeks). Cell lines derived from these double knockout mice were also about 10 times more resistant to gamma radiation than lines from scid mice. It was suggested that p53 may normally limit the survival of cells with broken DNA molecules and therefore that p53 loss promotes genetic instability [G23, N14].

107. Many of these p53-related responses may be associated with other functions of p53, in particular its role in radiation-induced apoptosis (programmed cell death, Section II.B.3). It is clear that DNA- break-inducing agents, including ionizing radiation, kill some cell types through a p53-dependent apoptotic pathway, although this pathway is not involved in the induction of apoptosis by other stimuli [C6, L13]. Alterations in the control of apoptosis may be linked to carcinogenesis, where a loss of p53 function may lead to the survival of precancerous cells, as has been shown in the induction of skin cancer by UV light [Z4].

108. It has also been known for many years that irradiated mammalian cells may arrest in G_2. The G_2 checkpoint is also controlled by specific cyclin/cdk proteins, and gamma radiation rapidly inhibits the kinase component (p34^{cdc2}) at

doses resulting in G_2 arrest [L12]. Conversely, p34^{cdc2} activation accompanies release of a radiation-induced G_2 block by drugs such as caffeine [H14]. In cultured cells, caffeine-induced release of the arrest decreases survival [B49] and increases the proportion of cells dying from apoptosis [B40]. However, it has also been found that caffeine treatment leads to an increased frequency of chromosomal aberrations [L40, T26], although the link between these two effects has been questioned [H18]. These observations suggest that the arrest promotes survival and may reduce the probability of genetic alterations. This view is supported by experiments with arrest-defective $rad9$ yeast cells, where the imposition of an artificial G_2 delay promotes survival [W6]. However, in some radiosensitive mammalian cell lines, including ataxia-telangiectasia, enhanced delay in G_2 is commonly linked to enhanced sensitivity [B3, B6]. It seems likely that because of repair defects these cells never progress further in the cycle and die (that is, the cell-cycle arrest is irreversible, because the damage signal is not reduced). The signal for this response could be unrepaired DNA breaks; recent experiments in recombination-deficient yeast carrying a single unrepaired double-strand break in an inessential DNA molecule show that this can be lethal to a majority of cells [B7].

109. Initially there was little evidence of a role for the p53 protein in G_2 arrest (paragraph 100), but more recent data with p53-deficient mice and cells have shown that p53 is required for correct control of entry into mitosis following DNA damage [B60, S74, S75]. Similarly, cells established from patients with germ-line $TP53$ mutations (Li-Fraumeni syndrome) have a consistent defect in G_2 response [G36].

3. Apoptosis: an alternative to repair?

110. Early studies of irradiated mammalian cells recognized that there are different forms of cell death [O1]. Attempts to correlate cell death with the formation of chromosomal aberrations concluded that a major cause of death in fibroblastic cells is genetic damage resulting in the loss of chromosome fragments [B22, C26, D8]. The most conclusive experiments were those of Revell and co-workers, showing a close correlation between loss of chromosome fragments, observed as micronuclei, and cell death in individual diploid Syrian hamster BHK21 cells [J13]. It is envisaged in these cells that genetic imbalance, perhaps due to the loss of specific essential genes, results in death once the pre-irradiation levels of their gene products have decayed. However, cells from a number of different developmental lineages are known to undergo rapid cell death, often termed interphase death, after irradiation [H16, P18, Y4].

111. It is now recognized that the balance between cell proliferation and cell death is crucial to the correct development of organisms, and that cell deaths in many tissue lineages are programmed by a genetically-controlled process known as apoptosis. This process was first described as a characteristic breakdown of cellular structures, including DNA degradation [K14, W33]. Apoptosis is initiated through specific cell surface receptors in response to external developmental signals but

may also be induced by DNA damage. Radiation-induced interphase death is considered to be an example of apoptosis [A3, U14, Y3], although the occurrence of apoptosis has more recently been reported in cell lines dying at various times after irradiation [R8, Y5].

112. Initiation of rapid apoptosis in normal cells by radiation appears to be dependent on the induction of the p53 protein [C6, L13], although delayed apoptosis may occur in cell lines deficient in wild-type p53 protein [B27, H37]. While p53-dependent apoptosis does not involve the same receptors as developmentally-regulated apoptosis, thereafter the molecular pathways converge to a common cell death programme. The mechanism of signalling by p53 is not well understood, although p53 can bind DNA at specific sites ("response elements"), and this binding is strongly stimulated by DNA ends, short regions of single-stranded DNA, and short mismatched DNA segments [J9, L28]. The importance of this DNA binding is demonstrated by the fact that the vast majority of TP53 mutations identified in tumour cells occur in the part of the gene specifying DNA-binding activity [H25]. Relatively stable binding to DNA (half-life >2 h) can allow a variety of subsidiary processes to occur, such as the activation of specific genes. Because p53 has the properties of a transcriptional activator, it may initiate apoptosis by switching on a specific set of genes. Alternatively there is evidence that in some situations p53 may act in the opposite way, repressing genes concerned with cell survival [C23, C32].

113. A second pathway of radiation-induced apoptosis is proposed to work through a membrane-associated signalling system responding to a variety of extracellular stresses. Sphingomyelin and possibly other sphingolipids at the membrane respond to stress by rapidly releasing ceramide, which accumulates and activates protein kinases to initiate cell-cycle arrest and apoptosis. It has been suggested that ceramide acts as a "biostat", measuring and initiating response to cell stresses in the same way as a thermostat measures and regulates temperature [H34]. Radiation can activate components of the ceramide pathway in isolated membranes, suggesting that DNA damage is not required [H35]; however, this suggestion has been challenged [R39], and recently it has been found that another function of the ATM gene is to modulate ceramide synthesis following radiation-induced DNA damage [L67]. Lymphoblasts from patients with the Niemann-Pick disorder have an inherited deficiency in acid sphingomyelinase, an enzyme hydrolysing sphingomyelin to give ceramide, and fail to show ceramide accumulation and apoptosis following irradiation. Comparison of mice with a knockout in the gene for acid sphingomyelinase with Tp53-knockout mice shows that the initial stages of the sphingomyelin-dependent apoptotic pathway are distinct from those of p53-mediated apoptosis. These two pathways may be prevalent in different tissues; sphingomyelin-dependent apoptosis prevails in the endothelium of the lung and heart and the mesothelium of the pleura and pericardium, while thymic apoptosis is especially dependent on p53. The acid-sphingomyelinase-deficient mice develop normally, showing also that this apoptotic pathway is distinct from develop-

mentally-regulated apoptosis [S55]. In tumour cell lines where TP53 is mutated, it has also been shown that resistance to radiation correlates to loss of ceramide accumulation [M1].

114. Whether or not the cell goes into apoptosis in response to a given radiation dose is thought to depend on the availability of other proteins that promote or inhibit further steps in the programme. The details of these steps are still being worked out, but it is clear that a diversity of factors influences apoptosis. For example, cells may become susceptible to radiation-induced apoptosis through functional loss of the retinoblastoma (Rb) protein, a cell-cycle regulator, by either mutation of the RB gene or the expression of viral oncoproteins that inactivate Rb [S42]. Conversely, the gene product of the BCL-2 oncogene, first identified at a site of chromosome translocation in B-cell lymphomas, can block radiation-induced apoptosis [R10]. It has been found that BCL-2 and related proteins localize to cellular membranes, especially of mitochondria, where permeability transitions are important in regulating the apoptotic process [G25]. The competitive formation of protein complexes by members of the BCL-2 family is suggested to control susceptibility to apoptosis; some of these proteins are antagonists, like BCL-2, while others are agonists, and the relative proportions of these proteins determine whether a cell will respond to an apoptotic signal [K5, K19, W34].

115. The extent of apoptosis for a given radiation dose does not appear to be affected by changes in the dose rate or by dose fractionation [H38, L26, L32], although there has been a recent report of a reduction in apoptosis at very low dose rates from gamma rays (<1.5 mGy min^{-1}) [B66]. There is an effect of cell-cycle stage on the extent of apoptosis; in contrast to the "classical" pattern of resistance (Section II.B.2), mammalian cells appear to be resistant in G_1 and sensitive in S phase [L26, L33]. In cells that are relatively resistant to apoptosis, the oxygen enhancement ratio was found to be similar for apoptosis and for clonogenic cell survival [H36]. The relative biological effectiveness (RBE) of fast neutrons for apoptosis in thymocytes was found to be 1 [W16], in accordance with earlier measurements of interphase death in thymocytes and lymphocytes [G15, H19]. However, for intestinal crypt cells of mice, where there is a highly sensitive subpopulation of cells in the stem-cell zone, the RBE for apoptosis was 4 for 14.7 MeV neutrons and 2.7 for 600 MeV neutrons [H38, H39]. It is not known why the RBE varies for different tissues; both thymocytes and intestinal cells show p53-dependent apoptosis [L13, M39], but it is possible that their relative abilities to repair radiation damage differ (see paragraph 87). The proportion of cells dying from apoptosis, relative to other forms of death, was shown to be constant at different doses (range 2-6 Gy) in cells where the main cause of death was apoptosis [L26]. This result does not suggest, at least for doses in excess of 2 Gy, that cells switch on the apoptotic pathway because their repair systems are unable to cope with the level of damage. In human lymphocytes, significant induction of apoptosis by gamma rays could be measured at doses as low as 50 mGy, and some evidence was found that individuals vary in their apoptotic response [B41].

116. It has been suggested that alterations in the p53-dependent apoptotic pathway play an important part in the sensitivity of ataxia-telangiectasia [M38] and Fanconi's anaemia [R20, R34], but this view has not been supported by other studies [B51, D15, E15, K35, T43]. It has been found that p53-mediated apoptosis is normal in irradiated *Atm* knockout mice [B50], but that *ATM* is involved in the regulation of the ceramide-dependent pathway [L67]. Further study of apoptotic responses, especially p53-independent pathways, are required to clarify these responses, preferably in primary cells from these disorders.

4. Summary

117. A link between radiation sensitivity and immune system defects has been revealed from studies of DNA repair genes; the same gene products are used to assemble functional immune genes and to repair radiation-induced DNA breaks by non-homologous end joining. Cells and animals defective in genes in this pathway are extremely sensitive to ionizing radiation and lack low-dose-rate sparing. While non-homologous end joining seems to be the main mechanism in mammalian cells for DNA double-strand break rejoining, at least two other pathways exist, and in lower organisms double-strand breaks are mostly repaired by the homologous recombination pathway.

118. An important component of the non-homologous end-joining pathway, the *XRCC7* gene encoding the catalytic subunit of DNA-dependent kinase, is related in structure to the ataxia-telangiectasia gene product and has been shown to behave as a tumour suppressor. The mechanism of rejoining by this pathway and the relationship of the components of the pathway to other regulatory processes (such as cell-cycle control) are at present the subject of intense research activity.

119. It is clear that the fidelity of rejoining by the different repair pathways is an important determinant of the mutagenic consequences of DNA damage (see also Figure III). Homologous recombination repair is likely to be the only error-free pathway. However, since even this pathway depends on copying information from another DNA strand, the fidelity must relate to the quality of both the information copied and the enzyme (polymerase) used for the copying. Pathways of double-strand break repair based on illegitimate recombination (non-homologous end joining, repeat-driven rejoining) will commonly be error-prone, although in principle the non-homologous end joining pathway can be error-free where the nature of the damage allows this (e.g. repair of a double-strand break formed by the overlap of two single-strand breaks sufficiently separated to allow for templated repair). Repeat-driven rejoining leads to deletion formation (mis-rejoining), and may act as a back-up for damage that is not (or cannot be) repaired by the other pathways. It is still not clear which factors determine the

likelihood that a specific damage site will be repaired by a particular pathway [T44]. The complexity of damage may influence this question (Section I.B), as well as the possibility that radiation damage induces the activity of specific repair enzymes or pathways (Section III.B). In future work it will be important to establish the relative efficiencies of these different pathways at different doses and dose rates and in different cell types, so that mutagenic consequences can be predicted. The potential consequences of error-prone repair pathways at low radiation doses are explored further in Annex G, *"Biological effects at low radiation doses"*.

120. When growing cells are irradiated, they arrest their cell-cycle processes, apparently to allow time for repair to be completed satisfactorily. This arrest is part of a checkpoint function that monitors the physical state of DNA at different stages of the cycle. Cells that have lost a checkpoint may be as radiation sensitive as cells that have lost DNA repair capability. Many genes are involved in controlling the cell cycle and determining checkpoints. In mammalian cells, the p53 protein is important in response to DNA breakage and controls both arrest in the G_1 phase and one pathway of programmed cell death (apoptosis). Animals and humans (Li-Fraumeni syndrome) deficient in p53 show elevated levels of cancer; irradiation of p53-deficient mice has a marked effect on the latency period for tumour formation and gives a high incidence of thymic lymphomas. The ataxia-telangiectasia disorder has defects in cell-cycle checkpoint functions, and there is evidence suggesting that ATM and p53 operate in a common pathway of cell-cycle control in response to DNA damage.

121. Apoptotis (programmed cell death) is also induced by ionizing radiation, both through the p53-dependent pathway and a membrane-associated signalling pathway. The relative importance of these pathways varies in different tissues; also, the importance of apoptosis as a mechanism of cell death in response to radiation varies with the cell type and developmental stage. The relationship between apoptotic death and radiation injury differs from that for genetic death (loss of essential genes through damage to the genome). Changes in dose rate or dose fractionation do not appear to affect apoptotic responses, and the response of cell-cycle stage is the reverse of the pattern found for genetic death. While it is attractive to consider that in the face of excessive damage, radiation-induced apoptosis is an alternative to DNA repair, the evidence for this possibility is not convincing. Loss of ability to respond to apoptotic stimuli will allow the accumulation of cells that may carry genetic damage and can therefore be a cancer-promoting event [W8]. This phenomenon is seen, for example, in *Tp53*-knockout mice (paragraph 104) and in mice overexpressing the BCL-2 protein [M21]. In this respect, the correct functioning of apoptotic pathways can be viewed as a complementary mechanism to the repair of DNA damage, removing damaged cells from the population.

III. HUMAN RADIATION RESPONSES

A. CONTRIBUTION OF MUTANT GENES TO HUMAN RADIOSENSITIVITY

122. The term radiosensitivity is used to indicate an abnormally increased response to ionizing radiation of both the whole body (see Section II.A.1) and cells derived from body tissues. Further, in cells the radiation response may be measured in different ways. For example, the ability of cells to grow and form clones (cell survival) following irradiation is commonly used as a measure of sensitivity, but more sensitive assays based on chromosome damage may also be used. In the comparative measurement of cell survival, the terms D_0 (mean lethal dose calculated as the inverse of the slope for exponential survival curves represented semilogarithmically), D_{bar} (mean lethal dose calculated as the area under the survival curve in linear representation [K34]), or D_{10} (dose to give 10% survival) are commonly used.

123. The range of sensitivity of cells from supposed normal individuals has been measured in survival experiments following multiple doses of acute low-LET radiation. In two major studies with human diploid fibro-blasts, the sensitivity to dose was found to vary by a factor of about 2: D_0 range = 0.98-1.6 Gy (mean = 1.22±0.17 Gy) in 34 cell lines tested [C8] and D_0 range = 0.89-1.75 Gy (mean = 1.23 ± 0.23 Gy) in 24 cell lines [L8]. The degree of interexperimental variation for a given cell line was generally small ($\leq 20\%$), although in one of these studies cell lines derived from a mother and daughter showed survival curves that could also vary by a dose factor of about 2 on repeated testing [L8]. However, it was concluded that neither cell culture conditions (including cloning efficiency and population doublings) nor age and sex of the donor were correlated to the observed differences in radiosensitivity [C8, L8]. This suggests that unknown genetic factors that affect radiosensitivity vary in the cells of normal individuals.

124. More recently it has also been possible to measure cell survival with peripheral blood T-lymphocytes, irradiated as resting (G_0 phase) cells, to assess the range of radiosensitivity in humans. G_0 lymphocytes tend to have a more curvilinear response to dose than cycling fibroblasts and, on average, a higher survival at low doses than fibroblasts. Early-passage fibroblast survival data generally show a good fit to a simple exponential, while the G_0 lymphocyte data are better fitted with models including a dose-squared term, such as the quadratic ($-\ln S = \alpha D + \beta D^2$). Using D_0 as a measure, the range of sensitivity to dose of lymphocytes from different individuals can also vary by a factor of about 2 [E7]; however, given the shape of the survival curves, D_0 is difficult to measure accurately, and the range is generally much less than 2 using measures such as D_{bar} or D_{10} [E7, G17, N4].

125. No correlation was found between the measured sensitivities of fibroblasts and lymphocytes from the same individual [G17, K23]. A similar lack of correlation was found between lymphocyte and fibroblast responses of pretreatment cancer patients, whether the fibroblasts were cycling or in the plateau phase, suggesting that these differences in response were not cell-cycle-dependent [G11]. The differences between fibroblasts and lymphocytes may reflect several other possibilities: modes of death may differ, cell-type-specific factors may affect the expression of genes that modify radiation response (especially after growth in culture), and/or there may be unknown variables in assay conditions. In support of the last possibility, Nakamura et al. [N4], using donors from one ethnic group, found a similar (relatively small) variation in the mean x-ray dose to kill 28 samples of T-lymphocytes from one individual (D_{10} = 3.66±0.21 Gy), very similar to results with samples from 31 different individuals (D_{10} = 3.59±0.18 Gy). Elyan et al. [E7] reported similar data with donors from three ethnic groups, although with a greater overall range of variability. In support of differences in cell-type-specific factors, other cell types such as keratinocytes also show survival that is significantly different from that of fibroblasts taken from the same individuals [S39].

126. The use of experimental conditions promoting recovery from irradiation, such as low-dose-rate irradiation, will, it is suggested, expand the range of sensitivities to dose shown by normal human cells. The exposure of non-growing fibroblasts from 14 different normal individuals to tritiated water (dose rate = 8.5-100 mGy h^{-1}) expanded the range of sensitivity to dose by a factor of more than 3 [L35]. Similarly, Elyan et al. [E7] found that low-dose-rate irradiation (9.8 mGy min^{-1}) of human G_0 lymphocytes from 19 individuals expanded the range of sensitivity to dose by a factor of about 4. When a higher dose rate (28.5 mGy min^{-1}) was used, however, there was little difference between the range of sensitivities measured for fibroblasts and lymphocytes and the range with a dose rate of 4.55 Gy min^{-1} [G11]. The latter result probably reflects incomplete recovery during the low-dose-rate irradiation period [E7, G11].

127. Other assays of cellular sensitivity have been developed based on the measurement of responses such as the growth of cells after single doses of radiation. A "growback" assay following gamma irradiation of lymphoblastoid cells, developed by Gentner and Morrison [G4], detected a wide range of sensitivity in the normal (asymptomatic) population. Of 270 lymphoblastoid lines tested, about 5% of normal lines showed hypersensitivity after acute irradiation, as measured by their overlap with the responses of ataxia-telangiectasia homozygotes, while on the same criterion about 12% were hypersensitive when assayed using chronic irradiation [G4, G13]. However, a wider range of sensitivity was detected using chronic irradiation, and rigorous statistical tests found that the proportion of lines showing a significantly hypersensitive response (p = 0.05) was 5%-6% under both irradiation conditions [G14].

128. There have been claims that a number of disorders with an increased incidence of cancer are hypersensitive to x rays, including retinoblastoma, basal-cell naevus syndrome, Gardner's syndrome, and Down's syndrome [A6, M2, W3]. The main method of determining sensitivity differences has been cell survival assays with cultured fibroblasts (as shown in Figure II), in which small numbers of cell cultures derived from normal individuals are compared with those of one or more derived from the patient(s). Unfortunately, such studies have not always given consistent sensitivity differences for the same disorder tested in different laboratories. In part this may be due to the large range of sensitivity found among normal individuals; if the normal lines used in a given laboratory are in the higher part of the survival range, a disorder showing relatively low survival may be classified as sensitive even though it falls within the overall normal range. However, it has been concluded from a statistical analysis of many published survival curves of human fibroblasts that the radiosensitivity of cells from some disorders, calculated as D_{bar}, can be discriminated from that of normal cells [D7]. The disorders include ataxia-telangiectasia homozygotes and heterozygotes, Cockayne's syndrome, Gardner's syndrome, Fanconi's anaemia, and 5-oxoprolinuria homozygotes and heterozygotes.

129. A number of studies had previously suggested that cells from individuals heterozygous for the disorder ataxia-telangiectasia (carriers with one mutated gene, such as the parents of those showing the full symptoms) are more sensitive than normal to the lethal effects of irradiation [A6, A14, C33, K17, P11]. Additionally, ataxia-telangiectasia heterozygotes show an average level of dose-rate sparing that is intermediate between that of lines from ataxia-telangiectasia homozygotes (paragraph 49) and normal persons [P12, W18, W21]. However, even under low-dose-rate conditions, where a number of carriers have been tested, it is evident that their sensitivity range overlaps that of the most sensitive normal individuals [L35, W18]. Survival assays on cells derived from human sources cannot therefore unequivocally distinguish carriers of a mutated AT gene from normal individuals. Similarly, it has been found that protracted radiation exposure will not readily distinguish other disorders suggested to have minor degrees of hypersensitivity [L35]. Inbred mouse strains may show less variation in response from individual to individual, and mice heterozygous for a defective *Atm* gene were found to show a significant reduction in lifespan following irradiation (4 Gy at 2-4 months of age), as well as premature greying of coat colour, compared to normal littermates [B61].

130. The finding of a range of sensitivities to radiation among supposedly normal persons implies that in addition to those with increased sensitivity, there are individuals with greater radiation resistance than average. There is little evidence, however, for specific disorders associated with radiation resistance. It has been noted that cell lines derived from mice deficient in the p53 tumour suppressor are more resistant to radiation than normal (paragraph 104), although this response may be cell-type-dependent [B43]. Similar

studies on fibroblasts derived from patients with the Li-Fraumeni syndrome, defective in p53, highlighted a family carrying a germ-line *TP53* gene mutation in codon 245. Cells from family members showed a radioresistance (D_0 value) that was increased by a factor of 1.2 [B44], compared to an approximately 1.5-fold increase for the p53-deficient mice [L6].

131. There is evidence from cell survival assays that radiation sensitivity at the cellular level may correlate with tissue response of cancer therapy patients. In a study of 811 patients treated with radiotherapy for breast cancer, five showed severe skin responses. The survival of fibroblast cell lines derived from these five patients was compared to six lines derived from women with normal radiotherapy responses (also from the same group of 811 patients). The sensitivity of the severe-response patients ($D_0 = 0.97\pm0.11$ Gy) was significantly higher than that of the normal group ($D_0 = 1.16\pm0.08$ Gy; see paragraph 123) [L39]. In a prospective study of 21 head-and-neck cancer patients it was found that the sensitivity of their cultured fibroblasts, but not their lymphocytes, correlated with whole-body late effects [G12]. Similarly, a study of fibroblasts derived from 10 radiotherapy patients showed a correlation between cellular sensitivity and late tissue reactions [B42]. The survival of lymphocytes from breast cancer patients suffering severe reactions to radiotherapy has also been studied in conjunction with ataxia-telangiectasia heterozygotes and homozygotes [W21]. Using high-dose-rate irradiation (1.55 Gy min^{-1}), the survival of lymphocytes from severe-response patients could not be distinguished from that of normal-response patients (or other normal controls), but with low-dose-rate irradiation (9.8 mGy min^{-1}) the cell survival of severe-response patients and ataxia-telangiectasia heterozygotes was similar and significantly different from that of normal individuals. Using the growback assay (paragraph 127), lymphoblastoid cell lines derived from newly presented, non-selected cancer patients showed a similar range of radiosensitivities to the normal lines, while patients already known to show some adverse reaction to radiotherapy showed a higher proportion with extreme sensitivity [G14].

132. Assays based on chromosomal damage have commonly shown a higher degree of sensitivity in the detection of differences between individuals. For example, the genetic disorders ataxia-telangiectasia, Bloom's syndrome, and Fanconi's anaemia were described as "chromosomal breakage syndromes" in early studies, showing significantly increased frequencies of chromosomal breaks and exchanges after irradiation [H24]. Chromosomal responses after low-dose-rate irradiation have also been used to examine the sensitivity of ataxia-telangiectasia heterozygotes [W13] and breast cancer patients who react severely to radiotherapy [J12]. Lymphocytes from 5 out of 16 patients (31%) sustaining severe reactions to radiotherapy had dicentric yields outside the control range [J12].

133. An assay based specifically on the chromosomal sensitivity of lymphocytes to irradiation in G_2 has been successful in detecting significant differences between

individuals carrying various cancer-prone disorders. The work of Sanford et al. is based on the measurement of chromatid breaks and gaps a short time after irradiation of lymphocyte cultures; this assay detected enhanced sensitivity in fibroblasts from a number of disorders, including ataxia-telangiectasia [T25], Fanconi's anaemia, Bloom's syndrome, Gardner's syndrome, basal-cell naevus syndrome, xeroderma pigmentosum, and familial polyposis [P2, S5]. In this G_2 assay, cells from cancer patients and tumour cells showed an even higher frequency of chromatid damage [H21, P3]. This assay has some technical difficulties; extensive tests in another laboratory using identical conditions failed to reproduce many of the results, which led to a modified assay that appears to be more encouraging [S28].

134. The modified G_2 chromosomal assay devised by Scott et al. [S27] has been used to assay the sensitivity of lymphocytes from 74 normal individuals, 28 ataxia-telangiectasia heterozygotes, and 50 breast cancer patients. The ataxia-telangiectasia heterozygotes were clearly distinguished in sensitivity from the majority of normal individuals, but about 9% of the normal individuals showed hypersensitivity. Human fibroblast cultures show a similar result, with four ataxia-telangiectasia hetero-zygotes distinguished from seven normal individuals [M47]. The range of sensitivities of lymphocytes from breast cancer patients was much greater than the range from normal individuals (42% overlapped the ataxia-telangiectasia heterozygote range). It was proposed that in addition to those mutated genes already identified with familial breast cancer, a number of other genes of lower penetrance may also predispose to breast cancer. It was suggested that these genes may be involved in the processing of DNA damage [H21, S27]. Family studies showed that the lymphocytes of first-degree relatives of sensitive individuals were significantly more sensitive to x rays than expectation; segregation analysis of family members (95 individuals in 20 families) suggested that a single gene accounts for most of the variability, but that a second rarer gene with an additive effect on radiosensitivity may also be present [R36].

135. It has also been found that a subset of patients with common variable immune deficiency (CVID) showed enhanced sensitivity to G_2 chromosomal damage [V1]. Among disorders with primary immune deficiency, CVID and ataxia-telangiectasia have the highest reported incidence of tumours, and both disorders show similar types of tumours, suggesting a common risk factor. This finding may relate to the link noted above (Section II.B.1) between immune system dysfunction and radiosensitivity. However, it is clear that the G_2 chromatid assay will detect enhanced levels of damage in cells showing a normal response in survival assays (e.g. xeroderma pigmentosum cells), and it may be argued that it is primarily a means of detecting genetic instability.

136. Another method based on alterations in the sedimen-tation of human lymphocyte nuclei is claimed to detect particularly sensitive individuals [S4]. When the repair of damage is measured by return to normal sedimentation properties after irradiation, it was possible to detect individuals sensitive to radiotherapy and patients with autoimmune disease [H5]. Further, patients showing post-therapy complica-tions were found to be more sensitive by this sedimentation assay than those without complications [D2].

137. It has been suggested that individuals heterozygous for the ataxia-telangiectasia defect are also at increased risk of cancer. Swift et al. [S13] compared prospectively the incidence of cancer in 1,599 relatives of ataxia-telangiectasia patients in 161 families in the United States and found an increased relative risk of 3.8 in men and 3.5 in women for all cancers, and an increased relative risk of 5.1 for breast cancer in women. If borne out, this predisposition would contribute significantly to the cancer incidence in the general population: heterozygotes are much more frequent than ataxia-telangiectasia homozygotes (both genes mutant) and from these data could constitute as much as 5%-8% of all adults with cancer [S12]. However, commentary following the publication of the work of Swift et al. [S13] questioned two aspects of their findings. First, bias in the control group was suggested [B24, K22, W14], although this was contested by Swift et al. [S41]. Second, against a background radiation level of 1 mGy a^{-1}, it was considered surprising that an increase in breast cancer could be detected following diagnostic x-ray procedures of ≤ 10 mGy [B24, L25]. Differences in dose rate might influence the effectiveness of these two sources, and in other studies an acute dose as low as 16 mGy gave a significant increase in breast cancer in normal women aged 5-9 years at exposure [M26]. However, it is clear that a relatively large difference in sensitivity of the ataxia-telangiectasia heterozygotes to the acute diagnostic procedures would be required to attribute the observed cancer incidence to deleterious effects of one mutated copy of the ataxia-telangiectasia gene.

138. Genotyping of 99 ataxia-telangiectasia families using markers tightly linked to the *ATM* gene showed that 25 of 33 women with breast cancer were heterozygous compared with an expectation of 15 of 33 (relative risk = 3.8; 95% CI: 1.1-7.6) [A18]. Two smaller European studies of cancer incidence among ataxia-telangiectasia heterozygotes tend to support the data from the United States, giving an overall relative risk for breast cancer of 3.9 (CI: 2.1-7.2) [E4]. The average relative risk for other cancers was lower at 1.9, with the European studies showing no statistically significant increase over controls. Using these combined data, the proportion of breast cancer cases due to ataxia-telangiectasia carriers would be about 4%. This small proportion is consistent with an inability in two studies [C36, W31] to detect linkage to AT gene markers in breast cancer families (i.e. on the basis of the risk estimates and gene frequency, most cases of familial breast cancer will be caused by other genes) [E4].

139. The identification of the gene mutated in ataxia-telangiectasia (*ATM*, paragraph 56) has led to further attempts to correlate this gene defect with breast cancer incidence or with severity of response to radiotherapy. In a

study of *ATM* mutations in 38 sporadic breast cancers, only rare polymorphisms were detected, none of which would lead to truncation of the gene (paragraph 57), giving no evidence for an increased proportion of ataxia-telangiectasia heterozygotes in breast cancer patients [V8]. The same workers also looked at 88 unrelated primary breast cancer cases from families previously associated with cancer susceptibility and found three *ATM* mutations (3.4%); this number was considered to be higher than expected by chance, but these mutations did not necessarily segregate with cancer incidence in the families [V9]. Screening a further 100 breast cancer cases from families with a history of breast cancer, leukaemia, and lymphoma revealed only one mutation, consistent with minimal involvement of the *ATM* gene [C65]. Fitzgerald et al. [F21] detected heterozygous mutations in 2 of 202 (1%) healthy women with no history of cancer, compared to 2 of 410 (0.5%) patients with early-onset breast cancer, consistent with a lack of association. Similarly, in a study of 18 families associated with a high incidence of breast and gastric cancers, only one *ATM* mutation was found, and this did not cosegregate with the gastric cancer in the family [B58]. In a study of 41 breast cancer patients showing marked normal tissue response to radiotherapy, one was found with a heterozygous mutation in the *ATM* gene (2.4%), compared with none in a comparable number of control patients [S69]. The conclusion that mutation of *ATM* is not a major cause of radiotherapy complications was supported by studies of smaller numbers (15-16) of breast cancer patients showing severe normal tissue damage following radiotherapy, in whom no *ATM* mutations typical of those in heterozygotes were found [A22, R31]. In contrast, a relatively high frequency of *ATM* mutations has been found for a group of prostate cancer patients (3 of 17, 17.5%) with severe late responses to radiotherapy [H47]. The majority of these studies suggest that heterozygosity for the *ATM* gene is not an important cause of breast cancer susceptibility or severe response to radiotherapy, but most do not have sufficient numbers of patients to exclude completely a role for ataxia-telangiectasia heterozygotes.

140. The association of certain forms of lymphocytic leukaemia with mutations in the *ATM* gene has already been noted (paragraph 59). Individuals with less severe mutations in the *ATM* gene, allowing some expression of ATM protein (paragraph 57), may also be at greater risk of developing breast cancer [S78]. It has also been observed that loss of heterozygosity (Section IV.B) at chromosome 11q23 is a frequent occurrence in breast carcinomas [C27, H15]. At least two regions of 11q23 are commonly deleted, and one of these includes the *ATM* gene. Other common human malignancies such as lung, cervical, colon, and ovarian carcinomas, as well as neuroblastoma and melanoma, also show an association with loss of heterozygosity at 11q23 and may include the *ATM* gene region [R9]. In a recent study, 40% loss of heterozygosity was found for markers of *ATM* in cases of sporadic invasive ductal breast carcinoma; in the same tumours, markers for the *BRCA1* and *BRCA2* genes showed 31% and 23% loss of heterozygosity, respectively. Loss of heterozygosity of *ATM* correlated with higher grade tumours and a younger age at diagnosis [R32].

141. It is clearly important to detect individuals in the human population who are hypersensitive to radiation and to understand the connection between radiosensitivity and cancer-proneness. While there are a number of indicators of this hypersensitivity, it has been difficult on the basis of cell survival assays to distinguish normal individuals from those carrying mutant genes giving intermediate levels of sensitivity, when the full sensitivity range of normal cells is taken into account (paragraph 128). In the case of ataxia-telangiectasia heterozygotes, newer assays based on G_2 chromosomal sensitivity (paragraphs 133-134), DNA damage levels measured by the Comet assay [D22], enhanced arrest in G_1 [N2, N3], or enhanced arrest in G_2 [H6, L4, L5] all show more promise, but it is still not clear whether any of the assays will detect these heterozygotes exclusively. The discovery of the *ATM* gene (paragraph 56) has made the task of detecting heterozygotes simpler, but because of the large size of this gene it is still a difficult task to screen large numbers of individuals for defects (see paragraph 139). Since many other genes are known to influence cellular radiosensitivity [Z13], it is likely that the molecular cloning of these genes will have a substantial impact on the ability to determine the importance of genetic predisposition to human radiation risk. Indeed, preliminary studies on nine recently-cloned repair genes including *XRCC1* and *XRCC3* (paragraph 78) showed that relatively common alterations in gene sequence exist in normal individuals. An average of 14 percent polymorphic sites yielding protein sequence variations was found in samples of 12-36 individuals, including some individuals who were homozygous for the variant site [M53, S84]. While the functional significance of these variations has yet to be established, their potential to reduce DNA repair capacity may influence individual response to radiation sensitivity and cancer susceptibility.

B. INFLUENCE OF REPAIR ON RADIATION RESPONSES

142. It is known that DNA repair processes influence the sensitivity of mammalian cells and organisms to radiation. If recovery from radiation damage is compromised, as in cell lines defective in double-strand break repair, the slope of the cellular survival curve is increased. Also, while normal cells show a "shouldered" curve, the repair-defective cells commonly show a loss of this feature. In simplistic terms, it is envisaged that repair processes in normal cells increase the chances of survival, but that as an acute dose increases, the amount of damage temporarily saturates the repair capacity of the cell [G6]. There is very good evidence in yeast cells that these aspects of survival are associated primarily with the repair of DNA double-strand breaks [F3]. It has been possible to make this correlation in yeast, because DNA double-strand breaks can be measured accurately, and mutant cells that are temperature-sensitive for double-strand break repair are available. Thus, as the post-irradiation temperature is altered, it is possible to see a concomitant alteration in double-strand break repair and survival. Unfortunately, it

has proved technically difficult to establish similar quantitative correlations in mammalian cells, which have a much larger genome than yeast.

143. Operationally defined measures of cellular recovery, termed potentially-lethal damage repair or sublethal damage repair, have been used for some years in cellular "repair" studies. Potentially-lethal damage repair is measured by the recovery found when cells are held in a non-proliferative state after irradiation, before respreading into fresh growth medium. Sublethal damage repair defines the recovery seen when survival is measured after splitting the dose over a suitable interval. While it has not been clear whether potentially-lethal damage repair and sublethal damage repair measure the same underlying repair processes, it is notable that both are absent in yeast single-gene mutants lacking DNA double-strand break repair (*rad52* series) [R2, R3]. It has also been argued from kinetic data that double-strand break repair is responsible for potentially-lethal damage repair in yeast [F3]. Similarly, with low-dose-rate irradiation, there is good evidence in yeast cells that DNA double-strand break repair is responsible for the dose sparing observed [F3]. Mammalian cell lines with defective double-strand break repair as well as ataxia-telangiectasia cells have also been shown to lack potentially-lethal damage repair (xrs, XR-1, ataxia-telangiectasia) or low-dose-rate sparing (xrs, ataxia-telangiectasia, irs2) (see Section II.A.2) [T30]. A correlation has recently been found for human normal and ataxia-telangiectasia cells, irradiated at high or low dose rates, between the amount of DNA double-strand breaks remaining after repair and radiosensitivity (measured as cell survival) [F18].

144. Mammalian cells irradiated with densely ionizing radiation, such as alpha particles, show both a more sensitive response (increased slope of survival curves) and, commonly, the loss of the curve shoulder region. Thus, characteristic increases in RBE are found as the density of ionization (LET) increases. This may well arise from the LET-dependence of damage complexity [A23, G28, G39] (see Section I.B), and refined measurement of DNA breakage following both x irradiation and high-LET particles has shown that there is a comparative excess of small fragments for the high-LET radiations [L53, R26]. However, several studies of mammalian cells using conventional biochemical techniques show that there is little or no difference in the numbers of DNA double-strand breaks induced by low- and high-LET radiations (RBE = ~1) but that fewer of the breaks induced by high-LET radiations are repaired [B52, C53, J4, L52, P27, P28, P37, R27]. Using these methods, a good correlation has been found between the relative number of non-rejoined DNA breaks and the RBE for mammalian cell inactivation over a wide range of LET [G32, R25]. Plasmid DNA, irradiated to give the same amount of damage by gamma rays or by alpha particles, showed much less repair of the alpha-particle damage when exposed to mammalian cell extracts [H41]. It is suggested that damage induced by high-LET radiations is more complex, because of the increased local clustering of ionizations, and therefore less repairable than low-LET radiation damage (see Section I.B). This finding is supported

by measurement of the rejoining kinetics of large fragments of cellular DNA, where rejoining was generally slower after irradiation with high-LET particles than with x rays [L62]. Additionally, using this method the proportion of mis-rejoined fragments increased with x-ray dose (10%-50% over the dose range 5-80 Gy), while similar experiments with high-LET radiation gave a constant 50% of fragments mis-rejoined at all doses tested [L72, K47].

145. In radiosensitive cells, specifically ataxia-telangiectasia cells [C10] and the xrs series [T7], the increase in RBE with LET is very much reduced. This result would be expected if the damage caused by densely ionizing radiation is intrinsically more difficult for normal cells to repair than the damage caused by sparsely ionizing radiation. On this basis, radiosensitive cells show little alteration in effectiveness as the LET increases, because they are already inefficient at repairing low-LET radiation damage. This analysis suggests that the RBE-LET relationship is largely caused by the normal cell's ability to repair low-LET radiation damage.

146. Extensive studies in bacteria have shown that some repair processes are inducible by treatment with DNA-damaging agents [W11]. That is, DNA damage causes an increase in the expression or activity of repair enzymes, mitigating the effects of the damage. This process is part of a much wider series of so-called stress responses by which cells adapt to their environment. Knowledge of such inducible processes in mammalian cells is fragmentary, and in some cases the data are controversial. Evidence for the existence of inducible repair processes in response to ionizing radiation damage comes from several different types of experiment. These may be broadly categorized as (a) adaptive response of pre-exposed cells (see also [U2]), (b) refined analysis of survival in low-dose regions, and (c) direct molecular evidence for the inducibility of specific gene products.

147. Pre-exposure to low doses of tritiated thymidine (18-37 Bq ml^{-1}) or x rays (5-10 mGy) was found to decrease the frequency of chromosomal breaks in proliferating human lymphocytes irradiated subsequently with a higher dose (1.5 Gy) of x rays [O2, S34]. The aberration frequency is reduced to about 60% of cells not receiving pre-exposure [S30]. Pre-exposure to low concentrations of radiomimetic chemicals, such as hydrogen peroxide and bleomycin, can also reduce the effect of a subsequent high dose of x rays [C35, V4, W29]. This adaptive response is, however, stopped by 3-aminobenzamide, an inhibitor of poly(ADP-ribose) polymerase (paragraph 77), which is itself induced in response to DNA breakage by radiation [S34, W23]. These features of the adaptive response have led to the suggestion that low levels of DNA breakage act as a signal for a response mechanism leading to accelerated repair of radiation damage in mammalian cells.

148. Although increased resistance following low radiation doses has been observed in a number of laboratories, it occurs only in the lymphocytes of some individuals [B21, B26, H13, S22, S26]. Also, lymphocytes from several persons with

Down's syndrome [K16] and fibroblasts from homocystinuria patients [Z11] did not show the adaptive response. The adaptive response is usually found when lymphocytes are proliferating, not resting (G_0) [S31, W15], although there are data suggesting that pre-irradiation in G_0, followed by a challenge dose in G_2, gives increased resistance [C63, K16]. These observations have led to suggestions that the increased resistance is a result of cell-cycle perturbation in the lymphocyte populations of some individuals following the pre-exposure (i.e. the cells sampled following the second dose are in a more resistant phase of the cycle than if no pre-dose had been given). This idea has received some experimental support [A9, W27], but recently an adaptive response for chromosomal and cell-killing responses has been found in cells other than lymphocytes when irradiated in proliferating or non-proliferating states [I1, L27, M24, S23]. Thus, the causes of the adaptive response remain controversial, but further work with different cell types has outlined the responses affected and the conditions under which the response is observed.

149. Measuring chromatid breaks or chromosome-type exchanges in lymphocytes, the maximum effect on resistance to a second dose occurs within 5-6 h following the pre-exposure [S34, W15]; the phenomenon may persist for up to three cell cycles [C24, S34]. While pre-exposure to relatively high acute x-ray doses (e.g. >200 mGy) does not give the adaptive response [C46], when 500 mGy was given at low dose rate (e.g. 10 mGy min^{-1}) the adaptive response was found. Conversely, there appears to be a minimum dose rate for the response induction by low doses (e.g. for 10 mGy pre-dose, the dose rate must be >50 mGy min^{-1} for full effect) [S33]. A pretreatment with an acute dose of 20 mGy from x rays will confer resistance to chromosome breakage by 150 mGy alpha particles from radon [W28]; however, when acute high-LET radiation was used to give the pre-exposure, there was no increased resistance to a second dose [K16, W24].

150. Pre-exposure to low doses from x or gamma rays has also been shown in several different cell types to decrease the frequency of cell killing [L27, M24, S32], mutation [K13, R12, S20, U17, Z8], apoptosis [F26], and morphological transformation [A15, R33]. However, the last result is still controversial. Using a near-diploid mouse skin cell line (m5S), Sasaki [S23] showed that x-ray-induced cell killing and chromosomal and mutation responses, but not morphological transformation, are mitigated by acute pre-exposure to 20 mGy from x rays. A malignant derivative of m5S cells lacked the adaptive response, but this was restored along with morphological reversion by transfer into these cells of human chromosome 11. The study also showed that chemicals that either activate or inhibit protein kinase C, which has an important role in signal transduction, either mimicked or abolished, respectively, the adaptive response [S23]. There have been reports suggesting that capacity to rejoin radiation-induced DNA double-strand breaks is greater following low-dose pre-exposure [I4, Z8], as well as data implicating activation of antioxidant metabolism in the adaptive response {B65, Z14}.

151. The repair of DNA base damage is also a potentially relevant process; a newly developed ultrasensitive assay for thymine glycols (paragraph 17) has shown that pre-exposure to 0.25 Gy from gamma rays increases the repair of this type of base damage from a subsequent dose of 2 Gy. The initial rate of removal of thymine glycols in human cellular DNA was found to be increased by a factor of 2 following pre-exposure [L60].

152. New methods to measure with improved accuracy the survival response of mammalian cells have led to the discovery that the dose-response curve may initially be steeper at low doses (<1 Gy) than predicted from the curve found at higher doses. This response has been characterized as an initial hypersensitivity at doses up to 0.5 Gy, followed by increased radioresistance of the cell population [L23, M17, W32]. The possibility that the hypersensitive response to low doses is the result of differentially sensitive fractions in cell populations has been rejected on the basis of two observations. First, survival curves for radiosensitive tumour cell lines and an ataxia-telangiectasia line appear to show no changes in sensitivity in the low-dose region. Preliminary data for two further radiosensitive lines, one with reduced DNA double-strand break repair and the other reduced excision repair, also show no increased radioresistance at low doses in comparison to their respective parental lines [S35]. Second, it is argued that the hypersensitivity seen would require one fraction of the cell population to have an unreasonably high level of sensitivity (e.g. about 7% of the cells to be >10 times more sensitive than ataxia-telangiectasia cells) [L24]. In parallel to the adaptive response of lymphocytes, recent data also show that pre-exposure to low doses from x rays or low concentrations of hydrogen peroxide increases radiation resistance in the hypersensitive region. Further, the additional resistance from pre-exposure was transitory, requiring time for development and diminishing after two or three cell cycles [M18]. Again, it has been found that low-dose hypersensitivity is absent with high-LET radiation [M17, M19].

153. Multi-laboratory experiments attempting to measure the dose response for chromosomal aberration induction in lymphocytes from two normal donors at low x-ray doses (down to 4 mGy) also appear to identify departures from fitted dose responses that have been attributed to the induction of repair processes [P15]. It was then suggested that the control value in these experiments was excessively high, biasing the result towards non-linearity [E5]. However, analysis of lymphocytes from individuals in Austria before and after exposure to fallout from the Chernobyl accident, where a peak twofold increase in radiation exposure was recorded, also showed dose-response curves departing from linearity (i.e. decreasing or levelling off) at annual doses between 0.3 and 0.5 mGy [P17]. Further *in vitro* studies with x rays and larger numbers of donors have not confirmed a significant departure from linearity at low acute doses (down to 20 mGy), although statistical variation in the small numbers of aberrations detected at lower doses do not allow conclusive statements on dose response [L37, L38]. These further studies did show an excess of multiply-damaged cells in some donors after

irradiation, leading to the possibility that a small subset of lymphocytes is especially sensitive to aberration induction by very low doses. Similar studies with 15 MeV D-T neutrons did not reveal significant departures from linearity at low doses [P16]. Similarly, a study of chromosomal aberration induction in Syrian hamsters injected with ^{137}Cs to give a whole-body dose of about 0.4 mGy, to mimic the Chernobyl fallout exposure, failed to reveal a significant increase above background values [L61]. Thus, the possibility that very low doses give a higher yield of chromosomal aberrations than expected on the basis of a linear extrapolation from high-dose data remains contentious.

154. Reports suggesting that low doses of ionizing radiation induce repair processes *in vivo* pre-date the cellular studies described above. Irradiation of mice with low doses of x or gamma rays has produced evidence of increased resistance to subsequent higher doses to both somatic and germ cells [C24, F6, W26]. Recent data show that the marked cytogenetic adaptive response of mouse germ cells does not influence the response of the somatic or germ cells of the offspring [C25]. Examination of chromosomal aberrations in the lymphocytes of people exposed occupationally to higher-than-average doses of alpha particles from radon (0.01-16 mGy) plus gamma rays (1-3 mGy) showed that the dose-effect curve was not related to that found at higher doses. Individuals subjected to the lowest doses gave a steep increase in aberration frequency with dose, while at higher doses the curve flattened out [P14]. Tuschl et al. [T33] showed that the people exposed to the higher doses (8-16 mGy a^{-1}) of alpha particles gave higher levels of repair (unscheduled DNA synthesis after UV light damage) in lymphocytes, compared to controls. Hospital workers exposed to low levels of x and gamma rays (maximum annual dose = 28 mSv) showed a reduction in the frequency of chromosome aberrations induced by a dose of 2 Gy to their blood lymphocytes, compared to non-exposed controls [B20]. However, it has been found that the lymphocytes of children living in areas contaminated by the Chernobyl accident showed no evidence of increased resistance to x rays [P9, U16], and separate reports suggests that fewer people from the region of the accident showed the adaptive response in lymphocytes compared with a control group [P19, M52].

155. A growing number of specific genes and proteins have been shown to be induced or repressed following irradiation, mostly using relatively high doses from x or gamma rays (2-6 Gy). While these do not as yet form coherent pathway(s) coordinating response to radiation, they do implicate genes and proteins involved in a variety of important molecular processes (see also Section II.B). Induced proteins include oncogenes/ transcription factors (c-jun, c-fos, interleukin-1, and egr-1), proteins involved in cell-cycle regulation (p53 and cyclins A and B), growth factors, and DNA-metabolizing proteins [PCNA, β-polymerase, and poly(ADP-ribose) polymerase], as well as the products of a number of unknown genes [B25, K15, W20]. The development of methods for rapidly screening hundreds or thousands of genes (micro-

arrays) in one experiment, for changes in levels of gene expression, is beginning to show the extent of transcriptional response to radiation damage. For example, using an array of more than 600 genes, gamma irradiation of a myeloid cell line showed induction of 48 genes by factors of two or more, and many of these genes had not been previously reported as radiation-inducible [A25]. Considering those genes which respond differentially in the presence or absence of p53, it was found using micro-arrays that several genes encoding secreted proteins with growth inhibitory functions were upregulated in a p53-dependent fashion following gamma irradiation. Thus, the p53 response to radiation may also be involved in growth inhibitory effects on surrounding cells, as observed in "bystander" effects [K44]. Methods are being developed similarly for assaying large numbers of proteins, to reveal translational and post-translational responses, using 2-dimensional gel systems coupled to mass spectrometry [B64, C69]. These methods will also revolutionize the classification of normal tissue and disease states, in particular cancer, by permitting a molecular description of the complete profile of gene products present [G40]. Such classifications will be invaluable in understanding the mechanistic basis of cancer induction by agents such as ionizing radiation.

156. Increased or decreased levels of certain gene products have also been found after low doses of radiation, within the range inducing the adaptive response [B18]. Doses of less than 500 mGy gamma rays have been found to induce the expression of a variety of stress-response genes [P35, P36], and dose-responses of several of these genes were shown to be approximately linear between 20 and 500 mGy gamma rays [A26]. Some candidate gene products with a potential role in induced radiation resistance have been identified, including a member of the heat shock protein 70 family (PBP74 [S79]), a heat-shock related immunophilin protein (DIR1 [R37]), ribonucleotide reductase [S81], and the MAPK and PKC protein kinases [S80].

C. SUMMARY

157. When the cells of normal individuals are examined for radiosensitivity, as shown by their survival in culture, variation by a factor of 2 is seen; this factor may be extended to 3 or 4 with the use of low-dose-rate irradiation conditions. It has been suggested that this variation has a genetic basis, but it has been technically difficult to establish this. Additionally, a fraction of cancer therapy patients suffers from severe skin reactions, and cells from these patients commonly show a slightly elevated radiation sensitivity.

158. Assays based on chromosomal damage in G$_2$ cells have also been used to estimate radiation sensitivity in normal individuals and in breast cancer patients, relative to individuals with known radiosensitivity disorders. The response of a relatively large fraction of breast cancer patients overlapped that of carriers of the ataxia-telangiectasia defect, suggesting that a number of genes involved in response to radiation damage may predispose to this form of cancer.

159. It has been suggested that individuals carrying one defective copy of the ataxia-telangiectasia gene (*ATM* heterozygotes) are at increased risk from cancer, especially breast cancer. A number of studies have recently tested this idea and come to the conclusion that *ATM* heterozygosity is not an important cause of breast cancer susceptibility (or severe response to radiotherapy), but commonly too few patients have been tested to exclude completely a role for *ATM*. Specific types of tumours, such as T-cell prolymphocytic leukaemia, are associated with high levels of mutation of the *ATM* gene (paragraph 59). Additionally, recent evidence suggests that the genes involved in familial susceptibility to breast and ovarian cancers (the *BRCA* genes, paragraph 72) are involved in DNA repair processes and lead to radiation sensitivity when defective in mice.

160. Repair processes affect the shape of survival curves, especially through differences in the processing of DNA double-strand breaks, as seen in specific repair-defective lines. The response to double-strand breaks probably also underlies operationally-defined measures of cellular recovery such as potentially-lethal damage repair and sublethal damage repair. High-LET radiations may not induce a higher frequency of DNA double-strand breaks than low-LET radiations, but high-LET radiation damage is much less repairable. The increase in RBE with LET, found for normal cells, is largely determined by the cell's ability to repair low-LET damage.

161. Evidence for the inducibility of repair processes in mammalian cells is fragmentary. Data from pre-exposures of cells to low radiation doses, as well as refined survival analyses at low doses (<1 Gy), show an altered response suggestive of an induction process but to date have failed to link this response to known inducible processes. However, recent experiments have shown that certain genes in at least two repair pathways are up-regulated following radiation damage (paragraphs 84 and 151). Additionally, refined methods for looking at protein structure and activity are beginning to reveal insights into more subtle modifications following irradiation. Some damage-response proteins may be activated by post-translational modifications (e.g. phosphorylation, paragraph 102), without any change in the amounts of these proteins, as part of a signalling mechanism promoting the repair of DNA damage. Other proteins have been shown to accumulate in the cell to form discrete foci following irradiation (paragraphs 71 and 74); these foci may represent repair protein complexes accumulating at sites of damage. While there has been little exploration of the dose dependence of these events, the data indicate that modification of cellular response mechanisms can occur following irradiation, raising the possibility that dose responses are altered as a result. Methods for data acquisition on the inducibility of gene products are presently being revolutionized by the introduction of micro-array and complementary techniques, so that rapid progress in this research area is to be expected. The potential importance of inducible repair processes in determining responses to low doses of ionizing radiation is considered further in Annex G, "*Biological effects at low radiation doses*".

IV. MECHANISMS OF RADIATION MUTAGENESIS

A. MUTATION AS A REPAIR-RELATED RESPONSE

162. The cellular processing of radiation-induced damage to DNA by enzymes may result in a return to normal sequence and structure (correct repair). Alternatively, the processing may fail or may cause alterations in DNA, with the consequence of lethality or inherited changes (mutations). It is also possible that some subtle forms of damage may be tolerated by the cell, particularly if it is non-replicating, and lead to persistent lesions in DNA. These lesions would have to be both chemically stable and not be substrates for repair enzymes and may include some minor types of damage such as methylated bases generated by non-enzymatic alkylation [L68].

163. It is likely that simple base damage or loss in the mammalian genome will commonly lead to base-pair substitutions. In recent years the defined production of single types of damage, at specific sites in DNA molecules *in vitro*, has given insights into their consequences. For example, DNA molecules carrying a single site of base loss (abasic site) have been shown to give rise to base-pair substitutions at that site, either when introduced on shuttle vectors [G8] or when the *RAS* gene was transfected and stable transformed clones selected [K7]. These substitutions (a form of point mutation) will often give rise to alterations in a gene product so that it works less efficiently or not at all.

164. Breaks in DNA, especially double-strand breaks, are thought to lead to larger alterations such as deletions and rearrangements. Some alterations are very large and are seen as chromosomal aberrations. While there is little formal proof of these relationships between breaks and mutation type, it has been shown that the transfer of restriction endonucleases into mammalian cells, causing site-specific DNA double-strand breaks, gives rise to mutations and chromosomal aberrations [T24]. Additionally, the processing of isolated double-strand breaks in defined DNA molecules by human cell extracts show that these may lead to large deletions of surrounding sequence [T10].

165. Reduction in the efficiency or fidelity of damage repair may lead to an increase in genetic change. This is seen strikingly in the recent discovery that loss of DNA mismatch repair capacity is involved in specific forms of cancer.

Mismatch repair is a form of base-excision repair (Section I.B) that removes bases that have been altered or incorrectly placed (by a polymerase) so that the two strands of DNA do not match in base sequence. The cancer connection was first noted as a high frequency of mutation of short-repeat-sequence (microsatellite) DNA in colon cancers, including hereditary non-polyposis colon cancer. It was rapidly established that the genes determining hereditary non-polyposis colon cancer co-localize with mismatch repair genes on human chromosomes 2 and 3 [B12, L15]. More genes with homology to these mismatch repair genes have since been identified, and at least two of these have also been shown to be mutated in the germ line of hereditary non-polyposis colon cancer patients [N10]. Colorectal cancer is one of the most common human cancers and perhaps the most frequent form of hereditary neoplasia; hereditary non-polyposis colon cancer accounts for as much as 5% of all cases of colon cancer, and the involvement of several genes in the phenotype may account for its prevalence [M20].

166. The high frequency of repeat-sequence mutation was recognized as characteristic of loss of the mismatch repair system (MutHLS), known for many years in bacteria and involving the concerted action of three mismatch enzymes to correct errors caused primarily during DNA replication (such as slippage of strands at repeat sequences) [M27]. Patients with hereditary non-polyposis colon cancer are therefore thought to have a mutator phenotype similar to mutant strains of bacteria with defects in mismatch repair. The role of inactive mismatch repair genes in humans is expected to be similar to the part played by tumour-suppressor genes; a germ-line mutation in one of the mismatch repair genes is followed by somatic mutation of the second gene copy during tumour development [H22, L36]. However, the mismatch repair genes are not thought to be directly involved in cancer; rather, they increase genetic instability and the probability that random mutations will affect those genes critical to cancer formation. This idea does not necessarily mean that mutations alone are sufficient for cancer formation; it is interesting to note that a subset of hereditary non-polyposis colon cancer patients carry numerous mutations in their cells and the expected defect in mismatch repair but have unexpectedly few tumours [P10]. Given the proposed mechanism, it is curious also that the increased cancer risk in hereditary non-polyposis colon cancer patients is selective; while other sites are affected, there is, for example, no increased risk for cancers of the breast and lung [W17]. Recent success at breeding transgenic mice defective in mismatch repair genes has shown that they are viable, but a large proportion develop lymphomas and sarcomas at an early age [B16, D9, R11]. While many of these mice die early, commonly succumbing to T-cell lymphoma, those that survive for more than 6 months develop gastrointestinal tumours, suggesting that these mice may be used as a model for human colon cancer [H50].

167. Cell lines lacking mismatch repair also show an increase in mutations at sites other than microsatellites. In colorectal carcinoma lines, shown to have high frequencies of microsatellite variation, the *HPRT* gene mutation rate was also found to be increased by a factor of more than 100 over the rate in normal human cells [B23, E8]. In one of these lines, known to have a defective human *MLH1* mismatch repair gene, about one quarter of the mutants had point mutations (frameshifts) at a hotspot within a run of guanine bases in the *HPRT* gene [B23]. An extensive study of one mismatch repair-defective line using the *APRT* gene similarly showed an elevation of frameshift mutations at sites of repeat base sequence, as well as AT → TA transversions at sites of secondary DNA structure [H23]. A potentially relevant site for mutations of this type is in the *BAX* gene, involved in promoting apoptosis (Section II.B.3); more than half of 41 colon carcinomas with microsatellite instability also had frameshift mutations in a run of eight guanines in the *BAX* gene [R19]. These findings are consistent with a role for mismatch repair in correcting base misalignments generated during DNA replication in normal cells. Interestingly, lack of mismatch repair also allows cells to become tolerant of certain forms of induced DNA damage. Alkylation of DNA bases normally produces cell-cycle arrest and/or death and can lead to cancer, but mismatch-repair-defective cell lines are highly resistant to the effects of alkylating agents [B28, K12]. Mismatch-repair-deficient mammalian cell lines also show a small but significant increase in resistance to gamma rays [F19], suggesting that repair of certain types of radiation-induced base damage is recognized by the mismatch repair system (e.g. 8-oxoguanine, which mispairs with adenine, paragraph 24). Mouse embryonic stem cells, heterozygous for a defect in the mismatch repair gene *Msh2*, showed resistance to low-dose-rate gamma radiation (0.004 Gy min^{-1}) but not to acute radiation (1 Gy min^{-1}). On the basis of this result, it was speculated that heterozygosity for mismatch repair may also contribute to tumorigenesis in a direct manner, without loss of the other gene copy [D20]. The mechanism of resistance is thought to follow from a reduction in abortive "repair": mismatch repair normally removes a base that is incorrectly incorporated opposite a damaged base but this repair is abortive, because the damaged base remains in place and may lead to lethality.

168. It is to be expected that the genes involved in other functions required for the maintenance of genome stability would lead to a similar effect of increasing mutation frequency, thereby affecting cancer rates. These functions would include the "proofreading" of DNA synthesis by polymerases and the regulation of DNA precursor synthesis. Indeed, it has been shown that mutations in the exonuclease domains of the DNA polymerase δ gene correlate with high mutation rates in some colorectal carcinoma cell lines lacking changes in mismatch repair genes [D6].

169. Repair enzymes may cause mutations by virtue of their imperfect response to damaged DNA, but in the last few years it has become apparent that many organisms have also retained specific enzymes that introduce mutations into damaged DNA. Thus the loss of certain repair functions can be antimutagenic, and indeed mutant strains of bacteria showing no increase in mutant frequency after treatment with DNA-damaging agents have been known for many years [W11]. Recent studies have defined these mutagenic or error-prone DNA repair processes in bacteria and yeast, and have

revealed that similar processes occur in other organisms, including humans [J23]. It has been found that genes coding for special types of DNA polymerases are responsible for many of the small mutations (base substitutions, frameshifts) occurring spontaneously and after treatment with DNA-damaging agents. In bacteria, for example, when DNA damage blocks the normal replication process, an "SOS" response is activated and more than 20 genes are induced [S83]. Among these genes are polymerases (e.g., DNA polymerase V) with a high affinity for damaged DNA, which are able to continue to synthesise DNA for a few bases in the presence of damage (translesion synthesis) and which commonly put in incorrect bases to give a mutation [R35, T45]. Similarly, in yeast, three genes (*REV1, REV3, REV7*) are required for much DNA-damage-induced mutagenesis, and these have been found to specify mutagenic DNA polymerase activity required for translesion synthesis [N23, N24]. A human gene named *hREV3* has recently been cloned through its homology to the yeast *REV3* gene and been found to have the properties of a mutagenic polymerase [G41]. As well as this error-prone activity, yeast cells have another specialized DNA polymerase (encoded by the *RAD30* gene) that can perform translesion synthesis in an error-free way. Strikingly, a human gene homologous to *RAD30 (hRAD30)* has been found mutated in a variant form of the human sunlight-sensitivity disorder xeroderma pigmentosum (see paragraph 26). Since individuals with this disorder suffer a high frequency of skin cancer, this result suggests that error-free translesion synthesis is important to protect against sunlight-induced cancers [J24, M49]. While much remains to be learned about the operation of these specialized polymerases, it seems clear that maintaining a balance between error-free and error-prone pathways of translesion synthesis is important in the determination of rates of (point) mutation in cells.

B. THE SPECTRUM OF RADIATION-INDUCED MUTATIONS

170. Ionizing radiation can induce many types of mutation, from small point changes to very large alterations encompassing many genes. From recent studies of mammalian cells, in which a few genes have been examined in some detail, it is clear that radiation is most effective at inducing large genetic changes (large deletions and rearrangements). This mutation spectrum (proportion of different types of mutations) differs from that found spontaneously or the spectra induced by many other DNA-damaging agents (e.g. ultraviolet light, alkylating chemicals). These other agents tend to induce mostly point mutations, independent of the genomic region assessed. However, it has also been found that the proportion of radiation-induced large genetic changes can vary with the genomic region assessed.

171. The size of genetic changes in a given region of the genome is limited by the extent to which that region can tolerate change. The region may contain genes that are essential for the viability of the cell and organism; if an essential gene is altered or lost, the changes incurred will

usually be lethal [E3, T16, T21]. Thus, although the initial occurrence of radiation-induced genetic changes may be similar at different sites in the genome, lethality will limit both the frequency and the apparent size of mutations recovered. Mutations will be recovered in some genes at low frequency, because they or the region they reside in can tolerate little change (certain genes have been found that, within experimental limits, show almost no radiation-induced mutation [T15]), while mutations in other genes may be detected at relatively high frequency because of cellular tolerance to large changes. Knowing this, it is possible to devise mutation detection systems that tolerate very large changes that would be inconsistent with survival in normal diploid cells. The most extreme example of high-frequency mutation detection is in a system devised by Waldren et al. [W12] that uses as the mutational target an accessory human chromosome 11 introduced into a Chinese hamster cell line. Since no genes on the human chromosome are essential for cell survival, very large changes including whole chromosome loss are tolerated, and mutant frequencies are about 100 times greater than for endogenous genes such as *HPRT*.

172. Much of the mutagenic response to ionizing radiation has been measured, for experimental ease, in regions of the genome that are present at the level of one copy per cell. This situation (monosomy) occurs naturally in parts of the genome such as the X and Y chromosomes in males; also, functional monosomy appears to be quite extensive through mechanisms that switch off gene expression in one copy (as in one copy of the X chromosome in females and in imprinted chromosomal regions). However, most chromosomes and genes are present as two copies per cell (disomic; see paragraph 7). Where the loss of function in one copy of a disomic gene is tolerated without harm (e.g. because of a recessive point mutation), it has been found that the other gene copy is mutated at relatively high frequency by ionizing radiation. Since the two gene copies differ in their functionality (that is, they are heterozygous), this mutation process is commonly termed loss of hetero-

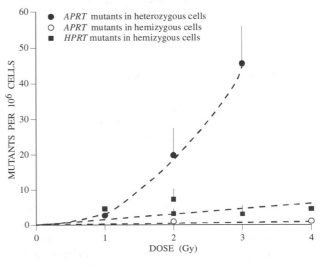

Figure V. Effect of gene location (for hemizygous genes) and copy number on mutant frequency in hamster cell lines carrying two copies of the gene (heterozygous) or single copy (hemizygous) [B11].

zygosity. The frequency of mutation in this disomic heterozygous situation has been shown experimentally to be higher than for the same gene in a monosomic situation [E3, M14, Y2], as illustrated in Figure V for the *APRT* gene, where the mutation frequency difference is about 20-fold [B11]. The reason for this higher mutation frequency in the disomic situation appears again to relate to tolerance of the frequent large genetic changes caused by radiation.

Thus, the first copy, despite carrying a point mutation, still has the remainder of that genomic region intact (including any linked essential genes), and large deletions in the second copy can be tolerated because these do not lead to a complete loss of linked essential genes. In a monosomic chromosomal region, however, there is no other copy (or no functioning copy) present, and the mutant cell will not survive large genetic changes (Figure VI).

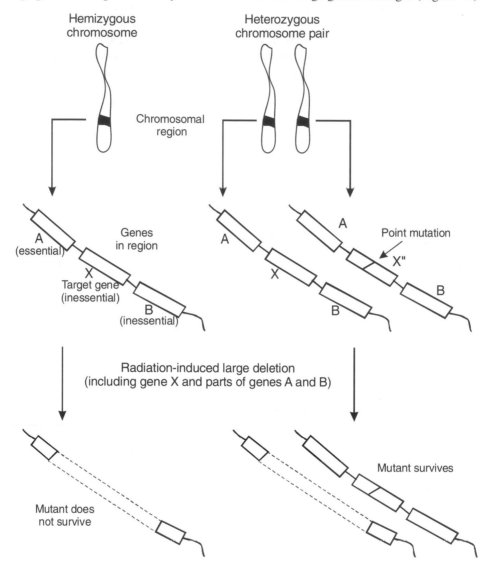

Figure VI. Consequences of radiation-induced deletion mutation in a hemizygous (monosomic) and heterozygous (disomic) gene.
Loss of function in gene X is selected for, but the mutation also removes parts of adjacent genes, one of which is essential for cell viability. The mutation results in death of the cell carrying the mutation when the genetic region is hemizygous, but not when it is present in a heterozygous state.

173. Where disomic genes are used to study radiation-induced mutation, the mutation frequency will also vary depending on which of the two copies (alleles) is used as the target. It is to be expected that in addition to the target gene, other linked genes will vary in their functional state (heterozygosity). Thus, if an active allele of the target gene is linked on one chromosome to the active allele of an essential gene, deletions of that chromosome will be severely limited. Conversely, linkage of the target gene to an inactive copy of the essential gene will not be limiting if the homologous

chromosome carries the active essential gene and will allow the cell to survive a large deletion. Thus, a 10-fold difference in mutant frequency has been recorded for the two alleles of a heterozygous *TK* gene in lymphoblastoid cells [A11]. This difference in mutant frequency was associated with a class of slow-growth mutants carrying large genetic changes at the site of the mutable allele of the *TK* gene [A12].

174. High mutation frequencies may therefore mainly reflect tolerance to large genetic changes, although higher

frequencies may also arise from additional mutation mechanisms available in heterozygous cells (paragraph 183). Since ionizing radiation is relatively good at inducing large genetic changes, it will be very effective as a mutagen for disomic genes compared with its effectiveness for monosomic genes (also, in general, ionizing radiation is less effective than potent chemical mutagens when used to mutate monosomic genes). It may be that, for example, an individual inherits a point mutation in one copy of a tumour-suppressor gene, which in itself does not lead to deleterious effects because of the "good" copy still present. However, should this person be subjected to an agent like radiation, which is very effective at inducing large genetic changes, the other gene copy can be readily mutated along with much of the surrounding chromosomal region, because that region is shielded by the presence of the other gene copies. In this way radiation can be seen as a very effective mutagen for specific types of mutational change; indeed it could be argued that for recessive gene mutation, individuals are susceptible to agents such as radiation if they already carry point mutations in disomic genes. This phenomenon is an aspect of predisposition of individuals to specific genetic changes and should not be confused with radiosensitivity (paragraph 122).

175. The prediction that large genetic changes would be found in disomic target genes has been borne out in studies of mice [C19]. Screening progeny after x irradiation of male germ cells revealed that very large chromosomal changes (deletions, rearrangements, and complex changes) occur at high frequency in one copy of disomic regions of the genome; these large changes were consistent with reasonable viability and fertility in the progeny. As expected, when these mutant regions are not shielded by a second normal copy of the genes involved, they are invariably lethal to the mice.

C. MOLECULAR ANALYSIS OF RADIATION-INDUCED MUTATIONS

176. While it can be argued that ionizing radiation has a particular mutation spectrum, it has not been found that any one type of mutation is induced specifically by ionizing radiation. However, detailed analysis of mutations (and to some extent of chromosomal aberrations) has given some possible indicators of differences between spontaneously-occurring and radiation-induced mutations, and of differences between densely and sparsely ionizing radiation.

177. Sequence analysis of x- or gamma-ray-induced point mutations in mammalian cells has shown that a variety of types occur, from base-pair substitutions and frameshifts to small deletions. Analysis of substitutions in the *HPRT* and *APRT* genes has shown that all of the 6 possible types occur, although at different frequencies (Table 3) [G10, M12, N11]. A majority of the radiation-induced substitutions were transversions (alteration of the base from a purine to a pyrimidine or vice versa), while many spontaneously-occurring substitutions were G:C → A:T transitions. A larger proportion of frameshift mutations also occurred in the radiation-induced mutants than spontaneously, and a few more of the radiation-induced point mutations were multiple substitutions (more than one base change in close proximity). Using a small gene target transferred into mouse cells, little difference was found in the mutation spectra for spontaeously-occurring and x-ray-induced mutations [K18]. The frequency of small rearrangements and deletions was increased in both spontaneous and radiation-induced mutants, indicating differences in the balance of mutagenic mechanisms at the site of integration of the transgene compared with those at endogenous genes.

Table 3
Comparison of spontaneous and radiation-induced point mutation spectra

Type of mutation	APRT *gene / hamster CHO cells [M12]*		HPRT *gene / human TK6 cells [N11]*	
	Spontaneous	*Induced by gamma rays (2.5–4 Gy)*	*Spontaneous*	*Induced by x rays (2 Gy)*
Base substitution	55 (71%)	19 (66%) [a]	10 (55%)	19 (54%)
Transition	31 (40%)	6 (21%)	6 (33%)	7 (20%)
Transversion	24 (31%)	13 (45%)	4 (22%)	12 (34%) [a]
Frameshift	6 (8%)	5 (17%)	3 (17%)	8 (23%)
Small deletion or rearrangement	16 (21%)	5 (17%)	5 (28%)	8 (23%)
Total	77	29	18	35

a Includes two tandem substitutions.

178. Radiation-induced point mutations were found at sites widely distributed within the gene, while spontaneously-occurring point mutations tended to cluster at certain sites [G10, M12, N11]. Differences in the types of base damage responsible and in the randomness of damage induction are likely to explain these differences.

179. Smaller numbers of large genetic changes have been sequenced because of the difficulty of locating and cloning the breakpoints of large deletions and rearrangements. No new or specific mechanism has been found for the induction of these larger changes. In the hamster *APRT* gene, present in the hemizygous state, the large gamma-ray-induced deletions tend

to have short direct or inverted repeat sequences around their breakpoints and to fall into regions rich in adenine-thymine base pairs. Insertions, however, involved repetitive sequences and were accompanied by short deletions [M13]. It was notable in these studies that large changes extending downstream of the *APRT* gene are not found, suggesting that an essential gene is present in this region.

180.   In the human *HPRT* gene, also in the hemizygous state, large deletions are found to extend both upstream and downstream of the gene over a region of about 2 Mb [M16, N9]. Radiation-induced large genetic changes tend to eliminate the whole *HPRT* gene, making their analysis difficult. Therefore a series of flanking markers has been identified to delimit these changes and ultimately to map the positions of the breakpoints. Sequence analysis of a few deletion mutations has again shown the presence of short direct repeats at some breakpoints induced by both x rays and alpha particles, as well as other sequence features in adjacent regions [M15, S18]. The large deletions induced by radiation frequently include sequences adjacent to *HPRT*, and it has been suggested that this may be a signature of radiation-induced events as opposed to spontaneous events [N17]. This type of mutation was equally common with gamma-ray doses of 0.2 or 2 Gy d^{-1}. Use of methods amplifying specific regions of the genome to measure rapidly the frequency of rearrangements following x- or gamma-irradiation, followed by sequencing the breakpoints, has given little indication of the involvement of specific sequence features or clustering of breakpoints [F22, F23].

181.   Shuttle vector systems consist of small defined DNA molecules that may be mutated in mammalian cells but rescued into bacteria for rapid analysis of sequence changes. While these facilitate the molecular analysis of induced mutations, they generally have the drawback that the target genes are very small and are flanked by essential sequence. This means that only point mutations are detectable, and the major class of radiation-induced mutations (large deletions and rearrangements) cannot be analysed. However, Lutze and Winegar [L19] devised a large shuttle vector based on Epstein-Barr viral sequence that could be maintained episomally in human lymphoblastoid cells and could detect changes of up to 8 kb or so. Using this vector, the mutation spectra of both x rays (150-600 Gy) and alpha particles (3 Gy) from radon gas were studied [L19, L20]. A larger proportion of deletions was found among alpha-particle mutations (64%) than among x-ray mutations (13%), but in a subsequent study with lower doses of x rays, more deletions were found (41% at 100 Gy, 33% at 20 Gy) [L21]. For both x rays and alpha particles, these deletions were large (>2.4 kb), and their breakpoints were clustered in specific regions of the shuttle vector. The breakpoints were commonly associated with short direct sequence repeats of up to six base pairs. Use of the same vector system in a lymphoblastoid cell line defective in double-strand-break repair gave a slightly higher frequency of deletions after x irradiation, in association with an additional class of small deletion mutations (<2.4 kb) [L21].

182.   The presence of sequence features such as short direct repeats of a few base pairs at large deletion junctions suggests that illegitimate recombination (see Section I.B) has driven the mutation process. Recently it has been possible to reconstruct the process of illegitimate recombination in cell-free conditions to show that the process can be associated with a DNA double-strand break. DNA molecules are broken at a specific site, using an endonuclease, and exposed to extracts from human cells for a brief period. Analysis of the extract-treated DNA shows that while the majority of broken molecules are correctly rejoined, a small fraction (about 0.5%) are mis-rejoined to give a deletion. The mis-rejoin mechanism involves the pairing of short direct sequence repeats, situated either side of the break, so that the intervening sequence (including one of the repeats) is deleted [T10]. This mechanism has been found in both radiation-induced mutations and in germinal and somatic-cell mutations in humans. A model for this process is shown in Figure IV. In addition to simple deletion, this cell-free system detected a small fraction of more complex changes, e.g. a large deletion associated with the insertion of several hundred base pairs.

183.   In heterozygotes, in addition to the tolerance of large changes, additional mechanisms of mutation may occur. The two gene copies may undergo mitotic recombination or non-disjunction, leading to the appearance of mutations. However, where loss of heterozygosity is measured simply by the presence or absence of the active (wild-type) gene, it is difficult to distinguish mutations occurring by these additional mechanisms from large deletions. In a study of mutation in the human *TK* gene, using both linked marker analysis and densitometry to assess gene copy number, about 50% of the spontaneous mutants appeared to involve recombination, while x-ray-induced mutants were mainly deletions [L29]. Recombination was similarly found to be involved in spontaneous mutation of the *APRT* gene in human cells, but this mechanism was also detected in a number of gamma-ray-induced mutants [F8].

184.   While it is difficult to compare somatic-cell mutation data with animal germ-cell data, some recent molecular analyses of the genes used in mouse specific-locus tests indicate possible similarities in the types and mechanisms of mutation. For example, with some dependence on both cell stage and radiation quality, a large proportion of *albino* (*c*) locus mutations on mouse chromosome 7 recovered from germ-cell irradiations are large deletions [R14]. Molecular analysis of radiation-induced mutations at the *c* locus shows that a genomic region of 1.5-2 Mb around the tyrosinase gene may be deleted in mutants without leading to inviability in homozygotes [R13]. This target size is very similar to that of the human *HPRT* gene region used in somatic-cell mutation studies. Also, mutation frequencies per unit dose of low-LET radiation at the *c* locus in mouse germ cells and at the *HPRT* locus in human somatic cells are in order-of-magnitude agreement [T19], given that mutations at the *c* locus occur with about average induced frequency relative to other loci in specific-locus tests [S29]. Comparisons of this type emphasize that the mechanisms of radiation mutagenesis are similar in

somatic and germ cells, and that if measurements are made under similar conditions of mutation tolerance, then the mutation frequencies will be comparable.

185. Comparison of two human lymphoblastoid cell lines derived from the same original cell line has given further insights into the mechanisms of radiation mutagenesis. Line TK6 is more sensitive to the lethal effects of x rays than line WIL2-NS, but the converse is true for mutation response to x rays (although not to chemical mutagens). The increased frequency in WIL2-NS at the hemizygous *HPRT* gene was modest (factor of 4 at 2 Gy), while the increase for the heterozygous *TK* gene is by a factor of 20-50, depending on the *TK* allele used as the target [A13]. Molecular studies have revealed that all *TK* mutants in WIL2-NS arose by loss of the active allele and linked markers (over a 5 Mb region), while point mutations and less extensive deletions were common in the TK6 line. Two copies of the *TK* gene were present on a majority of karyotypes in the WIL2-NS mutants, as expected if these arose from mitotic recombination rather than deletion, while the converse was true for TK6 mutants [X1]. It seems likely, therefore, that the WIL2-NS line has a higher frequency of recombination, leading to a greater ability to survive x-ray damage but incurring a concomitant increase in mutation. Analysis of the p53 status (see paragraph 112) of these two cell lines has shown that p53 protein levels are four times higher in WIL2-NS than in TK6, because of a mutation in exon 7 of the *TP53* gene in WIL2-NS (TK6 is wild-type) and that apoptotic death is substantially delayed in WIL2-NS [X2, Z7].

186. In a study of B-cell precursors from *Tp53*-knockout mice, it was also found that high frequencies of *HPRT* mutations occurred following x irradiation but that these resulted from a preferential survival of mutant clones rather than from a p53-dependent increase in mutation rate [G26]. It was concluded that loss of p53 function allows mutated cells to survive that would otherwise be eliminated by apoptosis. Other recent studies with cells from p53-deficient mice have similarly concluded that they do not have an intrinsically higher-than-average mutation rate [C68, N12, S21].

187. It has been suggested that a relatively unique feature of radiation mutagenesis is the induction of complex genetic changes [M11]. For example, a radiation-induced deletion may be associated with a rearrangement at the same site, and the rearrangement may be an exchange, inversion, or insertion. Additionally, complex changes may be seen as chromosomal exchanges involving several sites in the genome following a single acute radiation treatment [S15]. However, because of their large size and complexity, these changes have been difficult to analyse in cellular genes at the molecular level. Further molecular analyses will be necessary to establish whether or not large mutations and chromosomal aberrations are formed by similar mechanisms. It is clear from studies with one or two genetic regions, and especially the region containing the human *HPRT* gene (Xq26), that genetic changes

identified at the molecular level as mutations can extend to sizes that are visible in the light microscope [S17]. However, the breakpoints of such very large mutations and chromosomal aberrations have not been sequenced.

D. EFFECT OF RADIATION QUALITY

188. In general, high-LET radiation induces a higher frequency of mutants in rodent and human cells, per unit dose, than low-LET radiation [B17, B19, C22, C30, F7, G18, H20, L31, M22, M25, S88, S89, T16, T18, T32, W22]. The RBE varies with LET, peaking at 100-200 keV μm^{-1}, with values as high as 7-10 found for alpha particles and heavy ions in this LET range [C22, T16]. In some mouse cell lines, a relatively small RBE of 2-3 has been found for *HPRT* mutant frequency with alpha particles in the peak range, and when cell survival is taken into account, there is no difference in the effectiveness of alpha particles and x rays. This result appears to be due to the very high effectiveness of low-LET radiation on these cells, as reflected in increased cell killing and mutagenesis for a given dose compared with other rodent lines [B19, I2]. That is, the mutagenic effectiveness of low-LET radiation is more variable than that of high-LET radiation, presumably because cell lines have different abilities to repair low-LET radiation damage (while high-LET damage is less repairable; see Section III.B). An example of this variability is shown in Figure VII [T28].

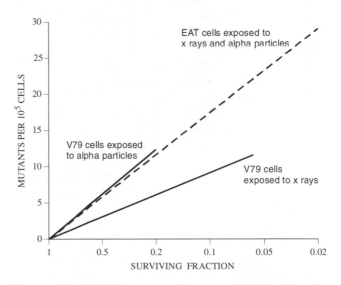

Figure VII. Fitted curves of mutant frequencies in Chinese hamster V79 cells [T18] and mouse ascites (EAT) cells [I20] induced by x rays and ^{238}Pu alpha particles (see paragraph 172).

189. A study of 19 non-smoking people living in houses with a range of radon concentrations (21-244 Bq m^{-3} in the United Kingdom suggested that there was a correlation of *HPRT* mutant frequency in blood T-lymphocytes with radon concentration. These data suggested that alpha particles in this LET range have a considerably higher effectiveness than found previously in high-dose experiments with human somatic cells. While the estimation of

alpha-particle dose to circulating blood lymphocytes is difficult, the authors estimated an annual doubling dose of 2.1 mSv from their results [B29]. A number of problems attend such measurements, not the least of which is the sample size. In a follow-up study by the same authors with a total of 65 persons from 41 houses in the same town, no significant correlation was found between mutant frequency and radon levels, even for people living in the same house [C34]. In a study in Belgium of 24 people living in houses with radon concentrations exceeding 100 Bq m^{-3}, a negative association between radon concentration and *HPRT* mutation frequency was found [A10]. This result led the authors to suggest the induction of repair processes by low levels of radon.

190. The *TP53* tumour-suppressor gene is commonly mutated in many different types of cancers, with a frequency ranging from <10% to >50% depending on tumour type. Analysis of mutations in the *TP53* gene in lung tumours from uranium miners initially gave some hope that these would provide a means of fingerprinting high-LET radiation damage. Vahakangas et al. [V2] sequenced exons 5-9 of the *TP53* gene in 19 lung tumours from miners working in New Mexico. They found several point mutations, although these lacked a class of base substitution (G:C → T:A transversions) and were not at the hotspots described for lung cancer. In contrast, in a larger study of 52 lung tumours from Colorado uranium miners exposed to a dose that was, on average, five times higher, Taylor et al. [T12] sequenced the same exons and found that about half of the substitutions were of the G:C → T:A type in codon 249 (codon = base-pair triplet encoding one subunit of the p53 protein). This transversion is very rare among *TP53* mutations in lung tumours (many of which are presumed to be smoking-associated) and was suggested to be a potential marker for radon-associated lung cancer. At present, apart from the difference in average dose, it is difficult to see why these two *TP53* mutation studies gave such contradictory results. One possibility is that the transversion mutations seen are a result of some other agent in the environment of the mine; if the mine is damp, then fungal toxins similar to aflatoxin B1, known to induce this type of mutation, might be present [V3]. Follow-up experiments looking specifically for the codon 249 transversion in lung cancers from individuals exposed to high domestic radon levels in the United Kingdom [L48] and from German uranium miners [B45] failed to find any examples of this mutation.

191. In a study of thymic lymphomas in RF/J mice induced by gamma rays or by neutrons (0.44 MeV), a similar frequency of lymphomas was associated with mutations activating *RAS* oncogenes (24% for gamma rays, 17% for neutrons). Of the gamma-ray set, 89% were K-*RAS*-activated, with the majority (7 of 8) having a GAT → GGT mutation in codon 12. However, in the neutron set no particular mutation site predominated, and one lymphoma contained a K-*RAS* gene activated by a point mutation in codon 146, a site not previously associated with any human or animal tumour [S36].

192. There are conflicting data on possible differences in the spectrum of large genetic changes induced by high-LET radiation relative to those induced by low-LET radiation. Early cytogenetic studies with diploid human fibroblasts suggested that a greater proportion of *HPRT* mutants induced by high-LET radiation carried large genetic changes than did x-ray-induced mutants [C37, T19]. Studies of specific-locus mutations in mouse spermatogonial stem cells had also concluded, based on a number of criteria, that ^{239}Pu-induced mutants carried more severe genetic damage than mutants induced by low-LET radiation [R16]. However, classification of *HPRT* mutations by molecular analysis in both diploid human cells and hamster cells has not shown differences in the proportions of large deletions to point mutations at doses of low- or high-LET radiation giving about 20% cell survival [A8, G16, S18, T20]. A confounding factor (commonly underestimated) in determining mutation spectra can be the non-independence of mutants when grown in bulk culture after irradiation [T22], but in the most recent studies [A8, S18] the experimental design ensured the independence of all mutants. Other studies have reported differences: irradiation with 190 keV μm^{-1} Fe ions gave 82% large deletions of the *HPRT* gene in human TK6 cells relative to 54% for x rays [K21]. High doses of radiation from incorporated ^{125}I, giving 1% survival of human TK6 cells, were reported to increase the frequency of deletions or rearrangements when compared with lower doses of ^{125}I or with x rays [W22]. It was noted that these high-dose mutations commonly showed loss of part of the *HPRT* gene rather than the entire gene. It has similarly been reported in recent studies with the same cell line that the average size of radon-induced deletions of the *HPRT* gene is not as great as that induced by x rays [B19, C29]. This feature has also been noted for mutations induced at the hemizygous DHFR gene in hamster cells; that is, the frequency of total-gene deletions was much higher for gamma-ray- than for alpha-particle-induced mutations, while the reverse was true for intragenic deletions [J11]. In the latter study, it was also found that some alpha-particle-induced deletions shared common breakpoint sites, indicating a non-random distribution. Use of the heterozygous *TK* gene as a target, also in TK6 cells, showed again that the proportion of deletions was similar for x rays and high-LET radiation (4.2 MeV neutrons and argon ions), but that high-LET radiation induced a category of large rearrangements not found for x rays [K20]. It has also been claimed that high LET radiations induce non-contiguous deletions that are not found with x or gamma rays (see also paragraph 208) [S90]. A number of these studies, therefore, suggest there could be differences in the types of large genetic changes induced by high-LET as opposed to low-LET radiation. Much more careful analysis is required before any general statement can be made about the extent or nature of these possible differences.

193. The mutagenic effects of a single alpha-particle traversal were measured with a high-frequency detection system (paragraph 171). Using a microbeam to localize 90 keV μm^{-1} alpha particles to cell nuclei, a dose-dependent increase in mutant frequency was found for traversals of 1-8 particles, with a twofold increase for 1 alpha particle traversal (cell surviving fraction = 0.8) [H42]. Molecular analysis of

these mutants gave some support for the view that higher doses of high-LET radiation induce larger genetic changes, but this system is unusual in being able to sustain loss of the whole target chromosome without lethality.

194. It has been proposed that the ratio of specific forms of chromosomal aberration may be a fingerprint of exposure to high-LET radiation [B13, S71]. A low ratio (about 6) of interchromosomal to intrachromosomal exchanges was found in several studies of aberration induction by high-LET radiation, while a value of ≥ 15 was found for x or gamma rays. However it was subsequently demonstrated [S54], using a simple two-dimensional chromosome model allowing the prediction of exchange sites relative to track structure, that more densely clustered tracks are not predicted to give a significantly lower ratio of aberration types than random scattering of tracks. More critical assessment of experimental data also does not support this theory [L54, S66]; in particular, reexamination of extensive plant data has ruled out both this theory and any LET-dependence in the ratio of interstitial deletions to inter-arm exchanges [S72]. A different high-LET fingerprint has more recently been proposed by Lucas [L55], who suggested that the proportion of unrejoined ("incomplete") chromosome aberrations will be greater with high-LET radiation, because the local density of DNA double-strand breaks is greater than for low-LET radiation and competition between broken ends leads to a greater chance of incomplete exchange events. Published data are cited that give a ratio of incomplete exchanges to complete exchanges (translocations) of about 9 for low-LET radiations and about 2 for high-LET radiations, with "mixed-LET" radiations giving intermediate values. While this theory has yet to receive critical appraisal in the literature, it may suffer from the same problems as the earlier theory, insofar as only certain sets of data give large differences between high- and low-LET radiations, partly because of difficulties in correctly defining incomplete aberrations. There may be further difficulties with the idea that competition between broken ends will necessarily lead to more incomplete aberrations and with the simple assumption that incomplete aberrations represent unrejoined breaks.

E. NOVEL MECHANISMS OF GENETIC CHANGE

195. Most of the processing of DNA damage by cellular repair enzymes is completed within a few hours of irradiation, including the fixation of mutations. However, there is evidence that cellular responses continue to occur for much longer periods, over many cell generations. These responses include delayed cell death and genetic changes. While these responses may in part be attributed to the time taken for the cell to recover from irradiation, there are in principle several reasons why persistent effects could occur. These include persistence of the damaging agent; persistence of certain forms of DNA damage, i.e. lack of repair; the repair of damage leading to rearrangements of the genome, which themselves upset the correct functioning of the cell (e.g. "position effects" on blocks of genes); and the induction of a long-lived metabolic disturbance in somatic cells, such that enzymatic activities (e.g. DNA polymerases) involved in the fidelity of maintaining the genome do not function properly.

196. It has been known for many years that cells may take some time to die following irradiation [E2, J7], but more recently emphasis has been placed on lethal mutations that may take effect many cell generations after irradiation [B53, G9, S16]. These two phenomena may in part be aspects of the same response. Delay in cell death in fibroblastic cell lines may be explained by the time taken for cells to show the effects of loss of essential genes through chromosome fragmentation (see paragraph 110). As cells divide, they segregate broken chromosomes (often seen as micronuclei), and the encoded gene products are diluted out of daughter cells, eventually causing death [J7]. In the same way, lethal mutations may cause late-onset death in cells; however, they can arise through more subtle effects on the genome than simple chromosomal fragmentation and loss. Thus, all types of mutation, from point mutation to large genetic changes, may represent lethal mutations if they lead to loss of essential gene products. However, to account for the high frequency of lethal mutations at long times after irradiation, it has also been proposed that some form of persistent genetic instability can be induced by radiation.

197. In addition to delayed death, the mutation frequency in inessential genes such as *HPRT* has been found to be persistently elevated in a large fraction of clones surviving irradiation [C20]. In these experiments, cells irradiated with x rays (12 Gy) were grown as separate clones and examined for both the proportion of cells surviving and for mutant frequency at different times after irradiation. It was found that clones showing reduced survival also commonly had elevated mutant frequencies at the *HPRT* gene for as many as 50-100 cell generations following irradiation. The mutant frequency in individual clones was highly variable but sometimes exceeded 1×10^{-3}. These delayed mutations appeared to be predominantly point mutations [L34].

198. Chromosomal aberrations have also been found to persist following irradiation. One-cell mouse embryos irradiated with x rays or neutrons showed an approximately linear increase in the frequency of chromosomal aberrations per cell in the first, second, and third mitoses post-irradiation. The relatively high frequency of aberrations, especially for neutrons, and the occurrence of chromatid-type aberrations on the third mitosis following irradiation suggested that new aberrations were being produced in post-irradiation cell cycles [W40, W41]. Similar results were obtained with x irradiation of two-cell mouse embryos [W42]. Delayed chromosomal aberrations were not observed in embryos treated with restriction endonucleases, suggesting that lesions other than DNA double-strand breaks are responsible for this effect [W48]. A significant increase in chromosome- and chromatid-type aberrations was also found in cell cultures derived from foetal skin biopsies of mice x irradiated as zygotes [P25]. The study of chromosomal aberrations in lymphocyte and fibroblast clones surviving x irradiation has similarly given evidence of persistent genetic effects [H30, M34, P33]. Analysis of clones

for two months following irradiation showed that >20% had sporadic aberrations as well as transmissable clonal karyotype alterations [H31].

199. Genetic instability in clones of cells surviving irradiation has also been described for alpha-particle irradiation of cultured haematopoietic stem cells from CBA/H mice [K8]. In this case, chromosomal aberrations were measured in cell clones surviving 3 Gy from x rays or 0.25-1 Gy from alpha particles (0.5 Gy from ^{238}Pu alpha particles corresponds to an average of one track per cell). About 50% of the clones surviving alpha-particle irradiation carried aberrations; these were mostly non-identical chromatid-type aberrations, suggesting that they had arisen many generations after irradiation. The frequency of such delayed aberration induction in x-ray survivors was only about 2% [K8]. This form of chromosomal instability in bone marrow cells was also found to be transmissible *in vivo*, by transplanting male cells irradiated with alpha particles into female recipients [W38]. The repopulated haemopoietic system showed instability persisting for up to one year. Alpha particles have also been shown to induce similar delayed chromosomal effects in the bone marrow of two out of four normal humans [K9]. It was suggested that the lack of effect in some individuals reflects genetic determinants that vary in the human population, and additional studies of other inbred mouse strains have also been found to show varying levels of this form of genetic instability. The alpha-particle-induced instability was, however, found to be independent of the p53 status of the cell [K32].

200. Heavy ion (neon, argon, or lead) irradiations have also been found to induce chromosomal instability in cultured human fibroblasts [M35, S53]. Analysis of mass cell cultures for up to 25 passages following irradiation showed that the frequency of aberrations declined at first but then increased until >60% of the cells showed aberrations. In contrast to clonal cell populations, these aberrations showed that the telomeric regions of specific chromosomes (1, 13, and 16) were involved. In a study of alpha particles of different LET, the frequency of micronuclei was found to be increased over that in unirradiated hamster cells seven days after irradiation, even at a dose giving one alpha-particle traversal per cell nucleus [M36]. Calculations indicate that the target size for this effect exceeds the size of the nucleus, suggesting that direct DNA damage by radiation tracks is not causal.

201. It has been speculated that these events indicate that ionizing radiation may induce an "untargeted" mechanism of mutagenesis in cells, as a result of the epigenetic alteration of enzymatic pathways controlling genomic stability [L18]. The idea that DNA may not require a direct hit from radiation to have an increased frequency of genetic changes has received support from the measurement of non-mutational responses. With reference to the process of carcinogenesis in particular, the work of Kennedy et al. [K10, K11] suggests that radiation may induce high-frequency events that predispose the cell towards further (spontaneously-occurring) changes in the process of malignant cell transformation. Furthermore, a study of the induction of sister-chromatid exchanges in

immortalized hamster cells by very low doses of ^{238}Pu alpha particles claimed that induction could be measured at doses (0.3 mGy) where <1% of the cell's nuclei were traversed by an alpha-particle track [N8]. Subsequent studies of primary human fibroblasts confirmed this finding; low doses of alpha particles give three times as many sister-chromatid exchanges [D13] or a five times higher *HPRT* gene mutation frequency [N25] than predicted from the number of nuclei traversed by the particle tracks, giving a considerably larger target size for this effect than expected from nuclear dimensions [D13]. Similarly, the use of a physical barrier to protect one part of a cell population from alpha-particle irradiation showed the expected reduction in killing of mouse bone-marrow stem cells but did not reduce chromosomal instability [L63]. These data are considered as evidence for the existence of a "bystander" effect; that is, damage signals may be transmitted from irradiated to neighbouring unirradiated cells. In a refinement of this type of experiment, an alpha-particle microbeam was used to give precise irradiation of the cytoplasm of cells, without damaging their nuclei. Under these conditions, an average increase of three-fold was found in the mutation frequency in a sensitive human-hamster hybrid cell line (paragraph 171), and the mutation spectrum was similar to that occurring spontaneously [W51]. One explanation of this phenomenon is that reactive oxygen species are generated by radiation in the whole cell and perhaps also in the surrounding medium, and these species or stable reaction products (e.g. lipid peroxides) diffuse into the nucleus and persist to cause chromosomal effects. Evidence for this idea has been found with x or neutron irradiation of bone marrow cells; various indicators of persistent oxygen radical activity were found in cell cultures seven days after irradiation [C40]. It has also been shown that alpha particles can produce factors in culture medium or in cells that cause increases in sister-chromatid exchanges, as noted above; these factors are inhibited by superoxide dismutase, an enzyme that catalyses the conversion of superoxide ions produced in water radiolysis [L17]. Similarly, in the microbeam experiments, reactive oxygen species were implicated in the mutagenic effects of cytoplamsic irradiation [W51]. Further, in human fibroblasts, the modulation of proteins involved in the p53-dependent pathway (Section II.B.2) following very low doses of alpha particles occurs in more cells than have been traversed by an alpha-particle track, apparently through cell-cell contact [A24]. The potential consequences of bystander effects are discussed further in Annex G, *"Biological effects at low radiation doses"*.

202. In an attempt to link the delayed appearance of chromosomal aberrations to cancer-proneness, Ponnaiya et al. [P24] measured this form of instability in strains of mice differing in their sensitivity to radiation-induced mammary cancer. Strikingly, cells from the more sensitive strain (BALB/c) showed a marked increase in the frequency of chromatid aberrations after 16 population doublings, while the less sensitive strain (C57BL/6) showed no increase in aberrations over the control level.

203. Higher-than-average frequencies of specific types of chromosomal aberrations have been found after irradiation of

strains of mice that are prone to certain types of cancer. These aberrations are not necessarily the delayed effects of irradiation but nonetheless may indicate that radiation damage to specific parts of the genome contributes disproportionately to cancer induction. As an example, interstitial deletions of mouse chromosome 2 are consistently associated with radiation-induced acute myeloid leukaemia in several inbred strains of mice. The chromosome breakpoints involved in these chromosome 2 deletions are non-randomly distributed and can be detected at high frequency at early times following irradiation, suggesting that they may be an early event in leukaemogenesis [B54, B55, H43, H45, R28, T23]. Investigation of two mouse strains showing large differences in susceptibility to radiation-induced thymic lymphomas showed that at early times following x irradiation the incidence of specific chromosomal aberrations, especially trisomy 15, was much higher in the cancer-prone strain than in the resistant strain [C21]. These examples suggest that certain chromosomal regions may be more sensitive to radiation-induced genetic changes that are associated with cancer, although it is difficult to quantify this contribution precisely without a better understanding of the development of aberrant clones of cells in animal tissues.

204. In some mutation systems very high frequencies of mutation occur both spontaneously and after irradiation. With radiation induction, the amount of initial damage (e.g. the numbers of DNA double-strand breaks per unit dose) in the mutational target may not in these cases be sufficient to account for the numbers of mutations induced. While it is possible to consider that every type of damage (paragraph 15, Table 1) may sometimes lead to mutation, these examples of high-frequency mutation have usually been proposed to arise from a form of genetic instability. For instance, somatic mutation frequencies of coat colour in *Pink-eyed unstable* mice are linear with x-ray dose down to 10 mGy and occur at least 100 times more frequently per unit dose than germ-line mutations in the mouse specific-locus tests [S57]. The *Pink-eyed unstable* male mice have a reduction in the pigment of the coat and eyes, caused by gene duplication interfering with normal pigment production; deletion of this duplication is scored as the mutation, giving normal pigmentation seen as black spots on the grey coat.

205. The spontaneous frequency of mutation at tandem-repeat (VNTR or "minisatellite") DNA sequences in the germ line of mice is also very high (1%-10% in offspring). The mechanism of mutation at these hypervariable loci does not appear to involve unequal exchange between homologous chromosomes [W30] but rather some form of complex gene conversion process [J10]. The incorporation of human tandem-repeat sequences into yeast cells, to study the mechanisms of variability, has confirmed that such sequences are destabilized in meiosis and that this process depends on the initiation of homologous recombination at a nearby DNA double-strand break [D23]. Irradiation of mouse spermatogonial stem cells by x rays increased the mutation frequency in offspring relative to controls [D10]. More extensive experiments at different stages of spermatogenesis using both single- and multi-locus probes showed no increase in mutation

frequency in post-meiotic spermatids but gave a doubling dose of 0.33 Gy for premeiotic spermatogonial and stem cells. The dose-response for mutation induction by 0.5-1 Gy x rays, combined for spermatogonial and stem cells, was linear [D19]. In contrast, in a separate study using one of the same probes at different germ-cell stages, the irradiation of spermatogonia gave a non-significant increase in mutation frequency, while the frequency in irradiated spermatids was significant. The increase in mutation frequency was not linear with dose, showing little increase above 1 Gy [F5, S19]. Similar data were reported for ^{252}Cf irradiations (35% gamma rays, 65% neutrons): spermatids again showed the highest induced mutation frequency, and a single dose (1 Gy) to spermatogonia was significantly mutagenic. The RBE for these mutations with ^{252}Cf was 5.9 for spermatogonia, 2.6 for spermatids, and 6.5 for spermatogonial stem cells [N15]. More work is required to reconcile the discrepancies that are seen for different germ-cell stages. Recently a striking result has been found from tandem-repeat mutation studies following irradiation of male mice with 0.5 Gy ^{252}Cf neutrons, and the mating of these mice to unirradiated mice through two generations to yield second-generation (F_2) progeny. Remarkably, the F_2 mice showed an elevated frequency of mutation in the repeat sequences inherited from both male and female lines (6-fold and 3.5-fold, respectively), suggesting that genetic instability can be transmitted through the germline [D24]. The induced mutation frequency at these hypervariable sequences seems to be too high, by two orders of magnitude, for direct damage by radiation at the sites of mutation; it was proposed that some indirect mechanism of mutation induction is responsible [D19, F5, S19, D24].

206. Human spontaneous germ-line mutation frequencies at tandem-repeat loci can also be high (1%-7% per gamete). In a pilot study of children of survivors of the atomic bombings (mean gonad dose, generally to only one of the parents = 1.9 Sv) and matched controls, mutation frequency was measured at six tandem-repeat loci. The average mutation frequency was similar in the two groups, at 1.5% per gamete per locus for the exposed gametes and 2% for the unexposed gametes [K26]. A further study of the same children using a probe to detect multiple-repeat loci again showed no increase in mutation rate for the exposed group [S67]. However, a twofold increase in the frequency of mutation at tandem-repeat loci has been reported for children of parents resident in the Mogilev district of Belarus, which was heavily contaminated in the Chernobyl accident [D14]. Four tandem-repeat loci were tested, three of which showed increases by a factor of 1.7-2.0, while one locus (in which only one mutation was found) showed a reduction in the exposed group. The 79 families tested were compared to a control Caucasian population from the United Kingdom, which showed a similar overall distribution of repeat lengths at these loci. In a follow-up study using five additional tandem-repeat loci and including a further 48 families from Mogilev, the same general twofold increase was found when compared with the same (United Kingdom) control population [D1]. The mean dose to the exposed population was calculated to be 27.6±3.3 mSv from ^{137}Cs; some evidence for a dose-response relationship was obtained by dividing the exposed group into

those receiving >20 mSv (mutation rate = 0.024) and those receiving <20 mSv (rate = 0.018), compared with the control rate of 0.011. While it is clear that there is a general increase in the mutation rate for several different genomic sites in the exposed population, difficulties of interpretation exist because of the geographical disparity of the control group and the possibility that other environmental agents may be responsible for the increased mutation rate [S67]. As a direct test of the involvement of tandem-repeat loci in radiation-induced carcinogenesis, normal and tumour DNA from post-Chernobyl thyroid carcinomas was examined for mutations in three loci [N22]. Mutations were found in 3 of 17 tumours (18%), with one of these having mutations in all three loci, while none of 20 sporadic thyroid cancers from patients without a history of radiation exposure showed tandem-repeat mutations.

207. Recent studies of "hypermutation" in non-dividing cultures of bacteria placed under stress have also suggested that recombination processes are involved in the formation of small deletions at high frequency [R29]. It is suggested that when bacteria are starved of nutrients, for example, mutations arise at a very high rate in order to survive (if sufficient mutations occur, one of these may be sufficiently favourable to allow the cells to adapt to the prevailing conditions). Evidence supports a model in which cells enter a transient hypermutable state in which DNA double-strand breaks initiate homologous recombination activity (see Figure Id), priming error-prone DNA synthesis. It seems likely that the errors (mutations) arise as a result of a down-regulation of the mismatch repair system (paragraph 166) [H44]. Similarly, specific types of oxidative base damage (e.g. 8-oxoguanine, paragraph 24) may lead to small deletion mutations when mismatch correction is compromised [B56]. The significance of these findings is that the intrinsic mutability of cells is modifiable, and this principle may apply equally to many types of cell, perhaps including those involved in carcinogenesis [R30].

208. Other reported phenomena may also be connected to the induction of genetic instability following irradiation. One of these is coincident mutation: in cells selected for radiation-induced mutation at one genomic site, a high frequency of mutations occurs at other sites. Li et al. [L16] found two mutations at tandem-repeat loci among 50 x-ray-induced TK gene mutants, a frequency of $4 \cdot 10^{-2}$ and far in excess of expectation based on current knowledge of radiation-induced gene mutation frequencies. No second-site mutations were found in 70 unirradiated clones. Perhaps the clearest example of coincident mutations was found in an extensive study of x-ray-induced mutation in the fungus Neurospora crassa [D5]. In a chromosomal region containing 21 genes, about 10% of the radiation-induced mutants had mutations in more than one gene following a single acute dose. However, a yeast cell study has revealed a possible mechanistic basis for non-targeted mutations that occur relatively close to the initial site of damage, without invoking genetic instability [S52]. A DNA double-strand break was placed at a single site in one yeast chromosome, using a site-specific endonuclease, and mutations were measured in a gene situated adjacent to the break site. After the break had undergone recombination repair, it was found that the adjacent gene had sustained a 300-fold increase in the frequency of point mutations. This result suggests that the break repair process is error-prone, presumably because of a lack of fidelity in DNA repair synthesis, and extends over distances of several hundred DNA base pairs.

209. It is difficult to know how the reports of delayed genetic effects and instability following both low- and high-LET radiation may apply to humans. As indicated above (paragraphs 201-203), a high frequency of radiation-induced genetic or epigenetic changes may contribute to cancer incidence, at least in those instances where an accumulation of somatically-stable changes is required. However, it is still not clear whether genetic instability or a higher-than-normal mutation rate is a necessary for the development of tumours [T38]. In the case of inherited genetic alterations, where germ cells are the target, measurements of mutation frequency made at short intervals after irradiation could underestimate the induced mutation frequency. However, in the case of alpha-particle irradiation, mice injected with ^{239}Pu and subsequently mated to tester stocks (specific-locus method) over many weeks were not found to show increasingly high levels of mutation [R16]. It is, of course, possible that some potential increase in mutations may be balanced by selection against sperm carrying an increased load of genetic damage due to instability. The type of mutation induced may be important; if these are point mutations, they could have a greater chance of transmission and could lead to dominant genetic effects in offspring. The potential importance of genetic instability in carcinogenesis, especially at low radiation doses, is discussed further in Annex G, "Biological effects at low radiation doses".

F. MUTATION FREQUENCIES AND CONSEQUENCES

210. Radiation-induced mutations are always measured against a background of spontaneously-occurring mutations. The mechanisms of spontaneous mutation are numerous; the chemical reactivity of DNA leads to instability, and there are inherent errors in replicating a very large molecule. Some of these mechanisms will overlap those of radiation damage; for example, oxidative damage from metabolic processes in aerobic organisms will give both base damage and strand breaks (paragraph 19). This indicates why many of the types of mutation that occur spontaneously are similar to those formed by ionizing radiation. The cell requires efficient repair processes to cope with endogenous damage; if unrepaired, the damage will lead to base-pair substitutions as well as some larger changes. The increase in mutant frequency found after exposure to ionizing radiation is likely to come from both the additional load of damage similar to that occurring spontaneously (such as DNA base damage and loss) and more complex radiation-induced damage that cannot be handled easily by the cell's battery of repair enzymes (Section I.B).

211. As was stressed in Section IV.B, an important influence on the observed frequency of mutation is the tolerance of the genome site to large changes. When large changes are tolerated, the spectrum of mutations induced by ionizing radiation shows that at least half (and commonly more than half) of the mutations measured shortly after irradiation are large deletions and rearrangements. This spectrum may change with time after irradiation, but as yet there are too few data to comment meaningfully on this possibility. Mutation frequencies have been measured for a few target genes, which have been chosen for their ease of use and especially for their presence in a hemizygous state. While these measurements seem unlikely to represent the genome as a whole (see Section IV.E), they do give some idea of the variation in frequencies and the reasons for this variation. An interesting comparison can be made of the hamster *APRT* and *HPRT* genes; since both of these code for inessential enzymes of similar function (purine salvage) they should both detect all types of mutant. However, the *APRT* gene is thought to have an essential gene situated downstream, which limits the possibility of detecting very large deletions (paragraph 179). Table 4 shows that the frequencies of both spontaneous and radiation-induced *APRT* mutations are lower than those for *HPRT* by more than an order of magnitude.

Table 4
Mutant frequency and spectrum in two different genes of hamster cells

Agent	Mutant frequency (10^{-6} cells)	Number analysed	Deletions/rearrangements (%)
APRT (autosomal hemizygous) [M13]			
None	0.13	125	7.2
EMS [a]	430	48	0
Gamma rays	1.5–3.0 [b]	85	22.3
HPRT (X-linked hemizygous) [T20]			
None	6.2	44	18
EMS [a]	690	56	0
Gamma rays [c]	38.8	48	71

a   Ethylmethane sulphonate, an alkylating agent.
b   Doses 2.5 and 4 Gy.
c   Dose 5 Gy (both EMS and gamma-ray doses gave 20% cell survival).

212. Whether available mutation systems are sensitive enough to yield useful data on genetic damage in situations of practical importance is questionable. There has been some controversy over attempts to measure, for example, the frequency of *HPRT* gene mutations in the blood lymphocytes of radiotherapy technicians and patients. It was concluded that some of these studies are better at revealing the variables involved in measuring mutation at low doses than at giving reliable data on mutation frequencies (reviewed in [T22]). One study [N16], however, included molecular analysis of the mutations found in lymphocytes of radioimmune therapy patients and showed that a higher proportion with large genetic changes occurred than in controls, with some dose dependence for the fraction of mutants with large changes due to cumulative ^{131}I activity. Studies of the survivors of the atomic bombings [H32, H33] reported a slight increase, about 10% per Gy, in *HPRT* mutant frequency as estimated dose increased. This frequency increase is considerably lower than found in freshly-irradiated lymphocytes, suggesting that *HPRT* mutants are selected against over the long time period involved. However, it has recently been found that a significant increase in *HPRT* mutant frequency could be detected in combined data from 142 liquidators involved in the Chernobyl accident (24% increase in the liquidator group relative to Russian controls) after adjustment for age and smoking [T49].

213. The problem of selection against mutant cells seems to be less severe in some more recently developed mutation assays. An assay based on the loss of one copy of the cell-surface marker glycophorin A, encoded by the *GPA* gene, in human erythrocytes gave dose-dependent increases in mutant frequency in survivors of the atomic bombings, and the frequencies per unit dose were similar to those found in human cells irradiated with low-LET radiation under laboratory conditions [K1, K33, L44, L45]. In contrast to the *HPRT* data, there was a positive correlation between the frequencies of chromosomal aberrations and *GPA* mutants in exposed individuals; additionally, survivors with malignant solid tumours showed a significantly higher mutant frequency than those without cancer. This assay will measure chromosome loss (by, for example, non-disjunction) as well as mutation and recombination events in individuals already heterozygous for the *GPA* gene (about half of the human population), but it has the drawback that no molecular analysis of the mutations is possible (mature erythrocytes lack nuclei) [G24]. The *GPA* mutation system has also shown dose-dependent mutation induction in individuals exposed to doses of up to 6 Gy during or following the Chernobyl accident, and the slope of the dose response was very similar for these measurements and the *GPA* mutants measured in survivors of the atomic bombings (at about 25 10^{-6} mutants Gy^{-1}) [J18]. This result has received support from subsequent measurements of *GPA* mutation in liquidators [W53] or

people living in the vicinity of Chernobyl [L71] at the time of the accident. However, use of the *GPA* system to measure mutation in more than 700 Estonian and Latvian workers involved in the Chernobyl clean-up, with estimated median doses of about 100 mGy, did not detect a consistent increase in mutant frequency [B39]. The mutagenic effects of 5.3 MeV alpha particles (average LET = 140 keV μm^{-1}) in a human exposed to ^{232}Th over a 43-year period following thorotrast injection were measurable using the *GPA* gene assay [L11]. These data showed a fivefold increase in *GPA* mutants and correlated to large increases in chromosomal aberration frequency in lymphocytes from the same individual. Other data associated with the consequences of the Chernobyl accident are considered in the Annex J, *"Exposures and effects of the Chernobyl accident"*.

214. Where enough measurements have been made, the average frequency of gene mutation induced by low-LET radiation is similar in cells from different somatic tissues and species. This can be illustrated by a plot of induced *HPRT* mutant frequency against surviving fraction (Figure VIIIa), where differences in intrinsic radiosensitivity are taken into account [T13, T14]. The plot also indicates that there is some consistency in the relationship between mutation and killing, suggesting that these responses derive from similar types of damage, a constant fraction of which is converted to mutations [T17]. It should be noted that this relationship does not say anything about the absolute mutant frequencies for different tissues or organisms; these may differ substantially, as seen, for example, in the response of different germ-cell stages in the mouse [S29].

215. The mutation frequency/survival plot can also be used to show that both the type of radiation and cellular parameters can influence the effectiveness of mutation induction. Thus, for *HPRT* mutation, densely ionizing radiation shows an increase in the effectiveness by a factor of no more than 2; i.e. the RBE for mutation is about twice that for cell killing (Figure VIIIb) [C22, T16, T18]. Similarly, x rays vary in effectiveness with the phase of the cell cycle; an increase in the effectiveness of mutation induction by factors of 2-3 is found relative to cell killing, especially in G_1/S phase [B14, J6]. It is tempting to speculate that the repair systems operational at this point in the cycle (see Section II.B.1) are more prone to recombinational errors.

216. Mutation induction is also subject to dose-rate effects; in general, the effectiveness of low-LET radiations is reduced by factors of 2-4 at low dose rates [T22]. However, it has also been found in specific conditions that the mutagenic effectiveness may remain the same or may increase at low dose rates relative to high dose rates [T22]. For example, if TK6 lymphoblastoid cells are exposed to low dose rates, no change in mutagenic effectiveness is found [K36], while their more radiation-resistant counterpart WIL2-NS (paragraph 185) shows an approximately twofold reduction in effectiveness [F11]. Lack of alteration in mutagenic effectiveness with dose rate has also been seen for radiation-sensitive (DNA-repair-deficient) cell lines [F12]. In addition, at dose rates of low-LET radiation of ≤0.5 mGy min^{-1}, rodent cell lines may

show no dose-rate effect [E10, E11, F13] or an increased (inverse) dose-rate effect [C44, C45]. Also, inverse dose-rate effects have been seen with low dose rates of high-LET radiation [K37, N18].

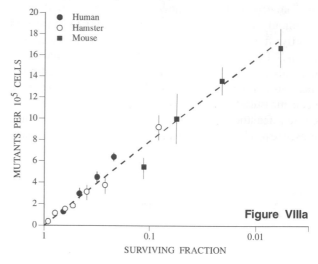

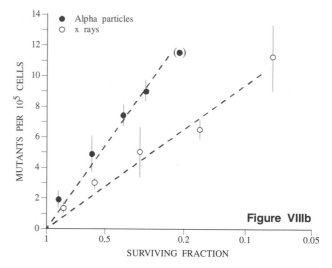

Figure VIII. Mutation-survival relationship for *HPRT* genes.
VIIIa: Similarity of induced mutant frequency in cell lines from human, mouse and hamster [T17];
VIIIb: The increased effectiveness of densely ionizing radiation (^{238}Pu alpha particles) compared to x rays in hamster cells [T18].

217. Some of these dose-rate data point to specific cellular processes influencing the response; in particular, it appears that the response is dependent on the capacity of cells to repair damage to DNA [T22]. It has long been thought that cellular repair capacity is responsible for the reduction in mutagenic effectiveness with dose rate, as seen for example in classical mouse germ-cell data [R6, R7]. Where little or no dose-rate effect is found, the repair capacity of the genetic region or of the whole cell is therefore likely to be impaired. This situation may also apply in some germ-cell stages; spermatozoa of both *Drosophila* and the mouse show no dose-rate effect [R6, T31]. The finding that very low dose rates give little relative effect, or an inverse effect, in some somatic cell lines may indicate that the full repair capacity of a cell requires a minimum level of insult (damage induced per unit

time) before it is brought to bear on that damage. However, it should be noted that chronic irradiation of mouse spermatogonial stem cells has shown no further reduction in specific-locus mutation frequency below 8 mGy min^{-1} with dose rates down to 7 µGy min^{-1} [R17].

G. SUMMARY

218. DNA damage caused by ionizing radiations leads to various types of mutation: the smaller mutations, such as base-pair substitutions, appear to result from damage to single bases, while larger changes, such as large deletions and rearrangements, probably arise from DNA double-strand breakage. The spectrum of radiation-induced mutations is dominated by the larger molecular changes if the genetic site assessed will tolerate these without lethality. In view of the propensity to induce large genetic changes, it may be supposed that the average radiation-induced mutation has greater consequences in terms of extensive alterations of the genome than the average spontaneous mutation or those induced by other agents. As yet no distinct mutational fingerprint has been identified for ionizing radiations, although there is a suggestion that the more complex types of molecular change may be overrepresented in the spectrum. High-LET radiations induce a higher mutation frequency per unit dose, but there are insufficient data to establish whether the mutation spectrum is different from that for low-LET radiations.

219. Attempts by cellular enzymes to repair DNA damage are intimately linked to mutation formation. The molecular mechanisms of radiation-induced mutation include illegitimate recombination, but where genes are heterozygous there is evidence that homologous recombination is involved (giving loss of heterozygosity, commonly seen in some cancers as loss of tumour-suppressor gene function). Some forms of radiation-induced base damage will be repaired by mismatch repair pathways; loss of mismatch repair function can lead to hypermutability, which is linked to specific forms of cancer such as hereditary non-polyposis colon cancer. The involvement of repair processes in mutation formation suggests that the intrinsic mutability of cells is not fixed but will vary with their repair capacity; this concept may have important consequences for the process of carcinogenesis.

220. There are several reports of persistent genetic effects in cells and animals following irradiation, suggesting that genetic instability is induced. A possible explanation for this phenomenon is that stable oxidative reaction products persist and can continue to give genetic damage in subsequent cell generations, although other explanations are possible (paragraph 195). In some mutation systems very high frequencies of mutation (hypermutability) occur spontaneously and after irradiation; on the basis of target size, it is unlikely that these mutations are induced directly by radiation damage at the sites of mutation. Hypermutability also suggests that some form of genetic instability is induced by radiation treatment, but at present it is difficult to know how these data relate to responses such as cancer induction.

221. Methods have been established for measuring mutations with some accuracy in human somatic cells, such as blood lymphocytes, but most mutation systems lack the sensitivity to be used as indicators of genetic damage in cases of low-dose radiation exposure. Some success has been found in measuring mutant frequencies in cells of individuals exposed to relatively high doses of radiation, such as atomic bomb survivors and Chernobyl recovery operation workers.

CONCLUSIONS

222. DNA repair processes have evolved in all biological organisms to combat the deleterious effects of damage to their genetic material. Attempts to repair DNA damage also cause many types of mutation, through inability to properly restore the DNA sequence. Many of the genes involved in the repair of DNA damage in human cells have now been cloned, and their functional analysis has led to considerable progress in our understanding of the ways in which cells and organisms respond to radiation damage. Several different pathways of repair are required to cope with damage from ionizing radiation (Figure III), and the consequences of losing repair capacity can be drastic. The importance of the repair gene function has been revealed in particular by the development of methods to knock out specific genes in experimental animals. Using these methods, it has been found that the loss of repair gene activity is often lethal in early stages of development (Table 2). Where the knockout animals survive the loss of repair capacity, they are commonly very prone to cancer. These findings reinforce more limited studies with rare individuals in the human population, which have suggested for some time that radiosensitivity is linked to cancer proneness. The converse finding, that mice lacking homologues of the human breast-cancer-predisposing (*BRCA*) genes are radiosensitive, is a further important illustration of this link. Loss of repair gene function is considered to lead to cancer proneness primarily through an increase in genetic instability, although the mechanistic details of this process remain to be elucidated. The precise contribution of loss of repair gene function to the risk of radiation-induced cancer in the population is unknown at present; if calculations are restricted to those rare individuals with recognized repair syndromes, then this contribution will be small, but it is already known that a more subtle variation in these genes occurs widely in the human population.

223. Cellular survival assays indicate a twofold variation in the response of individuals in the general population to acute radiation exposure. This variation may become three- or fourfold when irradiation is given at low dose rates. Specific groups of individuals may show a consistently elevated sensitivity; for example, individuals heterozygous for the ataxia-telangiectasia-mutated (*ATM*) gene, who constitute

about 1 percent of the population, show enhanced radio-sensitivity especially when tested in a chromosome-damage assay. This finding has recently been reinforced by similar data with transgenic mice heterozygous for ATM deficiency. Strikingly, using the chromosome-damage assay, one study has shown that about 40 percent of breast cancer patients show a similar enhancement of radiation sensitivity. Recent evidence shows that the enhanced radiosensitivity of breast cancer patients has a genetic basis; it has been suggested that the *ATM* gene is involved in predisposition to breast cancer, but the evidence for this is controversial. It is possible that small reductions in the efficiency of any one of a number of genes involved in radiation response, due to subtle mutations or polymorphisms, will account for the existence of radio-sensitive groups of individuals.

224. The analysis of DNA repair processes has revealed that they are part of a complex response system in our cells, which includes genes involved in recognizing and signalling the presence of damage and genes operating checkpoints to ensure that cells do not progress through the cell cycle if they carry DNA damage. The functional analysis of repair genes involved in radiation response has also revealed a link between radiosensitivity and immune dysfunction, because some of the gene products involved in the repair of radiation-induced DNA double-strand breaks also assemble functional immune genes. Thus, radiosensitivity in humans can arise from the loss of a broader spectrum of gene functions than was initially recognized.

225. A reduction in DNA repair capacity may have several consequences for the cellular radiation response, including alteration in the shapes of dose-response curves, the loss of low-dose-rate sparing, and a loss of relative effectiveness for high-LET radiations. It is probable that these consequences arise mainly from an alteration in the ability to repair complex forms of DNA damage, particularly those involving double-strand breaks, since this type of damage is extremely hazardous to the cell. DNA repair capacity is therefore an important component of radiation dose-response, and it is clear that it needs to be considered as a variable in modelling radiation action, especially when attempting to extrapolate to low dose exposure. However, before the effects of repair pathways can be modelled accurately, there is still consider-ably more to learn about the way in which different pathways contribute to the overall response to radiation. At present it is not known how cells control the use of different repair path-ways in responding to damage, and in particular how a balance is achieved between correct (error-free) repair and repair leading to mutation induction. Additionally, more information is needed on the relationship between DNA repair and apoptosis in different cells and tissues to predict the outcome of radiation exposure.

226. There is experimental evidence that pre-irradiation with low doses (5-10 mGy) can increase the resistance of cells to a subsequent higher dose, and that a more sensitive cellular response to radiation exists at low doses than at higher doses. These observations have led to some controversy; they suggest that under some circumstances a resistance factor can be induced by low radiation doses, but the mechanistic basis of

the observations has proved difficult to establish. It has also been found in separate studies that a number of different genes and proteins are induced or repressed by radiation, although few of these appear to be involved directly in the repair of DNA damage. To date little of this characterization has been carried out in a systematic fashion, because it has been conditioned by the availability of cloned genes and proteins. However, new technology based on large-scale gene sequencing coupled to micro-arraying of sequences is beginning to revolutionize this type of study by enabling the assay of hundreds or thousands of gene products at one time. By examining the levels of many gene products from cells before and after irradiation, as well as with time elapsed following irradiation, it will be possible to see how the damage response is coordinated. These methods also permit a molecular description of the differences in levels of gene products present in normal and diseased tissues, including tumour tissue, facilitating an understanding of the mechanistic basis of cancer induction by agents such as ionizing radiation.

227. Many different types of DNA damage are caused by ionizing radiations, ranging from isolated single-strand damage at sites of single ionizations to complex DNA altera-tions at sites of clustered ionizations. The more complex forms of damage may be unique to the interaction of ionizing radiations with DNA, compared with damage occurring spon-taneously or that caused by other DNA-damaging agents. Several attempts have been made to establish whether radia-tion damage can lead to a distinct fingerprint of genetic chan-ges, but this has proved elusive. It is clear, for example, that the spectrum of gene mutation arising from radiation damage has differences from the spontaneous mutation spectrum, but the overlap in the two spectra is considerable. Again, new technologies based on fluorescent *in situ* hybridization, the use of reporter genes with fluorescent tags, and comparative genomic hybridization will allow a refined view of radiation-induced genetic changes and the hope of future distinctions.

228. Many genetic changes caused by radiation occur within a few hours of giving the dose, but there is experimental evidence to show that some forms of change may occur after much longer times and following many cell divisions. In addition, at some sites in the human genome, the frequency of radiation-induced genetic changes is much higher than would be expected based on direct damage to DNA by radiation tracks. These observations broaden our knowledge of the mechanisms of radiation action, but elucidation is required further before the consequences of these mechanisms for radiation risk can be understood.

229. It is clear that much more knowledge of the structure of the human genome, in particular the disposition of the genes within it and their responses to radiation, is required before it will be possible to predict the average frequency of mutation induced by a given dose of radiation. Additionally, the range of repair capacities present in the human population, including carriers of defective repair genes, will have to be considered to predict mutability on an individual basis. A more complete knowledge of these response mechanisms will allow greater accuracy in the prediction of radiation-induced carcinogenic and hereditary effects.

References

A1 Alberts, B.M. Protein machines mediate the basic genetic processes. Trends Genet. 1: 26-30 (1985).

A2 Arlett, C.F., S.A. Harcourt, A.R. Lehmann et al. Studies of a new case of Xeroderma pigmentosum (XP3BR) from complementation group G with cellular sensitivity to ionising radiation. Carcinogenesis 1: 745-751 (1980).

A3 Afanasyev, V.N., B.A. Korol, Y.A. Mantsygin et al. Flow cytometry and biochemical analysis of DNA degradation characteristic of two types of cell death. FEBS Lett. 194: 347-350 (1986).

A4 Anderson, C.W. DNA damage and the DNA-activated protein kinase. Trends Biochem. 18: 433-437 (1993).

A5 Alt, F.W., E.M. Oltz, F. Young et al. VDJ recombination. Immunol. Today 13: 306-314 (1992).

A6 Arlett, C.F. and S.A. Harcourt. Survey of radiosensitivity in a variety of human cell strains. Cancer Res. 40: 926-932 (1980).

A7 Aurias, A. and B. Dutrillaux. Probable involvement of immunoglobulin superfamily gene in most recurrent chromosomal rearrangements from ataxia telangiectasia. Hum. Genet. 72: 210-214 (1986).

A8 Aghamohammadi, S.Z., T. Morris, D.L. Stevens et al. Rapid screening for deletion mutations in the hprt gene using the polymerase chain reaction: X-ray and α-particle mutant spectra. Mutat. Res. 269: 1-7 (1992).

A9 Aghamohammadi, S.Z. and J.R.K. Savage. A BrdU pulse double-labelling method for studying adaptive response. Mutat. Res. 251: 133-141 (1991).

A10 Albering, H.J., J.J. Engelen, L. Koulischer et al. Indoor radon, an extrapulmonary genetic risk? Lancet 344: 750-751 (1994).

A11 Amundson, S.A. and H.L. Liber. A comparison of induced mutation at homologous alleles of the tk locus in human cells. Mutat. Res. 247: 19-27 (1991).

A12 Amundson, S.A. and H.L. Liber. A comparison of induced mutation at homologous alleles of the tk locus in human cells. II. Molecular analysis of mutants. Mutat. Res. 267: 89-95 (1992).

A13 Amundson, S.A., F. Xia, K. Wolfson et al. Different cytotoxic and mutagenic responses induced by x-rays in two human lymphoblastoid cell lines derived from a single donor. Mutat. Res. 286: 233-241 (1993).

A14 Arlett, C.F., M.H.L. Green, A. Priestley et al. Comparative human cellular radiosensitivity. I. The effect of SV40 transformation and immortalization on the gamma-irradiation survival of skin derived fibroblasts from normal individuals and from ataxia-telangiectasia patients and heterozygotes. Int. J. Radiat. Biol. 54: 911-928 (1988).

A15 Azzam, E.I., G.P. Raaphorst and R.E.J. Michel. Radiation-induced adaptive response for protection against micronucleus formation and neoplastic transformation in C3H 10T½ mouse embryo cells. Radiat. Res. 138: S28-S31 (1994).

A16 Aurias, A., J-L. Antonie, R. Assathiany et al. Radiation sensitivity of Bloom''s syndrome lymphocytes during S and G_2 phase. Cancer Genet. Cytogenet. 16: 131-136 (1985).

A17 Antoccia, A., R. Ricordy, P. Maraschio et al. Chromosomal sensitivity to clastogenic agents and cell cycle perturbations in Nijmegen breakge syndrome lymphoblastoid cell lines. Int. J. Radiat. Biol. 71: 41-49 (1997).

A18 Athma, P., R. Rappaport and M. Swift. Molecular genotyping shows that ataxia-telangiectasia heterozygotes are predisposed to breast cancer. Cancer Genet. Cytogenet. 92: 130-134 (1996).

A19 Araki, R., M. Itoh, K. Hamatani et al. Normal D-JH rearranged products of the IgH gene in SCID mouse bone marrow. Int. Immuno. 8: 1045-1053 (1996).

A20 Araki, R., A. Fujimori, K. Hamatani et al. Nonsense mutation at Tyr-4046 in the DNA-dependent protein kinase catalytic subunit of severe combined immune deficiency mice. Proc. Natl. Acad. Sci. U.S.A 94: 2438-2443 (1997).

A21 Agarwal, M.L., W.R. Taylor, M.V. Chernov et al. The p53 network. J. Biol. Chem. 273: 1-4 (1998).

A22 Appleby, J.M., J.B. Barber, E. Levine et al. Absence of mutations in the ATM gene in breast cancer patients with severe responses to radiotherapy. Br. J. Cancer 76: 1546-1549 (1997).

A23 Andreev, S.G., I.K. Khvostunov, D.M. Spitkovsky et al. Clustering of DNA breaks in chromatin fibre: dependence on radiation quality. p. 133-136 in: Microdosimetry: An Interdisciplinary Approach (D.T. Goodhead et al., eds.). Roy. Soc. Chem., Cambridge, 1997.

A24 Azzam, E.I., S.M. de Toledo, T. Gooding et al. Intercellular communication is involved in the bystander regulation of gene expression in human cells exposed to very low fluences of alpha particles. Radiat. Res. 150: 497-504 (1998).

A25 Amundson, S.A., M. Bittner, Y. Chen et al. Fluorescent cDNA microarray hybridization reveals complexity and heterogeneity of cellular genotoxic stress responses. Oncogene 18: 3666-3672 (1999).

A26 Amundson, S.A., K.T. Do and A.J. Fornace. Induction of stress genes by low doses of gamma rays. Radiat. Res. 152: 225-231 (1999).

A27 Au, K.G., S. Clark, J.H. Miller et al. Escherichia coli mutY gene encodes an adenine glycosylase active on G-A mispairs. Proc. Natl. Acad. Sci. U.S.A. 86: 8877-8881 (1989).

A28 Aspinwall, R., D.G. Rothwell, T. Roldan-Arjona et al. Cloning and characterization of a functional human homoog of Escherichia coli endonuclease III. Proc. Natl. Acad. Sci. U.S.A. 94: 109-114 (1997).

B1 Bootsma, D. and J.H.J. Hoeijmakers. Engagement with transcription. Nature 363: 114-115 (1993).

B2 Baer, R., A. Heppell, A.M.R. Taylor et al. The breakpoint of an inversion of chromosome 14 in a T-cell leukemia: sequences downstream of the immunoglobulin heavy chain locus are implicated in tumorigenesis. Proc. Natl. Acad. Sci. U.S.A. 84: 9069-9073 (1987).

B3 Bates, P.R. and M.F. Lavin. Comparison of γ-radiation induced accumulation of ataxia telangiectasia and control cells in G_2 phase. Mutat. Res. 218: 165-170 (1989).

B4 Barbi, G., J.M.J.C. Scheres, D. Schindler et al. Chromosome instability and x-ray hypersensitivity in a microcephalic and growth-retarded child. Am. J. Med. Genet. 40: 44-50 (1991).

B5　Barnes, D.E., A.E. Tomkinson, A.R. Lehmann et al. Mutations in the DNA ligase-I gene of an individual with immunodeficiencies and cellular hypersensitivity to DNA-damaging agents. Cell 69: 495-503 (1992).

B6　Beamish, H. and M.F. Lavin. Radiosensitivity in ataxia-telangiectasia - anomalies in radiation-induced cell cycle delay. Int. J. Radiat. Biol. 65: 175-184 (1994).

B7　Bennett, C.B., A.L. Lewis, K.K. Baldwin et al. Lethality induced by a single site-specific double-strand break in a dispesible yeast plasmid. Proc. Natl. Acad. Sci. U.S.A. 90: 5613-5617 (1993).

B8　Biedermann, K.A., J. Song, A.J. Giaccia et al. scid mutation in mice confers hypersensitivity to ionising radiation and a deficiency in DNA double-strand break repair. Proc. Natl. Acad. Sci. U.S.A. 88: 1394-1397 (1991).

B9　Blocher, D., D. Sigut and M.A. Hannan. Fibroblasts from ataxia telangiectasia (AT) and AT heterozygotes show an enhanced level of residual DNA double-strand breaks after low dose-rate gamma-irradiation as assayed by pulsed field gel electrophoresis. Int. J. Radiat. Biol. 60: 791-802 (1991).

B10　Boder, E. Ataxia-telangiectasia: an overview. p. 1-63 in: Ataxia-telangiectasia: Genetics, Neuropathology and Immunology of a Degenerative Disease of Childhood (R.A. Gatti and M. Swift, eds.). Liss, New York, 1985.

B11　Bradley, W.E.C., A. Belouchi and K. Messing. The aprt heterozygote/hemizygote system for screening mutagenic agents allows detection of large deletions. Mutat. Res. 199: 131-138 (1988).

B12　Bronner, C.E., S.M. Baker, P.T. Morrison et al. Mutation in the DNA mismatch repair gene homologue Hmlh1 is associated with hereditary non-polyposis colon cancer. Nature 368: 258-261 (1994).

B13　Brenner, D.J. and R.K. Sachs. Chromosomal 'finger-prints' of prior exposure to densely ionizing radiation. Radiat. Res. 140: 134-142 (1994).

B14　Burki, H.J. Ionizing radiation-induced 6-thioguanine resistant clones in synchronous CHO cells. Radiat. Res. 81: 76-84 (1980).

B15　Badie, C., G. Iliakis, N. Foray et al. Defective repair of DNA double-strand breaks and chromosome damage in fibroblasts from a radiosensitive leukemia patient. Cancer Res. 55: 1232-1234 (1995).

B16　Baker, S.M., C.E. Bronner, L. Zhang et al. Male mice defective in the DNA mismatch repair gene PMS2 exhibit abnormal chromosome synapsis in meiosis. Cell 82: 309-319 (1995).

B17　Ban, S., S. Iida, A.A. Awa et al. Lethal and mutagenic effects of ^{252}Cf radiation in cultured human cells. Int. J. Radiat. Biol. 52: 245-251 (1987).

B18　Boothman, D.A., M. Meyers, E. Odegaard et al. Altered G1 checkpoint control determines adaptive survival responses to ionizing radiation. Mutat. Res. 358: 143-153 (1996).

B19　Bao, C-Y., A-H. Ma, H.H. Evans et al. Molecular analysis of HPRT gene mutations induced by α- and X-radiation in human lymphoblastoid cells. Mutat. Res. 326: 1-15 (1995).

B20　Barquinero, J.F., L. Barrios, M.R. Caballin et al. Occupational exposure to radiation induces an adaptive response in human lymphocytes. Int. J. Radiat. Biol. 67: 187-191 (1995).

B21　Bauchinger, M., E. Schmid, H. Braselmann et al. Absence of adaptive response to low-level irradiation from tritiated thymidine and X-rays in lymphocytes of two individuals examined in serial experiments. Mutat. Res. 227: 103-107 (1989).

B22　Bedford, J.S., J.B. Mitchell, H.G. Griggs et al. Radiation-induced cellular reproductive death and chromosome aberrations. Radiat. Res. 76: 573-586 (1978).

B23　Bhattacharyya, N.P., A. Skandalis, A. Ganesh et al. Mutator phenotypes in human colorectal carcinoma cell lines. Proc. Natl. Acad. Sci. U.S.A. 91: 6319-6323 (1994).

B24　Boice, J.D. and R.W. Miller. Risk of breast cancer in ataxia-telangiectasia. Lancet 326: 1357-1358 (1992).

B25　Boothman, D.A., G. Majmudar and T. Johnson. Immediate X-ray-inducible responses from mammalian cells. Radiat. Res. 138: S44-S46 (1994).

B26　Bosi, A. and G. Olivieri. Variability in the adaptive response to ionizing radiations in humans. Mutat. Res. 211: 13-17 (1989).

B27　Bracey, T.S., J.C. Miller, A. Preece et al. Gamma-radiation-induced apoptosis in human colorectal adenoma and carcinoma cell lines can occur in the absence of wild type p53. Oncogene 10: 2391-2396 (1995).

B28　Branch, P., G. Aquilina, M. Bignami et al. Defective mismatch binding and a mutator phenotype in cells tolerant to DNA damage. Nature 362: 652-654 (1993).

B29　Bridges, B.A., J. Cole, C.F. Arlett et al. Possible association between mutant frequency in peripheral lymphocytes and domestic radon concentrations. Lancet 337: 1187-1189 (1991).

B30　Beamish, H., R. Williams, P. Chen et al. Defect in multiple cell cycle checkpoints in ataxia-telangiectasia post-irradiation. J. Biol. Chem. 271: 20486-20493 (1996).

B31　Byrd, P.J., C.M. McConville, P. Cooper et al. Mutations revealed by sequencing the 5' half of the gene for ataxia telangiectasia. Hum. Mol. Genet. 5: 145-149 (1996).

B32　Bentley, N.J., D.A. Holzman, K.S. Keegan et al. The Schizosaccharomyces pombe rad3 checkpoint gene. EMBO J. 15: 6641-6651 (1996).

B33　Barlow, C., S. Hirotsune, R. Paylor et al. Atm-deficient mice: a paradigm of ataxia telangiectasia. Cell 86: 159-171 (1996).

B34　Benson, F.E., A. Stasiak and S.C. West. Purification and characterization of the human Rad51 protein, an analogue of E. coli RecA. EMBO J. 13: 5764-5771 (1994).

B35　Boulton, S.J. and S.P. Jackson. Saccharomyces cerevisiae Ku70 potentiates illegitimate DNA double-strand break repair and serves as a barrier to error-prone DNA repair pathways. EMBO J. 15: 5093-5103 (1996).

B36　Boulton, S.J. and S.P. Jackson. Identification of a Saccharomyces cerevisiae Ku80 homologue: roles in DNA double strand break rejoining and in telomeric maintenance. Nucleic Acids Res. 24: 4639-4648 (1996).

B37　Blunt, T., D. Gell, M. Fox et al. Identification of a non-sense mutation in the carboxy-terminal region of DNA-dependent protein kinase catalytic subunit in the scid mouse. Proc. Natl. Acad. Sci. U.S.A. 93: 10285-10290 (1996).

B38　Bosma, G.C., M. Fried, R.P. Custer et al. Evidence of functional lymphocytes in some (leaky) scid mice. J. Exp. Med. 167: 1016-1033 (1988).

B39　Bigbee, W.L., R.H. Jensen, T. Veidebaum et al. Bio-dosimetry of Chernobyl cleanup workers from Estonia and Latvia using the glycophorin A in vivo somatic cell mutation assay. Radiat. Res. 147: 215-224 (1997).

B40　Bernhard, E.J., R.J. Muschel, V.J. Bakanauskas et al. Reducing the radiation-induced G_2 delay causes HeLa cells to undergo apoptosis instead of mitotic death. Int. J. Radiat. Biol. 69: 575-584 (1996).

B41　Boreham, D.R., K.L. Gale, S.R. Maves et al. Radiation-induced apoptosis in human lymphocytes: potential as a biological dosimeter. Health Phys. 71: 685-691 (1996).

B42 Burnet, N.G., J. Nyman, I. Turesson et al. The relationship between cellular radiation sensitivity and tissue response may provide the basis for individualizing radiotherapy schedules. Radiother. Oncol. 33: 228-238 (1994).

B43 Biard, D.S.F., M. Martin, Y. Le Rhun et al. Concomitant *p53* gene mutation and increased radiosensitivity in rat lung embryo epithelial cells during neoplastic development. Cancer Res. 54: 3361-3364 (1994).

B44 Bech-Hansen, T., W.A. Blattner, B.M. Sell et al. Transmission of in vitro radioresistance in a cancer-prone family. Lancet i: 1335-1337 (1981).

B45 Bartsch, H., M. Hollstein, R. Mustonen et al. Screening for putative radon-specific p53 mutation hotspot in German uranium miners. Lancet 346: 121 (1995).

B46 Beckman, K.B. and B.N. Ames. Oxidative decay of DNA. J. Biol. Chem. 272: 19633-19636 (1997).

B47 Barlow, C., M. Liyanage, P.B. Moens et al. Partial rescue of prophase I defects of *Atm*-deficent mice by *p53* and *p21* null alleles. Nature Genet. 17: 462-466 (1997).

B48 Baskaran, R., L.D. Wood, L.L. Whitaker et al. Ataxia-telangiectasia mutant protein activates c-Able tyrosine kinase in a response to ionizing radiation. Nature 387: 516-519 (1997).

B49 Busse, P.M., S.K. Bose, R.W. Jones et al. The action of caffeine on X-irradiated HeLa cells. II Synergistic lethality. Radiat. Res. 71: 666-677 (1977).

B50 Barlow, C., K.D. Brown, C-X. Deng et al. Atm selectively regulates distinct p53-dependent cell-cycle checkpoint and apoptotic pathways. Nature Genet. 17: 453-456 (1997).

B51 Brown, J.M. Correspondence re: M.S. Meyn, Ataxia telangiectasia and cellular responses to DNA damage. Cancer. Res. 57: 2313-2315 (1997).

B52 Blocher, D. DNA double-strand break repair determines the RBE of alpha-particles. Int. J. Radiat. Biol. 54: 761-771 (1988).

B53 Beer, J.Z. and I. Szumiel. Slow clones, reduced clonogenicity and intraclonal recovery in X-irradiated L5178Y-S cell cultures. Radiat. Environ. Biophys. 33: 125-139 (1994).

B54 Bouffler, S.D., G. Breckon and R. Cox. Chromosomal mechanisms in murine radiation acute myeloid leukaemogenesis. Carcinogenesis 17: 655-659 (1996).

B55 Breckon, G., D. Papworth and R. Cox. Murine radiation myeloid leukaemogenesis: a possible role for radiation-sensitive sites on chromosome 2. Genes Chromosomes Cancer 3: 367-375 (1991).

B56 Bridges, B.A. and A.R. Timms. Mutation in Escherichia coli under starvation conditions: a new pathway leading to small deletions in strains defective in mismatch correction. EMBO J. 16: 3349-3356 (1997).

B57 Berdal, K.G., R.F. Johansen and E. Seeberg. Release of normal bases from intact DNA by a native DNA repair enzyme. EMBO J. 17: 363-367 (1998).

B58 Bay, J.O., M. Grancho, D. Pernin et al. No evidence for constitutional ATM mutation in breast/gastric cancer families. Int. J. Oncol. 12: 1385-1390 (1998).

B59 Banin, S., L. Moyal, S-Y. Shieh et al. Enhanced phosphorylation of p53 by ATM in response to DNA damage. Science 281: 1674-1677 (1998).

B60 Bouffler, S.D., C.J. Kemp, A. Balmain et al. Spontaneous and ionizing radiation-induced chromosomal abnormalities in *p53*-deficient mice. Cancer Res. 55: 3883-3889 (1995).

B61 Barlow, C., M.A. Eckhaus, A.J. Schaffer et al. *Atm* haploinsufficiency results in increased sensitivity to sublethal doses of ionizing radiation in mice. Nature Genetics 21: 359-360 (1999).

B62 Bullrich, F., D. Rasio, S. Kitada et al. ATM mutations in B-cell chronic lymphocytic leukemia. Cancer Res. 59: 24-27 (1999).

B63 Barlow, C., P.A. Dennery, M.K. Shigenaga et al. Loss of the ataxia-telangiectasia gene product causes oxidative damage in target organs. Proc. Natl. Acad. Sci. U.S.A. 96: 9915-9919 (1999).

B64 Blackstock, W.P. and M.P. Weir. Proteomics: quantitative and physical mapping of cellular proteins. Trends Biotechnol. 17: 121-127 (1999).

B65 Bravard, A., C. Luccioni, E. Moustacchi et al. Contribution of antioxidant enzymes to the adaptive response to ionizing radiation of human lymphoblasts. Int. J. Radiat. Biol. 75: 639-645 (1999).

B66 Boreham, D.R., J.A. Dolling, S.R. Maves et al. Dose-rate effects for apoptosis and micronucleus formation in gamma-irradiated human lymphocytes. Radiat. Res. 153: 579-86 (2000).

C1 Caporossi, D., B. Porfirio, B. Nicoletti et al. Hypersensitivity of lymphoblastoid lines derived from ataxia telangiectasia patients to the induction of chromosomal aberrations by etoposide (VP-16). Mutat. Res. 290: 265-272 (1993).

C2 Caldecott, K.W., C.K. McKeown, J.D. Tacker et al. An interaction between the mammalian DNA repair protein XRCC1 and DNA ligase III. Mol. Cell. Biol. 14: 68-76 (1994).

C3 Chen, P., F.P. Imray and C. Kidson. Gene dosage and complementation analysis of ataxia telangiectasia lymphoblastoid lines assayed by induced chromosome aberrations. Mutat. Res. 129: 165-172 (1984).

C4 Chessa, L., P. Petrinelli, A. Antonelli et al. Heterogeneity in ataxia-telangiectasia: classical phenotype associated with intermediate cellular radiosensitivity. Am. J. Med. Genet. 42: 741-746 (1992).

C5 Chen, F., A. Nastasi, Z. Shen et al. Cell-cycle dependent expression of mammalian homologs of yeast DNA double-strand break repair genes *Rad51* and *Rad52*. Mutat. Res. 384: 205-211 (1997).

C6 Clarke, A.R., C.A. Purdie, D.J. Harrison et al. Thymocyte apoptosis induced by p53-dependent and independent pathways. Nature 362: 849-852 (1993).

C7 Cox, R., G.P. Hosking and J. Wilson. Ataxia telangiectasia: evaluation of radiosensitivity in cultured skin fibroblasts as a diagnostic test. Arch. Dis. Childhood 53: 386-390 (1978).

C8 Cox, R. and W.K. Masson. Radiosensitivity in cultured human fibroblasts. Int. J. Radiat. Biol. 38: 575-576 (1980).

C9 Cox, R., W.K. Masson, R.R. Weichselbaum et al. The repair of potentially lethal damage in X-irradiated cultures of normal and ataxia-telangiectasia human fibroblasts. Int. J. Radiat. Biol. 39: 357-365 (1981).

C10 Cox, R. A cellular description of the repair defect in ataxia-telangiectasia. p.141-153 in: Ataxia-telangiectasia - A Cellular and Molecular Link between Cancer, Neuropathology and Immune Deficiency (B.A. Bridges and D.G. Harnden, eds.). Wiley, Chichester, 1982.

C11 Cornforth, M.N. and J.S. Bedford. On the nature of a defect in cells from individuals with ataxia-telangiectasia. Science 227: 1589-1591 (1985).

C12 Conley, M.E., N.B. Spinner, B.S. Emanuel et al. A chromosome breakage syndrome with profound immunodeficiency. Blood 67: 1251-1256 (1986).

C13 Cox, R., P.G. Debenham, W.K. Masson et al. Ataxia-telangiectasia: a human mutation giving high-frequency misrepair of DNA double-stranded scissions. Mol. Biol. Med. 3: 229-244 (1986).

C14   Cornforth, M.N. and J.S. Bedford. A quantitative comparison of potentially lethal damage repair and the rejoining of interphase chromosome breaks in low passage normal human fibroblasts. Radiat. Res. 111: 385-405 (1987).

C15   Cole, J. and T.R. Skopek. Somatic mutant frequency, mutation rates and mutational spectra in the human population in vivo. Mutat. Res. 304: 33-105 (1994).

C16   Collins, A.R. Mutant rodent cell lines sensitive to ultraviolet light, ionizing radiation and cross-linking agents: a comprehensive survey of genetic and biochemical characteristics. Mutat. Res. 293: 99-118 (1993).

C17   Costa, N.D. and J. Thacker. Response of radiation-sensitive human cells to defined DNA breaks. Int. J. Radiat. Biol. 64: 523-529 (1993).

C18   Curry, C.J.R., J. Tsai, H.T. Hutchinson et al. AT_{FRESNO}: a phenotype linking ataxia telangiectasia with the Nijmegen breakage syndrome. Am. J. Hum. Genet. 45: 270-275 (1989).

C19   Cattanach, B.M., M.D. Burtenshaw, C. Rasberry et al. Large deletions and other gross forms of chromosome imbalance compatible with viability and fertility in the mouse. Nature Genet. 3: 56-61 (1993).

C20   Chang, W.P. and J.B. Little. Persistently elevated frequency of spontaneous mutations in progeny of CHO clones surviving X-irradiation: association with delayed reproductive death phenotype. Mutat. Res. 270: 191-199 (1992).

C21   Chen, Y., E. Kubo, T. Sado et al. Cytogenetic analysis of thymocytes during early stages after irradiation in mice with different susceptibilities to radiation-induced lymphomagenesis. J. Radiat. Res. 37: 267-276 (1996).

C22   Cox, R. and W.K. Masson. Mutation and inactivation of cultured mammalian cells exposed to beams of accelerated heavy ions. III. Human diploid fibroblasts. Int. J. Radiat. Biol. 36: 149-160 (1979).

C23   Caelles, C., A. Helmberg and M. Karin. p53-dependent apoptosis in the absence of transcriptional activation of p53-target genes. Nature 370: 220-223 (1994).

C24   Cai, L. and S.Z. Liu. Induction of cytogenetic adaptive response of somatic and germ cells in vivo and in vitro by low dose X-irradiation. Int. J. Radiat. Biol. 58: 187-194 (1990).

C25   Cai, L. and P. Wang. Induction of a cytogenetic adaptive response in germ cells of irradiated mice with very low-dose rate of chronic gamma-irradiation and its biological influence on radiation-induced DNA or chromosomal damage and cell killing in their male offspring. Mutagenesis 10: 95-100 (1995).

C26   Carrano, A.V. Chromosome aberrations and radiation-induced cell death. II. Predicted and observed cell survival. Mutat. Res. 17: 355-366 (1973).

C27   Carter, S., M. Negrini, D.R. Gillum et al. Loss of heterozygosity at 11q22-23 in breast cancer. Cancer Res. 54: 6270-6274 (1994).

C28   Cavazzana-Calvo, M., F. La Deist, G. de Saint Basil et al. Increased radiosensitivity of granulocyte macrophage colony-forming units and skin fibroblasts in human autosomal recessive severe combined immunodeficiency. J. Clin. Invest. 91: 1214-1218 (1993).

C29   Chaudhry, M.A., Q. Jiang, M. Ricanati et al. Characterization of multilocus lesions in human cells exposed to X radiation and radon. Radiat. Res. 145: 31-38 (1996).

C30   Chen, D.J., G.F. Strinste and N. Tokita. The genotoxicity of α-particles in human embryonic skin fibroblasts. Radiat. Res. 100: 321-327 (1984).

C31   Chen, J.W., A.E. Tompkinson, W. Ramos et al. Mammalian DNA ligase. III: molecular cloning, chromosomal localization, and expression in spermatocytes undergoing meiotic recombination. Mol. Cell. Biol. 15: 5412-5422 (1995).

C32   Chen, M., J. Quintans, Z. Fuks et al. Suppression of Bcl-2 messenger RNA production may mediate apoptosis after ionizing radiation, tumor necrosis factor α, and ceramide. Cancer Res. 55: 991-994 (1995).

C33   Cole, J., C.F. Arlett, M.H.L. Green et al. Comparative human cellular radiosensitivity. II. The survival following gamma-irradiation of unstimulated (G_0) T-lymphocytes, T-lymphocyte lines, lymphoblastoid cell lines and fibroblasts from normal donors, from ataxia-telangiectasia patients and from ataxia-telangiectasia heterozygotes. Int. J. Radiat. Biol. 54: 929-943 (1988).

C34   Cole, J., M.H.L. Green, B.A. Bridges et al. Lack of evidence for an association between the frequency of mutants or translocations in circulating lymphocytes and exposure to radon gas in the home. Radiat. Res. 145: 61-69 (1996).

C35   Cortes, F., I. Dominguez, J. Pinero et al. Adaptive response in human lymphocytes conditioned with hydrogen peroxide before irradiation with x-rays. Mutagenesis 5: 555-557 (1990).

C36   Cortessis, V., S. Ingles, R. Millikan et al. Linkage analysis of DRD2, a marker linked to the ataxia-telangiectasia gene, in 64 families with premenopausal bilateral breast cancer. Cancer Res. 53: 5083-5086 (1993).

C37   Cox, R. and W.K. Masson. Do radiation-induced thioguanine-resistant mutants of cultured mammalian cells arise by HGPRT gene mutation or X-chromosome rearrangement? Nature 276: 629-630 (1978).

C38   Cimprich, K.A., T.B. Shin, C.T. Keith et al. cDNA cloning and gene mapping of a candidate human cell cycle checkpoint protein. Proc. Natl. Acad. Sci. U.S.A. 93: 2850-2855 (1996).

C39   Caldecott, K.W., S. Aoufouchi, P. Johnson et al. XRCC1 polypeptide interacts with DNA polymerase β and possibly poly(ADP-ribose) polymerase, and DNA ligase III is a novel molecular 'nick-sensor' in vitro. Nucleic Acids Res. 24: 4387-4394 (1996).

C40   Clutton, S.M., K.M.S. Townsend, C. Walker et al. Radiation-induced genomic instability and persisting oxidative stress in primary bone marrow cultures. Carcinogenesis 17: 1633-1639 (1996).

C41   Cooper, P.K., T. Nouspikel, S.G. Clarkson et al. Defective transcription-coupled repair of oxidative base damage in cockayne syndrome patients from XP group G. Science 275: 990-993 (1997).

C42   Carr, A.M. and M.F. Hoekstra. The cellular responses to DNA damage. Trends Cell Biol. 5: 32-40 (1995).

C43   Cox, L.S. and D.P. Lane. Tumour suppressors, kinases and clamps: how p53 regulates the cell cycle in response to DNA damage. Bioessays 17: 501-508 (1995).

C44   Crompton, N.E.A., F. Zoelzer, E. Schneider et al. Increased mutant induction by very low dose-rate gamma-irradiation. Naturwissenschaften 72: S439 (1985).

C45   Crompton, N.E.A., B. Barth and J. Kiefer. Inverse dose-rate effect for the induction of 6-thioguanine resistant mutants in Chinese hamster V79-S cells by ^{60}Co τ-rays. Radiat. Res. 124: 300-308 (1990).

C46   Chernikova, S.V., V.Y. Gotlib and I.I. Pelevina. Influence of low dose irradiation on the radiosensitivity. Radiat. Biol. Radioecol. 33: 537-541 (1994).

C47 Charlton, D.E., H. Nikjoo and J.L. Humm. Calculation of initial yields of single- and double-strand breaks in cell nuclei from electrons, protons and alpha particles. Int. J. Radiat. Biol. 56: 1-19 (1989).

C48 Cliby, W.A., C.J. Roberts, K.A. Cimprich et al. Over-expression of a kinase-inactive ATR protein causes sensitivity to DNA-damaging agents and defects in cell cycle checkpoints. EMBO J. 17: 159-169 (1998).

C49 Cleaver, J.E. and W.F. Morgan. Poly(ADP-ribose) polymerase: a perplexing participant in cellular responses to DNA breakage. Mutat. Res. 257: 1-18 (1991).

C50 Custer, R.P., G.C. Bosma and M.J. Bosma. Severe combined immunodeficiency (SCID) in the mouse: pathology, reconstitution, neoplasms. Am. J. Pathol. 120: 464-477 (1985).

C51 Critchlow, S.E., R.P. Bowater and S.P. Jackson. Mammalian DNA double-strand break repair protein XRCC4 interacts with DNA ligase IV. Curr. Biol. 7: 588-598 (1997).

C52 Carroll, A.M. and M.J. Bosma. T-lymphocyte development in scid mice is arrested shortly after the initiation of T-cell receptor δ gene recombination. Genes Dev. 5: 1357-1366 (1991).

C53 Coquerelle, T.M., K.F. Weibezahn and C. Lucke-Huhle. Rejoining of double-strand breaks in normal human and ataxia-telangiectasia fibroblasts after exposure to ^{60}Co gamma-rays, ^{241}Am alpha-particles or bleomycin. Int. J. Radiat. Biol. 51: 209-218 (1987).

C54 Carney, J.P., R.S. Maser, H. Olivares et al. The hMre11/Rad50 protein complex and Nijmegen breakage syndrome: linkage of double-strand break repair to the cellular DNA damage response. Cell 93: 477-486 (1998).

C55 Cadet, J., T. Douki and J-L. Ravenat. Artifacts associated with the measurement of oxidised DNA bases. Environ. Health Perspect. 105: 1034-1039 (1997).

C56 Chaudhry, M.A. and M. Weinfeld. The action of Escherichia coli endonuclease III on multiply damaged sites in DNA. J. Mol. Biol. 249: 914-922 (1995).

C57 Chaudhry, M.A. and M. Weinfeld. Reactivity of human apurinic/apyrimidinic endonuclease and Escherichia coli exonuclease III with bistranded abasic sites in DNA. J. Biol. Chem. 272: 15650-15655 (1997).

C58 Canman, C.E., D-S. Lim, K.A. Cimprich et al. Activation of the ATM kinase by ionizing radiation and phosphorylation of p53. Science 281: 1677-1679 (1998).

C59 Connor, F., D. Bertwistle, P.J. Mee et al. Tumorigenesis and a DNA repair defect in mice with a truncating Brca2 mutation. Nature Genet. 17: 423-430 (1997).

C60 Cunningham, R.P., S.M. Saporito, S.G. Spitzer et al. Endonuclease IV (nfo) mutant of Escherichia coli. J. Bacteriol. 168: 1120-1127 (1986).

C61 Chester, N., F. Kuo, C. Kozak et al. Stage-specific apoptosis, developmental delay, and embryonic lethality in mice homozygous for a targeted disruption of the murine Bloom's syndrome gene. Genes Dev. 12: 3382-3393 (1998).

C62 Clarke, A.A., N.J. Philpott, E.C. Godon-Smith et al. The sensitivity of fanconi anaemia group C cells to apoptosis induced by mitomycin C is due to oxygen radical generation not DNA crosslinking. Br. J. Haematol. 96: 240-247 (1997).

C63 Cotes, F., I. Dominguez, M.J. Flores et al. Differences in the adaptive response to radiation damage in Go lymphocytes conditioned with hydrogen peroxide or low-dose x-rays. Mutat. Res. 311: 157-163 (1994).

C64 Cedervall, B., R. Wong, N. Albright et al. Methods for the quantification of DNA double-strand breaks determined from the distribution of DNA fragment sizes measured by pulsed-field gel electrophoresis. Radiat. Res. 143: 8-16 (1995).

C65 Chen, J., G.G. Birkholtz, P. Lindblom et al. The role of ataxia-telangiectasia heterozygotes in familial breast cancer. Cancer Res. 58: 1376-1379 (1998).

C66 Cartwright, R., C.E. Tambini, P.J. Simpson et al. The XRCC2 repair gene from human and mouse encodes a novel member of the recA/RAD51 family. Nucleic Acids Res. 26: 3084-3089 (1998).

C67 Cortez, D., Y. Wang, J. Qin et al. Requirement of ATM-dependent phosphorylation of Brca1 in the DNA damage response to double-strand breaks. Science 286: 1162-1166 (1999).

C68 Chuang, Y.Y.E., Q. Chen et al. Radiation-induced mutations at the autosomal thymidine kinase locus are not elevated in p53-null cells. Cancer Res. 59: 3073-3076 (1999).

C69 Celis, J.E., M. Ostergaard, N.A. Jensen et al. Human and mouse proteomics databases: novel resources in the protein universe. FEBS Lett. 430: 64-72 (1998).

D1 Dubrova, Y.E., V.N. Nesterov, N.G. Krouchinsky et al. Further evidence for elevated human minisatellite mutation rate in Belarus eight years after the Chernobyl accident. Mutat. Res. 381: 267-278 (1997).

D2 Deeley, J.O.T. and J.L. Moore. Nuclear lysate sedimentation measurements of peripheral blood lymphocytes from radiotherapy patients. Int. J. Radiat. Biol. 56: 963-973 (1989).

D3 Donehower, L.A., M. Harvey, B.L. Slagle et al. Mice deficient for p53 are developmentally normal but susceptible to spontaneous tumours. Nature 356: 215-221 (1992).

D4 Dulic, V., W.K. Kaufmann, S.J. Wilson et al. p53-dependent inhibition of cyclin-dependent kinase activities in human fibroblasts during radiation-induced G_1 arrest. Cell 76: 1013-1023 (1994).

D5 de Serres, F.J. X-ray induced specific locus mutations in the ad-3 region of two-component heterokaryons of Neurospora crassa II. More extensive genetic tests reveal an unexpectedly high frequency of multiple locus mutations. Mutat. Res. 210: 281-290 (1989).

D6 da Costa, L.T., B. Liu, W. el-Deiry et al. Polymerase δ variants in RER colorectal tumours. Nature Genet. 9: 10-11 (1995).

D7 Deschavanne, P.J., D. Debieu, B. Fertil et al. Re-evaluation of in vitro radiosensitivity of human fibroblasts of different genetic origins. Int. J. Radiat. Biol. 50: 279-293 (1986).

D8 Dewey, W.C., S.C. Furman and H.H. Miller. Comparison of lethality and chromosomal damage induced by X-rays in synchronized Chinese hamster cells in vitro. Radiat. Res. 43: 561-581 (1970).

D9 Dewind, N., M. Dekker, A. Berns et al. Inactivation of the mouse MSH2 gene results in mismatch repair deficiency, methylation tolerance, hyperrecombination and predisposition to cancer. Cell 82: 321-330 (1995).

D10 Dubrova, Y.E., A.J. Jeffreys and A.M. Malashenko. Mouse minisatellite mutations induced by ionising radiation. Nature Genet. 5: 92-94 (1993).

D11 Danska, J.S., D.P. Holland, S. Mariathasan et al. Biochemical and genetic defects in the DNA-dependent protein kinase in murine scid lymphocytes. Mol. Cell. Biol. 16: 5507-5517 (1996).

D12 Danska, J.S., F. Pflumio, C. Williams et al. Rescue of T-cell specific V(D)J recombination in SCID mice by DNA-damaging agents. Science 266: 450-455 (1994).

D13 Deshpande, A., E.H. Goodwin, S.M. Bailey et al. Alpha-particle-induced sister chromatid exchange in normal human lung fibroblasts: evidence for an extranuclear target. Radiat. Res. 145: 260-267 (1996).

D14 Dubrova, Y.E., V.N. Nestorov, N.G. Krouchinsky et al. Human minisatellite mutation rate after the Chernobyl accident. Nature 380: 683-686 (1996).

D15 Duchaud, E., A. Ridet, D. Stoppa-Lyonnet et al. Deregulated apoptosis in ataxia-telangiectasia - association with clinical stigmata and radiosensitivity. Cancer Res. 56: 1400-1404 (1996).

D16 Dianov, G., A. Price and T. Lindahl. Generation of single-nucleotide repair patch following excision of uracil residues from DNA. Mol. Cell. Biol. 12: 1605-1612 (1992).

D17 Dolganov, G.M., R.S. Maser, A. Novikov et al. Human Rad50 is physically associated with hMre11: identification of a conserved multiprotein complex implicated in recombinational DNA repair. Mol. Cell. Biol. 16: 4832-4841 (1996).

D18 de Winter, J.P., Q. Waisfisz, M.A. Rooimans et al. The Fanconi anaemia group G gene *FANCG* is identical with *XRCC9*. Nature Genet. 20: 281-283 (1998).

D19 Dubrova, Y.E., M. Plumb, J. Brown et al. Stage specificity, dose response, and doubling dose for mouse minisatellite germ-line mutation induced by acute radiation. Proc. Natl. Acad. Sci. U.S.A. 95: 6251-6255 (1998).

D20 DeWeese, T.L., J.M. Shipman, N.A. Larrier et al. Mouse embryonic stem cells carrying one or two defective *Msh2* alleles respond abnormally to oxidative stress inflicted by low-level radiation. Proc. Natl. Acad. Sci. U.S.A. 95: 11915-11920 (1998).

D21 di Fagagna, F.d'A., M.P. Hande W-M. Tong et al. Functions of poly(ADP-ribose) polymerase in controlling telomere length and chromosomal stability. Nature Genet. 23: 76-80 (1999).

D22 Djuzenova, C.S., D. Schindler, H. Stopper et al. Identification of ataxia telangiectasia heterozygotes, a cancer-prone population, using the single-cell gel electrophoresis (Comet) assay. Lab. Invest. 79: 699-705 (1999).

D23 Debrauwere, H., J. Buard, J. Tessier et al. Meiotic instability of human minisatellite *CEB1* in yeast requires DNA double-strand breaks. Nature Genetics 23: 367-371 (1999).

D24 Dubrova, Y.E., M. Plumb, B. Gutierrez et al. Transgenerational mutation by radiation. Nature 405:37 (2000).

E1 Ejima, Y. and M.S. Sasaki. Enhanced expression of X-ray and UV-induced chromosome aberrations by cytosine arabinoside in ataxia telangiectasia cells. Mutat. Res. 159: 117-123 (1986).

E2 Elkind, M.M., A. Han and K.W. Volz. Radiation response of mammalian cells grown in culture. IV. Dose dependence of division delay and postirradiation growth of surviving and non-surviving Chinese hamster cells. J. Natl. Cancer Inst. 30: 705-721 (1963).

E3 Evans, H.H., J. Mencl, M.-F. Horng et al. Locus specificity in the mutability of L5178Y mouse lymphoma cells: the role of multilocus lesions. Proc. Natl. Acad. Sci. U.S.A. 83: 4379-4383 (1986).

E4 Easton, D.F. Cancer risks in A-T heterozygotes. Int. J. Radiat. Biol. 66: S177-S182 (1994).

E5 Edwards, A.A., D.C. Lloyd and J.S. Prosser. Chromosome aberrations in human lymphocytes: a radiobiological review. p. 423-432 in: Low Dose Radiation (K.F. Baverstock and J.W. Stather, eds.). Taylor and Francis, London, 1989.

E6 Ellis, N.A., J. Groden, T-Z. Ye et al. The Bloom"s syndrome gene product is homologous to RecQ helicases. Cell 83: 655-666 (1995).

E7 Elyan, S.A.G., C.M.L. West, S.A. Roberts et al. Use of low-dose-rate irradiation to measure the intrinsic radiosensitivity of human T-lymphocytes. Int. J. Radiat. Biol. 64: 375-383 (1993).

E8 Eshleman, J.R., E.Z. Lang, G.K. Bowerfind et al. Increased mutation rate at the *HPRT* locus accompanies micro-satellite instability in colon cancer. Oncogene 10: 33-37 (1995).

E9 Elson, A., Y. Wang, C.J. Daugherty et al. Pleiotropic defects in ataxia-telangiectasia protein-deficient mice. Proc. Natl. Acad. Sci. U.S.A. 93: 13084-13089 (1996).

E10 Evans, H.H., M-F. Horng, J. Mencl et al. The influence of dose rate on the lethal and mutagenic effects of X-rays in proliferating L5178Y cells differing in radiation sensitivity. Int. J. Radiat. Biol. 47: 553-562 (1985).

E11 Evans, H.H., M. Nielsen, J. Mencl et al. The effect of dose rate on X-radiation induced mutant frequency and the nature of DNA lesions in mouse lymphoma L5178Y cells. Radiat. Res. 122: 316-325 (1990).

E12 Essers, J., R.W. Hendriks, S.M.A. Swagemakers et al. Disruption of the mouse *RAD54* reduces ionizing radiation resistance and homologous recombination. Cell 89: 195-204 (1997).

E13 Engelward, B.P., G. Weeda, M.D. Wyatt et al. Base excision repair-deficient mice lacking the Aag alkyladenine DNA glycosylase. Proc. Natl. Acad. Sci. U.S.A. 94: 13087-13092 (1997).

E14 Eilam, R., Y. Peter, A. Elson et al. Selective loss of dopaminergic nigro-striatal neurons in brains of Atm-deficient mice. Proc. Natl. Acad. Sci. U.S.A. 95: 12653-12656 (1998).

E15 Enns, L., R.D.C. Barley, M.C. Paterson et al. Radiosensitivity in ataxia telangiectasia fibroblasts is not associated with deregulated apoptosis. Radiat. Res. 150: 11-16 (1998).

F1 Friedberg, E.C. Deoxyribonucleic acid repair in the yeast *Saccharomyces cerevisiae*. Microbiol. Rev. 52: 70-102 (1988).

F2 Fiorilli, M., A. Antonelli, G. Russo et al. Variant of ataxia-telangiectasia with low level radiosensitivity. Hum. Genet. 70: 274-277 (1985).

F3 Frankenberg-Schwager, M. and D. Frankenberg. DNA double-strand breaks: their repair and relationship to cell killing in yeast. Int. J. Radiat. Biol. 58: 569-575 (1990).

F4 Fritsche, M., C. Haessler and G. Brandner. Induction of nuclear accumulation of the tumor-suppressor protein p53 by DNA-damaging agents. Oncogene 8: 307-318 (1993).

F5 Fan, Y.J., Z.W. Wang, S. Sadamoto et al. Dose response of radiation induction of germline mutation at a hypervariable mouse minisatellite locus. Int. J. Radiat. Biol. 68: 177-183 (1995).

F6 Fomenko, L.A., I.K. Kozhanovskaia and A.I. Gaziev. Formation of micronuclei in bone marrow cells of chronically irradiated mice and subsequent acute exposure to gamma radiation. Radiobiologia 31: 701-715 (1991).

F7 Fuhrman Conti, A.M., G. Francone, M. Volonte et al. Induction of 8-azaguanine resistant mutants in human cultured cells exposed to 31 MeV protons. Int. J. Radiat. Biol. 53: 467-476 (1988).

F8   Fujimori, A., A. Tachibana and K. Tatsumi. Allelic losses at the *aprt* locus of human lymphoblastoid cells. Mutat. Res. 269: 55-62 (1992).

F9   Fanconi Anaemia/Breast Cancer Consortium. Positional cloning of the Fanconi anaemia group A gene. Nature Genet. 14: 324-328 (1996).

F10   Foray, N., C.F. Arlett and E.P. Malaise. Dose-rate effect on induction and repair rate of radiation-induced DNA double-strand breaks in a normal and an ataxia-telangiectasia human fibroblast cell line. Biochimie 77: 900-905 (1995).

F11   Furuno-Fukushi, I., K. Tatsumi, M. Takahagi et al. Quantitative and qualitative effect of γ-ray dose rate on mutagenesis in human lymphoblastoid cells. Int. J. Radiat. Biol. 70: 209-217 (1996).

F12   Furuno-Fukushi, I. and H. Matsudaira. Mutation induction by different dose rates of γ-rays in radiation-sensitive mutants of mouse leukaemia cells. Radiat. Res. 120: 370-374 (1989).

F13   Furuno-Fukushi, I., A.M. Ueno and H. Matsudaira. Mutation induction by very low dose rate gamma-rays in cultured mouse leukemia cells L5178Y. Radiat. Res. 115: 273-280 (1988).

F14   Fields, C., M.D. Adams, O. White et al. How many genes in the human genome? Nature Genet. 7: 345-346 (1994).

F15   Friedberg, E.C., G.C. Walker and W. Siede. DNA Repair and Mutagenesis. ASM Press, Washington, 1995.

F16   Foray, N., A. Priestley, G. Alsbeih et al. Hypersensitivity of ataxia telangiectasia fibroblasts to ionizing radiation is associated with a repair deficiency of DNA double-strand breaks. Int. J. Radiat. Biol. 72: 271-283 (1997).

F17   Flaggs, G., A.W. Plug, K.M. Dunks et al. Atm-dependent interactions of a mammalian Chk1 homolog with meiotic chromosomes. Curr. Biol. 7: 977-986 (1997).

F18   Foray, N., A. Priestley, G. Alsbeih et al. Hypersensitivity of ataxia telangiectasia fibroblasts to ionizing radiation is associated with a repair deficiency of DNA double-strand breaks. Int. J. Radiat. Biol. 72: 271-283 (1997).

F19   Fritzell, J.A., L. Narayanan, S.M. Baker et al. Role of DNA mismatch repair in the cytotoxicity of ionizing radiation. Cancer Res. 57: 5143-5147 (1997).

F20   Frank, K.M., J.M. Sekiguchi, K.J. Seidl et al. Late embryonic lethality and impaired V(D)J recombination in mice lacking DNA ligase IV. Nature 396: 173-177 (1998).

F21   Fitzgerald, M.G., J.M. Bean, S.R. Hegde et al. Heterozygous ATM mutations do not contribute to early onset of breast cancer. Nature Genet. 15: 307-310 (1997).

F22   Forrester, H.B., I.R. Radford et al. Selection and sequencing of interchromosomal rearrangements from gamma-irradiated normal human fibroblasts. Int. J. Radiat. Biol. 75: 543-551 (1999).

F23   Forrester, H.B., R.F. Yeh et al. A dose response for radiation-induced intrachromosomal DNA rearrangements detected by inverse polymerase chain reaction. Radiat. Res. 152: 232-238 (1999).

F24   Friedl, A.A., M. Kiechle, B. Fellerhof et al. Radiation-induced chromosome aberrations in *Saccharomyces cerevisiae*: influence of DNA repair pathways. Genetics 148: 975-988 (1998).

F25   Fukumura, R., R. Araki, A. Fujimori et al. Murine cell line SX9 bearing a mutation in the *dna-pkcs* gene exhibits aberrant V(D)J recombination not only in the coding joint but also the signal joint. J. Biol. Chem. 273: 13058-13064 (1998).

F26   Filippovich, I.V., N.I. Sorokina, N. Robillard et al. Radiation-induced apoptosis in human tumor cell lines:

adaptive response and split-dose effect. Int. J. Cancer 77: 76-81 (1998).

G1   Game, J.C. DNA double-strand breaks and the *RAD50-RAD57* genes in *Saccharomyces*. Cancer Biol. 4: 73-83 (1993).

G2   Gatti, R.A., I. Berkel, E. Boder et al. Localization of an ataxia-telangiectasia gene to chromosome 11q22-23. Nature 336: 577-580 (1988).

G3   Ganesh, A., P. North and J. Thacker. Repair and misrepair of site-specific DNA double-strand breaks by human cell extracts. Mutat. Res. 299: 251-259 (1993).

G4   Gentner, N.E. and D.P. Morrison. Determination of the proportion of persons in the population-at-large who exhibit abnormal sensitivity to ionizing radiation. p.253-262 in: Low Dose Radiation (K.F. Baverstock and J.W. Stather, eds.). Taylor and Francis, London, 1989.

G5   Giaccia, A., R. Weinstein, J. Hu et al. Cell-cycle-dependent repair of double-strand breaks in a γ-ray sensitive Chinese hamster cell. Somat. Cell Mol. Genet. 11: 485-491 (1985).

G6   Goodhead, D.T. Saturable repair models of radiation action in mammalian cells. Radiat. Res. 104: S58-S67 (1985).

G7   Gottlieb, T.M. and S.P. Jackson. The DNA-dependent protein kinase: requirement for DNA ends and association with ku antigen. Cell 72: 131-142 (1990).

G8   Gentil, A., J.B. Cabralneto, R. Mariagesamson et al. Mutagenicity of a unique apurinic/apyrimidinic site in mammalian cells. J. Mol. Biol. 227: 981-984 (1992).

G9   Gorgojo, L. and J.B. Little. Expression of lethal mutations in progeny of irradiated mammalian cells. Int. J. Radiat. Biol. 55: 619-630 (1989).

G10   Grosovsky, A.J., J.G. de Boer, P.J. de Jong et al. Base substitutions, frameshifts and small deletions constitute ionising radiation-induced point mutations in mammalian cells. Proc. Natl. Acad. Sci. U.S.A. 85: 185-188 (1988).

G11   Geara, F.B., L.J. Peters, K.K. Ang et al. Intrinsic radiosensitivity of normal human fibroblasts and lymphocytes after high- and low-dose-rate irradiation. Cancer Res. 52: 6348-6352 (1992).

G12   Geara, F.B., L.J. Peters, K.K. Ang et al. Prospective comparison of *in vitro* normal cell radiosensitivity and normal tissue reactions in radiotherapy patients. Int. J. Radiat. Oncol. Biol. Phys. 27: 1173-1179 (1993).

G13   Gentner, N.E. Increased definition of abnormal radiosensitivity using low dose rate testing. Adv. Radiat. Biol. 16: 293-302 (1992).

G14   Gentner, N.E. Communication to the UNSCEAR Secretariat (1995).

G15   Gerachi, J.P., P.D. Thrower, K.L. Jackson et al. The RBE of cyclotron fast neutrons for interphase death in rat thymocytes *in vitro*. Int. J. Radiat. Biol. 25: 403-405 (1974).

G16   Gibbs, R.A., J. Camakaris, G.S. Hodgson et al. Molecular characterization of ^{125}I decay and X-ray induced HPRT mutants in CHO cells. Int. J. Radiat. Biol. 51: 193-199 (1987).

G17   Green, M.H.L., C.F. Arlett, J. Cole et al. Comparative human cellular radiosensitivity: III. γ-radiation survival of cultured skin fibroblasts and resting T-lymphocytes from the peripheral blood of the same individual. Int. J. Radiat. Biol. 59: 749-765 (1991).

G18   Griffiths, S.D., S.J. Marsden, E.G. Wright et al. Lethality and mutagenesis of B lymphocyte progenitor cells following exposure to α-particles and X-rays. Int. J. Radiat. Biol. 66: 197-205 (1994).

G19 Gilad, S., R. Khosravi, D. Shkedy et al. Predominance of null mutations in ataxia telangiectasia. Hum. Mol. Genet. 5: 433-439 (1996).

G20 German, J. Bloom syndrome: a mendelian prototype of somatic mutational disease. Medicine 72: 393-406 (1993).

G21 Gschwend, M., O. Levran, L. Kruglyak et al. A locus for Fanconi anaemia on 16q determined by homozygosity mapping. Am. J. Hum. Genet. 59: 377-384 (1996).

G22 Greenwell, P.W., S.L. Kronmal, S.E. Porter et al. *TEL1*, a gene involved in controlling telomere length in S. cerevisiae, is homologous to the human ataxia telangiectasia gene. Cell 82: 823-829 (1995).

G23 Guidos, C.J., C.J. Williams, I. Grandal et al. V(D)J recombination activates a p53-dependent DNA damage checkpoint in *scid* lymphocyte precursors. Genes Dev. 10: 2038-2054 (1996).

G24 Grant, S.G. and W.L. Bigbee. In vivo somatic mutation and segregation at the human glycophorin A (*GPA*) locus: phenotypic variation encompassing both gene-specific and chromosomal mechanisms. Mutat. Res. 288: 163-172 (1993).

G25 Golstein, P. Controlling cell death. Science 275: 1081-1082 (1997).

G26 Griffiths, S.D., A.R. Clarke, L.E. Healy et al. Absence of p53 permits propagation of mutant cells following genotoxic damage. Oncogene 14: 523-531 (1997).

G27 Goodhead, D.T. Track structure considerations in low dose and low dose rate effects of ionizing radiation. Adv. Radiat. Biol. 16: 7-44 (1992).

G28 Goodhead, D.T. Initial events in the cellular effects of ionizing radiations: clustered damage in DNA. Int. J. Radiat. Biol. 65: 7-17 (1994).

G29 Goodhead, D.T. and H. Nikjoo. Clustered damage in DNA: estimates from track-structure simulations. Radiat. Res. 148: 485-486 (1997).

G30 Grawunder, U., M. Wilm, X. Wu et al. Activity of DNA ligase IV stimulated by complex formation with XRCC4 protein in mammalian cells. Nature 388: 492-495 (1997).

G31 Gu, Y., S. Jin, Y. Gao et al. Ku70-deficient embryoinc stem cells have increased ionizing radiosensitivity, defective DNA end-binding activity, and inability to support V(D)J recombniation. Proc. Natl. Acad. Sci. U.S.A. 94: 8076-8081 (1997).

G32 Goodhead, D.T., J. Thacker and R. Cox. Non-rejoining DNA breaks and cell inactivation. Nature 272: 379-380 (1978).

G33 Gao, Y., Y. Sun, K.M. Frank et al. A critical role for DNA end-joining proteins in both lymphogenesis and neurogenesis. Cell 95: 891-902 (1998).

G34 Gowen, L.C., A.V. Avrutskaya, A.M. Latour et al. BRCA1 required for transcription coupled repair of oxidative DNA damage. Science 281: 1009-1012 (1998).

G35 Gurley, K.E., K. Vo and C.J. Kemp. DNA double-strand breaks, p53, and apoptosis during lymphomagenesis in *scid/scid* mice. Cancer Res. 58: 3111-3115 (1998).

G36 Goi, K., M. Takagi, S. Iwata et al. DNA damage-associated dysregulation of the cell cycle and apoptosis control in cells with germ-line p53 mutation. Cancer Res. 57: 1895-1902 (1997).

G37 Gayther, S.A., W. Warren, S. Mazoyer et al. Germline mutations in the *BRCA1* gene in breast and ovarian cancer families provide evidence for a genotype-phenotype correlation. Nature Genet. 11: 428-433 (1995).

G38 Gayther, S.A., J. Mangion, P. Russell et al. Variation of risks of breast and ovarian cancer associated with different mutations of the *BRCA2* gene. Nature Genet. 15: 103-105 (1997).

G39 Goodhead, D.T., J. Thacker and R. Cox. Effects of radiations of different qualities on cells: molecular mechanisms of damage and repair. Int. J. Radiat. Biol. 63: 543-556 (1993).

G40 Golub, T.R., D.K. Slonim, P. Tamayo et al. Molecular classification of cancer: class discovery and class prediction by gene expression monitoring. Science 286: 531-537 (1999).

G41 Gibbs, P.E., W.G. McGregor, V.M. Maher et al. A human homolog of the Saccharomyces cerevisiae REV3 gene, which encodes the catalytic subunit of DNA polymerase zeta. Proc. Natl. Acad. Sci. U.S.A. 95: 6876-6880 (1998).

H1 Hanawalt, P.C. Heterogeneity of DNA repair at the gene level. Mutat. Res. 247: 203-211 (1991).

H2 Hozak, P., A.B. Hassan, D.A. Jackson et al. Visualization of replication factories attached to a nucleoskeleton. Cell 73: 361-373 (1993).

H3 Hoeijmakers, J. Nucleotide excision repair II from yeast to mammals. Trends Genet. 9: 211-217 (1993).

H4 Hutchinson, F. Chemical changes induced in DNA by ionising radiation. Prog. Nucleic Acid Res. Mol. Biol. 32: 115-154 (1985).

H5 Harris, G., W.A. Cramp, J.C. Edwards et al. Radio-sensitivity of peripheral blood lymphocytes in autoimmune disease. Int. J. Radiat. Biol. 47: 689-699 (1985).

H6 Hannan, M.A., M. Kunhi, M. Einspenner et al. Post-irradiation DNA synthesis inhibition and G_2 phase delay in radiosensitive body cells from non-Hodgkin's lymphoma patients: an indication of cell cycle defects. Mutat. Res. 311: 265-276 (1994).

H7 Henner, W.D. and M.E. Blaska. Hypersensitivity of cultured ataxia telangiectasia cells to etoposide. J. Natl. Cancer Inst. 76: 1107-1011 (1986).

H8 Hendrickson, E.A., D.G. Schatz and D.T. Weaver. The *scid* gene encodes a transacting factor that mediates the rejoining event of Ig gene rearrangement. Genes Dev. 2: 817-829 (1988).

H9 Hecht, F. and B.K. Hecht. Cancer in ataxia-telangiectasia patients. Cytogenet. Cell Genet. 46: 9-19 (1990).

H10 Hendrickson, E.A., X-Q. Qin, E.A. Bump et al. A link between double-strand break related repair and V(D)J recombination: the scid mutation. Proc. Natl. Acad. Sci. U.S.A. 88: 4061-4065 (1991).

H11 Hickson, I.D. and A.L. Harris. Mammalian DNA repair - use of mutants hypersensitive to cytotoxic agents. Trends Genet. 4: 101-106 (1988).

H12 Hsieh, C.L., C.F. Arlett and M.R. Lieber. V(D)J recom-bination in ataxia telangiectasia, Bloom's syndrome and a DNA ligase I-associated immunodeficiency disorder. J. Biol. Chem. 268: 20105-20109 (1993).

H13 Hain, J., R. Janussi and W. Burkart. Lack of adaptive response to low doses of ionising radiation in human lymphocytes from 5 different donors. Mutat. Res. 283: 137-144 (1992).

H14 Hain, J., N.E.A. Crompton, W. Burkart et al. Caffeine release of radiation induced S and G_2 phase arrest in V79 hamster cells: increase of histone messenger RNA levels and p34^{cdc2} activation. Cancer Res. 53: 1507-1510 (1994).

H15 Hampton, G.M., A. Mannermaa, R. Winqvist et al. Loss of heterozygosity in sporadic human breast carcinoma: a common region between 11q22 and 11q23. Cancer Res. 54: 4586-4589 (1994).

H16 Harris, A.W. and J.W. Lowenthal. Cells of some cultured lymphoma lines are killed rapidly by X-rays and bleomycin. Int. J. Radiat. Biol. 42: 111-116 (1982).

H17 Hartley, K.O., D. Gell, G.C.M. Smith et al. DNA-dependent protein kinase catalytic subunit: a relative of phosphatidylinositol 3-kinase and the ataxia telangiectasia gene product. Cell 82: 849-856 (1995).

H18 Harvey, A.N. and J.R.K. Savage. A case of caffeine-medited cancellation of mitotic delay without enhanced breakage in V79 cells. Mutat. Res. 304: 203-209 (1994).

H19 Hedges, M.J. and S. Hornsey. The effect of X-rays and neutrons on lymphocyte death and transformation. Int. J. Radiat. Biol. 33: 291-300 (1978).

H20 Hei, T.K., E.J. Hall and C.A. Waldren. Mutation induction and relative biological effectiveness of neutrons in mammalian cells. Radiat. Res. 115: 281-291 (1988).

H21 Helzlsouer, K.J., E.L. Harris, R. Parshad et al. Familial clustering of breast cancer: possible interaction between DNA repair proficiency and radiation exposure in the development of breast cancer. Int. J. Cancer 64: 14-17 (1995).

H22 Hemminki, A., P. Peltomaki and J.P. Mecklin. Loss of the wild type MLH1 gene is a feature of hereditary non-polyposis colorectal cancer. Nature Genet. 8: 405-410 (1994).

H23 Hess, P., G. Aquilina, E. Dogliotti et al. Spontaneous mutations at aprt locus in a mammalian cell line defective in mismatch recognition. Somat. Cell Mol. Genet. 20: 409-421 (1994).

H24 Higurashi, M. and P.E. Conen. In vitro chromosomal radiosensitivity in 'chromosomal breakage syndromes. Cancer 32: 380-383 (1973).

H25 Hollstein, M., K. Rice, M.S. Greenblatt et al. Database of p53 gene somatic mutations in human tumors and cell lines. Nucleic Acids Res. 22: 3551-3555 (1994).

H26 Huo, Y.K., Z. Wang, J-H. Hong et al. Radiosensitivity of ataxia-telangiectasia, X-linked agammaglobulinemia, and related syndromes using a modified colony survival assay. Cancer Res. 54: 2544-2547 (1994).

H27 Hays, S.L., A.A. Firmenich and P. Berg. Complex formation in yeast double-strand break repair: participation of Rad51, Rad52, Rad55 and Rad57 proteins. Proc. Natl. Acad. Sci. U.S.A. 92: 6925-6929 (1995).

H28 Hendry, J.H. and T-N. Jiang. Differential radiosensitizing effect of the scid mutation among tissues, studied using high and low dose rates; implications for prognostic indicators in radiotherapy. Radiother. Oncol. 33: 209-216 (1994).

H29 Huang, L-C., K.C. Clarkin and G.M. Wahl. Sensitivity and selectivity of the DNA damage sensor responsible for activaing p53-dependent G_1 arrest. Proc. Natl. Acad. Sci. U.S.A. 93: 4827-4832 (1996).

H30 Holmberg, K., S. Falt, A. Johansson et al. Clonal chromosome aberrations and genomic instability in X-irradiated human T-lymphocyte cultures. Mutat. Res. 286: 321-330 (1993).

H31 Holmberg, K., A.E. Meijer, G. Auer et al. Delayed chromosomal instability in human T-lymphocyte clones exposed to ionizing radiation. Int. J. Radiat. Biol. 68: 245-255 (1995).

H32 Hakoda, M., M. Akiyama, S. Kyoizumi et al. Measurement of in vivo HGPRT-deficient mutant cell frequency using a modified method for cloning human peripheral blood T-lymphocytes. Mutat. Res. 197: 161-169 (1988).

H33 Hirai, Y., Y. Kusunoki, S. Kyoizumi et al. Mutant frequency at the HPRT locus in peripheral blood T-lymphocytes of atomic bomb survivors. Mutat. Res. 329: 183-196 (1995).

H34 Hannun, Y.A. Functions of ceramide in coordinating cellular responses to stress. Science 274: 1855-1859 (1996).

H35 Haimovitz-Friedman, A., C. Kan, D. Ehleiter et al. Ionizing radiation acts on cellular membranes to generate ceramide and initiate apoptosis. J. Exp. Med. 180: 525-535 (1994).

H36 Hopcia, K.L., Y.L. McCarey, F.C. Sylvester et al. Radiation-induced apoptosis in HL60 cells: oxygen effect, relationship between apoptosis and loss of clonogenicity and dependence of time of apoptosis on radiation dose. Radiat. Res. 145: 315-323 (1996).

H37 Han, Z., D. Chatterjee, D.M. He et al. Evidence for a G_2 checkpoint in p53-independent apoptosis induction by X-irradiation. Mol. Cell. Biol. 15: 5849-5857 (1995).

H38 Hendry, J.H., C.S. Potten, C. Chadwick et al. Cell death (apoptosis) in the mouse small intestine after low doses: effects of dose-rate, 14.7 MeV neutrons and 600 MeV (maximum energy) neutrons. Int. J. Radiat. Biol. 42: 611-620 (1982).

H39 Hendry, J.H., C.S. Potten and A. Merritt. Apoptosis induced by high- and low-LET radiations. Radiat. Environ. Biophys. 34: 59-62 (1995).

H40 Hakem, R., J.L. de la Pomba, C. Sirard et al. The tumour suppressor gene Brca1 is required for embryonic cellular proliferation in the mouse. Cell 85: 1009-1023 (1996).

H41 Hodgkins, P., P. O'Neill, D. Stevens et al. The severity of alpha-particle induced DNA damage is revealed by exposure to cell-free extracts. Radiat. Res. 146: 660-667 (1996).

H42 Hei, T.K., L-J. Wu, S-X. Liu et al. Mutagenic effects of a single and an exact number of α particles in mammalian cells. Proc. Natl. Acad. Sci. U.S.A. 94: 3765-3770 (1997).

H43 Hayata, I., M. Seki, K. Yoshida et al. Chromosomal aberrations observed in 52 mouse myeloid leukaemias. Cancer Res. 43: 367-373 (1983).

H44 Harris, R.S., G. Feng, C. Thulin et al. Mismatch repair protein MutL becomes limiting during stationary-phase mutation. Genes Dev. 11: 2426-2437 (1997).

H45 Hayata, I., T. Ichikawa and Y. Ichikawa. Specificity in chromosomal abnormalities in mouse bone marrows induced by the difference of the condition of irradiation. Proc. Jpn. Acad. 63: 289-292 (1987).

H46 Haber, J.E. The many faces of Mre11. Cell 95: 583-586 (1998).

H47 Hall, E.J., P.B. Schiff, G.E. Hanks et al. A preliminary report: frequency of A-T heterozygotes among prostate cancer patients with severe late responses to radiation therapy. Cancer J. Sci. Am. 4: 385-389 (1998).

H48 Herzog, K-H., M.J. Chong, M. Kapsetaki et al. Requirement for Atm in ionizing radiation-induced cell death in the developing central nervous system. Science 280: 1089-1091 (1998).

H49 Harrison, L., Z. Hatahet, A.A. Purmal et al. Multiply damaged sites in DNA: interactions with Escherichia coli endonucleases III and VIII. Nucleic Acids Res. 26: 932-941 (1998).

H50 Heyer, J., K. Yang, M. Lipkin et al. Mouse models for colorectal cancer. Oncogene 18: 5325-5333 (1999).

H51 Harada, Y-N., N. Shiomi, M. Koike et al. Postnatal growth failure, short life span, and early onset of cellular senescence and subsequent immortalization in mice lacking the xeroderma pigmentosum Group G gene. Mol. Cell. Biol. 19: 2366-2372 (1999).

I1 Ikushima, T. Radio-adaptive response: characterization of a cytogenetic repair induced by low-level ionising radiation in cultured Chinese hamster cells. Mutat. Res. 227: 241-246 (1989).

I2   Iliakis, G. The mutagenicity of α-particles in Ehrlich ascites tumor cells. Radiat. Res. 99: 52-58 (1984).

I3   Iliakis, G. and M. Nuesse. Evidence that repair and expression of poentially lethal damage cause the variations in cell survival after X irradiation observed through the cell cycle in Ehrlich ascites tumour cells. Radiat. Res. 95: 87-107 (1983).

I4   Ikushima, T., H. Aritomi and J. Morisita. Radioadaptive response: efficient repair of radiation-induced DNA damage in adapted cells. Mutat. Res. 358: 193-198 (1996).

J1   Jaspers, N.G.J., R.D. Taalman and C. Baan. Patients with an inherited syndrome characterized by immunodeficiency, microcephaly and chromosomal instability: genetic relationship to ataxia-telangiectasia. Am. J. Hum. Genet. 42: 66-73 (1988).

J2   Jaspers, N.G.J., R.A. Gatti, C. Baan et al. Genetic complementation analysis of ataxia telangiectasia and Nijmegen breakage syndrome: a survey of 50 patients. Cytogenet. Cell Genet. 49: 259-263 (1988).

J3   Jeggo, P.A., J. Tesmer and D.J. Chen. Genetic analysis of ionising radiation sensitive mutants of cultured mammalian cell lines. Mutat. Res. 254: 125-133 (1990).

J4   Jenner, T.J., C.M. Delara, P. O'Neill et al. Induction and rejoining of DNA double-strand breaks in V79-4 mammalian cells following gamma-irradiation and alpha-irradiation. Int. J. Radiat. Biol. 64: 265-273 (1993).

J5   Jones, N.J., R. Cox and J. Thacker. Isolation and cross-sensitivity of X-ray sensitive mutants of V79-4 hamster cells. Mutat. Res. 183: 279-286 (1987).

J6   Jostes, R.F., K.M. Bushnell and W.C. Dewey. X-ray induction of 8-azaguanine-resistant mutants in synchronous Chinese hamster ovary cells. Radiat. Res. 83: 146-161 (1980).

J7   Joshi, G.P., W.J. Nelson, S.H. Revell et al. Discrimination of slow growth from non-survival among small colonies of diploid Syrian hamster cells after chromosome damage induced by a range of X-ray doses. Int. J. Radiat. Biol. 42: 283-296 (1982).

J8   Jackson, S.P. Ataxia-telangiectasia at the crossroads. Curr. Biol. 5: 1210-1212 (1995).

J9   Jayaraman, L. and C. Prives. Activation of p53 sequence-specific DNA binding by short single strands of DNA requires the p53 C-terminus. Cell 81: 1021-1029 (1995).

J10   Jeffreys, A.J., K. Tamaki, A. MacLeod et al. Complex gene conversion events in germline mutation at human minisatellites. Nature Genet. 6: 136-145 (1994).

J11   Jin, Y., T-A. Yie and A.M. Carothers. Non-random deletions at the dihydrofolate reductase locus of Chinese hamster ovary cells induced by α-particles simulating radon. Carcinogenesis 16: 1981-1991 (1995).

J12   Jones, L.A., D. Scott, R. Cowan et al. Abnormal radiosensitivity of lymphocytes from breast cancer patients with excessive normal tissue damage after radiotherapy: chromosome aberrations after low-dose-rate irradiation. Int. J. Radiat. Biol. 67: 519-528 (1995).

J13   Joshi, G.P., W.J. Nelson, S.H. Revell et al. X-ray induced chromosome damage in live mammalian cells, and improved measurements of its effects on their colony-forming ability. Int. J. Radiat. Biol. 41: 161-181 (1982).

J14   Johnson, R.D. and L.S. Symington. Functional differences and interactions among the putative RacA homologs Rad51, Rad55 and Rad57. Mol. Cell. Biol. 15: 4843-4850 (1995).

J15   Joenje, H., C. Mathew and E. Gluckman. Fanconi's anaemia research: current status and prospects. Eur. J. Cancer 31A: 268-272 (1995).

J16   Joenje, H., J.R. Lo Ten Foe, A. Oostra et al. Classification of Fanconi anaemia patients by complementation analysis: evidence for a fifth genetic subgroup. Blood 86: 2156-2160 (1995).

J17   Joenje, H., F. Arwert, A.W. Eriksson et al. Oxygen-dependence of chromosomal aberrations in Fanconi's anaemia. Nature 290: 142-143 (1981).

J18   Jensen, R.H., R.G. Langlois, W.L. Bigbee et al. Elevated frequency of glycophorin A mutations in erythrocytes from Chernobyl accident victims. Radiat. Res. 141: 129-135 (1995).

J19   Joenje, H., A.B. Oostra, M. Wijker et al. Evidence for at least eight Fanconi anemia genes. Am. J. Hum. Genet. 61: 940-944 (1997).

J20   Jhappan, C., H.C. Morse, R.D. Fleischmann et al. DNA-PKcs: a T-cell tumour suppressor encoded at the mouse *scid* locus. Nature Genet. 17: 483-486 (1997).

J21   Jongmans, W., M. Vuillaume, K. Chrzanowska et al. Nijmegen Breakage Syndrome cells fail to induce the p53-mediated DNA damage response following exposure to ionizing radiation. Mol. Cell. Biol. 17: 5016-5022 (1997).

J22   Jongmans, W., M. Artuso, M. Vuillaume et al. The role of ataxia telangiectasia and the DNA-dependent protein kinase in the p53-mediated cellular response to ionising radiation. Oncogene 13: 1133-1138 (1996).

J23   Johnson, R.E., M.T. Washington, S. Prakash et al. Bridging the gap: a family of novel DNA polymerases that replicate faulty DNA. Proc. Natl. Acad. Sci. U.S.A. 96: 12224-12226 (1999).

J24   Johnson, R.E., C.M. Kondratick, S. Prakash et al. HRAD30 mutations in the variant form of xeroderma pigmentosum. Science 285: 263-265 (1999).

K1   Kyoizumi, S., M. Akiyama, J.B. Cologne et al. Somatic cell mutations at the glycophorin A locus in erythrocytes of atomic bomb survivors: implications for radiation carcinogenesis. Radiat. Res. 146: 43-52 (1996).

K2   Kastan, M.B., O. Onyekwere, D. Sidransky et al. Participation of p53 protein in the cellular response to DNA damage. Cancer Res. 51: 6304-6311 (1991).

K3   Kastan, M.B., Q.M. Zhan, W.S. Eldeiry et al. A mammalian cell cycle checkpoint pathway utilizing p53 and GADD45 is defective in ataxia-telangiectasia. Cell 71: 587-597 (1992).

K4   Kaufmann, W.K. and S.J. Wilson. G_1 arrest and cell-cycle-dependent clastogenesis in UV-irradiated human fibroblasts. Mutat. Res. 314: 67-76 (1994).

K5   Kroemer, G. The proto-oncogene Bcl-2 and its role in regulating apoptosis. Nature Med. 3: 614-620 (1997).

K6   Khanna, K.K. and M.F. Lavin. Ionizing radiation and UV induction of p53 protein by different pathways in ataxia-telangiectasia cells. Oncogene 8: 3307-3312 (1993).

K7   Kamiya, H., M. Suzuki, Y. Komatsu et al. An abasic site analogue activates a c-Ha-ras gene by a point mutation at modified and adjacent positions. Nucleic Acids Res. 20: 4409-4415 (1992).

K8   Kadhim, M.A., D.A. Macdonald, D.T. Goodhead et al. Transmission of chromosomal instability after plutonium α-particle irradiation. Nature 355: 738-740 (1992).

K9   Kadhim, M.A., S.A. Lorimore, M.D. Hepburn et al. Alpha-particle induced chromosomal instability in human bone marrow cells. Lancet 344: 987-988 (1994).

K10   Kennedy, A.R., M. Fox, G. Murphy et al. Relationship between X-ray exposure and malignant transformation in C3H10T½ cells. Proc. Natl. Acad. Sci. U.S.A. 77: 7262-7266 (1980).

K11 Kennedy, A.R. and J.B. Little. Evidence that a second event in X-ray induced oncogenic transformation *in vitro* occurs during cellular proliferation. Radiat. Res. 99: 228-248 (1984).

K12 Kat, A., W.G. Thilly, W.H. Fang et al. An alkylation-tolerant, mutator cell line is deficient in strand-specific mismatch repair. Proc. Natl. Acad. Sci. U.S.A. 90: 6424-6428 (1993).

K13 Kelsey, K.T., A. Memisoglu, D. Frenkel et al. Human lymphocytes exposed to low doses of X-rays are less susceptible to radiation-induced mutagenesis. Mutat. Res. 263: 197-201 (1991).

K14 Kerr, J.F.R., A.H. Wyllie and A.R. Currie. Apoptosis: a basic biological phenomenon with wide-ranging implication in tissue kinetics. Br. J. Cancer 26: 239-257 (1972).

K15 Keyse, S.M. The induction of gene expression in mammalian cells by radiation. Cancer Biol. 4: 119-128 (1993).

K16 Khandogina, E.K., G.R. Mutovin, S.V. Zvereva et al. Adaptive response in irradiated human lymphocytes: radiobiological and genetical aspects. Mutat. Res. 251: 181-186 (1991).

K17 Kidson, C., P. Chen and P. Imray. Ataxia-telangiectasia heterozygotes: dominant expression of ionising radiation-sensitive mutants. p. 141-153 in: Ataxia-telangiectasia - A Cellular and Molecular Link between Cancer, Neuropathology and Immune Deficiency (B.A. Bridges and D.G. Harnden, eds.). Wiley, Chichester, 1982.

K18 Kimura, H., H. Higuchi, H. Iyehara-Ogawa et al. Sequence analysis of X-ray-induced mutations occurring in a cDNA of the human *hprt* gene integrated into mammalian chromosomal DNA. Radiat. Res. 134: 202-208 (1993).

K19 Korsmeyer, S.J. Regulators of cell death. Trends Genet. 11: 101-105 (1995).

K20 Kronenberg, A. and J.B. Little. Molecular characterization of thymidine kinase mutants of human cells induced by densely ionising radiation. Mutat. Res. 211: 215-224 (1989).

K21 Kronenberg, A., S. Gauny, K. Criddle et al. Heavy ion mutagenesis: linear energy transfer effects and genetic linkage. Radiat. Environ. Biophys. 34: 73-78 (1995).

K22 Kuller, L.H. and B. Modan. Risk of breast cancer in ataxia-telangiectasia. Lancet 326: 1357 (1992).

K23 Kushiro, J., N. Nakamura, S. Kyoizumi et al. Absence of correlations between radiosensitivities of human lymphocytes and skin fibroblasts from the same individuals. Radiat. Res. 122: 326-332 (1990).

K24 Kubota, Y., R.A. Nash, A. Klungland et al. Reconstitution of DNA base-excision repair with purified human proteins: interaction between DNA polymerase β and the XRCC1 protein. EMBO J. 15: 6662-6670 (1996).

K25 Komatsu, K., N. Kubota, M. Gallo et al. The *scid* factor on human chromosome 8 restores V(D)J recombination in addition to double-strand break repair. Cancer Res. 55: 1774-1779 (1995).

K26 Kodaira, M., C. Satoh, K. Hiyamam et al. Lack of effects of atomic bomb radiation on genetic instability of tandem-repetitive elements in human germ cells. Am. J. Hum. Genet. 57: 1275-1283 (1995).

K27 Keegan, K.S., D.A. Holzman, A.W. Plug et al. The ATR and ATM protein kinases associate with different sites along meiotically pairing chromosomes. Genes Dev. 10: 2423-2437 (1996).

K28 Kanaar, R., C. Troelstra, S.M.A. Swagemakers et al. Human and mouse homologs of the *Saccharomyces cerevisiae RAD54* DNA repair gene: evidence for functional conservation. Curr. Biol. 6: 828-838 (1996).

K29 Kubota, Y., R.A. Nash, A. Klungland et al. Reconstitution of DNA base excision repair with purified human proteins: interaction between DNA polymerase β and the XRCC1 protein. EMBO J. 15: 6662-6670 (1996).

K30 Kuhn, E.M. Effects of X-irradiation in G_1 and G_2 on Bloom's syndrome and normal chromosomes. Hum. Genet. 54: 335-341 (1980).

K31 Kataoka, M., M. Kawamura, K. Hamamoto et al. Radiation-induced oesophageal stricture in a case of Bloom's syndrome. Clin. Oncol. 1: 47-48 (1989).

K32 Kadhim, M.A., C.A. Walker, M.A. Plumb et al. No association between p53 status and α-particle-induced chromosomal instability in human lymphoblastoid cells. Int. J. Radiat. Biol. 69: 167-174 (1996).

K33 Kyoizumi, S., N. Nakamura, M. Hakoda et al. Detection of somatic mutations at the glycophorin A locus in erythrocytes of atomic bomb survivors using a single beam flow sorter. Cancer Res. 49: 581-588 (1989).

K34 Kellerer, A.M., E.J. Hall, H.H. Rossi et al. RBE as a function of neutron energy. II. Statistical analysis. Radiat. Res. 65: 172-186 (1976).

K35 Kruyt, F.A.E., L.M. Dijkmans, T.K. van den Berg et al. Fanconi anemai genes act to suppress a cross-linker-inducible p53-independent apoptosis pathway in lymphoblastoid cell lines. Blood 87: 938-948 (1996).

K36 Koenig, F. and J. Kiefer. Lack of dose-rate effect for mutation induction by gamma-rays in human TK6 cells. Int. J. Radiat. Biol. 54: 891-897 (1988).

K37 Kronenberg, A. and J.B. Little. Mutagenic properties of low doses of X-rays, fast neutrons and selected heavy ions in human cells. p. 554-559 in: Low Dose Radiation (K.F. Baverstock and J.W. Stather, eds.). Taylor and Francis, London, 1989.

K38 Kolodner, R.D. Mismatch repair: mechanisms and relationship to cancer susceptibility. Trends Biochem. Sci. 20: 397-401 (1995).

K39 Karow, J.K., R.K. Chakraverty and I.D. Hickson. The Bloom's syndrome gene product is a 3'-5' DNA helicase. J. Biol. Chem. 272: 30611-30614 (1997).

K40 Kupfer, G.M., D. Naf, A. Suliman et al. The Fanconi anaemia proteins, FAA and FAC, interact to form a nuclear complex. Nature Genet. 17: 487-490 (1997).

K41 Kooistra, R., K. Vreeken, J.B.M. Zonneveld et al. The *Drosophila melanogaster RAD54* homolog, *DmRAD54*, is involved in the repair of radiation damage and recombination. Mol. Cell. Biol. 17: 6097-6104 (1997).

K42 Klungland, A. and T. Lindahl. Second pathway for completion of human DNA base excision repair: reconstitution with purified proteins and requirement for DNase IV (FEN1). EMBO J. 16: 3341-3348 (1997).

K43 Khanna, K.K., K.E. Keating, S. Kozlov et al. ATM associates with and phosphorates p53: mapping the region of interaction. Nature Genet. 20: 398-400 (1998).

K44 Komarova, E.A., L. Diatchenko, O.W. Rokhlin et al. Stress-induced secretion of growth inhibitors: a novel tumour suppressor function of p53. Oncogene 17: 1089-1096 (1998).

K45 Kobayashi, S., M. Nishimura, Y. Shimada et al. Increased sensitivity of *scid* heterozygous mice to ionizing radiation. Int. J. Radiat. Biol. 72: 537-545 (1997).

K46 Klungland, A., M. Hoss, D. Gunz et al. Base excision repair of oxidative damage activated by XPG protein. Mol. Cell. 3: 33-42 (1999).

K47 Kuhne, M., K. Rothkamm and M. Lobrich. No dose dependence of DNA double-strand break misrejoining following alpha-particle irradiation. Int. J. Radiat. Biol. (2000, in press).

L1   Lehmann, A.R., B.A. Bridges, P.C. Hanawalt et al. Workshop on processing of DNA damage. Mutat. Res. 364: 245-270 (1996).

L2   Lindahl, T. Instability and decay of the primary structure of DNA. Nature 362: 709-715 (1993).

L3   Lindahl, T. and D.E. Barnes. Mammalian DNA ligases. Annu. Rev. Biochem. 61: 251-281 (1992).

L4   Lavin, M.F., P. Le Poidevin and P. Bates. Enhanced levels of G_2 phase delay in ataxia-telangiectasia heterozygotes. Cancer Genet. Cytogenet. 60: 183-187 (1992).

L5   Lavin, M.F., I. Bennett, J. Ramsay et al. Identification of a potentially radiosensitive subgroup among patients with breast cancer. J. Natl. Cancer Inst. 86: 1627-1634 (1994).

L6   Lee, J.M. and A. Bernstein. p53 mutations increase resistance to ionizing radiation. Proc. Natl. Acad. Sci. U.S.A. 90: 5742-5746 (1993).

L7   Little, J.B. Delayed initiation of DNA synthesis in irradiated human diploid cells. Nature 218: 1064-1065 (1968).

L8   Little, J.B., J. Nove, L.C. Strong et al. Survival of human diploid fibroblasts from normal individuals after X-irradiation. Int. J. Radiat. Biol. 54: 899-910 (1988).

L9   Lieber, M.R., J.E. Hesse, S. Lewis et al. The defect in murine severe combined immune deficiency: joining of signal sequences but not coding segments in V(D)J recombination. Cell 55: 7-16 (1988).

L10   Lipkowitz, S., M-H. Stern and I.R. Kirsch. Hybrid T-cell receptor genes formed by interlocus recombination in normal and ataxia-telangiectasia lymphocytes. J. Exp. Med. 172: 409-418 (1990).

L11   Littlefield, L.G., L.B. Travis, A.M. Sayer et al. Cumulative genetic damage in hematopoietic stem cells in a patient with a 40-year exposure to alpha particles emitted by thorium dioxide. Radiat. Res. 148: 135-144 (1997).

L12   Lock, R.B. and W.E. Ross. Inhibition of $p34^{cdc2}$ kinase activity by etoposide or irradiation as a mechanism of G_2 arrest in Chinese hamster ovary cells. Cancer Res. 50: 3761-3766 (1990).

L13   Lowe, S.W., E.M. Schmitt, S.W. Smith et al. p53 is required for radiation-induced apoptosis in mouse thymocytes. Nature 362: 847-849 (1993).

L14   Lu, X. and D.P. Lane. Differential induction of transcriptionally active p53 following UV or ionizing radiation - defects in chromosome instability syndromes? Cell 75: 765-778 (1993).

L15   Leach, F.S., N.C. Nicolaides, N. Papadopoulos et al. Mutations of a mutS homolog in hereditary nonpolyposis colorectal cancer. Cell 75: 1215-1225 (1993).

L16   Li, C-Y., D.W. Yandell and J.B. Little. Evidence for coincident mutations in human lymphoblast clones selected for functional loss of a thymidine kinase gene. Mol. Carcinog. 5: 270-277 (1992).

L17   Lehnert, B.E. and E.H. Goodwin. Extracellular factor(s) following exposure to α particles can cause sister chromatid exchange in normal human cells. Cancer Res. 57: 2164-2171 (1997).

L18   Little, J. DNA damage, recombination and mutation. p. 123-129 in: Molecular Mechanisms in Radiation Mutagenesis and Carcinogenesis (K.H. Chadwick, R. Cox, H.P. Leenhouts et al., eds.). European Commission, Brussels, 1994.

L19   Lutze, L. and R.A. Winegar. pHAZE: a shuttle vector system for the detection and analysis of ionising radiation-induced mutations. Mutat. Res. 245: 305-310 (1990).

L20   Lutze, L.H., R.A. Winegar, R. Jostes et al. Radon-induced deletions in human cells - role of nonhomologous strand rejoining. Cancer Res. 52: 5126-5129 (1992).

L21   Lutze, L.H., J.E. Cleaver and R.A. Winegar. Factors affecting the frequency, size and location of ionising-radiation-induced deletions in human cells. p. 41-46 in: Molecular Mechanisms in Radiation Mutagenesis and Carcinogenesis (K.H. Chadwick, R. Cox, H.P. Leenhouts et al., eds.). European Commission, Brussels, 1994.

L22   Lane, D.P. p53, guardian of the genome. Nature 358: 15-16 (1992).

L23   Lambin, P., B. Marples, B. Fertil et al. Hypersensitivity of a human tumour cell line to very low radiation doses. Int. J. Radiat. Biol. 63: 639-650 (1993).

L24   Lambin, P., B. Fertil, E.P. Malaise et al. Multiphasic survival curves for cells of human tumor cell lines: induced repair or hypersensitive subpopulation? Radiat. Res. 138: S32-S36 (1994).

L25   Land, C.E. Risk of breast cancer in ataxia-telangiectasia. Lancet 326: 1359-1360 (1992).

L26   Langley, R.E., S.G. Quartuccio, P.T. Keannealey et al. Effect of cell-cycle stage, dose rate and repair of sublethal damage on radiation-induced apoptosis in F9 terato-carcinoma cells. Radiat. Res. 144: 90-96 (1995).

L27   Laval, F. Pretreatment with oxygen species increases the resistance of mammalian cells to hydrogen peroxide and γ-rays. Mutat. Res. 201: 73-79 (1988).

L28   Lee, S., B. Elenbaas, A. Levine et al. p53 and its 14 kDa c-terminal domain recognize primary DNA damage in the form of insertion/deletion mismatches. Cell 81: 1013-1020 (1995).

L29   Li, C-Y., D.W. Yandell and J.B. Little. Molecular mechanisms of spontaneous and induced loss of heterozygosity in human cells in vitro. Somat. Cell Mol. Genet. 18: 77-87 (1992).

L30   Li, Z., T. Otrevel, Y. Gao et al. The *XRCC4* gene encodes a novel protein involved in DNA double-strand break repair and V(D)J recombination. Cell 83: 1079-1089 (1995).

L31   Liber, H.L., P.K. LeMotte and J.B. Little. Toxicity and mutagenicity of X-rays and [^{125}I]dUrd or [^{3}H]TdR incorporated in the DNA of human lymphoblast cells. Mutat. Res. 111: 387-404 (1983).

L32   Ling, C.C., C.H. Chen and W.X.L. Li. Apoptosis induced at different dose rates: implications for the shoulder regions of the cell survival curves. Radiother. Oncol. 32: 129-136 (1994).

L33   Ling, C.C., M. Guo, C.H. Chen et al. Radiation-induced apoptosis: effects of cell age and dose fractionation. Cancer Res. 55: 5207-5212 (1995).

L34   Little, J.B. Changing views of cellular radiosensitivity. Radiat. Res. 140: 299-311 (1994).

L35   Little, J.B. and J. Nove. Sensitivity of human diploid fibroblast cell strains from various genetic disorders to acute and protracted radiation exposure. Radiat. Res. 123: 87-92 (1990).

L36   Liu, B., R.E. Parsons, S.R. Hamilton et al. hMSH2 mutations in hereditary nonpolyposis colorectal cancer kindreds. Cancer Res. 54: 4590-4594 (1994).

L37   Lloyd, D.C., A.A. Edwards, A. Leonard et al. Frequencies of chromosomal aberrations induced in human blood lymphocytes by low doses of X-rays. Int. J. Radiat. Biol. 53: 49-55 (1988).

L38   Lloyd, D.C., A.A. Edwards, A. Leonard et al. Chromosomal aberrations in human lymphocytes induced in vitro by very low doses of X-rays. Int. J. Radiat. Biol. 61: 335-343 (1992).

L39   Loeffler, J.S., J.R. Harris, W.K. Dahlberg et al. *In vitro* radiosensitivity of human diploid fibroblasts derived from women with unusually sensitive clinical responses to definitive radiation therapy for breast cancer. Radiat. Res. 121: 227-231 (1990).

L40 Lucke-Huhle, C., L. Hieber and R.D. Wegner. Caffeine-mediated release of alpha-irradiation induced G_2 arrest increase the yield of chromosome aberrations. Int. J. Radiat. Biol. 43: 123-132 (1983).

L41 Lakin, N.D., P. Weber, T. Stankovic et al. Analysis of the ATM protein in wild-type and ataxia telangiectasia cells. Oncogene 13: 2707-2716 (1996).

L42 Lim, D-S. and P. Hasty. A mutation in mouse *rad51* results in an early embryonic lethal that is suppressed by a mutation in *p53*. Mol. Cell. Biol. 16: 7133-7143 (1996).

L43 Lo Ten Foe, J.R., M.A. Rooimans, L. Bosnoyan-Collins et al. Expression cloning of a cDNA for the major Fanconi anaemia gene, *FAA*. Nature Genet. 14: 320-323 (1996).

L44 Langlois, R.G., W.L. Bigbee, S. Kyoizumi et al. Evidence for increased somatic cell mutations at the glycophorin A locus in atomic bomb survivors. Science 236: 445-448 (1987).

L45 Langlois, R.G., M. Akiyama, Y. Kusunoki et al. Analysis of somatic cell mutations at the glycophorin A locus in atomic bomb survivors: a comparative study of assay methods. Radiat. Res. 136: 111-117 (1993).

L46 Liebetrau, W., A. Budde, A. Savoia et al. p53 activates Fanconi anemia group C gene expression. Hum. Mol. Genet. 6: 277-283 (1997).

L47 Lonn, U., S. Lonn, U. Nylen et al. An abnormal profile of DNA replication intermediates in Bloom's syndrome. Cancer Res. 50: 3141 3145 (1990).

L48 Lo, Y.M.D., S. Darby, L. Noakes et al. Screening for codon 249 p53 mutation in lung cancer associated with domestic radon exposure. Lancet 345: 60 (1995).

L49 Levran, O., T. Erlich, N. Magdalena et al. Sequence variation in the Fanconi anemia gen *FAA*. Proc. Natl. Acad. Sci. U.S.A. 94: 13051-13056 (1997).

L50 Liu, C.Y., A. Flesken-Nitikin, S. Li. et al. Inactivation of the mouse Brca1 gene leads to failure in the morphogenesis of the egg cylinder in early post-implantational development. Genes Dev. 10: 1835-1843 (1996).

L51 Lee, S.E., R.A. Mitchell, A. Cheng et al. Evidence for DNA-PK-dependent and independent DNA double-strand break repair pathways in mammalian cells as a function of the cell cycle. Mol. Cell. Biol. 17: 1425-1433 (1997).

L52 Lobrich, M., B. Rydberg and P.K. Cooper. DNA double-strand breaks induced by high-energy neon and iron ions in human fibroblasts. II. Probing individual *Not*I fragments by hybridization. Radiat. Res. 139: 142-151 (1994).

L53 Lobrich, M., P.K. Cooper and B. Rydberg. Non-random distribution of DNA double-strand breaks induced by particle irradiation. Int. J. Radiat. Biol. 70: 493-503 (1996).

L54 Lucas, J.N., F.S. Hill, A.M. Chen et al. A rapid method for measuring pericentric inversions using fluorescence *in situ* hybridization (FISH). Int. J. Radiat. Biol. 71: 29-33 (1997).

L55 Lucas, J.N. Cytogenetic signature for ionizing radiation. Int. J. Radiat. Biol. 73: 15-20 (1998).

L56 Liu, N. and P.E. Bryant. Response of ataxia telangiectasia cells to restriction endonuclease induced DNA double-strand breaks 1. Cytogenetic characterization. Mutagenesis 8: 503-510 (1993).

L57 Li, G., H. Ouyang, X. Li et al. *Ku70*: a candidate tumor suppressor gene for murine T cell lymphoma. Mol. Cell 2: 1-8 (1998).

L58 Luo, L., F. Lu, S. Hart et al. Ataxia-telangiectasia and T-cell leukemias: no evidence for somatic *ATMi* mutation in sporadic T-ALL or for hypermethylation of the *ATM-NPAT/E14* bidirectional promoter in T-PLL. Cancer Res. 58: 2293-2297 (1998).

L59 Liu, N., J.E. Lamerdin, J.D. Tucker et al. The human *XRCC9* gene corrects chromosomal instability and mutagen sensitivities in CHO UV40 cells. Proc. Natl. Acad. Sci. U.S.A. 94: 9232-9237 (1997).

L60 Le, X.C., J.Z. Xing, J. Lee et al. Inducible repair of thymine glycol detected by an ultrasensitive assay for DNA damage. Science 280: 1066-1069 (1998).

L61 Lloyd, D.C., P. Finnon, A.A. Edwards et al. Chromosome aberrations in syrian hamsters following very low radiation doses in vivo. Mutat. Res. 377: 63-68 (1997).

L62 Lobrich, M., P.K. Cooper and B. Rydberg. Joining of correct and incorrect DNA ends at double-strand breaks produced by high-linear energy transfer radiation in human fibroblasts. Radiat. Res. 150: 619-626 (1998).

L63 Lorimore, S.A., M.A. Kadhim, D.A. Pocock et al. Chromosomal instability in the descendants of unirradiated surviving cells after α-particle irradiation. Proc. Natl. Acad. Sci. U.S.A. 95: 5730-5733 (1998).

L64 Lee, H., A.H. Trainer, L.S. Friedman et al. Mitotic checkpoint inactivation fosters transformation in cells lacking the breast cancer susceptibility gene, *Brca2*. Mol. Cell 4: 1-10 (1999).

L65 Liu, N., J.E. Lamerdin, R.S. Tebbs et al. XRCC2 and XRCC3, new human RAD51-family members, promote chromosome stability and protect against DNA cross-links and other damages. Mol. Cell 1: 783-793 (1998).

L66 Liao, M.J. and T. van Dyke. Critical role for Atm in suppressing V(D)J recombination-driven thymic lymphoma. Genes Dev. 13: 1246-1250 (1999).

L67 Liao, W.C., A. Haimovitz-Friedman, R.S. Persaud et al. Ataxia telangiectasia-mutated gene product inhibits DNA damage-induced apoptosis via ceramide synthase. J. Biol. Chem. 274: 17908-17917 (1999).

L68 Lindahl, T. Repair of intrinsic DNA lesions. Mutat. Res. 238: 305-311 (1990).

L69 Lindahl, T. and R.D. Wood. Quality control by DNA repair. Science 286: 1897-1905 (1999).

L70 Lehmann, A.R. Nucleotide excision repair and the link with transcription. Trends Biochem. Sci. 20: 402-405 (1995).

L71 Livingston, G.K., R.H. Jensen, E.B. Silberstein et al. Radiobiological evaluation of immigrants from the vicinity of Chernobyl. Int. J. Radiat. Biol. 72: 703-713 (1997).

L72 Lobrich, M., M. Kuhne, J. Wetzel et al. Joining of correct and incorrect DNA double-strand break ends in normal human and ataxia telangiectasia fibroblasts. Genes, Chromosomes Cancer 27: 59-68 (2000).

M1 Michael, J.M., M.F. Lavin and D.J. Watters. Resistance to radiation-induced apoptosis in burkitt's lymphoma cells is associated with defective ceramide signaling. Cancer Res. 57: 3600-3605 (1997).

M2 Mullenders, L.H.F., H. Vrieling, J. Venema et al. Hierarchies of DNA repair in mammalian cells: biological consequences. Mutat. Res. 250: 223-228 (1991).

M3 Murray, A.W. Creative blocks - cell-cycle checkpoints and feedback controls. Nature 359: 599-604 (1992).

M4 Malkin, D., F.P. Li, I.C. Strong et al. Germ line p53 mutations in a familial syndrome of breast cancer, sarcomas and other neoplasms. Science 250: 1233-1238 (1990).

M5 McKinnon, P.J. Ataxia-telangiectasia: an inherited disorder of ionising radiation sensitivity in man. Hum. Genet. 75: 197-208 (1987).

M6 Morrison, C., G.C.M. Smith, L. Stingl et al. Genetic interaction between PARP and DNA-PK in V(D)J recombination and tumorigenesis. Nature Genet. 17: 479-482 (1997).

M7 McWhir, J., J. Selfridge, D.J. Harrison et al. Mice with DNA repair gene (*ERCC1*) deficiency have elevated levels of p53, liver nuclear abnormalities and die before weaning. Nature Genet. 5: 217-223 (1993).

M8 Meyn, M.S. High spontaneous intrachromosomal recombination rates in ataxia-telangiectasia. Science 260: 1327-1330 (1993).

M9 Menissier de Murcia, J.M., C. Niedergang, C. Trucco et al. Requirement of poly(ADP-ribose) polymerase in recovery from DNA damage in mice and in cells. Proc. Natl. Acad. Sci. U.S.A. 94: 7303-7307 (1997).

M10 Muris, D.F.R., O. Bezzubova, J.M. Buerstedde et al. Cloning of human and mouse genes homologous to RAD52: a yeast gene involved in DNA repair and recombination. Mutat. Res. 315: 295-305 (1994).

M11 Meuth, M. The structure of mutation in mammalian cells. Biochim. Biophys. Acta 1032: 1-17 (1990).

M12 Miles, C. and M. Meuth. DNA sequence determination of gamma-radiation-induced mutations of the hamster *aprt* locus. Mutat. Res. 227: 97-102 (1989).

M13 Miles, C., G. Sargent, G. Phear et al. DNA sequence analysis of gamma radiation-induced deletions and insertions at the *aprt* locus of hamster cells. Mol. Carcinog. 3: 233-242 (1990).

M14 Moore, M.M., A. Amtower, G.H.S. Strauss et al. Geno-toxicity of gamma-irradiation in L5178Y mouse lymphoma cells. Mutat. Res. 174: 149-154 (1986).

M15 Morris, T. and J. Thacker. Formation of large deletions by illegitimate recombination in the HPRT gene of primary human fibroblasts. Proc. Natl. Acad. Sci. U.S.A. 90: 1392-1396 (1993).

M16 Morris, T., W. Masson, B. Singleton et al. Analysis of large deletions in the *HPRT* gene of primary human fibroblasts using the polymerase chain reaction. Somat. Cell Mol. Genet. 19: 9-19 (1993).

M17 Marples, B. and M.C. Joiner. The response of Chinese hamster V79 cells to low radiation doses: evidence of enhanced sensitivity of the whole cell population. Radiat. Res. 133: 41-51 (1993).

M18 Marples, B. and M.C. Joiner. The elimination of low dose hypersensitivity in Chinese hamster V79-379A cells by pretreatment with X-rays or hydrogen peroxide. Radiat. Res. 141: 160-169 (1995).

M19 Marples, B., G.K.Y. Lam, H. Zhou et al. The response of Chinese hamster V79-379A cells exposed to negative pi-mesons: evidence that increased radioresistance is dependent on linear energy transfer. Radiat. Res. 138: S81-S84 (1994).

M20 Marra, G. and C.R. Boland. Hereditary non-polyposis colorectal cancer: the syndrome, the genes, and historical perspectives. J. Natl. Cancer Inst. 87: 1114-1125 (1995).

M21 McDonnell, T.J. and S.J. Korsmeyer. Progression from lymphoid hyperplasia to high-grade malignant lymphoma in mice transgenic for the t(14;18). Nature 349: 254-256 (1991).

M22 Metting, N.F., S.T. Palayoor, R.M. Macklis et al. Induction of mutations by bismuth-212 α-particles at two genetic loci in human B-lymphoblasts. Radiat. Res. 132: 339-345 (1992).

M23 Matsuoka, S., M. Huang and S.J. Elledge. Linkage of ATM to cell cycle regulation by the Chk2 protein kinase. Science 282: 1893-1897 (1998).

M24 Meyers, M., R.A. Schea, A.E. Petrowski et al. Role of X-ray inducible genes and proteins in adaptive survival response. p. 263-270 in: Low Dose Irradiation and Biological Defence Mechanisms (T. Sugahara et al., eds.). Elsevier, 1992.

M25 Miyazaki, N. and Y. Fujiwara. Mutagenic and lethal effects of [5-^{125}I]iodo-2'-deoxyuridine incorporated into DNA of mammalian cells and their RBEs. Radiat. Res. 88: 456-465 (1981).

M26 Modan, B., A. Chetrit, E. Alfandary et al. Increased risk of breast cancer after low-dose irradiation. Lancet 1: 629-631 (1989).

M27 Modrich, P. Mechanisms and biological effects of mismatch repair. Annu. Rev. Genet. 25: 229-253 (1991).

M28 Maser, R.S., K.J. Monsen, B.E. Nelms et al. hMre11 and hRad50 nuclear foci are induced during the normal cellular response to DNA double-strand breaks. Mol. Cell. Biol. 17: 6087-6096 (1997).

M29 Metcalfe, J.A., J. Parkhill, L. Campbell et al. Accelerated telomere shortening in ataxia telangiectasia. Nature Genet. 13: 350-353 (1996).

M30 McConville, C.M., T. Stankovic, P.J. Byrd et al. Mutations associated with variant phenotypes in ataxia-telangiectasia. Am. J. Hum. Genet. 59: 320-330 (1996).

M31 Morrow, D.W., D.A. Tagle, Y. Shiloh et al. *TEL1*, an S. cerevisiae homolog of the human gene mutated in ataxia telangiectasia, is functionally related to the yeast checkpoint gene *MEC1*. Cell 82: 831-840 (1995).

M32 Milne, G.T., S. Jin, K.B. Shannon et al. Mutations in two Ku homologs define a DNA end-joining repair pathway in *Saccharomyces cerevisiae*. Mol. Cell. Biol. 16: 4189-4198 (1996).

M33 Mason, R.M., J. Thacker and M.P. Fairman. The joining of non-complementary DNA double-strand breaks by mam-malian extracts. Nucleic Acids Res. 24: 4946-4953 (1996).

M34 Marder, B.A. and W.F. Morgan. Delayed chromosomal instability induced by DNA damage. Mol. Cell. Biol. 13: 6667-6677 (1993).

M35 Martins, M.B., L. Sabatier, M. Ricoul et al. Specific chromosome instability induced by heavy ions: a step towards transformation of human fibroblasts? Mutat. Res. 285: 229-237 (1993).

M36 Manti, L., M. Jamali, K.M. Prise et al. Genomic instability in Chinese hamster cells after exposure to X-rays or alpha particles of different mean linear energy transfer. Radiat. Res. 147: 22-28 (1997).

M37 Morita, T., Y. Yoshimura, A. Yamamoto et al. A mouse homolog of the *Escherichia coli recA* and *Saccharomyces cerevisiae RAD51* genes. Proc. Natl. Acad. Sci. U.S.A. 90: 6577-6580 (1993).

M38 Meyn, M.S. Ataxia-telangiectasia and cellular responses to DNA damage. Cancer Res. 55: 5991-6001 (1995).

M39 Merritt, A.J., C.S. Potten, C.J. Kemp et al. The role of *p53* in spontaneous and radiation-induced apoptosis in the gastrointestinal tract of normal and *p53*-deficient mice. Cancer Res. 54: 614-617 (1994).

M40 Mullaart, E., P.H.M. Lohman, F. Berends et al. DNA damage metabolism and aging. Mutat. Res. 237: 189-210 (1990).

M41 Matsuura, S., C. Weemaes, D. Smeets et al. Genetic mapping using microcell-mediated chromosome transfer suggest a locus for Nijmegen breakage syndrome at chromo-some 8q21-24. Am. J. Hum. Genet. 60: 1487-1494 (1997).

M42 Mizuta, R., J.M. LaSalle, H-L. Cheng et al. RAB22 and RAB163/mouse BRCA2: proteins that specifically interact with the RAD51 protein. Proc. Natl. Acad. Sci. U.S.A. 94: 6927-6932 (1997).

M43 Myung, K., D.M. He, S.E. Lee et al. KARP-1: a novel leucine zipper protein expressed from the Ku86 autoantigen locus is implicated in the control of DNA-dependent protein kinase activity. EMBO J. 16: 3172-3184 (1997).

M44 Myung, K., C. Braastad, D.M. He et al. KARP-1 is induced by DNA damage in a p53- and ataxia telangiectasia mutated-dependent fashion. Proc. Natl. Acad. Sci. U.S.A. 95: 7664-7669 (1998).

M45 Matsuura, S., H. Tauchi, A. Nakamura et al. Positional cloning of the gene for Nijmegen breakage syndrome. Nature Genet. 19: 179-181 (1998).

M46 Marmorstein, L.Y., T. Ouchi and S.A.A. Aaronson. The BRCA2 gene product functionally interacts with p53 and Rad51. Proc. Natl. Acad. Sci. U.S.A. 95: 13869-13874 (1998).

M47 Mitchell, E.L. and D. Scott G2 chromosomal radio-sensitivity in fibroblasts of ataxia-telangiectasia hetero-zygotes and a Li-Fraumeni syndrome patient with radio-sensitive cells. Int. J. Radiat. Biol. 72: 435-438 (1997).

M48 Morimatsu, M., G. Donoho and P. Hasty. Cells deleted for BRCA2 COOH terminus exhibit hypersensitivity to gamma-radiation and premature senescence. Cancer Res. 58: 3441-3447 (1998).

M49 Masutani, C., R. Kusumoto, A. Yamada et al. The XPV (xeroderma pigmentosum variant) gene encodes human DNA polymerase eta. Nature 399: 700-704 (1999).

M50 Moynahan, M.E., J.W. Chiu, B.H. Koller et al. Brca1 controls homology-directed DNA repair. Mol. Cell 4: 511-518 (1999).

M51 Maki, H. and M. Sekiguchi. MutT protein specifically hydrolyses a potent mutagenic substrate for DNA synthesis. Nature 355: 273-275 (1992).

M52 Makedonov, G.P., L.V. Tshovrebova, S.V. Unzhakov et al. Radioadaptive response in lymphocytes from the children from polluted Bryansk regions after Chernobyl accident. Radiat. Biol. Radioecol. 37: 640-644 (1997).

M53 Mohrenweiser, H.W. and I.M. Jones. Variation in DNA repair is a factor in cancer susceptibility: a paradigm for the promises and perils of individual and population risk estimation. Mutat. Res. 400: 15-24 (1998).

M54 Martin, S.G., T. Laroche, N. Suka et al. Relocalization of telomeric Ku and SIR proteins in response to DNA strand breaks in yeast. Cell 97: 621-633 (1999).

N1 Nagasawa, H., J.B. Robertson, C.S. Arundel et al. The effect of X-irradiation on the progression of mouse 10T½ cells released from density-inhibited cultures. Radiat. Res. 97: 537-545 (1984).

N2 Nagasawa, H., S.A. Latt, M.E. Lalande et al. Effects of X-irradiation on cell-cycle progression, induction of chromosomal aberrations and cell killing in ataxia telangiectasia fibroblasts. Mutat. Res. 148: 71-82 (1985).

N3 Nagasawa, H., K.H. Kraemer, Y. Shiloh et al. Detection of ataxia telangiectasia heterozygous cell lines by post-irradiation cumulative labelling index: measurements with coded samples. Cancer Res. 47: 398-402 (1987).

N4 Nakamura, N., R. Sposto, J.-I. Kushiro et al. Is inter-individual variation of cellular radiosensitivity real or artifactual? Radiat. Res. 125: 326-330 (1991).

N5 Nelson, W.G. and M.B. Kastan. DNA strand breaks - the DNA template alterations that trigger p53-dependent DNA damage response pathways. Mol. Cell. Biol. 14: 1815-1823 (1994).

N6 Nove, J., J.B. Little, P.J. Mayer et al. Hypersensitivity of cells from a new chromosomal-breakage syndrome to DNA-damaging agents. Mutat. Res. 163: 255-262 (1986).

N7 North, P., A. Ganesh and J. Thacker. The rejoining of double-strand breaks in DNA by human cell extracts. Nucleic Acids Res. 18: 6205-6210 (1990).

N8 Nagasawa, H. and J.B. Little. Induction of sister chromatid exchanges by extremely low doses of α-particles. Cancer Res. 52: 6394-6396 (1992).

N9 Nicklas, J.A., T.C. Hunter, J.P. O'Neill et al. Fine structure mapping of the human hprt gene region on the X-chromosome (Xq26). Am. J. Hum. Genet. 47 (Suppl.): A193 (1990).

N10 Nicolaides, N.C., N. Papadopoulos, B. Liu et al. Mutations of two PMS homologues in hereditary nonpolyposis colon cancer. Nature 371: 75-80 (1994).

N11 Nelson, S.L., C.R. Giver and A.J. Grosovsky. Spectrum of X-ray-induced mutations in the human hprt-gene. Carcinogenesis 15: 495-502 (1994).

N12 Nishino, H., A. Knoll, V.L. Buettner et al. p53 wild-type and p53 nullizygous Big Blue transgenic mice have similar frequencies and patterns of observed mutation in liver, spleen and brain. Oncogene 11: 263-270 (1995).

N14 Nacht, M., A. Strasser, Y.R. Chan et al. Mutations in p53 and SCID genes cooperate in tumorigenesis. Genes Dev. 10: 2055-2066 (1996).

N15 Niwa, O., Y-J. Fan, M. Numoto et al. Induction of a germ-line mutation at a hypervariable mouse minisatellite locus by ^{252}Cf radiation. J. Radiat. Res. 37: 217-224 (1996).

N16 Nicklas, J.A., M.T. Falta, T.C. Hunter et al. Molecular analysis of in vivo hprt mutations in human T lymphocytes. V. Effects of total body irradiation secondary to radio-immunoglobulin therapy. Mutagenesis 5: 461-468 (1990).

N17 Nelson, S.L., K.K. Parks and A.J. Grosovsky. Ionizing radiation signature mutations in human cell mutants induced by low-dose exposures. Mutagenesis 11: 275-279 (1996).

N18 Nakamura, N. and S. Sawada. Reversed dose-rate effect and RBE of 252-californium radiation in the induction of 6-thioguanine resistant mutations in mouse L5178Y cells. Mutat. Res. 201: 65-71 (1988).

N19 Nikjoo, H., P. O'Neill, D.T. Goodhead et al. Computational modelling of low-energy electron-induced DNA damage by early physical and chemical events. Int. J. Radiat. Biol. 71: 467-483 (1997).

N20 Nevaldine, B., J.A. Longo and P.J. Hahn. The scid defect results in a much slower repair of DNA double-strand breaks but not high levels of residual breaks. Radiat. Res. 147: 535-540 (1997).

N21 Nussenzweig, A., K. Sokol, P. Burgman et al. Hyper-sensitivity of Ku80-deficient cell lines and mice to DNA damage: the effects of ionizing radiation on growth, survival, and development. Proc. Natl. Acad. Sci. U.S.A. 94: 13588-13593 (1997).

N22 Nikiforov, Y.E., M. Nikiforova and J.A. Fagin. Prevalence of minisatellite and microsatellite instability in radiation-induced post-Chernobyl pediatric thyroid carcinomas. Oncogene 17: 1983-1988 (1998).

N23 Nelson, J.R., C.W. Lawrence and D.C. Hinckle. Thymine-thymine dimer bypass by yeast DNA polymerase zeta. Science 272: 1646-1649 (1996).

N24 Nelson, J.R., C.W. Lawrence and D.C. Hinckle. Deoxycytidyl transferase activity of yeast REV1 protein. Nature 382: 729-731 (1996).

N25 Nagasawa, H. and J.B. Little. Unexpected sensitivity to the induction of mutations by very low doses of alpha-particle radiation: evidence for a bystander effect. Radiat. Res. 152: 552-557 (1999).

N26 Naito, T., A. Matsuura and F. Ishikawa. Circular chromosome formation in a fission yeast mutant defective in to ATM homologues. Nature Genet. 20: 203-206 (1998).

N27 Nakane, H., S. Takeuchi, S. Yuba et al. High incidence of ultraviolet-B or chemical-carcinogen-induced skin tumours in mice lacking the xeroderma pigmentosum group A gene. Nature 377: 165-168 (1995).

O1 Okada, S. Radiation-induced death. p. 247-307 in: Radiation Biochemistry, Volume 1 (K.I. Altman et al., eds.). Academic Press, New York, 1970.

O2 Olivieri, G., J. Bodycote and S. Wolff. Adaptive response of human lymphocytes to low concentrations of radioactive thymidine. Science 223: 594-597 (1984).

O3 Olive, P.L. The role of DNA single and double-strand breaks in cell killing by ionizing radiation. Radiat. Res. 150: S42-S51 (1998).

O4 Oliver, F.J., J. Menissier-de Murcia and G. de Murcia. Poly(ADP-ribose) polymerase in the cellular response to DNA damage, apoptosis, and disease. Am. J. Hum. Genet. 64: 1282-1288 (1999).

O5 Osman, F. and S. Subramani. Double-strand break-induced recombination in eukaryotes. Prog. Nucleic Acid Res. Mol. Biol.58: 263-299 (1998).

P1 Perrino, F.W. and L.A. Loeb. Animal cell DNA poly-merases in DNA repair. Mutat. Res. 236: 289-300 (1990).

P2 Parshad, R., K.K. Sanford and G.M. Jones. Chromatid damage after G_2 phase X-irradiation of cells from cancer-prone individuals implicates deficiency in DNA repair. Proc. Natl. Acad. Sci. U.S.A. 80: 5612-5616 (1983).

P3 Parshad, R., K.K. Sanford and G.M. Jones. Chromosomal radiosensitivity during the G_2 cell-cycle period of skin fibroblasts from individuals with familial cancer. Proc. Natl. Acad. Sci. U.S.A. 82: 5400-5403 (1985).

P4 Painter, R.B. Inhibition of mammalian cell DNA synthesis by ionising radiation. Int. J. Radiat. Biol. 49: 771-781 (1986).

P5 Pandita, T.K. and W.N. Hittelman. The contribution of DNA and chromosome repair deficiencies to the radiosensitivity of ataxia telangiectasia. Radiat. Res. 131: 214-223 (1992).

P6 Pergola, F., M.Z. Zdzienicka and M.R. Lieber. V(D)J recombination in mammalian cell mutants defective in DNA double-strand break repair. Mol. Cell. Biol. 13: 3464-3471 (1993).

P7 Plowman, P.N., B.A. Bridges, C.F. Arlett et al. An instance of clinical radiation morbidity and cellular radiosensitivity, not associated with ataxia-telangiectasia. Br. J. Radiol. 63: 624-628 (1990).

P8 Pippard, E.C., A.J. Hall, D.J.P. Barker et al. Cancer in homozygotes and heterozygotes of ataxia-telangiectasia and xeroderma pigmentosum in Britain. Cancer Res. 48: 2929-2932 (1988).

P9 Padovani, L., M. Appolloni, P. Anzidei et al. Do human lymphocytes exposed to the fallout of the Chernobyl accident exhibit an adaptive response? Challenge with ionising radiation. Mutat. Res. 332: 33-38 (1995).

P10 Parsons, R., G-M. Li, M. Longley et al. Mismatch repair deficiency in phenotypically normal human cells. Science 268: 738-740 (1995).

P11 Paterson, M.C., A.K. Anderson, B.P. Smith et al. Enhanced radiosensitivity of cultured fibroblasts from ataxia telangiectasia heterozygotes manifested by defective colony-forming ability and reduced DNA repair replication after hypoxic X-irradiation. Cancer Res. 39: 3725-3734 (1979).

P12 Paterson, M.C., S.J. McFarlane, N.E. Gentner et al. Cellular hypersensitivity to chronic γ-radiation in cultured fibroblasts from ataxia-telangiectasia heterozygotes. p. 73-87 in: Ataxia-telangiectasia: Genetics, Neuropathology, and Immunology of a Degenerative Disease of Childhood (R.A Gatti & M. Swift, eds.). Liss, New York, 1985.

P13 Pegg, A.E. Properties of O^6-alkylguanine-DNA transferases. Mutat. Res. 233: 165-175 (1990).

P14 Pohl-Ruling, J. and P. Fischer. The dose-effect relationship of chromosome aberrations to α and γ irradiation in a population subjected an increased burden of natural radioactivity. Radiat. Res. 80: 61-81 (1979).

P15 Pohl-Ruling, J., P. Fischer, O. Haas et al. Effect of low-dose acute X-irradiation on the frequencies of chromosomal aberrations in human peripheral lymphocytes in vitro. Mutat. Res. 110: 71-82 (1983).

P16 Pohl-Ruling, J., P. Fischer, D.C. Lloyd et al. Chromo-somal damage induced in human lymphocytes by low doses of D-T neutrons. Mutat. Res. 173: 267-272 (1986).

P17 Pohl-Ruling, J., O. Haas, A. Brogger et al. The effect on lymphocyte chromosomes of additional radiation burden due to fallout in Salzburg (Austria) from the Chernobyl accident. Mutat. Res. 262: 209-217 (1991).

P18 Potten, C.S. Extreme sensitivity of some intestinal crypt cells to X and gamma irradiation. Nature 269: 518-520 (1977).

P19 Pelevina, I.I., V.A. Nikolaev, V.Ya. Gotlib et al. Adaptive response in human blood lymphocytes after low dose chronic irradiation. Radiat. Biol. Ecol. 34: 805-817 (1994).

P20 Pronk, J.C., R.A. Gibson, A. Savoia et al. Localization of the Fanconi anaemia complementation group A gene to chromosome 16q24.3. Nature Genet. 11: 338-343 (1995).

P21 Porter, S.E., P.W. Greenwell, K.B. Ritchie et al. The DNA-binding protein Hdf1p (a putative Ku homologue) is required for maintaining normal telomere length in Saccharomyces cerevisiae. Nucleic Acids Res. 24: 582-585 (1996).

P22 Petrini, J.H.J., M.E. Walsh, C. Dimare et al. Isolation and characterization of the human MRE11 homologue. Genomics 29: 80-86 (1995).

P23 Plug, A.W., J. Xu, G. Reddy et al. Presynaptic association of Rad51 protein with selected sites in meiotic chromatin. Proc. Natl. Acad. Sci. U.S.A. 93: 5920-5924 (1996).

P24 Ponnaiya, B., M.N. Cornforth and R.L. Ullrich. Radiation-induced chromosomal instability in BALB/c and C57BL/6 mice: the difference is as clear as black and white. Radiat. Res. 147: 121-125 (1997).

P25 Pampfer, S. and C. Streffer. Increased chromosome aberration levels in cells from mouse fetuses after zygote X-irradiation. Int. J. Radiat. Biol. 55: 85-92 (1989).

P26 Pennycook, J.L., Y. Chang, J. Celler et al. High frequency of normal DJH joints in B cell progenitors in severe combined immunodeficiency mice. J. Exp. Med. 178: 1007-1016 (1993).

P27 Prise, K.M., S. Davies and B.D. Michael. The relationship between radiation-induced DNA double-strand breaks and cell kill in hamster V79 fibroblasts irradiated with 250 kVp X-rays, 2.3 MeV neutrons or ^{238}Pu alpha-particles. Int. J. Radiat. Biol. 52: 893-902 (1987).

P28 Peak, M.J., L. Wang, C.K. Hill et al. Comparison of repair of DNA double-strand breaks caused by neutron or gamma radiation in cultured human cells. Int. J. Radiat. Biol. 60: 891-898 (1991).

P29 Pfeiffer, P. and W. Vielmetter. Joining of nonhomologous DNA double strand breaks in vitro. Nucleic Acids Res. 16: 907-924 (1988).

P30 Pflaum, M., O. Will and B. Epe. Determination of steady-state levels of oxidative DNA base modifications in mammalian cells by means of repair endonucleases. Carcinogenesis 18: 2225-2231 (1997).

P31 Pouget, J-P., J-L. Ravenat, T. Douki et al. Measurement of DNA base damage in cells exposed to low doses of gamma radiation using the comet assay associated with DNA glycosylases. Int. J. Radiat. Biol. 55: 51-58 (1999).

P32 Patel, K.J., V.P.C.C. Yu, H. Lee et al. Involvement of Brca2 in DNA repair. Mol. Cell 1: 347-357 (1998).

P33 Pelevina, A.A., V.Y. Gotlieb, O.V. Kudryashova et al. The properties of irradiated cell progeny. Tsitologiia 40: 467-477 (1998)

P34 Paull, T.T. and M. Gellert. Nbs1 potentiates ATP-driven DNA unwinding and endonuclease cleavage by the Mre11/Rad50 complex. Genes Dev. 13: 1276-1288 (1999).

P35 Prasad, A.V., N. Mohan, B. Chandrasekar et al. Activation of nuclear factor κB in human lymphoblastoid cells by low-dose ionizing radiation. Radiat. Res. 138: 367-372 (1994).

P36 Prasad, A.V., N. Mohan, B. Chandrasekar et al. Induction of transcription of "immediate early genes" by low-dose ionizing radiation. Radiat. Res. 143: 263-272 (1995).

P37 Prise, K.M., G. Ahnstrom, M. Belli et al. A review of dsb induction data for varying quality radiations. Int. J. Radiat. Biol. 74: 173-184 (1998).

R1 Roth, D. and J. Wilson. Illegitimate DNA recombination in mammalian cells. p. 621-653 in: Genetic Recombination (R. Kucherlapati and G.R. Smith, eds.). Am. Soc. Microbiol., Washington, 1988.

R2 Rao, B.S., N.M.S. Reddy and U. Madhvanath. Gamma radiation response and recovery studies in radiation sensitive mutants of diploid yeast. Int. J. Radiat. Biol. 37: 701-705 (1980).

R3 Reddy, N.M.S., K.B. Kshiti and P. Subrahmanyam. Absence of a dose-rate effect and recovery from sub-lethal damage in rad52 strain of diploid yeast Saccharomyces cerevisae exposed to γ-rays. Mutat. Res. 105: 145-148 (1982).

R4 Russo, G., M. Isobe, L. Pegoraro et al. Molecular analysis of a t(7;14)(q35;q32) chromosome translocation in a T cell leukemia of a patient with ataxia telangiectasia. Cell 53: 137-144 (1988).

R5 Runger, T.M., M. Poot and K.H. Kraemer. Abnormal processing of transfected plasmid DNA in cells from patients with ataxia telangiectasia. Mutat. Res. 293: 47-54 (1992).

R6 Russell, W.L., L.B. Russell and E.M. Kelly. Radiation dose rate and mutation frequency. Science 128: 1546-1550 (1958).

R7 Russell, W.L. Repair mechanisms in radiation mutation induction in the mouse. Brookhaven Symp. Biol. 20: 179-189 (1968).

R8 Radford, I.R., T.K. Murphy, J.M. Radley et al. Radiation response of mouse lymphoid and myeloid cell lines. II. Apoptotic death is shown by all lines examined. Int. J. Radiat. Biol. 65: 217-227 (1994).

R9 Rasio, D., M. Negrini, G. Manenti et al. Loss of heterozygosity at chromosome 11q in lung adeno-carcinoma: identification of 3 independent regions. Cancer Res. 55: 3988-3991 (1995).

R10 Reed, J.C. Bcl-2 and the regulation of programmed cell death. J. Cell Biol. 124: 1-6 (1994).

R11 Reitmair, A.H., R. Schmits, A. Ewel et al. MSH2 deficient mice are viable and susceptible to lymphoid tumours. Nature Genet. 11: 64-70 (1995).

R12 Rigaud, O., D. Papadopoulo and E. Moustacchi. Decreased deletion mutation in radioadapted human lymphocytes. Radiat. Res. 133: 94-101 (1993).

R13 Rinchik, E.M., J.P. Stoye, W.N. Frankel et al. Molecular analysis of viable spontaneous and radiation-induced albino (c)-locus mutations in the mouse. Mutat. Res. 286: 199-207 (1993).

R14 Russell, L.B. Information from specific-locus mutants on the nature of induced and spontaneous mutations in the mouse. p. 437-447 in: Genetic Toxicology of Environ-mental Chemicals, Part B. Liss, New York, 1986.

R15 Russell, N.S., C.F. Arlett, H. Bartelink et al. Use of fluorescence in situ hybridization to determine the relationship between chromosome aberrations and cell survival in eight human fibroblast strains. Int. J. Radiat. Biol. 68: 185-196 (1995).

R16 Russell, W.L. Genetic effects from internally deposited radionuclides. p. 14-17: NCRP Report No. 89 (1987).

R17 Russell, W.L. and E.M. Kelly. Mutation frequencies in male mice and the estimation of genetic hazards of radiation in men. Proc. Natl. Acad. Sci. U.S.A. 79: 542-544 (1982).

R18 Roth, D.B. and J.H. Wilson. Nonhomologous recombina-tion in mammalian cells: role of short sequence homologies in the joining reaction. Mol. Cell. Biol. 6: 4295-4304 (1986).

R19 Rampino, N., H. Yamamoto, Y. Ionov et al. Somatic frameshift mutations in the BAX gene in colon cancers of the microsatellite mutator phenotype. Science 275: 967-969 (1997).

R20 Rosselli, F., A. Ridet, R. Soussi et al. p53-dependent pathway of radio-induced apoptosis is altered in Fanconi anemia. Oncogene 10: 9-17 (1995).

R21 Reuvers, A.P., C.L. Greenstock, J.Borsa et al. Studies on the mechanism of chemical radioprotection by dimethyl sulphoxide. Int. J. Radiat. Biol. 24: 533-536 (1973).

R22 Reardon, J.T., T. Bessho, H.C. Kung et al. In vitro repair of oxidative DNA damage by human nucleotide excision repair system: possible explanation for neurodegeneration in xeroderma pigmentosum patients. Proc. Natl. Acad. Sci. U.S.A. 94: 9463-9468 (1997).

R23 Rajan, J.V., M. Wang, S.T. Marquis et al. Brca2 is coordinately regulated with Brca1 during proliferation and differentiation in mammary epithelial cells. Proc. Natl. Acad. Sci. U.S.A. 93: 13078-13083 (1996).

R24 Ramsden, D.A. and M. Gellert. Ku protein stimulates DNA end joining by mammalian DNA ligases: a direct role for Ku in repair of DNA double-strand breaks. EMBO J. 17: 609-614 (1998).

R25 Ritter, M.A., J.E. Cleaver and C.A. Tobias. High-LET radiations induce a large proportion of non-rejoining DNA breaks. Nature 266: 653-655 (1977).

R26 Rydberg, B. Clusters of DNA damage induced by ionizing radiation: formation of short DNA fragments. II. Experimental detection. Radiat. Res. 145: 200-209 (1996).

R27 Rydberg, B., M. Lobrich and P.K. Cooper. DNA double-strand breaks induced by high-energy neon and iron ions in human fibroblasts. I. Pulsed-field gel electrophoresis method. Radiat. Res. 139: 133-141 (1994).

R28 Rithidech, K.N., V.P. Bond, E.P. Cronkite et al. Hyper-mutability of mouse chromosome 2 during development of X-ray induced murine myeloid leukaemia. Proc. Natl. Acad. Sci. U.S.A. 92: 1152-1156 (1995).

R29 Rosenberg, S.M. Mutation for survival. Curr. Opin. Genet. Dev. 7: 829-834 (1997).

R30 Radman, M., I. Matic, J.A. Halliday et al. Editing DNA replication and recombination by mismatch repair: from bacterial genetics to mechanisms of predisposition to cancer in humans. Philos. Trans. R. Soc. Lond. 347: 97-103 (1995).

R31 Ramsay, J., G. Birrell and M. Lavin. Testing for mutations of the ataxia telangiectasia gene in radiosensitive breast cancer patients. Radiother. Oncol. 47: 125-128 (1998).

R32   Rio, P.G., D. Pernin, J.O. Bay et al. Loss of heterozygosity of BRCA1, BRCA2 and ATM genes in sporadic invasive ductal breast carcinoma. Int. J. Oncol. 13: 849-853 (1998).

R33   Redpath, J.L. and R.J. Antoniono. Induction of an adaptive response against spontaneous neoplastic transformation in vitro by low-dose gamma radiation. Radiat. Res. 149: 517-520 (1998).

R34   Ridet, A., C. Guillouf, E. Dechaud et al. Deregulated apoptosis is a hallmark of the Fanconi anaemia syndrome. Cancer Res. 57: 1722-1730 (1997).

R35   Reuven, N.B., G. Arad, A. Maor-Shoshani et al. The mutagenesis protein UmuC is a DNA polymerase activated by UmuD', RecA, and SSB and is specialized for translesion replication. J. Biol. Chem. 274: 31763-31766 (1999).

R36   Roberts, S.A., A.R. Spreadborough, B. Bulman et al. Heritability of cellular radiosensitivity: a marker of low-penetrance predisposition genes in breast cancer? Am. J. Hum. Genet. 65: 784-794 (1999).

R37   Robson, T., M.C. Joiner, G.D. Wilson et al. A novel human stress response-related gene with a potential role in induced radioresistance. Radiat. Res. 152: 451-461 (1999).

R38   Riballo, E., S.E. Critchlow, S-H. Teo et al. Identification of a defect in DNA ligase IV in a radiosensitive leukaemia patient. Curr. Biol. 9: 699-702 (1999).

R39   Radford, I.R. Review: Initiation of ionizing radiation-induced apoptosis: DNA damage-mediated or does ceramide have a role? Int. J. Radiat. Biol. 75: 521-528 (1999).

R40   Riboni, R., A. Casati, T. Nardo et al. Telomeric fusions in cultured human fibroblasts as a source of genomic instability. Cancer Genet. Cytogenet. 95: 130-136 (1997).

R41   Radicella, J.P., C. Dherin, C. Desmaze et al. Cloning and characterization of hOGG1, a human homolog of the OGG1 gene of Saccharomyces cerevisiae. Proc. Natl. Acad. Sci. U.S.A. 94: 8010-8015 (1997).

S1   Sancar, A. and G.B. Sancar. DNA repair enzymes. Annu. Rev. Biochem. 57: 29-67 (1988).

S2   Selby, C.P. and A. Sancar. Mechanisms of transcription-repair coupling and mutation frequency decline. Microbiol. Rev. 58: 317-329 (1994).

S3   von Sonntag, C. The Chemical Basis of Radiation Biology. Taylor and Francis, London, 1987.

S4   Sabovljev, S.A., W.A. Cramp, P.D. Lewis et al. Use of rapid tests of cellular radiosensitivity in radiotherapeutic practice. Lancet ii: 787 (1985).

S5   Sanford, K.K., R. Parshad, R. Gantt et al. Factors affecting and significance of G_2 chromatin radiosensitivity in predisposition to cancer. Int. J. Radiat. Biol. 55: 963-981 (1989).

S6   Schlissel, M., A. Constantinescu, T. Morrow et al. Double-strand signal sequence breaks in V(D)J recombination are blunt, 5' phosphorylated, RAG-dependent and cell cycle regulated. Genes Dev. 7: 2520-2532 (1993).

S7   Seemanova, E.E., E. Passarge, D. Beneskova et al. Familial microcephaly with normal intelligence, immunodeficiency and risk for lymphoreticular malignancies. Am. J. Med. Genet. 20: 639-648 (1985).

S8   Seemanova, E. An increased risk for malignant neoplasms in heterozygotes for a syndrome of microcephaly, normal intelligence, growth retardation, remarkable facies, immunodeficiency and chromosomal instability. Mutat. Res. 238: 321-324 (1990).

S9   Siede, W., A.S. Friedberg and E.C. Freidberg. RAD9-dependent G_1 arrest defines a second checkpoint for damaged DNA in the cell cycle of Saccharomyces cerevisiae. Proc. Natl. Acad. Sci. U.S.A. 90: 7985-7989 (1993).

S10   Shinohara, A., H. Ogawa, Y. Matsuda et al. Cloning of human, mouse and fission yeast recombination genes homologous to RAD51 and recA. Nature Genet. 4: 239-243 (1993).

S11   Smith, P.J., T.A. Makinson and J.V. Watson. Enhanced sensitivity to camptothecin in ataxia-telangiectasia cells and its relationship with the expression of DNA topoisomerase I. Int. J. Radiat. Biol. 55: 217-231 (1989).

S12   Swift, M., D. Morrell, E. Cromartie et al. The incidence and gene frequency of ataxia-telangiectasia in the United States. Am. J. Hum. Genet. 39: 573-583 (1986).

S13   Swift, M., D. Morrell, R.B. Massey et al. Incidence of cancer in 161 families affected by ataxia-telangiectasia. N. Engl. J. Med. 325: 1831-1836 (1991).

S14   Stamato, T.D., A. Dipatri and A. Giaccia. Cell cycle dependent repair of potentially lethal damage in the XR-1 gamma-ray sensitive CHO cell. Radiat. Res. 115: 325-333 (1988).

S15   Savage, J.R.K. and P.J. Simpson. Fish painting patterns resulting from complex exchanges. Mutat. Res. 312: 51-60 (1994).

S16   Seymour, C.B., C. Mothersill and T. Alper. High yields of lethal mutations in somatic mammalian cells that survive ionising radiation. Int. J. Radiat. Biol. 50: 167-179 (1986).

S17   Simpson, P., T. Morris, J. Savage et al. High-resolution cytogenetic analysis of X-Ray induced mutations of the HPRT gene of primary human fibroblasts. Cytogenet. Cell Genet. 64: 39-45 (1993).

S18   Singleton, B. and J. Thacker. Communication to the UNSCEAR Secretariat (1995).

S19   Sadamoto, S., S. Suzuki, K. Kamiya et al. Radiation induction of germline mutation at a hypervariable mouse minisatellite locus. Int. J. Radiat. Biol. 65: 549-557 (1994).

S20   Sanderson, B.J.S. and A.A. Morley. Exposure of human lymphocytes to ionising radiation reduces mutagenesis in subsequent ionising radiation. Mutat. Res. 164: 347-351 (1986).

S21   Sands, A.T., M.B. Suraokar, A. Sanchez et al. p53 deficiency does not affect the accumulation of point mutations in a transgene target. Proc. Natl. Acad. Sci. U.S.A. 92: 8517-8521 (1995).

S22   Sankaranayanan, K., A. van Duyn, M.J. Loos et al. Adaptive response of human lymphocytes to low-level radiation from radioisotopes or X-rays. Mutat. Res. 211: 7-12 (1989).

S23   Sasaki, M.S. On the reaction kinetics of the radioadaptive response in cultured mouse cells. Int. J. Radiat. Biol. 68: 281-291 (1995).

S24   Savitsky, K., A. Bar-Shira, S. Gilad et al. A single ataxia telangiectasia gene with a product similar to PI-3 kinase. Science 268: 1749-1753 (1995).

S25   Savitsky, K., S. Sfez, D.A. Tagle et al. The complete sequence of the coding region of the ATM gene reveals similarity to cell cycle regulators in different species. Hum. Mol. Genet. 4: 2025-2032 (1995).

S26   Schmid, E., M. Bauchinger and U. Nahrstedt. Adaptive response after X-irradiation of human lymphocytes? Mutagenesis 4: 87-89 (1989).

S27   Scott, D., A. Spreadborough, E. Levine et al. Genetic predisposition in breast cancer. Lancet 344: 1444 (1994).

S28   Scott, D., A.R. Spreadborough, L.A. Jones et al. Chromosomal radiosensitivity in G_2-phase lymphocytes as an indicator of cancer predisposition. Radiat. Res. 145: 3-16 (1996).

S29 Searle, A.G. Germ-cell sensitivity in the mouse: a comparison of radiation and chemical mutagens. p. 169-177 in: Environmental Mutagens and Carcinogens (T. Sugimura et al., eds.). University of Tokyo, 1982.

S30 Shadley, J.D. Chromosomal adaptive response in human lymphocytes. Radiat. Res. 138: S9-S12 (1994).

S31 Shadley, J.D., V. Afzal and S. Wolff. Characterization of the adaptive response to ionising radiation induced by low doses of X-rays to human lymphocytes. Radiat. Res. 111: 511-517 (1987).

S32 Shadley, J.D. and G. Dai. Cytogenetic and survival adaptive responses in G_1 phase human lymphocytes. Mutat. Res. 265: 273-281 (1992).

S33 Shadley, J.D. and J.K. Wiencke. Induction of the adaptive response by X-rays is dependent on radiation intensity. Int. J. Radiat. Biol. 56: 107-118 (1989).

S34 Shadley, J.D. and S. Wolff. Very low doses of X-rays can cause human lymphocytes to become less susceptible to ionising radiation. Mutagenesis 2: 95-96 (1987).

S35 Skov, K., B. Marples, J.B. Matthews et al. A preliminary investigation into the extent of increased radioresistance or hyper-radiosensitivity in cells of hamster lines known to be deficient in DNA repair. Radiat. Res. 138: S126-S129 (1994).

S36 Sloan, S.R., E.W. Newcomb and A. Pellicer. Neutron radiation can activate K-*ras* via a point mutation in codon 146 and induces a different spectrum of *ras* mutations than does gamma radiation. Mol. Cell. Biol. 10: 405-408 (1990).

S37 Sinclair, W.K. Cyclic X-ray responses in mammalian cells in vitro. Radiat. Res. 33: 620-643 (1968).

S38 Sinclair, W.K. Cell cycle dependence of the lethal radiation response in mammalian cells. Curr. Top. Radiat. Res. Quart. 7: 264-285 (1972).

S39 Stacey, M., S. Thacker and A.M.R. Taylor. Cultured skin keratinocytes from both normal individuals and basal cell naevus syndrome patients are more resistant to γ-rays and UV light compared with cultured skin fibroblasts. Int. J. Radiat. Biol. 56: 45-58 (1989).

S40 Savitsky, K., M. Platzer, T. Uziel et al. Ataxia-telangiectasia: structural diversity of untranslated sequences suggests complex post-transcriptional regulation of ATM gene expression. Nucleic Acids Res. 25: 1678-1684 (1997).

S41 Swift, M., D. Morrell, R.B. Massey et al. Risk of breast cancer in ataxia-telangiectasia. Lancet 326: 1360 (1992).

S42 Symonds, H., L. Krall, L. Remington et al. p53-dependent apoptosis suppresses tumor growth and progression in vivo. Cell 78: 703-711 (1994).

S43 Sanchez, Y., B.A. Desany, W.J. Jones et al. Regulation of *RAD53* by the *ATM*-like kinases *MEC1* and *TEL1* in yeast cell cycle checkpoint pathways. Science 271: 357-360 (1996).

S44 Shen, Z.Y., K. Denison, R. Lobb et al. The human and mouse homologs of the yeast *RAD52* gene: cDNA cloning, sequence analysis, assignment to human chromosome 12p12.2-p13, and mRNA expression in mouse tissues. Genomics 25: 199-206 (1995).

S45 Sturzbecher, H-W., B. Donzelmann, W. Henning et al. p53 is linked directly to homologous recombination processes via RAD51/RecA protein interaction. EMBO J. 15: 1992-2002 (1996).

S46 Strathdee, C.A., A.M.V. Duncan and M. Buchwald. Evidence for at least 4 Fanconi anaemia genes including *FACC* on chromosome 9. Nature Genet. 1: 196-198 (1992).

S47 Strathdee, C.A., H. Gavish, W.R. Shannon et al. Cloning of cDNAs for Fanconi's anaemia by functional complementation. Nature 356: 763-767 (1992).

S48 Saito, H., A.T. Hammond and R.E. Moses. Hyper-sensitivity to oxygen is a uniform and secondary defect in Fanconi anaemai cells. Mutat. Res. 294: 255 (1993).

S49 Seeberg, E., L. Eide and M. Bjoras. The base excision repair pathway. Trends Biochem. Sci. 20: 391-397 (1995).

S50 Siede, W., A.A. Friedl, I. Dianova et al. The *Saccharomyces cerevisiae* Ku autoantigen homologue affects radiosensitivity only in the absence of homologous recombination. Genetics 142: 91-102 (1996).

S51 Scully, R., J. Chen, A. Plug et al. Association of BRCA1 with Rad51 in mitotic and meiotic cells. Cell 88: 265-275 (1997).

S52 Strathern, J.N., B.K. Shafer and C.B. McGill. DNA synthesis errors associated with double-strand break repair. Genetics 140: 965-972 (1995).

S53 Sabatier, L., B. Dutrillaux and M.B. Martin. Chromosomal instability. Nature 357: 548 (1992).

S54 Savage, J.R.K. and D.G. Papworth. Comment on the ratio of chromosome-type dicentric interchanges to centric rings for track-clustered compared with random breaks. Radiat. Res. 146: 236-240 (1996).

S55 Santana, P., L.A. Pena, A. Haimovitz-Friedman et al. Acid sphingomyelinase-deficient human lymphoblasts and mice are defective in radiation-induced apoptosis. Cell 86: 189-199 (1996).

S56 Satoh, M.S., C.J. Jones, R.D. Wood et al. DNA excision-repair defect of xeroderma pigmentosum prevents removal of a class of oxygen free radical-induced base lesions. Proc. Natl. Acad. Aci. U.S.A. 90: 6335-6339 (1993).

S57 Schiestl, R.H., F. Khogali and N. Carls. Reversion of the mouse *pink-eyed unstable* mutation induced by low doses of X-rays. Science 266: 1573-1576 (1994).

S58 Singhal, R.K., R. Prasad and S.H. Wilson. DNA polymerase β conducts the gap-filling step in uracil-initiated base excision repair in a bovine testis nuclear extract. J. Biol. Chem. 270: 949-957 (1995).

S59 Saar, K., K.H. Chrzanowska, M. Stumm et al. The gene for the ataxia-telangiectasia variant, Nijmegen breakage syndrome, maps to a 1-cm interval on chromosome 8q21. Am. J. Hum. Genet. 60: 605-610 (1997).

S60 Stewart, E., C.R. Chapman, F. Al-Khodairy et al. *rhq1*$^+$, a fission yeast gene related to the Bloom's and Werner's syndrome genes, is required for reversible S phase arrest. EMBO J. 16: 2682-2692 (1997).

S61 Sharan, S.K., M. Morimatsu, U. Albrecht et al. Embryonic lethality and radiation hypersensitivity mediated by Rad51 in mice lacking *Brca2*. Nature 386: 804-810 (1997).

S62 Satoh, M.S., G.G. Poirier and T. Lindahl. Dual function for poly(ADP-ribose) synthesis in response to DNA strand breakage. Biochemistry 33: 7099-7106 (1994).

S63 Schreiber, V., D. Hunting, C. Trucco et al. A dominant-negative mutant of human poly(ADP-ribose) polymerase affects cell recovery, apoptosis and sister-chromatid exchange following DNA damage. Proc. Natl. Acad. Sci. U.S.A. 92: 4753-4757 (1995).

S64 Schar, P., G. Herrmann, G. Daly et al. A newly identified DNA ligase of *Saccharomyces cerevisiae* involved in *RAD52*-independent repair of DNA double-strand breaks. Genes Dev. 11: 1912-1924 (1997).

S65 Shafman, T., K.K. Khanna, P. Kedar et al. Interaction between ATM protein and c-Abl in response to DNA damage. Nature 387: 520-523 (1997).

S66 Schmid, E. and M. Bauchinger. Comments on direct biological evidence for a significant neutron dose to survivors of the Hiroshima atomic bomb. Radiat. Res. 146: 479-482 (1996).

S67 Satoh, C. and M. Kodaira. Effects of radiation on children. Nature 383: 226 (1996).

S68 Sala-Trepat, M., J. Boyse, P. Richard et al. Frequencies of *HPRT* mutants in T-lymphocytes and of glycophorin A variants in erythrocytes of Fanconi anemia patients and control donors. Mutat. Res. 289: 115-126 (1993).

S69 Shayeghi, M., S. Seal, J. Regan et al. Heterozygosity for mutations in the ataxia telangiectasia gene is not a major cause of radiotherapy complications in breast cancer patients. Br. J. Cancer 78: 922-927 (1998).

S70 Starostik, P., T. Manshouri, S. O'Brien et al. Deficiency of the ATM protein expression defines an aggressive subgroup of B-cell chronic lymphocytic leukemia. Cancer Res. 58: 4552-4557 (1998).

S71 Sachs, R.K., D.J. Brenner, A.M. Chen et al. Intra-arm and inter-arm chromosome intrachanges: tools for probing geometry and dynamics of chromatin. Radiat. Res. 148: 330-340 (1997).

S72 Savage, J.R.K. and D.G. Papworth. An investigation of LET 'fingerprints' in *Tradescantia*. Mutat. Res. 422: 313-322 (1998).

S73 Siliciano, J.D., C.E. Canman, Y. Taya et al. DNA damage induces phosphorylation of the amino terminus of p53. Genes Dev. 11: 3471-3481 (1997).

S74 Schwartz, D., N. Almog, A. Peled et al. Role of wild type p53 in the G2 phase: regulation of the gamma-irradiation-induced delay and DNA repair. Oncogene 15: 2597-2607 (1997).

S75 Stewart, N., G.G. Hicks, F. Paraskevas et al. Evidence for a second cell-cycle block at G2/M by p53. Oncogene 10: 109-115 (1995).

S76 Stankovic, T., P. Weber, G. Stewart et al. Inactivation of ataxia telangiectasia mutated gene in B-cell chronic lymphocytic leukaemia. Lancet 353: 26-29 (1999).

S77 Suzuki, K., S. Kodama and M. Watanabe. Recruitment of ATM protein to double strand DNA irradiated with ionizing radiation. J. Biol. Chem. 274: 25571-25575 (1999).

S78 Stankovic, T., A.M.J. Kidd, A. Sutcliffe et al. *ATM* mutations and phenotypes in ataxia-telangiectasia families in the British Isles: expression of mutant *ATM* and the risk of leukemia, lymphoma, and breast cancer. Am. J. Hum. Genet. 62: 334-345 (1998).

S79 Sadekova, S., S. Lehnert and T.Y-K. Chow. Induction of PBP74/mortalin/Grp75, a member of the hsp70 family, by low doses of ionizing radiation: a possible role in induced radioresistance. Int. J. Radiat. Biol. 72: 653-660 (1997).

S80 Shimizu, T., T. Kato, A. Tachibana et al. Coordinated regulation of radioadaptive response by protein kinase C and p38 mitogen-activated protein kinase. Exp. Cell Res. 251: 424-432 (1999).

S81 Svistuneko, D.A., G.Z. Ju, J. Wei et al. EPR study of mouse tissues in search for adaptive responses to low level whole-body X-irradiation. Int. J. Radiat. Biol. 62: 327-336 (1992).

S82 Stewart, G.S., R.S. Maser, T. Stankovic et al. The DNA double-strand break repair gene *hMRE11* is mutated in individuals with an ataxia-telangiectasia-like disorder. Cell 99: 577-587 (1999).

S83 Smith, B.T. and G.C. Walker. Mutagenesis and more: umuDC and the Escherichia coli SOS response. Genetics 148: 1599-1610 (1998).

S84 Shen, M.R., I.M. Jones and H. Mohrenweiser. Non-conservative amino acid substitution variants exist at polymorphic frequency in DNA repair genes in healthy humans. Cancer Res. 58: 604-608 (1998).

S85 Sargent, R.G., R.L. Rolig, A.E. Kilburn et al. Recombination-dependent deletion formation in mammalian cells deficient in the nucleotide excision repair gene ERCC1. Proc. Natl. Acad. Sci. U.S.A. 94: 13122-13127 (1997).

S86 Scully, R., S. Ganesan, K. Vlasakova et al. Genetic analysis of BRCA1 function in a defined tumor cell line. Mol. Cell 4: 1093-1099 (1999).

S87 Sutherland, B.M., P.V. Bennett, O. Sidorkina et al. Clustered DNA damages induced in isolated DNA and in human cells by low doses of ionizing radiation. Proc. Natl. Acad. Sci. U.S.A. 97: 103-108 (2000).

S88 Stoll, U., A. Schmidt, E. Schneider et al. Killing and mutation of Chinese hamster V79 cells exposed to accelerated oxygen and neon ions. Radiat. Res. 142: 288-294 (1995).

S89 Stoll, U., B. Barth, N. Scheerer et al. HPRT mutations in V79 Chinese hamster cells induced by accelerated Ni, Au and Pb ions. Int. J. Radiat. Biol. 70: 15-22 (1996).

S90 Schmidt, P. and J. Kiefer. Deletion-pattern analysis of alpha-particle and X-ray induced mutations at the HPRT locus of V79 Chinese hamster cells. Mutat. Res. 421: 149-161 (1998).

T1 Taylor, A.M.R. Unrepaired DNA strand breaks in irradiated ataxia telangiectasia lymphocytes suggested from cytogenetic observations. Mutat. Res. 50: 407-418 (1978).

T2 Taalman, R.D.F.M., N.G.J. Jaspers, J.M. Scheres et al. Hypersensitivity to ionising radiation, in vitro, in a new chromosomal breakage disorder, the Nijmegen breakage syndrome. Mutat. Res. 112: 23-32 (1983).

T3 Taylor, A.M.R., E. Flude, B. Laher et al. Variant forms of ataxia-telangiectasia. J. Med. Genet. 24: 669-677 (1987).

T4 Taccioli, G.E., G. Rathbun, G. Oltz et al. Impairment of V(D)J recombination in double-strand break repair mutants. Science 260: 207-210 (1993).

T5 Taccioli, G.E., T.M. Gottlieb, T. Blunt et al. Ku80: product of the XRCC5 gene and its role in DNA repair and V(D)J recombination. Science 265: 1442-1445 (1994).

T6 Tatsumi-Miyajima, J., T. Yagi and H. Takebe. Analysis of mutations caused by DNA double-strand breaks produced by a restriction enzyme in shuttle vector plasmids propagated in ataxia telangiectasia cells. Mutat. Res. 294: 317-323 (1993).

T7 Thacker, J. and A. Stretch. Responses of 4 X-ray-sensitive CHO cell mutants to different radiations and to irradiation conditions promoting cellular recovery. Mutat. Res. 146: 99-108 (1985).

T8 Thacker, J. and A.N. Ganesh. DNA break repair, radio-resistance of DNA synthesis and camptothecin sensitivity in the radiation-sensitive *irs* mutants: comparisons to ataxia-telangiectasia cells. Mutat. Res. 235: 49-58 (1990).

T9 Thacker, J. and R. Wilkinson. The genetic basis of resistance to ionising radiation damage in cultured mammalian cells. Mutat. Res. 254: 135-142 (1991).

T10 Thacker, J., J. Chalk, A. Ganesh et al. A mechanism for deletion formation in DNA by human cell extracts: the involvement of short sequence repeats. Nucleic Acids Res. 20: 6183-6188 (1992).

T11 Thacker, J., A.N. Ganesh, A. Stretch et al. Gene mutation and V(D)J recombination in the radiosensitive *irs* lines. Mutagenesis 9: 163-168 (1994).

T12 Taylor, J.A., M.A. Watson, T.R. Devereux et al. p53 mutation hotspot in radon-associated lung cancer. Lancet 343: 86-87 (1994).

T13 Thacker, J. and R. Cox. Mutation induction and inactivation in mammalian cells exposed to ionizing radiation. Nature 258: 429-431 (1975).

T14 Thacker, J., A. Stretch and M.A. Stephens. The induction of thioguanine-resistant mutants of Chinese hamster cells by gamma-rays. Mutat. Res. 42: 313-326 (1977).

T15 Thacker, J., M.A. Stephens and A. Stretch. Mutation to ouabain-resistance in Chinese hamster cells: induction by ethyl methanesulphonate and lack of induction by ionising radiation. Mutat. Res. 51: 255-270 (1978).

T16 Thacker, J., A. Stretch and M.A. Stephens. Mutation and inactivation of cultured mammalian cells exposed to beams of accelerated heavy ions. II. Chinese hamster cells. Int. J. Radiat. Biol. 36: 137-148 (1979).

T17 Thacker, J. The involvement of repair processes in radiation-induced mutation of cultured mammalian cells. p. 612-620 in: Radiation Research (S. Okada et al. eds.). Japanese Association for Radiation Research, Tokyo, 1979.

T18 Thacker, J., A. Stretch and D.T. Goodhead. The mutagenicity of α-particles from plutonium-238. Radiat. Res. 92: 343-352 (1982).

T19 Thacker, J. and R. Cox. The relationship between specific chromosome aberrations and radiation-induced mutations in cultured mammalian cells. p. 235-275 in: Radiation-induced Chromosome Damage in Man (T. Ishihara and M.S. Sasaki, eds.). Liss, New York, 1983.

T20 Thacker, J. The nature of mutants induced by ionising radiation in cultured hamster cells. III. Molecular characterization of HPRT-deficient mutants induced by gamma-rays or alpha-particles showing that the majority have deletions of all or part of the hprt gene. Mutat. Res. 160: 267-275 (1986).

T21 Thacker, J. Molecular nature of ionising radiation-induced mutations of native and introduced genes in mammalian cells. p. 221-229 in: Ionising Radiation Damage to DNA: Molecular Aspects (S. Wallace and R. Painter, eds.). Wiley/Liss, New York, 1990.

T22 Thacker, J. Radiation-induced mutation in mammalian cells at low doses and dose rates. Adv. Radiat. Biol. 16: 77-124 (1992).

T23 Trakhtenbrot, L., R. Kranthgamer, P. Resnitzky et al. Deletion of chromosome 2 is an early event in the development of radiation-induced myeloid leukaemia in SJL/J mice. Leukaemia 2: 545-550 (1991).

T24 Thacker, J. The study of responses the 'model' DNA breaks induced by restriction endonucleases in cells and cell-free systems: achievements and difficulties. Int. J. Radiat. Biol. 66: 591-596 (1994).

T25 Takai, S., K.K. Sanford and R.E. Tarone. A procedure for carrier detection in ataxia-telangiectasia. Cancer Genet. Cytogenet. 46: 139-140 (1990).

T26 Tanzarella, C., R. de Salvia, F. Degrassi et al. Effect of post-treatments with caffeine during G_2 on the frequency of chromosome type aberrations produced by X-rays in human lymphocytes during G_0 and G_1. Mutagenesis 1: 41-44 (1986).

T27 Terasima, T. and L.J. Tolmach. X-ray sensitivity and DNA synthesis in synchronous populations of HeLa cells. Science 140: 490-492 (1963).

T28 Thacker, J. Radiation-induced mutations in mammalian cells: what we know and what we do not know. p. 107 in: 37th Annual Meeting of the Radiation Research Society, Seattle, 1989.

T29 Thacker, J. Cellular radiosensitivity in ataxia-telangiectasia. Int. J. Radiat. Biol. 66: S87-S96 (1994).

T30 Thacker, J. and R.E. Wilkinson. The genetic basis of cellular recovery from radiation damage: response of the radiosensitive irs lines to low-dose-rate irradiation. Radiat. Res. 144: 294-300 (1995).

T31 Timofeeff-Ressovsky, N.W. The experimental production of mutations. Biol. Rev. Camb. Philos. Soc. 9: 411-457 (1934).

T32 Tsuboi, K., T.C. Yang and D.J. Chen. Charged particle mutagenesis. I. Cytotoxic and mutagenic effects of high-LET charged iron particles on human skin fibroblasts. Radiat. Res. 129: 171-176 (1992).

T33 Tuschl, H., H. Altmann, R. Kovac et al. Effects of low-dose radiation on repair processes in human lymphocytes. Radiat. Res. 81: 1-9 (1980).

T34 Tsuzuki, T., Y. Fujii, K. Sakumi et al. Targeted disruption of the Rad51 gene leads to lethality in embryonic mice. Proc. Natl. Acad. Sci. U.S.A. 93: 6236-6240 (1996).

T35 Tambini, C.E., A.M. George, J.M. Rommens et al. The XRCC2 DNA repair gene: identification of a positional candidate. Genomics 41: 84-92 (1997).

T37 Tashiro, S., N. Kotomura, A. Shinohara et al. S-phase specific formation of the human Rad51 protein nuclear foci in lymphocytes. Oncogene 12: 2165-2170 (1996).

T38 Tomlinson, I.P.M., M.R. Novelli and W.F. Bodmer. The mutation rate and cancer. Proc. Natl. Acad. Sci. U.S.A. 93: 14800-14803 (1996).

T39 Thyagarajan, B. and C. Campbell. Elevated homologous recombination activity in Fanconi anemia fibroblasts. J. Biol. Chem. 272: 23328-23333 (1997).

T40 Teo, S-H. and S.P. Jackson. Identification of Saccharomyces cerevisiae DNA ligase IV: involvement in DNA double-strand break rejoining. EMBO J. 16: 4788-4795 (1997).

T41 Tsukamoto, Y., J. Kato and H. Ikeda. Silencing factors participate in DNA repair and recombination in Saccharomyces cerevisiae. Nature 388: 900-903 (1997).

T42 Takata, M., M.S. Sasaki, E. Sonoda et al. Homologous recombination and non-homologous end-joining pathways of DNA double-strand break repair have overlapping roles in the maintenance of chromosomal integrity in vertebrate cells. EMBO J. 17: 5497-5508 (1998).

T43 Takagi, M., D. Delia, L. Chessa et al. Defective control of apoptosis, radiosensitivity, and spindle checkpoint in ataxia telangiectasia. Cancer Res. 58: 4923-4929 (1998).

T44 Thacker, J. Repair of ionizing radiation damage in mammalian cells: alternative pathways and their fidelity. C.R. Acad. Sci. Paris 322: (1999, in press).

T45 Tang, M., X. Shen, E.G. Frank et al. UmuD'(2)C is an error-prone DNA polymerase, Escherichia coli pol V. Proc. Natl. Acad. Sci. U.S.A. 96: 8919-8924 (1999).

T46 Tutt, A., A. Gabriel, D. Bertwistle et al. Absence of Brca2 causes genome instability by chromosome breakage and loss associated with centrosome amplification. Curr. Biol. 9: 1107-1110 (1999).

T47 Thacker, J. A surfeit of RAD51-like genes? Trends Genet. 15: 166-168 (1999).

T48 Tchou, J., H. Kasai, S. Shibutani et al. 8-oxoguanine (8-hydroxyguanine) DNA glycosylase and its substrate specificity. Proc. Natl. Acad. Sci. U.S.A. 88: 4690-4694 (1991).

T49 Thomas, C.B., D.O. Nelson, P. Pleshanov et al. Elevated frequencies of hypoxanthine phosphoribosyltransferase lymphocyte mutants are detected in Russian liquidators 6 to 10 years after exposure to radiation from the Chernobyl nuclear power plant accident. Mutat. Res. 439: 105-119 (1999).

U2 United Nations. Sources and Effects of Ionizing Radiation. United Nations Scientific Committee on the Effects of Atomic Radiation, 1994 Report to the General Assembly, with scientific annexes. United Nations sales publication E.94.IX.11. United Nations, New York, 1994.

U3   United Nations. Sources and Effects of Ionizing Radiation. United Nations Scientific Committee on the Effects of Atomic Radiation, 1993 Report to the General Assembly, with scientific annexes. United Nations sales publication E.94.IX.2. United Nations, New York, 1993.

U14   Umansky, S.R., B.A. Korol and P.A. Nelipovitch. *In vivo* DNA degradation in thymocytes of gamma-irradiated or hydrocortisone-treated rats. Biochim. Biophys. Acta 665: 9-17 (1981).

U15   Utsumi, H. and M.S. Sasaki. Deficient repair of potentially lethal damage in actively growing ataxia telangiectasia cells. Radiat. Res. 97: 407-413 (1984).

U16   Unzhakov, S.V., I.M. Vasilieva, I.A. Meliksetova et al. Adaptive response formation in lymphocytes from the children exposed to small doses of radiation as a result of the Chernobyl accident. Radiat. Biol. Ecol. 34: 827-831 (1994).

U17   Ueno, A.M., D.B. Vannais, D.L. Gustafson et al. A low adaptive dose of gamma rays reduced the number and altered the spectrum of S1-mutants in human-hamster hybrid A_L cells. Mutat. Res. 358: 161-169 (1996).

U18   Usui, T., T. Ohta, H. Oshiumi et al. Complex formation and functional versatility of Mre11 of budding yeast in recombination. Cell 95: 705-716 (1998).

V1   Vorechovsky, I., D. Scott, M.R. Haeney et al. Chromosomal radiosensitivity in common variable immune deficiency. Mutat. Res. 290: 255-264 (1993).

V2   Vahakangas, K.H., J.M. Samet, R.A. Metcalf et al. Mutations of p53 and *ras* genes associated lung cancer from uranium miners. Lancet 339: 576-580 (1992).

V3   Venitt, S. and P.J. Biggs. Radon, mycotoxins, p53 and uranium mining. Lancet 343: 795 (1994).

V4   Vijayalaxmi and W. Burkart. Resistance and cross-resistance to chromosome damage in human blood lymphocytes adapted to bleomycin. Mutat. Res. 211: 1-5 (1989).

V5   Vorechovsky, I., L. Luo, M.J.S. Dyer et al. Clustering of missense mutations in the ataxia-telangiectasia gene in a sporadic T-cell leukaemia. Nature Genet. 17: 96-99 (1997).

V6   Varon, R., C. Vissinga, M. Platzer et al. Nibrin, a novel DNA double-strand break repair protein, is mutated in Nijmegen breakage syndrome. Cell 93: 467-476 (1998).

V7   van der Burgt, I., K.H. Chrzanowska, D. Smeets et al. Nijmegen breakage syndrome. J. Med. Genet. 33: 153-156 (1996).

V8   Vorechovsky, I., D. Rasio, L. Luo et al. The *ATM* gene and susceptibility to breast cancer: analysis of 38 breast tumours reveals no evidence for mutation. Cancer Res. 56: 2726-2732 (1996).

V9   Vorechovsky, I., L. Luo, A. Lindblom et al. *ATM* mutations in cancer families. Cancer Res. 56: 4130-4133 (1996).

W1   Wallace, S.S. AP endonucleases and DNA glycosylases that recognize oxidative DNA damage. Environ. Mol. Mutagen. 12: 431-477 (1988).

W2   Weichselbaum, R.R., J. Nove and J.B. Little. Deficient recovery from potentially lethal radiation damage in ataxia telangiectasia and xeroderma pigmentosum. Nature 271: 261-262 (1978).

W3   Weichselbaum, R.R., J. Nove and J.B. Little. X-ray sensitivity of 53 human diploid fibroblast cell strains from patients with characterized genetic disorders. Cancer Res. 40: 920-925 (1980).

W4   Weemaes, C.M.R., T.W.J. Hustinx, J.M.J.C. Scheres et al. New chromosome instability disorder: the Nijmegen breakage syndrome. Acta Paediatr. Scand. 70: 557-562 (1981).

W5   Wegner, R.D., M. Metzger, F. Hanefelt et al. A new chromosome instability disorder confirmed by complementation studies. Clin. Genet. 33: 20-32 (1988).

W6   Weinert, T.A. and L.H. Hartwell. The *rad9* gene controls the cell cycle response to DNA damage in *Saccharomyces cerevisae*. Science 241: 317-322 (1988).

W7   Weinert, T.A. and L.H. Hartwell. Characterization of *RAD9* of *Saccharomyces cerevisiae* and evidence that its function acts posttranslationally in cell cycle arrest after DNA damage. Mol. Cell. Biol. 10: 6554-6564 (1990).

W8   Williams, G.T. Programmed cell death - apoptosis and oncogenesis. Cell 65: 1097-1098 (1991).

W9   Woods, W.G., T.D. Byrne and T.H. Kim. Sensitivity of cultured cells to gamma radiation in a patient exhibiting marked *in vivo* radiation sensitivity. Cancer 62: 2341-2345 (1988).

W10   Woods, C.G., S.E. Bundey and A.M.R. Taylor. Unusual features in the inheritance of ataxia telangiectasia. Hum. Genet. 84: 555-562 (1990).

W11   Walker, G.C. Mutagenesis and inducible responses to deoxyribonucleic acid damage in *Escherichia coli*. Microbiol. Rev. 48: 60-93 (1984).

W12   Waldren, C., L. Correll, M.A. Sognier et al. Measurement of low levels of X-ray mutagenesis in relation to human disease. Proc. Natl. Acad. Sci. U.S.A. 83: 4839-4843 (1986).

W13   Waghray, M., S. Al-Sedairy, P.T. Ozand et al. Cytogenetic characterization of ataxia telangiectasia heterozygotes using lymphoblastoid cell lines and chronic γ-irradiation. Hum. Genet. 84: 532-534 (1990).

W14   Wagner, L.K. Risk of breast cancer in ataxia-telangiectasia. Lancet 326: 1358 (1992).

W15   Wang, Z.-Q., S. Saigusa and M.S. Sasaki. Adaptive response to chromosome damage in cultured human lymphocytes primed with low doses of X-rays. Mutat. Res. 246: 179-186 (1991).

W16   Warenius, H.M. and J.D. Down. RBE of fast neutrons for apoptosis in mouse thymocytes. Int. J. Radiat. Biol. 68: 625-629 (1995).

W17   Watson, P. and H.T. Lynch. Extracolonic cancer in hereditary nonpolyposis colorectal cancer. Cancer 71: 677-685 (1993).

W18   Weeks, D.E., M.C. Paterson, K. Lange et al. Assessment of chronic γ radiosensitivity as an in vitro assay for heterozygote identification of ataxia-telangiectasia. Radiat. Res. 128: 90-99 (1991).

W19   Wei, Y.F., P. Robins, K. Carter et al. Molecular cloning and expression of human cDNAs encoding a novel DNA ligase IV and DNA ligase III, an enzyme active in DNA repair and recombination. Mol. Cell. Biol. 15: 3206-3216 (1995).

W20   Weichselbaum, R.R., D.E. Hallahan, V. Sukhatme et al. Biological consequences of gene regulation after ionising radiation exposure. J. Natl. Cancer Inst. 83: 480-484 (1991).

W21   West, C.M.L., S.A.G. Elyan, P. Berry et al. A comparison of the radiosensitivity of lymphocytes from normal donors, cancer patients, individuals with ataxia-telangiectasia (A-T) and A-T heterozygotes. Int. J. Radiat. Biol. 68: 197-203 (1995).

W22   Whaley, J.M. and J.B. Little. Molecular characterization of hprt mutants induced by low- and high-LET radiations in human cells. Mutat. Res. 243: 35-45 (1990).

W23   Wiencke, J.K., V. Afzal, G. Olivieri et al. Evidence that the [^{3}H] thymidine-induced adaptive response of human lymphocytes to subsequent doses of X-rays involves the induction of a chromosomal repair mechanism. Mutagenesis 1: 375-380 (1986).

W24 Wiencke, J.K., J.D. Shadley, K.T. Kelsey et al. Failure of high-intensity x-ray treatments or densely-ionising fast neutrons to induce the adaptive response in human lymphocytes. p. 212 in: Proceedings of the 8th International Congress on Radiation Research, Volume 1 (E.M. Fielden et al., eds.). Edinburgh, 1987.

W25 Winans, L.F., W.C. Dewey and C.M. Dettor. Repair of sublethal and potentially lethal X-ray damage in synchronous Chinese hamster cells. Radiat. Res. 52: 333-351 (1972).

W26 Wojcik, A. and H. Tuschl. Indications of an adaptive response in C57BL mice pre-exposed in vivo to low doses of ionising radiation. Mutat. Res. 243: 67-73 (1990).

W27 Wojcik, A. and C. Streffer. Application of a multiple fixation regimen to study the adaptive response to ionising radiation in lymphocytes of two human donors. Mutat. Res. 326: 109-116 (1995).

W28 Wolff, S., R. Jostes, F.T. Cross et al. Adaptive response of human lymphocytes for the repair of radon-induced chromosomal damage. Mutat. Res. 250: 299-306 (1991).

W29 Wolff, S., J.K. Wiencke, V. Afzal et al. The adaptive response of human lymphocytes to very low doses of ionising radiation: a case of induced chromosomal repair with the induction of specific proteins. p. 446-454 in: Low Dose Radiation (K.F. Baverstock and J.W. Stather, eds.). Taylor and Francis, London, 1989.

W30 Wolff, R., R. Plaetke, A.J. Jeffreys et al. Unequal crossing over between homologous chromosomes is not the major mechanism involved in the generation of new alleles at VNTR loci. Genomics 5: 382-384 (1989).

W31 Wooster, R., D. Ford, J. Mangion et al. Absence of linkage to the ataxia-telangiectasia locus in familial breast cancer. Hum. Genet. 92: 91-94 (1993).

W32 Wouters, B. and L.D. Skarsgard. The response of a human tumor cell line to low radiation doses: evidence of enhanced sensitivity. Radiat. Res. 138: S76-S80 (1994).

W33 Wyllie, A.H., J.F.R. Kerr and A.R. Currie. Cell death: the significance of apoptosis. Int. Rev. Cytol. 68: 251-306 (1980).

W34 Wyllie, A.H. The genetic regulation of apoptosis. Curr. Opin. Genet. Dev. 5: 97-104 (1995).

W35 Watt, P.M., E.J. Louis, R.H. Borts et al. Sgs1: a eukaryotic homolog of E. coli ReqQ that interacts with topoisomerase II in vivo and is required for faithful chromosome segregation. Cell 81: 253-260 (1995).

W36 Weksberg, R., C. Smith, L. Anson-Cartwright et al. Bloom syndrome: a single complementation group defines patients of diverse ethnic origin. Am. J. Hum. Genet. 42: 816-824 (1988).

W37 Whitney, M., M. Thayer, C. Reifsteck et al. Microcell-mediated chromosome transfer maps the Fanconi anaemia group D gene to chromosome 3p. Nature Genet. 11: 341-343 (1995).

W38 Watson, G.E., S.A. Lorimore and E.G. Wright. Long-term in vivo transmission of α-particle-induced chromosomal instability in murine haemopoietic cells. Int. J. Radiat. Biol. 69: 175-182 (1996).

W39 Ward, J.F. DNA damage produced by ionizing radiation in mammalian cells: identities, mechanisms of formation and reparability. Prog. Nucleic Acid Res. Mol. Biol. 35: 95-125 (1988).

W40 Weissenborn, U. and C. Streffer. Analysis of structural and numerical chromosomal anomalies at the first, second and third mitosis after irradiation of one-cell mouse embryos with X-rays or neutrons. Int. J. Radiat. Biol. 54: 381-394 (1988).

W41 Weissenborn, U. and C. Streffer. The one-cell mouse embryo: cell cycle-dependent radiosensitivity and development of chromosomal anomalies in post-radiation cell cycles. Int. J. Radiat. Biol. 54: 659-674 (1988).

W42 Weissenborn, U. and C. Streffer. Analysis of structural and numerical chromosomal aberrations at the first and second mitosis after X irradiation of two-cell mouse embryos. Radiat. Res. 117: 214-220 (1989).

W43 Ward, J.F. Molecular mechanisms of radiation-induced damage to nucleic acids. Adv. Radiat. Biol. 5: 181-239 (1975).

W44 Ward, J.F. The complexity of DNA damage: relevance to biological consequences. Int. J. Radiat. Biol. 66: 427-432 (1994).

W45 Wilson, D.M. and L.H. Thompson. Life without DNA repair. Proc. Natl. Acad. Sci. U.S.A. 94: 12754-12757 (1997).

W46 Wilson, T.E., U. Grawunder and M.R. Lieber. Yeast DNA ligase IV mediates non-homologous DNA end joining. Nature 388: 495-498 (1997).

W47 Westphal, C.H., C. Schmaltz, S. Rowan et al. Genetic interactions between atm and p53 influence cellular proliferation and irradiation-induced cell cycle checkpoints. Cancer Res. 57: 1664-1667 (1997).

W48 Wojcik, A., K. Bonk, W-U. Muller et al. Do DNA double-strand breaks induced by AluI lead to development of novel aberrations in the second and third post-treatment mitosis? Radiat. Res. 145: 119-127 (1996).

W49 Wang, Y.A., A. Elson and P. Leder. Loss of p21 increases sensitivity to ionizing radiation and delays the onset of lymphoma in atm-deficient mice. Proc. Natl. Acad. Sci. U.S.A. 94: 14590-14595 (1997).

W50 Wright, J.A., K.S. Keegan, D.R. Herendeen et al. Protein kinase mutants of human ATR increase sensitivity to UV and ionizing radiation and abrogate cell cycle checkpoint control. Proc. Natl. Acad. Sci. U.S.A. 95: 7445-7450 (1998).

W51 Wu, L-J., G. Randers-Pehrson, A. Xu et al. Targeted cytoplasmic irradiation with alpha particles induces mutations in mammalian cells. Proc. Natl. Acad. Sci. U.S.A. 96: 4959-4964 (1999).

W52 Wood, R.D. Nucleotide excision repair in mammalian cells. J. Biol. Chem. 272: 23465-23468 (1997).

W53 Wishkerman, V.Y., M.R. Quastel, A. Douvdevani et al. Somatic mutations at the glycophorin A (GPA) locus measured in red cells of Chernobyl liquidators who immigrated to Israel. Environ. Health Perspect. 105: 1451-1454 (1997).

X1 Xia, F., S.A. Amundson, J.A. Nickoloff et al. Different capacities for recombination in closely related human lymphoblastoid cell lines with different mutational responses. Mol. Cell. Biol. 14: 5850-5857 (1994).

X2 Xia, F., X. Wang, Y-H. Wang et al. Altered p53 status correlates with differences in sensitivity to radiation-induced mutation and apoptosis in two closely related human lymphoblast lines. Cancer Res. 55: 12-15 (1995).

X3 Xu, Y., T. Ashley, E.B. Brainerd et al. Targeted disruption of ATM leads to growth retardation, chromosomal fragmentation during meiosis, immune defects, and thymic lymphoma. Genes Dev. 10: 2411-2422 (1996).

X4 Xu, Y. and D. Baltimore. Dual roles of ATM in the cellular response to radiation and in cell growth control. Genes Dev. 10: 2401-2410 (1996).

X5 Xanthoudakis, S., G. Miao, F. Wang et al. Redox activation of Fos-Jun DNA binding activity is mediated by a DNA repair enzyme. EMBO J. 11: 3323-3335 (1992).

X6 Xanthoudakis, S., R.J. Smeyne, J.D. Wallace et al. The redox/DNA repair protein, Ref-1, is essential for early embryonic development in mice. Proc. Natl. Acad. Sci. U.S.A. 93: 8919-8923 (1996).

X7 Xu, X., K-U. Wagner, D. Larson et al. Conditional mutation of *Brca1* in mammary epithelial cells results in blunted ductal morphogenesis and tumour formation. Nature Genet. 22: 37-43 (1999).

X8 Xu, X., Z. Weaver, S.P. Linke et al. Centrosome amplification and a defective G2-M cell cycle checkpoint induce genetic instability in BRCA1 exon 11 isoform-deficient cells. Mol. Cell 3: 389-395 (1999).

Y1 Ying, K.L. and W.E. Decoteau. Cytogenetic anomalies in a patient with ataxia, immune deficiency and high α-fetoprotein in the absence of telangiectasia. Cancer Genet. Cytogenet. 4: 311-317 (1981).

Y2 Yandell, D.W., T.P. Dryja and J.B. Little. Somatic mutations at a heterozygous autosomal locus in human cells occur more frequently by allele loss than by intragenic structural alterations. Somat. Cell Mol. Genet. 12: 255-263 (1986).

Y3 Yamada, T. and H. Ohyama. Radiation-induced interphase death of rat thymocytes is internal programmed (apoptosis). Int. J. Radiat. Biol. 53: 65-75 (1988).

Y4 Yamaguchi, T. Relationship between survival period and dose of irradiation in rat thymocytes *in vitro*. Int. J. Radiat. Biol. 12: 235-242 (1967).

Y5 Yanagihara, K., M. Nii, M. Numoto et al. Radiation-induced apoptotic cell death in human gastric epithelial tumour cells; correlation between mitotic death and apoptosis. Int. J. Radiat. Biol. 67: 677-685 (1995).

Y6 Yoshimura, Y., T. Morita, A. Yamamoto et al. Cloning and sequence of the human *RecA*-like gene cDNA. Nucleic Acids Res. 21: 1665 (1993).

Y7 Yamashita, T., D.L. Barber, Y. Zhu et al. The Fanconi anaemia polypeptide FACC is localized to the cytoplasm. Proc. Natl. Acad. Sci. U.S.A. 91: 6712-6716 (1994).

Y8 Yuille, M.A., L.J. Coignet, S.M. Abraham et al. ATM is usually rearranged in T-cell prolymphocytic leukaemia. Oncogene 16: 789-796 (1998).

Z1 Ziv, Y., A. Amiel, N.G.J. Jaspers et al. Ataxia-telangiectasia: a variant with altered in vitro phenotype of fibroblast cells. Mutat. Res. 210: 211-219 (1989).

Z2 Ziv, Y., G. Rotman, M. Frydman et al. The ATC (ataxia-telangiectasia complementation group C) locus localizes to 11q22-q23. Genomics 9: 373-375 (1991).

Z3 Ziv, Y., M. Frydman, E. Lange et al. Ataxia-telangiectasia - linkage analysis in highly inbred arab and Druze families

and differentiation from an ataxia-microcephaly-cataract syndrome. Hum. Genet. 88: 619-626 (1992).

Z4 Ziegler, A., A.S. Jonason, D.J. Leffell et al. Sunburn and p53 in the onset of skin cancer. Nature 372: 773-776 (1994).

Z5 Zdzienicka, M.Z. and J.W.I.M. Simons. Mutagen-sensitive cell lines are obtained with a high frequency in V79 Chinese hamster cells. Mutat. Res. 178: 235-244 (1987).

Z6 Zdzienicka, M.Z., N.G.J. Jaspers, G.P. van der Schans et al. Ataxia-telangiectasia-like Chinese hamster V79 cell mutants with radioresistant DNA synthesis, chromosomal instability and normal DNA strand break repair. Cancer Res. 49: 1481-1485 (1989).

Z7 Zhen, W., C.M. Denault, K. Loviscek et al. The relative radiosensitivity of TK6 and WI-L2-NS lymphoblastoid cells derived from a common source is primarily determined by their p53 mutational status. Mutat. Res. 346: 85-92 (1995).

Z8 Zhou, P.-K., W-Z. Sun, X-Y. Liu et al. Adaptive response of mutagenesis and DNA double-strand break repair in mouse cells induced by low dose of γ-ray. p. 271-274 in: Low Dose Irradiation and Biological Defense Mechanisms (T. Sugahara et al., eds.). Elsevier, 1992.

Z9 Zhu, C., M.A. Bogue, D-S. Lim et al. Ku-86 deficient mice exhibit severe combined immunodeficiency and defective processing of V(D)J recombination intermediates. Cell 86: 379-389 (1996).

Z10 Zhang, Q-M., S. Yonei and M. Kato. Multiple pathways for repair of oxidative damages caused by X rays and hydrogen peroxide in *Escherichia coli*. Radiat. Res. 132: 334-338 (1992).

Z11 Zasukhina, G.D., T.A. Sinel'shchikova, I.M. Vasil'eva et al. Radioadaptive response in repair-defective cells from patients with homocyctinuria. Genetika 32: 1592-1595 (1996).

Z12 Zhong, Q., C-F. Chen, S. Li et al. Association of BRCA1 with the hRad50-hMre11-p95 complex and the DNA damage response. Science 285: 747-750 (1999).

Z13 Zdzienicka, M.Z. Mammalian X ray sensitive mutants: a tool for the elucidation of the cellular response to ionizing radiation. Cancer Surv. 28: 281-293 (1996).

Z14 Zhang, H., R.L. Zheng, Z.Q. Wei et al. Effects of pre-exposure of mouse testis with low dose $^{16}O8^+$ ions or ^{60}Co gamma-rays on sperm shape abnormalities, lipid peroxidation and superoxide dismutase activity induced by subsequent high dose irradiation. Int. J. Radiat. Biol. 73: 163-167 (1998).

ANNEX G

Biological effects at low radiation doses

CONTENTS

INTRODUCTION

1. Biological effects of ionizing radiation in humans, due to physical and chemical processes, occur immediately following the passage of radiation through living matter. These processes will involve successive changes at the molecular, cellular, tissue and whole organism levels. For acute whole-body exposures above a few gray from radiation of low linear energy transfer (LET), damage occurs principally as a result of cell killing. This can give rise to organ and tissue damage and, in extreme cases, death. These effects, termed early or deterministic, occur principally above a threshold dose that must be exceeded before they are manifested as clinical damage, although damage to individual cells will occur at lower doses. Protracted delivery of such high doses over several hours or days will usually result in effects of lower severity. Information on the early effects of radiation in humans was reviewed in the UNSCEAR 1993 and 1982 Reports [U3, U6].

2. A second type of damage can occur at late times after exposure. This damage consists primarily of damage to the nuclear material in the cell, causing radiation-induced cancer to develop in a proportion of exposed persons or hereditary disease in their descendants. Although the probability of both cancer and hereditary disease increases with radiation dose, it is generally considered that their severity does not. They are termed stochastic effects and were reviewed in the UNSCEAR 1977, 1988, and 1994 Reports [U2, U4, U7].

3. Direct information on radiation-induced cancer is available from epidemiological studies of a number of human populations. These include the survivors of the atomic bombings in Japan and groups that have been exposed to external radiation or to incorporated radionuclides, either for medical reasons or occupationally. Such studies provide quantitative information on the risk of cancer at intermediate to high doses and are reviewed in Annex I, *"Epidemiological evaluation of radiation-induced cancer"*. At lower levels of exposure, however, quantitative estimates of risk are not so readily obtained, and inferences need to be made by downward extrapolation from the information available at higher doses.

4. In the case of radiation-induced hereditary disease, studies on human populations have not provided quantitative information, so risk estimates have to be based on the results of animal studies. There is again the difficulty that quantitative data are available only following exposures to intermediate to high doses. Information on radiation-induced hereditary disease has been reviewed previously by the Committee [U3, U4].

5. For the majority of situations in which human beings are exposed to ionizing radiation in the home, in the natural environment, and in many places of work, the principal concern is the consequence of exposure to low doses and low dose rates. For the purposes of radiation protection, the establishment of the expected incidence of cancer or heredi-

tary disease following radiation exposure is presently based on the hypothesis that the frequency of their induction increases proportionally with radiation dose. A linear, no-threshold dose-response relationship has generally been adopted by national and international bodies for assessing the risks resulting from exposures to low doses of ionizing radiation (see, e.g. [I2, U4]). This hypothesis implies that the risk of cancer increases (linearly) with increasing exposure and that there is no threshold, i.e. no dose below which there is absolutely no risk. As yet no definitive experimental data are available on this issue (see Chapter IV).

6. Experimental and epidemiological data on which quantitative evaluations of the risk of cancer following exposure to low-LET radiation are based come principally from studies involving exposures at moderate to high doses and dose rates. Most organizations have extrapolated linearly and then applied a reduction factor to estimate risks at low doses and low dose rates. This reduction factor has been variously termed a dose and dose rate effectiveness factor (DDREF) [I2], a dose-rate effectiveness factor (DREF) [N1], a linear extrapolation overestimation factor (LEOF), and a low dose extrapolation factor (LDEF) [P1, P14]. The basis for the application of such a reduction factor was described in the UNSCEAR 1993 Report [U3]. For high-LET radiations, such as neutrons and alpha particles, no reduction factor has generally been applied, because the dose response for radiation-induced cancer and hereditary disease is essentially linear between the lowest dose at which effects have been observed and that at which cell killing becomes a factor in the dose response [I2, U3, U4].

7. There has been extensive debate as to the shape of the dose-response relationship below the range at which effects can be directly measured. It has been argued that irradiating cells and tissues with small radiation doses can result in an adaptive response that reduces the amount of damage caused by subsequent radiation exposure [U2, W6] or even results in a beneficial effect, termed hormesis [A9, T11, W13]. There have been suggestions that, at very low doses, radiation may have no effect at all; these suggestions are based on the proposition that there could be a threshold for a response, in the same way as there is for clinically observed deterministic effects. This situation may arise, for example, if damage to a number of cells is needed before any adverse effect occurs or if interaction between cells is a prerequisite for an effect [K19, M34]. An apparent threshold may also arise if the latent period between exposure and the appearance of a cancer exceeds the normal lifespan of the individual [R1, R14].

8. Several mechanistic models have been proposed to describe the effects of radiation at the different levels of biological organization. There has been considerable effort in developing such models to quantitatively describe cellular survival, repair and transformation, based on the stochastic (probabilistic) process of energy deposition in radiosensitive

targets representing elements of cell structure, or employing track structure concepts. Other models have concentrated on representing the processes of repair and misrepair of damaged cell structures. In general, a mechanistic model should, apart from quantitative description of available data, have a predictive capability and offer crucial tests of its validity. Some mechanistic models support the linear no-threshold expressions used to fit epidemiological data, while others point to power law dose-effect relationships, implying a zero initial slope. There have been suggestions that the limits of dose-based quantities have been reached and that fluence and an action cross section are more appropriate concepts for assessing damage to cells. No quantitative attempt has, however, been made to apply these concepts in radiological protection [S5].

9.   It has been recognized by the Committee for some time that information is needed on the extent to which both total dose and dose rate influence the induction of cancer and hereditary disease. A number of considerations are important in determining the risks of exposures to radiation at low doses and low dose rates. These include (a) careful analysis of epidemiological studies to determine the lowest doses at which effects are statistically evident, (b) examination of the shape of the dose-response relationships in the low-dose region using available experimental and epidemiological data, and (c) assessment of the possibilities for extrapolation to lower levels of dose based on an understanding of the mechanisms involved in the radiation response of tissues. Extrapolation based on mechanistic considerations can, in principle, be made using information on relevant biological factors such as cellular/molecular targets for tumour initiation, the nature of radiation-induced damage to deoxyribonucleic acid (DNA) and the fidelity of its repair, together with information on adaptive responses and cellular surveillance. Many of these factors were discussed in the UNSCEAR 1993 and 1994 Reports [U2, U3].

10.   The objective of this Annex is to examine the sources of data that are available for assessing the risks of radiation-induced cancer and hereditary disease at low doses for both sparsely ionizing (low-LET) and densely ionizing (high-LET) radiation and their associated uncertainties. This Annex brings together information reviewed by the Committee in separate specialized Annexes, material from previous UNSCEAR reports, and additional data from dosimetric and cellular studies, epidemiological investigations, recent advances in molecular biology, and developments in mechanistic models. The aim is to provide an overview of the data available on the relationship between radiation exposure and the induction of cancer and hereditary disease, with emphasis on the extent to which radiation effects can be observed at low doses. This information, coupled with knowledge on the mechanisms of damage to cells and tissues, provides a basis for informed judgements to be made about the likely form of the dose response at exposures below those at which direct information is available.

11.   Dose-response relationships for radiation effects in cellular systems are reviewed in Chapter I. Considered first of all is the definition of a low dose and a low dose rate, as they may be described either physically or biologically. This will depend upon the level of biological organization considered. Also addressed are theoretical aspects of the interactions of radiation with cells and tissues; the influence of track structure on radiation response; the concept of dose as it applies to tissues, cells, or subcellular targets; and the possible implications for dose-response relationships. The results of cellular studies are then reviewed. The range of endpoints of these studies include cell killing, cell transformation, chromosome aberrations, and mutation, which occur principally as a consequence of damage to the nuclear material in individual cells.

12.   The results of animal studies related to radiation-induced cancer and hereditary disease are considered in Chapter II. For tumour induction, animal studies have demonstrated that dose-response relationships can be complex, depending on the age, gender, and species or strain of the animal, the sensitivity of individual tissues, the tumour type, and the dose rate. The results obtained for dose-response relationships for life-shortening and tumour induction with different animal models following exposure to external radiation or incorporated radionuclides are illustrated, and information is presented on the extent to which animal data can provide information on the risks of exposure at low doses.

13.   In the case of damage to germ cells, the mutational events resulting from DNA damage generally arise as a simple function of dose and dose rate and depend principally on the radiation sensitivity of the specific gene locus. Dose-response relationships are reviewed in Chapter II. Radiation-induced hereditary effects were comprehensively examined by the Committee in the UNSCEAR 1986, 1988, and 1994 Reports [U2, U4, U5].

14.   Epidemiological studies give information on dose-response relationships for tumour induction and provide the basis for quantitative risk estimates for human populations. The available data have been the subject of substantive reviews by the Committee [U2, U4, U5], and a further review is contained in Annex I, "*Epidemiological evaluation of radiation-induced cancer*". The information available on dose-response relationships is described in Chapter III, with emphasis on the extent to which data are available at low doses. These data relate to the consequences of exposure *in utero* as well as the exposure of infants, children, and adults.

15.   The direct information on tumour induction, both from experimental and epidemiological studies, is insufficient, on its own, to elucidate the shape of the dose-response relationship at low doses. In Chapter IV, present knowledge is examined on the mechanisms of radiation tumorigenesis that can be used to gain further insight into effects at low doses. Emphasis is placed on gaps in knowledge and the consequent uncertainties. This topic was last reviewed by the Committee in the UNSCEAR 1993 Report [U3], and other issues relevant to those discussed here are considered in

Annex F, "*DNA repair and mutagenesis*" and Annex H, "*Combined effects of radiation and other agents*".

16. As modern molecular methods are developed and applied, the understanding of the mechanisms of tumorigenesis has, in recent years, increased substantially. At the same time there has been an equivalent increase in knowledge of radiation action on cellular DNA; of control of the reproductive cell cycle; of the mechanisms of DNA repair, genomic maintenance, and mutagenesis; and of non-mutational mechanisms of stable cellular changes. All this information could be relevant to assessing the shape of the dose response for both radiation-induced cancer and hereditary disease at low doses and dose rates and the effects of radiation quality at exposures below those at which direct information is available.

17. An important aim of Chapter IV is, accordingly, to highlight the critical elements of the current understanding of the mechanisms of tumorigenesis in order to relate them to data on dose-effect relationships and permit extrapolation to doses beneath those at which quantitative information is available. In Chapter V, the judgements developed in Chapter IV are used to examine biologically based computational models that may in turn be used to assess the risk of radiation-induced cancer at low doses and low dose rates.

I. CELLULAR EFFECTS

18. Damage to DNA, which carries the genetic information in chromosomes in the cell nucleus, is considered to be the main initiating event by which radiation damage to cells results in the development of cancer and hereditary disease [U3]. Either one or both strands of the DNA helix in cells may be damaged or broken, resulting in cell death, damage to chromosomes, or mutational events. Radiation is thought to have an effect on DNA either through the direct interaction of ionizing particles with DNA molecules or through the action of free radicals or other chemical intermediates produced by the interaction of radiation with neighbouring molecules. Damage can also be caused to other cellular structures, resulting in death or sublethal damage in individual cells; such damage does not in general result in radiation-induced cancer or hereditary disease. An exception is damage to cells that results in fibrosis, as this seems to be a precursor to the development of some tumour types (see Chapter II). It is also possible that other more indirect mechanisms can influence tumour development.

19. This Annex is concerned with the examination of the biological effects of radiation at low radiation doses. It is appropriate, therefore, to consider first how these should be defined. The designation of low doses and low doses rates has been considered in earlier reports by the Committee [U3, U5] and is summarized here briefly. The following Sections then consider radiation damage to DNA, relative biological effectiveness (RBE) of radiations of different quality, and the influence of track structure on cellular response. Cellular studies related to the determination of dose-response relationships for chromosome aberrations, cell transformation, and mutation induction in somatic cells are then summarized.

A. DESIGNATION OF LOW DOSES AND LOW DOSE RATES

20. In interpreting the responses of cells and tissues to ionizing radiation, judgements need to be made as to the bounds for low and high doses of low-LET radiation. In the 1993 UNSCEAR Report [U3], the physical and biological factors that need to be considered in making these evaluations were examined in the context of the doses and dose rates below which it would be appropriate to apply a reduction factor when assessing risks (per unit dose) at low doses and low dose rates from information on risks obtained at high doses and dose rates.

21. The following Sections deal with physical and biological approaches to designating exposures that may be considered to be either low-dose or low-dose-rate and with experimental data that can give information on the dose-response relationship for stochastic effects in cells either *in vitro* or *in vivo*.

1. Physical factors

22. Various models have been developed to account for the features of dose-response relationships obtained in experimental studies. A common aspect of many of these models is that a single radiation track, for any radiation quality, is taken to be capable of producing the initial damage and hence the cellular effect. The fundamental physical quantity used to define the deposition of energy in organs and tissues from ionizing radiation is the absorbed dose. The tissue or organ absorbed dose, D_T, is generally taken to be the mean energy absorbed in the target organ or tissue divided by the mass, T. This definition of the absorbed dose does not, however, characterize the fluctuation of energy absorption resulting from the stochastic nature of the energy deposition events (tracks) in individual cells. The fluctuation in the energy deposition between cells in a tissue is generally disregarded but can be significant when the possible effects of ionizing radiation on cells at low doses are considered. The number of independent tracks within each cell follows a Poisson distribution, and thus the numbers of cells receiving zero or few tracks will depend on the fluence of tracks through the organ or tissue.

23. The physical factors that can influence the effect of radiation on cells and tissues are generally well understood as a result of advances that have taken place in recent years in

microdosimetry at the cellular and subcellular levels [B31, B32, G6, G12, P13, R18]. A microdosimetric argument for defining low doses and low dose rates can be based on statistical considerations of the occurrence of independent radiation tracks within cells or nuclei. For ^{60}Co gamma rays, for example, and a spherical cell or nucleus (taken to be the sensitive target) assumed to be 8 μm in diameter, there will be, on average, one track per nucleus when the averaged absorbed dose is about 1 mGy [B31, B32]. If the induction of damage in the nucleus depends on energy deposition in single nuclei, with no interaction between them, a departure from linearity is unlikely unless there have been at least two independent tracks within the cell nucleus. The number of tracks within cells follows a Poisson distribution, as illustrated in Table 1, with the mean number of tracks proportional to the

average absorbed dose. For average tissue absorbed doses of 0.2 mGy from low-LET ^{60}Co gamma rays, for example, spherical nuclei of say 8 μm diameter would each receive, on average, about 0.2 tracks. In this case, just 18% of cells would receive any radiation track at all and less than 2% of cells would receive more than one track. Halving the exposure would simply halve the fraction of the total cells affected, and so, at such low doses, the dose-effect should be linear. This microdosimetric argument for a low dose (taken here to be 0.2 tracks per cell) would apply to biological effects where the energy deposited in a cell produces effects in that cell and no other cell. It might apply, for example, to cell killing, the induction of chromosome aberrations, and mutations. Its applicability to cell transformation and cancer induction is less certain. It would need modification if, for example, the pro-

Table 1

Proportions of a cell population traversed by tracks for various mean doses [a] **from gamma rays and alpha particles**

Mean tracks per cell	Percentage of cells in population suffering						Percentage of hit cells with only one track
	0 track	1 track	2 tracks	3 tracks	4 tracks	>5 tracks	
0.1	90.5	9	0.5	0.015	-	-	95.1
0.2	81.9	16.4	1.6	0.1	-	-	90.3
0.5	60.7	30.3	7.6	1.3	0.2	-	77.1
1	36.8	36.8	18.4	6.1	1.5	0.4	58.2
2	13.5	27.1	27.1	18	9	5.3	31.3
5 [b]	0.7	3.4	8.4	14	17.5	56	3.4
10 [b]	0.005	0.05	0.2	0.8	1.9	97.1	0.05

a   Approximately 0.1 mGy for gamma rays, 300 mGy for alpha particles.
b   At these values appreciable proportions of the cell population will incur more than five tracks.

bability of an effect was influenced by a subsequent track at a later time, as could be the case for multi-stage carcino-genesis, or if there was interaction between cells in the development of a specific radiation effect, as, for example, has been suggested for so-called bystander effects. This is con-sidered further in Chapter IV.

24. To develop the microdosimetric argument for assessing a low dose, knowledge is required of the sensitive volume in the cell. A sphere of 8 μm diameter, as described above, is typical of the size of some cell nuclei, although they may be larger or smaller. If only a part of the nucleus responds autonomously to radiation damage and repair, then a smaller sensitive volume may be more appropriate, and the estimate of a low dose would increase. Figure I illustrates, for various volumes, the specific energy of low-LET radiation that would correspond to this microdosi-metric criterion of a low dose when less than 2% of cells receive more than 1 track. Thus for a nucleus of diameter 4 μm, a low dose (0.2 tracks per cell, on average) would be about 0.8 mGy, and for 32 μm it would be about 0.01 mGy. As described in the UNSCEAR 1993 Report [U3], this definition of a low dose could also take into account information on the time characteristics for DNA repair, which would give a low dose rate of 10^{-3} mGy min^{-1}, or be based on only a single track traversing a cell in a lifetime (say, 60 years), allowing essentially no scope for track interactions, which would give a low dose rate of 10^{-8} mGy min^{-1}.

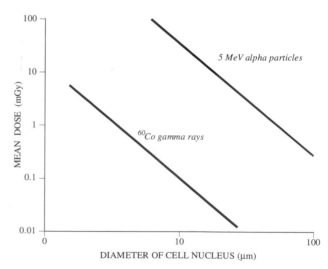

Figure I. Mean dose from an average of 0.2 radiation tracks per cell nucleus as a function of diameter (<2% of nuclei receive more than one track).

25.   The situation is quite different for exposures to high-LET radiation. When a tissue receives an average dose of 1 mGy from alpha particles, only about 0.3% of the nuclei are struck by a track at all; the remaining 99.7% are totally unirradiated. When a single track does strike, it delivers to the nucleus a very large dose, of about 370 mGy on average. In individual nuclei the dose may be any value up to about 1,000 mGy [G22, G23].

2. Biological approaches

26. *Biological approaches.* The definition of a low dose and a low dose rate can also be based on direct observation of damage in experimental systems or in epidemiological studies. One approach to assessing a low dose is based on parametric fits to observed dose-response data for cellular effects at low to intermediate doses, below those at which cell killing will become important. For the induction of cellular damage, the incidence, I, of an effect can then be related to the dose, D, by an expression of the form

$$I(D) = \alpha D + \beta D^2 \qquad (1)$$

in which α and β, the coefficients for the linear and quadratic terms fitted to the radiation response, are constants and are different for different endpoints. This equation has been shown to fit data on the induction of chromosome aberrations in human lymphocytes [U3]. It can also be extended to cover cell killing as described in Section I.B.1. For some types of unstable chromosome aberrations in human lymphocytes, the α/β quotient (which corresponds to the average dose at which the linear and quadratic terms contribute equally to the biological response) is about 200 mGy for ^{60}Co gamma rays [L34], and thus the response is essentially linear up to about 20 mGy, with the dose-squared term contributing only 9% of the total response. Even at 40 mGy, the dose-squared term still contributes only about 17% to the overall response. On this basis a low dose might be judged to be 20-40 mGy.

27. Another approach to assessing the low dose range can be based on animal studies. The results of studies designed to examine the effect of dose and dose rate on tumour induction were comprehensively reviewed in the UNSCEAR 1993 Report [U3]. The results obtained with experimental animals, predominantly mice, comparing the effect of various dose rates of low-LET radiation on the induction of leukaemia and solid tumours have suggested that, on average, a dose rate of about 0.06 mGy min^{-1} over a few days or weeks may be regarded as low. At lower dose rates no further reduction in tumour incidence, per unit

exposure, was obtained. The choice by the Committee in the UNSCEAR 1986 Report [U5] of a low dose rate to include values up to 0.05 mGy min^{-1} appears to have been based on dose rate studies in experimental animals.

28. The analysis of information from epidemiological studies, in particular the data from the survivors of the atomic bombings in Japan, can also be used for estimating a low dose. Analysis of the dose response for mortality from solid cancers in the range 0-4 Gy (adjusted for random errors) has suggested an α/β quotient from a minimum of about 1 Gy, with a central estimate of about 5 Gy [P1, P14]. An α/β quotient of 1 Gy suggests that at a dose of 100 mGy the dose-squared term contributes less than 10% to the response and at 200 mGy still less than 20%. It was suggested in the 1993 UNSCEAR report that for solid tumour induction in humans, a low dose could be taken to be less than 200 mGy [U3]. There is, in practice, little evidence of a departure from linearity up to about 3 Gy. In the case of leukaemia in the survivors of the atomic bombings, where there is a significant departure from linearity at doses above about 1.5 Gy, a central estimate of α/β has been calculated to be 1.7 Gy, with a minimum value less than 1 Gy [P1, P14]. On the basis of this central estimate, the dose-squared term would contribute about 10% to the response at a dose of 200 mGy and about 23% at 500 mGy. A low dose might therefore be considered to be any exposure up to about 200 mGy [U3].

29. On the basis of these various analyses of physical and biological data, the Committee concluded in the UNSCEAR 1993 Report [U3] that for the purpose of assessing the risk of tumour induction in humans at low doses and dose rates of low-LET radiation, a reduction factor (dose and dose rate effectiveness factor, DDREF) should be applied, either if the total dose is less than 200 mGy, whatever the dose rate, or if the dose rate is below 0.1 mGy per min^{-1} (when averaged over about an hour), whatever the total dose. The various approaches to assessing a low dose and low dose rate from low-LET radiation are summarized in Table 2.

Table 2
Alternative criteria for upper limits of low dose and low dose rate for assessing risks of cancer induction in humans (low-LET radiation)
[U3]

Basis of estimation	Low dose (mGy)	Low dose rate (mGy min^{-1})
UNSCEAR 1986 Report [U5]	200	0.05
UNSCEAR 1993 Report [U3]		0.1 ^{c}
Linear term dominant in parametric fits to single-cell dose responses	20-40	–
Microdosimetric evaluation of minimal multi-track coincidences in cell nucleus	0.2	10^{-8} (lifetime) 10^{-3} (DNA repair)
Observed dose-rate effects in animal carcinogenesis	–	0.06
Epidemiological studies of survivors of the atomic bombings in Japan	200	–

a Approximately 0.1 mGy for gamma rays, 300 mGy for alpha particles.
b At these values appreciable proportions of the cell population will incur more than five tracks.
c Averaged over about an hour.

30. For high-LET radiation, the experimental data suggested that little consistent effect of dose rate or dose fractionation on the tumour response at low to intermediate doses has been obtained. It was therefore concluded [U3] that there was no need to apply a reduction factor to risks calculated at high doses and dose rates.

B. DAMAGE TO DNA

31. Cells are able to repair both single- and double-strand breaks in DNA over a period of a few hours, but this repair can be imperfect, resulting in long-term cellular damage and mutation. It has been assumed in previous reports by the Committee that damage to DNA causing mutational events in germ cells is the result of a single biological event but that carcinogenesis is a multi-stage process in which the radiation can induce one or more of the stages involving damage to DNA and interference with cellular homeostatic mechanisms [U3].

32. The vast majority of endogenous DNA lesions take the form of DNA base damage, base losses, and breaks to one of the sugar-phosphate backbone strands of the double helix [A14, W7]. Such single-strand damage may be reconstituted rapidly in an error-free fashion by cellular repair processes since the enzyme systems involved will, for all these lesions, have the benefit of the DNA base sequence on the undamaged strand acting as a template on which to recopy the damaged or discontinuous code. Single ionizing tracks of radiation will induce a preponderance of such single-strand damage as a result of energy loss events occurring in close proximity to a single DNA strand in the double helix. A cluster of such events within the diameter of the DNA duplex (about 2 nm) has, however, a finite probability of simultaneously inducing coincident damage to both strands. In support of this, an approximately linear dose response for double-strand break induction by low-LET radiation has been observed [J6], confirming that breakage of both strands of the duplex may be achieved by the traversal of a single ionizing track and does not require multiple-track action. There is also evidence that a propor-tion of radiation-induced double-strand breaks are complex and involve locally, multiply damaged sites, LMDS [J6]. On the basis of a body of experimental evidence it may be judged that the ratio of low-LET radiation-induced single-strand DNA breaks plus base damages to double-strand breaks is around 50:1. The probability of a double-strand break per cell has been judged to be about 4 per cell per 100 mGy [G10].

33. A fraction of radiation-induced double-strand damage will be repaired efficiently and correctly, but error-free repair of all such damage, even at the low abundance expected after low-dose exposure, is not anticipated. Unlike damage to a single strand of the DNA duplex, a proportion of double-strand lesions, perhaps that component represented by LMDS, will result in the loss of DNA coding from both strands. Such losses are inherently difficult to repair correctly, and it is believed that misrepair of such DNA double-strand lesions is the critical factor

underlying the principal hallmarks of stable mutations induced by ionizing radiation of various qualities [T2, T14]. Double-strand DNA losses may in principle be repaired correctly by DNA repair recombination, but such damage may be subject to error-prone repair, which can result in the formation of gene and chromosomal mutations that are known to characterize malignant development [C23]. This interpretation would, however, be flawed if cellular repair processes were totally effective in repairing damage in the case of small numbers of double-strand breaks in the affected cells. In such a case, a threshold dose before any response could occur would be possible The most basic, although not necessarily sufficient, condition for a true dose threshold would be that any single track of the radiation should be unable to produce the effect. Thus, no biological effect would be observed in the true low-dose region, where cells are traversed only by single tracks. This is considered further in Chapter IV.

1. Dose response for low-LET radiation

34. The approach that has been frequently used to describe both the absolute and the relative biological effectiveness of a given radiation exposure from low-LET radiation is based on the assumption that the induction of an effect can be approximated by an expression of the following form:

$$I(D) = (\alpha D + \beta D^2) e^{-(p_1 D + p_2 D^2)} \qquad (2)$$

where α and β are coefficients of the linear and quadratic terms for the induction of stochastic effects and p_1 and p_2 are linear and quadratic terms for cell killing. This equation has been shown to fit much of the published data on the effects of radiation on cells and tissues resulting from damage to DNA in the cells, including the induction of chromosome aberrations, mutation in somatic and germ cells, and cell transformation.

35. The nature of the initial damage to DNA was considered in the UNSCEAR 1993 Report [U3]. The theoretical considerations were described in terms of the general features of target theory, because the insult of ionizing radiation is in the form of finite numbers of discrete tracks. On this basis, it was proposed that the nature of the overall dose response for low-LET radiation could be subdivided into a number of regions:

(a) *Low-dose region.* At the lowest doses, a negligible proportion of cells (or nuclei) would be intersected by more than one track, so the dose response for single-cell effects would be dominated by individual track events acting alone and would therefore be expected to increase linearly with dose and be largely independent of dose rate;

(b) *Intermediate-dose region.* In this dose region, where there may be several tracks per cell, it has been commonly assumed that tracks act independently if a linear term (α) can be obtained by curve-fitting to equations such as (1). For most of the experimental dose-response data used for curve-fitting, the lowest dose at which a significant effect is obtained is usually towards the higher end of this dose region, when individual cells may, in fact, have been traversed by considerable numbers of tracks. The assumption of one-track action for this region considers that the relevant metabolic processes of the cell are not influenced by the additional tracks in any way that could alter the expression of the ultimate biological damage of each individual track. On this assumption, it is conventional to interpolate linearly from this region to zero dose to deduce the effectiveness of low doses and low dose rates of radiation. Such interpolation is based on the coefficient α in equation (1) and on the assumption that it remains unchanged down to very low doses and very low dose rates. There are a number of radiobiological studies, mostly with cells *in vitro* but also from animals exposed at different dose rates, that suggest that this common assumption is not universally valid.

(c) *High-dose region.* In this region, where there are many tracks per cell, multi-track effects are clearly seen as non-linearity of dose response, with upward or downward curvature of the dose response. The simpler forms of the dose-response relationship that are observed experimentally can commonly be fitted by a general polynomial with terms for dose and dose rate. At high doses, a separate term is needed to account for the effects of cell killing.

36. Equation (1), or some modification of it, has been conventionally used to estimate the biological effectiveness of radiation at minimal doses, assuming a constant value of α from the intermediate-dose region down to zero dose, with independence of dose rate. There are, however, instances in which this may not apply.

37. The assumption of such a dose response for single-cell stochastic effects may not hold if there are significant multi-track events in the intermediate-dose region. Such events could include, for example, the induction of multiple independent steps in radiation carcinogenesis, cellular damage-fixation processes influencing repair of DNA damage; the induction of enhanced repair by small numbers of tracks; multiple tracks or enhancement of misrepair; and variations in cell sensitivity with time. The dose response may also be modified if the biological effect of interest required damage to more than one cell or if it was influenced by damage to additional or surrounding cells.

2. Dose response for high-LET radiation

38. There are extensive radiobiological data indicating that high-LET radiations (neutrons and alpha particles) have a greater biological effect, per unit of average absorbed dose, than low-LET radiation. The influence of radiation quality on a biological system is usually quantified in terms of its relative biological effectiveness (RBE). The RBE of a specific radiation, R, can be defined as the absorbed dose of reference radiation required to produce a specific level of response divided by the absorbed dose of radiation, R, required to produce an equal response, with all physical and biological variables, except radiation quality, being held constant as far as possible. This definition does not depend on the dose response for the two radiations being the same, it simply depends on comparing the dose to give a specific level of effect for a particular endpoint. Low-LET radiation (x rays or gamma rays) is normally used as the reference radiation. A particular form of RBE is RBE_m, which is the maximum RBE that would be obtained at low doses and low dose rates. Various authors and committees (see for example [M18, S35, U6]) have reviewed the relevant biological data. A comprehensive review of the literature relevant to the determination of values for RBE may be found in NCRP Report No. 104 [N6]. Maisin et al. [M53] reported information on tumour induction in 7- or 21-day old C57BL mice exposed to 3.1 MeV neutrons (0.5, 1 or 3 Gy) or 250 kVp x rays (0.125, 0.25, 0.5 or 1 Gy). When the incidence of all malignant tumours and of hepatocellular cancer was fitted to a linear of linear-quadratic function, an RBE in the range 5 to 8 was obtained.

39. It is apparent from the studies summarized above that the RBE for high-LET radiation is dependent on the biological response being studied. For early effects in tissues caused by cell killing (e.g. skin burns, cataracts, and sterility), an ICRP task group [I13] concluded that for a range of tissues and for both neutrons and alpha particles, the RBE was generally less than 10. For damage to the lung from inhaled alpha particles causing fibrosis and loss of fluid into the lung (pneumonitis), the RBE for rats and beagle dogs was estimated to be between 7 and 10. Similarly, for the induction of chromosome aberrations in human blood lymphocytes by alpha particles from ^{242}Cm, RBE values of about 6 have been obtained in comparison with x rays and 18 in comparison with gamma rays [E12]. For the induction of micronuclei (caused by fragmentation of chromosomes) in lymphocytes by alpha particles from plutonium, an RBE of 3.6 has been found at low doses (<1 Gy), and for DNA double-strand breaks in Ehrlich ascites tumour cells, RBEs in the range 1.6–3.8 have been reported [B40].

40. In a few experimental studies a biological effect has been obtained for alpha particle irradiation although a similar effect has not been found with low-LET radiation. Studies in which sister chromatid exchanges (SCE) have been measured in human lymphocytes in the G_0 stage of the cell cycle give a measurable frequency of SCEs following exposure to alpha particles from ^{241}Am, but no effect of x-ray irradiation was obtained. From the definition of RBE given above, this implies an infinite RBE, although it is solely a consequence of there having been no observable effect of x rays at low doses [A16]. Similar results have been reported for SCEs in Chinese hamster ovary cells irradiated in the G_1 phase of the cell cycle by ^{238}Pu alpha particles or x rays. High values of RBE, up to about 245, have also been reported [N10] for sperm head abnormalities in mice when the effect of external exposure to x rays was compared with the effects of tissue-incorporated ^{241}Am. This may partly be accounted for by the heterogeneous distribution of ^{241}Am incorporated in the testis; it is known that actinides such as ^{241}Am tend to concentrate in interstitial

tissue in the mouse testis, in close proximity to the developing sperm cells.

41. For tumour induction, a number of studies have demonstrated that both high- and low-LET radiation may induce cancer in a range of tissues. Data relevant to the choice of RBEs for neutrons and alpha particles are summarized below. Values of RBE obtained for long-term effects can be useful for transferring information on risks calculated following exposure to high-LET to assess risks in populations exposed to low-LET radiation, and vice versa.

(a) Neutrons

42. Values for RBE_m obtained for various biological endpoints in mammals and in mammalian cells for fission neutrons compared with gamma rays are summarized in Table 3. Similar reviews have been published by UNSCEAR [U5] and Sinclair [S35]. Information on the variation of RBE_m with neutron energy comes partly from data from cellular studies, in particular using point mutations, chromosomal aberrations, and cell transformation as endpoints.

Table 3
Estimated RBE_m values for fission neutrons compared with gamma rays
[N6]

Endpoint	RBE_m
Cytogenetic studies, human lymphocytes in culture	34–53
Cell transformation	3–80
Genetic endpoints in mammalian systems	5–70
Life shortening (mouse)	10–46
Tumour induction	16–59

---

43. There is uncertainty in the value of RBE_m for fission neutrons. This uncertainty comes principally from how the data for low-LET radiation, mainly for cancer induction and life shortening in mammals, are extrapolated to low doses and low dose rates. The derivation of values for RBE_m is illustrated in Figure II. The straight line A of slope α_H represents the dose-response relationship for high-LET radiation. The data points shown in the Figure are representative of data for low-LET radiation and can be extrapolated to low doses by the linear relationship B of slope α_{LH} or by curve C, based upon a linear-quadratic dose-response relationship. Curve D represents the extrapolated linear portion with slope α_{LL} of the low dose response of curve C. The ratio of the slopes of curves A (slope α_H) and D (slope α_{LL}) represents RBE_m.

44. There are few data in whole animals that measure the variation of RBE_m for specific tumour induction or life shortening with neutron energy. Knowledge of the variation of RBE_m with neutron energy is confined to cellular studies. Chromosomal aberrations in human lymphocytes [E12, E13] indicate a monotonic decrease by a factor of about four from 1 MeV to 14 MeV. Mutations in a human hamster hybrid cell line (A_L) indicated [H29] a monotonic decrease by a factor of about seven from 0.3 to 14 MeV. The oncogenic transformation of C3H10T½ cells showed a more erratic variation with neutron energy [M15], but an overall variation by a factor of three from 230 keV to 14 MeV. The cellular data suggest a decrease in RBE_m by a factor of about four from 100 keV with an increase of neutron energy. There are very few experimental data at lower neutron energies. Some cellular data observing chromosome aberrations in human lymphocytes [E14] suggest an RBE_m close to that for fission neutrons, whereas similar data from Sevan'kaev et al. [S36] suggest lower values.

(b) Alpha particles

45. Alpha particles have a very short range in tissue. For the highest energy natural alpha particles from ^{226}Ra and its decay products with an energy up to about 7.8 MeV, the maximum range is about 80 µm; for 5 MeV alpha particles from ^{239}Pu, the range is about 40 µm. These dimensions may be compared with the dimensions of the cell nucleus, which range from about 5 to 10 µm in diameter. The dose that a single alpha particle delivers crossing the cell nucleus, considered to be the radiosensitive target, is highly variable. It may range from very low doses for particles that graze the nucleus, to more than 1 Gy for particles crossing the diameter. Thus the concept of average tissue dose is a considerable simplification, and individual cells in a tissue will receive very different doses. Furthermore, alpha-emitting radionuclides

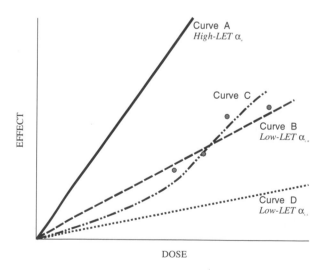

Figure II. Typical dose-effect relationship for low- and high-LET radiations [M18].

may be deposited on the surfaces of organs within the body; this is the case, for example, for radon decay products deposited in the lung and for plutonium isotopes accumulated by the skeleton. There can therefore be a very heterogeneous distribution of alpha particle dose within an organ (or tissue). The specified dose depends on whether an average organ dose or the mean dose to a particular localized tissue volume is calculated. In practice, average organ dose is usually calculated.

46.   When it comes to choosing an appropriate RBE to use for estimating the risk of tumour induction in organs and

tissues, there are rather few data available. The main difficulty is that comparable patterns of exposure are needed for both the alpha-emitting and the reference radiation (x rays or gamma rays). Although extensive data are available on tumour induction for both of these radiations on their own, their effects have been directly compared much less frequently. Published data relevant to the estimation of RBE_m are available for the induction of bone sarcomas and lung tumours in experimental animals, from studies on cells in culture and, to a limited extent, from epidemiological studies on human populations. Relevant data are summarized in Table 4.

Table 4
Estimated RBE$_m$ values for alpha particle irradiation and for gamma rays

Endpoint		RBE_m	Ref.
Bone tumours			
	Dogs	26	[N6]
	Mice	25	[N6]
	Dogs	5.4 (4.0–5.8)	[G16]
Lung tumours			
	Various species	30 (6–40)	[I5]
	Dogs	10–18	[B41]
	Rats	25	[H30]
	Dogs	36	[H30]
Cell transformation (C3H10T½)		10–25	[B42]
Cell mutation			
	Human lung cells HF19	up to 7.1 [a]	[C27]
	Chinese hamster cells V79	up to 18	[T16]
Chromosome aberrations		5–35	[E12, P20]
Germ cell mutations (chromosome fragments, chromosome translocations, dominant lethals)		22–24	[S37]

[a]   Compared with x rays (from [M18]).

3. Influence of track structure

47.   As described above, it is commonly observed that high-LET radiations (neutrons and alpha particles) are much more effective per unit dose than low-LET (electrons and photons), in producing cellular effects such as chromosome aberrations and mutations or for effects in animals such as cancer and life shortening. Despite this, the number of DNA breaks produced per gray is not very different for high- and low-LET radiations. Yet it is these breaks that lead to chromosomal and mutation events in cells and eventually to cancer. The explanation lies either in the difference in the efficiency/fidelity of double-strand break repair after high- and low-LET radiation, or in the difference between the spatial distribution of the initial physical events (ionizations and excitations) which lead, via double-strand breaks, to aberrations and mutations in the cell. If the second explanation is true, there is some biological relevance to the distribution of initial events of energy deposition around tracks of charged particles, i.e. to track structure.

48.   Computer programmes based on Monte Carlo techniques are now available to calculate on a scale of nanometres or smaller the exact position of ionizations and excitations in the track of charged particles [N9]. Examples for a 500 eV electron and a 4 MeV alpha particle are given in Figure III [G10]. Electrons meander by scattering and may travel in any direction. In contrast, heavy charged particles (from protons to much heavier ions) essentially travel in straight lines on a well-defined path. They pass their energy on to secondary electrons, which wander from the path of the ion. Generally, ions of higher velocity produce higher energy electrons that can travel further from the path of the ion. As an example, Figure IV shows calculations of the fraction of energy deposited within a distance, r, of a track for protons of energy from 0.3 to 20 MeV. For 0.3 MeV protons, at least 99% of the energy is deposited within 30 nm of the centre of the track. For a 20 MeV proton, some 2% of the energy is deposited more than 1 μm away from the path of the particle track.

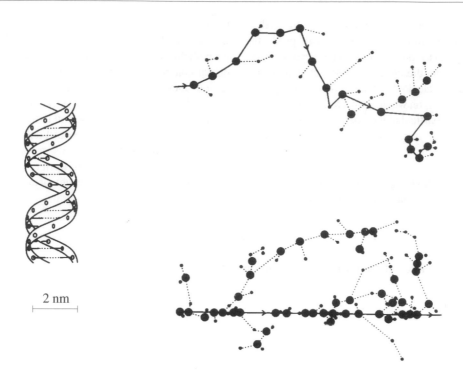

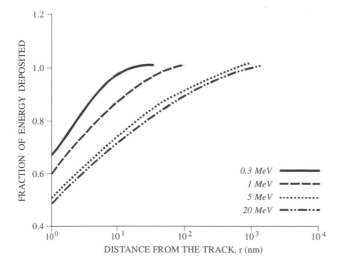

Figure III. Simulated low-energy track (upper panel: initial energy 500 eV) and simulated short portion of an alpha-particle track (lower panel: initial energy 4 MeV). A section of DNA is shown to give a perspective on dimensions [G10].

49.   By choosing particle type and energy, it is possible to select particles of the same LET but markedly different track structures, and when such experiments are done then different biological effects are found [C24]. Clearly, track structure is important for understanding how radiations of different quality cause differences in RBE, although opinions vary as to which particular features of track structure and which objects and characteristic distances in the cells relate to given biological endpoints. As an example, Savage [S32] considers that for producing chromosome aberrations, DNA breaks may exchange over distances up to one or two hundred nanometers. Thus, energy around a particle track deposited within volumes of the order of 100 nm may be important. Other dimensions

Figure IV.   Fraction of energy deposited at distances from a proton track less than a specified value, r [E16].

may apply to other biological effects. Some analytic models employing volume sizes as fitted parameters have been quite successful in representing quantitative relationships between RBE and LET for biological endpoints such as cell survival or transformation in irradiated mammalian cell cultures [C25, C29, K24, K28].

C. CELLULAR DAMAGE

50.   A range of assays has been developed for evaluating radiation damage to cells either *in vitro* or *in vivo*. These include survival curves, cell transformation, induction of mutations, and chromosome aberrations. Various models have been developed to describe these radiation-induced cellular effects and their dependence on dose, dose rate, and radiation quality. Some of these models have been summarized by Goodhead [G6, G7]. The near-consensus view of the critical damage to single cells that has resulted from these studies is that single radiation tracks from ionizing radiations can lead to cellular damage. The various models that have been developed assume *inter alia* that cellular damage arises from DNA double-strand breaks, either singly or in pairs; pairs of DNA single-strand breaks; localized clusters of radiation damage in unspecified molecular targets or in DNA; unspecified single or double lesions, probably in DNA, but qualitatively similar and independent of radiation quality; or damage to DNA and associated nuclear membranes [U3]. Such models indicate that a single track, even from low-LET radiation, has a finite probability of producing one, or more than one, double-strand break in DNA in a cell nucleus. Hence the cellular consequences of a double-strand break, or of interactions between them, should be possible even at the lowest of doses or dose rates. This would not, however, be the

outcome if cellular repair processes for single or small numbers of double-strand breaks were totally efficient, in which case a threshold for the response might be anticipated. There is no evidence for repair being 100% efficient, however, although experimental assays to test the hypothesis have limited resolution.

51. The following Sections illustrate dose-response relationships that have been obtained for chromosome aberrations, for cell transformation, and for mutagenesis in somatic cells. Emphasis is placed on assessing effects at low doses. Other cellular effects, including changes in gene expression, are examined in Annex F, *"DNA repair and mutagenesis"* and noted in Chapter IV of this Annex. In a number of cellular systems, an adaptive response has been described in which a small initial radiation dose can modify the effect of a subsequent larger dose. Some examples of studies demonstrating this response are given below.

1. Chromosome aberrations

52. The scoring of chromosome aberrations in human peripheral blood lymphocytes provides a sensitive method for biological dosimetry. It also provides a valuable approach to assessing dose-response relationships for chromosome mutations. By scoring dicentric aberrations in the full genome of about 1,000 cells, average whole-body doses of about 100 mGy from x rays or gamma rays may be detected and higher doses estimated. Calibration curves have been prepared by a number of laboratories for a wide range of radiations. All dose-effect curves for low-LET radiation conform to equation (2) up to 5–10 Gy. At higher doses, saturation of the curve can occur when yields of dicentrics approach five per cell [E8].

53. The difficulty of experimentally demonstrating the presence or absence of a threshold for single cellular events can be illustrated by work on the induction of chromosome aberrations in lymphocytes by x rays. In assessing radiation exposure by the analysis of aberrations in blood lymphocytes, measurements are normally made of the incidence of both dicentrics and total aberrations. These are unstable aberrations, and they will slowly disappear from peripheral blood. They are, however, more sensitive for the detection of effects at low doses than are stable aberrations.

54. The background incidence of dicentric aberrations in blood lymphocytes observed at metaphase is about 1 in 1,000 cells. As radiation-induced aberrations arise at the rate of about 4 per 100 cells per gray, the ability to detect a dose of 100 mGy would require about 1,000 lymphocytes to be scored, which would take about three man days. Radiation damage at lower doses can be detected, but this requires the assay of proportionately more cells. Investigations at doses much less than 100 mGy require very large numbers of cells to be assayed, which would be likely to exceed the scoring capacity of any single laboratory and cannot, as yet, be satisfactorily undertaken on a routine basis, even with automated scoring techniques.

55. A number of *in vitro* studies of chromosome aberrations published in the 1970s and 1980s gave data that could be fitted with a linear dose-response relationship, although other functions were also reported. Thus, Luchnik and Sevan'kaev [L13] reported a plateau in the dicentric response to gamma rays at doses between 100 and 300 mGy (low-LET), and Kučerová et al. [K9] produced dicentric data for x rays, which might have indicated a threshold at about 150 mGy, although the authors interpreted the data with a linear function. Lloyd et al. [L9] found a linear response in the lower dose region for x rays down to 50 mGy for both dicentrics and total aberrations. Wagner et al. [W1] found a linear-quadratic response using doses in the range 50–500 mGy from 220 kVp x rays. One study by Pohl-Rüling et al. [P15] reported that 4 mGy of 200 kVp x rays produced a significant reduction in aberration frequencies below the control value until doses of 20 mGy and above were received. This was interpreted as evidence for the stimulation of repair mechanisms at doses below a few tens of milligrays. It was only when repair processes were overwhelmed that aberration yields rose following a linear-quadratic response. It was subsequently noted that if the control aberration yield from a similar experiment designed to examine the effect of D-T neutrons (and in which one of the two controls was common to the x-ray study) [P23], then the yield for 4 mGy of x rays was no longer significant [E15].

56. To provide better data on the response at low doses, a coordinated project was carried out by scientists in six laboratories They collaborated in a study to examine the yield of unstable aberrations induced in peripheral blood lymphocytes *in vitro* by x rays [L8]. The study covered doses of 0, 3, 5, 6, 10, 20, 30, 50, and 300 mGy. Cells from 24 donors were examined, and a total of about 300,000 metaphases were scored. Aberration yields significantly in excess of control values were seen at doses down to 20 mGy, and the dose-response data at low doses were consistent with a linear extrapolation from higher doses. The overall dose response up to 290 mGy was fitted with a linear-quadratic dose-response relationship of the form $I = C + \alpha D + \beta D^2$, where I is the incidence of dicentrics, C is a constant equal to the spontaneous dicentric yield, and α and β are the coefficients for the linear and quadratic terms as a function of the dose D (in gray). Values of $C = 0.0012 \pm 0.002$, $\alpha = 0.027 \pm 0.012$, and $\beta = 0.044 \pm 0.042$ ($\chi^2 = 5.2$ for 5 degrees of freedom, df) were obtained as best fits to the data. At doses below 20 mGy, the observed dicentric yields were generally lower than background, but not significantly so. Excess acentric aberrations and centric rings, in contrast, were higher than controls, although the increase was not statistically significant. A number of uncertainties associated with this type of analysis were described in the paper, including differences in scoring by the participating laboratories, and it was concluded that the statistical uncertainties were such that it is unlikely that this technique would ever allow the response for aberrations to be directly measured at doses much below 20 mGy.

57. The complete set of dicentric data published in the paper have been subject to further analysis to determine the extent to which other models could fit the dose-response

information obtained in the study [E2]. A threshold-linear dose response of the form I = C + α(D – D$_0$) has been used for the analysis in which I is the incidence of dicentrics at dose D, in gray, and D$_0$ is the threshold dose. With values of C = 0.0013 and α = 0.040, the best estimate of the threshold, D$_0$, was 0.0097 ± 0.0045 Gy (±1 SE) (χ^2 = 4.0 for 5 df). It may be concluded, therefore, that while the data can be reasonably fitted with a simple linear-quadratic function, the possibility of a threshold for doses up to about 10 mGy cannot be excluded.

58.   It has been found that the yield of aberrations following a given radiation dose can be influenced by an earlier radiation exposure. Some of the earliest studies on this so-called adaptive response were carried out in human lymphocytes that had incorporated tritiated thymidine [O6]. The cells were exposed to chronic, low doses from tritium in culture and were subsequently exposed at the relatively high dose of 1.5 Gy from x rays. Approximately half as many chromosome aberrations were induced in cells that had incorporated thymidine as in those that had not. This observation was repeatable, and subsequent experiments showed that exposure to tritium need not be chronic [W14] and that pre-exposure to 10 mGy of x rays could also cause the lymphocytes to become adapted [S25]. Subsequent work showed that the response to low doses took several hours to fully manifest itself [S26] and that it depended on synthesis of proteins (possibly an enzyme), which was inhibited by the addition of cycloheximilidine 4–6 hours after the 10 mGy priming dose of x rays [W15]. It has been postulated that stimulation of the synthesis of enzymes responsible for DNA repair is the key factor in this response [U2]. This type of an adaptive response has been observed in other cellular systems, as for somatic mutations, but it has been most comprehensively studied in human lymphocytes [U2, W13]. This issue is considered further in Chapter IV.

59.   The measurement of unstable aberrations in blood lymphocytes has its limitations as a biological dosimeter, because the incidence of dicentrics, rings, and other aberrations decreases with time. In recent years stable chromosome aberrations have been extensively studied, as they provide a method for assessing exposures that occurred some years previously. Some data have been published on dose-response relationships for stable aberrations using fluorescent *in situ* hybridization techniques. However, these aberrations have a higher background yield, which increases and becomes more variable with age and lifestyle of the individual [R21]. Cumulative background radiation exposure accounts for only a small part of the increased frequency with age; clastogenic physiological processes of normal ageing are more important [H34, L50]. This higher and inherently more variable background of stable translocations means that it has not been possible to measure a significant increase in response at doses below 200–300 mGy [G13, L38, N7]. Stable aberrations are, therefore, of little value at present in obtaining information on the shape of the dose response at low doses.

60.   Recently chromosome aberrations have been used to examine the effects of radiation in a high-background-radiation area (HBRA) in China and in a control area [J9]. The level of radiation in the high-background-radiation area was 3 to 5 times higher than that in the control area. Overall the cumulative doses in 39 individuals ranged from 31 to 360 mGy for high-background-radiation area and 6.0 to 59 Gy for the controls. The frequency of dicentrics and ring chromosomes (unstable aberrations) increased in proportion to the cumulative dose in the high-background-radiation area group. Such a dose-response relationship was not clear for those in the control area. The increase in the frequency of these aberrations at such an extremely low dose rate suggested that there is no threshold dose for the induction of chromosome aberrations. In contrast, in the case of translocations any effect of radiation, at up to 3 times control levels, could not be detected against the background incidence [H33].

61.   For high-LET radiation the dose response obtained following exposure *in vitro* both to alpha particles and to neutrons is generally well fitted with a linear response up to doses of around 1.5 Gy [E8] (Figure V). The lowest doses used in these studies were about 50 mGy (high-LET).

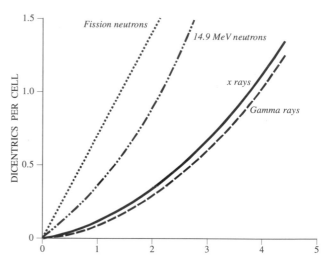

Figure V. Yield of dicentrics in human lymphocytes as a function of dose for some photon and neutron radiations [E8].

2. Cell transformation

62.   Cell transformation systems *in vitro* have been widely used to study the initial stages of oncogenesis. Cell transformation describes the cellular changes associated with loss of normal homeostatic control, particularly of cell division, which ultimately results in the development of a neoplastic phenotype. They are considered to be the closest *in vitro* model for carcinogenesis. The only biological endpoint generally accepted as being definitive of oncogenic transformation is the growth of malignant clones in "nude" or immunologically suppressed host animals. As it is not practicable to screen every transformed cell in this way, other endpoints are normally used, such as enhanced growth rate;

lack of contact inhibition and indefinite growth potential; anchorage-independent growth; and the ability to grow in less nutritious media. The specific criteria used to define oncogenic transformation depend on the particular system that is being used.

63. The most common cell transformation systems such as the BALB/c3T3 and the C3H10T½ mouse-embryo-derived lines are based on cell lines derived from rodent fibroblasts. There are disadvantages to using such models for carcinogenesis in humans, and these were reviewed by the Committee in the UNSCEAR 1993 Report [U3]. When transformed C3H10T½ cells are innoculated into suitable hosts, they form fibrosarcomas. These are not typical of the tumour types that arise in humans after exposure to radiation, which are mainly epithelial in origin. Ideally, more relevant cell lines based on human epithelial tissues are needed for studying the mechanisms of tumorigenesis and dose-response relationships, but they have proven to be much more difficult to develop.

64. Both carcinogenesis and transformation are multi-stage processes, although transformation *in vitro* is normally studied in cells that have already undergone one or more of the possible steps involved, in particular, immortalization. Too much reliance on studies of rodent cell lines can, however, lead to errors in interpretation, if directly applied to humans. Thus, a correlation between anchorage-independent growth and the tumorigenic phenotype has been established in rodent cells [F10, O1, S8], which has permitted the selection of neoplastically transformed cells by growth in soft agar. This does not, however, apply to cultured human cells, as normal human fibroblasts are capable of anchorage-independent growth when cultured in the presence of high concentrations of bovine serum.

65. Despite such limitations, a number of characteristics of *in vitro* cell transformation have allowed their use as model systems for studying the early stages of radiation carcino-genesis *in vivo*. These have been summarized by Little [L4] and include a high correlation between animals and cell transformation systems for carcinogenicity of many chemicals; the response of transformed cells to initiation and promotion similar to two-stage carcinogenesis in the tissues of experimental animals; and the provision of quantitative information on the conversion of normal to tumour cells. Cell-based transformation systems should be free of the influence of hormonal and immunological factors, although cell-cell interactions are still possible.

66. Both BALB/c3T3 and C3H10T½ cell lines have been used to measure the oncogenic effects of ionizing radiation (see, for example, [H1, H18, M3]) and chemical agents (see, for example, [B25, R13]), as well as to screen for possible carcinogenic agents (see, for example, [S9]). Dose-response relationships for cell transformation following exposure to low-LET radiation were reviewed in the UNSCEAR 1986 and 1993 Reports [U3, U5] and by Barendsen [B3]. The pattern of response is very dependent on cell-cycle kinetics; nevertheless, in carefully controlled experiments, the results from

transformation studies on dose and dose-rate effects agree closely with the results obtained with other cellular effects. There are, however, limitations to the sensitivity at low doses and dose-response data for low-LET radiation are generally available down to doses of around 100 mGy (see, for example, [M35, M36]). Above 3 Gy, cell reproductive death starts to predominate over the transformation frequency per plated cell [B2, B3, H1].

67. For transformation by low-LET radiation, various dose-response relationships have been reported. A linear dose response has been described by a number of authors (see, for example, [B33, H19, H27]), while linear-quadratic or curvilinear relationships have been described by others (see, for example, [B30, H19, H24, M15]). Balcer-Kubiczek and Harrison [B29] reported a linear dose response for the induction frequency, IF, of cell transformation in C3H10T½ cells exposed to single doses of x rays (4 Gy min^{-1}) between 250 mGy and 2 Gy (Figure VI), described by IF = 2.50 ± 0.11 10^{-4} Gy^{-1}. The overall fits to the data were evaluated by comparing χ^2 values. The addition of a quadratic function was found not to be justified by least-squares fitting. For continuous exposures over 1 hour or 3 hours, linear responses were also obtained but with a reduced transformation frequency described by 1.5 ± 0.03 10^{-4} Gy^{-1} and 0.87 ± 0.5 10^{-4} Gy^{-1}, respectively. In this study, no transformation was observed in unirradiated cultures. The laboratory estimate for transformation was less than 0.81 ± 0.04 10^{-5} transformants per viable cell; thus it is reasonable to assume that even the lowest transformation frequency obtained at 250 mGy was due to radiation exposure.

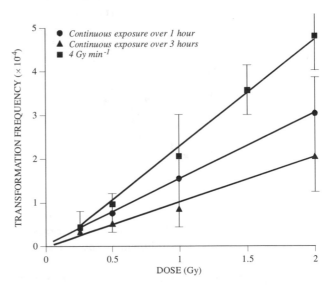

Figure VI. Transformation frequency in C3H10T½ cells following exposure to a single or protracted doses of x rays [B29]. *Lines are fits to the data with a linear dose-response function.*

68. Little [L4] compared results from BALB/c3T3 and the C3H10T½ cell lines. Following exposures between 100 mGy and 3 Gy, the dose response for the BALB/c3T3 cells was nearly linear but that for the C3H10T½ cells could be represented by a linear-quadratic or quadratic relationship (Figure VII).

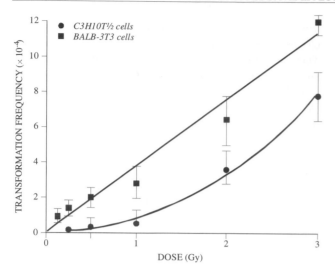

Figure VII.   Transformation frequency in C3H10T½ cells and BALB-3T3 cells following exposure to single doses of x rays [L4].

69.   Miller et al. [M35, M36] measured the effect on C3H10T½ cells of x-ray doses down to 100 mGy delivered just 24 hours after seeding. They found a plateau in the incidence of transformants per surviving cell between about 300 mGy and 1 Gy, which may have reflected the fact that the cells had not achieved asynchronous growth at the time of exposure.

70.   In a major study, six European laboratories collaborated in a study that was specifically designed to address the issue of the dose response at low doses of low-LET radiation with the C3H10T½ transformation system [M14]. Considerable effort went into standardizing techniques in the different laboratories and carrying out extensive intercomparison exercises. One laboratory carried out all the irradiations, and care was taken to ensure that transport conditions did not interfere with the assays. Dose-response data were obtained for exposure to 250 kVp x rays at dose intervals from 0.25 to 5 Gy, and a total of 51,000 petri dishes (of 55 cm^2) were scored. In total, 759 transformed loci were obtained, far in excess of the numbers reported in any other study involving low-LET radiation and the C3H10T½ cell transformation system.

71.   The combined data are shown graphically in Figure VIIIa. A regression fit to the data on transformation induction frequency, IF, between 250 mGy and 5 Gy gave a linear fit of the form IF = (0.83 ± 0.08) 10^{-4} Gy^{-1}. A fit using a linear-quadratic relationship resulted in a non-significant value for the dose-squared term. The authors concluded that the data supported a linear dose-response relationship for cell transformation *in vitro* and that there was no evidence for a threshold dose. A presentation of these data in terms of the numbers of cells at risk might be more relevant. The data from Figure VIIIa are therefore replotted in Figure VIIIb, showing the transformation frequency per cell at risk. This follows the standard bell-shaped curve with a fall in frequency at doses above about 2 Gy, reflecting the effect of cell killing. At the lower end of the curve the response is similar to that shown in Figure VIIIa.

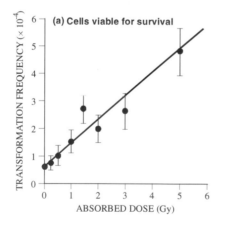

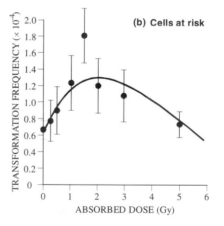

Figure VIII.   Transformation frequencies in C3H10T½ cells following exposure to 250 kVp x rays at 2 Gy min^{-1} [M14].

72.   These experiments were carried out under as near-identical conditions as possible in the participating laboratories and probably represent the optimum conditions that could be achieved for this type of experiment. The results support a linear dose-response relationship for cell transformation *in vitro* at low doses and do little to support the concept of either a threshold dose or an enhanced supralinear response. Nevertheless, the lowest dose at which effects could be detected was 250 mGy, clearly demonstrating the limitations of the technique for assessing dose-response relationships at low doses. Because of the large amount of scoring needed, it would not have been practicable to obtain information at appreciably lower doses.

73.   An adaptive response to low doses of gamma radiation that reduces the effectiveness of a subsequent challenge dose in inducing spontaneous neoplastic transformation has been reported for C3H10T1/2 cells [A20]. In a subsequent study the same group reported that doses of 1–100 mGy from gamma radiation resulted in a suppression of the transformation frequency of C3H10T1/2 cells to levels below that seen for spontaneous transformation of unirradiated cells [A21]. Similar results have been obtained with HeLa x skin

fibroblast human hybrid cells [R22]. In the latter study the frequency of transformation of unirradiated cultures was compared with that of cultures irradiated with 10 mGy from gamma radiation and either plated immediately or held for a further 24 hours at 37°C prior to plating. Pooled data from a number of studies indicated an adaptive response in the case of post-irradiation holding, although the results of four individual studies were quite variable.

74. Exposure to high-LET radiation results in a higher transformation frequency than exposure to low-LET radiation, with a general tendency towards a linear dose-response relationship, but with a tendency to plateau and then fall at high doses (see, for example, [H19, H25, M15, M16]). There is no tendency for the response per unit dose to decrease at low doses or low dose rates, although a number of studies have shown an enhanced effect. As described in the UNSCEAR 1993 Report [U3], the main evidence for this enhanced effect is restricted to 5.9 MeV or fission neutrons, and in more recent studies with monoenergetic neutrons of various energies (see, for example, [M37, M38]) the magnitude of this so-called inverse dose-rate effect has been reduced from a factor of around 9 to a factor of 2 or 3 and has been shown to be radiation-quality dependent. A model has been developed that can satisfactorily explain many experimental results showing this enhanced dose-rate effect [B26, H20].

75. Recently, Miller et al. [M40] carried out a detailed analysis of the effect of transformation of C3H10T½ cells by alpha particle irradiation using charged particle microbeams. Cells in a monolayer in a cell culture dish were irradiated in turn under a highly collimated shuttered beam of alpha particles. The technique permitted measurement of the oncogenic potential of a single or a fixed number of alpha particles passing through a nucleus. The nucleus of each cell was exposed to a predetermined exact number of alpha particles with energy similar to that of radon decay progeny. In parallel with these microbeam studies, "broad beam" alpha particle exposures were also carried out such that cell nuclei received different fluences of alpha particles with mean numbers of 0, 1, 2, 6, and 8. In this case the actual number of "hits" of each cell was determined by Poisson statistics. Thus, if the mean number of traversals of a cell is 1, then 37% are not traversed at all, 37% once, and the remainder are traversed by two or more alpha particles.

76. The authors reported that the measured oncogenicity from exactly one alpha particle was significantly less than from a Poisson distributed mean of one alpha particle, implying that cells traversed by multiple alpha particles were more likely to be subject to transformation. Transformation frequencies for an exact or mean number of one alpha particle per cell were 1.2 10^{-4} and 3.1 10^{-4}, respectively. The incidence of transformations in the exact single-traversal cells was not significantly different from that in the zero-dose (sham) irradiated cells (0.86 10^{-4}). The result was taken to imply that the majority of the yield of transformed cells following irradiation with a mean of one alpha particle per cell must come from the minority of cells subject to multiple

traversals. While these results suggest a non-linear response at low doses and that the risk is less than might be expected for single-track traversals on the basis of a linear dose response, these conclusion were based on only a single result for each exposure condition and need further replication before any confidence can be placed in them. Nevertheless, they have demonstrated a unique approach to examining the carcinogenic potential of alpha particles at low doses.

3. Mutagenesis in somatic cells

77. The principal mechanism resulting in a neoplastic initiating event is induced damage to DNA, which predisposes target cells to subsequent malignant development (see Chapter IV). There is also strong evidence linking a number of tumours to specific gene mutations. An understanding of the dose-response relationships for this initial mutational change is relevant to an assessment of the effect of low doses on tumour induction. The experimental data have been reviewed by Thacker [T2] and are considered in detail in Annex F, "DNA repair and mutagenesis". Considered here is information on dose-response relationships for somatic mutation induction resulting from exposure to both low- and high-LET radiations.

78. A range of mutation systems have been described in the literature, but only a few are sufficiently well defined for quantitative studies. There are also a number of difficulties in interpreting results from somatic cell systems. In particular, the mutation frequency of a given gene is to some extent modifiable, depending on the exact conditions of the experiment. It may also be that a period of time is needed for the mutation to manifest itself. Thus, the true mutation frequency may be difficult to determine, and this can present particular difficulties in studies of dose-response relationships. Several established cell lines, derived from mouse, hamster, or human tissue, have also been used to measure mutation frequencies at different doses and dose rates. Because the cells lines used experimentally can have sensitivities that depend on the stage of the cell cycle, to ensure as consistent a response as possible, it is preferable to use a stationary culture in plateau phase in which only a limited number of the cells will be cycling in the confluent monolayer.

79. The mutation of a single gene is a relatively rare event; the majority of experimental systems are therefore designed to select out cells carrying mutations. Commonly used systems employ the loss of function of a gene product (enzyme) that is not essential for the survival of cells in culture. Thus, cells may be challenged with a drug that they would normally metabolize with fatal consequences. If mutation renders the gene product producing the specific enzyme ineffective, the cell will survive, and thus the mutation frequency can be obtained by measuring the survivors. Frequently used examples of such a system are those employing the loss of the enzyme hypoxanthine-guanine phosphoribosyl transferase (HPRT), which renders cells resistant to the drug 6-thioguanine (6-TG), and of the enzyme thymidine kinase (TK), which gives resistance to trifluorothymidine (TFT). HPRT activity is specified by an X-linked gene,

hprt, while TK is specified by an autosomal gene *tk* and therefore has to be used in the heterozygous state.

80.   Mutation induction in a human lymphoblastoid cell line (TK$_6$) after acute x ray and continuous low-dose-rate gamma irradiation was investigated using the *hprt* and *tk* mutation assays [K20]. The TK$_6$ cells are radiosensitive, and increases in mutation rate for both 6-TK and HPRT were obtained at acute x-ray doses from 250 mGy to 1 Gy, with a response that could be fitted with a linear function down to zero dose. At high doses (1.5 and 2 Gy), mutation data were difficult to obtain because of a very low surviving fraction (1%–4%). Mutation frequency after continuous gamma irradiation could also be fitted with a linear response and with mutation rates at both 27 mGy h^{-1} and 2.7 mGy h^{-1} that did not differ significantly from those for acute exposure.

81.   Evans et al. [E7] examined the effect of dose and dose rate on the mutation frequency at both the *tk* and *hprt* loci in two variants (LY-S1 and LY-R16) of mouse lymphoma L5178Y cells. Mutation at the *tk* locus, resulting from x-ray exposure, was dose-rate-dependent in the LY-R16 variant but not in the LY-S1 variant. This was thought to reflect the deficiency of DNA double-strand break repair in LY-S1. In contrast, with the *hprt* locus, mutation was dose-rate-independent in both strains. The results suggested that mutation at the *hprt* locus is caused by single lesions, with dose-rate-independent repair, whereas for the *tk* locus, interaction of DNA damaged sites is important. In both cases, however, an increased incidence of mutations was obtained at doses down to about 500 mGy, the lowest dose tested. The data could be fitted with a linear dose response, with no threshold up to 3 Gy for LY-R16 cells and 2 Gy for LY-S1 cells.

82.   Induction of mutation to 6-TG resistance was examined in a radiation-sensitive mutant strain of mouse leukaemia cells following gamma irradiation at dose rates of 30 Gy h^{-1}, 200 mGy h^{-1}, and 6.2 mGy h^{-1} [F14]. The mutation frequency increased linearly with increasing dose for all dose rates, with no significant difference between the responses at any of the dose rates. The lowest dose tested was 250 mGy.

83.   A particularly sensitive mutation assay has been described using the pink-eyed unstable (p^{un}) mutation in the mouse [S23]. This causes a reduction in the pigment in coat colour and eye colour as a result of a gene duplication and reverts to wild type by deletion of one copy. Reversion events are assayed as black spots on the grey coat. The reversion frequency of p^{un} is at least five orders of magnitude greater than that of other recessive mutations at other coat colour loci. Female mice, homozygous for the reversion, were irradiated with various doses of x rays between 10 mGy and 1 Gy and the frequency of reversions measured. Even at a dose of 10 mGy, the incidence of black melanosome streaks was increased threefold. There was a linear dose response over the dose range examined, with no indication of a threshold (Figure IX).

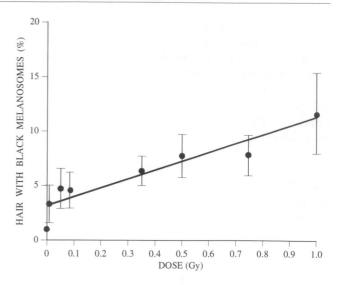

Figure IX.   Dose response for x-ray induced reversion of pink-eyed unstable mutations in mice [S23].

84.   Albertini et al. [A10] compared the effects of ^{137}Cs gamma rays and alpha irradiation from ^{222}Rn on *hprt* mutations induction in human T cells *in vitro*. For gamma-ray doses between 500 mGy and 4 Gy, the dose response could be represented by either a linear or linear-quadratic relationship, with a significant increase in mutations at the lowest dose. The doubling dose for mutation induction was calculated to be about 0.8 Gy. For alpha particle irradiation, a linear regression gave a best fit to the data for doses between about 0.25 and 0.9 Gy. In this case the doubling dose was calculated to be about 0.2 Gy. This would suggest a relative biological effectiveness (RBE) of about 4 for alpha particle irradiation compared with low-LET radiation.

85.   Further comparative data on mutagenesis in human lymphoblastoid cells have been published by Amundson et al. [A11, A12]. Despite being derived from the same donor, two cell lines, WTK1 and TK6, have very different responses to radiation. Alpha particles from a ^{238}Pu source produced about four to five times more *hprt* mutants per gray in WTK1 cells than did x rays. On the other hand, there was little difference between the dose-response curves for the TK6 cells, although alpha particles were somewhat more effective. In contrast, the *tk* locus was only slightly more sensitive to alpha particles than to x rays, although the WTK1 cells were considerably more sensitive than the TK6 cells. The authors considered that the results suggested that WTK1 cells have an error-prone repair pathway that is either missing or deficient in TK6, and they further suggested that this pathway might be involved in the processing of alpha-particle-induced damage. In all these studies the data could be fitted with a linear dose-response function, with no threshold, although the lowest exposure dose used was about 200 mGy for alpha particle irradiation.

86.   Data on the response of TK6 cells exposed to the alpha emitter ^{212}Bi have been reported by Metting et al. [M41]. The radionuclide was added directly to the cell suspension as a chelate complex. The incidence of mutations was a simple linear function of the dose from 200 mGy up to 800 mGy. Induced mutant frequencies were 2.5 10^{-5} Gy^{-1} at the *hprt* locus and 3.75 10^{-5} Gy^{-1} at the *tk* locus.

87. An adaptive response of the kind observed in blood lymphocytes [W6, W15] has also been demonstrated to have an influence on mutation induction. Thus, experiments with the *hprt* locus in human lymphocytes have shown that exposure to tritiated thymidine [S27] or 10 mGy from x rays [K21] can markedly decrease the number of mutations induced by subsequent high doses of radiation. The response to low doses from x rays eliminated about 70% of the effect of a challenging dose of 3 Gy.

88. Similarly, when SR-1 mammary carcinoma cells were irradiated with 10 mGy from x rays and subsequently challenged 18 or 24 hours later with 3 Gy from x rays, approximately one half as many mutations were induced as when the cells were irradiated only with 3 Gy [Z3]. Furthermore, the rate of repair of DNA double-strand breaks, which are the lesions responsible for chromosomal breaks, increased in cells that had been pre-exposed.

89. Wolff [W13] pointed out that the data on mutation induction illustrate two aspects of adaptation that are similar to that observed in blood lymphocytes: after the initial dose it takes time for the induction to occur, and once induced it disappears with time. The SR-1 data show that in this system the effect takes more than 6 hours to become effective and it then disappears if 48 hours elapse between the two doses.

90. In addition to being studied in mammalian systems, radiation-induced mutations have also been studied in plant cells. Mutations can occur in stamen hairs in tradescantia that result in the normal dominant blue colour being replaced by recessive pink. This is a sensitive system for detecting effects at low doses. Dose-response curves for pink mutations have shown for 250 kV x rays a linear response between 2.5 mGy and 50 mGy and for neutrons (0.43 MeV) a linear response between 0.1 mGy and 80 mGy [S28]. In neither case was evidence for a threshold obtained.

D. SUMMARY

91. Damage to DNA in the nucleus is considered to be the main initiating event by which radiation causes damage to cells that results in the development of cancer and hereditary disease. Information on the effects of radiation on individual cells can, therefore, provide insight into the fundamental damage that may ultimately give rise to cancer or hereditary disease. It can also provide information on the consequences of damage to other cellular structures, such as the cellular membrane and the cytoplasm, although damage here is less significant in terms of long-term health effects.

92. Double-strand breaks in DNA are generally regarded as the most likely candidate for causing the critical damage to the nucleus that can subsequently manifest itself as a mutation in somatic or germ cells. Single radiation tracks have the potential to cause double-strand breaks and in the absence of 100% efficient repair could result in long-term damage, even at the lowest doses.

93. In examining the effects of radiation at low doses it has been appropriate to consider how they should be defined. A number of physical and biological approaches have been examined for designating low doses and low dose rates. Microdosimetric arguments suggest low doses will be less than 1 mGy. However, radiobiological experiments on cells in culture suggest that acute doses of about 20 mGy are low, while epidemiological studies suggest that a low dose is of the order of 200 mGy, whatever the dose rate. In addition, studies of tumour induction in experimental animals suggest that dose rates of about 0.1 mGy min^{-1} are low, whatever the total dose.

94. A range of assays is available for evaluating radiation damage to cells occurring either *in vivo* or *in vitro* and the form of the dose-response relationships. In the low-dose region, damage from low-LET radiation can be considered to be due to single tracks acting independently, whereas at higher doses multi-track effects can occur, causing non-linearity in the dose response. In the case of high-LET radiation, the dose response is generally found to be linear.

95. The results of studies of the induction by low-LET radiation of chromosome aberrations in blood lymphocytes, of the transformation of cells in culture, and of somatic mutations in mammalian cell systems at low doses all give results that are somewhat variable. They depend on the experimental design and the effort that went into assessing risks at doses of less than about 1 Gy. In the most comprehensive studies the results are consistent with an increasing incidence with increasing dose at low to intermediate doses. Nevertheless, even very extensive studies, which have taken considerable resources, have demonstrated that it is not practical to obtain information on radiation effects at doses much below about 20 mGy for chromosome aberrations, 100 mGy for cell transformations, and 200 mGy for somatic mutations. The exact form of the response for cellular effects at low doses must therefore remain unclear. One exception is a particularly sensitive system based on the assay of reversion events in a gene mutation in pink-eyed unstable p^{un} mutations in the mouse, which cause a reduction in coat colour. A linear dose response, with no indication of a threshold, was obtained over the dose range for x rays from 10 mGy to 1 Gy. Similarly, pink mutations have been induced in tradescantia stamen hairs at doses down to 2.5 mGy from x rays and with a linear response up to about 50 mGy.

96. For high-LET radiation the experimental data again indicate a linear dose response down to the lowest doses that have been tested. In the case of chromosome aberrations, the lowest neutron dose used was about 50 mGy. For mutation induction, little information is available from mammalian cellular systems at doses much below 200 mGy.

97. The relative biological effectiveness (RBE) of high-LET radiation compared with low-LET radiation varies considerably depending on the biological damage and the dose range. In the case of deterministic effects, caused by cell killing, RBE values are generally less than 10. For stochastic effects, values of RBE depend on the dose, reflecting an essentially linear dose response for high-LET

radiation and a linear-quadratic response for low-LET radiation. The maximum value, RBE$_m$, occurs at low doses. For exposures to both neutrons and alpha particle irradiation, values of RBE$_m$ depend on the biological endpoint but for cytogenetic damage, cell transformation, and tumour induction are generally greater than 10.

98. A so-called adaptive response has been observed for a number of indicators of cellular damage: a small radiation dose reduces the amount of cellular damage caused by a later higher dose. For the induction of unstable chromosome aberrations and of mutation, it has been demonstrated that a

small priming dose of low-LET radiation of about 10 mGy can reduce the effect caused by a subsequent higher dose. This adaptive response seems to take a few hours to manifest itself and then lasts for up to about 40 hours. There are no indications that this would modify the shape of the dose response, although it could alter the magnitude of any effect.

99. The information on the lowest doses at which effects from low-LET radiation have been detected in cellular systems are summarized in Table 5. These are principally for endpoints arising from mutations.

Table 5
Lowest doses at which chromosome aberrations and mutations have been detected in experimental systems exposed to low-LET radiation

System	Endpoint	Radiation	Lowest dose [a] (mGy)	Paragraph [b]	Ref.
Human lymphocytes	Unstable chromosomal aberrations	x rays	20	56	[L8]
Human lymphocytes	Stable chromosomal aberrations	gamma rays	250	59	[L38]
C3H10T½ cells	Cells transformation	x and gamma rays	100	66/69	[M35]
Mouse	Pink-eye mutation	x rays	10	83	[S23]
TK$_6$ cells	*hprt* and *tk* mutation	x rays	250	80	[K20]
Tradescantia	Pink mutation	x rays	2.5	90	[S28]

a   Acute exposures (minutes).
b   Text paragraph in which endpoints are further discussed.

II. ANIMAL EXPERIMENTS

100. Studies with experimental animals are important for predicting the long-term effects of radiation in humans. Provided the limitations of such studies are acknowledged, considerable information can be obtained on both radiation-induced cancer and hereditary disease. The most directly relevant data come from studies with mammalian species, with the majority of work relevant to assessing effects at low doses and low dose rates having been carried out with rodents and beagle dogs.

101. Studies with experimental animals are valuable for examining the biological and physical factors that may influence tumour induction by radiation. They can be used to examine the form of dose-response relationships over a wide range of doses; the effect of spatial and temporal distribution of dose; and the influence of factors such as sensitivity of individual organs and tissues, age at exposure, radiation quality, and dose protraction or fractionation on the tumour response. Animal models are also of increasing value in understanding the molecular and cellular mechanisms underlying tumour response (Chapter IV). Quantitative risk coefficients for radiation-induced cancer in humans cannot, however, be based on the results of animal studies because there are differences in radiation sensitivity between different mammalian species.

102. Experimental studies of genetic damage in the offspring of irradiated animals, mainly mice, have been used to assess the hereditary effects of radiation. In the absence of any clear evidence from observations in humans on the risks of radiation-induced hereditary disease, animal studies provide information on dose-response relationships as well as quantitative risk estimates. Studies of germ-cell mutations are also relevant to understanding the dose-response relationship for the initial damage to DNA that could ultimately result in the development of cancer.

A. CANCER

103. Data on radiation-induced tumours in experimental animals were extensively reviewed in the UNSCEAR 1977 and 1986 Reports [U5, U7], by NCRP [N1], by Upton [U22], and in a comprehensive monograph on radiation carcinogenesis [U23]. The effect of dose rate on tumour response was examined by NCRP [N1] and in the UNSCEAR 1993 Report [U3]. As noted in the UNSCEAR 1993 Report, the experimental animals used in many studies are inbred strains with patterns of disease that can be very different from those found in humans. Very many studies have used rodents as the experimental animal. Different strains of mice and rats have varying susceptibilities

to both spontaneous and radiation-induced tumours; furthermore, within a given strain, there are frequently differences between the sexes and ages in the incidence and time of onset of specific tumour types. A number of tumour types for which information is available are either not found in humans (e.g. Harderian gland) or appear to require substantial cell killing for their development and thus may exhibit a threshold in the dose response (e.g. ovarian tumour, thymic lymphoma). For a number of other tumours there may be a human counterpart (e.g. myeloid leukaemia and tumours of the lung, breast, pituitary, and thyroid), but even here there can be differences in the cell types involved and in the development of the tumour. Although data for larger animals are not as extensive, broadly similar findings are found for tumour induction in dogs or other species.

104. There are also substantial variations in the rates of turnover of cells and in the lifespan of most experimental animals compared with humans. Furthermore, the development of tumours in both humans and animals is subject to the modifying influence of various internal and external environmental factors, all of which can potentially influence dose-response relationships. Their development will also depend on the genetic background, the physiological state, and the environmental conditions of the animals. All these factors make it difficult to interpret the results of animal studies and to apply them to humans. Nevertheless, most tumours in laboratory animals appear to arise as clonal growths and to

develop, as do most human tumours, through stages of initiation, promotion, and progression. They are therefore of considerable value for helping to understand the form of the dose response for tumour induction in humans and the potential for effects at very low doses.

105. While extensive data exist on tumour induction in laboratory animals exposed either to external radiation or to incorporated radionuclides, many of these studies were carried out in the 1960s and 1970s. The ability to detect radiation-induced cancer at low doses depends, as with epidemiological studies, on the number of animals in the study, the spontaneous incidence of the disease, and the radiation sensitivity of the particular tumour type(s). This is illustrated in Table 6. Thus, in the CBA strain of mouse with a very low spontaneous incidence of acute myeloid leukaemia ($1\ 10^{-4}$) and a high sensitivity to induction by radiation (about $1\ 10^{-1}\ Gy^{-1}$), only 300 exposed animals and a similar number of controls would be needed to detect a significant increase (p=0.05) in tumour incidence at a whole-body dose of 100 mGy (low-LET). It would also be possible to detect an effect of 10 mGy with groups of about 4,000 animals. In contrast, for the RFM mouse strain, which has a much higher spontaneous rate of the disease ($7\ 10^{-3}$) and a lower sensitivity ($7\ 10^{-3}\ Gy^{-1}$), the number of animals needed to detect a significant increase in acute myeloid leukaemia at 100 mGy would be $1.2\ 10^{5}$, an impractically high number of animals.

Table 6
Statistically determined sample sizes of irradiated and control mice needed to detect a significant increase in tumour risk [a]

Mouse strain	Tumour	Sample size			
		1 000 mGy	*100 mGy*	*10 mGy*	*1 mGy*
RFM	Thymic lymphoma [b]	1 300	$1.2\ 10^{5}$	$1.2\ 10^{7}$	$1.2\ 10^{9}$
RFM	Myeloid leukaemia [c]	1 700	$1.2\ 10^{5}$	$1.2\ 10^{7}$	$1.2\ 10^{9}$
CBA	Myeloid leukaemia [d]	30	300	4 000	$1.3\ 10^{5}$

a   p = 0.05.
b   Spontaneous incidence $1.3\ 10^{-1}$; risk of $3\ 10^{-2}\ Gy^{-1}$ assumed.
c   Spontaneous incidence $7\ 10^{-3}$; risk of $7\ 10^{-3}\ Gy^{-1}$ assumed.
d   Spontaneous incidence $1\ 10^{-4}$; risk of $1\ 10^{-1}\ Gy^{-1}$ assumed.

106. At low radiation doses the number of animals used and their sensitivity is thus important in determining the ability to detect any effect. Animal studies do not generally involve as many individuals as there are in the more extensive epidemiological studies. They do, however, have the advantage that they are planned; the groups of animals exposed are, in general, genetically homogeneous; and the numbers of animals allocated to various dose groups can be chosen to maximize the information obtained. Laboratory animals are exposed to sources of radiation under controlled conditions, and there is much greater certainty associated with the dosimetry. Information may also be available from studies of animals exposed at different dose rates. Data on irradiation of laboratory animals can thus give information for a range of tumour types on the shape of dose-response relationships and

provide an estimate of the lowest dose at which a significant effect on the induction of tumours from exposure to ionizing radiation can be observed.

107. Despite a substantial number of research studies on tumour induction in experimental animals potentially available for analysis, there are in practice only a limited number that can help to define the dose-response relationship for cancer induction down to low doses. The range of dose-response relationships that have been obtained in experimental animals and the effect of dose rate on tumour response were reviewed in the UNSCEAR 1993 Report [U3]. Accordingly, only illustrative examples of dose-response relationships are given here for tumour induction following exposure to external radiation and intakes of radionuclides.

1. Dose-response relationships

108. Tumour induction has been demonstrated in laboratory animals exposed to both low- and high-LET radiation. Information on dose-response relationships was previously examined in the UNSCEAR 1986 and 1993 Reports [U3, U5]. In the UNSCEAR 1986 Report, the Committee limited its analysis to those models that appeared to be supported by general knowledge of cellular and subcellular radiobiology. Because most readily induced human tumours, such as those of the breast, thyroid, and lung, as well as leukaemia, did not indicate the existence of a threshold [U5, Annex B, paragraph 108], analyses were confined to the linear no-threshold, the linear-quadratic and the quadratic dose-response relationships. It was considered that these three dose-response relationships provided a general envelope for observation of tumour induction in experimental animals as well as in human populations. In the UNSCEAR 1993 Report [U3], emphasis was placed on examining the effect of dose rate on the tumour response, although information was also given on the dose-response relationships.

109. Dose-response functions other than those adopted in 1986 have also been proposed [E4, G1, U2]. Models that incorporate a threshold assume that there is no response up to some level of exposure, and that thereafter the response increases with dose. Some animals models of tumour induction show this type of response. Dose-response models that incorporate an adaptive response have also aroused some interest. These consider the possibility that stimulated repair of radiation damage as a result of the effect of a toxic agent, including radiation, at low doses would reduce the influence of subsequent, higher doses. The evidence for such an adaptive response was reviewed by the Committee in the UNSCEAR 1994 Report [U2]. Much of the evidence for such a response that is presently available comes from observations of short-term effects in both plants and animals and from studies on cells in culture (Chapter I). Extensive data from animal experiments on dose-response relationships for cancer induction and limited human epidemiological data on low-level exposures have, however, provided no firm evidence that the adaptive response decreases the incidence of late effects such as cancer induction after exposure to low radiation doses. Molecular and cellular studies have shown that DNA damage in the form of double-strand breaks is repairable but that some degree of misrepair is to be expected (Chapter IV). On this basis, it may be concluded that the extent of damage caused by ionization events resulting from exposure to low radiation doses may be influenced by the stimulation of DNA repair mechanisms, but even so, such repair can only be partially effective and for many tumour types cannot entirely eliminate the risk of tumour development following radiation exposure.

110. Published reports of dose-response relationships obtained with various animal species have described responses for different tumour types or life shortening (as a surrogate for tumour induction) using a wide range of functions. Although in many studies dose-effect relationships can be defined by a linear, linear-quadratic, or quadratic response, the data are generally not well defined, particularly at low doses, and alternative fits to the data are also possible. Some animal models also indicate the presence of a threshold for a response. Extensive data are available on a wide range of tumour types including leukaemias and lymphomas arising in haematologic tissue as well as tumours of solid tissues (e.g. lung, liver, and bone). Examples of dose-response relationships for exposures to both external radiation and internally incorporated radionuclides are given below. The studies have been chosen to illustrate the various patterns of dose response that have been obtained with some emphasis on those that give information at low doses.

(a) External radiation

111. *Life shortening.* Extensive studies in laboratory animals have reported life shortening as a result of whole-body external irradiation. This is a precise biological endpoint and reflects the early onset of lethal diseases, an increased incidence of early occurring diseases, or a combination of the two. At radiation doses up to a few gray (low-LET), life shortening in experimental animals appears to be mainly the result of an increase in tumour incidence. There is little suggestion that there is a general increase in other non-specific causes of death [G14, M39, S38]. At higher doses, into the lethal range, a non-specific component of life shortening becomes apparent owing to cellular damage to the blood vasculature and other tissues. It was concluded in the UNSCEAR 1993 Report [U3] that life shortening at low to intermediate doses can be used as a basis for examining the factors that influence dose-response relationships for tumour induction.

112. The majority of comprehensive studies that give quantitative information on dose-response relationships for life shortening from exposure to low-LET radiation as well as on factors such as the effects of age, dose fractionation and dose rate have used the mouse as the experimental animal. Substantial differences in sensitivity have, however, been noted between strains and between the sexes. A review of 10 studies involving about 20 strains of mice given single exposures to x or gamma radiation showed that estimates of life shortening ranged from 15 to 81 days Gy^{-1}, although the majority of values (9 of 14 quoted in the review) were between 25 and 45 days Gy^{-1}, with an overall unweighted average of 35 days Gy^{-1} [G14]. In general, in the range from about 0.5 Gy to acutely lethal doses, the dose response was either linear or curvilinear upwards. In male BALB/c mice exposed to acute doses of ^{137}Cs gamma rays (4 Gy min^{-1}), life shortening was a linear function of dose between 0.25 and 6 Gy, with a loss of life expectancy of 46.2 ± 4.3 days Gy^{-1} [M39]. Similar data have been reported on B6CF3 mice irradiated at 17 days before birth or at various times up to 365 days after birth [S38], although the lowest dose used was 1.9 Gy. The effects of acute single doses on life shortening in other species were summarized in the UNSCEAR 1982 Report [U6], although they are not as comprehensive as the data for mice.

113. The effect of dose fractionation appears to be very dependent on the strain of mouse and the spectrum of dis-

eases contributing to the overall death rate. Overall no clear trend in the effect of dose fractionation on life shortening could be found [U3], and the results from a number of studies suggested that when compared with acute exposures, the effects of dose fractionation are small and in some studies have given either small increases or small decreases in life-span. When the effects in mice of acute exposures to low-LET radiation are compared with those of protracted irradiation given more or less continuously, the effectiveness of the radiation decreases with decreasing dose rate and increasing time of exposure. With lifetime exposures there is some difficulty assessing the total dose contributing to the loss of lifespan. The results available suggest, however, that with protracted exposures over a few months to a year, the effect on life shortening is reduced by factors of between about 2 and 5, compared with exposures at high dose rates.

114. The results of a number of studies on life shortening as a result of exposure to high-LET radiation were examined in the UNSCEAR 1993 Report [U3]. The data were all reasonably consistent and suggest that the dose response for life shortening is a linear function of dose, at least for total doses up to about 0.5 Gy, and that neither dose fractionation or dose protraction has much effect.

115. *Tumour induction.* In the late 1970s, Ullrich and Storer published a series of studies on tumour induction in mice (see, for example, [U16, U17, U18]). The data have provided comprehensive information on the effects of dose and dose rate on the induction of a range of neoplastic diseases, including myeloid leukaemia and solid tumours of the ovary, pituitary, lung, and thymus.

116. In a large study in female RFM mice, animals were exposed to acute doses from ^{137}Cs gamma rays (0.45 Gy min^{-1})

at 10 ± 0.5 weeks of age [U16]. Groups of animals received a range of doses (0, 0.1, 0.25, 0.5, 1.0, 1.5, 2.0, and 3.0 Gy), were followed for their lifespan, autopsied at death, and diagnosed for various types of neoplastic disease. Dose-response data were obtained for a range of tumour types. A significant increase in the incidence, I (%), of acute myeloid leukaemia was obtained at doses of 1.0 Gy and above. A linear dose response of the form I = 0.63 + 1.4D, where D is the dose in gray, adequately described the data, and the doses were not high enough for a cell-killing term to have become apparent. A linear-quadratic model, I = 0.69 + 0.86D + 0.00227D^2, also provided a fit to the data, although the dose-squared term was not significant. Ullrich and Storer [U18] published further data on myeloid leukaemia in female RFM mice exposed under similar conditions. The results were similar to those published earlier, but with fewer exposure points (0, 0.5, and 2.0 Gy).

117. The information on myeloid leukaemia induction in mice for these two data sets has been combined in Table 7. An analysis of the combined data carried out for this Annex indicates that the incidence of myeloid leukaemia is increased over controls at doses of about 0.5 Gy and above. The data have been fitted with linear, linear-quadratic, and threshold-linear dose responses. All three models give a good fit to the data, and in the case of the threshold-linear model, a threshold at about 0.22 Gy can be obtained (Figure X, Table 8). These studies by Ullrich and Storer [U16, U18] involved a total of nearly 18,000 mice, and yet the information at low doses is equivocal because of the small numbers of acute myeloid leukaemias occurring. Few other animal studies have been carried out on such a scale, and this clearly illustrates the limited ability of such animal studies to provide detailed information on the effects of whole-body radiation at low doses.

Table 7
Myeloid leukaemia incidence in female RFM mice exposed to acute doses of gamma rays [U16, U18]

Dose (Gy)	Number of animals	Incidence
0	4 763	0.72 ± 0.10
0.1	2 827	0.72 ± 0.15
0.25	965	0.84 ± 0.30
0.5	1 918	1.17 ± 0.26
1	1 100	1.60 ± 0.41
1.5	1 054	3.6 ± 0.76
2	1 099	3.22 ± 0.43
3	4 133	5.2 ± 0.51
Total	17 859	

Table 8
Model fits to data on myeloid leukaemia in mice exposed to ^{60}Co gamma rays

Function	C	α	β	D_0	χ^2	DF
I = C + αD [a]	0.64 ± 0.09	1.39 ± 0.13	-	-	4.7	6
I = C + αD βD^2 [b]	0.69 ± 0.09	0.87 ± 0.39	0.22 ± 0.15	-	2.7	5
I = C + α(D − D$_0$) [c]	0.72 ± 0.08	1.55 ± 0.18	-	0.22 ± 0.14	2.6	5

[a] Linear. [b] Linear-quadratic. [c] Linear-threshold.

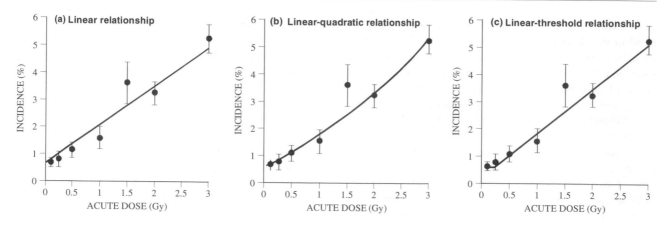

Figure X.   Dose-response relationships fitted to data on myeloid leukaemia in female RFM mice [U16, U18].

118. The induction of lung tumours has been compared in female BALB/c mice given doses from ^{60}Co gamma rays in the range 0.5–2 Gy at two dose rates (0.4 Gy min^{-1} and 0.06 mGy min^{-1}) [U18, U24]. Tumour induction was less at low dose rates than at high dose rates. After high-dose-rate exposure, the age-correlated incidence, I (%), could be represented by a linear function, [I(D) = 13.4 + 12D; p>0.5], where D is the absorbed dose in gray. At low dose rates, a linear function also gave a good fit to the data [I(D) = 12.5 + 4.3D; p>0.8]. The data were adjusted for differences in the distribution of ages at death among the treatment groups, and the authors indicated there were no changes with age over the period of irradiation. The differences in slope were taken to indicate variations in effectiveness for tumour induction at the two dose rates. The data were subsequently extended to provide additional information at the high dose rate (0.4 Gy min^{-1}) in the dose range from 0.1 to 2 Gy [U14]. Although the tumour incidence data could again be fitted with a linear dose response [I(D) = 10.9 + 11D; p>0.70], a linear-quadratic dose response [I(D) = 11.9 + 4D + 4D^2; p>0.70] would also give a fit. In this extended analysis, the linear term was very similar to that obtained in the low-dose-rate study, and it was concluded by the authors that the result was consistent with a linear-quadratic response in which the linear term is independent of dose rate, at least for the dose rates used in the study.

119. One of the most extensively studied tumours in the mouse is that arising in the thymus. The dose response for the induction of thymic lymphomas by acute whole-body irradiation found in a number of studies has been of the threshold type (Figure XI). Thus, Maisin et al. [M39] exposed 12-week-old male mice to single or fractionated doses of ^{137}Cs gamma rays (4 Gy min^{-1}) in the dose range from 0.25 to 6 Gy. The dose-response curve was of a threshold type; the incidence of thymic lymphomas rose above that in controls only following exposures at 4 Gy and above. Similarly, Ullrich and Storer [U18, U24] studied the dose-response relationship for thymic lymphoma in female RFM/Un mice. Exposures were at 0.45 Gy min^{-1} and 0.06 mGy min^{-1}. For the highest dose rate, the incidence of lymphoma up to 0.25 Gy increased with the square of the dose, although a threshold for a response up to about 0.1 Gy could not be excluded. Linearity was rejected over this limited dose range. From 0.5 to 3 Gy

the increase in incidence with dose was nearly linear. At the lower dose rate the response was best described by a linear-quadratic dose response with a shallow (perhaps zero) initial linear slope, again allowing the possibility of a threshold at low total doses.

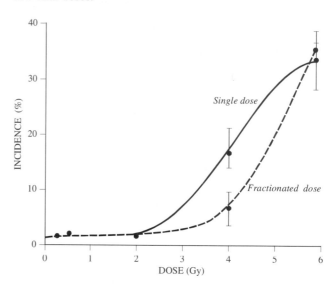

Figure XI.   Incidence of thymic lymphoma as a function of dose for single or fractionated x rays [M39].

120. The induction of ovarian tumours in mice exposed to x rays or gamma radiation has also indicated the presence of a threshold in the response for some strains, and this is reflected in a pronounced effect of dose and dose rate [U17, U18, U21, U24]. Thus, in SPF/RFM mice exposed at 0.45 Gy min^{-1}, a significant increase in tumour incidence was obtained at doses from 0.25 to 3 Gy [U18]. The data could be fitted with a linear-quadratic dose response with a negative linear component [I(D) = 2.3 + (−23)D + 1.8 D^2; p>0.25] or by a threshold plus quadratic model [I(D) = 2.2 + 2.3 (D−D*)^2; p>0.75], where the threshold dose, D*, was estimated to be 0.12 Gy. Linear and quadratic dose responses were rejected. This pattern of response is considered to reflect the fact that ovarian tumour development in the mouse seems to follow changes in hormonal status that occur after substantial killing of oocytes. For low-dose-rate exposures, cell killing is less effective, and as a consequence there is a substantial reduction, by a factor of about 6, of

effectiveness in inducing tumours at low dose rates. The results suggested the possibility of a threshold up to about 0.115 Gy.

121. A further substantial study in the mouse has demonstrated that dose-response relationships for tumour induction can vary in different organs and tissues [S39]. Groups of B6C3F$_1$ mice were exposed to various doses between 0.48 and 5.7 Gy low-LET radiation from ^{137}Cs gamma rays. The dose-response curves for tumour induction in the liver, pituitary, ovary, and lungs were convex upwards in the dose region examined, with a significant increase in numbers of tumour at 0.48 Gy. The data suggested a progressive increase with dose up to about 1 Gy. A subsequent gradual increase to the highest incidence obtained was seen and then a declining incidence at doses above about 1.5 to 3 Gy, depending on the tumour type. The results could be interpreted as showing an increasing risk with dose up to the maximum incidence, although the lack of data below 480 mGy limited the ability to elucidate the dose response at low doses. In contrast, the shape of the dose response for bone tumour induction was quite different from that for other solid tumours: the initial slope was concave upwards, with the highest incidence observed in the group given 3.8 Gy. Bone tumour incidence up to about 3 Gy was a function of the square of the dose, and the existence of a threshold could not be excluded because the incidence of bone tumours in groups irradiated with doses below 1.43 Gy was not significantly increased.

122. Variations in sensitivity to radiation-induced mammary cancers in different strains of mice and rats are well known, although the reasons underlying these differences are not well understood. Thus studies of mammary carcinogenesis in Sprague-Dawley, WAG/Rij, and BN/BiRij rats have shown that only in WAG/Rij rats was an appreciable number of carcinomas induced by radiation [V3]. Analysis of data on radiation-induced mammary tumours gave a linear dose-response function for fibroadenomas in Sprague-Dawley rats and for both fibroadenomas and carcinomas in WAG/Rij rats after irradiation with either 0.5 MeV neutrons or x rays. In the case of exposure to x rays, the lowest data point was at 200 mGy (Figure XII).

123. Studies of mammary tumours in mice by Adams et al. [A13] have demonstrated that irradiation resulted in many more transformed mammary cells than are ultimately expressed as tumours. A later study by Ullrich et al. [U26] examined possible reasons for differences in sensitivity in sensitive BALB/c and resistant C57BL and B6CF1 hybrid mice. They demonstrated that variations in sensitivity could be correlated with differences between strains in the sensitivity of the mammary epithelial cells to radiation-induced trans-formation. Differences in sensitivity could not, however, be accounted for by differences in the number of sensitive cells or by systemic or cellular influences on progression. This observation of inherent differences in sensitivity to radiation-induced tumour initiation may be one approach to understanding the mechanism by which radiation induces cancer in these different mouse strains and may have more general application.

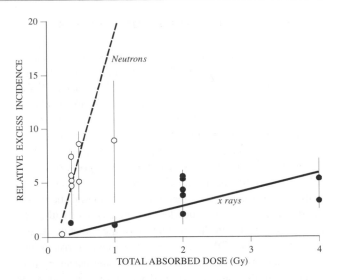

Figure XII. Relative excess Incidence of carcinomas in WAG/Rij rats after irradiation with x rays and 0.5 MeV neutrons [B34].

124. A unique experimental system has been described by Tanooka and Ootsuyama [O8, T13] in mice. The backs of female ICR mice were irradiated with beta particles from ^{90}Sr-^{90}Y three times a week throughout life. At radiation doses per exposure between 1 Gy and 11.8 Gy (low-LET), the tumour incidence was 100%. At 0.75 Gy per exposure, however, no tumours occurred in 31 mice over a period of 790 days from the start of irradiation. One osteosarcoma did arise at 791 days and one squamous cell carcinoma at 819 days. This was despite the fact that the cumulative dose was extremely high (305 Gy in 950 days). The appearance of tumours in irradiated mice depended on a fractionated regime: no tumours occurred following single exposures with doses up to 30 Gy. At such doses depilation and severe skin damage occurred. The authors proposed that at the lower dose fractionation regime efficient repair occurs, resulting in an apparent threshold in the tumour response. No histological information was reported, but it seems likely that at these high doses deterministic damage would occur, resulting in the development of a fibrotic response preceding tumour development.

125. Tumour induction in rats and mice exposed to high-LET neutron irradiation was described in the UNSCEAR 1993 Report [U3] and has also been summarized by Broerse [B34] and Fry [F15]. In general the experimental results reviewed indicated that there are differences between tissues in their tumorigenic response following either dose fractionation or reductions in dose rate as compared with acute radiation exposures. Taken together, however, the effects of dose rate and fractionation are small, and for the majority of studies a linear dose response would give a good fit to the data up to about 1 Gy. Exceptions are tumour types for which cell killing seems to play a significant role in tumour induction, as for thymic lymphomas, when a threshold dose may be found. In many cases, however, information is not available down to low levels of exposure. For life shortening at low doses, which has been shown to be the result of tumour induction, again little effect of fractionation or dose rate has been found [F15, U3].

(b) Internally incorporated radionuclides

126. In the case of intakes of radionuclides, many factors in addition to those for external radiation exposure may influence the dose response. For radionuclides such as ^{90}Sr or ^{239}Pu with a long physical half-life and a long biological half-time in the body, radiation exposure after an intake will generally be for the remaining lifespan of the animal, making it difficult to relate tumour incidence to radiation dose. Additional difficulties in interpreting dose-response data arise from the heterogeneous distribution of dose between and within body organs and tissues as well as temporal changes in the distribution of radionuclides, and hence dose, within the body. A key factor in the calculation of the radiation dose is the identification of the sensitive "target" cells at risk. When a radionuclide is uniformly distributed throughout an organ or tissue, as is the case for tritiated water or ^{137}Cs, then the calculation of average tissue dose is sufficient to assess the dose to these critical cells. In other cases, however, as with the bone-seeking radionuclides ^{239}Pu and ^{241}Am, the distribution of dose may be very heterogeneous, and then the calculation of dose to sensitive cells is essential in assessing dose-response relationships.

127. *Stem cells.* The International Commission on Radiological Protection (ICRP) has developed a comprehensive set of biokinetic and dosimetric models to enable the calculation of organ doses from inhaled or ingested radionuclides [B36, I9, I10]. These models take account of the distribution and retention of radionuclides in individual organs and the proportion of the energy of decay deposited in different organs. For penetrating photon radiation, it is necessary to take account of crossfire between organs, but in these cases the calculation of average energy absorbed in a tissue is sufficient. For non-penetrating alpha and beta radiations, energy is taken to be deposited in the organ in which the radionuclide is retained. For these radiation types it is necessary in some cases to take account of the distribution of the radionuclide within the organ relative to sensitive target cells. This consideration has been addressed in models developed by ICRP for the respiratory and gastrointestinal tracts, the skin, and the skeleton. In the case of other tissues (for example the liver, kidneys, and spleen), the average tissue dose is calculated on the assumption that sensitive cells are uniformly distributed throughout them. In relation to intakes of radionuclides, only the respiratory tract, the gastrointestinal tract and the skeleton are directly relevant. Radiation doses to the sensitive cells in the skin are, however, important in the case of radionuclides deposited on the surface of the body.

128. *Respiratory tract.* The ICRP model for the human respiratory tract [I11] takes account of the distribution of sensitive cells for cancer induction in the extrathoracic region and the bronchial and bronchiolar regions of the lung. For the region of the lung in which gaseous exchange occurs, the alveoli and terminal bronchioles, the dimensions of the structures are considered to be sufficiently small for doses to be calculated on the assumption that sensitive cells are uniformly distributed.

129. The extrathoracic region of the respiratory tract (the nose, oropharynx, and larynx) are lined mainly with stratified squamous epithelium. Excess nasal and laryngeal cancers have been observed in luminizer workers and patients receiving head and neck exposures [B37] but not in atomic bomb survivors or patients treated for spondylitis [D12]. Sinonasal cancers were described in humans as a result of systemic contamination with radium [E11, F13]. Radiation-induced tumours were mainly carcinomas, including basal cell, squamous cell, and epidermoid carcinomas, for which the cells at risk are assumed by ICRP [I11] to be the basal cells of the epithelial layer with their nuclei at average depths of 40–50 μm.

130. The trachea, bronchi, and bronchioles are lined by a pseudostratified, ciliated, columnar epithelium separated from the subepithelial connective tissue by a prominent basement membrane. Radiation-induced lung cancers have been documented in uranium miners, atomic bomb survivors, and therapeutically irradiated patients [B38, I12, P16]. Lung cancers occur predominantly in the bronchial region; there is no evidence that radiation induces tracheal cancer. There are four main classes of tumour observed: squamous cell carcinoma (most frequent), small-cell carcinoma, adenocarcinoma, and large-cell carcinoma. It appears that these tumour types share the same endodermic progenitors [M44, Y5]; the most likely candidate cells for tumour induction were considered by ICRP to be secretory cells [T15]. Basal cells may also be involved, although their role may be limited [J7]. ICRP therefore assumes for dosimetric purposes that the sensitive cells in the bronchial region are secretory and basal cells, with nuclei at average depths of 10–40 μm and 35–50 μm, respectively [I11]. The sensitive cells in the bronchiolar region are taken to be secretory cells, with nuclei at an average depth of 4–15 μm.

131. Estimates of dose to the lung from short-range emitters, particularly alpha emitters, depend on the assumptions made regarding the depth and thickness of the sensitive layer in the bronchi and bronchioles. For example, in a recent sensitivity analysis of doses from radon progeny, a dose range that varied by a factor of 2.6 resulted from consideration of sensitive cell parameters [M45].

132. *Gastrointestinal tract.* The current dosimetric model of the gastrointestinal tract makes only a simple generalized allowance for the position of sensitive cells relative to ingested radionuclides [I9]. Doses are calculated separately for the mucosal layer of each region modelled: the stomach, small intestine, upper large intestine, and lower large intestine. For penetrating radiations, the average dose to the wall of each region is used as a measure of the dose to the mucosal layer. For non-penetrating radiations, the fraction absorbed by the mucosal layer is taken to be equal to $0.5v/M$, where M is the mass of the contents of that section of the gastrointestinal tract and v is a factor (between 0 and 1) representing the proportion of energy that reaches sensitive cells. The factor of 0.5 is introduced because the dose at the surface of the contents will be approximately half that within the contents for non-penetrating radiations. For beta particles, v is taken to be 1.

For alpha particles, v is taken to be 0.01. This value is based on weak experimental evidence from an acute toxicity study in rats in which the LD_{50} for ingested [91]Y was estimated at about 12 Gy while a more than 100 times greater dose to the mucosal surface from [239]Pu had no effect [S33].

133. This model is currently being revised. The new model is expected to consider the location of sensitive cells in all regions of the alimentary tract, from the mouth to the large intestine. Radiation-induced cancer in human populations has been documented for the oesophagus, stomach, and colon; the small intestine is not a significant site for cancer induction [B39]. The sensitive cells in the oesophagus are assumed to be the basal layer of the stratified squamous epithelial lining. This epithelium is quite thick (300–500 μm) and is protected by a surface layer of mucus. At the gastro-oesophageal junction, it is abruptly succeeded by a simple columnar epithelium with gastric pits and glands. In the stomach, the sensitive cells for cancer induction are assumed to be the epithelial stem cells, located within but towards the top of the gastric pits, at a depth of about 75–100 μm from the surface. In the small intestine, stem cells are located above the paneth cells, towards the base of the crypts. In the large intestine, stem cells are situated at the very base of the crypts. These locations have been deduced from a variety of cell kinetic, mutational, and regeneration studies in mouse models [P17] and their positions are likely to be qualitatively similar in man. The number of stem cells per colonic crypt in mice has been estimated to be in the range 1–8, and as colonic crypts in man are around six times as large as in mice, it is possible that the number of stem cells per crypt may be greater in man. The depth of the stem cells, measured in human tissue samples, is about 100–150 μm in the small intestine and 200–400 μm in the large intestine [P18].

134. **Skin.** The skin is broadly divisible into two component layers: the outer epidermis and the underlying dermis. The epidermis arises from a single basal layer of cells, overlaid by layers of cells with dead layers on the outer surface. The basal layer is separated from the dermis by a basement membrane. This boundary is not flat but undulates, with discrete points known as rete pegs where the epidermis projects down into the dermis. In addition, the basal layer extends around the skin appendages, notably the shaft and base of the hair follicles, which project even deeper into the dermis. At some sites on the body, over 50% of the basal layer stem cells are associated with the hair follicles. Thus, the depth of the basal layer is highly variable. In most body areas it ranges from 20 to 100 μm in the interfollicular sites, but exceptionally (e.g. the finger tips), it can be over 150 μm deep because of increased outer cornification [L46]. The deeper projections associated with hair follicles result in basal cells being situated more than 200 μm deep.

135. There is substantial evidence linking the incidence of non-melanoma skin cancer (NMSC) with exposure to ionizing radiation, including studies on irradiated children and atomic bomb survivors [L46]. The two main types of non-melanoma skin cancer are squamous cell carcinoma and basal cell carcinoma, with the sensitive cells for cancer induction assumed to be the basal layer of cells in each case. This assumption is supported by animal data [A15, H28].

136. In calculating dose to the skin, ICRP has recommended that skin dose should be evaluated at an average depth of 70 μm [I2]. However, when assessing dose in cases of non-uniform exposure, it may be necessary to use skin thickness values appropriate to the area of interest.

137. **Skeleton.** Biokinetic and dosimetric models for the skeleton take account of the two main types of bone, cortical and trabecular, and the behaviour of different bone-seeking radionuclides as well as the location of sensitive cells for the induction of bone sarcoma and leukaemia [I9, I10]. Cortical bone is the hard, dense bone that forms the outer wall of bones and the whole of the shaft of long bones. Trabecular bone is a soft, spongy bone with a lattice-work structure that is found within flat bones and in the ends of the long bones. The endosteal layer of cells on the inner bone surfaces in cortical and trabecular bone is taken to be the sensitive cells for bone sarcoma and the red bone marrow is taken to be the sensitive cells for leukaemia. It is assumed that all haemopoietically active red marrow is confined to the spaces in trabecular bone in adults, with cortical bone containing inactive yellow marrow. In children, a proportion of cortical marrow is assumed to be haemopoietically active and therefore a target for leukaemia induction. In its 1979 Report, ICRP classified bone-seeking radionuclides into two groups: bone-surface seekers, including the actinide elements, and bone-volume seekers, including the alkaline earth elements [I9]. Thus, radionuclides were assumed to be retained either on endosteal bone surfaces or uniformly distributed throughout the volume of bone mineral. Absorbed fractions were calculated for the proportions of alpha and beta energy emitted in each case that would be deposited in the sensitive regions of the endosteal layer, taken to lie within 10 μm of bone surfaces and red marrow. More realistic biokinetic models have since been developed for the main bone-seeking radionuclides, isotopes of the actinides, alkaline earths and similar elements, which allow for initial deposition on bone surfaces, movement into bone owing to bone remodelling and chemical exchange, and loss from bone, principally owing to bone resorption [I10]. For the actinides, transfer from bone to marrow is also included.

138. An increased incidence of bone sarcomas has been observed in populations exposed to alpha-emitting radium isotopes, particularly in painters of luminuous dials, but also radium chemists and people treated with radium salts for the supposed therapeutic effect [M18]. Although the ICRP assumption [I8] that the sensitive cells constitute a 10-μm-thick layer on endosteal surfaces gives reasonable dose estimates, it has been suggested that all bone surfaces may not be equally sensitive [P19] and that the sensitive region may include cells at a greater depth into the marrow [G15]. Priest [P19] argued that the observed difference in toxicity between [226]Ra (half-life = 1,600 years) and [224]Ra (3.6 days) in animals and humans cannot be explained simply in terms of a greater wastage of alpha dose from the longer-lived [226]Ra within bone mineral. He suggests that a

greater proportion of alpha dose from ^{224}Ra may be delivered to active trabecular surfaces and that these regions have a greater than average sensitivity.

139. Gössner et al. [G15] have reviewed the histopathology of radiation-induced bone sarcomas, showing that there are of two main types, bone-producing osteosarcomas, and non-bone-producing sarcomas of the fibrous-histiocytic type. A trend to a greater proportion of fibrous-histiocytic tumours was identified at lower doses and shorter latency periods. The data suggest that cells at risk are not only those committed to bone formation on the bone surfaces but multipotent marrow stromal cells located at some distance from the bone surface.

140. Excess leukaemia has been recorded in patients exposed to the alpha-emitting contrast medium thorotrast and in the atomic bomb survivors, but it is not a feature of exposure to isotopes of radium [I9, M46]. Comparison of leukaemia induction by thorotrast and external low-LET irradiation suggests a low RBE for alpha-induced leukaemia. The inability of ^{226}Ra to induce leukaemia [R16] may be explained by a low alpha RBE, but the distribution of sensitive cells in the marrow may also be a contributory factor. While the colloidal thorium oxide preparation thorotrast was retained in macrophages throughout the marrow, radium on bone surfaces delivers a dose only to peripheral marrow, and it may be that sensitive cells are concentrated more towards the centre of marrow spaces. Some evidence for this was provided by studies using mice [L47]. It may be, therefore, that the ICRP assumption that sensitive cells for leukaemia induction are uniformly distributed throughout red marrow [I9] may overestimate the risk of leukaemia from bone-seeking radionuclides.

141. *Tumour induction.* A number of reviews and papers have examined dose-response relationships for tumour induction in animals exposed to either alpha emitters or beta/gamma emitters (see, for example, [I5, L27, M11, N6, Y6]). Most information is available on the induction of bone tumours following the entry of radionuclides into the blood or lung tumours after inhalation of radioactive materials in various chemical forms, although more limited data on other organs and tissues are also available. A wide range of dose-response relationships has been obtained. These encompass data that can be fitted with simple linear models up to intermediate levels of dose and other responses with clear evidence of a threshold. The results of studies on tumour induction from intakes of radionuclides are illustrated by data on tumour induction in the lungs and skeleton.

142. *Bone tumours.* The incidence of bone tumours in mice, rats, dogs, and pigs given graded doses of ^{90}Sr was examined by Mays and Lloyd [M11]. Although limited data were available at low doses, and the various species had different sensitivities to tumour induction, in all cases the incidence of bone tumours at the lowest doses examined was less than would have been predicted on the basis of a linear dose response. Thus, in beagle dogs with average skeletal doses from ^{90}Sr at 1 year before death of between 0.27 Gy and 111 Gy, no tumours were found in the three lowest dose

groups (1, 3.35, and 5.97 Gy), with an 8% incidence occurring at 21.7 Gy [N6]. The numbers of dogs in each group was, however, only about 12, and a small increase in incidence could not have been detected. Similar data have been reported for osteosarcoma induction by ^{90}Sr in female CF$_1$ mice. In groups of about 100 animals with average bone doses ranging from 0.26 to 120 Gy, no significant increase in tumour incidence was found in animals with average doses below about 10 Gy (1.3, 4.5, and 8.9 Gy) [M11, N6].

143. An extensive series of studies has examined tumour induction in animals given various alpha emitters. Lloyd et al. [L27] examined the occurrence of skeletal tumours in young adult beagle dogs given single intravenous injections of monomeric ^{239}Pu citrate. The relationship between the incidence of osteosarcoma and average dose to bone at the presumed time of tumour initiation, taken to be at 1 year before death, appeared to be linear below about 1.3 Gy (26 Sv, assuming an RBE for alpha radiation of 20) (Figure XIII). The observed tumour incidence, I (%), could be approximated by the expression I = 0.76 + 75D, where D is the average dose to bone in gray. Similar analyses of data from dogs given ^{226}Ra also gave a linear response with the expression I = 0.76 + 4.7D (for doses up to 20 Gy). The ratio of the coefficients (75/4.7 = 16±5) shows that ^{239}Pu is more effective in inducing osteosarcoma than ^{226}Ra. This is thought to be due to the tendency of plutonium to remain longer on bone surfaces and to more effectively irradiate the sensitive cells for tumour induction.

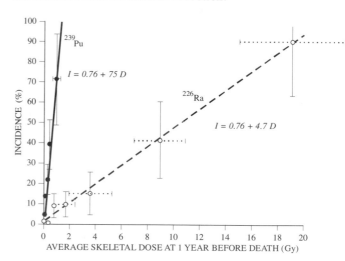

Figure XIII.   Bone cancer incidence in beagle dogs following a single injection of plutonium-239 or radium-226 citrate at about one year of age [L14].
The ratio of plutonium to radium dose coefficients (75:4.7) is 16.

144. Further data on bone tumour induction in animals given alpha emitters were analysed by Mays and Lloyd [M17]. They found that although the induction of bone tumours appeared to increase linearly with dose in some cases, in others it followed threshold or sigmoid relationships. In CF$_1$ female mice injected intravenously at 70 days of age with ^{239}Pu [F11], no tumours were observed in groups of mice with average bone doses of 0.01 Gy (N=99) and 0.22 Gy (N=96), whereas a linear response would have predicted some 5.3 cases. The

probability of observing zero cases, if 5.3 cases is the true number, is only 5%. At 0.4 Gy and above, bone tumour incidence increased linearly with dose.

145. A linear dose response was found for osteosarcoma induction in female CF_1 mice given ^{226}Ra by intravenous injection at 70 days of age. In 1,436 mice with average bone doses below 3 Gy (high-LET), 115 cases of osteosarcoma were observed, in good agreement with 92 cases predicted using a linear dose response [F12, M17]. In contrast, in beagle dogs given ^{228}Ra and ^{228}Th, the dose-response data suggested the presence of thresholds at about 2 Gy and 0.5 Gy (high-LET), respectively [M17].

146. More complex models have also been developed to interpret dose-response relationships for bone tumour induction. Raabe [R1] has described an example for predicting risks associated with protracted exposure to ionizing radiation from internally deposited radionuclides. For long-lived radionuclides such as ^{90}Sr, ^{226}Ra, or ^{239}Pu, the radiation dose will be delivered over the lifespan of the animal. Raabe et al. [R1, R14] have interpreted the data from various lifetime studies with beagle dogs exposed by injection, ingestion, and inhalation to either beta emitters or alpha emitters. The cumulative absorbed dose required to give a specified level of cancer risk was found to be less at lower dose rates than at the higher dose rates, and the induction time required for tumours to manifest themselves tended to be longer at lower dose rates and could exceed the normal lifespan of the animal. The authors interpreted the data to suggest that at the lowest dose rates there is an effective threshold for the induction of fatal radiation-induced cancer.

147. For example, beagle dogs given eight fortnightly injections of ^{226}Ra in amounts from 0.099 kBq kg^{-1} to 46.3 kBq kg^{-1} received average lifetime skeletal doses from 0.9 ± 0.2 Gy to 167 ± 44 Gy (±1 SD). Death in these dogs was considered to be a function of three effects: (a) spontaneous death arising from causes associated with the natural lifespan, (b) death associated with radiation-induced bone tumours, and (c) death from radiation-induced skeletal injury such as radiation osteodystrophy and bone fractions occurring at high doses (Figure XIV). Mathematical three-dimensional dose-rate/time response models with log-normal probability distributions were fitted to the lifespan data for the dogs. The data plots indicated that bone cancer predominates as a cause of death at intermediate doses and is infrequent at low dose rates (because of death associated with natural lifespan) and at high dose rates (because of deaths from acute radiation injury). The cumulative dose required to cause bone cancer is smaller at the low dose rates; however at lower dose rates it takes longer to reach any specified level of risk, perhaps longer than the natural lifespan of the animal. This results in a lifespan effective threshold for cancer induction similar to the "practical" threshold described by Rowland [R17] at a cumulative lifespan alpha dose of about 1 Gy in man (see Chapter III). In practice, the lack of a significant effect during the lifespan of the animals could also be taken to indicate a risk of cancer with a very low probability of occurrence at low doses.

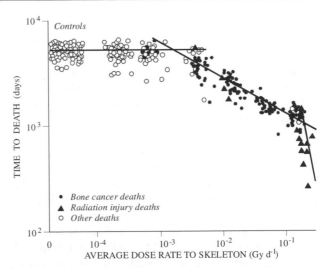

Figure XIV.   Deaths from non-neoplastic radiation injury, bone cancer and other causes in beagle dogs injected with ^{226}Ra. Initial intake occurred at 435 days of age [R1].

148. Raabe et al. [R14] have also compared data on bone tumour induction in humans and CF_1 mice with data obtained in beagle dogs. When time was normalized with respect to lifespan, the three species were found to have bone cancer dose-rate/time risk functions that were almost identical and could be represented by one median regression line.

149. *Lung tumours.* Extensive data have also been published on lung tumour induction in rodents and beagle dogs exposed to internally incorporated alpha and beta/gamma emitters. Studies conducted in the 1960s and 1970s were considered by a Task Group of Committee 1 of ICRP [I5]. The data reviewed were from laboratories around the world and from studies using a range of different protocols and methodologies. One specific aim of the analyses was to determine the relative effectiveness of alpha emitters and beta/gamma emitters in causing lung damage, including neoplastic development.

150. The Task Group commented on some of the difficulties in ascertaining the dose response for lung tumour induction. In studies with inhaled radionuclides it is impossible to deposit the same amount of activity in the pulmonary region of different animals in a group. As a consequence, authors have commonly shown dose ranges rather than a single value. Further, researchers do not agree on how to express dose to the types of tissue found in the lung. Cumulative doses may be estimated for individual animals at death, at time of the first tumour, or for the average lifespan of the group of animals. Other variants have also been used. In all cases considered by the Task Group, average lung dose was calculated.

151. The analyses of pooled data from studies with different species generally used a probit model, as had commonly been used in dose-response analyses, and the linear dose response, which the Task Group considered to reflect conservatism. Both linear and probit models gave an adequate description of the incidence data for alpha emitters over the range of

observed doses. For beta/gamma emitters, however, neither model gave a good fit to the incidence data, which were rather variable for similar doses. The linear dose response considerably overestimated incidence at low doses. In general the pooling of data from numerous species, although it is a comprehensive approach, does not readily permit detailed comparison of any of the individual studies.

152. The results of a number of separate studies, mainly of rodents exposed to both alpha and beta/gamma emitters have been published. Sanders and Lundgren [S11] compared lung cancer induction in F344 and Wistar rats exposed to $^{239}PuO_2$. In the F344 strain, significantly increased lung tumour incidences were found at lung doses of both 0.98 Gy (20%) and 37 Gy (34%) compared with 1.7% in controls. There were insufficient data to define a dose-response function, but there was no evidence for a threshold in the response. In contrast for the Wistar rats, there was no significant increase in lung tumour incidence in animals with an average lung dose of 0.75 Gy (0%) compared with controls (0.1%), but for animals with a lung dose of 34 Gy the incidence was 68%. These data suggested the presence of a threshold at doses somewhat above 0.75 Gy.

153. The data on Wistar rats [S11] were similar to those found in a more comprehensive lifespan study [S12]. In 3,157 female Wistar rats that had inhaled $^{239}PuO_2$ only three adenomas were found in 1,877 rats at lung doses <1.5 Gy, for an incidence of 0.16%; tumour incidence increased to 41% in 228 rats with lung doses >1.5 Gy. Pulmonary squamous metaplasia was not seen in controls and was first noted in exposed rats at lung doses >1 Gy. All tumour types induced by inhaled $^{239}PuO_2$ exhibited a threshold at lung doses >1 Gy. It was concluded that for lung tumours in Wistar rats resulting from inhaled $^{239}PuO_2$, plutonium particle aggregation is required to cause proliferation of initiated cells and to promote the formation of premalignant and malignant lesions.

154. Similar results in Wistar rats were obtained by Oghiso et al. [O5], although the study was not as extensive as that by Saunders et al. [S12]. Dose-response relationships were compared among primary tumours, classified by histological type, following a single inhalation exposure to $^{239}PuO_2$. In this study there were 130 controls and 310 animals, separated into seven groups, exposed to $^{239}PuO_2$. Initial lung contents in the different groups varied between about 97 and 1,670 Bq, giving average lung doses from 0.7 to 8.5 Gy. A differential tumour response was obtained. In general, metaplasia and benign adenomas were induced at lower doses (<1 Gy), whereas malignant carcinomas were induced at relatively high doses (>1.5 Gy) (Figure XV). The peak incidence of adenomas occurred at a dose of 0.7 Gy, of adenocarcinomas at 2.9 Gy, and adenosquamous and squamous cell carcinomas at 5.4–8.5 Gy. These results were considered by the authors to indicate a differential dose response for pulmonary carcinogenesis, in which metaplasia and benign adenomas were induced at lower doses (<1 Gy) and malignant carcinomas were induced at

higher doses (>1.5 Gy). It was also noteworthy that the lifespan of the 0.7 Gy group (871 ± 105 days, ±1 SD) was significantly longer than that of the control group (790 ± 144 days, p<0.01). In the higher exposure dose groups, lifespan was reduced.

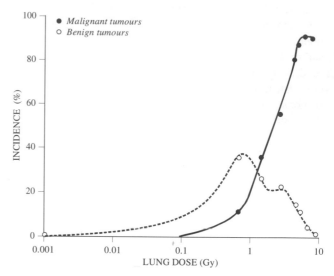

Figure XV. Benign and malignant lung tumours in rats after inhalation of $^{239}PuO_2$ aerosols [O5].

155. This threshold type of response did not seem to be found in Fisher 344 rats exposed to ^{244}Cm as the oxide. Groups of 100–200 male and female rats received average lung doses from $^{244}Cm_2O_3$ between 0.2 and 36 Gy [L28]. In general the prevalence of benign and malignant lung neoplasms increased with increasing average lung dose. For lung tumours, a linear dose-response function adequately fitted the data ($I = 0.38 ± 0.04 \ Gy^{-1}$). The response in rats exposed to $^{239}PuO_2$ ($I = 0.70 ± 0.07 \ Gy^{-1}$) was about twice the response following exposure to $^{244}Cm_2O_3$.

156. Some information is available on tumour induction in rats exposed to the beta/gamma emitter ^{144}Ce as the oxide [L39]. A total of 1,059 F344/N male and female rats (about 12 weeks of age) were exposed to graded levels of $^{144}CeO_2$, and a further 1,064 rats were maintained as controls (exposed to stable CeO_2). Groups of rats received mean lung doses of 3.6, 12, and 37 Gy. The incidence, I (%), of lung tumours increased with increasing lung dose and could be represented by a linear function of the form $I = 0.13 + 0.51D \ Gy^{-1}$, where D is the dose in gray. Because the data are limited in extent, more complex functions such as the linear-quadratic, exponential linear-quadratic, and Weibull functions also described the dose response adequately over the dose range of the study.

157. An extensive series of studies has been carried out in rodents exposed to radon and its decay products. They have demonstrated that exposures at high doses can give rise to radiation-induced lung cancers. Experimental animal studies have been valuable for understanding the consequences of exposure at varying dose rates and the influence of other environmental factors on the lung tumour response as the animals can be exposed to a variety of agents under carefully controlled conditions. Much of the information available on

experimental animals was reviewed by Cross [C22]. In animals, exposure to radon has resulted in respiratory tract tumours (adenomas, bronchiolar carcinomas or adenocarcinomas, epidermoid carcinomas, adenosquamous carcinomas, and sarcomas). In addition, pulmonary fibrosis, pulmonary emphysema, and life shortening have occurred at exposures above about 1,000 WLM^{-1} (3.5 J m^{-3}). Excess respiratory tract tumours have occurred in rats at exposure levels well below 100 WLM^{-1} (0.35 J m^{-3}), even at levels comparable to those found for typical lifespan exposures in homes. Further, tumours occurred in animals exposed to radon decay products alone; thus indicating that exposure to other environmental agents (uranium ore dust, cigarette smoke) is not necessary for carcinoma development. Most (~80%) radon-induced lung tumours in rats are considered to originate peripherally and to occur at the bronchiolar-alveolar junction. The remaining 20% are centrally located in association with the bronchi, the actual percent depending on exposure rate and possibly on

exposure level. [*Note: Working level (WL) is defined as any combination of short-lived radon decay products in 1 litre of air resulting in the ultimate emission of 1.3 10^5 MeV of potential alpha energy (1 WL = 2.08 10^{-5} J m^{-3}). Working-level month: exposure equivalent to 170 hours at 1 WL concentration (1 WLM = 3.5 10^{-3} J h m^{-3}).*]

158. A notable finding in these animal studies has been that longer duration of exposure at a lower dose rate induces more lung cancers than exposures for a shorter duration at a higher dose rate. Table 9 compares the incidence of lung tumours in rats exposed at either 50 or 500 WLM per week. With exposure levels between 320 and 5,120 WLM, in all except the lowest exposure group there is a higher incidence of tumours in the groups exposed at the lower dose rate. The decrease in exposure rate not only increased the tumour incidence but specifically increased the incidence of epidermoid carcinomas.

Table 9
Percentage incidence of primary and fatal lung tumours in rats exposed to radon and decay products [a]
[C22]

Cancer type	Exposure (WLM)				
	320	640	1 280	2 560	5 120
At 500 WLM per week					
Number of animals examined	131	70	38	38	41
Adenoma	5	3	0	3	2
Adenocarcinoma	8	7	26	24	44
Epidermoid carcinoma	1	0	0	3	2
Adenosquamous carcinoma	0	0	3	0	0
Sarcoma	0	0	0	3	2
Fatal lung tumours	2	1	5	11	15
Animals with lung tumours (%)	15	10	29	32	49
At 50 WLM per week					
Number of animals examined	127	64	32	32	32
Adenoma	5	3	(22)	9	(22)
Adenocarcinoma	5	(20)	41	41	53
Epidermoid carcinoma	1	3	(13)	(47)	(44)
Adenosquamous carcinoma	1	3	9	(9)	3
Sarcoma	1	2	3	0	0
Fatal lung tumours	2	6	(22)	(50)	(44)
Animals with lung tumours (%)	10	(28)	(66)	(69)	(75)

a   15 mg m^3 ore dust exposures accompanied radon and radon progeny exposures; data in parentheses at 50 WLM per week are significantly
      (p<0.05) higher than corresponding data at 500 WLM per week.

159. A series of studies has also been conducted in France on the effects of radon exposure [G18, M48]. In these experiments more than 2,000 rats were exposed to cumulative doses of up to 28,000 WLM of radon gas. There was an excess of lung cancer at exposures down to 25 WLM (80 mJ h m^{-3}). These exposures were carried out at relatively high concentrations of radon and its decay products (2 J m^3). Above 6,000 WLM, rats suffered increasingly from life shortening due to radiation-induced non-neoplastic causes, thus limiting tumour development. When the dose-response data were adjusted for these competing causes of death, the hazard function for the excess risk of developing pulmonary tumours was approximately linearly related to dose. This

suggests that the apparent reductions in tumour induction found at high doses may have been chiefly the result of acute damage. Later experiments, however, found that chronic exposure protracted over 18 months at an alpha energy of 2 WL (0.0042 mJ m^3) resulted in fewer lung tumours in rats (0.6%, 3/500 animals, 95% CI: 0.32–2.33) than similar exposures at a potential alpha energy of 100 WL (2 mJ m^{-3}) protracted over 4 months (2.2%, 11/500 animals, CI: 0.91–3.49) or over 6 months (2.4%, 12/500, CI: 1.06–3.74). The incidence of lung tumours in controls was 0.6% (5/800, CI: 0.20–1.49) [M48]. The confidence intervals are, however, wide, and the longer period of exposure (18 months) would in itself have been expected to result in fewer lung tumours.

160. It has been suggested by Moolgavkar et al. [M13, M29], based on a two-stage initiation-progression model for carcinogenesis, that extended duration of exposure allows time for proliferation of initiated cells and thus for a higher disease incidence. The findings are that the first mutation rate is very strongly dependent on the rate of exposure to radon progeny and consistent with *in vitro* rates measured experimentally. The second mutation rate is much less so, suggesting that the nature of the two mutational events is different. Furthermore, by incorporating cell killing into such a model, Luebeck et al. [L12] proposed that the model that gave the best fit to the data indicated that the initial increase in the proliferation rate of initiated cells depended on a second promotional step, which may be due principally to the presence of ore dust and not to radon decay products. The inverse exposure rate effect may thus be reduced in the absence of ore dust.

161. The main factors found to influence the tumorigenic potential of radon exposure in laboratory rats include cumulative exposure to radon progeny, exposure rate, unattached fraction, and associated cigarette smoke exposure. The respiratory tract cancer risk increases with the increase in cumulative exposure to radon progeny and in the magnitude of the unattached fraction. The increased risk with a high unattached fraction of radon progeny is particularly relevant to indoor radon exposure, where the unattached levels are generally much higher than in mines. The influence of cigarette smoke has been variable, depending in part on the temporal sequence of radon and cigarette smoke exposure.

162. Overall, the data on lung cancer risk resulting from exposure to radon and its decay products show an increasing risk with increasing exposure, although there are strong indications of an inverse dose-rate effect that is influenced by the presence of ore dust in the atmosphere. The data are broadly similar to those obtained from follow-up studies on uranium miners (see Chapter III).

2. Cancer risks at low doses

163. An essential input to the analysis of dose-response relationships is not only the shape of the dose response but the extent to which a statistically significant effect of radiation can be detected at low doses. It is informative to examine a number of studies that have been concerned with assessing risks at low doses.

(a) Studies

164. Laboratory animal studies that are most suitable for determining the lowest doses at which effects of radiation on tumour induction can be detected have been carried out predominantly with mice. Comprehensive data are, however, rather limited. Some of the more significant studies are briefly summarized below and analysed in the following Section.

165. Mole and Major [M3] and Mole et al. [M4] reported myeloid leukaemia incidence in male CBA-H mice acutely exposed to x rays (0.5 Gy min^{-1}) and ^{60}Co gamma rays (0.25 Gy min^{-1}) and chronically exposed to gamma rays

over a period of 28 days (0.4-0.11 mGy min^{-1}). This strain of mice is exceptional, in that no case of myeloid leukaemia has been observed in more than 1,400 unirradiated male mice, so that every case occurring in irradiated animals can be regarded as radiation-induced. Total acute doses were from 0.25 to 1.0 Gy for x rays and 1.5 to 3.0 Gy for gamma rays.

166. Upton et al. [U21] used RFM mice of both sexes. For acute exposures of female mice, a dose rate of 67 mGy min^{-1} from ^{60}Co gamma rays was used, giving doses between 0.25 and 4 Gy. For male mice, x rays at 800 mGy min^{-1} were used, with doses from 0.25 to 3 Gy. Male and female mice were also exposed chronically. Data are available on the induction of myeloid leukaemia, thymic lymphoma, and ovarian tumours.

167. Ullrich [U14] and Ullrich et al. [U15, U16, U17, U18, U19] carried out experiments similar to those of Upton et al. [U21] using RFM male and female mice acutely exposed (450 mGy min^{-1}) to ^{137}Cs gamma rays. Data were reported on the tumours examined by Upton et al. [U21], together with data on Harderian gland and pituitary tumours.

168. Ullrich [U14], Ullrich and Storer [U16], and Ullrich and Preston [U20] also used BALB-C female mice to obtain further data on dose-response relationships. Acute (450 mGy min^{-1}) and chronic (0.06 mGy min^{-1}) exposures from ^{137}Cs were given. Acute doses were between 0.01 and 1 Gy, and chronic doses were between 0.25 and 2 Gy. Tumours showing a positive increase with dose were ovarian tumours as well as mammary and lung adenocarcinomas.

169. Coggle [C6] reported data on the induction of lung adenocarcinomas in male and female SAS/4 mice acutely exposed to x rays at 0.6 Gy min^{-1}. The dose range used was 0.25-3.0 Gy.

170. Covelli et al. [C7, C8] reported tumour induction in male and female BC3F$_1$ mice. They observed various types of radiation-induced tumours following acute exposure of 113 mGy min^{-1} (dose range males, 0.04-2.5 Gy; females, 0.5-5.0 Gy). The authors gave age-adjusted incidences of tumours and described tests showing which doses gave significant increases in cancer yield.

(b) Analysis

171. To determine the lowest dose at which a significant increase in tumour yield occurred in the various studies, the following method was used. The tumour yield in control animals was tested against the yield at the lowest dose used in the study. If the difference in tumour incidence is statistically significant, then that dose is taken as the lowest dose at which a significant effect is found. If the difference is not significant, the data point with the next lowest dose is included and a weighted linear regression performed, either by weighted least squares or, where possible, by iteratively re-weighted least squares. This process is continued at progressively higher doses until the linear regression coefficient becomes signifi-

cantly different from zero (p=0.05). The last dose added is then taken to be the lowest dose to give a significant effect. When calculating statistical significance, any lack of fit to a straight line is taken into account in computing uncertainties.

172. The exposure levels at which significant increases in risks of leukaemia and solid cancers could be observed are

given in Table 10. The lowest dose at which a significant effect on tumour incidence could be determined is very different from study to study. It depends on factors that influence statistical power, such as the number of mice used and the spontaneous cancer rate, the cancer type, the level of radiation risk, the dose range used, and the period of follow-up.

Table 10
Lowest acute doses at which significant increases in cancers have been observed in mice

Cancer	Mouse strain	Sex	Irradiation	Dose (Gy)	Ref.
Myeloid leukaemia	RFM	Male	x rays	0.25	[U21]
		Male	Gamma rays	1	[U16, U17, U20]
		Female	Gamma rays	1	[U15, U16, U17, U18]
		Female	Gamma rays	2	[U21]
	CBA-H	Male	x rays	0.5	[M4]
	BC3F$_1$	Male	Gamma rays	1.5	[M3]
		Female	x rays	1	[C7, C8]
Thymic lymphoma	RFM	Male	Gamma rays	1	[U16, U17, U20]
		Male	x rays	3	[U21]
		Female	Gamma rays	1	[U15, U16, U17, U18]
		Female		2	[U21]
Lung adenocarcinoma	BALB-c	Female	Gamma rays	0.5	[U16]
	SAS/4	Both	x rays	2.5	[C6]
Mammary adenocarcinoma	BALB-c	Female	Gamma rays	0.2	[U14]
Ovarian tumour	BC3F$_1$	Female	x rays	0.16	[C7, C8]
	BALB-c	Female	Gamma rays	0.25	[U18]
	RFM	Female	Gamma rays	0.5	[U21]
Harderian gland tumour	RFM	Male	Gamma rays	3	[U16, U17, U20]
All solid tumours	BC3F$_1$	Female	x rays	1.3 $^{a\ b}$	[C7, C8]
				4	[C7, C8]

a   Excluding ovarian tumour.
b   p=0.05.

173. For leukaemia induction in mice there was little evidence for an increase in risk below 1 Gy, although two studies indicated statistically significant increases at 0.25 Gy [U21] and 0.5 Gy [M3]. Most of the dose-response data for acute exposures showed no significant departure from linearity. An exception was the study by Mole and Major [M3], which showed a reduced effectiveness of radiation, per unit dose, at 1 Gy. There was also a suggestion of a departure from linearity at high doses in the results reported by Ullrich et al. in RFM mice [U16, U17, U20].

174. For solid cancers the overall results (Table 10) are similar to those for leukaemias, with significant increases in tumour incidence occurring principally at acute doses of 1 Gy and above. Increases in risk were, however, seen at lower doses for ovarian tumours (0.16 and 0.25 Gy), mammary adenocarcinomas (0.2 Gy), and lung adeno-carcinomas (0.5 Gy). Although data are not given here, the use of a lower dose rate consistently resulted in a lower risk per unit of dose.

B. HEREDITARY DISEASE

175. In addition to inducing neoplastic changes in somatic tissues, ionizing radiation may produce transmissible (heritable) effects in irradiated populations by inducing muta-tions in the DNA of male or female germ cells. These muta-tions, while having no direct consequences for the exposed individual, may be expressed in subsequent generations as genetic disorders of widely differing types and severity.

176. Studies of germ-cell mutations *in vivo* are not only relevant for assessing dose-response relationships for hereditary effects but they also have value for assessing effects on the primary lesion in DNA likely to be involved in tumour initiation. As described in Chapter IV, subsequent tumour expression will depend on the influence of many other factors.

177. The evaluation of genetic hazards associated with the exposure of human populations to ionizing radiation is an important area in which the Committee has been active since its inception. To date, however, there has been a lack

of direct data that give quantitative information on genetic effects leading to disease states in humans. The substantial amount of data from other species indicates that radiation can give rise to mutations in humans that will be manifested as disease. So far there has been no alternative but to use data from experimental animals as the main basis for predicting quantitative effects in humans.

178. In the UNSCEAR 1988 Report [U4], the Committee summarized the principal assumptions thought to be necessary for extrapolating data on hereditary damage in mice and other animals to humans. The main considerations are the following:

(a)   the amount of genetic damage induced by a given type of radiation under a given set of conditions is the same in human germ cells and in those of the test species used as a model;

(b)   the various biological and physical factors affect the magnitude of the damage in similar ways and to a similar extent in the experimental species from which extrapolations are made and in humans; and

(c)   at low doses and low dose rates of low-LET radiation there is a linear relationship between dose and the frequency of genetic effects studied.

179. Studies in mice have provided the main basis for assessing the risks of hereditary disease in humans. The doubling dose for hereditary disease that has been adopted by most national and international organizations for chronic exposure is 1 Gy (e.g. [C1, M18, U4]). Reviews of experimental data from mice generally give values in the range from 1 to 4 Gy and would therefore suggest that the value of 1 Gy adopted for humans is conservative [M18, S13].

180. A series of studies have been reported on dose-response relationships for the induction of germ-cell mutations in mice. The most comprehensive information comes from studies in male mice in which specific locus mutations were measured. Russell et al. [R5, R6], for example, presented data on dose-response relationships for male mice exposed to 0.72–0.9 Gy min^{-1} for doses between 3 Gy and 6.7 Gy and ≤8 mGy min^{-1} for doses between 0.38 and 8.61 Gy. In both cases the data could be fitted by a linear dose-response relationship over the whole dose range examined. For chronic exposure, I = (8.04 10^{-6} ± 1.19 10^{-6}) + (7.34 10^{-6} ± 0.83 10^{-6})D; for acute exposure, I = (8.12 10^{-6} ± 1.19 10^{-6}) + (2.19 10^{-5} ± 0.19 10^{-5})D, where I is the mutation frequency per locus and D is the dose in gray. The difference in slope for the two exposure conditions, by a factor of about 3, reflects the difference in the dose rates and opportunity for repair of damage at lower dose rates (Figure XVI). It might be expected that if lower doses had been used in the high-dose-rate study (0.72–0.9 Gy min^{-1}), the slope of the response at lower total doses would approach that found for low-dose-rate exposures.

181. It was notable that although the incidence of mutations fell by a factor of about 3 for a reduction in dose rate from 800–900 mGy min^{-1} to 8 mGy min^{-1}, further reduction in the dose rate to 0.007 mGy min^{-1} failed to further reduce the yield

of mutations. This independence of dose rate was shown over a range of doses differing by rather more than a factor of 1,000, and it was concluded that it was unlikely that a further reduction in mutation frequency would be obtained at even lower dose rates [R6]. This suggests that a substantial fraction of the damage to DNA that results in the induction of heritable mutations is not amenable to effective repair.

182. Similar results for specific-locus mutations in male mice were obtained in studies by Lyon et al. [L40]. For a gamma ray dose from ^{60}Co of 6.3 Gy given in 60 equal daily fractions at 0.17 Gy min^{-1}, the mutation frequency (4.17 10^{-5} per locus) was very similar to that obtained in mice chronically exposed at 0.08 mGy min^{-1} to a total dose of 6.2 Gy (3.15 10^{-5} per locus). The mutation rate with fractionation was, however, about a third of that obtained for a single exposure to 6.4 Gy at 0.17 Gy min^{-1} (13.1 10^{-5} per locus).

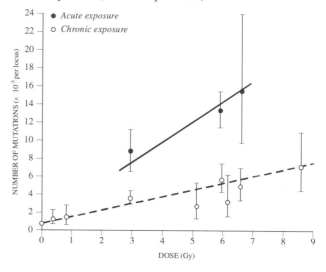

Figure XVI.   Specific locus mutations in mouse spermatogonia following radiation exposure [R6].

183. Searle [S10] reviewed data from a number of publications on specific-locus mutations in spermatogonia of mice after chronic exposure to gamma rays. Data points from a number of authors, including those by Russell et al. [R5, R6], were obtained for doses in the range 0.38–8.6 Gy. A linear relationship gave a good fit to the data on mutation frequency: I = 8.34 10^{-6} + 6.59 10^{-6}D (assuming 100 roentgens ≈ 1 Gy). This fit does not differ significantly from that obtained by Russell and Kelly [R6]. With acute exposures, a peak in the incidence of mutations was obtained with a decline in the incidence at doses between 6.7 and 10 Gy. The reduction at high dose rates may be attributed to more extensive killing of spermatogonia at doses above 6.7 Gy and to a lower mutational response in those more resistant spermatogonia that survive.

184. Searle [S10] also reviewed data on specific locus mutations in mice both acutely and chronically exposed to neutron irradiation with dose rates varying between 0.79 Gy min^{-1} and 0.01 Gy min^{-1}. A linear function fitted essentially all the data points between 0.5 Gy and 2.1 Gy, with the exception of a single value at 1.9 Gy. The fit had the form I = 8.30 10^{-6} + 1.25 10^{-4}D. There was no evidence

of a dose-rate effect for neutrons. The ratio between the slopes for male mice chronically exposed to low-and high-LET radiations gives an RBE of 19.

185. In female mice irradiated just before birth, there is a more pronounced dose-rate effect for mutational damage to oocytes than in those irradiated later. Using the specific-locus method, Selby et al. [S24] examined the effect of dose rate on mutation induction in mouse oocytes. Female mice were given 3 Gy of whole-body x-irradiation at dose rates of 0.73–0.93 Gy min^{-1} and 7.9 mGy min^{-1} at 18.5 days after conception. The frequency of specific-locus mutations measured in the offspring decreased from 8.7 10^{-6} to 6 10^{-7} mutations per gray per locus (a factor of about 14) between the high-dose-rate and low-dose-rate exposure. In practice, the mutation rate at the low dose rate (7.9 mGy min^{-1}) did not differ significantly from that in controls, indicating very effective repair. Similar calculations based on the results of irradiating mature and maturing oocytes at the same dose rates [R19, R20] suggest an approximately fourfold drop in the induced mutation frequency in the adult. These results suggested that females irradiated just before birth have a more pronounced dose-rate effect, although the confidence limits of magnitude of the dose-rate effect are too wide for firm conclusions to be drawn.

C. SUMMARY

186. The results of animal studies contribute to the database of information available for assessing the biological effects of low doses of ionizing radiation and dose-response relationship. Because of differences in radiosensitivity between animals and humans, the results obtained from animals cannot be used directly to obtain quantitative estimates of cancer risks for human populations. Animal studies are, however, valuable for determining the shape of dose-response relationships as well as for examining the biological and physical factors that may influence radiation responses. They are also of use for examining how factors such as age at exposure, radiation quality, and dose protraction or fractionation can influence the tumour response. Laboratory animals have the advantage that they are a homogeneous population with minimal biological variability, and the influence of confounding factors can be eliminated. Although studies with laboratory animals generally involve fewer individuals than epidemiological studies, they have the advantage that they are carried out under controlled conditions with good estimates of the radiation dose and with a known spontaneous cancer rate. In the case of radiation-induced hereditary disease animal studies provide the principal source of quantitative information.

187. Dose-response relationships for many tumour types in various animal models following exposure to both low- or high-LET radiation can be reasonably well represented by a linear or linear-quadratic function for the dose ranges analysed. In many cases, however, alternative fits to the data are also possible. Other model fits include the possibility of a threshold dose below which tumours will not occur, as well as more complex functions in which the time for the tumour to appear is much later at low dose rates than at high dose rates and thus can also suggest the presence of a threshold for a response. For some lung tumours it has been demonstrated that high local doses from alpha irradiation are required to cause proliferation of initiated cells and to promote the development of malignant lesions.

188. Analysis of a series of studies in mice has shown that the lowest dose at which a statistically significant (p=0.05) increase in tumour yield is observed varies from study to study. It depends on the number of animals used in each experiment, the radiation sensitivity of the species to specific cancers, and the spontaneous cancer rate. It also depends on the range of doses used. From the animal data reviewed, the lowest single (acute) dose to give a significant (p=0.05) effect on tumour yield is of the order of 100–200 mGy (low-LET). The higher values obtained in other studies can be attributed to lack of sensitivity, high control incidence, or to small numbers of animals. Values for the lowest dose to give a significant increase in risk following continuous (chronic) irradiation are generally higher than those for acute irradiation. It may be concluded that animal studies provide quantitative information on risks of radiation-induced tumour induction at low to intermediate doses but do not, and probably cannot, provide direct information at doses much less than about 100 mGy.

189. For radiation-induced hereditary disease, the most comprehensive information comes from measurements of specific-locus mutations in mouse spermatogonia. Data from a number of laboratories have demonstrated a dose-response relationship for low-dose exposures from low-LET radiation that is well fitted by a linear response. The lowest dose tested in these studies was 380 mGy. Data from both male and female mice indicate a significant effect of dose rate. It was notable that although the incidence of mutations in male mice fell by a factor of about 3 for a reduction in dose rate from 800–900 mGy min^{-1} to 8 mGy min^{-1}, a further reduction in the dose rate to 0.007 mGy min^{-1} failed to further reduce the yield of mutations. This independence of dose rate occurred over a range of doses differing by rather more than a factor of 1,000, and it was concluded that it was unlikely that a further reduction in mutation frequency would be obtained at even lower dose rates. This suggests that a substantial fraction of the damage to DNA that results in the induction of heritable mutations is not amenable to effective repair.

III. EPIDEMIOLOGY

190. The extent to which epidemiological studies can provide information on the effect of ionizing radiation on the induction of cancer at low doses is considered in this Chapter. Although the role of radiation in inducing cancer was recognized soon after the discovery of x rays by Röntgen in 1895, up to the early 1950s only high doses causing acute tissue damage were considered to be important. This view was reflected in the early recommendations of the International Commission on Radiological Protection (ICRP) and by national organizations. By 1959, however, the stated aim of ICRP in setting dose limits was to "prevent or minimize somatic injuries and to minimize the deterioration of the genetic constitution of the population" [I6]. These recommendations reflected an increasing awareness of the effect of radiation in inducing cancer, particularly leukaemia, at low doses and was largely the result of information becoming available from the follow-up of the survivors of the atomic bombings and groups exposed for medical reasons (see, for example, [L29]).

191. By the early 1970s it was known that radiation is capable of causing tumours in many tissues of the body, although the frequency of appearance following a unit dose varied markedly from one organ or tissue to another. Information on the dose-related frequency of tumour induction by radiation had become available from a number of epidemiological studies of persons exposed to external radiation or internally incorporated radionuclides. In the UNSCEAR 1972 Report [U8], the Committee gave preliminary estimates of the risk of leukaemia and some solid cancers based on the survivors of the atomic bombings and other groups exposed at high dose rates. It also pointed out that animal studies suggested that risks per unit dose at lower dose rates could be lower and that risk estimates based on groups exposed at high dose rates would be overestimates for doses and dose rates received from environmental sources.

192. The chief sources of information on the risks of radiation-induced cancer were the survivors of the atomic bombings exposed to whole-body irradiation at Hiroshima and Nagasaki; patients with ankylosing spondylitis and other patients who were exposed to partial body irradiation therapeutically, either from external radiation or from internally incorporated radionuclides; and various occupationally exposed populations, in particular uranium miners and radium dial painters. Follow-up of these populations had shown that there is a minimum period of time between irradiation and the appearance of a radiation-induced tumour, although this "latency period" varies with age and from one tumour type to another. Some types of leukaemia and bone cancer have latency periods of only a few years, with most of the risk being expressed within about 25 years of exposure. Many tumours of solid tissues (e.g. liver or lung) have latency periods of 10 years or more, and it was not clear whether their incidence passes through a maximum and subsequently declines with time

following exposure or whether the risk levels out or even increases indefinitely during the remainder of life.

193. To project the overall cancer risk for an exposed population, it is therefore necessary to use empirical models that extrapolate over time data based on only a limited portion of the lives of the individuals. Two such projection models have been generally considered:

(a) the additive (absolute) risk model, which postulates that radiation will induce cancer independently of the spontaneous rate after a period of latency and that variations in risk due to gender and age at exposure may occur; and

(b) the multiplicative (relative) risk model, in which the excess cancer risk (after latency) is given by a constant factor applied to the age-dependent incidence of natural cancers in the population.

Both models imply an increasing risk of cancer with increasing radiation dose. In addition to these two models, alternative fits to the epidemiological data to assess lifetime risks have also been used such as a model expressing excess relative risk as a function of attained age [K27]. Further information is given in Annex I, *"Epidemiological evaluation of radiation-induced cancer"*.

194. Most organizations assessing risks in the 1970s, including UNSCEAR in its 1977 Report [U7], adopted the additive model for the assessment of cancer risks, although the Committee on the Biological Effects of Ionizing Radiation (BEIR I) [C17] of the United States National Academy of Sciences used both models for risk assessment. In a major revision of its recommendations in 1977, ICRP, in its System of Dose Limitation, considered it necessary to limit the incidence of radiation-induced fatal cancers and severe hereditary disease to a level accepted by society [I8]. Implicit in this approach for stochastic effects was the necessity to use quantified risks of radiation-induced disease in setting limits on exposure. The risks of cancer and hereditary disease adopted by ICRP were derived mainly from reviews by UNSCEAR [U7]. Organ-specific risks were given for the red bone marrow, the lungs, cells on bone surfaces, and the thyroid and breast. No specific risks were given for the other organs and tissues of the body, which were pooled in a risk factor for all the "remainder" organs and tissues.

195. During the 1980s new information progressively became available from the Life Span Study in Japan, and this necessitated a revision of the earlier risk estimates by UNSCEAR. The data available from the extended follow-up of the survivors of the atomic bombings indicated that a multiplicative risk model now gave a best fit to data for most solid cancers [U4]. These new risk estimates, which also allowed for improvements in dosimetry, were taken into account by ICRP in its 1990 recommendations [I2].

196. Overall, the lifetime risks calculated in recent years by various national and international organizations are not too different (e.g. [C1, I2, M18, U4]). The estimates of risk have, however, been obtained by direct extrapolation from epidemiological studies. They are, therefore, appropriate for populations exposed at high doses and dose rates. To allow for a reduced effect of radiation in inducing cancer when exposures are at low doses or low dose rates, most organizations have recommended the use of a reduction factor to obtain risks for application in radiation protection. ICRP [I2] applied a reduction factor of 2 (it called the factor a dose and dose-rate effectiveness factor (DDREF)) to give a risk coefficient for radiation protection purposes. In the UNSCEAR 1993 Report [U3], the Committee reviewed epidemiological and experimental data relevant to the choice of a reduction factor. It recommended that for tumour induction, the DDREF adopted should, to be on the safe side, "have a low value, probably no more than 3". Insufficient data were available to make recommendations for specific tissues.

197. Epidemiological studies were recently reviewed in the UNSCEAR 1994 Report [U2], and further studies and results are reviewed in an accompanying Annex I, "*Epidemiological evaluation of radiation-induced cancer*". In this Chapter the statistical difficulties associated with obtaining quantitative estimates of the risk of radiation-induced cancer from epidemiological studies at low doses are first examined. The available data from groups exposed at high dose rates, from which dose-response relationships and quantitative risk estimates are generally obtained, and from groups exposed at low doses and dose rates are then reviewed. Also considered is the choice of an appropriate value of the reduction factor for assessing risks at low doses and doses rates from studies of groups exposed at high doses and high dose rates.

198. There are no human data so far that can be applied in determining quantitative dose-response relationships or risk estimates for hereditary disease. Risk factors for hereditary disease have been considered in previous UNSCEAR reports [U3, U4].

A. STATISTICAL CONSIDERATIONS

199. Making quantitative estimates of the risk of cancer associated with low doses of ionizing radiation is compli-cated. Small epidemiological studies often have insufficient statistical power to detect any increase in risks. If bias has arisen in a study through, for example, failure to follow up a large percentage of a cohort of persons or to allow for con-founding factors, then spurious positive or negative findings could occur. In low-dose studies where the excess risks are predicted to be small, it is particularly important to ensure that the potential for bias and confounding is kept as low as possible, as this can create spurious results.

200. It is not, at present, possible to distinguish cancers induced by ionizing radiation from those due to other causes. A particular result of an epidemiological study is normally considered to be "statistically significant" if, in the absence of an effect, the probability of its occurrence is less than 1 in 20. If a large number of disease outcomes (e.g. different cancer types) are examined, however, possibly for each of several age groups and time periods, it is quite likely that a "statistically significant" finding will arise simply by chance. It is therefore important to examine the results of any epidemiological study in the context of possible dose-response relationships, other epidemiological studies, and supporting experimental evidence.

201. As with animal studies, the statistical power of an epidemiological study to detect an excess risk associated with ionizing radiation exposure depends on a number of factors. In Annex I, "*Epidemiological evaluation of radiation-induced cancer*", a procedure is described for assessing the power of a study to detect an elevated risk of a disease before a study is conducted. The statistical precision of completed studies is also examined. An example illustrates how the power of a cohort study to detect an elevated risk depends on the relative sizes of both the exposed and control populations, their absolute numbers, and the total numbers of cancers. These, in turn, depend on the baseline cancer rates, the length of follow-up, the radiation dose, and the specific radiation sensitivity of the organ(s) or tissue(s). Thus a study based on a very large cohort may not be particularly informative if a rare cancer is under investigation and the follow-up is short. Conversely, a study based on a fairly small cohort may be quite informative if a common cancer is being investigated and the follow-up is long. The distribution of the population and the number of cancer cases between various exposure dose groups will also influence the ability of a study to define a dose-response relationship. More detailed information on statistical considerations is given in the above-mentioned Annex.

202. The limitations of statistical power and the possibility of bias or confounding will constrain not only the ability to detect small increases in the risk of cancer but also the determination of whether or not there is the potential for a dose threshold for radiation carcinogenesis in specific tissues. Some examples of dose-response relationships obtained from epidemiological studies and the ability to detect risks at low dose are illustrated below.

B. HIGH-DOSE AND HIGH-DOSE-RATE EXPOSURES

203. The primary basis for evaluating risks of cancer associated with radiation exposure is the epidemiological study of human health in populations that include groups exposed at high doses and generally at high dose rates. The main features of the major high-dose-rate epidemiological studies were considered in the UNSCEAR 1994 Report [U2] and are reviewed in Annex I, "*Epidemiological evaluation of radiation-induced cancer*". The Life Span Study of the survivors of the atomic bombings at Hiroshima and Nagasaki by the Radiation Effects Research Foundation (RERF) is of particular importance in risk estimation. As well as involving a population of all ages and both sexes, the Life Span Study is based on large numbers of persons with a wide range of

whole-body doses. Consequently, it has high statistical power to examine any variation in cancer risk with dose. The interpretation of the dose-response data is, however, complicated by the fact that exposure was to both gamma rays and neutrons. An RBE of 10 has generally been assumed when fitting the dose-response data.

204. Other high-dose, high-dose-rate epidemiological studies are more limited in terms of the sex and age structure of the exposed population or in terms of the organs irradiated. However, they do provide additional information on risks for particular organs or for exposures at particular ages.

205. As discussed in Annex I, "*Epidemiological evaluation of radiation-induced cancer*", the statistical power of epidemiological studies to assess risks depends on the range of doses received by the study population and the spontaneous cancer rate. Analyses based on a restricted set of data for exposures in the low-dose region would have much reduced statistical power to detect risks. However, it may still be possible to detect raised risks in some circumstances. Furthermore, analysis of the dose-response relationship over the whole range of doses can be informative in making inferences about risks at low doses, when interpreted in conjunction with the mechanistic and computational modelling approaches described in Chapters IV and V.

1. Dose-response relationships

(a) Survivors of atomic bombings

206. Various analyses of the dose-response data for the Life Span Study have been reported, and with increasing length of follow-up the quality of the information available has improved considerably. Pierce and Vaeth [P1] examined mortality data from the follow-up of the Life Span Study to 1985, based on the most recent published DS86 dosimetry. In their analyses, those persons with shielded kerma estimates in excess of 4 Gy were excluded, in view of an apparent levelling-off in the dose response that may be associated with errors in the estimates of such high doses or with cell killing. The authors concluded that for all cancers other than leukaemia the data could be well fitted by a linear dose-response model, although a linear-quadratic model would not be inconsistent with the data.

207. Shimizu et al. [S1] assessed the slopes of the dose-response curves for the survivors of the atomic bombings in various low-dose regions. Over the lowest dose range (0–0.49 Gy) with a statistically significant trend (p<0.05), the value of the excess relative risk per gray for all cancers, other than leukaemia, was 0.38. This is similar to the value obtained for the whole dose range (0.41), in line with the analysis by Pierce and Vaeth [P1], suggesting a linear dose-response relationship.

208. For leukaemia mortality, the data up to 1985 on survivors of the atomic bombings suggested that a linear dose-response model did not provide a good fit and that a linear-quadratic model would be preferred [P1]. In the analysis of Shimizu et al. [S1], the excess relative risk per

gray of leukaemia mortality in the dose region 0–0.49 Gy was 2.40 (p<0.05), which is about half of the value over the whole dose region (0–6 Gy) of 5.21 (p<0.001). This supported the conclusions of Pierce and Vaeth [P1] that a linear-quadratic dose-response model better fits the data.

209. Errors in the estimates of dose in the Life Span Study can substantially alter the shape of dose-response relationships. The problem of random dosimetry errors for the RERF data on the Life Span Study has been investigated by a number of authors [G2, J3, P1, P7]. Pierce and Vaeth [P1] found that after adjustment for dosimetric errors there were non-significant indications of upward curvature in the dose-response function for mortality from all solid cancers, while for leukaemia the evidence for curvilinearity became stronger.

210. The evidence for possible curvilinearity in the dose response for leukaemia and for solid cancers in the most recent cancer incidence data [P3] has been examined [L7, M19]. A variety of relative risk models have been fitted to the data, including those that allow for a possible dose threshold. Errors in estimates of doses were also allowed for, as these can substantially alter the shape of the dose-response relationship.

211. For solid cancers taken together, a variety of models provided little evidence for curvilinearity. A significant positive dose response was found for all survivors receiving doses less than 0.5 Sv but not for doses less than 0.2 Sv (assuming an RBE of 10 for neutrons). A threshold-linear relative risk model fitted to the data gives no support for a threshold above about 0.2 Sv, and the data are consistent with the absence of a threshold. For most solid cancers taken separately, the data on cancer incidence are also consistent with a linear dose-response relationship [T4] (also see Annex I, "*Epidemiological evaluation of radiation-induced cancer*"). These findings are in accord with previous analyses of the dose response for all solid cancers taken together.

212. In contrast, the latest data on non-melanoma skin cancer incidence indicate substantial curvilinearity, consistent with a possible dose threshold of about 1 Sv to the skin or with a dose response in which the excess relative risk (ERR) is proportional to the fourth power of dose, with a decrease in the response at high doses (>3 Sv) [L30] (Figure XVII). Supporting epidemiological data on the shape of the dose response for non-melanoma skin cancer are, however, limited. For example, Ron et al. [R15] found no evidence for curvilinearity in the dose response in a group of children in Israel who had been treated with large therapeutic doses of radiation for tinea capitis (ringworm of the scalp). However, the doses in this study were generally much higher than those received by survivors of the atomic bombings. Thus there are no patients with doses less than 5 Gy (low-LET) in the Israeli data set. The only other information on the shape of the dose response for skin cancer comes from animal experiments. Some evidence of a threshold has been obtained in studies with mice and rats [A5, B23, O3, P8], although a linear-exponential form of induction curve was obtained for beta-irradiated male SAS/4 mice [W8].

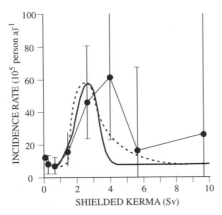
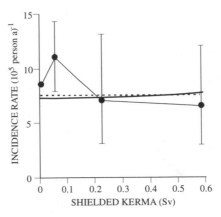

Figure XVII. Observed incidence of non-melanoma skin cancer in survivors of atomic bombings (CI: 90%) compared with fourth-power exponential (solid line) and linear-exponential threshold (dotted line) models of dose response [L30].
The diagram on the right shows the low-dose region in detail.

213. In contrast with solid cancers, the analysis by RERF and other groups of the dose-response relationship for leukaemia incidence in the Life Span Study cohort found quite a marked upward quadratic component, i.e. significant upward curvature [P2, P3], with the evidence for non-linearity being strongest for acute myeloid leukaemia [P3]. For the three main radiation-inducible leukaemia subtypes analysed together (acute lymphatic leukaemia, acute myeloid leukaemia, and chronic myeloid leukaemia), there is a significant increase in the risk of leukaemia if the dose responses for all survivors with doses less than 0.5 Sv are considered together [L7, M19]. This significance vanishes, however, if doses less than 0.2 Sv are considered.

214. Analysis of leukaemia incidence among the Japanese atomic bomb survivors by Little and Muirhead [L7] showed that incorporation of a threshold in the linear-quadratic dose response yielded an improvement in fit at borderline levels of statistical significance [best estimate of threshold for a linear-quadratic-threshold model was 0.12 Sv (95% CI: 0.01–0.28; two-sided p=0.04)]. This analysis takes account of random dosimetric errors, but not possible systematic errors in dose estimates. The fits of a linear-quadratic-threshold dose response to the recently released leukaemia mortality data [P2] and that takes account of random dosimetric errors, demonstrated that the threshold was not significantly different from zero [best estimate of threshold for a linear-quadratic-thresh-old model was 0.09 Sv (95% CI: <0.00–0.29; two-sided p=0.16)] [L44]. Similar findings have been reported by Hoel and Li [H26] in analyses that do not take account of dosimetric error. Comparison of the incidence and mortality data by Little and Muirhead [L44] and Little [L49] demonstrates the essential similarity of the leukaemia incidence and mortality data. Little and Muirhead [L44] concluded that the most probable reason for the difference between the findings in the incidence and mortality data sets was the finer subdivision of dose groups in the mortality data set. (There are 14 dose groups in the mortality data sets in their publicly available form, compared with 10 dose groups in the incidence data sets.)

215. Recent analyses by Kellerer and Nekolla [K25] and Little and Muirhead [L52] of the tumour incidence and mortality data demonstrate that if account is taken of possible systematic errors in the Hiroshima DS86 neutron dose estimates, then there is evidence of appreciable upward curvature in the dose response for solid tumours in the Life Span Study data. This is particularly marked if analysis is restricted to the 0–2 Gy dose range rather than the 0–4 Gy dose range that has been used for most analyses of dose response in the Life Span Study. Over the 0–2 Gy dose range, the low-dose extrapolation factor (LDEF) for all solid tumour incidence is 1.43 (95% CI: 0.97–2.72), and so is comparable with the LDEF for leukaemia incidence, 1.58 (95% CI: 0.90–10.58) [L52].

216. Recent data on the mortality of the atomic bomb survivors was reported by Pierce et al. [P2]. The follow-up covers the period to 1990 and includes an extra 10,500 survivors for whom DS86 dose estimates have been calculated. The total cohort comprises approximately 86,500 persons, 60% of whom received doses in excess of 5 mSv. Of the total population, 44% had died by 1990, including 8,827 who died of cancer. The shape of the dose-response curve for all solid cancers is essentially linear up to 3 Sv, beyond which there is an apparent decrease in risk. This may be attributed both to cell killing and to imprecision in the estimates of high doses (Figure XVIII).

217. As discussed in Annex I, *"Epidemiological evaluation of radiation-induced cancer"*, the dose-response relationships for mortality from many specific tumour types (stomach, colon, lung) are consistent with a linear response although generally based upon the analysis of a restricted number of cases. For leukaemia, the dose response over the range 0–3 Sv can be fitted with a linear-quadratic dose-response relationship (Figure XVIII).

218. While the Life Span Study provides information on cancer risks in a number of tissues, there are others for which there is either very little or no evidence for an effect. These include, for example, the bone, cervix, prostate, testes and rectum.

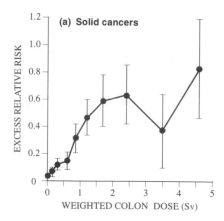 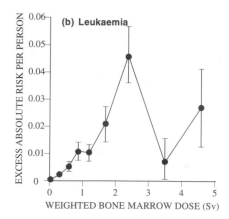

Figure XVIII. Dose response for mortality from solid cancer (males of 30 years of age at exposure) and leukaemia in survivors of atomic bombings in Japan [P2].

(b) Other groups exposed to low-LET radiation

219. Additional data on dose-response relationships for groups exposed to low-LET radiation are available from a number of other studies. For Canadian tuberculosis patients given fluoroscopies, Miller et al. [M2] showed that a linear dose-response relationship gave a good fit to the data on breast cancer among patients in Canadian provinces other than Nova Scotia. For patients from Nova Scotia, who generally received higher doses, the dose-response relationship was also consistent with linearity, but it had a steeper slope than for other Canadian provinces. Howe and McLaughlin [H31] have given further results from an extended follow-up of this population. The data on breast cancer mortality could again be fitted with a linear dose-response relationship. As before, the slope of the dose trend was greater for patients in Nova Scotia than for patients in other provinces.

220. Dose-response analyses have also been performed for some other groups with medical exposures. Boice et al. [B6] studied the relationship between the risk of breast cancer and dose for women in Massachusetts (United States) given multiple chest x-ray fluoroscopies. For this study, doses were mostly in the range 0–3 Gy. A linear dose-response model was found to provide as good a fit to these data as a linear-quadratic model, whereas a purely quadratic model did not fit well. Among women given radiotherapy for cervical cancer, the risk of leukaemia increased with dose up to 4 Gy, in a manner consistent with linearity, although the data were also consistent with a quadratic dose response; beyond 4 Gy the risk decreased, probably as a result of cell killing [B7]. At lower doses, Ron et al. [R8] found that the risk of thyroid cancer among children in Israel irradiated for tinea capitis was consistent with a linear dose-response relationship, based on doses that were mostly less than 0.15 Gy.

(c) Groups exposed to high-LET radiation

221. Information on dose-response relationships that depart from the conventional linear or linear-quadratic response has been obtained for bone tumours arising from alpha particle irradiation of bone following the deposition of isotopes of radium. Extensive epidemiological information is available on groups of persons exposed, principally by ingestion, to ^{226}Ra and ^{228}Ra in the 1920s and 1930s. The most comprehensive data relate to female radium dial painters. The data on tumour induction in this population have been the subject of extensive analysis over the last 40 years (e.g. [E1, F13, H21, R4]). After the radium programme at the Argonne National Laboratory finished in the early 1990s, Rowland brought together all the data collected in this long-term study [R16]. His most recent analysis considered all female radium dial painters with body content measurements and who had entered the study prior to 1950, a total of 1,530 women [R17]. In this cohort, 46 women had bone sarcomas and 19 had head sinus carcinomas; 3 women had both a bone sarcoma and a head sinus carcinoma. The analysis incorporated revised estimates of systemic intake, which took into account the magnitude of the original intake. This has been shown to influence the retention kinetics and hence the cumulative doses [K15, R7]. The intakes by the various members of the cohort covered several orders of magnitude. The 46 bone sarcomas had appearance times ranging from 7 to 63 years. The lowest systemic intake associated with a bone sarcoma was 3.7 MBq (100 μCi). This malignancy, diagnosed in 1981 and resulting from an intake in 1918, was thus detected 63 years later.

222. Various forms of a general incidence–systemic intake expression

$$I = (\alpha SI + \beta SI^2)\, e^{-\gamma SI} \qquad (3)$$

were fitted to the data and tested with a χ^2 statistic. In the equation, I is the incidence of bone tumours, α, β, and γ are constants, and SI is the systemic intake. No acceptable fit to the equation was found. However, when a constant, C, was included in a general function of the form

$$I = C + kSI^\beta\, e^{-\gamma SI} \qquad (4)$$

in which k and β are constants, a good fit to the data could be obtained with C = $-1.44\ 10^{-4}$, k = $2.14\ 10^{-15}$, β = 3.15, and γ = $7.06\ 10^{-5}$. With the incidence I equal to zero, this gives an intercept at 2,920 kBq. This fit to the data, shown in Figure XIX, gives good evidence for a threshold dose for the induction of bone tumours. The dose-response data were further analysed by Thomas [T12], who suggested the data were consistent with a threshold for tumour induction in the range 3.9–6.2 Gy high-LET (average bone dose). He proposed a rounded value of 10 Gy (average bone dose) as a "practical threshold" below which there should be little cause for concern.

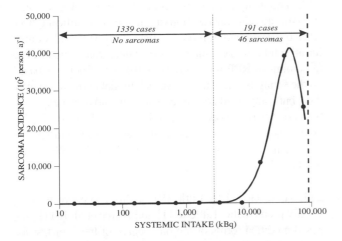

Figure XIX. Bone sarcoma incidence in female radium dial painters [R17].
Systemic intake is kBq ^{226}Ra plus 2.5 × ^{228}Ra activities.

223. Various forms of the general dose-response expression were also fitted to the data on head sinus carcinoma. In contrast to the data on bone sarcomas, linear, linear-exponential, and dose-squared-exponential functions all provided acceptable fits. Models that included a threshold would also fit the data, but the threshold value was not statistically significant.

224. It was concluded that the tumour induction data for osteosarcoma induction show a very steep dose response [R16]. Whether this actually demonstrated a threshold or simply showed a very low probability of osteosarcoma induction at intakes below about 3,000 kBq could not be determined. For head sinus carcinoma, the data did not suggest the presence of a threshold, although various model fits to the data were possible, reflecting the paucity of data, which prevented discrimination between alternative functions.

225. Further information on bone tumour pathology in persons exposed to external radiation or internally incorporated radionuclides may explain some of these observations. A review of bone tumour pathology in patients treated with ^{224}Ra revealed an unexpectedly high proportion of bone sarcomas of the fibrous connective tissue type, including the first case of malignant fibrous histiocytoma (MFH) of bone described after internal irradiation [G17]. Out of 46 bone tumours in the ^{224}Ra patients, osteosarcoma was the most common histological type (48% of cases), but 30% of these were fibrosarcoma-MFHs and the remainder were chondrosarcomas, malignant lymphomas, myelomas, and malignant chordomas. The 30% of fibrosarcoma-MFHs substantially exceeds the usual prevalence of this disease, which is 8%–11% in spontaneous bone

tumours. In a follow-up study, a similar spectrum of tumours was obtained in persons occupationally exposed to $^{226/228}$Ra, patients given external irradiation and other so-called secondary bone tumours arising at sites of pre-existing bone lesions as had been obtained in the ^{224}Ra patients [G17].

226. The authors of the review [G17] concluded that disturbance of the local cellular system caused by deterministic radiation damage and repair resulted in the unexpectedly high proportion of fibrosarcoma-MFHs. It was considered that the development of the tumours reflected the cell types involved in a disturbed remodelling process in the skeleton. The reactive proliferation of the predominantly fibroblastic tissue at the site of tissue damage could be the presumptive origin of this special type of radiation-induced bone sarcoma. As a fibrotic response would be likely to arise as a consequence of deterministic radiation damage, the fibrosarcoma-MFH type of tumour might well arise only at doses above a limiting threshold.

(d) Groups exposed to radon

227. Radon has been extensively studied as a human carcinogen. Epidemiological studies are reviewed in Annex I, *"Epidemiological evaluation of radiation-induced cancer"* and are summarized here only briefly. The results of a series of cohort studies of miners in countries throughout the world have provided the basis for estimates of the risk of lung cancer associated with exposure to radon and its decay products. These data, although subject to some uncertainties, have allowed characterization of exposure response relationships [L51]. The exposure response relationship in the various studies of radon exposed miners is consistent with linearity, but the slope appears to be higher at lower exposure rates (Figure XX). As discussed in Annex I, this apparent inverse exposure-rate effect does not imply that low exposures carry a greater risk than higher exposures; rather it suggests that for a given total exposure, the risk is higher if the exposure is received over a long rather than a shorter period of time. This

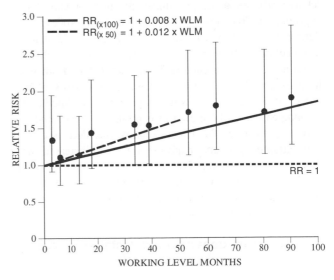

Figure XX. Relative risks of lung cancer from pooled data for miners, restricted to <100 WLM exposure and also to <50 WLM exposures [L53].

could reflect some cell killing at high exposure rates. Case-control studies of residential radon exposure and lung cancer have also been conducted in various countries. Although these have also been informative, the generally lower exposures of people and methodological difficulties have meant the power of these studies is less than that of the occupational studies. However, the estimates of lung cancer risk based upon a recent meta-analysis of these eight studies are in close agreement with the risk predicted on the basis of miner data [L53].

228. The Committee on Biological Effects of Ionizing Radiations (BEIR VI) considered the data on 11 miner cohorts exposed to radon previously analysed by Lubin et al. [L48]. The most recent data available were used in developing the Committee's risk models for radon exposure [C26]. The Committee recognized that care is needed in combining data from different cohorts of underground miners around the world. The levels of exposure to radon and other relevant covariates, such as arsenic and tobacco smoke, differed appreciably among groups of miners. The completeness and quality of the data available on relevant exposures also differed notably among the cohorts. Information on tobacco consumption was available for only 6 of the 11 cohorts; of these 6, only 3 had information on duration and intensity of exposure to tobacco smoke. Lifestyle and genetic factors that influence susceptibility to cancer might also account for heterogeneity among cohorts.

229. Despite those differences, the Committee concluded that the best possible estimate of lung cancer risk associated with exposure to radon and its decay products would be obtained by combining, in a judicious manner, the available information from all 11 cohorts. The Committee used statistical methods for combining data that both allowed for heterogeneity among cohorts and provided an overall summary estimate of the lung cancer risk. Confidence limits for the overall estimate of risk allowed for such heterogeneity.

230. The Committee's risk models described the excess relative risk as a simple linear function of cumulative exposure to radon, allowing for differential effects of exposure during the periods 5–14 years, 15–24 years, and 25 years or more before death from lung cancer. The most weight was given to exposures occurring 5–14 years before death from lung cancer. The Committee examined two types of risk models in which the excess relative risk was modified either by attained age and duration of exposure or by attained age and exposure rate. The excess relative risk decreased with both attained age and exposure rate and increased with duration of exposure. For cumulative exposures below 0.175 J h m^{-3} (50 WLM), a constant-relative-risk model without these modifying factors appeared to fit the data as well as the two models that allow for effect modification.

2. Minimum doses for a detectable increase in cancer risk

231. It is important to examine the lowest levels of dose at which a significantly elevated level of radiation-induced cancer has been observed in human populations. Relevant information from epidemiological studies is available from the follow-up of the atomic bomb survivors, from other studies of thyroid cancer in infants, children, and adults, and from studies of the risk of cancer in children following radiation exposure *in utero*.

(a) Survivors of atomic bombings

232. The analysis by Pierce et al. [P2] of the atomic bomb survivor mortality data set finds a statistically significant (two-sided p<0.05) trend in mortality risks in the 0–50 mSv range for all solid cancers combined, based upon follow-up to 1990 (assuming an RBE for neutrons of 10). This finding is based on the fitting of a linear relative risk model to the 0–50 mSv data, but using fixed adjustments for sex and age at exposure based on fits of a model to the full data set. Pierce et al. [P2] pointed out that without these adjustments for sex and age, the significance of the trend in dose in the 0–50 mSv group would be lost.

233. As discussed by Little [L11], this procedure is statistically problematic. Little [L11] and Pierce et al. [P11] proposed modified forms of the one-degree-of-freedom test for trend in the low-dose region, using nested models that incorporate sex and age adjustments but that do not rely on fixed modifications fitted to the whole dose range. When either of these modified tests is used, the finding of a significant increasing (two-sided p<0.05) trend with dose in the low-dose region (0–50 mSv) remains valid [P11].

234. Notwithstanding these statistical considerations, Pierce et al. [P2, P11] were cautious in their interpretation of this finding, which is at variance with the findings in the latest atomic bomb survivor solid tumour incidence data in which a significant excess risk of solid cancers is only seen down to doses of 200–500 mSv [T4]; they indicated that the finding in the 0–50 mSv group might be artefactual, resulting from the differential misclassification of cause of death in the lowest dose groups. Further information on recent data from the atomic bomb survivors is given in Annex I, *"Epidemiological evaluation of radiation-induced cancer"*.

(b) Thyroid cancer incidence

235. Information on the risks of radiation-induced thyroid cancer is described in Annex I, *"Epidemiological evaluation of radiation-induced cancer"*. Studies of thyroid cancer incidence following radiation exposure were reviewed by Shore [S6], and a combined analysis of seven studies was performed by Ron et al. [R9]. Among various cohorts with external low-LET exposures, the excess relative risk per gray tends to be higher for thyroid cancer than for most other solid cancers. Furthermore, the excess relative risk is higher for those irradiated at young ages than for adults. Studies of cohorts with low-dose, external irradiation of the thyroid in childhood are therefore of value for examining risks at low doses. The risks of thyroid cancer following exposure to ^{131}I are less well understood, as discussed in Annex I, *"Epidemiological evaluation of radiation-induced cancer"*.

236. A study of about 10,800 children in Israel given x-ray treatment for tinea capitis was reported by Ron et al. [R8]. The total dose was given in five daily fractions to five treatment fields on the scalp. While the dose to the scalp was of the order of several gray, the average total thyroid dose was calculated to be only about 100 mGy. An analysis over the range 0–0.5 Gy showed a statistically significant trend of increasing risk of thyroid cancer with increasing thyroid dose. In addition, the trend in relative risk per unit dose was greater for those irradiated at ages under five years than for those irradiated at older ages, in line with the general observation of an increasing relative risk with decreasing age at exposure, as well as being consistent with a study of thyroid cancer in young persons irradiated for enlarged tonsils [P4]. Among the tinea capitis patients less than five years old at exposure, the relative risk at about 0.1 Gy (100 mGy) was approximately 5 and was significantly greater than 1. This finding was, however, based on a fairly small number of cases, although it arose among those persons for whom the risk would be predicted to be greatest.

237. Thyroid cancer in a cohort of 2,657 infants in New York State given x-ray treatment for a purported enlarged thymus gland and followed for an average of 37 years has been reported by Shore et al. [S2]. Estimated thyroid doses ranged from 0.03 to more than 10 Gy, with 62% receiving less than 0.5 Gy. The dose-response relationship for thyroid cancer was fitted by a linear dose-response relationship, with no evidence of a quadratic dose component. An analysis restricted to the range 0–0.3 Gy showed a statistically significant trend with dose (p=0.002), although based on just four thyroid cancer cases with non-zero doses. The estimate of absolute excess risk per unit dose over this dose range was similar to that from the Israeli tinea capitis study [R8].

238. Ron et al. [R9] conducted a combined analysis of data from seven studies of thyroid cancer after exposure to external radiation. The range of doses varied considerably between the different studies. For exposure before age 15 years, linearity was considered to best describe the dose response, even down to 0.1 Gy. The estimated excess relative risk per gray of 7.7 (95% CI: 2.1–28.7) is one of the highest values found for any organ.

239. Thyroid cancer in a cohort of 4,404 children of whom 2,827 were given x-ray treatment for cancer in childhood and followed for an average of 15 years has been reported by de Vathaire [D14]. Estimated thyroid doses ranged from 0.001 to 75 Gy, with 41% receiving less than 0.5 Gy. The dose-response relationship for thyroid cancer was best fitted by a linear dose-response relationship, with no evidence for a quadratic component. A standardized incidence ratio of 35 (90% CI: 10–87, p<0.01) was found to be associated with a dose of 0.5 Gy to the thyroid.

(c) Exposures *in utero*

240. A number of studies have been published that have examined the risks of cancer in childhood following exposures *in utero*. These studies have particular advantages for detecting risks of cancer at low doses because of the low spontaneous cancer rate in childhood.

241. Information on cancer risks following radiation exposure *in utero* are available from studies of those with prenatal diagnostic x-ray exposures, as well as those irradiated as a consequence of the atomic bombings of Hiroshima and Nagasaki. The largest study of childhood cancer following prenatal x-ray exposure is the Oxford Survey of Childhood Cancers (OSCC), which is a national case-control study of childhood cancer mortality carried out in the United Kingdom. Information is also available from other studies of prenatal x-ray exposure that have been carried out in North America and elsewhere [B28].

242. The Oxford Survey of Childhood Cancers was started in the mid-1950s. Up to 1981, mothers of 15,276 cases and the same number of matched controls had been interviewed [K10]. During the late 1950s, the study investigators reported a doubling in the risk of childhood cancer associated with prenatal x-ray exposure [S7]. Later analysis covering a longer period indicated a falling risk with time and an average raised risk of about 40% (95% CI: 31–50) [B8, K10].

243. Data on doses to the embryo and fetus are available for the Oxford Survey of Childhood Cancers, although there is some uncertainty in these values. Table 11 shows the mean number of films per x-ray examination in the

Table 11
Doses from prenatal x rays in the Oxford Study of Childhood Cancers

Birth year	Mean number of films per examination	Mean dose per film according to reference (mGy)	
		[U8]	[S21]
1943-1949	1.9	18	4.6
1950-1954	2.2	10	4.0
1955-1959	1.9	5	2.5
1960-1965	1.5	2	2.0

Survey according to calendar period, together with estimates of the average dose per film made by the Committee in the UNSCEAR 1972 Report [U8] and by Stewart and Kneale

[S21]. UNSCEAR estimated that the average dose per examination was 10–20 mGy (low-LET) during the 1950s and decreased over time. Stewart and Kneale's dose estimates were about half of the UNSCEAR values. Based on the UNSCEAR dose estimates, Muirhead and Kneale [M8] estimated the absolute radiation-induced risk for the incidence of all cancers up to age 15 years to be about 0.06 Gy^{-1} (low-LET) (95% CI: 0.04–0.10). A similar risk estimate was calculated by Mole [M6] based on a national survey in the United Kingdom of doses from obstetric radiography performed in 1958, for which the average dose was about 6 mGy.

244. There has been concern that owing to the retrospective nature of the Oxford Survey of Childhood Cancers, which relied at least partially on mothers' memories, some bias may have been introduced. The results of the follow-up were supported by a study in the United States [M9, M10] of contemporary records of x-ray exposures of children born in hospitals in the north-eastern United States. In an initial study [M9] of 734,243 children born between 1947 and 1954, 556 children were identified as having died from cancer between 1947 and 1960. Prenatal x-ray exposure was associated with an increased risk of cancer, with relative risks for leukaemia of 1.58 and for solid cancers of 1.45. These increases were very similar to those in the Oxford Survey. In an extension of the study, however, with a further 695,157 children born up to 1960 and having 786 additional cases of cancer, the relative risk for leukaemia was similar to the first phase (1.48) but that for solid cancers was appreciably lower (1.06). The overall values of relative risks for leukaemia (1.52) and solid cancers (1.27) do not differ significantly when compared directly (p=0.4).

245. Bithell [B28] reviewed a number of studies that examined the risk of childhood cancer following *in utero* radiation exposure. None of the studies on their own had the statistical power of the Oxford Survey of Childhood Cancers, but a total of 12 studies, when taken together, gave a weighted average of an increase in relative risk of 1.37 (95% CI: 1.26–1.49). Including the Oxford Survey of Childhood Cancers data gave a relative risk of 1.39 (CI: 1.33–1.45). While the individual study designs and methods of analysis were very different, the overall finding lends support to the results of Childhood Cancers.

246. Doll and Wakeford [D3] reviewed the evidence from epidemiological studies on the risk of cancer in childhood from exposure of the fetus *in utero* from diagnostic radiology. They also considered the limited studies in experimental animals. They concluded that while information is available from a number of epidemiological studies, the most significant comes from the Oxford Survey. It was concluded that there is strong evidence for a causal relationship, with radiation doses to the fetus of the order of 10–20 mGy giving increases in the risk of childhood leukaemia and solid cancers of about 40%. Because of the low risk of cancer in childhood, the calculated absolute risk coefficient was approximately 6% Gy^{-1}. The analysis supports the view that small doses of radiation are potentially carcinogenic. The possibility still exists that there may be some as yet unidentified confounding factor in the Oxford Survey affecting both the probability of the fetus being irradiated *in utero* and the risk of subsequent cancer. A feature

of the data from the Oxford Survey that remains unexplained is that the increase in risk for both leukaemia and solid cancers following exposure *in utero* is essentially the same, with a relative risk of about 1.4. Most other human and animal studies consistently indicate different sensitivities of leukaemia and solid cancers [B45].

247. Several cohort studies of *in utero* exposures have not shown evidence of excess risk. Those studies, however, were small in size. Among those exposed to atomic bomb radiation *in utero* [J1], no childhood leukaemia cases have been observed. For 1,263 children irradiated *in utero* and followed from birth, two cases of cancer arose up to 15 years of age, compared with 0.73 expected from Japanese national rates [Y2]. The resulting upper limit on the 95% confidence interval for the absolute radiation-induced risk is 2.8 10^{-2} Gy^{-1} (low-LET). Continued follow-up showed an excess of adult cancers among those exposed to atomic bomb radiation *in utero*. Based on the follow-up to 1988, the relative risk at 1 Gy was estimated to be 3.77 [Y2], which is similar to that seen among survivors of the atomic bombings irradiated in the first 10 years of life [S1]. Further follow-up to the end of 1989 suggested a subsequent decrease in the relative risk [Y1], in line with the pattern indicated by the earlier follow-up of those exposed post-natally at ages under 10 years.

248. More recently, Delongchamp et al. [D2] reported cancer mortality data in atomic bomb survivors exposed *in utero* for the period October 1950 to May 1992. Only 10 cancer deaths were reported among persons exposed *in utero*. Although there were only two leukaemia deaths, this was higher than in a control group (p=0.054). Mortality from solid cancers at ages over 16 years was in excess of expected (ERR = 2.4 Sv^{-1}; 90% CI: 0.3–6.7); all the deaths occurred in females.

249. Thus, although there is some consistency in case-control studies in showing a raised risk of childhood cancer, the absence of clear confirmation in cohort studies leaves some uncertainty in establishing a risk estimate. For the Oxford Survey of Childhood Cancers, however, an increase in childhood cancer risk by about 40% is associated with doses of about 10–20 mGy (low-LET). A number of other studies, taken together, support this finding.

3. Effect of dose and dose rate

250. As explained above, quantitative information on the risk of cancer in human populations comes largely from epidemiological studies of population groups exposed at intermediate and high doses and dose rates. For the assessment of the risk of cancer from environmental and occupational exposure to radiation, a reduction factor, frequently termed a dose and dose-rate effectiveness factor (DDREF), has normally been used to assess risks at low doses and low dose rates. The choice of reduction factors was reviewed most recently in the UNSCEAR 1993 Report [U3] and has also been reviewed by ICRP [I2] and by a number of other international bodies. In the UNSCEAR 1993 Report [U3], the Committee examined cellular studies, data from experimental animal studies, and

information from epidemiological studies that would allow judgements to be made on an appropriate reduction factor. The judgements made in that report remain valid and are summarized here only briefly.

251. The dose-response information on cancer induction in the survivors of the atomic bombings in Japan provided, for solid tumours, no clear evidence for a reduction factor much in excess of 1 for low-LET radiation. For leukaemia, the dose response fits a linear-quadratic relationship, and a best estimate of the reduction factor is about 2. Analyses by Little and Muirhead [L52] of the latest cancer incidence data that take account of possible random errors and possible systematic errors in DS86 dose estimates show that there is little indication of upward curvature in the dose response for solid tumours over the 0–4 Gy dose range, although over the 0–2 Gy dose range and after adjustment of Hiroshima DS86 neutron dose estimates the upward curvature is more pronounced. There is marked upward curvature in the dose response for leukaemia over the 0–4 Gy dose range, which becomes less pronounced if attention is restricted to those receiving less than 2 Gy [L52]. If adjustments are made to the Hiroshima DS86 neutron dose estimates, then over the 0–2 Gy dose range the LDEF for all solid tumours is 1.43 (95% CI: 0.97–2.72), and so is comparable with the LDEF for leukaemia, 1.58 (95% CI: 0.90–10.58) [L52]. There is only limited support for the use of a reduction factor from other epidemiological studies of groups exposed at high dose rates, although for both thyroid cancer and female breast cancer some data suggest a value of about 3 may be appropriate.

252. The results of studies in experimental animals conducted over a dose range that was similar to, although generally somewhat higher than, the dose range to which the survivors of the atomic bombings in Japan were exposed, and at dose rates that varied by factors between about 100 and 1,000 or more, give reduction factors from about 1 to 10 or more, with a central value of about 4. Some of the tumour types for which information is available have a human counterpart (e.g. myeloid leukaemia and tumours of the breast and lung) while others do not (e.g. Harderian gland in the mouse) or require for their development substantial cell killing and/or changes in hormonal status (ovarian tumour, thymic lymphoma). Similar results to those obtained with animal tumour models have been obtained for somatic mutations and for transformation of cells in culture, although the reduction factors obtained have not been as large. In a number of the experimental studies on tumour induction, linear functions would give a good fit to both the high- and low-dose-rate data in the range from low to intermediate doses. This indicates that even if the cellular response can, in principle, be fitted by a linear-quadratic dose response, in practice it is not always possible to resolve a common linear term for exposures at different dose rates.

253. If human response is similar to that in experimental animals, then it can be envisaged that at lower dose rates than were experienced in Hiroshima and Nagasaki, a reduction factor greater than the value of about 1.5 that is suggested by analysis of the dose-response data could be obtained. However,

information from human populations exposed at low dose rates suggests risk coefficients that are not very different from those obtained for the atomic bomb survivors, although the risk estimates have wide confidence intervals.

254. In the UNSCEAR 1993 Report [U3], the Committee concluded that, when taken together, the available epidemiological and experimental data suggested that for tumour induction, the reduction factor adopted should, to be on the safe side, have a low value, probably no more than 3. Insufficient data were available to make recommendations for specific tissues [U3]. For high-LET radiation, a reduction factor of 1 was indicated on the basis of experimental data that suggested little effect of dose rate or dose fractionation on tumour response at low to intermediate doses. It was noted that a value of somewhat less than 1 is suggested by some studies, but the results are equivocal, and cell killing may be a factor in the tissue response [U3].

255. In the case of hereditary disease, the adoption of a reduction factor of 3 was supported by experimental data in male mice, although a somewhat higher value has been found with one study of female mice.

C. LOW-DOSE-RATE EXPOSURES

256. Information from studies of groups exposed to low dose rates is potentially of more direct relevance to risk estimates. However, studies of low-dose-rate exposure generally involve low doses and, because of the probably low excess risks, are likely to be hampered by a lack of statistical power and possibly also by confounding factors. Examination of the results of low-dose-rate studies can, however, provide a check on the risks derived by extrapolation from high-dose-rate studies.

1. Occupational exposures

257. Several studies have been conducted of nuclear industry workers. In the United States, Gilbert et al. [G3] performed a joint analysis of data for about 36,000 workers at the Hanford, Oak Ridge, and Rocky Flats weapons plants. Neither for the grouping "all cancers" nor for leukaemia was there any indication of an increasing trend in risk with dose. However, the upper limit of the 90% confidence interval for the excess relative risk per unit dose was several times greater than the corresponding value for the survivors of the atomic bombings in Japan in the case of all cancers other than leukaemia and slightly greater than the value from Japan in the case of leukaemia.

258. The first analysis of the National Registry for Radiation Workers (NRRW) in the United Kingdom examined cancer mortality in relation to dose in a cohort of over 95,000 workers [K3]. The mean lifetime dose received was 33.6 mSv; however, over 8,000 workers had a lifetime dose in excess of 100 mSv. For all malignant neoplasms, the trend in the relative risk with dose was positive but was not statistically significant (p=0.10). Based on a relative risk projection model,

the central estimate of the lifetime risk based on these data was 10% Sv^{-1}, which is 2½ times the value of 4% Sv^{-1} cited by ICRP [I2] for risks associated with the exposure of workers (based on applying a DDREF of 2 to the Japanese data). The 90% confidence interval for the NRRW-derived risk ranged from a negative value up to about six times the ICRP value. For leukaemia (excluding chronic lymphatic leukaemia, which does not appear to be radiation-inducible), the trend in risk with dose was statistically significant (p=0.03). Based on a projection model as used by BEIR V [C1], the central estimate of the corresponding lifetime leukaemia risk was 0.76% Sv^{-1}, which is 1.9 times the ICRP [I2] value for a working population (0.4% Sv^{-1}), with 90% confidence limits ranging from just above zero up to about six times the ICRP value.

259.   A second analysis of the NRRW cohort was published in 1999 [M47] and covered a total of 124,743 workers. For leukaemia, excluding chronic lymphatic leukaemia, there was a marginally significant increasing risk with dose. The central estimate of excess relative risk per sievert, 2.55 (90% CI: −0.03–7.16), is similar to that estimated for the Japanese atomic bomb survivors at low doses (2.15 Sv^{-1}, 90% CI : 0.43–4.68); the corresponding 90% confidence limits were tighter than in the first analysis, ranging from just under four times the risk estimated at low doses from the Japanese atomic bomb survivors to about zero. For all malignancies other than leukaemia, the central estimate of the trend with dose, 0.09 Sv^{-1} (90% CI: 0.28–0.52), was closer to zero than in the first analysis and smaller than the Japanese atomic bomb estimate of 0.24 Sv^{-1} (90% CI: 0.12–0.37) (without the incorporation of a dose-rate reduction factor). Also, the 90% confidence intervals were tighter than before and include zero. Overall, the second NRRW analysis provides stronger evidence than the first on occupational radiation exposure and cancer mortality; the 90% confidence interval for the risk per unit dose now excludes values that are more than four times those seen in the atomic bomb survivors, although they are also consistent with there being no risk at all.

260.   The NRRW therefore provides some evidence of an elevated risk of leukaemia associated with occupational exposure to radiation and, like the combined study of workers in the United States, is consistent with the risk estimates for low-dose/low-dose-rate exposures derived by ICRP [I2] from the data on the survivors of the atomic bombings in Japan.

261.   A cohort study of occupational radiation exposure has been conducted using the records of the National Dose Registry of Canada [A17]. The cohort consisted of 206,620 individuals monitored for radiation exposure between 1951 and 1983, with mortality followed up to the end of 1987. A total of 5,425 deaths were identified by computerized record linkage with the Canadian Mortality Database. A trend of increasing mortality with increasing cumulative radiation exposure was found for all causes of death in both males and females. In males, cancer mortality appeared to increase with radiation exposure without any relationship to specific types. Unexplained trends of increasing mortality due to cardiovascular diseases (males and females) and accidents (males) were also noted. The excess relative risk for radiation-

induced cancer was calculated to be 3.0% per 10 mSv (90% CI: 1.1–4.8) for all cancers combined and was significantly higher than the comparable risk estimate for survivors of the atomic bombings. However, the very low SMR for all-cause mortality suggests that record linkage procedures between the Canadian National Dose Registry and the Canadian Mortality Database may have been imperfect and that there could have been some confounding of the dose response.

262.   In the UNSCEAR 1994 Report [U2] information was given on the association of leukaemia and radiation exposure among workers at the Mayak facility in the Russian Federation, some of whom received substantial exposures several decades ago [K26]. Risk coefficients for radiation-induced leukaemia were similar to those given by the ICRP [I2] for workers, although no confidence interval was provided. Limitations in the study were that 15% of the original cohort had been lost to follow-up, and bone marrow doses from plutonium remained to be evaluated.

263.   An international study of cancer risk among radiation workers in the nuclear industry was coordinated by the International Agency for Research on Cancer (IARC) [C20, I1]. It consisted of a combined analysis of mortality data for nearly 96,000 workers in Canada, the United Kingdom, and the United States. The groups of workers studied were the subject of individual analyses that had been published in 1988 or earlier. The United Kingdom component of this study was the Nuclear Industry Combined Epidemiological Analysis (NICEA) [C3], based on workers at BNFL Sellafield, the United Kingdom Atomic Energy Establishment, and the Atomic Weapons Establishment. The other groups studied were workers at the three United States Department of Energy plants referred to earlier (Hanford, Oak Ridge, and Rocky Flats) [G3] and workers at Atomic Energy Canada Ltd. [G4].

264.   Analysis of the combined cohort of radiation workers showed a statistically significant trend in the risk of leukaemia (excluding chronic lymphatic leukaemia) with external dose. This finding is similar to that reported in the first analysis of the NRRW [K3], although the results are not independent, since many of the workers in the NRRW were also in the IARC study. The central estimate of risk per unit dose corresponded to 0.59 times the value estimated from the atomic bomb survivors based on a linear dose-response model and 1.59 times the value based on a linear-quadratic model fitted to the atomic bomb survivor data; the corresponding 90% confidence interval ranged from about zero up to four times the value from the linear-quadratic atomic bomb survivor model. The evidence for a trend with dose was particularly strong for chronic myeloid leukaemia, as was also reported in the large study of workers in the United Kingdom [M47], some of whom were included in the international study.

265.   For all cancers other than leukaemia, the central estimate of the trend in risks with dose was negative, but the upper 90% confidence limit corresponded to about twice the value arising from a linear extrapolation to low doses of

results for the survivors of the atomic bombings in Japan, i.e. about four times the estimate for low-dose-rate exposures based on a reduction factor of 2.

266. The authors of the IARC study concluded that their analysis provides little evidence that the risk estimates that form the basis of current radiation protection standards are appreciably in error. Since most of the workers studied are still alive, however, they recommended further follow-up of these and other workers to increase the precision of risk estimates. To further address the issue of effects at low doses, IARC is now coordinating an enlarged International Collaborative Study of Cancer Risk among Radiation Workers in the Nuclear Industry [C2]. This study will contain additional workers from countries such as France and Japan, and the combined cohort should number several hundred thousand.

2. Environmental exposures

267. Studies of exposures to natural background radiation (other than radon) or to environmental contamination from man-made sources have generally involved examining geographical correlations in cancer rates. Such studies can be difficult to interpret, owing to the effect of confounding factors such as sociodemographic variables and other factors that vary geographically, together with the lack of information on doses.

268. Sources of natural background radiation include terrestrial gamma rays and cosmic radiation, which vary considerably with geographical location. Many attempts have been made to correlate radiation exposure with cancer mortality or incidence in different populations. While this would in principle give information on exposures at relatively low radiation doses, such attempts are subject to considerable difficulties, as was described in the UNSCEAR 1994 Report [U2]. Interpreting the data is made difficult by uncertainties in the doses actually received, geographical variation in the accuracy of cancer diagnoses, and confounding with the numerous other environmental factors. Furthermore, when different geographical areas are compared, exact matching of control groups or groups exposed at different levels can be difficult. As a consequence, studies that have tried to compare cancer risks from natural background radiation in different geographical locations are subject to considerable uncertainty and must be interpreted with care [C1].

269. Darby [D10] has made some estimates of the proportion of deaths from various cancers that might be caused by exposure to natural background radiation based on models developed by the BEIR V Committee [C1]. These models were based on the data from the survivors of the atomic bombings in Japan. They predict that about 11% of deaths from leukaemia might be caused by post-natal exposure to natural background sources, excluding radon. For other cancers the estimate was 4% or less. The interval between exposure and the development of the disease is shorter for leukaemia than for most other tumours. There is a higher relative risk for leukaemia, and the influence of other environmental factors on leukaemia risk is less than for many other types of cancer. It might be expected, therefore, that any effect of variations in natural background would be more readily detectable for leukaemia than for other cancer types. As described in the UNSCEAR 1994 Report [U2], however, well designed studies conducted in a number of countries find no significant association between natural background radiation and leukaemia (excluding chronic lymphatic leukaemia) (e.g. [I7, T10, U2, W11]).

270. Few of the studies examining cancer incidence in relation to exposure to natural background radiation have tried to obtain realistic dose estimates that take into account differences between indoor and outdoor exposure and the effects of population movement. One exception is a Chinese study [W12] that compares leukaemia mortality in two neighbouring regions having quite different levels of exposure as a result of the high thorium content in monazite sands. Yangjiang is a high-background-radiation area and Taishan/ Enping is a control area. In both regions there was a highly stable population, and considerable effort went into measuring radiation exposure both indoors and outdoors. In the high-background area the radiation dose calculated to the red bone marrow by age 50 years would have been about 60 mSv greater than that for someone living in the low-background area. During 1970–1985, the age-adjusted mortality rates for leukaemia in females were 2.21 and 3.56 10^{-5} PY^{-1} in the high- and low-background areas, respectively, while in males the rates were 3.32 and 3.82 10^{-5} PY^{-1}, where PY stands for person-years. The differences were not significant; if anything, they suggested a lower risk in the more highly exposed population [W12]. The study had low statistical power to detect an effect, if one existed, as the relative risk expected was about 1.2 in the highly exposed group, and effects of this magnitude are very difficult to detect epidemiologically.

271. An extension to this study covering 1987 to 1990 has also been reported [T9]. The later study covered a fixed cohort with 78,614 persons in the high-background-radiation area and 27,903 in the control area at the start of 1987. Dose estimates were obtained by measurements using environmental gamma-ray dose-rate measurements and individual TLDs. The cohort was added to that monitored previously to give a total population of 64,070 subjects in the high-background-radiation area and 24,876 in the control area at the beginning of 1979. In total, the study covered 949,018 person-years (PY) during 1979–1990 (696,181 in the high-background-radiation area and 252,837 in the control area). The relative risks (the high-background-radiation area compared with the control area) for all cancers and for all cancers except leukaemia in each of three dose subgroups in the high-background-radiation area did not differ significantly from 1. The relative risks for site-specific cancers of the lungs, liver, and stomach were generally less than 1, while for nasopharyngeal cancer and leukaemia they were greater than 1. It is noteworthy that the result for leukaemia was the reverse of that found in the earlier study [W12]. The authors concluded that even for the combined data the sample size in each group was not large enough to come to any definite conclusions.

272. A further extension to the study has also been reported [T17] covering a total of 125,079 subjects with 1,698,350 PY (10,415 cancer deaths) followed from 1979 to 1995. The population was separated into controls and high, medium and low dose groups. Despite higher death rates in the males than in the females no significant difference was found between the persons from the high-background-radiation area and the controls; if anything, the death rates in the high-background-radiation area were lower.

273. It may be concluded from this and other studies reviewed in the UNSCEAR 1994 Report [U2] that comparative studies of groups exposed to differing levels of natural background gamma radiation have not demonstrated any significant effects on cancer incidence.

274. Some studies of environmental exposures have examined the temporal trends in cancer rates. For example, Darby et al. [D1] examined temporal trends in childhood leukaemia in the Nordic countries in relation to fallout from atmospheric nuclear weapons testing during the 1950s and the 1960s. They concluded that there was some evidence of a raised risk associated with the "high" exposure period, when children would have received a dose from fallout of about 1.5 mSv, compared with the adjacent "medium" exposure period, when the dose received would have been about 0.5 mSv (relative risk for ages 0–14 years is 1.07; 95% CI: 1.00–1.14). These data are consistent with a relative risk of 1.03 predicted with the BEIR V leukaemia model [C1], although the central estimate from this study is larger than the BEIR V value, a difference that may be explained by the different follow-up times on which the two values are based (0–7 years and 5–15 years, respectively).

275. Studies have been reported of a population in the East-Urals that was exposed to radioactive materials following an accident at the Mayak reprocessing plant in September 1957 [K29]. A total of 7,854 persons who received radiation doses estimated to be between 40 and 500 mSv have been followed. No statistically significant changes in causes of death, mortality or reproductive function have been found compared with control values from the province and USSR data. Although this study is to be continued, it illustrated the difficulties in conducting carefully controlled epidemiological studies, which require a defined control group and accurate dose estimates.

D. SUMMARY

276. Epidemiological studies provide direct quantitative data on the risks of cancer in humans following radiation exposure. The main source of information is the Life Span Study of survivors of the atomic bombings in Hiroshima and Nagasaki in 1945. Substantial information is also available from studies of people occupationally or medically exposed either to external radiation or to internally incorporated radionuclides.

277. The Life Span Study is important, as it gives information on the effects of whole-body irradiation following exposure at different ages. The interpretation of the dose-response data is, however, complicated by the fact that exposure was to both gamma rays and to neutrons. An RBE of 10 has generally been assumed when fitting the dose-response data. The data show a pattern of increasing risk with increasing dose for both leukaemia and most solid cancers. The most recent analyses of the data suggest that the numbers of solid cancers induced in the population depends on the spontaneous cancer rate, and that at least for those exposed in adulthood the absolute level of the radiation-induced risk increases with age over the period of follow up. The follow-up study indicates a significant (p=0.05) increase in the risk of radiation-induced fatal solid cancers over the dose range of 0–50 mSv (assuming an RBE for neutrons of 10). Caution is needed in interpreting this finding, however, as an increased incidence of solid cancers is seen only at doses down to 200–500 mSv, suggesting the possibility of bias at the lower dose range.

278. The data on mortality from leukaemia are best fitted by a linear-quadratic dose response, while for all solid cancers taken together, a linear dose response provides a best fit for dose-response data up to doses of about 3 Sv. However, while a linear dose response can also be fitted to the data for a number of individual tumour types, in the case of non-melanoma skin cancer there is substantial curvilinearity in the dose response, consistent with a possible dose threshold of about 1 Sv or with a dose response in which the excess relative risk is proportional to the fourth power of dose. It is notable that if analyses are restricted to the dose range up to 2 Gy and account taken of possible systematic errors in the Hiroshima DS86 data, then there is evidence of appreciable upward curvature of the dose response for solid tumours. It has become clear that further follow-up and improved information on the doses received will be needed before the shape of the dose response at low doses for both morbidity and mortality can be determined with confidence at doses below about 100–200 mSv. While the Life Span Study has shown elevated cancer risks in a number of tissues, there are others for which there is either very little or no evidence for an effect. These include, for example, the bone, cervix, prostate, testes and rectum.

279. Information on cancer risks is also available from a number of studies of patients irradiated for medical reasons. Many of the patients in these studies received high doses to particular organs, often 1 Gy or more, although some received much lower doses. Patients were generally given acute exposures, although women treated with fluoroscopy for tuberculosis were given highly fractionated doses. As with solid cancers in the Life Span Study, the dose-response data from many of these studies are generally consistent with a linear dose-response relationship at low to intermediate doses. Results from several studies have suggested a statistically significant increase in the risk of thyroid cancer at doses of about 100–300 mGy received in childhood.

280. In contrast, the best fit to the data on bone tumour induction in radium dial painters exposed to $^{226/228}$Ra can be obtained with a model indicating a "practical threshold" for a response at an average bone dose in the range 3.9–6.2 Gy (high-LET). This observation might also reflect the extent of the data available at low doses. For head sinus carcinomas in the radium dial painters, linear, linear-exponential, or dose-squared exponential functions all provided acceptable fits to the data. Data are also available on the risk of bone sarcomas in patients given ^{224}Ra. Recent analysis of the pathology of these tumours has shown that a high proportion of them (30%) are malignant fibrous histiosarcomas, which is higher than would have been expected in sarcomas occurring spontaneously (8%–11%). It has been proposed that these tumours can only be expected to arise in tissue with deterministic radiation damage and so would be expected to appear only above a threshold dose. Similar conclusions have been drawn for the bone tumours arising in the radium workers.

281. Extensive data are available on cohorts of miners occupationally exposed to radon and its decay products. These studies have provided information on the risk of radiation-induced lung cancer. The most recent analyses of the data examined a range of risk models. However, for cumulative exposures below 0.175 J h m^{-3} (50 WLM), a constant-relative-risk model without any modifying factors, such as attained age and exposure rate, appeared to fit the data well.

282. A number of studies have provided information on the risk of childhood cancer following obstetric radiography. In the Oxford Survey of Childhood Cancers, a statistically significant 40% increase in the childhood leukaemia rate (up to 15 years of age) has been seen following doses of 10–20 mGy (low-LET). Similar results have been obtained in a number of other, smaller studies of the effects of obstetric radiography. Although there may be some increase in sensitivity to radiation at this early stage of development, there is no reason to believe the mechanisms involved in tumour induction will be fundamentally different from those in adults. The number of cells at risk would, however, be different. The principal reasons for being able to determine this increase in risk, which in absolute terms is small, is the low background incidence of leukaemia in childhood and greater sensitivity to radiation. A feature of the data from the Oxford Survey that remains unexplained is that the increase in risk for both leukaemia and solid cancers following exposure *in utero* is essentially the same, with a relative risk of about 1.4. Most other human and animal studies consistently indicate different sensitivities of leukaemia and solid cancers.

283. More recently, direct information on the effects of low-dose, chronic exposure has become available from studies of radiation workers. The estimation of cancer risks associated with exposure to low doses poses particular problems. The predicted level of excess risk associated with such exposures is lower than that for high-dose exposures, and consequently the size of the study population required to detect a raised risk is usually much larger than that required for the high-dose studies. The information available to date is generally consistent with information on the risks of cancer obtained from the high-dose-rate studies, although having wide confidence intervals, and would also be consistent with there being no risk at all. A long period of follow-up and pooling of data from different studies will be necessary if statistically useful data are to be obtained.

284. A number of studies have been published that have examined the risks of cancer in areas of high natural background. Comparative studies on groups exposed to different levels of natural background radiation do not, however, have the statistical power to detect significant effects on cancer incidence. There are difficulties in interpreting the data as a result of uncertainties in the doses actually received, geographical variation in the accuracy of cancer diagnoses, and confounding by other environmental factors.

IV. MECHANISMS AND UNCERTAINTIES IN MULTI-STAGE TUMORIGENESIS

285. The development and application of modern molecular methods has, in recent years, substantially increased the understanding of the mechanisms of tumorigenesis. At the same time, there has been an equivalent increase in the understanding of the action of radiation on cellular DNA, control of the reproductive cell cycle, and the mechanisms of DNA repair and mutagenesis.

286. Mechanisms of radiation oncogenesis were reviewed by the Committee in the UNSCEAR 1993 Report [U3] and are considered further in Annex F, "*DNA repair and mutagenesis*"; and Annex H, "*Combined effects of radiation and other agents*". Accordingly, the aim of this Chapter is to provide an updated view of the mechanisms of tumorigenesis in order to relate them to data on dose-effect relationships. Emphasis will be placed on current uncertainties surrounding the mechanisms of radiation tumorigenesis, with a view to exploring their importance for the development of biologically based computational models that seek to describe radiation cancer risk at low doses and low dose rates (Chapter V).

A. MULTI-STAGE PROCESSES IN TUMORIGENESIS

287. In accord with earlier proposals on spontaneously arising neoplasia [F1, F5, V1], UNSCEAR supports a multi-stage model as a conceptual framework for describing radiation tumorigenesis [U3]. A generalized model of this form is illustrated in Figure XXI. In this model, radiation tumorigenesis is imprecisely subdivided into four phases: neoplastic initiation, promotion, conversion, and progression. This operational framework, while subject to considerable uncertainty, may be used to illustrate the critical cellular and molecular processes that direct neoplastic change.

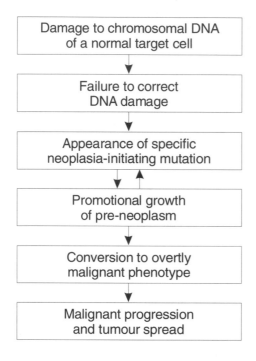

Figure XXI.   A simple generalized scheme for multi-stage oncogenesis.

1.   Initiation of neoplasia

288. Neoplastic initiation may be broadly defined as essentially irreversible changes to appropriate target somatic cells, driven principally by gene mutations that create the potential for neoplastic development [C9, U3]. Such tumour gene mutations can have profound effects on cellular behaviour and response, e.g. dysregulation of genes involved in biochemical signalling pathways associated with the control of cell proliferation and/or disruption of the natural processes of cellular communication, development, and differentiation. Although the full expression of such neoplasia-initiating mutations invariably requires interaction with other later-arising gene mutations and/or changes to the cellular environment, the initiating mutation creates the stable potential for pre-neoplastic cellular development in cells with proliferative capacity.

2.   Promotion of neoplasia

289. Neoplastic development is believed to be highly influenced by the intra- and extracellular environment, with the expression of the initial mutation being dependent not only on interaction with other endogenous mutations but also on factors that may transiently change the patterns of specific gene expression, e.g. cytokines, lipid metabolites, and certain phorbol esters. As a consequence, there may be an enhancement of cellular growth potential and/or an uncoupling of the intercellular communication processes that act to restrict cellular autonomy and thereby coordinate tissue maintenance and development [T5, U3]. In this way, tumour-initiated cells can receive a supranormal growth stimulus and begin to proliferate in a semi-autonomous manner, allowing for the clonal development of pre-neoplastic lesions in tissues, e.g. benign papillomas, adenomas, or haemopoietic dysplasias.

3.   Neoplastic conversion

290. Neoplastic conversion of pre-neoplastic cells to a state in which they are more committed to malignant development is believed to be driven by further gene mutations accumulating within the expanding pre-neoplastic cell clone. Evidence is accumulating that the dynamic cellular heterogeneity that is a feature of malignant development may in many instances be a consequence of the early acquisition of gene-specific mutations that destabilize the genome. Mutations of the *TP53* gene or one of a set of DNA mismatch repair genes provide examples of such destabilizing events in neoplasia [F3, H6, H17, L10]. There is also evidence that mutations resulting in enhanced chromosomal non-disjunction may also contribute to oncogenic change [L3].

291. An elevated mutation rate established relatively early in tumour development may, therefore, provide for the high-frequency generation of variant cells within a pre-malignant cell population. Such variant cells having the capacity to evade the constraints that act to restrict proliferation of aberrant cells will tend to be selected during tumorigenesis.

4.   Progression of neoplasia

292. The progression of neoplastic disease may be dependent on metastatic changes that facilitate (a) the invasion of local normal tissues, (b) the entry and transit of neoplastic cells in the blood and lymphatic systems, and (c) the subsequent establishment of secondary tumour growth at distant sites [H4, T1]. It is the metastatic process and tumour spreading that are mainly responsible for the lethal effects of many common human tumours. Again, it is believed that in many cases gene mutations are the driving force for tumour metastasis, with the development of tumour vasculature an important element in disease progression [F4].

B. MUTATIONAL EVENTS MEDIATING THE TUMORIGENESIS PROCESS

293. Although models such as that shown in Figure XXI are most useful in placing clinical, histopathological, and cellular/molecular experimental data in the context of a generalized mechanism of tumour development, important

uncertainties remain. Identification of these uncertainties should help to guard against over-interpretation of data relating to radiation tumorigenesis, particularly with respect to biological modelling.

294. The gene-specific determinants of the initiation process that is believed to operate to allow entry of normal somatic cells into a given neoplastic pathway are incompletely understood, although for some organs there are strong associations with specific tumour gene mutations [U3, V1]. A similar degree of uncertainty attaches to the molecular events that determine the other cellular transitions noted in Figure XXI. While it is accepted that, in general, target cells for tumorigenic initiation will reside in the stem cell compartment of most tissues, the specific identity and location of these cells is poorly understood. As noted earlier in the Annex, this represents a significant uncertainty in some areas of radiation tumorigenesis, particularly with respect to alpha particle irradiation.

295. In general, the concept of stepwise interaction between loss-of-function mutation of tumour-suppressor genes and gain-of-function mutations of proto-oncogenes [U3] is still believed to apply. Further tumour-specific gene mutations have been identified, and there is much new information on the biochemical interactions between tumour gene mutations, which may destabilize the genome, compromise control of cell signalling, proliferation, and differentiation, and interfere with the normal interaction of cells in tissues (see [K1, S3]).

296. In the UNSCEAR 1993 Report [U3], the Committee concluded that the somatic genetic changes to cells that inter-

mediate multi-stage tumour development potentially involve sequential mutation of different classes of genes, i.e. proto-oncogenes, tumour-suppressor genes, genes involved in cell-cycle regulation, and genes that play roles in maintaining normal genomic stability. It should be recognized, however, that the above classification serves principally as a framework for discussion and that there is substantial functional overlap between these classes.

1. Proto-oncogenes

297. Proto-oncogenes may be broadly defined as tumour-associated genes that can sustain productive gain-of-function mutations that result in over-expression or more subtle functional abnormalities in a wide range of cellular proteins. These proteins normally serve to control or effect cellular signalling and the temporal maintenance of growth and development [H9, L14, M20, W3]. Indeed, the known proto-oncogene proteins perform an extraordinary range of specific cellular functions, many of them interacting with each other in biochemical signalling cascades that, for example, target mitogenic processes, apoptotic activity, cell-to-cell inter-actions, and cytoskeletal functions. The capacity to effect transcriptional or post-translational activation of such pathways is a common theme for many such genes.

298. Thus, mutations resulting in altered proto-oncogene activity/specificity can lead to profound and constitutively expressed cellular effects; the close linkage between many cellular signalling pathways means that these effects are frequently pleiotropic. Table 12 gives a convenient scheme for classifying proto-oncogenes, along with a few examples.

Table 12
Classification scheme for proto-oncogene products

Designation	Product	Examples
Class 1	Growth factors	PDDGF-β chain (*sis*) and FGF-related growth factor (*hst*)
Class 2	Receptor and non-receptor protein tyrosine kinases (RPTK and NRPTK)	*src* (NRPTK) and *erbB* (RPTK); also *ret*
Class 3	Receptors lacking protein kinase activity	Angiotensin receptor (*mas*)
Class 4	Membrane-associated G proteins	*Ras* family
Class 5	Cytoplasmic protein-serine kinases	*raf-1* and *mos*
Class 6	Cytoplasmic regulators	SH2/SH3 protein (*crk*)
Class 7	Nuclear transcription factors	*myc, myb, jun, fos*
Class 8	Cell survival factors	*bcl-2*
Class 9	Cell cycle genes	*PRAD1* (cyclin D1)

299. Numerous and often multiple proto-oncogene activation events characterize different tumours; some of these were discussed in the UNSCEAR 1993 Report [U3]; others are noted in a series of reviews [B9, L14, M20, M21, V1] and are also mentioned later in this Annex.

300. In the context of this Annex, a very important issue is the nature of the mutational events that characterize proto-oncogene activation. Although there has been a large gain in the biochemical understanding of proto-oncogene action, little has changed since 1993 in respect of mutational activation mechanisms [U3]. In essence, human proto-oncogenes may be

activated by point mutation (e.g. *RAS*), by gene amplification (e.g. *MYC*), or by chromosomal rearrangement (e.g. *ABL*) [W3].

301. There has, however, been a rapid increase in knowledge of the range of proto-oncogene activation events via chromosomal rearrangement and their often early role in tumorigenesis. These advances have been reviewed [R2], and cytogenetic data relevant to human tumorigenesis have been subject to detailed analyses [M23].

302. In brief, an increasingly wide range of chromosomal rearrangements associated with many subtypes of human

lympho-haemopoietic neoplasms have been characterized at the molecular and biochemical levels. More than 30 gene activation events resulting from proto-oncogene juxtaposition with T-cell receptors and immunoglobulin loci are known in T- and B-cell neoplasms, respectively. In the case of specific gene fusion by chromosome translocation/inversion, more than 25 examples have been characterized, with myeloid neoplasms the predominant carriers. Table 13 provides examples of chromosome translocations in human lympho-haemopoietic neoplasms.

Table 13
Examples of human tumour-suppressor genes

Gene	Chromosome map location	Cancer type	Product location	Mode of action
APC	5q21	Colon carcinoma	Cytoplasm	Transcription regulator
DCC	18q21	Colon carcinoma	Membrane	Cell adhesion/signalling
NF1	17q21	Neurofibromas	Cytoplasm	GTPase-activator
NF2	22q12	Schwannomas and meningiomas	Inner membrane	Links membrane to cytoskeleton?
p53	17p13	Multiple	Nucleus	Transcription factor
RB1	13q14	Multiple	Nucleus	Transcription factor
VHL	3p25	Kidney carcinoma	Membrane	Transcription factor
WT-1	11p13	Nephroblastoma	Nucleus	Transcription factor
p16	9p21	Multiple	Nucleus	CDK inhibitor
BRCA-1	17q21	Breast carcinoma	Nucleus	Transcription factor/DNA repair
PTCH	9q	Skin (basal cell)	?	Signalling protein
TSC2	16p13	Multiple	?	?

303. It has also become apparent that although proto-oncogene juxtaposition/fusion is most commonly observed in lympho-haemopoietic tumours, such events are also charac-teristic of certain solid tumours. For example, Ewing's sarcoma frequently carries a chromosomally mediated *FLI/EWS* gene fusion [R2], and in some papillary thyroid cancers the *RET* proto-oncogene can be activated by a set of specific chromosome rearrangements [Z1].

304. Since the cytogenetics of solid tumours are often complex and difficult to resolve accurately, it may be that proto-oncogene activation via chromosomal rearrangement is being underestimated. New methods of cytogenetic analysis by FISH are now available to approach this problem [S14].

2. Tumour-suppressor genes

305. Tumour-suppressor genes are defined as genes that can act as negative regulators of cellular processes such as signal transduction, gene transcription, mitogenesis, and cell development/ differentiation [H10, H11, L15, W3]. As noted later, some genes that act to regulate cell-cycle progression, apoptosis, and various aspects of DNA processing may also be included in this category. Consequently, all cancer-associated genes that act via a loss-of-function mechanism may be described as tumour suppressors even though, universally, they may not have true tumour-suppressing activity [H11]. The loss of function of tumour-suppressor genes characterizes a broad range of human neoplasms; some examples discussed in this Annex are listed in Table 14.

Table 14
Examples of chromosome translocations in lympho-haemopoietic neoplasia

Chromosome translocation	Disease	Translocation	Genes involved
Involving T-cell receptors	T-cell acute lymphatic leukaemia	t(1;7)(p32;q34)	TCRβ-TCL5
		t(1;14)(p32;q11)	TCRδ-TCL5
		t(1;7)(p34;q34)	TCRβ-LCK
		t(7;9)(q34;q32)	TCRβ-TAL2
		t(7;9)(q34;q34)	TCRβ-TAN1
Involving immunoglobulin	Burkitts lymphoma/B-cell acute lymphatic leukaemia	t(8;14)(q24;q32)	IgH-MYC
		t(2;8)(p11;q24)	Igk-MYC
		t(8;22)(q24;q11)	Igl-MYC
	B-cell chronic lymphatic leukaemia	t(2;14)(p13;q32)	IgH-REL
	Pre-B-cell acute lymphoma	t(5;14)(q31;q32)	IgH-IL-3
Involving fusion gene sequences	Pre-B-cell acute lymphoma	t(1;19)(q23;p13)	E2A-PBXσσ
	Acute myeloid leukaemia	t(6;9)(p23;q34)	DEK-CAN
		t(9;9)(q34;q34)	SET-CAN
		t(8;21)(q22;q22)	AML1-ETO
	Chronic myeloid leukaemia /B-cell acute lymphatic leukaemia	t(9;22)(q34;q11)	BCR-ABL

306. Unlike activated proto-oncogenes, which are functionally dominant, most tumour-suppressor genes require mutation of both autosomal copies to occur, often via intragenic point mutation of one copy and complete deletion of the other [U3]. Thus, the genomic location of potential tumour-suppressor genes is often revealed by the presence of consistent, region-specific DNA losses in a given tumour type. As noted in the UNSCEAR 1993 Report [U3], there are, however, examples where the mutation of one copy of such a gene can result in a change in cellular phenotype via intragenic mutations that result in so-called dominant negative effects. There are other examples where it seems that one gene copy is mutated conventionally and the other silenced by DNA methylation; there are also cases where effects from gene copy number have been found [H11]. Changes in chromosome complement (ploidy) are common during the development of many tumours, and it seems likely that some specific numerical chromosome changes in neoplasia relate to the loss of tumour-suppressor functions.

307. Overall, the loss of function that is characteristic of the role of tumour-suppressor genes means that the responsible mutational events can vary greatly, i.e. there can be intragenic point mutation/deletion, interstitial chromosome segment deletion, whole chromosome loss, or epigenetic silencing. Much will depend on the capacity of the target cell to remain viable, particularly with the deletion of large segments of DNA. This position contrasts with proto-oncogene activation, which demands relatively high DNA sequence specificity with respect to both intragenic point mutation and gene-specific juxtaposition or fusion. Few such gain-of-function mutations are expected to involve large DNA losses.

3. Genes involved in cell-cycle control and genomic stability

308. Abrogation of normal control of the cell cycle and maintenance of genomic stability is frequently observed in neoplasia. These phenotypes are sometimes closely linked, and recent advances have led to the consensus view that mutations leading to cell-cycle defects and mutator phenotypes can be critical for neoplastic development [H12, L10].

309. An example of the effect of tumour-suppressor mutations on cell-cycle control and genomic stability is provided by the *TP53* gene, which is mutated in a high proportion of tumours of various types [G8, L41]. The p53 protein is known to bind DNA and can act on a transcriptional regulator with potential effects on cell-cycle progression, DNA repair/recombination, and apoptosis [B43, H13, O1].

310. The half-life of the p53 protein in cells is short but increases in response to cellular stress, including DNA damage. Through mechanisms that remain uncertain, the increase in p53 protein serves to check the cell cycle in G_1/S or sometimes in G_2/M. It is believed that such cell-cycle checkpoints promote cellular recovery from stress, including the facilitation of DNA repair [H6, H14].

311. According to these proposals, when *TP53* is appropriately mutated, cell-cycle control and its checkpoints for repair are compromised, and during subsequent cellular development, errors of DNA replication and damage repair accumulate. Failure to adequately effect apoptotic death in damaged cells is also believed to be a feature of *TP53*-deficiency that contributes to neoplastic development [O1]. Other protein products of tumour-suppressor genes that impinge on cell-cycle control include pRb, p16, p27, and p85, and a complex series of cascade interactions involving tumour-suppressor and proto-oncogene proteins, together with cytokines, is believed to maintain close control of cell replication and apoptosis (see [K11, M22, N2]). It can be seen that many of the mutations that accumulate during neoplastic development do so because of the need for cooperation in order to fully compromise normal proliferative control.

312. In this context it has been argued for many years that the accumulation of the series of gene-specific clonal mutations that are believed to drive tumorigenesis would be improbable if normal genomic stability was maintained. Thus, recent findings regarding the spontaneous development of genomic instability in tumours has come as no great surprise. In addition to the *TP53*-mediated effects noted above, other aspects of somatically acquired genomic instability have been debated widely [H6, L10, L17]. The most important of these is the role of defects in DNA mismatch repair.

313. Mismatches in DNA base pairing occur at a relatively high spontaneous rate through replication errors (RER) and spontaneous oxidative/hydrolytic damage; these mismatches are corrected at high fidelity by a repair system that is highly conserved across species [F3]. Following the finding of a high frequency of replication errors in short microsatellite repeat sequences (the RER$^+$ phenotype) in a variety of tumours, some form of DNA mismatch repair defect was suspected.

314. Subsequently, many RER$^+$ human tumours, particularly those of the gastrointestinal (GI) tract, were shown to harbour mutations in DNA mismatch repair genes, principally *hMSH2* and *hMLH1* [A2, F3, H17, S15]. It is clear, however, that "instability" genes other than *TP53* and those associated with DNA mismatch repair are somatically mutated in human tumours. For example, the onset of aneuploidy is often a feature of the transition from pre-neoplastic to malignant phenotype, but the genes participating in the control of ploidy remain poorly understood. Recently, however, a dominantly expressing gene in this category has been revealed by a combination of FISH cytogenetics and somatic cell fusion techniques applied to a panel of colorectal cell lines [L3]; this gene appears to be functionally independent of *TP53* status.

315. Some progress is also being made with respect to somatic tumour genes that have a known or suspected role in DNA damage recognition and processing. The 11q22-encoded *ATM* gene of human ataxia-telangiectasia (A-T) and its role in the cellular and biochemical response to radiation damage are described in Annex F, *"DNA repair and mutagenesis"*. With the knowledge that ataxia-telangiectasia patients are genetically predisposed to the development of neoplasms of the T-lymphoid lineage, a

search has been conducted for somatic *ATM* mutation in sporadic T-prolymphocytic leukaemia (T-PLL) [S16]. This investigation revealed that 11q22 losses and biallelic *ATM* mutations were present in a high proportion of sporadic T-PLL, suggesting a tumour-suppressor-like role for this gene in target T-cell precursors. It may be speculated that this is associated with its role in controlling genomic stability, particularly with respect to T-cell receptor sequences. Also noted in Annex F, *"DNA repair and mutagenesis"*, is the growing recognition that the breast cancer suppressor genes *BRCA1* and *BRCA2* play a role in the recognition/repair of damage to cellular DNA. Although the specific functions of these genes with respect to genomic stability remain to be resolved, their importance to heritable and sporadic breast cancer is well established.

4. Early events in multi-stage tumorigenesis

316. In the UNSCEAR 1993 Report [U3], the Committee recognized the difficulties of identifying the specific genes that, in mutant form, act at the initiation phase of tumorigenesis. For some lympho-haemopoietic neoplasms, specific chromosomally mediated proto-oncogene events were suggested to occur early in neoplastic development, and the *Mll1* gene data outlined later strengthen this view. Equally, however, many human myeloid neoplasms are characterized by region-specific chromosome deletions [M23], some of which are believed to arise early.

317. In the case of human solid tumours of certain tissues, there is growing evidence that those genes that act early are also represented as rare germ-line determinants of heritable cancer; the principal examples of this association are the *RET* gene in thyroid cancer, the *APC* gene in colorectal cancer, the *VHL* gene in renal cancer, the *PTCH(patched)* gene in basal cell carcinoma, and the *RB1* gene in retinoblastoma/osteosarcoma [H11, S17, W3].

318. Although the *BRCA1* and *BRCA2* genes of breast cancer may be exceptions, a concept of early tumour development is evolving from the above associations. The concept requires a relatively tissue-specific "gatekeeper" gene to be mutated in order for stem-like cells to enter a phase of inappropriate clonal expansion [K12, S17]; this expansion then allows for the accumulation of further mutations. According to the concept, the accumulation of other mutations in the neoplastic pathway in the absence of gatekeeper defects will only infrequently result in the clonal development of recognizable tissue lesions. In essence, the temporal order of mutational events is likely to be important for productive neoplastic growth with loss of specific gatekeeper genes as critical early events.

319. In the UNSCEAR 1993 Report [U3], the Committee drew heavily on evolving models of colorectal carcinogenesis to support its views on the mechanisms that drive the genesis of solid tumours. In the same way, further data relating to this tumour type, while not necessarily fully representative of all solid cancers, may also be used to support the gatekeeper hypothesis.

320. A key element in this hypothesis as it relates to colorectal cancer is that the first consistent mutation in tissue lesions should be monoclonal mutation of the *APC* gatekeeper gene, which acts as a transcriptional regulator [N4]. In the main, the data to be discussed later [K12] support this, but a recent investigation of the temporal sequence of gene mutations adds considerable weight to the argument.

321. Using tumour microdissection and allelotyping methods, the sequence and tempo of allelic losses in a series of colorectal cancers at different stages of development was followed [B10]. The principal losses that were tracked were those associated with deletion of *APC* (5q21), *TP53* (17p13), and *DCC* (18q21). In brief, loss of heterozygosity (LOH) via allelic loss was not recorded in normal tissue surrounding colorectal tumours. However, 5q but not 17p losses arose abruptly and consistently at the transition from normal tissue to benign adenoma; a proportion of adenomas also showed 18q losses. Losses to 17p occurred equally abruptly and consistently at the adenoma to carcinoma transition border, and in highly advanced and invasive carcinomas, there was a high level of allelic variation indicative of clonal heterogeneity due to genomic instability.

322. Thus, commencing with *APC* loss from cells in normal tissue, the development of colonic tumours is characterized by abrupt waves of clonal expansion, with *TP53* loss and chaotic allelic variation being critical watersheds in the evolution of the fully malignant phenotype. Considering these and other molecular genetic observations with colorectal cancer, a temporal model of neoplastic initiation and malignant development has been proposed [B10]. This is illustrated in Figure XXII.

323. Although critical evidence in support of the gatekeeper gene hypothesis remains to be gathered, the hypothesis does account for many key observations made with respect to tumour genetics. Assuming for the moment that the hypothesis realistically reflects the processes of tumour initiation and subsequent development, then the spontaneous or induced mutation of rate-limiting and tissue-specific genes will be critical. Albeit less forcefully, the data discussed in this Chapter also imply that induced mutation of other genes in the neoplastic pathway for a given tissue will tend to be of less importance. This would be particularly true if a mutator phenotype, as described earlier, were to arise relatively early in the development of a malignancy; evidence for such early development of genomic instability is accumulating [S34]. With such mutator phenotypes, secondary mutations might be expected to arise in a developing neoplastic clone at a sufficiently elevated spontaneous rate for exogenous DNA-damaging agents at low doses to have no great effect on subsequent tumour development. Some caution is needed, however, before concluding firmly that the majority of induced neoplasms will spontaneously acquire genomic instability during malignant development. In this context it has been argued that current models of tumorigenesis place too much emphasis on the elevation of mutation rates and that cellular selection of evolving clones is more critical [T8].

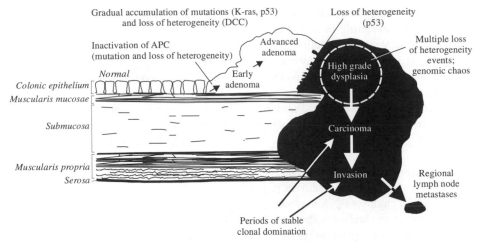

Figure XXII.   A model of the sequence of genetic events in neoplastic development in the human colon [B10].

Initial mutations of loss of heterogeneity at the APC locus of a colonic epithelial cell is followed by adenoma development involving k-ras mutation and DCC loss.

Loss of p53 from advanced adenoma marks the transition between benign and malignant disease characterized in turn by the development of genomic instability, multiple gene losses and invasive behaviour/metastasis to regional lymph nodes.

5.   Non-mutational stable changes in tumorigenesis

324. It has been recognized for some years that non-mutational but stable changes to cellular genomes can contribute to neoplastic development [C10, F2, U3]. In the light of knowledge of the role of specific genes in the tumorigenic process, the central questions are whether the activation/silencing of such genes can be identified in neoplasms and what mechanisms are involved. Such non-mutational mechanisms are broadly termed epigenetic and are believed to involve DNA methylation, genomic imprinting, and changes in DNA-nucleoprotein structure. As will be seen, these mechanisms are not mutually exclusive.

325. The DNA methylation status is believed to be one of the principal determinants of gene expression, and numerous studies have revealed widespread changes in the methylation patterns of the genomes of neoplastic cells [B11]. According to one theory, these changes contribute to the epigenetic modulation of gene expression, while another theory states that increased abundance of 5-methylcytosine serves to elevate spontaneous mutation rates in affected genomic domains. There is some evidence that both processes can occur, but attention will be given here to gene expression effects.

326. The promotor regions of genes are often rich in islands of CpG dinucleotides. These islands are normally free of methylation, irrespective of the state of expression of the genes in question [C10]. Studies with a wide range of neoplastic cells have revealed that *de novo* methylation of CpG islands is frequently acquired, e.g. [D4, J2, J3, M24]. Such effects have been recorded, for example, in a significant fraction of sporadic retinoblastoma and renal tumours for the *RB1* and *VHL* genes, respectively. Recently the *p16* gene (various cancers) and oestrogen receptor gene (colonic cancers) have been shown to be similarly methylated, sometimes at an early stage of tumorigenesis [I3, I4, M1]. In essence, methylation-mediated epigenetic changes in somatic gene expression appear to be an alternative route to mutation for the inactivation of tumour-suppressor genes.

327. Cytosine methylation of CpG dinucleotides is also known to be involved in the process of genomic imprinting, whereby specific genes are marked during gametogenesis for subsequent differential somatic expression [B12, C10]. These imprints, which inactivate one gene copy of sets of autosomal genes throughout the genome, are retained throughout development in spite of a wave of genome-wide demethylation during embryogenesis. Since certain genes involved in neoplastic development are believed to lie within imprinted genomic regions, the remaining active copy will be exposed, and a single somatic mutation can therefore result in the full expression of a mutant cellular phenotype [F2, U3].

328. Evidence of the involvement of this gametic form of imprinting on tumorigenesis was outlined in the UNSCEAR 1993 Report [U3] and has been discussed in depth elsewhere [F2]. Although the original gene-inactivating role ascribed to gametic imprinting with respect to the *RB1, IGF2,* and *H19* genes may be correct in certain instances, an alternative process may also operate. According to this second hypothesis, genomic imprinting serves principally to repress the expression of one somatic copy of growth-promoting genes. Loss of imprinting (LOI) during tumorigenesis acts to de-repress this normally silent copy, thereby increasing gene dosage and contributing to the deregulation of cellular growth and development. The data that support this second hypothesis include *N-MYC* gene amplification in neuroblastoma, loss of imprinting in colorectal cancer, and certain aspects of *H19* gene activation and *BCR-ABL* gene fusion, together with the overall picture of DNA demethylation in neoplasia [F2, R10, M31].

329. However, a considerable degree of uncertainty attaches to the contribution of genomic imprinting in tumorigenesis. Overall it appears that classical region-specific genomic imprinting established during gametogenesis may not play a large role in the development of common tumours. On the other hand, somatic changes in gene expression that do involve changes in the methylation status of critical genes may be widespread in common tumours and contribute significantly to their development.

330. The third stream of knowledge concerning epigenetic changes in gene expression derives from relatively recent findings in yeast concerning the nature of DNA-nucleoprotein interactions and its relationship to chromatin structure. So-called mating-type switches in yeast depend on the silencing of *HM* mating loci by trans-acting factors [R11]. The silencing of *HM* loci has been shown to occur via the action of silent information regulator (Sir) proteins; these also act on silent genes close to chromosome termini. Current evidence favours a role for a complex of Sir and other regulators in sequence-specific binding to silent target genes, with the acetylation of neighbouring histone proteins as a critical factor; CpG island methylation may also be important.

331. Overall it seems that gene silencing demands the formation of tightly folded nucleoprotein configurations in chromatin (heterochromatization), with the Sir proteins playing a role in establishing the necessary pattern of histone deacetylation. In subsequent studies, a mutant form of the yeast gene *SAS2* was found to enhance the loss of gene silencing; the yeast and mammalian forms of this gene have been shown to have sequence homology with several known acetyltransferase genes. Again, a role in the formation of heterochromatin structure is implied [R11]. These and other mechanisms that link gene expression with chromatin structure have been considered in depth [E4]. Knowledge of the biochemistry and genetics of Sir proteins also provides evidence that their diverse functions include regulation of the cell cycle and repair of DNA double-strand breaks. It has been suggested that such regulation may involve the provision of heterochromatic sites for the storage of DNA repair and replication proteins [G19]. DNA strand break repair is considered in depth in Annex F, *"DNA repair and mutagenesis"*.

332. An association between these gene-silencing observations in yeast and tumorigenic processes in man was established by the finding that *MOZ*, a human homologue of yeast *SAS2*, was a partner in a fusion gene (*MOZ-CBP*) generated by the primary t(8;16) chromosome translocation in human myeloid leukaemia [B13]. Although critical evidence is lacking, it seems feasible that the fusion protein could act to redirect MOZ acetylation function to an inappropriate set of genomic domains. In this way normal patterns of heterochromatization and gene activation/silencing would be compromised. Further to this, there is now evidence of a synergy between DNA demethylation and inhibition of histone deacetylase in the re-expression of genes silenced during tumorigenesis [C18].

333. Thus it is becoming clearer that region-specific changes in the heterochromatic state of chromosome regions can have profound effects on gene expression. Given the accepted role of gene expression changes in neoplasia, it would be surprising if the acetylation-related *MOZ-CBP* fusion noted above proved to be an isolated example of oncoprotein involvement, and recent studies point towards a more general role in neoplasia of the genes involved in modifying chromatin [K18].

6. Summary

334. Proto-oncogenes and tumour-suppressor genes control a complex array of biochemical pathways involved in cellular signalling and interaction, growth, mitogenesis, apoptosis, genomic stability, and differentiation. Mutation of these genes can, in an often pleiotropic fashion, compromise these controls and contribute to the multi-stage development of neoplasia. Mutant proto-oncogenes disturb cellular homeostasis in a dominant gain-of-function manner, whereas for tumour-suppressor genes sequential loss-of-function mutation of both autosomal copies is usually, but not always, required. Thus, proto-oncogene mutations are invariably subtle, while the mutations of tumour-suppressor genes can range up to gross DNA deletion.

335. On the basis of accumulating knowledge it is argued that early proto-oncogene activation by chromosomal translocation is often associated with the development of lympho-haemopoietic neoplasia. In contrast, for many solid tumours there is a requirement that tissue-specific tumour-suppressor genes that act as gatekeepers to the neoplastic pathway must undergo mutation; some of these mutations directly or indirectly affect control of the cell cycle and apoptosis. On the basis that solid tumour initiation is most frequently associated with tumour-suppressor gene mutation, it has also been proposed that the subsequent onset of spontaneous genomic instability via further clonal mutation is a critical event in neoplastic conversion from a benign to a malignant phenotype. Loss of apoptotic control is believed to be an important feature of neoplastic development and is described in more detail later in this Annex. Apoptosis as a response to radiation is also discussed in Annex F, *"DNA repair and mutagenesis"*.

336. In spite of continuing gains in knowledge, it is important to recognize that much of the information available on multi-stage tumorigenesis remains incomplete, thus limiting the predictive power of mechanistic models that seek to describe these complex cellular processes. Although the concept of sequential and interacting gene mutations as the driving force is more firmly established, there is a lack of understanding of the complex physiological interplay between these events and its consequences for cellular behaviour and tissue homeostasis. It is also important to stress that the concepts outlined in this Section derive from detailed studies in a somewhat limited set of tumour types; there is an inherent danger in applying a single mechanistic concept to all or many tumour types.

337. Uncertainty also surrounds the degree to which non-mutational (epigenetic) changes to the genomes of neoplastic cells contribute to tumorigenesis. Increases in the methylation status of critical tumour-suppressor genes is known to be an alternative to mutational inactivation in a range of neoplasms, and loss of methylation imprints may also serve to increase the activity of some growth-promoting genes. DNA methylation is also believed to be involved in the genomic imprinting processes occurring during gametogenesis, but these may not make a major contribution to tumorigenesis. New evidence also implicates histone acetylation in genomic hetero-chromatization and gene silencing. It is suggested that such gene silencing may make a potentially important contribution to epigenetic change.

338. An important feature of recent studies has been the clarification of the role of specific gene mutations in tumours that serve to destabilize the genome, thereby allowing for the rapid spontaneous development of clonal heterogeneity and tumour progression. Although critical evidence is lacking, it is possible to envisage that after this transition point is reached, tumour development may be relatively independent of exogenously induced DNA damage.

C. CELLULAR AND MOLECULAR TARGETS FOR TUMOUR INITIATION

339. In the UNSCEAR 1993 Report [U3], the Committee reviewed data for appraising the cellular targets that are or might be involved in tumour initiation. The critical question posed was whether the mutation of single genes in a single normal target cell in tissue could, in principle, divert that mutant cell into a potentially neoplastic pathway. At the time of the UNSCEAR 1993 Report, the evidence available broadly supported this view. Uncertainties on this issue were, however, recognized, and since that time there have been further developments, which are summarized in the following paragraphs.

1. Monoclonal origin of tumours

340. The critical features of human and animal tumours that lend support to the single-cell (monoclonal) origin of tumours are that they exhibit (a) consistent and characteristic chromosomal and/or gene mutations in all neoplastic cells, (b) clonality with respect to the expression of X-chromosome-encoded genes in tumours of females, and (c) characteristic monoclonal restriction enzyme polymorphism of known and anonymous DNA sequences. It has also been noted that molecular analysis of human tumours associated with exposure to chemical carcinogens and ultraviolet radiation, together with that of tumours arising in genetically pre-disposed individuals, adds weight to the concept that the majority of tumours are of single-cell origin [U3].

341. It was recognized, however, that because such analyses are performed on macroscopic neoplasms, monoclonality might, in some circumstances, be due to cellular selection via proliferative advantage, i.e. initially neoplasms are pre-dominantly polyclonal but become increasingly monoclonal during early growth. Although this issue remains somewhat problematical, a number of recent observations allow further comment.

342. The first observation concerns tumours that are believed to have their origins *in utero*. Some leukaemias arising in monozygotic twin children have, in the past, been shown to share the same primary chromosomal anomaly, implying, but not proving, that they arose in a monoclonal fashion from an early precursor cell population present when the two fetuses shared a common (*in utero*) blood supply. This interpretation has been greatly strengthened by the finding that such leukaemia in monozygotic twins can have identical molecular rearrangements of a proto-oncogene termed *Mll1* [F6].

343. The monoclonality of childhood solid cancers is also strongly supported by the finding that a specific tumour-suppressor-gene-associated chromosome loss/reduplication event in early embryogenesis can lead not only to mosaicism in normal tissue but also to the development of monoclonal Wilms' tumour [C11].

344. A second line of evidence concerns further developments in the understanding of multi-stage colon carcinogenesis [U3]. Mutation/loss of the tumour-suppressor gene *APC* has for some time been believed to be a critical early event in the development of human colon cancer. Up to about 70% of early colonic adenomas show apparently monoclonal structural/functional loss of this gene [P5], and a critical role in tumour initiation seems likely [B10].

345. With use of a mouse (Min) model of intestinal carcino-genesis, this view of monoclonal tumour initiation has been strengthened. In essence, aberrant crypts, the earliest intestinal lesions detectable microscopically, have been microdissected from Min mice and shown to be monoclonal with respect to *Apc* loss [L18, L19] (but see also para. 284). This and another mouse model of myeloid leukaemia, described below, have also been used to provide evidence of early monoclonal events associated with radiation tumori-genesis. There are also data describing early events in thymic lymphomagenesis [M33].

346. Given the recent evidence outlined in this Annex and that previously reviewed by the Committee [U3], it seems likely that the vast majority of tumours in humans and animals arise from mutation of single target stem-like cells in tissues. This view continues to find the support of most commentators but has been debated widely [A3, F7, F8, R12].

347. The Committee also noted publications where tumour monoclonality has been questioned [U3], and a striking contribution to the debate was recently published [N3]. The basic finding was that a large proportion of intestinal adenomas in the gastrointestinal tracts of human familial adenomatous polyposis patients, who were also XO/XY in genotype and therefore mosaic for the Y-chromosome, was apparently polyclonal. In this study polyclonality was judged by the presence within single adenomas of a mixed

population of cells with respect to the Y-chromosome sequence. By this measure, up to 76% of adenomas were polyclonal. However, early Y-chromosome loss and field effects creating tumour clustering and collision [U3] might contribute to this finding.

348. Tumour clustering may also explain new data on the apparent polyclonality of a proportion of spontaneously arising intestinal adenomas in Min mice as assessed by genetic features other than *Apc* loss [D9]. Thus, polyclonality may be acquired during adenoma development rather than arising *de novo* at the time of initiation. For example, fusion of independent *Apc*-deficient microclones may allow for cooperative growth. These data illustrate some of the problems that remain in resolving the early molecular events and complex cellular interactions of tumour development. In spite of these uncertainties there remain experimental data on intestinal tumorigenesis that forcibly support monoclonality for induced neoplasms [G11].

2. Molecular targets for radiation tumorigenesis

349. Following its review of the mechanisms of mutagenesis, oncogenesis and the data available on molecular targets, the Committee suggested that loss of critical tumour-suppressor genes via DNA deletion might be the principal mechanism by which radiation damage might contribute to tumour development [U3]. It was suggested that proto-oncogene activation via point mutation or chromosomal rearrangement played a less critical role overall but might be important for certain tumours.

350. Direct human data relating to these issues remain, however, fragmentary. As noted in Annex F, *"DNA repair and mutagenesis"*, more data have emerged on *TP53* gene mutations in radiation-associated human tumours, particularly lung tumours. Unfortunately the interpretation of these data remains highly problematical, and at present it is not possible to judge whether intragenic mutation of this gene is an early radiation-associated event in any human tumour type. *TP53* gene mutations have also been studied in liver tumours arising in excess in patients treated with the radiographic contrast agent thorotrast, which contains alpha-emitting thorium oxide [I14]. These studies comment more on secondary *TP53* mutation than on early radiation-associated events in liver tumorigenesis.

351. In the case of human thyroid cancer, chromosomally mediated rearrangement of the *RET* proto-oncogene is a common but not invariate feature; such events are believed to occur early in the genesis of the papillary form of this tumour [Z1]. *RET* proto-oncogene rearrangements have been found in some cases of papillary childhood thyroid cancer arising in areas contaminated by the Chernobyl accident. Since three different forms of *RET* rearrangement are present, overall, among spontaneously arising papillary thyroid cancer cases, it is possible that in radiation-associated cases one particular form will predominate. A recent commentary [W4] on one data set suggests that the spectrum of these rearrangements in

Chernobyl-related papillary thyroid cancer is unremarkable, although in other studies [B14, K13] one type (*RET/PTC3*) appeared to be more frequent than expected. A causal relationship between *RET* rearrangement and radiation remains, therefore, a matter of some uncertainty. Nevertheless, the finding of *RET* rearrangement following experimental high-dose irradiation of human thyroid cells [M25] suggests that specific and rare *RET* proto-oncogene rearrangements associated with the genesis of these tumours can be induced by radiation.

352. The study of second cancers after radiotherapy [C12, C28] provides another direct approach to the problem. Investigations of gene-specific mutations in such tumours have yet to be particularly informative, and at this stage of knowledge cytogenetic approaches may prove to be more productive. Studies that include cytogenetic evaluation of therapy-related sarcoma, meningioma, and rectal carcinoma provide some evidence that chromosomally complex monoclonal tumours having hypodiploid karyotypes with multiple deletions may be most common [C13, C28]. The number of therapy-related tumours characterized in this way remains, however, too small to make these findings conclusive.

353. With respect to target DNA regions and genes for radiation tumorigenesis, more rapid progress is being made through the use of experimental models of tumorigenesis in rodents. Regarding as yet uncharacterized molecular targets, some studies with breast and thyroid clonogens provide evidence of an apparently high frequency of tumour-initiating events that, it is argued, may reflect the involvement of non-mutational processes [C19, U3]. Uncertainties attaching to the status of these events are discussed later in this Annex, and here attention will focus on mouse models that more specifically suggest genomic targets for tumorigenesis. Three examples of this work are given below.

354. In a mouse genetic model of germ-line *p53*-deficiency (*p53*$^{+/-}$), quantitative studies of tumorigenesis (principally lymphomas and sarcomas) showed these mice to be extremely sensitive to tumour induction by an acute dose of 4 Gy from gamma rays [K2]. Of particular note was the shortening of the tumour latency period after irradiation. Molecular studies of these tumours revealed that complete loss of wild type *p53* was a consistent event; these data, together with those from quantitative studies, provided good evidence that *p53* and surrounding sequences could act as a direct target for radiation.

355. A somewhat unexpected finding was that in almost all cases of *p53* loss from induced tumours there had been duplication of the mutant *p53* gene. A likely reason for this was provided by subsequent *in vivo* analyses of post-irradiation cytogenetic damage in the haemopoietic system of murine *p53* genotypes [B4]. These studies showed that although *p53*-deficiency had only marginal effects with respect to the frequency of structural chromosomal rearrangements after radiation, there were substantial effects on whole chromosome loss and gain (aneuploidy). This enhancement of radiation-induced aneuploidy appeared to be driven by a *p53*-associated

defect in a G$_2$/M cell-cycle checkpoint. Thus, it was suggested that loss of wild type *p53* occurred through loss of the whole of the encoding chromosome (chromosome 11). For the purposes of establishing genetic balance and viability, there was strong selection for those cells that had duplicated the remaining chromosome 11, which accounts for the duplication of mutant *p53*. On this basis it may be seen that in some circumstances the molecular target for radiation oncogenesis may be as large as a whole chromosome.

356. Somewhat similar studies have been undertaken using the Min mouse (*Apc*$^{+/-}$) model of intestinal carcinogenesis [S18]. Using F1 hybrid mice carrying the *Apc* mutation, a 2 Gy whole-body dose from x rays has been shown to double the spontaneous incidence of intestinal adenomas [E5]. In hybrid genetic backgrounds it is possible to determine through the use of polymorphic microsatellites whether spontaneous and radiation-associated adenomas arise through early loss of wild type *Apc* and, if so, the type/extent of the mutation involved. Published studies [H32, L20] reveal that complete loss of wild type *Apc* is characteristic of the majority of both spontaneously arising and radiation-associated early adenomas. These mutational events may involve whole chromosome loss or interstitial deletions, but deletion events tend to predominate in radiation-associated adenomas [H32]. Again, therefore, a tumour-suppressor gene appears to be acting as a direct target for radiation, with gene losses usually being associated with gross deletion events.

357. The third example of informative animal data concerns the induction by radiation of acute myeloid leukaemia (AML) in certain strains of mice. It has been known for some years that these acute myeloid leukaemias are consistently associated with early arising deletions from chromosome 2 [B15, H15]. More recently, however, cytogenetic studies of bone marrow cells of irradiated mice [B16] have revealed that characteristic chromosome 2 deletions are apparent within the first few days following *in vivo* irradiation. Carrier cells of stem-like origin remain, however, relatively indolent in bone marrow for many months until unknown secondary events trigger them into rapid clonal expansion prior to the development of overt monoclonal acute myeloid leukaemia. The identity of the gene loss from mouse chromosome 2 that initiates acute myeloid leukaemia development remains unknown, but the critical chromosomal region encoding an acute myeloid leukaemia suppressor gene has been narrowed to around 1 cM (~10^6 base pairs) [C5, S40]. Thus, data on the mechanisms of radiation-induced murine acute myeloid leukaemia point to tumour-initiating loss of gene function from stem-like cells in bone marrow, followed by the accumulation of spontaneous secondary events that trigger initiated cells into a pathway leading to monoclonal leukaemia development.

3. Summary

358. On the basis of a large body of data it may be judged that, in the main, tumours appear to have their origin in gene/chromosomal mutations affecting single target stem-like cells in tissues. It is recognized, however, that there are circumstances where early phases of tumour development may be bi- or even polyclonal and that monoclonal selection occurs later.

359. Direct evidence on the nature of radiation-associated initiating events in human tumours is sparse, and rapid progress in this area should not be anticipated. By contrast, good progress is being made in resolving early events in radiation-associated tumours in mouse models. In the case of tumours induced in *p53* and *Apc* heterozygously deficient mice, radiation appears to target the remaining wild type tumour-suppressor gene via gross chromosomal deletion. Radiation-induced deletion of a specific chromosomal segment also appears to act as an initiating event for mouse acute myeloid leukaemia. These molecular observations lend further support to the view expressed in the UNSCEAR 1993 Report [U3] that radiation-induced tumorigenesis will tend to proceed via gene-specific losses from target stem cells.

D. CELLULAR FACTORS THAT COUNTER ONCOGENIC DEVELOPMENT

360. Somatic cells employ a series of measures to protect against the development of abnormal and potentially neoplastic phenotypes. In essence, a certain proportion of these barriers has to be breached by the cell before it becomes committed to malignant development. Thus, the process of multi-stage oncogenesis may be viewed as the stepwise acquisition of cellular properties that allow evasion of these protective functions [U3].

1. Control of cellular proliferation and genomic stability

361. The ordered replication of DNA during the reproductive cell cycle, the equal sharing of the replicated genome to the daughter cells, and the close control of mitotic activity is an essential element of normal tissue development and maintenance [U3]. Under normal circumstances cells can respond to specific mitogenic stimuli, continue proliferation while that stimulus is maintained, and fall to a resting state when it is removed. Such normal somatic cells are also believed to have a finite lifespan, and as a consequence of an internal genetic programme, after completing a given series of reproductive cycles, they cease proliferative activity and enter a degenerative senescence phase. There is also a strong requirement for phase controls within the reproductive cycle itself, such that DNA replication is appropriately initiated and completed before genomic segregation to daughter cells and that in the event of non-optimal cellular conditions, the cell cycle is checked until the problem is rectified. Cell-cycle checkpoints may be particularly sensitive to induced DNA damage, and there is some evidence that the presence of very few DNA double-strand breaks can lead to cell-cycle arrest [H23].

362. In recent years much has been learned of the control of cellular proliferation [N2], the process of cellular senescence [H3, H5, S4], and the importance of cell-cycle checkpoint

control [H7] for maintenance of genomic stability. As this information accumulates, it has become evident that all of the above normal controls are potentially subject to mutational change during multi-stage oncogenesis and therefore require some consideration in the modelling of tumour development.

363. The information discussed previously by the Committee [U3] and in Section B of this Chapter includes a number of examples of how gene/chromosomal mutations and sometimes epigenetic events in tumours can compromise control of the cell cycle (e.g. *RB1*, *TP53*, and *p16*) and/or decrease genomic stability, e.g. *TP53*, DNA mismatch repair genes, and *ATM*. As noted earlier, the resulting abnormal patterns of cellular proliferation and the generation of clonal heterogeneity are sentinel features of neoplastic growth, representing the escape from normal cellular constraints. Thus, precise control of the cell cycle and high-fidelity DNA damage recognition/repair are clearly important protective factors against tumour development. Also associated with proliferative control and genomic stability arc the DNA sequences present at chromosome termini (telomeres).

364. The characteristic hexamer repeat sequences (TTAGGG) at mammalian telomeres erode via incomplete replication during each cell cycle. Since the majority of human somatic cells lack expression of the enzyme telomerase that adds these hexamer repeats to telomeres, it has been suggested that the process of erosion acts as a "molecular clock" (see [H16, K14]). In this way the replication potential of somatic cells is limited, and there is direct evidence that the senescence of human cells is, in some part, determined by the absence of telomerase [B35, J4].

365. During the senescence process as measured *in vitro*, there is a tendency for cells to become chromosomally unstable, with many of the resulting aberrations being centred on chromosome termini [C14, K14]. Thus, telomeric erosion during senescence renders chromosomes prone to end-to-end association and subsequent cycles of breakage and fusion.

366. It is believed that telomere-sequence-mediated cellular senescence is one of the means whereby cells may be eliminated from neoplastic pathways. It follows, therefore, that in human cells the stabilization of telomeres and the generation of immortal or lifespan extended phenotypes is likely to be a critical step in tumorigenesis [G9, K14]. In accord with this proposition many, but not all, human tumours have been shown to carry stabilized telomeres via reactivation of telomerase or utilization of alternative pathways of telomere maintenance (see [B17, K14]).

367. Although good progress continues to be made in this whole area, a simple relationship between senescence, immortalization, telomerase, and tumorigenesis should not be inferred [B17, J4, L21].

2. Programmed cell death and gene expression

368. Programmed cell death, also termed apoptosis, plays an important role in restricting the growth of many normal cell lineages and is an important element in the regulation of organ development and maintenance [R3]. Apoptotic processes are believed to be controlled by the interaction of intra- and extracellular factors with the signalling machinery of the cell. These signals, or in some cases their absence, can trigger the recipient cell into a characteristic biochemical suicide pathway that usually involves genomic degradation. Importantly, apoptotic responses can also accompany exogenous insult, induced by ionizing radiation, genotoxic chemicals, and other sources of stress; in some cellular systems these responses have been associated with prior perturbation of the cell cycle. The biochemistry and genetics of apoptosis are becoming much better understood, and advances in the whole area have been reviewed extensively [C16, H22, K6] and have received comment with respect to radiation protection [S22]. A detailed description of these mechanisms is beyond the scope of this Annex, but a brief outline is appropriate.

369. The process of cellular apoptosis may be divided conveniently into three phases: initiation, effector, and degradation (nucleolytic and cellular). The initiation phase differs according to cell type and the source of stress, while the effector and degradation phases, although regulated, tend to be more uniform [K7]. As noted in Section B, a range of proto-oncogenes and tumour-suppressor genes participate in the intracellular signal cascades that can initiate apoptosis. Here, information on two apoptosis-related genes, *TP53* [E3] and *Bcl2* [K7], will be presented.

370. Biochemical studies indicate that almost all productive mutations of *TP53* compromise the ability of the protein to bind to gene-specific DNA sequences and to regulate transcription; in general, the cellular consequences are alterations in growth arrest or apoptosis. Although p53 protein response is apparent under a range of stresses, recent studies suggest a common root. A series of findings (see [K5]) imply that intracellular oxidative stress is a critical trigger for p53-mediated apoptosis. Together with p85 and perhaps Abl protein, p53 is believed to regulate the redox state of the cell [K5, Y3], and it is this state that may be a common determinant of apoptosis or survival. Other aspects of p53-dependent and developmentally regulated apoptosis including the role of ceramide are discussed in Annex F, "*DNA repair and mutagenesis*".

371. The second example concerns the genes in the *Bcl2* family [K6, K7]. These include *Bcl X_L*, *Bcl-w*, and *Blf-1* (death antagonists) and *Bax, Bak, Bcl X_S*, and *Bad* (death promotors). The protein products of these genes participate in a complex network of biochemical reactions that differ between cell types. These pathways may be linked with *Raf*, *MEK*, and *Jun* amino terminal kinase (JNK) protein and also, via *Ras* protein, with the p85 pathway noted above [K7, P6].

372. In a broad sense it is believed that it is the balance between pro- and anti-apoptotic factors that determines cell fate [K7]. Thus, apoptotic signals via, for example, cell surface receptors, redox changes, reactive oxygen species, and Ca^{++} ion concentration (initiation phase) are sensed by

the Bcl-2 regulatory complex, resulting in changes in mitochondrial permeability (effector phase). According to current proposals, the degredative and nucleolytic phases then proceed as a consequence of the release of directly apopto-genic factors, e.g. caspases, superoxide anions, and endo-nucleases from mitochondria into the cellular cytosol [K7].

373. Overall, it may be seen that cells possess a highly developed system for detecting stress, eliciting biochemical responses, and, in essence, deciding on the basis of bio-chemical balance whether to survive or to proceed towards cell death.

374. The potential of these stress-related apoptotic pathways to reduce tumorigenic risk, although not formally established, is strongly indicated. There appear to be at least two principal

points of action of apoptosis during tumorigenesis. There is evidence for at least three stress-related pathways in cells that respond to genotoxic insult, including that from ionizing radiation, i.e. those pathways centred on Abl, JNK, and p53 proteins [C4]; the p53 pathway has been judged to be the "universal sensor" of damage in normal cells. A variety of other cellular genes have also been shown to be up- or down-regulated in response to radiation. While a comprehensive review of such studies is beyond the scope of this Annex, Table 15 provides, by way of example, a summary of the data obtained after neutron or gamma-ray exposure of Syrian hamster cells [W10]. These results were obtained after exposures of 0.21–2.0 Gy of neutrons or 0.96–3.0 Gy of gamma rays at low and high dose rates. These data should not, however, be taken as representative of mammalian cells in general, and cell type dependency in induced gene expression should be expected.

Table 15
Radiation effects on gene expression in Syrian hamster cells
[W10]

Gene	Effect on expression [a]		Function
	Neutrons	Gamma rays	
Interleukin-1	Increase	Increase	Cytokine
β-actin	Decrease	Decrease	Cytoskeleton
γ-actin	Increase	Increase	Cytoskeleton
β-PKC	No change	Increase	Signal transduction
Rp-8	Increase	No change	Apoptosis
c-fos	Decrease	Increase	Transcription factor
c-myc	No change	No change	Nuclear protein
α-tubulin	Increase	Increase	Cytoskeleton
fibronectin	Decrease	Increase	Cellular matrix
Interleukin-6	Increase	-	Cytokine
Proliferating cell nuclear Ag (PCNA)	Increase	Increase	Transcription factor/repair
Superoxide dismutase	-	Increase	Scavenger
c-jun	Increase	Increase	Transcription factor
Rb	Decrease	Increase	Nuclear protein
H4-histone	Decrease	Increase	Nuclear protein
p53	No change	No change	Nuclear protein

a   All changes evident within the first four hours following radiation exposure. Neutron dose rates: 1 mGy min^{-1} and 140 mGy min^{-1}; gamma ray dose rates: 10 mGy min^{-1} and 120 mGy min^{-1}.

375. The development of high-throughput screening technologies promises to greatly increase the power of resolution of studies on such radiation-associated changes in gene expression in mammalian cells. For example, using these new techniques a linear non-threshold dose response for the transcriptional induction of the stress-related genes *CIP1/WAF1* and *GADD45* has been demonstrated for gamma ray doses in the range 20–500 mGy [A18]. The consequences of such induced stress responses for low-dose tumorigenesis remain a matter for speculation. Nevertheless, an association between the *in vitro* induction of *PBP 74* gene transcription by 250 mGy radiation and human cell hypersensitivity to cell inactivation might be explained by some form of damage threshold for the enhancement of DNA repair [A18, S41]. One speculation is that if such a hypersensitive mechanism for cell inactivation were to dominate at low doses (say, up to around 100 mGy), mutation induction rates would be

depressed, leading to a non-linear and, perhaps, a threshold-type relationship for radiation cancer risk [J8]. If, however, this increased sensitivity to cell inactivation were to be accompanied by increased cell mutation rates, no such threshold would be expected. The data available do not allow these two possibilities to be distinguished, although some of the data discussed in Annex F, *"DNA repair and mutagenesis"*, suggest a direct dose-effect relationship between cell inactivation and gene mutation. Stress-related cellular responses are also discussed in Annex F, *"DNA repair and mutagenesis"*, which draws attention to new work that associates the appearance of specific novel proteins with cellular stress.

376. In the absence of complete DNA repair fidelity, the whole organism gains a large advantage by promoting the death of damaged and potentially neoplastic cells. However,

the true effectiveness of apoptotic pathways in removing such aberrant cells cannot be judged at this time. The mere fact that the frequency of gene/chromosomal mutations increases in cell populations surviving genotoxic insult argues against an extremely high capacity for apoptotic surveillance of mutagenic damage in all cell types. In the context of ionizing radiation, the shape of the low dose response for the induction of apoptosis in different cell types remains very uncertain, and equal uncertainty surrounds the influence of dose rate. Accordingly, for the purposes of modelling tumorigenic risk, judgements on the balance between mutagenesis and apoptosis at low doses cannot be made with confidence. The radiobiological factors that influence the induction of apoptosis vary with cell type, and there is also some dependency on the mechanisms involved [B27, S29]. In general, doses greater than 0.5 Gy of low-LET are necessary to obtain statistically significant increases in apoptotic rates; a plateau in the dose response is frequently seen at doses >5 Gy. In the well-studied human lymphocyte system there is evidence that the induction of apoptosis is largely independent of LET and dose rate, implying that in these cells, initial DNA damage is more important than its repair [V4]. DNA double-strand lesions are believed to be one of the determinants of apoptotic response, but some have suggested that damage to plasma membranes may also act as an apoptotic signal [O4]. There is also evidence of linkage of the signalling of apoptosis and cell-cycle arrest; for example, a protein known as survivin has been implicated in the control of both apoptosis and a mitotic spindle checkpoint [L37]. Additional aspects of apoptotic response are discussed in Annex F, *"DNA repair and mutagenesis"*.

377. Apoptosis is also believed to play an important role in tumour cell survival during post-initiation clonal expansion. At a critical point during clonal expansion, the oxygen supply to the neoplasm begins to become limiting [F4, U3]. It has been proposed that under these circumstances the redox stress placed on tumour cells triggers an apoptotic response, with cell death being most pronounced in the regions most distant from vascular supply [K5, N5]. Thus, during this phase in tumorigenesis, apoptosis will be playing a crucial role in limiting *in situ* growth and invasive behaviour.

378. Given the scenarios noted above, it is not surprising that a broad array of tumour types carry a variety of mutant genes that directly or indirectly uncouple stress response and apoptosis. Resistance to apoptosis may be viewed as the means whereby cell survival is favoured over cell death, and under conditions of *in vivo* stress, this phenotype will tend to be strongly selected. The p53 pathway appears to be the universal sensor of cellular stress, and it is this feature that may make loss-of-function *TP53* mutations so prominent in human tumorigenesis.

379. Overall, it is judged that apoptotic suicide of cells provides an important protective mechanism against aberrant cell growth and neoplasia. However, via gene-specific mutation, a number of potential bypasses or mechanisms of tolerance are available.

3. Cellular differentiation and other cellular interactions

380. The stepwise accumulation of genetic/epigenetic events demands continuing growth potential in cells that have sustained a neoplasia-initiating event. Running counter to this is the normal process of terminal cellular differentiation, whereby uncommitted progenitor cells assume specialized functions in tissues and no longer retain proliferative potential. Thus, a developing subpopulation of cells may carry a tumour-initiating mutation that, for example, deregulates cellular proliferation but in the absence of further phenotypic change will complete a quasi-normal programme of terminal differentiation mediated by cellular interactions.

381. In this way neoplasia-initiated cells will, in the absence of other changes, exit the pathway to malignancy. Thus, the antiproliferative process of terminal differentiation will tend to be rate-limiting with respect to overt malignancy and may be evidenced by the accumulation of benign lesions in tissues. There are numerous examples of associations between proto-oncogene/tumour-suppressor gene functions and cellular differentiation/development [K1, L1, R2, S3]; here it will be sufficient to give only a few examples.

382. In the case of the lympho-haemopoietic system, the development of the different cell lineages is known to depend on a complex interplay between cell-cell interaction, cytokines, and intracellular signalling cascades [O2]. The genes *AML1* and *tal/SCL* have been implicated in haemopoietic stem-cell differentiation and, in mutant form, are known to contribute to the genesis of certain types of leukaemia. Other examples of leukaemia/lymphoma-associated genes with roles in normal differentiation processes include genes of the *Hox* and *Pax* families, *RBTN2*, *RARA*, and *Mll1* [O2, R2]. As noted in Section B, the functional development of the T- and B-haemopoietic cell lineages is highly dependent on the recombination of immune gene sequences, and specific mis-recombination of these sequences makes a major contribution to T- and B-cell neoplasia. Recent evidence also links downstream signals from these recombinogenic processes with subsequent clonal growth and differentiation; that is, normal growth and differentiation of cells may be blocked if recom-bination does not proceed normally [W5]. In general, it may be concluded that many mutations in lympho-haemopoietic neoplasia serve to compromise the closely controlled process of cell-lineage-dependent differentiation.

383. Similar evidence exists with respect to solid tumours [L14]. For example, via its interaction with *Ras* proteins, the protein product of the *NF1* tumour-suppressor gene plays a role in regulating the normal growth and differentiation of neural cells, and the Ras/Raf/MAP kinase intracellular signalling pathway appears to play a more general role in the regulation of cellular differentiation. In addition, there are data that support a role for tumour-associated catenin/APC, Rb1, and DCC proteins in transcriptional regulation/cellular signalling processes that control cell-lineage-dependent growth and differentiation.

384. Thus, many forms of proto-oncogene/tumour-suppressor gene mutations, often in combination and via perturbation of cellular signalling, will have dual effects on cellular growth and differentiation. The maintenance of constitutive growth of stem-like cells having differentiation defects may be viewed as an important element in the early phases of tumour development. In essence, the homeostatic imbalance created by these mutations will tend to promote the clonal expansion of cells having limited potential for terminal differentiation. Alone, however, such clonal growth may not be productive, since the cells will be potentially subject to senescence and apoptosis, both of which serve to limit the opportunity to accumulate the further mutations necessary for overtly malignant development. Nonetheless, as noted earlier, each of the processes of senescence and apoptosis may itself be compromised by gene mutation, e.g. by telomerase deregulation and *TP53* gene mutation, respectively. In principle, therefore, extended clonal growth can be achieved. Further to this, and in accord with previous discussion, the early appearance of genome-destabilizing mutations may dramatically accelerate neoplastic development.

385. Overall, the frequency with which the genes involved in normal cellular differentiation are mutated in tumours testifies to the protective function offered by terminal cellular differentiation to a non-proliferative state. It is judged, however, that alone, such aberrant differentiation is usually insufficient for full malignancy and that cooperating mutations that further extend clonal lifespan and/or destabilize the genome are likely to be required.

386. Intercellular transmission of biological signals followed by intracellular biochemical cascades is believed to be an integral component of the differentiation of cells. Not surprisingly, cell-to-cell communication has also been implicated in the expression of neoplastic phenotypes, and more recently, cellular communication has been shown to influence radiation response.

387. The UNSCEAR 1993 Report [U3] reviewed the role of cellular communication via gap junctions in neoplastic development. In summary, it is believed that the establishment of such intercellular communication can lead to the suppression of neoplastic features by neighbouring normal cells. During clonal evolution, however, many tumour cells lose the capacity to communicate with normal cells and in this way become less receptive to tissue regulation, i.e. they become increasingly autonomous. Mechanistic links between gap junction processes, tumour promotion, and cell cycle control were also discussed in the UNSCEAR 1993 Report [U3]. In respect of radiation response, the term "bystander effect" has been coined to describe a range of *in vitro* responses occurring in unirradiated cells that are close neighbours of others receiving a given radiation dose, usually from single ionizing particles.

388. The effects seen in bystander cells include changes in gene expression [A19], lethality [M49], sister chromatid exchange [D13], chromosome breakage [P21], and gene mutation [N11]. The mechanisms involved are not well understood but are believed to involve the transfer of factors from irradiated cells via the extracellular medium or via intercellular communication [A19, M50]. Such effects have yet to be demonstrated *in vivo*, and their consequences for tumour risk cannot be judged. In the context of this Annex, the most important data set concerns apparent alpha-particle-induced bystander effects on gene mutation [N11]. These studies imply that at a low fluence of alpha particles, the frequency of gene mutations arising in bystander cells exceeds by up to fivefold that in cells intersected by single particles. At higher particle fluences, the bystander contribution to mutation rates decreases, and in this way the dose response for mutation induction is supralinear, with a steep rise at doses below ~50 mGy. Assuming that there is a direct relationship between mutation rate and tumour risk, the data noted above imply that per unit dose of alpha particles, tumour risk at doses below ~50 mGy is substantially greater than that at higher doses. Whether these data represent an important source of uncertainty in high-LET radiation risk estimates must await replication of the study and the establishment of its generality, particularly in the *in vivo* situation. Other data on radiation response implicate cellular interactions in the induction of genomic instability [G20, M51] and in adaptive responses [I15]. Some studies in this general area also imply that cellular DNA may not always be the principal target for radiation effects, particularly those that may have transient epigenetic components. Although it remains difficult to integrate such data into a mechanistic framework for assessing tumour risk at low doses, the findings noted above caution against a dogmatic view in the modelling of dose-response data on the basis of DNA damage alone.

4. Cellular surveillance

389. Following review of epidemiological, animal, and cellular studies, the Committee concluded that conventional T- and B-lymphocyte-mediated immune response was not a particularly effective protective mechanism against the development of most human tumours [U3]. Although these immune responses appear to be able to target the specific non-self antigens presented by oncogenic viruses or their associated neoplasms, the common radiogenic tumours seem unable to effectively initiate timely immune responses or are capable of efficiently evading surveillance [B18, U3]. The Committee did, however, recognize some uncertainty surrounding the potential protective role of certain classes of cytotoxic T-lymphocytes (CTL), including natural killer (NK) cells [U3]. A number of novel approaches have been used to resolve some of these uncertainties.

390. One area of recent study [L23] has been to determine whether the poor immunogenicity of most tumour types is due to the lack of signals for co-stimulation of full CTL activity; this is believed to be mediated by specialized antigen-presenting cells. A study with mice using an antibody to block CTLA-4, a negative regulator of CTL activation, showed that such a blockade resulted in the rejection of transplanted and pre-existing human tumour cells and also the development of resistance to a second challenge. It seems, therefore, that with fully malignant

cells, an effective anti-tumour response can be elicited provided that specific immune regulators are manipulated.

391. Another approach to the problem has been to seek mutational signatures in tumours indicative of evasion of immune surveillance. In one such study [B19], it was argued that the RER$^+$ mutator phenotype of certain colorectal carcinoma cell lines might generate a sufficiently diverse array of mutant protein neo-antigens to elicit a strong CTL response. If this is the case, inefficient antigen presentation via the loss of beta-2-microglobulin (β2M) would be strongly selected in the resulting tumour cell population. Such correlation was observed in a study of 37 cell lines, where the four mutator lines were the only examples in which β2M expression was lost.

392. Beta-2-microglobulin associates with polymorphic heavy-chain glycoproteins in cell membranes for the purposes of antigen presentation by the resulting class I major histocompatibility complex (MHC). Intracellular antigens are believed to be transported to the MHC by proteins of the TAP family [E6]. Not only do a substantial fraction of human tumours lack expression of the class I MHC, but there is some evidence of the involvement of TAP gene mutation in tumours. Other strategies adopted by tumour cells in order to evade immune surveillance include the expression of decoy receptors, the Fas-mediated inactivation of CTL/NK cells at membrane surfaces, and the secretion of factors that inhibit or inactivate CTL [V5].

393. In general, these observations, while providing some correlative support for the view that evasion of CTL-mediated surveillance may be of some importance in tumour development, provide no information on the effectiveness of this surveillance in reducing tumour risk. It is, however, intriguing to note emerging evidence of a possible two-way interaction between genomic instability in neoplasia and CTL response. On the one hand, the RER$^+$ mutator phenotype may serve to generate sufficiently strong tumour antigen signals for CTL response; on the other, it can provide an enhanced mutational capacity to evade the resulting T-cell surveillance. For potentially anti-tumorigenic CTL response, there is also some evidence to support a model whereby normal cells that engulf apoptotic tumour cells can migrate to lymph nodes, where, in principle, they can invoke a response to tumour antigens [A8].

394. With respect to NK cells, it is now well established that this class of cytotoxic cells can, in principle, exert a degree of anti-tumour activity via the release of factors such as interferon γ, tumour necrosis factor, and Fas ligand. There is also some evidence of an additional antitumour mechanism involving NK attack on tumour vasculature [B24, F9]. In spite of numerous studies, there is, however, no convincing evidence of a correlation between NK abundance/function in vivo and tumour development or prognosis (see [F9]). In general, this area of study remains most controversial, and with current knowledge it is not possible to judge the extent to which NK cells act to protect against non-viral human cancers.

395. Overall, the role of immune surveillance in protecting against common neoplasms has yet to be adequately described, and some studies tend to argue against this proposition [B18, U3]. Gains in fundamental knowledge will probably contribute to the debate. For example, the complement protein system is an important determinant of humoral immune surveillance and is believed to target certain malignant cells. In accord with this, a novel stress-related protein has been revealed that appears to participate in the discrimination of malignant cells by homologous complement [M26]. The recent observation that the proto-oncogene PML of human myeloid leukaemia plays a role in the regulation of antigen presentation in cells also implies the need for some developing haemopoietic neoplasms to evade cellular surveillance [Z2]. Equally, however, the tumorigenic expression of Apc-deficiency in Min mice is not enhanced by a defect (scid) in immune function [S18]. Thus, recent findings can be used to both support and question the true role of cellular surveillance in tumour defence. Studies of low-dose stimulation of immune functions, e.g. [L58, M54] have previously been reviewed by the Committee [U2] and a few additional studies have been published (e.g. [H36, K30, S42, S43]) in more recent years. Doubts were expressed as to whether the immune system plays a significant role in any cancer-related adaptive processes at low doses.

5. Summary

396. Through a better understanding of the processes that mediate multi-stage tumorigenesis it has become evident that neoplastic development is subject to a number of cellular constraints. The main constraints are control of cellular proliferation/genomic stability, the induction of programmed cell death, tumorigenic suppression by cell-cell communication, and terminal differentiation to a non-proliferative cellular state. In addition, for at least certain tumour types there is evidence that immunosurveillance mechanisms can recognize and restrict the growth of neoplastic cells.

397. These protective mechanisms are believed to provide a high level of protection against neoplastic growth and development. In spite of this, there is growing evidence that during the evolution of tumours, resistance to or tolerance of all these countermeasures can be developed via gene-specific mutation. Thus a substantial proportion of consistent mutations in tumours may be linked directly with cellular strategies aimed at maintaining viability and growth, avoiding terminal differentiation and immune recognition, and promoting genomic instability such that a wide range of clonal variants are available for the full development of malignancy. On the basis of current molecular genetic knowledge, there seems no good reason to suppose that different modes of in vivo constraint apply to spontaneously arising and carcinogen-induced tumours. Evidence is also accumulating in support of the view that cellular communication can also influence early in vitro radiation responses, with possible effects on cellular recovery, genomic stability, and mutation rates. The present state of knowledge does not allow for extrapolation

of these findings to tumorigenesis *in vivo*, but some recent alpha-particle mutation data, if confirmed, may be of importance.

E. DNA REPAIR AND TUMORIGENESIS

398. For the purposes of relating mechanisms of radiation tumorigenesis to mechanisms that are believed to apply to spontaneously arising disease, it is important to consider in greater depth the evidence on the role of DNA repair and the uncertainties that attach to this association.

1. DNA repair as a determinant of oncogenic response

399. Data relating to the influence of DNA repair on mutagenic and other cellular radiation responses are discussed in Annex F, *"DNA repair and mutagenesis"*. Critical to the role of DNA repair in radiation tumorigenesis is the now unambiguous evidence that heritable human deficiency in genes controlling DNA repair and maintenance of genomic stability is frequently associated with an increased incidence of spontaneously arising neoplasms.

400. Thus, such DNA processing functions in normal somatic human cells must play a critical role in protecting against spontaneous neoplastic development. As discussed in the UNSCEAR 1993 Report [U3], these data also provide important support for the mutational origin of neoplasia via failures in repair of DNA damage.

401. With respect to tumours associated with human exposure to exogenous genotoxic agents, studies with two categories of genetic disorders, xeroderma pigmentosum (XP) [K8] and Li-Fraumeni syndrome (LFS) [H8], provide evidence that defects in DNA damage processing are also important to oncogenic development after ultraviolet and ionizing radiation, respectively.

402. In the case of xeroderma pigmentosum, there is unambiguous evidence that the inherited deficiency in repair of DNA photoproducts is associated with an excess of cancer in regions of skin receiving significant solar exposure [K8]. Unexposed skin of XP patients shows an unremarkable frequency of these neoplasms, indicating the critical importance of DNA photoproduct induction for skin carcinogenesis and the high level of protection afforded by high-fidelity DNA repair processes.

403. The cancer-prone genetic disorder Li-Fraumeni syndrome (LFS) is frequently, although not always, characterized by a deficiency in the *TP53* tumour-suppressor gene that normally plays a role in DNA damage sensing, cell-cycle control, and apoptosis (see Sections IV.B and IV.D). Although the data are less compelling than those for ultraviolet radiation exposure of XP patients, there is evidence that LFS and LFS-like patients exhibit, in childhood, an elevated risk of tumour induction after radiotherapy [H8, S19]. Thus, inherited human defects in

DNA damage processing can be reflected in an increased risk of carcinogen-induced as well as spontaneously arising tumours.

404. In addition to these important human studies, experimental animal data relating to the role of DNA repair in radiation tumorigenesis are also beginning to emerge, largely through studies of mice that have been genetically manipulated to be deficient in specific genes involved in DNA repair and genomic stability [W2].

405. These animal data are discussed in Section IV.C. In brief, studies with *p53*-deficient mice give evidence of enhanced tumorigenic radiosensitivity associated with abrogation of a G_2/M cell-cycle checkpoint for chromosomal repair; radiation-induced *p53* gene loss via increased sensitivity to the induction of aneuploidy appeared to be the principal mechanism involved.

406. The *ATM* gene of human ataxia-telangiectasia is discussed in Annex F, *"DNA repair and mutagenesis"*, together with recent data relating to the radiosensitivity of mice manipulated to be deficient in this critical DNA damage response gene. In brief, *Atm*-deficient mice are highly radiosensitive, and while radiation tumorigenesis studies have yet to be reported, the animals, like their human counterparts, are prone to the spontaneous development of lymphoma and specific lymphoma-associated chromosomal rearrangement in haemopoietic cells [B1]. It may be anticipated that radiation tumorigenesis studies with these and other relevant genetically manipulated animals, e.g. those deficient in *BRCA1*, *BRCA2*, and *Rad51*, will be informative on the further relationships between DNA damage repair and tumorigenesis.

2. Implications and uncertainties

407. The Annex F, *"DNA repair and mutagenesis"*, provides evidence that certain forms of DNA double-strand lesions are the principal biologically relevant event induced by radiation in mammalian cells. Since current data imply that these lesions are usually repaired via a process of illegitimate rather than homologous recombination, there will be an inherent degree of error proneness in DNA repair after radiation exposure. Such misrepair events may be represented by gross chromosomal abnormality (deletion or rearrangement) or subchromosomal and intragenic events. Judging from molecular analyses of radiation-induced somatic cell mutants, these misrepair events take the form of DNA base-pair substitutions, frameshift mutations, or, more frequently, DNA deletions of varying size. Some data are suggestive of changes in mutational spectra with radiation dose rate and LET, but this issue remains controversial.

408. Also noted in Annex F, *"DNA repair and mutagenesis"*, is the growing evidence that DNA repair functions are important determinants of dose, dose rate, and radiation quality effects in mammalian cells. In brief, there is evidence that the extent and fidelity of repair

strongly influence the initial and final slopes of low-LET dose-effect relationships and the progressively steeper slopes of these relationships with increasing LET. It is concluded that RBE-LET effects are largely dependent on the repairability of initial DNA damage. To what extent are these *in vitro* data reflected in current knowledge of radiation tumorigenesis *in vivo*?

409. The data discussed in the UNSCEAR 1993 Report [U3] and in this Annex are broadly consistent with the single-cell mutational origin of most tumours. Loss-of-function mutations of critical gatekeeper genes appear to be early events in the genesis of many human solid tumours, and gain-of-function chromosomal events occur early in leukaemias and lymphomas (see Section B). There is evidence from animal studies, described earlier, that radiation-associated gene and chromosomal loss/deletion can act as an initiating event for tumorigenesis and that DNA damage processing is a crucial protective factor for *in vivo* radiation response. Given the obvious parallels between *in vitro* and *in vivo* data, it becomes possible to consider a general mechanistic framework within which to model dose-effect/RBE-LET relationships for radiation tumorigenesis and the protective functions that may operate. These modelling approaches will be discussed and developed later in this Annex.

410. There are, however, aspects of the data discussed in Annex F, *"DNA repair and mutagenesis"*, that caution against seeking oversimplistic correlations between *in vitro* cellular response data and tumorigenesis *in vivo*. First is the issue of novel mechanisms of genetic change in mammalian cells. In addition to epigenetic changes such as imprinting and gene silencing noted in this Annex, it has also been speculated that radiation may induce unknown cellular pathways that promote untargeted mutation. In some studies the activity of these pathways has been shown to persist for many post-irradiation cell generations, leading to an apparent elevation of the spontaneous mutation rate. As discussed in Annex F, *"DNA repair and mutagenesis"*, such findings have been made with respect to lethal mutation, gene mutation, and unstable chromosomal damage.

411. Given the emphasis in this Annex on the early development of genomic instability in tumour development, it is possible to speculate that any persistent genomic instability induced by radiation in target somatic cells *in vivo* might make a significant, late-expressing contribution to tumorigenic risk [L24]. The cellular processes underlying this induced instability remain uncertain, however, and no single mechanism seems capable of explaining the various and sometimes inconsistent manifestations referred to in Annex F, *"DNA repair and mutagenesis"*. A collection of recent papers [M32] addresses various aspects of this developing field; of particular note is the finding of a possible genetic association in mouse strains between the post-irradiation expression of persistent chromosomal instability and susceptibility to radiation tumorigenesis [U25]. The authors were, however, cautious about the implications of these initial findings. In follow-up studies [O7], late-expressing chromosomal instabi-

lity in the mammary-tumour-susceptible BALB/c mouse was shown to be genetically associated with changes in expression of repair-related DNA PK protein as well as with reduced DNA double-strand break repair. From these studies it is possible to indirectly implicate late-expressing genomic instability in radiation tumorigenesis in certain genetic settings, but whether a causal relationship applies is uncertain.

412. Overall, a general link between such induced instability and radiation tumour risk remains to be established. Indeed, the cytogenetic findings of *in vivo* studies relating to mechanisms of radiation-induced lymphomagenesis in *p53*-deficient mice and myeloid leukaemogenesis in CBA mice tend to argue against a significant contribution from induced and persistent genome-wide instability [B4, B20]. However, a specific and persistent clonal feature of chromosomal instability that has been closely associated with the development of human neoplasia is the so-called segmental jumping translocation. These events are not uncommon in spontaneous human lymphohaemopoietic tumours [U3] and were reported recently in myeloid neoplasms arising in irradiated survivors of the atomic bombings [N8]. It may be concluded that certain elements of radiation-associated persistent genomic instability probably do play a role in tumorigenic processes, but there is much uncertainty as to the overall importance of that role and therefore as to how it might be taken into account in the modelling of radiation risk.

413. A second source of uncertainty discussed in the UNSCEAR 1994 Report [U2] and in Annex F, *"DNA repair and mutagenesis"*, is the cellular phenomenon of adaptive response to DNA damage; the data describing such responses were reviewed recently [W6]. In brief, in certain experimental systems a small priming dose of radiation (or of some other genotoxic agents) can result in the development of partial resistance to a challenge by a second, higher dose. The radiobiological endpoint most frequently employed in cellular and animal systems has been cytogenetic damage, but there are also some data with respect to gene mutation and cell survival. In addition, some mechanistic studies have been undertaken of the possible role of activation/induction of novel proteins and cell-cycle perturbation. In principle, inducible DNA repair might underlie some manifestations of adaptive response, but as noted in Annex F, *"DNA repair and mutagenesis"*, evidence for the up-regulation of relevant repair genes is fragmentary. There is, however, growing evidence for subtle post-irradiation modification of repair-related protein complexes and the induction of stress-response genes. More detailed information on adaptive responses to cytogenetic and lethal cellular damage in mammalian cells are provided in Annex F, *"DNA repair and mutagenesis"*.

414. These studies lend credibility to the true existence of adaptive responses, but they also draw attention to the fact that the response is transient, not usually robust, and frequently lacking a clear mechanistic basis. For example, at the cellular level, adaptive responses have rather uncertain dose and dose-rate dependency and when expressed lead to only a modest decrease in sensitivity to

the subsequent radiation challenge. In addition, at the cytogenetic level this response is not consistently expressed, with cells from some humans and mouse strains failing, for unknown reasons, to show adaptive responses. Although some novel proteins have been detected in "adapted" cells, cell-cycle-related changes are not obvious, and the relationships between cytogenetic adaptive responses and known stress-related biochemical signalling pathways remain to be clarified [W6].

415. In the UNSCEAR 1994 Report [U2], the Committee reviewed animal studies of life-shortening and tumour induction that were relevant to the possible role of adaptive responses, but these studies did not report convincing evidence of such effects. Subsequently there have been reports of possible adaptive effects in relation to radiation lymphomagenesis [B21] and life-shortening in mice [Y4]. Of particular note is a recent report on a possible gamma-ray-induced adaptive response in respect of acute myeloid leukaemia (AML) induction in the mouse [M52]. This study reported that a chronic priming dose of 100 mGy did not influence the incidence of acute myeloid leukaemia following a chronic challenge dose of 1 Gy; the lifespan of mice without acute myeloid leukaemia was similarly unaffected. However, the animals receiving the priming dose showed modestly increased latency for induced acute myeloid leukaemia, implying that the later stages of the leukaemogenic process had been modified. The mechanistic basis of this unexpected result remains highly uncertain, but the authors speculate that the increased tumour latency might reflect the triggering of some form of persistent stress response.

416. Although the relevance of adaptive responses to human tumorigenesis should not be discounted, in the absence of a consistent body of *in vivo* tumorigenesis data and with current uncertainties on cellular mechanisms, it would be most difficult to include adaptive response parameters in mechanistic models of low-dose radiation tumorigenesis.

417. A third area of uncertainty surrounding DNA damage and repair concerns the relationships between spontaneous and radiation-induced damage and their implications for low-dose tumorigenic risk. Debate on these issues has been conducted for some years [A4, B22, C15, W7], and the main areas of contention may be outlined as follows.

418. In spite of its critical information-carrying role in the cell, genomic DNA has limited chemical stability. Via hydrolysis, oxidative attack, and chemical methylation processes, cellular DNA is constantly modified by its endogenous environment irrespective of the influence of exogenous agents such as electrophilic chemicals, ultraviolet radiation, and ionizing radiation [L25].

419. Endogenous damage to the mammalian genome may take the form of hydrolytic depurination and deamination of DNA bases, oxidative attack on DNA bases and the sugar-phosphate backbone, and non-enzymatic DNA methylation of certain bases. For largely technical reasons, estimates of the rate of accumulation and abundance of such endogenous DNA lesions vary considerably [L25].

420. Less uncertainty surrounds the general form that this DNA damage takes [L25]. By their very nature, hydrolytic, oxidative, and methylation events are random and unclustered, affecting chemical moieties on one or the other strands of the DNA duplex; examples are the formation of abasic sites due to hydrolytic depurination, 8-hydroxyguanine formation via hydroxyl radical attack, and uracil formation via deamination of 5-methylcytosine. DNA single-strand breaks as a consequence of base loss, oxidative attack, and as repair intermediates also arise spontaneously.

421. The evidence concerning the type, abundance, and repair of endogenously arising spontaneous DNA damage is summarized in Annex F, *"DNA repair and mutagenesis"*. The Annex emphasizes the technical uncertainties surrounding estimates of the abundance of such DNA damage.

422. The general conclusion that may be reached from these data is that while it is difficult to make precise quantitative comparisons, there are differences between the spectra of DNA damage types arising spontaneously and those induced by radiation; there are also differences in their repair characteristics (see also [C15, L25, W7]).

423. This view of endogenous damage and its consequences may be set against the following theoretical proposition: since the cell is able to repair a very high level of endogenous DNA damage without frequent mutagenic consequences, a further small increment of DNA damage from low doses of radiation will not impose significant risk; that risk only becomes significant at relatively high doses, when, at a given level of genomic damage, DNA repair capacity is exceeded.

424. The fundamental scientific uncertainty surrounding this proposition is that it assumes that the nature and reparability of spontaneous and radiation-induced DNA damage are essentially equivalent [C15, W7]. The data in Annex F, *"DNA repair and mutagenesis"*, provide evidence that although there are some similarities between the DNA lesion types arising spontaneously and those induced by radiation, DNA double-strand breaks almost certainly make a substantially greater relative contribution after radiation exposure. More important, however, is the evidence accumulating on the chemical nature of radiation-induced DNA double-strand breaks and other double-strand lesions.

425. Through a combination of cellular, biophysical, biochemical, and molecular approaches, it has become apparent that a high proportion of radiation-induced DNA double-strand breaks and related lesions are chemically complex and/or part of multiply damaged DNA sites. This feature stems from the requirement for local clustering of energy loss events from a given radiation track to effect coincident damage to both sugar-phosphate backbones of the DNA duplex [G5, G10]. This chemical complexity of

DNA double-strand breaks is apparent after low-LET radiation but will increase with LET.

426. As noted in Annex F, *"DNA repair and mutagenesis"*, the correct repair of such complex damage is difficult because of multiple and coincident damage to coding on both DNA strands. In most instances such repair is believed to proceed via illegitimate recombination, which is inherently error-prone, and it is this reparability factor that will principally distinguish spontaneous and radiation-induced DNA lesions. Accordingly, excess dicentric chromosomes have been recorded in human lymphocytes *in vitro* at low-LET doses of around 20 mGy, while the spontaneous rate of generation of these events is very low (~1 per 1,000 cell generations) [L8].

427. Stated simply, the relative abundance of complex and poorly repairable DNA lesions after radiation exposure is judged to be very much greater than that of lesions that arise spontaneously. Therefore, it is this feature rather than lesion abundance overall that should guide judgements on the role of DNA repair in low-dose response and radiation quality effects. Accordingly, the proposition stated in paragraph 423 runs counter to advances in fundamental knowledge and therefore has no obvious role in the modelling of tumorigenic risk.

3. Summary

428. A large body of information points to the crucial importance of DNA repair and other damage-response functions in tumorigenesis. Not only do these DNA damage-processing functions influence the appearance of initial events in the multi-stage process, but they also serve to reduce the probability that a benign neoplasm will spontaneously acquire the secondary mutations necessary for full malignant development. Thus, mutations of DNA damage-response genes in tumours play important roles in the spontaneous development of genomic instability. Various forms of radiation-induced persistent genomic instability have been recorded in experimental cellular systems. With the possible exception of instability at certain chromosomal translocation junctions, these phenomena are not well understood, and their association with *in vivo* tumorigenesis has yet to be established.

429. With respect to radiation damage to DNA, it is concluded that the repair of sometimes complex DNA double-strand lesions is inherently error-prone and is most likely to be an important determinant of dose, dose rate, and radiation quality effects for the induction of tumorigenic lesions. Uncertainties remain on the significance for tumorigenesis of adaptive responses to DNA damage; the mechanistic basis of such responses has yet to be clarified. In contrast, recent scientific advances provide clear evidence of the differences in complexity and reparability between spontaneously arising and radiation-induced DNA lesions. In the modelling of radiation tumorigenesis these data argue against basing judgements about low-dose response on uncritical comparisons between overall lesion abundance and repair capacity. Overall, the general concepts linking DNA damage

repair and tumorigenesis that were summarized previously in the UNSCEAR 1993 Report [U3] remain valid. However, the data discussed here and in Annex F, *"DNA repair and mutagenesis"*, provide a far more robust scientific framework to support these concepts than was available to the Committee in 1993.

F. BIOLOGICAL MODELLING OF TUMORIGENIC RESPONSES

1. Implications of current data

430. As knowledge of the fundamental basis of multi-stage tumorigenesis continues to advance, it becomes possible to identify critical features of the process and the uncertainties that may attach to the development of biological models to describe risk at low doses.

431. The earliest phase of tumour development (initiation) appears, in the main, to involve loss- or gain-of-function mutation of single genes in single target stem-like cells in tissue. In the case of solid tumours, a set of tissue-/cell-type-specific gatekeeper genes in the tumour-suppressor category may be the principal loss-of-function gene targets. For lympho-haemopoietic tumours, both loss-of-function mutations and gain-of-function chromosomal translocations are likely to be important. In biological modelling, these two types of tumorigenic event may require different forms of treatment. Not only do mutational mechanisms differ but also gain-of-function mutations in leukaemia/lymphoma can have more profound cellular effects than loss of a single tumour-suppressor gene.

432. An important source of uncertainty is provided, however, by the non-mutational (epigenetic) events that can, for certain genes, substitute for gene mutation. The overall contribution of, for example, gene silencing and loss of imprinting to human tumorigenesis, although very difficult to quantify, seems likely to be significant.

433. Some uncertainties about the probability of accumulation of multiple genetic events in tumorigenesis have been reduced by the characterization of the spontaneous development of genomic instability at a relatively early point in neoplastic development. Some of the mutator genes that serve to drive tumorigenesis have been characterized, and from a modelling standpoint it seems most appropriate to view the development of the mutator phenotype as marking the transition between benign and malignant disease. Although critical evidence is lacking, the early appearance of a strongly expressing mutator phenotype with respect to spontaneously arising DNA damage may mean that low doses of exogenous carcinogens such as radiation will make a relatively small contribution to later phases of tumour development compared with those phases occurring before the spontaneous onset of genomic instability. This might serve as a mechanistic explanation for the observation that radiation usually acts only weakly on tumour promotion and progression [U3].

434. Evidence is growing, however, that in some circumstances, the expression of radiation-induced DNA/chromosomal damage in cells may be a persistent phenomenon. Secondary chromosomal change centred on primary exchange junctions, e.g. jumping translocations, is not unexpected and has been recorded in cellular systems and human/animal lymphohaemopoietic neoplasms [B44, G21, N8, U3]. The other cellular features of induced genomic instability discussed in this Annex and in Annex F, *"DNA repair and mutagenesis"* are more difficult to relate directly to tumorigenesis, so for this reason it may be premature to attempt to integrate them specifically into mechanistic models of tumour risk. The same general problem applies to bystander effects of radiation on cell inactivation and mutagenesis, although the alpha-particle data [N11] discussed earlier deserve some attention. That said, the Committee recognizes recent signs of progress in resolving the mechanistic uncertainties associated with the role of stress-related processes in cellular response to radiation and anticipates that much better informed judgements will be possible within a few years.

435. By contrast, many sources of data on tumorigenesis point toward the crucially important protective role played by high-fidelity DNA repair. In the light of information on DNA repair capacity in relation to the high flux of spontaneous DNA damage in mammalian cells, it has been suggested by some that low doses of radiation would be expected to contribute little to tumour risk. If true, this would have important implications for biological modelling. Uncertainties on this contentious issue have been reduced by the growing evidence that an important fraction of the DNA double-strand lesions induced by radiation is chemically complex, extremely difficult to repair correctly, and only very rarely occurs spontaneously. In the context of biological models of tumorigenesis, it is judged that overly simplistic analysis of data on DNA lesion abundance and repair can be most misleading. New data on the role of DNA repair in cellular dose and dose-rate effects and the associations between DNA lesion complexity and RBE-LET effects do, however, have implications for mechanistic models.

436. Other important protective factors in tumorigenesis include apoptotic cell death, cellular senescence, terminal differentiation of cells to a non-proliferating state, and the elimination of neoplastic cells by immunosurveillance mechanisms. Through their capacity to remove cells from neoplastic pathways, these processes collectively serve to provide a high level of protection against tumours arising spontaneously or induced by carcinogens. Critically, however, molecular genetic studies have provided compelling evidence of gene-specific neoplastic mutations that serve to block these processes. Thus, a given overtly malignant tumour will have succeeded via mutation or epigenetic change in evading or gaining tolerance to each of the protective challenges it has faced during development.

437. There are few quantitative data on which to base judgements on the relative magnitude or efficiency of the protective factors noted above. Indeed, it seems likely that this relative magnitude will vary from one tumour type to another, and the differences between virally associated and other tumour types with respect to immunological protection may be evidence of this. Thus, with limited knowledge, the modelling of protective factors in tumorigenesis can, at best, be only empirical.

438. In spite of this, a critically important question is whether the extent or effectiveness of such protective factors might be different for tumours arising spontaneously and those induced by radiation. As noted in the UNSCEAR 1993 Report [U3], exposure to high doses of radiation, where cell killing becomes important, would be expected to influence final tumour yield not only by initially reducing target cell numbers but also by subsequently mobilizing quiescent stem cells for tissue repopulation. High-dose suppression of immune functions might also be important for certain tumour types, principally those associated with oncogenic viruses.

439. At low doses and dose rates, where cell killing is not significant, there is, however, no specific reason to anticipate profound and long-term effects of radiation on the function of protective mechanisms. Transient changes in the activity of these systems resulting from stress-related effects on cellular biochemical signalling might be anticipated, but with current knowledge it is not possible to relate these to final tumour yields.

440. The possible exception to this are the adaptive responses to radiation noted in this Annex and discussed in depth in the UNSCEAR 1994 Report [U2]. These would include not only adaptive DNA damage responses but also possible stimulating effects on immune system and other potentially protective functions.

441. Stated simply, if low doses of radiation could be shown to enhance profoundly, over an extended period, the anti-tumour activities outlined in this Annex, then radiation-induced tumours would be expected to be subject to greater suppression than those arising spontaneously. Under these theoretical circumstances the shape of the dose-effect relationship for tumorigenesis would not be expected to be simple and might well depart from those for related radiobiological endpoints in single cells, i.e. the induction of chromosome aberrations, gene mutation, or cell transformation.

442. Although current knowledge does not exclude the possibility of this scenario, the data available on adaptive responses in cells or animals are judged to be insufficiently well developed for the purposes of biological modelling. Accordingly, the existence or otherwise of an adaptive response for radiation tumorigenesis remains a continuing source of debate.

2. Basic premises

443. Although much knowledge has been gained on the cytogenetic, molecular, and biochemical processes involved in the development of neoplasia, considerable uncertainty remains, particularly with respect to the quantitative

aspects of multi-stage tumour development. For this reason any attempts to include such data in the modelling of radiation tumorigenesis demands a set of simplifying judgements.

444. In full recognition of the uncertainties discussed in earlier Sections of this Annex, the following premises, based on the weight of current evidence, may be stated:

(a) The principal role of radiation in tumour development is to generate the DNA damage that can give rise to gene-specific mutation in critical target cells. Repair of that damage may be enhanced by cell-cycle checkpoints and, possibly, adaptive repair processes, but there is no specific expectation of wholly error-free repair, even at low doses and dose rates. Equally, the elimination of radiation-damaged cells by apoptotic processes is very unlikely to be complete;

(b) The vast majority of spontaneous and induced tumours have their origin in single specific mutations in single target cells in tissues. The probability of such a mutated cell progressing to overt malignancy is, however, very low because of the defences afforded by protective processes such as apoptosis, terminal differentiation, senescence, and cellular surveillance. Further mutation during pre-neoplastic clonal development can serve to bypass these defensive measures, and none are likely to be wholly protective or to be consistently enhanced by low doses of radiation;

(c) Although both loss-of-function (tumour-suppressor) and gain-of-function (proto-oncogene) gene mutations can contribute to multi-stage tumorigenesis, the DNA deletion mechanism characteristic of radiation will tend to make loss-of-function events the predominant process at all doses of radiation; and

(d) Radiation acts principally at the early stages of tumorigenesis by inducing specific mutations in normal stem-like cells. During protracted radiation exposures a contribution to the later stages of tumour development is possible, but during these later phases the acquisition of a mutator phenotype and/or one associated with epigenetic gene silencing/activation may be the primary driving force for neoplastic selection and progression.

445. Assuming these premises to be correct, it would seem that the dose-response parameters for radiation tumorigenesis at low doses are determined principally by factors that apply to the induction of the specific gene/chromosomal mutations in the target cells in question; the abundance and kinetics of these target cells will also be important determinants of the response. Stated simply, these radiation-induced mutations would be adding in a dose-dependent manner to the *in vivo* pool of tumour-initiating mutational events contributed by spontaneous processes and other genotoxic exposures. Thereafter, it seems reasonable to assume that all such events will be subject to the same variable sets of cellular and humoral factors that serve to suppress or enhance malignant development. On this basis, significant departure from a simple dose-response relationship would demand a dose-

dependent change in the kinetics of one or more of these post-irradiation modifying processes. For example, if there were to be persistent post-irradiation elevation of error-free DNA repair, apoptosis, terminal differentiation, cellular senescence, or immunosurveillance, then the radiation cancer risk might be depressed. Conversely, if post-irradiation mutation rates are persistently high (as a form of induced genomic instability), tumour development might be enhanced.

446. Although it is possible to speculate on the roles that the above processes might play in determining tumorigenic responses, any such hypotheses currently lack critical experimental support and plausible mechanisms that might operate after low doses. It is not, however, difficult to envisage substantial modification of cellular/tissue behaviour during the long-term cellular repopulation required for tissue regeneration after high doses; these processes may well have consequences for local tumour development.

447. At low doses, transient changes in biochemical equilibria and cell kinetics should be expected, but these are probably part of the normal cellular damage response pathways associated with the cell-cycle checkpoint, DNA repair, and mutagenesis functions already discussed. There is, however, no reason to believe that induced transient changes are unique to radiation.

448. The overall judgement that, on mechanistic grounds, cancer risk at low doses will increase as a simple function of dose is, however, subject to a number of important caveats. Some of these have already been rehearsed in this Annex, but two deserve additional attention.

3. Error-free DNA repair at low doses

449. Recombinational repair involving fully homologous DNA sequences may be regarded as the sole source of potentially error-free repair, particularly with respect to DNA double-strand breaks/lesions that involve complexity of DNA damage at single sites, so-called multiply damaged sites. It has been argued that such homologous recombination is not the predominant mode of repair after ionizing radiation and that the majority of the relevant double-strand lesions are processed via error-prone pathways involving non-homologous recombination. The data that underpin this judgement have, however, been obtained largely through cellular and molecular studies conducted after relatively high radiation doses. For example, as discussed earlier, the formation of dicentric chromosome aberrations probably reflects such error-prone repair processes, and for these aberrations, the lowest dose at which an excess has been reproducibly obtained is around 20 mGy (low-LET). Human epidemiological studies, also outlined in this Annex, reveal evidence of excess cancer risk at somewhat higher doses. Below these doses there must be complete dependence on an understanding of mechanistic processes and, critically, on the contention that error-prone DNA repair processes remain in place in cells down to single-track intersections of DNA.

450. Although there is no specific reason to depart from this position, it remains possible that at low doses, below, say, 10 mGy (low-LET), where radiogenic DNA lesions are few, error-free homologous recombination predominates. In this hypothetical situation, error-prone repair would be a secondary response that applied only when the abundance of DNA lesions increased above some critical level. Under these conditions the form of the dose-response relationship for mutational/tumorigenic risk would be expected to have a threshold-like component at very low doses.

451. Formal experimental approaches to this problem are beyond the resolution of current quantitative techniques in cellular radiobiology. One particular set of observations argues, however, against the proposition of wholly error-free repair at very low abundance of radiogenic DNA lesions.

452. As noted in this Annex and in Annex F, *"DNA repair and mutagenesis"*, spontaneously arising DNA double-strand breaks have a relatively low abundance in mammalian cells. In spite of this, dicentric chromosome aberrations, a manifestation of DNA break misrepair, occur at a reproducibly measurable spontaneous rate of about 1 per 1,000 cell generations [L8]. These observations argue that such DNA misrepair processes are not solely a product of a high incidence of DNA lesions and therefore that error-free repair at low doses is unlikely. Although somewhat less certain, a proportion of the DNA deletions/rearrangements that characterize some spontaneously arising gene mutations in mammalian cells may also arise as a consequence of DNA break misrepair mechanisms (see Annex F, *"DNA repair and mutagenesis"*).

453. Finally, although in a repair context homologous recombination is potentially error-free, it may carry its own risk in that it can serve to duplicate genes from the undamaged homologous chromosome of a given target cell. These recombinational processes are well-recognized mechanisms for the unmasking of variant heterozygous genes that can contribute to tumorigenic development, i.e. homologous recombination as well as DNA deletion results in the loss of heterozygosity in DNA that characterizes many human and animal tumours [V2]. Thus, it may be that there is no risk-free way in which complex DNA double-strand lesions can be processed in the cells of genetically heterozygous organisms such as humans.

4. Epigenetic events in tumorigenesis

454. Stable epigenetic effects on gene activity such as gene silencing via heterochromatization and gene activation via loss of imprinting have been described in this Annex as being involved in tumour development. Whereas there is a wealth of information on dose-response relationships for the induction of gene/chromosomal mutations, it remains most uncertain whether radiation damage contributes directly to the establishment of these stable epigenetic events and whether there is any form of dose response. In general, the implications for low-dose cancer risk remain a matter of speculation, but a number of issues may be raised.

455. First and most simply, in some instances the translocation of intact genetic material from one genomic location to another can lead to changes in gene activity, so-called position effects (see [P12]). The dose response for such effects should follow that of chromosomal exchange, so there should be no major uncertainty. Cellular targets for loss of imprinting and/or changes in the status of DNA methylation and/or heterochromatin *in situ* remain obscure, however.

456. Second, some *in vivo/in vitro* studies with rodents noted in the UNSCEAR 1993 Report [U3] and in this Annex imply that the rate of induction by radiation of early tumour-associated events in clonogens of thyroid and breast tissue is too high to be explained by conventional mutational mechanisms. These observations, together with those made in certain cellular transformation systems, have led to suggestions that induced epigenetic processes may play an important role in radiation tumorigenesis [C19].

457. The cellular and molecular basis of these high-frequency events remains unresolved. Nevertheless, the finding of induced frequencies for tumour-initiated clonogens of around 10^{-2} cannot possibly be explained by a gene-specific deletion mechanism such as that which applies to *HPRT* in cultured cells, where induced frequencies after low-LET irradiation rarely exceed a value of 10^{-4}. As noted in Annex F, *"DNA repair and mutagenesis"*, induced mutation frequency is, however, influenced strongly by genetic context, with tolerance of DNA loss an important factor; recent observations concerning tumorigenic mechanisms in mice, outlined below, suggest that extreme forms of such tolerance may not be unusual and may be misleading with respect to gene targets.

458. In studies of radiation tumorigenesis in *p53*-deficient (+/−) mice, Kemp et al. [K2] noted the very high frequency at which tumour-initiating events appeared to be induced and suggested that radiation-induced persistent genomic instability with respect to wild type *p53* might be responsible. Follow-up studies [B4] suggest, however, that this high frequency of initial events can be explained by a mechanism involving whole chromosome gain and loss (aneuploidy), where the target size for *p53* gene loss from critical cells may be orders of magnitude greater than the gene itself. In the same way, loss of wild-type *Apc* during the early development of intestinal neoplasia in *Apc*+/− mice frequently involves loss of the whole of the encoding chromosome [L20, S18] and is also likely to be a much more frequent event than single gene deletion. Thus, alone, frequency of tumour initiation may not be a reliable indicator of epigenetic involvement in radiation tumorigenesis. In this respect the extrapolation of mechanistic data from rodent experimental systems to human tumorigenesis should be undertaken with caution.

459. Nevertheless, these findings indicate that in seeking to model multi-stage radiation tumorigenesis, it is not always necessary to be constrained by conventional values of induced mutation frequency that apply to single genes in cells irradiated *in vitro*. Very large chromosomal deletions

and specific forms of aneuploidy can be tolerated and selected during *in vivo* tumour development, and the rate of many forms of gene/chromosomal mutation is often enhanced following the spontaneous development of mutator phenotypes. In addition, the suppression of apoptosis during tumorigenesis may allow the proliferation of genomically aberrant cells that would otherwise have been eliminated. In effect, the loss of apoptotic processes can further enhance the rate at which viable mutant cells appear within evolving neoplastic clones.

G. SUMMARY

460. Current evidence suggests that the biological modelling of radiation tumorigenesis might best proceed on the initial assumption that at low doses radiation acts primarily as a mutational initiator of neoplasia. The situation regarding protracted low-dose irradiation is biologically more complex, and mechanistic studies have yet to comment specifically upon the extent to which radiation may influence the later stages of tumorigenesis. The possible existence of error-free DNA repair in target cells that might generate a low-dose threshold for tumour induction is recognized but judged on fundamental grounds to be unlikely. Other cellular and humoral factors would need to be modulated in a dose-dependent fashion to specifically change the initial slope of the dose response for tumour induction. At present, conclusive evidence for these radiation-specific modulations operating at low doses is lacking, but such effects would not be unexpected after high-dose irradiation. An additional uncertainty is the balance between mutational and epigenetic contributions to induced neoplasia, particularly the role and possible dose response for epigenetic effects. Epigenetic effects following radiation could, in principle, impact all stages of tumour development. Transient biochemical stress responses and bystander effects are most likely to influence the mutagenic and apoptotic aspects of tumour initiation, while

induced and persistent genomic instability may be envisaged to impact pre-neoplastic clonal evolution and malignant development. Although fundamental knowledge is increasing rapidly, the extent to which such processes specifically determine low-dose tumorigenic response remains largely a matter of speculation.

461. Overall, evidence on the fundamental aspects of radiation action and its relationship to tumour induction provide no firm scientific reasons to believe that at low doses the form of cellular dose response is related in a complex fashion to increasing dose. Employing the principle of parsimony, it is therefore suggested that low-dose cellular mutagenic risk and, by implication, that for tumorigenesis rises from the zero-dose baseline as a simple function of dose. The linear form is the simplest of these responses and is not inconsistent with the majority of the quantitative data discussed in this Annex. Irrespective of future scientific developments, however, it may well be impossible to provide formal scientific proof of linearity or any other form of low-dose radiation response for tumorigenesis *in vivo*.

462. In addition, for a complex multifactorial response such as *in vivo* tumorigenesis, the expression of initial *in vivo* cellular events may in some circumstances be subject to high dose modification. Accordingly, caution needs to be exercised in interpreting dose-response data for *in vivo* tumorigenesis that encompass wide dose ranges, say, 0–5 Gy low-LET.

463. The observation of apparently simple forms of *in vivo* dose response over such a dose range can, in principle, disguise competing dose-dependent elements in the tumorigenic process. For example, at high doses, the suppressive effects of initial inactivation of target cells may compete with subsequent tumour promotion in damaged normal tissue. For the purposes of modelling the biological elements of radiation tumorigenesis, it is not possible at present to quantify such competing effects, but as knowledge accumulates these problems will demand increasing attention.

V. BIOLOGICALLY BASED MODELLING OF RADIATION CARCINOGENESIS

464. The principal aims of this Chapter are (a) to review general aspects of the computational models that seek to interpret epidemiological data on cancer risk, (b) to describe the empirical and mechanistic models that have been developed, and (c) to illustrate and compare the predictive features of empirical and mechanistic models with emphasis on risk at low doses. Computational models of cancer risk can also play a role in describing the possible interactions between radiation and other agents. This complex issue is explored in Annex H, *"Combined effects of radiation and other agents"*.

465. In Chapter IV the review of fundamental data on radiation tumorigenesis allowed proposals to be made on how evolving knowledge might guide the development of mechanistic models of radiation risk. At this early stage of

understanding, much of the guidance remains to be implemented, but three main principles can be considered: (a) radiation will tend to act at the earliest stage of tumorigenesis (initiation), (b) in general, no low-dose threshold should be expected, and (c) time-constant relative risk is suggested on the basis that radiation-induced and spontaneously arising tumorigenic events will be subject to the same host and environmental modifications although this is recognized as being somewhat simplistic.

466. The biological uncertainties noted in earlier Chapters suggest, however, that dose-dependent differences between tumour types with respect to their induction and the mechanisms involved should be expected. Accordingly, general comment will be provided on the predictive value of

biologically based models with respect to the projection of risk with time and dose response for radiation tumorigenesis. At the outset it is however important to stress that although a number of valuable mathematical and statistical tools have been developed, the outcome of cancer risk modelling is often dependent on the initial biological assumptions made. Even using the same data sets, different groups of workers can arrive at different optimal mathematical/statistical solutions depending on these assumptions. This is clearly a significant source of uncertainty.

A. GENERAL ASPECTS OF THE PROBLEM

467. One of the principal uncertainties that surround the calculation of cancer risks from epidemiological data results from the fact that few radiation-exposed cohorts have been followed up to extinction. For example, 50 years after the atomic bombings of Hiroshima and Nagasaki, about half of the survivors were still alive [P2]. In attempting to calculate lifetime population cancer risks it is therefore important to predict how risks might vary as a function of time after radiation exposure, in particular for that group for whom the uncertainties in projection of risk to the end of life are most uncertain, namely those who were exposed in childhood.

468. One way to model the variation in risk is to use empirical models incorporating adjustments for a number of variables (e.g. age at exposure, time since exposure, sex) and indeed this approach was used by the BEIR V Committee [C1] in its analyses of data on the Japanese atomic bomb survivors and various other irradiated groups. Recent analyses of solid cancers for these groups have found that the radiation-induced excess risk can be described fairly well by a relative risk model [I2]. The time-constant relative risk model assumes that if a dose of radiation is administered to a population, then, after some latent period, there is an increase in the cancer rate, the excess rate being proportional to the underlying cancer rate in an unirradiated population. For leukaemia, this model provides an unsatisfactory fit, consequently a number of other models have been used for this group of malignancies, including one in which the excess cancer rate resulting from exposure is assumed to be constant i.e. the time-constant additive risk model [U4].

469. It is well known that for all cancer subtypes (including leukaemia) the excess relative risk (ERR) diminishes with increasing age at exposure [U2]. For those irradiated in childhood there is evidence of a reduction in the excess relative risk of solid cancer 25 or more years after exposure [L6, L33, P2, T4]. For solid cancers in adulthood the excess relative risk is more nearly constant, or perhaps even increasing over time [L32, L33], although there are some indications to the contrary [W9]. Clearly then, even in the case of solid cancers various factors have to be employed to modify the excess relative risk.

470. Associated with the issue of projection of cancer risk over time is that of projection of cancer risk between two populations with differing underlying susceptibilities to cancer. Analogous to the relative risk time projection model one can employ a multiplicative transfer of risks, in which the ratio of the radiation-induced excess cancer rates to the underlying cancer rates in the two populations might be assumed to be identical. Similarly, akin to the additive risk time projection model one can use an additive transfer of risks, in which the radiation-induced excess cancer rates in the two populations might be assumed to be identical. The data that are available suggests that there is no simple solution to the problem [U2]. For example, there are weak indications that the relative risks of stomach cancer following radiation exposure may be more comparable than the absolute excess risks in populations with different background stomach cancer rates [U2]. The breast cancer relative risks observed in the most recent analysis of the Japanese atomic bomb survivor incidence data [T4] are rather higher than those seen in various other data sets, particularly for older ages at exposure [B6, M2, S20]. The observation that sex differences in solid tumour excess relative risk are generally offset by differences in sex-specific background cancer rates [U2] might suggest that absolute excess risks are more alike than excess relative risks. Taken together, these considerations suggest that in various circumstances relative or absolute transfers of risk between populations may be advocated, or indeed, the use of some sort of hybrid approach such as that which has been employed by Muirhead and Darby [M12] and Little et al. [L56].

471. The exposed populations that are often used for deriving cancer risks e.g. the Japanese atomic bomb survivors, were exposed to ionizing radiation at high doses and high dose rates. However, it is the possible risks arising from low dose and low dose-rate exposure to ionizing radiation which are central to the setting of standards for radio-logical protection. The ICRP [I2] recommended application of a dose and dose-rate effectiveness factor of 2 to scale cancer risks from high dose and high dose-rate exposure to low dose and low dose-rate exposure on the basis of animal data, the shape of the cancer dose response in the bomb survivor data and other epidemiological data. Although the linear-quadratic dose-response model (with upward curvature) found for leukaemia is perhaps the most often employed departure from linearity in analyses of cancer in radiation-exposed groups [P1, P2], other shapes are possible for the dose-response curve [U3]. While for most tumour types in the Japanese data linear-quadratic curvature adequately describes the shape of the dose-response curve, for non-melanoma skin cancer (NMSC) there is evidence for departures from linear-quadratic curvature. The non-melanoma skin cancer dose response in the Japanese cohort is consistent with a dose threshold of 1 Sv [L7, L30] or with an induction term proportional to the fourth power of dose, with, in each case, an exponential cell sterilization term to reduce non-melanoma skin cancer risk at high doses (>3 Sv).

472. Arguably, models which take account of the biological processes leading to the development of cancer can provide insight into these related issues of projection of cancer risk over time, transfer of risk across population and extrapolation of risks from high doses and dose-rates to low doses and dose-

rates. For example, Little and Charles [L32] have demonstrated that a variety of mechanistic models of carcinogenesis predict an excess relative risk which reduces with increasing time after exposure for those exposed in childhood, while for those exposed in adulthood the excess relative risk might be approximately constant over time. Mechanistic considerations also imply that the interactions between radiation and the various other factors that modulate the process of carcinogenesis may be complex [L2], so that in general one would not expect either relative or absolute risks to be invariant across populations. Some of the general features of interaction between radiation and other factors are described in Annex H, *"Combined effects of radiation and other agents"*.

B. EMPIRICAL AND MECHANISTIC MODELS

1. Armitage-Doll multi-stage model

473. Mechanistic models of carcinogenesis were originally developed to explain phenomena other than the effects of ionizing radiation. One of the more commonly observed patterns in the age-incidence curves for epithelial cancers is that the cancer incidence rate varies approximately as Ct^β for age t and some constants C and β. At least for most epithelial cancers in adulthood, the exponent β of age seems to lie between 4 and 6 [D5]. The so-called multi-stage model of

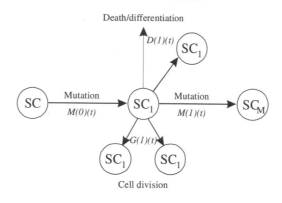

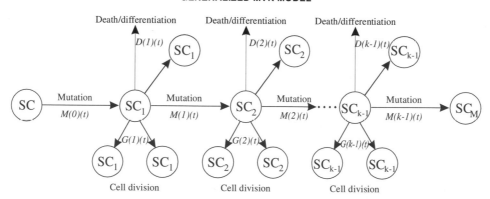

Figure XXIII. Empirical/mechanistic models of multi-stage carcinogenesis.

SC: Initial stem cell
M(i)(t): Mutation rate
at stage i and age t

SC_i: Stem cell of stage i
G(i)(t): Stem cell rate
at stage i and age t

SC_M: Malignant stem cell
D(i)(t): Death/differentiation rate
at stage i and age t

carcinogenesis of Armitage and Doll [A1] was developed in part as a way of accounting for this approximately log-log variation of cancer incidence with age. The model supposes that at age t an individual has a population of X(t) completely normal (stem) cells and that these cells acquire one mutation at a rate M(0)(t). The cells with one mutation acquire a second mutation at a rate M(1)(t), and so on until at the (k − 1)th

stage the cells with (k − 1) mutations proceed at a rate M(k − 1)(t) to become fully malignant. The model is illustrated schematically in Figure XXIII. It can be shown that when X(t) and M(i)(t) are constant, a model with k stages predicts a cancer incidence rate that is approximately given by the expression Ct^{k-1}, with C = M(0) M(1)...M(k − 1)/(1 × 2 ... (k − 1)) [A1, M27].

474. In developing their model, Armitage and Doll [A1] were driven largely by epidemiological findings and, in particular, by the age distribution of epithelial cancers. In the intervening 30 years, substantial biological evidence has been gathered that cancer is a multi-step process involving the accumulation of a number of genetic and epigenetic changes in a clonal population of cells. This evidence was reviewed in the UNSCEAR 1993 Report [U3], and subsequent data and concepts are outlined in Chapter IV of this Annex. However, there are certain problems with the model proposed by Armitage and Doll [A1] associated with the fact that to account for the observed age-incidence curve Ct^β with β between 4 and 6, between five and seven stages are needed. For colon cancer there is evidence that six stages might be required [F1, U3]. However, for other cancers there is, at present, insufficient evidence to conclude that there are as many rate-limiting stages as this. BEIR V [C1] surveyed evidence for all cancers and found that two or three stages might be justifiable, but not a much larger number. To this extent, the large number of stages predicted by the Armitage-Doll model appears to be verging on the biologically unlikely. Related to the large number of stages required by the Armitage-Doll multi-stage model are the high mutation rates predicted by the model. Moolgavkar and Luebeck [M28] fitted the Armitage-Doll multi-stage model to data sets describing the incidence of colon cancer in a general population and in patients with familial adenomatous polyposis. Moolgavkar and Luebeck [M28] found that Armitage-Doll models with five or six stages gave good fits to these data sets, but that both of these models implied mutation rates that were too high by at least two orders of magnitude. The discrepancy between the predicted and experimentally measured mutation rates might be eliminated, or at least significantly reduced, if account is taken of the fact that the experimental mutation rates are locus-specific. A "mutation" in the sense in which it is defined in this model might result from the "failure" of any one of a number of independent loci, so that the "mutation" rate would be the sum of the failure rates at each individual locus.

475. Notwithstanding these problems, much use has been made of the Armitage-Doll multi-stage model as a framework for understanding the time course of carcinogenesis, particularly for the interaction of different carcinogens [P10]. Thomas [T3] has fitted the Armitage-Doll model with one and two radiation-affected stages to the solid cancer data in the Japanese Life Span Study 11 cohort of bomb survivors. Thomas [T3] found that a model with a total of five stages, of which either stages one and three or stages two and four were radiation-affected, fitted significantly better than models with a single radiation-affected stage. Little et al. [L5, L35] also fitted the Armitage-Doll model with up to two radiation-affected stages to the Japanese Life Span Study 11 data set and also to data on various medically exposed groups, using a slightly different technique to that of Thomas [T3]. Little et al. [L5, L35] found that the optimal solid cancer model for the Japanese data had three stages, the first of which was radiation affected, while for the Japanese leukaemia data the best fitting model had three stages, the first and second of which were radiation affected. A version of the Armitage-Doll has also been fitted to the Life Span Study solid tumour incidence data by Pierce and Mendelsohn [P22], who found that a model with five or six stages gave the best fit to this data.

476. Both the paper of Thomas [T3] and those of Little et al. [L5, L35] assumed the ith and the jth stages or mutation rates [M(i - 1), M(j - 1)] (j > i) in a model with k stages to be (linearly) affected by radiation and the transfer coefficients (other than M(i - 1) and M(j - 1)) to be constant, as is the stem-cell population X(t). In these circumstances, it can be shown [L5] that if an instantaneously administered dose of radiation d is given at age a, then at age t (>a) the cancer rate is approximately as follows:

$$\mu\, t^{k-1} + \alpha\, d\, a^{i-1}\, t^{k-i-1} + \tag{5}$$
$$\beta\, d\, a^{i-1}\, t^{k-j-1} + \gamma\, d^2\, a^{i-1}\, t^{k-j-1}$$

for some positive constants μ, α and β, and where γ is given by

$$\gamma = \frac{\alpha\, \beta\, \Gamma(k-i)\, \Gamma(j)}{2\, \mu\, \Gamma(k)} \quad \text{if } j = i + 1 \tag{6}$$
$$and \qquad = 0 \quad \text{if } j > i + 1$$

and $\Gamma(i)$ is the gamma function [A6].

477. The first term (μt^{k-1}) in expression (5) corresponds to the cancer rate that would be observed in the absence of radiation, while the second term ($\alpha da^{i-1}t^{k-i-1}$) and the third term ($\beta da^{i-1}t^{k-j-1}$) represent the separate effects of radiation on the ith and jth stages, respectively. The fourth term ($\gamma d^2 a^{i-1}t^{k-j-1}$), which is quadratic in dose, d, represents the consequences of interaction between the effects of radiation on the ith and the jth stages and is only non-zero when the two radiation-affected stages are adjacent (j = i + 1). Thus if the two affected stages are adjacent, a quadratic (dose plus dose-squared) relationship will occur, whereas the relationship will be approximately linear if the two affected stages have at least one intervening stage. Another way of considering the joint effects of radiation on two stages is that for a brief exposure, unless the two radiation-affected stages are adjacent, there will be insignificant interaction between the cells affected by radiation in the earlier and later of the two radiation-affected cell compartments. This is simply because very few cells will move between the two compartments in the course of the radiation exposure. If the ith and the jth stages are radiation-affected, the result of a brief dose of radiation will be to cause some of the cells that have already accumulated (i -1) mutations to acquire an extra mutation and move from the (i - 1)th to the ith compartment. Similarly, it will cause some of the cells that have already acquired (j - 1) mutations to acquire an extra mutation and so move from the (j - 1)th to the jth compartment. It should be noted that the model does not require that the same cells be hit by the radiation at the ith and jth stages, and in practice for low total doses, or whenever the two radiation-affected stages are separated by an additional unaffected stage or stages, an insignificant

proportion of the same cells will be hit (and mutated) by the radiation at both the ith and the jth stages. The result is that unless the radiation-affected stages are adjacent, for a brief exposure the total effect on cancer rate is approximately the sum of the effects, assuming radiation acts on each of the radiation-affected stages alone. One interesting implication of models with two or more radiation-affected stages is that as a result of interaction between the effects of radiation at the various stages, protraction of dose, in general, results in an increase in cancer rate, i.e. an inverse dose-rate effect [L5]. However, it can be shown that in practice the resulting increase in cancer risk is likely to be small [L5].

478. The variant of the Armitage-Doll model fitted by Pierce and Mendelsohn [P22] is unusual in that it assumes that radiation equally affects all k mutation rates in the model except the last. (In the last stage, radiation is not assumed to have any effect.) This assumption distinguishes their use of this model from the approaches of Little et al. [L5] or Thomas [T3], both of which assumed that radiation affected at most two of the mutation rates (and neither of which constrained the effects of radiation to be equal in these stages). There are some technical problems with the paper of Pierce and Mendelsohn [P22] arising from the authors' failure to take account of interactions between the effects of radiation on the (k – 2) pairs of adjacent stages. These interactions contribute significantly by adding a quadratic term in the dose response and cannot be ignored, even to a first-order approximation. The fact that in general there is little evidence for upward curvature in the solid cancer dose response in the Life Span Study [P1, L7, L44] argues that if proper account had been taken of these interaction terms, the model of Pierce and Mendelsohn [P22] would not fit the data well. Moreover, one implication of the model of Pierce and Mendelsohn [P22] is that the excess relative risk will be proportional to 1/a, i.e. the inverse of attained age. However, this is known to provide a poor description of the excess relative risk of solid cancer, even within the Life Span Study cohort [L56, L57].

479. The optimal leukaemia model found by Little et al. [L5, L35], having adjacent radiation-affected stages, predicts a linear-quadratic dose response, in accordance with the significant upward curvature which has been observed in the Japanese data set [P1, P2, P3]. This leukaemia model, and also that for solid cancer, predicts the pronounced reduction of excess relative risk with increasing age at exposure which has been seen in the Japanese atomic bomb survivors and other data sets [U2]. The optimal Armitage-Doll leukaemia model predicts a reduction of excess relative risk with increasing time after exposure for leukaemia. At least for those exposed in childhood, the optimal Armitage-Doll solid cancer model also predicts a reduction in excess relative risk with time for solid cancers. These observations are consistent with the observed pattern of risk in the Japanese and other data sets [L33, U2]. Nevertheless, there are indications that the Armitage-Doll model may not provide an adequate fit to the Japanese data [L6]. For this reason, and because of the other problems with the Armitage-Doll model discussed above, a slightly different class of models need to be considered.

2. Two-mutation models

480. In order to reduce the biologically implausible number of stages required by their first model, Armitage and Doll [A7] developed a further model of carcinogenesis, which postulated a two-stage probabilistic process whereby a cell following an initial transformation into a pre-neoplastic state (initiation) was subject to a period of accelerated (exponential) growth. At some point in this exponential growth a cell from this expanding population might undergo a second transformation (promotion) leading directly to the development of a neoplasm. Like their previous model, it satisfactorily explained the incidence of cancer in adults, but was less successful in describing the pattern of certain childhood cancers.

481. The two-mutation model developed by Knudson [K16] to explain the incidence of retinoblastoma in children took account of the process of growth and differentiation in normal tissues. Subsequently, the stochastic two-mutation model of Moolgavkar and Venzon [M7] generalized Knudson's model, by taking account of cell mortality at all stages as well as allowing for differential growth of intermediate cells. The two-stage model developed by Tucker [T7] is very similar to the model of Moolgavkar and Venzon but does not take account of the differential growth of intermediate cells. The two-mutation model of Moolgavkar, Venzon, and Knudson (MVK) supposes that at age t there are X(t) susceptible stem cells, each subject to mutation to an intermediate type of cell at a rate M(0)(t). The intermediate cells divide at a rate G(1)(t); at a rate D(1)(t) they die or differentiate; at a rate M(1)(t) they are transformed into malignant cells. The model is illustrated schematically in Figure XXIII. In contrast to the case of the (first) Armitage-Doll model, there is a considerable body of experimental biological data supporting this initiation-promotion type of model (see, e.g. [M5, T6]). The model has been developed to allow for time-varying parameters at the first stage of mutation [M30]. An additional generalization of this model (to account for time-varying parameters at the second stage of mutation) was presented by Little and Charles [L32], who also demonstrated that the excess relative risk predicted by the model when the first mutation rate was subject to instantaneous perturbation decayed at least exponentially for a sufficiently long time after the perturbation. Moolgavkar et al. [M29] and Luebeck et al. [L12] and Heidenreich et al. [H2] used the two-mutation model to describe the incidence of lung cancer in rats exposed to radon, and in particular to describe the inverse dose-rate effect that has been observed in these data. Other groups have also modelled lung tumour risk in rats exposed to radon [H35, L2], and in this experimental system the modelling suggests effects of radiation on the later stages of tumorigenesis. Moolgavkar et al. [M13] and Luebeck et al. [L16] also applied the model to describe the interaction of smoking and radiation as causes of lung cancer in the Colorado Plateau uranium miner cohort. More recently the two-mutation model has been utilized to model lung, stomach, and colon cancer in the atomic bomb survivor incidence data [K17].

482. Moolgavkar and Luebeck [M28] have used models with two or three mutations to describe the incidence of colon cancer in a general population and in patients with familial adenomatous polyposis. They found that both models gave good fits to both data sets, but that the model with two mutations implied biologically implausibly low mutation rates. The three-mutation model, which predicted mutation rates more in line with biological data, was therefore somewhat preferable. The problem of implausibly low mutation rates implied by the two-mutation model is not specific to the case of colon cancer, and is discussed at greater length by Den Otter et al. [D6] and Derkinderen et al. [D7], who argue that for most cancer sites a model with more than two stages is required.

3. Generalized Moolgavkar-Venzon-Knudson (MVK) multi-stage models

483. A number of generalizations of the Armitage-Doll and two- and three-mutation models have been developed [L31, T6]. In particular, two closely related models have been developed, whose properties have been described by Little [L31]. The first model is a generalization of the two-mutation model of Moolgavkar, Venzon, and Knudson and so will be termed the *generalized MVK model*. The second model generalizes the multi-stage model of Armitage and Doll and will be termed the *generalized multi-stage model*.

For the generalized MVK model it may be supposed that at age t there are X(t) susceptible stem cells, each subject to mutation to a type of cell carrying an irreversible mutation at a rate of M(0)(t). The cells with one mutation divide at a rate G(1)(t); at a rate D(1)(t) they die or differentiate. Each cell with one mutation can also divide into an equivalent daughter cell and another cell with a second irreversible mutation at a rate M(1)(t). For the cells with two mutations there are also assumed to be competing processes of cell growth, differentiation, and mutation taking place at rates G(2)(t), D(2)(t), and M(2)(t), respectively, and so on until at the (k − 1)th stage the cells that have accumulated (k − 1) mutations proceed at a rate M(k − 1)(t) to acquire another mutation and become malignant. The model is illustrated schematically in Figure XXIII. The two-mutation model of Moolgavkar, Venzon, and Knudson corresponds to the case k = 2. The generalized multi-stage model differs from the generalized MVK model only in that the process whereby a cell is assumed to split into an identical daughter cell and a cell carrying an additional mutation is replaced by the process in which only the cell with an additional mutation results, i.e. an identical daughter cell is not produced. The classical Armitage-Doll multi-stage model corresponds to the case in which the intermediate cell proliferation rates G(I)(t) and the cell differentiation rates D(i)(t) are all zero.

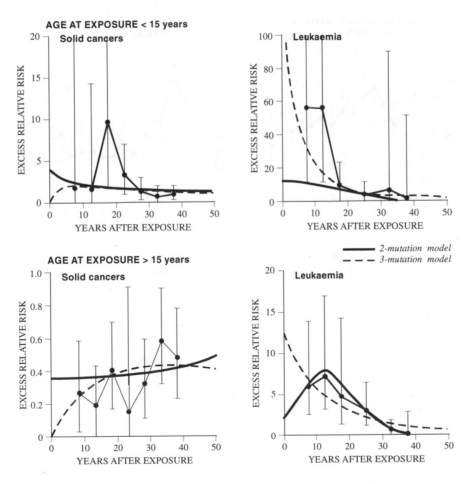

Figure XXIV. Comparison of generalized MVK models fitted to the observed excess relative risk Sv⁻¹ (CI: 90%) in survivors of the atomic bombings in Japan [L36].

484. It can be shown [L31] that the excess relative risk for either model following a perturbation of the parameters will tend to zero as the attained age tends to infinity. One can also demonstrate that perturbation of the parameters M(k − 2), M(k − 1), G(k − 1), and D(k − 1) will result in an almost instantaneous change in the cancer rate [L31].

485. Generalized MVK models have been fitted to the atomic bomb survivor mortality data [L36]. For both leukaemia and solid cancers, the only models with a single radiation-affected parameter that give even slightly satisfactory fits are those in which radiation is assumed to affect M(0) [L36]. For both leukaemia and solid cancer, generalized two- and three-mutation MVK models fit equally well. For leukaemia, the three-mutation model provides a satisfactory fit only when M(0) and M(1) are assumed affected by radiation. For solid cancer and leukaemia there are indications of lack of fit to the youngest age-at-exposure group for the three-mutation model; there is also some lack of fit of the optimal solid cancer three-mutation model to this age-at-exposure group (Figure XXIV).

486. For solid cancer, only M(0) is (linearly) affected by radiation for two- or three-mutation generalized MVK models. In contrast to the solid cancer models, both leukaemia models assume a linear-quadratic dose dependence of the M(i). The non-linearity found in the leukaemia M(i) dose response reflects known curvature in the leukaemia dose response in the atomic bomb survivor data [C1, P1]. There is some evidence, e.g. for chromosome aberrations, that the mutation induction curve is linear-quadratic at least for low-LET radiation, although linearity is generally observed for high-LET radiation [L34].

487. Despite the indications of lack of fit discussed above, the variation of excess relative risk with time since exposure and age at exposure predicted by the optimal two- and three-mutation models for solid cancer (Figure XXIV) is in qualitative agreement with the variation seen in the Japanese bomb survivors and in other irradiated groups [U2]. In particular the optimal models demonstrate the progressive reduction in excess relative risk with increasing age at exposure seen in many data sets [U2], together with the marked reduction in excess relative risk with increasing time since exposure observed in various groups exposed in childhood [L42, P2].

488. Figure XXIV reinforces the theoretical predictions of an earlier paper by Little [L31] and shows that immediately after perturbing M(0) in the two-mutation model, the excess relative risk for solid cancers and leukaemia quickly increases. However, there are no data in the first five years of follow-up in the survivor cohort [P2], so that it is difficult to test the predictions [L31] in respect of the variation in risk shortly after exposure using that data set.

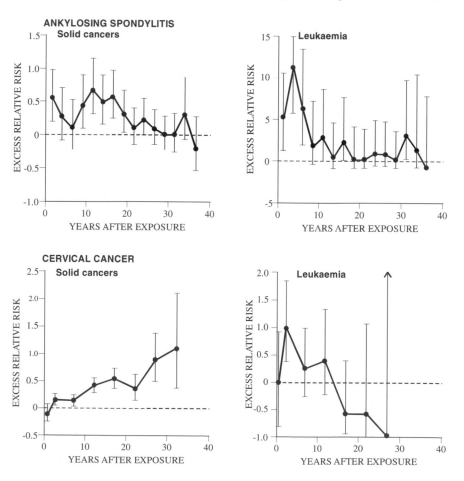

Figure XXV.   Excess relative risk (CI: 90%) of solid cancers and leukaemia in Ankylosing Spondylitis Study in United Kingdom and the International Study of Cervical Cancer [B5, D8].

489. There is a suggestive increase in the excess relative risk of cancers other than leukaemia and colon cancer in the UK ankylosing spondylitis patients <5 years after first treatment (the first two datapoints in the top-left panel of Figure XXV), but the authors caution against interpreting this as the effect of the x-irradiation [D8]. There are no strong indications of an elevation in risk in the first five years after radiotherapy for cancers other than leukaemia and of the reproductive organs in a study of women followed up for second cancer after radiotherapy for cervical cancer [B5]. This corresponds to the first two datapoints in the bottom panel of Figure XXV. (Lung cancers are also excluded from the International Radiation Study of Cervical Cancer (IRSCC) data shown in the lower left panel of Figure XXV because of indications of above-average smoking rates in this cohort [B5].) In general there are no strong indications of an elevation in solid cancer risk soon after irradiation in other exposed groups [U2]. To this extent there are indications of inconsistency for solid cancers between the predictions of the two-mutation model and the observed variation in risks shortly after exposure.

490. In their analysis of the Colorado uranium miners data, Moolgavkar et al. [M13] partially overcame the problem posed by this instantaneous rise in the hazard after perturbation of the two-mutation model parameters by assuming a fixed period (3.5 years) between the appearance of the first malignant cell and the clinical detection of malignancy. However, the use of such a fixed latency period only translates a few years into the future the sudden step-change in the hazard. To achieve the observed gradual increase in excess relative risk shortly after exposure, a stochastic process must be used to model the transition from the first malignant cell to detectable cancer; such a process is provided by the final stage(s) in the three- or four-mutation generalized MVK models used in the analysis of Little [L36]. In particular, an exponentially growing population of malignant cells could be modelled by a penultimate stage with $G(k - 1) > 0$ and $D(k - 1) = 0$, the probability of detection of the clone being determined by $M(k - 1)$. In their analysis of lung, stomach, and colon cancer in the atomic bomb survivor incidence data, Kai et al. [K17] did not assume any such period of latency, perhaps because of the long time after the bombings (12.4 years) before solid cancer incidence follow-up began in the Life Span Study.

491. The evidence with respect to the variation in excess relative risk shortly after exposure for leukaemias is rather different from that for solid cancers. In the United Kingdom ankylosing spondylitis patients [D8] there is significant excess risk even in the period <2.5 years after first treatment (first datapoint in top-right panel of Figure XXV). The IRSCC data [B5] shows a significant excess risk for acute non-lymphocytic leukaemia in the period 1–4 years after first treatment (the second datapoint in the lower-right panel of Figure XXV), and this pattern is observed in many other groups [U2]. More detailed analysis of UK leukaemia incidence data indicate that the age-incidence curves for all subtypes of lymphocytic leukaemia can be adequately modelled by two- and three-mutation generalized MVK models [L26, L36]. However, the two-mutation models for acute lymphocytic leukaemia (ALL) imply a very small number of stem cells (<10^4 cells) if the model is not to yield implausibly low mutation rates [L26].

492. Little [L55] fitted various generalized MVK models to the three main radiogenic leukaemia subtypes, namely acute myeloid leukaemia (AML), chronic myeloid leukaemia (CML) and acute lymphocytic leukaemia (ALL) in two incidence data sets, one relating to a subset of the population of the United Kingdom recently assembled by the Leukaemia Research Fund (LRF) [C21] and the second the Japanese atomic bomb survivors [P3]. The results of this model fitting are illustrated by Figures XXVI and XXVII. Figure XXVI shows that the optimal two-mutation models adequately describe the background incidence of all three leukaemia subtypes in the United Kingdom Leukaemia Research Fund data [L55]. The optimal two-mutation model for AML assumes a step change in the numbers of susceptible stem cells and a simultaneous change in the intermediate cell proliferation parameters, $G(1)(t)$ and $D(1)(t)$. The optimal two-mutation model for CML assumes a step change in the numbers of susceptible stem cells and a simultaneous change in the number of susceptible stem cells and a simultaneous change in the intermediate cell growth parameter, $G(1)(t)$, although the cell death or differentiation rate, $D(1)(t)$, is constant. The optimal two-mutation model for ALL assumes a susceptible stem cell population of the form $X = X_0 \exp [X_1 t + 1_{[1>T]} X_+]$ and a step change in the intermediate cell proliferation parameters, $G(1)(t)$ and $D(1)(t)$. As can be seen from Figure XXVI, three-mutation models provide a rather worse fit for all leukaemia subtypes, particularly for ALL [L55]. For ALL, two-mutation models which assumed ionizing radiation acts to elevate mutation rates for life fitted the Japanese atomic bomb survivor incidence data rather worse than models which assumed ionizing radiation acts to elevate the first mutation rate instantaneously [L55] (see Figure XXVII, lower panel). For CML, two-mutation models which assumed ionizing radiation acts to elevate mutation rates for life fitted the Japanese atomic bomb survivor incidence data rather better than models which assumed ionizing radiation acts to elevate the first mutation rate instantaneously [L55] (see Figure XXVII, center panel). For AML (Figure XXVII, upper panel), both sorts of two-mutation models fitted equivalently well [L55].

4. Multiple pathway models

493. Little et al. [L6] fitted a generalization of the Armitage-Doll model to the Japanese atomic bomb survivor and IRSCC leukaemia data which allowed for two cell populations at birth, one consisting of normal stem cells carrying no mutations, the second a population of cells each of which has been subject to a single mutation. The leukaemia risk predicted by such a model is equivalent to that resulting from a model with two pathways between the normal stem cell

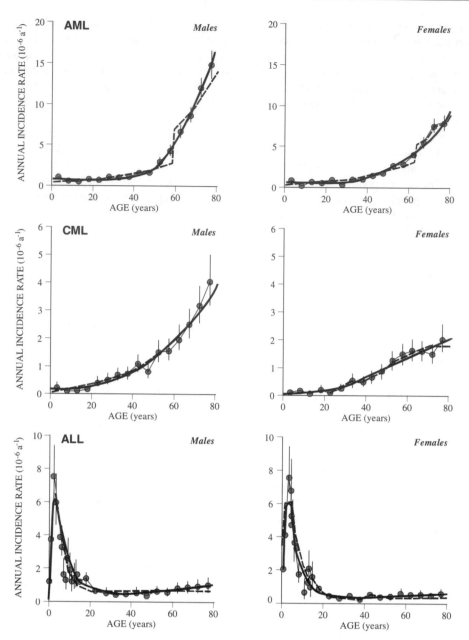

**Figure XXVI. Fit of optimal two-stage and three-stage generalized MVK models
to Leukeamia Research Fund data for acute myeloid leukaemia (AML), chronic myeloid leukaemia (CML),
and acute lymphocytic leukaemia (ALL) [L55].**
Observed risks are shown with 95% CI.
●————● *Observed* ———— *Two-stage model* ----- *Three-stage model*

compartment and the final compartment of malignant cells, the second pathway having one fewer stage than the first. This model fitted the Japanese and IRSCC leukaemia data sets significantly better, albeit with biologically implausible parameters, than a model which assumed just a single pathway [L6]. The findings of Kadhim et al. [K4], namely that the exposure of mammalian haemopoietic stem cells to alpha particles could generally elevate mutation rates to very much higher than normal levels, imply (if they are at all relevant to tumorigenesis) that there might be multiple pathways in the progression from normal stem cells to malignant cells (discussed in Chapter IV). The mutation rates and indeed the number of rate-limiting stages might be substantially different in these two or more pathways. A number of such models are described by Tan [T6], who also

discusses at some length the biological and epidemiological evidence for such models.

C. DOSE-RESPONSE RELATIONSHIPS

494. The shape of the cancer dose response is largely driven by assumptions made about the shape of the dose-response curve for the initiating lesion or lesions. In particular, if a lesion induced by a single acutely delivered dose D of ionizing radiation at age α has a dose response given by the function $F(D,a)$ and this is assumed to act on a single stage (not necessarily the first) in the multi-stage process of carcinogenesis and assuming also that the dose is low enough to avoid saturation effects, then it can be shown [L5, L31] that

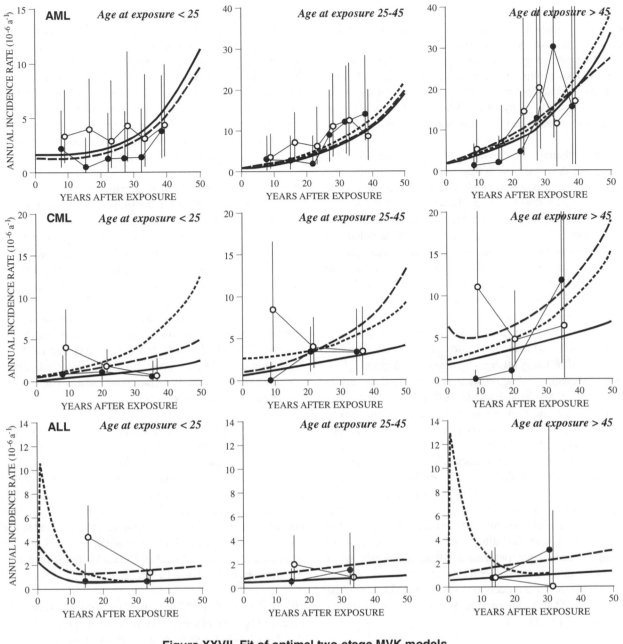

**Figure XXVII. Fit of optimal two-stage MVK models
to Japanese data for acute myeloid leukaemia (AML), chronic myeloid leukaemia (CML),
and acute lymphocytic leukaemia (ALL) [L55].**
Observed risks are shown with 95% CI.

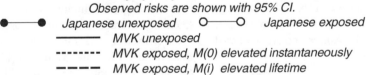

Japanese unexposed   Japanese exposed
——————— MVK unexposed
---------- MVK exposed, M(0) elevated instantaneously
– – – – – MVK exposed, M(i) elevated lifetime

the dose-response curve for carcinogenesis i years after exposure is given by $w_0(t,a) + F(D,a) - w_1(t,a)$ for some functions $w_0(t,a)$ and $w_1(t,a)$. In other words, the dose response for cancer has the same shape as the dose response for initial lesion production. In particular, if the initial lesion production is a linear-quadratic function of dose D, $F(D) = \sigma_0 + \sigma_1 - D + \sigma_2 - D^2$, then the cancer dose response will also be linear-quadratic, with the same ratio of quadratic to linear coefficients.

495. It is a crucial assumption underlying this invariance in "shape" of the dose-response curves for the production of initial lesions and cancer that the radiation-induced lesions act only at a single "stage" in the carcinogenic process. If, for example, there is a quadratic term in the dose response resulting from interactions between the (linear) effects of radiation on different "stages" (e.g. adjacent "stages" in the classical multistage model of carcinogenesis) then the ratio of the quadratic to the linear coefficients for the cancer dose response would change with time after exposure [L5]. It has been hypothesized that the quadratic term in the dose-response curve for chromosome aberrations results from interactions between the effects of two radiation tracks through the cell nucleus [E10, M42].

496. Comparison of the shape of dose-response relationships for tumour induction with that of *in vitro* cellular endpoints such as chromosome aberration induction is not straightforward. Chapter IV provides evidence of the cellular complexity of multistage tumour development, and some distortion of dose-response parameters for initial events in single cells seems likely. Accordingly, such dose-response comparisons need to be made with some caution.

497. Nevertheless, the assumption made here is that tumorigenic dose response is determined largely by the dose response for the production of initial lesions. The shape of the dose response for the various candidate lesions that might be associated with cancer has been discussed in earlier Chapters of this Annex. There is some information on this question that can be obtained from the epidemiological data, although this is generally obtained from moderately high dose studies (with total dose up to 5 Gy). There is very little reliable epidemiological data relating to total doses less than 20 mGy. In most analyses of epidemiological data, linear dose-response models are used, and give satisfactory fit. In particular this is the case for most solid tumours in the Japanese atomic bomb survivors [P2, T4] and in many other radiation-exposed groups [U2]. The linear-quadratic dose response (with upward curvature) that is found for leukaemia is perhaps the most often employed departure from linearity. However, in analyses of the shape of the cancer dose-response curve in radiation-exposed groups [P1, S1], there are various other possible shapes to the dose-response curve [U3]. For the class of deterministic effects defined by the ICRP [I2], it is assumed that there is a threshold dose, below which there is no effect [E9]. Such a form of dose response has also been employed in analyses of brain damage among those exposed *in utero* to the atomic bombings in Hiroshima and Nagasaki [O9, O10]. Arguments have been put forward that sufficiently small doses of radiation induce either no increase in cancer risk (i.e. a dose threshold), or a reduction in cancer risk (i.e. hormesis) [L45, K19, K22, P9], although these interpretations have been challenged [C15, U3].

498. Recently there have appeared a number of assessments of possible threshold-type departures from linear-quadratic curvature in the cancer dose-response curve in the Japanese atomic bomb survivor tumour incidence and mortality data. These data are noted earlier in the Annex and are discussed in detail below.

499. Analysis by Little and Muirhead [L7, L43] and by Hoel and Li [H26] of the Japanese atomic bomb survivor incidence data demonstrated a significant improvement in fit to the leukaemia incidence data when a threshold is incorporated in a linear-quadratic relative risk model, albeit at borderline levels of statistical significance (two-sided p=0.04). Little and Muirhead [L7, L43] examined the three radiogenic leukaemia subtypes (AML, ALL, CML), as well as the principal solid cancer sites in the incidence data, and found that, apart from leukaemia, only for non-melanoma skin cancer was there evidence of a threshold (at about 1 Sv); this last finding was reinforced by a more detailed examination of this cancer type [L30]. The evidence for there being a significant excess risk of non-melanoma skin cancer at relatively low doses (<1 Gy) in other (Caucasian) populations [S31] may indicate the limited relevance of these findings in the Japanese data to Western populations with respect to this cancer type. Paralleling the analysis of Little and Muirhead [L7, L43]. Hoel and Li [H26] also examined the fit of linear-threshold models to a number of solid cancer sites in both the Japanese incidence and mortality data and found that for none of the cancer sites was there evidence that incorporation of a threshold significantly improved the fit.

500. It is well recognized that errors in the estimates of dose can substantially alter the shape of the dose-response relationship and hence the evidence both for a dose response and also for any possible curvature in that dose response. The problem of random dosimetric errors for the Radiation Effects Research Foundation (RERF) data has been previously investigated by Jablon [J5], Gilbert [G2], Pierce et al. [P7] and Pierce and Vaeth [P1]. Such random errors in doses were taken into account in the analysis of the tumour incidence and mortality data by Little and Muirhead [L7, L32, L44], but not by Hoel and Li [H26]. The issue of dosimetric errors in epidemiological data is considered in Annex I, *"Epidemiological evaluation of radiation-induced cancer"*.

501. There are certain technical problems associated with use of threshold models. In general, the asymptotic (χ^2) distribution of the deviance difference statistic used for significance tests is not guaranteed, because of a lack of sufficient smoothness in the likelihood as a function of the dose threshold parameter [S30]. This problem is circumvented by the likelihood-averaging techniques used to take account of dosimetric errors in the analysis by Little and Muirhead [L7, L43, L44]; this problem is not addressed in the analyses of Hoel and Li [H26].

502. Other subtle problems affect the interpretation of the results of both Hoel and Li [H26] and Little and Muirhead [L7, L43, L44]. These problems are connected with the use of the grouped form of the data, and in particular the grouped dose categories, in the publicly available forms of both the Japanese incidence and mortality data sets. A likelihood-averaging technique used to take account of dosimetric errors [L7, L43, L44] is one possible way around this problem. Little and Muirhead [L7, L43, L44], following the methodology of Pierce et al. [P7], evaluated the average, for a given "nominal" dose, d, of the relative risk, RR(i,D), evaluated at the "true" dose D: Avg[RR(i,D)|d]. The data set used was in grouped form, the strata being defined in each case by the variables city, sex, age at exposure, time since exposure, and dose. (In the mortality data set [P2], there is additional stratification by attained age.) For each such stratum, i, the average "nominal" dose was available for the persons in that stratum $Avg_i[d]$. Ideally one should calculate for each strata $Avg_i[Avg[RR(i,D)|d]]$, i.e. the average of Avg[RR(i,D)|.] over all individuals in the stratum i. It was not possible to calculate this quantity using the grouped data that were publicly available, so that in the analyses of Little and Muirhead [L7, L43, L44] the quantity $Avg[RR(i,D)|Avg_i[d]]$

Table 16
Leukaemia cases and deaths in various dose groups in the Japanese atomic bomb survivor incidence and mortality data

Bone marrow dose group [a]	Number of cases	
	Incidence [b] [P3]	Deaths [c] [P2]
0.0-0.1	128	125
0.1-0.2	8	11
0.2-0.5	27	24
0.5-1.0	24	22
1.0-1.5	19	16
1.5-2.0	8	9
2.0-3.0	17	18
3.0-4.0	2	7
≥4.0	4	4
Total	237	236

a   In the incidence data a neutron relative biological effectiveness (RBE) of 1 is used to calculate the neutron component of bone marrow dose
   (in Gy), for the mortality data a neutron RBE of 10 is used to calculate dose (in Sv).
b   All leukaemia cases over the years 1950-1987.
c   All leukaemia deaths over the years 1950-1987.

was evaluated, i.e. the value of Avg[RR(i,D)|.] evaluated at the average "nominal" dose (Avg$_i$[d]) within stratum i. Even for linear dose-response models there can be differences between Avg[RR(i,D)|Avg$_i$[d]] and Avg$_i$[Avg[RR(i,D)|d]]. As shown by Little and Muirhead [L43], at least when 35% dosimetric errors were assumed, this approximation did not introduce appreciable errors for the optimal linear-quadratic-threshold model for leukaemia; errors were at most 5% [L43].

503.   Although the errors introduced by this approximation were small, nevertheless they may be sufficiently great to question the analyses of Little and Muirhead [L7, L43] and Hoel and Li [H26]. As discussed previously, one of the main findings of the analysis of leukaemia incidence among the Japanese atomic bomb survivors by Little and Muirhead [L7] was that incorporation of a threshold in the linear-quadratic model yielded an improvement in fit at

borderline levels of statistical significance (best estimate of threshold for a linear-quadratic-threshold model was 0.12 Sv, 95% CI: 0.01-0.28; two-sided p=0.04). In contrast, the fits of a linear-quadratic-threshold model to the mortality data by the same authors demonstrated that the threshold was not significantly different from zero (best estimate of threshold for a linear-quadratic-threshold model was 0.09 Sv, 95% CI: <0.00-0.29; two-sided p=0.16) [L44]. Comparison of the leukaemia incidence and mortality data in Figure XXVIII and Table 16 demonstrates their similarity. Little and Muirhead [L44] concluded that the most probable reason for the difference between the reported findings in the incidence and mortality data sets was the finer subdivision of dose groups in the mortality data set. (There are 14 dose groups in the mortality data sets in a publicly available form, compared with 10 dose groups in the incidence data sets.)

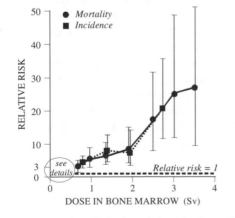
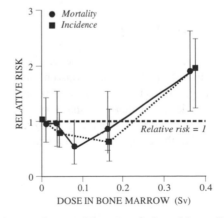

Figure XXVIII.   Relative risk of leukaemia in survivors of the atomic bombings [L44].
The diagram on the right shows the low-dose region in detail.

504.   Substantial low-dose curvilinearity in dose response has been observed for skin cancer in some human populations [L30] (although not in all [S31, R15]) and in CBA/CaH mice [P8], which is consistent either with a threshold or with a

power of dose substantially greater than 2. Low-dose curvature, consistent with a quadratic-exponential dose response, has also been observed for myeloid leukaemia in CBA mice [M4, D11]. For a specific endpoint a threshold dose response, or something approximating to it (e.g. a dose response proportional to some high power (>2) of dose), might be expected if radiation must hit a large number of targets relevant to that endpoint, for example in order to inactivate a critical number of cells in a particular tissue [E9]. In particular, there are grounds for believing that this might be the case for induction of cataract, sterility, *in utero* severe mental retardation and various other deterministic effects [E9]. There is substantial evidence that oncogenesis arises from damage to a single cell, and in particular from damage to the genetic material in a cell, as reviewed by UNSCEAR [U3], in Chapter IV of the present Annex and discussed by Little and Muirhead [L7]. Given the evidence that single tracks of all types of ionizing radiation can induce a variety of damage, including DNA double-strand breaks [C15, G10, G12], a dose threshold for cancer induction is judged to be unlikely but cannot be excluded formally. These mechanistic issues have been discussed in depth in Chapter IV. The finding of a significant excess leukaemia risk in various occupationally exposed groups [C20], in which total doses generally are administered in an episodic manner, and also among those exposed to small doses (<0.02 Sv) of x-irradiation *in utero* [K23], provide further evidence that argues against a low dose threshold in the leukaemia dose response.

505. In conclusion, although there is evidence at borderline levels of statistical significance for threshold departures from linear-quadratic curvature for leukaemia incidence in the Japanese atomic bomb survivor data, the grouped nature of the Japanese data make inferences on a possible dose threshold problematic. Based on the most current analysis of the mortality data [L44], there is no evidence for a threshold departure from linear-quadratic curvature for leukaemia in the Japanese atomic bomb survivor data, nor is there for any other cancer type, with the possible exception of non-melanoma skin cancer. In arriving at these conclusions the Committee recognizes the uncertainties that attach to current modelling approaches to cancer risk and the shape of the dose-response relationship.

D. SUMMARY

506. The classical multi-stage model of Armitage and Doll and the two-mutation model of Moolgavkar, Venzon, and Knudson, and various generalizations of them also, are capable of describing, at least qualitatively, many of the observed patterns of excess cancer risk following ionizing radiation exposure. However, there are certain inconsistencies

with the biological and epidemiological data for both the multi-stage and two-mutation models. In particular, there are indications that the two-mutation model is not totally suitable for describing the pattern of excess risk for solid cancers that is often seen after exposure to ionizing radiation, although leukaemia may be better fitted by this type of model. Generalized MVK models which require three or more mutations are easier to reconcile with biological and epidemiological data relating to solid cancers. Firm statements on the relative validity of different biologically based models of radiation tumorigenesis must await further developments. In general, however, it appears that those models retaining biologically realistic parameters while providing satisfactory fits to the data tend to require radiation action at the early stage of tumorigenesis. This feature is consistent with the conclusions reached in Chapter IV following review of mechanistic data. At the same time it is recognized that the optimal solutions in such modelling can often depend on initial assumptions made on the role of radiation-induced damage in complex multi-stage tumorigenic processes. Some influence of radiation on the later stages of tumorigenesis should be anticipated, particularly perhaps with respect to protracted exposures.

507. Although there is evidence at borderline levels of statistical significance for threshold departures from linear-quadratic curvature for leukaemia incidence data in the Japanese atomic bombings, the grouped nature of the Japanese data make inferences on a possible dose threshold problematic. Based on the most current analysis of the mortality data, there is no evidence for a threshold departure from linear-quadratic curvature for leukaemia in the Japanese atomic bomb survivor data, nor is there for any other cancer type, with the possible exception of non-melanoma skin cancer. Thus, as is the case for data from animal, cellular and molecular studies, evidence from the modelling of epidemiological data tends to favour the view that, in general, cancer risk at low doses rises as a simple function of dose. It is recognized, however, that, at present, the descriptions of dose-effect relationships that have been published are principally qualitative in nature and the choice of models for the quantitative estimation of risk remains to be satisfactorily resolved. Substantial uncertainties attach to the true form of these dose-response relationships and the extent to which they are determined by biological assumptions. The Committee is supportive of further work aimed at the further development and validation of these biologically-based models; it is believed that they will have an important role in the future work of the Committee.

SUMMARY AND CONCLUSIONS

508. A number of considerations are important in determining the risks of exposures to radiation at low doses and low dose rates. These include (a) analysis of epidemiological and experimental studies to determine the lowest doses at which, for statistical and methodological reasons, radiation effects are directly observable; (b) examination of the shape of the dose-response relationships in the low-dose region using available epidemiological and experimental data; and (c) assessment of the possibilities for extrapolation to lower levels of dose based on an understanding of the mechanisms involved in radiation response. The aim of this Annex has been to provide an overview of the data available on the relationship between radiation exposure and the induction of cancer and hereditary disease, with emphasis on the limits of detection of effects at low doses of low-LET radiation and the associated uncertainties. This information, coupled with the understanding to date of mechanisms of damage to cells and tissues, provides a basis for reasoned judgements to be made about the likely form of the dose response at exposures below those at which direct information is available.

509. *DNA damage.* It is generally recognized that damage to DNA in the nucleus is the main initiating event by which radiation causes long-term damage to organs and tissues of the body. Double-strand breaks in DNA are generally regarded as the most likely candidate for causing the critical damage. Single radiation tracks have the potential to cause double-strand breaks and in the absence of 100% efficient repair could result in long-term damage, even at the lowest doses, although with a low probability. Damage to other cellular components (epigenetic changes) may influence the functioning of the cell and progression to the malignant state.

510. *Direct observations.* Studies of cellular systems, animal experiments and human epidemiological investigations provide direct and relatively consistent evidence of linear or linear-quadratic dose-response relationships at high to intermediate levels of dose and dose rate. However, all such studies are hampered by statistical limitations in providing clear indications of effects at acute doses much less than about 100 mGy (low-LET). Epidemiological studies at low doses are also subject to uncertainties due to methodological issues related to bias and confounding that can limit interpretation of the data. Some exceptions are the induction of cancer following irradiation *in utero* for which an increase in risk has been observed at doses of about 10–20 mGy, experimental data on mouse hair mutations at 10 mGy, and unstable chromosomal aberrations at 20 mGy. In the case of high-LET radiation, the experimental data on cellular damage generally indicate a linear dose-response relationship.

511. For the induction of unstable chromosome aberrations and mutations, a small priming dose of low-LET radiation can sometimes reduce the effect caused by a subsequent higher dose. This adaptive response seems to be a consequence of stimulating the expression/production of genes/proteins in cells involved in DNA damage response and takes a few hours to become effective. Such adaptive responses appear to be transient and have also been observed for cell transformation.

512. Animal studies are valuable for determining the shapes of dose-response relationships and examining how the biological and physical conditions of exposure may influence radiation responses. For many tumour types, the dose response following exposure to both low- or high-LET radiation can be reasonably well represented by a linear or linear-quadratic function. In many cases, however, alternative fits to the data are also possible. Other model fits include the possibility of a threshold dose below which tumours do not occur, as well as more complex functions in which the time for the tumour to appear is much later at low dose rates, which can also suggest the presence of a threshold for response. Animal studies do not, and probably cannot, provide direct information at acute doses much less than about 100 mGy. Values for the lowest doses to give a significant increase in tumour yield following chronic irradiation are generally higher than those for acute irradiation.

513. For radiation-induced hereditary disease, the most comprehensive information comes from measurements of specific locus mutations in mouse spermatogonia. The dose-response relationship for low-dose exposures from low-LET radiation is well fitted by a linear response. The lowest dose tested in these studies was 380 mGy (low-LET). The incidence of mutations in male mice falls by a factor of about three for a reduction in dose rate from $800–900$ mGy min^{-1} to 0.007 mGy min^{-1}. This suggests that a substantial fraction of the damage to DNA that results in the induction of heritable mutations is not amenable to effective repair.

514. Epidemiological studies provide a substantial amount of direct quantitative data on the risks of cancer in humans following radiation exposure. The main source of information is the Japanese Life Span Study (LSS), which gives information on the effects of whole-body irradiation following exposure at different ages. The follow-up study indicates a significant (p=0.05) increase in the risk of radiation-induced fatal solid cancers in the 0–50 mSv dose range.

515. The dose-response relationship for mortality from leukaemia has been fitted by a linear-quadratic function, while for all solid cancers taken together, a linear dose response provides a best fit for the data for doses up to about 3 Sv. A linear dose response can also be fitted to the data for a number of individual tumour types. There are a number of cancers that have not been significantly increased, including those of the rectum, bone, prostate and testes. Further follow-up and better information on the doses received will be needed before the shape of the dose response for both morbidity and mortality can be determined with confidence at doses below about 100 mSv.

516. Dose-response data from a number of other epidemiological studies can also be fitted with a linear or

linear-quadratic dose response at doses up to a few gray, but alternative relationships have also been obtained. Thus, in radium dial painters exposed to the alpha emitters $^{226/228}$Ra, the best fit to the data on bone tumour induction can be obtained with a model indicating a "practical threshold". Data are also available on an increased risk of bone sarcomas in patients given ^{224}Ra. It has been proposed that some of these tumours would be expected to arise only in tissue with deterministic radiation damage and only above a threshold dose. Similar conclusions have been drawn for the bone tumours arising in the radium dial painters. For exposure to radon and its decay products a constant-relative-risk model without any modifying factors, such as attained age and exposure rate, appear to give a good fit to the data at low doses.

517. Data on patients irradiated for medical reasons are generally consistent with a linear dose-response relationship at doses below a few gray. Results suggest a statistically significant increase in the risk of thyroid cancer at external radiation doses above about 100 mGy received in childhood.

518. A number of studies provide information on the risk of childhood cancer following obstetric radiography at low doses. A statistically significant, 40% increase in the relative risk of leukaemia and other childhood cancers (up to 15 years of age) has been seen following doses in the 10–20 mGy (low-LET) range. The principal reason for being able to determine this increase in risk, which in absolute terms is modest, is the low background incidence of cancer in childhood.

519. Data on the effects of low-dose, chronic exposure in radiation workers are generally consistent with results obtained from the high-dose-rate studies on leukaemia induction, although having wide statistical uncertainties. A longer period of follow-up and pooling of data from different studies will, however, be necessary if information on the slope of the dose-response relationship is to be obtained.

520. Some data are available on the risks of cancer in areas of high natural background. Comparative studies of groups exposed to different levels of natural background radiation do not have the statistical power to detect predicted effects on cancer incidence. Generally, there are substantial difficulties in interpreting the data because of uncertainties in the doses actually received, geographical variation in the accuracy of cancer diagnoses, and confounding by environmental factors.

521. *Mechanistic considerations.* Proto-oncogenes and tumour-suppressor genes control a complex array of biochemical pathways involved in cellular signalling and interaction, growth, mitogenesis, apoptosis, genomic stability, and differentiation. Mutation of these genes can, in an often pleiotropic fashion, compromise these controls and contribute to the multi-stage development of neoplasia.

522. On the basis of accumulating knowledge it is argued that early gain-of-function proto-oncogene activation by chromosomal translocation is often associated with the development of human lympho-haemopoietic neoplasia, although gene loss is not infrequent. For many solid tumours there is a requirement for loss of function mutation of tissue-specific tumour-suppressor genes that act as cellular gatekeepers. It has also been proposed that the subsequent onset of spontaneous genomic instability via further clonal mutation is a critical event in neoplastic conversion from a benign to a malignant phenotype. Loss of apoptotic control is also believed to be an important feature throughout neoplastic development.

523. Much information on multi-stage tumorigenesis still remains to be learned. Although the concept of sequential and interacting gene mutations as the driving force for neoplasia is more firmly established, there is insufficient understanding of the complex physiological interplay between these events and the consequences for cellular behaviour and tissue homeostasis.

524. Uncertainty also surrounds the degree to which non-mutational (epigenetic) changes to the genomes of neoplastic cells contribute to tumorigenesis. Increases in the methylation status of critical tumour-suppressor genes is known to be an alternative to mutational inactivation in a range of neoplasms, and loss of methylation imprints may also serve to increase the activity of some growth-promoting genes. DNA methylation is also believed to be involved in genomic imprinting processes. Loss of such imprinting may be important in a number of tumour types. New evidence also implicates histone acetylation in genomic heterochromatization and gene silencing; this process is suggested to be a potentially important contributor to epigenetic change. Epigenetic processes (bystander effects and induced genomic instability) have been shown to influence certain aspects of cellular response *in vitro*. The relevance of these poorly understood processes to *in vivo* tumour induction at low doses of radiation remains to be established.

525. Studies have clarified the role of specific gene mutations in tumours that serve to destabilize the genome, thereby allowing for the accelerated spontaneous development of clonal heterogeneity and tumour progression. Although critical evidence is lacking, it is possible to envisage that after this transition point is reached, tumour development may be relatively independent of exogenously induced DNA damage. Cellular selection during neoplastic development is judged to be of crucial importance at all stages of tumorigenesis. Overall it is judged that most tumours have their origin in gene/chromosomal mutations affecting single target stem-like cells in tissues.

526. Direct evidence on the nature of radiation-associated initiating events in human tumours is sparse, and rapid progress in this area should not be anticipated. By contrast, good progress is being made in resolving early events in radiation-associated tumours in mouse models. These molecular observations strengthen the view expressed in the UNSCEAR 1993 Report [U3] that radiation-induced tumorigenesis will tend to proceed via gene-specific losses; a contribution from early arising epigenetic events should not, however, be discounted.

527. Neoplastic development is subject to a large number of cellular constraints, which provide a high level of protection against neoplastic growth and development. Principal of these are control of cellular proliferation/genomic stability, the induction of apoptosis, and terminal differentiation to a non-proliferative cellular state. For at least certain tumour types there is evidence that immunosurveillance mechanisms can recognize and restrict the growth of neoplastic cells. In spite of these constraints, resistance to or tolerance of all these countermeasures can be developed via gene-specific mutation. On the basis of current molecular genetic knowledge, different modes of *in vivo* constraint are unlikely to apply to spontaneously arising and radiation-induced tumours.

528. Much information points to the crucial importance of DNA repair and other damage-response functions in tumorigenesis. DNA damage response functions influence the appearance of initial events in the multi-stage process, and reduce the probability that a benign neoplasm will spontaneously acquire the secondary mutations necessary for full malignant development. Thus, mutations of DNA damage-response genes in tumours play an important role in the spontaneous development of genomic instability.

529. The repair of sometimes complex DNA double-strand lesions is largely error-prone, and is an important determinant of dose, dose rate, and radiation quality effects in cells. Uncertainties continue to surround the significance to tumorigenesis of adaptive responses to DNA damage; the mechanistic basis of such responses has yet to be well characterized although associations with the induction of biochemical stress responses seems likely. Recent scientific advances highlight the differences in complexity and reparability between spontaneously arising and radiation-induced DNA lesions. These data argue against basing judgements concerning low-dose response on comparisons of overall lesion abundance rather than their nature.

530. *Biological uncertainties and dose-response models.* Evidence suggesting the predominance of error-prone repair of radiation damage to cellular DNA has grown, implying that mutational/ tumorigenic risk should be expected at low doses. Important uncertainties remain, however, on whether error-free DNA repair might apply at very low doses, although there are some arguments against it.

531. There are also uncertainties about whether radiation-induced non-mutational (epigenetic) events, such as induced genomic instability, contribute significantly to tumour risk. The dose-response characteristics of such events are obscure, and there is no way to judge the ensuing risk at low doses, if indeed it exists. The involvement of such processes cannot be inferred solely on the basis of the frequency of phenotypic effects after radiation.

532. Since tumorigenic processes are highly complex, attention is drawn to the problems of judging the shape of the low-dose response on data sets that are over-reliant on high-dose estimates of effect. Apparently simple dose-response relationships may disguise competing processes that have different dose dependencies.

533. In spite of these uncertainties the weight of evidence from fundamental studies favours the mutagenic action of radiation acting primarily at a very early stage of tumorigenesis (initiation), with risk rising as a function of dose. Thus the risk of developing malignant tumours should follow the dose response for initiating lesions unless there are dose-dependent effects on the later phases of tumorigenesis.

534. *Computational modelling of tumorigenesis.* The different characteristics of empirical and biologically based models of radiation tumorigenesis have been considered by the Committee. The classical multi-stage model of Armitage and Doll and the two-mutation model of Moolgavkar, Venzon, and Knudson, and various generalizations of both, are capable of describing, at least qualitatively, many of the observed patterns of excess cancer risk following ionizing radiation exposure. However, different solutions have been obtained by different investigators and there are certain inconsistencies with the biological and epidemiological data for both the multi-stage and two-mutation models. Generalized MVK models that require three or more mutations are easier to reconcile with biological and epidemiological data relating to solid cancers.

535. Although there is evidence at borderline levels of statistical significance for threshold departures from linear-quadratic curvature for leukaemia incidence in the Japanese atomic bomb survivor data, the grouped nature of the Japanese data make inferences on a possible dose threshold problematic. The most current analysis of the mortality data provides no evidence for a threshold departure from linear-quadratic curvature for leukaemia in the Japanese atomic bomb survivor data, nor is there evidence of this for any other cancer type, with the possible exception of non-melanoma skin cancer.

536. *Conclusions.* DNA is the principal target for the initiation of radiation-induced cancer and for radiation-induced hereditary disease. Experimental studies of the effects of ionizing radiation on cellular systems, including the induction of chromosome aberrations, cell transformation and somatic mutations are of value for providing information on damage to DNA. The data obtained have been generally consistent with a linear or linear-quadratic dose response at exposures below those at which cell killing becomes significant (a few gray). In general, significant radiation effects can be detected at doses of about 100 mGy (low-LET) and above, although there are some experimental systems for which effects at lower doses have been observed. In the case of high-LET radiation, the experimental data generally indicate a linear dose-response relationship in the absence of cell killing.

537. For most tumour types in experimental animals and in man a significant increase in risk is only detectable at doses above about 100 mGy. An exception is for human exposures *in utero* when a significant increase in tumour induction in children has been found for doses in the 10–20 mGy range (low-LET). No such excess was observed in the studies of Japanese atomic bomb survivors irradiated *in utero*.

538. In both experimental animals and in humans the dose-response data for tumour induction can be frequently fitted by a linear or linear-quadratic dose response at doses below a few gray. There is evidence though that for some cancer types this form of response does not apply and there may be a practical threshold for a response. Other forms of dose response can also be fitted for the induction of some tumour types.

539. With respect to direct observations of radiation effects, which all carry statistical and/or methodological uncertainty, there are no circumstances where it is scientifically valid to equate the absence of an observable biological effect with the absence of risk.

540. Although mechanistic uncertainty remains, studies on DNA repair and the cellular/molecular processes of radiation tumorigenesis provide no good reason to assume that there will be a low-dose threshold for the induction of tumours in general. However, curvilinearity of the dose response in the low-dose region, perhaps associated with biochemical stress responses and/or changing DNA repair characteristics, cannot be excluded as a general feature. The mechanistic modelling of radiation tumorigenesis is at a relatively early stage of development, but the data available tend to argue against a dose threshold for most tumour types.

541. Until the above uncertainties on low-dose response are resolved, the Committee believes that an increase in the risk of tumour induction proportionate to the radiation dose is consistent with developing knowledge and that it remains, accordingly, the most scientifically defensible approximation of low-dose response. However, a strictly linear dose response should not be expected in all circumstances.

542. The dose response for the induction of heritable disease carries fewer low-dose biological uncertainties than that of multi-stage tumorigenesis, but the same uncertainties surrounding DNA damage response remain; an increase in the risk of germ-cell mutation that is proportionate to radiation dose is judged to be a scientifically reasonable approximation for the induction of heritable effects at low doses.

543. The Committee recognizes that ongoing and future studies in epidemiology and animal sciences, while remaining of great importance for quantitative risk assessment, will not resolve the uncertainties surrounding the effects in humans of low-dose radiation. Accordingly, there will be an increasing need for weight-of-evidence judgements based on largely qualitative data from cellular/molecular studies of the biological mechanisms that underlie health effects; the provision of such judgements demands strong support from biologically validated computational models of risk. With ever-improving experimental technology, fundamental knowledge will continue to grow. On this basis, the Committee emphasizes the need for further work on the mechanisms of DNA damage response/cellular stress and studies of the consequences of these responses for neoplastic development. Current uncertainties on the role of epigenetic factors, such as bystander effects and induced genomic instability, are expected to be reduced, but it may remain difficult to estimate their overall contribution to risk. However, the development of mechanistic models of radiation risk demands more than a simple improvement in the understanding of cellular/molecular processes. Issues such as target-cell identity/multiplicity; the kinetics of pre-neoplastic clonal development; and rates of cell mutation, clonal proliferation, differentiation, and apoptosis, as well as the pattern of energy deposition in critical cellular targets, all need to be better understood in order to define biological parameters for use in the biological modelling of tumorigenesis. These are difficult areas of research, and it is not easy to anticipate the rate of progress. In spite of such experimental difficulties, the Committee believes that advances in computational modelling of the physical and biological aspects of radiation tumorigenesis will provide an essential tool for estimating radiation risk at low doses and low dose rates.

References

A1   Armitage, P. and R. Doll. The age distribution of cancer and a multi-stage theory of carcinogenesis. Br. J. Cancer 8: 1-12 (1954).

A2   Arnheim, N. and D. Shibata. DNA mismatch repair in mammals: role in disease and meiosis. Curr. Opin. Genet. Dev. 3: 364-370 (1997).

A3   Alexander, P. Do cancers arise from a single transformed cell or is monoclonality of tumours a late event in carcinogenesis? Br. J. Cancer 51: 453-457 (1985).

A4   Abelson, P.H. Risk assessments of low-level exposures. Science 265: 1507 (1994); *also* Correspondence. Science 266: 114-113 (1994).

A5   Albert, R.E., F.J. Burns and P. Bennett. Radiation-induced hair follicle damage and tumour formation in mouse and rat skin. J. Natl. Cancer Inst. 49: 1131-1137 (1972).

A6   Abramowitz, M. and I.A. Stegun. Handbook of Mathematical Functions. National Bureau of Standards, Washington, D.C., 1964.

A7   Armitage, P. and R. Doll. A two-stage theory of carcinogenesis in relation to the age distribution of human cancer. Br. J. Cancer 9: 161-169 (1957).

A8   Albert, M.L., J. Darnell, A. Bender et al. Tumour-specific killer cells in paraneoplastic cerebellar degeneration. Nature Med. 4: 1321-1325 (1998).

A9   Académie des sciences. Problèmes liés aux effets des faibles doses de radiations ionisantes. Rapport No. 34, Tcc Doc, Paris (1995).

A10   Albertini, R.J., L.S. Clark, J.A. Nicklas et al. Radiation quality affects the efficiency of induction and the molecular spectrum of *HPRT* mutations in human T cells. Radiat. Res. 148: S76-S86 (1997).

A11   Amundson, S.A., F. Xia, K.B. Wolfson et al. Different cytotoxic and mutagenic responses induced by X-rays in two human lymphoblastoid cell lines derived from a single donor. Mutat. Res. 286: 233-241 (1993).

A12   Amundson, S.A., D.J. Chen and R.T. Okinaka. Alpha particle mutagenesis of human lymphoblastoid cell lines. Int. J. Radiat. Biol. 70(2): 219-226 (1996).

A13   Adams, L.M., S.P. Ethier and R.L. Ullrich. Enhanced *in vitro* proliferation and *in vivo* tumorigenic potential for mammary epithelium from BALB/c mice exposed *in vivo* to γ-radiation and/or 7,12-dimethylbenz(a)anthracene. Cancer Res. 47: 4425-4431 (1987).

A14   Ames, B.N. Endogenous DNA damage as related to cancer and ageing. Mutat. Res. 214: 41-46 (1989).

A15   Albert, R.E., F.J. Burns and R.D. Heimbach. The effect of penetration depth of electron radiation on skin tumour formation in the rat. Radiat. Res. 30: 515-524 (1967).

A16   Aghamohammadi, S.Z, D.T. Goodhead and J.R.K. Savage. Induction of sister chromatid exchanges (SCE) in G lymphocytes by plutonium-238 α-particles. Int. J. Radiat. Biol. 53: 909-915 (1988).

A17   Ashmore, J.P., D. Krewski, J.M. Zielinski et al. First analysis of mortality and occupational radiation exposure based on the National Dose Registry of Canada. Am. J. Epidemiol. 148: 564-574 (1998).

A18   Amundson, S.A., K.T. Do and A.J. Fornace. Induction of stress genes by low doses of gamma rays. Radiat. Res. 152: 225-231 (1999).

A19   Azzam, E., S. de Toledo, T. Gooding et al. Inter-cellular communication is involved in the bystander regulation of gene expression in human cells exposed to very low fluences of alpha particles. Radiat. Res. 159: 497-504 (1998).

A20   Azzam, E.I., G.P. Raaphorst and R.E.J. Mitchel. Radiation-induced adaptive response for protection against micronucleus formation and neoplastic transformation in C3H 10T1/2 mouse embryo cells. Radiat. Res. 138: S28-S31 (1994).

A21   Azzam, E.I., S.M. de Toledo, G.P. Raaphorst et al. Low-dose ionizing radiation decreases the frequency of neo-plastic transformation to a level below the spontaneous rate in C3H 10T1/2 cells. Radiat. Res. 146: 369-373 (1996).

B1   Barlow, C., S. Hirotsume, R. Paylor et al. *Atm*-deficient mice: a paradigm of ataxia-telangiectasia. Cell 86: 159-171 (1996).

B2   Barendsen, G.W. Do fast neutrons at low dose rate enhance cell transformation *in vitro*? A basic problem of micro-dosimetry and interpretation. Int. J. Radiat. Biol. 47: 731-734 (1985).

B3   Barendsen, G.W. Physical factors influencing the frequency of radiation induced transformation of mammalian cells. p. 315-324 in: Cell Transformation and Radiation Induced Cancer (K.H. Chadwick et al. eds.). Adam Hilger, Bristol, 1989.

B4   Bouffler, S.D., C.J. Kemp, A. Balmain et al. Spontaneous and ionizing radiation-induced chromosomal abnormalities in *p53*-deficient mice. Cancer Res. 55: 3883-3889 (1995).

B5   Boice, J.D., N.E. Day, A. Andersen et al. Second cancers following radiation treatment for cervical cancer. An international collaboration among cancer registries. J. Natl Cancer Inst. 74: 955-975 (1985).

B6   Boice, J.D., D.L. Preston, F.G. Davis et al. Frequent chest x-ray fluoroscopy and breast cancer incidence among tuberculosis patients in Massachusetts. Radiat. Res. 125: 214-222 (1991).

B7   Boice, J.D., M. Blettner, R.A. Kleinerman et al. Radiation dose and leukemia risk in patients treated for cancer of the cervix. J. Natl. Cancer Inst. 79: 1295- 1311 (1987).

B8   Bithell, J.F. and A.M. Stewart. Prenatal irradiation and childhood malignancy: a review of British data from the Oxford Survey. Br. J. Cancer 35: 271-287 (1975).

B9   Bishop, J.M. Molecular themes in oncogenesis. Cell 64: 235-248 (1991).

B10   Boland, C.R., J. Sato, H.D. Appelman et al. Microallelo-typing defines the sequence and tempo of allelic losses at tumour suppressor gene loci during colorectal cancer progression. Nature Med. 1: 902-909 (1995).

B11   Balmain, A. Exploring the bowels of DNA methylation. Curr. Biol. 5: 1013-1016 (1995).

B12   Barlow, D.P. Gametic imprinting in man. Science 270: 1610-1613 (1995).

B13   Borrow, J., V.P. Stanton, J.M. Andersen et al. The translocation t(18;16) (p11;p13) of acute myeloid leukaemia fuses a putative acetyltransferase to the CREB-binding protein. Nature Genet. 14: 33-41 (1996).

B14   Bongarzone, I., M.G. Butti, L. Fugazzola et al. Comparison of the breakpoint regions of *ELE1* and *RET* genes involved in the generation of *RET/PTC3* oncogene in sporadic and in radiation-associated papillary thyroid carcinomas. Genomics 42: 252-259 (1997).

B15   Breckon, G., D. Papworth and R. Cox. Murine radiation myeloid leukaemogenesis: a possible role for radiation sensitive sites on chromosome 2. Genes, Chromosome, Cancer 3: 367-375 (1991).

B16   Bouffler, S.D., E.I.M. Meijne, D.J. Morris et al. Chromosome 2 hypersensitivity and clonal development in murine radiation acute myeloid leukaemia. Int. J. Radiat. Biol. 72: 181-189 (1997).

B17   Bryan, T.M., A. Englezou, L. Dalla-Pozza et al. Evidence for an alternate mechanism for maintaining telomere length in human tumours and tumour-derived cell lines. Nature Med. 3: 1271-1274 (1997).

B18   Beverley, P. Cell mediated immune responses to cancer. p. 311-329 in: Introduction to the Cellular and Molecular Biology of Cancer (L.M. Franks and N.M. Teich, eds.). Oxford University Press, Oxford, 1997.

B19   Branch, P., D.C. Bicknell, A. Rown et al. Immune surveillance in colorectal carcinoma. Nature Genet. 9: 231-232 (1995).

B20   Bouffler, S.D., E.I.M. Meijne, R. Huiskamp et al. Chromosomal abnormalities in neutron-induced acute myeloid leukaemias in CBA/H mice. Radiat. Res. 146: 349-352 (1996).

B21   Bhattarcharjee, D. Role of radio-adaptation on radiation-induced thymic lymphoma in mice. Mutat. Res. 358: 231-235 (1996).

B22   Billen, D. Spontaneous DNA damage and its significance for the 'negligible' dose controversy in radiation protection. Radiat. Res. 124: 242-245 (1990).

B23   Burns, F.J. and R.E. Albert. Radiation carcinogenesis in rat skin. p. 199-214 in: Radiation Carcinogenesis (A.C. Upton, R.E. Albert, F.J. Burns et al., eds.). Elsevier, New York, 1986.

B24   Berke, G. The function and mechanism of action of cytolytic lymphocytes. p. 965-1014 in: Fundamental Immunology, 3rd edition (W.E. Paul, ed.). Raven Press, New York, 1993.

B25   Benedict, W.F., A. Banerjee, A. Gardner et al. Induction of morphological transformation in C3H 10T½ cells clone 8 cells and chromosomal damage in hamster A(T1) Cl-3 cells by cancer chemotherapeutic agents. Cancer Res. 37: 2202-2208 (1977).

B26   Brenner, D.J. and E.J. Hall. The inverse dose-rate effect for oncogenic transformation by neutrons and charged particles: a plausible interpretation consistent with published data. Int. J. Radiat. Biol. 58: 745-758 (1990).

B27   Blank, K.R., M.S. Rudoltz, D.G. Kao et al. Review: the molecular regulation of apoptosis and implications for radiation oncology. Int. J. Radiat. Biol. 71: 455-466 (1997).

B28   Bithell, J.F. Epidemiological studies of children irradiation in utero. p. 77-87 in: Low Dose Radiation: Biological Bases of Risk Assessment (K.F. Baverstock and J.W. Stather, eds.). Taylor and Francis, London, 1989.

B29   Balcer-Kubiczek, E.K. and G.H. Harrison. Survival and oncogenic transformation of C3H10T½ cells after extended x irradiation. Radiat. Res. 104: 214-223 (1985).

B30   Balcer-Kubiczek, E.K. and G.H. Harrison. Effect of x-ray dose protraction and a tumour promoter on transformation induction in vitro. Int. J. Radiat. Biol. 54: 81-89 (1988).

B31   Bond, V.P., L.E. Feinendegen and J. Booz. What is a "low dose" of radiation? Int. J. Radiat. Biol. 53(1): 1-12 (1988).

B32   Booz, J. and L.E. Feinendegen. A microdosimetric understanding of low-dose radiation effects. Int. J. Radiat. Biol. 53(1): 13-21 (1988).

B33   Borsa, J., M.D. Sargent, M. Einspenner et al. Effects of oxygen and misonidazole on cell transformation and cell killing in C3H10T½ cells by x-rays in vitro. Radiat. Res. 100: 96-103 (1984).

B34   Broerse, J.J. Influence of physical factors on radiation carcinogenesis in experimental animals. p. 181-194 in: Low Dose Radiation: Biological Bases of Risk Assessment (K.F. Baverstock and J.W. Stather, eds.). Taylor and Francis, London, 1989.

B35   Bodnar, A.G., M. Onellette, M. Frolkis et al. Extension of life-span by introduction of telomerase into normal human cells. Science 279: 349-352 (1998).

B36   Boecker, B.B. Development and use of biokinetic models for incorporated radionuclides. Radiat. Prot. Dosim. 79: 223-228 (1998).

B37   Batsakis, J.G. Tumours of the Head and Neck. Williams and Wilkins, Baltimore, Maryland, 1996.

B38   Boice, J.D. Carcinogenesis - a synopsis of human experience with external exposure in medicine. Health Phys. 55: 621-630 (1988).

B39   Boice, J.D. and R.J.M. Fry. Radiation carcinogenesis in the gut. p. 291-306 in: Radiation and Gut (C.S. Potten and J.H. Hendry, eds.). Elsevier, Oxford, 1995.

B40   Blocher, D. DNA double-strand break repair determines the RBE of α-particles. Int. J. Radiat. Biol. 54: 761-771 (1988).

B41   Boecker, B.B., F.F. Hahn, B.A. Muggenburg et al. The relative effectiveness of inhaled alpha and beta-emitting radionuclides in producing lung cancer. p. 1059-1062 in: Proceedings IRPA7: 7th International Congress of the International Radiation Protection Association, Sydney, April 1988. Volume 2. Pergamon Press, London, 1988.

B42   Brenner, D.J. The microdosimetry of radon daughters and its significance. Radiat. Prot. Dosim. 31: 399-403 (1990).

B43   Bertrand, P., D. Rouillard, A. Boulet et al. Increase of spontaneous intrachromosomal recombination in mammalian cells expressing a mutant p53 protein. Oncogene 14: 1117-1122 (1997).

B44   Bouffler, S.D., G. Breckon and R. Cox. Chromosomal mechanisms in murine radiation acute myeloid leukaemo-genesis. Carcinogenesis 17: 101-105 (1996).

B45   Boice, J.D. Jr. and R.W. Miller. Childhood and adult cancer after intrauterine exposure to ionizing radiation. Teratology 59: 227-233 (1999).

C1   Committee on the Biological Effects of Ionizing Radiations (BEIR V). Health Effects of Exposure to Low Levels of Ionizing Radiation. United States National Academy of Sciences, National Research Council. National Academy Press, Washington, 1990.

C2   Cardis, E., J. Estève and B.K. Armstrong. Meeting recommends international study of nuclear industry workers. Health Phys. 63: 405-406 (1992).

C3   Carpenter, L., C. Higgins, A. Douglas et al. Combined analysis of mortality in three United Kingdom nuclear industry workforces. Radiat. Res. 138: 224-238 (1994).

C4   Canman, C.E. and M.B. Kastan. Three paths to stress relief. Nature 384: 213-214 (1996).

C5   Clark, D.J., E.I.M. Meijne, S.D. Bouffler et al. Micro-satellite analysis of recurrent chromosome 2 deletions in acute myeloid leukaemia induced by radiation in F1 hybrid mice. Genes, Chromosome Cancer 16: 238-246 (1996).

C6   Coggle, J.E. Lung tumour induction in mice by x-rays and neutrons. Int. J. Radiat. Biol. 53: 585-598 (1988).

C7   Covelli, V., M. Coppola, V. DiMajo et al. Tumour induction and life shortening in BC3F$_1$ mice at low doses of fast neutrons and x-rays. Radiat. Res. 113: 362-374 (1988).

C8   Covelli, V., V. DiMajo, M. Coppola et al. The dose-response relationships for myeloid leukaemia and malignant lymphoma in Bc3F$_1$ mice. Radiat. Res. 119: 553-561 (1989).

C9   Cox, R. Mechanisms of radiation oncogenesis. Int. J. Radiat. Biol. 65: 57-64 (1994).

C10   Cross, S.H. and A.P. Bird. CpG islands and genes. Curr. Opin. Genet. Dev. 5: 309-314 (1995).

C11   Chao, L-Y., V. Huff, G. Tomlinson et al. Genetic mosaicism in normal tissues of Wilms' tumour patients. Nature Genet. 3: 127-131 (1993).

C12   Cosset, J-M. Secondary cancers after radiotherapy. Radioprotection 32: C1241-C1248 (1997).

C13   Chauveinc, L., M. Ricoul, L. Sabatier et al. Radiation-induced malignant tumours: a specific cytogenetic profile? Radioprotection 32: C1249-C1250 (1997).

C14   Counter, C.M., A.A. Avillon, C.R. Lefeuvre et al. Telomere shortening associated with chromosome instability is arrested in immortal cells which express telomere activity. EMBO J. 11: 1921-1929 (1992).

C15   Cox, R., C.R. Muirhead, J.W. Stather et al. Risk of radiation-induced cancer at low doses and low dose rates for radiation protection purposes. Doc. NRPB 6(1): (1995).

C16   Chinnaiyan, A. and V. Dixit. The cell-death machine. Curr. Biol. 6: 555-562 (1996).

C17   Committee on the Biological Effects of Ionizing Radiations (BEIR I). The Effects on Populations of Exposure to Low Levels of Ionizing Radiation. National Academy of Sciences, National Research Council. National Academy Press, Washington, 1972.

C18   Cameron, E.A., K.E. Bachman, S. Myohanen et al. Synergy of demethylation and histone deacetylase inhibition in the re-expression of genes silenced in cancer. Nature Genet. 21: 103-107 (1999).

C19   Clifton, K.H. Comments on the evidence in support of the epigenetic nature of radiogenic initiation. Mutat. Res. 350: 77-80 (1996).

C20   Cardis, E., E.S. Gilbert, L. Carpenter et al. Effects of low doses and low dose rates of external ionizing radiation: cancer mortality among nuclear industry workers in three countries. Radiat. Res. 142: 117-132 (1995).

C21   Cartwright, R.A., R.J.Q. McNally, D.J. Rowland et al. The Descriptive Epidemiology of Leukaemia and Related Conditions in Parts of the United Kingdom 1984-1993. Leukaemia Research Fund, London, 1997.

C22   Cross, F.T. A review of experimental animal radon health effects data. p. 476-481 in: Radiation Research: A Twentieth-Century Perspective, Vol. II (J.D. Chapman et al., eds.). Academic Press, San Diego, 1992.

C23   Chadwick, K.H., R. Cox, H.P. Leenhouts et al. Molecular mechanisms in radiation mutagenesis and carcinogenesis. EUR 13294 (1994).

C24   Cera, F., R. Cherubini, M. Dalla Vecchia et al. Cell inactivation, mutation and DNA damage induced by light ions: Dependence on radiation quality. p. 191-194 in: Microdosimetry: An Interdisciplinary Approach (D.T. Goodhead, P. O'Neill and H.G. Menzel, eds.). Royal Society of Chemistry, Cambridge, 1997.

C25   Cucinotta, F.A., J.W. Wilson, M.R. Shavers et al. Effects of track structure and cell inactivation on the calculation of heavy ion mutation rates in mammalian cells. Int. J. Radiat. Biol. 69: 593-600 (1995).

C26   Committee on the Biological Effects of Ionizing Radiations (BEIR VI). The Health Effects of Exposure to Indoor Radon. National Academy of Sciences, National Research Council. National Academy Press, Washington, 1999.

C27   Cox, R. and W.K. Masson. Mutation and activation of cultured mammalian cells exposed to beams of accelerated heavy ions. III. Human diploid fibroblasts. Int. J. Radiat. Biol. 36: 149-160 (1979).

C28   Chauveinc, L., M. Ricoul, L. Sabatier et al. Dosimetric and cytogenetic studies of multiple radiation-induced meningiomas for a single patient. Radiother. Oncol. 43: 285-288 (1997).

C29   Chadwick, K.H., G. Moschini and M.M. Varma (eds.). Biophysical Modelling of Radiation Effects. Adam Hilger, Bristol, Philadelphia, New York, 1992.

D1   Darby, S.C., J.H. Olsen, R. Doll et al. Trends in childhood leukaemia in the Nordic countries in relation to fallout from atmospheric nuclear weapons testing. Br. Med. J. 304: 1005-1009 (1992).

D2   Delongchamp, R.R., K. Mabuchi, Y. Yasuhiko et al. Cancer mortality among atomic bomb survivors exposed *in utero* or as young children, October 1950-May 1992. Radiat. Res. 147: 385-395 (1997).

D3   Doll, R. and R. Wakeford. Risk of childhood cancer from fetal irradiation. Br. J. Radiol. 79: 130-139 (1997).

D4   de Bustros, A., D.D. Nelkin, A. Silverman et al. The short arm of chromosome 11 is a "hot spot" for hypermethylation in human neoplasia. Proc. Natl. Acad. Sci. U.S.A. 85: 5693-5697 (1988).

D5   Doll, R. The age distribution of cancer: implications for models of carcinogenesis. J. R. Stat. Soc., Ser. A 132: 133-166 (1971).

D6   Den Otter, W., J.W. Koten, B.J.H. van der Vegt et al. Oncogenesis by mutations in anti-oncogenes: a view. Anticancer Res. 10: 475-488 (1990).

D7   Derkinderen, D.J., O.J. Boxma, J.W. Koten et al. Stochastic theory of oncogenesis. Anticancer Res. 10: 497-504 (1990).

D8   Darby, S.C., R. Doll, S.K. Gill et al. Long term mortality after a single treatment course with X-rays in patients treated for ankylosing spondylitis. Br. J. Cancer 55: 179-190 (1987).

D9   Dove, W.F., R.T. Cormier, K.A. Gould et al. The intestinal epithelium and its neoplasms: genetic, cellular and tissue interactions. Philos. Trans. R. Soc. Lond. 353: 915-923 (1998).

D10   Darby, S.C. The contribution of natural ionizing radiation to cancer mortality in the United States. p. 183-190 in: The Origins of Human Cancer (J. Brugge et al., eds.). Cold Spring Harbour Laboratory Press, New York, 1991.

D11   Di Majo, V., M. Coppola, S. Rebessi et al. Dose-response relationship of radiation-induced harderian gland tumors and myeloid leukemia of the CBA/Cne mouse. J. Natl. Cancer Inst. 76: 955-966 (1986).

D12   Darby, S.C., E. Nakashima and H. Kato. A parallel analysis of cancer mortality among atomic bomb survivors and patients ankylosing with spondylitis given x-ray therapy. J. Natl. Cancer Inst. 75: 1-21 (1985).

D13   Deshpande, A., E.H. Goodwin, S.M. Bailey et al. Alpha particle induced sister chromatid exchange in normal human lung fibroblasts. Evidence for an extracellular target. Radiat. Res. 145: 260-267 (1996).

D14   de Vathaire, F., C. Hardiman, A. Shamsaldin et al. Thyroid carcinoma following irradiation for a first cancer during childhood. Arch. Intern. Med. 159: 2713-2719 (2000).

E1 Evans, R.D. The effect of skeletally deposited alpha-ray emitters in man. Br. J. Radiol. 39: 881-895 (1966).

E2 Edwards, A. Communication to the UNSCEAR Secretariat (1998).

E3 Elledge, R. and W.H. Lee. Life and death by p53. BioEssays 17: 923-930 (1995).

E4 Elgin, S.C.R. and S.P. Jackson (eds). Chromosomes and expression mechanisms. Curr. Opin. Genet. Dev. 7: (1997).

E5 Ellender, M., S.M. Larder, J.D. Harrison et al. Radiation-induced intestinal neoplasia in a genetically-predisposed mouse (*min*). Radioprotection 32: C1-287 (1997).

E6 Elliot, T. Tapping into tumours. Nature Genet. 13: 139-140 (1996).

E7 Evans, H.H., M. Nielsen, J. Mencl et al. The effect of dose rate on x-radiation-induced mutant frequency and the nature of DNA lesions in mouse lymphoma L5178Y cells. Radiat. Res. 122: 316-325 (1990).

E8 Edwards, A.A. The use of chromosomal aberrations in human lymphocytes for biological dosimetry. Radiat. Res. 148: S39-S44 (1997).

E9 Edwards, A.A. and D.C. Lloyd. Risk from deterministic effects of ionising radiation. Doc. NRPB 7(3): 1-31 (1996).

E10 Edwards, A.A., V.V. Moiseenko and H. Nikjoo. On the mechanism of formation of chromosomal aberrations by ionising radiation. Radiat. Environ. Biophys. 35: 25-30 (1996).

E11 Evans, R.D., A.T. Keane, R.J. Kolenkow et al. Radiogenic tumours in the radium and mesothorium cases studied at M.I.T. p. 157-194 in: Delayed Effects of Bone-Seeking Radionuclides (C.W. Mays et al., eds.). University of Utah Press, Salt Lake City, 1969.

E12 Edwards, A.A., R.J. Purrott, J.S. Prosser et al. The induction of chromosome aberrations in human lymphocytes by alpha-radiation. Int. J. Radiat. Biol. 38: 83-91 (1980).

E13 Edwards, A.A., D.C. Lloyd and J.S. Prosser. The induction of chromosome aberrations in human lymphocytes by accelerated charged particles. Radiat. Prot. Dosim. 13: 205-209 (1985).

E14 Edwards, A.A., D.C. Lloyd and J.S. Prosser. The induction of chromosome aberrations in human lymphocytes by 24 keV neutrons. Radiat. Prot. Dosim. 31: 265-268 (1990).

E15 Edwards, A.A., D.C. Lloyd and J.S. Prosser. 38 chromosome aberrations in human lymphocytes - a radiobiological review. p. 423-432 in: Low Dose Radiation: Biological Bases for Risk Assessment (K.F. Baverstock and J.W. Stather, eds.). Taylor and Francis, London, 1989.

E16 Edwards, A.A. Communication to the UNSCEAR Secretariat based on calculations by H. Nikjoo, Medical Research Council (MRC), Radiobiology Unit, Chilton, United Kingdom (2000).

F1 Fearon, E.R. and B. Vogelstein. A genetic model for colorectal tumourigenesis. Cell 61: 759-767 (1990).

F2 Feinberg, A.P. Genomic imprinting and gene activation in cancer. Nature Genet. 4: 110-113 (1993).

F3 Fishel, R. and R.D. Kolodner. The identification of mismatch repair genes and their role in the development of cancer. Curr. Opin. Genet. Dev. 5: 382-395 (1995).

F4 Folkman, J. Angiogenesis in cancer, vascular rheumatoid and other disease. Nature Med. 1: 27-31 (1995).

F5 Foulds, L. Neoplastic Development, Volume 2. Academic Press, New York, 1975.

F6 Ford, A.M., S.A. Ridge, M.E. Cabrera et al. *In utero* rearrangements in the trithorax-related oncogene in infant leukaemias. Nature 363: 358-360 (1993).

F7 Fialkow, P.J. Clonal origin of human tumours. Annu. Rev. Med. 30: 135-143 (1979).

F8 Fearon, E.R., K.R. Cho, J.M. Nigro et al. Identification of a chromosome 18q gene that is altered in colorectal cancers. Science 247: 49-56 (1991).

F9 Ferrara, N. Natural killer cells, adhesion and tumour angiogenesis. Nature Med. 2: 971-972 (1996).

F10 Freedman, V.H. and S. Shin. Cellular tumorigenicity in male mice: correlation with cell growth in semi solid medium. Cell 3: 355-359 (1974).

F11 Finkel, M.P. and B.O. Biskis. Toxicity of plutonium in mice. Health Phys. 8: 565-579 (1962).

F12 Finkel, M.P., B.O. Biskis and P.B. Jinkins. Toxicity of ^{226}Ra in mice in: Radiation Induced Cancer (A. Ericson, ed.). IAEA, Vienna, 1971.

F13 Finkel, A.J., C.E. Miller and R.J. Hasterlik. Radium-induced malignant tumours in man. p. 195-225 in: Delayed Effects of Bone-Seeking Radionuclides (C.W. Mays et al., eds.). University of Utah Press, Salt Lake City, 1969.

F14 Furuno-Fukushi, I. and H. Matsudaira. Mutation induction by different dose rates of γ rays in radiation-sensitive mutants of mouse leukaemia cells. Radiat. Res. 120: 370-374 (1989).

F15 Fry, R.J.M. Time-dose relationship and high-LET radiation. Int. J. Radiat. Biol. 58: 866-870 (1990).

G1 Groupe de travail de l'Académie des Sciences. Problèmes liés aux effets des faibles doses de radiations ionisantes. Rapport No. 34. Académie des Sciences, Paris (1995).

G2 Gilbert, E.S. Some effects of random dose measurement errors on analyses of atomic bomb survivor data. Radiat. Res. 98: 591-605 (1984).

G3 Gilbert, E.S., D.L. Cragle and L.D. Wiggs. Updated analyses of combined mortality data on workers at the Hanford site, Oak Ridge National Laboratory and Rocky Flats weapons plant. Radiat. Res. 136: 408-421 (1993).

G4 Gribben, M.A., J.L. Weeks and G.R. Howe. Cancer mortality (1956-85) among male employees of Atomic Energy of Canada Limited with respect to occupational exposure to external low-linear-energy-transfer ionising radiation. Radiat. Res. 133: 375-380 (1993).

G5 Goodhead, D.T., P. O'Neill and H.G. Menzel (eds.). Microdosimetry: An Interdisciplinary Approach. Royal Society of Chemistry, Cambridge, 1997.

G6 Goodhead, D.T. Biophysical models of radiation-action - introductory review. p. 306-311 in: Proceedings of the 8th International Congress of Radiation Research (Abstract), Edinburgh, 1987, Volume 2 (E.M. Fielden et al. eds.). Taylor and Francis, London, 1987.

G7 Goodhead, D.T. Physical basis for biological effect. p. 37-53 in: Nuclear and Atomic Data for Radiotherapy and Related Radiobiology. STI/PUB/741. IAEA, Vienna, 1987.

G8 Greenblatt, M.S., W.P. Bennett, M. Hollstein et al. Mutations in the p53 tumour suppressor gene: clues to cancer etiology and molecular pathogenesis. Cancer Res. 54: 4855-4878 (1994).

G9 Greider, C.W. Telomere length regulation. Annu. Rev. Biochem. 65: 337-365 (1996).

G10 Goodhead, D.T. Initial events in the cellular effects of ionizing radiations: clustered damage in DNA. Int. J. Radiat. Biol. 65: 7-17 (1994).

G11 Griffiths, D.F.R., P. Sacco and D. Williams. The clonal origin of experimental large bowel tumours. Br. J. Cancer 59: 385-387 (1989).

G12 Goodhead, D.T. and H. Nikjoo. Track structure analysis of ultrasoft x-rays compared to high- and low-LET radiations. Int. J. Radiat. Biol. 55(4): 513-529 (1989).

G13 Granath, F., F. Darroudi, A. Auvinen et al. Retrospective dose estimates in Estonian Chernobyl clean-up workers by means of FISH. Mutat. Res. 369: 7-12 (1996).

G14 Grahn, D. and G.A. Sacher. Fractionation and protraction factors and the late effects of radiation in small mammals. p. 2.1-2.27 in: Proceedings of a Symposium on Dose Rate in Mammalian Radiation Biology (D.G. Brown et al., eds.). CONF-680410 (1968).

G15 Gössner, W., R.R. Wick and H. Spiess. Histopathological review of ^{224}Ra induced bone sarcomas. p. 255-260 in: Health Effects of Internally Deposited Radionuclides: Emphasis on Radium and Thorium (G. van Kaick, A. Karaoglou and A.M. Kellerer, eds.). World Scientific, London, 1995.

G16 Griffith, W.C., B.B. Boecker, N.A. Gillett et al. Comparison of risk factors for bone cancer induced by inhaled ^{90}SrCl$_2$ and ^{238}PuO$_2$. in: Proceedings EULEP/DoE Joint Bone Radiobiology Workshop, Toronto, July 1991. USDOE Report UCD-472-136 (1991).

G17 Gössner, W. Pathology of radium-induced bone tumours: new aspects of histopathology and histogenesis. Radiat. Res. 152: S12-S15 (1999).

G18 Gray, R.G., J. Lafuma and S.E. Paris. Lung tumours and radon inhalation in over 2000 rats: approximate linearity across a wide range of doses and potentiation by tobacco smoke. p. 592-607 in: Lifespan Radiation Effects Studies in Animals: What Can They Tell Us? (R.C. Thompson and J.A. Mahaffey, eds.). CONF-830951 (1986).

G19 Guarente, L. Diverse and dynamic functions of the Sir silencing complex. Nature Genet. 23: 281-285 (1999).

G20 Grosovsky, A.J. Radiation-induced mutations in unirradiated DNA. Proc. Natl. Acad. Sci. (USA) 96: 5346-5347 (1999).

G21 Grosovsky, A.J., K.K. Parks and S.L. Nelson. Clonal analysis of delayed karyotypic abnormalities and gene mutations in radiation-induced genetic instability. Mol. Cell Biol. 16: 6252-6262 (1996).

G22 Goodhead, D.T. Spatial and temporal distribution of energy. Health Phys. 55(2): 231-240 (1988).

G23 Goodhead, D.T. Microscopic features of dose from radionuclides, particularly emitters of α-particles and Auger electrons. Int. J. Radiat. Biol. 60(3): 550-553 (1991).

H1 Han, A. and M.M. Elkind. Transformation of mouse C3H10T½ cells by single and fractionated doses of x-rays and fission-spectrum neutrons. Cancer Res. 39: 123-130 (1979).

H2 Heidenreich, W.F., P. Jacob, H.G. Paretzke et al. Two-step model for the risk of fatal and incidental lung tumors in rats exposed to radon. Radiat. Res. 151: 209-217 (1999).

H3 Holliday, R. Endless quest. BioEssays 18: 3-5 (1996).

H4 Hart, I.R. and I. Saini. Biology of tumour metastasis. Lancet 339: 1453-1457 (1992).

H5 Holliday, R. Understanding Ageing. Cambridge University Press, Cambridge, 1995.

H6 Hartwell, L.H. and M.B. Kastan. Cell cycle control and cancer. Science 266: 1821-1828 (1994).

H7 Hartwell, L.H., T. Weinert, L. Kadyk et al. Cell cycle checkpoints, genomic integrity, and cancer. Cold Spring Harbor Symp. Quant. Biol. 59: 259-263 (1994).

H8 Heyn, R., V. Haeberlen, W.A. Newton et al. Second malignant neoplasms in children treated for rhabdomyo-sarcoma. J. Clin. Oncol. 11: 262-270 (1993).

H9 Hall, A. Ras-related proteins. Curr. Opin. Cell Biol. 5: 63-69 (1993).

H10 Hinds, P.W. and R.A. Weiberg. Tumour suppressor genes. Curr. Opin. Genet. Dev. 4: 135-141 (1994).

H11 Haber, D. and E.D. Harlow. Tumour-suppressor genes: evolving definitions in the genomic age. Nature Genet. 16: 320-322 (1997).

H12 Hartwell, L. Defects in a cell cycle checkpoint may be responsible for the genomic instability of cancer cells. Cell 71: 543-546 (1992).

H13 Haffner, R. and M. Oren. Biochemical properties and biological effects of p53. Curr. Opin. Genet. Dev. 5: 84-90 (1995).

H14 Hartwell, L., T. Weinert, L. Kadyk et al. Cell cycle checkpoints, genomic integrity and cancer. Cold Spring Harbor Symp. Quant. Biol. 59: 259-263 (1994).

H15 Hayata, I., M. Seki, K. Yoshida et al. Chromosomal aberrations observed in 52 mouse myeloid leukaemias. Cancer Res. 43: 367-373 (1983).

H16 Harley, C.B. Telomere loss: mitotic clock or genetic time bomb? Mutat. Res. 256: 271-282 (1991).

H17 Hemminki, A., P. Peltomaki and J.P. Mecklin. Loss of the wild type MLH1 gene is a feature of hereditary non-polyposis colorectal cancer. Nature Genet. 8: 405-410 (1994).

H18 Hall, E.J. Finding a smoother pebble: a workshop summary. p. 401-402 in: Cell Transformation and Radiation Induced Cancer (K.H. Chadwick et al., eds.). Adam Hilger, Bristol, 1989.

H19 Hei, T.K., K. Komatsu, E.J. Hall et al. Oncogenic transformation by charged particles of defined LET. Carcinogenesis 9: 747-750 (1988).

H20 Hall, E.J., R.C. Miller and D.J. Brenner. Neoplastic transformation and the inverse dose rate effect for neutrons. Radiat. Res. 128: S75-S80 (1991).

H21 Hasterlik, R.J. The delayed toxicity of radium deposited in the skeleton of human beings. p. 149-155 in: Proceedings of the International Conference on the Peaceful Uses of Atomic Energy, Volume 11, Geneva, 1956.

H22 Hockenbery, D.M. Bcl-2, a novel regulator of cell death. BioEssays 17: 631-638 (1995).

H23 Huang, L.C., K.C. Clarkin and G.M. Wahl. Sensitivity and selectivity of the DNA damage sensor responsible for activating p53-dependent G1 arrest. Proc. Natl. Acad. Sci. U.S.A. 93: 4827-4832 (1996).

H24 Hall, E.J. and R.C. Miller. The how and why of in vitro oncogenic transformation. Radiat. Res. 87: 208-223 (1981).

H25 Hill, C.K. and L. Zhu. Energy and dose-rate dependence of neoplastic transformation and mutations induced in mammalian cells by fast neutrons. Radiat. Res. 128: S53-S59 (1991).

H26 Hoel, D.G. and P. Li. Threshold models in radiation carcinogenesis. Health Phys. 75: 241-250 (1998).

H27 Hill, C.K., A. Han and M.M. Elkind. Promotion, dose-rate, and repair processes in radiation-induced neoplastic transformation. Radiat. Res. 109: 347-351 (1987).

H28 Heimbach, R.D., F.J. Burns and R.E. Albert. An evaluation by alpha-particle Bragg peak radiation of the critical depth in rat skin for tumour induction. Radiat. Res. 39: 332-344 (1969).

H29 Hei, T.K., E.J. Hall and C.A. Waldren. Neutron risk assessment based on low dose mutation data. p. 481-490 in: Low Dose Radiation: Biological Bases for Risk Assessment (K.F. Baverstock and J.W. Stather, eds.). Taylor and Francis, London, 1989.

H30 Hahn, F.F., W.C. Griffith and B.B. Boecker. Comparison of the effects of inhaled ^{238}PuO$_2$ and β-emitting radionuclides on the incidence of lung carcinomas in laboratory animals. p. 916-919 in: Proceedings IRPA8: 8th International Congress of the International Radiation Protection Association, Montreal, 1991. Volume 1. Pergamon Press, London, 1991.

H31   Howe, G.R. and J. McLaughlin. Breast cancer mortality between 1950 and 1987 after exposure to fractionated moderate-dose-rate ionizing radiation in the Canadian fluoroscopy cohort study and a comparison with breast cancer mortality in the atomic bomb survivors study. Radiat. Res. 145: 694-707 (1996).

H32   Haines, J., R. Dunford, J. Moody et al. Loss of heterozygosity in spontaneous and x-ray-induced intestinal tumors arising in F1 hybrid *Min* mice: evidence for sequential loss of *Apc*+ and *Dpc4* in tumour development. Genes Chrom. Cancer 28(4): 387-394 (2000).

H33   Hayata, I., C. Wang, W. Zhang et al. Chromosome translocation in residents of the high background radiation area in Southern China. J. Radiat. Res. 41: (2000, in press).

H34   Hayata, I. Insignificant risk at low dose (rate) radiation predicted by cytogenetic studies. p. 268 (T-17-3) in: Proceedings of the 10th International Congress of the International Radiation Protection Association, Hiroshima, Japan, 14-19 May 2000.

H35   Harms, M.D., H.P. Leenhouts and P.A.M. Uijt deHaag. A two mutation model of carcinogenesis: application to lung tumours using rat experimental data. RIVM Report No. 610065.006 (1998).

H36   Hyun, S-J., M-Y. Yoon, T-I. Kum et al. Enhancement of mitogen-stimulated proliferation of low dose radiation-adapted mouse splenocytes. Anticancer Res. 17: 225-230 (1997).

I1   International Agency for Research on Cancer Study Group on Cancer Risk among Nuclear Industry Workers. Direct estimates of cancer mortality due to low doses of ionising radiation: An international study. Lancet 344: 1039-1043 (1994).

I2   International Commission on Radiological Protection. 1990 Recommendations of the International Commission on Radiological Protection. Annals of the ICRP 21(1-3). ICRP Publication 60. Pergamon Press, Oxford, 1991.

I3   Issa, J-P., Y.L. Ottaviano, P. Celano et al. Methylation of the oestrogen receptor CpG island links ageing and neo-plasia in human colon. Nature Genet. 7: 536-540 (1994).

I4   Issa, J-P. and S.B. Baylin. Epigenetics and human disease. Nature Med. 2: 281-282 (1996).

I5   International Commission on Radiological Protection. Biological Effects of Inhaled Radionuclides. ICRP Publication 31. Annals of the ICRP 4(1-2). Pergamon Press, Oxford, 1980.

I6   International Commission on Radiological Protection. Recommendations of the International Commission on Radiological Protection. ICRP Publication 1. Pergamon Press, Oxford, 1959.

I7   Iwasaki, T., M. Minowa, S. Hashimoto et al. Non-cancer mortality and life expectancy in different natural background radiation levels of Japan. p. 107-112 in: Low Dose Irradiation and Biological Defense Mechanisms (T. Sugahara et al., eds.). Elsevier Science Publishers, Amsterdam, 1992.

I8   International Commission on Radiological Protection. Recommendations of the International Commission on Radiological Protection. ICRP Publication 26. Pergamon Press, Oxford, 1977; reprinted (with additions) in 1987.

I9   International Commission on Radiological Protection. Limits on Intakes of Radionuclides by Workers. ICRP Publication 30, Part 1. Annals of the ICRP 2(3/4). Pergamon Press, Oxford, 1979

I10   International Commission on Radiological Protection. Age-dependent Doses to Members of the Public from Intake of Radionuclides. Part 2: Ingestion Dose Coefficients. ICRP Publication 67. Annals of the ICRP 23(3/4). Pergamon Press, Oxford, 1993.

I11   International Commission on Radiological Protection. Human Respiratory Tract Model for Radiological Protection. ICRP Publication 66. Annals of the ICRP 24(1/3). Pergamon Press, Oxford, 1994.

I12   International Commission on Radiological Protection. Lung Cancer Risk from Indoor Exposures to Radon Daughters. ICRP Publication 50. Annals of the ICRP 17(1). Pergamon Press, Oxford, 1987.

I13   International Commission on Radiological Protection. RBE for Deterministic Effects. ICRP Publication 58. Annals of the ICRP 20(4). Pergamon Press, Oxford, 1989.

I14   Iwamoto, K.S., S. Fujii, A. Kurata et al. p53 mutations in tumour and non-tumour tissues of Thorotrast recipients: a model for cellular selection during radiation carcinogenesis in the liver. Carcinogenesis 20: 1283-1291 (1999).

I15   Ishii, K. and M. Watanabe. Participation of gap-junctional cell communication on the adaptive response in human cells induced by low doses of X-rays. Int. J. Radiat. Biol. 69: 291-299 (1996).

J1   Jablon, S. and H. Kato. Childhood cancer in relation to prenatal exposure to atomic bomb radiation. Lancet i: 1000-1003 (1970).

J2   Jones, P.A., W.M.I. Rideout, J-C. Shen et al. Methylation, mutation and cancer. Bioessays 14: 33-36 (1992).

J3   Jones, P.A., M.J. Wolkewicz, W.M. Ridout et al. *De novo* methylation of the MyoD1 CpG island during the establishment of immortal cell lines. Proc. Natl. Acad. Sci. U.S.A. 87: 6117-6121 (1990).

J4   Jiang, X-R., G. Jiminez, E. Chang et al. Telomerase expression in human somatic cells does not induce changes associated with a transformed phenotype. Nature Genet. 21: 111-114 (1999).

J5   Jablon, S. Atomic bomb radiation dose estimation at ABCC. ABCC Technical Report 23-71 (1971).

J6   Jenner, T.J., C.M. DeLara, P. O'Neill et al. Induction and rejoining of double strand breaks in V79-4 mammalian cells following γ- and α-irradiation. Int. J. Radiat. Biol. 64: 265-273 (1993).

J7   Johnson, N.F., A.F. Hobbs and D.G. Thomassen. Epithelial progenitor cells in the rat respiratory tract. p. 88-98 in: Biology, Toxicology and Carcinogenesis of Respiratory Epithelium (D.G. Thomassen and P. Nettesheim, eds.). Hemisphere Publishing Co., Washington D.C., 1990.

J8   Joiner, M.C., P. Lamblin and B. Marples. Adaptive response and induced resistance. C.R. Acad. Sci. Ser. 3 (322): 167-175 (1999).

J9   Jiang, T., I. Hayata, C. Wang et al. Dose-effect relationship of dicentric and ring chromosomes in lymphocytes of individuals living in high background radiation area in China. J. Radiat. Res. 41: (2000, in press).

K1   Karp, J.E. and S. Broder. Molecular foundations of cancer: new targets for intervention. Nature Med. 1: 309-320 (1995).

K2   Kemp, C.J., T. Wheldon and A. Balmain. *p53*-deficient mice are extremely susceptible to radiation-induced tumorigenesis. Nature Genet. 8: 66-69 (1994).

K3 Kendall, G.M., C.R. Muirhead, B.H. MacGibbon et al. Mortality and occupational exposure to radiation: first analysis of the National Registry for radiation workers. Br. Med. J. 304: 220-225 (1992).

K4 Kadhim, M.A., D.A. Macdonald, D.T. Goodhead et al. Transmission of chromosomal instability after plutonium α-particle irradiation. Nature 355: 738-740 (1992).

K5 Kinzler, K.W. and B. Vogelstein. Life and death in a malignant tumour. Nature 379: 19-20 (1996).

K6 Korsmeyer, S.J. Regulators of cell death. Trends Genet. 11: 101-105 (1995).

K7 Kroemer, G. The proto-oncogene Bcl-2 and its role in regulating apoptosis. Nature Med. 3: 614-620 (1997).

K8 Kraemer, K.H., D.D. Levy, C.N. Parris et al. Xeroderma pigmentosum and related disorders: examining the linkage between defective DNA repair and cancer. J. Invest. Dermatol. 103: 96-101 (1994).

K9 Kučerová, M., A.J.B. Anderson, K.E. Buckton et al. X-ray-induced chromosome aberrations in human peripheral blood leucocytes: the response to low levels of exposure in vitro. Int. J. Radiat. Biol. 21: 389-396 (1972).

K10 Knox, E.G., A.M. Stewart, G.W. Kneale et al. Prenatal irradiation and childhood cancer. J. Soc. Radiol. Prot. 7: 177-189 (1987).

K11 Kastan, M.B. On the TRAIL from p53 to apoptosis. Nature Genet. 17: 130-131 (1997).

K12 Kinzler, K.W. and B. Vogelstein. Lessons from hereditary colorectal cancer. Cell 87: 159-170 (1996).

K13 Klugbauer, S., E. Lengerfelder, E.P. Demidchik et al. High prevalence of ret rearrangements in thyroid tumours of children from Belarus after the Chernobyl reactor accident. Oncogene 11: 2459-2467 (1995).

K14 Kipling, D. The Telomere. Oxford University Press, Oxford, 1995.

K15 Keane, A.T., J. Rundo and M.A. Essling. Postmenopausal loss of Ra acquired in adolescence or young adulthood: quantitative relationship to radiation-induced skeletal damage and dosimetric implications. Health Phys. 54: 517-527 (1988).

K16 Knudson, A.G. Mutation and cancer: statistical study of retinoblastoma. Proc. Natl. Acad. Sci. U.S.A. 68: 820-823 (1971).

K17 Kai, M., E.G. Luebeck and S.H. Moolgavkar. Analysis of the incidence of solid cancer among atomic bomb survivors using a two-stage model of carcinogenesis. Radiat. Res. 148: 348-358 (1997).

K18 Kouzarides, T. Histone acetylases and deacetylases in cell proliferation. Curr. Opin. Genet. Dev. 9: 40-48 (1999).

K19 Kondo, S. Health Effects of Low-Level Radiation. Kinki University Press, Osaka, Japan and Medical Physics Publishing, Madison, WI USA, 1993.

K20 König, F. and J. Kiefer. Lack of dose-rate effect for mutation induction by γ-rays in human TK_6 cells. Int. J. Radiat. Biol. 54(6): 891-897 (1988).

K21 Kelsey, K.T., A. Memisoglu, D. Frenkel et al. Human lymphocytes exposed to low doses of X-rays are less susceptible to radiation-induced mutagenesis. Mutat. Res. 263: 197-201 (1991).

K22 Kathren, R.L. Pathway to a paradigm: the linear non-threshold dose-response model in historical context: the American Academy of Health Physics 1995 Radiology Centennial Hartman oration. Health Phys. 70: 621-635 (1996).

K23 Kneale, G.W. and A.M. Stewart. Age variation in the cancer risks from foetal irradiation. Br. J. Cancer 35: 501-510 (1977).

K24 Katz, R., R. Zachariah, F.A. Cucinotta et al. Survey of radiosensitivity parameters. Radiat. Res. 14: 356-365 (1994).

K25 Kellerer, A.M. and E. Nekolla. Neutron versus γ-ray risk estimates. Inferences from the cancer incidence and mortality data in Hiroshima. Radiat. Environ. Biophys. 36: 73-83 (1997).

K26 Koshurnikova, N.A., G.D. Bysogolov, M.G. Bolotnikova et al. Mortality among personnel who worked at the Mayak complex in the first years of its operation. Health Phys. 71: 90-93 (1996).

K27 Kellerer, A.M. and D. Barclay. Age dependencies in the modelling of radiation carcinogenesis. Radiat. Prot. Dosim. 41: 273-281 (1992).

K28 Kiefer, J. Quantitative Mathematical Models in Radiation Biology. page 208. Springer Verlag, Berlin, Heidelberg, New York, 1987.

K29 Kostyuchenko, V.A. and L.Yu. Krestinina. Long-term irradiation effects in the population evacuated from the East-Urals radioactive trace area. Sci. Total Environ. 142: 119-125 (1994).

K30 Kusunoki, Y., S. Kyoizumi, Y. Hirai et al. Flow of cytometry measurements of subsets of T, B and NK cells in peripheral blood lymphocytes of atomic bomb survivors. Radiat. Res. 150: 227-236 (1998).

L1 Lanfrancone, L., G. Pelicci and P.E. Pelicci. Cancer genetics. Curr. Opin. Genet. Dev. 4: 109-119 (1994).

L2 Leenhouts, H.P. and K.H. Chadwick. A two-mutation model of radiation carcinogenesis: application to lung tumours in rodents and implications for risk evaluation. J. Radiol. Prot. 14: 115-130 (1994).

L3 Lengauer, C., K.W. Kinzler and B. Vogelstein. Genetic instability in colorectal cancers. Nature 386: 623-626 (1997).

L4 Little, J.B. The relevance of cell transformation to carcinogenesis in vivo. p. 396-413 in: Low Dose Radiation: Biological Bases of Risk Assessment (K.F. Baverstock and J.W. Stather, eds.). Taylor and Francis, London, 1989.

L5 Little, M.P., M.M. Hawkins, M.W. Charles et al. Fitting the Armitage-Doll model to radiation-exposed cohorts and implications for population cancer risks. Radiat. Res. 132: 207-221 (1992).

L6 Little, M.P., C.R. Muirhead, J.D. Boice et al. Using multistage models to describe radiation-induced leukaemia. J. Radiol. Prot. 15: 315-334 (1995).

L7 Little, M.P. and C.R. Muirhead. Evidence for curvilinearity in the cancer incidence dose-response in the Japanese atomic bomb survivors. Int. J. Radiat. Biol. 70: 83-94 (1996).

L8 Lloyd, D.C., A.A. Edwards, A. Leonard et al. Chromosomal aberrations in human lymphocytes induced in vitro by very low doses of x-rays. Int. J. Radiat. Biol. 61: 335-343 (1992).

L9 Lloyd, D.C., A.A. Edwards and J.S. Prosser. Chromosome aberrations induced in human lymphocytes by in vitro acute x and gamma radiation. Radiat. Prot. Dosim. 15: 83-88 (1986).

L10 Loeb, L. Mutator phenotype may be required for multi-stage carcinogenesis. Cancer Res. 51: 3075-3079 (1991).

L11 Little, M.P. Comments on the article "Studies of the mortality of atomic bomb survivors. Report 12, Part I. Cancer: 1950-1990" by D.A. Pierce et al. (Radiat. Res. 146: 1-27 (1996)). Radiat. Res. 148: 399-401 (1997).

L12 Luebeck, E.G., S.B. Curtis, F.T. Cross et al. Two stage model of radiation-induced malignant lung tumours in rats: effects of cell killing. Radiat. Res. 145: 163-173 (1996).

L13   Luchnik, N.V. and A.V. Sevan'kaev. Radiation-induced chromosomal aberrations in human lymphocytes. I. Dependence on the dose of gamma-rays and an anomaly at low doses. Mutat. Res. 36: 363-378 (1976).

L14   Levine, A.J. and J.R. Broach (eds). Oncogenes and cell proliferation. Curr. Opin. Genet. Dev. 5(1): (1995).

L15   Levine, A.J. The tumour suppressor genes. Annu. Rev. Biochem. 62: 623-651 (1993).

L16   Luebeck, E.G., W.F. Heidenreich, W.D. Hazelton et al. Biologically based analysis of the data for the Colorado uranium miners cohort: age, dose and dose-rate effects. Radiat. Res. 152: 339-351 (1999).

L17   Loeb, L.A. Microsatellite instability: marker of a mutator phenotype in cancer. Cancer Res. 54: 5059-5063 (1994).

L18   Levy, D.B., K.J. Smith, Y. Beazer-Barclay et al. Inactivation of both *Apc* alleles in human and mouse tumours. Cancer Res. 54: 5953-5958 (1994).

L19   Luongo, C., A.R. Moser, S. Gledhill et al. Loss of *Apc*⁺ in intestinal adenomas from Min mice. Cancer Res. 54: 5947-5952 (1994).

L20   Luongo, C. and W.F. Dove. Somatic genetic events linked to the *Apc* locus in intestinal adenomas of the Min mouse. Genes, Chromosome Cancer 17: 194-198 (1996).

L21   Lundblat, V. and W.E. Wright. Telomeres and telomerase: a simple picture becomes complex. Cell 87: 369-375 (1996).

L22   Little, M.P., C.R. Muirhead and C.A. Stiller. Modelling lymphocytic leukaemia incidence in England and Wales using generalisations of the two-mutation model of carcinogenesis of Moolgavkar, Venzon and Knudson. Stat. Med. 15: 1003-1022 (1996).

L23   Leach, D.R., M.F. Krummel and J.P. Allison. Enhancement of antitumour immunity by CTLA-4 blockade. Science 271: 1734-1736 (1996).

L24   Little, J.B. Induced genetic instability as a critical step in radiation carcinogenesis. p. 597-601 in: Radiation Research 1895-1995, Vol. 2 (U. Hagen, D. Harder, H. Jung et al., eds.). H. Sturtz AG, Wurzburg, 1995.

L25   Lindahl, T. Instability and decay of the primary structure of DNA. Nature 362: 709-715 (1993).

L26   Little, M.P., C.R. Muirhead and C.A. Stiller. Modelling acute lymphocytic leukaemia using generalizations of the MVK two-mutation model of carcinogenesis: implied mutation rates and the likely role of ionising radiation. p. 244-247 in: Microdosimetry. An Interdisciplinary Approach (D.T. Goodhead, P. O'Neill and H.G. Menzel, eds.). Royal Society of Chemistry, Cambridge, 1997.

L27   Lloyd, R.D., G.N. Taylor, W. Angus et al. Bone cancer occurrence among beagles given ²³⁹Pu as young adults. Health Phys. 64: 45-51 (1993).

L28   Lundgren, D.L., F.F. Hahn and W.W. Carlton. Dose response from inhaled monodisperse aerosols of ²⁴⁴Cm₂O₃ in the lung, liver and skeleton of F344 rats and comparison with ²³⁹PuO₂. Radiat. Res. 147: 598-612 (1997).

L29   Lewis, E.B. Leukaemia and ionizing radiation. Science 125: 965-972 (1957).

L30   Little, M.P. and M.W. Charles. The risk of non-melanoma skin cancer incidence in the Japanese atomic bomb survivors. Int. J. Radiat. Biol. 71(5): 589-602 (1997).

L31   Little, M.P. Are two mutations sufficient to cause cancer? Some generalizations of the two-mutation model of carcinogenesis of Moolgavkar, Venzon and Knudson and of the multistage model of Armitage and Doll. Biometrics 51: 1278-1291 (1995).

L32   Little, M.P. and M.W. Charles. Time variations in radiation-induced relative risk and implications for population cancer risks. J. Radiol. Prot. 11: 91-110 (1991).

L33   Little, M.P. Risks of radiation-induced cancer at high doses and dose rates. J. Radiol. Prot. 13: 3-25 (1993).

L34   Lloyd, D.C. and A.A. Edwards. Chromosome aberrations in human lymphocytes: effect of radiation quality, dose and dose rate. p. 23-49 in: Radiation-induced Chromosome Damage in Man (T. Ishihara and M.S. Sasaki, eds.). Alan Liss, New York, 1983.

L35   Little, M.P., M.M. Hawkins, M.W. Charles et al. Letter to the Editor. Corrections to the paper "Fitting the Armitage-Doll model to radiation-exposed cohorts and implications for population cancer risks". Radiat. Res. 137: 124-128 (1994).

L36   Little, M.P. Generalisations of the two-mutation and classical multi-stage models of carcinogenesis fitted to the Japanese atomic bomb survivor data. J. Radiol. Prot. 16: 7-24 (1996).

L37   Li, F., G. Ambrosini, E.Y. Chu et al. Control of apoptosis and mitotic spindle checkpoint by survivin. Nature 396: 580-582 (1998).

L38   Lloyd, D.C. New developments in chromosomal analysis for biological dosimetry. Radiat. Prot. Dosim. 77 (1/2): 33-36 (1998).

L39   Lundgren, D.L., F.F. Hahn, W.C. Griffith et al. Pulmonary carcinogenicity of relatively low doses of beta-particle radiation from inhaled ¹⁴⁴CeO₂ in rats[1,2]. Radiat. Res. 146: 525-535 (1996).

L40   Lyon, M.F., R.J.S. Phillips and H.J. Bailey. Mutagenic effect of repeated small radiation doses to mouse spermatogonia. I. Specific locus mutation rates. Mutat. Res. 15: 185-190 (1972).

L41   Lutz, W.K., T. Fekete and S. Vamvakas. Position and base pair specific comparison of p53 mutation spectra in human tumours: elucidation of relationships between organs for cancer etiology. Environ. Health Perspect. 106: 207-211 (1998).

L42   Little, M.P., M.M. Hawkins, R.E. Shore et al. Time variations in the risk of cancer following irradiation in childhood. Radiat. Res. 126: 304-316 (1991).

L43   Little, M.P. and C.R. Muirhead. Curvilinearity in the dose-response curve for cancer in Japanese atomic bomb survivors. Environ. Health Perspect. 105 (Suppl. 6): 1505-1509 (1997).

L44   Little, M.P. and C.R. Muirhead. Curvature in the cancer mortality dose response in Japanese atomic bomb survivors: absence of evidence of threshold. Int. J. Radiat. Biol. 74: 471-480 (1998).

L45   Luckey, T.D. Radiation Hormesis. CRC Press, Boca Raton, 1991.

L46   Little, M., M.W. Charles, J.W. Hopewell et al. Assessment of skin doses. Doc. NRPB 8(3) (1997).

L47   Lord, B.I. The architecture of bone marrow cell populations. Int. J. Cell Cloning 8: 317-331 (1990).

L48   Lubin, J.H., J.D. Boice Jr., C. Edling et al. Radon and lung cancer risk: a joint analysis of 11 underground miners studies. NIH Publication No. 94-3644 (1994).

L49   Little, M.P. Comments on the article "Threshold models in radiation carcinogenesis" by D.G. Hoel and P. Li (Health Phys. 75: 241-250 (1998)). Health Phys. 76: 432-434 (1999).

L50   Lucas, J.N., W. Deng, D. Moore et al. Background ionizing radiation plays a minor role in the production of chromosome translocations in a control population. Int. J. Radiat. Biol. 75: 819-827 (1999).

L51 Lubin, J.H., J.D. Boice Jr., C. Edling et al. Lung cancer risk in radon-exposed miners and estimation of risk from indoor exposure. J. Natl. Cancer Inst. 87: 817-827 (1995).

L52 Little, M.P. and C.R. Muirhead. Derivation of low dose extrapolation factors from analysis of curvature in the cancer incidence dose response in the Japanese atomic bomb survivors. Int. J. Radiat. Biol. 76: 939-953 (2000).

L53 Lubin, J.H. and J.D. Boice. Lung cancer risk from residential radon: meta analysis of eight epidemiologic studies. J. Natl. Cancer Inst. 89: 49-57 (1997).

L54 Lloyd, D.C., R.J. Purrott, G.W. Dolphin et al. Chromosome aberrations induced in human lymphocytes by neutron irradiation. Int. J. Radiat. Biol. 29: 169-182 (1976).

L55 Little, M.P. Modelling leukaemia risk in the Japanese atomic bomb survivors and in a UK population using multistage generalizations of the two-mutation model of carcinogenesis of Moolgavkar, Venzon and Knudson. Br. J. Cancer (2000, in press).

L56 Little, M.P., C.R. Muirhead and M.W. Charles. Describing time and age variations in the risk of radiation-induced solid tumour incidence in the Japanese atomic bomb survivors using generalized relative and absolute risk models. Stat. Med. 18:17-33 (1999).

L57 Little, M.P., F. de Vathaire, M.W. Charles et al. Variations with time and age in the relative risks of solid cancer incidence after radiation exposure. J. Radiol. Prot. 17: 159-177 (1997).

L58 Liu, S.Z., W.H. Liu and J.B. Sun. Radiation hormesis: its expression in the immune system. Health Phys. 52: 579-583 (1987).

M1 Merlo, A., J.E. Herman, D.J. Lee et al. 5'CpG island methylation is associated with transcriptional silencing of the tumour suppressor (p16/CDKN2/MTS) in human cancers. Nature Med. 1: 686-692 (1995).

M2 Miller, A.B., G.R. Howe, G.J. Sherman et al. Mortality from breast cancer after irradiation during fluoroscopic examinations in patients being treated for tuberculosis. N. Engl. J. Med. 321: 1285-1289 (1989).

M3 Mole, R.H. and I.R. Major. Myeloid leukemia frequency after protracted exposure to ionising radiation: experimental confirmation of the flat dose-response found in ankylosing spondylitis after a single treatment course with x-ray. Leuk. Res. 7: 295-300 (1983).

M4 Mole, R.H., D.G. Papworth and M.J. Corp. The dose response for x-ray induction of myeloid leukaemia in male CBA/H mice. Br. J. Cancer 47: 285-291 (1983).

M5 Moolgavkar, S.H. and A.G. Knudson. Mutation and cancer: a model for human carcinogenesis. J. Natl. Cancer Inst. 66: 1037-1052 (1981).

M6 Mole, R.H. Fetal dosimetry by UNSCEAR and risk coefficients for childhood cancer following diagnostic radiology in pregnancy. J. Radiol. Prot. 10: 199-203 (1990).

M7 Moolgavkar, S.H. and D.J. Venzon. Two-event models for carcinogenesis: incidence curves for childhood and adult tumours. Math. Biosci. 47: 55-77 (1979).

M8 Muirhead, C.R. and G.W. Kneale. Prenatal irradiation and childhood cancer. J. Radiol. Prot. 9: 209-212 (1989).

M9 MacMahon, B. Prenatal x-ray exposure and childhood cancer. J. Natl. Cancer Inst. 28: 1173-1191 (1962).

M10 Monson, R.R. and B. MacMahon. Prenatal x-ray exposure and cancer in children. p. 97-105 in: Radiation Carcinogenesis: Epidemiology and Biological Significance (J.D. Boice and J.F. Fraumeni, eds.). Raven Press, New York, 1984.

M11 Mays, C.W. and R.D. Lloyd. Bone sarcoma risk from [90]Sr. p. 352-375 in: Biomedical Implications of Radiostrontium Exposure (M. Goldman and L.K. Bustad, eds.). CONF-710201 (1972).

M12 Muirhead, C.R. and S.C. Darby. Modelling the relative and absolute risks of radiation-induced cancers. J. R. Stat. Soc., Ser. A 150: 83-118 (1987).

M13 Moolgavkar, S.H., E.G. Luebeck, D. Krewski et al. Radon, cigarette smoke, and lung cancer: a re-analysis of the Colorado plateau uranium miners' data. Epidemiology 4: 204-217 (1993).

M14 Mill, A.J., D. Frankenberg, D. Bettega et al. Transformation of C3H 10T½ cells by low doses of ionising radiation: a collaborative study by six European laboratories strongly supporting a linear dose-response relationship. J. Radiol. Prot. 18: 79-100 (1998).

M15 Miller, R.C., C.R. Geard, D.J. Brenner et al. Neutron-energy-dependent oncogenic transformation of C3H 10T½ mouse cells. Radiat. Res. 117: 114-127 (1989).

M16 Miller, R.C., C.R. Geard and S.G. Martin. Neutron induced cell cycle-dependent oncogenic transformation of C3H 10T½ cells. Radiat. Res. 142: 270-275 (1995).

M17 Mays, C.W. and R.D. Lloyd. Bone sarcoma incidence vs alpha particle dose. p. 409-430 in: Radiobiology of Plutonium (B.J. Stover and W.S.S Jee, eds.). The J.W. Press, University of Utah, 1972.

M18 Muirhead, C.R., R. Cox, J.W. Stather et al. Estimates of late radiation risks to the UK population. Docs. NRPB 4(4): 15-157 (1993).

M19 Muirhead, C.R. and M.P. Little. Evidence for curvature in the cancer incidence dose-response curve in the Japanese atomic bomb survivors. p. 156-159 in: Health Effects of Low Dose Radiation. British Nuclear Energy Society, London, 1997.

M20 Marshall, C.J. and E. Nigg (eds.). Oncogenes and cell proliferation. Curr. Opin. Genet. Dev. 7(7): (1998).

M21 McCormick, F. Activators and effectors of ras p21 proteins. Curr. Opin. Genet. Dev. 4: 71-76 (1994).

M22 Murray, A.W. Cell cycle checkpoints. Curr. Opin. Cell Biol. 6: 872-876 (1994).

M23 Mitelman, F., F. Mertens and B. Johansson. A breakpoint map of recurrent chromosomal rearrangements in human neoplasia. Nature Genet. 15: 417-474 (1997).

M24 Makos, M., B.D. Nelkin, M.I. Weman et al. Distinct hypermethylation patterns occur at altered chromosome loci in human lung and colon cancer. Proc. Natl. Acad. Sci. U.S.A. 89: 1929-1933 (1992).

M25 Mizuno, T., S. Kyoizumi, T. Suzuki et al. Continued expression of a tissue specific activated oncogene in the early steps of radiation-induced human thyroid carcinogenesis. Oncogene 15: 1455-1460 (1997).

M26 Matsumoto, M., J. Takeda, N. Inoue et al. A novel protein that participates in non-self discrimination of malignant cells by homologous complement. Nature Med. 3: 1266-1269 (1997).

M27 Moolgavkar, S.H. The multistage theory of carcinogenesis and the age distribution of cancer in man. J. Natl. Cancer Inst. 61: 49-52 (1978).

M28 Moolgavkar, S.H. and E.G. Luebeck. Multistage carcinoenesis: population-based model for colon cancer. J. Natl. Cancer Inst. 84: 610-618 (1992).

M29 Moolgavkar, S.H., F.T. Cross, G. Luebeck et al. A two-mutation model for radon-induced lung tumors in rats. Radiat. Res. 121: 28-37 (1990).

M30 Moolgavkar, S.H., A. Dewanji and D.J. Venzon. A stochastic two-stage model for cancer risk assessment. I. The hazard function and the probability of tumor. Risk Anal. 8: 383-392 (1988).

M31 Miyaki, M. Imprinting and colorectal cancer. Nature Med. 4: 1236-1237 (1998).

M32 Mothersill, M. (ed). Genomic Instability. Int. J. Radiat. Biol. 74: 661-770 (1998).

M33 Muto, M., Y. Chen, E. Kubo et al. Analysis of early initiating events in radiation-induced thymic lymphomagenesis. Jpn. J. Cancer Res. 87: 247-257 (1996).

M34 Muckerheide, J. Low Level Radiation Health Effects: Compiling The Data. Radiation, Science and Health Inc., 1998.

M35 Miller, R.C. and E.J. Hall. X-ray dose fractionation and oncogenic transformations in cultured mouse embryo cells. Nature 272: 58-60 (1978).

M36 Miller, R.C., E.J. Hall and H.H. Rossi. Oncogenic transformation of mammalian cells in vitro with split doses of x-rays. Proc. Natl. Acad. Sci. U.S.A. 76: 5755-5758 (1979).

M37 Miller, R.C., D.J. Brenner, C.R. Geard et al. Oncogenic transformation by fractionated doses of neutrons. Radiat. Res. 114: 589-598 (1988).

M38 Miller, R.C., C.R. Geard, D.J. Brenner et al. The effects of temporal distribution of dose on neutron-induced transformation. p. 357-362 in: Cell Transformation and Radiation Induced Cancer (K.H. Chadwick et al., eds.). Adam Hilger, Bristol, 1989.

M39 Maisin, J.R., A. Wambersie, G.B. Gerber et al. The effects of a fractionated gamma irradiation on life shortening and disease incidence in BALB/c mice. Radiat. Res. 94: 359-373 (1983).

M40 Miller, R.C., G. Randers-Pehrson, C.R. Geard et al. The oncogenic transforming potential of the passage of single α particles through mammalian cell nuclei. Proc. Natl. Acad. Sci. U.S.A. 96: 19-22 (1999).

M41 Metting, N.F., S.T. Palayoor, R.M. Macklis et al. Induction of mutations by Bismuth-212 α particles at two genetic loci in human b-lymphoblasts. Radiat. Res. 132: 339-345 (1992).

M42 Moiseenko, V.V., A.A. Edwards and H. Nikjoo. Modelling the kinetics of chromosome exchange formation in human cells exposed to ionising radiation. Radiat. Environ. Biophys. 35: 31-35 (1996).

M43 Moiseenko, V.V., A.A. Edwards, H. Nikjoo et al. The influence of track structure on the understanding of relative biological effectiveness for induction of chromosomal exchanges in human lymphocytes. Radiat. Res. 147: 208-214 (1997).

M44 McDowell, E., J.S. McLaughlin, D.K. Merenyl et al. The respiratory epithelium histogenesis of lung carcinoma in the human. J. Natl. Cancer Inst. 2: 587-606 (1978).

M45 Marsh, J.W. and A. Birchall. Sensitivity analysis of the weighted equivalent lung dose per unit exposure from radon progeny. Radiat. Prot. Dosim. (2000, in press).

M46 Mole, R.H. Leukaemia induction in man by radionuclides and some relevant experimental and human observations. p. 1-13 in: The Radiobiology of Radium and Thorotrast (W. Gossner et al., eds.). Urban and Schwarzenberg, Munich, 1986.

M47 Muirhead, C.R., A.A. Goodill, R.G.E. Haylock et al. Occupational radiation exposure and mortality: second analysis of the National Registry for Radiation Workers. J. Radiol. Prot. 19: 3-26 (1999).

M48 Morlier, J.P., M. Morin, J. Chameaud et al. Importance du rôle du débit de dose sur l'apparition des cancers chez le rat après inhalation de radon. C.R. Acad. Sci., Ser. 3 (315): 436-466 (1992).

M49 Mothersill, C. and C. Seymour. Medium from irradiated human epithelial cells but not human fibroblasts reduces the clonogenic survival of unirradiated cells. Int. J. Radiat. Biol. 71: 421-427 (1997).

M50 Mothersill, C. and C. Seymour. Cell to cell contact during gamma irradiation is not required to induce a bystander effect in normal human keratinocytes: evidence for release during irradiation of a signal controlling survival into the medium. Radiat. Res. 149: 256-262 (1998).

M51 Manti, L., M. Jamali, K.M. Prise et al. Genetic instability in Chinese hamster cells after exposure to X-rays or alpha particles of different mean linear energy transfer. Radiat. Res. 147: 22-28 (1997).

M52 Mitchel, R.E.J., J.S. Jackson, R.A. McCann et al. The adaptive response modifies latency for radiation-induced myeloid leukaemia in CBA/H mice. Radiat. Res. 152: 273-279 (1999).

M53 Maisin, J.R., G.B. Gerber, J. Vankerkom et al. Survival and diseases in C57BL mice exposed to X rays or 3.1 MeV neutrons at an age of 7 or 21 days. Radiat. Res. 146: 453-460 (1996).

M54 Makinodan, T. and S.J. James. T cell potentiation by low dose ionizing radiation: Possible mechanisms. Health Phys. 59: 29-34 (1990).

N1 National Council on Radiation Protection and Measurements. Influence of dose and its distribution in time on dose-response relationships for low-LET radiation. NCRP Report No. 64 (1980).

N2 Nasmyth, K. Viewpoint: putting the cell cycle in order. Science 274: 1643-1645 (1996).

N3 Novelli, M.R., J.A. Williamson, I.P.M. Tomlinson et al. Polyclonal origin of colonic adenomas in an XO/XY patient with FAP. Science 272: 1187-1190 (1996).

N4 Nakamura, Y. Cleaning up on β-catenin. Nature Med. 3: 499-500 (1997).

N5 Naik, P., J. Karrim and D. Hanahan. The rise and fall of apoptosis during multistage tumorigenesis. Genes Dev. 10: 2105-2116 (1996).

N6 National Council on Radiation Protection and Measurements. The relative biological effectiveness of radiations of different quality. NCRP Report No. 104 (1990).

N7 Natarajan, A.T., S.J. Santos, F. Darroudi et al. [137]Cesium-induced chromosome aberrations analyzed by fluorescence in situ hybridization: eight years follow up of the Goiânia radiation accident victims. Mutat. Res. 400: 299-312 (1998).

N8 Nakanishi, M., K. Tanaka, T. Shintani et al. Chromosomal instability in actue myelocytic leukaemia and myelodysplastic syndrome patients among atomic bomb survivors. J. Radiat. Res. 40: 159-167 (1999).

N9 Nikjoo, H., S. Uehara and D.J. Brenner. Track structure calculations in radiobiology: How can we improve them and what can they do? p. 3-10 in: Microdosimetry: An Interdisciplinary Approach (D.T. Goodhead, P. O'Neill and H.G. Menzel, eds.). Royal Society of Chemistry, Cambridge, 1997.

N10 Nagasawa, H. and J.B. Little. Induction of sister chromatid exchanges by extremely low doses of α-particles. Cancer Res. 52: 6394-6396 (1992).

N11 Nagasawa, H. and J.B. Little. Unexpected sensitivity to the induction of mutations by very low doses of alpha particle radiation: evidence for a bystander effect. Radiat. Res. 152: 552-557 (1999).

O1 Oren, M. Relationship of p53 to the control of apoptotic cell death. Semin. Cancer Biol. 5: 221-227 (1994).

O2 Orkin, S.H. Development of the haemopoietic system. Curr. Opin. Genet. Dev. 6: 597-02 (1996).

O3 Ootsuyama, A. and H.A. Tanooka. One hundred percent tumour induction in mouse skin after repeated β irradiation in a limited dose range. Radiat. Res. 115: 488-494 (1988).

O4 Ojeda, F., H.A. Diehl and H. Folch. Radiation-induced membrane damage and programmed cell death: possible interrelationships. Scanning Microsc. 8: 645-651 (1994).

O5 Oghiso, Y., Y. Yamada, I. Haruzo et al. Differential dose responses of pulmonary tumor types in the rat after inhalation of plutonium dioxide aerosols. J. Radiat. Res. 39: 61-72 (1998).

O6 Olivieri, G., J. Bodycote and S. Wolff. Adaptive response of human lymphocytes to low concentrations of radioactive thymidine. Science 223: 594-597 (1984).

O7 Okayasu, R., K. Suetomi, Y. Yu et al. A defect in DNA repair associates with the high incidence of solid tumour formation in Balb/c mice. Cancer Res. (2000, in press).

O8 Ootsuyama, A. and H. Tanooka. Threshold-like dose of local β irradiation repeated throughout the life span of mice for induction of skin and bone tumours. Radiat. Res. 125: 98-101 (1991).

O9 Otake, M. and W.J. Schull. Radiation-related small head sizes among prenatally exposed A-bomb survivors. Int. J. Radiat. Biol. 63: 255-270 (1993).

O10 Otake, M., W.J. Schull and H. Yoshimaru. A review of radiation-related brain damage in the prenatally exposed atomic bomb survivors. RERF CR 4-89 (1990).

P1 Pierce, D.A. and M. Vaeth. The shape of the cancer mortality dose-response curve for atomic bomb survivors. Radiat. Res. 126: 36-42 (1991).

P2 Pierce, D.A., Y. Shimizu, D.L. Preston et al. Studies of the mortality of A-bomb survivors, Report 12, Part I. Cancer: 1950-1990. Radiat. Res. 146: 1-27 (1996).

P3 Preston, D.L., S. Kusumi, M. Tomonaga et al. Cancer incidence in atomic bomb survivors. Part III: leukaemia, lymphoma and multiple myeloma, 1950-1987. Radiat. Res. 137: S68-S97 (1994).

P4 Pottern, L.M., M. Kaplan, P. Larsen et al. Thyroid nodularity after childhood irradiation for lymphoid hyperplasia: a comparison of questionnaire and clinical findings. J. Clin. Epidemiol. 43: 449-460 (1990).

P5 Powell, S.M., N. Zilz, Y. Beazer-Barclay et al. APC mutations occur early during colorectal tumorigenesis. Nature 359: 235-237 (1992).

P6 Pritchard, C. and M. McMahon. Raf revealed in life or death decisions. Nature Genet. 16: 214-215 (1997).

P7 Pierce, D.A., D.O. Stram and M. Vaeth. Allowing for random errors in radiation dose estimates for the atomic bomb survivor data. Radiat. Res. 123: 275-284 (1990).

P8 Papworth, D.G. and E.V. Hulse. Dose-response models for the radiation-induction of skin tumours in mice. Int. J. Radiat. Biol. 44: 423-431 (1983).

P9 Patterson, H.W. Setting standards for radiation protection: the process appraised. Health Phys. 72: 450-457 (1997).

P10 Peto, R. Epidemiology, multistage models, and short-term mutagenicity tests. p. 1403-1428 in: Origins of Human Cancer (H.H. Hiatt and J.A. Winsten, eds.). Cold Spring Harbor Laboratory, Cold Spring Harbor, 1977.

P11 Pierce, D.A., Y. Shimizu, D.L. Preston et al. Response to the letter of M.P. Little (letter). Radiat. Res. 148: 400-401 (1997).

P12 Pirrotta, V. PcG complexes and chromatin silencing. Curr. Opin. Genet. Dev. 7: 249-258 (1997).

P13 Paretzke, H.G. Physical aspects of radiation quality. p. 514-522 in: Low Dose Radiation: Biological Bases of Risk Assessment (K.F. Baverstock and J.W. Stather, eds.). Taylor and Francis, London, 1989.

P14 Pierce, D.A. and M. Vaeth. Cancer risk estimation from the A-bomb survivors: extrapolation to low doses, use of relative risk models and other uncertainties. p. 54-69 in: Low Dose Radiation: Biological Bases of Risk Assessment (K.F. Baverstock and J.W. Stather, eds.). Taylor and Francis, London, 1989.

P15 Pohl-Rüling, J., P. Fischer, O. Haas et al. Effect of low-dose acute X-irradiation on the frequencies of chromosomal aberrations in human peripheral lymphocytes in vitro. Mutat. Res. 110: 71-82 (1983).

P16 Preston, D.L., H. Kato, K. Kopecky et al. Life Span Study, Cancer mortality among A-bomb survivors in Hiroshima and Nagasaki 1950-1982, Report 10, Part 1. RERF TR 1-86 (1986).

P17 Potten, C.S. Stem cells in gastrointestinal epithelium: numbers, characteristics and death. Philos. Trans. R. Soc. Lond., Ser. B: Biol. Sci. 353: 821-830 (1998).

P18 Potten, C.S. Communication to the UNSCEAR Secretariat (1999).

P19 Priest, N.D. Risk estimates for high LET alpha-irradiation of skeletal tissues: problems with current methods? p. 423-429 in: Health Effects of Internally Deposited Radionuclides: Emphasis on Radium and Thorium (G. van Kaick, A. Karaoglou and A.M. Kellerer, eds.). World Scientific, London, 1995.

P20 Purrott, R.J., A.A. Edwards, D.C. Lloyd et al. The induction of chromosome aberrations in human lymphocytes by in vitro irradiation with α-particles from plutonium-239. Int. J. Radiat. Biol. 38: 277-284 (1980).

P21 Prise, K.M., O.V. Belyakov, M. Folkard et al. Studies of bystander effects in human fibroblasts using a charged particle microbeam. Int. J. Radiat. Biol. 74: 793-798 (1998).

P22 Pierce, D.A. and M.L. Mendelsohn. A model for radiation-related cancer suggested by atomic bomb survivor data. Radiat. Res. 152: 642-654 (1999).

P23 Pohl-Rüling, J., P. Fischer, D.C. Lloyd et al. Chromosomal damage induced in human lymphocytes by low doses of D-T neutrons. Mutat. Res. 173: 267-272 (1986).

R1 Raabe, O.G. Three-dimensional models of risk from internally deposited radionuclides. Internal radiation dosimetry. Health Physics Society, 1994 Summer School (1994).

R2 Rabbitts, T.H. Chromosomal translocations in human cancer. Nature 372: 143-149 (1994).

R3 Raff, M.C. Social controls on cell survival and cell death. Nature 356: 397-399 (1992).

R4 Rowland, R.E., A.F. Stehney and H.F. Lucas. Dose-response relationships for female radium dial workers. Radiat. Res. 76: 368-383 (1978).

R5 Russell, W.L., L.B. Russell and E.M. Kelly. Radiation dose rate and mutation frequency. Science 128: 1546-1550 (1958).

R6 Russell, W.L. and E.M. Kelly. Mutation frequencies in male mice and the estimation of genetic hazards of radiation in man. Proc. Natl. Acad. Sci. U.S.A. 79: 542-544 (1982).

R7 Rundo, J., A.T. Keane and M.A. Essling. Long-term Retention of Radium in Female Former Dial Workers. pages 77-85. CMTP Press Ltd., Lancaster, 1985.

R8 Ron, E., B. Modan, D.L. Preston et al. Thyroid neoplasia following low-dose radiation in childhood. Radiat. Res. 120: 516-531 (1989).

R9 Ron, E., J.H. Lubin, R.E. Shore et al. Thyroid cancer after exposure to external radiation: a pooled analysis of seven studies. Radiat. Res. 141: 259-277 (1995).

R10 Ranier, S., A. Johnson, C.J. Dobry et al. Relaxation of imprinting in human cancer. Nature 362: 749-751 (1993).

R11 Roth, S.Y. Something about silencing. Nature Genet. 14: 3-4 (1996).

R12 Rubin, H. Cancer as a dynamic development disorder. Cancer Res. 45: 2935-2942 (1985).

R13 Reznikoff, C.A., J.S. Bertram, D.W. Brankow et al. Quantitative and qualitative studies of chemical transformation of cloned C3H mouse embryo cells sensitive to post confluence inhibition of cell division. Cancer Res. 33: 3239-3249 (1973).

R14 Raabe, O.G., M.R. Culbertson, R.G. White et al. Lifetime radiation effects in beagles injected with ^{226}Ra as young adults. p. 313-318 in: Health Effects of Internally Deposited Radionuclides: Emphasis on Radium and Thorium (G. van Kaick, A. Karaoglou and A.M. Kellerer, eds.). World Scientific, London, 1995.

R15 Ron, E., B. Modan, D. Preston et al. Radiation-induced skin carcinomas of the head and neck. Radiat. Res. 125: 318-325 (1991).

R16 Rowland, R.E. Radium in humans: a review of US studies. ANL/ER-3, UV-408 (1994).

R17 Rowland, R.E. Dose-response relationships for female radium dial workers: a new look. p. 135-143 in: Health Effects of Internally Deposited Radionuclides: Emphasis on Radium and Thorium (G. van Kaick, A. Karaoglou and A.M. Kellerer, eds.). World Scientific, London, 1995.

R18 Rossi, H.H. Microscopic energy distribution in irradiated matter. p. 43 in: Radiation Dosimetry, Volume 1 (F.H. Atix and W.C. Roesch, eds.). Academic Press, New York, 1968.

R19 Russell, W.L. and E.M. Kelly. Mutation frequencies in female mice and the estimation of genetic hazards or radiation in women. Proc. Natl. Acad. Sci. U.S.A. 74: 3523-3527 (1977).

R20 Russell, W.L. The genetic effects of radiation. p. 487-500 in: Peaceful Uses of Atomic Energy. STI/PUB/300, Volume 13. IAEA, Vienna, 1972.

R21 Ramsey, M.J., D.H. Moore II, J.F. Briner et al. The effects of age and lifestyle factors on the accumulation of cytogenetic damage as measured by chromosome painting. Mutat. Res. 338: 95-106 (1995).

R22 Redpath, J.L. and R.J. Antoniono. Induction of an adaptive response against spontaneous neoplastic transformation in vitro by low-dose gamma radiation. Radiat. Res. 149: 517-520 (1998).

S1 Shimizu, Y., H. Kato and W.J. Schull. Studies of the mortality of A-bomb survivors. 9. Mortality, 1950-1985: Part 2: Cancer mortality based on the recently revised doses (DS 86). Radiat. Res. 121: 120-141 (1990).

S2 Shore, R.E., N. Hildreth, P. Ovoretsky et al. Thyroid cancer among persons given x-ray treatment in infancy for an enlarged thymus gland. Am. J. Epidemiol. 137: 1068-1080 (1993).

S3 Skuse, G.R. and J.W. Ludlow. Tumour suppressor genes in disease and therapy. Lancet 345: 902-906 (1995).

S4 Smith, J.R. and O.M. Pereira-Smith. Replicative senescence: implications for in vivo ageing and tumour suppression. Science 273: 63-67 (1996).

S5 Simmons, J.A. and D.E. Watt. Radiation Protection Dosimetry - A Radical Reappraisal. Medical Physics Publishing, Madison, Wisconsin, U.S.A., 1999.

S6 Shore, R.E. Issues and epidemiological evidence regarding radiation-induced thyroid cancer. Radiat. Res. 131: 98-111 (1992).

S7 Stewart, A.M., J. Webb and D. Hewitt. A survey of childhood malignancy. Br. Med. J. 1: 1495-1508 (1958).

S8 Shin, C., L.C. Padhy and R.A. Weinberg. Transforming genes of carcinomas and neuroblastomas introduced into mouse fibroblasts. Nature 290: 261-264 (1981).

S9 Schechtman, L.M., E. Kiss, J. McCorvill et al. A method for the amplification of chemically induced transformation in C3H 10T½ clone 8 cells: its use as a potential screening assay. J. Natl. Cancer Inst. 79: 487-498 (1987).

S10 Searle, A.G. Mutation induction in mice. Adv. Radiat. Biol. 4: 131-207 (1974).

S11 Sanders, C.L. and D.L. Lundgren. Pulmonary carcinogenesis in the F344 and Wistar rats after inhalation of plutonium dioxide. Radiat. Res. 144: 206-214 (1995).

S12 Saunders, C.L., K.E. Lauhala and K.E. McDonald. Lifespan studies in rats exposed to ^{239}PuO$_2$. III. Survival and lung tumours. Int. J. Radiat. Biol. 64: 417-430 (1993).

S13 Sankaranarayanan, K. Ionising radiation and genetic risks IV. Current methods, estimates of risk of Mendelian disease, human data and lessons from biochemical and molecular studies of mutations. Mutat. Res. 258: 99-122 (1991).

S14 Speicher, M.R., S.G. Ballard and D.C. Ward. Karyotyping human chromosomes by combinational multi-fluor FISH. Nature Genet. 12: 368-375 (1996).

S15 Shibata, D., M.A. Peinado, Y. Ionov et al. Genomic instability in repeated sequences is an early somatic event in colorectal tumorigenesis that persists after transformation. Nature Genet. 6: 273-281 (1994).

S16 Stilgenbaur, S., C. Schuffner, A. Litterst et al. Biallelic mutations in the ATM gene in T-prolymphocytic leukaemia. Nature Med. 3: 1155-1159 (1997).

S17 Sidransky, D. Is human patched the gatekeeper of common skin cancers? Nature Genet. 14: 7-8 (1996).

S18 Shoemaker, A.R., K.A. Gould, C. Luongo et al. Studies of neoplasia in the Min mouse. Biochim. Biophys. Acta 1332: F25-F48 (1997).

S19 Strong, L.C. and W.R. Williams. The genetic implications of long-term survival of childhood cancer. Am. J. Pediatr. Hematol./Oncol. 9: 99-103 (1987).

S20 Shore, R.E., N. Hildreth, E. Woodard et al. Breast cancer among women given X-ray therapy for acute postpartum mastitis. J. Natl. Cancer Inst. 77: 689-696 (1986).

S21 Stewart, A.M. and G.W. Kneale. Radiation dose effects in relation to obstetric x-rays and childhood cancers. Lancet i: 1185-1188 (1970).

S22 Sugahara, T. A radiation protection system aimed at cancer prevention: a proposal. in: Proceedings of DAE Symposium on Recent Advances in Genetic Epidemiology and Population Monitoring, Madras, India, 1998.

S23 Schiestl, R.H., F. Khogali and N. Carls. Reversion of the mouse pink-eyed unstable mutation induced by low doses of x-rays. Science 266: 1573-1576 (1994).

S24 Selby, P.B., S.S. Lee, E.M. Kelly et al. Specific-locus experiments show that female mice exposed near the time of birth to low-LET ionizing radiation exhibit both a low mutational response and a dose rate effect. Mutat. Res. 249: 351-367 (1991).

S25 Shadley, J.D. and S. Wolff. Very low doses of X-rays can cause human lymphocytes to become less susceptible to ionizing radiation. Mutagenesis 2: 95-96 (1987).

S26 Shadley, J.D., V. Afzal and S. Wolff. Characterization of the adaptive response to ionizing radiation induced by low doses of X-rays to human lymphocytes. Radiat. Res. 111: 511-517 (1987).

S27 Sanderson, B.J.S. and A.A. Morley. Exposure of human lymphocytes to ionizing radiation reduces mutagenesis by subsequent ionizing radiation. Mutat. Res. 164: 347-351 (1986).

S28 Sparrow, A.H., A.G. Underbrink and H.H. Rossi. Mutations induced in tradescania by small doses of x rays and neutrons: analysis of dose response curves. Science 176: 916-918 (1972).

S29 Szumiel, I. Review: ionising radiation-induced cell death. Int. J. Radiat. Biol. 66: 329-341 (1994).

S30 Schervish, M.J. Theory of Statistics. Springer-Verlag, New York, 1995.

S31 Shore, R.E., R.E. Albert, M. Reed et al. Skin cancer incidence among children irradiated for ringworm of the scalp. Radiat. Res. 100: 192-204 (1984).

S32 Savage, J.R.K. Sites of radiation induced chromosome exchanges. Curr. Top. Radiat. Res. VI: 129-196 (1970).

S33 Sullivan, M.F., P.L. Hackett, L.A. George et al. Irradiation of the intestine by radioisotopes. Radiat. Res. 13: 343-355 (1960).

S34 Schmutte, C. and R. Fishel. Genomic instability: First step to carcinogenesis. Anticancer Res. 19: 4665-4696 (1999).

S35 Sinclair, W.K. Experimental RBE value of high LET radiations at low doses and the implications for quality factor assignment. Radiat. Prot. Dosim. 13: 319-326 (1985).

S36 Sevan'kaev, A.V., E.A. Zherbin, N.V. Luchnik et al. Cytogenetic effects produced by neutrons in lymphocytes of human peripheral blood in vitro. 1. Dose-response dependence of neutrons of different energies for different types of chromosomal aberrations. Genetika 15: 1046-1060 (1979). (In Russian).

S37 Searle, A.G., C.V. Beechey, D. Green et al. Cytogenetic effects of protracted exposures to alpha particles from ^{239}Pu and to gamma rays from ^{60}Co compared in male mice. Mutat. Res. 41: 297-310 (1976).

S38 Sasaki, S. Influence of the age of mice at exposure to radiation on life-shortening and carcinogenesis. J. Radiat. Res. 2 (Suppl.): 73-85 (1991).

S39 Sasaki, S. and N. Fukuda. Dose-response relationship for induction of solid tumours in female B6C3F$_1$ mice irradiated neonatally with a single dose of gamma rays. J. Radiat. Res. 40: 229-241 (1999).

S40 Silver, A., J. Moody, R. Dunford et al. Molecular mapping of chromosome 2 deletions in murine radiation-induced AML localises a putative tumour suppressor gene to a 1.0 cM region homologous to human chromosome segment 11p 11-12. Genes Chrom. Cancer 24: 95-104 (1999).

S41 Sadekova, S., S. Lehnert, B. Chandrasekar et al. Induction of a PBP74/mortalin/Grp75, a member of the hsp 70 family, by low doses of ionizing radiation. A possible role in induced radioresistance. Int. J. Radiat. Biol. 72: 653-660 (1997).

S42 Sambani, C., H. Thomou and P. Kitsiou. Stimulatory effect of low dose x-irradiation on the expression of the human T lymphocyte CD2 surface antigen. Int. J. Radiat. Biol. 70: 711-717 (1996).

S43 Shankar, B., S. Premachandran, P. Bharambe et al. Modification of immune response by low dose ionizing radiation: role of apoptosis. Immunol. Lett. 68: 237-245 (1999).

T1 Takeichi, M. Cadherins in cancer: implications for invasion and metastasis. Curr. Opin. Cell Biol. 5: 806-811 (1993).

T2 Thacker, J. Radiation-induced mutation in mammalian cells at low doses and dose rates. Adv. Radiat. Biol. 16: 77-117 (1992).

T3 Thomas, D.C. A model for dose rate and duration of exposure effects in radiation carcinogenesis. Environ. Health Perspect. 87: 163-171 (1990).

T4 Thompson, D.E., K. Mabuchi, E. Ron et al. Cancer incidence in atomic bomb survivors. Part II: solid tumors, 1958-1987. Radiat. Res. 137: S17-S67 (1994).

T5 Trosko, J.E., C.L. Chang, B.V. Madhukar et al. Intercellular communication: a paradigm for the interpretation of the initiation/promotion/progression model of carcinogenesis. in: Chemical Carcinogenesis: Modulation and Combination Effects (J.C. Arocs, ed.). Academic Press, New York, 1992.

T6 Tan, W-Y. Stochastic Models of Carcinogenesis. Marcel Dekker, New York, 1991.

T7 Tucker, H.G. A stochastic model for a two-stage theory of carcinogenesis. p. 387-403 in: Fifth Berkeley Symposium on Mathematical Statistics and Probability. University of California Press, Berkeley, 1967.

T8 Tomlinson, I. and W. Bodmer. Selection, the mutation rate and cancer: ensuring that the tail does not wag the dog. Nature Med. 5: 11-12 (1999).

T9 Tao, Z.-F., H. Kato, Y.-R. Zha et al. Study on cancer mortality among the residents in high background radiation area of Yangjiag, China. p. 249-254 in: High Levels of Natural Radiation 96: Radiation Dose and Health Effects (L. Wei et al., eds.). Elsevier, Amsterdam. 1997.

T10 Tirmarche, M., A. Rannou, A. Mollié et al. Epidemiological study of regional cancer mortality in France and natural radiation. Radiat. Prot. Dosim. 24: 479-482 (1988).

T11 Tubiana, M. The report of the French Academy of Sciences: Problems associated with the effects of low doses of ionising radiation. J. Radiol. Prot. 18(4): 243-248 (1998).

T12 Thomas, R.G. Tumorigenesis in the US radium luminizers: how unsafe was this occupation? p. 145-148 in: Health Effects of Internally Deposited Radionuclides: Emphasis on Radium and Thorium (G. van Kaick, A. Karaoglou, A.M. Kellerer, eds.). World Scientific, Singapore, 1995.

T13 Tanooka, H. and A. Ootsuyama. Radiation carcinogenesis in mouse skin and its threshold-like response. J. Radiat. Res. 2 (Suppl.): 195-201 (1991).

T14 Thacker, J. The nature of mutants induced by ionizing radiation in cultured hamster cells. Mutat. Res. 160: 267-275 (1986).

T15 Trump, B.F., E.M. McDowell, F. Glavin et al. The respiratory epithelium III. Histogenesis of epidermoid metaplasia and carcinoma in situ in the human. J. Natl. Cancer Inst. 61: 563-575 (1978).

T16 Thacker, J., A. Stretch and M.A. Stephens. Mutation and inactivation of cultured mammalian cells exposed to beams of accelerated heavy ions. II. Chinese hamster V79 cells. Int. J. Radiat. Biol. 36: 137-148 (1979).

T17 Tao, Z.F., S. Akiba, Y.R. Zha et al. Analysis of data from investigation of cancer mortality in high background radiation area of Yangjiang, China (1987-1995). Chin. J. Radiol. Med. Prot. 19(2): 75-82 (1999).

U2 United Nations. Sources and Effects of Ionizing Radiation. United Nations Scientific Committee on the Effects of Atomic Radiation, 1994 Report to the General Assembly, with scientific annexes. United Nations sales publication E.94.IX.11. United Nations, New York, 1994.

U3 United Nations. Sources and Effects of Ionizing Radiation. United Nations Scientific Committee on the Effects of Atomic Radiation, 1993 Report to the General Assembly, with scientific annexes. United Nations sales publication E.94.IX.2. United Nations, New York, 1993.

U4 United Nations. Sources, Effects and Risks of Ionizing Radiation. United Nations Scientific Committee on the Effects of Atomic Radiation, 1988 Report to the General Assembly, with annexes. United Nations sales publication E.88.IX.7. United Nations, New York, 1988.

U5 United Nations. Genetic and Somatic Effects of Ionizing Radiation. United Nations Scientific Committee on the Effects of Atomic Radiation, 1986 Report to the General Assembly, with annexes. United Nations sales publication E.86.IX.9. United Nations, New York, 1986.

U6 United Nations. Ionizing Radiation: Sources and Biological Effects. United Nations Scientific Committee on the Effects of Atomic Radiation, 1982 Report to the General Assembly, with annexes. United Nations sales publication E.82.IX.8. United Nations, New York, 1982.

U7 United Nations. Sources and Effects of Ionizing Radiation. United Nations Scientific Committee on the Effects of Atomic Radiation, 1977 Report to the General Assembly, with annexes. United Nations sales publication E.77.IX.1. United Nations, New York, 1977.

U8 United Nations. Ionizing Radiation: Levels and Effects. Volume I: Levels, Volume II: Effects. United Nations Scientific Committee on the Effects of Atomic Radiation, 1972 Report to the General Assembly, with annexes. United Nations sales publication E.72.IX.17 and 18. United Nations, New York, 1972.

U14 Ullrich, R.L. Tumour induction in BALB/c female mice after fission neutron or γ-irradiation. Radiat. Res. 93: 506-515 (1983).

U15 Ullrich, R.L., M.C. Jernigan, G.E. Cosgrove et al. The influence of dose and dose rate on the incidence of neoplastic disease in RFM mice after neutron irradiation. Radiat. Res. 68: 115-131 (1976).

U16 Ullrich, R.L. and J.B. Storer. Influence of γ irradiation on the development of neoplastic disease in mice. I. Reticular tissue tumours. Radiat. Res. 80: 303-316 (1979).

U17 Ullrich, R.L. and J.B. Storer. Influence of γ irradiation on the development of neoplastic disease in mice. II. Solid tumours. Radiat. Res. 80: 317-324 (1979).

U18 Ullrich, R.L. and J.B. Storer. Influence of γ irradiation on the development of neoplastic disease in mice. III. Dose-rate effects. Radiat. Res. 80: 325-342 (1979).

U19 Ullrich, R.L., M.C. Jernigan, L.C. Satterfield et al. Radiation carcinogenesis: time-dose relationships. Radiat. Res. 111: 179-184 (1987).

U20 Ullrich, R.L. and R.J. Preston. Myeloid leukaemia in male RFM mice following irradiation with fission spectrum neutrons or γ-rays. Radiat. Res. 109: 165-170 (1987).

U21 Upton, A.C., M.L. Randolph and J.W. Conklin. Late effects of fast neutrons and gamma-rays in mice as influences by the dose rate of irradiation: induction of neoplasia. Radiat. Res. 41: 467-491 (1970).

U22 Upton, A.C. Radiological effects of low doses. Implications for radiological protection. Radiat. Res. 71: 51-74 (1977).

U23 Upton, A.C., R.E. Albert, F.J. Burns et al. Radiation Carcinogenesis. Elsevier, New York, 1986.

U24 Ullrich, R.L. and J.B. Storer. Influence of dose, dose rate and radiation quality on radiation carcinogenesis and life shortening in RFM and BALB/c mice. p. 95-113 in: Late Effects of Ionizing Radiation, Volume II. IAEA, Vienna, 1978.

U25 Ullrich, R.L. and B. Ponnaiya. Radiation-induced instability and its relation to radiation carcinogenesis. Int. J. Radiat. Biol. 74: 747-754 (1998).

U26 Ullrich, R.L., N.D. Bowles, L.C. Satterfield et al. Strain-dependent susceptibility to radiation-induced mammary cancer is a result of differences in epithelial cell sensitivity to transformation. Radiat. Res. 146: 353-355 (1996).

V1 Vogelstein, B. and K.W. Kinzler. The multistep nature of cancer. Trends Genet. 9: 138-141 (1993).

V2 Vogelstein, B. and K.W. Kinzler. The Genetic Basis of Human Cancer. McGraw-Hill, New York, 1998.

V3 van Zwieten, M.J. The Rat as Animal Model in Breast Cancer Research. Martinus Nijhoff, Boston, 1984.

V4 Vral, A., H. Cornelisen, H. Thierens et al. Apoptosis induced by fast neutrons versus ^{60}Co gamma rays in human peripheral blood lymphocytes. Int. J. Radiat. Biol. 73: 289-295 (1998).

V5 Villunger, A. and A. Strasser. The great excape: is immune evasion required for tumour progression? Nature Med. 5: 874-875 (1999).

W1 Wagner, R., E. Schmid and M. Bauchinger. Application of conventional and FPG staining for the analysis of chromosome aberrations induced by low levels of dose in human lymphocytes. Mutat. Res. 109: 65-71 (1983).

W2 Williams, B.O. and T. Jacks. Mechanisms of carcino-genesis and the mutant mouse. Curr. Opin. Genet. Dev. 6: 65-70 (1996).

W3 Weinberg, R.A. Oncogenes and tumour suppressor genes. CA Cancer J. Clin. 44: 160-170 (1994).

W4 Williams, E.D. Thyroid cancer and the Chernobyl accident. in: Health Effects of Low Dose Radiation. British Nuclear Energy Society, London, 1997.

W5 Willerford, D.M., S. Wojciech and F. Alt. Developmental regulation of V(D)J recombination and lymphocyte differentiation. Curr. Opin. Genet. Dev. 6: 603-609 (1996).

W6 Wolff, S. Adaptive responses. p. 103 in: Low Doses of Ionizing Radiation: Biological Effects and Regulatory Control. IAEA, Vienna, 1998.

W7 Ward, J.F. Response to commentary by D. Billen. Radiat. Res. 126: 38-57 (1991).

W8 Williams, J.P., J.E. Coggle, M.W. Charles et al. Skin carcinogenesis in the mouse following uniform and non-uniform beta irradiation. Br. J. Radiol. (Suppl.) 19: 61-64 (1986).

W9 Weiss, H.A., S.C. Darby and R. Doll. Cancer mortality following X-ray treatment for ankylosing spondylitis. Int. J. Cancer 59: 327-338 (1994).

W10 Woloschak, G.E., T. Pauneska, C-M. Chang-Liu et al. Changes in gene expression associated with radiation exposure. p.545-547 in: Radiation Research 1895-1995 (U. Hagen, D. Harder, H. Jung et al., eds). H. Sturtz AG, Wurzburg, 1995.

W11 Walter, S.D., J.W. Meigs and J.F. Heston. The relationship of cancer incidence to terrestrial radiation and population density in Connecticut, 1935-1974. Am. J. Epidemiol. 123: 1-14 (1986).

W12 Wang, J. Statistical analysis of cancer mortality data of high background radiation areas in Yiangjiang. Chin. J. Radiol. Med. Prot. 13: 291-294 and 358 (1993).

W13 Wolff, S. Adaptive responses. Environ. Health Perspect. 106: 277-283 (1998).

W14 Wiencke, J.K., V. Afzal, G. Olivieri et al. Evidence that the [3H]thymidine-induced adaptive response of human lymphocytes to subsequent doses of X-rays involves the induction of a chromosomal repair mechanism. Mutagenesis 1: 375-380 (1986).

W15 Wolff, S., J.K. Wiencke, V. Afzal et al. The adaptive response of human lymphocytes to very low doses of ionizing radiation: A case of specific proteins. p. 446-454 in: Low Dose Radiation: Biological Bases of Risk Assessment (K.F. Baverstock and J.W. Stather, eds.). Taylor and Francis, London, 1989.

Y1 Yoshimoto, Y., R. Delongchamp and K. Mabuchi. *In-utero* exposed atomic bomb survivors: cancer risk update. Lancet 344: 345-346 (1994).

Y2 Yoshimoto, Y., H. Kato and W.J. Schull. Risk of cancer among children exposed *in utero* to A-bomb radiations, 1950-84. Lancet ii: 665-669 (1988).

Y3 Yin, Y., Y. Terauchi, G. Solomon et al. Involvement of p85 in p53-dependent apoptotic response to oxidative stress. Nature 391: 707-710 (1998).

Y4 Yonezawa, M., J. Misonoh and Y. Hosokawa. Two types of x-ray-induced radioresistance in mice - Presence of four dose ranges with distinct biological effects. Mutat. Res. 358: 237-243 (1996).

Y5 Yessner, R. The dynamic histopathologic spectrum of lung cancer. Yale J. Biol. Med. 54: 447-456 (1981).

Y6 Yamamoto, O., T. Seyama, H. Itoh et al. Oral administration of tritiated water (HTO) in mouse. III. Low dose-rate irradiation and threshold dose-rate for radiation risk. Int. J. Radiat. Biol. 73: 535-541 (1998).

Z1 Zimmerman, D. Thyroid neoplasia in children. Curr. Opin. Pediatr. 9: 413-418 (1997).

Z2 Zheng, P., Y. Guo, Q. Niu et al. Proto-oncogene *PML* controls genes devoted to MHC class 1 antigen presentation. Nature 396: 373-376 (1998).

Z3 Zhou, P.K., X.Y. Liu, W.Z. Sun et al. Cultured mouse SR-1 cells exposed to low dose of gamma-rays become less susceptible to the induction of mutagenesis by radiation as well as bleomycin. Mutagenesis 8(2): 109-111 (1993).

ANNEX H

Combined effects of radiation and other agents

CONTENTS

INTRODUCTION

1.   Living organisms are exposed to numerous natural and man-made agents that interact with molecules, cells, and tissues, causing reversible deviations from homeostatic equilibrium or irreversible damage. Many aspects of aging and many diseases are thought to stem from exogenous and endogenous deleterious agents acting on key components of cells within the body. Because of the worldwide proliferation of a number of man-made agents and the increasing release of natural agents due to human activities into the environment, the assessment of toxicity, carcinogenicity, and mutagenicity of a specific chemical, physical, or biological agent is, in fact, a study of combined exposures [G10]. Although this has been recognized for a long time, risk assessment is generally performed with the simplifying assumption that the agent under study acts largely independently of other substances. Studies of interactions have indicated, however, that, at least at high exposures, the action of one agent can be influenced by simultaneous exposures to other agents. The combined effects may be greater or smaller than the sum of the effects from separate exposures to the individual agents. The action at low levels of exposure, which are commonly encountered in occupational and environmental situations, is less clear. Continued, critical review of studies on the effects of combined exposures to radiation and other toxic agents is necessary, particularly at the lowest levels of exposure, to be sure that any modifications of the radiation effects caused by other environmental or occupational agents are recognized and, as far as possible, taken into account in risk assessments.

2.   In the UNSCEAR 1982 Report [U6], the Committee discussed the problem of the combined action of radiation with other agents. In reviewing the approaches and the many reports in which synergisms were claimed, the Committee noted that, in general, an adequate conceptual framework was lacking. Despite many reports showing the potential importance of interactions between different agents under specific conditions, mostly occupational, information on the mechanisms of action was largely missing, and the methodologies for data analysis in different branches of the biological sciences were based on different approaches. The UNSCEAR 1982 Report concluded that it was not possible to document clear cases of interaction that could justify substantial modifications to the existing radiation risk estimates. The Committee felt that systematic investigations of combined effects were needed to allow this field to move forward from its early stage of development.

3.   The objective of this Annex is to update the Committee's previous review of this subject [U6] and to reconsider whether interactions of radiation and one or more other agents should be taken into account in evaluating radiation risks at low doses. To achieve this objective, the following subjects are considered:

(a)   the concepts of doses, targets, and detriments currently used in risk assessments of radiation and chemical agents;

(b)   recent developments from research on the possible mechanisms of combined effects from low-level exposures to radiation and other agents;

(c)   results and evaluations of data from experimental and epidemiological studies;

(d)   mechanistic models applied to experimental and epidemiological results, with generalizations and extrapolations that might be pertinent to low and chronic exposures;

(e)   concepts and approaches in other areas of biological science (for example, molecular biology and toxicology) that could suggest ways to develop databases and to identify and assess the effects of interactions important for human populations.

4.   Combined effects must be viewed in the light of the considerable insights gained from wider studies of cancer induction (see Annex E, "Mechanisms of radiation oncogenesis", in the UNSCEAR 1993 Report [U3] and Annex G, *"Biological effects at low radiation doses"*), heritable defects (see Annex G, "Hereditary effects of radiation" in the UNSCEAR 1993 Report [U3]), and DNA integrity (see Annex F, *"DNA repair and mutagenesis"*). Where necessary, the following text refers to these and other Annexes.

5.   Since at low levels of exposure, the main endpoints from ionizing radiation alone and from its interaction with other agents are stochastic in nature, this Annex will mainly focus on this type of effect and consider cancer induction, mutation and the possibility of prenatal effects. Several specific areas where the combined action of high doses of radiation and chemical agents are known to lead to considerable deviation from additivity will also be considered but only in so far as they help to elucidate the mechanisms of combined exposures. These areas include the interaction of chemotherapeutic compounds and sensitizers to enhance radiation effects in clinical radiotherapy, the effects of protective agents on acute radiation exposure, and stimulatory responses to radiation (reviewed in the UNSCEAR 1994 Report [U2]). The endpoints of interest in these situations of high-dose exposure are deterministic effects.

6.   The Annex begins by introducing the problem of combined effects, considering the additivity or non-additivity of biological effects and the possible differences between radiation and chemical carcinogenesis. This is followed by concepts and definitions of physical and biological dosimetry for radiation and other agents. Interactions of other agents in the development of radiation-induced cancer are then considered from a mechanistic point of view. A very important part of the Annex is a review of data on the effects of specific combined exposures on carcinogenesis. This is followed by a chapter on interactions in humans that produce effects other than cancer. Finally, conclusions are drawn and recommendations are offered. A detailed account of the combined effects of radiation and specific physical, chemical, and biological agents is provided in the Appendix.

I. IDENTIFYING INTERACTIONS AND COMBINED EFFECTS

A. SCOPE OF THE PROBLEM

7.   When discussing combined effects, it is of utmost importance to provide clear definitions and terminology. Multiple-agent toxicology uses many concepts the nomenclature for which is not unambiguous. Different names are sometimes used for the same phenomenon, and sometimes the same name is used for different mechanisms. The confusion arises in part because the concepts were developed in different disciplines, such as pharmacology, toxicology, biology, statistics, epidemiology, and radiation biology. Starting from different basic assumptions and with different aims in mind, attempts are made to describe the effects of combined exposures to chemical and physical agents. The confusing terminology inhibits clear understanding and thwarts the comparison of different investigations and results. In this Chapter some basic problems concerning combined exposures are discussed.

1. Additivity and deviations from additivity

8.   One of the basic questions surrounding the combined effects of two agents is the question of whether the effect of a combined exposure to two or more agents is the same as or different from the sum of the effects of each agent separately. Many terms and synonyms are used to indicate the result (Table 1). They are, in general, based on deviations from the expected outcome (additivity). On a descriptive level, two classes of combined effects can be considered. In the first case, both ionizing radiation and the other agent (or agents) are deleterious on their own and combine to produce an effect not directly predictable from the single exposures. In the second case, only ionizing radiation produces an effect, but its nature or severity may be modified by the other agent, which is non-toxic by itself.

Table 1
Terms and synonyms for combined effects

Effect smaller than anticipated	*Effect as anticipated*	*Effect larger than anticipated*
Antagonism	Additivity	Augmentation
Antergism	Additivism	Enhancement
Depotentiation	Independence	Positive interaction
Desensitation	Indifference	Potentiation
Inhibition	Non-interaction	Sensitation
Infra-additivity	Summation	Superadditivity
Negative interaction	Zero-interaction	Supra-additivism
Negative synergism		Synergism
Subadditivity		Synergy

---

9.   On a mechanistic level, insights gained in more recent years indicate that a much more refined classification may be needed. The main classes of genotoxic and non-genotoxic agents must be considered in relation to specific targets of action. For example, a chemical may act specifically at the site of a radiation-induced lesion, modifying DNA repair fidelity, or it may modify cell growth, strongly influencing the clonal expansion of precancerous cells. The many possibilities for interaction are related to the complexity of the development of the radiation effect and the many steps involved in carcinogenesis. These steps are prone to the influence of many classes of agents, both endogenous and environmental. The multi-step process and the many levels of interaction to be considered are schematically depicted in Figure I. In view of this complexity, it is not surprising that many models, both descriptive and mechanistic, have been developed to describe the combined effects of exposures to different agents [B11, L2, L8, L28, M16, S15, S16, S23, S25, Z1]. In the UNSCEAR 1982 Report [U6], the Committee reviewed these approaches.

10.   Although classical epidemiology is important in identifying critical combined effects, it has little potential for dissecting such interactions from the complex interplay possible among the undocumented (and sometimes unknown) exposures that the individuals in these studies incur during their lifetimes. In epidemiological studies, effects that may be associated with exposures to specific agents or circumstances may be the result of interactions among components of a mixture of agents and may have resulted from, or been influenced by, previous exposures. The emerging field of molecular epidemiology may be able to address such questions in the near future.

11.   Most knowledge of interaction effects has been provided by experimental studies. These studies have an advantage over epidemiological studies: they retain control of

(a)   the population (e.g. selection of systems ranging from DNA to intact animals and of species, strain, age, gender and previous exposure history);

(b)   the exposure (e.g. precise knowledge of the type, dose, dose rate and timing of exposure); and

(c)   the endpoints (e.g. selection of sampling time and frequency, use of invasive and destructive tests, consistency and completeness of health status evaluations).

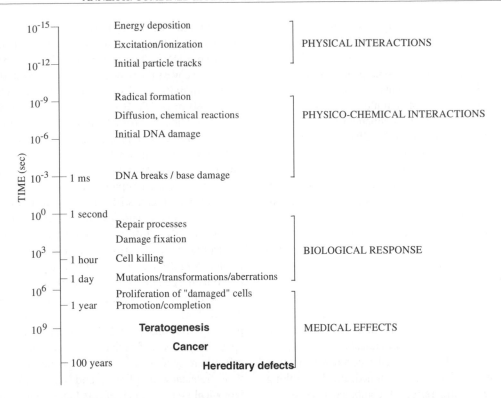

Figure I. Schematic development of the events leading to stochastic radiation effects.

Moreover, in experimental studies one can relate exposures to effects more directly than is typically possible in human studies. This is due, in part, to the fact that both the history of the subject and the exposures under study are known and controlled, making cause-effect linkage easier. In addition, experimental scientists can often determine that an exposure actually results in a dose to the tissue manifesting an effect.

12. Experiments with animals or cells have the disadvantage that the results and conclusions have to be extrapolated to humans. Additionally, conclusions drawn from high-level exposures of animals and cells have to be extrapolated to the low levels of human exposures. The greatest uncertainty is largely a problem of not knowing the shape of the dose-effect relationship at low exposure levels and whether there are effect thresholds. A well balanced conclusion on the combined action of two agents can only be given if the dose-effect relationships of both agents separately and of the combined exposure are known and can be analysed using a (mathematical) model in which the interaction can be consistently and quantitatively defined. The majority of studies on combined effects, including those with radiation, do not meet these conditions.

13. For the basic case of a single agent acting on a biological system, the resulting effect will be dependent on the dose of the agent and will follow some kind of functional dose-effect relationship. The effect level in the absence of the agent is termed the spontaneous or background effect. The simplest relationship between dose and effect is linear. In the realm of linear dose-effect relationships, the three most commonly considered types of interactions between two agents are additivity, synergism and antagonism, giving a combined effect equal to, greater or less than the effects of independent actions, respectively (reviewed in [M16]).

14. For combined effects of agents with non-linear dose-effect relationships, the analysis is complicated, and more precise definitions of the terms antagonism, additivity, and synergism must be provided [S25, S49]. For example, for an upward-bending dose-effect relationship (Figure II), an additional increment of dose from a single agent will result in a non-linear increase in response, even in the case of additivity. The term synergism has sometimes been erroneously used for such situations [Z3]. Although correct on a descriptive and mathematical level, such a broad definition would render the term synergism practically useless in the study of combined effects. With such a definition, different agents with the same action spectrum, i.e. fully independent agents, would

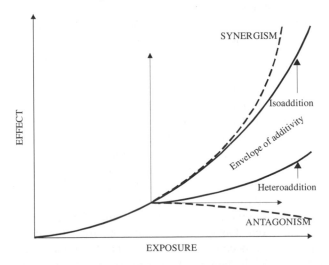

Figure II. Interaction of two agents having non-linear dose-effect relationships. Isoaddition results for mechanically similar agents, heteroaddition for independently acting agents [I3, S49].

produce an apparent synergism in any combination of concentrations as long as the single dose-effect relationships are bent upwards, as is often the case in the dose range of interest. From a mechanistic point of view, synergism can be defined more narrowly to imply that agents combine by acting at different rate-limiting steps of a multi-step process or at different sites of a molecule, thereby enhancing the chance for a negative outcome, such as cancer, by different mechanisms [B23]. Such an assessment is often hindered by insufficient knowledge of the underlying mechanism of action and therefore can rarely be made. Clearly, deviation from additivity is a poor indicator of synergism or antagonism, since non-linear dose-effect relationships and threshold phenomena are the rule rather than the exception for most endpoints in biological systems, and interaction in the statistical-mathematical sense does not define an interaction in a biological-mechanistic sense [B69].

15. In this Annex the term synergism will be used in a narrow sense. The most important question is whether data on combined effects do show some modification of stochastic radiation effects as a result of combined exposure with another agent. If not, no interaction will be assumed, and the resulting effect is additive; if the result of combined exposure is different, some form of interaction has to be assumed, and the resulting effect will be called sub- or supra-additive, depending on whether the effect is lesser or greater, respectively, than the sum of the single-agent effects separately.

2. Radiation effects and effects of other agents

16. As far as carcinogenesis is concerned, the primary effects of ionizing radiation are on DNA, compromising cell survival, cell proliferation, and proper physiological cell functioning. Although the deposition of energy along the track of ionizing radiation can directly affect DNA, most of the damage to DNA from low-LET radiation comes from the formation of radical intermediates stable enough to diffuse several nanometers and interact with critical cellular constituents (for details see Annex F, "*DNA repair and mutagenesis*"). Only a small fraction of the radiation-induced molecular modifications occur in the DNA of the cell nucleus, but practically all experimental and theoretical evidence indicates that DNA, the main carrier of genetic information in living matter, is the critical target. Especially at the doses under consideration in this Annex, damage to structural and functional proteins and lipids has not been shown to contribute noticeably to the detriment from ionizing radiation. To protect the integrity of the genetic information, most cells have highly intricate enzyme systems to repair DNA damage efficiently and effectively based on information contained within the undamaged complementary DNA strand. Despite that, residual fixed damage may result even from low-dose exposures, especially when both DNA strands are damaged. Such damage may lead to reproductive cell death, and therefore possible deterministic effects; to somatic cell mutation, enhancing the risk of cancer; or to mutations in germ cells, with possible deleterious effects in offspring.

17. Longer wavelength radiation, such as ultraviolet (UV) light, although not ionizing itself, still acts mainly by modifying DNA. The UV portion of the electromagnetic spectrum covers the wavelengths between 200 and 400 nm. Conventionally, a distinction is made between UV-C (200–280 nm), UV-B (280–320 nm), and UV-A (320–400 nm). The effects of UV light depend on the wavelength and the absorption properties of the target. Ultraviolet radiation mainly causes the formation of pyrimidine dimers and 6–4 photoproducts, which may also lead to residual DNA damage after repair. Apart from visible light up to 525 nm, which can still interact with photosensitizers to generate reactive species and, subsequently, oxidative damage to DNA, infrared, microwave, and low-frequency electromagnetic radiation have no direct genotoxic effects of their own. Indirect effects might arise from local heating or from charge effects across membranes activating signal transduction pathways and neurons. Such cellular changes may be long-lasting or even be passed from one cell to its progeny. Sugahara and Watanabe [S10] reviewed the epigenetic aspects of radiation carcinogenesis. Studies using cell culture systems show that magnetic fields, depending on their frequency, amplitude, and wave form, interact with biological systems. Such effects have been seen on enzymes related to growth regulation, on intracellular calcium balance, on gene expression, and on peripheral levels of the oncostatic hormone melatonin [H45]. These effects are potentially related to tumour promotion. However, the considerable research conducted thus far has not elucidated critical mechanisms or revealed important health risks from non-thermal exposures. Other than crude effects present only at high exposures, for example strong irritations or protein denaturation, cellular perturbations resulting from non-ionizing radiation cannot be labelled harmful per se.

18. Chemical agents may act as genotoxicants by, for example, forming direct covalent links, by transferring reactive molecular subgroups to DNA, by inducing DNA-DNA or DNA-protein cross-links, or by generating strand breaks. The mode of action may be direct, by the formation of small or bulky DNA adducts as well as strand breaks, or indirect, by the formation of radicals in the vicinity of DNA, leading to strand breaks or small adducts. On the epigenetic or non-genotoxic level, chemicals may interfere with DNA synthesis or repair or may prevent radical scavenging, thereby promoting DNA damage. Non-genotoxic agents may also influence a broad spectrum of other cellular events. Of concern in cancer induction is any interference with cell proliferation, cell differentiation, cell senescence, and apoptosis or with the regulation of these processes.

19. Biological agents may also act at the genetic and epigenetic levels, i.e. they may be genotoxic or non-genotoxic, respectively. Viruses are effective transport vectors for genome fragments and may activate or block the expression of endogenous genetic information. Viral involvement in many animal tumours and also in human malignancies is well established, e.g. the DNA tumour viruses of the papilloma family in cervical carcinoma and the retroviruses HTLV-1 in adult T-cell leukaemia (reviewed in [H13]). In addition,

biological-agent-induced influences on immune responses, inflammation, fever, and endogenous radicals may lead to cytotoxic and/or growth stimulatory responses that are co-carcinogenic, as described later.

B. EXPOSURE ASSESSMENT

20. The most important prerequisite for a comparative assessment of biological effects of different agents, and also of their possible interactions, is the characterization of the exposure or the dosimetry of both agents that may be related to subsequent effects. Some of the main concepts used in toxicology and radiation biology to convert exposures into meaningful measures of dose and health impact are introduced in the following paragraphs.

21. The toxicity of an agent can be defined as its inherent ability to adversely affect living organisms. The spectrum of undesired effects is very wide, ranging from local, reversible effects to irreversible changes leading to the failure of critical organ systems and then to death. The objective of dosimetry is to relate the amount of agent presented to the organism in a way that is relevant to the effects observed and that is measurable in a physical, chemical, or biological manner. Identification of processes occurring at the molecular level, i.e. at a mechanistic level of the effect, would give the most basic indication of a dosimetric measure. The present approaches and possibilities are discussed below. In Section I.B.1, dosimetry based on the measurement of physical or chemical parameters of the agent itself, the physical or chemical dosimetry, is considered. In Section I.B.2, measurement of immediate biological damage caused by the agent (biochemical monitoring) is discussed; this damage may or may not be directly related to the biological effect being considered.

22. Sometimes, when physical, chemical, or biochemical measurements are not possible or cannot be made accurately enough, certain biological effects may be detectable. Such effects may serve as indicators of the exposure to biologically active agents. These "biological markers" reflect damage resulting from toxic interaction, either at the target or at an analogous site that is known or believed to be pathogenically linked to health effects. A wide variety of biological markers fall into this category, including gene mutation; alterations in oncogenes and tumour-suppressor genes; DNA single- and double-strand breaks; and unscheduled DNA synthesis, sister chromatid exchanges, chromosomal aberrations; and micronuclei. None of these markers is highly agent- or exposure-specific, and other factors (lifestyle and environment) that affect these endpoints can act as confounding variables in molecular studies. Some possibilities for assay systems to measure biological markers such as specific gene mutations and cytogenetic damage in exposed humans are presented in Sections I.B.3 and I.B.4, respectively.

1. Dose concepts for physical and chemical agents

23. Ionizing radiation exposure is generally measured in terms of absorbed dose, i.e. the average energy deposited per unit mass. The unit of absorbed dose is the gray (Gy), with 1 Gy equal to 1 J kg^{-1} [I3]. At the level of a cell or cell nucleus, the minimal dose is determined by the ionization density of a single track. Averaged over the volume of a cell nucleus, a single event amounts to between one and several milligray (mGy) for electrons and about 300 mGy for an alpha particle [U3]. Below these dose levels, the probability of a cell being hit varies but the absorbed dose per cell nucleus does not. For internally deposited radionuclides, their location and fate in the organism are used to calculate the absorbed dose in the organs of interest, and usually the average absorbed dose in the organ is taken as the relevant dose that causes the biological effect, assuming a rather homogeneous distribution of energy absorption in the tissue.

24. The definition of exposure or dose for non-ionizing radiation and for most chemical and biological agents is more difficult than for radiation. Ultraviolet radiation can penetrate into tissue at most only for several millimetres, depending on wavelength. The energy absorbed in the tissue of interest, and thus the effectiveness of UV, cannot be easily estimated. Exposure to a toxic agent may be estimated by environmental monitoring (referred to as external dose evaluation in toxicology), internal monitoring (internal dose evaluation), and biochemical effect monitoring (tissue dose or biologically effective dose determination) [E1].

25. For chemical and biological agents, the dose can be based on the time integral of concentration, as for internal exposures with radionuclides. However, in addition to the common important question of defining the critical cellular targets, it is the activation and biodegradation of a chemical agent in the different compartments of the organism that will determine the degree of genetic damage or strength of an epigenetic signal. Although the local concentrations of receptors or reactants could possibly be estimated or determined, these vary considerably in their response to endogenous and environmental factors, which can lead to different sensitivities to the physical or chemical agent. This may restrict the use of biochemical markers somewhat, because their concentrations in body fluids will depend on the mechanisms of uptake, the formation of reactive molecular species, and their breakdown. Somewhat like the dose concept for ionizing radiation, exposure can be related to the number of primary chemical events on DNA leading to the effect under consideration. The above-mentioned quantitative link between DNA alkylation and the product of concentration and time for ethylene oxide may serve as an example [E3] (see also paragraph 34). However, only rarely is the nature of such events known or quantifiable.

26. To give exposure (or dose) its full biological meaning, the concentration-time product at the level of the cellular target structure should be known. Even this is difficult to determine owing to the many membranes and other barriers to be crossed between the intake port and the place of action. Many chemicals also undergo modifications by detoxification in the liver, lung, and other organs, which change both their toxicity and their biokinetics. One of the best known carcino-

gens, benzo(a)pyrene, becomes toxic only after metabolic activation, leading to the ultimate reactive electrophilic carcinogen. Such transformations, called metabolic activation, may differ considerably among species and even between male and female subjects, making inferences from experimental systems still more difficult. The induction of kidney cancer in male rats by a group of chemicals (n-1,4-dichlorobenzene, hexachloroethane, isophorone, tetrachloroethylene, and unleaded gasoline) may serve as an example. It took a great deal of research [B29, E4] to show that the risk for the endpoint under consideration, namely kidney tumours in male rats, is a species- and sex-specific finding that relates to male rats but not to female rats, mice of either sex, or humans. Mechanistic studies showed that male rats have a specific circulating protein, alpha-2u-globulin, that binds the chemicals under consideration and leads to renal accumulation, with subsequent kidney damage and the development of kidney tumours. This protein was shown to be absent in humans. Only such detailed molecular information allows a reasonable risk estimate to be generated for humans [B29]. Unfortunately, the species-specific detection and quantification of toxic agents formed in biochemical pathways are rarely achieved.

2. Biochemical monitoring

27.   For chemical agents, internal dose evaluation involves the measurement of the amount of a carcinogen or its metabolites present in cells, tissues, or body fluids. Analysis of internal dose takes into account individual differences in absorption or bioaccumulation of the compound in question. It may be relatively easy to measure the concentrations of the compound in body fluids. However, doing so does not provide data on the interactions of the compound with critical cellular targets. Examples of this type of monitoring include organic compounds or metals (e.g. lead) in the diet, cigarette smoke, or industrial exposures that can be detected in blood or urine [P3]. The binding of chemicals with cell constituents may be measured directly with radioactive labels. Even *in vivo*, correlations between the administered amount of the toxicant, the number of molecules bound to critical targets, and the biological effect can be established [P6].

28.   From the energy deposition pattern of ionizing radiation in the tissue constituents and from some critical biochemical parameters, such as oxygen pressure and the local concentrations of radical scavengers, the primary damage, i.e. the number of primary DNA lesions, can be estimated. A few of these parameters are even stable enough to be used as biological indicators such as cytogenetic changes in peripheral blood lymphocytes to assess exposures retrospectively.

29.   In toxicology, the tissue dose or biologically effective dose reflects the amount of carcinogen that has directly interacted with cellular macromolecules at a target site. It can be assessed from the amount of DNA and protein damage (strand breaks, DNA adducts, protein adducts) in the target tissue or by extrapolating from damage levels found in surrogate tissues, such as white blood cells. Experiments have shown that, in general, DNA damage levels in target tissues and non-target cells are proportional to the external dose. This

class of markers is more mechanistically relevant to carcinogenesis than internal dose, since it takes into account differences in metabolism (activation vs. detoxification) of the compound in question, as well as the extent of repair of carcinogen-altered DNA. Perera and Santella [P3] provided examples of compounds and exposures that might be analysed using this type of biologically effective dosimetry, as well as the populations that have been studied.

30.   DNA and protein adducts are measures of exposure to carcinogenic compounds [E1]. They are mechanistically linked to cancer, as they cause DNA damage and mutations in important genes, such as genes coding for growth control or damage repair enzymes. Adducts have been used to estimate cancer risk by comparing their mutagenicity relative to that of x rays. In the same way that the unit cancer risk of x rays is defined, the relative mutagenicity is used to estimate the cancer risk of a chemical exposure that causes adducts (gray-equivalent approach) [E1, E3].

31.   In the case of agents binding covalently to different cellular macromolecules, the degree of alkylation of proteins can be used as a surrogate measure for their effects on DNA. Ehrenberg et al. [E3, E10] showed in the mouse that the tissue dose of ethylene oxide, i.e. the concentration of the alkylating agent integrated over time, correlated well with the alkylation pattern. In male mice, the authors were able to show with this method that the tissue dose for ethylene oxide, an agent rapidly distributed to all organs after inhalation, was about 0.5 mM h per ppm h for most organs, including the testes. On the basis of dose-effect curves of ethylene oxide and x rays in barley, the same authors [E3] set a tissue dose of ethylene oxide in humans of 1 mM h equal to 0.8 Gy of low-LET radiation. Such an approach facilitates the comparison and combination of risks of various agents.

32.   Despite their relevance as dosimeters of biological effects, the limitations of the current methods should be noted. Most available assays provide information on total or multiple adducts and are rarely capable of pinpointing the critical adducts on DNA. Only for a few target organs, such as the lung or bladder, are epithelial cells available for routine analysis. For other organs, DNA is not readily accessible; many studies therefore use surrogate tissues (e.g. peripheral blood cells and placentas). However, the relationship between adducts in the target and those in surrogate tissues has not been well characterized in humans, although for certain carcinogens it has been characterized in experimental animals [S17]. Again, it must be considered that there are species- and sex-dependent differences in the absorption and metabolism of chemicals in their various forms.

33.   By definition, all types of ionizing radiations generate ions. Ionizing radiation can directly induce ionizations in DNA, causing direct damage. However, the majority of damage from low-LET radiation occurs in an indirect manner via the formation of free radicals and H_2O_2, which are precursors of oxidative damage [B8, S34]. When living cells or organisms are irradiated, OH radicals are generated in cells or tissue, which leads to many DNA

lesions, including oxidative DNA base products. Both ionizing radiation and oxidative stress generate free radicals near DNA. Most of these radicals (–R) interact with oxygen, forming peroxyl intermediates (–ROO) and final products (–P). Most of the products are eliminated by nucleotide excision repair and glycosylases [F10], while a small fraction remain in the DNA [S35]. Critical are lesions leading to double-strand breaks or even more complex local damage.

34. Free radicals are difficult to detect, identify, and monitor because of their short half-life, particularly in living organisms. Such detection and monitoring can be achieved only by detecting and measuring the products of their reaction with endogenous bio-components or exogenous components selectively added to a biosystem. Specific products of such reactions or their metabolites may qualify as markers of a particular process or specific free radical. In biosystems, these products are called molecular markers, a subclass of bio-markers [G9]. For a product to qualify as a molecular marker, there must be unequivocal proof of an exclusive origin of the product. First, a comprehensive understanding of the kinetics, energetics, and mechanisms of product generation is required. Then other possible sources of the product must be excluded [S39].

35. Although a molecular marker can be quantified by measurement *in vivo*, quantification of oxidative stress is considerably more complex. The reactivity of all five bases, adenine, cytosine, guanine, thymine and uridine, with OH radicals is extremely high, whereas that of deoxyribose is about five times lower [B34]. The distribution of damage will therefore be governed by the relative abundance and reactivity of DNA and RNA components. Each DNA and RNA base contains more than one site of attack. For example, OH adds to the double bond of thymine at C-5 (56%) and C-6 (35%) and removes hydrogen from the methyl group (9%) [J7]. The 5-hydroxythymidine intermediate leads to formation of thymine glycol. The 6-hydroxythymidine intermediate is an oxidizing radical that gives rise to unstable hydroxy-hydrothymine. The radical on the methyl group of thymine, however, is a reducing radical that yields 5-hydroxymethyl-uracil as the final product (reviewed in [S39]). Addition of OH to the C-8 position of guanine yields a well-known product, 8-hydroxyguanine or 8-oxoguanine, which was discovered by Kasai et al. [K1, K47] and described in detail [J4, S39].

Numerous other products have been identified, and the kinetics and mechanisms of their formation have been described [B8, S12, S34, S36].

36. On the basis of extensive studies in radiation chemistry and radiation biology of the kinetics and mechanisms of OH radical reaction with DNA components, it was suggested that detection of thymine glycol, thymidine glycol, and 5-hydroxy-methyluracil indicated endogenous OH generation in rats and humans [C3, H18, W8]. Because thymine glycol can be absorbed through the gastrointestinal tract and 5-hydroxy-methyluracil may be generated by enzymatic hydroxylation of thymine, these products may not always qualify as biomarkers for oxidative damage in organisms. Thymidine glycol is less prone to such confounders and qualifies as one of the best endogenous markers of OH [S39]. It was suggested that 8-hydroxyguanine could be another OH marker in biosystems [B12, F3, K1, R9, S30, W12]. Enzymatic hydroxylation of guanine, however, has not been ruled out unequivocally. Hence it is prudent to monitor more than one marker for each specific free radical under investigation. 8-Hydroxyguanosine was analysed in the DNA of peripheral blood leukocytes of patients exposed to therapeutic doses of ionizing radiation [W12]. Radiation-generated oxidative DNA base products were also measured in the DNA of irradiated cells [N2].

37. The chemical reaction products in DNA are excised from damaged DNA over a certain period of time by repair mechanisms and eventually appear in the cell medium or urine. Some oxidative DNA base products have been measured in the urine of irradiated humans and mice. The radiation yields of these markers, i.e. the increments per unit of energy (mass × dose), were obtained from the level one day after irradiation minus the level before irradiation and are shown in Table 2. In contrast to the metabolic levels of these markers, the irradiation yields per unit energy are the same for both mouse and human, as expected, because the same number of OH radicals is generated in both cases [S39]. The metabolic rate plays an important role in the variability of relative rates of oxidative DNA damage. A high metabolic rate, as in rodents, generates a high yield of urinary markers, i.e. higher rates of DNA damage. The rate of DNA damage, however, is not always proportional to the specific metabolic rate because the efficacy of inhibition and scavenging of oxygen radicals and peroxides as well as of DNA repair systems varies in different species.

Table 2
Yield in urine of biological markers of oxidative DNA damage [B12]

Species	Specific metabolic rate $(kJ\ kg^{-1}\ d^{-1})$	Metabolic yield $(nmol\ kg^{-1}\ d^{-1})$		Increment induced by radiation $(nmol\ kg^{-1}\ Gy^{-1})$	
		Thymidine glycol	8-Hydroxy-guanine	Thymidine glycol	8-Hydroxy-guanine
Human	100	0.3±0.1	0.3±0.1	3.1±0.8	6.7±1.5
Mouse	750	7.3±1	11±2	3.0±0.6	6.9±1.3

3. Gene mutation analysis

38. The analysis and quantification of genomic changes are important steps in monitoring and in the elucidation of mechanisms leading to critical health effects. Functional changes, i.e. changes in the phenotype of oncogenes and tumour-suppressor genes, may be of direct relevance to the process of carcinogenesis. The ultimate effects of ionizing radiation and other genotoxic agents are genetic changes, which are heritable, i.e. which can be passed in a clonal fashion from one somatic cell to the following cell genera-tions, or from a germ cell to the offspring. Therefore critical studies of combined effects should include gene mutation analysis as one very important biological endpoint for stochastic health effects. Several methods are available for the study of gene mutations arising in human somatic cells *in vivo*. These methods allow determination of the frequency of mutant lymphocytes or erythrocytes or characterization of mutations at the molecular level in lymphocytes. The study of types, frequencies, and mechanisms of human somatic muta-tions *in vivo* is valuable in its own right and may also improve the understanding of individual variation in sensiti-vity to environmental exposures, the influence of DNA repair and metabolism, and the relationship between mutagenesis and carcinogenesis [L7, M21]. Genetic changes are key events in carcinogenesis. Most human tumours contain more or less specific mutations that are directly or indirectly related to the carcinogenic process. A description of mutations in human tumours and the scientific background of many of the concepts and methods addressed in this and the following Chapter are presented in more detail in Annex F, "*DNA repair and mutagenesis*".

39. Early and probably single-step biological end points, such as morphological changes in *in vitro* cell lines, might also serve as indicators of genetic changes. The development of cell culture systems has made it possible to assess the oncogenic potential of a variety of agents at the cellular level. Many assays for oncogenic transformation have been developed, ranging from those in established rodent cell lines, where morphological alteration is scored (e.g. loss of contact inhibition in 10T½ cells), to those in human cells growing in nude mice, where tumour invasiveness is determined. The mutational changes involved are rarely defined. In general, simple *in vitro* systems that deliver reproducible results are the least relevant in terms of human carcinogenesis and human risk estimation. The most important potential of these systems lies in the opportunity they offer to identify and quantify factors and conditions that prevent or enhance cellular transformation by radiation and chemicals [H11].

(a) Mutation frequencies

40. Five systems for biomonitoring humans exposed to carcinogenic agents have been developed in which gene muta-tion is the endpoint. Two of these use as markers haemo-globin variants (Hb) [S18, T4] and loss of the cell-surface glycoprotein glycophorin A (GPA) in donors heterozygous at the MN locus in erythrocytes [L1, L3, L6]. The other three involve detection of mutations in T lymphocytes in the X-linked locus for the purine salvage pathway enzyme, hypoxanthine phosphoribosyltransferase (*hprt*) [A5, A6, A8, M22, R7, R8, T10], in the autosomal locus for human leukocyte antigen-A (HLA-A) [J3, M18, T7, T8], and in the autosomal T-cell receptor genes (*TCR*) [K23, K24, N3, U15].

41. Mean background mutation frequencies in human cells *in vivo,* as analysed by the five mutation assays, differ by about four orders of magnitude. In summary, the relative order of background mutation frequency values from normal adults for the five markers are Hb ($5 \cdot 10^{-8}$) < *hprt* ($5 \cdot 10^{-6}$) < GPA ($1 \cdot 10^{-5}$) < HLA-A ($>1 \cdot 10^{-5}$) < *TCR* ($>1 \cdot 10^{-4}$) (reviewed in [C23]). For at least three of these mutation systems, sufficient numbers of donors have been tested to show that, as a general rule, the mutant frequency in normal, non-exposed donors is low at birth, increases with age, is often elevated in smokers, and is increased in people who have been exposed to known mutagens and carcinogens. Despite the great variation in mutant frequency among individuals at each of the loci studied, these findings show the potential relevance of muta-tional analysis in the assessment of combined environmental exposures. More recently, the polymerase chain reaction (PCR) has also been applied in the analysis of mutational spectra. A fairly complete database has been compiled by Cariello et al. [C51, C52].

42. The frequencies of *hprt* mutant cells in healthy adults range from <0.5 to $112 \cdot 10^{-6}$. In most cases the frequency of *hprt* mutant cells is significantly increased after smoking [C17, C19, H2, T4, T11]. There seems to be no effect of sex on the *hprt* mutant frequency. In most studies, an age-related increase in mutant frequency is seen at the *hprt* locus, estimated to be 1%–5% per year in adult donors. Radio-chemotherapy for various malignant disorders, including breast cancer, hepatoma, other solid tumours, and lymphoma increased the frequency of *hprt* mutant T cells by a factor of 3–10 [D5, M20, N8]. Cole et al. [C18, C20] examined factory workers exposed to styrene or to nitrogen mustard. In contrast to styrene, nitrogen mustard significantly increased the number of mutant *hprt* cells in these donors. Tates et al. [T9] described a significantly increased mutant frequency in a group of factory workers exposed to ethylene oxide.

(b) Mutational spectrum

43. The spectrum of mutational changes that arise spontaneously or that may be induced by a physical or chemical agent in human cells is broad. At the DNA level it encompasses, at one extreme, single-base events, and at the other, chromosomal rearrangements involving small to large deletions or translocations. In addition, an important category of mutational events in humans involves losses or gains of whole chromosomes. The mutation spectrum in the mammalian genome is reviewed in Annex F, "*DNA repair and mutagenesis*".

44. Many known mutagens form covalent DNA adducts that are released from DNA either spontaneously or by bio-logical repair processes [H16]. Mutations induced by a large number of compounds, e.g. alkylating agents, arylating

agents, and radiation, have been scored and characterized using shuttle vectors. These experiments elucidate the sequence specificity of adduct formation and, subsequently, the mutations and the mutational efficiencies of different adducts [D9, I6, M10].

45. About 15%–20% of *hprt* mutations in normal adults result from gross structural alterations [A8, B31, H8, N7, T10], as detected by Southern blot analysis. These include deletions, insertions, and rearrangements. The break points or alterations are distributed randomly within the gene, with no hot spots having thus far been identified [A8]. The remaining 80% of the background *in vivo hprt* mutations in adults consist of point mutations or small deletions, insertions, and frameshifts beyond the resolution of Southern analysis. Considering only the *in vivo hprt* mutations (46 Lesch-Nyhan germinal, 51 normal adult somatic, 86 exposed adult somatic), several hot spots of point mutations were observed. In particular, four base-pair sites have been observed to be mutated in all groups [C13].

46. Ionizing radiation is known to induce gross structural alterations in *hprt* and other reporter genes in cultured human cells. After exposure to radionuclides for diagnostic purposes, an increase in the frequency of mutants with gross structural alterations on Southern blots was observed to be 33%, compared with 13% before receiving radionuclides [B31]. Mutations from post-radioimmunotherapy patients showed clearly greater frequencies of gross structural alterations than mutations from pre-radioimmunotherapy patients or normal individuals. The latter two frequencies are quite similar, suggesting that cancer per se does not produce this sort of damage at *hprt* [A9]. Taken *in toto,* the data from Albertini et al. [A9] on *in vivo hprt* T-cell mutations indicate that ionizing radiation produces deletions, particularly large deletions.

47. The yield of mutations caused by ionizing radiation may be influenced strongly by adaptive responses to other toxicants or earlier exposures to the same agent. This topic was reviewed in Annex B, "Adaptive responses to radiation in cells and organisms" of the UNSCEAR 1994 Report [U2]. A 70% reduction in *hprt* mutant frequency in radioadapted human lymphoblastoid cells has been reported, as analysed by Southern blot analysis and multiplex polymerase chain reaction assay [R10, R12]. The treatment was 4 Gy from gamma rays alone or in addition to an adaptive dose of 0.02 Gy. The proportion of deletion-type mutations was decreased in adapted cells (42%) compared with that in mutants treated with the high dose alone (77%).

48. Using a shuttle vector system, Kimura et al. [K12] analysed mutational spectra of the human cDNA *hprt* gene, a recombinant DNA copy of the *hprt* RNA, arising spontaneously or induced by the mutagens methylnitrosourea (MNU); 3-amino-1-methyl-5H-pyridol[4,3-b-]-indole (Trp-P2), a tryptophan pyrolysate; and acetylaminofluorene (AAF). Most mutations induced by MNU are G:C to A:T transitions. This can be predicted by the major premutagenic lesion in DNA produced by MNU, namely O^6-methyl-guanine that specifi-

cally mispairs with thymine [S41]. Mutations that arise spontaneously or are caused by x rays, Trp-P2, or AAF give rise to a similar mutation spectrum of c-*hprt*. Base substitutions account for about one third of all mutations. Mutations other than base substitutions make up some two thirds of all mutations. The main mutational event in these cases is deletion. A noticeable feature of these deletion mutations is the frequent presence of short, direct repeats at the site of the deletion.

49. Mutational alterations in *p53*, a tumour-suppressor gene, are mostly (more than 85%) missense mutations, while those of *APC*, another tumour-suppressor gene, and *hprt* are largely composed of nonsense, frameshift, deletion, and insertion mutations, resulting in truncated gene products or loss of genes. The mutational spectrum in *p53* is therefore clearly different from that of other genes. Mutations in the *p53* gene detected in tumours seem to be the result of a functional selection process for mutant *p53* protein that gives growth advantages to the cell. On the other hand, large deletions in the *p53* region may not be compatible with cell survival. This suggests that the mutations of *p53* observed in tumours may reflect only those mutations of the initial events that are compatible with cell proliferation and may even reflect those that give the transformed cell a growth advantage over the surrounding cells. Mutational selectivity in tumour-suppressor genes is discussed in detail in Annex F, "*DNA repair and mutagenesis*".

50. With respect to interaction mechanisms leading to combined effects, present knowledge indicates that the mutational spectrum found in tumours often reflects not the agent responsible for the primary DNA damage but rather growth selection based on specific changes in the phenotype or general chromosome instability emerging during carcinogenesis. Analysis of marker cells in peripheral lymphocytes may overcome this problem, albeit at the expense of losing the direct link to human disease.

4. Cytogenetic analysis

51. The main conceptual basis for using cytogenetic assays for biological monitoring is that genetic damage in easily available cells, such as peripheral blood lymphocytes, reflects comparable events in target cells. The fact that chromosomal abnormalities are often a characteristic feature in malignant cells points to the direct relevance of such markers for clastogenic agents to be considered in combined exposures. In addition, long-term follow-up of populations screened for chromosomal aberrations shows a clearly higher cancer risk for the subgroup with an elevated level of chromosome damage [B20, H5]. Microscopically recognizable chromosomal damage includes numerical aberrations and structural chromosomal aberrations, in which a gross change in the morphology of a chromosome has occurred. Chromosome and chromatid breaks, dicentrics, and ring chromosomes are important examples of this class of damage [N11]. The yield of sister chromatid exchanges, which represent apparently symmetrical intrachromosomal exchanges between the two identical sister chromatids and which are already quite

frequent in unexposed cells, is also increased. Micronuclei arising either from acentric chromosome fragments or from a lagged whole chromosome with centromere [W15] are also important markers, although the second production pathway points to a mechanism driven partially by epigenetic factors.

(a) Chromosomal aberrations

52.   Induced chromosomal aberrations can be divided into two main classes: chromosome-type aberrations, involving both chromatids of a chromosome, and chromatid-type aberrations, involving only one of the two chromatids. Ionizing radiation induces chromosome-type aberrations in the G_0 or G_1 stage of the cell cycle (e.g. prior to replication), while chromatid-type aberrations are produced during the S or G_2 stage (e.g. during or after replication of the affected chromatid segment). In peripheral lymphocytes, most of which are in the G_0 stage of the cell cycle, ionizing radiation induces mainly chromosome-type aberrations.

53.   Most chemical mutagens are S-dependent clastogens and therefore produce mainly chromatid-type aberrations. S-dependent compounds have no direct effect on the chromosomes of peripheral lymphocytes *in vivo*, because they replicate only after stimulation in cell culture. Peripheral lymphocytes can, however, carry unrepaired/misrepaired, long-lived lesions that may lead to aberrations during replication of DNA *in vitro* [S28].

54.   The classical chromosome aberration assay for measuring dicentrics is a reasonably good measure of dose down to 100 mGy whole-body exposure [L31] or, with much effort, even lower. However, it is based on a genetic change that considerably impairs the survival of indicator cells and their stem cells, so that the signal fades with time. Reciprocal translocations are considered less disruptive to the proliferative future of affected cells. It is possible to score translocations with G-banding or FISH (fluorescent *in situ* hybridization) techniques, with the latter technique having a higher detection limit, about 500 mGy. In such systems, the preferential loss of affected cells may still be a minor problem; in addition, clonal expansion of cells carrying translocations conferring a growth advantage may lead to an overestimation of the dose with time. Biological dosimetry using cytogenetic parameters will be discussed later. It seems that all agents that apparently induce single-base changes (i.e. base deletions, transversions, or transitions) also induce gross chromosomal changes that are visible under the microscope. However, the number of agents clearly shown to induce cytogenetic changes in humans is still relatively limited [A24, S31]. From known or suspected carcinogenic agents, mixtures, or complex exposures to humans, cytogenetic data are available for 27 compounds in Group 1 of the IARC classification (known carcinogens to humans), for 10 compounds in Group 2A (probable carcinogens to humans), and 15 compounds in Group 2B (possible carcinogens to humans) [I1, I2]. Chromosome damage in humans was found in 19/27, 6/10, and 5/15 cases in these groups, respectively.

55.   Most of the informative data on induced chromosomal aberrations in humans arise from high-exposure occupational situations. The comparisons of experimental animal data and human data for the endpoint of chromosomal aberration are generally in good agreement. However, in a few cases there are discrepancies between animal and human data. High occupational exposure to radon induces chromosomal aberrations in humans. Animal experiments with comparable exposures are negative. The most likely explanation is a confounding by other clastogenic exposures in humans, e.g. smoking.

56.   Unlike radiation exposure, chemical exposures have been considered in very few cytogenetic follow-up studies. Studies on the induction of chromosomal aberrations after exposure to alkylating agents expressed in peripheral lymphocytes show, like studies after radiation exposure, that damage can be conserved over several months or even years after treatment [G3]. The persistence of chromosome damage, however, varies with the type of exposure and the cytogenetic endpoint examined.

(b) Sister chromatid exchange

57.   The induction of sister chromatid exchange can be observed in cells that have undergone two rounds of DNA replication in the presence of bromodeoxyuridine (BrUdR), which results in chromosomes having sister chromatids that are chemically different from one another: one is unifilarly labeled with BrUdR and the other bifilarly labeled. Such sister chromatids stain differently from one another, and any exchanges that occur between the sister chromatids can be clearly seen and counted [W7]. A number of studies confirmed the ability of low-LET radiation to induce sister chromatid exchanges in rodent cells [G4, L22, R5, U14] and human lymphocytes [G14]. However, in other studies, when normal human lymphocytes in G_0 were assessed for their ability to express sister chromatid exchanges following low-LET radiation exposure, they failed to do so, in contrast to the quantifiable induction of chromosomal aberrations [L21, M28, P2]. This difference could possibly be attributed to the presence of BrUdR, a known radiosensitizer, at the time of irradiation in the rodent cell studies [L25]. Nevertheless, low-LET ionizing radiation and radiomimetic chemicals are not very effective at inducing sister chromatid exchanges, contrary to S-dependent agents such as UV light [W11], alkylating agents [T1, Y4], and cross-linking agents [S4]. High-LET radiation (neutrons and alpha particles), however, induces sister chromatid exchanges in normal human peripheral lymphocytes exposed in G_0. This suggests that the relative biological effectiveness for sister chromatid exchange induction is very large, since there is little low-LET response [A2, S11]. The induction of sister chromatid exchange as a function of charged-particle LET in Chinese hamster cells was recently described [G7]. At each LET examined there was a dose-dependent increase in the frequency of sister chromatid exchanges. In contrast to the majority of biological endpoints, however, where relative biological effectiveness increases as LET increases up to a maximum and then declines, it was found that sister chromatid exchange

induction already declined as LET changed from 10 to 120 keV mm^{-1} [G7]. These observations can be explained on the basis of repair differences for DNA damage induced by radiations of different LET, i.e. the faster the repair, the less likelihood there will be of unrepaired DNA damage at the time of replication when sister chromatid exchanges are formed.

(c) Micronuclei induction

58.   Micronuclei can be formed from entire chromosomes or chromosome fragments [M36]. They result from chromosome breakage and/or damage to the mitotic spindle and are used as a measure of genotoxicity [H15]. Techniques to block cytokinesis in mitogen-stimulated lymphocytes [F4, F5, M24, P19] allow these micronuclei to be observed in binucleated cells found after the abortive attempt of the cell to divide. There is, however, a large and variable background frequency of some 5–12 micronuclei per 10^3 binucleated cells [F5, Y2]. The background frequency increases with age from about 4 per 10^3 among those $\geq$20 years, to 8 per 10^3 for those $\geq$30 years, and nearly 12 per 10^3 for those $\geq$40 years [Y2]. The increase is about 4% per year [F5]. The range of variability increases with age as well. Farooqi and Kesavan [F18] also found that the yield of radiation-induced micronuclei in mouse polychromatic erythrocytes was strongly influenced by small conditioning doses (25 mGy). Micronuclei assays are faster and have a greater potential for automation than the scoring of chromosome aberrations [M36].

59.   Caffeinated and alcoholic beverages have no significant effects on *in vivo* mean micronuclei frequency in binucleated lymphocytes. Even the intraperitoneal (ip) injection of large amounts of caffeine (15 mg kg^{-1} body weight) did not induce chromosomal aberrations in mice [F19]. However, the estimated number of diagnostic x-ray examinations to an individual in the year prior to measurement was significantly correlated to micronuclei frequency [Y2, Y5]. The effect of age and x rays on lymphocyte micronuclei has been shown repeatedly [A11, E2, F5, I1, I2]. Tobacco smoke and tobacco-related exposures are listed in the IARC Monograph series [I1, I2] as micronuclei-inducing agents.

60.   In an analysis of micronuclei frequency in survivors of the atomic bombings, Ban et al. [B4] confirmed the age dependency of background micronuclei levels in peripheral lymphocytes. Females showed a somewhat higher frequency of binucleated cells. Age and sex were independently acting factors. There is no evidence for an effect of radiation dose on present-day background micronuclei frequency in the survivors.

5. Summary

61.   The primary molecular and cellular effects of the many agents potentially involved in combined effects are extremely diverse. No unifying concept of dose can therefore be applied. However, comparisons of toxicity may be based on relevant experimental and clinical endpoints with sometimes only loose and enigmatic links to primary lesions and interactions. A large number of quantitative and semi-quantitative indicators of exposure are presently available. On the level of genotoxicity, DNA damage can be measured up to the functional level of single genes, thus allowing a comparison of the biological activity of different agents and an assessment of possible interactions on a directly relevant level. The accessibility of critical cells and tissues to standard analysis remains a problem. Qualitative and quantitative monitoring of biological effects at the different levels of organization, from molecules to organisms, not only might allow an assessment of the exposure to the different agents involved but could also form the basis for a better understanding of the mechanisms of combined effects and for the elucidation of dose-effect functions for cellular and clinical endpoints.

II. MECHANISTIC CONSIDERATIONS

62.   In view of the many different agents that may be involved in combined exposures with radiation and the complexity of the possible interactions, it is necessary to gain some insight from the mechanistic point of view. This Chapter will give a qualitative insight into the interaction processes by describing important steps in the development of the radiation effect and by suggesting how the radiation effect might be influenced by other agents. For a quantitative insight, various models have been developed to describe the biological effects. Examples of such models will be briefly discussed, in so far as they serve to improve understanding of the mechanisms involved in combined effects. However, it should be kept in mind that models have limited applicability, and agents do not always have only a single mode of interaction.

63.   Since cancer is the most important health effect for radiation at low doses, the review presented in this Chapter deals mainly with mechanisms that are central to the emergence of malignant growth. An in-depth review of the scientific background of some of the concepts discussed here was presented in Annex E, "Mechanisms of radiation oncogenesis", of the UNSCEAR 1993 Report [U3], Annex F, "*DNA repair and mutagenesis*" and Annex G, "*Biological effects at low radiation doses*".

64.   The timescale of events for the various stages of radiation-induced cancer ranges from less than a second to tens of years. Schematically, three crude time-scale-based phases can be defined on the molecular, the cellular, and the tissue/organ level. The molecular phase ranges from the early interaction of the radiation track until initial damage in biologically important molecules has occurred (of the order of seconds). The cellular phase follows and lasts until the biological reactions of the cells involved have occurred and biological cellular effects are induced (of the order of a few

days). Ultimately, on the tissue/organ level, cellular damage may progress in due time, with or without cooperation from other damage, to clinically detectable cancer, which can occur up to 40 or more years after the initial irradiation. These phases are described below. A schematic representation of the processes is given in Figure I. The separation into these phases is arbitrary; it is time-scale-motivated and serves here only to describe the possible interactions of the radiation effect with other agents. In reality, the processes are not separated that rigorously, and interactions with another agent may occur on more than one level or phase.

65.   Radiation-induced effects other than cancer, such as deterministic and teratogenic effects, involve similar phases in the development of the radiation damage. For conciseness, these effects are not explicitly mentioned and considered here, but the data in humans are reviewed briefly in Chapter V.

A.   EFFECTS ON THE MOLECULAR LEVEL

66.   Following the primary interaction of a radiation track with biological matter, an avalanche of events occurs, and various reactive species are left after passage of an ionizing particle or photon: molecules are excited and ionized, radicals are formed, and secondary electrons progress through the material. Most of these species are chemically very reactive and produce other molecular species. These initial processes develop in a very short time (of the order of microseconds) and at short distances from the radiation track. The processes are dependent on the physical and chemical characteristics of the material, the type of radiation, and the conditions in the immediate environment of the target molecule, such as the availability of oxygen, the presence of sensitizing or protecting agents, the ambient temperature, and the ionization density of the radiation. The processes involved in the interaction of radiation at the molecular level are extensively studied in radiation biochemistry and microdosimetry, the concepts of which have been described by the International Commission on Radiation Units and Measurements (ICRU) [I7].

67.   The biological effects of radiation arise mainly from damage induced in DNA molecules. Important types of DNA damage are DNA single- and double-strand breaks, base damage, intra- and intermolecular cross-links, and multiply damaged sites (mds) (see Annex F, "*DNA repair and mutagenesis*"). A review of special models with emphasis on the importance of the DNA damage is given by Goodhead et al. [G17]. As far as epigenetic damage or modifications of other cell constituents are concerned, cytoplasmic changes and mitochondrial or membrane damage may also play a role in certain types of radiation effects, but the importance of these for radiation-induced cancer is disputed. Indirect effect modifiers such as growth stimulation as a result of stem cell killing may become important at higher doses.

68.   The possibility of another agent interacting with the radiation effect in this early phase is dependent on changes in the DNA environment. The direct environment of the DNA defines the fate of radiation-induced reactive species, such as water radicals, and the possibility for direct or indirect damage to the DNA. Interaction leads to changes in the dose-effect relationship for DNA damage and consequently to changes in the dose-effect relationship for cellular effects (see Section II.B). A well known modification of the radiation effect is caused by a change in the oxygen content. Anoxic cells, in general, are more resistant to radiation than well oxygenated cells. Typical agents interacting with the radiation effect at this level are electrophilic compounds, such as N_2O, NO_2, NO, CO_2, SO_2, and SO_3, and nucleophilic agents, such as cysteamine and cysteine [G17, O11]. For interaction with the radiation effect, the agents should, in general, be present in the DNA environment during irradiation. They may modify radiation effects by a factor of up to 3. More indirect effects may result from vasodilators and constrictors modulating oxygen pressure in irradiated tissue.

69.   An important class of agents are hypoxic cell radiosensitizers, also called oxygen-mimetic agents, which have potential use in radiotherapy to enhance the effectiveness of the radiation treatment in anoxic or poorly oxygenated parts of the tumour. These sensitizers must be present at the instant of irradiation. The mechanisms are free-radical-based: the compounds, in general, have increased electron affinity and are believed to involve fast electron transfer processes in DNA [A1]. Well-known agents include nitroheterocyclic compounds, such as metronidazole, misonidazole, and related compounds, metal-based compounds containing Pt, Rh, Fe, Co, and other metals, and nitro-compounds, such as nitrosoureas [S2].

70.   Other chemicals protect healthy cells against the radiation effect. They may also be used in radiotherapy. These radioprotectors are mainly sulphur-containing compounds. They act, in part, as radical scavengers and have to be present at the time of irradiation to produce their protective effect. The radioprotective effect is a factor of 3 or less. Typical compounds of this type are cysteine, cysteamine, aminoethyl-isothiourea (AET), mercaptoethylamine (MEA), and other sulfhydryl-group-containing agents [M4].

B.   EFFECTS ON THE CELLULAR LEVEL

71.   When the radiation has induced molecular damage, the cell reacts by attempting to remove the damage and restore normal cellular function. The reaction depends on the type of damage. For simplicity, only damage to the DNA is considered here, which may be characterized as single-strand or double-strand damage. Single-strand damage, such as breaks or base damage, may be readily and effectively repaired. Complex localized damage, such as a double-strand break, is more difficult to repair and may lead to a biologically different behaviour of the cell. Repair depends on the cell's genotype. It takes place within a few hours after the irradiation. Some of the damage may be persistent and lead to a radiation effect at the cellular level. The most important cellular effects are chromosomal aberrations, mutations and cell inactivation, killing, and apoptosis. Changes leading to

malignant transformation, which can be considered a specific class of somatic mutations or chromosomal aberrations, are particularly important for radiation carcinogenesis.

72. Attempts to characterize the initial biological effect of a radiation exposure and its dose-effect relationship have led to the development of mechanistic biophysical models of radiation action. The aim of these models is to present a mathematical description of radiation action based on realistic assumptions related to basic mechanisms [G12]. Broadly, a common characteristic of these models is that they describe the cellular radiation effect E(D) by a linear quadratic dose-effect relationship:

$$E(D) = E_0 + \alpha D + \beta D^2 \tag{1}$$

where E(D) is the cellular effect from a dose D, E_0 is the effect without radiation (D = 0), α is the contribution to the effect per unit dose and β is the contribution to the effect per unit dose squared.

73. The interpretation of the linear and quadratic dose terms depends on the underlying assumptions of the model. The linear term has a single-track nature, sometimes called intratrack damage. The quadratic term has a dual or multitrack nature, involving the accumulation of sublethal damage or sublesions [C16, K5, K8]. Some models do not account for repair; in other models, repair is considered essential for development of the radiation effect. Most models do not specify the initial type of damage [C16, K4], while others are more specific [C45]. In general, double-strand breaks in DNA play an essential role in the radiation effect.

74. Equation (1) broadly describes the dose-effect relationships for exposures within one cell cycle and is generally used to analyse cellular experimental data, such as chromosomal aberrations, mutations, cellular transformation, and cell killing [I14]. The dose coefficients α and β depend on the effect considered, the cell type, the type of radiation, and the development of the radiation damage during the molecular phase [L11]. For instance, α is particularly dependent on the type of radiation and, in general, is larger for densely ionizing radiation than for sparsely ionizing radiation. The coefficient β, in general, tends to decrease with higher-LET radiation. As far as irradiation time is concerned, α hardly changes and is mostly invariable, but β changes markedly: it reaches a maximum for acute irradiation, decreases for lower dose rates, and for irradiation times of more than a few hours is negligible or zero. This implies that for chronic irradiation a linear dose-effect relationship for cellular effects is anticipated.

75. The mechanism of interaction of another agent with the radiation effect at the cellular level is broadly based on three types of action: (a) the accumulation of sublesions and lesions; (b) interference with cellular repair; and (c) changes in cell-cycle kinetics. All types of interaction are most effective when the potentially interacting agent is present in the cell at the time of irradiation or within a few hours later, roughly as long as the radiation effect is not fixed and repair is still possible.

1. Accumulation of (sub)lesions

76. An important category of combined exposures involving accumulation of sublesions is that of combined exposures to different types of ionizing radiation. For cellular effects such as cell killing, mutations, and chromosomal aberrations, it is well known that the combined exposure to two types of radiation can lead to a larger than additive effect. Understanding how cellular damage produced by densely ionizing radiation (high-LET radiation) interacts with that produced by low-LET radiation is important both in radiation therapy and in evaluating risk.

77. With similarity in the underlying radiation mechanism, interaction between different types of ionizing radiation can be shown to be, in general, of the so-called isoadditive type. Modellers of cellular radiation effects tend to describe the larger effect of combined radiation exposures in terms of accumulation of and interaction between sublethal damaged sites, which may lead to an extra contribution to the radiation effect (increase of the quadratic term of the linear-quadratic dose-effect relationship) [B35, C15, L10, Z14].

78. In general, if the (additional) radiation effect E_i of radiation type i is linear-quadratic with dose D_i,

$$E_i(D_i) = \alpha_i D_i + \beta_i D_i^2 \tag{2}$$

then the combined exposure to radiation types 1 and 2 will lead to effect E_c, given by

$$E_c(D_1, D_2) = \alpha_1 D_1 + \alpha_2 D_2 + (\sqrt{\beta_1} D_1 + \sqrt{\beta_2} D_2)^2 \tag{3}$$

In the absence of interaction, the effect would be given by

$$E_a(D_1, D_2) = \alpha_1 D_1 + \alpha_2 D_2 + \beta_1 D_1^2 + \beta_2 D_2^2 \tag{4}$$

The extra effect is expressed in the difference between equations (3) and (4) and can be calculated to be

$$E_c(D_1, D_2) - E_a(D_1, D_2) = 2\sqrt{\beta_1 \beta_2} D_1 D_2 \tag{5}$$

Equation (5) indicates that the extra effect is dependent on β_1 and β_2. Experimental evidence [B5, C6] shows that β is practically independent of radiation type (i.e. low- or high-LET radiation), so that interaction of sublethal damage can be expected. Using this assumption, the radiation effect of combined exposures of acute high- and low-LET radiation could well be described by the equations given here [L10].

79. Considering this interaction process, one has to keep in mind the following restrictions:
(a) sublethal damage can be repaired by the cell, so that when there is time between the two exposures, the extra effect will decrease;
(b) the quadratic term for each radiation type separately is dependent on dose rate, i.e. irradiation time, which implies that the extra term for combined exposures

also vanishes for dose rates below a certain value (i.e. less than 10 mGy min^{-1});

(c) the interaction process described here is, strictly speaking, proven only for exposures occurring within one cell cycle. Deviations may be expected when exposures occur over more than one cell cycle; and

(d) for practical applications in risk analysis, deviations from additivity are generally not very large, with the most significant deviations being expected for acute irradiation exposures such as are used in radiation therapy; additivity is virtually expected for combined chronic exposures.

80. As Lam [L47] has shown, the interaction of two types of radiation can also be described using the linear isobolic relationship, which is usually used for the combined action of two toxic agents. The reverse also applies: the interaction with radiation of a toxic chemical that has a supralinear or quadratic exposure-effect relationship for cellular effects can be similarly described as the interaction of two types of radiation. As described above, if the radiation effect after a dose D is given by equation (1) and the effect after an exposure X of a second agent is given by

$$E_s(X) = \sigma X + \varepsilon X^2 \qquad (6)$$

then the effect of a combined exposure will be

$$E_c(D, X) = \alpha D + \beta D^2 + \sigma X + \varepsilon X^2 + \eta DX \qquad (7)$$

This means that the effect of the combined exposure to radiation and the second agent is given by the sum of the effects of the two agents separately and an extra effect (ηDX), which is proportional to the dose D of radiation and exposure X of the second agent. This extra term is the result of the interaction of sublethal damage of radiation with sublethal damage of the second agent.

81. This description of the effect of combined exposures can be used for a number of compounds with radiation [L51]. In this analysis it is assumed that the cellular effect of physical and chemical agents can be described as a linear-quadratic function of exposure X. Examples of such agents are ultraviolet radiation (UV) [L52]; alkylating agents such as the nitrosouric compounds ethylnitrosourea (ENU), 1,3-bis(2-chloroethyl)-1-nitrosourea (BCNU), and 1-(2-chloroethyl)-3-cyclohexyl-1-nitrosourea (CCNU) [L48]; benzo[a]pyrene (BP); ethylmethane sulphonate (EMS) [C7] and many more. The conditions mentioned in paragraph 84 concerning the interaction of two types of radiation should be kept in mind for this interaction of radiation with another physical or chemical agent as well. Repair of the sublesions from the first agent before the other agent becomes effective can lessen the enhancement effect of the combined exposure and lead to an effect more nearly like additivity. In general, the analysis can be applied to different cellular endpoints, such as cell killing [L48, L50, L51], chromosomal aberrations, and mutations [C7].

2. Cellular repair

82. The speed and fidelity of DNA repair is one of the main determinants of the yield of fixed damage. Most molecular damage to DNA is subject to a sequential series of enzymatic reactions that constitutes the repair process. This topic has been the subject of much recent study, and a spectrum of analytical procedures, operative at both the molecular and cellular level, has been developed to monitor DNA repair [F10]. DNA damage may include altered bases, the covalent binding of bulky adducts, intrastrand or interstrand cross-links and the generation of strand breaks. Altered bases may be generated by spontaneous reactions, most importantly deamination of cytosine to form uracil, of adenine to form hypoxanthine, and of 5-methylcytosine to form thymine. A range of alkylated products is formed in DNA as a consequence of exposure to nitroso compounds and other alkylating agents. Bulky adducts are formed as a consequence of the covalent binding, to purines in particular, of polycyclic hydrocarbons, aromatic amines, aflatoxins, and similar substances. Two types of pyrimidine dimer are induced by exposure to UV radiation: cyclobutyl pyrimidine dimers are most common, and the so-called 6-4 photoproducts are also produced. Cross-linking of DNA strands may occur following exposure to bifunctional alkylating agents and chemicals such as cis-diaminedichloroplatinum. Strand breakage may be caused by ionizing radiation, heavy metals, chemicals such as bleomycin, and endogenously generated active oxygen species (reviewed in [S9]).

83. Efficient repair of DNA damage is necessary to retain genomic stability and to prevent somatic and genetic disease in humans and other organisms as well. There are several modes of repair, and these may also be affected themselves by mutagenic agents. Failure of repair may thus be as much a cause of disease as the initial DNA damage. To safeguard the genome, cells are able to block cell-cycle progression in response to DNA damage at specific transition points to allow DNA repair. Most prominent are the so-called checkpoint control mechanisms at the G_1/S phase and G_2/M phase transition. The subject of DNA repair is reviewed in Annex F, "*DNA repair and mutagenesis*".

84. Programmed cell death, known as apoptosis, obviates the risks from error-prone repair in heavily damaged cells and is, accordingly, another important defence mechanism of the cell, preventing the survival of aberrant cells and, hence, tumour development [D13]. Apoptosis can become activated under physiological conditions and also after damage to DNA [H44, T17]. *p53* plays an important role in DNA damage-induced apoptosis [L32], so the clonal selection of cells with non-functional *p53* by hypoxia [G19] or by UV radiation [Z5] is potentially an important mechanism to increase tumour yield. This was also shown for radiation teratogenesis in mice. Norimura et al. [N18] found that p53-mediated apoptosis strongly reduced fetal malformations after *in utero* exposure to ionizing radiation (2 Gy), whereas p53$^{-/-}$ strains displayed a 70% incidence of anomalies. Such effects may lead to an apparent threshold in the dose-effect relationship for malformations after *in utero* irradiation [N19]. Several other

types of cell loss or irreversible growth arrest occur in mammalian systems in addition to apoptosis; these include terminal differentiation, senescence, and necrosis. Necrosis, in contrast to apoptosis, is not an orderly cellular process but rather the disorganized death of a cell. Several recent reviews of this topic have been published [S5, T17, W6]. Apoptosis is discussed further in Annex F, *"DNA repair and muta-genesis"*.

85. A second class of agents that can interact with radiation and cause changes in the radiation effect at the cellular level are agents that modify the repair capacity of cells. Repair inhibitors often influence the DNA structure and may be immunosuppressive [S1]. These agents might have toxic effects themselves. Examples are the intercalating agents actinomycin D, adriamycin, and quinacrine. The xanthine derivatives (caffeine, theobromine, and theophylline) also belong to this type of agent. The different effects reported for these agents may be due to the different kinetics of repair in the cell cycle and the presence of the drugs during different phases of the cell cycle [B3, T16]. Depending on the drugs, the repair of sublethal damage or of potentially lethal damage might be involved in the interaction process.

3. Cytokinetics

86. Another important class of agents are chemicals that change the behaviour of the cells in the cell cycle. These agents are indirectly related to those that interfere with repair, because cells that are irradiated tend to move more slowly through the cell cycle in order to have more time for repair. Some cytokinetic agents inhibit changes in the cell cycle. Caffeine is an agent known to remove or alter the cell's capacity to induce a G_2 block or shorten the S phase after irradiation [S1]. The result is that caffeine enhances radiation-induced cell killing and chromosomal aberrations. Effects of cytokinetic agents are investigated for different purposes, among which is to study the mechanism of radiation-induced cellular response and to answer questions such as, in which phase of the cell cycle is the radiation damage fixed? These effects are also investigated for their possible application in radiotherapy. Cytokinetic agents are not normally considered important for environmental risks of stochastic radiation effects. However, some chemicals, for example those with hormonal side effects such as environmental estrogens, have been shown to be effective even at environmental concentrations [S81].

4. Toxicological analysis

87. The cellular effects of combined exposures to radiation and other agents are part of the broad, classical field of toxicology, in which the effects of exposures to two agents are analysed using the isobolic method. The method is primarily useful for agents with isoadditive effects, but it is used for other agents as well. It has been applied to radiation effects [L47, S1].

88. The use of an isobolic diagram to describe the combined effect of two agents is shown in Figure III. The

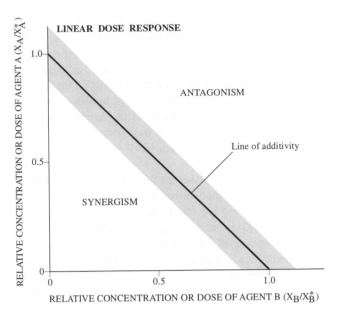

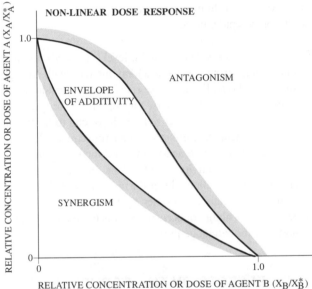

Figure III. Isobolic diagrams for a given level of response in two agents, both acting with linear (upper diagram) and non-linear (lower diagram) exposure response [U6].
The axes are normalized to values of 1.0 for each agent acting separately, i.e. X_A and X_B.

exposures are indicated on the two ordinates, usually with the single-agent exposures yielding the same effect normalized to one. The case of additivity is described by a straight isoeffect line for any combination of two agents with linear dose-response relationships for separate action (Figure III, upper diagram). If the points deviate significantly to the left of the isobolic line, the interaction is synergistic. An antagonistic interaction is postulated when the experimental points lie to the right of the isobolic line. Even in such a simple theoretical case, to assess the combined action of two agents, several combinations of the two agents leading to the given effect, E_{AB}, have to be tested. Although there are some important biological endpoints, such as frequency of point mutations, that show a linear or nearly linear increase after separate exposures to genotoxic chemicals or radiation, the dose-effect relationships for health impairments caused by complex multi-stage changes in biological systems are often better described by exponential or sigmoid functions of dose. For these more realistic circumstances, the line of additivity in the isobolic diagram becomes curved and transforms into an envelope of additivity (Figure III, lower diagram). In general, the order of exposure to agents with differing dose-response relationships then becomes important as well [R4].

89.   Such isobolic analyses are important tools, for example in optimizing combination therapy [L16, R4], but are of less value in evaluating the effects of chronic exposures in the workplace and in non-occupational settings [B69]. An extended review of this approach and its mathematical background was presented in the UNSCEAR 1982 Report [U6]. This approach is based on producing equal effects with different combinations of the two agents under restricted conditions of time. Owing to the general lack of such ranges of exposures in human populations, this method is not applicable in epidemiology.

90.   The interaction at the cellular level is restricted in time, so that damage from one agent seldom interacts with damage from a second one. For low dose rates of radiation and long-term exposures of other agents, the supralinear or quadratic dose terms of the dose-effect relationships tend to diminish, and only a linear dose-effect relationship remains. In these cases, the possible interaction in combined exposure also decreases, and additivity results. This implies that since interaction at the cellular level during low-dose, long-term exposures to radiation and other agents can be expected to have a low probability of occurrence, it is therefore of limited importance for carcinogenesis.

C.   EFFECTS ON THE TISSUE/ORGAN LEVEL

91.   After fixation of the radiation effect at the cellular level, which occurs within a few days, a much longer time is needed before an effect at the organ level occurs, i.e. before a stochastic radiation effect is evident. The period of occurrence of a stochastic effect is dependent on the type of effect. For example, hereditary defects may occur when a germ cell, after having been irradiated, forms the origin of an organism of the next generation. For radiation-induced cancer, it is the time between the initiation event, or possibly one of the following steps of the carcinogenesis process, and the detection of a tumour. Full consideration of the mechanistic aspects of cancer development is given in Annex G, "Biological effects at low radiation doses". The events occurring after the initiation event in the development of tumorigenesis are considered to take place on the tissue and organ level and may occur years or decades later.

92.   It is generally accepted that carcinogenesis is a multi-step process. The usual chain of events is considered to be initiation of damage, tumour promotion, possibly with activated proto-oncogenes or deactivated tumour-suppressor genes, and malignant progression. Each of these processes can be related to effects at the cellular level. The basic aspects of these processes were reviewed in Annex E, "Mechanisms of radiation oncogenesis", of the UNSCEAR 1993 Report [U3]. The concepts of multi-stage carcinogenesis have evolved over many years of cancer research [A17, B10, B14, B15, C8, F7, M27, R6]. Several lines of evidence that support the multi-stage model of cancer derive from studies of pathology, epidemiology, chemical and radiation carcinogenesis in animals, cell biology, molecular biology, and human genetics [K28, M5]. Germ-line mutations, somatic genetic events, and epigenetic stimulation by the host organism may all play important roles in neoplastic development. The definition of two broad classes of genes, proto-oncogenes with growth-enhancing functions and tumour- suppressor genes with growth-inhibiting functions, brought a biological basis and a unifying concept to the multi-stage theory of cancer [V2]. Owing to the functional diversity of the products of these genes involving cell surface receptors, protein kinases, phosphatases, and DNA-binding proteins, to mention only a few, this concept does not lend itself directly to a better understanding quantitatively. However, in this area the modifications of the cancer process after exposure to external agents may be investigated.

93.   The number of genetic changes involved in the evolution of a specific malignant neoplasm is not known with certainty. In some cancers that occur early in life, soon after exposure, or in genetically susceptible individuals, there may be only one rate-limiting change needed for malignant disease. Certain forms of leukaemia, e.g. those resulting from reciprocal translocations [B64] or cancer induction in retinoblastoma heterozygotes, seem to follow this course. Multi-hit models developed on the basis of specific incidence rates of solid cancers from epidemiological data often show an exponential increase in the incidence of specific cancers with the fifth to seventh power of age [K11]. Most colorectal cancers have three or more altered genes, [F6, V1, V2], and estimates of as many as 10 or more mutational changes have been proposed to occur in adult human cancers [B17]. Basically, all these genetic changes might be induced by ionizing radiation, other genotoxic agents, and the inherent instability of DNA alone.

94.   The distinction between proto-oncogenes and tumour-suppressor genes has important repercussions for dose-effect models, because the former class would generally express its function dominantly, whereas the latter could fulfill its protective function as long as one allele is functionally intact, i.e. the tumour-suppressor function would be a recessive trait. However, the probability of developing cancer is in many cases higher in heterozygotes than a pure recessive trait would predict, indicating the importance of penetrance in the genetics of the different tumour-suppressor genes. Moreover, mutations in some tumour-suppressor genes like p53 and WT1 may be of the dominant negative type, in which the mutated protein overrides the action of the suppressor wild-type allele [H4, M25].

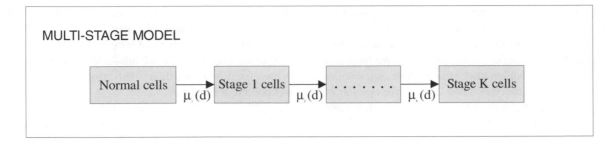

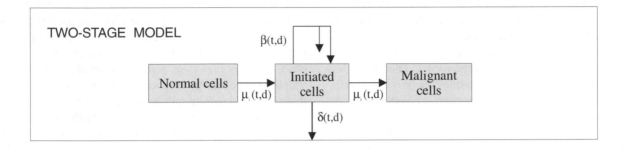

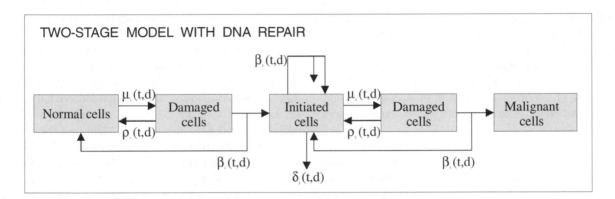

Figure IV. Models of carcinogenesis.
*Parameters: μ is the probability of transformation, β_i is the birth or replication rate,
δ_i is the death rate, ρ_i is the repair rate, i is the stage, d the dose, and t the time.*

95. Multi-stage cancer models are described in Annex G, *"Biological effects at low radiation doses"*. The multi-stage model proposed by Armitage and Doll [A16, A17, A18] represents one of the first attempts to develop a biological model of carcinogenesis. They postulated that cancer develops from a single cell that must pass sequentially through a particular series of transformations to become a malignant cell. The multi-stage Armitage-Doll model is illustrated in the upper portion of Figure IV. This model assumes that a normal cell must pass through k sequential stages before becoming fully malignant. This model has k + 1 types of cells: normal cells, stage 1 cells, stage 2 cells, ..., and stage k (malignant) cells. The model supposes that at age t an individual has a population $N_0(t)$ of completely normal cells and that these cells acquire a first mutation at a rate $\lambda_1(t)$. The cells with one mutation acquire a second mutation at a rate $\lambda_2(t)$, and so on until at the (k − 1) stage the cells with (k − 1) mutations proceed at a rate $\lambda_k(t)$ to become fully malignant.

96. The instantaneous tumour incidence rate h(t) at time t in the multi-stage model is therefore approximately of the form

$$dN_k(t)/dt = \lambda_k(t) \int_0^t \int_0^{x_{k-1}} \int_0^{x_2} N_0(x_1)\lambda_1(x_1) \dots$$
$$\lambda_{k-1}(x_{k-1}) dx_1 \dots dx_{k-1} \tag{8}$$

where N is the number of cells in the target tissue and $\lambda_i(t)$ is the instantaneous rate of the i-th cellular change (i = 1, ..., k). For simplicity, it is often assumed that the transition rates are linearly related to the dose $d_i(t)$ of the carcinogen at time t for the i-th stage. Therefore the transition rate from one stage to the next is given by $\lambda_i(t) = a_i + b_i d_i(t)$. Here, a_i denotes the transition rate in the absence of exposure and b_i reflects the effects of the carcinogen on the transition rate into stage i = 1, ..., k. With similar values for spontaneous transition rates (a_i to a_k), this model predicts that the age-specific tumour incidence rate will be proportional to the (k − 1)st power of time and provides a good description of human cancer incidence data with 2 < k < 6 stages [A20, A22].

97. To encompass the growing biological evidence that the process of carcinogenesis involves intermediate cells having

a growth advantage over normal cells, Armitage and Doll [A17] modified their initial model. The initial model is generally viewed as not biologically plausible, because it does not account for cell kinetics, more specifically the birth and death of cells. The modified model that includes cell kinetics must sometimes assume very small and, in the opinion of Armitage and Doll [A17], unlikely values for the growth rate of intermediate cells to fit the data.

98.　Moolgavkar and Venzon [M29] and Moolgavkar and Knudson [M23] proposed a two-stage birth-death-mutation model to describe the process of carcinogenesis in adults. By incorporating both cell kinetics and tissue growth, this model can be used to describe a broader class of tumour incidence data than the classical multi-stage model. This model has three cell types: normal cells, intermediate or initiated cells, and malignant cells. The middle portion of Figure IV displays the general two-stage model of carcinogenesis in which for a normal cell to become malignant, it must pass from the normal state through the intermediate state and into the malignant state. The simplicity of this model allows classifying external agents in three categories of carcinogen: initiators, which stimulate the first transition of a normal stem cell into the intermediate stage; completers, which transform an intermediate cell into a malignant cell by the second transition; and promoters, which enhance cell division and the net increase of intermediate cells with time [K46]. Ionizing radiation and other genotoxic agents may be both initiators and completers.

99.　The Moolgavkar-Venzon-Knudson model was later extended to account for more than two mutational stages [L23, L24, M14]. For the generalized Moolgavkar-Venzon-Knudson model it may be supposed that at age t there are N(t) susceptible stem cells, each subject to mutation to a type of cell carrying an irreversible mutation at a rate of $\mu_0(t)$. The cells with one mutation divide into two such cells at a rate $\gamma_1(t)$. At a rate $\delta_1(t)$ they die or differentiate. Each cell with one mutation can also divide into an equivalent daughter cell and another cell with a second irreversible mutation at a rate $\mu_1(t)$. For the cells with two mutations there are also assumed to be competing processes of cell growth, death and differentiation, and mutation taking place at rates $\gamma_2(t)$, $\delta_2(t)$, and $\mu_2(t)$, respectively. This continues until at the (k-1) stage the cell will have accumulated (k-1) mutations. It will eventually acquire another mutation and become fully malignant.

100.　With the advent of more sophisticated experimental techniques and a growing understanding of the process of carcinogenesis, more refined mathematical models have been developed and continue to be developed to embody the current scientific knowledge and mechanisms of cancer. Mutation is the result of DNA damage and the subsequent fixation and propagation of the damage by DNA replication. This process is included as a single rate constant in the models described in the previous paragraphs. However, agents can affect DNA damage rates, cellular replication rates, and/or the processes of DNA repair. Kopp-Schneider and Portier [K19] expanded the modelling of the mutation process to account explicitly for the process of cellular damage to DNA, DNA repair, and

DNA replication. The two-stage damage-fixation model has five types of cells: normal cells, damaged normal cells and damaged initiated cells both of which are subject to DNA repair, initiated cells, and malignant cells (in which damage has been fixed by replication). This model is shown in the lower portion of Figure IV.

101.　It is clear that research on quantitative multi-stage models is still in progress and that the complexity of the carcinogenic process inhibits a choice of a universally accepted and applicable model. However, it is also clear that multi-stage models have a biological basis and could describe tumour incidence quantitatively and as such have a future in improving radiation risk estimates and estimates of combined effects. Always important is the question of complexity vs. simplicity. The biology of cancer formation is so complicated that an ever-increasing number of parameters are needed to cover all possibilities of tumour formation mathematically. On the other hand the available data are limited, so the number of parameters that can be fixed is limited as well. As far as the mathematics and statistics are concerned, it is preferable that the number of unknown parameters be as low as possible.

102.　Most multi-stage models are used to describe the age dependence of tumour incidence and the influence of chemical carcinogens in animal experiments. In a few cases, they have been used to describe radiation-induced tumours. As far as human data are concerned, the induction of lung tumours by radon in miners [L9, M39] and the lifespan studies of the Japanese atomic bomb survivors [H1, K45, L5] were used to test multi-stage models. In general, ionizing radiation acts mainly as an initiator, although it has some influence on other coefficients. An important conclusion of the use of multi-stage models for radiation carcinogenesis is that radiation generally seems to affect only one step in the carcinogenesis process; in other words, it is a co-factor of background tumour incidence. This implies that the radiation effect is dependent on background tumour incidence as well as on other agents or factors that produce tumours or cancer.

103.　The timescale for effects at the tissue or organ level is long and can last for years. The implication is that interaction with another agent is possible even when the exposures of the two agents are separated in time for up to several years. A comprehensive treatment of the carcinogenic effect of combined exposures and the implications for dose-effect relationships using a two-mutation carcinogenesis model is given in Krewski et al. [K46]. They classify carcinogenic agents as initiators, completers, and promoters and conclude that the joint effect of two compounds that both affect the same stage in the carcinogenic process will be described well by the additive risk model; however, the effect of combined exposure to two carcinogens that influence different stages will not necessarily result in a multiplicative model. Short exposures that occur close together in time and do not occur at either very young or very old ages can produce a nearly additive relative relationship. Synergism would, however, arise when the contribution to different transitions by different carcinogenic agents is large compared with the spontaneous rate and when the time course of exposure is penalizing if the sequence of steps matters. Brown and Chu [B9] concluded that the observation of a multiplicative relative risk relationship in studies of joint exposure to two carcinogens is evidence of action at two different stages of the carcinogenesis process. These examples illustrate the importance of a full understanding of the timescale of exposure for both agents.

104. An overview of interactions for simple binary exposures to agents with specific effects is given in Table 3. According to the terminology used by Krewski et al. [K46], many interactions leading to considerable deviations from additivity are possible although hardly predictable. The effectiveness of a carcinogenic agent depends not only on the exposure but also on the time of exposure, age at exposure, time since exposure, and duration of exposure. This time dependence is completely different from and should not be confused with the

time involved in the cellular dose-rate effect. It is therefore not possible to quantitatively predict the dose-effect relationship for tumour induction after combined exposure.

105. This assessment is based on the evaluation of carcinogenesis data using a two-mutation model. Deviations of the carcinogenesis process in other ways, e.g. by disturbing organ functions in a crude way, are ignored. The assumption of otherwise undisturbed functioning of the organ or organism is probably reasonable for low exposures but may complicate analysis for high doses and exposures. The long-term development of tumours implies a long period of time over which the process can be influenced. For interaction in the genesis of radiation-induced tumours, it implies that exposure to a different agent at a time that is completely separated from the time of irradiation may influence the radiation effect in often poorly predictable ways.

Table 3
Anticipated interaction response of two single-agent carcinogens [a]
[K29]

Carcinogen A	Carcinogen B	Interaction response
Initiator	Initiator	Additive
Completer	Completer	Additive
Initiator	Completer	Multiplicative
Initiator	Promoter	Multiplicative to supra-multiplicative
Initiator	Promoter and completer	Multiplicative to supra-multiplicative
Initiator	Initiator and completer	Supra-additive to sub-multiplicative
Initiator	Initiator and promoter	Supra-additive to supra-multiplicative
Promoter	Promoter and completer	Supra-multiplicative

a   See glossary for definition of terms.

D.   DOSE MODIFIERS AND OTHER INDIRECT INTERACTIONS

106. Apart from the direct interference with the development of the radiation effect, further indirect interaction mechanisms are possible when an agent changes the retention of the radioactive substance following inhalation or ingestion and consequently changes the organ dose. A well-known case is blockage of ^{131}I uptake to the thyroid by stable iodine, which is used in nuclear medicine and envisaged in future radiological emergencies to greatly reduce the dose to the thyroid gland. Other drugs, such as ethylenediaminetetracetic acid (EDTA), are used for therapeutic reasons, to stimulate the metabolic transfer of inhaled or ingested radionuclides and, consequently, to reduce the relevant organ dose. Natural chelators such as citrate may also modulate the biological half-lives of metal and actinide ions. Synergistic effects on this level are known from the inhibition of mucocilliary clearance by a second agent [F28]. Several examples of dose-modifying agents are given in the Appendix, which covers specific interactions. Mechanistic considerations indicate that the irritants and cytotoxicants implicated here generally display an effect threshold and are therefore of little concern for combined effects at low exposure levels.

107. Other modifications of the radiation effect are possible when the physiological condition of the organism is changed, either intentionally or by chance. Examples are changes in hormone levels or in the immunological system. Such changes may also be induced by radiation (e.g. UV radiation). Also, novel mechanisms of genetic change such as radiation- or chemical-induced genetic instability, which leads to new genetic damage many cell generations after exposure (see Annex F, "*DNA repair and mutagenesis*"), may be prone to more-than-additive effects. The results of this type of interaction are dependent on the conditions of the change and are, in general, poorly predictable.

E.   SUMMARY

108. Carcinogenesis and, consequently, also the development of radiation-induced tumours is a long-term process. Mechanistically, three levels can be distinguished in the development of the radiation effect on cancer: the molecular, cellular, and tissue/organ levels. On each level, ionizing radiation induces changes and processes, and these may be influenced by combined exposures to other agents. A summary of the levels, processes involved in cancer development, and examples of the many classes of substances and agent with a potential to interfere at different

levels of radiation-induced carcinogenesis is presented in Figure V. The biological effect of combined exposures to radiation and other agents at low doses is, in general, expected to be additive, especially in the case of chronic exposures. Deviations from additivity may primarily be expected from interactions on the tissue/organ level. In this phase, exposure to other agents may take a long time, up to tens of years, and for interaction to occur, the exposures to radiation and the other agent need not be simultaneous.

Thus, interaction can last a relatively long time, and the radiation effect can be influenced to a significant extent by the interactions that take place during this phase. Studies using multi-stage models show that classifying the agents involved in terms of their action as initiators, completers, and/or promoters may help to predict the result of their interaction with radiation and other agents. These studies also indicate that the radiation effect depends on the background tumour incidence.

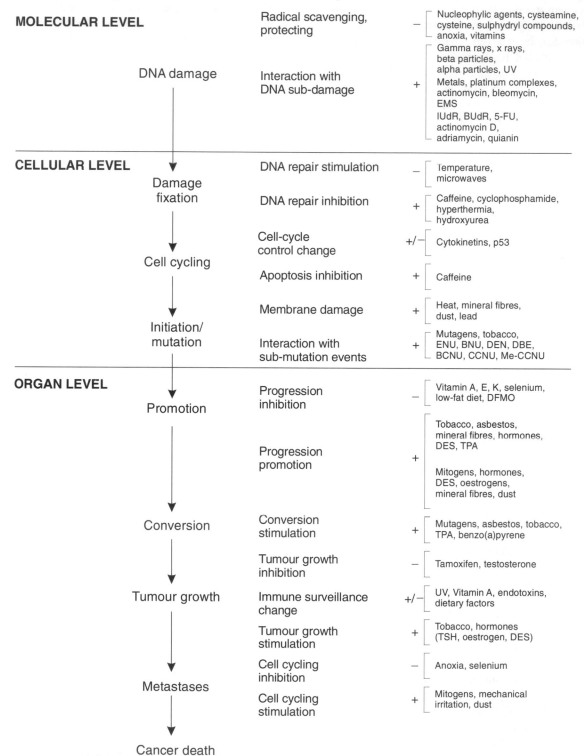

Figure V. Schematic representation of radiation-induced carcinogenesis and possible interaction mechanisms with examples of agents having shown a potential for more (+) or less (−) than additive effects for at least one tumour site.

III. SPECIFIC COMBINED EXPOSURES

109. Recent data on combined exposures to radiation and specific physical, chemical, or biological agents are reviewed in this Chapter. For each type of interaction, data from epidemiological studies of the adverse health effects in humans are generally reviewed first. Studies involving experimental animal models are considered next. Lastly, various effects observed in *in vitro* systems are reviewed. Carcinogenesis is the principal endpoint of interest; non-neoplastic endpoints are viewed mainly in relation to mechanistic considerations.

110. Only a minor fraction of the interacting agents described below are found in the human environment at potentially critical levels. An overview of agents known or suspected to affect human health on their own is given in Table 4. Details of the experimental and epidemiological conditions and results of studies on specified combined exposures may be found in the Appendix and are summarized in Table A.1.

Table 4
Exposure conditions and characteristics of prominent environmental and occupational agents and substances that may produce combined effects with them
[B6]

Agent	Typical environmental exposure (E)	Occupational limit for chronic exposure (O)	Major health endpoint	Estimated contribution to total incidence/ mortality of endpoint (%)	Estimated lifetime risk [a] (%)	Substance with known or suspected combined effect
Ionizing radiation	1.5-4 mSv a^{-1}	20 mSv a^{-1}	Cancer	3-8 (E) ~10 (O)	0.4-1 (E) ~4 (O)	Smoking, asbestos, hormones, arsenic?
UV radiation [W31]	Noontime intensity: UV A: 40 W m^{-2} UV-B: 3 W m^{-2}	350 nm: 150 kJ m^{-2} in 8 h 300 nm: 100 J m^{-2} in 8 h	Skin aging Skin cancer Melanoma	Important (E) >50 (E) ?	- >20 (E) ?	Phototoxicity, allergy with UV sensitizers
Asbestos [I12]		Crocidolite: 0.2 fibers cm^{-3} Other forms: 2 fibres cm^{-3}	Lung cancer Mesothelioma	Low >50%		Smoking
Benzene [M60, W34]	14 μg m^{-3} in indoor air	3.2 mg m^{-3}	Leukaemia	2.5 (E)	0.01 (E)	Substrates for activation/ detoxification systems
Carbon tetra-chloride [W34]	3 μg m^{-3} in indoor air	65 mg m^{-3}	Cancer	0.05 (E)	0.01 (E)	Chloroform
Chloroform [I2]	1.2 μg kg^{-1} d^{-1} from tap water and showering	50 mg m^{-3}	Cancer	0.15 (E)	0.03 (E)	Carbon tetra-chloride
Dioxins/furans [F8]	1.3 μg kg^{-1} d^{-1} dietary intake	50 pg m^{-3}	Cancer	0.1 (E)	0.02 (E)	
Ethylenebisdithio-carbamates (EBDCs) [L49]	n.a.	n.a.	Cancer, adverse reproductive outcomes	0.17 (E)	0.034 (E)	
Polychlorinated biphenyls (PCBs) [G8]	14 ng kg^{-1} d^{-1}	1 mg m^{-3} (42% CI)	Cancer	0.06 (E)	0.01 (E)	

a   Assuming 80 and 40 years of exposure for environmental (E) and occupational (O) levels, respectively. Risks are generally upper bound estimates based on linear extrapolations from high exposures.

A. RADIATION AND PHYSICAL AGENTS

1. Combinations of ionizing radiation

111. Many experiments have been undertaken to investigate the cellular effects of combined exposures of two types of ionizing radiations. In view of their potential radiotherapeutic applications, a wealth of data on the combined action of neutrons, heavy ions, and gamma or x rays was accumulated in the 1970s and early 1980s. This information was reviewed in Annex L, *"Biological effects of radiation in combination with other physical, chemical and biological agents"*, in the UNSCEAR 1982 Report [U6]. The results generally indicate additive effects of combined exposures characterized by the so-called isoaddition of these agents at the cellular level [L10, L47], as described in Chapter II. A few data [E5, L26] were reported on radiation-induced tumours; these, in general, include exposure to internal emitters. The results indicate additive to slightly supra-additive effects for combined exposures, mainly because of the lower dose rates that are involved in the internal exposure to alpha radiation in these experiments. For estimating the risk of carcinogenesis, the effects of combined exposures to more than one type of ionizing radiation are expected to be isoadditive, i.e. to add up in the same way as effects of increments of the same agents, when at least one of the radiations is delivered at a low dose rate (chronic irradiation), as is generally the case for occupational and environ-mental exposure levels.

2. Ultraviolet radiation

112. Ultraviolet (UV) radiation is recognized as the most important initiator and co-factor for human skin carcinogenesis [S29]. It is mainly the skin that is exposed to UV radiation. A study of combined exposures to gamma radiation and UV radiation was presented by Shore, who analysed 12 studies on the incidence of skin cancer in populations irradiated with known skin doses [S29]. In the absence of a proper control (skin exposed to ionizing radiation but not to UV), it was concluded that, at least for combined exposures, the data are compatible with a linear dose-response relationship for ionizing radiation [S29] but that the interaction is unclear. The question of whether relative risk or absolute risk models are more appropriate remains open. From the mechanistic point of view, interaction at the cellular level may be expected, which results in a more or less additive effect for low exposure rates [L52]. The considerable variations in skin cancer among different populations and subgroups seem to reflect the large differences in UV exposures due to latitude and lifestyle and the differences in genetic predisposition to skin cancer due to skin type. The overwhelming dependence of skin cancer on extended exposures to UV prevents conclusive epidemiological data on the interaction of UV with ionizing radiation. Another important factor to take into account in possible interactions is UV-induced suppression of the immune system [B76, L58, N26]

3. Electromagnetic radiation

113. Neither low- nor high-frequency electromagnetic radiation have enough single photon energy to directly damage DNA and therefore cannot be cancer initiators. However, strong electromagnetic fields may modify and stimulate growth [K16, S33], and this has led to the hypothesis that electromagnetic fields may influence cancer development. However, no straightforward inferences from experimental results to exposure situations in occupational or environmental settings have been found at this stage for the combination of electromagnetic and ionizing radiation [B77, B78, B79, U19]. Moreover, there is at present little indication from a mechanistic standpoint for potentially harmful interactions between electromagnetic fields and ionizing radiation at controlled exposure levels in the workplace or the clinic. The possible modulation of radiation effects by heating produced by strong electromagnetic fields is considered in the next Section.

4. Temperature

114. Heat can kill mammalian cells in a predictable and stochastic way [D3]. Elevated temperature is used as a modifier of radiation sensitivity in many therapies to control tumour growth. In combination with ionizing radiation, heat can act synergistically on cell survival, cell proliferation, and cytogenetic damage by, for example, interfering with DNA repair. However, extremely high temperatures, which are generally not found in the workplace or in environmental conditions, are needed in the cells at risk, so heat is not considered as potentially enhancing radiation risk.

5. Ultrasound

115. Ultrasound has achieved widespread use in medical diagnostic and therapeutic procedures. Studies have shown that at the intensities used for diagnostic purposes, ultrasound does not interact with ionizing radiation to cause cytogenetic damage in treated cells, although the yield of sister chromatid exchanges was observed to be slightly increased in one study [K20]. Because cavitation-induced mechanical damage by ultrasound shows high thresholds, this mechanism is of little concern for environmental exposures. Such damage has to be prevented in other situations already caused by single-agent effects.

6. Dust, asbestos, and other mineral fibres

116. Mineral dust and fibres such as asbestos generally act through non-genotoxic mechanisms such as mechanical irritation and cell killing [B13]. The combination of radiation exposure and exposure to dusts and fibres is quite common in industrial settings and in the environment, and these agents are reported in both animal studies and *in vitro* studies to act synergistically at high exposures [B38, H11]. Silicosis was shown to be a risk factor for human lung cancer in metal miners in the 1940s [H9] and is implicated as a modifier of lung cancer risk in radon-exposed underground workers [K49]. Combined exposure to phosphate ore dust, gamma

radiation, and radon daughter products also resulted in elevated lung cancer risks in earlier practices [B74]. Although exposures and thus risks are considerably lower today, mineral dust and fibres still deserve attention because they may interact with radiation, including densely ionizing radiation such as alpha radiation in mining environments, to enhance the risk of cancer.

7. Space flight

117. A special form of combined exposure is experienced in space flights, where a multitude of stressors act in combination on astronauts. This problem has been investigated using animals [A13, A14, V3]. The most important environmental parameter is microgravity. Space radiation effects were comprehensively reviewed by Kiefer et al. [K54], the interaction of microgravity and radiation at the cellular level [H52, K55]. No synergistic actions were found. A very important aspect is a possible reduction of the immune response [S86], which could have an influence on cancer development. The changes of many parameters that are normally stable in experimental work on earth make well designed studies in space potentially important in addressing combined effects of physical agents.

B. RADIATION AND CHEMICAL AGENTS

118. A multitude of natural and man-made chemicals with cancer initiating and promoting potential are present in the human environment and may interact with radiation. Classification based on their mode of action is often difficult, as many have more than one type of action, but at least a crude separation can be made into substances that mainly act by damaging DNA (genotoxic substances) and those that act in other ways (non-genotoxic substances) [C48]. The former group includes chemically active species, or substances that can be activated, bind to or modify DNA directly, or indirectly via radicals. The non-genotoxic substances range from nonspecific irritants and cytotoxins to natural hormones, growth factors, and their analogues. They interact with the regulatory systems of cells and organs and cannot always be considered toxic by themselves. Some are clearly protective, e.g. they scavenge reactive species before they interact with DNA.

1. Genotoxic chemicals

119. Numerous examples of combined exposures to radiation and chemical genotoxic agents can be found in the literature, including studies on the improvement of radiation therapy by simultaneous treatment with a chemical (see Chapter IV). In many cases, supra-additive effects are reported, caused by interaction in the cellular phase and by the high exposure levels involved. The agents include 1,2-dimethylhydrazine (DMH) [S27], N-methyl-N-nitrosourea (MNU), butyl-nitrosourea (BNU), N-ethyl-N-nitrosourea (ENU) [H6, K15, S13, S20, S21, S22], diethylnitrosamine (DEN) [M8, P26], N-2-fluorenylacetamide (FAA), 4-nitroquinoline-1-oxide (4NQO) [H39], bleomycin [D6], and 1,2-dibromoethane

(DBE) [L13]. The effects are dependent on the species, exposure conditions, time of exposure, etc., and sometimes the same chemical is involved in a supra-additive and a sub-additive result. In general, for short exposures to high concentrations and for low chronic concentrations, deviations from additivity are small, if at all existent. In most epidemiological and experimental studies, effects exceeding a level predicted from isoaddition have not specifically been demonstrated.

2. Non-genotoxic chemicals

120. Many chemicals in the human environment or their metabolites do not specifically attack DNA but influence cell proliferation and cell differentiation on an epigenetic level. These include the tumour promoter 12-O-tetradecanoylphorbol-13-acetate (TPA) [H3, L24], carbon tetrachloride, and α-difluoromethylornithine (DFMO) [O12]. These agents act in combination with radiation on the cellular and tissue level of cancer development and can significantly enhance the induction of tumours. Specific mitogens may interfere with regulatory mechanisms and cell-cell signalling, but many substances with a high chemical reactivity act as non-specific irritants or toxicants via membranes or proteins. For example, toxin-induced cell death will induce proliferation in neighbouring cells, which may enhance the progression of premalignant cells. Substances acting in a non-specific manner, for example lipophilic solvents, quite often show highly non-linear dose-response relationships with apparent thresholds. Other agents may interfere with critical cellular processes involved in repairing damage to cellular constituents such as DNA. The assessment of possible synergistic effects at the exposure levels relevant for risk estimation remains very difficult because of the high exposures used in experimental systems and the apparent threshold levels. One important group of chemicals, which includes cysteamine and mexamine, has radioprotective effects; these chemicals scavenge radicals formed by ionizing radiation [M4]. A considerable number of agents may have both genotoxic and epigenetic functionalities such as the base analogue 5-bromo-2'-deoxyuridine (BrUdR) [A15].

3. Tobacco

121. Given the large collective dose from radon and its decay products in non-occupational and occupational settings and the prevalence of active smoking, the combined effect of these two exposures on human health deserves special attention [B27]. A large body of epidemiological evidence from uranium miner studies allows, at least for higher radiation doses, calculating risks and interaction coefficients directly from human data [C46]. However, the fact that tobacco smoke is itself a complex mixture of genotoxic and non-genotoxic substances and even contains some natural radionuclides (the long-lived radon progeny ^{210}Po and ^{210}Pb) makes a mechanistic assessment difficult.

122. Because of the complex composition of tobacco smoke, the issues surrounding combined exposures to radiation and

tobacco smoke are even more difficult to elaborate than for binary combinations. Some 4,000 individual chemical components of cigarette smoke have been identified, and there are probably a number of additional important but unidentified components, for example, extremely reactive, short-lived compounds or those present in very low concentrations [G1]. The complexity of tobacco smoke means that the action in combined exposures with radiation can take place in both the cellular and organ phase of cancer development. Tobacco smoke contains only relatively small amounts of DNA-reactive carcinogens such as nitrosamines, polycyclic aromatic hydrocarbons, and pyrolysis products such as carbolines. Hence enhancing and promoting factors, e.g. catechols, other phenols and terpenes, are important. Discontinuation of smoking progressively reduces the relative risk of cancer development as time since withdrawal increases, probably because of reduced pressure from the action of promoters [W1].

123. In the last few years, joint analyses of original data sets [C1, L18] and meta-analyses of published results [T14] have yielded a detailed assessment of risk patterns and have allowed investigators to test risk models. The most comprehensive and complete analysis of radon-induced health risks was published by Lubin et al. [L18]. The review contains a joint analysis of original data from 11 studies of male underground miners. Data on smoking were available for 6 of the 11 cohorts, but the assessments were limited by incomplete data on lifetime tobacco consumption patterns and the sometimes exotic forms of tobacco use, such as water pipes in the Chinese study [L18]. Single studies for which smoking data could be analysed were generally not informative enough to allow choosing between an additive or a multiplicative joint relationship between radon progeny and smoking. The Chinese cohort seemed to suggest an association more consistent with additivity, while the Colorado cohort suggested a relationship more consistent with a multiplicative interaction. For all studies taken together, the combined influence of smoking and radon progeny exposures on lung cancer was clearly more than purely additive but less than multiplicative and compatible with isoadditivity [B69]. The most recent analyses of the BEIR VI Committee [C46], which were based on an update of these data, suggested synergism between the two agents that is statistically most consistent with a slightly sub-multiplicative interaction. A best estimate from miner data indicates that the lung cancer risk for smokers expressed in absolute terms is higher by a factor of at least 3. To further characterize the association, more detailed data on tobacco use would be needed. Age of starting to smoke, amount and duration of smoking, and type of tobacco were recognized as important determinants of risk. A further handicap of present studies is that the sub-cohorts of lifetime non-smokers exposed only to radon are very small. The statistical power of the conclusions on the radon-tobacco smoke interaction is correspondingly low. Data are available from a study by Finch et al. [F28] of smoke exposure and alpha-particle lung irradiation over the lifespan of exposed rats. The pulmonary retention of inhaled ^{239}Pu was higher, increasing with the concentration of the ^{239}Pu, in smoke-exposed rats than in sham-smoke-exposed rats. This effect on retention resulted in

increased alpha-radiation doses to the lung. Assuming an approximately linear dose-response relationship between radiation dose and lung neoplasm incidence, approximate increases of 20% and 80% in tumour incidence over controls would be expected in rats exposed to ^{239}PuO$_2$ + low-level cigarette smoke and ^{239}PuO$_2$ + high-level cigarette smoke, respectively.

124. Hypotheses on the mechanistic interaction between tobacco smoke and radon were tested by applying the two-mutation clonal expansion model of carcinogenesis of Moolgavkar to data from the Colorado plateau miners [M39]. No interaction between radon and tobacco smoke in any of the three steps (the two mutation steps and clonal expansion) is needed to fit the data, which are clearly supra-additive for radon and smoking combined. The model, however, shows a significant dependence on age at exposure. Quantitatively similar results were obtained by Leenhouts and Chadwick [L9, L57]. A highly significant decrease in excess relative risk with time since exposure is found in miner studies in contrast to findings on lung cancer in survivors of the atomic bombings. This may be explained by microdosimetric considerations. In the case of high-LET alpha radiation from radon progeny, the minimal local dose from one single alpha track averaged over a cell nucleus is already in the range of several hundred milligray, whereas one electron track yields a dose to the nucleus of only 1–3 mGy. This means that even at the lowest possible nuclear dose from alpha exposure, stem cells that are hit carry a multitude of DNA lesions, which may considerably impair long-term cell survival and maintenance of proliferative capacity [B25, B27].

125. Smoking is also of great importance for non-occupational radon exposures in the indoor environment. Until now, little quantitative evidence has come from indoor radon studies. Most of the case-control studies published are inconclusive [A28, K53, P11]. Only one larger study [P11] was indicative of an indoor radon risk and its modification by tobacco that is comparable to what is predicted from miner studies. It remains doubtful whether the results from the many case-control studies under way will in the near future allow narrowing the uncertainties that surround indoor radon risk and the possible interactions with smoking. Emerging study results from Europe based on much longer residence times may offer better statistical power. Several large indoor case-control studies under way will narrow uncertainties in the next few years. First results from the United Kingdom [D33] and Germany [K53, W35] are indicating a lung cancer risk in the range of ICRP projections. However, confidence intervals are relatively large and still include zero risk in most analyses. Because of the limitation of the indoor radon studies, risk estimates based on miner data remain the main basis for predicting lung cancer from indoor radon exposure. A best linear estimate of the risk coefficients found in the joint analysis of Lubin et al. [L18, L35] for the indoor environment indicates that in the United States, some 10%–12%, or 10,000 cases, of the lung cancer deaths among smokers and 28%–31%, or 5,000 cases, of the lung cancer deaths among never-

smokers are caused by radon progeny. About half of these 15,000 lung cancer deaths traceable to radon would then be the result of overadditivity, i.e. synergistic interactions between radon and tobacco. Based on the same risk model, Steindorf et al. [S47] estimated an attributable risk for indoor radon of 4%–7% for smokers and 14%–22% for non-smokers. Because of the many differences between exposed persons and exposure situations in mines and homes and the additional carcinogens such as arsenic, dust, and diesel exhaust in mine air, these figures should be interpreted with caution.

4. Metals

126. Toxic metals are important trace pollutants in the human environment (Table 5). They interact in many ways with cellular constituents and may produce oxidative DNA damage or influence enzyme activity at low concentrations, e.g. by competing with essential metal ions [H38]. Carcinogenic transition metals are capable of causing promutagenic damage, such as DNA base modifications, DNA-protein cross-links, and strand breaks [K7]. The underlying mechanism seems to involve active oxygen and other radicals arising

Table 5
Metals in the environment and effects on humans
[M38, N14, S48]

Metal	Release ^{a} (10^9 g a^{-1})	Main sources of intake and typical levels in the body	Characteristics affecting health
Arsenic	31 (61)	*Source:* food (seafood up to 120 mg kg^{-1}) and drinking water *Concentration in body:* 0.3 mg kg^{-1}	Mutagenic, teratogenic, co-carcinogenic, As^{3+} causes skin cancer
Cadmium	8.9 (85)	*Source:* inhalation (2 µg cigarette^{-1}) and food (0.025 mg kg^{-1})	Mutagenic, teratogenic, co-carcinogenic, causes cancer at multiple sites
Mercury	6.1 (59)	*Source:* metal vapours, food (up to 1 mg kg^{-1} MeHg$^+$ naturally in fish), tooth fillings *Intake:* by inhalation and ingestion 0.2 and 25 µg d^{-1}, respectively	Mutagenic, teratogenic (brain damage), co-carcinogenic, causes sarcomas and renal tumours
Nickel	86 (65)	*Source:* food intake (0.2 mg d^{-1}) *Concentration in body:* 0.007 mg kg^{-1}	Essential element; allergenic, comutagenic, cocarcinogenic, causes nasal sinus cancer
Lead	12 (96)	*Source:* Food, dust, air (0.15 mg d^{-1}) *Amount in body:* steady increase to about 200 mg at age of 60 years	Substitutes for Ca^{2+}, neurobehavioural deficits (decrease in fertility, abortifacient) Low mutagenic and carcinogenic potential (may cause renal adenocarcinoma)
Antimony	5.9 (59)	*Source:* food and tobacco (0.005 mg d^{-1})	Mutagenic as Sb^{3+}, organic antimony compounds used as emetics
Vanadium	114 (75)	*Source:* food (0.01-0.05 mg d^{-1})	Essential element Inhibits Na$^+$/K$^+$ ATPase and drug detoxification enzymes at low concentration Mutagenic, teratogenic, carcinogenic
Zinc	177 (66)	*Source:* food intake (10-50 mg d^{-1})	Essential element with small window of tolerance Clastogenic (causes chromosome aberrations) Causes growth of some tumours at elevated concentrations

a Global values; percentage of anthropogenic contribution is given in parentheses.

from metal-catalysed redox reactions. Cadmium, nickel, cobalt, lead, and arsenic may also disturb DNA repair processes [H48]. Only a few data are available from combined exposures of radiation and metals in human populations; no firm evidence of interactions was observed. However, metals and ionizing radiation have been shown to produce combined effects in many other biological systems (see the Appendix). Especially in underground mining, possible effects from the epidemiologically proven lung carcinogens arsenic, cadmium, chromium, nickel, and antimony [M65] have to be assessed together with high-LET radiation from radon. Arsenic in particular is a major risk factor in combined exposures to

mineral dust, radon, metals, and diesel fumes [K48, T5]. The risk-enhancing effects of iron dust seem to be limited to very high dust concentrations, leading to changes in lung function [B74]. The significance of these data for radiation risk estimation at low dose levels remains unclear.

5. Mitogens and cytotoxicants

127. Although many mitogenic and cytotoxic compounds could have been considered above with genotoxic or non-genotoxic agents, they should be mentioned separately because of their potential to interact with radiation,

principally by virtue of their ability to stimulate cell proliferation. From a mechanistic standpoint, they can be expected to interact in the organ phase of radiation carcinogenesis, but the resulting interaction (sub- or supra-additivity) is not always predictable. Examples of such agents include N-methylformamide (NMF) [L15], caffeine [M11], theobromine, theophylline [Z4], 2-aminopurine, and tributyl phosphate. Many studies assessing deviations from additivity in combined exposures of mitogens/cytotoxicants and ionizing radiation are found in the literature (see Appendix), but the high exposure levels applied and the biological endpoint studied generally do not allow directly transferring the results to carcinogenesis in humans. However, any endogenous or dietary levels of agents influencing stem-cell population size or kinetics will have the potential to modulate response to radiation.

6. Antioxidants, vitamins, and other dietary factors

128. Diet can modify the effectiveness of chemical carcinogens, sometimes by a large factor, and interactions with radiation are found as well [B24, C26, H11, W29]. All classes of substances described in the five preceding Sections III.B.1–5 are found in human food supplies. Actions ranging from subadditive to supra-additive may occur, depending on the specific agent. The radiation risk may be reduced when growth stimuli are reduced as a result of nutritional deficiency or when repair possibilities are optimized. Synergism can be expected where lower levels of radical scavengers or the coenzymes needed for repair increase the yield of effective damage from ionizing radiation or impair the speed and accuracy of cellular recovery from damage. Some of the underlying mechanisms of specific agents have been identified in animal experiments. Tumour-incidence-enhancing effects have been noticed with elevated consumption of, for example, riboflavin, ethanol, and marihuana. Tumour-incidence-reducing effects are found for low-caloric diets, vitamins A, C, K, and E, retinoic acid derivatives (but enhancing effects of artificial beta carotin in some smoker cohorts), selenium, and 3-aminobenzamide. Very important in view of population health are behavioural changes and a tendency to malnutrition in alcohol addicts, which may increase the susceptibility to toxicants in the environment or at the workplace [U18]. In general, the combined action is not specific for radiation but is also found for other carcinogens, and the interaction is dependent on the dosage.

129. In summary, dietary factors are proven modifiers of risk from diverse agents at levels found in human populations and probably also influence the production and repair of endogenously arising lesions. Absence or deficiency of important coenzymes and nutrients on the one side and high levels of directly or indirectly acting mitogens on the other interfere with molecular, cellular, and tissue responses to ionizing radiation. A modulation in the radiation risk may occur in situations where growth stimuli are reduced or increased, owing to nutritional deficiency or surplus or where the number of stem cells at risk is changed. Synergisms are also to be expected where reduced levels of radical scavengers or

coenzymes needed for repair increase the yield of primary damage from ionizing radiation or impair the speed and accuracy of cellular responses to damage. In general, these mechanisms apply to most deleterious agents in the human environment.

C. RADIATION AND BIOLOGICAL AGENTS

130. Many hormones are potent growth stimulators, and there is considerable evidence that they may modify cancer risk. They include thyroid-stimulating hormone (TSH), oestradiol-17 beta (E_2), prolactin, diethylstilbestrol (DES), and androgens in general. Their effect is dependent on tissue, type of hormone, and dosage and is important enough to be kept in mind when analysing radiation risks. Tamoxifen, a synthetic anti-oestrogen, has both cancer risk-enhancing functions (endometrium) and protective properties (breast), depending on the organ [J9]. An important consequence of interaction with hormones is the sex difference in tumour sensitivity, mainly of organs of the reproductive system.

131. Viruses, bacteria and microbial genetic sequences have been shown to play an important role in the development of tumours. Cancer viruses may interact with radiation by mutation or translocation of dormant viral sequences. Experiments so far give no clear indication of any interaction with radiation that influences cancer development. Viruses may induce genotypic and functional changes, i.e. they may act as highly site-specific genotoxic agents in multi-step mechanisms. Highly synergistic effects due to increased sensitivity may arise for some endpoints. Little information is available at present on the mechanism of the induction of gastric cancer by bacteria (Helicobacter pylori).

132. The interaction of several miscellaneous factors with radiation exposure and its role in carcinogenesis has been investigated. Some of these factors are reviewed in the Appendix. The role of others, such as psychosocial factors, remains unclear and is outside the scope of this Annex.

D. SUMMARY

133. Combinations of different types of ionizing radiation show mainly isoadditive effects. For decreasing doses and chronic exposure, the quadratic terms of the dose-effect relationships tend to vanish and the linear terms to prevail, indicating additivity for low-level exposures. Also, for the combination of UV radiation and ionizing radiation, additive effects are expected for low exposure levels. Temperature and ultrasound are not considered to significantly modify radiation risk. The temperature range and the ultrasound intensities necessary for an interaction with radiation are too high to be of relevance for environmental or occupational settings. Mineral dust and fibres, including asbestos, tend to show supra-additive interaction with radiation at high exposure levels. These levels were reached in workplaces in the 1950s and earlier. Today the occupational exposures are lower, but these agents still deserve attention for their potential to enhance risks after combined exposure.

134. At high exposures, a wealth of supra-additive effects between genotoxic chemicals (e.g. alkylating agents) and radiation were recorded. For low-level exposures, there is no mechanistic evidence of combined effects at the cellular level greater than those predicted from isoadditivity. Nor are these agents expected to show a more-than-additive effect at the organ level. However, non-genotoxic agents with mitogenic, cytotoxic, or hormonal activity may interact with radiation in an additive to highly supra-additive manner. High exposures clearly have a considerable potential for enhancing radiation risk during the organ phase of radiation-induced cancer. Since most of these substances show highly non-linear dose-effect relationships with sometimes considerable thresholds, the combined effects with radiation at low concentrations could be ex-

pected not to deviate much from additivity, i.e. to be additive to slightly supra-additive. Special attention has to be given to the combined effects of radiation and tobacco smoke. Tobacco smoke itself is a complex mixture of different genotoxic and non-genotoxic chemicals. Combined exposures to radiation and tobacco smoke show clearly supra-additive effects. Heavy metals and arsenic may generate free radicals or disturb DNA repair mechanisms and therefore may also cause more-than-additive effects. Many human cancers show considerable dependence on lifestyle, nutrition, and other dietary factors. Tumour- incidence-enhancing effects have been reported for riboflavin, ethanol, and high fat diets and incidence-reducing effects for low fat diets and some vitamins. In general, these combination effects have been found not just for radiation but also for other carcinogens.

IV. COMBINED EXPOSURES IN CANCER THERAPY

135. Many modern cancer treatment regimens combine surgery, radiotherapy, chemotherapy, and/or immunotherapy. Generally, combining the different treatments does not mean that the different therapeutic agents interact in a mechanistic manner. Central to the discussion of cancer therapy is, therefore, the distinction between non-interactive and interactive combinations, with the latter being of interest in this Annex. From the rapidly emerging understanding of the action and interaction mechanisms of different agents in combined modality therapy, information relevant to possible interaction mechanisms in environmental and occupational exposure situations may be obtained.

136. Central to cancer therapy is the relationship between the desired and undesired effects of the therapies chosen. This relationship is defined as the therapeutic index ratio or gain [G21, H41]. The gap between the sigmoid curves of tumour cure (tumour control probability) and dose-limiting toxicity to normal tissue (normal tissue complication probability) is the therapeutic index (Figure VI). The goal of cancer therapy is to increase the therapeutic index by separating the two curves. The therapeutic index is increased when the tumour control probability curve is displaced to the left of the normal tissue complication probability curve. This can be achieved in radiotherapy by altering the exposure schedule. Important techniques are hyperfractionation, accelerated fractionation, split-course techniques, interstitial irradiation, manipulation of target volumes, shrinking field techniques and others. Another approach to increasing the therapeutic index is to combine radiotherapy with chemotherapy. Drug-ionizing radiation interaction in therapy is useful only when it leads to a further separation of the curves, not just to their displacement [K43].

137. It should, however, be clearly noted here that the final goal of tumour therapy is tumour control and therefore cell death (apoptosis, necrosis) or blockage of cellular growth (loss of proliferative capacity, differentiation, senescence). These effects are mostly deterministic and often mechanistically

different from the stochastic radiation effects that are of concern in radiation protection. Emphasis is therefore placed on the mechanisms of interaction between the drugs and radiation that reveal possible mechanisms of interaction between chemical agents and radiation under environmental and normal occupational settings. Clinical results will be mentioned only if mechanistic information with relevance for low dose effects can be provided.

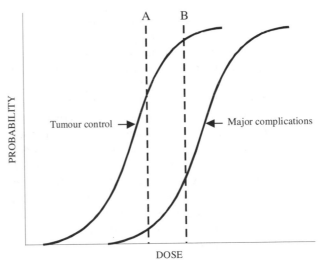

Figure VI. Sigmoid curves of tumour control and complications [H41].
A: Dose for tumour control with minimum complications.
B: Maximum tumour dose with significant complications.

A. MECHANISMS OF INTERACTIONS

138. Publications on the mechanisms of interaction between radiation and drugs are numerous but often lack precise and quantitative information. Factors on which the interaction of these two treatment modalities depends include the type of tumour and normal tissue involved, the endpoints studied, the drug and its dose level, the radiation dose, dose rate, and

fractionation, and the intervals between and sequencing of the combined treatments. Chemotherapy could slow the process of cell repopulation after radiotherapy or it could synchronize the cell cycle. Moreover, tumour reduction by chemotherapy could improve tumour oxygenation, thus increasing the effect of radiotherapy. At the cellular level, inhibition of repair of sublethal and potentially lethal radiation damage by anticancer drugs is probably the most important mechanism of radiosensitization. Exploitable mechanisms in combined chemotherapy and radiotherapy treatment can be described under four headings, as was originally done by Steel and Peckham [S1, S23, S46]: spatial cooperation, independent cytotoxicity, protection of normal tissues, and enhancement of tumour response.

139. Spatial cooperation describes a non-interactive combination of radiotherapy, chemotherapy, surgery, and other therapeutic strategies that act at different anatomical sites. The commonest situation is where surgery and/or radiation is used to treat the primary tumour and chemotherapy is added as an adjuvant to attack remaining local tumour cells and distant metastases. There is an analogous situation in the treatment of leukaemia, where chemotherapy is the mainline treatment and radiotherapy is added to deal with the disease in anatomical sites, e.g. the brain, protected from chemotherapeutic attack by vascular constraints or by blood barriers. Spatial cooperation still appears to be one of the main clinical benefits of combination modality treatment. This mechanism does not require interactive processes between drugs and radiation.

140. Independent cytotoxicity describes another form of non-interactive combination of therapeutic modalities. If two modalities can both be given at full dose, then even in the absence of interactive processes the tumour response should be greater than that achieved with either modality alone. The cost of this improvement on tumour response is that the patient has to tolerate a wider range of toxic reactions in normal tissues (within and outside of the radiation field). As with spatial cooperation, the mechanism of independent cytotoxicity does not require interactive processes between drugs and radiation. Independent cytotoxicity can even tolerate a subadditive interaction of the modalities and still produce an increase in therapeutic gain. The relative extent of reduction in toxicity to normal tissue within the radiation field is the critical parameter of this mechanism.

141. The protection of normal tissues requires an antagonistic interaction of the combined modalities. Since two toxic agents usually tend to produce more damage than either agent alone, it would seem rather unlikely that chemotherapy in conjunction with radiation could reduce the damage to dose-limiting normal tissue. However, there are well-documented situations in which certain cytotoxic drugs increase the resistance of normal tissue to radiation or to a second cytotoxic treatment.

142. Studies of this seemingly contradictory mechanism have concentrated on the bone marrow and the intestinal epithelium. It has been shown by Millar et al. [M51, M52] that in the bone marrow, the most effective cytotoxic agent, cytara-

bine, does not modify stem-cell radiosensitivity; instead, it stimulates enhanced repopulation by the surviving stem cells. This phenomenon is highly dependent on the timing of the two modalities. Maximal radioprotection is achieved when the drug is given two days before radiation. In the small intestine, microcolony survival was increased when cytosine arabinoside was given 12 hours before irradiation [P24]. Other cytotoxic drugs with radioprotective action are cyclophosphamide, chlorambucil, and methotrexate [M51]. Recently, a topoisomerase II inhibitor (etoposide) was shown to increase the radioresistance of the bone marrow when given one day before whole-body irradiation [Y1].

143. Normal tissue can be protected from radiation effects by radioprotective agents. Increasing the differential between tumour and normal tissue radiosensitivities would give a therapeutic advantage. Radioprotectors can thus be used as selective protectors against radiation damage to normal tissue, allowing higher curative doses of radiation to be delivered to tumours.

144. Chemical radioprotectors target the detoxifying mechanisms of the cell, in particular the antioxidant enzymes that are available for removal or detoxification of the reactive oxygen species and their products formed by the action of ionizing radiation. By far the most widely studied class of radioprotective agents is the thiols, and the most important non-protein thiol present in cells is glutathione. Other classes of agents conferring radioresistance to normal tissue are the eicosanoids, which are biologically active compounds derived from arachidonic acid, the lipoic acids, and calcium antagonists (reviewed in [M51, M56]). The effects of biological response modifiers such as the cytokines IL-1 and TNF_α as radioprotectors in normal tissue have been discussed in recent reviews [M53, M54, N12, N13, Z11, Z12].

145. Relative enhancement of tumour response is commonly perceived to be the principal aim of adding chemotherapy to radiotherapy. A wide variety of biological mechanisms have been proposed to explain interactions between radiation and therapeutic agents. In the context of this Annex, this kind of interaction is the most important mechanism with respect to environmental and normal occupational settings.

146. DNA adduct repair regularly involves strand scissions by repair enzymes. Conversion of repairable into lethal DNA damage may occur if a DNA-repair-associated single-strand break combines with a radiation-induced single-strand break to produce new DNA double-strand breaks. This mechanism has been suggested for the interaction of cisplatin and radiation. A similar mechanism, the production of double-strand breaks by combining single-strand breaks, may occur when topoisomerase I or II inhibitors and radiation are combined.

147. Many drugs inhibit the repair of radiation damage. Antitumour antibiotics (e.g. dactinomycin and doxorubicin), antimetabolites (e.g. hydroxyurea, cytarabine, and arabinofuranosyl-adenine), and alkylating agents and platinum analogues (e.g. cisplatin) have been shown to inhibit radiation-

induced DNA damage repair. Repair inhibition has been detected in a number of ways, including removal of the shoulder on the cell survival curve, inhibition of split-dose recovery, and inhibition of delayed plating recovery.

148. Cell-cycle synchronization exploits the fact that many cytotoxic drugs and radiation show some degree of selectivity in cell killing at certain phases of the cell cycle. Antimetabolites show a maximum effect on cells undergoing the S phase. Radiation sensitivity is highest in the G_2/M phase. There is, therefore, an attractive possibility of complementary action between drugs and radiation. The most attractive possibility seem to be the interaction between microtubule and topoisomerase poisons and primary DNA-damaging agents such as radiation.

149. Activation of apoptosis by differential pathways increases cell killing during tumour therapy and is therefore another possibility for combined action of radiation and chemotherapeutic drugs. Ionizing radiation may activate the apoptotic process by a DNA damage-p53 dependent pathway, whereas taxoids like paclitaxel may activate a pathway downstream of p53 by phosphorylation of Bcl-2. There is, therefore, a possibility that radiation-induced cell killing can increase, even in p53-deficient tumours. The involvement of apoptosis in radiation-induced cell killing has recently been studied extensively [B75, D34, H44, H49, M69, O19].

150. Reduction of the hypoxic fraction by bioreductive drugs targeted at hypoxic tumour cells increases tumour radiosensitivity. Most promising here is the development of dual-function drugs specific to hypoxic cells and with intrinsic cytotoxic activity (e.g. alkylating activity).

B. SECONDARY CANCERS FOLLOWING COMBINED MODALITY TREATMENT

151. The successful treatment of cancers involves radiation therapy and/or multi-agent chemotherapy, each of which is used either as primary therapy or as an adjunct to therapy of the primary tumour, and it often includes surgery. With further improvements in modern cancer therapy, the duration of survival and the curability of many patients has increased up to 45%. However, along with this progress has come a recognition of the long-term complications of therapy, such as secondary cancers (reviewed in [T30]). Although other clinical consequences in non-target tissues are known, the main focus in this Section is on secondary cancers after combined modality treatments. Secondary cancers resulting from the combined effect of radiotherapy and tobacco smoke are discussed in the Appendix, Section B.3.

152. No one specific type of secondary cancer is seen after therapeutic irradiation. Secondary cancers can occur after any initial cancer, when survival surpasses the latent period. Radiation-induced leukaemias begin to appear after 3–5 years. Solid cancers typically emerge more than 10 years after treatment but may occur earlier in particularly susceptible individuals [F9, G32, T31, V10]. When the risk

of secondary solid cancer is elevated, it rises with increasing radiation dose to the site and with increasing time since treatment and persists as long as 20 years.

153. The predominant secondary cancer associated with chemotherapy is acute non-lymphocytic leukaemia (ANL). Most ANLs have occurred after treatment with alkylating agents or nitrosoureas. The findings are similar for Hodgkin's disease, paediatric cancers, ovarian cancer, multiple myeloma, polycythemia vera, gastrointestinal cancers, small-cell lung cancer, and breast cancer [B22, B36, B63, B65, C14, F9, G32, G33, R22, T31, T32, V10]. The risk for leukaemia rises with increasing cumulative dose of the alkylating agent or nitrosourea. A few ANL cases were reported following combination chemotherapy, including teniposide or etoposide. The leukaemias differ from those that follow alkylating agents in that they occur sooner and that specific chromosomal abnormalities are induced [P9, P29, P30, W32]. Few solid tumours have been linked to chemotherapy. Bladder cancer has been associated with cyclophosphamide treatment, and risk is dependent on the cumulative dose of cyclophosphamide [P4, T33, T36, W33]. Excess bladder cancer risk following treatment with both radiotherapy and cyclophosphamide was as expected from a summation of the individual risks. Bone sarcomas have also followed treatment with alkylating agents [T34]. In general, the risk for solid tumours after chemotherapy alone has been difficult to evaluate because too few patients survived long enough after treatment by chemotherapy alone. At present, several cohort studies are under way to assess this risk.

154. Earlier reports indicated a distinctive pattern of secondary cancers after treatment of childhood malignancies [M1]. The most common secondary cancer was bone sarcoma, followed by soft tissue sarcomas, leukaemias, and cancer of the brain, thyroid, and breast. The cancers showing the highest increases compared with the usual distribution of childhood cancers were retinoblastoma, followed by Hodgkin's disease, soft tissue sarcomas, Wilms' tumour, and brain cancer. This difference may reflect both the genetic predisposition to develop multiple tumours in the case of heritable retinoblastoma, soft tissue sarcoma of Li-Fraumeni syndrome, and possibly the immune dysfunction associated with Hodgkin's disease. To summarize the findings from recent study results [B68, O17, S6, S7], there is little indication that heritable sensitivity to treatment is a significant component of secondary cancer, but intensive multiple agent therapy used in childhood cancer treatment acts as an independent aetiological factor for a second tumour. The risk for a second malignant neoplasm after cancer in childhood is considerable. Absolute risks up to 7% over 15 years following diagnosis of the primary cancer were found for Hodgkin's disease [B68]. This amounts to an excess relative risk (ERR) of about 17, with breast cancer contributing most. A follow-up study in the Nordic countries showed a significant increase in the ERR from a low of 2.6 in patients first diagnosed in the 1940s and 1950s to 5.9 for cohort members included in the late 1970s and 1980s, indicating that newer treatments are not only more successful but also carry a higher long-term risk [O17].

155. In patients with bone sarcomas as the secondary tumour following childhood cancer therapy, the effects of radiation therapy, chemotherapy, and combined modality treatment have been analysed [H43, T34]. The risk for bone sarcoma rose dramatically with increasing doses of radiation with a linear trend. Patients with heritable retinoblastoma had a much higher risk for secondary bone sarcoma, but their response to radiation was similar to that of patients with other childhood cancers. In addition to the radiation dose, the exposure to chemotherapy was evaluated. There was an independent effect of exposure to alkylating agents in the risk for bone cancer, i.e. radiation and alkylating agents acted additively. The risk rose with increasing cumulative dose of the alkylating agents. The effect of alkylating agents was much smaller than that of radiation, and in the presence of radiation at the site of the bone sarcoma, the alkylating agents added little to the risk.

156. Thyroid cancer risk after treatment of childhood cancer is increased 53-fold compared with general population rates [T35]. The risk for thyroid cancer rose with increasing radiation dose. There was no increased risk of thyroid cancer associated with alkylating-agent chemotherapy.

157. There was a sevenfold increased risk of secondary cancers after treatment of acute lymphoblastic leukaemia (ALL) [N22]. Most of this risk was due to a 22-fold increase in brain cancers. The brain cancers occurred in patients diagnosed with ALL before the age of five years and who received cranial or whole-body irradiation.

158. Among 29,552 patients with Hodgkin's lymphoma, 163 cases of secondary leukaemia after treatment of the primary disease indicated a considerable risk [K44]. There was no difference in the relative risk of secondary leukaemias from chemotherapy alone (MOPP regimen) and chemotherapy plus radiotherapy. A relatively small risk for leukaemia was seen after radiation alone, and this risk increased with radiation dose. The risk did not vary significantly or consistently across radiation doses for any given number of chemotherapy cycles but increased consistently with more cycles of chemotherapy in each radiation dose range.

159. Significantly elevated risks for secondary solid tumours (lung, non-Hodgkin's lymphoma, stomach, melanoma, bone, and connective tissue) were reported in patients treated for Hodgkin's disease [T31]. The pattern of secondary tumours was distinctive and was similar to the distribution of cancers seen in immunosuppressed populations, such as renal transplantation patients or patients with non-Hodgkin's lympho-ma. All cancers of the stomach, bone, and connective tissue occurred within areas previously treated with radiation therapy. All those who developed lung cancer had received radiation therapy and smoked. For breast cancer, a fourfold elevated risk was reported in Hodgkin's disease patients after 15 years of follow-up. The highest risk was in women irradiated before the age of 30 [H19]. Comparable results were reported from a Dutch study [V11]. These authors reported an overall relative risk of 3.5 for secondary cancers after Hodgkin's disease. Significant increases in relative risk of

34.7, 20.6, 8.8, 4.9, 3.7, 2.4, and 2.0 were reported for leukaemia, non-Hodgkin's lymphoma, soft tissue sarcoma, melanoma, lung cancer, urogenital cancers, and gastrointestinal cancers, respectively. Risk factors for leukaemia were chemotherapy and host factors; for non-Hodgkin's lymphoma they were combined modality treatment rather than single modality treatment or host factors. For lung cancer the risk factors were strongly related to radiation therapy, while an additional role for chemotherapy could not be demonstrated.

160. Significant excesses of ANL followed therapy for non-Hodgkin's disease with either prednimustine, a derivative of nitrogen mustard, or with regimens containing mechlor-ethamine and procarbazine, for example MOPP therapy, (nitrogen mustard, vincristine, and procarbazine prednisone) [T6]. Chlorambucil and cyclophosphamide were associated with smaller increased risk of ANL. In this study, radio-therapy did not add to the leukaemogenicity of alkylating agents. This finding should be interpreted cautiously, however, because of the small number of patients and the large number of parameters evaluated.

161. Few studies have evaluated the late effects of adjuvant chemotherapy and radiotherapy in breast cancer. The inter-action of alkylating agents with radiation in producing leukaemia in women treated for breast cancer was investigated in a cohort of 82,700 patients in the United States [C29]. Based on 74 cases, the risk of ANL was significantly increased after radiotherapy alone (relative risk = 2.4, 7.5 Gy mean dose to the active marrow) and alkylating agents (melphalan and cyclophosphamide) alone (relative risk = 10). Combined therapy resulted in a more-than-additive relative risk of 17.4. The most common solid cancer that occurs after breast cancer is contralateral breast cancer, but fewer than 3% of these tumours could be attributed to radiation [B2]. The risk was highest in women treated at young age (under 45 years). The usefulness of such studies is still hampered by the fact that an important proportion of patients developing primary tumours might already belong to a genetically more sensitive subpopulation [E11]. In addition, combined treatments might be more often used in more advanced stages of tumours needing higher total doses or more cycles for cure.

C. SUMMARY

162. A large number of chemotherapeutic drugs are used in clinical cancer therapy in combination with radiation. The main ones in use or proposed for use are described in the Appendix, with emphasis on the mechanisms of interaction between the drugs and radiation that may have relevance for combined effects between chemical agents and radiation even at the low exposure levels found in controlled environmental and occupational settings. The main findings on modes of action and combined effects are summarized in Table 6.

163. The predominant secondary cancer associated with chemotherapy is ANL and, to a lesser degree, bladder cancer. No one specific type of secondary cancer follows

Table 6
Combined modality treatments in tumour therapy

Class of agent	Agent(s)	Mode of action	Critical target(s)	Main effects	Combined effects with radiation
Adduct forming agents [a]					
Alkylating agents	Nitrogen mustards (mechlorethamine, melphalan, chlorambucil, cyclophosphamide)	Addition of alkyl group to nucleophilic sites in biomolecules	DNA, proteins	DNA adducts, cross-links	Mainly additive (isoadditive) to borderline supra-additive
Nitrosoureas (Chloroethylnitrosoureas)	BCNU, CCNU, MeCCNU	Bifunctional with two reactive intermediates: isocyanate reacts with amine groups (carbamoylation reaction) and chloroethyl carbonium ion with nucleophilic sites (alkylation)	DNA	DNA adducts, cross-links, glutathione depletion	Additive to borderline Supra-additive for alkylation Supra-additive for glutathione effects
Platinum coordination complexes	Cisplatin [cisdiamino-dichloroplatinum (II)], carboplatin [diaminecyclobutane-dicarboxylatoplatinum (II)]	Bifunctional cross-linker	DNA, RNA, proteins	Cross-links, DNA adducts	Supra-additive, double-strand breaks result from the combination of strand breaks associated with DNA-platinum repair and radiation-induced strand breaks
Antimetabolites [b]					
Antifolates	Methotrexate	Depletion of intracellular nucleotide pools	DNA	Impaired DNA repair and synthesis	Supra-additive
Pyrimidine analogs and precursors	BUdR, IUdR (thymidine analogs)	Replacement of thymidine in DNA	DNA	Impaired DNA repair, synthesis and transcription	Supra-additive
	5-FU, precursor for FUdR	Depletion of nucleotide pools by thymidylate synthase inhibition	DNA, RNA	Depletion of dTTP pool	Supra-additive
	FUdR (uridine analog)	Replacement of uridine in RNA and thymidine in DNA	DNA, RNA	Impaired DNA repair and synthesis, impaired transcription	Supra-additive
Hydroxyurea		Depletion of nucleotide pools by ribo-nucleoside diphosphate reductase inhibition	DNA	Inhibition of DNA synthesis and repair	Supra-additive
Natural products					
Antitumour antibiotics	Anthracyclines (doxorubicin, daunomycin, epirubicin, idarubicin)	Free-radical formation and/or topoisomerase II inhibition Increased tumour oxygenation by inhibition of respiration (doxorubicin)	DNA, mitochondria	Protein-associated DNA breaks	Additive for free radical formation; supra-additive for topoisomerase II inhibition; supra-additive for respiratory chain inhibition

Table 6 (continued)

Class of agent	Agent(s)	Mode of action	Critical target(s)	Main effects	Combined effects with radiation
Antitumour antibiotics (continued)	Bleomycin	Produces active oxygen species after intercalating with DNA	DNA	Radiomimetic drug	Isoaddition
	Mitomycin C	Bifunctional alkylating activity after reduction of quinone entity; highest activity in reducing environment	DNA	DNA adducts, cross-links	Supra-additive when given before radiation; pronounced cytotoxic effect of the drug against radioresistant hypoxic cells
	Actinomycin D	Intercalates into DNA	DNA	Inhibition of transcription	Additive to supra-additive
Microtubule poisons	Vinca alkaloids (vincristine, vinblastine, desacetyl-vinblastine)	Inhibition of tubulin polymerization, induction of microtubule disassembly	Spindle apparatus	Cells blocked in mitotic phase	Additive to supra-additive when given before radiation
Topoisomerase poisons	Epipodophyllotoxins (etoposide, teniposide)	Inhibition of topoisomeras-II	DNA	Protein-associated DNA breaks	Supra-additive when applied together, repairable radiation-induced DNA damage is transformed into lethal damage
	Camptothecin, topotecan	Inhibition of topoisomeras-I	DNA	Blocking of replication forks (S-phase specific), single-strand breaks	Supra-additive
Oxygen	Hyperbaric oxygen	Increased oxygenation of tumour			Additive to supra-additive
Bioreductive drugs	Quinone alkylating agents (EO9, porfiromycin)	Metabolic reduction in anoxic tumour cells to alkylating agents, mechanism similar to mitomycin C	DNA	DNA adducts, cross-links	Supra-additive when given before radiation, pronounced cytotoxic effect of the drug against radio-resistant hypoxic cells
	Nitroimidazoles (metronidazole, misonidazole, etanidazole, RSU 1096)	Metabolic reduction in anoxic tumour cells to cytotoxic agents; additional alkylating function in RSU 1069	DNA	Oxidative stress under aerobic conditions, DNA-covalent reaction products under hypoxic conditions	Supra-additive
	Benzotrizine di-N-oxides (tirapazamine)	Metabolic reduction in anoxic tumour cells to cytotoxic agents with production of free radicals	DNA	DNA damage by free radicals	Supra-additive

a   Effects cell-cycle independent.
b   Effects mainly cell-cycle dependent.

radiotherapy. In general, there are independent effects of exposure to alkylating agents and radiotherapy. For secondary solid tumours, radiation is the main risk factor, while a role for chemotherapy has been demonstrated in some cases. For lung cancer, an additional role of smoking was reported. Host factors, for example, age of diagnosis and treatment for breast cancer, are additional risk factors. For secondary leukaemias the main risk factors are chemotherapy and host factors. There is only a small increase in this risk due to radiation. The effect on secondary cancers of increasingly used adjuvant treatments with topoisomerase I or II inhibitors, microtubule poisons (discussed in the Appendix), and hormone treatment is as yet unknown. In summary, secondary, treatment-related cancers are observed increasingly because of the long-term success of the initial treatments. At present, no important synergistic effects between ionizing radiation and other agents are known. Further investigations are needed to assess and to develop strategies to reduce this potential complication.

V. EFFECTS OTHER THAN CANCER

164. Given that there is an overlap in the development of radiation-induced cancer and other biological effects, such as cellular effects, deterministic effects, and teratogenic effects (see Chapter II), results of combined exposures to radiation and other agents for these other effects might give some information on the mechanistic aspects of possible interactions for radiation carcinogenesis. In this Chapter effects other than cancer following combined exposures are reviewed with the aim of concluding whether the agents and low-level radiation interact. It must be kept in mind that most deterministic effects and many aspects of teratogenesis are a result of cytotoxicity and cytolethality having apparent threshold levels in tissue. Qualitatively, the results reported in this Chapter can be considered as interactions occurring at the cellular and tissue/organ level of radiation-induced cancer. In view of the many data available and the aim of this Annex, only effects in humans and mammalian organisms are reviewed.

165. Especially in earlier occupational situations, concomitant exposures to other agents may have caused pathological changes in organs such as the lung, with considerable implications for exposure-dose conversion coefficients and possibly also for target sensitivity towards stochastic effects from ionizing radiation. For example, in the studies of miners, reduced pulmonary function and early onset of silicosis from exposure to dust is also correlated with end points of interest in the context of this Annex [K21, K49, N25]. Although these combined exposures have little relevance at present, they contribute to the uncertainties involved in drawing inferences from historic occupational risks and applying them to modern-day working environments and non-occupational settings.

A. PRE- AND POST-NATAL EFFECTS

166. The effects of x-irradiation and hyperthermia at 43°C both individually and in combination on mouse embryos were investigated by Nakashima et al. [N1]. Cultured eight-day B6C3F$_1$ embryos were exposed to 0.3–2 Gy from x rays, 5–20 minutes of heating, or 5 minutes of heating and irradiation at 0.3, 0.6, and 0.9 Gy. Irradiation alone at 0.3 Gy showed no apparent effect on embryonic development, but irradiation at 0.6–2 Gy caused a dose-dependent increase in malformed embryos. Heating alone for 5 minutes produced no malformed embryos, while heating for 10–20 minutes caused malforma-

tions as a function of heating time. Combined treatments produced higher frequencies (22%–100%) of malformations than would have been expected from considering the sum of the separate treatments (0%–42%). The malformations observed were primarily microphthalmia, microcephaly, and open neural tubes. The results indicate that in cultured mouse embryos irradiation combined with a non-teratogenic dose of hyperthermia increases the formation of malformed embryos. The interaction is most probably in the cellular phase of effects development (see Section II.B).

167. The interaction of exposures to heavy metals with radiation was studied during the pre-implantation stage in mice by Müller and Streffer [M3]. At this stage, placental protection against chemical attack is lacking, and low cell numbers limit replacement of damaged cells. Of the metals arsenic, cadmium, lead, and mercury tested in micromolar concentrations, arsenic showed no interaction with radiation [M59] and cadmium and lead showed supra-additivity only for single endpoints: morphological development for cadmium and micronuclei formation for lead. Mercury, however, showed considerable interaction for morphological development and cell proliferation. A classical construction of the envelope of additivity in the range 0–3 Gy from x rays and 0–8 µM of mercury chloride showed synergism, i.e. an interaction effect exceeding isoadditivity. However, there was no effect on micronuclei formation by mercury. Time factors were shown to play an important role in these experiments [M57, M58]. For an enhancement of radiation risk, exposure to mercury (3 µM) had to start immediately after irradiation and to last for an extended time period afterwards (112 hours). The interaction is probably in the cellular phase, but the fact that a 24-hour exposure has little effect speaks against inhibition of repair of radiation-induced DNA damage as the only mechanism of mercury toxicity in this system.

168. The interaction of ionizing radiation with cadmium, which at higher concentrations is teratogenic by itself, was studied by Michel and Balla [M37]. Metal exposure (2 mg kg^{-1}) on day 8 of gestation significantly increased exencephaly and eye anomalies. They found considerable antagonistic effects on survival, growth retardation, and developmental malformations for combined exposures to CdCl$_2$ and x rays (0.5 and 1 Gy) in NMRI mouse embryos. Since the metal exposure had to precede radiation for an

antagonistic effect, induction of maternal metallothionein was proposed as a protective mechanism. However, application of metallothionein shortly before $CdCl_2$ exposure exerted no protective effect. $HgCl_2$ alone induced a low rate of exencephaly, and combined treatment with x rays resulted in additivity of single-exposure effects in the range tested (0.5 and 1 Gy, 2 mg kg^{-1}). No conclusion on possible implications for chronic, low-level exposures and radiation carcinogenesis is possible.

B. GENETIC AND MULTI-GENERATION EFFECTS

169. An understanding of mutations in germ cells and of carcinogenic and teratogenic effects from germ-cell exposure is of great importance for assessing health risks in future generations [N19, N24]. The same holds true for potential combined effects in these exposure situations. The hereditary effects of radiation have been considered in most previous reports of the Committee. Experimental studies in animals and emerging human evidence from epidemiology considered here deal mainly with combined modalities in tumour therapy. The combined effects of cytotoxic substances used in tumour therapy on mouse stem cells and gamma-ray doses of 5 and 9 Gy were studied using the spermatocyte test [D7]. Most of the chemicals tested showed additive effects when combined with doses in the ascending part of the dose-response curve and potentiating effects when combined with doses in the curve's descending part. This has generally been considered additional confirmation that any kind of spermatogonia depletion is sufficient to modify the genetic response of stem cells. The chemicals mitomycin C and N,N',N''-triethylen-ethiophosphoramide (thiotepa) induced very low yields of translocations after single treatments. In combined treatments with a dose of 5 Gy, mitomycin C was found to have a subadditive effect and thiotepa, an additive effect. Combined with a dose of 9 Gy, the compounds potentiated the effect of radiation.

170. Based on the generally accepted hypothesis that most cancers are multifactorial in origin, perinatal and multi-generation carcinogenesis should be considered in depth. Nevertheless, the consequences of prenatal exposures and of prenatal events are often ignored [T15], partially because it is not possible at present to quantify the role of prenatal exposures to carcinogens/mutagens in determining or modulating the risk of cancer in humans. Tomatis [T15] listed prenatal events important to the occurrence of cancer as the consequence of one of the following:

(a) the direct exposure of embryonal or fetal cells to a carcinogenic agent;

(b) a prezygotic exposure of the germ cells of one or both parents to a carcinogen/mutagen before mating; or

(c) a genetic instability and/or a genetic rearrangement resulting from selective breeding, which may favour a deregulation of cellular growth and differentiation.

Because they involve both germ and somatic cells, studies of prenatal carcinogenesis are sometimes difficult to interpret but are essential for a more accurate estimation of the risks attributable to environmental agents. At the same time they may contribute to an understanding of some of the mechanisms underlying individual variability in the genetic predisposition to cancer. With regard to combined effects, no new or additional mechanisms are apparent for genetic and multi-generation effects.

C. DETERMINISTIC EFFECTS

171. Deterministic effects of ionizing radiation are the result of exposures that cause sufficient cell damage or loss of proliferative capacity in stem cells to impair function in the irradiated tissue or organ. For a given deterministic effect, a large proportion of cells must generally be affected, so that in most cases there are considerable thresholds in the range from tenths of a sievert to several sievert. Deterministic effects were reviewed by the Committee in Annex I, "Late deterministic effects in children", of the UNSCEAR 1993 Report [U3]. Although deterministic effects are practically excluded in controlled settings, side effects in tissue adjacent to treated tumours and localized effects in skin, eyes, and lungs must still be considered when assessing human health risks. Since loss of the ability to divide is also a result of DNA damage, many of the molecular mechanisms that modulate combined effects in carcinogenesis also modulate deterministic effects. In this context, the scavenging of radiation-induced radicals by scavengers such as cysteamine will also exert antagonistic effects for deterministic endpoints. Because this field is of limited relevance for this Annex, only a few examples from this poorly explored field are given below.

172. There are suggestions from clinical findings that pre-existing diabetes exacerbates radiation injury to the retinal vasculature. Gardiner et al. [G2] studied this phenomenon in streptozotocin-induced diabetic rats. In both diabetic and control rats, the right eye was irradiated with 90 kVp x rays to 10 Gy and the prevalence of acellular capillaries in trypsin digests of the retinal vasculature was quantified 6.5 months after irradiation. Diabetes as well as irradiation led to a statistically significant higher prevalence of acellular capillaries. The net increase in acellular capillaries following irradiation was much greater in rats with an eight-month term of pre-existing diabetes (180%) than in those that had been diabetic for only three months (36%). These results suggest a synergistic relationship between pre-existing diabetes and ionizing radiation in the development of retinal vasculopathy that seems to depend on the duration of diabetes before radiation exposure.

173. Ivanitskaia [I4] studied the reduction of spermatogenesis and of activities of key enzymes as a result of single or combined action of ionizing radiation and mercury in rats. The combined biological effects seemed to be close to the sum of the effects caused by the single agents.

174. Higher acute radiation doses are known to impair immune functions at least temporarily. Generally, immune

deficiencies are the condition being considered. However, overstimulation of the immune functions or the emergence of new antigenic sites as a secondary effect of radiation damage have to be considered as well. A stimulation effect at high doses was described by Lehnert et al. [L14] in inbred C57BL mice. The mice were irradiated with 10 Gy delivered to the thorax 24 hours prior to the induction of graft-versus-host disease by the injection of allogeneic lymphoid cells (2 10^7 cells). In mice only irradiated or only injected, survival was 100% at 250 days. In contrast, a combination of the two treatments, graft-versus-host disease and partial-body irradiation, resulted in a mortality of 83% and a mean survival time of only 29 days, indicating strong synergy between graft-versus-host disease and partial-body irradiation. From histological studies of the lung, it appeared that about 40% of the deaths occurring after combined graft-versus-host disease and partial-body irradiation (PBI) treatment might be attributable to pneumonia. The cause of death in the remaining mice that received combined treatment is unknown. Mice receiving combined PBI/lymphoid cell treatment also develop a characteristic skin lesion that is not seen in non-irradiated mice and that is confined to the irradiated area. A first indication of the mechanism involved is the fact that the amplifying effect of pre-induction partial-body irradiation on the timing and severity of graft-versus-host disease is similar to the effect that would be produced by an increase in the number of effector cells. Such a proliferative response should display a highly non-linear dose-response relationship with an apparent threshold similar to immune deficiencies based on widespread stem-cell killing. Therefore, no direct relevance of these findings for much lower occupational or environmental exposures is apparent.

175. Guadagny et al. [G18] described an increase in immunogenicity of murine lymphoma cells following exposure to gamma rays *in vivo*. On the basis that mutagenic compounds such as 5' (3,3'-dimethyl-1-triazeno)-imidazole-4-carboxamide (DTIC) cause a marked increase in immunogenicity in murine lymphoma cells *in vivo* or *in vitro*, they then conducted further experiments to test whether ionizing radiation would be able to affect the immunogenic properties of cancer cells in a mouse leukaemia model. Male $CD2F_1$ mice were inoculated with histocompatible L1210 Ha leukaemia cells and treated with 4 Gy of whole-body irradiation. A number of transplant generations were carried out with leukaemic cells collected from irradiated donors, generating a radiation-treated line. The immunogenicity of radiation-treated cell lines increased significantly compared with that of the L1210 Ha line as early as after three passages *in vivo*. However, no strong transplantation antigens comparable to those elicited by treatment with DTIC were found in radiation-treated cell lines, even after a number of transplant generations. The combination of bis-chloroethyl-nitrosourea and the weakly antigenic radiation-treated cell line elicited a strongly synergistic immune response of the host. Moreover, lymphoma induced with radiation-treated cell lines acquired strong immunogenic properties after a single cycle of DTIC treatment *in vivo*. Again, these results may well provide an experimental model for the exploitation of a radiation-induced increase of tumour cell immunogenicity for combined radioimmunochemotherapy in cancer treatment, but no direct relevance for the risk of radiation carcinogenesis is evident. As with other combined modalities in tumour therapy that also enhance and modulate deterministic effects (described in depth in the Appendix), the above examples do not indicate mechanisms leading to marked supra-additivity for effects from low-level exposures to multiple agents.

EXTENDED SUMMARY

176. In this Annex, the effects of combined exposures to radiation and other agents are considered particularly with respect to the induction of stochastic effects at low doses. A large amount of information on the combined exposures of radiation and other physical, chemical, and biological agents is reviewed. In many situations agents can interact with radiation and may significantly modify the biological processes and outcomes. The implications for radiation risk assessment and limitation of individual and collective health risks are considered.

177. For ionizing radiation, the main potential risk to humans from exposures at low doses, i.e. at the level of background radiation or a few times that level, is the enhanced incidence of stochastic effects, i.e. carcinogenesis and heritable genetic effects. In this Annex the effects of combined exposures to radiation and other agents are considered, particularly with respect to the possibility of enhanced radiation carcinogenesis caused by chronic low doses. Many radiobiological experiments, however, used

acute, high radiation doses and high exposures to the other agent. It is usually not clear how these results might be extrapolated to low and chronic irradiation conditions and to humans. Such data may, however, be informative on the possible mechanisms of the interactions between various agents and radiation, and in that sense they may be of relevance for combined, low-dose exposures leading to carcinogenesis.

178. For assessing the effects of chemical agents, the situation is somewhat more complex than for ionizing radiation. For genotoxic chemicals, the main biological effects from low and chronic exposures are comparable to effects from low levels of radiation, i.e. stochastic in nature. However, as with radiation, most experimental data are from high, acute exposures. The wealth of epidemiological data and risk estimates based on such information, is much greater for radiation protection than for toxicology. With only a few exceptions (e.g. asbestos, smoking, and arsenic) for which human data are available, chemical

carcinogenesis data are based solely on biochemical, cell biological, and/or animal data. The general assumption about the dose-effect relationships for radiation and genotoxic chemicals in the low-dose region for chronic exposures and for stochastic endpoints is, however, the same. In general, it is assumed that these relationships are linear from high-dose ranges with observed effects down to zero dose. For non-genotoxic chemicals, non-linear dose-effect relationships are the norm, since higher order enzyme reactions are involved in most cases (uptake of agents, incorporation, metabolization, cell physiological reactions, etc.). These reactions are dominated by sigmoidal dose-effect relationships with apparent thresholds, related to the biochemical Michaelis-Menten kinetics.

179. The starting point for an analysis of combined exposures is to specify the doses of the agents at the site of interaction. For ionizing radiation, absorbed dose is the quantity most generally applied to characterize the exposure, and methods have been developed to calculate the dose in target cells from the irradiation conditions. For other agents, different methods are used, depending on the characteristics of the agent involved. Unfortunately, the exposure of the cells at risk for carcinogenesis is not always clear. No unifying concept of dose exists. In this Annex, methods of biochemical and biological monitoring and dosimetric evaluation are reviewed. The conclusion is that physiologically based parameters, such as concentration of toxic agents in blood or urine, are often not specific and sensitive enough to be generally applicable for the analysis of the biological action of a chemical agent. Therefore, biological endpoints at the cellular level, which are more directly related to stochastic health effects, have been developed and used as measures for genetic changes in somatic as well as germ cells. To these endpoints belong biochemical parameters such as DNA adducts, and gene mutation parameters such as mutation frequency and spectrum, and stable chromosomal alterations. A generally applicable method for analysing combined effects is still lacking, but taking these endpoints as a measure of the genotoxic burden of radiation or of chemical agents, a unifying risk concept based on genetic burden can be envisaged.

180. Carcinogenesis is, in general, a slowly developing process extending over years and even decades. From a mechanistic standpoint, different phases in the development of the effects of an agent can be considered. These may be characterized broadly as changes on the molecular, cellular, and tissue/organ levels. Agents can interact with radiation in each phase to produce an effect. Radiobiological research has turned up numerous agents potentially capable of influencing the progression of early radiation effects towards adverse health effects. General conclusions are hindered by the multitude and complexity of the possible interactions and the dependence of the combined effect on the sequence of the exposures. More explicitly, because of the long time period between the initial radiation event and the final effect, a combined exposure to radiation and another agent may occur after simultaneous exposure but also from exposures hours or even years apart.

181. In the early, molecular phase of the development of the radiation effect, interactions of chemicals with the primary radiation process can occur that are important for the fixation of the primary molecular radiation damage. For an interaction to occur, the active agents must be in close proximity to the DNA, which is the most important molecule for radiation carcinogenesis, at the time of irradiation or during repair and in a sufficiently high concentration. Molecular interactions are studied particularly to investigate the early radiation mechanism, and changes in the radiation effect from interactions have been seen. However, because of the high concentrations needed to observe a significant effect, the results from these investigations are not of direct relevance for the low levels of exposure found in occupational or environmental settings.

182. An impressive amount of information concerning interactions in the cellular phase can be found in the literature. For acute exposures, many agents can interact with radiation in this phase, including physical (e.g. UV) and chemical (e.g. alkylating and other genotoxic) agents. Toxic chemicals have been evaluated, using the isobolic method of analysis. Different interaction mechanisms are involved, ranging from an accumulation of DNA (sub)lesions, sometimes enhanced by repair inhibitors, to modulation of cell-cycle kinetics. The results of the interactions range from subadditive to supra-additive; however, for interaction to occur, the agents must generally be present during or shortly after irradiation, and the interaction effects decrease at low doses and dose rates. This mode of interaction may have implications for radiation carcinogenesis, but at low doses and for chronic irradiation, deviations from additivity are expected to be small.

183. In the organ phase of cancer development, interactions in combined exposures can be significant. The long duration of this phase creates many opportunities for interaction with other agents. As is also concluded for chemical carcinogenesis, these interactions are potentially important, but only a few data from human epidemiological studies are suitable for quantitative analysis. Radiation has been found to interact with physical agents such as UV radiation and mineral fibres; with chemical agents such as alkylating chemicals, tumour promoters, dietary factors, arsenic, and heavy metals; and with biological agents such as hormones and viruses. Well-defined effects are summarized in Table 7. Observation and identification of combination effects in this phase is difficult, because the duration of the interaction with the radiation damage may be long. In view of the many possibilities, it may well be that different interactions in the organ phase are largely responsible for the variations in background cancer incidence between populations.

184. A very important combined effect is the interaction of smoking and exposure to radon, although even in this case there is still no unambiguous conclusion on the interaction mechanism. Epidemiological data clearly indicate that combined exposure to radon and cigarette smoke leads to more-than-additive effects on lung cancer. These results warrant special consideration in estimating the radiation risks because a large proportion of the world's population

is exposed concomitantly to considerable levels of indoor radon and smoking. The combined analysis of 11 miner studies [L18] indicates that the effect of radon may be enhanced by a factor of about 3 by being combined with smoking.

185. Since the Committee's previous review of this subject [U6], there have been advances in modelling the multi-stage processes involved in carcinogenesis. The development and application of mechanistically based, multi-stage carcinogenesis models promise to give new insights into the interaction processes, especially because with these models it is possible to analyse interactions at the tissue/organ level of carcinogenesis. The results indicate, for example, that the

effect of radiation is dependent on the background tumour incidence; they also show how the interaction of radiation with other agents might influence carcinogenesis.

186. Information is scarce on combined exposures of radiation and specific agents that might alter the radiation health risks caused by ambient exposures in the human environment. The possible relevance of the interaction of other agents with the radiation effect is obscured by the many sometimes poorly known or unknown sources of uncertainty surrounding radiation-induced carcinogenesis, such as variations in background cancer incidences, population characteristics and genetics, diet, and individual susceptibility.

Table 7
Agents that interact with ionizing radiation of importance in radiation carcinogenesis

Interacting agent	Interaction	Endpoint
Physical agents		
External ionizing radiation with internal emitters	Supra-additive	Bone cancer
Ultraviolet radiation (UV)	Possibly supra-additive	Skin cancer
Alpha emitters with mineral fibres, including asbestos	Supra-additive	Lung cancer
Chemical agents		
Nitroso compounds, such as MNU, DEN, 4NQO	Supra-additive	
Tumour promoters, such as TPA	Supra-additive	Effects shown only in animal experiments
Smoking	Supra-additive	Lung cancer
Vitamins	Subadditive	
Diet/fat	Sub- to supra-additive	Interaction dependent on comparing level
Arsenic	Supra-additive	Extrapolated from chemical carcinogenesis
Biological agents		
DES	Supra-additive	Breast cancer
Testosterone	Supra-additive	Prostate cancer

CONCLUSIONS

187. Combined exposures are a characteristic of life. The environment in which organisms reside and the organisms themselves are complex systems in which a multitude of interactions between physical, chemical, and biological factors occur. The specific agents involved in exposures in the environment and in occupational settings vary widely, but almost all physical and chemical agents, both natural and man-made, are capable of producing adverse effects under some exposure conditions, although individual agents differ considerably in their capacity to do so. In general, for many agents essential for life, there is a spectrum of effects associated with exposure, ranging from deficiency through sufficiency to adverse effects with increasing levels of exposure.

188. Although both synergistic and antagonistic combined effects are common at high exposures, there is no firm evidence for large deviations from additivity at controlled occupational or environmental exposures. This holds for

mechanistic considerations, animal studies, and epidemiology-based assessments. Therefore, in spite of the potential importance of combined effects, results from assessments of the effects of single agents on human health are generally deemed applicable to exposure situations involving multiple agents.

189. With the exception of radiation and smoking, there is little indication from epidemiological data for a need to adjust for strong antagonistic or synergistic combined effects. The lack of pertinent data on combined effects does not imply per se that interactions between radiation and other agents do not occur. Indeed, substances with tumour promoter and/or inhibitor activities are found in the daily diet, and cancer risk therefore depends on lifestyle, particularly eating habits. Not only can these agents modify the natural or spontaneous cancer incidence, but they may also modify the carcinogenic potential of radiation. Such modifications would influence the outcome particularly when radiation

risks are projected relative to the spontaneous cancer incidence.

190. The analysis of small effects of combined exposures of other agents with radiation is also inhibited by the lack of well defined and pertinent harmonized measures of the exposures to radiation and the other agent. A generally applicable method for use in the analysis of combined effects is still lacking, but taking end points such as measures of the genotoxic effects of radiation or chemical agents, a unifying risk concept based on genetic burden could be envisaged for combinations of genotoxicants. At this stage, the uncertainties in the data permit the possible interactions from combined exposures of radiation with other agents to be only qualitatively recognized. A quantitative assessment of the radiation risks at low doses is not yet possible. In other words, even possible deviations from additivity at lower exposures are generally too low to show up in experimental studies or population cohorts.

191. The extent to which the effects of combined exposures can be elucidated is highly dependent on clarification of the carcinogenesis process itself and its dependence on environmental and lifestyle factors. Interactions of agents with radiation can be broadly grouped into three different levels (molecular, cellular, and tissue/organ levels) of the radiation effect. The molecular phase lasts only a fraction of a second until the primary radiation damage to DNA has occurred. For interactions during this phase, the other agent has to be present concomitantly and in a high enough concentration.

192. The cellular phase of the radiation effect lasts for one or a few cell cycles until the primary radiation damage in DNA has been repaired or the remaining damage has been fixed into heritable genetic damage (mutation in somatic and germ cells). For low doses and dose rates of radiation and low doses and chronic exposures to genotoxic chemical agents, the supralinear or quadratic terms of dose-effect relationships tend to vanish, and the linear terms dominate for single-agent effects. In the absence of target specificity, this implies that interaction at the cellular level during long-term low-level exposures to radiation and chemicals is of limited importance.

193. During the tissue/organ phase of radiation-induced carcinogenesis, which lasts from the first fixed genetic alteration to the clinically manifested tumour and which may include several genetic and epigenetic changes, combined effects can occur from exposures of two and more agents spread over days or decades, giving a large potential for combined effects. Besides genotoxic chemicals, many non-genotoxic agents may interact during the organ phase. Tumour promotion, mitogenic stimulation, and hormonal activation are a few of the important examples of processes with the potential for more-than-additive effects. Also, radiation- or chemical-induced genetic instability, which leads to new genetic damage after many cell generations, may be prone to more-than-additive effects. An overview of possible interaction processes and groups of agents involved in these processes is given in Figure V.

194. Within the framework of the multi-stage mechanism of carcinogenesis, the following general conclusions can be drawn for the combined action of different carcinogenic and co-carcinogenic agents:

(a) genotoxic agents with similar biological and mechanistic behaviour and acting at the same time will interact in an isoadditive or concentration-additive manner. This means that concurrent exposures to ionizing radiation and other DNA-damaging agents with no specific affinity to those DNA sequences that are critically involved in carcinogenesis will generally result in effects not far from isoadditive. Isoadditivity at this point includes "apparent synergisms" or "autosynergisms" resulting from non-linear dose-effect relationships of the single-agent effects. Supra-additivity of this quality generally does not exceed the expectation value derived from high-exposure, single-agent effects combined with linear dose-effect models;

(b) for genotoxic agents acting on different rate-limiting steps of multi-stage stochastic diseases like cancer, strong deviations from additivity might result. Deviation from additivity can depend on the specificity of the agents for the different steps, sequence specificity, and the sequence of exposures. Highly synergistic effects are, however, only to be expected in cases where both agents are responsible for a large fraction of the total transitions through the respective rate-limiting steps;

(c) in combinations of radiation and non-genotoxic agents in which the second agent causes promotion, i.e. the multiplication of premalignant cells, highly synergistic effects may arise. This combined effect is dependent on the exposure schedule. Thresholds for such combined effects are generally implicated from the highly non-linear dose-effect relationship for the non-genotoxic agent acting alone;

(d) for agents acting independently and through different mechanisms and pathways, heteroadditivity or effect additivity is predicted. Apparent thresholds will not interfere with each other, and possible conservatisms in linear dose-effect extrapolations from high exposures will not be affected;

(e) in combinations of agents, in which one agent induces adaptive mechanisms, e.g. increased DNA repair capacity or increased radical scavenger function, and the other agent induces DNA damage, antagonistic effects may arise. Owing to the generally short half-times of adaptive mechanisms, the exposures have to occur concurrently or nearly concurrently.

195. In summary, the following parameters need to be considered to address and assess potential combined effects: the mode of action of the agent (genotoxic or non-genotoxic); the shape of the dose-effect relationship for single-agent effects; the dose or concentration involved (low or high); the type of exposure (chronic or acute); and the sequence and time interval between exposures (simultaneous or before or after radiation exposure).

196. There has been little systematic research on possible interactions of radiation with other agents. An exception is

the use of combined modalities in tumour therapy, but the high doses and deterministic effects involved cannot be easily related to stochastic, low-level combined effects. Considerable progress in the biological sciences and the many radiological and toxicological disciplines involved will be needed to allow predicting the potential presence of combined effects at low exposure levels and negative

health outcomes. It can be stated, however, that the conclusion of the Committee's previous review on combined effects [U6] still holds: except for radiation and smoking, there is no evidence that low-level exposures to multiple agents yield combined effects far from additivity, or above the estimates resulting from linear extrapolation of single agent effects to lower doses.

FURTHER RESEARCH NEEDS

197. The lack of systematic mechanistic understanding and quantitative assessments of combined exposures and the resulting possible interactions urgently needs to be resolved. This elucidation of interactions of agents in combined effects critically depends on both a qualitative (mechanistic) and a quantitative knowledge of the action of any single exogenous or endogenous agent involved. The basic tenets of experimental toxicology and radiobiology will have to be applied in studies of the adverse health effects of combinations of exposures. In view of convincing evidence that the critical stochastic endpoint, cancer, is multifactorial, the many studies concentrating on single carcinogenic agents and attempting to quantify cancer risks as if they were due to single factors have to be supplemented and extended to address potential modifications from joint effects.

198. Present knowledge of the many qualitatively different interactions already found in biological systems speaks against the emergence of simple unifying concepts to predict modifications of risk from combined exposures. However, mechanistically based classifications of interactions may be helpful in predicting effects. At present, relevant knowledge is being gained on interaction mechanisms in different parts of the long process of radiation carcinogenesis. A better understanding of how important these separate physical, chemical, and biological interaction mechanisms are for the ultimate endpoint will help to create a basis for risk assessment, and a better understanding of the carcinogenesis process itself and the rate-limiting steps involved may contribute to an understanding of the interactions as well. The development of mechanistically based cancer models could greatly improve the estimation of quantitative risks.

199. Individual genetic susceptibility is already a concern for the assessment of radiation risks. In addition to those parts of the genome susceptible to radiation-induced effects, e.g. repair and proofreading genes, heterozygosity for oncogenes and tumour suppressor genes, many additional gene products determining the biokinetics and biotransformation of chemical agents will have to be considered in the individual response to combined exposures involving chemicals.

200. Progress in the analysis of interactions between ionizing radiation and toxicants is often hampered by a lack of scientific data that quantitatively relates chemical exposure to health risk or experimental endpoints. The implementation of standard protocols and dosimetry to harmonize reported research is urgently needed to allow comparison of data from

studies on different agents. Data will also have to be extended over a sufficiently large exposure range to allow extrapolating to the doses relevant in environmental health.

201. Epidemiological studies have already revealed important combined effects for carcinogenesis, particularly for the joint effect of cigarette smoke with either radiation or asbestos. New tools in molecular biology point the way to the field of molecular epidemiology and will provide investigators with markers of exposure and damage that are much more sensitive than the cruder incidence measures of clinical diseases. Such approaches, if successful, can be expected to yield significant new information on interactions between agents at the cellular and molecular levels. Markers of this kind can also be used in more classical human epidemiological studies of cancer, some of which may help in probing potential interactions between agents that may have induced the disease.

202. A mechanistic assessment of combined effects is dependent on progress in the scientific understanding of other generic issues, such as extrapolation from high to low levels, transfer of data from laboratory animals to humans, and age dependence of the radiation risk, that are central to the general risk assessment process. Current approaches to the risk assessment of complex exposures rely heavily on linear dose-effect relationships and additivity models. However, in the dose and concentration range of interest for human exposures, dose-effect relationships other than linear (sigmoidal and even U-shaped curves in the case of partially stimulatory or essential agents) are reported as well. This issue must be fully addressed in assessing the risks of combined effects.

203. Finally, a comprehensive approach for the study and quantitative assessment of combined effects must be developed. The gap between different conceptual approaches in the assessment of risk in chemical toxicology and radiological protection has to be bridged urgently. Multidisciplinary approaches to research (radiobiology, toxicology, cell and molecular biology, biostatistics, epidemiology) have to be forged. In some instances, recasting and combining results from recent studies and on-going work into more refined models that take into account additional mechanisms of responses and that take advantage of the multidisciplinary approach may improve the understanding and quantification of specific interactions. This, together with the application of refined multi-stage models, will help to reduce uncertainties at the low exposure levels found in the human environment.

APPENDIX

Combined effects of specific physical, chemical, and biological agents with ionizing radiation

1.	Studies of the combined effects of specific physical, chemical, and biological agents in association with radiation exposure are reviewed in detail in this Appendix. The intention is to provide an overview of the available literature in support of the more general findings, summaries, and conclusions of this Annex. First, data from epidemiological studies of the adverse health effects in humans of each group of agents are reviewed. Studies involving experimental animal models are then considered. Finally, various effects observed using *in vitro* systems are reviewed. Carcinogenesis is the principal endpoint of interest, but non-neoplastic endpoints are also discussed. Only a minor fraction of the interacting agents described below are found in the human environment at potentially critical levels. A few such critical agents already known or suspected to affect human health on their own are listed in Table 4. The findings of specific combined exposures and effects described in depth in this Appendix are summarized in Table A.1.

## A.	RADIATION AND PHYSICAL AGENTS

### 1.	Combinations of different types of ionizing radiation

2.	Understanding how cellular damage produced by high linear-energy-transfer (LET) radiation interacts with that produced by low-LET radiation is important both in radiation therapy and in evaluating risk. In view of the possible radiotherapeutic applications, a wealth of data on the combined action of neutrons, heavy ions, and gamma or x rays was accumulated in the 1970s and early 1980s. This information was reviewed in Annex L, "Biological effects of radiation in combination with other physical, chemical and biological agents", in the UNSCEAR 1982 Report [U6]. Because the underlying damage mechanism is similar, the interaction between different types of ionizing radiation is, in general, of the isoadditive type in the case of cell killing. Deviations from additivity are generally not very large and can be explained in most cases by concomitant changes in exposure rates and in exposures of critical target structures. For example, survival of Chinese hamster V79 cells *in vitro* after irradiation first with neon ions (LET = 180 keV μm^{-1}) and then with x rays (225 kVp) was additive, as predicted from independent action. The system also showed no dependence on the order of application [N5].

3.	In extreme emergency situations, localized exposures to the skin or other organs in the presence of elevated external radiation fields is of considerable concern. Randall and Coggle [R3] studied the deterministic effects of concomitant whole-body irradiation and localized radiation trauma from

beta activity on the skin. They modelled the immunosuppressive effects of whole-body gamma radiation in the sublethal to lethal range (1–11 Gy) on skin reactions produced by 50 Gy from superficial beta radiation from ^{171}Tm in male mice. For gamma doses below 4 Gy, no interaction effects were detectable. For gamma doses in the range 4–8 Gy, the skin reaction developed more slowly, but it was not much more severe. The overall time for the resolution of the skin reaction, about 45 days, was also unaffected by high-dose whole-body irradiation. The authors ascribed the absence of any considerable deviation from additivity in this system to the mismatch in time between maximal immunosuppression and localized severe beta burns ranging from 2 to 10 days and 10 to 25 days, respectively. Although such beneficial mismatches in time are species-specific, these mechanisms may also be important in humans. Deterministic combined effects in Chernobyl power plant staff and emergency workers are discussed in Annex J, "*Exposures and effects of the Chernobyl accident*".

4.	A strong antagonistic combined effect was found in an experimental study of deterministic effects in rats. When the animals received high external gamma (about 6 Gy) or beta (about 24 Gy surface dose) radiation, lethality was lower by a factor of 5 when the animals received a concomitant exposure to the thyroid gland of 0.3 kBq g^{-1} from ^{131}I given orally [M13]. The protective influence of the combined treatment was attributed to ^{131}I-induced changes in the hormonal state in the course of acute radiation sickness. In another study by the same author with lower sublethal external doses of up to 3 Gy and additional orally administered ^{131}I, there was an increased yield of mammary tumours in the combined treatment group receiving low exposures from iodine (0.04–0.8 kBq g^{-1}). For higher iodine exposures, however, the reverse was true [M30]. The combined effects observed seem to be deterministic and can be attributed mainly to different organ doses rather than different radiation qualities.

5.	Changes in the haematopoietic bone marrow, i.e. in the number of colony-forming units (CFU), of rats were observed by Brezani et al. [B1] after a single whole-body neutron dose of 2 Gy and combined single neutron (2 Gy) and continuous gamma irradiation (6 Gy, daily dose rate of 0.57 Gy). Neutron irradiation alone significantly reduced the number of karyocytes, including CFU-S in the bone marrow and induced extensive cytogenetic damage. When followed by continuous gamma irradiation, the primary damage from neutrons was not enhanced, however CFU-S remained at a decreased level for the whole time of irradiation. Recovery from damage began only after termination of the continuous irradiation; its course was similar to that after single neutron irradiation. A long-lasting supra-additive influence of the combined exposure to neutrons and gamma rays is nevertheless manifested in later periods after irradiation by a reduction in the total CFU-S number in the bone marrow.

6.   Most studies of stochastic effects from combined irradiations are undertaken in connection with cancer. For bone-seeking radionuclides, a synergistic effect was found in mice for osteosarcomas after combined exposure to short-lived ^{227}Th (190 Bq g^{-1}, corresponding to about 10 Gy mean skeletal alpha dose) and to longer-lived ^{227}Ac (1.9 Bq g^{-1}) bone-seeking radionuclides. The beta emitter ^{227}Ac produces protracted internal alpha exposures through ingrowth of the decay product ^{227}Th. At 700 days after intraperitoneal (ip) injection of pure ^{227}Th or ^{227}Th contaminated with 1% ^{227}Ac (combined exposure) in the form of citrate, an osteosarcoma incidence higher than additive was found for the combined exposure. With incidences of less than 1% for controls, 8% for ^{227}Ac alone, and 36% for ^{227}Th alone, the combined effect of 62% amounted to an interaction factor of 1.7 [L26]. In terms of the time for 50% tumour appearance, the interaction factor was reduced to a barely significant 1.3. The authors speculated that the increased oncogenetic effectiveness of ^{227}Th contaminated with 1% ^{227}Ac may be caused by the continuous stimulation of cell proliferation or by the activation of retroviruses by protracted low-level alpha irradiation from ^{227}Ac/^{227}Th.

7.   Bukhtoiarova and Spirina [B33] studied the combined effect of external gamma irradiation and ^{239}Pu on the incidence of osteosarcomas in inbred male rats. Osteosarcomas occurred more frequently and at earlier times and displayed a more pronounced multicentric pattern of growth and metastatic spreading than the malignancies induced by exposure to only one of the two agents. The differences resulted from increased development of tumours and decreased osteogenesis. A quantitative evaluation of the combined effect of the same radiation mix on biochemical parameters of the rat immune system was undertaken by Elkina and Lumpov [E5]. The combined effect of external gamma radiation (^{137}Cs, 1–4 Gy) and incorporated alpha radiation (^{239}Pu nitrate, 9.3–93 kBq kg^{-1} body mass) was estimated by determining changes in nucleic acid metabolism and the number of cells in rat thymus, spleen, and bone marrow. The data obtained for the lower end of exposures were consistent with an additive model. The same researchers also studied aminotransferase and lactate dehydrogenase activity in the blood of dogs exposed to the joint action of external gamma and internal alpha radiation [E6]. After the effect of external gamma radiation (0.25–2 Gy) and inhaled ^{239}Pu submicron oxide containing 25% ^{241}Am (approximately 7–10 kBq kg^{-1}) delivered separately and in combination, activities of alanine-aspartate aminotransferase and lactate dehydrogenase changed in an undulatory manner, tending to increase at later times. The change was a function of type and level of radiation as well as time elapsed from the onset of exposure. Even at the relatively high exposures used in these experiments, the combined effect of gamma and alpha radiation did not exceed the additive effect of the two factors delivered separately. In view of the deterministic nature of the endpoints studied, no inferences for controlled exposures are apparent.

8.   Several authors have shown that large radiation doses influence biokinetics and hence exposure from incorporated radionuclides. The influence of external gamma radiation on ^{239}Pu redistribution in pregnant and lactating rats was described by Ovcharenko and Fomina [O1]. A quite high dose range of acute external gamma radiation, from 0.5 to 4 Gy, was investigated. Transplacental transfer of ^{239}Pu to the embryo increased with dose to a maximum at 1 Gy and then declined. However, transfer of ^{239}Pu via milk to newborn rats was decreased by external gamma irradiation of lactating rats with the dose of 0.5 Gy. The nature of the biological mechanisms responsible for the changes in biokinetics remains elusive. There are no suggestions by the authors that these radiation effects on metabolism are stochastic in nature and would extend to low doses and dose rates.

9.   Lundgren et al. [L19, L20] examined the carcinogenicity of a single, acute pernasal inhalation exposure of 3,201 male and female F344 rats to ^{239}PuO$_2$ followed one and two months later by whole-body x-irradiation. Plutonium lung burdens were 56 or 170 Bq, and the x-ray exposure was fractionated into two exposures totalling either 3.8 or 11.5 Gy. Other groups of rats received control (sham) exposures. Minor x-ray-dependent differences in ^{239}Pu lung retention were observed; however, exposure to x rays significantly reduced the median survival times in rats of both sexes [L19]. For a given level of x-ray exposure (0, 3.8, or 11.5 Gy), the level of ^{239}Pu exposure (0, 56, or 170 Bq) had no effect on median survival time. A preliminary histological evaluation of primary lung tumours produced has been reported for approximately two thirds of the rats in this study. The authors noted an apparently antagonistic interaction between the two agents in producing lung tumours; for example, crude tumour incidences were 10.8% in rats receiving 11.5 Gy x-irradiation alone, 9.2% in rats receiving a 170 Bq lung burden of ^{239}PuO$_2$ alone, but only 11.7% in rats receiving a combined exposure at these levels [L20]. The authors cautioned, however, that a simple evaluation of the crude tumour incidence is insufficient because of the effect of exposure on lifespan. They further state that analysis of this study is not yet complete.

10.   An apparent synergism was described in an *in vitro* study of the combined effect of alpha particles and x rays on cell killing and micronucleus induction in rat lung epithelial cells (LEC) [B39]. The cells were grown on Mylar films and exposed to both x rays and alpha particles, separately or simultaneously. X rays and alpha particles given separately caused dose-related increases in cell cycle time, with alpha particles producing greater mitotic delay than x rays. Damage from alpha particles and x rays given simultaneously did not interact to further alter the cell cycle. Cell survival data following exposure to x rays and alpha particles, combined or individually, were fitted by linear-quadratic models. Survival curves following exposure to alpha particles only, or to 1 Gy from alpha particles plus graded x-ray doses, were adequately described using only the linear (alpha) terms with values of the coefficients of 0.9±0.04 and 1.03±0.18 Gy^{-1}, respectively. Survival following exposure to x rays only or to 0.06 Gy from alpha particles combined with x rays was

best fitted using both alpha and beta terms $(0.12\pm0.03)D + (0.007\pm0.002)D^2$ and $(0.57\pm0.08)D + (0.3\pm0.02)D^2$, respectively. The numbers of micronuclei in binucleated cells produced by exposure to alpha particles or x rays alone increased linearly with dose, with slopes of 0.48 ± 0.07 and 0.19 ± 0.05 micronuclei per binucleated cell per Gy for alpha particles and x rays, respectively. Simultaneous exposure to graded levels of x rays and a constant alpha dose of either 1.0 or 0.06 Gy increased micronuclei frequency, with a slope of 0.74 ± 0.05 or 0.58 ± 0.04 micronuclei per binucleated cell and Gy, respectively. These slopes are similar to that produced by alpha particles alone. These studies demonstrated that both cell killing and the induction of micronuclei were greater with combined exposures than with separate exposures.

11. A refined model for the combined effects of mixtures of ionizing radiations was recently published by Lam [L4]. Assuming that ionizing radiation is a special group of toxic agents whose general interaction can be calculated, the model postulates the existence of a common intermediate lesion and the relative action of lesions before, at, and after this common stage. General quantitative dose-effect relationships of mixed radiations can be derived from the dose-effect relationships of the components in the mixture. Again, only small deviations from isoadditivity are predicted by this damage function, which allows treating mixed irradiation as two different increments of dose from the same radiation source.

12. A unifying concept to predict the expected combined stochastic radiobiological effects of different ionizing radiations was presented by Scott [S15]. Additive-damage dose-effect models were developed for predicting the radiobiological effects of sequential and simultaneous exposures. These additive-damage dose-effect models assume that

(a)   each type of radiation in the combined exposure produces initial damage, called critical damage, that could lead to the radiobiological effect of interest; and

(b)   doses of different radiations that lead to the same level of radiobiological effect (or risk) can be viewed as producing the same amount of critical damage, which is indistinguishable as far as the effects of subsequently administered radiation are concerned.

The methodologies allow the use of known radiation-specific risk functions to derive risk functions for the combined effects of different radiations, called global risk functions. For sequential exposures to different ionizing radiations, the global risk functions derived depend on how individual radiation doses are ordered. Global risk functions can also differ for sequential and simultaneous exposures. The methodologies are used to account for some previously unexplained radiobiological effects of combined exposures to high- and low-LET radiations. Since all radiation effects are traced to a common initial damage mainly occurring in DNA, the model is basically additive.

13. At doses lower than those that induce deterministic effects, no large deviations from additivity are found in the interaction of different radiation qualities (see also Table A.1). Although the mathematical modelling of mixed radiation showing non-linear dose-response relationships with a single radiation quality yields apparent synergistic interactions when the analysis of endpoints like survival is based on some current definitions [Z2], these definitions are clearly inappropriate for the approach used in this Annex. This point is also made by Suzuki, who stressed the need for definitions based on biological mechanisms [S3].

14. In summary, it can be stated that when dose rates and other possible confounders are taken into account, practically all the results from mixed radiation yielding more than the sum of the single agents can be explained by isoaddition, so that general quantitative dose-response relationships for mixed radiations can be derived from the dose-response relationships of the components in the mixture [L4, L47]. There is no indication that the influence of external radiation on the biokinetics of radionuclides found at high doses is relevant at occupational or environmental exposure levels.

2. Ultraviolet radiation

15. Ultraviolet (UV) radiation is recognized as an important initiator and co-factor for human skin carcinogenesis. Genetic predisposition, i.e. skin type, age at exposure, and duration of exposures are important determinants of risk for UV radiation-induced skin cancer. Shore analysed 12 studies on the incidence of skin cancer in irradiated populations with known skin doses [S29]. In the absence of a proper control (skin exposed to ionizing radiation but not to UV), it was concluded that at least for combined exposures, there was no evidence of a dose threshold for radiation-induced skin cancer. The data are compatible with a linear dose-response relationship [S29]. The question whether relative risk or absolute risk models are more appropriate remains open. Considerable variations in sensitivity to skin cancer induction among demographic and genetic subgroups may be mainly a reflection of the large differences in UV exposures because of lifestyle, skin type, and tanning.

16. Combined exposure to UV and x rays leads to synergistic interaction in killing mammalian cells [H53], confirming previous studies in yeast [S87]. Only a small interaction was found for mutations at the *hprt* locus in Chinese hamster cells [K56]. A recent study by Spitkovsky et al. [S82] in human peripheral lymphocytes on the interaction between x-ray doses of $5-250$ mGy and 20 J m^{-2} of 254 nm UV light in DNA repair, measured by unscheduled DNA synthesis (UDS), indicated that the repair of UV-induced damage was modulated by previous x-ray exposures. For radiation alone, UDS was highest for $20-30$ mGy and $150-200$ mGy and lowest at 100 mGy. For combined exposures, i.e. ionizing radiation followed by UV, UDS was highest in cells previously exposed to 100 mGy and lower than in UV-only controls for cells previously exposed to $20-30$ or $150-200$ mGy. The mechanism of this proposed adaptive response remains to be elucidated.

Table A.1
Combined effects of ionizing radiation and other agents

Radiation	Combining agent	Study system	Endpoint studied	Nature of effect	Comments	Ref.
Physical agents						
Beta to skin	Gamma, whole body	Male mice	Deterministic effect	Additive		[R3]
Beta, ^{131}I to thyroid	Gamma, whole body	Rats	Deterministic effect; change in hormone status	Subadditive		[M13]
Alpha	Gamma	Rat lung epithelial cells *in vitro*	Cell killing and micronuclei formation	Supra-additive		[B39]
Alpha, ^{227}Th	Beta, ^{227}Ac	Mice	Osteosarcoma	Supra-additive		[L26]
Alpha, ^{239}Pu	Gamma	Male rats	Osteosarcoma	Supra-additive	Additive at low exposure	[E5]
Low LET	UV radiation	Humans	Skin cancer	?	Meta-analysis of 12 epidemiological studies; controls with ionizing radiation only were not available	[S29]
Alpha	Laser (633 nm)	Bacteria (*E. Coli*)	Survival	Subadditive	Mechanism only found in bacteria	[V6]
X rays (5.5 Gy)	Microwave field (200 µW cm^{-2})	Rats	Survival	Subadditive		[G13, G16]
X rays	Static field (58 mT)	Cell culture	Growth and survival rate	Supra-additive		[K16]
Gamma (3-9 Gy)	Static field (10-400 mT)	Mice	Survival	Subadditive	Field has to precede irradiation by 7-30 days for full protection	[S33]
X rays	Temperature	Human, mammals	Cell killing, suppression of proliferation	Supra-additive	Complex adaptive processes, important for tumour control	[G6, M15, Z3]
X rays	Ultrasound	C3HT½ cells	Transformation	None		[H10]
Alpha (radon, 6 000 WLM)	Mineral fibres (intrapleurally)	Rats	Lung carcinomas and mesotheliomas	Supra-additive		[B38]
Alpha (radon)	Mineral fibres	Rats	Thoracic tumours	Supra-additive		[B38]
Alpha (radon)	Mineral fibres	C3HT½ cells	Transformation	Supra-additive		[H11]
Chemical toxicants (genotoxic chemicals)						
X rays (7.5 Gy)	Alkylating agents	Human	Secondary non-lymphocytic leukaemia after breast cancer therapy	Supra-additive		[C29]

Table A.1 (continued)

Radiation	Combining agent	Study system	Endpoint studied	Nature of effect	Comments	Ref.
X rays	Cyclophosphamide	Mice	Mutations in germ cells	Supra-additive	Interference of cyclophosphamide with DNA repair	[E7]
X rays (9 Gy)	DMH, 1,2-dimethylhydrazine (0.15 g kg⁻¹)	Rats	Colon carcinogenesis	Supra-additive		[S27]
X rays (0.25-3 Gy)	Methyl, ethyl, butylnitrosourea (10 mg kg⁻¹)	Rats Mice Rats	Gastrointestinal tumours T-cell lymphomas Neural tumours	Supra-additive Supra-additive Subadditive	Deterministic effect suggested by enhanced risk in generating lympho-haemopoiesis Killing of stem cells? (Radiation suppresses schwannoma induction by ENU and also spontaneous squamous-cell carcinomas)	[M32] [S22] [H6, K15]
		Rats	Brain tumours	Subadditive	Probably due to radiation-induced alkyl-transferase	[S13]
		C57Bl/6N mice	Lymphomas	Supra-additive	Effect traced to cellular kinetics and clonal expansion	[S21]
		BDF₁ mice	Thymic lymphomas	Subadditive to supra-additive	Lower radiation doses enhanced, higher doses delayed leukaemogenesis	[S20]
Gamma (0.75-3 Gy)	Diethylnitrosamine (DEN) (100 mg kg⁻¹)	Rats	Liver foci	Supra-additive	Large sex differences, different histo-chemically characterized foci show divergent response	[P26]
Neutrons (0.125-0.5 Gy)	Diethylnitrosamine (DEN) (100 mg kg⁻¹)	Mice	Liver carcinomas and foci	Supra-additive	Effect mainly of increased foci appearance	[M8]
⁹⁰Sr/⁹⁰Y (27 Gy, skin surface)	4-Nitroquinoline 1-oxide	ICR female mice	Skin tumour	Supra-additive	No tumours from single-agent exposure despite acute effects	[H39]
Gamma (1, 9 Gy)	Bleomycin (60 mg kg⁻¹)	Mouse germ cells	Reciprocal translocations	Additive to supra-additive		[D6]
X rays	1,2-Diethylnitrosamine (DEN)	Tradescantia	Stamen hair cell mutations	Supra-additive	Interaction of single-strand lesions, covered by isoaddition	[L13]
X rays (3 Gy)	Paraquat (superoxide generating agent)	C3HT½ cells	Survival, transformation	Supra-additive	Additive for sister chromatid exchange	[G5]
Chemical toxicants (non-genotoxic chemicals)						
X rays	TPA	see Table A.3		Synergism	TPA not present in natural environment	
X rays	Sulfhydryl-carrying radioprotectors	Male BALB/c and C57Bl mice	Survival, tumour induction	Antagonism	Many systems and endpoints studied with similar results	[M6, M7]
⁹⁰Sr/⁹⁰Y (3 × 3 Gy per week)	alpha-Difluoromethylornithine (DFMO) antipromoter (1% in drinking water)	ICR mice	Skin and bone tumours	Antagonism	Tumour emergence delayed	[O12]

Table A.1 (continued)

Radiation	Combining agent	Study system	Endpoint studied	Nature of effect	Comments	Ref.
Neutrons (1.7-3.3 Gy)	CCl_4	C57Bl6 mice	Liver carcinomas	Supra-additive	No effect with chloroform	[B16]
Alpha (PuO_2)	CCl_4	Rats and hamsters	Modification of biokinetics		Results not yet known	[B16]
X rays (1.5 Gy)	Bromo-2'-deoxyuridine (BrdUrd) (3.2 mg per rat)	Rats	Total tumour yield. latency	Supra-additive	Potential clinical problem	[A15]
X rays (4-10 Gy)	Nicorandil, inhibitor of free radical production	Chinese hamster V79 cells	Survival	Subadditive		[N20]
X rays (2-8 Gy)	Calyculin A, protein phosphatase inhibitor (2.5-10 mM)	BHK21 cells	Cell killing	Supra-additive	Disruption of protein kinase - mediated signal transduction?	[N21]

Tobacco

Radiation	Combining agent	Study system	Endpoint studied	Nature of effect	Comments	Ref.
Alpha (^{222}Rn)	Active smoking [a]	Humans (miners)	Lung cancer	Supra-additive	Confounders: other air pollutants; transfer to modern workplaces and indoor exposure	[C1, L18]
Alpha (^{222}Rn)	Active smoking [b]	Humans (miners)	Lung cancer	Additive to supra-additive	More-than-multiplicative only for smoking preceding radon	[T18]
Alpha (^{222}Rn)	Active smoking [a]	Humans (home)	Lung cancer	Supra-additive	Little statistical power	[P11]
Alpha (^{222}Rn)	Active smoking [a]	Humans (thorotrast patients)	Lug cancer	None	Proper lung dosimetry lacking, therefore little statistical power	[I5]
Alpha (^{222}Rn)	Mainstream smoke [b]	Rats	Lung cancer	Supra-additive	Interaction only for radon	[C9]
Alpha (^{222}Rn)	Mainstream smoke [b]	Beagle dogs	Lung cancer	Subadditive	Reduced dose to critical cells by thickening of bronchial mucus?	[C21]
Alpha (^{222}Rn)	Diesel fumes [b]	Sprague-Dawley rats	Lung cancer	None	Diesel fumes alone had no effect	[M61]
Alpha (^{239}Pu)	Mainstream smoke [a]	Mice	Lung dose, histopathology	Supra-additive	Lung clearance reduced	[T2]
Alpha (^{239}Pu)	Mainstream smoke [a]	F344 rats	Lung lesions, lung cancer	Supra-additive	Lung clearance reduced	[F17]
X rays (>9 Gy to target)	Active smoking [b]	Humans (Hodgkin's disease patients)	Lung cancer	Supra-additive	Clinical problem	[V7]
X rays	Active smoking [b]	Humans (breast cancer patients)	Lung cancer	Subadditive to supra-additive	Possible bias in positive study	[I9, I10, N4]

Metals

Radiation	Combining agent	Study system	Endpoint studied	Nature of effect	Comments	Ref.
Alpha (^{222}Rn)	Arsenic	Humans (miners)	Lung cancer	None	Adjustment for arsenic exposure reduced the radon risk estimate (ERR WLM^{-1})	[L18, T5]

Table A.1 (continued)

Radiation	Combining agent	Study system	Endpoint studied	Nature of effect	Comments	Ref.
Alpha (150-6 700 Bq ^{239}Pu)	Beryllium (1-91 µg lung burden)	Rats	Lung dose Lung tumours	Supra-additive None	Lung clearance reduced	[S42]
Alpha (60-170 Bq ^{239}Pu)	Beryllium (50-450 µg lung burden)	F344 rats	Lung dose Lung tumours	Supra-additive Supra-additive	Lung clearance reduced	[F12]
X rays (1, 2 Gy)	Beryllium (0.2, 1 mM)	CHO cells	Cell-cycle delay Chromosome aberrations	Supra-additive Supra-additive	Multiplicative interaction probably limited to S and G_2 cells	[B37]
Gamma	Cadmium (1 mg kg^{-1})	Mice	DNA damage in peripheral lymphocytes	Subadditive	Cadmium exposure has to precede radiation by several hours	[P27]
^{134,137}Cs (~2 500 Bq kg^{-1})	Lead (16-320 mg kg^{-1})	Arabidopsis thaliana	Mutations	Subadditive to supra-additive		[K42]
^{137}Cs (-0.012 Gy)	Zinc (10-100 mg kg^{-1}) cadmium (0.5-16 mg kg^{-1})	Soil microbes	Nitrogen fixation, dentri-fication, CO_2 flux	Supra-additive	Catalase activity reduced	[E19, E20]

Mitogens and cytotoxicants

Radiation	Combining agent	Study system	Endpoint studied	Nature of effect	Comments	Ref.
X rays (0.24, 0.94 Gy)	Caffeine (1-7 mM)	Pre-implantation mouse embryos	Micronuclei induction Embryonal development	Supra-additive Supra-additive	Apparent threshold below 1 mM	[M11]
X rays	Caffeine	In vitro chemistry	Oxygen radical concentrations	Subadditive		[K29]
X rays	Caffeine, theobromine, theophylline	T lymphoma cells TKG cells	Apoptosis Apoptosis	Supra-additive Subadditive		[Z4]
X rays	N-methylformamide (0-170 mM)	Human colon tumour line	Survival	Supra-additive	Effect on alpha term, i.e. stronger at lower exposures	[L15]

Antioxidants, vitamins and other dietary factors

Radiation	Combining agent	Study system	Endpoint studied	Nature of effect	Comments	Ref.
X rays	Restricted food intake [a]	Animals	Tumour incidence	Subadditive	Probably via reduced growth stimuli	[C26]
X rays (10 Gy)	NaCl (1% in diet), ethanol (10%) [b]	CD(SD):Crj rats	Intestinal metaplasia	Supra-additive	No effect on incidence of gastric tumours	[W29]
Gamma rays (1.5-4 Gy)	Tetrahydrocannabinol [b]	Rats	Tumour incidence	Supra-additive	Breast adenocarcinoma increased by a factor 5	[M24]
Gamma rays (4 Gy)	Vitamin A, E, selenium, 3-aminobenzamide [a]	C3H10T½ cells	Transformation	Subadditive	Epidemiological studies indicate importance of dietary form	[B24, H11]

Biological agents (hormones, viruses, other)

Radiation	Combining agent	Study system	Endpoint studied	Nature of effect	Comments	Ref.
Gamma, neutrons	Oestradiol-17 beta [b]	Rats	Mammary tumours	Questionable	High hormone levels applied	[B32]

Table A.1 (continued)

Radiation	Combining agent	Study system	Endpoint studied	Nature of effect	Comments	Ref.
Neutrons	Prolactin [a]	Rats	Mammary tumours	Supra-additive		[Y6]
Neutrons (0.064 Gy)	Diethylstilbestrol (DES) (12.5 mg kg^{-1}) [a]	Rats	Mammary tumours	Supra-additive	Increased time span between radiation and hormone did not reduce interaction	[S24]
Gamma rays (2.6 Gy)	DES pellet (release 1 µg d^{-1}) [a]	Rats	Mammary tumours	Supra-additive	Irradiation in late pregnancy or during late lactation clearly more effective	[I11, S40]
Beta rays (1.5 MBq ^{131}I)	Castration ± testosterone [b]	Male rates	Thyroid neoplasms	Supra-additive		[H12]
X rays (2 × 10 Gy)	Testosterone and DES dimethyl-estradiol (0.2–2.5 mg implant) [b]	Normal and gonadecto-mized CD(SD);Cjr rats	Intestinal metaplasia	Subadditive / Supra-additive	DES in females / Testosterone in females; DES in males	[W28]
X rays	Retrovirus T1223/B [a]	C57Bl mice	Leukaemogenesis	Supra-additive	Recombinant viral DNA found in the genome of every tumour	[A25]
X rays	Microbial substances [a]	Mammals	Survival	Subadditive	Stimulation of immune system?	[A12]
Gamma rays (1–4 Gy)	Urethane [a]	Athymic nude and euthymic mice	Radiation enhancement of urethane-induced lung tumours	Additive	Immunosurveillance by T cells not important in this system	[K41]
X rays	Dimethylhydrazine-induced inflammation of bowel [a]	BALB/c mice	Colon cancer	Additive		[W2]

a   Relative for low doses.
b   Uncertain whether relative for low doses.

17. In the early study of molecular genetics, a wealth of data were accumulated on interactions between UV and ionizing radiation in bacterial systems. For a review, see Annex L, "Biological effects of radiation in combination with other physical, chemical, and biological agents", in the UNSCEAR 1982 Report [U6]. Recently, laser applications have become important in industrial settings. A sparing effect from visible light on irradiated bacteria was reported by Voskanian et al. [V6]. The study measured the combined effect of laser (helium-neon laser, 633 nm) and alpha radiation on the survival of *Escherichia coli* K-12 cells of different genotypes. Pre- and post-irradiation exposures to laser radiation diminished the damaging effect of alpha particles. The increase in survival was more pronounced for post-irradiation exposure. There is a well-known molecular basis for enhanced DNA repair and hence for survival: photoreactivation with visible light after UV irradiation. The protective mechanism involved in the repair of damage from alpha irradiation, especially the one involved in pre-irradiation exposure to laser light, remains to be elucidated. However, at this stage there are no such mechanisms known in mammalian cells. Despite large human populations with considerable combined exposures to ionizing and UV radiation to parts of the skin, no indications of a critical interaction are apparent.

3. Low- and high-frequency electromagnetic radiation

18. The photon energies of all frequencies of electro-magnetic radiation below infrared are clearly too low to produce direct chemical damage to DNA. However, there is a large body of published data suggesting the presence of effects at exposure levels below those from critical thermal effects, i.e. local heating by several degrees Celsius. (Heat stress is discussed in Section A.4.) The epigenetic influences of heat stress could only act on later stages of cancer development. Whether so-called athermic levels of high-frequency non-ionizing radiation may interfere with cell signalling, levels of cellular calcium, or systemic melatonin remains disputed on the level of the single agent.

19. Tyndall undertook an investigation [T3] to ascertain the combined effects of magnetic resonance imaging fields and x-irradiation on the developing eye in mice from the strain C57Bl/6J. Dams in groups were subjected to absorbed doses of 50, 150, and 300 mGy. Other dams were exposed to T2 spin-echo magnetic resonance imaging fields under clinically realistic conditions following exposure to 300 mGy from x-irradiation. It was found that the 300 mGy dose had significant teratogenic effects on the eye of C57BL/6J mice. Groups exposed to both types of radiation fields demonstrated malformation levels similar to those in animals irradiated only with 300 mGy from ionizing radiation. The results confirmed the teratogenic effects of low-level x rays but gave no evidence for an enhancement of the teratogenicity of x-irradiation on eye malformations in the mouse system tested.

20. Somewhat unexpected results of combined effect of microwave exposures of non-thermal intensity and ionizing radiation were reported in rats and chicken embryos by Grigor'ev et al. [G13, G16]. Rats were pre-exposed to electromagnetic radiation of power flux density (PFD) 200 μW cm^{-2} 30 minutes daily for 8 days, followed the next day by single whole-body gamma irradiation at 5.5 Gy. Pre-exposure to microwave radiation reduced the mortality rate of the test animals by 33% compared with the controls. Immuno-biological examinations revealed a significant increase in the stimulation index in mitogen (phytohemagglutinin, PHA) induced lymphocytes. The imprinting of chicks was disrupted when they were irradiated in early embryogenesis for 5 minutes with microwaves (PFD = 40 μW cm^{-2}) and then with gamma rays at a dose of 0.36 Gy.

21. The same group also described changes in humoral immunity and in autoimmune processes under the combined action of microwave, infrasonic, and gamma irradiation [G13, G16]. The exposure regimens for rats and rabbits were 9.3 GHz and 0.1 GHz (200 and 1,530 μW cm^{-2}, respectively), infrasound (8 Hz, 115 db), and gamma radiation (cumulative dose of 5.5 Gy). It was shown that pre-irradiation with microwaves increased the resistance of the animal to gamma radiation, but microwaves combined with infrasound enhanced the biological effect of gamma radiation. Since no hypotheses on possible mechanisms are suggested, no inferences applicable to controlled human environments can be drawn at this stage from the extremely high exposure levels in this study.

22. A very strong radioprotective effect of static magnetic fields of 10, 120, and 350 mT on the survival of mice (CBA × C57Bl/6) after acute ^{60}Co irradiation with a dose of 9 Gy was described by Schein [S33]. The adaptive effect of an exposure of 6 hours in a static field increased with time and was strongest in animals irradiated 30 days later. The weakest field, 10 mT, led to a survival of up to 60% of the animals, whereas controls had survival rates of only 0%–4%. The mechanisms behind this antagonism are speculated to be unspecific stress-induced stimulation of endocrine systems by magnetic fields, an increase in surviving stem cells after ionizing radiation, or a faster proliferation and differentiation of bone marrow stem cells in adapted animals.

23. Growth and survival rates of cultured cells (FM3A) were investigated in a static gradient magnetic field with a strength of 58 mT at the center and a mean gradient of 0.6 T m^{-1} [K16]. The magnetic field alone reduced the growth rate by 5% and survival by 20%. The combined effect of ^{60}Co irradiation followed by exposure to the magnetic field showed synergism.

24. Magnetic fields have been shown under certain reaction conditions to perturb the rates at which radical pairs recombine. An example is catalase-catalysed decom-position of H_2O_2, which is increased by 20% in an extremely high magnetic field of 0.8 T [M35]. In theory, this could lead to changes in the kinetics of free-radical production and recombination [S60]. To measure the interaction potential of this indirect genotoxic effect of magnetic fields with ionizing radiation, the exposures

would have to be simultaneous and not sequential, as described in the preceding paragraphs.

25. In assessing the association between exposure to electromagnetic fields and cancer, Koifman [K17] defined the elements necessary for quantitative analysis. Obtaining more accurate measurements of exposure to electromagnetic fields is a key to understanding any possible association. In certain circumstances, strong electromagnetic fields may stimulate growth and hence fulfill the characteristics of a cancer-promoter in biomechanistic models of carcinogenesis. This leads to the hypothesis that electromagnetic fields do not act alone to affect health, as is assumed in many epidemiological studies, but only where their action is combined with that of other initiator agents.

26. In summary, no straightforward inferences from experimental results to exposures in occupational settings are possible at this stage for the combination of electromagnetic and ionizing radiation. From the standpoint of mechanistic considerations, there is little evidence for potentially harmful interactions between the two radiation modes for controlled exposure levels in the workplace or in the clinic.

4. Temperature

27. Heat kills mammalian cells in a predictable and stochastic way [D3]. Heat stress at the cell and tissue level may disrupt energy metabolism (local depletion of oxygen and ATP) as a result of the enhanced reactivity of most enzymes, the production of heat shock proteins, and finally denaturation and cell death. Critical changes leading to a loss of proliferative capacity involve cell membrane blebbing, probably owing to detachment of the cytoskeleton from the plasma membrane [R23]. A slow mode of cell killing by hyperthermia in CHO cells involves the formation of multi-nucleated cells from damage to centrioles [D3]. Above 42.5°C, cell-survival curves for Chinese hamster ovary cells in culture where the abscissa is the duration of heat treatment are similar to the curves for x rays. At 42°C and below, the survival curves tend to flatten out with time as tolerance to the elevated temperature develops. The cell-cycle dependence of sensitivity to heat contrasts with that of x rays, with late S-phase cells being the most sensitive to hyperthermia treatment. Cells at low pH or deficient in nutrients also show elevated heat sensitivity. Temperature is therefore an important modifier of radiation sensitivity in many therapies to control tumour growth. In general, hyperthermia increases the relative susceptibility of tumour cells to radiation compared with healthy tissue. Very hot or very cold ambient temperatures are rarely encountered in the modern workplace and the temperatures that do prevail generally do not change the body core temperature. No correlation with elevated radiation exposure is apparent in such workplace settings. The same is true for recreational settings and even for hot spas with elevated radon levels. Therefore the combined action of high and low tempera-

tures remains in the realm of clinical research, and the following paragraphs give only some cursory remarks on recent *in vitro* work.

28. At the mechanistic level, it is important to note that the large effects found in hyperthermia treatments cannot be attributed solely to changes in blood flow and concomitant changes in local oxygen pressure alone. The disruption of energy metabolism due to considerably accelerated biochemical reactions and a decrease in molecular stability are important far below the threshold of protein denaturation. Dauncey and Buttle [D2] found a tendency towards elevated plasma concentrations of growth hormone and prolactin in 14-week-old pigs acclimated to 35° or 10°C, respectively. In mammalian cell culture (L5178Y), protease inhibitors such as phenylmethylsulfonyl fluoride were shown to potentiate hyperthermic cell killing [Z13]. It is suggested that protease inhibitors sensitize by inhibiting the proteases that are needed to degrade denatured proteins induced by heat. In response to heat, cells and tissue produce proteins of mainly 70 and 90 kilodaltons. These proteins are called heat-shock proteins, although many other agents such as arsenite and ethanol also induce them. Their appearance coincides with the development of thermotolerance, an important effect that can influence the slope of the survival curve by a factor of up to 10. The development of thermotolerance and the production of heat-shock proteins occur during heating at temperatures up to 42°C (CHO cells) but are delayed by several hours for heat treatment with higher temperatures [H36].

29. Skin is the only tissue whose temperature might differ considerably from the core temperature. Therefore, Zölzer et al. [Z3] studied the influence of radiation and/or hyperthermia on the proliferation of human melanoma cells *in vitro*. DNA synthesis and content were both determined with two-parameter flow cytometry. In controls, most of the S-phase cells showed incorporation of BrUdR. The fraction of quiescent S-phase cells increased after irradiation (up to 8 Gy from x rays) and/or hyperthermia (up to 6 hours at 42°C or up to 2 hours at 43°C). There was a clear dose dependence for radiation and hyperthermia alone or in combination. In general, the combined effect seemed to be additive.

30. Combination effects of radiation and hyperthermia were found, however, in several other *in vitro* cell systems. Matsumoto et al. [M15] treated cultured human retinal pigment epithelial cells by radiation, hyperthermia, or a combination of the two. The effect on cell proliferation was evaluated by counting the cell number and measuring the uptake of bromodeoxyuridine. x-irradiation with a dose of 1 Gy or 3 Gy was not effective in suppressing proliferation of the retinal pigment epithelial cells. Similarly, heat treatment at 42°C for 30 minutes did not suppress proliferation. However, combining hyperthermia at 42°C for 30 minutes with 3 Gy irradiation suppressed cellular growth of the retinal pigment epithelial cells to 36% of the control, as estimated by cell counting, and to 48% by the bromodeoxyuridine uptake assay. The effect of radiation combined with heat on three human prostatic carcinoma cell lines was investigated by Kaver et al. [K6]. Cells were exposed to different radiation

doses followed by heat treatment at 43°C for 1 hour. Heat treatment given 10 minutes after radiation significantly reduced the survival rate of all the cell lines studied. The combined effect of radiation and heat produced greater cytotoxicity than predicted from the additive effects of the two individual treatment modalities alone. Impairment of DNA repair with elevated temperature is considered an important mechanism [W36].

31.   Growth, cell proliferation, and morphological alterations *in vivo* in mammary carcinomas of C57 mice exposed to x rays and hyperthermia were followed by George et al. [G6]. Radiation doses of 10, 20, or 30 Gy from x rays or heating to 43°C for 30 minutes preceded or not by exposure to 10 Gy were studied. Tumour growth, cell proliferation kinetics, induction of micronuclei, and morphological changes in necrosis and vascular density were simultaneously determined. These showed very complex adaptive responses. Treatment with radiation and/or hyperthermia produced only a delay in tumour growth of between 1 and 3.8 days. However, the effects of the treatments became more apparent when the amounts of muscle and necrosis were deducted from the originally measured tumour volume. Radiation-induced G_2 block of the cells was observed 12 hours after radiation alone. After combined treatment, however, the G_2 block was delayed beyond 12 hours. Whereas the amount of necrosis was markedly enhanced five days after treatment with 10 Gy plus heat, as well as after 30 Gy, no changes in the density of small blood vessels could be observed during this period. These results clearly demonstrate that the apparent changes in tumour volume after x rays and hyperthermia do not truly reflect the response of the constituent cells and that there are many other factors, for instance cell proliferation and morphological alterations, that influence the combined effects of radiation and hyperthermia.

32.   Heat shock before, during, or immediately after exposure to ionizing radiation can increase cell killing in a supra-additive manner [B70]. The heat-shock treatment was shown to inactivate the Ku auto-antigen binding to DNA, and this binding capacity of Ku was directly related to the hyperthermic radiosensitizing effect. The Ku auto-antigen is the regulatory subunit of the DNA-dependent protein kinase and is directly involved in DNA double-strand break repair and V(D)J recombination.

33.   In general, it can be said that because of the high temperatures and exposures needed to produce enhanced cell killing in poorly oxygenated tissue, combined effects from hyperthermia and ionizing radiation are not relevant outside the realm of tumour therapy. Temperature in combination with ionizing radiation can act synergistically on cell survival, cell proliferation, and cytogenetic damage. However, temperatures higher than those found in the human body are needed to cause these effects.

5. Ultrasound

34.   Possible effects from ultrasound exposures alone or in combination with ionizing radiation are of some concern because ultrasound is so widely used in diagnostic procedures. Above a threshold level, ultrasound by itself may induce cavitation, leading to mechanical damage to cellular structures and to microlesions. Kuwabara et al. [K20] studied the effects of ionizing radiation and ultrasound at exposure levels typical for diagnostic purposes on the induction of chromosomal aberrations and sister chromatid exchanges in peripheral lymphocytes. No statistically significant increases in the frequencies of dicentric and ring chromosomes or sister chromatid exchanges were discovered after ultrasound exposure alone at the diagnostic level (Table A.2). Nor could elevated frequencies of these phenomena be found following exposure to ultrasound before or after ionizing radiation, compared with the frequencies found after the same dose of ionizing radiation alone. However, simultaneous exposure to ultrasound and ionizing radiation seemed to induce a slight enhancement of sister chromatid exchanges, although no significant changes were noted in the yields of dicentric and ring chromosomes.

Table A.2
Effects of combined exposures to ionizing radiation and ultrasound in peripheral human lymphocytes
[K20]

Exposure		Dicentrics and rings	Sister chromatid exchanges
Radiation	*Ultrasound*		
None (control)	None (control)		6.64±0.40
3 Gy	None	0.61±0.08	7.92±0.54
	40 min (immediately following)		6.31±0.53
	80 min (immediately following)		7.00±0.47
	30 min (simultaneous)	0.52±0.07	9.80±0.91
4 Gy	None	1.12±0.11	
	30 min (simultaneous)	1.10±0.11	9.96±0.50

35.   Continuous-wave ultrasound and neoplastic transformation was assayed *in vitro* by Harrison and Balcer-Kubiczek [H10] in C3H10T½ cells in suspension. An initiation-promotion protocol for neoplastic transformation induced by continuous-wave ultrasound was used. Cells were insonated at 1.8 MHz for 40 minutes. Two ultrasonic

intensities were used: 1.3 and 2.6 W cm^{-2} spatial average. The first intensity was found to be non-cytotoxic; the second was above the threshold level for cavitation and resulted in immediate lysis of 20% of the cells (cavitation-induced cell killing), followed by the clonogenic survival of 64% of the remaining cells. Ultrasound was delivered alone or in combination with x rays (2 Gy, 240 kVp given before ultrasound) and/or TPA (0.1 µg ml^{-1} after irradiation). Under all treatment conditions, ultrasound had no effect on transformation at the 95% confidence level. The effects of high-energy shock waves, i.e. therapeutic levels of ultrasound generated by a lithotripter in combination with ^{137}Cs gamma rays were shown to act additively or slightly supra-additively in colony-forming assays and cell-cycle analysis [F29]. Both pellets of single cells and multicellular spheroids of the bladder cancer cell line RT4 gave similar results.

36.   In conclusion, it can be said that the ultrasound intensities used for diagnostic purposes and ionizing radiation did not interact to cause cytogenetic damage in treated cells. However, sister chromatid exchanges were slightly increased in one study. *In vitro* transformation rates caused by ionizing radiation were not changed by ultrasound.

6. Dust, asbestos, and other mineral fibres

37.   The combination of radiation exposure and exposure to dusts and fibres is quite common in important industrial environments such as mining, metallurgical industries, and power plants. Some dusts and fibres are pathogenic or carcinogenic by themselves. Both experimental results from mammals and epidemiological evidence are available [B9, B13, C22, K13, P1, P5]. In cases where the main biological effect results from soluble toxicants that dissolve from the surface of dust particles to interact with biological structures, the interaction is basically between radiation and a chemical, which is dealt with in Section B of this Appendix.

38.   Silica is often considered to be a co-carcinogen through the route of silicosis. Harlan and Costello [H9] studied 9,912 metal miners (369 silicotics and 9,543 non-silicotics) to investigate the association between silicosis and lung cancer mortality. When lung cancer mortality in silicotics and non-silicotics was compared, the age-adjusted rate ratio was 1.56 (95% CI: 0.91–2.68). Further adjustment for smoking yielded a rate ratio of 1.96 (95% CI: 0.98–3.67), and the value for employment in mines with low levels of radon was 2.59 (95% CI: 1.44–4.68). The statistical power of the study was too weak to quantify single contributions and interactions between metal, radon, silica, and smoking. For high dust loads and concomitant exposures to gamma radiation and radon in earlier times, there is indication for an increased lung cancer risk (standardized mortality ratio = 2.5 with 20 years of employment and hired before 1960) in the phosphate industry [B74].

39.   The molecular mode of action of mineral fibres is quite distinct from radiation and genotoxic chemicals interacting directly with nuclear DNA. They are relatively ineffective as mutagens but quite powerful inducers of human meso-theliomas and bronchial cancers. Fibre dimensions, fibre durability, and surface characteristics are important properties affecting their carcinogenicity. In the case of asbestos, there is clear evidence for the induction of chromosomal aberrations and aneuploidy [B13]. A possible mechanism of asbestos cell toxicity is phagocytosis and accumulation of the fibres in the perinuclear region of cells. During mitosis, the fibres would then interfere with chromosome segregation, and chromosomal abnormalities would result. In addition, mechanical irritation and cell killing may lead to growth stimulation and transcellular epigenetic promotion. The production of active oxygen species on fibre surfaces was proposed as a directly acting genotoxic mechanism; however, the relatively long diffusion length from the site of radical production outside the nucleus to the target structures argues against the importance of this pathway.

40.   Recent reviews of mortality and cancer morbidity in asbestos worker cohorts with large cumulative exposures showed an ERR for pleural mesothelioma of about 1 for each fibre-year ml^{-1} of air [A3]. For lung cancer, an ERR from 0.0009 to 0.08 per fibre-year ml^{-1} has been found [N9], which, in absolute terms, is considerably higher than the mesothelioma risk. The ratio of the number of meso-theliomas to the excess number of cases of lung cancer ranges from 0.06 to 0.78.

41.   Few epidemiological data exist describing potential interactions between mineral fibres and radiation. In a case-control analysis of deaths from lung cancer among persons employed at the Portsmouth Naval Shipyard at Kittery, Maine, in the United States, elevated odds ratios for exposures to ionizing radiation, asbestos, and welding by-products were found in a first crude assessment. Further analysis of data on radiation exposure, controlling for exposures to asbestos and welding, found no evidence for a risk related to radiation exposure. The low cumulative radiation doses and the absence of data on cigarette smoking and socioeconomic status precluded an assessment of possible interactions among the three toxic agents [R1].

42.   The synergistic effects of the combined exposures to asbestos and smoking in the causation of human lung cancer was one of the first examples of a supra-additive interaction of importance for protection in the workplace [S19]. In most studies, very high risk ratios were observed in asbestos-exposed subjects who were heavy smokers. The interaction observed in most cases conforms more closely to a multiplicative model than an additive one. Brown et al. [B7] were able to show in organ cultures derived from Fischer F344 rats that the ability to metabolize benzo(a)pyrene was significantly reduced after *in vivo* exposure to crocidolite, thus suggesting possible mechanisms leading to a departure from linearity. Work by Fasske [F2] showed that after the combined instillation of 1 mg chrysotile and 0.5 mg benzo(a)pyrene, lung tumours arose much earlier than after the instillation of only one of the carcinogens.

43. Regarding animal experimentation, Bignon et al. [B38] inoculated radon-exposed Sprague-Dawley rats intrapleurally with asbestos fibres, glass fibres, or quartz. In rats given mineral materials, bronchopulmonary carcinomas and mixed carcinomas were observed, as well as typical mesotheliomas and combined pulmonary pleural tumours, whereas in rats inhaling radon alone, only bronchopulmonary carcinomas occurred. A clear co-carcinogenic effect of the insult from the minerals was established for malignant thoracic tumours. Significant differences in survival time were found for exposures to different types of dust, depending on the additional tumour types induced. The same group also studied whether similar co-carcinogenic effects would take place over longer distances, i.e. from subcutaneous injection of chrysotile fibres. Neither mesotheliomas nor evidence of co-carcinogenic effects were found in the animals treated with both radon and asbestos fibres [M19]. Three groups of animals were used: 109 rats that inhaled radon only (dose = 1,600 working-level months [WLM]); 109 rats given a subcutaneous injection in the sacrococcygeal region of 20 mg of chrysotile fibres after inhalation of radon resulting in the same dose; and 105 rats injected with fibres only. As already stated, no mesotheliomas occurred in any of the three groups. The incidence of lung cancer was 55% in the second group, 49% in the first, and 1% in the third group. Statistical analysis using the Pike model showed that the carcinogenic insult was slightly higher in the second group than in the first group. Electron microscopy analysis of fibre translocation from the injection site showed that less than 1% of injected fibres migrated to the regional lymph nodes and only about 0.01% to the lungs. After injection, the mean length of the fibres recovered in lung parenchyma increased with time, suggesting that short fibres are cleared by pulmonary macrophages, whereas long fibres remain trapped in the alveolar walls. Kushneva [K49] studied pathological processes in the lungs of white rats exposed intratrachealy to 50 mg of finely dispersed quartz dust and to 3 hours of $3 \cdot 10^8$ Bq m^{-3} radon. Supra-additivity is clearly implied but only described qualitatively.

44. To assess the possible co-carcinogenic effects of mineral dust in radon-prone mines, five groups of 30 Sprague-Dawley rats received minerals typically found in metal mines (nemalite; biotite, present in many granites; iron pyrite; chlorite) by intratracheal instillations one month after the end of a 1,000 WLM radon exposure. No or only slight co-carcinogenic effects were found [M62]. In earlier work with the same experimental system to investigate the effect of intrapleural injection of asbestos fibres (chrysotile), glass fibres, and quartz on the yield of radon-induced thoracic tumours, a clear promoting effect was noted [B38].

45. Densely ionizing alpha particles, similar to those emitted by radon progeny, are highly effective in inducing transformations in cell cultures such as CH310T½ cells. The yield of foci from combined alpha/asbestos exposure is clearly greater than would be predicted from the sum of the effects found with single-agent exposures. Figure A.I shows a clearly supra-additive interaction with asbestos fibres [H11].

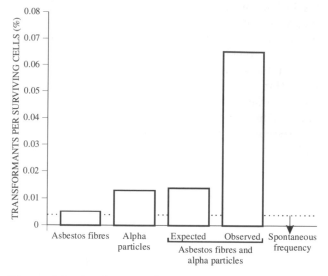

Figure A.I. *In vitro* **transformation of C3H10T½ cells exposed to asbestos fibres and alpha particles alone and in combination [H11].**

46. In an experimental study, Donham et al. [D10] studied possible combined effects of asbestos ingestion and localized x-irradiation of the colon in rats based on the hypothesis that the mucous produced by goblet cells that normally coats the normal bowel surface protects against tissue penetration by ingested asbestos. X-ray treatment results in localized damage to the colonic mucosa and theoretically disrupts the normal mucous coating, allowing increased tissue penetration by the fibres. To study this, segments of the colons of laboratory rats were exposed to x-irradiation. The animals were then divided into three groups, which were fed a diet containing 10% chrysotile asbestos, a diet containing 10% non-nutritive cellulose fibre, or a standard laboratory diet. Autopsies and histopathology were performed on all animals that died spontaneously and those that were killed at 350 days. Various types of inflammatory and degenerative lesions were commonly seen, but there was little difference in frequency between the diet groups. Five adenocarcinomas and two sarcomas were seen in the fibre groups (three tumours in the asbestos group and four tumours in the cellulose group), but no tumours were seen in animals on the standard diet. There was no significant difference in tumour rates between the asbestos and cellulose groups, nor was there a significant difference between the combined fibre groups and the standard diet group. Ingested asbestos did not increase the risk of tumour development and does not, therefore, seem to be co-carcinogenic or to promote tumours by disrupting the mucous coating.

47. In summary, it can be stated that mineral dust and fibres such as asbestos generally act through non-genotoxic mechanisms. These include mechanical irritation and cell killing. However, chromosomal aberrations, especially aneuploidy, can be induced by interfering with the spindle apparatus of mitotic cells. At exposure levels found in

workplaces until the early 1940s, there was a clearly supra-additive interaction between asbestos and tobacco smoke exposure in the causation of lung cancers, with a concomitant shift in the cancer spectrum from mesotheliomas to broncho-pulmonary carcinomas. A similar supra-additive interaction and shift in the cancer spectrum was observed in animals exposed to both asbestos and radon. The much lower occupational exposures experienced today considerably decrease the risk for potential detrimental interactions between dust/fibres and radiation. However, in view of the proven interaction effects in humans, any stochastic and/or genotoxic effects of these agents merit further consideration.

7. Space flight

48. In space flight, which involves an extreme situation of controlled exposures, a multitude of stressors act in combination on astronauts, the most important being microgravity. Its biological and medical role has been extensively reviewed [M71]. Microgravity effects may occur at all levels of biological organization, and in principle can also lead to modifications of radiation action. From an experimental point of view there are no clear-cut results at the organ and tissue level. With simple organisms, a synergistic action of microgravity and radiation has been reported for teratogenic effects [B80]. Antipov et al. [A14] analysed structural and functional changes in the central nervous system of experimental animals exposed to the isolated and combined effects of space flights. They evaluated the significance of ionizing and non-ionizing radiation, hyperoxia, hypoxia, acceleration, vibration, and combined effects of some of these factors for anatomic and physiological changes in the rat brain. Neuronal functions were found to be sensitive to ionizing radiation and hypoxia, but these synapses were shown to be highly resistant to short-term hyperoxia and electromagnetic radiation [A13]. Along with radiation, the investigated stressors had additive, synergistic, and antagonistic effects on the central nervous system. However, as significant effects and deviations from the sum of effects from exposure to isolated stressors were always linked to high exposures and exposure rates, they have little relevance for exposure situations on the ground.

49. In radiobiological experiments in space, a more-than-additive interaction between microgravity and radiation was reported in several cases (reviewed in [H47]). Insect embryos in particular appear to be susceptible. Conflicting results were reported for cellular systems. In human lymphocytes that were exposed to ^{32}P-irradiation in space, chromosomal aberrations were significantly increased compared with ground controls [B18]. However, the follow-up experiment by the same authors did not show this interaction [B19]. More recently, experiments on the interaction of space microgravity and DNA repair were performed by Hornek et al. [H14]. Microgravity had no measurable effect on strand rejoining of x-ray-induced DNA strand breaks in *Escherichia coli* (120 Gy) and in human fibroblasts (5 and 10 Gy) or on the induction of SOS reponse in *E. coli* (300 Gy). In yeast no microgravity

related effects on the repair of DNA double-strand breaks were found both for cells irradiated previously on ground [P31] or during flight using a ^{63}Ni beta source [P32]. Therefore, repair of radiation-induced DNA damage seems not to be disturbed by microgravity, and other mechanisms must be involved in the reported interaction between radiation and space gravity.

50. At similarly high exposures, Vasin and Semenova [V3] showed synergistic effects for combined stress from radiation and vibration or normobaric hyperoxia. A study was made of the combined effect of normobaric hyperoxia and vibration on the sensitivity of hybrid mice (CBA × C57Bl)F_1 and F_2(CBWA) to gamma radiation. Both single and protracted (for five days, daily) vibration before irradiation aggravated acute radiation sickness. Hyperoxia also enhanced the development of the intestinal form of radiation sickness. The combined effect of the two additional factors aggravated the intestinal syndrome of acute radiation sickness. These deterministic effects have no direct implication for present-day controlled exposure situations. Nevertheless, the changes of many parameters that are normally stable in experimental work on earth make well-designed studies in space potentially important in addressing the combined effects of physical agents.

B. RADIATION AND CHEMICAL TOXICANTS

51. A multitude of natural and man-made chemicals with cancer-initiating and -promoting potential are present in the human environment and may interact with radiation. Classification based on their mode of action is often difficult, but at least a crude separation can be made into substances that mainly act by damaging DNA directly (genotoxic substances) and non-genotoxic substances [C50]. The former group includes chemically active species (activation-independent chemicals) or species dependent on biotransformation and their active metabolites (activation-dependent chemicals). The mode of action is either direct, by forming covalent links with DNA, or indirect, via radical attack of DNA. The latter group comprises chemicals ranging from nonspecific irritants and cytotoxins to natural hormones and growth factors and their analogues that interact with the regulatory systems of cells and organs. At this point, chemicals that protect against ionizing radiation should also be mentioned. Many endogenous and exogenous sulfhydryl-carrying molecules as well as other radical-scavenging agents considerably reduce the primary damage and hence the clinical effects caused by radiation [M6, M7]. A wealth of experimental data is available to describe the action of single chemical agents, but the literature on interactions between these substances and other agents is far more sketchy. It is important to note that recent efforts to quantify tissue doses of chemical toxicants and their metabolites showed the decisive importance of interactions in activation and deactivation/excretion processes. For example, an assessment of the toxicity of benzene and its metabolites was

shown to depend crucially on the presence of other toxicants such as toluene, and this effect extended to concentrations found in human exposures [M60].

1. Genotoxic chemicals

52.   The large group of genotoxic chemicals may be further subdivided on the basis of their need to be activated by metabolism. Most chemicals require metabolic activation through the generation of highly reactive electrophiles, which form DNA adducts by binding covalently to nucleic acids. The metabolism of any individual chemical can be very complex, because the chemical can be the substrate of several metabolizing enzymes. Genotoxic chemicals can also be subdivided based on whether the reactive compound acts directly by covalent binding to DNA or indirectly by the generation of free radicals. In the latter case, effects similar to those of radiation can be envisaged.

(a) Activation-independent alkylating agents

53.   Modern cancer therapy involves many combined treatments using radiation and genotoxic drugs. Although exposures are well known and strong interactions exist, this human experience is of limited direct importance for risk assessment at low doses, because with therapy, cell killing is the main endpoint envisaged. Therefore this subject is considered separately in Section D of this Appendix. The occurrence of second primary tumours in healthy tissue adjacent to treated tumours is of great direct relevance.

54.   Morishita et al. [M32] examined the effects of x rays on N-methyl-N-nitrosourea (MNU)-induced multi-organ carcinogenesis in both sexes of ACI rats. Rats were treated with MNU (25 or 50 mg kg^{-1}) at 6 weeks of age and/or with x rays (3 Gy) at 10 weeks of age. The incidence of adenocarcinomas in the small and large intestines of male rats treated with 50 mg kg^{-1} MNU and x-irradiation (small intestine, 48%; large intestine, 32%) was significantly higher than the sum of the incidences resulting from 50 mg kg^{-1} MNU alone (small intestine, 17%; large intestine, 8%) and with radiation only (small intestine, 0%; large intestine, 0%) and also higher than the frequency of adenocarcinomas in the large intestine of males treated with 25 mg kg^{-1} MNU alone (0%). Strongly synergistic effects in these high-exposure studies were restricted to the gastrointestinal system. When MNU or 1,2-dimethyl-hydrazine (DMH) treatment was started two months after x-irradiation, no induction of gastric tumours was observed with MNU [W3], and only a low incidence was observed with DMH [A7]. Surprisingly, an inverse relationship between incidences of gastric tumours and intestinal metaplasias was apparent. These findings again indicate the importance of the order and timing of the exposures in the induction of combined effects. It comes as a further surprise that the presence of intestinal metaplasia, long considered a basis for further malignant growth, does not exert a positive influence on the induction of gastric neoplasia by MNU in the rat.

55.   Seidel [S22] studied the effects of radiation on chemically induced T-cell lymphomas (thymomas) in BDF$_1$ mice. N-methyl-N-nitrosourea or butylnitrosourea (BNU) were the main inducers, and x rays in various dose schedules were applied. The radiation was seen to shorten the latency period between induction and lymphoma emergence in protocols of 12 exposures of 0.25 Gy. This effect was most pronounced compared with chemically induced non-irradiated controls with a prolonged median induction time as a result of a dose reduction of the chemical (median induction time 27–36 weeks instead of 16–18 weeks under optimal conditions using 50 mg kg^{-1} of MNU). Irradiation 2–5 weeks before administrating 40 mg kg^{-1} of MNU also enhanced leukaemogenesis. Again, mice with regenerating lymphohaemopoiesis after lethal irradiation and bone marrow transplantation were more sensitive to both chemicals than were the controls. Combined effects from radiation and N-ethyl-N-nitrosourea (ENU) on neural tumours in Wistar rats were reported by Hasgekar et al. [H6]. The animals received 2 Gy whole-body irradiation, followed immediately by 10 mg kg^{-1} of ENU on the day of birth. Of 33 rats given ENU alone, 14 developed 22 tumours of the nervous system, of which 15 (68%) were gliomas and 7 (32%) were schwannomas. Of 34 rats given both irradiation and ENU, 12 were found to harbour 15 neural tumours, of which 14 (93%) were gliomas and 1 (7.1%) was a schwannoma. The pretreatment with irradiation seems to have resulted in selective suppression of schwannoma induction. Whether this antagonistic relationship is a result of overkill or whether it may be relevant for lower radiation doses remains to be elucidated.

56.   The combined effects of radiation and BNU on murine T-cell leukaemogenesis was studied by Seidel and Bischof [S20] in BDF$_1$ mice. The animals were exposed to BNU (0.02% in drinking water) for 12 weeks, and they died of thymic lymphomas with median latency periods of 12–20 weeks. Groups of mice received weekly radiation doses of 0.06–1.0 Gy in addition to BNU. Lower doses (12 × 0.25 Gy) enhanced leukaemogenesis, high doses (12 × 0.75 Gy) delayed it, and intermediate doses (12 × 0.50 Gy) had no effect. Doses lower than 12 × 0.25 Gy had marginal enhancing effects. After a dose of 12 × 1.0 Gy, the mice died earlier than after treatment with BNU alone, and as with the dose of 12 × 0.75 Gy, some extrathymic lymphomas were observed. The numbers of CFU-S in the femur and the spleen showed a dose-dependent depression, in addition to the decrease from BNU alone. In lymphocyte stimulation assays with Con A and LPS and also in the mixed lymphocyte reaction, a reduced proliferation was found, again dependent on the radiation dose. Thus, there was an inverse correlation between leukaemogenesis and the degree of stem-cell reduction or depression of these immune parameters.

57.   Stammberger et al. [S13] analysed the activity of O^6-alkylguanine-DNA alkyltransferase (AT) in the fetal brain and liver and made long-term observations of Wistar rats that were treated in utero either with x-irradiation

(1 or 2 Gy), with ENU (50 mg kg^{-1}), or with both in combination. They hoped to reveal any relationship between the O^6-alkylguanine repair capability and tumour incidence in the organs of the offspring. The AT activity in the brain was affected to the same extent in the fetuses as in the dams. There was a 61% decrease in AT activity in fetuses 24 hours after ENU treatment. This correlated with a significant increase in the incidence of brain tumours in the treated offspring (44%) compared with control animals. The inductive effects of x-irradiation on AT activity (131% for 1 Gy and 202% for 2 Gy) corresponded with a reduction in the incidence of tumours after the combined treatment (27% and 8.3% tumour incidence, 103% and 158% AT activity). Comparing biochemical and morphological results suggests that this antagonistic effect may be the result of the AT induction by x rays.

58.   Yokoro et al. [Y6] found that whole-body irradiation facilitates chemically initiated T-cell lymphomagenesis in mice. This was attributed to the amplification of the cell population susceptible to a chemical carcinogen in the target tissues, bone marrow, and thymus during the recovery phase after irradiation. Split administration of ENU showed different effects in the different phases of carcinogenesis leading to T-cell lymphomas. Once more the authors emphasized that after a cell has been initiated by a genotoxic agent, its fate is determined by the presence of promoters and inhibitors and that modifiers of target cells play a crucial role in the induction yield of tumours. The possibility of synergistic effects in carcinogenesis due to changes in cellular kinetics brought about by combined treatment with radiation and ENU was studied by Seyama et al. [S21]. Lymphomas in female C57Bl/6N mice were used as a model system. A single intragastric administration of 5 mg (about 200 mg kg^{-1} body weight) of ENU was only slightly lymphomagenic, inducing thymic lymphomas in 20% of mice; the incidence was elevated to 92% if the ENU treatment was preceded (five days earlier) by 4 Gy from whole-body x-irradiation, which alone is seldom lymphomagenic. A high yield of lymphoma (84%–93%) was also obtained when 5 mg (about 200 mg kg^{-1}) of ENU was delivered in two split doses four days apart of 4 mg and 1 mg (160 and 40 mg kg^{-1}), indicating that cellular kinetics or clonal expansion, but not two agent-specific different initiation events in the combined treatment, is at the root of this apparent synergism. Drastic injury to both the thymus and bone marrow caused by either 4 Gy whole-body x-irradiation or the first dose of ENU (4 mg, or about 160 mg kg^{-1}) was followed by a vigorous regeneration within a few days. The maximum induction rate of lymphoma was obtained when the subsequent dose of ENU (1 mg, or 40 mg kg^{-1}) was given at the peak of DNA synthesis in the bone marrow and thymus following the first treatment. The data indicate that the principal effect of irradiation or the first dose of ENU was to provide a susceptible cell population, and that a high yield of lymphomas was brought about by the action of the subsequent dose of ENU on a larger number of potentially radiation-modified target cells engaged in heightened DNA synthesis.

59.   A clear antagonistic effect of ENU and x-irradiation was observed by Knowles [K14, K15] for neurogenic tumours in neonatal rats. After neonatal injection of rats with 10 mg kg^{-1} of ENU, whole-body x-irradiation with 1.25 Gy caused a reduction in induced neurogenic tumours, which was greatest when radiation was given 1 day after ENU and progressively decreased with irradiation at 5 and 30 days. Although x-irradiation did not affect the range of histological appearances in the tumours, malignant schwannomas, particularly those of the trigeminal nerve, were significantly reduced by 1.25 Gy given after ENU (10 mg kg^{-1}). The mean latency for clinical signs of tumour appearance was not affected by radiation. Another important finding in this study also points to the importance of the size of stem cell pools in interactions: a significant reduction in the high spontaneous incidence of squamous-cell carcinomas of the mouth in the inbred strain used after 1.25 Gy from x-irradiation. The reduction was greater after irradiation at 5 days of age than at 30 days. A large study on the incidence rates of neural, pituitary, and mammary tumours in Sprague-Dawley rats treated with x-irradiation and ENU during the early post-natal period was undertaken by Mandybur et al. [M2]. These late effects of early post-natal treatment with ENU, preceded by x-irradiation to the head, were studied in 226 neonatal CD rats. The animals were divided into six groups, each receiving one of the following treatments: x-irradiation with 5 Gy to the head on the third post-natal day; ip injection with 30 mg kg^{-1} ENU on the fourth post-natal day; ip injection with 30 mg kg^{-1} ENU on the seventh post-natal day; a combination of x-irradiation with 5 Gy to the head on the third post-natal day, followed by ip 30 mg kg^{-1} ENU on the fourth post-natal day; a combination of x-irradiation of 5 Gy to the head on the third post-natal day, followed by ip 30 mg kg^{-1} ENU on the seventh post-natal day; and untreated controls. The results indicated that (a) x-irradiation to the head alone significantly extended the lifespan of females compared with that of control females and did not affect the survival of males; (b) x-irradiation did not influence the latency period or mortality from neurogenic tumours when ENU was given 1 or 3 days afterwards; (c) ENU itself was a factor in shortening latency periods for mammary tumours; (d) x-irradiation alone did not increase the incidence of mammary tumours and revealed no protective effect on the ENU-induced mammary carcinogenesis; (e) x-irradiation increased the prevalence of pituitary tumours in the females; (f) no enhancement of pituitary tumours by ENU was observed; and (g) there was a statistically significant association of pituitary and mammary tumours in females. Again, these widely divergent findings speak against the possibility of simple concepts for the interaction of different genotoxic agents.

60.   Post-natal development and cancer patterns in NMRI mice after combined treatment with ENU and x-irradiation on different days of the fetal period were studied by Wiggenhauser and Schmahl [W10]. When mice were irradiated to 1 Gy on day 14, 15, or 16 of gestation, this did not result in an increased tumour frequency in the offspring until 12 months. Mice treated with ENU (45 mg kg^{-1}) on day 15 of gestation developed a significantly increased tumour frequency in the lungs and liver and in the ovaries. After

combined treatment in the sequence x rays plus ENU with an interval of 4 hours, a significantly increased incidence of animals with tumours was observed in the offspring treated on gestation day 14 or 16. Moreover, the treatment on day 16 exhibited the highest tumour frequency per examined animal (5.7) of all treatment groups. Although the result was due to a relatively uniform increase of all tumour types, the frequency of liver tumours was most marked. In the reverse sequence (ENU plus x rays), the total tumour outcome was not significantly altered compared with the effects of ENU alone. However, detailed analysis also showed a significant augmentation of the liver tumour frequency with treatment on day 15.

(b) Metabolism-dependent alkylating agents

61.   Maisin et al. [M8] studied the effects of x rays alone or combined with the initiator diethylnitrosamine (DEN) on liver cancer induction in infant C57Bl/Cnb mice. The number of induced liver foci and carcinomas was found to depend essentially on the dose of DEN. X rays did not produce any combined effect on the induction of foci or carcinomas when given seven days before or after administration of DEN [M34]. Using the same system for exposures to DEN and neutrons (average energy = 3.1 MeV), it was shown that even high-LET irradiation (0.125–0.5 Gy) initiated only small numbers of nodular lesions, whereas DEN alone increased liver nodules significantly and proportional to dose (0.3–2.5 µg). A supra-additive interaction between the two initiating agents was found mainly in the increased rate of foci appearance after 1.25 µg of DEN and 0.125 Gy of neutrons, both given seven days before or after DEN exposure [M33]. Peraino et al. [P26] studied three altered hepatocyte foci (elevated gamma-glutamyl transpeptidase [GG+] and/or iron-exclusion [Fe−]) in Sprague-Dawley rats exposed to DEN (0.15 µmol g^{-1}) and/or gamma rays (0.75, 1.5, and 3 Gy) shortly after birth. The exposure was followed by a phenobarbital (0.05%) diet to promote focus expression. Radiation alone was a weak hepatocarcinogen. A strong synergism was seen at the lower radiation doses for the induction of [GG+] foci but not for other focus phenotypes. A qualitatively different type of genetic damage for DEN (point mutations) and for radiation (rearrangements) is postulated from the result. Large sex differences in the yield of DEN-induced [GG+/Fe−] foci by a factor of up to 10 are additional indicators of the complexity of this system.

62.   The potential for pulmonary carcinogenic interactions between $^{239}PuO_2$ and the tobacco-specific nitrosamine 4-(N-methyl-N-nitrosamino)-1-(3-pyridyl)-1-butanone (NNK), a genotoxic lung carcinogen, was studied in 740 male rats [L17]. The animals received $^{239}PuO_2$ by inhalation to result in lung burdens of 0 or 470 Bq. The NNK was administered by multiple ip injection at doses of 0, 0.3, 1.0, or 50 mg kg^{-1}. The highest dose of NNK markedly reduced the median lifespan of the rats, whereas in the other treatment groups survival was minimally reduced in comparison with the controls. Results on carcinogenicity are not yet available from this study.

63.   An apparent synergism between low-LET ionizing radiation and the carcinogen 1,2-dimethylhydrazine (DMH) in the induction of colonic tumours in rats has been described by Sharp and Crouse [S27]. They evaluated the interaction of radiation (9 Gy to the abdomen only) and DMH (150 mg kg^{-1}) with respect to colon carcinogenesis in male Fischer 344 rats. Radiation was administered 3.5 days before the DMH. At eight months post-treatment, the incidence of DMH-induced colon tumours was doubled by prior radiation exposure. When the protocol of radiation plus DMH was repeated three times at monthly intervals, a 15-fold increase in tumour incidence (from 5% to 74%) was observed at six months post-treatment. This finding demonstrated an apparent synergy between radiation and the chemical carcinogen. Throughout the study, the appearance of carcinomas was associated with pre-existing colonic lymphoid nodules. The reproducibility of tumour induction as well as the range of tumour incidence generated by treatment variations in this system appeared to be sensitive enough to allow the examination of combined effects of much lower doses of radiation and/or chemical carcinogens. The model could be used to evaluate the relationship between existing lymphoid aggregates, which alter local epithelial cell kinetics and are associated with fenestrations in the basement membrane. The quantification of the development of colon cancer in congruent sites may assist in defining dose-response curves for combined agents and may also provide a system for evaluating the mechanisms underlying their interactions. When DMH treatment was started two months after x-irradiation, only a slight increase in gastric tumour incidence was recorded [A7]. These tumours occurred on top of a background of radiation-induced gastrointestinal metaplasia.

64.   Ehling and Neuhäuser-Klaus [E7] studied the induction of specific-locus and dominant-lethal mutations by combined cyclophosphamide (see also Section D.1.a for uses in combined modalities in tumour therapy) and radiation treatment in male mice. Unlike radiation, this widely used antineoplastic agent, used alone, induced recessive mutations in spermatozoa and spermatids but not in spermatocytes and spermatogonia. Pretreatment (with 60 mg kg^{-1}) 24 hours before radiation, however, enhanced the frequency of specific-locus mutations in spermatogonia. The mutational spectrum among seven loci remained the same as in animals treated only with radiation. The synergistic interaction was mechanistically explained by the interference of cyclophosphamide, a strong inhibitor of DNA and RNA synthesis, with repair of radiation-induced damage.

65.   The effect of radiation on chemical hepatocarcinogenesis has also been examined in male ACI/N rats [M26]. The number of neoplastic nodules or hepatocellular carcinomas in rats given N 2-fluorenylacetamide (FAA) (0.02% in diet for 16 weeks) followed by x-irradiation (3 Gy) was significantly greater than in rats given FAA alone (p < 0.001). In addition, the incidence of hepatocellular carcinomas in rats given the combined treatment was also higher than in rats given FAA alone (p < 0.003). No liver lesions

were found in animals receiving only an x-ray dose of 3 Gy. The authors suggested that these highly supra-additive results indicate that ionizing radiation acts as a promoter in this model.

66.   An inhibition of urethane(ethyl carbamate)-induced pulmonary adenomas by inhaled ^{239}Pu in random-bred male A2G mice was reported as far back as 1973 by Brightwell and Heppleston [B66, B67]. This early study of combined exposure to alpha radiation and a genotoxic chemical comprised four groups, each of 32 animals, receiving plutonium inhalation followed by urethane (PU), plutonium followed by saline (PS), mock inhalation followed by urethane (MU), and mock inhalation followed by saline (MS). Exposures consisted of initial lung burdens of 925 Bq ^{239}Pu and ip urethane injections of 1 mg g^{-1} body weight two weeks later. Eight weeks after the injections, PS-treated animals showed no increase in pulmonary tumours over control animals (MS), whereas practically all animals in the PU and MU groups had multiple tumours. The number of tumours per animal 8, 16, and 24 weeks after urethane treatment was clearly lower in the PU group, which had 4.2, 11.4, and 13 as compared with 8, 24.4, and 38 in the MU group. An earlier hypothesis, that this finding is the result of alpha irradiation counteracting immuno-suppression by urethane, is rejected on the basis of ultrastructural evidence. Severe morphological changes in mouse type-II cells in the vicinity of alpha particles indicate that functional impairment of the initiated cells is the main cause of the effect. The authors said, however, that this apparent antagonism needs to be viewed with caution; it remains to be determined, they concluded, if much smaller local plutonium doses would augment urethane tumorigenesis.

67.   The transgenerational combined effects of x rays (2.2 Gy) and urethane were studied by Nomura [N23, N24] in three different mice strains (ICR, LZ, and N5). Urethane treatment of F$_1$ offspring of either irradiated males or females yielded an 18% incidence of tumour nodule clusters in the lung compared with only 2.8% in offspring of non-irradiated controls. Tumour clusters were defined as having 12 or more nodules. The transgenerational effect of radiation alone resulted in lung tumours (at least one tumour nodule) in 7.5% of the animals, whereas the value in unexposed controls was 4.7%.

68.   The interaction of gamma rays with urethane in lung tumorigenesis in mice in relation to the immune status has been studied by Kobayashi et al. [K41]. Male athymic nude mice (nu/nu) and their female heterozygous litter mates (nu/+) were treated with 1–4 Gy of ^{137}Cs gamma rays and 0.5 mg g^{-1} of urethane. Gamma-ray exposure alone caused relatively few lung tumours (in up to 10% of animals); urethane alone caused tumours in 70%–80%. The combined effect was supra-additive. There was a tendency towards higher yields in nu/+ mice, suggesting that impaired immunosurveillance from T-cell deficiency does not increase lung tumorigenesis in this system. Since relatively radiation-resistant macrophages and natural killer cells had higher

activities in nu/nu mice, the authors concluded that the influence of immunological status on tumorigenesis remained unresolved.

69.   A strong synergism was found by Hoshino and Tanooka [H39] for skin tumours in beta-irradiated ICR mice painted later with 4-nitroquinoline 1-oxide (4NQO); 27 Gy of ^{90}Sr/^{90}Y radiation or 20 applications of 5 mg ml^{-1} 4NQO in benzene to the skin alone did not produce any skin tumours in groups of 50 mice. Radiation followed by 4NQO painting with an interval of 11–408 days between the two treatments resulted in an incidence of malignant skin tumours (squamous-cell carcinomas and papillomas) of up to 17%. There was no significant decrease of the synergistic effect with increasing interval, the greatest effect being seen with an interval of 234 days.

70.   A notable finding indicating the considerable uncertainties and misinterpreting the results of experimental animal studies was described by Little et al. [L55], who studied the potential synergistic interactions between ^{210}Po (185 Bq, resulting in a lifetime lung dose of about 3 Gy) and benzo[a]pyrene (0.3 mg) in the induction of lung cancer in Syrian golden hamsters. It was shown that simultaneous administration by intratracheal instillation led to additive effects. A significant apparent synergism was found when benzo(a)pyrene was given 4 months after the ^{210}Po. Most of this effect could be ascribed, however, to a potentiating effect of the seemingly innocuous 0.9% NaCl instillation solution alone.

71.   The effects of repeated low exposures at high dose rates such as used in some diagnostic radiologic procedures at the time of the study were published by Lurie and Cutler in 1979 [L56]. The induction of lingual tumours by 7,12-dimethylbenz[a]anthracene (DMBA) and radiation to the head and neck was studied in Syrian golden hamsters. Treatment schedules were topical application of 0.5% DMBA in acetone on the lateral middle third of the tongue three times a week for 15 consecutive weeks, about 200 mGy radiation exposures (x rays with 100 kV peak) of the head and neck once a week for 15 consecutive weeks, or concurrent radiation and DMBA treatments for 15 consecutive weeks. Histopathology was performed 35 weeks after the start of the treatment. Animals receiving radiation alone had no detectable changes. The combined treatment led to an excess of lingual papillomas compared with animals receiving only DMBA (35% versus 15%). In addition, an excess of non-lingual oral tumours (lip, gingiva, and floor of the mouth) was found in the animals receiving the combined treatment compared with the DMBA-treated animals. Whether this radiation enhancement of DMBA-induced tumorigenesis has implications for the lower combined exposures found for cigarette-smoke-derived carcinogens in the bucal cavity of humans and dental x rays, remains to be elucidated.

72.   Studies on chromosome aberrations from the combined effect of gamma rays and the mutagen thiotepa on unstimulated human leukocytes showed no significant

difference from the sum of their separately induced effects. The sequence of treatment and the interval between them (up to 4 hours) did not affect the frequency of chromosome aberrations [B21].

73. Leenhouts et al. [L13] investigated the combined effect of 1,2-dibromoethane (DBE) and x rays on the induction of somatic mutations in the stamen hair cells of tradescantia KU 9. At low radiation doses, a synergistic interaction was found between the two agents for both DBE exposure followed by acute x rays and chronic simultaneous exposures. The synergism was considered to result from an interaction of single-strand lesions in the DNA. It was concluded that this type of interaction would not be too important for radiological protection. However, it could be of significance in evaluating the effects of chemicals at low exposure rates.

(c) Free-radical-generating chemicals

74. Superoxide (O_2^-) generating agents such as the dipyridilium compound paraquat might also interact directly with the fixation or repair of radiation-induced damage. Geard et al. [G5] investigated the combined effects of paraquat and radiation on mouse C3H10T½ cells. Effects on oncogenic transformation, chromosome alteration, cytokinetics, or cellular survival were the endpoints measured. Paraquat alone is a cytotoxic agent and is also a weak radiosensitizer. Treatment with 0.1 mM for 24 hours results in about 30% cell survival and enhances the cell-killing effects of ^{137}Cs gamma rays by a factor of about 1.2. The drug appears to function lethally by initiating interphase cell death and also by slowing cell cycling. In combination with radiation (3 Gy), paraquat acted either additively (sister chromatid exchanges) or with a greater-than-additive effect (cell survival and oncogenic transformation).

75. De Luca et al. [D6] studied the induction of reciprocal translocations in mouse germ cells (BALB/c) by bleomycin alone or combined with radiation (see also Section D.3 for bleomycin used in combined modalities in tumour therapy). The dose-response relationships after treatments with doses of 20, 40, and 60 mg kg^{-1} of bleomycin as well as the combined effect of bleomycin and gamma rays were studied. A positive, significant correlation between the dose of bleomycin and the frequency of translocations was found. Both potentiation and additivity were found when the yields of translocations induced after combined treatments, separated by a lapse of 24 hours, were compared with the sum of translocation frequencies induced after the corresponding single treatments. Potentiation occurred in the treatments with 1 Gy plus 9 Gy and 60 mg kg^{-1} of bleomycin plus 9 Gy, while additivity occurred in the treatments with 60 mg kg^{-1} of bleomycin plus 1 Gy and 1 Gy plus 60 mg kg^{-1} of bleomycin. In mice irradiated with 1 Gy plus 9 Gy and mice treated with 60 mg kg^{-1} of bleomycin plus 9 Gy, similar translocation yields were found. The potentiating effect of bleomycin was found to be similar to that obtained with non-radiomimetic compounds such as triethylenemelamine, cyclophosphamide, and adriamycin. The high doses involved and the erratic changes from

synergistic to additive relationships preclude extending inferences from these experiments beyond cancer therapy to occupational or non-occupational settings.

76. In summary, there are many examples of strong deviations from hetero- and isoadditivity in the interactions between genotoxic chemicals and ionizing radiation (Table A.1). Owing to generally high exposures to both agents under study, deterministic effects were shown or suspected to be the cause of strong deviations from additivity in many studies. Thus, the several cases of synergism found seem to be mostly the result of modifications of the biokinetics and the metabolism of the chemical rather than of agent-specific genotoxicity at different stages of the pathological processes. Similar considerations hold for antagonistic effects, where depletion of stem cells and inhibition of cellular growth may be a factor in the high dose range. Additional risks, beyond the level predicted from isoaddition, from the combined effects of ionizing radiation and genotoxic chemicals at low exposures levels are, accordingly, not specifically demonstrated by the many epidemiological and experimental studies reviewed in this Section.

2. Non-genotoxic chemicals

77. Many chemicals in the human environment or their metabolites do not specifically attack DNA but influence cell proliferation and cell differentiation on an epigenetic level. Specific mitogens may interfere with regulatory mechanisms and cell-cell signaling, but many substances with a high chemical reactivity act as unspecific irritants or toxicants via membranes or proteins. Toxin-induced cell death will induce proliferation in neighbouring cells, which may enhance the progression of premalignant cells. Substances acting in a non-specific manner, for example lipophilic solvents, quite often show highly non-linear dose-response relationships with apparent thresholds. Other agents may interfere with the critical cellular processes involved in repairing damage to cellular constituents such as DNA. The assessment of possible synergistic effects at the exposure levels relevant to this Annex is very difficult, because of the high exposures used in experimental systems and the apparent threshold levels.

78. The tumour promoter 12-O-tetradecanoylphorbol-13-acetate (TPA) has the potential to enhance the yield of radiation-induced tumours. This has been well documented *in vitro* and in animal systems. The combined effects of paternal x-irradiation and TPA on skin tumours in two generations of descendants of male mice was studied by Vorobtsova et al. [V5]. Progeny of outbred SHR male mice non-irradiated or exposed to a single dose of whole-body x-irradiation (4.2 Gy) were skin-painted twice a week for 24 consecutive weeks from the age of four months onwards with acetone or with TPA in acetone (6.15 µg ml^{-1}). The incidence and number of skin papillomas were monitored between week 2 and week 20 after the last application of the promoter (TPA). Exposure to acetone was never followed by skin tumour development in the progeny of either irradiated or non-irradiated males. Two weeks after

TPA treatment, the incidence of skin tumours in the progeny of non-irradiated mice was 21% in males and 37% in females, and 20 weeks later it was 12% in males and 15% in females. The skin tumour incidence in the progeny of the irradiated male mice 2 and 20 weeks after the last painting was clearly elevated: 75% and 68% in males and 50% and 43% in females, respectively. Some of the F_1 offspring of irradiated male mice were mated before the start of TPA treatment, and F_2 progeny were exposed to acetone or TPA as F_1. The incidence of skin papilloma 2 weeks after the last TPA painting was 58% in males and 40% in females, whereas at 20 weeks after the last exposure to the promoter it was 53% and 36%, respectively. In the progeny of irradiated male mice there were more animals with multiple (>4) skin papillomas than in the progeny of non-irradiated mice. The incidence of other than skin tumours in offspring was also clearly increased in TPA-treated progeny from irradiated male mice. The authors suggested that irradiation of males before mating increases the susceptibility of progeny in at least two generations to promoters of carcinogenesis as a result of persisting genomic instability. On the other hand, Brandner et al. [B30] found no influence of ip-administered TPA on the incidence of radiation lymphomas in C57Bl/6 mice. Female C57Bl/6 mice, given four x-irradiations each with 1.7 Gy, developed lethal lymphomas in more than 90% of animals 270 days after irradiation. Intraperitoneal application of TPA, 30 ng g^{-1} twice weekly for 240 days, had no influence on survival of the animals or on incidence of the malignant lymphomas. However, the incidence in radiation-only treated animals was already so high that this test was highly insensitive to the promoting effects of TPA.

79. Jaffe et al. [J2] studied the effect of proliferation and promotion time on radiation-initiated tumour incidence in Sencar mice. In this system, a single subcarcinogenic dose of ionizing radiation followed by 60 weeks of TPA treatment led to the formation of squamous-cell carcinomas. Even TPA pretreatment before irradiation seemed to result in an overall increase in total tumour incidence, including both epidermal and non-epidermal tumours [J1]. Based on these findings, the effect of the proliferative state of the skin before irradiation and the promotion duration after irradiation on tumour incidence was further investigated in CD-1 mice. To examine the influence of the proliferative state of the skin, a 17 nmol TPA solution was applied to one half of the mice 24 hours before irradiation. The skin was irradiated using 4 MeV x rays at a dose rate of 0.31 Gy min^{-1}. Animals received a single dose of x rays of 0.5 or 11.3 Gy, followed by twice weekly applications of TPA (8 nmol). The animals were then promoted for either 10 or 60 weeks. All animals promoted with TPA for the same duration had a similar incidence of papillomas regardless of radiation or TPA pretreatment. Increasing the promotion duration did not significantly alter the incidence of squamous-cell carcinomas at either initiation dose. At the lower initiation dose, only animals that were promoted for 60 weeks developed squamous-cell carcinomas. TPA pretreatment at the higher dose resulted in a slight decrease

in tumour incidence; however, this was not statistically significant. The incidence of basal-cell carcinomas was radiation-dose-dependent and appeared to be independent of TPA promotion. Again, as in many other cases, no common pattern emerged for the different tumour types.

80. The interaction between ionizing radiation and TPA has been studied using a three stage model of initiation, promotion, and progression. Ionizing radiation is well established as an initiator, whereas its potential for promotion and progression is less well known. Therefore, Jaffe and Bowden [J1] performed a three-stage experiment using ionizing radiation in the third stage of mouse skin carcinogenesis. CD-1 mice were initiated with N-methyl-N'-nitro-N-nitrosoguanidine (MNNG), followed by biweekly promotion with TPA. After 20 weeks of promotion, the animals were treated with either acetone, TPA (twice a week for two weeks), or eight fractions of 1 MeV electrons (1 Gy per fraction over a period of 10 days). The conversion of papillomas to squamous-cell carcinomas was 80% for animals treated with ionizing radiation compared with 25% for tumour-bearing animals treated with TPA. Ionizing radiation increased the number of cumulative carcinomas per group. The absence of an increase in the number of cumulative papillomas per group due to late exposure to ionizing radiation suggests that the dose and fractionation protocol used in this study enhanced the progression of pre-existing papillomas.

81. The tumour-initiating and -promoting effects of ionizing radiation in mouse skin was also studied with TPA by Ootsuyama and Tanooka [O2]. Neither single 24 Gy ^{90}Sr/^{90}Y beta irradiation followed by repetitive treatment with TPA nor single pretreatment with 7,12-dimethylbenz-(alpha)-anthracene (DMBA), followed by repetitive 4.7 Gy beta irradiation, produced tumours above the level of significance within a period of 210 days, while a positive control, DMBA + TPA, yielded a high incidence of papilloma in a shorter period. In this system, DMBA seemed to exert an action antagonistic to beta particles in the induction of malignant tumours. It was concluded that the tumour-enhancing activity of repetitive radiation is qualitatively different from the promoting activity of TPA.

82. Nomura et al. [N6] were able to show that *in utero* irradiation at early stages of embryogenesis, which was not visibly carcinogenic by itself in a tester strain of mice (PT × HT F_1), followed by post-natal application of TPA, led to a high incidence of skin tumours. Radiation doses in this system were 0.3 and 1.0 Gy of 180 kVp x rays, respectively, at about 10.5 days after fertilization. Two dose rates, 0.54 and 0.0043 Gy min^{-1}, were used. The incidence of both embryonic mutations, determined as spots of different coat color, and tumours increased with *in utero* doses. Low-dose-rate irradiation led to a large (about 80%) reduction in tumour incidence.

83. TPA also causes enhanced transformation of irradiated mouse 10T½ cells (Figure A.II). For the loss-of-contact inhibition, two genetic steps and modulation by epigenetically acting substances were proposed by Little

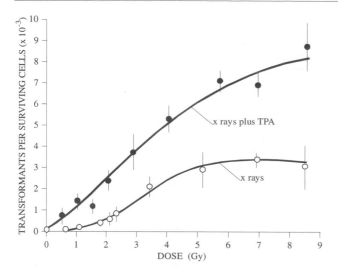

Figure A.II. *In vitro* transformation of C3H10T½ cells exposed to x rays (50 kV) with and without post-irradiation incubation in TPA (0.1 µg ml⁻¹) [H3].

[L24]. TPA promotes following exposure to x rays or to fission-spectrum neutrons without any effect on cell survival [H3]. However, treatment of unirradiated cells with 0.1 µg ml⁻¹ of TPA resulted in a small increase in transformation frequency above background (i.e. from $1.1 \cdot 10^{-5}$ to $1.0 \cdot 10^{-4}$). Thus, besides being a promoter, TPA seems to be also a weak initiator. The enhancement factor of TPA for radiation-induced transformation was greater after low doses than high doses of either radiation. In addition, TPA caused the RBE of neutrons as compared to x rays to increase with increasing dose. For x-ray doses from zero to approximately 1.2 Gy, TPA raised transformations to frequencies approximately equal to those due to neutrons alone. Analysis of TPA enhancement in the context of the combined effect of two inducing agents, TPA plus radiation, indicates that with either x rays or neutrons, TPA acts synergistically. The main mechanism of action of TPA is suggested by the finding that the dependence of transformation frequency on the density of viable cells is also altered by the tumour promoter. In contrast to the constant frequency of transformants per surviving (or viable) cell, which was observed after a fixed dose of x rays or neutrons for a range of cell inocula, the increase in the frequency of transformation caused by TPA and radiation was dependent on cell inocula. The frequency of transformation from combined treatment decreased with increasing size of the inoculum, from approximately 20 to 6,000 viable cells per 90-mm Petri dish, a result that the authors interpreted as an interference with cell-to-cell communication by TPA plus the fading of initiation events caused by radiation.

84. DNA base analogues are another group of substances with the potential to modify the effects of radiation and other genotoxic agents (see also Section D.2). 5-Bromo-2'-deoxyuridine (BrUdR) is an analogue for thymidine and widely used in tumour diagnosis, cytogenetics, and flow cytometry. Important examples of epigenetic and (indirect) genetic effects are the inhibition of differentiation in cultured myoblasts and photosensitivity of patients,

respectively. Anisimov and Osipova [A15] investigated carcinogenesis induced by combined neonatal exposure to BrUdR and subsequent whole-body x-irradiation of rats. Outbred LIO rats at 1, 3, 7, and 21 days of post-natal life were exposed to subcutaneous injections of 3.2 mg of BrUdR per animal and/or at the age of 3 months to single whole-body x-irradiation at a dose of 1.5 Gy. In males, treatment with BrUdR alone decreased the latency of all tumours and increased the incidence of malignant tumours and the number of tumours per rat compared with controls. Combined exposure to BrUdR and x-irradiation increased total and malignant tumour yield and multiplicity over that in all other groups. More testicular Leydigomas, tumours of prostata, kidney, and adrenal cortex, and leukaemia were seen in male rats exposed to BrUdR plus x rays, compared with male rats treated with BrUdR or x-irradiation alone. In female rats, treatment with BrUdR alone decreased the latency for the total number of tumours and increased their incidence and number per rat, in comparison with controls. Combined exposure of females to BrUdR and x rays did not increase total tumour incidence in comparison with females that had only been irradiated; however, it shortened tumour latency. The incidence and multiplicity of malignant tumours and incidences of pituitary adenomas, mammary adenocarcinomas, and uterine polyps were significantly increased, whereas the latency of kidney tumours was decreased in females exposed to BrUdR plus x rays, compared with all other groups. The data from this experimental model provide, together with other studies, evidence that perturbation of DNA induced by the nucleoside analogue BrUdR contributes substantially to the spontaneous development of tumours and enhances the sensitivity of target cells to carcinogenesis induced by x-irradiation as well as by chemicals or hormones.

85. Information on the effects of the interaction of thorium and phenobarbital, an anticonvulsive drug inducing liver detoxification functions and showing promoting activity, may be available from earlier epileptic patients. Thorium exposure (thorotrast) resulting from angiographic procedures correlated with the use of anticonvulsive drugs. Olsen et al. [O15, O16] found considerably increased risks for liver cancer, but since thorotrast exposure was considered a confounder in both studies, no definitive quantitative information on combined effects from thorium and phenobarbital was given.

86. The potentially important interaction of phenobarbital, a widely used anticonvulsant and sedative, with x-irradiation was studied by Kitagawa et al. [K18]. Male newborn Wistar-Ms rats received whole-body x-irradiation of 0.5, 1, and 4 Gy at 8 or 22 days. After weaning they were fed either a basal diet or a diet containing 0.05% phenobarbital. The x rays induced numerous adenosine-triphosphatase-deficient islands appearing in the liver by week 22 of age. However, no hepatic tumours were observed by 22 months after radiation, even in phenobarbital-treated animals.

87. Supra-additivity was also found for a combination of fast-neutron irradiation and subcutaneously applied carbon

tetrachloride in male and female C57Bl6 mice. The animals received a single whole-body dose of 1.7 or 3.3 Gy from fast neutrons, followed nine weeks later by a single subcutaneous injection of carbon tetrachloride. Carbon tetrachloride markedly increased the incidence of radiation-induced liver carcinomas, whereas chloroform, which was also tested in this system, did not influence the incidence of radiation-induced tumours [B16].

88. The potential for carbon tetrachloride to modify the biokinetics of an inhaled, soluble form of plutonium is also being examined in both F344 rats and Syrian hamsters [B16]. Groups of animals were exposed to carbon tetrachloride in whole-body chambers at concentrations of 0, 5, 20, or 100 ppm for 6 hours per day, 5 days per week, for a total of 16 weeks. After 4 weeks of exposure, approximately one half of the animals were exposed by a single pernasal inhalation exposure to ^{239}Pu nitrate. Serial sacrifices of groups of animals were conducted at 4 hours and 2, 4, 6, or 13 weeks after plutonium exposure for the quantification of ^{239}Pu in lung, liver, kidney, and bone (femur) and for the evaluation of histologic changes in various tissues. Results describing possible carbon tetrachloride effects on plutonium disposition are not yet available from this study. Another subgroup of rats and hamsters was exposed to a radioactively labelled insoluble tracer particle. Tracer particle clearance was analysed for 13 weeks following exposure, and no significant clearance differences were observed between carbon-tetrachloride-treated and control groups.

89. Since ionizing radiation and tumour-promoting agents increase the level of ornithine decarboxylase (ODC) involved in polyamine biosynthesis, the effect of alpha-difluoromethyl-ornithine (DFMO), an inhibitor of ODC, on tumour yield from beta radiation was tested in female ICR mice [O12]. The chronic radiation exposure consisted of three times 3 Gy ^{90}Sr/^{90}Y surface dose per week to the back. DFMO was added to the drinking water in a final concentration of 1%. It significantly delayed the time of tumour emergence from 245 days with radiation exposure only to 330 days in animals also given DFMO. The antagonistic effect of DFMO was also observed for bone tumours.

90. Monchaux et al. [M61] addressed the important question of possible synergistic contributions from diesel fumes present in mine air to radon-induced lung tumours. Three groups of 50 male Sprague-Dawley rats were exposed to radon (1,000 WLM) and/or diesel exhaust (300 hours; 22-25 ppm CO and 4-5 mg m^{-3} diesel particles), with the diesel exposure succeeding the radon exposure by one month. Contrary to the strong synergistic effect of cigarette smoke found in this system (discussed under tobacco), exhausts had only a slight, non-significant effect on the risk for thoracic tumours from radon. Diesel exhausts alone were not carcinogenic.

91. Since phosphorylation and dephosphorylation of proteins play an important role in cellular metabolism, Nakamura and Antoku [N21] studied the effect of calyculin A (CL-A), a specific inhibitor of protein phosphatase 1 and 2A isolated from the marine sponge *Discodermia calyx*, on x-ray-induced cell killing in cultured mammalian cells (BHK21). At concentrations above 2.5 nM, CL-A enhanced the radiation effect considerably. As also shown in another cell culture system with the inhibition of protein kinases [H40], agents that interfere with protein-kinase-mediated signal transduction after radiation exposure may enhance damage and represent a new class of radiosensitizers.

92. Many non-genotoxic agents clearly produce strong synergistic effects with ionizing radiation. The combined effects of this class of agent are summarized in Table A.1. Table A.3 lists more detailed effects of TPA, probably the best-studied modifier of genotoxic agents, on several endpoints. These studies are of great importance for the elucidation of mechanisms affecting expression of risk. At this stage, however, no functional analogues of potent experimental enhancers of radiation risk, such as TPA or DNA bases, are known to exist in critical concentrations in the human environment.

3. Tobacco

93. The important interaction of tobacco smoke and radiation was introduced in the main text of this Annex. Epidemiological studies of uranium miners have allowed the risks and interaction coefficients to be quantified, at least for higher radiation doses. The complex composition of tobacco smoke makes the interaction not simply a binary combination, however. Some 4,000 individual chemical components of cigarette smoke have been identified, and a number of additional unidentified components surely exist (for example, extremely reactive, short-lived compounds or those present in very low concentrations) [G1]. Identified compounds in smoke include several known carcinogens of the polycyclic aromatic hydrocarbon and nitrosamine classes.

94. The studies reviewed below refer to mainstream smoke, sidestream smoke, or environmental tobacco smoke. Mainstream smoke is defined as the smoke originating from the butt end of a cigarette; it is generated during the active puffing process. Sidestream smoke is the smoke released at the burning tip of a cigarette, whether the cigarette is being puffed or simply smoldering. Lastly, environmental tobacco smoke is a mixture of sidestream smoke and exhaled mainstream smoke. This term most accurately describes the smoke that would be found within an enclosed space with a smoker present. Tobacco smoke contains relatively small amounts of DNA-reactive carcinogens, such as nitrosamines, polycyclic aromatic hydrocarbons, and pyrolysis products, such as carbolines. Hence enhancing and promotional factors, e.g. catechols, other phenols, and terpenes, are an important component. Probably because it reduces pressure from the action of promoters, discontinuation of smoking progressively reduces the risk of cancer development with time since withdrawal [W1].

Table A.3
TPA as a modulator of transformation and cancer yield from ionizing radiation

Endpoint	Experimental system	Interaction	Proposed mechanism	Outcome	Ref.
Transformations in surviving cells	10T½ cell culture	x rays, TPA	Initiation, promotion	Higher linear yield Loss of threshold (see Figure V)	[H3]
Transformations in surviving cells	10T½ cell culture	x rays/neutrons, TPA	Initiation, promotion	Enhancement factor greater at lower exposures RBE of neutrons enhanced at higher doses	[H3]
Transformations in surviving cells	10T½ cell culture	Radiation, TPA	Two genetic steps, epigenetic modulation	Genetic effect fading with culture time TPA interferes with cell-cell interaction	[L24]
Squamous-cell carcinoma	CD-1 mice	Beta radiation, TPA; MNNG, TPA, beta radiation	Initiation, promotion, progression	High papilloma yield with TPA only Progression to carcinoma by radiation	[J1]
Skin papilloma	Mice	Beta radiation, TPA; DMBA, beta radiation	Initiation, promotion	Promotion by repetitive irradiation different from TPA	[O2]
Skin papilloma	SHR mice	Radiation (4.2 Gy) to father TPA to offspring F_1 and F_2	Genetic modification, promotion	Skin tumours elevated in TPA-treated offspring Weaker effect in female offspring	[V5]

(a) Epidemiological studies

95.   In the last few years, joint analyses of original data sets [C1, L18] and meta-analyses of published results [T14] have yielded detailed assessments of risk patterns from combined exposure to high-LET alpha radiation from radon and its short-lived decay products and tobacco smoke, and have allowed investigators to test risk models. The most comprehensive and complete analysis of radon-induced health risks was published by Lubin et al. [L18]. The review contains a joint analysis of original data from 11 studies of male underground miners; 2,736 lung cancer deaths among 67,746 miners were observed in 1,151,315 person-years. A linear relationship was found for the ERR of lung cancer with the cumulative exposure to radon progeny, estimated in working level months (WLM). This coefficient (ERR/WLM) was strongly influenced by various factors. Contrary to the low-LET experience from Hiroshima and Nagasaki, ERR/WLM decreased significantly with attained age and time after cessation of exposure to radon progeny. A stronger decline of risk with time since exposure than in survivors of the atomic bombings was also found. A considerably higher lung cancer risk was initially found for exposures received at low rates as compared with high rates. Depletion of stem cells at risk in high dose rate exposures was implied. However, the epidemiological database was said to be too weak to project

this indication of an inverse dose-rate effect to non-occupational settings, i.e. to typical indoor radon exposures and exposure rates [L18]. Also, a recent reassessment of the Beaverlodge cohort, which earlier on gave the strongest indication of such an effect, no longer does so. Revised exposure estimates of this study of miners with relatively low exposures now bring the modifying effects of risk with time since exposure and age at risk in line with those from other studies [H46]. The highly significant decrease in ERR with time since exposure may be explained with microdosimetric considerations. In the case of high-LET alpha radiation from radon progeny, the minimal local dose from one single alpha track averaged over a cell nucleus is already in the range of several hundred milligray, whereas one electron track yields a dose to the nucleus in the range of only 1–3 mGy. This means that even at the lowest possible nuclear dose from alpha exposure, stem cells that are hit carry a multitude of DNA lesions, which may considerably impair long-term cell survival and maintenance of proliferative capacity [B25, B27].

96.   In the joint analysis by Lubin et al. [L18], data on smoking were available for 6 of the 11 cohorts, but assessments were limited by incomplete data on lifetime tobacco consumption patterns and sometimes exotic tobacco use, such as in water pipes in the Chinese study. Most studies for which smoking data could be analysed were generally not informa-

tive enough to allow deciding between an additive or a multiplicative joint relationship for radon progeny and smoking. The Chinese cohort seemed to suggest an association more consistent with additivity, while the Colorado cohort suggested a relationship more consistent with a multiplicative interaction. For all studies combined, the joint relationship of smoking and radon progeny exposures with lung cancer was stable over the different age groups and deviated quite clearly from either a purely additive or a multiplicative relationship. The most recent analyses of the BEIR VI Committee [C46], which were based on an update of these data, suggest that the joint effect is statistically closer to a multiplicative than an additive interaction. To further characterize the association, more detailed data on tobacco use would be needed. Age at onset of smoking, amount and duration of smoking, and type of tobacco were recognized as important determinants of risk. Such a refined analysis of smoking patterns is possible only in the prospective part of ongoing studies and is subject, furthermore, to potential bias in the affected individuals owing to the rapidly decreasing public acceptance of smoking. In general, the single-exposure subcohorts of lifetime non-smokers are very small in all studies. The statistical power of the conclusions on the interaction between radon and tobacco smoke is correspondingly small. Applying the two-mutation clonal expansion model of carcinogenesis of Moolgavkar et al. to data from the Colorado plateau miners shows no interaction between radon and tobacco smoke in any of the three steps [M39], but the predicted lung cancer incidence caused by radon and smoking remains more than additive and less than multiplicative, an indication of isoadditivity.

97. Microdosimetric considerations are also important in extrapolating the inverse dose-rate effect found for oncogenic endpoints caused by alpha radiation in general and for lung cancer in miners [L36]. Brenner [B40] postulated that protraction enhancement is a mechanism limited to cells receiving multiple hits over a human lifespan. Since a typical domestic exposure to radon progeny of 14 WLM yields a very small probability of multiple traversals in a cell nucleus (<1% for the most highly exposed stem cells in the tracheobronchial epithelium), dose-rate effects are probably of no relevance, and lung cancer risk per unit exposure will not increase further at low radon levels.

98. Two recent analyses by Yao et al. [Y7] and Thomas et al. ([T18] with erratum) on the radon-smoking interaction showed a considerable influence of timing of exposures. The former study found a higher lung cancer risk for exposure to radon progeny and tobacco use occurring together as compared to radon exposure preceding tobacco use. The second study on Colorado uranium miners found a significantly more-than-multiplicative effect for smoking followed by radon, whereas radon exposure followed by tobacco use produced an essentially additive effect. These findings are in conflict with earlier notions based on experimental results in rats, whereby radon is an initiator and tobacco smoke, a promoter [G20]. However the relevance of this animal system is questionable, because tobacco smoke alone does not produce lung tumours in this system.

99. Despite the remaining uncertainties, it is quite clear that the joint effect of radon progeny exposure and smoking is greater than the sum of each individual effect. The combined analysis [L18] shows that a linear exposure-response estimate for radon and lung cancer is compatible with the data and gives a relative risk that is about three times higher in non-smokers than in smokers. Assuming a 10-fold difference in the tobacco-caused lung cancer risk between smokers and non-smokers, this means that the lung cancer risk for smokers expressed in absolute terms is higher by a factor of about 3. Such a supra-additive effect, if also demonstrated to hold for present occupational and non-occupational exposure settings, would be of great importance for the regulation of smoking and radon progeny in the human environment. Until now, little quantitative evidence has come from indoor radon studies. The few case-control studies published are inconclusive [A28, P11]. Only one larger study [P11] was indicative of an indoor radon risk and its modification by tobacco that is comparable to what is predicted from miner studies. It remains doubtful whether the results from the many case-control studies under way will in the near future allow narrowing of the uncertainties that surround indoor radon risk and possible interactions with smoking. Based on inconclusive results from 1,000 computer-simulated large case-control studies assuming an ERR of 0.015 WLM^{-1}, Lubin et al. [L33] questioned the assumption that epidemiological studies, even when pooled in meta-analyses, will produce reliable estimates of risk from residential radon exposure. Errors in exposure assessment, migration, and confounding by smoking are at the root of this pessimistic assessment. At least for the second confounder, studies in Europe based on much longer mean residence times may offer better statistical power. Several large indoor case-control studies under way will narrow uncertainties in the next few years. First results from the United Kingdom [D33] and Germany [W35] are indicative of a lung cancer risk in the range of ICRP projections. However, confidence intervals are relatively large and include zero risk in most analyses.

100. Because of the limitation of the indoor radon studies, risk estimates based on miner data remain the main basis for predicting lung cancer from indoor radon exposure. A best linear estimate of the risk coefficients found in the joint analysis of Lubin et al. [L18, L35] for the indoor environment indicates that in the United States, some 10%–12%, or 10,000 cases, of the lung cancer deaths among smokers and 28%–31%, or 5,000 cases, of the lung cancer deaths among never-smokers are caused by radon progeny. About half of these 15,000 lung cancer deaths traceable to radon would then be the result of overadditivity, i.e. synergistic interactions between radon and tobacco. Based on the same risk model, Steindorf et al. [S47] predicted that about 7% of all lung cancer deaths in the western part of Germany are due to residential radon. This corresponds to 2,000 deaths per year, 1,600 in males and 400 in females. The attributable risk estimate was 4%–7% for smokers and 14%–22% for non-smokers. The most recent central estimates for the proportion of radon-attributable lung cancer deaths in the United States in 1995 was recently provided by the BEIR VI Committee [C46] in

1998, based on an updated data set of the miners studies reported by Lubin et al. [L18]. The Committee applied a sub-multiplicative relation to model the joint effect of tobacco smoking and radon. Depending on two different models (exposure-age-concentration model or exposure-age-duration model) about 14% or 9% of all lung cancer deaths among ever-smokers and 27% or 19% among never-smokers were estimated to be attributable to radon. Because of the many differences between mines and homes and the additional carcinogens such as arsenic, dust and diesel exhaust in mine air, these figures should be interpreted with caution. A population-based case-control study of incident lung cancers among women in Missouri who where lifetime non-smokers or long-term ex-smokers yielded a very low and non-significant estimate of the attributable lung cancer risk from radon in non-smokers [A4].

101. It has been questioned whether toxicants other than joint exposures to radon progeny and cigarette smoke contribute considerably to the high lung cancer risk found in miners [I5]. Heavy exposures to mine dust containing silicates, diesel exhausts, and fumes from explosives may add to or combine with the two main lung carcinogens, radon and cigarette smoke. Patients who received thorotrast continuously exhale the very short-lived ^{220}Rn derived from ^{232}Th deposits in the body and therefore provide a model for lung carcinogenesis by radon without concomitant dust exposure. Ishikawa et al. [I5] studied the lung cancer incidence in a Japanese thorotrast cohort and found 11 lung cancer cases in 359 thorotrast autopsy cases. The analysis revealed that while the proportion of small-cell lung cancer considered to be related to alpha radiation was significantly increased, the overall lung cancer incidence was not significantly higher than in controls, in spite of the high levels of ^{220}Rn in the patients' breath. The authors took this as an indication that the risk for radon-induced lung cancer is not as high as expected from risk coefficients deduced from miner studies. To substantiate this hypothesis, the build-up of ^{220}Rn decay products in the lung air space before exhalation and the resulting exposure to critical stem cells would have to be quantified.

102. Owing to the generally good linear correlation between radon progeny exposure and lung cancer in the major miner studies, few additional carcinogens in mine dust were considered in depth. Toxic metals are, however, of special concern. Results from the Chinese [X1], Canadian (Ontario), [K21] and Czech [T41] cohorts showed arsenic to be an important additional risk factor for lung cancer. Adjustment for arsenic exposure reduced the radon risk estimate in these cohorts considerably. Even in the most recent joint analysis by Lubin et al. [L18], other mine exposures were difficult to interpret, since the information was quite limited and of poor quality. In most cases these concomitant exposures to suspected carcinogens or promoters are typically highly correlated with radon progeny exposures in a given study and therefore difficult to assess independently (see also following Section B.4).

103. The mechanism of interaction between DNA lesions caused by radon progeny and those caused by chemical toxicants contained in tobacco smoke is not known. There is clear evidence that the prevalence of mutations in critical genes is dependent on the type of insult. The most common known gene mutations in lung cancer cells are found in the tumour-suppressor gene *p53*, which is thought to be crucial in the initiation of this and many other types of cancer. Several groups analysed the molecular changes in the conserved regions of the *p53* gene in lung cancer tissue and reported differences between non-smokers (survivors of the atomic bombings and unirradiated controls), Japanese smokers, and uranium miners with high radon exposures [T12, T13] (see also Annex F, "*DNA repair and mutagenesis*"). The non-smokers from Hiroshima showed mainly transition mutations (all G:C to A:T) but no G:C to T:A transversions. By contrast, the changes in 77 Japanese smokers showed a predominance of G:C to T:A transversions in which the guanine residues occur in the non-transcribed DNA [T12]. In 16 of 52 lung cancers of miners, a specific transversion AGG to ATG at codon 249 was reported [T13]. The prevalence of 31% for this mutation in miners was compared with only 1 reported case in 241 published *p53* mutations from lung cancers in the general population (mainly smokers). Such a marker might help to define a causal relationship, but even in the first study, only a minor fraction of the *p53* genes from lung cancer tissue of miners, all of whom had a unique genotoxic exposure, showed the specific change. However, later studies were not able to confirm the initial finding [B73, L53]. As was pointed out, a multitude of different primary lesions can lead to the same cellular and clinical endpoints, in this case a non-functional repressor protein and lung cancer, respectively, and highly specific molecular markers of single agents in all affected individuals are not to be expected.

104. A difficult matter of some concern for the protection of the public is the combined exposure to indoor radon progeny and environmental tobacco smoke. The presence of environmental tobacco smoke in homes has been implicated in the causation of lung cancer. In the absence of direct epidemiological information, the clearly higher-than-additive combined effects of smoking and radon progeny in mine air may lead to the application of a multiplicative model for risk assessment. While of interest in its own right, environmental tobacco smoke also influences the risk imposed by radon and its decay products through its strong influence on aerosol characteristics. The interaction between radon progeny and environmental tobacco smoke alters the exposure, intake, uptake, biokinetics, dosimetry, and radiobiology of those progeny. Crawford-Brown [C10] developed model predictions of the various influences of environmental tobacco smoke on these factors in the population of the United States and provided estimates of the resulting change in the dose from average levels of radon progeny. It was predicted that environmental tobacco smoke produces a very small, non-measurable increase in the risk of radiation-induced tracheobronchial cancer in homes with initially very high

particle concentrations for both active and never-smokers but that it significantly lowers the dose in homes with initially lower particle concentrations for both groups when generation 4 of the tracheobronchial tree is considered the target site. For generation 16, the presence of environmental tobacco smoke generally increases the lung dose from radon progeny, although the increase should be unmeasurable at high initial particle concentrations. Although the author shows that the dose-modifying effects of environmental tobacco smoke are negligible, the main problem, a potential synergism between environmental tobacco smoke and radon progeny, was not assessed.

105. A smaller but still considerable cohort may be at risk from the combined effects of low-LET radiation and tobacco smoke, namely cigarette-smoking women who underwent breast cancer radiation therapy. Ionizing radiation has already been shown to be a lung carcinogen after breast cancer radiation therapy. Neugut et al. [N4] used a case-control study to explore whether cigarette smoking and breast cancer radiation therapy have a multiplicative effect on the risk of subsequent lung cancer. Case and control women were persons registered with primary breast cancer in the Connecticut Tumour Registry who developed a second malignancy between 1986 and 1989. Cases, i.e. those diagnosed with a subsequent primary lung cancer, were compared with controls diagnosed with a subsequent non-smoking, non-radiation-related second malignancy, and age-adjusted odds ratios were calculated with logistic regression. No effects from radiation therapy were observed within 10 years of initial primary breast cancer. Among both smokers and non-smokers diagnosed with second primary cancers more than 10 years after an initial primary breast cancer, radiation therapy was associated with a threefold increased risk of lung cancer. A multiplicative effect was observed, with women exposed to both cigarette smoking and breast cancer radiation therapy having a relative risk of 32.7 (95% CI: 6.9–154) (Figure A.III). Further evidence for a direct causal relationship was the observation that the carcinogenic effect of radiation was seen only for the ipsilateral lung and not for the contralateral lung in both smokers and non-smokers. The authors concluded that breast cancer radiation therapy, as delivered before 1980, increased the risk of lung cancer after 10 years in non-smokers, and a multiplicative effect was observed in smokers. The significance of the findings is, however, strongly reduced by the fact that the study also indicates a large difference in the incidence of ipsilateral and contralateral lung tumours for smokers who had no radiation therapy (Figure A.III), resulting in concerns about unidentified bias [I10]. A similar case-control investigation was based on 61 lung cancer cases from the Connecticut Tumour Registry who had received radiation therapy for the treatment of breast cancer [I9]. The authors of this study found no indication of a strong positive association between smoking and radiotherapy in the 27 cases where information on cigarette use was available. Therefore, it is not possible at this stage to decide whether current treatment practices involving much lower radiation doses to the lung may need to be reassessed in view of the detriment (late stochastic effects) for young breast cancer patients who smoke.

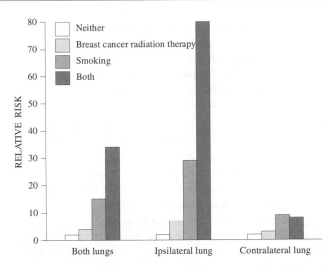

Figure A.III. Age-adjusted relative risk of lung cancer for separate or combined exposures to radiation and cigarette smoke [N4].

106. Long-term survivors of Hodgkin's disease display an increased lung cancer risk. Van Leeuwen et al. [V7] conducted a case-control study with 30 lung cancer cases from a cohort of 1,939 patients treated for Hodgkin's disease from 1966 through 1986 in the Netherlands to investigate the effects of radiation, chemotherapy, and smoking. Comparing patients who had a radiation dose of more than 9 Gy to the area where malignant growth developed with those who had less than 1 Gy, the relative risk was 9.6 (95% CI: 0.98–98, p for trend = 0.01). Patients smoking more than 10 pack-years (number of years with more than 1 pack per day) after diagnosis had a sixfold higher risk than patients with less than 1 pack-year. A multiplicative interaction was observed between the lung cancer risk from smoking and from increasing levels of radiation. On the other hand, no such trend was found with the drugs mechlorethamine or pro-carbazine, either in relation to the number of cycles of chemotherapy or to cumulative dose. It was suggested that Hodgkin's disease patients should be dissuaded from smoking after radiotherapy [V7].

(b) Animal studies

107. Although there are no well-suited animal model systems in which to examine potential carcinogenic inter-actions between environmental tobacco smoke and radiation, the issue of interactions between exposure to mainstream cigarette smoke and either radon or ^{239}PuO$_2$ has been examined. Relationships between increased risk for lung cancer in animals and exposure to radon and/or radon progeny [G11] or to ^{239}PuO$_2$ [C2] have recently been reviewed.

108. Studies conducted in France involved the whole-body exposure of rats to diluted mainstream cigarette smoke administered either before or after exposure to radon [C9]. Rats received high-level exposures to smoke for ten 15-minute periods four times weekly for one year. Smoke exposures given before the exposures to radon did not influence radon-induced tumour incidence, but smoke

exposures given after radon exposure increased the tumour incidence by a factor of 2-3 over rats receiving radon alone. These data indicated that cigarette smoke may have acted to promote radon-induced carcinogenesis, as reviewed in the UNSCEAR 1982 Report [U6].

109. In contrast, studies conducted on dogs exposed to the smoke from 10 cigarettes per day for 4-5 years combined with radon suggested that the incidence of lung tumours was less than that in dogs receiving radon alone [C21]. Lung tumours were produced in 8 of 20 dogs receiving radon alone, whereas tumours developed in only 2 of 20 dogs receiving both agents. The investigators speculated that increased mucus flow may have led to a reduced radiation dose to target cells in the smoke-exposed dogs; however, the small number of animals made interpretation of these results difficult.

110. Thus, despite the directly relevant epidemiological data on smoking and albeit high exposure to radon progeny, a significant problem remains, for example, for extrapolations to low exposures, in that the epidemiological and animal data related to lung cancer are in agreement for rats [C9] but in disagreement for dogs [C21]. Archer [A21] tried to explain this disagreement by advancing a hypothesis based on an additive interaction of the two agents at the level of initiation and on temporal differences of cancer expression. The hypothesis is that among cigarette smokers a given radiation exposure induces a finite number of lung cancers that have shorter latency periods as a result of the cancer-promoting activity of smoke.

111. In a study with hamsters exposed to ^{210}Po, benzo(a)pyrene was used as a substitute for tobacco smoke [L24]. As compared with animals exposed only to ionizing radiation (lung cancers incident in about 3% of the animals) or only to benzo(a)pyrene (no incident cases in over 280 treated animals), animals receiving benzo(a)-pyrene instillations after exposure to ionizing radiation were at a much higher risk (about 50%) of developing a lung tumour. It is noteworthy that the instillation of saline after radiation exposure also induced lung tumours in about 30% of the animals.

112. Douriez et al. [D30] investigated the role of cytochrome P-450 1A1 (CYP1A1) inducers on radon-induced lung cancers in rats. CYP1A1 is the member of the cytochrome P-450 gene family producing the most mutagenic activation products from polycyclic hydro-carbons. All three inducers tested (methylcholanthrene, 5,6-benzoflavone, 2,3,7,8-tetrachlorodibenzo-p-dioxin) increased the incidence of epidermoid carcinoma to 100%, independent of whether the inducer itself was converted to a powerful carcinogen or not. Depletion of retinoid acid in CYP1A1-stimulated rats is implicated as a further step leading to increased susceptibility to lung cancer. Since tobacco smoke is a powerful inducer of CYP1A1, this mechanism could account for the supra-additive effects in radiation-exposed smokers.

113. Preliminary studies on an interaction between ^{239}PuO$_2$ and cigarette smoke were reported by Talbot et al. [T2]. The experiments were designed to show whether exposure to cigarette smoke for 12 months enhances the incidence of lung tumours in mice that had previously inhaled ^{239}PuO$_2$. The main difference found was a reduced growth rate in both smoke- and sham-exposed mice relative to that of cage controls. After 3 months of treatment, histopathology and morphometry of lung sections found only slight smoke-induced changes. On a per-unit-area basis, these changes included a reduced proportion of alveolar space and an increased number of pulmonary alveolar macrophages that were larger than those from sham-exposed or control mice and had an increased proportion of binucleated cells. All mice in a second study were initially exposed to ^{239}PuO$_2$, then subsequently divided into three treatment groups as above. Cigarette smoke exposure was shown to increase lung weight and inhibit clearance of ^{239}Pu from the lung. The authors pointed out a dosimetric problem: the group receiving ^{239}PuO$_2$ and subsequently tobacco smoke would receive a higher radiation dose to the lung than those receiving ^{239}PuO$_2$ alone. Although this aspect is important for elucidating the mechanisms by which synergism or antagonism occur, for radiation protection, an apparent combined effect traced to a modification of exposure/dose conversion factors by one agent would still be considered synergism or antagonism.

114. A cigarette-smoke-induced reduction in the lung clearance of inhaled ^{239}PuO$_2$ was also observed in a study in rats [F15, F28]. Animals were first exposed by a whole-body inhalation mode to diluted mainstream cigarette smoke at a concentration of 100 or 250 mg m^{-3} of total particulate matter for six hours per day, five days per week. Control rats received filtered air alone. After three months, all groups of rats received a single pernasal exposure to radioactively labelled insoluble tracer particles; then the rats were returned to their respective cigarette smoke or filtered air exposure. External whole-body counting of the tracer was continued for six months, and substantial smoke-induced clearance inhibition was found. Lifetime radiation doses were 3.8 Gy, 4.4 Gy, or 6.7 Gy for the control, 100 and 250 mg m^{-3} total particulate matter groups, respectively [F28]. The results for the highest level of cigarette smoke exposure suggested that the radiation dose increased by a factor between 1.6 and 1.7 by this effect, compared with the group of rats receiving filtered air alone. It should be noted that cigarette smoking has been shown to reduce the lung clearance of relatively insoluble particles in humans as well as in animals [C5].

115. The study described above is part of a carcinogenicity experiment in which 2,170 male and female F344 rats received exposures to cigarette smoke and/or ^{239}PuO$_2$ [F17]. Groups of animals were exposed for up to 30 months to filtered air or to low or high concentrations of cigarette smoke. For each of these groups, approximately one half of the rats also received a single pernasal inhalation exposure to ^{239}PuO$_2$ that resulted in an initial lung burden of approximately 400 Bq. Cigarette smoke exposure did not

markedly influence survival, but it did result in decreased weight gain and a variety of lung lesions such as alveolar macrophage hyperplasia, interstitial fibrosis, chronic-active inflammation, hyperplasia of the alveolar epithelium, and bronchial mucous-cell hyperplasia. A preliminary evaluation of lung cancer in females indicated that crude lung tumour incidences were approximately 7% in rats exposed to high concentrations of smoke, 20% in rats exposed to $^{239}PuO_2$, and 74% in groups receiving both agents (Figure A.IV). Thus, the interaction was clearly synergistic. This study illustrates the manner in which a dose from one agent can be markedly affected by exposure to a second agent, leading to a clear synergism in carcinogenic response. Less certain, however, is the extent to which the interaction resulted strictly from the impaired clearance (and associated increased radiation dose) in the combined exposure groups rather than a more fundamental interaction between the radiation and cigarette smoke constituents at the molecular or cellular level. Another mechanism by which synergism could occur might relate to the localized radiation dose rather than the dose to the whole organ. For example, the synergistic interaction between smoking and radiation in this example could result from the alpha radiation dose delivered at the site of smoke-induced lung lesions, where the processes of cell hyperplasia, fibrosis, and activated phagocytes were already occurring.

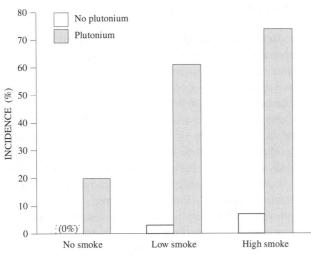

Figure A.IV. Incidence of lung tumours in female rats exposed to plutonium dioxide in combination with varying levels of cigarette smoke [F17].

116. In an early study by Cowdry et al. [C49], the carcinogenicity of ^{90}Sr beta irradiation to the skin of Swiss mice applied twice weekly and 3 tar paintings per week to distributed skin areas was studied. Surface doses were about 2 Gy per fraction and 200 Gy totally. Skin tumour incidences after 30 months were 12.3% for radiation alone, 42.9% for tar paint alone, and 61.3% for the combined treatment. It was concluded that there is no synergism between the two carcinogens in the study system. It is noteworthy that a monthly skin irradiation of several Gy surface dose did not produce any skin tumours, whereas in the control group only painted with acetone (the solvent for cigarette tar) an incidence of 6.8% was seen after 30 months.

117. In view of the many active substances contained in cigarette smoke, possible interactions are very numerous. It is outside the scope of this Annex to cover this fully, but the many reported interactions of caffeine with cigarette smoke components may merit mention, especially because caffeine at higher concentrations also modifies the effects of ionizing radiation. Rothwell [R13] found an inhibition of cigarette-smoke-induced carcinogenesis in mouse skin by caffeine. Other recent reports showed an inhibition of tobacco-specific nitrosamine-induced lung tumorigenesis in A/J mice by polyphenols extracted from green tea and cruciferous vegetables [C24, X2]. It is believed that these dietary compounds act as antioxidants (see also Section B.6).

(c) Cellular studies

118. To examine the interaction between radiation exposure and smoking, Piao and Hei [P8] studied the toxicity and oncogenic transforming incidence of alpha-particle irradiation with and without concurrent exposure to cigarette smoke condensate on C3H10T½ cells *in vitro*. In this system, additive modes of interaction between cigarette smoke condensate and ionizing radiation were observed for the oncogenic transforming potential of both gamma rays and alpha particles. In a recent study made possible by the development of charged-particle microbeams, it was shown that even for radon alone, induction of transformation in C3H10T½ cells *in vitro* from exactly one alpha particle was significantly lower than for a Poisson-distributed mean of one alpha particle through a cell nucleus [M66]. This implies that cells traversed by multiple alpha particles contribute most of the risk, and that a linear extrapolation from high exposures may overestimate the transforming potential of high-LET radiation in low-level exposures. If generally applicable, such results would speak against the potential of combined effects at low exposures to surpass values expected from linear dose-effect relationships and additivity.

119. The combined genotoxic effect of cigarette smoke condensate and gamma radiation was also studied in a simple eukaryotic organism [S8]. The induction of gene conversion in diploid yeast (*Saccharomyces cerevisiae*) strains was investigated following exposure to cigarette smoke condensate and gamma radiation. Cells exposed to a combination of cigarette smoke condensate and low-LET radiation showed an additive response irrespective of the order of treatments. The system also showed large differences in sensitivity depending on growth status, with log-phase cells being 2–3 times more sensitive than stationary cells. The relevance of these findings is limited by the fact that critical toxicants in tobacco smoke require activation by biotransformation, a mechanism that is highly species- and tissue-specific.

(d) Summary

120. In summing up the many results from the well-studied interaction between tobacco smoking and high levels of radon exposure, it can be stated that this important combined exposure leads to clearly overadditive effects for lung cancer in humans. Some of the more

important findings are summarized in Table A.1. The quite different dose-effect relationships for the two agents, apparently linear for radon and clearly non-linear for tobacco smoke, speaks against iso-addition and for a true synergism. However, large uncertainties remain with regard to quantifying the health effects of these important agents at prevailing levels of combined exposures in the present-day workplace and in non-occupational settings. In view of the complexities involved in the toxicological assessment of tobacco smoke, which is itself a combination of genotoxic and non-genotoxic agents, it is not possible at this time to extend inferences from mechanistic considerations to low combined exposures. Several large case-control studies under way involving non-occupational exposures may help to solve this enigma by creating better estimates and reducing the uncertainties surrounding synergistic effects from smoking and radon.

4. Metals

121. Toxic metals are important trace pollutants in the human environment. They interact in many ways with cell constituents and may produce oxidative gene damage or influence enzyme activity at low concentrations, e.g. by competing with essential metal ions [H38]. Carcinogenic transition metals are capable of causing promutagenic damage such as DNA base modifications, DNA-protein cross-links, and strand breaks [K7]. The underlying mechanism seems to involve active oxygen and other radicals arising from metal-catalysed redox reactions. Cadmium, nickel, cobalt, lead, and arsenic may also disturb DNA repair processes [H48]. Lead neurotoxicity, an example of an important non-genotoxic metal effect, is a result of intracellular regulatory dysfunction caused by this heavy metal. Lead activates calmodulin-dependent phosphodiesterase, calmodulin-sensitive potassium channels, and calmodulin-independent protein kinase C(PKC) [G31]. The latter effect already occurs at picomolar concentrations and indicates second messenger metabolism as a potential sensitive site for the disruptive action of lead. Epidemiologically proven metal lung carcinogens are arsenic, cadmium, chromium, nickel, and antimony [M65]. In the critical field of underground mining, possible metal effects have to be assessed together with high-LET radiation from radon. Arsenic in particular has been shown to be a major risk factor in combined exposures to mineral dust, radon, metals, and diesel fumes [K48, T5]. The risk-enhancing effects of iron dust seems to be limited to very high dust concentrations, leading to changes in lung function [B74]. An elevated stomach cancer risk in Ontario gold miners was statistically associated with chronium exposures but not with arsenic, mineral fibres, or diesel emissions [K50].

122. Multiple exposures to radon, arsenic, and tobacco smoke were common in several uranium mines (see also Section B.3). An assessment of 107 living tin miners who had lung cancer and an equal number of age-matched controls from tin miners without lung cancer provided no evidence for synergism between radon and arsenic or between arsenic and smoking [T5]. That there is no obvious synergism between this heavy metal and radon

progeny exposure is implied by the fact that the risk of lung cancer among workers exposed to arsenic (and radon) in mining only is slightly less than for miners whose exposure came from smelting operations. In a study on gold miners with quite low radon exposures, linear regressions indicated that exposure to 1 WLM of radon decay products increases lung cancer mortality rates by 1.2%, a finding comparable to other studies, and that each year of employment in a poorly ventilated mine (before 1946) with an arsenic content of the host rock of 1% is associated with a 31% increase in lung cancer mortality rates [K21, K51]. Adding an interaction term to allow for a deviation from additivity for the combined effect of arsenic and radon decay products did not improve the fit. Noteworthy is the fact that the duration of the arsenic exposure seems to be more important than its intensity [T5].

123. The induction of radical-scavenging metallothionein by higher concentrations of heavy metals may confer protection against ionizing radiation. Single ip injection of cadmium (1 mg kg^{-1}) two hours before radiation exposure increased the yield of DNA lesions in peripheral blood lymphocytes of mice, but cadmium injection 24–48 hours in advance of ionizing radiation reduced DNA damage in lymphocytes in comparison with untreated animals [P27]. In this study the protective effect was due to reduced levels of initial DNA damage per unit dose of radiation as well as accelerated DNA repair measured in a single-cell gel assay.

124. Beryllium is another metal that has been examined for potential interactions with radiation. Although not considered to be a genotoxic metal [A10], beryllium is a known animal carcinogen and has recently been classified as a demonstrated human carcinogen [W9]. The potential for carcinogenic interactions between inhaled beryllium oxide and ^{239}PuO$_2$ was examined in rats [S42]. The agents were administered alone or in combination at initial lung burdens of 1–91 μg beryllium and 0.15–6.7 kBq ^{239}Pu. Beryllium oxide exposure induced few tumours and did not markedly influence Pu-induced lung tumorigenicity, despite the fact that beryllium oxide exposure decreased the clearance of ^{239}Pu from the lung and thus served to increase the total radiation dose to the lung.

125. Another ongoing study investigates the potential carcinogenic interactions between inhaled beryllium metal and ^{239}PuO$_2$ in some 5,456 male and female F344 rats [F1]. Preliminary results from the study demonstrate that inhaled beryllium metal is a potent rat lung carcinogen; over 90% of rats that survived at least 12 months after inhaling an initial lung burden of 450 μg beryllium developed benign and/or malignant lung tumours [F12]. At lower lung burdens of approximately 50 μg beryllium, some 65% of exposed rats (39 of 60 rats) developed at least one malignant lung tumour. When this level of beryllium was combined with a lung burden of 60 and 170 Bq ^{239}PuO$_2$, which alone caused crude malignant lung tumour incidences of 8% (6 of 60 rats) and 7% (2 of 27 rats), respectively, crude tumour incidences ranged from 57% (16 of 28 rats) to 90% (54 of 60 rats) [F16, F28]. Thus, indications of a more-than-additive response were observed.

As was the case for the cigarette smoke or beryllium oxide exposures described above, inhalation of beryllium metal was found to markedly decrease the clearance of ^{239}PuO$_2$ from the lung [F1], serving to increase the radiation dose in groups receiving combined exposures and to leave in question the role of beryllium-induced increased radiation dose vs. a more fundamental interaction between the two agents at the molecular or cellular level. In addition, the investigators reported that exposure to beryllium markedly reduced the median lifespan of exposed animals, and they noted that a complete analysis of the combined carcinogenic effects of the two agents would require an analysis more sophisticated than an examination of crude tumour incidence [F15]. Specifically, the authors noted that the age-specific tumour incidence for the two agents alone and in combination should be analysed, and it was noted that this analysis is under way.

126. The potential for beryllium and x radiation administered alone and in combination to affect cell-cycle kinetics, cell killing, and induction of chromosomal aberrations was examined in mammalian cell culture (Chinese hamster ovary cells, Figure A.V) [B37]. Beryllium was administered in a soluble form (BeSO$_4$) at 0.2 or 1 mM concentrations, and x rays at levels of 1 or 2 Gy. It was found that exposure to beryllium significantly inhibited the capacity of the cells to repair DNA damage induced by x rays. The combined exposures were characterized by a multiplicative model when total chromosomal aberrations were examined hours after exposure. Both agents caused an accumulation of cells in the G$_2$/M stage of the cell cycle, and an analysis using varying times between exposures suggested that the multiplicative interaction observed may have been limited to cells in the S and G$_2$ stages of the cell cycle.

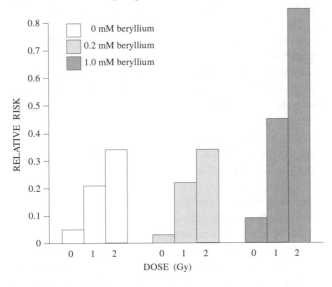

Figure A.V. Induction of chromosome aberrations in Chinese hamster ovary cells from exposures to x rays and beryllium [B37].

127. Micronucleus formation in mouse bone marrow polychromatocytes as a measure of the modulation of the mutagenic action of gamma rays by chromium and lead salts was used by Vitvitskii et al. [V12]. Chromium (VI)

ions enhanced radiation effects in acute and chronic experiments. Acute exposures of lead (II) ions below 15 mg kg^{-1} body weight had an antagonistic effect, i.e. they decreased the number of gamma-ray-induced micronuclei, whereas higher doses increased it. Chronic combined action of lead (III) ions and gamma rays resulted in a lower yield of micronuclei. For an extrapolation to environmental exposures and humans, an elucidation of the underlying mechanisms, i.e. heavy metal influence on cell kinetics and/or on DNA damage and repair, will be necessary.

128. The combined effect of ^{134,137}Cs and lead (Pb^{2+}) at the soil concentrations found in highly contaminated habitats in the Russian Federation on the mutation rate in the plant *Arabidopsis thaliana (L:) Heynh* has been investigated [K42]. At concentrations of 220–2,500 Bq kg^{-1} and 16–320 mg kg^{-1}, respectively, both antagonistic and synergistic effects were seen. The radiation-induced mutation rate was significantly reduced in the presence of 16 mg kg^{-1} Pb^{2+}, whereas higher lead concentrations increased the rate in plants grown in soil with up to 1,000 Bq kg^{-1} radiocaesium. At the highest radiation level and 32–320 mg kg^{-1} Pb^{2+}, an apparent decrease in the mutation rate was linked to a large number of sterile seeds. In an ecological study, the combined effect of zinc or cadmium and external radiation on microbial activity in soil was determined by measuring nitrogen fixation, dentrification, and CO$_2$ flux [E20]. At metal concentrations in soil of 10–100 mg kg^{-1} for Zn^{2+} and 0.5–16 mg kg^{-1} for Cd^{2+}, small radiation doses ranging from 3.6 to 12 mGy led to a supra-additive effect in the inhibition of microbial activity in soddy-podzolic soil. It was further shown that the enzyme level of invertase increased in combined exposures, whereas catalase and dehydrogenase activities were lower [E19].

129. A sensitive assay in spring barley (*Hordeum vulgare L.*) leaf meristem to record effects from ionizing radiation and/or heavy metals was developed by Gerask'in et al. [G34]. The radiation-induced frequency of cells with aberrant chromosomes in the intercalary meristem allows doses to be registered in the range of a few tens of milligray [G35]. Irradiations were performed at the shoot stage and involved doses of 40, 80, and 200 mGy at a dose rate of 2 Gy h^{-1}. Lead (II) and cadmium (II) were applied as nitrates in two concentrations of 40 and 200, and 3 and 20 mg kg^{-1} of soil, respectively. The authors claimed that in this system, radiation and heavy metals alone exhibit clearly non-linear relationships, i.e. supralinearity, with a higher slope for aberrations at lower doses than at higher doses. Combined exposures show an antagonism for low doses of ionizing radiation (40 mGy) and for all lead concentrations. At doses of 80 and 200 mGy, a slightly supra-additive effect is reported. For cadmium, supra-additivity is found at low metal concentrations of 4 mg kg^{-1} for 80 and 200 mGy but not for 40 mGy. At high metal concentration, less-than-additive effects were found. Although these findings may be important for environmental assessments and potentially extendable to mechanistic studies, no direct inferences to humans are warranted at this stage.

130. Metals and ionizing radiation have been shown to produce combined effects in many biological systems (Table A.1). Because metals cause many biological effects with no or very low thresholds, possible interactions would potentially extend to very low exposures. In the case of relatively unspecific damage to DNA, such as oxidative attack, iso-addition would be predicted. As an example of a synergistic effect at high exposure levels, a threshold phenomenon, decreased lung clearance of internal radionuclide content by high metal concentrations, was found to be the cause of the combined effect. No supra-additive effects are seen in the albeit weak database on combined occupational exposure to radon and arsenic. The relative importance of different damage-inducing mechanisms of metals for combined exposures in human remains to be elucidated.

5. Mitogens and cytotoxicants

131. Although mitogenic and cytotoxic compounds are generally non-genotoxic agents and could have been included in Section B.2, they are considered here separately, principally because of their ability to stimulate cell proliferation. The combination of mitogens or differentiation-inducing agents with radiation has some potential as a cancer therapeutic strategy. Experiments in this area employ high doses, but studies intended to elucidate the mechanisms of interaction may still be relevant outside the clinic. Leith and Bliven [L15] investigated the x-ray responses of a human colon tumour cell line after exposure to the differentiation-inducing agent N-methylformamide (NMF). A human colon tumour line was exposed for three passages to varying concentrations (0–170 mM) of NMF and the change in sensitivity to ionizing radiation was examined in vitro. The linear-quadratic formalism of survival with two constants (alpha and beta) was used to characterize the single graded dose-survival curves. As the NMF concentration increased, the alpha parameter increased and the beta parameter decreased, yielding a concentration-dependent radiosensitization that was most marked in the low-dose region of the survival curve. Upon removal of NMF, the original radioresistance was regained within two or three cell culture doubling times.

132. Müller et al. [M11] studied the formation of micronuclei in preimplantation mouse embryos in vitro after combined treatment with x rays and caffeine. The exposures to caffeine were 0.1 or 2 mM and to x rays, 0.2 or 0.9 Gy. X rays as well as caffeine induced micronuclei. The dose-effect curve after irradiation was linear for the dose range measured (0–3.8 Gy). Caffeine only induced micronuclei at concentrations higher than 1 mM; between 1 mM and 7 mM, however, there was a linear increase in the number of micronuclei. A considerable enhancement of the number of radiation-induced micronuclei was observed when irradiation of the embryos was followed by treatment with caffeine. The sum of the single effects was clearly exceeded by the combination effects. An earlier study in the same laboratory [M12] was on the effects of a combination of x rays (0.2, 0.9, or 1.9 Gy) and caffeine (0.1, 1, or 2 mM) on the formation of blastocysts (96 hours post-conception), hatching of blastocysts (144 hours post-conception), and on the cell numbers of embryos at different times (48, 56, 96, and 144 hours post-conception). The embryos were irradiated in the G_2 phase of the two-cell stage (28 or 32 hours post-conception), either 1 hour after or immediately before application of caffeine. Caffeine was present during the whole incubation period (until 144 hours post-conception). Specific conditions under which caffeine markedly enhanced the radiation risk, i.e. under which the combination effect exceeded the sum of the single effects, were described. This was the case, in particular, for embryonal development, for which the risk was almost doubled, whereas the enhancement of risk was smaller for the proliferation of cells. The amount of caffeine necessary for supra-additivity, however, is so high (at least 1 mM caffeine for rather long times) that it is clearly above the range achievable in vivo by consumption of caffeine-containing beverages. At physiological levels, caffeine also displays antioxidant properties and inhibits carcinogenesis induced in rats and mice by various known carcinogens. Examples are the inhibition of smoke-condensate-induced carcinogenesis in mouse skin [R13] and gastric tumour promotion by NaCl in rats [N10]. Based on these and other findings, Devasagayam et al. [D1] suggest that at lower concentrations, the potency of the antioxidant action of caffeine far outweighs the deleterious effects, if any, from its inhibition of DNA repair.

133. Besides the interference of caffeine with repair processes as a consequence of its effect on cell-cycle blocks at high concentrations, this ubiquitous substance also scavenges oxygen species induced by radiation and genotoxic chemicals [K27, K28]. The chemical basis of this effect was shown to be the removal of free electrons and hydroxyl radicals by caffeine. The reaction rate constants for these two reactions were shown to be about $1.5 \ 10^{10} \ M^{-1} \ s^{-1}$ and $6.9 \ 10^9 \ M^{-1} \ s^{-1}$, respectively [K27]. The former value is high enough to compete with oxygen for the scavenging of free electrons and therefore may reduce oxidative damage involving superoxide anion (O_2^-), hydroperoxyl radical (HO_2^-), and hydrogen peroxide (H_2O_2). This mode of action is backed by recent findings in barley seeds that caffeine affords protection only at high oxygen concentrations but potentiates radiation damage (albeit less damage) at low oxygen pressures [K29].

134. Both a reduction of the radiation-induced G_2/M phase arrest and the antioxidant effect of caffeine may indirectly influence apoptosis and modulate survival and expansion of cells with a modified genome. In different systems, an enhancement of the degree of DNA fragmentation by caffeine, theobromine, theophylline, and 2-aminopurine was found in murine T-lymphoma cells [P10], whereas in TKG cells, 2 mM caffeine eliminated the degradation of DNA entirely [Z4]. At this stage, it is doubtful whether these findings have any meaning for risk assessments at controlled exposure levels.

135. Caffeine, which may potentiate radiation damage at higher concentrations owing to its release of protective cell-

cycle blocks, seems also to influence the clastogenic effects of radiation and other genotoxic agents (see also Section B.5). Several studies found an inhibition of oxic radiation damage [K25, K26]. Stoilov et al. [S43] found both potentiation and protection against radiation-induced chromosomal damage in human lymphocytes. Temperature and concentration were shown to be decisive for the direction of the effect.

136. Kalmykova et al. [K3] evaluated the effectiveness of joint exposure to ^{239}Pu and tributyl phosphate on the induction of leukopenia in Wistar rats. It was shown in this system that the additive effect of the two agents delivered simultaneously was exceeded only at high doses, i.e. acute levels. With levels ranging from subacute effective to minimum effective, the effect of the combined treatment was less than projected from additivity.

137. Cattanach and Rasberry [C27] reviewed the literature on the genetic effects of combined treatments with cytotoxic chemicals and x rays. Some pretreatments clearly enhanced the yield of genetic damage. With spermatogonial cells, chemicals that kill cells can substantially modify the genetic response to subsequent radiation exposure over several days or weeks. Both enhancement and reduction in the genetic yield were found, and the modifications also depended on the type of genetic damage scored, with specific-locus-mutation response differing from that for translocations. Selective killing of rapidly dividing cells in the areas most heavily damaged by radiation was a suggested explanation [C28]. In general, such interactions based on perturbations of cell kinetics should be of little relevance for lower exposure levels.

138. Cyanate (KOCN)-induced modification of the effect of gamma radiation and benzo(a)pyrene was studied by Serebryanyi et al. [S80] in cultured CHO-AT3-2 cells. Sensitizing effect was found for radiation and benzo(a)pyrene effects such as cell viability, micronuclei induction, and mutations in the thymidinekinase and Na$^+$/K$^+$-ATPase loci. The authors suggested that repair inhibition and/or changes in the cell chromatin structure produced by KOCN is responsible for these sensitizing effects. The proposed mechanisms as well as the concentration and dose ranges used in the experiment preclude direct transfer to occupational or environmental levels.

139. In summary, many studies assessing deviations from additivity in combined exposures between mitogens/cyto-toxicants and ionizing radiation are found in the literature (Table A.1). In most cases, the high exposure levels applied and the biological endpoints studied do not allow the transfer of results to humans. However, any endogenous or dietary levels of agents influencing stem-cell population size or kinetics will have the potential to modulate response to radiation.

6. Antioxidants, vitamins, and other dietary factors

140. The genetic effects of combined treatments of radio-protecting agents and x rays were reviewed by Cattanach and Rasberry [C27]. Chemicals such as cysteamine, mexamine, and glutathione given in advance of radiation were not always protective but gave contradictory results, with significant protection of specific germ-cell stages being restricted to different dose ranges. This might be attributable to the different radiation sensitivities and cell-cycle kinetics of the germ-cell stages tested. Some pretreatments clearly enhanced the yield of genetic damage.

141. Dietary caloric intake and type of food are important variables affecting the rate of spontaneous DNA damage, as was discovered recently in humans [D11, S37, S38]. These findings are supported by similar findings of reduced oxidative damage to mitochondrial and nuclear DNA in food-restricted rats and mice [C12]. It is known from experiments with rats that caloric restriction of food is correlated with a lower incidence of cancer, an increased lifespan, and less free-radical damage to lipids, proteins, and DNA [W4, Y3, Y5, Y14]. Dietary fat is associated with increased breast cancer risk. In a study involving 21 women at high risk, the level of the oxidized thymine (5-hydroxymethyluracil) per 10^4 thymine was 9.3 ± 1.9 in the nucleated peripheral blood cells of women consuming 57 g of dietary fat per day compared with 3 ± 0.6 for women consuming 32 g per day [D11, F3].

142. Diet can also modify the effectiveness of chemical carcinogens, sometimes by a large factor. Some of the underlying mechanisms have been identified. Rats with a deficiency of riboflavin in their diet become highly sensitive to liver tumour formation when treated with 4-dimethyl-aminoazobenzene, because reduced levels of a flavin adenine dinucleotide-dependent azo dye reductase increase the effective dosage of the carcinogen [C25]. On the other hand, a protein-free diet prevents liver toxicity of dimethylnitro-samine in rats, and a fat-restricted diet decreases tumour induction in mammary glands of rats. Silverman et al. [S32] studied the effect of dietary fat on mammary cancer induction in Sprague-Dawley rats given 3.5 Gy whole-body x-irradiation at 50 days of age. Rats on a high-fat diet (20% lard) from 30 days of age had more tumours than rats on a low-fat diet (5% lard) and a higher multiplicity of carcinomas per rat. Rats on the low-fat diet exhibited longer median tumour latency periods than did those on the high-fat diet. Spontaneous breast cancer incidence in humans is also influenced by the level and type of fat intake. Potential mechanisms in dietary-fat-dependent mammary tumorigenesis were reviewed by Welsch [W5]. Yoshida et al. [Y14] reported that caloric restriction significantly reduced the incidence of x-ray-induced myeloid leukemia in C3H mice. Again, in this system, caloric restriction either before or after irradiation also significantly prolonged the lifespan of the animals.

143. In some instances, the degree of tumour formation depends on the amount of food provided during the pro-moting phase and not on the nutritional status at time of exposure. Polyunsaturated oils are potent promoters, probably also for humans [W14]. It is now generally accepted that restricted food intake, particularly during development phases, reduces the incidence of neoplasms

and increases longevity. An epigenetic effect, namely a general decrease in cell duplication rates, especially in endocrine-sensitive organs, is at the root of this finding [C26].

144. Several vitamins and many food constituents display radical scavenging activities and antioxidant properties. There is considerable scientific and economic interest in the still unresolved question whether diets enriched in vitamins, antioxidants, carotinoids, and selenium reduce the risk of cancer [W13]. Vitamin A and retinoic acid derivatives are considered important micronutrients involved in the modulation of cancer risk in humans. Vitamin A seems to affect the incidence of lung cancer in smokers and tobacco chewers positively. Hence, clinical trials in Finland and the United States randomized the use among smokers of artificial beta-carotene (precursor of vitamin A) and, in the Finnish study, the use of artificial alpha-tocopherol (vitamin E) [H7, O14]. Surprisingly, these two studies found significant increases in lung cancer risk related to beta-carotene use. Whether this finding is due to the dietary form of the provitamin remains to be elucidated. Human cervix and bladder cancer are somewhat more frequent in individuals with low vitamin intake [S45]. These beneficial effects are thought to arise from differentiation of epithelial tissues and from improved cell-cell communication. Vitamins E and K are benzo- and naphthoquinones and therefore potential antioxidants. Reduction of tumour induction by the former in animal systems was shown only at levels much higher than are found in the human organism.

145. Selenium also reduces tumour risk in animal systems. Its salts are indicated as a co-factor for glutathione peroxidase. Vitamins C, E, and K, the latter two in the lipid phase and its boundary, prevent the formation of nitrosamines and nitrosamides and seem to be important in the protection of the gastro-intestinal linings, the liver, and the respiratory tract [M31]. Although any molecule with antioxidant and radical scavenger activity is also a potential radioprotector, the extreme speed of the interaction of reactive species formed by radiation with DNA would require high concentrations to make a difference. For combined effects, the available information indicates that micronutrients are important. The sizeable influence of vitamin A, vitamin E, selenium, and 3-aminobenzamide as radioprotectors in the C3H10T½ transformation assay is shown in Figure A.VI [H11].

146. Borek et al. [B24] studied the anticarcinogenic action of selenium and vitamin E. The single and combined effects of these chemicals were examined on cell transformations induced in C3H10T½ cells by x rays and benzo(a)pyrene and on the levels of cellular scavenging and peroxide destruction. Incubation of C3H10T½ cells with 2.5 µM Na$_2$SeO$_3$ (selenium) or with 7 µM alpha-tocopherol succinate (vitamin E) 24 hours prior to exposure to x rays or the chemical carcinogens resulted in an inhibition of transformation by each of the antioxidants with an additive-inhibitory action when the two nutrients were combined. Cellular pretreatment with selenium resulted in increased

levels of cellular glutathione peroxidase, catalase, and non-protein thiols (glutathione) and in an enhanced destruction of peroxide. Cells pretreated with vitamin E did not show these biochemical effects, and the combined pretreatment with vitamin E and selenium did not augment the effect of selenium on these parameters. These results support the notion that free-radical-mediated events play a role in radiation and chemically induced transformation. They indicate that selenium and vitamin E act alone and in additive fashion as radioprotecting and chemopreventing agents. The results further suggest that selenium confers protection in part by inducing or activating cellular free-radical scavenging systems and by enhancing peroxide breakdown, while vitamin E appears to confer its protection through another, complementary mechanism.

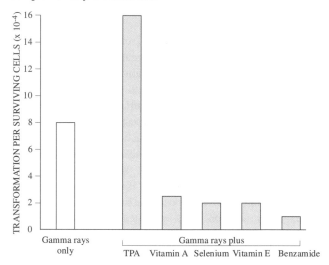

Figure A.VI. Effect of chemical and dietary factors on response of C3H10T½ cells in combined exposures with gamma rays giving a dose of 4 Gy [H11].

147. The importance of sulfhydryl groups as an antagonist of or protector against radiation-induced radical attack on DNA is well known from molecular and *in vivo* studies [M4, M6, M7, V4]. In view of the fast and localized action of ionizing radiation, these substances have to be small enough to reach the target to be protective and to be present there in considerable concentrations for noticeable effects. Even for small water-soluble substances displaying sulfhydryl groups, such as cysteamines, this is difficult to achieve in humans. Therefore the use of sulfhydryl radioprotectors is limited by their toxicity and the short period during which they are active [M4]. At environmental levels, no protective effect is to be envisaged. However, for lipophilic substances such as vitamins A, E, and K or for coenzymes with high-affinity binding to active centres, local concentrations in specific compartments might become high enough for a protective effect, even with low dietary intakes. In addition to these directly acting protectors, immunomodulators such as endotoxins and bacterial or yeast polysaccharides are known to protect against the deleterious effects of radiation [M4]. Their mode of action, stimulation of the reticulo-endothelial system, is probably irrelevant for stochastic effects.

148. Ethanol consumed in alcoholic beverages is known to increase the incidence of several cancers of the oral cavity and the oesophagus, especially in combination with active cigarette smoking [I8]. No data from human studies on ethanol and ionizing radiation are available, but the irritating effect of higher concentrations of ethanol makes this form a potential tumour promoter. Mechanistic studies also suggest that ethanol modifies the biochemical activation in the oral cavity and the oesophagus of some tobacco-specific carcinogens [S44], a mechanism of no direct relevance to radiation. Acetaldehyde is a toxic reactive metabolite of ethanol in tissue where biotransformation occurs. Very important in view of population health are behavioural changes and a tendency to malnutrition in alcohol addicts, which may increase their susceptibility to toxicants in the environment or at the workplace. The many direct and indirect effects of alcohol consumption on radiation-induced changes at the cellular, organ, and behavioral levels were discussed in depth in a review by Ushakov et al. [U18] assessing experiences in human populations affected by the Chernobyl accident.

149. Iodine as a constituent of thyroxine, the hormone of the thyroid gland, is often deficient in inland areas, where geological factors and the absence of seafood produce a diet low in iodine. Since its fission yield, relative volatility, and half-life make ^{131}I one of the critical fission products that may be present in environmental exposures, the potential increase of thyroid dose per unit uptake in humans with iodine-deficient diets is a major concern in radiation protection. It is still not known whether the higher stimulation of the gland in iodine deficiency by endogenous hormones will also alter the radiosensitivity of the stem cells and the risk coefficient for thyroid carcinoma. The wealth of data, mainly from therapeutic procedures in nuclear medicine, showing little to no carcinogenic potential for ^{131}I, even at high exposures [U2], is somehow contradicted by recent results showing large increases in thyroid carcinoma in children affected by the Chernobyl accident. Initial measurements of iodine in urine from Belarus indicated that areas most heavily affected by iodine deposition are also deficient in a dietary supply of iodine and are endemic goitre areas.

150. The modifying influence of sodium chloride (NaCl), miso (Japanese soybean paste), and ethanol on the development of intestinal metaplasia after x-ray exposure was examined by Watanabe et al. [W29] in CD(SD):Crj rats. Intestinal metaplasia in the glandular stomach is considered a precursor lesion for differentiated gastric adenocarcinoma. Five-week-old rats were treated with two doses of 10 Gy from x rays to the gastric region at a three-day interval. After exposure, the rats were given NaCl (1% or 10% in diet), miso (10% in diet), or ethanol (10% in drinking water) for 12 months. The number of alkaline phosphatase-positive foci of intestinal metaplasia in rats given a 1% NaCl diet after x rays was significantly elevated compared with that in rats given x rays alone or x rays with a 10% NaCl diet. In the pyloric gland mucosae, total numbers of metaplastic foci in rats given x rays and

1% NaCl diet were much higher than other combined-treatment groups. The incidence of atypical hyperplasia was less than 6% in all treatment groups, and no promoting effect on gastric tumorigenesis was observed. These results demonstrated that the occurrence of intestinal metaplasia induced by x rays can be significantly modified by basic and common food constituents, but this is not associated with any influence on gastric neoplasia.

151. A potentially important interaction was investigated by Montour et al. [M24], who studied the modification of radiation carcinogenesis by marihuana (tetrahydrocannabinol, delta(9)-tetrahydro-cannabinolic acid). Male, female, and ovariectomized female Sprague-Dawley rats were irradiated with doses of 1.5, 3, or 4 Gy, respectively, from ^{60}Co gamma rays at between 40 and 50 days of age. The animals were injected three times weekly with either marihuana extract or with alcohol-emulphor carrier. Mean survival time in males was significantly shorter in the 4 Gy plus marihuana group compared with the three other groups, whose mean survival times did not differ. Throughout the 546-day period in which the male rats were observed, the total number of tumours other than fibrosarcomas was significantly greater following radiation and marihuana administration (22) than following irradiation alone (6). Fifteen of the tumours originated in breast or endocrine tissues. No differences were seen in the unirradiated groups. In the females, which were observed for 635 days, the total number of breast tumours was significantly higher in the combined treatment group (38) compared with the group treated with radiation alone (22). This was entirely due to a marked difference in the adenocarcinoma incidence, which was 21 (radiation plus marihuana) compared with 4 (radiation alone). The number of adenofibromas was similar in the two groups. In the unirradiated female groups, the breast adenocarcinoma incidence was 8 in the marihuana group and 2 in the control group. Ovariectomy resulted in a lower breast tumour incidence in all groups. Non-breast tumours were more frequent in the ovariectomized-irradiated groups. Radiation plus marihuana produced more non-breast tumours (25) than radiation alone (17) in the ovariectomized females.

152. Dietary factors are proven modifiers of risk from diverse agents at levels found in human populations and probably also influence the production and repair of endogenously arising lesions. Absence or deficiency of important coenzymes and nutrients on one side and high levels of directly or indirectly acting mitogens on the other interfere with molecular, cellular, and tissue responses to ionizing radiation. In view of the many mechanisms involved, the full spectrum of interactions from antagonisms to synergisms must be expected (see also Table A.1). A reduction in the radiation risk may occur in situations where growth stimuli are reduced owing to nutritional deficiency or where the number of stem cells at risk are reduced. Synergisms are to be expected where reduced levels of radical scavengers or coenzymes needed for repair increase the yield of primary damage from

ionizing radiation or impair the speed and accuracy of cellular responses to damage. In general, the health risks not only from ionizing radiation but also those from most other deleterious agents in the human environment will be affected by deviations from an optimum diet.

C. RADIATION AND BIOLOGICAL AGENTS, MISCELLANEOUS

1. Hormones

153. Many hormones are potent growth stimulators. Considerable evidence is available for the modulation of cancer risk by hormones. Animal experiments have shown that increased levels of thyroid-stimulating hormone (TSH) can enhance tumour growth and increase the risk of cancer [D31]. Thyroid stimulating hormone is increased during puberty and pregnancy as a result of increased levels of female sex hormones [H37, P25]. There is epidemiological evidence suggesting that the development of thyroid cancer after high-dose radiation exposure in females can be potentiated by subsequent child bearing. Marshall Islanders who were exposed to radioactive fallout from a nuclear weapons test in 1954 received high thyroid doses from radioiodines. Women who later became pregnant were at higher risk of thyroid cancer than exposed women who remained nulliparous [C43]. The numbers, however, were small.

154. The same effect was found in a population-based case-control study in Connecticut in the United States involving 159 subjects with thyroid cancer and 285 controls [R11]; 12% of the cases but only 4% of the controls reported prior radiotherapy to the head and neck. Among women, this risk appeared to be potentiated by subsequent live births (RR = 2.7). The risk for ever parous alone was, however, higher (1.6) than for prior radiotherapy (1.1). Another case-control study, carried out in Washington in the United States, linked a 16.5-fold increased risk of thyroid cancer to prior radiotherapy of the head and neck among 282 females and 394 controls [M17]. Overall, 20.2% of the cases but only 1.5% of the controls reported earlier radiotherapy (RR = 16.5; 95% CI: 8.1–33.5). In this study, pregnancy following radiotherapy was associated with only a small additional risk (RR = 1.3), which was far from statistically significant (95% CI: 0.1–15.7) [M9]. Combined with similar findings from Sweden [W30], these studies suggest that TSH-mediated tissue proliferation in adolescence and pregnancy may be a risk factor in radiation-induced thyroid cancer.

155. The long-term use of tamoxifen, a synthetic anti-oestrogen that has been shown to reduce mortality from breast cancer and the occurrence of contralateral breast cancer, increases the risk for endometrial cancer. In a case-control study of woman treated for breast cancer, Sasco et al. [S26] showed that tamoxifen or radiation castration (which included high doses to the uterus as well as the ovaries) considerably increased the risk for subsequent endometrial cancer. The odds ratios for tamoxifen use for more than five years and radiotherapeutic castration were 3.5 and 7.7, respectively. Women who had undergone combined treatment had an odds ratio of only 7.1. Since the study was based on small numbers (43 cases and 177 controls), the power is not sufficient to postulate an antagonism, but there is enough evidence to reject an enhancement of risk between the two carcinogenic factors.

156. One well-studied interaction is that between radiation and the natural hormone oestradiol-17 beta (E_2) in mammary carcinogenesis. In a publication by Broerse et al. [B32], the combined effects of irradiation and E_2 administration on the mammary gland in different rat strains were investigated. Three rat strains, Sprague-Dawley, Wistar WAG/Rij, and Brown Norway, with different susceptibilities to the induction of mammary cancer, were irradiated with x rays and mono-energetic neutrons; increased hormone levels were obtained by subcutaneous implantation of pellets with E_2. Mean plasma levels were 100–300 pg ml^{-1} plasma, while normal levels in these rat strains were about 50 pg ml^{-1}. The latency period for the hormone-treated animals was shown to be considerably shorter than for animals with normal endocrinological levels. Administration of the hormone alone also appreciably increased the proportion of rats with malignant tumours. At the high levels of hormones applied in the study, there was little indication that radiation and hormones produced any supra-additive effect, but the single-agent effect levels in this study might have been too high to properly assess this other effect. The effect of hormone administration and irradiation on mammary tumorigenesis was the same for hormone administration one week prior to or 12 weeks after irradiation. The RBE values for induction of mammary carcinomas after irradiation with 0.5 MeV neutrons have a maximum value of 20 and are not strongly dependent on hormone levels.

157. The carcinogenic and co-carcinogenic effects of radiation on rat mammary carcinogenesis and mouse T-cell lymphomagenesis were studied by Yokoro et al. [Y6]. For both experimental models, the study clearly showed the importance of the promotion stage and of the physiological condition of target cells at the time of initiation. In rat mammary carcinogenesis, prolactin was shown to be a powerful promoter regardless of the initiating agent. The authors also suggested that an enhancer like prolactin might be useful in detecting the carcinogenicity of small doses of carcinogens; for example, a high RBE of 2.0 MeV fission spectrum neutrons was demonstrated by the application of prolactin to radiation-initiated mammary carcinogenesis in rats. Because cellular reactions are somewhat different for different LET values, it remains to be proven that the sensitizing effect of a hormone is independent of radiation quality.

158. Shellabarger et al. [S24] investigated the influence of the interval between neutron irradiation and diethylstilbestrol (DES) on mammary carcinogenesis in female ACI rats. Both radiation and DES are carcinogens for the

mammary gland of ACI female rats and act in a synergistic fashion, particularly with regard to the number of mammary adenocarcinomas per rat when DES is given at about the same time as radiation. DES, in the form of a compressed pellet containing a mixture of cholesterol and DES, formulated to average 1.25 mg of DES per 100 g body weight, was given to groups of approximately 28 rats at 2 days before or 50, 100, or 200 days after 0.064 Gy from 0.43 MeV neutron irradiation. Every time DES was given to irradiated rats, it was also given to non-irradiated rats. When the total number of mammary adenocarcinomas 375 days after administration of DES was analysed as a percentage of 24 sites per rat at risk, DES and radiation always produced a response that was larger than the sum of the responses of DES alone and radiation alone. The supra-additive interaction between radiation and DES did not decline as the time interval between irradiation and DES was lengthened, which suggests that neutron-initiated mammary carcinogenesis is not subject to repair, since DES promotion continues to be effective for long times.

159. Irradiation of pregnant Wistar rats at days 7, 14, and 20 of pregnancy, followed by DES treatment after nursing for one year, showed a strong correlation of mammary gland tumours with the hormonal status of the gland during radiation exposure [I11]. Irradiation alone (2.6 Gy) resulted in a 23% incidence of mammary gland tumours. The additional implantation of a DES pellet (releasing about 1 mg d^{-1}) increased this value to 35% and 93% for radiation exposure at days 14 and 20, respectively. The data suggest that the initiation of tumorigenesis by gamma rays is critically dependent on the developmental status of the gland at exposure. Since no group with DES exposure only was included, no direct assessment of the combined effects is possible from this study. When the radiation exposure was delayed to day 21 of lactation or day 5 post-weaning, combined treatment with gamma rays and DES resulted in an incidence of mammary gland tumours of 94% and 73%, respectively. Since the value from combined treatment in virgin animals was only 24%, it is suggested that the differentiation state of the radiation-exposed tissue is more relevant than the hormonal and proliferative state of the cell populations at risk [S40]. Rats with weaning experience receiving only a gamma dose at day 21 of lactation or DES had tumour incidences of 35% and 27%. When compared with the combined effect of 94%, a synergistic effect has to be postulated.

160. The effect of age and oestrogen treatment on radiation-induced mammary tumours in rats was analysed by Bartstra et al. [B71, B72]. The excess normalized risk of mammary carcinoma was 0.9 for 1 Gy and 2.2 for 2 Gy in the age groups 8, 12, 16, 22, and 36 weeks, with no significant differences between the age groups. However, irradiation at 64 weeks yielded fewer carcinomas than in the controls, the excess normalized risk being 0.7 and -0.3 for 1 and 2 Gy, respectively. After oestradiol-17 beta2 treatment, the excess normalized risk for carcinomas was 7.7 for both 1 and 2 Gy in the age groups 8, 10, 12, and 15 weeks, with no significant differences between the age groups. However, in the age groups 22, 36, and 64 weeks, the excess normalized risk decreased with increasing age at exposure. Irradiation at 64 weeks yielded fewer carcinomas than in controls, with an excess normalized risk of -0.6 for both 1 and 2 Gy. The excess normalized risk was 10-80 in oestrogen-treated controls compared with untreated animals. The findings indicated that administration of oestrogen increased the radiation sensitivity of the mammary gland in young animals considerably. Administration of oestrogen influenced the shape of the dose-response curve for radiation-induced mammary cancer in young rats. In untreated animals there was a linear dose-effect relationship, whereas in oestrogen-treated ones the relationship could only be described by a quadratic function. In older rats, radiation dose-effect relationships in oestrogen-treated and non-treated animals were best fitted by linear relationships. The reduced risk of radiation exposure at mid-life was observed in oestrogen-treated and control rats.

161. The influence of androgens in the development of radiation-induced thyroid tumours in male Long-Evans rats was investigated by Hofmann et al. [H12]. When eight-week-old male rats were treated with radiation (1.5 MBq $Na^{131}I$), thyroid follicular adenomas and carcinomas were observed at 24 months of age with a high incidence, 94%. Castration of males prior to irradiation significantly reduced this tumour incidence to 60%. When testosterone was replaced in castrated, irradiated male rats, differentially increased incidences of thyroid tumours occurred, depending on the time interval for hormone replacement. Immediate (age 2-6 months) or early (age 6-12 months) testosterone replacement at approximate physiological levels led to thyroid follicular tumour incidences of 100% and 82%, respectively, whereas intermediate (12-18 months) or late (18-24 months) testosterone treatment led to only 70% and 73% incidences, respectively. Continuous testosterone replacement (2-24 months) in castrated, irradiated male rats raised the thyroid tumour incidence to 100%. Only the two 100% values are significantly different from the value of 60% in castrated irradiated animals not receiving testosterone replacement. Since elevated TSH is a reported requisite for the development of radiation-associated thyroid tumours, the effects of testosterone on serum thyroid-stimulating hormone levels were examined. Mean serum thyroid-stimulating hormone values in all irradiated animal groups were significantly elevated and well above those in age-matched, non-irradiated animals at 6, 12, 18, and 24 months. Serum thyroid-stimulating hormone levels were higher in continuous testosterone-replaced irradiated castrates than in intact, irradiated males but lower in irradiated castrates without testosterone treatment. Interval testosterone replacement in castrated male rats was generally associated with increased serum thyroid-stimulating hormone levels during the treatment interval and with lowered thyroid-stimulating hormone levels after discontinuation of testosterone treatment, particularly in irradiated rats. However, when irradiated, castrated males received late (age 18-24 months) testosterone replacement, there was no elevation of thyroid-stimulating hormone at

the end of the treatment interval. Thus an indirect effect of testosterone via early stimulation of thyroid-stimulating hormone may be at least partly responsible for the high incidence of radiation-induced thyroid tumours in male rats.

162. Watanabe et al. [W28] examined the influence of sex hormones on the induction of intestinal metaplasia by x rays in five-week-old Crj:CD(SD) rats of both sexes. At the age of four weeks, the animals were gonadectomized and given testosterone or DES in the form of subcutaneous implants containing 0.25–2.5 mg hormone. One week later, they were irradiated with x rays to give two doses of 10 Gy to the gastric region at a three-day interval, for a total of 20 Gy. Six months after radiation exposure, the incidence of intestinal metaplasia with alkaline phosphatase (ALP) positive foci in males was significantly higher than in females, in orchidectomized males, or orchidectomized plus DES-treated rats. The incidence of intestinal metaplasia with ALP-positive foci in normal females appeared lower than in ovariectomized females and was increased by treatment with testosterone or decreased by DES. Numbers of foci of intestinal metaplasia with Paneth cells and total numbers appeared to increase in males treated with DES. These results suggest a promoting role for testosterone in the development of radiation-induced ALP positive lesions and also indicate considerable differences among intestinal metaplasia subtypes in their response to hormone stimulation.

163. Rat prostate tumours after androgen ablation by castration showed an increase, from 0.4% to 1.0%, of the apoptotic index as determined by the TUNEL assay. The apoptotic index did not vary significantly over time after castration. Irradiation of intact rats to 7 Gy resulted in an apoptotic response of 2.3%. When castration was initiated three days prior to irradiation, peak levels of 10.1% for the apoptotic response were recorded. Androgen restoration with testosterone implants restored the intact animal response [J10].

164. In conclusion, it can be said that many hormones are powerful regulators of cell proliferation and programmed cell death in specific tissues and organs. The resulting influence on radiation risk per unit dose is well proven (see also Table A.1). An important part of differences in risks linked to gender or age may be traced to hormones acting as endogenous growth factors.

2. Viruses, bacteria, and genetic sequences

165. Viruses, bacteria, and microbial genetic sequences have been shown to play an important role in the pathogenesis of animal tumours. Human malignancies such as Burkitt lymphoma and T-cell leukaemia are caused by the Epstein-Barr virus [L12] and the retrovirus HTLV-1 [D4], respectively, and a variety of carcinomas, including cervical, skin, anal, and others, by papilloma viruses [H13]. Hepatitis type B and C virus, the bacterium Heliobacter pylori, some parasites such as Opistorchis viverrini and Schistosoma haematobium are proven or

putative causes of hepatoma, gastric cancer, cholangio-carcinoma and urinary bladder cancer, respectively [M64]. One mechanism of interaction might be the inhibition of DNA repair by viral proteins. The HBV protein HBx was shown to interact with cellular DNA repair capacity in a p53-independent manner after ultraviolet C irradiation [G37]. Interaction of cancer viruses with radiation may also occur by mutation or translocation of dormant viral sequences. In a multi-stage process, virally infected organisms may also be much more susceptible to radiation-induced cancer if a virus is causing or facilitating one of the genetic transformations leading to the outbreak of malignancy. Astier-Gin et al. [A25] investigated the role of retroviruses in murine radioleukaemogenesis in C57B1 mice. The protocol associated the injection of a non-pathogenic retrovirus (T1223/B virus) and a dose from x rays (2 × 1.75 Gy), which alone was non-leukaemogenic in this system. Thymic lymphomas induced by the combined effect of virus and irradiation or irradiation alone were analysed for MuLV proviral organization and RNA expression with the Southern or Northern blotting techniques, respectively. The active involvement of the retrovirus was shown by the detection of a recombinant provirus in the chromosomal DNA of every tumour induced by the combined treatment with virus and radiation. No specific site in the genome was found for provirus integration and no relationship was observed between viral RNA expression and tumour induction. Trisomy 15 was observed in all metaphases irrespective of the protocol of tumour induction. The G-banding technique revealed an extra band in several thymic lymphomas induced by irradiation and T1223/B virus injection. This complex pattern of viral behaviour may pose great obstacles for diagnosis and for the elucidation of risk from combined exposures.

3. Miscellaneous factors

166. Many other sometimes poorly defined biological materials have also been shown to influence the response of organisms to ionizing radiation. For example, the modulating effect of microbial substances on survival after acute radiation doses in mammals (mice, rats, dogs, sheep, and monkeys) was studied by Andrushenko et al. [A12]. The highest protection was found for some vaccines containing inactivated bacteria and given before the radiation exposure. Polysaccharides, lipopolysaccharides, and protein-lipopolysaccharide complexes were also able to increase the radioresistance. The mechanisms involved in the modulation of the status and the number of stem cells of the immune system remains to be elucidated. Such effects might also be of importance at low exposure levels, e.g. for malignancies of the haematopoietic system.

167. To test the hypothesis that low-dose radiation, such as is used for diagnosis, may act as a co-carcinogen in inflammatory bowel disease, Weinerman et al. [W2] induced inflammation with DMH in a mouse system to study potential sensitization towards the radiation exposure (see Section B.1 for genotoxic action of DMH). Four

groups of BALB/c mice (a control, DMH, DMH plus low-dose radiation, and low-dose radiation) were studied. No protective or carcinogenic effects of the radiation in combination with DMH were found compared with DMH alone. This type of negative experimental finding is directly important for radiation protection of the patient, in that individuals with inflammatory bowel disease undergo many diagnostic x-ray examinations throughout life.

168. A strong effect was, however, found from the interaction of ionizing radiation with surgical procedures on the stomach. Griem et al. [G15] followed patients with peptic ulcer who had received radiotherapy to control excessive gastric acid secretions, a method used between 1937 and 1965 (mean dose to the stomach = 14.8 Gy). The mortality study involved 3,609 patients; 1,831 were treated with radiation and 1,778 were treated by other means. Compared with the general population, patients treated with or without radiation were at significantly increased risk of dying of cancer and non-malignant diseases of the digestive system. Radiotherapy was linked to significantly elevated relative risk for all cancers combined (RR = 1.53; 95% CI: 1.3–1.8). Radiotherapy and surgery together increased the rate of stomach cancer (RR = 10) above the sum of individual effects. There is no specific information on co-carcinogenic mechanisms in the post-surgical reaction of stomach tissue or on tumour location.

169. The influence of pre-immunization with a rectal extract on radiation-induced carcinoma of the rectum was studied by Terada et al. [T37] in 4–7-week-old A/HeJ mice. The animals received 40 Gy (20 Gy per week from x rays) in the pelvic region with or without two prior injections of rectal extract from adult animals of the same strain emulsified with complete Freud's adjuvant. After eight months, rectal adenocarcinomas were observed in significantly higher numbers in pre-immunized mice compared with non-immunized animals (62% vs. 18%). The results indicate that local immunological reactions sensitize to the carcinogenic action of ionizing radiation.

170. Finally, the effect of psychosocial factors such as fear, anguish, and chronic stress on the health status of individuals and populations, both in psychosomatic expressions and in the subjective perception of radiation-exposed persons, is clearly an important problem during and after accidents and cases of environmental contamination, such as seen in areas affected by the Chernobyl accident [I13]. However, a review of these aspects is beyond the scope of this Annex and involves professional disciplines outside the realm of UNSCEAR. Despite the attention given by the media to the potential deleterious effects of ionizing radiation in combination with conventional industrial pollutants in such instances, little scientific information is available on specific exposure situations. Some potentially important modifying factors are discussed in connection with dietary factors in Section B.6.

D. COMBINED MODALITIES IN RADIATION THERAPY

171. A large number of chemotherapeutic drugs are used in clinical cancer therapy in combination with radiation. The main ones in use or about to be used are described in this Section, with emphasis on the mechanisms of interaction between the drugs and radiation to reveal possible mechanisms of interaction between chemical agents and radiation under environmental and normal occupational settings. The main findings relating to modes of action and combined effects are summarized in Table 6. However, it should be clearly noted here that the final goal of tumour-therapy-related studies is tumour control and therefore cell death (apoptosis, necrosis) or cell inactivation (loss of proliferative capacity, differentiation, senescence). These effects are mostly deterministic and often mechanistically different from the stochastic radiation effects that are of concern in radiation protection. Therefore, highly sigmoidal dose-effect relationships and considerable threshold doses are found for the contribution of many of these agents to the interaction with radiation. Several groups of agents are also covered in the preceding sections, e.g. alkylating agents under the heading "genotoxic chemicals".

1. Alkylating agents, nitrosoureas, and platinum coordination complexes

172. Alkylating agents were among the first compounds found to be useful in cancer chemotherapy, and because of their variety and relative tumouricidal selectivity, they remain important components of many modern chemotherapeutic regimens. Although the alkylating agents are a diverse series of chemical compounds, they all have the common property of displaying a positively charged, electrophilic alkyl group capable of attacking negatively charged, electron-rich nucleophilic sites on most biologic molecules, thereby adding alkyl groups to oxygen, nitrogen, phosphorus, or sulphur atoms. Their chemotherapeutic usefulness derives from their ability to form a variety of DNA adducts that sufficiently alter DNA structure or function, or both, so as to have a cytotoxic effect [L37]. Many of the pharmacologically useful agents undergo a complex activation process.

173. The most common site of DNA alkylation is the N-7 position of guanine. Alterations at this position are relatively silent in their effect on DNA function, because these adducts do not interfere with the base-pairing scheme. In contrast, adducts at the N-3 position of cytosine, the O-6 position of guanine, and the O-4 position of thymidine interfere with the Watson-Crick base-pairing scheme and are therefore likely to interfere with fidelity of replication and transcription, leading to mutagenicity and cytotoxicity. In addition to direct interference with replication and transcription, the formation of DNA adducts leads to a variety of structural lesions, including ring openings, base deletions, and strand scissions [B41, F20, H20]. Many of the DNA adducts and lesions are further acted on by repair enzymes that can restore the integrity of the DNA, or if the repair process is only partially

completed, they can cause additional DNA damage, such as the creation of apurinic or apyrimidinic sites or DNA strand breaks. Bifunctional alkylating agents with the capacity to generate two electrophilic groups and to form two adducts are capable of forming DNA-interstrand and DNA-protein cross-links that interfere directly with DNA replication, repair, and transcription [L38].

174. Alkylating agents are cell-cycle-dependent but not cell-cycle-specific. They exert their cytotoxic effects on cells throughout the cell cycle but have quantitatively greater activity against rapidly proliferating cells, possibly because these cells have less time to repair damage before entering the vulnerable S phase of the cell cycle [T19]. Cells in which cross-links occur accumulate and die in the G_2 phase of the cell cycle. Persistent DNA strand breaks may result in lethal chromosomal damage in the mitotic phase of the cell cycle.

(a) Nitrogen mustards (mechlorethamine, melphalan, chlorambucil, cyclophosphamide)

175. Nitrogen mustard, originally studied for its potential as a vesicant in chemical warfare, is a highly reactive analogue of sulphur mustard and was the first alkylating agent introduced into clinical therapy [G22]. Exposure to these alkylating agents results in the formation of simple DNA adducts, DNA-interstrand cross-links, and DNA-protein cross-links [E12].

176. Experimental investigations of interactions between derivatives of nitrogen mustard and radiation *in vitro* showed that these interactions are additive, independent of sequence of treatment with the two agents, and not markedly influenced by the interval between treatments [D14, H21]. An isobolic analysis confirms the additivity, although when radiation precedes the mustard by 4 hours, the effect is on the borderline of supra-additivity, indicating that the two agents may share a common mechanism of cell killing [D14]. Neither radiosensitization nor interference with sublethal damage repair has been implicated in these interactions. Hetzel et al. [H21], examining the effects of combined treatment on V79 cell spheroids, put forth the interesting proposal that the enhancement seen with the nitrogen mustard derivative chlorambucil in combination with irradiation may be related to its ability to alter the internal oxygen profile in spheroids, resulting in partial reoxygenation.

177. The main use of nitrogen mustards was in the treatment of lymphomas, breast and ovarian cancer, and cancers of the central nervous system. A prospective randomized study examined whether MOPP (nitrogen mustard, vincristine, procarbacine prednisone) therapy alone is superior to combined modality treatment of extended field radiation and MOPP in patients with Hodgkin's disease [O13]. No significant differences were noted between the combined modality therapy and therapy with MOPP alone. However, overall toxicity was different. Viral and fungal infections occurred more frequently in the combined modality. In an overview by Cuzick et al. [C44]

of post-operative radiation therapy of breast cancer, no difference was seen in mortality over the first 10 years between patients treated with radiation therapy. After 10 years, however, there was a lower survival associated with radiation therapy. In recent years, chemotherapy has been favoured for breast cancer treatment. However, the use of post-operative radiation therapy needs to be reconsidered in patients who receive adjuvant chemotherapy and in whom drug resistance develops, leading to failure of chemotherapy. By decreasing the local tumour burden, adjuvant radiation therapy may decrease the probability of drug resistance and increase the probability of cure in those patients [H42].

(b) Nitrosoureas

178. The chloroethylnitrosoureas, including 1,3-bis(2-chloroethyl)-1-nitrosourea (BCNU), 1-(2-chloroethyl)-3-cyclohexyl-1-nitrosourea (CCNU), and methyl-CCNU (Me-CCNU), are highly lipophilic and chemically reactive compounds that are clinically active against a variety of tumours (reviewed in [L39]). Chemical decomposition of these agents in aqueous solution yields two reactive intermediates, a chloroethyldiazohydroxide and an isocyanate group [C30, M40]. The latter react with amine groups in a carbamoylation reaction. The isocyanates are believed to deplete glutathione, inhibit DNA repair, and alter maturation of RNA. The chloroethyldiazohydroxide undergoes further decomposition to yield reactive chloroethyl carbonium ions that form a variety of adducts with all four DNA bases and the phosphate groups of DNA. Of major importance in the antitumour effects of nitrosoureas is the formation of DNA interstrand cross-links, as demonstrated by the close correlation between cross-link formation and cytotoxicity [E13, K30, L40]. Alkylation seems to be the more important feature of direct nitrosourea action.

179. Additive or greater-than-additive responses have been recorded in *in vitro* and animal studies, with the greatest enhancement associated with the presence of the drug in some experiments before irradiation and in others after irradiation. Deen and Williams [D15] provided an isobolic analysis of the effects of combined BCNU-radiation treatment of 9L rat brain tumour that suggested some concentration dependence of these interactions. At two levels of BCNU (1 and 7.5 mg ml^{-1}) all data points fell within the additivity envelope, indicating similar mechanisms of action for the drug and radiation, but at other levels (3 and 5 mg ml^{-1}) supra-additivity was noted, suggesting that alternative mechanisms might be operating. CCNU resulted in less interaction than did BCNU [D16], a finding confirmed by the study of Kann et al. [K31] on L1210 cells. In experiments comparing the radiosensitizing effects of four nitrosoureas, the compound without alkylating activity, 1,3-bicyclohexyl-1-nitrosourea (BCyNU), was the most effective sensitizer [K31]. BCyNU was reported to selectively inhibit glutathione reductase activity [M41]. Kann et al. [K31] concluded that because the agent without alkylating activity was the most potent radiation synergist, alkylation was not involved in the enhancing effect, which

may relate instead to differential repair-inhibiting activity. However, even if the enhancement of the cytotoxic effects of ionizing radiation could be ascribed to repair inhibition, details of the mechanisms of cell killing are still not clear, because inhibition of DNA repair by the nitrosoureas was not complete and permanent [K32]. Rather, their effect was to slow the rate of strand rejoining, prolonging the period when numerous unjoined breaks are present, and lethality was considered a consequence of this prolongation.

180. Controlled clinical trials have demonstrated the efficacy of nitrosoureas combined with irradiation as adjuvant therapy for glioblastoma and anaplastic astrocytoma (reviewed in [L39]).

(c) Platinum coordination complexes

181. Cisplatin and its analogues are an important group of agents now in use for cancer therapy [R15]. Cisplatin (cis-diaminodichloroplatinum (II)) can bind to all DNA bases, but in intact DNA, there appears to be preferential binding to the N7 positions of guanine and adenine [B42, M42, P12]. Cisplatin binds to RNA more extensively than to DNA, and to DNA more than to protein [P13]. In the reaction of cisplatin with DNA or other macromolecules, the two chloride ligands can react with two different sites to produce cross-links [E14, E15, F21, F22]. Studies of the effects of platinum DNA binding on the three-dimensional structure of the DNA double helix revealed that the platinum lesions cause bending of the DNA double helix, suggesting that the stereochemistry of the platinum molecule is maintained and that DNA is modified in its three-dimensional conformation [R14]. The cytotoxicity of cisplatin against cells in culture has been found to be related directly to total platinum binding to DNA and to interstrand and intrastrand cross-links. Intrastrand guanine-guanine cross-linkage inhibits DNA replication [G23, P14]. Diaminocyclobutane-dicarboxylatoplatinum (II), carboplatin, and other cisplatin analogues appear to have subcellular mechanisms of action similar to cisplatin. They form lesions with DNA that are recognized by antibodies reacting with cisplatin-DNA lesions [P15].

182. More than two decades ago, Zak and Drobnik [Z6] reported an apparent interaction between cisplatin and ionizing radiation after whole-body irradiation of mice. Since then, cisplatin has been reported to enhance the cytotoxicity of radiation in a number of studies in both cell culture and tumour-bearing animals (reviewed in [B26, B43, C31, D17, D18, D19, H22, H23]). Isobolic analysis provides some evidence that this interaction can be supra-additive [D20]. A pronounced inhibition of repair of both radiation-induced potentially lethal damage and sublethal damage by platinum drugs has been demonstrated in several cell lines [B44, C32, D18, D21, O3, Y8]. The survival curves of cells exposed to platinum compounds have either a reduction or no shoulder, and this effect is interpreted as evidence for the inhibition of sublethal damage repair by platinum because of the role of sublethal repair in the formation of the shoulder of the radiation

survival curve. The enhanced killing of irradiated cells by platinum compounds may be due to an enhanced production of DNA double-strand breaks. Repair of DNA-platinum adducts results in a gap that, in association with radiation-induced DNA single-strand breaks (rejoining of which is retarded by platinum compounds), produces new DNA double-strand breaks [Y9, Y10].

183. DNA-protein cross-links and the binding of high-mobility-group proteins to DNA-platinum lesions seem to play a role in the radiosensitizing mechanism of cisplatin at moderate doses in hypoxic cells [K33, S50, S51, W16]. However, comparable in vivo experiments with RIF-1 tumours in mice failed to show the preferential radio-sensitization of hypoxic cells at low radiation doses by cisplatin [S52]. Herman et al. [H24] provided evidence that intracellular pH is an important variable in the action of cisplatin as a radiosensitizer of hypoxic cells using murine fibrosarcoma cells in vitro. Radiosensitization of cancer cells in vitro and as spheroids was observed when platinum drugs were delivered before and during irradiation. In addition, enhanced cell killing was demonstrated when these drugs were added immediately after irradiation (reviewed in [B26, H22, H23, S53]).

184. A number of animal in vivo studies have reported sequence-dependent positive interactions between the two modalities. Increased lifespan was reported when cisplatin was administered before whole abdominal irradiation of Krebs II ascitic carcinoma-bearing mice compared with cisplatin after irradiation [J8]. Supra-additivity was reported when cisplatin was given before x rays in SCCVII [T20, Y11] and RIF-1 [L41] or simultaneously in SCCVII and RIF-1 carcinoma-bearing mice [K34].

185. The platinum coordination complexes are the most important group of agents now in use for cancer treatment. They are curative in combination therapy for testicular cancer and ovarian cancer and play a central role in the treatment of lung [A29, K52, S54, S84, T21], head and neck [A27, B45, C33, C34, H25, O4, S55], brain [S56, S57], and bladder cancers [C35].

2. Antimetabolites

(a) Antifolates

186. Despite the clinical importance of antifolates in cancer therapy, there are only a limited number of reports of experimental data relating to interactions of methotrexate and radiation in vitro. Early studies of Berry [B47, B48] suggested that methotrexate might be useful as a radiosensitizer, with the greatest enhancement occurring with a cytotoxic drug concentration or in hypoxic cells. Enhancement was influenced by the proliferation status of the cells, and although stationary-phase cells showed an enhanced response to radiation, this was accompanied by a decreased response to methotrexate, which cancelled any gain [B49]. The synergistic effects between methotrexate and radiation can be explained by impaired DNA repair

owing to depleted intracellular pyrimidine and purine pools [A19]. Methotrexate cytotoxicity may also result from drug-induced single- and double-strand breakage of DNA [B46]. These breaks appear come from the methotrexate-induced depletion of intracellular nucleotide pools, with impairment of the ability to repair DNA damage. Synergistic effects were observed only when drug and radiation were given at the same time. Radioprotective effects of methotrexate were observed when it was administered hours before radiation treatment (see paragraph 119).

187. Effects of intracerebral injections of methotrexate, whole-brain radiation, or a combination of both were analysed on intracerebrally implanted RT-9 gliosarcoma in male CD-Fisher rats. Methotrexate alone and radiation alone each prolonged survival moderately. Combined methotrexate and radiation caused a significant prolongation of survival in all animals [W17].

188. In acute lymphoblastic leukaemia, current treatment is divided into four phases: remission induction by chemotherapy; central nervous system preventive therapy by radiation or combined modality treatment (radiation plus methotrexate); consolidation; and maintenance with chemotherapy. However major adverse effects of central nervous system preventive therapy have been documented, including CT-detected brain abnormalities, impaired intellectual and psychomotor function, and neuroendocrine dysfunction. These adverse effects have been attributed mainly to radiation therapy [R16, S58]. Several approaches have been tested to decrease adverse effects, including reduction of cranial irradiation from 24 to 18 Gy in regimens using cranial radiation plus intrathecal chemotherapy with methotrexate or the use of triple intrathecal chemotherapy with methotrexate alone or with methotrexate, cytarabine, and hydrocortisone (reviewed in [P28]).

(b) Pyrimidine analogues and precursors

189. Deoxyuridine analogues that increase radiosensitivity include 5-bromodeoxyuridine (BrUdR), 5-fluoro-2'-deoxyuridine (FUdR), and 5-iododeoxyuridine (IUdR). In these compounds, a halogen atom replaces the hydrogen at the 5 position on the pyrimidine ring of deoxyuridine. Because the van der Waals radii of bromine (1.95 Å) and iodine (2.15 Å) resemble closely the methyl group of deoxythymidine (2.00 Å), BrUdR and IUdR are more accurately referred to as thymidine analogues. FUdR is considered a uridine analogue because the van der Waals radius of the fluorine atom (1.35 Å) most closely resembles hydrogen (1.20 Å) [S59]. The biological effects of FUdR are significantly different from those of BrUdR/IUdR and will be discussed separately.

190. 5-Fluorouracil (5-FU) and FUdR are the fluoropyrimidines of greatest clinical interest. The fluoropyrimidines require intracellular activation to exert their cytotoxic effects. They are converted by multiple alternative biochemical pathways to one of several active cytotoxic forms. Incorporation of 5-FU into DNA inhibits

DNA replication and alters DNA stability by producing DNA single-strand breaks and DNA fragmentation [C36]. Fluoropyrimidines may also induce DNA strand breaks without being directly incorporated into DNA, possibly through the inhibition of DNA repair as a result of dTTP (deoxy-thymidine-triphosphate) depletion [Y12].

191. The synthesis and antitumour activity of 5-FU was initially described by Heidelberger et al. in 1957 [H26]. Complete tumour regression was observed in mice bearing sarcoma tumours after 5-FU and radiation, not observed after each treatment alone [H27]. In mice with a transplanted leukaemic cell line, 5-FU and radiation interacted synergistically when the drug was given before and after radiation, with the effects being most noticeable in the latter situation [V8]. Squamous-cell carcinoma responses in mice from combined exposures were dependent on total drug doses; however, the response was independent of the schedule of drug administration and consistent only with an additive effect [W18].

192. *In vitro* studies of Nakajima et al. [N15] with mouse L cells showed an enhanced effect of combined treatment of 5-FU and radiation on cell survival. The results suggest that maximum enhancement occurred when drug-treated cells were irradiated in the S phase and also confirmed the importance of post-irradiation drug treatment. Enhancement was dependent on drug concentration, increasing with increased dosage, and on treatment duration. The prolonged temporal requirement and the cytotoxic dose of 5-FU for the induction of sensitization following x-ray exposure implicates incorporation of 5-FU into RNA as an important mechanism involved in the combined effect.

193. A series of *in vitro* combined treatments using ionizing radiation and 5-FU on the human adenocarcinoma cell lines HeLa and HT-29 were performed by Byfield et al. [B50]. Based on these experiments they concluded that (a) sensitization occurred only with post-irradiation drug treatment, with prior exposure to 5-FU being strictly additive; (b) enhanced cell killing could not be explained by drug-induced additional acute damage or inhibition of sublethal damage repair; (c) the effect is maximized if cells are exposed to 5-FU for prolonged periods following irradiation; and (d) the concentration of 5-FU required for these effects is associated with dose-limiting toxicity in clinical studies.

194. Attempts to define more clearly the mechanism of interaction of 5-FU have used the derivative FUdR, which may limit the complex effects of 5-FU. Radiosensitization by FUdR in human colon cancer cells (HT-29) was critically dependent on the timing of exposure, being most marked when irradiation occurred 8–12 hours after exposure to a clinically achievable drug concentration, with no effect resulting when the cells were irradiated first [B51]. FUdR impaired sublethal damage repair in a dose-dependent manner but had no effect on the induction of double-strand breaks [H28]. Sensitization correlated with thymidylate synthase inhibition [B51] and depletion of

dTTP pools [H28] and was blocked by co-incubation with thymidine [B51]. These findings strongly suggest that FUdR acts by inhibiting thymidylate synthase.

195. In more recent studies by Miller and Kinsella [M43], a 2-hour exposure to low doses of FUdR resulted in extended thymidylate synthase inhibition after the drug was removed (up to 30 hours after treatment). Although the enzyme was nearly completely inhibited (>90%), an increase in radiosensitivity of cells was not evident until 16 hours after removal of the drug. Therefore, no direct correlation between thymidylate synthase inhibition and radiosensitization was observed. Parallel analysis of cell-cycle kinetics showed that cells accumulated during the early S phase after drug exposure and the rise and fall of radiosensitivity of the entire cell population over time followed the change of proportion of cells in early S phase [M44], a relatively radiosensitive phase of the cell cycle [T22]. These data suggest that radiosensitization by FUdR is in part caused by alterations in cell kinetics and a redistribution of cells through the cell cycle.

196. Concomitant radiotherapy with 5-FU has been evaluated in patients with cancers of the oesophagus, rectum, anus, bladder, and advanced laryngeal tumours (reviewed in [K35, M45, O5]). A recent consensus conference at the National Institute of Health reviewed the data from clinical trials and has recommended combined post-operative 5-FU and radiotherapy as the most effective management of patients with stage II or III surgically resected rectal cancer [N16]. In general, 5-FU and radiation in combined modality treatment is superior to radiation alone in the treatment of intestinal tumours.

197. Differential sensitization of tumours with bromated or iodinated pyrimidines, including BrUdR and IUdR, has been observed. These analogues influence only proliferating cells and may therefore preferentially sensitize rapidly growing tumours surrounded by more slowly proliferating normal tissue. BrUdR and IUdR are readily incorporated into the DNA of mammalian cells. The incorporation follows the thymidine salvage pathway. The extent of thymidine replacement in DNA, however, is not simply a function of competition within the salvage pathway, because the preferred pathway for thymidine incorporation is through the *de novo* synthesis of pyrimidine nucleotides.

198. Steric hindrance resulting from analogue incorporation into DNA appears minimal. In contrast, the physicochemical properties of altered DNA are influenced by thymidine replacement. Incorporation of BrUdR increases the forces that bind the strands of DNA together [P16]. This may alter DNA transcription and replication. The affinity of chromosomal proteins for BrUdR- and IUdR-substituted DNA is increased. This increased affinity has been associated with the repression or induction of cellular proteins, receptors, and growth factors (reviewed in [M45]). BrUdR and IUdR cause a dose-dependent delay of cells in the S and G_2 phases of the cell cycle, as demonstrated in human ileal and spleen cells *in vitro* [P17].

199. The physicochemical properties of IUdR- and BrUdR-containing DNA have been implicated in its increased sensitivity to radiation. The large, highly electronegative halogen atoms greatly increase the cross-sectional area available for trapping radiation-produced electrons. In addition, migration of absorbed energy to a halogenated base has been demonstrated [F23, L43]. Highly reactive uracilyl radicals may result from these reactions.

200. Erikson and Szybalski [E16, E17] reported radiosensitization of human cells exposed to BrUdR and IUdR. These studies revealed that the incorporation of halogenated pyrimidine radiosensitizes the cell through a direct effect on DNA [S59]. It was demonstrated that BrUdR resulted in greater thymidine replacement than did IUdR. However, IUdR was a more effective sensitizer to x rays, even at lower levels of incorporation [E18, M46, M47]. The distribution between the two DNA strands was not a critical factor in radiosensitization. Sensitization was also shown to be independent of the presence of oxygen [H29].

201. Recent analysis of radiosensitization by IUdR and BrUdR in two exponentially growing human colon cancer cell lines (HCT116 and HT29) using the linear-quadratic model revealed that an increase in the initial slope of the cell survival curve is the predominant mode of radiosensitization [M46, M47]. This suggests that the radiosensitizing effect may be the result of an increase in the amount of initial DNA damage. However, other recent *in vitro* studies with plateau-phase cells (CHO cells) suggest that IUdR and BrUdR are, in fact, potentially lethal damage repair inhibitors [F27, W19]. These different proposed mechanisms of radiosensitization of BrUdR and IUdR in exponentially growing and plateau-phase cells are not inconsistent and may reflect a bimodal mechanism.

202. Significant systemic toxicity was noted in animals [B52, G24] and humans, suggesting minimal tumour selectivity for these analogues. For clinical investigations in humans, therefore, tumours were selected that were surrounded by practically non-proliferating normal tissue (brain, bone, and muscle), thereby limiting the incorporation of BrUdR and IUdR into normal cells within the irradiated volume (reviewed in [M44]).

203. Gemcitabine (2',2'-difluorodeoxycytidine) is a new antimetabolite. It is a pyrimidine analogue and appears to prevent the addition of other nucleotides by DNA polymerase (masked chain termination) and to impair DNA repair. Gemcitabine has been shown to be a potent radiosensitizer in a variety of tumour cell lines, including HT-29 colorectal carcinoma, pancreatic cancer, breast, non-small-cell lung and head and neck cancer cell lines. It was most effective when administered prior to radiation. For most cell lines, sensitization was evident at non-cytotoxic concentrations of gemcitabine. For most cell lines, the primary radiosensitizing effect seems to be associated with depletion of endogeneous nucleotide pools [L54, S83, S85]. Radiosensitization by gemcitabine was observed in mice bearing tumours *in vivo* [M70]. In

clinical trials gemcitabine seems to be a powerful radiation enhancer in the treatment of non-small-cell lung cancer [H51, V13].

(c) Hydroxyurea

204. Hydroxyurea, a relatively simple compound, is a representative of a group of compounds that have as their primary site of action the enzyme ribonucleoside diphosphate reductase. This enzyme, which catalyses the reactive conversion of ribonucleotides to deoxyribonucleotides, is a crucial and rate-limiting step in the biosynthesis of DNA. The mechanism of cytotoxicity from hydroxyurea is related to direct inhibition of DNA synthesis and repair. Hydroxyurea causes cells to arrest at the G_1/S phase transition [S61]. It selectively kills cells synthesizing DNA at concentrations that have no effect on cells in other stages of the cell cycle [S62].

205. Additive or greater-than-additive responses have generally been reported for combined treatment with hydroxyurea and ionizing radiation in vitro. Phillips and Tolmach [P7], using synchronized HeLa cells, reported that enhancement occurred only when the drug was present post-irradiation. They demonstrated that hydroxyurea inhibits potentially lethal damage repair. In synchronized V79 cells, hydroxyurea treatment was necessary before and after irradiation to be effective as a radiosensitizer [S61]. Sensitizing by hydroxyurea resulted from its inhibitory action at the G_1/S-phase transition or its lethal action during the S phase. Kimler and Leeper [K36] showed that the enhancement of radiation-induced lethality observed when hydroxyurea was present after irradiation was specific for G_1 and S phase cells, but that the drug did not interfere with recovery from radiation-induced division delay in the G_2 phase. Non-cytotoxic doses of hydroxyurea significantly increased the early S-phase population in a human bladder cancer cell line (647V) [K37]. Exposure to these non-toxic concentrations of hydroxyurea before irradiation resulted in radiosensitization. In the human cervix carcinoma cell line Caski, the radiosensitizing effect of hydroxyurea was mainly due to a significantly longer G_2 block, indicating effects on DNA repair [K2].

206. Hydroxyurea has been shown to inhibit the repair of radiation-induced single-strand breaks in HeLa cells. The time course for repair of radiation-induced, single-strand DNA breaks is partially inhibited by exposure to hydroxyurea before and after irradiation [F24]. The effects of hydroxyurea on DNA repair after UV irradiation have also been studied, and depletion of the triphosphate pools (except dTTP) appears to be responsible for the observed alterations in DNA repair and enhanced cytotoxicity [C38].

207. Piver et al. [P18] showed a significant reduction in the radiation dose needed to control mammary tumours in mice when hydroxyurea was given with fractionated radiation exposure. However, a similar study on implanted squamous cells of cervix carcinoma in nude mice showed no radiosensitizing effect [X3].

208. Clinical trials of hydroxyurea radiosensitization have involved patients with head and neck malignancies [R17, S63] and primary brain tumours [L27]. The most convincing trials, suggesting radiosensitization and improved local control with hydroxyurea, involved patients with uterine cervical carcinoma [P19, P20, S64]. Although these studies suggested improved results, none of the trials considered cell cycle times of the tumour and normal tissue or hydroxyurea concentrations in the relevant tissues [S65].

(d) Other antimetabolites

209. Other antimetabolites of clinical relevance are arabinose nucleosides, including cytarabine (cytosine arabinoside, ara-C); ara-C, a cytidine analogue, has important clinical activity against acute myelocytic leukaemia. It is an inhibitor of DNA synthesis and kills cells selectively during the S phase of the cell cycle. Synergism between cytarabine and a number of antitumour agents, including alkylating agents, platinum coordination complexes, and topoisomerase II inhibitors, has been observed in vitro and in animal models. ara-C enhances the activity of these compounds by inhibiting the repair of strand breaks associated with these agents.

210. Another class of antimetabolites is the purine analogues, including 6-mercaptopurine and 6-thioguanine, which act as guanine analogues, and adenosine analogues, including arabinofuranosyladenine (9-ß-arabinofuranosyladenine, ara-A). All these compounds have antileukaemic activity and are used in combination chemotherapy. Their activity is directed against DNA replication and repair.

211. The potential of the thymidylate synthase inhibitor tomudex to interact with ionizing radiation was assessed by Teicher et al. [T38]. Tomudex (1 μM) decreased the shoulder of the radiation survival curve in both oxygenated and hypoxic HT-29 cells (human colon carcinoma) and SCC-25 cells (squamous-cell carcinoma of the head and neck), respectively. The effect was more significant in oxygenated cells. In tumour-bearing animals, tomudex in combination with radiation showed an additive to supra-additive effect on tumour control. The interaction effect was dependent on the fractionation schedule of drug and radiation. In each assay, the results obtained with tomudex were equal to or exceeded the results of comparable experiments with 5-fluorouracil.

3. Antitumour antibiotics

212. Anthracyclines such as doxorubicin, daunomycin, epirubicin, and idarubicin cause a range of biochemical effects in tumour cells. The antitumour activity and toxicity are the result of free-radical formation and/or triggering of topoisomerase-II-dependent DNA fragmentation. The enzyme is prevented from finishing its cycle with the religation of the broken strands. In addition, the alteration of the DNA helical structure that occurs on DNA intercalation by anthracyclines may trigger enhanced topoisomerase II activity. The net result is that addition of

anthracyclines to tumour cells dramatically increases protein-associated breaks. There is strong evidence that the topoisomerase II mechanism is the means by which doxorubicin and other anthracyclines kill leukaemia and lymphoma cells.

213. As a second mechanism, anthracyclines are able to form oxygen radicals. Evidence suggests a role for anthra-cycline-induced radical formation by virtue of its killing of ovary, breast, and colon tumour cells. Much of this evidence depends on the key roles that gluthathione and gluthathione peroxidase play in detoxifying hydrogen pero-xide and organic peroxides. Doxorubicin is an inhibitor of mitochondrial and cell respiration and reduces oxygen consumption by cells in the outer layers of the tumour. This may lead to improved oxygenation and radiosensi-tivity of hypoxic areas of the tumour [D23]. On the other hand, aclarubicin, an anthracycline differing in its sugar moiety from doxorubicin, was shown to exert its enhance-ment effect on x-ray-induced cell killing in HeLa cells only when given after radiation exposure (5 μg ml^{-1} for one hour) [M63]. The authors hypothesized that this potentia-tion, which is visible through 10 cell divisions, is due to the interaction between radiation and drug damage, a mechanism probably relevant only for very high acute exposures.

214. Bleomycin is another important antitumour anti-biotic. Its action has been associated primarily with its ability to produce single- and double-strand breaks in DNA. The sequence of events leading to DNA breakage begins with the metabolic activation of bleomycin. The activated agent binds to DNA as the result of intercalation. Highly toxic oxygen intermediates, such as the superoxide or hydroxyl radicals, are then formed that attack DNA. There is indirect evidence that the same processes required to repair ionizing radiation damage also are used in bleomycin repair [C39]. The lesions caused by bleomycin include chromosome breaks and deletions, very similar to the action of ionizing radiation. Bleomycin is therefore called a radiomimetic drug. There is, however, some base sequence specificity for the site of DNA cleavage. Bleo-mycin binds preferentially to the DNA strand opposing the sequences GpT and GpC to attack and cleave the strand at the 3' side of G [P21, S66]. A primary point of attack in non-mitotic cells is considered to be the link regions of DNA between nucleosomes [K38].

215. The effects of the interaction between bleomycin and ionizing radiation on cell survival have been reported to range from additive to greater than additive [B26, H30, T28, T29]. These effects are schedule-dependent, with maximum interaction occurring when there is only a short time interval between administration of the two agents or when they are administered simultaneously, possibly reducing the extent of repair of any induced damage or similar lesions by these two agents. Although both ionizing radiation and bleomycin induce G$_2$ arrest, their damage is independent and purely additive [K39]. Thus, in contrast to the sometimes greater-than-additive effects observed for cell lethality, bleomycin and radiation do not interact in the induction of cell-cycle blocks.

216. Bleomycin and radiation have been combined frequently in the treatment of head and neck cancer. There are several randomized clinical trials (reviewed in [S67]), some of which showed a benefit in response rate and/or survival; others, however, including the largest trials, did not reveal any benefit from the use of bleomycin and radiotherapy.

217. Mitomycin C is a bioreductive alkylating agent that is inactive in its original form but is activated to an alkylating species by reduction of the quinone and sub-sequent loss of the methoxy group. Recent studies indicate that bifunctional alkylation by mitomycin C occurs preferentially in a reducing environment [T23]. In an aerobic environment, the reduction of mitomycin C initiates a chain of electron transfers that leads ultimately to the formation of toxic hydroxyl and superoxide radicals [B54].

218. Mitomycin C significantly reduced the radiation-resistant subpopulation of KHT carcinomas growing intra-muscularly in C3H/HeJ mice when administered 24 hours before radiation. Isobolic analysis indicated that this treat-ment combination led to supra-additive cell killing in the tumour [S68]. Combined treatment with mitomycin C and radiation of C3H mouse mammary carcinoma *in vivo* showed that the drug significantly enhanced the radiation-induced growth delay when administered before radiation [G25]. Isobolic analysis revealed that pre-irradiation treatment with mitomycin C resulted in a supra-additive response, whereas post-irradiation treatment resulted in only an additive response. The enhancement appeared to be related to both a direct radiosensitization and a pronounced cytotoxic effect of the drug against radioresistant hypoxic cells.

219. A randomized trial using mitomycin C with radio-therapy for head and neck cancer showed a disease-free survival benefit [W21]. However, a high incidence of pulmonary complications was reported. Mitomycin C is included in many multimodality therapy regimens for gastrointestinal tumours in combination with 5-FU and radiation. For cancers of the anal region, chemotherapy with 5-FU and mitomycin C plus irradiation have been widely accepted as the conventional treatment for most patients, and surgery may not be required in many cases (reviewed by [S14]).

220. Actinomycin D (dactinomycin) binds to DNA by intercalation. The intercalation depends on a specific interaction between the polypeptide chains of the antibiotic and deoxyguanosine and blocks the ability of DNA to act as a template for RNA and DNA synthesis [R17]. The predominant effect is selective inhibition of DNA-dependent RNA synthesis. In addition to these effects, actinomycin D causes single-strand breaks in a manner similar to doxorubicin.

221. Experimental investigations of interactions between actinomycin D and radiation were reviewed by Hill and Bellamy [B26, H30]. The overall conclusion of this review was that irrespective of the sequence employed, the two agents are at least additive and that a shorter rather than a longer interval between the two agents is the most beneficial. Because actinomycin D generally leads to a decreased D_q and a decrease in split-dose recovery, it inhibits sublethal damage repair but does not effect potentially lethal damage repair in exponentially growing cells. This increased radiation damage expression, resulting from residual non-repaired single- and double-strand breaks in DNA, has been proposed as a possible mechanism for interaction between actinomycin D and ionizing radiation.

222. Actinomycin D is effective in the treatment of Wilms' tumour, Ewing's sarcoma, embryonal rhabdomyosarcoma, and gestational choriocarcinoma. It enhances radiation effects in clinical therapy when both are given simultaneously. When given after radiation therapy, actinomycin D, like doxorubicin, can recall the irradiation volumes by erythema of the skin or by producing pulmonary reactions [D32]. It is not known whether this is due to interaction between the damage done by radiation and that by the drug or whether it represents only additivity of the effects. The recall effect can be observed even after a period of several months between radiation and drug treatments.

4. Microtubule poisons

223. Many antineoplastic agents currently in use are biosynthetic products and were initially isolated from plants [D24]. In this Section, the mode of action alone and in combination with radiation of the vinca alkaloids, epipodophyllotoxins, and taxanes is reviewed.

224. Vinca alkaloids are present naturally in minute quantities in the common periwinkle plant, *Catharanthus roseus*. Vincristine (VCR), vinblastine (VBL), desacetyl vinblastine (vindesine), and vinorelbine are in clinical use. Vinca alkaloids exert their cytotoxic effects by binding to a specific site on tubulin and preventing polymerization of tubulin dimers, disrupting the formation of microtubules [M55]. The binding occurs at sites that are distinct from binding sites of other antimicrotubule agents, such as colchicine, podophyllotoxin, and paclitaxel [B55].

225. The effect of single doses of VCR on mice spermatogonia was investigated by Hansen and Sorensen [H31], and the influence of these drugs on the radiation response of murine spermatogonial stem cells was examined. VCR significantly reduced the survival in the differentiated spermatogonia and to a lesser extent in the stem cells. VCR radiosensitized spermatogonial stem cells, with the effect being most prominent when it was administered after irradiation. Grau et al. [G26] evaluated the interaction between VCR and x rays in a murine C3H mammary carcinoma and its surrounding skin. VCR caused a temporary blockage of cells in the mitotic phase. The

tumour control studies, however, showed a lack of correlation between the VCR-induced accumulation of cells in the G_2/M cell-cycle phase and enhancement of tumour radiation response. Nevertheless, pre-irradiation VCR caused radiosensitization in both tumours and skin, whereas post-irradiation VCR mostly resulted in responses equal to radiation only.

226. The effect of combining VBL and ionizing radiation on tumour response was investigated in CDF1 mice bearing the MO4 mouse fibrosarcoma [V9]. Different treatment schedules for the combination of VBL and radiation all resulted in additive tumour responses. The maximum percentage of tumour cells that could be accumulated in mitosis by a single intravenous bolus of VBL was around 13%. The results show that this will probably be insufficient for significant radiation enhancement.

227. Paclitaxel (commercial name Taxol), another microtubule poison, was first isolated from the Pacific yew, *Taxus brevifolia*. Paclitaxel promotes microtubule assembly *in vitro* and stabilizes microtubules in mouse fibroblast cells exposed to the drug [S69, S70]. It binds preferentially to microtubules rather than to tubulin dimers [P22]. Although the binding site for paclitaxel on microtubules is distinct from the binding site for exchangeable guanosine triphosphate (GTP) and for colchicine, podophyllotoxin, and VBL, the specific binding site for paclitaxel on microtubules has not been identified. Unlike other antimicrotubule agents such as colchicine and the vinca alkaloids, which induce microtubule disassembly, paclitaxel shifts the equilibrium towards microtubule assembly and stabilizes microtubules. Distinct morphological effects suggest that paclitaxel adversely affects critical microtubule functions during interphase and mitosis. Paclitaxel belongs to the group of taxanes, microtubuli stabilizing agents containing a taxane ring. Microtubuli stabilizing agents without taxane ring are called taxoids.

228. Choy et al. [C40] evaluated the possible radiosensitizing effects of paclitaxel on the human leukaemic cell line (HL-60). When HL-60 cells were treated with paclitaxel, up to 70% of them were blocked in the G_2/M phase. Isobolic analysis of the data revealed that the combined effects of ionizing radiation and paclitaxel fell within the range between additivity and synergism. Reasoning that paclitaxel could function as a cell-cycle-selective radiosensitizer, Tishler et al. [T24, T25] examined the consequences of combined drug/radiation exposures on the radioresistant human grade 3 astrocytoma cell line, G18, under oxic conditions. Survival curve analysis showed a dramatic interaction between paclitaxel and ionizing radiation, with the degree of enhanced cell killing dependent on paclitaxel concentration and on the fraction of cells in the G_2 or M phases of the cell cycle.

229. Three human ovarian cancer cell lines were used to examine the radiosensitizing effects of paclitaxel: BG-1, SKOV-3, and OVCAR-3 [S71]. Paclitaxel was found to have a significant radiosensitizing effect on all cell lines. Proliferating cells were more sensitive to paclitaxel, radiation, and the combination than confluent cells.

Treatment of proliferating cells with paclitaxel 48 hours prior to irradiation had a greater radiosensitizing effect than treatment 24 hours prior to irradiation.

230. Liebmann et al. [L44] examined the radiosensitizing effects of paclitaxel in four cell lines: MCF-7, A549, OVG-1, and V79. All cell lines developed a G_2/M block after paclitaxel exposure. Paclitaxel acted as a radiosensitizer in human breast cancer cells (MCF-7), in human ovary adenocarcinoma cells (OVG-1), and in Chinese hamster lung fibroblast cells (V79). However, paclitaxel was unable to enhance the radiation sensitivity of human lung adenocarcinoma cells (A549). Paclitaxel increased the linear component of the radiation survival curves in all cell lines. The quadratic component was unaffected by paclitaxel in the rodent cell line. The cells that were sensitized to radiation by paclitaxel had a relatively small baseline linear component, while A549 cells had a large linear component. Asynchronous and synchronous cells from carcinomas of the human uterine cervix were irradiated alone and after paclitaxel treatment [G27, T39]. Irradiating paclitaxel-treated cells resulted in a strictly additive response, like the response in lung adenocarcinoma cells and in contrast to the earlier supra-additive results with astrocytoma cells, breast cancer cells, and ovarian cancer cells. Paclitaxel affected the cervical carcinoma cells at stages of the cell cycle other than G_2/M. This may explain the failure to observe paclitaxel radiosensitization with these cells, and it may indicate that paclitaxel has a multiplicity of actions, with differences in effectiveness likely between cells of different origins. Similar cell-line-specific results on the cell-cycle specificity of the combined paclitaxel radiation effects were reported for other tumour cell lines. In non-synchronized and synchronized human fibroblasts, however, the combined effect was additive to even subadditive [G36]. Subadditive effects on cell survival between radiation and paclitaxel were reported for the human laryngeal squamous-cell carcinoma cell line SCC-20 [I15].

231. Besides having cell-cycle effects, paclitaxel is able to induce apoptosis by a p53-independent mechanism. On a molecular level, paclitaxel effects primarily involve phosphorylation of the product of the bcl-2 gene downstream of p53 [M68].

232. Hei et al. [H32, H33] assessed the potential oncogenic effects of paclitaxel either alone or in combination with gamma irradiation in C3H10T½ cells. In contrast to human cells *in vitro*, the mitotic block induced by paclitaxel in 10T½ cells was only partial. While paclitaxel was ineffective in transformant induction, it enhanced the oncogenic transforming potential of gamma rays in a supra-additive manner.

233. *In vivo* experiments with animal tumours showed that enhanced tumour radiosensitivity after paclitaxel treatment was attributable to two distinct mechanisms. Paclitaxel was able to enhance the radioresponse of apoptosis-sensitive and -resistant tumours but not the normal tissue radioresponse, thus providing true therapeutic gain. Tumour reoxygenation and antiangiogenic properties occurring as a result of paclitaxel-induced apoptosis in apoptosis-sensitive tumours

and mitotic arrest after paclitaxel treatment in apoptosis-resistant tumours are two distinct radiosensitizing mechanisms of paclitaxel [M67]. In mice bearing spontaneous mammary carcinoma, paclitaxel and radiation interacted in a supra-additive manner in controlling tumour growth. However, no supra-additive response has been observed in normal tissue, indicating a favourable therapeutic gain [C48].

234. Antitumour activity of paclitaxel has been observed in advanced ovarian cancer and metastatic breast cancer. The initial activity reported in refractory ovarian cancer has now been confirmed in three subsequent studies with response rates ranging from 21% to 40% (reviewed in [Y13]). Significant activity (56%–62%) has also been observed in metastatic breast cancer [H34]. Docetaxel, a paclitaxel derivative, has been shown to be 100-fold more potent than paclitaxel in bcl-2 phosphorylation and apoptotic cell death [H50]. The radiosensitizing activity of docetaxel has been reported in clinical trials with head and neck cancer [S79] and non-small-cell lung cancer [G38, O20].

5. Topoisomerase poisons

235. Epipodophyllotoxins from extracts of the mandrake plant (*Podophyllum pelatum*) have been used for medical purposes for centuries as cathartics or as treatment for parasites or venereal warts. Podophyllotoxin, an antimitotic agent that binds to a site on tubulin distinct from that occupied by the vinca alkaloids or paclitaxel, was identified as the main constituent possessing cytostatic activity. These early tubulin-binding podophyllotoxins possessed a prohibitively high degree of clinical toxicity. For example, a considerable risk of pneumonitis was observed following irinotecan and radiotherapy for lung cancer [Y15]. However, two glycosidic derivatives of podophyllotoxin, etoposide (VP-16) and teniposide (VM-26), have very significant clinical activity against a wide variety of neoplasms. Their main target is DNA topoisomerase II.

236. Epipodophyllotoxins were found to arrest cells in the late S or early G_2 phase of the cell cycle rather than the G_2/M border that would have been expected of an antimicrotubule agent [K40]. It was noted that these agents had no effect on microtubule assembly at concentrations that were highly cytotoxic [L45]. It was subsequently found that these drugs produced DNA strand breaks in intact cells but that these effects were not seen when the epipodophyllotoxins were incubated *in vitro* with purified DNA, suggesting that direct chemical cleavage in DNA was not occurring [W20]. The epipodophyllotoxins exert their cytotoxic effects by interfering with the scission-reunion reaction of the enzyme DNA topoisomerase II [Y10]. The enzyme binds to DNA covalently and forms single-strand, protein-associated breaks. On a molar basis, teniposide is approximately 10 times more effective than etoposide at inducing DNA strand breaks [L46]. In addition, the epipodophyllotoxins inhibit the catalytic or "strand-passing" activity of topoisomerase II that permits the enzyme to catenate DNA circles and disentangle topologically constrained DNA.

237. Isobolic analysis of the combined modality treatment of etoposide and radiation on asynchronous growing V79 fibroblasts showed that considerable potentiation occurs upon concomitant radiation/drug exposure [G28]. Synergistic cell killing was observed as radiation was applied before or concomitantly with etoposide. Rapidly repairable radiation-induced DNA damage was fixed into lethal lesions by etoposide, giving rise to supra-additive interaction under concomitant radiation/drug exposure. The shoulder of the radiation survival curve was eliminated. A second interaction mechanism was that cells arrested in the G_2 phase of the cell cycle by irradiation were hypersensitive to the cytotoxic effects of the drug. Recently, Goswami et al. [G29] reported that the synthesis of topoisomerase II is suppressed as cells accumulate in G_2 following irradiation. Ng et al. [N17] investigated the ability of etoposide to potentiate the x-irradiation response and to inhibit the repair of potentially lethal damage and sublethal damage in confluent cultures of a radioresistant (Sk-Mel-3) and a radiosensitive (HT-144) human melanoma cell line. In both cell lines, etoposide inhibited sublethal damage repair; however, in contrast to camptothecin, a topoisomerase I inhibitor, it also inhibited potentially lethal damage repair in HT-144 cells but not in the radioresistant cell line Sk-Mel-3.

238. Non-cytotoxic concentrations of etoposide (1.7 mM) caused little or no effect in V79 cells when combined with radiation [S72]. Even at highly toxic doses of etoposide, human bladder carcinoma cells were not radiosensitized by the drug [M48]. Etoposide and teniposide have demonstrated highly significant clinical activity against a wide variety of neoplasms, including non-Hodgkin's lymphomas, germ-cell malignancies, leukaemias, and small-cell lung carcinoma [B56, O6, W22].

239. Camptothecin, a heterocyclic alkaloid, and its analogues are inhibitors of topoisomerase I and possess antitumour activity. Camptothecin was first isolated from the stem wood of *Camptotheca acuminata*, a tree native to northern China. Characterization of the molecular structure of camptothecin critical for antitumour activity has led to the development of the camptothecin analogue topotecan and others with greater solubility and improved therapeutic indices in preclinical models.

240. DNA topoisomerase I is the unique target for camptothecin [S73]. Topoisomerase I transiently breaks a single strand of DNA, thereby reducing torsional strain and unwinding DNA ahead of the replication fork. Human DNA topoisomerase I binds to its nucleic acid substrate non-covalently. The bound enzyme then creates a transient break in one strand and concomitantly binds covalently to the 3'-phosphoryl end of the broken DNA strand. Topoisomerase I then allows passage of the unbroken DNA strand through the break site and religates the cleaved DNA. Camptothecin blocks the topoisomerase I in the form that is covalently bound to DNA [C41]. Camptothecin-induced DNA strand breaks have been detected frequently at replication forks close to growth points. The cytotoxicity

of camptothecin, a highly S-phase-specific agent, may be explained by the collision of drug-stabilized topoisomerase I-DNA complexes with moving replication forks, leading to replication arrest and conversion of topoisomerase-I-bound transient DNA strand breaks into persistent breaks [H35]. A direct stereospecific interaction between camptothecin and DNA topoisomerase is essential for the radiosensitizing effect of the inhibitor [C47].

241. Exposure to camptothecin under conditions of low-dose-rate irradiation (1 Gy h^{-1}) induced the accumulation of cells in the S phase in V79 and HeLa cells. Isobolic analysis of survival data consistently showed supra-additivity of cell killing in both cell lines upon concomitant exposure to camptothecin and low-dose-rate irradiation. Cytokinetic cooperation appears to be the main determinant of cell survival in treatments associating camptothecin and radiation in growing cells. Non-cytotoxic concentrations of camptothecin produced a reproducible effect at x-ray doses of up to 2 Gy; however, like cells treated with etoposide at non-toxic concentrations, the radiation survival curves for drug-treated and untreated V79 cells were comparable at higher radiation doses [S72]. X-irradiation of camptothecin-treated SV40 transformed normal (MRC5CVI) and ataxia-telangiectasia (AT5BIVA) fibroblast cells resulted in additive prolongation of S-phase delay in MRC5CVI cultures and additive effects for cell killing in both cell lines [F11]. Hypersensitivity of AT5BIVA to camptothecin was not attributable to elevated levels of complex trapping.

242. HT-29 human colon adenocarcinoma cells growing in spheroids were more resistant to both SN-38, a metabolite of a derivative of camptothecin (irinotecan: CPT-11), and radiation than HT-29 monolayers. SN-38 at a subtoxic concentration (2.5 µg ml^{-1}) increased the lethal effects of radiation on spheroids in a supra-additive manner but only acted additively on monolayers. The mechanism of radiosensitization of SN-38 is due to the inhibition of potentially lethal damage repair in spheroids [O18]. In both small-cell lung cancer and small-cell/large-cell lung carcinoma xenografts, combination treatment with SN-38 and radiation resulted in a significant tumour regression compared with the use of SN-38 or radiation alone [T40].

243. Gamma-ray irradiation of AS-30D rat hepatoma cells followed by a 2-hour exposure to camptothecin *in vitro* was found to act additively at low radiation doses and synergistically at higher radiation doses, as shown by isobolic analysis [R18]. Treatment of established ascites tumours in rats with either camptothecin or ^{131}I-labelled monoclonal antibody RH1, which specifically localizes in hepatoma ascites, prolonged rat survival but was ineffective at curing animals of tumours. In contrast, combined therapy consisting of camptothecin followed by the injection of ^{131}I-labelled monoclonal antibody RH1 cured 86% of animals. These results suggest that topoisomerase I inhibitors may be useful for increasing the efficacy of radioimmunoconjugates for the treatment of cancer.

244. Subtoxic concentrations of topotecan potentiated radiation-induced killing of exponentially growing Chinese hamster ovary or P388 murine leukaemia cultured cells [M49]. Survival curve shoulders were reduced; the slopes of the exponential portions of the curves were slightly decreased. Potentiation of radiation-induced cell killing by topotecan was absolutely dependent on the presence of the topoisomerase I inhibitor during the first few minutes after irradiation. A dose-dependent reduction in cell survival was obtained with a 4-hour exposure of topotecan following irradiation of human carcinoma cells in culture and murine fibrosarcoma in mice [K22]. No enhancement of cell killing was seen when cells were treated with the drug before irradiation. *In vivo* tumour studies showed a significant radiosensitizing effect of topotecan that was dependent on both drug dose and time sequence (before irradiation). There was no enhanced skin reaction following the combined treatments [K22].

6. Bioreductive drugs

245. The oxygenation status of clonogenic cells in solid tumours is believed to be one of the main factors adversely affecting tumour response in radiotherapy. In totally hypoxic cells, the radiation dose must be raised by a factor as great as 3 to achieve the effects obtained in fully oxic cells. The presence of 2%–3% of such resistant cells may double the total radiation dose required for eradication of all tumour cells [G30, T26]. It appears that solid tumours can contain two distinct classes of hypoxic cells: chronically and transiently hypoxic cells [C42, T26]. However, in clinical radiotherapy, treatment is usually sufficiently protracted to allow a significant re-oxygenation.

246. Results of clinical studies on the use of hyperbaric oxygen in combination with radiotherapy to increase oxygenation of hypoxic tumour cells have been conflicting. Nine randomized trials have been reported, of which only three gave statistically significant positive results for the use of hyperbaric oxygen, particularly in tumours of the head and neck region and advanced carcinoma of the cervix [D25, D26, F13, W23]. A second approach towards increased delivery of oxygen to tumours involved the use of erythrocyte transfusions. Retrospective studies of cancer patients with anaemia showed some indications for a negative correlation between anaemia and the outcome of radiotherapy [B57, D27, D28]. Use of the perfluoro-chemical oxygen-carrying emulsion Fluosol-DA and 100% oxygen as an adjunct to radiotherapy is a third approach to increased oxygen delivery. Clinical trials in the treatment of head and neck cancer showed a benefit of this combined modality treatment [L42, R19].

247. Clinical trials with hypoxic cell radiosensitizers rely on a different approach [R2]. Drugs that replace oxygen in chemical reactions that lead to radiation-induced DNA damage are used as adjuncts to radiotherapy. These drugs sensitize hypoxic tumour cells to radiation but do not sensitize normal tissue, which is already maximally sensitized by oxygen. Hypoxia-directed drugs would have limited use as single agents, because they would not destroy the normally oxygenated tumour cells; however, they could be extremely valuable in combination with radiotherapy or drugs that selectively kill aerobic cells. Optimal use of hypoxia-directed drugs would therefore require the development of regimens in which concomitant therapies with agents attacking each cell population were combined effectively to eradicate all the different cell populations within the tumour. Drugs that are selectively toxic to hypoxic cells should be relatively non-toxic to healthy normal tissue, which is generally well perfused and well oxygenated.

248. Bioreductive drugs are activated by metabolic reduction in tumour cells to form highly effective cyto-toxins. Tumour selectivity exploits the presence of hypoxia in tumours, since oxygen can reverse the activating step, thereby greatly reducing drug activity in most normal tissues. Selectivity can also depend on the level of expression in tumour cells of the particular reductase for which the drug can act as a substrate. These include DT-diaphorase, various P450 isozymes, cytochrome P450 reductase, xanthine oxidase, and doubtless other enzymes as well.

(a) Quinone alkylating agents

249. Quinone alkylating agents, as well as various nitro compounds and the benzotriazine di-N-oxides, have the ability to undergo metabolic reduction in such a way as to selectively kill hypoxic cells. When quinones are reduced under normal aerobic conditions, the cell is placed in oxidative stress due to a process known as redox cycling [P23, T27]. Although oxidative stress due to cycling is considered important in the toxicity of quinones and other redox labile agents to normal oxic cells, this pathway is in fact less damaging than the highly toxic metabolites that predominate in hypoxic cells. This is partly because of the protective enzymes that detoxify superoxide, that is, superoxide dismutase and catalase. Another pathway that protects oxic cells from the toxic action of quinones is direct reduction by DT-diaphorase. Unlike other reduct-ases, DT-diaphorase catalyses a concerted two-electron reduction step, which is therefore not reversible by oxygen. Radiosensitizing effects of EO9, an analogue of mito-mycin C, and porfiromycin, another quinone alkylating agent, were reported in experimental animal tumour models [A26, R20].

(b) Nitroimidazoles

250. Nitroimidazoles are reduced intracellularly, but in the absence of adequate supplies of oxygen they undergo further reduction to more reactive products [E9]. The formation of these products is initiated by an enzyme-mediated single-electron reduction of the nitro group to a free radical that is an anion at neutral pH. The reduction pathway can proceed in successive steps past the nitro-radical anion (one electron addition), the nitroso (two electrons), and the hydroxylamine (four electrons) to

terminate at the relatively inactive amine derivative (six electrons). In aerobic conditions the predominant reaction is redox cycling through radical anion analogous to quinone bioreduction, and oxidative stress may result from this pathway. The precise molecular nature of the covalent reaction products that predominate under hypoxia have not been identified, but these products almost certainly derive from the nitroso- or hydroxylamine reduction level or their ring cleavage products such as glyoxal [W24].

251. The first compound tested in clinical trials was the 5-nitroimidazole, metronidazole [D29, U16, U17]. It was selected because of its known activity both *in vitro* [C37, F25] and in experimental murine tumours [B58, R21, S74]. Misonidazole was the first in a series of 2-nitro-imidazole compounds to be used in the clinic. Because the 2-nitroimidazole compounds are more electron-affinitive than metronidazole, they are more efficient as hypoxic cell sensitizers. Misonidazole was shown to be more efficient as a radiosensitizer in experimental tumour systems than metronidazole [F26]. Clinical experience with misoni-dazole as a radiosensitizer showed some benefit of the drug in some head and neck cancer and pharyngeal cancer studies [D12, D22, F14, O7]. However, clinical use is limited because it induces cumulative peripheral neuro-pathy.

252. Neurotoxicity is linked to the lipophilic properties of the compound [B59, B60, B61]. The less lipophilic misonidazole analogue SR 2508 (etanidazole), with radiosensitizing activity comparable to that of misoni-dazole, was subsequently used. Adding etanidazole to conventional radiotherapy was beneficial for patients who had squamous-cell carcinoma of the head and neck without regional lymph node metastasis [L34]. Nimorazole, a weakly basic 5-nitroimidazole with an electron affinity lower than that of the 2-nitroimidazoles, was evaluated in a randomized trial in patients with squamous-cell carcinoma of the larynx and pharynx [O8]. Results demonstrated a statistically significant improvement in locoregional control. Nimorazole was much less toxic than etanidazole, and the toxicity was reversible.

253. RSU 1069 is the leading compound of dual-function hypoxic cell radiosensitizers. It is a 2-nitroimidazole containing a monofunctional, alkylating aziridine ring. RSU 1069 has radiosensitizing properties and can be up to 100 times more toxic to hypoxic cells than to aerobic cells [S75]. The increased differential toxicity compared with that of other simple nitroimidazoles is due to the alkylating function in the molecule [W25]. Following bioreduction, therefore, the drug is converted into a bifunctional agent that can cause both DNA strand breaks and cross-links [J5, O9, S76, W26]. In mice injected with RSU 1069, aerobic cells exhibited large numbers of DNA single-strand breaks, while toxic DNA interstrand cross-links were produced only in hypoxic cells. Cells from bone marrow and spleen showed extensive numbers of DNA single-strand breaks but minimal cross-linking compared with tumours [O10]. However, clinical testing revealed severe gastrointestinal

toxicity at doses below those needed for therapeutic benefit [H17].

254. A series of pro-drugs (e.g. RB 6145) have been developed that release RSU 1069 spontaneously under physiological conditions [J6]. *In vitro* and *in vivo* animal data showed that the hypoxic cell specificity and cytotoxic activity are retained but that at the same time the acute toxicity is reduced in animal models [A23, C4, C11, C44, S77, S78]. The efficacy of the combined treatment of SCCVII transplantable tumours is significantly higher than that of treatment with radiation alone.

(c) Benzotriazine di-N-oxides

255. Brown and collaborators [M50, Z7, Z8] introduced the benzotriazine di-N-oxide tirapazamine (SR 4233) and analogues into the field of bioreductive drugs. Like the nitro compounds and quinones, the benzotriazine di-N-oxides are reduced to one-electron reduced free radicals [B62, K10, L30, Z9]. Tirapazamine is highly efficient in killing hypoxic cells *in vitro* and *in vivo* [B53, K9, Z7, Z8]. Unlike the toxicity of other bioreductive drugs studied, the toxicity of SR 4233 does not level off at normal oxygen concentrations but continues to decrease as the oxygen concentration increases.

256. The drug appears to induce DNA strand breaks by means of an oxidative damage to pyrimidines [E8]. Analysis of DNA and chromosomal breaks after hypoxic exposure to SR 4233 suggests that DNA double-strand breaks are the primary lesion causing cell death [B28]. More DNA single-strand breaks and a greater heterogeneity in DNA damage were observed in tumour cells than in spleen and marrow cells of mice exposed to tirapazamine, consistent with the presence of hypoxic cells and the greater bioreductive capacity of tumours [O10].

257. SR 4233 is also extremely active when used in combination with fractionated radiation schedules [B62, E8, Z9]. This enhancement is seen when SR 4233 is given before and after irradiation [Z10]. In two animal tumour models (KHT and SCCVII), SR 4233 with radiation produced a significantly greater enhancement than did nicotinamide with carbogen, a combination that has been shown to improve tumour oxygenation. In RIF-1 tumour, which has the lowest hypoxic fraction of the three, the response was comparable for the two modalities [D8]. SR 4233 was able to enhance the tumour growth delay produced by radioimmunotherapy in severe combined immunodeficient phenotype mice with human cutaneous T-cell lymphoma xenografts [W27]. In a study by Lartigau and Guichard [L29], the pO_2 dependence of the survival of three human cell lines (HRT 18, Na11+, and MEWO) exposed to tirapazamine (SR 4233) alone or combined with ionizing radiation, was studied *in vitro*. There was a marked increase in cell killing when tirapazamine was combined with radiation, compared with either tirapazamine or radiation given alone.

Glossary

Absolute risk	Excess risk attributed to an agent and usually expressed as the numeric difference between exposed and unexposed populations (e.g. five cancer deaths over a lifetime per 100 people, each irradiated with 1 Sv).
Additivity	Effect of a combined exposure equaling the sum of the effects from single-agent exposures.
Absorbed dose	*Chemicals* The amount of an applied dose of chemical absorbed into the body or into organs and tissues of interest.
	Radiation The average energy imparted to matter in an element of volume by ionizing radiation divided by the mass of that element of volume. The SI unit for absorbed dose is joule per kilogramme ($J\ kg^{-1}$) and its special name is gray (Gy).
Alkylating agents	compounds that transfer an alkyl group to DNA.
Antagonism	*General* A combined effect of two or more interacting agents that is smaller than the addition of the single-agent effects with known dose-effect relationships.
	Chemical antagonism or inactivation Chemical reaction between two compounds to produce a less toxic product. Example: toxic metal and chelator.
	Dispositional antagonism Alteration of absorption, biotransformation, distribution, or excretion of one agent in such a way that the time-concentration product in the target organ is diminished. Example: prevention of absorption with charcoal.
	Functional antagonism Two agents balance each other by producing opposite effects on the same function. Example: drug with vasodepressing side effect and vasopressor.
	Receptor antagonism Competitive binding to the same receptor producing a smaller effect. Examples: oxygen in carbon monoxide poisoning, ethanol in methanol poisoning.
Antioxidants	substances preventing oxidation.
Biochemical effect monitoring	Monitoring of biochemical and molecular effects, i.e. changes in sequence, structure, and/or function of biologically relevant molecules caused by an exposure to an agent or a mixture of agents. Biochemical effect monitoring determines tissue dose or biologically effective dose. Examples are direct measurement of DNA adducts and strand breaks. Biochemical effect monitoring takes into account individual differences such as genetic background and deficiencies in DNA repair. A disadvantage is the difficulty of directly monitoring changes in target cell populations. Most analyses are therefore done on surrogate tissue such as blood cells.
Biological indicator	Measurable biological effect that is clearly, specifically, and quantifiably related to an exposure.
Biological effective dose	Biological effect in cells or tissues at risk with direct relevance to the initiation or progression of a disease; see also *Biochemical effect monitoring.*
Biological monitoring	Continuous or repeated monitoring of potentially toxic agents or their metabolites in cells, tissues, body fluids, or excretions (internal dose). Biological monitoring takes into account individual differences in absorption or bioaccumulation of agents in question. It has the advantage of being comparatively easy to monitor.
Biomonitoring	monitoring the environment or a population with biological markers.
Carcinogen	An agent, chemical, physical, or biological, that can act on living tissue in such a way as to cause a malignant neoplasm.
	Solitary or complete The agent does not need additional action of further exogenous cancer risk factors to cause a neoplasm.
	Indirect or precarcinogen The agent has to be transformed to its active molecular form (ultimate carcinogen) in the metabolism.

Co-factor	A substance or agent that acts with another substance to bring about certain effects; e.g. coenzyme, a low-molecular entity needed for enzymatic activity of the apoenzyme.
Combined effect	The joint effects of two or more agents on the level of molecules, cells, organs, and organisms in the production of a biological effect.
Concentration additivity	Combined effect is predicted by addition of concentrations of different agents on a normalized concentration-effect graph; valid in the case of isoaddition. In this case, a combined effect can arise even when all single-agent concentrations are below their threshold for the endpoint under study (see also effect additivity).
Confounder	A variable that can cause or prevent the outcome of interest, is not an intermediate variable, and is not associated with the factor under investigation. Such a variable must be controlled in order to obtain an undistorted estimate of the effect of the study factor(s) on risk.
Confounding	A situation in which the effects of two processes are not separated. The distortion of the apparent effect of an exposure on risk brought about by the association with other factors that can influence the outcome. Distortion of a measure of the effect because of the association of exposure with other factor(s) that influence the outcome under study (WHO).
Dependent action	Action of two and more agents, in which the effect of a second agent depends on the effect of a first agent. Dependent action leads to combined effects different from heteroadditivity.
Deterministic effect	Effect on sufficient proportion of cells to disrupt tissue or organ function. The probability of causing observable damage will be zero at small doses but will increase steeply to unity above a threshold. Above the threshold, the severity of damage will also increase with dose. Examples include cataracts, skin erythema, and stem-cell depression in bone marrow or the small intestine.
Dose	*Radiation* See *Absorbed dose radiation*. *Chemicals* The amount of a chemical administered to an organism. See also *Absorbed dose chemicals*.
Dose-response relationship	The relationship between the magnitude of exposure to a chemical, biological, or physical agent (dose) and the magnitude or frequency and/or severity of associated adverse effects (response).
Effect additivity	Combined effect is predicted by adding the effects of different agents; valid in the case of heteroaddition. In this case, the combined effect is zero as long as all single-agent concentrations are below their threshold.
Environmental monitoring	Quantitative determination of a potentially detrimental agent in the environment (external dose).
Epigen, epigenetic	Changes in an organism brought about by alterations in the expression of genetic information without any change in the genome itself; the genotype is unaffected by such a change but the phenotype is altered.
Exposure	Concentration, amount, or intensity of a particular physical or chemical or environmental agent that reaches the target population, organism, organ, tissue, or cell, usually expressed in numerical terms of substance concentration, duration, and frequency (for chemical agents or microorganisms) or intensity (for physical agents such as radiation). In the radiation field, exposure may also denote the electrical charge of ions caused by x or gamma rays per unit mass of air; however the term is used in its more general sense as described here.
External dose	*Radiation* Dose from an external radiation source; obtained from being within a radiation field. *Chemicals* Concentration of an agent in an exposure medium, i.e. air or water; see also *Environmental monitoring*.
Genotoxicity	Ability to cause damage to genetic material. Such damage may be mutagenic and/or carcinogenic.

Hazard	Set of inherent properties of a substance, mixture of substances, or a process involving substances that, under production, usage, or disposal conditions, make it capable of having adverse effects on organisms or the environment, depending on the degree of exposure; in other words, a source of danger.
Heteroadditivity	Additive effect from two independently acting agents with different modes of action and therefore different dose-effect relationships. See also *Effect additivity*.
Initiator	In the multi-stage model of carcinogenesis, initiators are defined by their ability to induce persistent changes (probably due to genotoxic effects) in the cell (initiation). If there is subsequent promotion, these changes may result in tumour formation.
Independent action	Action of two and more agents in which the effect of one agent is independent of the effect of the other agent. Independent action leads to combined effects defined as heteroadditive.
Interaction	Combined, mutual effects between agents on a molecular and/or cellular level within a short time.
Internal dose	*Radiation* Dose from radioactive material deposited in the body.
	Chemicals (a) Amount of a chemical recently absorbed; measured, e.g. as metal concentration in blood; (b) amount of chemical stored in one or several body compartments or in the whole body (body burden); used mainly for cumulative toxicants; (c) in the case of ideal biological monitoring, amount of active chemical species bound to the critical sites of action (target dose; e.g. carbon monoxide binding to haemoglobin).
Isoadditivity	Additive effect from two similarly acting agents or from two increments of the same agent on an upward bent dose-effect relationship. See also *Concentration additivity*. On a descriptive level without detailed information about dose-effect relationships, isoadditivity is sometimes indistinguishable from supra-additivity or synergism.
Mitogens	substances with a mitogenic effect on cells.
Multiplicative response	Effect of two agents for which the single-agent response coefficients or relative risks have to be multiplied to describe the combined response.
Mutagen	A substance that can induce heritable changes (mutations) of the genotype in a cell as a consequence of alteration or loss of genes or chromosomes (or parts thereof).
Mutation	A hereditary change in genetic material. A mutation can be a change in a single base (point mutation) or a single gene or it can involve larger chromosomal rearrangements such as deletions and translocations.
Non-genotoxic effect	Effect of an agent at the cellular, organ, or organism level without direct effects on the genome such as DNA damage.
Non-stochastic effect	see *Deterministic effect*.
No Observed Adverse Effect Level (NOAEL)	The greatest concentration or amount of a substance, found by experiment or observation, that causes no detectable adverse alteration of morphology, functional capacity, growth, development, or lifespan of the target organism under defined conditions of exposure. Alterations of morphology, functional capacity, growth, development, or lifespan of the target can be detected at this level but may be judged not to be adverse.
No Observed Effect Level (NOEL)	The greatest concentration or amount of a substance, found by experiment or observation, that causes no detectable adverse alteration of morphology, functional capacity, growth, development, or lifespan of the target organisms distinguishable from those observed in normal (control) organisms of the same species and strain under the same defined conditions of exposure.
Potentiation	Synergism.
Precursor	Substance from which another, usually more biologically active, substance is formed.

Progression	Increase in autonomous growth and malignancy; used in particular to describe the transition from benign to malignant tumours and the progression of malignancy. There are probably numerous stages of progression during neoplastic development. The process of progression features in the general model of carcinogenesis as well as in the multi-stage model.
Promoter	Risk factors of cancer that are capable of triggering preferential multiplication of a cell changed by initiation. Often, following initiation, a long-term action on the target tissue is necessary. Promoters often cause enzyme induction, hyperplasia, and/or tissue damage. The essential primary effects are considered to be reversible. As a rule, promoters do not bind covalently to cell components and do not exert an immediate genotoxic action.
Relative risk (RR)	Ratio between the cancer cases in the exposed population to the number of cases in the unexposed population. A relative risk of 1.5 indicates a 50% increase in cancer due to the agent under consideration. Excess relative risk (ERR) is RR - 1.
Sensitizer	An agent or substance that is capable of causing a state of abnormal responsiveness in an individual. In most cases, initial exposure results in a normal response, but repeated exposures lead to progressively strong and abnormal responses.
Stochastic effect	Effect of an agent on a cell of a random or statistical nature in which the cell is modified rather than killed. If this cell is able to transmit the modification to later cell generations, any resulting effect, of which there may be many different kinds and severity, are expressed in the progeny of the exposed cell. The probability of such a transmittable effect resulting from an exposure to a genotoxic agent increases with increments of dose, at least for doses well below the threshold for deterministic effects. The severity of the damage is not affected by the dose. When the modified cell is a germ cell, the stochastic effect is called a hereditary effect.
Subadditivity	Less than additive; effect of a combined exposure being less than the sum of effects from single-agent exposures.
Supra-additivity	More than additive; effect of a combined exposure exceeding the sum of the effects from single-agent exposures.
Synergism	A combined effect of two or more interacting agents that is greater than the addition of the single-agent effects with known dose-effect relationships.
Target (biological)	Any organism, organ, tissue, or cell that is subject to the action of a pollutant or other chemical, physical, or biological agent.
Threshold dose	The minimum dose that will produce a biological effect. Dose below which no effects occur ("true", mechanistically derived threshold) or are measurable (apparent threshold). For a given agent there can be multiple threshold doses, in essence one for each definable effect.
Tissue dose	Local dose in an organ or a functional or structural entity of an organ. See also *Absorbed dose* and *Internal dose-chemicals*.
Topoisomerase	ubiquitous enzymes that alter DNA configuration or topology.
Toxicity	Capacity of an agent to cause injury to a living organism. Toxicity can only be defined in quantitative terms with reference to the quantity of substance administered or absorbed, the way in which this quantity is administered (e.g. inhalation, ingestion, or injection) and distributed in time (e.g. single or repeated doses), the type and severity of injury, and the time needed to produce the injury.

References

A1 Adams, G.E. Redox, radiation and bioreductive bioactivation. Radiat. Res. 132: 129-139 (1992).

A2 Aghamolammadi, S.Z., D.T. Goodhead and J.R.K. Savage. Induction of sister chromatid exchanges (SCE) in G_0 lymphocytes by plutonium-238 alpha particles. Int. J. Radiat. Biol. 53: 909-915 (1988).

A3 Albin, M., K. Jakobsson, R. Attewell et al. Mortality and cancer morbidity in cohorts of asbestos cement workers and referents. Br. J. Ind. Med. 47(9): 602-619 (1990).

A4 Alvanja, M.C.R., R.C. Brownson, J. Benichou et al. Attributable risk of lung cancer in lifetime nonsmokers and long-term ex-smokers (Missouri, United States). Cancer Cases and Control 6: 209-216 (1995).

A5 Albertini, R.J., K.L. Castle and W.R. Borcherding. T-cell cloning to detect the mutant 6-thioguanine-resistant lymphocytes present in human peripheral blood. Proc. Natl. Acad. Sci. U.S.A. 79(21): 6617-6621 (1982).

A6 Albertini, R.J., J.P. O'Neill, J.A. Nicklas et al. Alterations of the hprt gene in human in vivo-derived 6-thioguanine-resistant T lymphocytes. Nature 316: 369-371 (1985).

A7 Ando, Y., H. Watanabe and M. Tatematsu. Gastric tumorigenicity of 1,2-dimethylhydrazine on the background of gastric intestinal metaplasia induced by X-irradiation in CD (SD) rates. Jpn. J. Cancer Res. 87: 433-436 (1996).

A8 Albertini, R.J., J.P. O'Neill, J.A. Nicklas et al. HPRT mutations in vivo in human T-lymphocytes, frequencies, spectra and clonality. p. 15-20 in: Mutation and the Environment, Fifth International Conference on Environmental Mutagens, Part C: Somatic and Heritable Mutation, Adduction and Epidemiology (M.L. Mendelsohn and R.J. Albertini, eds.). Wiley/Liss, New York, 1990.

A9 Albertini, R.J., J.A. Nicklas, J.C. Fuscoe et al. In vivo mutations in human blood cells: biomarkers for molecular epidemiology. Environ. Health Perspect. 99: 135-141 (1993).

A10 Ashby, J., M. Ishidate, G.D. Stoner et al. Studies on the genotoxicity of beryllium sulphate in vitro and in vivo. Mutat. Res. 240: 217-225 (1990).

A11 Amos, N., D. Bass and D. Roe. Screening human populations for chromosome aberrations. Mutat. Res. 143: 155-160 (1985).

A12 Andrushenko, V.N., A.A. Ivanov and V.N. Maltsev. The radioprotective effect of microbial substances. J. Radiationbiol. Radioecol. 36(2): 195-208 (1996). (In Russian).

A13 Antipov, V.V., V.S. Tikhonchuk, I.B. Ushakov et al. Status of the synapses of the end brain of rats exposed to the factors of space flight. Kosm. Biol. Aviakosm. Med. 22(4): 54-61 (1988). (In Russian).

A14 Antipov, V.V., B.I. Davydov, I.B. Ushakov et al. The effects of space flight factors on the central nervous system. Structural-functional aspects of radiomodifying action. Probl. Kosm. Biol. 66: 1-328 (1989). (In Russian).

A15 Anisimov, V.N. and G.Y. Osipova. Carcinogenesis induced by combined neonatal exposure to 5-bromo-2'-deoxyuridine and subsequent total-body X-ray irradiation in rats. Cancer Lett. 70(1-2): 81-90 (1993).

A16 Armitage, P. and R. Doll. The age distribution of cancer and a multistage theory of carcinogenesis. Br. J. Cancer 8: 1-12 (1954).

A17 Armitage, P. and R. Doll. A two-stage theory of carcinogenesis in relation to the age distribution of human cancer. Br. J. Cancer 11: 161-169 (1957).

A18 Armitage, P. and R. Doll. Stochastic models for carcinogenesis. p. 19-37 in: Proceedings of the Fourth Berkeley Symposium on Mathematical Statistics and Probability (J. Neyman, ed.). Berkeley, University of California, 1961.

A19 Allegra, C.J. Antifolates. p. 110-153 in: Cancer Chemotherapy: Principles and Practice (B.A. Chabner and J.M. Collins, eds.). J.B. Lippincott Co., Philadelphia, 1990.

A20 Armitage, P. The assessment of low-dose carcinogenicity. Biometrics 28 (Suppl.): 119-129 (1982).

A21 Archer, V.E. Enhancement of lung cancer by cigarette smoking in uranium and other miners. Carcinog. Compr. Surv. 8: 23-37 (1985).

A22 Armitage, P. Multistage models of carcinogenesis. Environ. Health Perspect. 63: 195-201 (1985).

A23 Adams, G.E. and I.J. Stratford. Bioreductive drugs for cancer therapy: the search for tumor specificity. Int. J. Radiat. Oncol. Biol. Phys. 29: 231-238 (1994).

A24 Ashby, J. and C.R. Richardson. Tabulation and assessment of 113 human surveillance cytogenetic studies conducted 1965-1984. Mutat. Res. 154(2): 111-133 (1985).

A25 Astier-Gin, T., M. Galiay, E. Legrand et al. Murine thymic lymphomas after infection with a B-ecotropic murine leukemia virus and/or X-irradiation: proviral organization and RNA expression. Leuk. Res. 10(7): 809-817 (1986).

A26 Adams, G.E., I.J. Stratford, H.S. Edwards et al. Bioreductive drugs as post-irradiation sensitizers: comparison of dual function agents with SR 4233 and the mitomycin C analogue EO9. Int. J. Radiat. Oncol. Biol. Phys. 22: 717-720 (1992).

A27 Al Sarraf, M., T.F. Pajak, V.A. Marcial et al. Concurrent radiotherapy and chemotherapy with cisplatin in inoperable squamous cell carcinoma of the head and neck. An RTOG Study. Cancer 59: 259-265 (1987).

A28 Alavanja, M.C., R.C. Brownson, J.H. Lubin et al. Residential radon exposure and lung cancer among nonsmoking women [see comments]. J. Natl. Cancer Inst. 86(24): 1829-1837 (1994).

A29 Amsari, R., R. Tokars, W. Fisher et al. A phase III study of thoracic irradiation with or without concomitant cisplatin in locoregional unresectable non-small cell lung cancer: a Hoosier Oncology Group Protocol. Proc. Am. Soc. Clin. Oncol. 10: 241 (Abstract) (1991).

B1 Brezani, P., I. Kalina and A. Ondrussekova. Changes in cellularity, CFU-S number and chromosome aberrations in bone marrow and blood of rats after neutron and continuous gamma irradiation. Radiobiol. Radiother. 30: 431-438 (1989).

B2 Boice, J.D. Jr., E.B. Harvey, M. Blettner et al. Cancer in the contralateral breast after radiotherapy of breast cancer. N. Engl. J. Med. 326: 781-785 (1992).

B3 Busse, P.M., S.K. Bose, R.W. Jones et al. The action of caffeine on x-irradiated HeLa cells. III. Enhancement of x-ray-induced killing during G_2 arrest. Radiat. Res. 76: 292-307 (1978).

B4 Ban, S., J.B. Cologne, S. Fujita et al. Radiosensitivity of atomic bomb survivors as determined with a micronucleus assay. Radiat. Res. 134: 170-178 (1993).

B5 Blakely, E.A., C.A. Tobias, T.C.H. Yang et al. Inactivation of human kidney cells by high-energy monoenergetic heavy-ion beams. Radiat. Res. 80: 122-160 (1979).

B6   Burkart, W., G.L. Finch and T. Jung. Quantifying health effects from the combined action of low-level radiation and other environmental agents: Can new approaches solve the enigma? Sci. Total Environ. 205: 51-70 (1997).

B7   Brown, R.C., A. Poole and G.T. Flemming. The metabolism of benzo(a)pyrene by the lungs of asbestos exposed rats. Exp. Pathol. 26(2): 101-105 (1984).

B8   Bensasson, R.V., E.J. Land and T.G. Truscott. Flash Photolysis and Pulse Radiolysis. Pergamon Press, New York, 1983.

B9   Brown, C.C. and K.C. Chu. Additive and multiplicative models and multistage carcinogenesis theory. Risk Anal. 9: 99-105 (1989).

B10   Boweri, T. Zur Frage der Entstehung maligner Tumoren. Gustav Fischer, Jena, 1914.

B11   Berenbaum, M.C. Antagonism between radiosensitizing agents. Br. J. Radiol. 61: 975-976 (1988).

B12   Bergtold, D.S., M.G. Simic, H. Alessio et al. Urine biomarkers for oxidative DNA damage. p. 483-490 in: Oxygen Radicals in Biology and Medicine (M.G. Simic et al., eds.). Plenum Press, New York, 1988.

B13   Barrett, J.C., P.W. Lamb and R.W. Wiseman. Multiple mechanisms for the carcinogenic effects of asbestos and other mineral fibers. Environ. Health Perspect. 81: 81-89 (1989).

B14   Bishop, J.M. Molecular themes in oncogenesis. Cell 64: 235-248 (1991).

B15   Berenblum, I. The mechanism of carcinogenic action related phenomena. Cancer Res. 1: 807-814 (1941).

B16   Benson, J.M., E.B. Barr, D.L. Lundgren et al. Pulmonary retention and tissue distribution of ^{239}Pu nitrate in F344 rats and syrian hamsters inhaling carbon tetrachloride. p. 146-148 in: Annual Report 1993-1994, Inhalation Toxicology Research Institute (S.A. Belinsky et al., eds.). ITRI-144 (1994).

B17   Boyd, J.A. and J.C. Barret. Genetic and cellular basis of multistep carcinogenesis. Pharmacol. Ther. 46: 469-486 (1990).

B18   Bender, M.A., P.C. Gooch and S. Kondo. The Gemini-3 S-4 space flight-radiation interaction experiment. Radiat. Res. 31: 91-111 (1967).

B19   Bender, M.A., P.C. Gooch and S. Kondo. The Gemini XI S-4 space flight-radiation interaction experiment: The human blood experiment. Radiat. Res. 34: 228-238 (1968).

B20   Bonassi, S., A. Abbondandolo, L. Camurri et al. Are chromosome abberations in circulating lymphocytes predictive of future cancer onset in humans? Cancer Genet. Cytogenet. 79: 133-135 (1995).

B21   Bochkov, N.P., K.N. Yakovenko and N.I. Voskoboiynik. Dose and concentration dependence of chromosome aberrations in human cells and the combined action of radiation and chemical mutagens. Cytogenet. Cell Genet. 33(1): 42-47 (1983).

B22   Boice, J.D. Jr., M.H. Greene, J.Y. Killen Jr. et al. Leukemia and preleukemia after adjuvant treatment of gastrointestinal cancer with semustine (methyl-CCNU). N. Engl. J. Med. 309: 1079-1084 (1983).

B23   Burkart, W. Molecular mechanisms in radiation-induced cancer. p. 141-160 in: Epidemiology of Childhood Leukemia (J. Michaelis, ed.). Gustav Fischer Verlag, Stuttgart/New York, 1993.

B24   Borek, C., A. Ong, H. Mason et al. Selenium and vitamin E inhibit radiogenic and chemically induced transformation in vitro via different mechanisms. Proc. Natl. Acad. Sci. U.S.A. 83(5): 1490-1494 (1986).

B25   Burkart, W., P. Heusser and V. Vijayalaxmi. Micro-dosimetric constraints on specific adaptation mechanisms to reduce DNA damage caused by ionizing radiation. Radiat. Prot. Dosim. 31: 69-274 (1990).

B26   Bellamy, A.S. and B.T. Hill. Interactions between clinically effective antitumor drugs and radiation in experimental systems. Biochim. Biophys. Acta 738: 125-166 (1984).

B27   Burkart, W. Radiation biology of the lung. Sci. Total Environ. 89: 1-230 (1989).

B28   Biedermann, K.A., J. Wang, R.P. Graham et al. SR 4233 cytotoxicity and metabolism in DNA repair-competent and repair-deficient cell cultures. Br. J. Cancer 63: 358-362 (1991).

B29   Borghoff, S.J. α2u-Globulin-mediated male rat nephropathy and kidney cancer: relevance to human risk assessment. CIIT Activities 13(4): 1-8 (1993).

B30   Brandner, G., A. Luz and S. Ivankovic. Incidence of radiation lymphomas in C57BL/6 mice is not promoted after intra-peritoneal treatment with the phorbolester tetra-decanoylphorbol acetate. Cancer Lett. 23(1): 91-95 (1984).

B31   Bradley, W.E., J.L. Gareau, A.M. Seifert et al. Molecular characterization of 15 rearrangements among 90 human in vivo somatic mutants shows that deletions predominate. Mol. Cell. Biol. 7(2): 956-960 (1987).

B32   Broerse, J.J., L.A. Hennen, W.M. Klapwijk et al. Mammary carcinogenesis in different rat strains after irradiation and hormone administration. Int. J. Radiat. Biol. Relat. Stud. Phys. Chem. Med. 51(6): 1091-1100 (1987).

B33   Bukhtoiarova, Z.M. and S.S. Spirina. Characteristics of osteosarcomas and morphological changes in rat thymus during the combined action of gamma-radiation and 239 Pu. Radiobiologiya 29(6): 804-808 (1989). (In Russian).

B34   Buxton, G.V., C.L. Greenstock, W.P. Helman et al. Critical review of rate constants for reactions of hydrated electrons, hydrogen atoms and hydroxyl radicals. Phys. Chem. Ref. Data 17: 513-517 (1988).

B35   Barendsen, G.W. Interpretation of the LET dependence of radiation induced lethal and sublethal lesions in mammalian cells. p. 13-20 in: Biophysical Modelling of Radiation Effects (K.H. Chadwick, G. Moschini and M.N. Varma, eds.). Adam Hilger, Bristol, 1992.

B36   Berk, P.D., J.D. Goldberg, M.N. Silverstein et al. Increased incidence of acute leukemia in polycythemia vera associated with chlorambucil therapy. N. Engl. J. Med. 304: 441-447 (1981).

B37   Brooks, A.L., W.C. Griffith, N.F. Johnson et al. The induction of chromosome damage in CHO cells by beryllium and radiation given alone and in combination. Radiat. Res. 120: 494-507 (1989).

B38   Bignon, J., G. Monchaux, J. Chameaud et al. Incidence of various types of thoracic malignancy induced in rats by intrapleural injection of 2mg of various mineral dusts after inhalation of ^{222}Rn. Carcinogenesis 4(5): 621-628 (1983).

B39   Brooks, A.L., G.J. Newton, L.J. Shyr et al. The combined effects of alpha-particles and X-rays on cell killing and micronuclei induction in lung epithelial cells. Int. J. Radiat. Biol. 58(5): 799-811 (1990).

B40   Brenner, D.J. The significance of dose rate in assessing the hazards of domestic radon exposure. Health Phys. 67(1): 76-79 (1994).

B41   Bohr, V.A., D.H. Phillips and P.C. Hanawalt. Heterogeneous DNA damage and repair in the mammalian genome. Cancer Res. 47: 6426-6436 (1987). *[published erratum appeared in Cancer Res. 48: 1377 (1988)].*

B42 Bau, R., R.W. Gellert and S.M. Lehovec. Crystallographic studies on platinum-nucleoside and platinum-nucleotide complexes. J. Clin. Hematol. Oncol. 7: 51-63 (1977).

B43 Begg, A.C. Cisplatin and radiation: interaction probabilities and therapeutic possibilities. Int. J. Radiat. Oncol. Biol. Phys. 19: 1183-1189 (1990).

B44 Begg, A.C., P.J. van der Kolk, L. Dewit et al. Radio-sensitization by cisplatin of RIF1 tumour cells in vitro. Int. J. Radiat. Biol. Relat. Stud. Phys. Chem. Med. 50: 871-884 (1986).

B45 Bachaud, J.M., J.M. David, G. Boussin et al. Combined postoperative radiotherapy and weekly cisplatin infusion for locally advanced squamous cell carcinoma of the head and neck: preliminary report of a randomized trial. Int. J. Radiat. Oncol. Biol. Phys. 20: 243-246 (1991).

B46 Borchers, A.H., K.A. Kennedy and J.A. Straw. Inhibition of DNA excision repair by methotrexate in Chinese hamster ovary cells following exposure to ultraviolet irradiation or ethylmethanesulfonate. Cancer Res. 50: 1786-1789 (1990).

B47 Berry, R.J. A reduced oxygen enhancement ratio for x-ray survival of HeLa cells in vitro, after treatment with 'Methotrexate'. Nature 208: 1108-1110 (1965).

B48 Berry, R.J. Some observations on the combined effects of x-rays and methotrexate on human tumor cells in vitro with possible relevance to their most useful combination in radiotherapy. Am. J. Roentgenol., Radium Ther. Nucl. Med. 102: 509-518 (1968).

B49 Berry, R.J. and J.M. Huckle. Radiation and methotrexate effects on stationary phase cells. Br. J. Radiol. 45: 710-711 (1972).

B50 Byfield, J.E., P. Calabro Jones, I. Klisak et al. Pharma-cologic requirements for obtaining sensitization of human tumor cells in vitro to combined 5-Fluorouracil or florafur and X rays. Int. J. Radiat. Oncol. Biol. Phys. 8: 1923-1933 (1982).

B51 Bruso, C.E., D.S. Shewach and T.S. Lawrence. Fluoro-deoxyuridine-induced radiosensitization and inhibition of DNA double strand break repair in human colon cancer cells. Int. J. Radiat. Oncol. Biol. Phys. 19: 1411-1417 (1990).

B52 Brown, J.M., D.R. Goffinet, J.E. Cleaver et al. Preferential radiosensitization of mouse sarcoma relative to normal skin by chronic intra-arterial infusion of halogenated pyrimidine analogs. J. Natl. Cancer Inst. 47: 75-89 (1971).

B53 Baker, M.A., E.M. Zeman, V.K. Hirst et al. Metabolism of SR 4233 by Chinese hamster ovary cells: basis of selective hypoxic cytotoxicity. Cancer Res. 48: 5947-5952 (1988).

B54 Bachur, N.R., S.L. Gordon and M.V. Gee. A general mechanism for microsomal activation of quinone anticancer agents to free radicals. Cancer Res. 38: 1745-1750 (1978).

B55 Bryan, J. Definition of three classes of binding sites in isolated microtubule crystals. Biochemistry 11: 2611-2616 (1972).

B56 Blume, K.G., G.D. Long, R.S. Negrin et al. Role of etoposide (VP-16) in preparatory regimens for patients with leukemia or lymphoma undergoing allogeneic bone marrow transplantation. Bone Marrow Transplant (Suppl. 4) 14: 9-10 (1994).

B57 Bush, R.S. The significance of anemia in clinical radiation therapy. Int. J. Radiat. Oncol. Biol. Phys. 12: 2047-2050 (1986).

B58 Begg, A.C., P.W. Sheldon and J.L. Foster. Demonstration of radiosensitization of hypoxic cells in solid tumours by metronidazole. Br. J. Radiol. 47: 399-404 (1974).

B59 Brown, J.M. and W.W. Lee. Pharmacokinetic considera-tions in radiosensitizer development. p. 2-13 in: Radiation Sensitizers - Their Use in the Clinical Management of Cancer (L.M. Brady, ed.). Masson, New York, 1980.

B60 Brown, J.M. and P. Workman. Partition coefficient as a guide to the development of radiosensitizers which are less toxic than misonidazole. Radiat. Res. 82: 171-190 (1980).

B61 Brown, J.M., N.Y. Yu, D.M. Brown et al. SR-2508: a 2-nitroimidazole amide which should be superior to misonidazole as a radiosensitizer for clinical use. Int. J. Radiat. Oncol. Biol. Phys. 7: 695-703 (1981).

B62 Brown, J.M. and M.J. Lemmon. Potentiation by the hypoxic cytotoxin SR 4233 of cell killing produced by fractionated irradiation of mouse tumors. Cancer Res. 50: 7745-7749 (1990).

B63 Bergsagel, D.E., A.J. Bailey, G.R. Langley et al. The chemotherapy of plasma-cell myeloma and the incidence of acute leukemia. N. Engl. J. Med. 301: 743-748 (1979).

B64 Boice, J.D. Jr. and P.D. Inskip. Radiation-induced leukemia. p. 195-209 in: Leukemia, 6th edition (E.S. Henderson, T.A. Lister and M.F. Greaves, eds.). W.B. Saunders Co., Philadelphia, 1996.

B65 Boice, J.D. Jr., M.H. Greene, J.Y. Killen Jr. et al. Leukemia after adjuvant chemotherapy with semustine (methyl-CCNU) - evidence of a dose-response effect. N. Engl. J. Med. 314: 119-120 (1986).

B66 Brightwell, J. and A.G. Heppleston. Inhibition of urethane-induced pulmonary adenomas by inhaled plutonium-239. Br. J. Radiol. 46: 180-182 (1973).

B67 Brightwell, J. and A.G. Heppleston. Effect of inhaled plutonium dioxide on development of urethane-induced pulmonary adenomas. Br. J. Cancer 35: 433-438 (1977).

B68 Bhatia, S., L.L. Robison, O. Oberlin et al. Breast cancer and other second neoplasms after childhood Hodgkin's disease. N. Engl. J. Med. 334: 745-751 (1996).

B69 Burkart, W. and T. Jung. Health risks from combined exposures: mechanistic considerations on deviations from additivity. Mutat. Res. 411: 119-128 (1998).

B70 Burgaman, P., H. Ouyang, S. Peterson et al. Heat inactivation of Ku autoantigen: possible role in hyper-thermic radiosensitization. Cancer Res. 57(14): 2847-2850 (1997).

B71 Bartstra, R.W., P.A. Bentvelzen, J. Zoetelief et al. Induction of mammary tumours in rats by single-dose gamma irradiation at different ages. Radiat. Res. 150(4): 442-450 (1998).

B72 Bartstra, R.W., P.A. Bentvelzen, J. Zoetelief et al. The influence of estrogen treatment on induction of mammary carcinoma in rats by single-dose gamma irradiation at different ages. Radiat. Res. 150(4): 451-458 (1998).

B73 Bartsch, H., M. Hollstein, R. Mustonen et al. Screening for putative radon-specific p53 mutation hotspot in German uranium miners. Lancet 346(8967): 121 (1995).

B74 Block, G., G.M. Matanoski, R. Seltser et al. Cancer morbidity and mortality in phosphate workers. Cancer Res. 48: 7298-7303 (1988).

B75 Brown, J.M. and B.G. Wouters. Apoptosis, p53, and tumor cell sensitivity to anticancer agents. Cancer Res. 59(7): 1391-1399 (1999).

B76 Beissert, S. and T. Schwarz. Mechanisms involved in ultraviolet light-induced immunosuppression. J. Invest. Dermatol. 4(1): 61-64 (1999).

B77 Boorman, G.A., R.D. Owen, W.G. Lotz et al. Evaluation of in vitro effects of 50 and 60 Hz magnetic fields in regional EMF exposure facilities. Radiat. Res. 153: 648-657 (2000).

B78   Boorman, G.A., C.N. Rafferty, J.M. Ward et al. Leukemia and lymphoma incidence in rodents exposed to low-frequency magnetic fields. Radiat. Res. 153: 627-636 (2000).

B79   Boorman, G.A., D.L. McCormick, J.M. Ward et al. Magnetic fields and mammary cancer in rodents: a critical review and evaluation of published literature. Radiat. Res. 153: 617-626 (2000).

B80   Bücker, H., R. Facius, G. Horneck et al. Embryogenesis and organogenesis of *Carausius morosus* under spaceflight conditions. Adv. Space Res. 6: 115-124 (1986).

C1   Committee on the Biological Effects of Ionizing Radiations (BEIR V). Health Effects of Exposure to Low Levels of Ionizing Radiation. United States National Academy of Sciences, National Research Council. National Academy Press, Washington, 1990.

C2   Committee on the Biological Effects of Ionizing Radiations (BEIR IV). Health Risks of Radon and Other Internally Deposited Alpha-Emitters. United States National Academy of Sciences, National Research Council. National Academy Press, Washington, 1988.

C3   Cathcart, R., E. Schwiers, R.L. Saul et al. Thymine glycol and thymidine glycol in human and rat urine: a possible assay for oxidative DNA damage. Proc. Natl. Acad. Sci. U.S.A. 81: 5633-5637 (1984).

C4   Cole, S., I.J. Stratford, E.M. Fielden et al. Dual function nitroimidazoles less toxic than RSU 1069: selection of candidate drugs for clinical trial (RB 6145 and/or PD 130908. Int. J. Radiat. Oncol. Biol. Phys. 22: 545-548 (1992).

C5   Cohen, D., S.F. Arai and J.D. Brain. Smoking in pairs long-term dust clearance from the lung. Science 204: 514-517 (1979).

C6   Chapman, J.D., E.A. Blakely, K.C. Smith et al. Radio-biological characterization of the inactivating events produced in mammalian cells by helium and heavy ions. Int. J. Radiat. Oncol. Biol. Phys. 3: 97-102 (1977).

C7   Cebulska-Wasilewska, A., H.P. Leenhouts and K.H. Chadwick. Synergism between EMS and X-rays for the induction of somatic mutations in Tradescantia. Int. J. Radiat. Biol. 40: 163-173 (1981).

C8   Case, R.A.M., M.E. Hosker, D.B. McDonald et al. Tumors of the urinary bladder in workmen engaged in the manufacture and use of certain dyestuf intermediates in the British chemical industry. I. The role of aniline, benzidine, alpha-naphtylamine and beta-naphtylamine. Br. J. Ind. Med. 11: 75-104 (1954).

C9   Chameaud, J., R. Perraud, J. Chretien et al. Lung carcinogenesis during in vivo cigarette smoking and radon daughter exposure in rats. Cancer Res. 82: 11-20 (1982).

C10   Crawford-Brown, D.J. Modeling the modification of the risk of radon-induced lung cancer by environmental tobacco smoke. Risk Anal. 12(4): 483-493 (1992).

C11   Cole, S., I.J. Stratford, J. Bowler et al. Oral (po) dosing with RSU 1069 or RB 6145 maintains their potency as hypoxic cell radiosensitizers and cytotoxins but reduces systemic toxicity compared with parenteral (ip) administration in mice. Int. J. Radiat. Oncol. Biol. Phys. 21: 387-395 (1991).

C12   Chung, M.H., H. Kasei, S. Nishimura et al. Protection of DNA damage by dietary restriction. Free Radic. Biol. Med. 12: 523-525 (1992).

C13   Craft, T.R., N.F. Cariello and T.R. Skopek. Mutational profile of the human hprt gene: summary of mutations observed to date. Environ. Mol. Mutagen. 17(19): 18-25 (1991).

C14   Chak, L.Y., B.I. Sikic, M.A. Tucker et al. Increased incidence of acute nonlymphocytic leukemia following therapy in patients with small cell carcinoma of the lung. J. Clin. Oncol. 2: 385-390 (1984).

C15   Curtis, S.B. Application of the LPL model to mixed radiations. p. 21-28 in: Biophysical Modelling of Radiation Effects (K.H. Chadwick, G. Moschini and M.N. Varma, eds.). Adam Hilger, Bristol, 1992.

C16   Curtis, S.B. Lethal and potentially lethal lesions induced by radiation - a unified repair model. Radiat. Res. 106: 252-270 (1986).

C17   Cole, J., M.H.L. Green, S.E. James et al. A further assessment of factors influencing measurements of thioguanine-resistant mutant frequency in circulating T-lymphocytes. Mutat. Res. 204(3): 493-507 (1988).

C18   Cole, J., C.F. Arlett, M.H.L. Green et al. Measurement of mutant frequency to 6-thioguanine resistance in circulating T-lymphocytes for human population monitoring. p. 175-203 in: New Trends in Genetic Risk Assessment (G. Jolles and A. Cordier, eds.). Academic Press, New York, 1989.

C19   Cole, J., M.H.L. Green, G. Stephens et al. HPRT somatic mutation data. p. 25-35 in: Mutation and the Environment, Fifth International Conference on Environmental Mutagens, Part C: Somatic and Heritable Mutation, Adduction and Epidemiology (M.L. Mendelsohn and R.J. Albertini, eds.). Wiley/Liss, New York, 1990.

C20   Cole, J., C.F. Arlett, M.H.L. Green et al. An assessment of the circulating T-lymphocyte 6-thioguanine resistance system for population monitoring. in: Biomonitoring and Carcinogen Risk Assessment (R.C. Garner, ed.). Oxford University Press, Oxford, 1991.

C21   Cross, F.T., R.F. Palmer, R.E. Filipy et al. Carcinogenic effects of radon daughters, uranium ore dust and cigarette smoke in Beagle dogs. Health Phys. 42: 33-52 (1982).

C22   Cross, F.T., R.F. Palmer, R.E. Filipy et al. Study of the combined effects of smoking and inhalation of uranium ore dust, radon daughters and diesel oil exhaust fumes in hamsters and dogs. PNL-2744 (1978).

C23   Cole, J. and T.R. Skopek. International Commission for Protection Against Environmental Mutagens and Carcinogens. Working paper no. 3. Somatic mutant frequency, mutation rates and mutational spectra in the human population in vivo. Mutat. Res. 304(1): 33-105 (1994).

C24   Chung, F.L., M.A. Morse, K.I. Eklind et al. Inhibition of tobacco-specific nitrosamine-induced lung tumorigenesis by compounds derived from cruciferous vegetables and green tea. Ann. N.Y. Acad. Sci. 686: 186-201 (discussion 201-202) (1993).

C25   Conney, A.H. Induction of microsomal enzymes by foreign chemicals and carcinogenesis by polycyclic aromatic hydrocarbons: G.H.A. Clowes Memorial Lecture. Cancer Res. 42(12): 4875-4917 (1982).

C26   Clayson, D.B., E.A. Nera and E. Lok. The potential for the use of cell proliferation studies in carcinogen risk assessment. Regul. Toxicol. Pharmacol. 9(3): 284-295 (1989).

C27   Cattanach, B.M. and C. Rasberry. Genetic effects of combined chemical X-ray treatments in male mouse germ cells. Int. J. Radiat. Biol. Relat. Stud. Phys. Chem. Med. 51(6): 985-996 (1987).

C28   Cattanach, B.M. and J.H. Barlow. Evidence for the re-establishment of a heterogeneity in radiosensitivity among spermatogonial stem cells repopulating the mouse testis following depletion by X-rays. Mutat. Res. 127(1): 81-91 (1984).

C29 Curtis, R.E., J.D. Boice Jr., M. Stovall et al. Risk of leukemia after chemotherapy and radiation treatment for breast cancer. N. Engl. J. Med. 326(26): 1745-1751 (1992).

C30 Colvin, M., R.B. Brundrett, W. Cowens et al. A chemical basis for the antitumor activity of chloroethylnitrosoureas. Biochem. Pharmacol. 25: 695-699 (1976).

C31 Coughlin, C.T. and R.C. Richmond. Biologic and clinical developments of cisplatin combined with radiation: concepts, utility, projections for new trials and the emergence of carboplatin. Semin. Oncol. 16: 31-43 (1989).

C32 Carde, P. and F. Laval. Effect of cis-dichlorodiammine platinum II and X rays on mammalian cell survival. Int. J. Radiat. Oncol. Biol. Phys. 7: 929-933 (1981).

C33 Choi, K.N., M. Rotman, H. Aziz et al. Locally advanced paranasal sinus and nasopharynx tumors treated with hyperfractionated radiation and concomitant infusion cisplatin. Cancer 67: 2748-2752 (1991).

C34 Crispino, S., G. Tancini, S. Barni et al. Simultaneous cisplatinum (CDDP) and radiotherapy in patients with locally advanced head and neck cancer. Proc. Am. Soc. Clin. Oncol. 6: 123 (Abstract) (1987).

C35 Coppin, C. Improved local control of invasive bladder cancer by concurrent cisplatin and preoperative or radical radiation. Proc. Am. Soc. Clin. Oncol. 11: 198 (Abstract) (1992).

C36 Cheng, Y.C. and K. Nakayama. Effects of 5-fluoro-2'-deoxyuridine on DNA metabolism in HeLa cells. Mol. Pharmacol. 23: 171-174 (1983).

C37 Chapman, J.D., A.P. Reuvers and J. Borsa. Effectiveness of nitrofuran derivatives in sensitizing hypoxic mammalian cells to x rays. Br. J. Radiol. 46: 623-630 (1973).

C38 Collins, A. and D.J. Oates. Hydroxyurea: effects on deoxyribonucleotide pool sizes correlated with effects on DNA repair in mammalian cells. Eur. J. Biochem. 169: 299-305 (1987).

C39 Chafouleas, J.G., W.E. Bolton and A.R. Means. Potentiation of bleomycin lethality by anticalmodulin drugs: a role for calmodulin in DNA repair. Science 224: 1346-1348 (1984).

C40 Choy, H., F.F. Rodriguez, S. Koester et al. Investigation of taxol as a potential radiation sensitizer. Cancer 71: 3774-3778 (1993).

C41 Covey, J.M., C. Jaxel, K.W. Kohn et al. Protein-linked DNA strand breaks induced in mammalian cells by camptothecin, an inhibitor of topoisomerase I. Cancer Res. 49: 5016-5022 (1989).

C42 Chaplin, D.J., P.L. Olive and R.E. Durand. Intermittent blood flow in a murine tumor: radiobiological effects. Cancer Res. 47: 597-601 (1987).

C43 Conard, R.A. Late radiation effects in Marshall Islanders exposed to fallout 28 years ago. p. 57-71 in: Radiation Carcinogenesis: Epidemiology and Biological Significance (J.D. Boice Jr. and J.R. Fraumeni Jr., eds.). Raven Press, New York, 1984.

C44 Cuzick, J., H. Stewart, R. Peto et al. Overview of randomised trials of postoperative adjuvant radiotherapy in breast cancer. Cancer Treat. Rep. 71: 15-29 (1987).

C45 Chadwick, K.H. and H.P. Leenhouts. The Molecular Theory of Radiation Biology. Springer Verlag, Berlin, 1981.

C46 Committee on the Biological Effects of Ionizing Radiations (BEIR VI). The Health Effects of Exposure to Indoor Radon. United States National Academy of Sciences, National Research Council. National Academy Press, Washington, 1998.

C47 Chen, A.Y., P. Okunieff, Y. Pommier et al. Mammalian DNA topoisomerase I mediates the enhancement of radiation cytotoxicity by camptothecin derivatives. Cancer Res. 57(8): 1529-1536 (1997).

C48 Cividalli, A., G. Arcangeli, G. Cruciani et al. Enhancement of radiation response by paclitaxel in mice according to different treatment schedules. Int. J. Radiat. Oncol. Biol. Phys. 40(15): 1163-1170 (1998).

C49 Cowdry, E.V., A. Pild, M.A. Croninger et al. Combined action of cigarette tar and beta radiation on mice. Cancer 2(14): 344-352 (1961).

C50 Cohen, S.M. and L.B. Ellwein. Cell proliferation in carcinogenesis. Science 249: 1007-1011 (1990).

C51 Cariello, N.F., G.R. Douglas, N.J. Gorelick et al. Databases and software for the analysis of mutations in the human p53 gene, human hprt gene and both the lacI and lacZ gene in transgenic rodents. Nucleic Acids Res. 26(1): 198-199 (1998).

C52 Cariello, N.F., T.R. Craft, H. Vrieling et al. Human HPRT mutant database: software for data entry and retrieval. Environ. Mol. Mutagen. 20: 81-83 (1992).

D1 Devasagayam, T.P.A., J.P. Kamat, H. Mohan et al. Caffeine as an antioxidat: inhibition of lipid peroxidation induced by reactive oxygen species. Biochim. Biophys. Acta 1282: 63-70 (1996).

D2 Dauncey, M.J. and H.L. Buttle. Differences in growth hormone and prolactin secretion associated with environmental temperature and energy intake. Horm. Metab. Res. 22(10): 524-527 (1990).

D3 Dewey, W.C. The search for critical cellular targets damaged by heat. Radiat. Res. 120: 191-204 (1989).

D4 Dalgleish, A.G. Human T cell leukaemia/lymphoma virus and blood donation. Br. Med. J. 307: 1224-1225 (1993).

D5 Dempsey, J.L., R.S. Seshadri and A.A. Morley. Increased mutation frequency following treatment with cancer chemotherapy. Cancer Res. 45(6): 2873-2871 (1985).

D6 De Luca, J.C., F.N. Dulot and J.M. Andrieu. The induction of reciprocal translocations in mouse germ cells by chemicals and ionizing radiations. I. Dose-response relationships and combined effects of bleomycin with thio-tepa and gamma-rays. Mutat. Res. 202(1): 65-70 (1988).

D7 De Luca, J.C., F.N. Dulot, M.A. Ulrich et al. The induction of reciprocal translocations in mouse germ cells by chemicals and ionizing radiations. II. Combined effects of mitomycin C or thio-tepa with two doses of gamma-rays. Mutat. Res. 232(1): 11-16 (1990).

D8 Dorie, M.J., D. Menke and J.M. Brown. Comparison of the enhancement of tumor responses to fractionated irradiation by SR 4233 (tirapazamine) and by nicotinamide with car-bogen. Int. J. Radiat. Oncol. Biol. Phys. 28: 145-150 (1994).

D9 Dixon, K., E. Roilides, J. Hauser et al. Studies on direct and indirect effects of DNA damage on mutagenesis in monkey cells using an SV40-based shuttle vector. Mutat. Res. 220(2-3): 73-82 (1989).

D10 Donham, K.J., L.A. Will, D. Denman et al. The combined effects of asbestos ingestion and localized X-irradiation of the colon in rats. J. Environ. Pathol., Toxicol. Oncol. 5(4-5): 299-308 (1984).

D11 Djuric, Z., L.K. Heilbrun, B.A. Reading et al. Effects of low-fat diet on levels of oxidative damage to DNA in human peripheral nucleated blood cells. J. Natl. Cancer Inst. 83: 766-769 (1991).

D12 Dische, S. Chemical sensitizers for hypoxic cells: a decade of experience in clinical radiotherapy. Radiother. Oncol. 3: 97-115 (1985).

D13 Deng, C., P. Zhang, J.W. Harper et al. Mice lacking p21CIP1/WAF1 undergo normal development, but are defective in G1 checkpoint control. Cell 82: 675-684 (1995).

D14 Deen, D.F., P.M. Bartle and M.E. Williams. Response of cultured 9L cells to spirohydantoin mustard and X-rays. Int. J. Radiat. Oncol. Biol. Phys. 5: 1663-1667 (1979).

D15 Deen, D.F. and M.E. Williams. Isobologram analysis of X-ray-BCNU interactions in vitro. Radiat. Res. 79: 483-491 (1979).

D16 Deen, D.F., M.E. Williams and K.T. Wheeler. Comparison of the CCNU and BCNU modification of the in vitro radiation response in 9L brain tumor cells of rats. Int. J. Radiat. Oncol. Biol. Phys. 5: 1541-1544 (1979).

D17 Dewit, L. Combined treatment of radiation and cisdiamminedichloroplatinum (II): a review of experimental and clinical data. Int. J. Radiat. Oncol. Biol. Phys. 13: 403-426 (1987).

D18 Douple, E.B. Interactions between platinum coordination complexes and radiation. p. 171-190 in: Interactions Between Antitumor Drugs and Radiation (B.T. Hill and A. Bellamy, eds.). CRC Press, Boca Raton, 1990.

D19 Douple, E.B. Platinum chemotherapy for cancer treatment. p. 349-376 in: Noble Metals and Biological Systems (R.R. Brooks, ed.). CRC Press, Boca Raton, 1993.

D20 Double, E.B. and R.C. Richmond. A review of interactions between platinum coordination complexes and ionizing radiation: implications for cancer therapy. p. 125-147 in: Cisplatin - Current Status and New Developments (A.W. Prestayko, S.T. Crooke and S.K. Carter, eds.). Academic Press, New York, 1980.

D21 Dritschilo, A., A.J. Piro and A.D. Kelman. The effect of cis-platinum on the repair of radiation damage in plateau phase Chinese hamster (V-79) cells. Int. J. Radiat. Oncol. Biol. Phys. 5: 1345-1349 (1979).

D22 Dische, S., M.I. Saunders, I.R. Flockhart et al. Misonidazole – a drug for trial in radiotherapy and oncology. Int. J. Radiat. Oncol. Biol. Phys. 5: 851-860 (1979).

D23 Durand, R.E. and S.L. Vanderbyl. Sequencing radiation and adriamycin exposures in spheroids to maximize therapeutic gain. Int. J. Radiat. Oncol. Biol. Phys. 17: 345-350 (1989).

D24 Donehower, R.C. and E.K. Rowinsky. Anticancer drugs derived from plants. p. 409-417 in: Cancer: Principles and Practice of Oncology, 4th edition (V.T. DeVita Jr., S. Hellman and S.A. Rosenberg, eds.). J.B. Lippincott Co., Philadelphia, 1993.

D25 Dische, S., P.J. Anderson, R. Sealy et al. Carcinoma of the cervix-anaemia, radiotherapy and hyperbaric oxygen. Br. J. Radiol. 56: 251-255 (1983).

D26 Dische, S. A review of hypoxic cell radiosensitization. Int. J. Radiat. Oncol. Biol. Phys. 20: 147-152 (1991).

D27 Dische, S., M.I. Saunders and M.F. Warburton. Hemoglobin, radiation, morbidity and survival. Int. J. Radiat. Oncol. Biol. Phys. 12: 1335-1337 (1986).

D28 Dische, S. Radiotherapy and anaemia - the clinical experience. Radiother. Oncol. 20 (Suppl. 1): 35-40 (1991).

D29 Deutsch, G., J.L. Foster, J.A. McFadzean et al. Human studies with "high dose" metronidazole: a non-toxic radio-sensitizer of hypoxic cells. Br. J. Cancer 31: 75-80 (1975).

D30 Douriez, E., P. Kermanach, P. Fritsch et al. Cocarcinogenic effect of cytochrome P-450 1A1 inducers for epidermoid lung tumour induction in rats previously exposed to radon. Radiat. Prot. Dosim. 56(1-4): 105-108 (1994).

D31 Doniach, I. Effects including carcinogenesis of I-131 and x-rays on the thyroid of experimental animals - a review. Health Phys. 9: 1357-1362 (1963).

D32 Donaldson, S.C., J.M. Click and J.R. Wilburn. Adriamycin activating a recall phenomenon after radiation therapy. Ann. Intern. Med. 81: 407-408 (1974).

D33 Darby, S., E. Whitley, P. Silocks et al. Risk of lung cancer associated with residential radon exposure in south-west England: a case-control study. Br. J. Cancer 78(3): 394-408 (1998).

D34 Dewey, W.C., C.C. Ling and R.E. Meyn. Radiation-induced apoptosis: relevance to radiotherapy. Int. J. Radiat. Oncol. Biol. Phys. 33(4): 781-796 (1995).

E1 European Chemical Industry Ecology and Toxicology Centre. DNA and Protein Adducts: Evaluation of their Use in Exposure Monitoring and Risk Assessment. Monograph No. 13. European Chemical Industry, Brussels, 1989.

E2 Eastmond, D.A. and J.D. Tucker. Identification of aneuploidy-inducing agents using cytokinesis-blocked human lymphocytes and an antikinetochore antibody. Environ. Mol. Mutagen. 13(1): 34-43 (1989).

E3 Ehrenberg, L., K.D. Hiesche, S. Osterman-Golkar et al. Evaluation of genetic risks of alkylating agents: tissue doses in the mouse from air contaminated with ethylene oxide. Mutat. Res. 24(2): 83-103 (1974).

E4 Environmental Protection Agency. Alpha-2u-globulin: association with chemically-induced renal toxicity and neoplasia in the male rat. EPA-625/3-91/019F (1991).

E5 Elkina, N.I. and A.I. Lumpov. Quantitative evaluation of the combined effect of external gamma radiation and internal alpha radiation from ^{239}Pu based on biochemical indices of the organs of the rat immune system. Radiobiologiya 28(5): 695-699 (1988). (In Russian).

E6 Elkina, N.I. and A.I. Maksutova. Aminotransferase and lactate dehydrogenase activity in the blood of dogs exposed to the joint action of external gamma- and internal alpha-radiation. Radiobiologiya 28(5): 657-659 (1988). (In Russian).

E7 Ehling, U.H. and A. Neuhäuser-Klaus. Induction of specific-locus and dominant-lethal mutations by cyclo-phosphamide and combined cyclophosphamide-radiation treatment in male mice. Mutat. Res. 199: 21-30 (1988).

E8 Edwards, D.I. and N.S. Virk. Repair of DNA damage induced by SR 4233. p. 229-230 in: Proceedings of the Seventh International Conference on Chemical Modifiers of Cancer Treatment. Clearwater, FLorida, 1991.

E9 Edwards, D.I. Nitroimidazole drugs - action and resistance mechanisms. I. Mechanisms of action. J. Antimicrob. Chemother. 31: 9-20 (1993).

E10 Ehrenberg, L., E. Moustacchi and S. Osterman-Golkar. International Commission for Protection Against Environmental Mutagens and Carcinogens. Dosimetry of genotoxic agents and dose-response relationships of their effects. Mutat. Res. 123(2): 121-182 (1983).

E11 Eng, C., F.P. Li, D.H. Abramson et al. Mortality from second tumors among long-term survivors of retinoblastoma [see comments]. J. Natl. Cancer Inst. 85(14): 1121-1128 (1993).

E12 Ewig, R.A. and K.W. Kohn. DNA damage and repair in mouse leukemia L1210 cells treated with nitrogen mustard, 1,3-bis(2-chloroethyl)-1-nitrosourea, and other nitrosoureas. Cancer Res. 37: 2114-2122 (1977).

E13 Ewig, R.A. and K.W. Kohn. DNA-protein cross-linking and DNA interstrand cross-linking by haloethylnitro-soureas in L1210 cells. Cancer Res. 38: 3197-3203 (1978).

E14 Eastman, A. Characterization of the adducts produced in DNA by cis-diamminedichloroplatinum(II) and cis-dichloro(ethylene-diamine)platinum(II). Biochemistry 22: 3927-3933 (1983).

E15 Eastman, A. Reevaluation of interaction of cis-dichloro (ethylene-diamine)platinum(II) with DNA. Biochemistry 25: 3912-3915 (1986).

E16 Erikson, R.L. and W. Szybalski. Molecular radiobiology of human cells. I: comparative sensitivity of x-ray and ultra-violet light of cells containing halogen-substituted DNA. Biochem. Biophys. Res. Commun. 4: 258-261 (1961).

E17 Erikson, R.L. and W. Szybalski. Molecular radiobiology of human cell lines. V: comparative radiosensitizing properties of 5-halodeoxycytidines and 5-halodeoxy-uridines. Radiat. Res. 20: 252-262 (1963).

E18 Erikson, R.L. and W. Szybalski. Molecular radiobiology of human cell lines. III: radiation-sensitizing properties of 5-iododeoxyuridine. Cancer Res. 23: 122-130 (1963).

E19 Egorova, E.I. and S.M. Polyakova. Enzyme activity of soil for the condition of combined γ-irradiation and heavy metal contamination. J. Radiationbiol. Radioecol. 36(2): 227-233 (1996). (In Russian).

E20 Egorova, E.I. Nitrogen fixation, dentrification and CO_2 flux under condition of combined γ-irradiation and heavy metal contamination in soil. J. Radiationbiol. Radioecol. 36(2): 219-226 (1996). (In Russian).

F1 Finch, G.L., P.J. Haley, M.D. Hoover et al. Interactions between inhaled beryllium metal and plutonium dioxide in rats: effects on lung clearance. p. 49-51 in: Proceedings of the Fourth International Conference on the Combined Effects of Environmental Factors (L. Fechter, ed.). John Hopkins University, Baltimore, 1991.

F2 Fasske, E. Experimental lung tumours following specific intrabronchial application of chrysotile asbestos. Longitudinal light and electron microscopic investigation in rats. Respiration 53(2): 111-127 (1988).

F3 Floyd, R.A. The role of 8-hydroxydeoxyguanosine in carcinogenesis. Carcinogenesis 11: 1447-1450 (1990).

F4 Fenech, M. and A.A. Morley. Measurement of micronuclei in lymphocytes. Mutat. Res. 147: 29-36 (1985).

F5 Fenech, M. and A.A. Morley. Cytokinesis-block micronucleus method in human lymphocytes: effect of in vivo ageing and low dose X-irradiation. Mutat. Res. 161(2): 193-198 (1986).

F6 Fearon, E. and B. Vogelstein. A gentic model for colorectal tumorgenesis. Cell 61: 759-767 (1990).

F7 Foulds, L. The experimental study of tumor progression: a review. Cancer Res. 14: 327-339 (1954).

F8 Fingerhut, M.A., W.E. Halferin, D.A. Marlow et al. Cancer mortality in workers exposed to 2,3,7,8-tetrachlorodi-benzo-p-dioxin. N. Engl. J. Med. 324: 212-496 (1991).

F9 Fisher, B., H. Rockette, E.R. Fisher et al. Leukemia in breast cancer patients following adjuvant chemotherapy or postoperative radiation: the NSABP experience. J. Clin. Oncol. 3: 1640-1658 (1985).

F10 Friedberg, E.C. DNA Repair. W.H. Freeman & Co., New York, 1985.

F11 Falk, S.J. and P.J. Smith. DNA damaging and cell cycle effects of the topoisomerase I poison camptothecin in irradiated human cells. Int. J. Radiat. Biol. 61: 749-57 (1992).

F12 Friedberg, E.C. and P.C. Hanawalt. DNA Repair. A Laboratory Manual of Research Procedures. Marcel Dekker Inc., New York, 1981.

F13 Fischer, J.J., S. Rockwell and D.F. Martin. Perfluoro-chemicals and hyperbaric oxygen in radiation therapy. Int. J. Radiat. Oncol. Biol. Phys. 12: 95-102 (1986).

F14 Fazekas, J., T.F. Pajak, T. Wasserman et al. Failure of misonidazole-sensitized radiotherapy to impact upon out-come among stage III-IV squamous cancers of the head and neck. Int. J. Radiat. Oncol. Biol. Phys. 13: 1155-1160 (1987).

F15 Finch, G.L., F.F. Hahn, W.W. Carlton et al. Combined exposure of F344 rats to beryllium metal and $^{239}PuO_2$ aerosols. p. 81-84 in: Annual Report 1993-1994, Inhalation Toxicology Research Institute (S.A. Belinsky et al., eds.). ITRI-144 (1994).

F16 Fishel, R., M.K. Lescoe, M.S.R. Rao et al. The human mutator gene homolog MSH2 and its assiciation with hereditary nonpolyposis colon cancer. Cell 75: 1027-1038 (1993).

F17 Finch, G.L., K.J. Nikula, E.B. Barr et al. Exposure of F344 rats to aerosols of $^{239}PuO_2$ and chronically inhaled cigarette smoke. p. 75-77 in: Annual Report 1993-1994, Inhalation Toxicology Research Institute (S.A. Belinsky et al., eds.). ITRI-144 (1994).

F18 Farooqi, Z. and P.C. Kesavan. Low-dose radiation-induced adaptive response in bone marrow cells of mice. Mutat. Res. 302(2): 83-89 (1993).

F19 Farooqi, Z. and P.C. Kesavan. Radioprotection by caffeine pre- and post-treatment in the bone marrow chromosomes of mice given whole-body gamma-irradiation. Mutat. Res. 269: 225-230 (1992).

F20 Friedberg, E.C., G.C. Walker and W. Siede. DNA Repair And Mutagenesis. American Society for Microbiology Press, Washington, D.C., 1995.

F21 Fichtinger-Schepman, A.M., A.T. van Oosterom, P.H. Lohman et al. cis-diamminedichloroplatinum(II)-induced DNA adducts in peripheral leukocytes from seven cancer patients: quantitative immunochemical detection of the adduct induction and removal after a single dose of cis-diamminedichloroplatinum(II). Cancer Res. 47: 3000-3004 (1987).

F22 Fichtinger-Schepman, A.M., F.J. van Dijk, W.H. De Jong et al. In vivo cis-diamminedichloroplatinum (II)-DNA adduct formation and removal as measured with immuno-chemical techniques. p. 32-46 in: Platinum and Other Metal Coordination Compounds in Cancer Chemotherapy (M. Nicolini, ed.). Martinum Nijhoff, Boston, 1988.

F23 Fielden, E.M., S.C. Lillicrap and A.B. Robins. The effect of 5-bromouracil on energy transfer in DNA and related model systems: DNA with incorporated 5-BUdR. Radiat. Res. 48: 421-431 (1971).

F24 Fram, R.J. and D.W. Kufe. Effect of 1-beta-D-arabino-furanosyl cytosine and hydroxyurea on the repair of X-ray-induced DNA single-strand breaks in human leukemic blasts. Biochem. Pharmacol. 34: 2557-2560 (1985).

F25 Foster, J.L. and R.L. Willson. Radiosensitization of anoxic cells by metronidazole. Br. J. Radiol. 46: 234-235 (1973).

F26 Fowler, J.F., G.E. Adams and J. Denekamp. Radio-sensitizers of hypoxic cells in solid tumors. Cancer Treat. Rev. 3: 227-256 (1976).

F27 Franken, N.A., C.V. Van-Bree, J.B. Kipp et al. Modifi-cation of potentially lethal damage in irradiated Chinese hamster V79 cells after incorporation of halogenated pyrimidines. Int. J. Radiat. Biol. 72(1): 101-109 (1997).

F28 Finch, G.L., D.L. Lundgren, E.B. Barr et al. Chronic cigarette smoke exposure increases the pulmonary retention and radiation dose of ^{239}Pu inhaled as $^{239}PuO_2$ by F344 rats. Health Phys. 75(6): 597-609 (1998).

F29 Fickweiler, S., P. Steinbach et al. The combined effects of high-energy shock waves and ionising radiation on a human bladder cancer cell line. Ultrasound Med. Biol. 22(8): 1097-1102 (1996).

G1 Guerin, M.R. The Chemistry of Environmental Tobacco Smoke: Composition and Measurements. Lewis Publishing, Boca Raton, 1992.

G2 Gardiner, T.A., W.M. Amoaku and D.B. Archer. The combined effect of diabetes and ionising radiation on the retinal vasculature of the rat. Curr. Eye Res. 12(11): 1009-1014 (1993).

G3 Gebhart, E., J. Losing and F. Wopfner. Chromosome studies on lymphocytes of patients under cytostatic therapy. I. Conventional chromosome studies in cytostatic interval therapy. Hum. Genet. 55(1): 53-63 (1980).

G4 Geard, C.R., M. Rutledge-Freeman, R.C. Miller et al. Antipain and radiation effects on oncogenic transformation and sister chromatid exchanges in Syrian hamster embryo and mouse C3H10T½ cells. Carcinogenesis 2: 1229-1233 (1981).

G5 Geard, C.R., C.M. Shea and M.A. Georgsson. Paraquat and radiation effects on mouse C3H10T½ cells. Int. J. Radiat. Oncol. Biol. Phys. 10(8): 1407-1410 (1984).

G6 George, K.C., D. van Beuningen and C. Streffer. Growth, cell proliferation and morphological alterations of a mouse mammary carcinoma after exposure to X-rays and hyperthermia. Recent results. Cancer Res. 107: 113-117 (1988).

G7 Geard, C.R. Induction of sister chromatid exchange as a function of charged-particle linear energy transfer. Radiat. Res. 134: 187-192 (1993).

G8 Gartrell, M.J., J.C. Craun, D.S. Podrebarac et al. Pesticides, selected elements, and other chemicals in adult total diet samples, October 1980 - March 1982. J. Assoc. Off. Anal. Chem. 69: 146-159 (1986).

G9 Gledhill, B.L. and F. Mauro (eds.). New Horizons in Biological Dosimetry. Wiley/Liss, New York, 1991.

G10 German Radiation Protection Commission. Synergism and radiological protection. Comment of the Radiological Protection Commission, September 1977. Bundesanzeiger 29: 1-3 (1977).

G11 Guilmette, R.A., N.F. Johnson, G.J. Newton et al. Risks from radon progeny exposure: what we know, and what we need to know. Annu. Rev. Pharmacol. Toxicol. 31: 569-601 (1991).

G12 Goodhead, D.T. Biophysical models of radiation action; introductory review. p. 306-311 in: Proceedings of the 8th International Congress on Radiation Research (E.M. Fielden et al., eds.). Taylor and Francis, London, 1987.

G13 Grigor'ev, I.G., L.I. Beskhlebnova, Z.I. Mitiaeva et al. Combined effect of microwaves and gamma-rays on the imprinting of chickens, irradiated in early embryogenesis. Radiobiologiya 24(2): 204-207 (1984). (In Russian).

G14 Grundy, S., L. Varga and M.A. Bender. Sister chromatid exchange frequency in human lymphocytes exposed to ionizing radiation during in vivo and in vitro. Radiat. Res. 100: 47-54 (1984).

G15 Griem, M.L., R.A. Kleinermann, J.D. Boice Jr. et al. Cancer following radiotherapy for peptic ulcer. J. Natl. Cancer Inst. 86(11): 824-829 (1994).

G16 Grigor'ev, I.G., V.S. Stepanov, G.V. Batanov et al. Combined effect of microwave and ionizing radiation. Kosm. Biol. Aviakosm. Med. 21(4): 4-9 (1987). (In Russian).

G17 Goodhead, D.T., H.P. Leenhouts, H.G. Paretzke et al. Track structure approaches to the interpretation of radiation effects of DNA. Radiat. Prot. Dosim. 52: 217-223 (1994).

G18 Guadagny, F., M. Roselli, M.P. Fuggetta et al. Increased immunogenicity of murine lymphoma cells following exposure to gamma rays in vivo. Chemiotheraoia 3(6): 358-364 (1984).

G19 Graeber, T.G., C. Osmanian, T. Jacks et al. Hypoxia-mediated selection of cells with diminished apoptotic potential in solid tumours. Nature 379: 88-91 (1996).

G20 Gray, R., J. LaFuma, S.E. Parish et al. Lung tumors and radon inhalation in over 2000 rats: Approximate linearity across a wide range of doses and potention by tobacco smoke. p. 592-607 in: Life-Span Radiation Effects In Animals: What Can They Tell Us? (R.C. Thompson and J.A. Mahaffey, eds.). Department of Energy, Washington, D.C., 1986.

G21 Goodman, L.S. and A. Gilman. The Pharmacological Basis of Therapeutics. Macmillan, London, 1970.

G22 Goodman, L.S., M.M. Wintrobe, W. Dameshek et al. Nitrogen mustard therapy: use of methylbis(B-chlorethyl)-aminohydrochloride for Hodgkin's disease, lymphosarcoma, leukemia and certain allied and miscellaneous disorders. J. Am. Med. Assoc. 132: 126-132 (1946).

G23 Gralla, J.D., S. Sasse Dwight and L.G. Poljak. Formation of blocking lesions at identical DNA sequences by the nitrosourea and platinum classes of anticancer drugs. Cancer Res. 47: 5092-5096 (1987).

G24 Goffinet, D.R. and J.M. Brown. Comparison of intravenous and intra-arterial pyrimidine infusion as a means of radiosensitizing tumors in vivo. Radiology 124: 819-822 (1977).

G25 Grau, C. and J. Overgaard. Radiosensitizing and cytotoxic properties of mitomycin C in a C3H mouse mammary carcinoma in vivo. Int. J. Radiat. Oncol. Biol. Phys. 20: 265-269 (1991).

G26 Grau, C., M. Hoyer and J. Overgaard. The in vivo interaction between vincristine and radiation in a C3H mammary carcinoma and the feet of CDF1 mice. Int. J. Radiat. Oncol. Biol. Phys. 30: 1141-1146 (1994).

G27 Geard, C.R. and J.M. Jones. Radiation and taxol effects on synchronized human cervical carcinoma cells. Int. J. Radiat. Oncol. Biol. Phys. 29: 565-569 (1994).

G28 Giocanti, N., C. Hennequin, J. Balosso et al. DNA repair and cell cycle interactions in radiation sensitization by the topoisomerase II poison etoposide. Cancer Res. 53: 2105-2111 (1993).

G29 Goswami, P.C., M. Hill, R. Higashikubo et al. The suppression of the synthesis of a nuclear protein in cells blocked in G2 phase: identification of NP-170 as topoisomerase II. Radiat. Res. 132: 162-167 (1992).

G30 Gray, L.H., A.O. Conger, M. Ebert et al. The concentration of oxygen dissolved in tissues at the time of irradiation as a factor in radiotherapy. Br. J. Radiol. 36: 638-648 (1953).

G31 Goldstein, G.W. Evidence that lead acts as calcium substitute in second messenger metabolism. Neurotoxicology 14(2-3): 97-102 (1993).

G32 Greene, M.H., E.L. Harris, D.M. Gershenson et al. Melphalane may be a more potent leukemogen than cyclophosphamide. Ann. Intern. Med. 105: 360-367 (1986).

G33 Greene, M.H., J.D. Boice Jr., B.E. Greer et al. Acute non-lymphocytic leukemia after therapy with alkylating agents for ovarian cancer. N. Engl. J. Med. 307: 1416-1421 (1982).

G34 Gerask'in, S.A., V.G. Dikarrev, A.A. Udalova et al. The combined effect of ionizing irradiation and heavy metals on the frequency of chromosome abberations in spring barley leaf meristem. Russian J. Gen. 32(2): 246-254 (1996).

G35 Gerask'in, S.A., V.G. Dikarrev, N.S. Dikareva et al. Effect of ionizing or heavy metals on the frequency of chromosome abberations in spring barley leaf meristem. Russian J. Gen. 32(2): 240-245 (1996).

G36 Gorodetsky, R., L. Levdansky, I. Ringel et al. Paclitaxel-induced modification of the effect of radiation and alterations in the cell cycle in normal and tumour mammalian cells. Radiat. Res. 150: 283-291 (1998).

G37 Groisman, I.J., R. Koshy, F. Henkler et al. Downregulation of DNA excision repair by the hepatitis B virus-x protein occurs in p53-proficient and p53-deficient cells. Carcinogenesis 20: 479-483 (1999).

G38 Gordon, G.S. and E.E. Vokes. Chemoradiation for locally advanced, unresectable NSCLC. New standard of care, emerging strategies. Oncology 13(8): 1075-1088 (1999).

H1 Heidenreich, W.F., P. Jacob and H.G. Paretzke. Exact solutions of the clonal expansion model and their application to the incidence of solid tumors of atomic bomb survivors. Radiat. Environ. Biophys. 36: 45-58 (1997).

H2 Huttner, E., U. Mergner, R. Braun et al. Increased frequency of 6-thioguanine-resistant lymphocytes in peripheral blood of workers employed in cyclophosphamide production. Mutat. Res. 243(2): 101-107 (1990).

H3 Han, A. and M.M. Elkind. Enhanced transformation of mouse 10T½ cells by 12-0-tetradecanoylphorbol-13-acetate following exposure to X-rays or to fission-spectrum neutrons. Cancer Res. 42(2): 477-483 (1982).

H4 Hann, B.C. and D.P. Lane. The dominating effect of mutant p53. Nature Genet. 9: 221 (1991).

H5 Hagmar, L., A. Brogger, I.L. Hansteen et al. Cancer risk in humans predicted by increased levels of chromosomal aberrations in lymphocytes: Nordic Study Group on the Health Risk of Chromosome Damage. Cancer Res. 54: 2919-2922 (1994).

H6 Hasgekar, N.N., A.M. Pendse and V.S. Lalitha. Effect of irradiation on ethyl nitrosourea induced neural tumors in Wistar rat. Cancer Lett. 30(1): 85-90 (1986).

H7 Heinonen, O.P., D. Albanes, P.R. Taylor et al. Alpha-tocopherol and beta-carotene supplements and lung-cancer incidence in the alpha-tocopherol, beta-carotene cancer prevention study - effects of base-line characteristics and study compliance. J. Natl. Cancer Inst. 88(N21): 1560-1570 (1996).

H8 Hakoda, M., Y. Hirai, S. Kyoizumi et al. Molecular analyses of in vivo hprt mutant T cells from atomic bomb survivors. Environ. Mol. Mutagen. 13(1): 25-33 (1989).

H9 Harlan, A. and J. Costello. Silicosis and lung cancer in U.S. metal miners. Arch. Environ. Health 46(2): 82-86 (1991).

H10 Harrison, G.H. and E.K. Balcer-Kubiczek. Continuous-wave ultrasound and neoplastic transformation in vitro. Ultrasound Med. Biol. 15: 335-340 (1989).

H11 Hall, E.J. and T.K. Hei. Modulation factors in the expression of radiation-induced oncogenetic transformation. Environ. Health Perspect. 88: 149-155 (1990).

H12 Hofmann, C., R. Oslapas, R. Nayyar et al. Androgen-mediated development of irradiation-induced thyroid tumors in rats: dependence on animal age during interval of androgen replacement in castrated males. J. Natl. Cancer Inst. 77(1): 253-260 (1986).

H13 Hesketh, R. The Oncogene Handbook. Academic Press, London, 1994.

H14 Hornek, G., P. Rettberg, S. Kozubek et al. The influence of microgravity on repair of radiation-induced DNA damage in bacteria and human fibroblasts. Radiat. Res. 147: 376-384 (1997).

H15 Heddle, J.A., A.B. Kreijnsky and R.R. Marshall. Cellular sensitivity to mutagens and carcinogens in the chromosome-breakage and other cancer-prone syndroms. p. 203-234 in: Chromosome Mutation and Neoplasia (J. German, ed.). A.R. Liss, New York, 1983.

H16 Hemminki, K. Nucleic acid adducts of chemical carcinogens and mutagens. Arch. Toxicol. 52(4): 249-285 (1983).

H17 Horwich, A., S.B. Holliday, J.M. Deacon et al. A toxicity and pharmacokinetic study in man of the hypoxic-cell radiosensitiser RSU-1069. Br. J. Radiol. 59: 1238-1240 (1986).

H18 Hegi, M.E., P. Sagelsdorff and W.K. Lutz. Detection by ^{32}P-postlabeling of thymidine glycol in γ-irradiated DNA. Carcinogenesis 10: 43-47 (1991).

H19 Hancock, S.L., M.A. Tucker and R.T. Hoppe. Breast cancer after treatment for Hodgkin's disease. J. Natl. Cancer Inst. 85: 25-31 (1993).

H20 Hanawalt, P.C., P.K. Cooper, A.K. Ganesan et al. DNA repair in bacteria and mammalian cells. Annu. Rev. Biochem. 48: 783-836 (1979).

H21 Hetzel, F.W., M. Brown, N. Kaufman et al. Radiation sensitivity modification by chemotherapeutic agents. Cancer Clin. Trials 4: 177-182 (1981).

H22 Hill, B.T. Overview of experimental laboratory investigations of antitumor drug-radiation interactions. p. 225-246 in: Antitumor Drug Radiation Interactions (B.T. Hill and A.S. Bellamy, eds.). CRC Press, Boca Raton, 1990.

H23 Hill, B.T. Interactions between antitumour agents and radiation and the expression of resistance. Cancer Treat. Rev. 18: 149-190 (1991).

H24 Herman, T.S., B.A. Teicher, M.R. Pfeffer et al. Effect of acidic pH on radiosensitization of FSaIIC cells in vitro by misonidazole, etanidazole, or cisdiamminedichloroplatinum (II). Radiat. Res. 124: 28-33 (1990).

H25 Haselow, R.E., M.G. Warshaw, M.M. Oken et al. Radiation alone versus radiation with weekly low dose cisplatinum in unresectable cancer of the head and neck. p. 279-281 in: Head and Neck Cancer, Volume 2 (W.E. Fee Jr., H. Goepfert, M.E. Johns et al., eds.). B.C. Decker, Philadelphia, 1990.

H26 Heidelberger, C., N.K. Chaudhuari, P. Danenberg et al. Fluorinated pyrimidines: a new class of tumor inhibitory compounds. Nature 179: 663-666 (1957).

H27 Heidelberger, C., L. Griesbach, B.J. Montag et al. Studies on fluorinated pyrimidines. II: effects on transplantable tumors. Cancer Res. 18: 305-317 (1958).

H28 Heimburger, D.K., D.S. Shewach and T.S. Lawrence. The effect of fluorodeoxyuridine on sublethal damage repair in human colon cancer cells. Int. J. Radiat. Oncol. Biol. Phys. 21: 983-987 (1991).

H29 Humphrey, R.M., W.C. Dewey and A. Cork. Effect of oxygen in mammalian cells sensitized by incorporation of 5-bromo-deoxyuridine into DNA. Nature 198: 268-269 (1963).

H30 Hill, B.T. and A.S. Bellamy. An Overview of experimental investigations of interaction between certain antitumor drugs and X-irradiation in vitro. Adv. Radiat. Biol. 11: 211-267 (1984).

H31 Hansen, P.V. and D. Sorensen. Effect of vincristine or bleomycin on radiation-induced cell killing of mice spermatogonial stem cells: the importance of sequence and time interval. Int. J. Radiat. Oncol. Biol. Phys. 20: 339-341 (1991).

H32 Hei, T.K. and E.J. Hall. Taxol, radiation and oncogenic transformation. Cancer Res. 53: 1368-1372 (1993).

H33 Hei, T.K., C.Q. Piao, C.R. Geard et al. Taxol and ionizing radiation: interaction and mechanisms. Int. J. Radiat. Oncol. Biol. Phys. 29: 267-271 (1994).

H34 Holmes, F.A., R.S. Walters, R.L. Theriault et al. Phase II trial of taxol, an active drug in the treatment of metastatic breast cancer. J. Natl. Cancer Inst. 83: 1797-1805 (1991).

H35 Hsiang, Y.H., M.G. Lihou and L.F. Liu. Arrest of replication forks by drug-stabilized topoisomerase. I. DNA cleavable complexes as a mechanism of cell killing by camptothecin. Cancer Res. 49: 5077-5082 (1989).

H36 Hall, E.J. Radiobiology for the Radiologist, Third edition. J.B. Lippincott Co., Philadelphia, 1988.

H37 Hempelmann, L.H. and J. Furth. Etiology of thyroid cancer. p. 37-49 in: Thyroid Cancer (L.D. Greenfield, ed.). CRC Press, West Palm Beach, FL., 1978.

H38 Hartwig, A. Current aspects in metal genotoxicity. Biometals 8(1): 3-11 (1995).

H39 Hoshino, H. and H. Tanooka. Interval effect of β-irradiation and subsequent 4-nitroquinoline 1-oxide painting on skin tumor induction in mice. Cancer Res. 35: 3663-3666 (1975).

H40 Hallahan, D.E., S. Virudachalam, J.L. Schwartz et al. Inhibition of protein kinases sensitizes human tumor cells to ionizing radiation. Radiat. Res. 129: 345-350 (1992).

H41 Hellman, S. Principles of radiation therapy. p. 248-275 in: Cancer: Principles and Practice of Oncology, 4th edition (V.T. DeVita Jr., S. Hellman and S.A. Rosenberg, eds.). J.B. Lippincott Co., Philadelphia, 1993.

H42 Harris, J.R., M. Morrow and G. Bonadonna. Cancer of the breast. p. 1264-1332 in: Cancer: Principles and Practice of Oncology, 4th edition (V.T. DeVita Jr., S. Hellman and S.A. Rosenberg, eds.). J.B. Lippincott Co., Philadelphia, 1993.

H43 Hawkins, M.M., L.M. Wilson, H.S. Burton et al. Radiotherapy, alkylating agents, and risk of bone cancer after childhood cancer. J. Natl. Cancer Inst. 88: 270-278 (1996).

H44 Harms-Ringdahl, M., P. Nicotera and I.R. Radford. Radiation induced apoptosis. Mutat. Res. 366: 171-179 (1996).

H45 Holmberg, B. Magnetic fields and cancer: animal and cellular evidence an overview. Environ. Health Perspect. (Suppl. 2) 103: 63-67 (1995).

H46 Howe, G.R. and R.H. Stager. Risk of lung cancer mortality after exposure to radon decay products in the Beaverlodge cohort based on revised exposure estimates. Radiat. Res. 146(1): 37-42 (1996).

H47 Hornek, G. Radiobiological experiments in space: a review. Nucl. Tracks Radiat. Meas. 20: 185-205 (1992).

H48 Hartwig, A. Carcinogenicity of metal compounds: possible role of DNA repair. Toxicol. Lett. 102/103: 235-239 (1998).

H49 Hendry, J.H. and C.M. West. Apoptosis and mitotic cell death: their relative contributions to normal-tissue and tumour radiation response. Int. J. Radiat. Biol. 71(6): 709-719 (1997).

H50 Hortobagyi, G.N. Recent progress in clinical development of decetaxel (Taxotere). Semin. Oncol. 26 (3 Suppl. 9): 32-36 (1999).

H51 Herbst, R.S. and R. Lilenbaum. Gemcitabine and vinorelbine combinations in the treatment of non-small cell lung cancer. Semin. Oncol. 26 (5 Suppl. 16): 67-70 (1999).

H52 Horneck, G. Impact of microgravity on radiobiological processes and efficiency of DNA repair. Mutat. Res. 430: 221-228 (1999).

H53 Han, A. and M.M. Elkind. Ultraviolet light and x-ray damage interaction in Chinese hamster cells. Radiat. Res. 74: 88-100 (1978).

I1 International Agency for Research on Cancer. Genetic and related effects: An updating of selected IARC Monographs, Volumes 1-42. IARC Monographs on the Evaluation of Carcinogenic Risks to Humans, Suppl. 6 (1987).

I2 International Agency for Research on Cancer. Overall evaluations of carcinogenicity: an updating of IARC Monographs, Volumes 1-42. IARC Monographs on the Evaluation of Carcinogenic Risks to Humans, Suppl. 7 (1987).

I3 International Commission on Radiation Units and Measurements. Quantitative concepts and dosimetry in radiobiology. ICRU Report 30: 14-15 (1979).

I4 Ivanitskaia, N.F. Evaluation of combined effects of ionizing radiation and mercury on the reproductive function of animals. Gig. Sanit. 12: 48-52 (1991).

I5 Ishikawa, Y., T. Mori, Y. Kato et al. Lung cancers associated with thorotrast exposure: high incidence of small-cell carcinoma and implications for estimation of radon risk. Int. J. Cancer 52(4): 570-574 (1992).

I6 Ingle, C.A. and N.R. Drinkwater. Mutational specificities of 1'-acetoxysafrole, N-benzoyloxy-N-methyl-4-aminoazobenzene, and ethyl methanesulfonate in human cells. Mutat. Res. 220: 133-142 (1989).

I7 International Commission on Radiation Units and Measurements. Microdosimetry. ICRU Report 36 (1983).

I8 International Agency for Research on Cancer. Cancer: causes, occurrence and control (L. Tomatis, A. Aitio, N.E. Day et al., eds.). IARC Scientific Publication 100 (1990).

I9 Inskip, P.D., M. Stovall and J.T. Flannery. Lung cancer risk and radiation dose among women treated for breast cancer. J. Natl. Cancer Inst. 86(13): 983-988 (1994).

I10 Inskip, P.D. and J.D. Boice Jr. Radiotherapy-induced lung cancer among women who smoke. Cancer 73(6): 1541-1543 (1994).

I11 Inano, H., K. Suzuki, H. Ishii-Ohba et al. Pregnancy-dependent initiation in tumorigenesis of Wistar rat mammary glands by ^{60}Co-irradiation. Carcinogenesis 12(6): 1085-1090 (1991).

I12 International Agency for Research on Cancer. Monographs on the evaluation of carcinogenic risk to humans, man-made fibers and radon. IARC Publication 43 (1988).

I13 Institute of Biophysics. Problems in Establishing Norms for Ionizing Radiation in the Presence of Modifying Factors. Ministry of Health, Moscow, 1991. (In Russian).

I14 International Commission on Radiological Protection. 1990 Recommendations of the International Commission on Radiological Protection. ICRP Publication 60. Annals of the ICRP 21(1-3). Pergamon Press, Oxford, 1991.

I15 Ingram, M.L. and J.L. Redpath. Subadditive interaction of radiation and taxol in vitro. Int. J. Radiat. Oncol. Biol. Phys. 37(5): 1139-1144 (1997).

J1 Jaffe, D.R. and G.T. Bowden. Ionizing radiation as an initiator: effects of proliferation and promotion time on tumor incidence in mice. Cancer Res. 47: 6692-6696 (1987).

J2 Jaffe, D.R., J.F. Williamson and G.T. Bowden. Ionizing radiation enhances malignant progression of mouse skin tumors. Carcinogenesis 8(11): 1753-1755 (1987).

J3 Janatipour, M., K.J. Trainor, R. Kutlaca et al. Mutations in human lymphocytes studied by an HLA selection system. Mutat. Res. 198(1): 221-226 (1988).

J4 Jovanovic, S.V. and M.G. Simic. The DNA-guanyl radical: kinetics and mechanisms of generation and repair. Biochim. Biophys. Acta 1008: 39-44 (1989).

J5 Jenner, T.J., P. O'Neill, M.A. Naylor et al. The repair of DNA damage induced in V79 mammalian cells by the nitroimidazole-aziridine, RSU-1069. Implications for radiosensitization. Biochem. Pharmacol. 42: 1705-1710 (1991).

J6 Jenkins, T.C., M.A. Naylor, P. O'Neill et al. Synthesis and evaluation of alpha-[[(2-haloethyl)amino]methyl]-2-nitro-1H-imidazole-ethanols as prodrugs of alpha-[(1-aziridinyl)methyl]-2-nitro-1H-imidazole-1-ethanol (RSU-1069) and its analogues which are radiosensitizers and bioreductively activated cytotoxins. J. Med. Chem. 33: 2603-2610 (1990).

J7 Jovanovic, S.V. and M.G. Simic. Mechanisms of OH radical reaction with thymine and uracil derivatives. J. Am. Chem. Soc. 108: 5968-5972 (1986).

J8 Julia, A.M., P. Canal, D. Berg et al. Concomitant evaluation of efficiency, acute and delayed toxicities of combined treatment of radiation and CDDP on an in vivo model. Int. J. Radiat. Oncol. Biol. Phys. 20: 347-350 (1991).

J9 Jordan, V.C. Designer estrogens. Sci. Am. 279(Oct): 60-67 (1998).

J10 Joon, D.L., M. Hasegawa, C. Sikes et al. Supraadditive apoptotic response of R3327-G rat prostate tumours to androgen ablation and radiation. Int. J. Radiat. Oncol. Biol. Phys. 38(5): 1071-1077 (1997).

K1 Kasai, H., P.F. Crain, Y. Kuchino et al. Formation of 8-hydroxyguanine moiety in cellular DNA by agents producing oxygen radicals and evidence for its repair. Carcinogenesis 7: 1849-1851 (1986).

K2 Kuo, M.L., K.A. Kunugi, M.J. Lindstrom et al. The interaction of hydroxyurea and ionizing radiation in human cervix carcinoma cells. Cancer J. Sci. Am. 3(3): 163-173 (1997).

K3 Kalmykova, Z.I., V.A. Chudin and Z.S. Men'shikh. Quantitative evaluation of the effectiveness of exposure of the joints to ^{239}Pu and tributyl phosphate by the degree of leukopenia in rats. Radiobiologiya 28(3): 420-424 (1988). (In Russian).

K4 Kellerer, A.M. and H.H. Rossi. A generalized formulation of dual radiation action. Radiat. Res. 75: 471-488 (1978).

K5 Kellerer, A.M. and H.H. Rossi. The theory of dual action. Curr. Top. Radiat. Res. 8: 85-158 (1972).

K6 Kaver, I., J.L. Ware, J.D. Wilson et al. Effect of radiation combined with hyperthermia on human prostatic carcinoma cell lines in culture. Urology 38(1): 88-92 (1991).

K7 Kasprzak, K.S. Possible role of oxidative damage in metal-induced carcinogenesis. Cancer Invest. 13(4): 411-430 (1995).

K8 Katz, R., B. Ackerson, M. Homayoonfar et al. Inactivation of cells by heavy ion bombardment. Radiat. Res. 47: 402-425 (1971).

K9 Koch, C.J. Unusual oxygen concentration dependence of toxicity of SR-4233, a hypoxic cell toxin. Cancer Res. 53: 3992-3997 (1993).

K10 Khan, S. and O.B. PJ. Molecular mechanisms of tirapazamine (SR 4233, Win 59075)-induced hepatocyte toxicity under low oxygen concentrations. Br. J. Cancer 71: 780-785 (1995).

K11 Kaldor, J.M. and N.E. Day. Interpretation of epidemiological studies in context of the multistage model of carcinogenesis. p. 21-57 in: Mechanisms of Environmental Carcinogenesis: Multistep Models of Carcinognesis, Volume 2 (J.C. Barret, ed.). CRC Press, Boca Raton, 1987.

K12 Kimura, H., H. Iyehara-Ogawa and T. Kato. Slippage-misalignment: to what extent does it contribute to mammalian cell mutagenesis. Mutagenesis 9(5): 395-400 (1994).

K13 Knizhnikov, V.A., V.A. Grozovskaya, N.N. Litvinov et al. Appearance of malignant tumors in the lungs of mice after inhalation of shistous ash, 3,4-benzopyzene and ^{210}Po. in: Hygienic Problems of Radiation and Chemical Carcinogenesis (L.A. Iljin and V.A. Knizhnikov, eds.). Institute of Biophysics, Moscow, 1979. (In Russian).

K14 Knowles, J.F. The effect of x-radiation given after neonatal administration of ethylnitrosourea on incidence of induced nervous system tumors. Neoropathol. Applied Neurobiol. 8(4): 265-276 (1982).

K15 Knowles, J.F. Changing sensitivity of neonatal rats to tumorigenic effects of N-nitroso-N-ethylurea and x-radiation, given singly or combined. J. Natl. Cancer Inst. 74(4): 853-857 (1985).

K16 Kobayashi, H. and S. Sakuma. Biological effects of static gradient magnetic field on cultured mammalian cells and combined effects with ^{60}Co gamma rays. Nippon Acta Radiologica 52(12): 1679-1685 (1992).

K17 Koifman, S. Electromagnetic fields: a cancer promoter? Med. Hypotheses 41(1): 23-27 (1993).

K18 Kitagawa, T., K. Nomura and S. Sasaki. Induction by X-irradiation of adenosine triphosphatase-deficient islands in the rat liver and their characterization. Cancer Res. 45: 6078-6082 (1985).

K19 Kopp-Schneider, A. and C. Portier. Distinguishing between models of carcinogenesis: the role of clonal expansion. Fundam. Appl. Toxicol. 17: 601-613 (1991).

K20 Kuwabara, Y., S. Matsubara, S. Yoshimatsu et al. Combined effects of ultrasound and ionizing radiation on lymphocyte chromosomes. Acta Radiol. Oncol. 25(4-6): 291-294 (1986).

K21 Kusiak, R.A., J. Springer, A.C. Ritchie et al. Carcinoma of the lung in Ontario gold miners: possible aetiological factors. Br. J. Ind. Med. 48: 808-817 (1991).

K22 Kim, J.H., S.H. Kim, A. Kolozsvary et al. Potentiation of radiation response in human carcinoma cells in vitro and murine fibrosarcoma in vivo by topotecan, an inhibitor of DNA topoisomerase I. Int. J. Radiat. Oncol. Biol. Phys. 22: 515-518 (1992).

K23 Kyoizumi, S., M. Akiyama, Y. Hirai et al. Spontaneous loss and alteration of antigen receptor expression in mature CD4+ T cells. J. Exp. Med. 171(6): 1981-1999 (1990).

K24 Kyoizumi, S., S. Umeki, M. Akiyama et al. Frequency of mutant T lymphocytes defective in the expression of the T-cell antigen receptor gene among radiation-exposed people. Mutat. Res. 265(2): 173-180 (1992).

K25 Kesavan, P.C., G.J. Sharma and S.M.J. Afzal. Differential modification of oxic and anoxic radiation damage by chemicals (1) Simulation of the action of caffeine by certain inorganic radical scavengers. Radiat. Res. 75: 18-30 (1978).

K26 Kesavan, P.C. and A.T. Natarajan. Protection and potentiation of radiation clastogenesis by caffeine: nature of possible initial events. Mutat. Res. 143: 61-68 (1985).

K27 Kesavan, P.C. and E.L. Powers. Differential modification of oxic and anoxic components of radiation damage in Bacillus megaterium spores by caffeine. Int. J. Radiat. Biol. Relat. Stud. Phys. Chem. Med. 48: 223-233 (1985).

K28 Kesavan, P.C. Protection by caffeine against oxic radiation damage and chemical carcinogens: mechanistic considerations. Curr. Sci. 62: 791-797 (1992).

K29 Kesavan, P.C., S.P. Singh and N.K. Sah. Chemical modification of postirradiation damage under varying oxygen concentrations in barley seeds. Int. J. Radiat. Biol. 59(3): 729-737 (1991).

K30 Kohn, K.W. Interstrand cross-linking of DNA by 1,3-bis(2-chloroethyl)-1-nitrosourea and other 1-(2-haloethyl)-1-nitrosoureas. Cancer Res. 37: 1450-1454 (1977).

K31 Kann, H.E. Jr., M.A. Schott and A. Petkas. Effects of structure and chemical activity on the ability of nitrosoureas to inhibit DNA repair. Cancer Res. 40: 50-55 (1980).

K32 Kann, H.E. Jr., B.A. Blumenstein, A. Petkas et al. Radiation synergism by repair-inhibiting nitrosoureas in L1210 cells. Cancer Res. 40: 771-775 (1980).

K33 Korbelik, M. and K.A. Skov. Inactivation of hypoxic cells by cisplatin and radiation at clinically relevant doses. Radiat. Res. 119: 145-156 (1989).

K34 Kallman, R.F., D. Rapacchietta and M.S. Zaghloul. Schedule-dependent therapeutic gain from the combination of fractionated irradiation plus c-DDP and 5-FU or plus c-DDP and cyclophosphamide in C3H/Km mouse model systems. Int. J. Radiat. Oncol. Biol. Phys. 20: 227-232 (1991).

K35 Kelsen, D.P. and D.H. Ilson. Chemotherapy and combined-modality therapy for esophageal cancer. Chest 107: 224s-232s (1995).

K36 Kimler, B.F. and D.B. Leeper. Effect of hydroxyurea on radiation-induced division delay in CHO cells. Radiat. Res. 72: 265-276 (1977).

K37 Kuo, M.L., K.A. Kunugi, M.J. Lindstrom et al. The interaction of hydroxyurea and iododeoxyuridine on the radiosensitivity of human bladder cancer cells. Cancer Res. 55: 2800-2805 (1995).

K38 Kuo, M.T. and T.C. Hsu. Bleomycin causes release of nucleosomes from chromatin and chromosomes. Nature 271: 83-84 (1978).

K39 Kimler, B.F. The effect of bleomycin and irradiation on G2 progression. Int. J. Radiat. Oncol. Biol. Phys. 5: 1523-1526 (1979).

K40 Krishan, A., K. Paika and E. Frei III. Cytofluorometric studies on the action of podophyllotoxin and epipodophyllotoxins (VM-26, VP-16-213) on the cell cycle traverse of human lymphoblasts. J. Cell Biol. 66: 521-530 (1975).

K41 Kobayashi, S., H. Otsu, Y. Noda et al. Comparison of dose-dependent enhancing effects of γ-ray irradiation on urethan-induced lung tumorigenesis in athymic nude (nu/nu) mice and euthymic (nu/+) littermates. J. Cancer Res. Clin. Oncol. 122: 231-236 (1996).

K42 Kryukov, V.I., V.A. Shishkin and S.F. Sokolenko. Chronical influence of radiation and lead on mutation rates in plants of arabidopsis thaliana (L.) heyhn. J. Radiationbiol. Radioecol. 36(2):209-218 (1996). (In Russian).

K43 Kallman, R.F. The importance of schedule and drug dose intensity in combinations of modalities. Int. J. Radiat. Oncol. Biol. Phys. 28: 761-771 (1994).

K44 Kaldor, J.M., N.E. Day, A. Clarke et al. Leukemia following Hodgkin's disease. N. Engl. J. Med. 322: 7-13 (1990).

K45 Kai, M., E.G. Luebeck and S.H. Moolgavkar. Analysis of solid cancer incidence among atomic bomb survivors using a two-stage model of carcinogenesis. Radiat. Res. 148: 348-358 (1997).

K46 Krewski, D., M.J. Goddard and J.M. Zielinski. Dose-response relationships in carcinogenesis. p. 579-599 in: Mechanisms of Carcinogenesis in Risk Identification (H. Vainio, P.N. Magee, D.B. McGregor et al., eds.). IARC Scientific Publication 116 (1992).

K47 Kasai, H. and S. Nishimura. Hydroxylation of deoxyguanosine at C-8 position by ascorbic acid and other reducing agents. Nucleic Acids Res. 12: 2137-2145 (1984).

K48 Kusiak, R.A., A.C. Ritchie et al. Mortality from lung cancer in Ontario uranium miners. Br. J. Ind. Med. 50: 920-928 (1993).

K49 Kushneva, V.S. On the problem of the long-term effects of the combined injury to animals of silicon dioxide and radon. p. 21-28 in: Long-term Effects of Injuries Caused by the Action of Ionizing Radiation (D.I. Zatukinski, ed.). AEC-tr-4473 (1959).

K50 Kusiak, R.A., A.C. Ritchie et al. Mortality from stomach cancer in Ontario miners. Br. J. Ind. Med. 50: 117-126 (1993).

K51 Kusiak, R.A. Lung cancer mortality in Ontario gold miners. Chron. Dis. Canada 13(6): 23-26 (1992).

K52 Komaki, R., C. Scott, J.S. Lee et al. Impact of adding concurrent chemotherapy to hyperfractionated radiotherapy for locally advanced non-small cell lung cancer (NSCLC): comparison of RTOG 83-11 and RTOG 91-06. Am. J. Clin. Oncol. 20(5): 435-440 (1997).

K53 Kreienbrock, L., M. Kreuzer, M. Gerken et al. Case-control study on lung cancer and residential radon in West Germany. Am. J. Epidemiol. (2000, in press).

K54 Kiefer, J., M. Kost and K. Schenk-Meuser. Radiation biology. p. 300-367 in: Biological and Medical Research in Space (D. Moore, P. Bie and H. Oser, eds.). Springer-Verlag, New York, 1996.

K55 Kiefer, J. and H.D. Pross. Space radiation effects and microgravity. Mutat. Res. 430: 307-313 (1999).

K56 Kiefer, J., A. Schreiber, F. Gutermuth et al. Mutation induction by different types of radiation at the HPRT locus. Mutat. Res. 431: 429-448 (1999).

L1 Langlois, R.G., W.L. Bigbee and R.H. Jensen. Measurements of the frequency of human erythrocytes with gene expression loss phenotypes at the glycophorin A locus. Hum. Genet. 74(4): 353-362 (1986).

L2 Lam, G.K.Y. The survival response of a biological system to mixed radiations. Radiat. Res. 110: 232-243 (1987).

L3 Langlois, R.G., W.L. Bigbee, S. Kyoizumi et al. Evidence for increased somatic cell mutations at the glycophorin A locus in atomic bomb survivors. Science 236: 445-448 (1987).

L4 Lam, G.K.Y. The combined effects of mixtures of ionizing radiations. J. Theor. Biol. 134(4): 531-546 (1988).

L5 Little, M. and C. Muirhead. Mechanistic models of carcinogenesis. Radiol. Prot. Bull. 164: 10-19 (1995).

L6 Langlois, R.G., N. Bigbee and R.H. Jensen. The glycophorin A assay for somatic cell mutations in humans HPRT somatic mutation data. p. 47-56 in: Mutation and the Environment, Fifth International Conference on Environmental Mutagens, Part C: Somatic and Heritable Mutation, Adduction and Epidemiology (M.L. Mendelson and R.J. Albertini, eds.). Wiley/Liss, New York, 1990.

L7 Lambert, B. Biological markers in exposed humans: gene mutation. p. 535-542 in: Mechanisms of Carcinogenesis in Risk Identification (H. Vainio, P.N. Magee, D.B. McGregor et al., eds.). IARC Scientific Publication 116 (1992).

L8 Loewe, S. The problem of synergism and antagonism of combined drugs. Arzneim.-Forsch. 3: 285-290 (1953).

L9 Leenhouts, H.P. and K.H. Chadwick. Use of a two-mutation carcinogenesis model for analysis of epidemiological data. p. 145-149 in: Health Effects of Low Dose Radiation: Challenges of the 21st Century. British Nuclear Energy Society, London, 1997.

L10 Leenhouts, H.P. and K.H. Chadwick. Analysis of the interaction of different radiations on the basis of DNA damage. p. 29-36 in: Biophysical Modelling of Radiation Effects (K.H. Chadwick, G. Moschini and M.N. Varma, eds.). Adam Hilger, Bristol, 1992.

L11 Leenhouts, H.P. and K.H. Chadwick. The influence of dose rate on the dose-effect relationship. J. Radiol. Prot. 10: 95-102 (1990).

L12 Lenoir, G.M., G.T. O'Conor and C.L.M. Olweny. Burkitts Lymphoma: A Human Cancer Model. WHO/IARC, 1985.

L13 Leenhouts, H.P., M.J. Sijsma, A. Cebulska-Wasilewska et al. The combined effect of DBE and X-rays on the induction of somatic mutations in Tradescantia. Int. J. Radiat. Biol. Relat. Stud. Phys. Chem. Med. 49(1): 109-119 (1986).

L14 Lehnert, S., W.B. Rybka and T.A. Seemayer. Amplification of the graft-versus-host reaction by partial body irradiation. Transplantation 41(6): 675-679 (1986).

L15 Leith, J.T. and S.F. Bliven. X ray responses of a human colon tumor cell line after exposure to the differentiation-inducing agent N-methylformamide: concentration dependence and reversibility characteristics. Int. J. Radiat. Oncol. Biol. Phys. 14(6): 1231-1237 (1988).

L16 Lystsov, V.N. and I.I. Samolinenko. Quantitative assessments of synergism. Radiobiologiya 25(1): 43-46 (1985). (In Russian).

L17 Lundgren, D.L., S.A. Belinsky, K.J. Nikula et al. Effects of combined exposure of F344 rats to inhaled $^{239}PuO_2$ and a chemical carcinogen. p. 78-80 in: Annual Report 1993-1994, Inhalation Toxicology Research Institute (S.A. Belinsky et al., eds.). ITRI-144 (1994).

L18 Lubin, J.H., J.D. Boice Jr., C. Edling et al. Radon and lung cancer risk: A joint analysis of 11 underground miners studies. NIH Publication No. 94-3644 (1994).

L19 Lundgren, D.L., F.F. Hahn, W.C. Griffith et al. Effects of combined exposure of F344 rats to $^{239}PuO_2$ and whole-body x-irradiation. p. 61-63 in: Annual Report 1992-1993, Inhalation Toxicology Research Institute (K.J. Nikula et al., eds.). ITRI-140 (1993).

L20 Lundgren, D.L., F.F. Hahn, W.C. Griffith et al. Effects of combined exposure of F344 rats to $^{239}PuO_2$ and whole-body x-irradiation. p. 115-117 in: Annual Report 1991-1992, Inhalation Toxicology Research Institute (G.L. Finch et al., eds.). LMF-138 (1992).

L21 Littlefield, L.G., S.P. Colyer, E.E. Joiner et al. Sister chromatid exchanges in human lymphocytes exposed to ionizing radiation during G_0. Radiat. Res. 78: 514-521 (1979).

L22 Livingston, G.K. and L.A. Dethlefsen. Effects of hyperthermia and x irradiation on sister chromatid exchange (SCE) frequency in Chinese hamster ovary (CHO) cells. Radiat. Res. 77: 512-520 (1979).

L23 Little, J.B. Influence of noncarcinogenic secondary factors on radiation carcinogenesis. Radiat. Res. 87: 240-250 (1981).

L24 Little, J.B. Low-dose radiation effects: interactions and synergism. Health Phys. 59(1): 49-55 (1990).

L25 Luchnik, N.V., N.A. Projadkova, T.V. Kondrashova et al. Production of sister chromatid exchange by irradiation during the G_1 stage. The probable role of 5-bromo-deoxyuridine. Mutat. Res. 190: 149-152 (1987).

L26 Luz, A., W.A. Linzner, V. Müller et al. Synergistic osteosarcoma induction by incorporation of the short-lived alpha-emitter ^{227}Th and low-level activity of the long-lived ^{227}Ac. p. 141-151 in: Biological Implications of Radio-nuclides Released from Nuclear Industries, Volume 1. IAEA Vienna, 1979.

L27 Levin, V.A., C.B. Wilson, R. Davis et al. A phase III comparison of BCNU, hydroxyurea, and radiation therapy to BCNU and radiation therapy for treatment of primary malignant gliomas. J. Neurosurg. 51: 526-532 (1979).

L28 Loewe, S. Die quantitativen Probleme der Pharmakologie. Ergebn. Physiol. 27: 47-187 (1928).

L29 Lartigau, E. and M. Guichard. Does tirapazamine (SR-4233) have any cytotoxic or sensitizing effect on three human tumour cell lines at clinically relevant partial oxygen pressure? Int. J. Radiat. Biol. 67: 211-216 (1995).

L30 Laderoute, K., P. Wardman and A.M. Rauth. Molecular mechanisms for the hypoxia-dependent activation of 3-amino-1,2,4-benzotriazine-1,4-dioxide (SR 4233). Biochem. Pharmacol. 37: 1487-1495 (1988).

L31 Lloyd, D. and R. Purrot. Chromosome aberration analysis in radiological protection dosimetry. Radiat. Prot. Dosim. 1(1): 19-28 (1981).

L32 Lane, D.P., C.A. Midgley, T.R. Hupp et al. On the regulation of the p53 tumour suppressor, and its role in the cellular response to DNA damage. Philos. Trans. R. Soc. Lond., Ser. B: Biol. Sci. 347(1319): 83-87 (1995).

L33 Lubin, J., J.D. Boice Jr. and J. Samet. Errors in exposure assessment, statistical power and the interpretation of residential radon studies. Radiat. Res. 144: 329-341 (1995).

L34 Lee, D.J., D. Cosmatos, V.A. Marcial et al. Results of an RTOG phase III trial (RTOG 85-27) comparing radiotherapy plus etanidazole with radiotherapy alone for locally advanced head and neck carcinomas [see com-ments]. Int. J. Radiat. Oncol. Biol. Phys. 32: 567-576 (1995).

L35 Lubin, J., J.D. Boice Jr., C. Edling et al. Lung cancer in radon-exposed miners and estimation of risk from indoor exposure. J. Natl. Cancer Inst. 87(11): 817-827 (1995).

L36 Lubin, J.H. and K. Steindorf. Cigarette use and the estimation of lung cancer attributable to radon in the United States. Radiat. Res. 141(1): 79-85 (1995).

L37 Ludlum, D.B. Alkylating agents and the nitrosoureas. p. 285-307 in: Cancer: A Comprehensive Treatise, Volume 5 (F.F. Becker, ed.). Plenum Press, New York, 1977.

L38 Lawley, P.D. and P. Brookes. Molecular mechanism of the cytotoxic action of difunctional alkylating agents and of resistance to this action. Nature 206: 480-483 (1965).

L39 Levin, V.A., P.H. Gutin and S. Leibel. Neoplasms of the central nervous system. p. 1679-1737 in: Cancer: Principles and Practice of Oncology, 4th edition (V.T. DeVita Jr., S. Hellman and S.A. Rosenberg, eds.). J.B. Lippincott Co., Philadelphia, 1993.

L40 Ludlum, D.B., B.S. Kramer, J. Wang et al. Reaction of 1,3-bis(2-chloroethyl)-1-nitrosourea with synthetic polynucleotides. Biochemistry 14: 5480-5485 (1975).

L41 Lelieveld, P., M.A. Scoles, J.M. Brown et al. The effect of treatment in fractionated schedules with the combination of X-irradiation and six cytotoxic drugs on the RIF-1 tumor and normal mouse skin. Int. J. Radiat. Oncol. Biol. Phys. 11: 111-121 (1985).

L42 Lustig, R., N. McIntosh Lowe, C. Rose et al. Phase I/II study of Fluorosol-DA and 100% oxygen as an adjuvant to radiation in the treatment of advanced squamous cell tumors of the head and neck. Int. J. Radiat. Oncol. Biol. Phys. 16: 1587-1593 (1989).

L43 Lillicrap, S.C. and E.M. Fielden. The effect of 5-bromouracil on energy transfer in DNA and related model systems. Purine: pyrimidine crystal complexes. Radiat. Res. 48: 432-446 (1971).

L44   Liebmann, J., J.A. Cook, J. Fisher et al. Changes in radiation survival curve parameters in human tumor and rodent cells exposed to paclitaxel (Taxol). Int. J. Radiat. Oncol. Biol. Phys. 29: 559-564 (1994).

L45   Loike, J.D. and S.B. Horwitz. Effects of podophyllotoxin and VP-16-213 on microtubule assembly in vitro and nucleoside transport in HeLa cells. Biochemistry 15: 5435-5443 (1976).

L46   Long, B.H. and M.G. Brattain. The activity of etoposide (VP-16-213) and teniposide (VM-26) against human lung tumor cells in vitro: cytotoxicity and DNA breakage. p. 63-86 in: Etoposide (VP-16): Current Status and New Developments (B.F. Issell, F.M. Muggia and S.K. Carter, eds.). Academic Press, New York, 1984.

L47   Lam, G.K.Y. An isoeffect approach to the study of combined effects of mixed radiations - the nonparametric analysis of in vivo data. Radiat. Res. 119: 424-431 (1989).

L48   Leenhouts, H.P., K.H. Chadwick and D.F. Deen. An analysis of the interaction between two nitrosourea compounds and X-radiation in rat brain tumour cells. Int. J. Radiat. Biol. 37: 169-181 (1980).

L49   Lu, M.-H. and G.L. Kennedy. Teratogenic evaluation of mancozeb in the rat following inhalation exposure. Toxicol. Appl. Pharmacol. 84: 355-368 (1986).

L50   Leenhouts, H.P. and K.H. Chadwick. A molecular model for the cytotoxic action of UV and ionizing radiation. p. 71-81 in: Photobiology (E. Riklis, ed.). Plenum Press, New York, 1991.

L51   Leenhouts, H.P. and K.H. Chadwick. A quantitative analysis of the cytotoxic action of chemical mutagens. Mutat. Res. 129: 345-357 (1984).

L52   Leenhouts, H.P. and K.H. Chadwick. Fundamental aspects of the dose-effect relationship for ultraviolet radiation. p. 21-25 in: Human Exposure to Ultraviolet Radiation: Risks and Regulations (W.F. Passchier and B.F. M. Bosnjakovic, eds.). Elsevier Science Publishing B.V. (Biomedical Division), Amsterdam, 1987.

L53   Lo, Y.M., S. Darby, L. Noakes et al. Screening for codon 249 p53 mutation in lung cancer associated with domestic radon exposure. Lancet 345(8941): 60 (1995).

L54   Lawrence, T.S., A. Eisbruch and D.S. Shewach. Gecitabine-mediated radiosensitization. Semin. Oncol. 24 (2 Suppl. 7): 24-28 (1997).

L55   Little, J.B., R.B. McGandy and A. Kennedy. Interactions between polonium-210 a-radiation, benzo(a)pyrene, and 0.9% NaCI solution instillations in the induction of experimental lung cancer. Cancer Res. 38: 1929-1935 (1978).

L56   Lurie, A.G. and L.S. Cutler. Effects of low-level x-radiation on 7,12-dimethylbenz[a]anthracene-induced lingual tumors in Syrian golden hamsters. J. Natl. Cancer Inst. 63(1): 147-152 (1979).

L57   Leenhouts, H.P. Radon-induced lung cancer in smokers and non-smokers: risk implications using a two-mutation carcinogenesis model. Radiat. Environ. Biophys. 38: 57-71 (1999).

L58   Longstreth, J., F.R. de Gruijl, M.L. Kripke et al. Health risks. J. Photochem. Photobiol. B 46(1-3): 20-39 (1998).

M1   Meadows, A.T. Risk factors for second malignant neoplasms: report from the Late Effects Study Group. Bull. Cancer (FR) 75: 125-130 (1988).

M2   Mandybur, T.I., I. Ormsby, S. Samuels et al. Neural, pituitary and mammary tumors in Sprague-Dawley rats treated with X irradiation to the head and N-Ethyl-N-nitrosourea (ENU) during the early postnatal period: a statistical study of tumor incidence and survival. Radiat. Res. 101(3): 460-472 (1985).

M3   Müller, W.-U. and C. Streffer. Risk to preimplantation mouse embryos of combinations of heavy metals and radiation. Int. J. Radiat. Biol. 51(6): 997-1006 (1987).

M4   Maisin, J.R. Chemical radioprotection: past, present and future prospects. Int. J. Radiat. Biol. 73: 443-450 (1998).

M5   Mihara, K., L. Bal, Y. Kano et al. Malignant transformation of human fibroblasts previously immortalized with ⁶⁰Co gamma rays. Int. J. Cancer 50: 639-643 (1992).

M6   Maisin, J.R., G. Mattelin and M. Lambiet-Collier. Chemical protection against the long-term effects of a single whole-body exposure of mice to ionizing radiation. I. Life shortening. Radiat. Res. 71: 119-131 (1977).

M7   Maisin, J.R., A. Declève, G.B. Gerber et al. Chemical protection against the long-term effects of a single whole-body exposure of mice to ionizing radiation. II. Causes of death. Radiat. Res. 74: 415-435 (1978).

M8   Maisin, J.R., L. De Saint-Georges, M. Janowski et al. Effect of X-rays alone or combined with diethylnitrosamine on cancer induction in mouse liver. Int. J. Radiat. Biol. Relat. Stud. Phys. Chem. Med. 51(6): 1049-1057 (1987).

M9   McTiernan, A.M., N.S. Weiss and J.R. Daling. Incidence of thyroid cancer in women in relation to reproductive and hormonal factors. Am. J. Epidemiol. 120: 423-435 (1984).

M10   Maher, V.M., J.L. Yang, M.C. Mah et al. Comparing the frequency and spectra of mutations induced when an SV-40 based shuttle vector containing covalently bound residues of structurally-related carcinogens replicates in human cells. Mutat. Res. 220(2-3): 83-92 (1989).

M11   Müller, W.U., C. Streffer and R. Wurm. Supraadditive formation of micronuclei in preimplantation mouse embryos in vitro after combined treatment with X-rays and caffeine. Teratog. Carcinog. Mutagen. 5(2): 123-131 (1985).

M12   Müller, W.U., C. Streffer and C. Fischer-Lado. Effects of a combination of X-rays and caffeine on preimplantation mouse embryos in vitro. Radiat. Environ. Biophys. 22(2): 85-93 (1983).

M13   Moskalev, Y.I., I.K. Reitarovsky and I.K. Petrovich. The influence of ¹³¹I on the development of lesions brought by combined action of external and internal irradiation. p. 224-227 in: Distribution, Kinetic of Metabolism and Biological Action of Radioactive Isotopes of Iodine (Y.I. Moskalev, ed.). Medizina Moscow, 1970. (In Russian).

M14   Moolgavkar, S.H. and E.G. Luebeck. Multistage carcinogenesis: population-based model for colon cancer. J. Natl. Cancer Inst. 84: 610-618 (1992).

M15   Matsumoto, M., H. Takagi and N. Yoshimura. Synergistic suppression of retinal pigment epithelial cell proliferation in culture by radiation and hyperthermia. Invest. Ophthalmol. Visual Sci. 34(6): 2068-2073 (1993).

M16   Mauderly, J.L. Toxicological approaches to complex mixtures. Environ. Health Perspect. 101 (Suppl.): 155-165 (1993).

M17   McTiernan, A.M., N.S. Weiss and J.R. Daling. Incidence of thyroid cancer in women in relation to previous exposure to radiation therapy and history of thyroid disease. J. Natl. Cancer Inst. 73: 575-581 (1984).

M18   McCarron, M.A., A. Kutlaca and A.A. Morley. The HLA-A mutation assay: improved technique and normal results. Mutat. Res. 225(4): 189-193 (1989).

M19   Monchaux, G., J. Chameaud, J.P. Morlier et al. Translocation of subcutaneously injected chrysotile fibres: potential cocarcinogenic effect on lung cancer induced in rats by inhalation of radon and its daughters. p. 161-166 in: IARC Scientific Publication 90 (1989).

M20 Messing, K. and W.E. Bradley. In vivo mutant frequency rises among breast cancer patients after exposure to high doses of gamma-radiation. Mutat. Res. 152(1): 107-112 (1985).

M21 Mohrenweiser, H.W. and I.M. Jones. Review of the molecular characteristics of gene mutations of the germline and somatic cells of the human. Mutat. Res. 231(1): 87-108 (1990).

M22 Morley, A.A., K.J. Trainor, R. Seshadri et al. Measurement of in vivo mutations in human lymphocytes. Nature 302: 155-156 (1983).

M23 Moolgavkar, S. and A. Knudson. Mutation and cancer: a model for human carcinogenesis. J. Natl. Cancer Inst. 66: 1037-1052 (1981).

M24 Montour, J.L., W. Dutz and L.S. Harris. Modification of radiation carcinogenesis by marihuana. Cancer 47(6): 1279-1285 (1981).

M25 Milner, J. Conformational hypothesis for the suppressor and promotor functions of p53 in cell growth control and in cancer. Proc. R. Soc. London 245: 139-145 (1991).

M26 Mori, H., H. Itawa, Y. Morishita et al. Synergistic effect of radiation on N-2-fluorenylacetamide-induced hepato-carcinogenesis in male ACI/N rats. Jpn. J. Cancer Res. 81: 975-978 (1990).

M27 Mottram, J.C. The origin of tar tumours in mice, whether from single cells or many cells. J. Pathol. 40: 407-409 (1935).

M28 Morgan, W.F. and P.E. Crossen. X-irradiation and sister chromatid exchanges in cultured human lymphocytes. Environ. Mutagen.2: 149-155 (1980).

M29 Moolgavkar, S. and D. Venzon. Two-event models for carcinogenesis: incidence curves for childhood and adult tumors. Math. Biosci. 47: 55-77 (1979).

M30 Moskalev, Y.I. and S.V. Kalistratova. Late effects for the yield of mammary tumors after incorporation of iodine-131 and combined treatment. Government Report of the USSR to UNSCEAR 78-4 (1978). (In Russian).

M31 Mirvish, S.S. Effects of vitamins C and E on N-nitroso compound formation, carcinogenesis and cancer. Cancer 58 (Suppl): 1842-1850 (1986).

M32 Morishita, Y., T. Tanaka, H. Mori et al. Effects of X-irradiation on N-methyl-N-nitrosourea-induced multiorgan carcinogenesis in rats. Jpn. J. Cancer Res. 84: 26-33 (1993).

M33 Maisin, J.R., J. Vankerkom, L. De Saint-Georges et al. Effect of neutrons alone or combined with diethyl-nitrosamine on tumor induction in the livers of infant C57BL mice. Radiat. Res. 142: 78-84 (1995).

M34 Maisin, J.R., J. Vankerkom, L. De Saint-Georges et al. Effect of X rays alone or combined with diethyl-nitrosamine on tumor induction in infant mouse liver. Radiat. Res. 133: 334-339 (1993).

M35 Molin, Y.N., R.Z. Sagdeev, T.V. Leshina et al. Magnetic Resonance and Related Phenomena (E. Kundla, E. Lippmaa, T. Saluvere, eds.). Springer Verlag, Berlin, 1979.

M36 Müller, W. and C. Streffer. Micronucleus assays. p. 1-134 in: Advances in Mutagenesis Research, Volume 5 (G. Obe, ed.). Springer Verlag, Berlin, 1994.

M37 Michel, C. and I. Balla. Interaction between radiation and cadmium or mercury in mouse embryos during organogenesis. Int. J. Radiat. Biol. Relat. Stud. Phys. Chem. Med. 51(6): 1007-1019 (1987).

M38 Merian, E. Metals and their Compunds in the Environment.VCH Verlagsgesellschaft, Weinheim/New York/Basel, 1991.

M39 Moolgavkar, S.H., E.G. Luebeck, D. Krewski et al. Radon, cigarette smoke and lung cancer: a re-analysis of the Colorado Plateau uranium miners' data [see comments]. Epidemiology 4(3): 204-217 (1993).

M40 Montgomery, J.A., R. James, G.S. McCaleb et al. The modes of decomposition of 1,3-bis(2-chloroethyl)-1-nitrosourea and related compounds. J. Med. Chem. 10: 668-674 (1967).

M41 Miller, A.C. and W.F. Blakely. Inhibition of glutathione reductase activity by a carbamoylating nitrosourea: effect on cellular radiosensitivity. Free Radic. Biol. Med. 12: 53-62 (1992).

M42 Macquet, J.P., K. Jankowski and J.L. Butour. Mass spectrometry study of DNA-cisplatin complexes: perturbation of guanine-cytosine base-pairs. Biochem. Biophys. Res. Commun. 92: 68-74 (1980).

M43 Miller, E.M. and T.J. Kinsella. Radiosensitization by fluorodeoxyuridine: effects of thymidylate synthase inhibition and cell synchronization. Cancer Res. 52: 1687-1694 (1992).

M44 McGinn, C.J., E.M. Miller, M.J. Lindstrom et al. The role of cell cycle redistribution in radiosensitization: implications regarding the mechanism of fluorodeoxy-uridine radiosensitization. Int. J. Radiat. Oncol. Biol. Phys. 30: 851-859 (1994).

M45 McGinn, C.J. and T.J. Kinsella. The clinical rationale for S-phase radiosensitization in human tumors. Curr. Probl. Cancer 17: 273-321 (1993).

M46 Miller, E.M., J.F. Fowler and T.J. Kinsella. Linear-quadratic analysis of radiosensitization by halogenated pyrimidines. I. Radiosensitization of human colon cancer cells by iododeoxyuridine. Radiat. Res. 131: 81-89 (1992).

M47 Miller, E.M., J.F. Fowler and T.J. Kinsella. Linear-quadratic analysis of radiosensitization by halogenated pyrimidines. II. Radiosensitization of human colon cancer cells by bromodeoxyuridine. Radiat. Res. 131: 90-97 (1992).

M48 Musk, S.R. and G.G. Steel. The inhibition of cellular recovery in human tumour cells by inhibitors of topoisomerase. Br. J. Cancer 62: 364-367 (1990).

M49 Mattern, M.R., G.A. Hofmann, F.L. McCabe et al. Synergistic cell killing by ionizing radiation and topoisomerase I inhibitor topotecan (SK&F 104864). Cancer Res. 51: 5813-5816 (1991).

M50 Minchinton, A.I., M.J. Lemmon, M. Tracy et al. Second-generation 1,2,4-benzotriazine 1,4-di-N-oxide bioreductive anti-tumor agents: pharmacology and activity *in vitro* and *in vivo*. Int. J. Radiat. Oncol. Biol. Phys. 22: 701-705 (1992).

M51 Millar, J.L., N.M. Blackett and B.N. Hudspith. Enhanced post-irradiation recovery of the haemopoietic system in animals pretreated with a variety of cytotoxic agents. Cell Tissue Kinet. 11: 543-553 (1978).

M52 Millar, J.L. and T.J. McElwain. Combinations of cytotoxic agents that have less than expected toxicity on normal tissues in mice. Antibiot. Chemother. 23: 271-282 (1978).

M53 Michalowski, A. On radiation damage to normal tissues and its treatment. I. Growth factors. Acta Oncol. 29: 1017-1023 (1990).

M54 Michalowski, A.S. On radiation damage to normal tissues and its treatment. II. Anti-inflammatory drugs. Acta Oncol. 33: 139-157 (1994).

M55 Madoc Jones, H. and F. Mauro. Interphase action of vinblastine and vincristine: differences in their lethal action through the mitotic cycle of cultured mammalian cells. J. Cell Physiol. 72: 185-196 (1968).

M56 Murray, D., A. Prager, R.E. Meyn et al. Radioprotective agents as modulators of cell and tissue radiosensitivity. Cancer Bull. 44: 137-144 (1992).

M57 Müller, W.-U. and C. Streffer. Enhancement of radiation effects by mercury in early postimplantation mouse embryos in vitro. Radiat. Environ. Biophys. 25: 213-217 (1986).

M58 Müller, W.-U. and C. Streffer. Time factors in combined exposures of mouse embryos to radiation and mercury. Radiat. Environ. Biophys. 27: 115-121 (1988).

M59 Müller, W.-U., C. Streffer and C. Fischer-Lahdo. Toxicity of sodium arsenite in mouse embryos in vitro and its influence on radiation risk. Arch. Toxicol. 59: 172-175 (1986).

M60 Medinsky, M.A., P.M. Schlosser and J.A. Bond. Critical issues in benzene toxicity and metabolism: the effect of interactions with other organic chemicals on risk assessment. Environ. Health Perspect. 102(9): 119-124 (1994).

M61 Monchaux, G., J.P. Morlier, M. Morin et al. Carcinogenic effects on rats of exposure to mixtures of diesel exhausts, radon and radon daughters. Am. Occup. Hyg. 38(1): 281-288 (1994).

M62 Monchaux, G., J.P. Morlier, M. Morin et al. Carcinogenic effects in rats of exposure to different minerals from metallic mine ores, radon and radon daughters. p.159-164 in: Cellular and Molecular Effects of Mineral and Synthetic Dusts and Fibres (J.M.G. Davis and M.-C. Jaurand, eds.). Nato ASI Series, Volume H85. Springer Verlag, Berlin-Heidelberg, 1994.

M63 Miyamoto, T., M. Wakabayashi and T. Terasima. Aclarubicin (aclacinomycin A) and irradiation: evaluation using HeLa cells. Radiology 149: 835-839 (1983).

M64 Maeda, H. Carcinogenesis via microbial infection. Gan to Kagaku Ryoho 25(10): 1474-1485 (1998).

M65 Magos, L. Epidemiological and experimental aspects of metal carcinogenesis: physiochemical properties, kinetics and the active species. Environ. Health Perspect. 95: 157-189 (1991).

M66 Miller, R.C., G. Randers-Person et al. The oncogenic transforming potential of the passage of single α particles through mammalian cell nuclei. Proc. Natl. Acad. Sci. U.S.A. 96: 19-22 (1999).

M67 Milross, C.G., K.A. Mason, N.R. Hunter et al. Enhanced radioresponse of paclitaxel-sensitive and -resistant tumours in vivo. Eur. J. Cancer 33(8): 1299-1308 (1997).

M68 Milas, L., M.M. Milas and K.A. Mason. Combination of taxanes with radiation: preclinical studies. Semin. Radiat. Oncol. 9 (2 Suppl. 1): 12-26 (1999).

M69 Muschel, R.J., D.E. Sato, W.G. McKenna et al. Radio-sensitization and apoptosis. Oncogene 17(25): 3359-3363 (1998).

M70 Milas, L., T. Fujii, N. Hunter et al. Enhancement of tumor radioresponse in vivo by gemcitabine. Cancer Res. 59(1): 107-114 (1999).

M71 Moore, D., P. Bie and H. Oser (eds.). Biological and Medical Research in Space. Springer-Verlag, New York, 1996.

N1 Nakashima, K., A. Kawamata, I. Fujiki et al. The individual and combined effects of X-irradiation and hyperthermia on early somite mouse embryos in culture. Teratology 44(6): 635-639 (1991).

N2 Nackerdien, Z., R. Olinski and M. Dizdaroglu. DNA base damage in chromatin of γ-irradiated cultured human cells. Free Radic. Res. Commun. 16: 259-273 (1992).

N3 Nakamura, N., S. Umeki, Y. Hirai et al. Evaluation of four somatic mutation assays for biological dosimetry of radiation-exposed people including atomic bomb survivors. p. 341-350 in: Progress in Clinical and Biological Research, Volume 372 (B.L. Gledhill and F. Mauro, eds.). Wiley/Liss, New York, 1991.

N4 Neugut A.I., T. Murray, J. Santos et al. Increased risk of lung cancer after breast cancer radiation therapy in cigarette smokers. Cancer 73(6): 1615-1620 (1994).

N5 Ngo, F.Q.H., E.A. Blakely and C.A. Tobias. Sequential exposures of mammalian cells to low- and high-LET radiations. I. Lethal effects following X-ray and Neon-ion irradiation. Radiat. Res. 87: 59-78 (1981).

N6 Nomura, T., H. Nakajiama, T. Hatanaka et al. Embryonic mutation as a possible cause of in utero carcinogenesis in mice relevated by postnatal treatment with 12-O.tetra-decanoylphorbol-13-acetate. Cancer Res. 50: 2135-2138 (1990).

N7 Nicklas, J.A., T.C. Hunter, L.M. Sullivan et al. Molecular analyses of in vivo hprt mutations in human T-lympho-cytes. I. Studies of low frequency 'spontaneous' mutants by Southern blots. Mutagenesis 2(5): 341-347 (1987).

N8 Nicklas, J.A., M.T. Falta, T.C. Hunter et al. Molecular analysis of in vivo hprt mutations in human T lymphocytes. V. Effects of total body irradiation secondary to radio-immunoglobulin therapy (RIT). Mutagenesis 5(5): 461-468 (1990). [published erratum appeared in Mutagenesis 6(1): 101 (1991)].

N9 Nicholson, W.I. Comparative dose-response relationships of asbestos fiber types: magnitudes and uncertainties. Ann. N.Y. Acad. Sci. 643: 74-84 (1991).

N10 Nishikawa, A., F. Furukawa, T. Imazawa et al. Effects of caffeine on glandular stomach carcinogenesis induced in rats by N-methyl-N-nitro-N-nitrosoguanidine and sodium chloride. Food Chem. Toxicol. 33: 21-26 (1995).

N11 Natarajan, A.T., J.D. Tucker and M.S. Sasaki. Monitoring cytogenetic damage in vivo. in: Methods to Assess DNA Damage and Repair, Chapter 8 (R.G. Tardiffet, ed.). John Wiley, London, 1994.

N12 Neta, R. and J.J. Oppenheim. Radioprotection with cytokines: a clarification of terminology. Cancer Cells 3: 457 (1991).

N13 Neta, R. and J.J. Oppenheim. Radioprotection with cytokines-learning from nature to cope with radiation damage. Cancer Cells 3: 391-396 (1991).

N14 Nriagu, J. A global assessment of natural sources of atmospheric trace metals. Nature 338: 47-49 (1989).

N15 Nakajima, Y., T. Miyamoto, M. Tanabe et al. Enhancement of mammalian cell killing by 5-fluorouracil in combination with X-rays. Cancer Res. 39: 3763-3767 (1979).

N16 National Institute of Health Consensus Conference (NIH). Adjuvant therapy for patients with colon and rectal cancer. J. Am. Med. Assoc. 264: 1444-1450 (1990).

N17 Ng, C.E., A.M. Bussey and G.P. Raaphorst. Inhibition of potentially lethal and sublethal damage repair by camptothecin and etoposide in human melanoma cell lines. Int. J. Radiat. Biol. 66: 49-57 (1994).

N18 Norimura, T., S. Nomoto, M. Katsuki et al. p53-dependent apoptosis suppresses radiation-induced teratogenesis. Nat. Med. 2(5): 577-580 (1996).

N19 Nomura, T. Quantitative studies on mutagenesis, terato-genesis and carcinogenesis in mice. p. 27-34 in: Problems of Threshold in Chemical Mutagenesis (Y. Tazima et al., eds.). The Environmental Mutagen Society, Japan, 1984.

N20 Nakamura, K. and S. Antoku. Radioprotective effects of nicorandil in cultured mammalian cells. Med. Sci. Res. 21: 501-502 (1993).

N21 Nakamura, K. and S. Antoku. Enhancement of x-ray cell killing in cultures mammalian cells by the protein phosphatase inhibitor calyculin A. Cancer Res. 54: 2088-2090 (1994).

N22 Neglia, N.P., A.T. Meadows, L.L. Robinson et al. Second neoplasms after acute lymphoblastic leukemia in childhood. N. Engl. J. Med. 325: 1330-1336 (1991).

N23 Nomura, T. X-ray induced germ-line mutation leading to tumors. Its manifestation in mice given urethane postnatally. Mutat. Res. 121: 59-65 (1983).

N24 Nomura, T. Role of radiation induced mutations in multigeneration carcinogenesis. p. 375-387 in: Perinatal and Multigeneration Carcinogenesis (N.P. Napalkov and J.M. Rice, eds.). IARC, Lyon, 1989.

N25 National Council on Radiation Protection and Measurements. Evaluation of occupational and environmental exposures to radon and radon daughters in the United States. NCRP Report No. 78 (1984).

N26 Nishigori, C., D.B. Yarosh, C. Donawho et al. The immune system in ultraviolet carcinogenesis. J. Invest. Dermatol. 1(2): 143-146 (1996).

O1 Ovcharenko, E.P. and T.P. Fomina. Combined effect of ^{239}Pu and external gamma radiation on pregnant and lactating rats. Radiobiologiya 23(2): 275-277 (1983).

O2 Ootsuyama, A. and H. Tanooka. The tumor-initiating and promoting effects of ionizing radiations in mouse skin. Jpn. J. Cancer Res. 78(11): 1203-1206 (1987).

O3 O'Hara, J., E.B. Douple and R.C. Richmond. Enhancement of radiation-induced cell kill by platinum complexes (carboplatin and iproplatin) in V79 cells. Int. J. Radiat. Oncol. Biol. Phys. 12: 1419-1422 (1986).

O4 Osoba, D., A.D. Flores, J.H. Hay et al. Phase I study of concurrent carboplatin and radiotherapy in previously untreated patients with stage III and IV head and neck cancer. Head Neck 13: 217-222 (1991).

O5 O'Connel, M., J. Martenson, T. Rich et al. Protracted venous infusion (PVI) 5-fluorouracil (5FU) as a component of effective combined modality post-operative surgical adjuvant therapy for high-risk rectal cancer. Proc. Am. Soc. Clin. Oncol. 12: 193 (1993).

O6 O'Dwyer, P., B. Leyland Jones, M.T. Alonso et al. Etoposide (VP-16-213): current status of an active anticancer drug. N. Engl. J. Med. 312: 692-700 (1985).

O7 Overgaard, J., H.S. Hansen, A.P. Andersen et al. Misonidazole combined with split-course radiotherapy in the treatment of invasive carcinoma of larynx and pharynx: report from the DAHANCA 2 study. Int. J. Radiat. Oncol. Biol. Phys. 16: 1065-1068 (1989).

O8 Overgaard, J., H.S. Hansen, B. Lindelov et al. Nimorazole as a hypoxic radiosensitizer in the treatment of supraglottic larynx and pharynx carcinoma. First Report from the Danish Head and Neck Cancer Study (DAHANCA) Protocol 5-85. Radiother. Oncol. 20 (Suppl. 1): 143-149 (1991).

O9 O'Neill, P., S.S. McNeil and T.C. Jenkins. Induction of DNA crosslinks in vitro upon reduction of the nitroimidazole-aziridines RSU-1069 and RSU-1131. Biochem. Pharmacol. 36: 1787-1792 (1987).

O10 Olive, P.L. Use of the comet assay to detect hypoxic cells in murine tumours and normal tissues exposed to bioreductive drugs. Acta Oncol. 34: 301-305 (1995).

O11 O'Neill, P. and E.M. Fielden. Primary free radical processes in DNA. Adv. Radiat. Biol. 17: 53-120 (1993).

O12 Ootsuyama, A. and H. Tanooka. Effect of an inhibitor of tumor promotion, α-difluoromethylornithine, on tumor induction by repeated beta irradiation in mice. Jpn. J. Cancer Res. 84: 34-36 (1993).

O13 O'Dwyer, P.J., P.H. Wiernik, M.B. Steward et al. Treatment of early stage Hodgkin's disease: a randomised trial of radiotherapy plus chemotherapy versus chemotherapy alone. p. 329-336 in: Malignant Lymphomas and Hodgkin's Disease: Experimental and Therapeutic Advances (F. Cavalli, G. Bonadonna, M. Rozencweig, eds.). Martinus Nijhoff, Boston, 1985.

O14 Omenn, G.S., G.E. Goodman, M.D. Thornquist et al. Effects of a combination of beta carotene and vitamin A on lung cancer and cardiovascular disease. N. Engl. J. Med. 334(18): 1150-1155 (1996).

O15 Olsen, J.H., J.D. Boice Jr., J.P.A. Jensen et al. Cancer among epileptic patients exposed to anticonvulsant drugs. J. Natl. Cancer Inst. 81(10): 803-808 (1989).

O16 Olsen, J.H., G. Schulgen, J.D. Boice Jr. et al. Antiepileptic treatment and risk for hepatobiliary cancer and malignant lymphoma. Cancer Res. 55(2): 294-297 (1995).

O17 Olsen, J.H., S. Garwicz, H. Hertz et al. Second malignant neoplasms after cancer in childhood or adolescence. Br. Med. J. 307: 1030-1036 (1993).

O18 Omura, M., S. Torigoe and N. Kubota. SN-38, a metabolite of the camptothecin derivative CPT-11, potentiates the cytotoxic effect of radiation in human colon adenocarcinoma cells grown as spheroids. Radiother. Oncol. 43(2): 197-201 (1997).

O19 Olive, P.L. and R.E. Durand. Apoptosis: an indicator of radiosensitivity in vitro? Int. J. Radiat. Biol. 71(6): 695-707 (1997).

O20 Ornstein, D.L., A.M. Nervi and J.R. Rigas. Docetexel (Taxotere) in combination chemotherapy and in association with thoracic radiotherapy for the treatment of non-small cell lung cancer. Thoracic Oncology Program. Ann. Oncol. 10 (Suppl. 5): S35-S40 (1999).

P1 Panov, D., V. Višnjić, Lj. Novak et al. Radiobiological effects of joint action of alpha radiation (Po-210) and quartz dust in the rats respiratory and renal system. p. 1165-1168 in: Proceedings of the Fourth Congress of International Radiation Protection Association, Vol. 4 (1977).

P2 Painter, R.B. and W.F. Morgan. SCE induced by ionizing radiation are not the result of exchanges between homologous chromosomes. Mutat. Res. 121: 205-210 (1983).

P3 Perera, F.P. and R.M. Santella. Carcinogenesis. p. 227-300 in: Molecular Epidemiology: Principles and Practices (P.A. Schulte and F.P. Perera, eds.). Academic Press, San Diego, 1993.

P4 Pedersen-Bjergaard, J., J. Ersboll, V.L. Hansen et al. Carcinoma of the urinary bladder after treatment with cyclophosphamide for non-Hodgkin's lymphoma. N. Engl. J. Med. 318: 1028-1032 (1988).

P5 Ponomareva, V.L., P.P. Lyarskii, L.N. Burykina et al. The results of experimental studies of the late effects of dust combined with radiation and the problems of hygienic standardization. in: Proceedings of the Radiobiological Conference of Socialist Countries, Varna, 1978. (In Russian).

P6 Pereira, M.A., F.J. Burns and R.E. Albert. Dose response for benzo(a)pyrene adducts in mouse epidermal DNA. Cancer Res. 39: 2556-2559 (1979).

P7 Phillips, R.A. and L.J. Tolmach. Repair of potentially lethal damage in x-irradiated HeLa cells. Radiat. Res. 29: 413-432 (1966).

P8   Piao, Ch.Q. and T.K. Hei. The biological effectiveness of radon daughter alpha particles. I. Radon, cigarette smoke and oncogenic transformation. Carcinogenesis 14(3): 497-501 (1993).

P9   Pedersen-Bjergaard, J., P. Philip, S.O. Larson et al. Chromosome aberrations and prognostic factors in therapy-related myelodysplasia and acute nonlymphocytic leukemia. Blood 76: 1083-1091 (1990).

P10   Palayoor, S.T., R.M. Macklis, E.A. Bump et al. Modulation of radiation-induced apoptosis and G2/M block in murine T-lymphoma cells. Radiat. Res. 141(3): 235-243 (1995).

P11   Pershagen, G., G. Akerblom, O. Axelson et al. Residential radon exposure and lung cancer in Sweden. N. Engl. J. Med. 330(3): 159-164 (1994).

P12   Pinto, A.L. and S.J. Lippard. Binding of the antitumor drug cis-diamminedichloroplatinum(II) (cisplatin) to DNA. Biochim. Biophys. Acta 780: 167-180 (1985).

P13   Pascoe, J.M. and J.J. Roberts. Interactions between mammalian cell DNA and inorganic platinum compounds. I. DNA interstrand cross-linking and cytotoxic properties of platinum(II) compounds. Biochem. Pharmacol. 23: 1359-1365 (1974).

P14   Pinto, A.L. and S.J. Lippard. Sequence-dependent termination of in vitro DNA synthesis by cis- and transdiammine-dichloro-platinum (II). Proc. Natl. Acad. Sci. U.S.A. 82: 4616-4619 (1985).

P15   Poirier, M.C., M.J. Egorin, A.M. Fichtinger Schepman et al. DNA adducts of cisplatin and carboplatin in tissues of cancer patients. p. 313-320 in: IARC Scientific Publication 89 (1988).

P16   Prusoff, W.H. A review of some aspects of 5-iododeoxy-uridine and azauridine. Cancer Res. 23: 1246-1259 (1963).

P17   Post, J. and J. Hoffman. The effects of 5-iodo-2'-deoxyuridine upon the replication of ileal and spleen cells in vivo. Cancer Res. 29: 1859-1863 (1969).

P18   Piver, M.S., A.E. Howes, H.D. Suit et al. Effect of hydroxyurea on the radiation response of C3H mouse mammary tumors. Cancer 29: 407-412 (1972).

P19   Piver, M.S., J.J. Barlow, V. Vongtama et al. Hydroxyurea: a radiation potentiator in carcinoma of the uterine cervix. A randomized double-blind study. Am. J. Obstet. Gynecol. 147: 803-808 (1983).

P20   Piver, M.S., M. Khalil and L.J. Emrich. Hydroxyurea plus pelvic irradiation versus placebo plus pelvic irradiation in nonsurgically staged stage IIIB cervical cancer. J. Surg. Oncol. 42: 120-125 (1989).

P21   Povirk, L.F., Y.H. Han and R.J. Steighner. Structure of bleomycin-induced DNA double-strand breaks: predominance of blunt ends and single-base 5' extensions. Biochemistry 28: 5808-5814 (1989).

P22   Parness, J. and S.B. Horwitz. Taxol binds to polymerized tubulin in vitro. J. Cell Biol. 91: 479-487 (1981).

P23   Powis, G. Metabolism and reactions of quinoid anticancer agents. Pharmacol. Ther. 35: 57-162 (1987).

P24   Phelps, T.A. and N.M. Blackett. Protection of intestinal damage by pretreatment with cytarabine (cytosine arabino-side). Int. J. Radiat. Oncol. Biol. Phys. 5: 1617-1620 (1979).

P25   Pacchiarotti, A., E. Martino, L. Bartalena et al. Serum thyrotropin by ultra sensitive immunoradiometric assay and serum free thyroid hormones in pregnancy. J. Endocrinol. Invest. 9: 185-189 (1986).

P26   Peraino, C., D.J. Grdina and B.A. Carnes. Synergistic induction of altered hepatocye foci by combined gamma radiation and diethylnitrosamine administered to neonatal rats. Carcinogenesis 7(3): 445-448 (1986).

P27   Privezentzev, C.V., N.P. Sirota and A.I. Gaziev. The influence of combined treatment of Cd and γ-irradiation on DNA damage and repair in lymphoid tissues of mice. J. Radiationbiol. Radioecol. 36(2): 234-240 (1996).

P28   Poblack, D.G., I.T. Magrath, L.E. Kun et al. Leukemias and lymphomas of childhood. p. 1792-1818 in: Cancer: Principles and Practice of Oncology, 4th edition (V.T. DeVita Jr., S. Hellman and S.A. Rosenberg, eds.). J.B. Lippincott Co., Philadelphia, 1993.

P29   Pui, C.H., R.C. Ribeiro, M.L. Hancock et al. Acute myeloid leukemia in children treated with epipodophyllotoxins for acute lymphoblastic leukemia. N. Engl. J. Med. 325: 1682-1687 (1991).

P30   Pedersen-Bjergaard, J., G. Daugaard, S.W. Hansem et al. Increased risk of myelodysplasia and leukemia after etoposide, cisplatin and bleomycin for germ-cell tumours. Lancet 338: 359-363 (1991).

P31   Pross, H.D. and J. Kiefer. Repair of cellular radiation damage in space under microgravity conditions. Radiat. Environ. Biophys. 38: 133-138 (1999).

P32   Pross, H.D., A. Casares and J. Kiefer. Induction and repair of DNA double-strand breaks under irradiation and microgravity. Radiat. Res. 153: 521-525 (2000).

R1   Rinsky, R.A., J.M. Melius, R.W. Hornung et al. Case-control study of lung cancer in civilian employees at the Portsmouth Naval Shipyard, Kittery, Maine. Am. J. Epidemiol. 127(1): 55-64 (1988).

R2   Rockwell, S. Use of hypoxia-directed drugs in the therapy of solid tumors. Semin. Oncol. 19: 29-40 (1992).

R3   Randall, K. and J.E. Coggle. The effect of whole-body gamma-irradiation on localized beta-irradiation-induced skin reactions in mice. Int. J. Radiat. Biol. 62(6): 729-733 (1992).

R4   Redpath, J.L. Mechanisms in combination therapy: isobologram analysis and sequencing. Int. J. Radiat. Biol. 38: 355-356 (1980).

R5   Renault, G., A. Gentil and I. Chouroulinkov. Kinetics of induction of sister-chromatid exchanges by X-rays through two cell cycles. Mutat. Res. 94: 359-368 (1982).

R6   Rous, P. and J.W. Beard. The progression to carcinoma of virus-induced rabbit papillomas (shape). J. Exp. Med. 62: 523-548 (1935).

R7   Rossi, A.M., J.C. Thijssen, A.D. Tates et al. Mutations affecting RNA splicing in man are detected more frequently in somatic than in germ cells. Mutat. Res. 244(4): 353-357 (1990).

R8   Recio, L., J. Cochrane, D. Simpson et al. DNA sequence analysis of in vivo hprt mutation in human T lymphocytes. Mutagenesis 5(5): 505-510 (1990).

R9   Richter, C., J.W. Park and B.N. Ames. Normal oxidative damage to mitochondrial and nuclear DNA is extensive. Proc. Natl. Acad. Sci. U.S.A. 85: 6465-6467 (1988).

R10   Rigaud, O., D. Papadopoulo and E. Moustacchi. Decreased deletion mutation in radioadapted human lymphoblasts. Radiat. Res. 133: 94-101 (1993).

R11   Ron, E., R.A. Kleinermann, J.D. Boice Jr. et al. A population-based case-control study of thyroid cancer. J. Natl. Cancer Inst. 79: 1-12 (1987).

R12   Rigaud, O. and E. Moustacchi. Radioadaptation to the mutagenic effect of ionizing radiation in human lymphoblasts: molecular analysis of HPRT mutants. Cancer Res. 54 (Suppl.): 1924-1928 (1994).

R13   Rothwell, K. Dose-related inhibition of chemical carcinogenesis in mouse skin by caffeine. Nature 252: 69-70 (1974).

R14 Rice, J.A., D.M. Crothers, A.L. Pinto et al. The major adduct of the antitumor drug cis-diamminedichloroplatinum(II) with DNA bends the duplex by approximately equal to 40 degrees toward the major groove. Proc. Natl. Acad. Sci. U.S.A. 85: 4158-4161 (1988).

R15 Reed, E. Platinum analogs. p. 390-400 in: Cancer: Principles and Practice of Oncology, 4th edition (V.T. DeVita Jr., S. Hellman and S.A. Rosenberg, eds.). J.B. Lippincott Co., Philadelphia, 1993.

R16 Roman, D.D. and P.W. Sperduto. Neuropsychological effects of cranial radiation: current knowledge and future directions. Int. J. Radiat. Oncol. Biol. Phys. 31: 983-998 (1995).

R17 Reich, E., R.M. Franklin, A.J. Shatkin et al. Action of actinomycin D on animal cells and viruses. Proc. Natl. Acad. Sci. U.S.A. 48: 1238-1245 (1962).

R18 Roffler, S.R., J. Chan and M.Y. Yeh. Potentiation of radio-immunotherapy by inhibition of topoisomerase I. Cancer Res. 54: 1276-1285 (1994).

R19 Rose, C., R. Lustig, N. McIntosh et al. A clinical trial of Fluosol DA 20% in advanced squamous cell carcinoma of the head and neck. Int. J. Radiat. Oncol. Biol. Phys. 12: 1325-1327 (1986).

R20 Rockwell, S., S.R. Keyes and A.C. Sartorelli. Preclinical studies of porfiromycin as an adjunct to radiotherapy. Radiat. Res. 116: 100-113 (1988).

R21 Rauth, A.M. and K. Kaufman. In vivo testing of hypoxic radiosensitizers using the KHT murine tumour assayed by the lung-colony technique. Br. J. Radiol. 48: 209-220 (1975).

R22 Reimers, R.R., R.N. Hoover, J.F. Fraumeni Jr. et al. Acute leukemia after alkylating-agent therapy for ovarian cancer. N. Engl. J. Med. 307: 1416-1421 (1982).

R23 Roti Roti, J.L. and A. Laszlo. The effects of hyperthermia on cellular macromolecules. p. 13-56 in: Hyperthermia and Oncology, Volume 1 (M. Urano and E. Douple, eds.). VSP BV, The Netherlands, 1988.

S1 Steel, G.G. The search for therapeutic gain in the combination of radiotherapy and chemotherapy. Radiother. Oncol. 11: 31-53 (1988).

S2 Stratford, I.J. Concepts and developments in radiosensitization of mammalian cells. Int. J. Radiat. Oncol. Biol. Phys. 22: 529-532 (1992).

S3 Suzuki, S. The 'synergistic' action of mixed irradiation with high-LET and low-LET radiation. Radiat. Res. 138: 297-301 (1994).

S4 Sahar, E., C. Kittrel, S. Fulghum et al. Sister-chromatid exchange induction in Chinese hamster ovary cells by 8-methoxypsoralen and brief pulses of laser light. Assessment of the relative importance of 8-methoxypsoralen-DNA monoadducts and crosslinks. Mutat. Res. 83: 91-105 (1981).

S5 Steller, H. and S.J. Korsmeyer. Regulators of cell death. Trens. Genet. 11: 101-105 (1995).

S6 Scaradavou, A., G. Heller, C.A. Sklar, et al. Second malignant neoplasms in long-term survivors of childhood rhabdomyosarcoma. Cancer 76(10): 1860-1867 (1995).

S7 Sankila, R. et al. Risk of subsequent malignant neoplasms among 1641 Hodgkin's disease patients diagnosed in childhood and adolescence: a population-based cohort study in the five Nordic countries. J. Clin. Oncol. 14: 1442-1446 (1996).

S8 Sankara Narayanan, N. and B.S. Rao. Interaction between cigarette smoke condensate and radiation for the induction of genotoxic effects in yeast. Mutat. Res. 208: 45-49 (1988).

S9 Stewart, B.W. Role of DNA repair in carcinogenesis. p. 307-320 in: Mechanisms of Carcinogenesis in Risk Identification (H. Vainio, P.N. Magee, D.B. McGregor et al., eds.). IARC Scientific Publication 116 (1992).

S10 Sugahara, T. and M. Watanabe. Epigenetic nature of radiation carcinogenesis at low doses. Int. J. Occup. Med. Toxicol. 3: 129-136 (1994).

S11 Savage, J.R.K. and M. Holloway. Induction of sister-chromatid exchanges by d(42 MeV)-Be neutrons in unstimulated human-blood lymphocytes. Br. J. Radiol. 61: 231-234 (1988).

S12 Steenken, S. Purine bases, nucleosides and nucleotides: aqueous solution redox chemistry and transformation of their radical cations e⁻ and OH adducts. Chem. Rev. 89: 503-520 (1989).

S13 Stammberger, I., W. Schmahl and L. Nice. The effects of x-irradiation, N-ethyl-N-nitrosourea or combined treatment on O^6-alkylguanine-DNA alkyltransferase activity in fetal rat brain and liver and the induction of CNS tumours. Carcinogenesis 11(2): 219-222 (1990)

S14 Shank, B., A.M. Cohen and D. Kelsen. Cancer of the anal region. p. 1006-1022 in: Cancer: Principles and Practice of Oncology, 4th edition (V.T. DeVita Jr., S. Hellman and S.A. Rosenberg, eds.). J.B. Lippincott Co., Philadelphia, 1993.

S15 Scott, B.R. Methodologies for predicting the expected combined stochastic radiobiological effects of different ionizing radiations and some applications. Radiat. Res. 98(1): 182-197 (1984).

S16 Streffer, C. and W.-U. Müller. Dose-effect relationships and general mechanisms of combined exposures. Int. J. Radiat. Biol. 51: 961-969 (1987).

S17 Stowers, S.J. and M.W. Anderson. Formation and persistence of benzo(a)pyrene metabolite-DNA adducts. Environ. Health Perspect. 62: 31-39 (1985).

S18 Stamatoyannapoulos, G., P.E. Nute, D. Lindsley et al. Somatic cell mutation monitoring system based on human hemoglobin mutants. p. 1-35 in: Single Cell Monitoring Systems (Topics in Chemical Mutagenesis) (A.A. Ansari and F.J. deSerres, eds.). Plenum Press, New York, 1984.

S19 Selikoff, I.J., E.C. Hammond and H. Seidman. Mortality experience of insulation workers in the United States and Canada, 1943-1976. Ann. N.Y. Acad. Sci. 330: 91-116 (1979).

S20 Seidel, H.J. and S. Bischof. Effects of radiation on murine T-cell leukemogenesis induced by butylnitrosourea. J. Cancer Res. Clin. Oncol. 105(3): 243-249 (1983).

S21 Seyama, T., T. Kajitani, A. Inoh et al. Synergistic effect of radiation and N-nitrosoethylurea in the induction of lymphoma in mice: cellular kinetics and carcinogenesis. Jpn. J. Cancer Res. 76(1): 20-27 (1985).

S22 Seidel, H.J. Effects of radiation and other influences on chemical lymphomagenesis. Int. J. Radiat. Biol. Relat. Stud. Phys. Chem. Med. 51(6): 1041-1048 (1987).

S23 Steel, G.G. and M.J. Peckham. Exploitable mechanisms in combined radiotherapy-chemotherapy: the concept of additivity. Int. J. Radiat. Oncol. Biol. Phys. 5: 85-91 (1979).

S24 Shellabarger, C.J., J.P. Stone and S. Holtzman. Effect of interval between neutron radiation and diethylstilbestrol on mammary carcinogenesis in female ACI rats. Environ. Health Perspect. 50: 227-232 (1983).

S25 Streffer, C. and W.-U. Müller. Radiation risk from combined exposures to ionizing radiations and chemicals. Adv. Radiat. Biol. 11: 173-209 (1984).

S26 Sasco, A.J., G. Chaplain, E. Amoros et al. Endometrial cancer following breast cancer: effect of tamoxifen and castration by radiotherapy. Epidemiology 7(1): 9-13 (1996).

S27 Sharp, J.G. and D.A. Crouse. Apparent synergism between radiation and the carcinogen 1,2-dimethyl-hydrazine in the induction of colonic tumors in rats. Radiat. Res. 117(2): 304-317 (1989).

S28 Sorsa, M., J. Wilbourn and H. Vainio. Human cytogenetic damage as a predictor of cancer risk. p. 543-554 in: Mechanisms of Carcinogenesis in Risk Identification (H. Vainio, P.N. Magee, D.B. McGregor et al., eds.). IARC Scientific Publication 116 (1992).

S29 Shore, R.E. Overview of radiation-induced skin cancer in humans. Int. J. Radiat. Biol. 57(4): 809-827 (1990).

S30 Shigenaga, M.K. and B.N. Ames. Assays for 8-hydroxy-2'-deoxyguanosine: a biomarker of in vivo oxidative DNA damage. Free Radic. Biol. Med. 10: 211-216 (1991).

S31 Sorsa, M. and J.W. Yager. Cytogenetic surveillance of occupational exposures. p. 345-360 in: Cytogenetics (G. Obe and A. Basler, eds.). Springer Verlag, Berlin, 1987.

S32 Silverman, J., C.J. Shellaberger, S. Holtzman et al. Effect of dietary fat on X-ray-induced mammary cancer in Sprague-Dawley rats. J. Natl. Cancer Inst. 64(3): 631-634 (1980).

S33 Schein, V.I. Kombinierte Wirkung eines konstanten Magnetfeldes und ionisierender Strahlung. Radiobiologiya 28: 703-706 (1988). (In Russian).

S34 Sonntag, von C. The Chemical Basis of Radiation Biology. Taylor and Francis, New York, 1987.

S35 Simic, M.G. Introduction to mechanisms of DNA damage and repair. p. 1-8 in: Mechanisms of DNA Damage and Repair (M.G. Simic, L. Grossman and A.C. Upton, eds.). Plenum Press, New York, 1986.

S36 Simic, M.G. DNA damage, environmental toxicants and rate of aging. J. Environ. Carcinog. Ecotoxicol. Rev. C9: 113-153 (1991).

S37 Simic, M.G. and D.S. Bergtold. Urinary biomarkers of oxidative DNA base damage and human caloric intake. p. 217-225 in: Biological Effects of Dietary Restriction (L. Fishbein, ed.). Springer Verlag, Berlin, 1991.

S38 Simic, M.G. and D.S. Bergtold. Dietary modulation of DNA damage in human. Mutat. Res. 250: 17-24 (1991).

S39 Simic, M.G. DNA markers of oxidative processes in vivo: relevance to carcinogenesis and anticarcinogenesis. Cancer Res. 54 (Suppl.): 1918-1923 (1994).

S40 Suzuki, K., H. Ishii-Ohba, H. Yamanouchi et al. Susceptibility of lactating rat mammary glands to gamma-ray-irradiation-induced tumorigenesis. Int. J. Cancer 56: 413-417 (1994).

S41 Snow, E.T., R.S. Foote and S. Mitra. Base-pairing properties of O6-methylguanine in template DNA during in vitro DNA replication. J. Biol. Chem. 259(13): 8095-8100 (1984).

S42 Sanders, C.L., W.C. Cannon and G.J. Powers. Lung carcinogenesis induced by inhaled high-fired oxides of beryllium and plutonium. Health Phys. 35: 193-199 (1978).

S43 Stoilov, L.M., L.H. Mullenders and A.T. Natarajan. Caffeine potentiates or protects against radiation-induced DNA and chromosomal damage in human lymphocytes depending on temperature and concentration. Mutat. Res. 311(2): 169-174 (1994).

S44 Seitz, H.K. and U.A. Simanowski. Alcohol and carcinogenesis. Annu. Rev. Nutr. 8: 99-119 (1988).

S45 Stich, H.F., K.D. Brunnemann, B. Mathew et al. Chemopreventive trials with vitamin A and beta-carotene: some unresolved issues. Prev. Med. 18(5): 732-739 (1989).

S46 Steel, G.G. Principles for the combination of radiotherapy and chemotherapy. p. 14-22 in: Combined Radiotherapy and Chemotherapy in Clinical Oncology (A. Horwich, ed.). Edward Arnold, London, 1992.

S47 Steindorf, K., J. Lubin, H.E. Wichmann et al. Lung cancer deaths attributable to indoor radon exposure in West Germany. Int. J. Epidemiol. 24(3): 485-492 (1995).

S48 Seiler, H., H. Sigel and A. Sigel. Handbook on Toxicity of Inorganic Compounds (H. Seiler, ed.). Marcel Dekker Inc., New York/Basel, 1988.

S49 Sinclair, W.K. Radiation, chemicals and combined effects. in: The Future of Human Radiation Research (G.B. Gerber, D.M. Taylor, E. Cardis et al., eds.). British Institute of Radiology, Report 22 (1991).

S50 Skov, K. and S. MacPhail. Interaction of platinum drugs with clinically relevant x-ray doses in mammalian cells: a comparison of cisplatin, carboplatin, iproplatin, and tetraplatin. Int. J. Radiat. Oncol. Biol. Phys. 20: 221-225 (1991).

S51 Skov, K.A., H.S. MacPhail and B. Marples. The effect of radiosensitizers on the survival response of hypoxic mammalian cells: the low X-ray dose region, hypersensitivity and induced radioresistance. Radiat. Res. 138: S113-S116 (1994).

S52 Sun, J.R. and J.M. Brown. Lack of differential radiosensitization of hypoxic cells in a mouse tumor at low radiation doses per fraction by cisplatin. Radiat. Res. 133: 252-256 (1993).

S53 Schwachöfer, J.H., R.P. Crooijmans, J. Hoogenhout et al. Effectiveness in inhibition of recovery of cell survival by cisplatin and carboplatin: influence of treatment sequence. Int. J. Radiat. Oncol. Biol. Phys. 20: 1235-1241 (1991).

S54 Schaake-Koning, C., W. van den Bogaert, O. Dalesio et al. Effects of concomitant cisplatin and radiotherapy on inoperable non-small-cell lung cancer. N. Engl. J. Med. 326(8): 524-530 (1992).

S55 Schnabel, T., N. Zamboglou, H. Pape et al. Phase II trial with carboplatin and radiotherapy in previously untreated advanced squamous cell carcinoma of the head and neck (SCCHN). Proc. Am. Soc. Clin. Oncol. 9: 176 (Abstract) (1990).

S56 Stewart, D.J. The role of chemotherapy in the treatment of gliomas in adults. Cancer Treat. Rev. 16: 129-160 (1989).

S57 Stewart, D.J., J.M. Molepo, L. Eapen et al. Cisplatin and radiation in the treatment of tumors of the central nervous system: pharmacological considerations and results of early studies. Int. J. Radiat. Oncol. Biol. Phys. 28: 531-542 (1994).

S58 Stehbens, J.A., T.A. Kaleita, R.B. Noll et al. CNS prophylaxis of childhood leukemia: what are the long-term neurological, neuropsychological, and behavioral effects? Neuropsychol. Rev. 2: 147-177 (1991).

S59 Szybalski, W. Properties and applications of halogenated deoxyribonucleic acids. p. 147-171 in: The Molecular Basis of Neoplasia; Program and Abstracts of Papers; Fifteenth Annual Symposium on Fundamental Cancer Research, Houston, 1961. University of Texas Press, Austin, 1962.

S60 Steiner, U.E. and T. Ulrich. Magnetic effects in chemical kinetics and related phenomena. Chem. Rev. 89: 51-147 (1989).

S61 Sinclair, W.K. The combined effect of hydroxyurea and x-rays on Chinese hamster cells in vitro. Cancer Res. 28: 198-206 (1968).

S62 Sinclair, W.K. Hydroxyurea: differential lethal effects on cultured mammalian cells during the cell cycle. Science 150: 1729-1731 (1965).

S63 Stefani, S., R.W. Eells and J. Abbate. Hydroxyurea and radiotherapy in head and neck cancer. Radiology 101: 391-396 (1971).

S64 Stehman, F.B., B.N. Bundy, G. Thomas et al. Hydroxyurea versus misonidazole with radiation in cervical carcinoma: long-term follow-up of a Gynecologic Oncology Group trial. J. Clin. Oncol. 11: 1523-1528 (1993).

S65 Sinclair, W.K. Hydroxyurea revisited: a decade of clinical effects studies. Int. J. Radiat. Oncol. Biol. Phys. 7: 631-637 (1981).

S66 Sugiyama, H., C. Xu, N. Murugesan et al. Chemistry of the alkali-labile lesion formed from iron(II) bleomycin and d(CGCTTTAAAGCG). Biochemistry 27: 58-67 (1988).

S67 Schantz, S.P., L.B. Harrison and W.K. Hong. Tumors of the nasal cavity and paranasal sinuses, nasopharynx, oral cavity and oropharynx. in: Cancer: Principles and Practice of Oncology, 4th edition (V.T. DeVita Jr., S. Hellman and S.A. Rosenberg, eds.). J.B. Lippincott Co., Philadelphia, 1993.

S68 Siemann, D.W. and P.C. Keng. Responses of tumor cell subpopulations to single modality and combined modality therapies. Natl. Cancer Inst. Monogr. 6: 101-105 (1988).

S69 Schiff, P.B., J. Fant and S.B. Horwitz. Promotion of microtubule assembly in vitro by taxol. Nature 277: 665-667 (1979).

S70 Schiff, P.B. and S.B. Horwitz. Taxol stabilizes microtubules in mouse fibroblast cells. Proc. Natl. Acad. Sci. U.S.A. 77: 1561-1565 (1980).

S71 Steren, A., B.U. Sevin, J. Perras et al. Taxol as a radiation sensitizer: a flow cytometric study. Gynecol. Oncol. 50: 89-93 (1993).

S72 Skov, K., H. Zhou and B. Marples. The effect of two topoisomerase inhibitors on low-dose hypersensitivity and increased radioresistance in Chinese hamster V79 cells. Radiat. Res. 138: S117-S120 (1994).

S73 Schneider, E., Y.H. Hsiang and L.F. Liu. DNA topoisomerases as anticancer drug targets. Adv. Pharmacol. 21: 149-183 (1990).

S74 Stone, H.B. and H.R. Withers. Metronidazole: effect on radiosensitivity of tumor and normal tissues in mice. J. Natl. Cancer Inst. 55: 1189-1194 (1975).

S75 Stratford, I.J., P. O'Neill, P.W. Sheldon et al. RSU 1069, a nitroimidazole containing an aziridine group. Bioreduction greatly increases cytotoxicity under hypoxic conditions. Biochem. Pharmacol. 35: 105-109 (1986).

S76 Stratford, I.J., J.M. Walling and A.R. Silver. The differential cytotoxicity of RSU 1069: cell survival studies indicating interaction with DNA as a possible mode of action. Br. J. Cancer 53: 339-344 (1986).

S77 Sebolt-Leopold, J.S., P.W. Vincent, K.A. Beningo et al. Pharmacologic/pharmacokinetic evaluation of emesis induced by analogs of RSU 1069 and its control by antiemetic agents. Int. J. Radiat. Oncol. Biol. Phys. 22: 549-551 (1992).

S78 Sebolt-Leopold, J.S., W.L. Ellicot, H.D. Showater et al. Rationale for selection of PD 144872, the R isomer of RB 6145, for clinical development as a radiosensitizer. Proc. Am. Assoc. Cancer Res. 34: 362 (Abstract 2155) (1993).

S79 Schrijvers, D. and J.B. Vermorken. Update on the taxoids and other new agents in head and neck cancer therapy. Curr. Opin. Oncol. 10(3): 233-241 (1998).

S80 Serebryanyi, A.M., A.L.E. Salnikov, A.L.M. Bakhitov et al. Potassium Cyanate-induced modification of toxic and mutagenic effects of γ-radiation and benzo(A)pyrene. Wiss. Akad. CCCP Radiologia 29(2): 235-240 (1989).

S81 Sumpter, J.P. Reproductive effects from oestrogen activity in polluted water. Arch. Toxicol. (Suppl.) 20: 143-150 (1998).

S82 Spitkovsky, D.M., A.V. Ermakov, A.I. Gorin et al. Peculiarities of unscheduled DNA synthesis and human lymphocyte nuclear parameter alterations induced by x-rays and with following uv-irradiation. Radiat. Biol. Radioecol. 34(1): 23-30 (1994).

S83 Shewach, D.S. and T.S. Lawrence. Radiosensitization in human solid tumor cell lines with gemcitabine. Semin. Oncol. 23 (5 Suppl. 10): 65-71 (1996).

S84 Schaake-Koning, C., W. van den Bogaert, O. Dalesio et al. Radiosensitization by cytotoxic drugs. The EORTC experience by the Radiotherapy and Lung Cancer Cooperative Group. Lung Cancer 10 (Suppl. 1): S263-S270 (1994).

S85 Shewach, D.S. and T.S. Lawrence. Gemcitabine and radiosensitization in human tumor cells. Invest. New Drugs 14 (3): 257-263 (1996).

S86 Sonnenfeld, G. Immune responses in space flight. Int. J. Sports Med. 19 (Suppl. 3): S195-S202 (1998).

S87 Schneider, E. and J. Kiefer. Interaction of ionizing and ultraviolet light in diploid yeast strains of different sensitivity. Photochem. Photobiol. 24: 573-578 (1976).

T1 Takehisa, S. and S. Wolff. Induction of sister chromatide exchanges in Chinese hamster cells by carcinogenic mutagens requiring metabolic activation. Mutat. Res. 45: 263-270 (1977).

T2 Talbot, R.J., A. Morgan, S.R. Moores et al. Preliminary studies of the interaction between ^{239}PuO$_2$ and cigarette smoke in the mouse lung. Int. J. Radiat. Biol. Relat. Stud. Phys. Chem. Med. 51(6): 1101-1110 (1987).

T3 Tyndall, D.A. MRI effects on the teratogenecity of x-irradiation in the C57BL/6J mouse. Magn. Reson. Imaging 8(4): 423-433 (1990).

T4 Tates, A.D., L.F. Bernini, A.T. Natarajan et al. Detection of somatic mutants in man: HPRT mutations in lymphocytes and hemoglobin mutations in erythrocytes. Mutat. Res. 213(1): 73-82 (1989).

T5 Taylor, P.R., Y.L. Quiao, A. Schatzkin et al. Relation of arsenic exposure to lung cancer among tin miners in Yunnan Province, China. Br. J. Ind. Med. 46: 881-886 (1989).

T6 Travis, L.B., R.E. Curtis, M. Stovall et al. Risk of leukemia following treatment for non-Hodgkin's lymphoma. J. Natl. Cancer Inst. 19: 1450-1457 (1994).

T7 Turner, D.R. and A.A. Morley. Human somatic mutation at the HLA-A locus. p. 25-35 in: Mutation and the Environment, Fifth International Conference on Environmental Mutagens (M.L. Mendelson and R.J. Albertini, eds.). Wiley/Liss, New York, 1990.

T8 Turner, D.R., S.A. Grist, M. Janatipour et al. Mutations in human lymphocytes commonly involve gene duplication and resemble those seen in cancer cells. Proc. Natl. Acad. Sci. U.S.A. 85(9): 3189-3192 (1988).

T9 Tates, A.D., T. Grummt, M. Tornqvist et al. Biological and chemical monitoring of occupational exposure to ethylene oxide. Mutat. Res. 250(1-2): 483-497 (1991). *[published erratum appeared in Mutat. Res. 280(1): 73 (1992)].*

T10   Turner, D.R., A.A. Morley, M. Haliandros et al. In vivo somatic mutations in human lymphocytes frequently result from major gene alterations. Nature 315(6017): 343-345 (1985).

T11   Tates, A.D., F.J. van Dam, H. van Mossel et al. Use of the clonal assay for the measurement of frequencies of HPRT mutants in T-lymphocytes from five control populations. Mutat. Res. 253(2): 199-213 (1991).

T12   Takeshima, Y., T. Seyama, W.P. Bennett et al. p53 mutations in lung cancers from non-smoking atomic-bomb survivors. Lancet 342: 1520-1521 (1993).

T13   Taylor, J.A., M.A. Watson, T.R. Devereux et al. p53 mutation hotspot in radon-associated lung cancer. Lancet 343: 86-87 (1994).

T14   Thomas, D.C., K.G. McNeill and C. Dougherty. Estimates of lifetime lung cancer risks resulting from Rn progeny exposures. Health Phys. 49: 825-846 (1985).

T15   Tomatis, L. Overview of perinatal and multigeneration carcinogenesis. p. 1-15 in: IARC Scientific Public. 96 (1989).

T16   Tolmach, L.J. and P.M. Busse. The action of caffeine on x-irradiated HeLa cells. IV. Progression delays and enhanced cell killing at high caffeine concentrations. Radiat. Res. 82: 374-392 (1980).

T17   Thompson, C.B. Apoptosis in the pathogenesis and treatment of disease. Science 267: 1456-1462 (1995).

T18   Thomas, D., J. Pogoda, B. Langholz et al. Temporal modifiers of the radon-smoking interaction. Health Phys. 66(3): 257-262 (1994).
       [published erratum in Health Phys. 67(6): 675 (1994)].

T19   Tannock, I. Cell kinetics and chemotherapy: a critical review. Cancer Treat. Rep. 62: 1117-1133 (1978).

T20   Tanabe, M., D. Godat and R.F. Kallman. Effects of fractionated schedules of irradiation combined with cis-diamminedichloro-platinum II on the SCCVII/St tumor and normal tissues of the C3H/KM mouse. Int. J. Radiat. Oncol. Biol. Phys. 13: 1523-1532 (1987).

T21   Trovò, M.G., E. Minatel, G. Franchin et al. Radiotherapy versus radiotherapy enhanced by cisplatin in stage III non-small cell lung cancer: randomized cooperative study. Lung Cancer 7: 158 (Abstract) (1991).

T22   Terasima, T. and L.J. Tolmach. Variations in several responses of HeLa cells to X-irradiation during the devision cycle. Biophys. J. 3: 11-33 (1963).

T23   Tomasz, M., A.K. Chawla and R. Lipman. Mechanism of monofunctional and bifunctional alkylation of DNA by mitomycin C. Biochemistry 27: 3182-3187 (1988).

T24   Tishler, R.B., P.B. Schiff, C.R. Geard et al. Taxol: a novel radiation sensitizer. Int. J. Radiat. Oncol. Biol. Phys. 22: 613-617 (1992).

T25   Tishler, R.B., C.R. Geard, E.J. Hall et al. Taxol sensitizes human astrocytoma cells to radiation. Cancer Res. 52: 3495-3497 (1992).

T26   Thomlinson, R.H. and L.H. Gray. The histological structure of some human lung cancers and the possible implications for radiotherapy. Br. J. Cancer 9: 539-549 (1955).

T27   Thor, H., M.T. Smith, P. Hartzell et al. The metabolism of menadione (2-methyl-1,4-naphthoquinone) by isolated hepatocytes. A study of the implications of oxidative stress in intact cells. J. Biol. Chem. 257: 12419-12425 (1982).

T28   Terasima, T., Y. Takabe and M. Yasakuwa. Combined effect of x-ray and bleomycin on cultured mammalian cells. GANN Monogr. Cancer Res. 66: 701-703 (1975).

T29   Takabe, Y., T. Miyamoto, M. Watanabe et al. Synergysm of x-rays and bleomycin on ehrlich ascites tumour cells. Br. J. Cancer 36: 196-200 (1977).

T30   Tucker, M.A. Secondary cancers. p. 2407-2416 in: Cancer: Principles and Practice of Oncology, 4th edition (V.T. DeVita Jr., S. Hellman and S.A. Rosenberg, eds.). J.B. Lippincott Co., Philadelphia, 1993.

T31   Tucker, M.A., N.C. Coleman, R.S. Cox et al. Risk of secondary malignancies following Hodgkin's disease after 15 years. N. Engl. J. Med. 318: 76-81 (1988).

T32   Tucker, M.A., A.T. Meadows, J.D. Boice Jr. et al. Leukemia after therapy with alkylating agents of childhood cancer. J. Natl. Cancer Inst. 78: 459-464 (1987).

T33   Travis, L.B., R.E. Curtis, J.D. Boice Jr. et al. Bladder cancer after chemotherapy for non-Hodgkin's lymphoma. N. Engl. J. Med. 321: 544-545 (1989).

T34   Tucker, M.A., G.J. D'Angio, J.D. Boice Jr. et al. Bone sarcomas linked to radiotherapy and chemotherapy in children. N. Engl. J. Med. 317: 588-593 (1987).

T35   Tucker, M.A., P.H. Morries Jones, J.D. Boice Jr. et al. Therapeutic radiation at a young age is linked to secondary thyroid cancer. Cancer Res. 51: 2885-2888 (1991).

T36   Travis, L.B., R.E. Curtis, B. Glimelius et al. Bladder and kidney cancer following cyclophosphamide therapy for non-Hodgkin's lymphoma. J. Natl. Cancer Inst. 87: 524-530 (1995).

T37   Terada, Y., H. Watanabe and N. Fujimoto. Enhancing effects of immunization with rectal extract on rectal carcinogenesis by local X-irradiation caused in male A/HeJ mice. Oncol. Rep. 3: 707-712 (1996).

T38   Teicher, B.A., G. Ara, Y.N. Chen et al. Interaction of tomudex with radiation in vitro and in vivo. Int. J. Oncol. 13(3): 437-442 (1998).

T39   Talwar, N. and J.L. Redpath. Schedule dependence of the interaction of radiation and taxol in HeLa cells. Radiat. Res. 148(1): 48-53 (1997).

T40   Tamura, K., M. Takada, I. Kawase et al. Enhancement of tumor radio-response by irinotecan in human lung tumor xenografts. Jap. J. Cancer Res. 88(2): 218-223 (1997).

T41   Tomásek, L., S.C. Darby, T. Fearn et al. Patterns of lung cancer mortality among uranium miners in West Bohemia with varying rates of exposure to radon and its progeny. Radiat. Res. 137: 251-261 (1994).

U2   United Nations. Sources and Effects of Ionizing Radiation. United Nations Scientific Committee on the Effects of Atomic Radiation, 1994 Report to the General Assembly, with Scientific Annexes. United Nations sales publication E.94.IX.11. United Nations, New York, 1994.

U3   United Nations. Sources and Effects of Ionizing Radiation. United Nations Scientific Committee on the Effects of Atomic Radiation, 1993 Report to the General Assembly, with scientific annexes. United Nations sales publication E.94.IX.2. United Nations, New York, 1993.

U6   United Nations. Ionizing Radiation: Sources and Biological Effects. United Nations Scientific Committee on the Effects of Atomic Radiation, 1982 Report to the General Assembly, with annexes. United Nations sales publication E.82.IX.8. United Nations, New York, 1982.

U14   Uggla, A.H. and A.T. Natarajan. X-ray-induced SCEs and chromosomal aberrations in CHO cells. Influence of nitrogen and air during irradiation in different stages of the cell cycle. Mutat. Res. 122: 193-200 (1983).

U15   Umeki, S., S. Kyoizumi, Y. Kusunoki et al. Flow cytometric measurements of somatic cell mutations in Thorotrast patients. Jpn. J. Cancer Res. 82(12): 1349-1353 (1991).

U16   Urtasun, R.C., P. Band, J.D. Chapman et al. Radiation and high-dose metronidazole in supratentorial glioblastomas. N. Engl. J. Med. 294: 1364-1367 (1976).

U17 Urtasun, R.C., J. Sturmwind, H. Rabin et al. "High-dose" metronidazole: a preliminary pharmacological study prior to its investigational use in clinical radiotherapy trials. Br. J. Radiol. 47: 297-299 (1974).

U18 Ushakov, I.B., V.E. Lapaev, Z.A. Vorontsova et al. Radiation and alcohol (the essays of radiation narcology or alcoholic „Chernobyl"). Voronezh „Istoki" 248 (1998).

U19 United States National Research Council. Possible Health Effects of Exposure to Residential Electric and Magnetic Fields. National Academy Press, Washington, 1997.

V1 Vogelstein, B., E.R. Fearon, S.R. Hamilton et al. Genetic alterations during colorectal-tumor development. N. Engl. J. Med. 319: 525-532 (1988).

V2 Vogelstein, B., E.R. Fearon, S.E. Kern et al. Allelotype of colorectal carcinomas. Science 244: 207-211 (1989).

V3 Vasin, M.V. and L.A. Semenova. Combined effect of normobaric hyperoxia and vibration on the radiosensitivity of animals. Radiobiologiya 27(5): 704-705 (1987). (In Russian).

V4 Vasin, M.V., V.V. Antipov, G.A. Chernov et al. Investigation of radioprotective effect of indraline on hematopoietic system in different species of animal. J. Radiationbiol. Radioecol. 36(2): 168-189 (1996). (In Russian).

V5 Vorobtsova, I.E., L.M. Aliyakparova and V.N. Anisimov. Promotion of skin tumors by 12-O-tetradecanoylphorbol-13-acetate in two generations of descendants of male mice exposed to X-ray irradiation. Mutat. Res. 287(2): 207-216 (1993).

V6 Voskanian K.S., N.V. Simonian, T.M. Avakian et al. Modification of the damaging effect of alpha-particles on Escherichia coli K-12 by low intensity laser irradiation. Radiobiologiya 26(3): 375-377 (1986). (In Russian).

V7 van Leeuwen, F., W. Klokmann, M. Stovall et al. Roles of radiotherapy and smoking in lung cancer following Hodgkin's disease. J. Natl. Cancer Inst. 87(20): 1530-1537 (1995).

V8 Vietti, T., F. Eggerding and F. Valeriote. Combined effect of x radiation and 5-fluorouracil on survival of transplanted leukemic cells. J. Natl. Cancer Inst. 47: 865-870 (1971).

V9 Van Belle, S., L. Fortan, M. De Smet et al. Interaction between vinblastine and ionizing radiation in the mouse MO4 fibrosarcoma in vivo. Anticancer Res. 14: 1043-1048 (1994).

V10 Valgussa, P., A. Santoro, F. Bellani-Fossati et al. Second acute leukemia and other malignancies following treatment for Hodgkin's disease. J. Clin. Oncol. 4: 830-837 (1986).

V11 van Leeuwen, F.E., W.J. Klokman, A. Hagenbeek et al. Second cancer risk following Hodgkin's disease: a 20-year follow-up study. J. Clin. Oncol. 12: 312-325 (1994).

V12 Vitvitskii, V.N., L.M. Bakhitova, L.S. Soboleva et al. Modification of γ-rays mutagenic effects by heavy metal salts. Russ. Acad. Sci. 4: 495-498 (1996).

V13 Vokes, E.E., A. Gregor and A.T. Turrisi. Gemcitabine and radiation therapy for non-small cell lung cancer. Semin. Oncol. 25 (4 Suppl. 9): 66-69 (1998).

W1 Williams, G.M. and J.H. Weisburger. Chemical carcinogenesis. p. 127-194 in: Casaret & Doull's Toxicology. Pergamon Press, New York, 1992.

W2 Weinerman, B.H., K.B. Orr, L. Lu et al. Low-dose (diagnostic-like) x-ray as a cocarcinogen in mouse colon carcinoma. J. Surg. Oncol. 31(3): 163-165 (1986).

W3 Watanabe, H., Y. Ando and K. Yamada. Lack of any positive effect of intestinal metaplasia on induction of gastric tumors in Wistar rats with N-methyl-N-nitrosourea in their drinking water. Jpn. J. Cancer 85: 892-896 (1994).

W4 Weindruch, R. and R.L. Walford. The Retardation of Aging and Disease by Dietary Restriction. Charles C. Thomas Publishers, Springfield, 1988.

W5 Welsch, C.W. Enhancement of mammary tumorigenesis by dietary fat: review of potential mechanisms. Am. J. Clin. Nutr. 45: 192-202 (1987).

W6 Wyllie, A.H. The genetic regulation of apoptosis. Curr. Opin. Genet. Dev. 5: 97-104 (1995).

W7 Wolff, S. Biological dosimetry with cytogenetic endpoints. Prog. Clin. Biol. Res. 372: 351-362 (1991).

W8 Weinfeld, M. and K.-J.M. Soderlind. ^{32}P-postlabelling detection of radiation-induced DNA damage: identification and estimation of thymine glycols and phosphoglycolate termini. Biochemistry 30: 1091-1097 (1991).

W9 World Health Organization and International Agency for Research on Cancer (WHO/IARC). Beryllium and beryllium compounds. Beryllium, cadmium, mercury and exposures in the glass manufacturing industry, Working Group views and Expert opinions. p. 41-117 in: IARC Scientific Publication 58 (1993).

W10 Wiggenhauser, A. and W. Schmahl. Postnatal development and neoplastic disease pattern in NMRI mice after combined treatment with ethylnitrosourea and x-irradiation on different days of the fetal period. Int. J. Radiat. Biol. Relat. Stud. Phys. Chem. Med. 51(6): 1021-1029 (1987).

W11 Wolff, S., J. Bodycote and R.B. Painter. Sister chromatide exchanges induced in chinese hamster cells by UV irradiation of different stages of the cell cycle: the necessity of cells to pass through S. Mutat. Res. 25: 73-81 (1974).

W12 Wilson, V.L., B.G. Taffe, P.G. Shields et al. Detection and quantification of 8-hydroxydeoxyguanosine adducts in peripheral blood of people exposed to ionizing radiation. Environ. Health Perspect. 99: 261-263 (1993).

W13 Weisburger, J.H. Nutritional approach to cancer prevention with emphasis on vitamins, antioxidants and carotenoids. Am. J. Clin. Nutr. 53(Suppl. 1): 226s-237s (1991).

W14 Weisburger, J.H. and C. Horn. Causes of cancer. in: American Cancer Society Textbook on Clinical Oncology, 6th edition, Chapter 7 (A. Holleb and D. Fink, eds.). American Cancer Society, Atlanta, GA, 1990.

W15 Weissenborn, U. and C. Streffer. Micronuclei with kinetochores in human melanoma cells and rectal carcinomas. Int. J. Radiat. Biol. 59(2): 373-383 (1991).

W16 Walter, P., M. Korbelik, I. Spadinger et al. Investigations into mechanisms of the interaction between platinum complexes and irradiation at low (2 Gy) doses in hypoxic cells. I. The role of single-strand breaks. Radiat. Oncol. Invest. 1: 137-147 (1993).

W17 Wilkinson, H.A., T. Fujiwara and S. Rosenfeld. Synergistic effect between intraneoplastic methotrexate and radiation on experimental intracerebral rat gliosarcoma. Neurosurgery 34: 665-668 (1994).

W18 Weinberg, M.J. and A.M. Rauth. 5-Fluorouracil infusions and fractionated doses of radiation: studies with a murine squamous cell carcinoma. Int. J. Radiat. Oncol. Biol. Phys. 13: 1691-1699 (1987).

W19 Wang, Y. and G. Iliakis. Effects of 5'-iododeoxyuridine on the repair of radiation induced potentially lethal damage interphase chromatin breaks and DNA double strand breaks in Chinese hamster ovary cells. Int. J. Radiat. Oncol. Biol. Phys. 23: 353-360 (1992).

W20 Ward, J.F., W.F. Blakeley and E.I. Jones. Effects of inhibitors of DNA strand break repair of HeLa cell radiosensitivity. Int. J. Radiat. Oncol. Biol. Phys. 8: 811-819 (1982).

W21 Weissberg, J.B., Y.H. Son, R.J. Papac et al. Randomized clinical trial of mitomycin C as an adjunct to radiotherapy in head and neck cancer. Int. J. Radiat. Oncol. Biol. Phys. 17: 3-9 (1989).

W22 Warde, P. and D. Payne. Does thoracic irradiation improve survival and local control in limited-stage small-cell carcinoma of the lung? A meta-analysis. J. Clin. Oncol. 10: 890-895 (1992).

W23 Watson, E.R., K.E. Halnan, S. Dische et al. Hyperbaric oxygen and radiotherapy: a Medical Research Council trial in carcinoma of the cervix. Br. J. Radiol. 51: 879-887 (1978).

W24 Workman, P. Bioreductive mechanisms. Int. J. Radiat. Oncol. Biol. Phys. 22: 631-637 (1992).

W25 Walton, M.I., C.R. Wolf and P. Workman. Molecular enzymology of the reductive bioactivation of hypoxic cell cytotoxins. Int. J. Radiat. Oncol. Biol. Phys. 16: 983-986 (1989).

W26 Whitmore, G.F. and S. Gulyas. Studies on the toxicity of RSU-1069. Int. J. Radiat. Oncol. Biol. Phys. 12: 1219-1222 (1986).

W27 Wilder, R.B., J.K. McGann, W.R. Sutherland et al. The hypoxic cytotoxin SR 4233 increases the effectiveness of radioimmuno-therapy in mice with human non-Hodgkin's lymphoma xenografts. Int. J. Radiat. Oncol. Biol. Phys. 28: 119-126 (1994).

W28 Watanabe, H., T. Okamoto, M. Matsuda et al. Effects of sex hormones on induction of intestinal metaplasia by X-irradiation in rats. Acta Pathol. Jpn. 43: 456-463 (1993).

W29 Watanabe, H., T. Okamoto, T. Takahashi et al. The effects of sodium, chloride, miso or ethanol on development of intestinal metaplasia after x-irradiation of the rat glandular stomach. Jpn. J. Cancer Res. 83: 1267-1272 (1992).

W30 Wingren, G., T. Hatschek and O. Axelson. Determinants of papillary cancer of the thyroid. Am. J. Epidemiol. 138: 482-491 (1993).

W31 World Health Organization. Ultraviolet radiation. Environmental Health Criteria 160 (1994).

W32 Whitlock, J.A., J.P. Greer and J.N. Lukens. Epipodo-phyllotoxin-related leukemia: identification of a new subset of secondary leukemia. Cancer 68: 600-604 (1991).

W33 Wall, R.L. and K. Clausen. Carcinoma of the urinary bladder in patients receiving cyclophosphamide. N. Engl. J. Med. 293: 271-273 (1975).

W34 Wallace, L. Personal exposures, indoor and outdoor air concentrations, and exhaled breath concentrations of selected volatile organic compounds measured for 600 residents of New Jersey, North Dakota, North Carolina, and California. Toxicol. Environ. Chem. 12: 215-236 (1986).

W35 Wichmann, H.E., L. Kreienbrock, M. Kreuzer et al. Lungenkrebsrisiko durch Radon in der Bundesrepublik Deutschland (West). in: Fortschritte in der Umweltmedizin (H.E. Wichmann, H.W. Schlipköter und G. Fülgraf, eds.). Ecomed Verlagsgesellschaft, Landsberg, 1998.

W36 Woudstra, E.C., A.W. Konings, P.A. Jeggo et al. Role of DNA-PK subunits in radiosensitization by hyperthermia. Radiat. Res. 152(2): 214-218 (1999).

X1 Xuan, X.Z., J.H. Lubin, J.Y. Li et al. A cohort study in southern China of workers exposed to radon and radon decay products. Health Phys. 64: 120-131 (1993).

X2 Xu, Y., C.T. Ho, S.G. Amin et al. Inhibition of tobacco-specific nitrosamine-induced lung tumorigenesis in A/J mice by green tea and its major polyphenol as antioxidants. Cancer Res. 52(14): 3875-3879 (1992).

X3 Xynos, F.P., I. Benjamin, R. Sapiente et al. Adriamycin and hydroxyurea as radiopotentiators in the treatment of squamous cell carcinoma of the cervix implanted in nude mice: a preliminary report. Gynecol. Oncol. 9: 170-176 (1980).

Y1 Yamada, S., K. Ando, S. Koike et al. Accelerated bone marrow recovery from radiation damage in etoposide-pretreated mice. Int. J. Radiat. Oncol. Biol. Phys. 29: 621-625 (1994).

Y2 Yager, J.W. The effect of background variables on human peripheral lymphocyte micronuclei. p. 147-150 in: IARC Scientific Publication 104 (1990).

Y3 Yu, B.P. Free radicals and modulation by dietary restriction. Age Nutr. 2: 84-89 (1991).

Y4 Yager, J.W., W.M. Paradisin, E. Symanski et al. Sister chromatid exchanges induced in peripheral lymphocytes of workers exposed to low concentrations of styrene. Prog. Clin. Biol. Res. 340C: 347-356 (1990).

Y5 Youngman, L.D., J.-Y.K. Park and B.N. Ames. Protein oxidation associated with aging is reduced by dietary restriction of proteins or calories. Proc. Natl. Acad. Sci. U.S.A. 89: 9112-9116 (1992).

Y6 Yokoro, K., O. Niwa, K. Hamada et al. Carcinogenic and co-carcinogenic effects of radiation in rat mammary carcinogenesis and mouse T-cell lymphomagenesis: a review. Int. J. Radiat. Biol. Relat. Stud. Phys. Chem. Med. 51(6): 1069-1080 (1987).

Y7 Yao, S.X., J.H. Lubin, Y.L. Qiao et al. Exposure to radon progeny, tobacco use and lung cancer in a case-control study in southern China. Radiat. Res. 138(3): 326-336 (1994).

Y8 Yang, L-X., E.B. Douple, J.A. O'Hara et al. Enhanced radiation-induced cell killing by carboplatin in cells of repair-proficient and repair-deficient cell lines. Radiat. Res. 144: 230-236 (1995).

Y9 Yang, L-X., E.B. Douple, J.A. O'Hara et al. Carboplatin enhances the production and persistence of radiation-induced DNA single-strand breaks. Radiat. Res. 143: 302-308 (1995).

Y10 Yang, L-X., E.B. Douple, J.A. O'Hara et al. Production of DNA double-strand breaks by interactions between carboplatin and radiation: a potential mechanism for radiopotentiation. Radiat. Res. 143: 309-315 (1995).

Y11 Yan, R.D. and R.E. Durand. The response of hypoxic cells in SCCVII murine tumors to treatment with cisplatin and x rays. Int. J. Radiat. Oncol. Biol. Phys. 20: 271-274 (1991).

Y12 Yoshioka, A., S. Tanaka, O. Hiraoka et al. Deoxyribo-nucleoside triphosphate imbalance 5-Fluorodeoxy-uridine-induced DNA double strand breaks in mouse FM3A cells and the mechanism of cell death. J. Biol. Chem. 262: 8235-8241 (1987).

Y13 Young, R.C., C.A. Perez and W.J. Hoskins. Cancer of the ovary. p. 1226-1263 in: Cancer: Principles and Practice of Oncology, 4th edition (V.T. DeVita Jr., S. Hellman and S.A. Rosenberg, eds.). J.B. Lippincott Co., Philadelphia, 1993.

Y14 Yoshida, K., T. Inoue and K. Nojima. Caloric restriction reduces the incidence of myeloid leukemia induced by a single whole-body radiation in C3H/He mice. Proc. Natl. Acad. Sci. 94: 2615-1619 (1997).

Y15 Yamada, M., S. Kudoh and K. Hirata. Risk factors of pneumonitis following chemoradiotherapy for lung cancer. Eur. J. Cancer 34: 71-75 (1998).

Z1   Zaider, M. and H.H. Rossi. The synergistic effect of different radiations. Radiat. Res. 83: 732-739 (1980).

Z2   Zaider, M. and D.J. Brenner. Comments on "V79 survival following simultaneous or sequential irradiation by 15-MeV neutrons and ^{60}Co photons" by Higgins et al. [Radiat. Res. 95: 45-56 (1983)]. Radiat. Res. 99: 438-441 (1984).

Z3   Zölzer, F., C. Streffer and T. Pelzer. Induction of quiescent S-phase cells by irradiation and/or hyperthermia. I. Time and dose dependence. Int. J. Radiat. Biol. 63(1): 69-76 (1993).

Z4   Zhen, W. and A.T. Vaughan. Effect of caffeine on radiation-induced apoptosis in TK6 cells. Radiat. Res. 141(2): 170-175 (1995).

Z5   Ziegler, A., A.S. Jonason, D.J. Leffell et al. Sunburn and p53 in the onset of skin cancer. Nature 372: 773-776 (1994).

Z6   Zak, M. and J. Drobnik. Effect of cisdichlorodiamine platinum (II) on the post-irradiation lethality in mice after irradiation with X-rays. Strahlentherapie 142: 112-115 (1971).

Z7   Zeman, E.M., J.M. Brown, M.J. Lemmon et al. SR-4233: a new bioreductive agent with high selective toxicity for hypoxic mammalian cells. Int. J. Radiat. Oncol. Biol. Phys. 12: 1239-1242 (1986).

Z8   Zeman, E.M., M.A. Baker, M.J. Lemmon et al. Structure-activity relationships for benzotriazine di-N-oxides. Int. J. Radiat. Oncol. Biol. Phys. 16: 977-981 (1989).

Z9   Zeman, E.M., V.K. Hirst, M.J. Lemmon et al. Enhancement of radiation-induced tumor cell killing by the hypoxic cell toxin SR 4233. Radiother. Oncol. 12: 209-218 (1988).

Z10   Zeman, E.M. and J.M. Brown. Pre- and post-irradiation radiosensitization by SR 4233. Int. J. Radiat. Oncol. Biol. Phys. 16: 967-971 (1989).

Z11   Zucali, J.R. Mechanisms of protection of hematopoietic stem cells from irradiation. Leuk. Lymphoma 13: 27-32 (1994).

Z12   Zucali, J.R., A. Suresh, F. Tung et al. Cytokine protection of hematopoietic stem cells. Prog. Clin. Biol. Res. 389: 207-216 (1994).

Z13   Zhu, W.-G., S. Antoku, K. Shinobu et al. Enhancement of hyperthermic killing in L5178Y cells by protease inhibitors. Cancer Res. 55: 739-742 (1995).

Z14   Zaider, M. The application of the principle of "dual radiation action" in biophysical modelling. p. 37-46 in: Biophysical Modelling of Radiation Effects (K.H. Chadwick, G. Moschini and M.N. Varma, eds.). Adam Hilger, Bristol, 1992.

ANNEX I

Epidemiological evaluation of radiation-induced cancer

CONTENTS

INTRODUCTION

1.   Epidemiological studies of the cancer risks associated with both external and internal exposure to ionizing radiation were the subject of an extensive review in the UNSCEAR 1994 Report [U2]. Covered in that review were studies of cancer mortality and incidence up to 1987 among the survivors of the atomic bombings at Hiroshima and Nagasaki, who received a single dose of radiation; patients exposed to radiation for diagnostic or therapeutic purposes, usually as multiple doses; and radiation workers and individuals exposed chronically to environmental radiation. Estimates of risks observed in the major epidemiological studies of external low linear energy transfer (low-LET) exposures were presented in a common format. Data from the Life Span Study of survivors of the atomic bombings, in particular, were used to estimate the lifetime risk of total cancer mortality following external exposure to low-LET radiation [U2]. Lifetime risks for specific cancer sites were also estimated, based on a Japanese population.

2.   Information from follow-up through the end of 1990 of mortality among the survivors of the atomic bombings has recently been published [P9]. The extended period of follow-up was not very informative for survivors over 40 years old at the time of the bombings, since many of these people had already died. On the other hand, the data for survivors exposed at younger ages, particularly in childhood, are highly valuable, because these people have only recently reached the ages at which baseline rates for most solid tumours begin to increase sharply. Methods used in the UNSCEAR 1994 Report [U2] to project risks beyond the period of follow-up assume that the relative risks for solid tumours either remain constant throughout life (following a minimum latency period) or decrease at long times following exposure. It was shown that lifetime risk estimates based on the latter approach were 20%–40% lower than estimates based on the former [U2]. This difference was larger for those exposed at young ages.

Further follow-up of this group is needed to reduce the uncertainties in lifetime risk projections.

3.   Although the Life Span Study of survivors of the atomic bombings is the single most informative study on the effects of low-LET exposure of humans, a considerable amount of data is available from many other epidemiological studies. For example, studies of people with partial-body exposures, such as those from medical examinations or treatments, provide valuable information on risks for specific cancers. Despite the extensive knowledge of radiation risks gained through epidemiological investigations, much still remains to be learned. For example, the effects of chronic low-level exposures and internal exposures are not well described. Further data are being obtained through updates of individual studies and parallel analyses for sites such as breast and thyroid. Information is also becoming available from *inter alia* further studies of occupational exposures, including workers at the Mayak nuclear facility in the Russian Federation and from past radiological events in the former Soviet Union, such as at Chernobyl and around the Techa River.

4.   In addition to individuals exposed to low-LET radiation, various groups with exposure to high-LET radiation have been studied. Some of these exposures have arisen in occupational settings (e.g. radon in mines, radium in dial painting, or plutonium in some nuclear facilities), some from medical interventions (e.g. injections with ^{224}Ra or thorotrast), and some environmentally (e.g. radon in homes). Combined analyses of existing data, as well as several studies of residential radon that are in progress, should provide additional information on the risks of high-LET radiation. A review of these data in a format similar to that for low-LET radiation may be helpful in comparing risks.

5. The mortality follow-up of the survivors of the atomic bombings yields little data on cancers that are usually non-fatal. However, comprehensive cancer incidence data are now available for the survivors of the atomic bombings in Japan [T1], and comparisons between the two types of endpoint have been reported [R1]. Data on cancer incidence from this and other studies will assume greater importance as the treatment of cancers improves.

6. While there is a need for estimates of the total risks of cancer mortality and incidence arising from radiation exposure, there are also situations in which risk estimates for specific cancer sites are of particular value. These include (a) evaluating the effects of partial-body irradiation arising either from external exposure or from internal exposure to radionuclides and (b) estimating the probability that a prior radiation exposure led to the development of cancer in an individual, i.e. the probability of causation [I12, N1]. Epidemiological studies carried out in countries with differing baseline rates for certain cancer sites may also assist in determining how to transfer radiation-induced risks from one population to another. This is an important topic in view of the differences in baseline rates for cancers such as breast, lung, and stomach between Japan and many other countries. Depending on the form of the model used to transfer radiation risks derived from data on the Japanese atomic bomb survivors to other populations, quite different estimates of radiation-induced cancer risks can arise for such sites [L12]. It was concluded in the UNSCEAR 1994 Report [U2] that the

epidemiological data available at that time provided no clear indication of how to transfer risks. Ongoing and future studies of genetic (host) susceptibility and interactions with other carcinogens have the potential to both increase knowledge and provide new information on radiation risks.

7. The UNSCEAR 1994 Report [U2] contained a comparison of risk estimates for specific cancer types derived from various epidemiological studies. The aim of this Annex is to provide a more detailed comparison of site-specific cancer risks. It incorporates more recent data, including the updated mortality follow-up for the survivors of the atomic bombings and additional analyses of cancer incidence data for this group. The methodology and findings for this and other studies are described and compared. The potential for bias or confounding, the impact of errors in dosimetry, and other sources of uncertainty are discussed. Among the general considerations addressed are the advantages or limitations of the various types of epidemiological studies, statistical power, the influence of factors that modify radiation-induced risks, and the approach to be taken in examining risks. This approach is applied to data for specific cancer sites, namely oesophagus, stomach, colon, liver, lung, bone and connective tissue, skin, female breast cancer, prostate, bladder cancer, brain and central nervous system tumours, thyroid cancer, non-Hodgkin's lymphoma, Hodgkin's disease, multiple myeloma, and leukaemia. Risk estimates for all cancers combined are then derived, although it should be recognized that cancer is a heterogeneous group of diseases.

I. FEATURES OF EPIDEMIOLOGICAL STUDIES

8. Epidemiology is the study of the distribution and determinants of disease in humans [M10]. One of the key facets of epidemiology is that it is observational rather than experimental in nature. In contrast to randomized clinical trials, there is the possibility that bias or confounding associated with the design and conduct of an epidemiological study may give rise to spurious results. Another difficulty, which may also arise in randomized trials, is the possibility that low statistical power can hinder the ability to detect, or to quantify with precision, an elevated risk. Bias, confounding, and statistical power are discussed in more detail below. It should be emphasized that not all epidemiological studies are equally informative or of equal quality. Some have such low statistical power that they provide very little information on risks; others are so susceptible to potential or actual biases that the findings have little or no validity. It, therefore, is important to consider such methodological issues when interpreting the evidence from different studies.

9. Epidemiological investigations of radiation effects are usually constructed around either a cohort study or a case-control study. In a cohort study, a defined population (preferably with a wide range of exposures) is followed forward in time to examine the occurrence of effects. Such a

study may be performed either prospectively (i.e. by following a current cohort into the future) or retrospectively (i.e. by constructing a cohort of persons alive at some time in the past and following it forward, possibly to the current time). In a case-control study, people with and without a specified disease (the cases and controls, respectively) are compared to examine differences in exposures. Some case-control studies are nested within a cohort study, in that the cases and controls are selected from the cohort. The nested case-control study design is often used when it is difficult to obtain estimates of radiation dose or other exposures for all members of a cohort, but possible to collect them for a smaller number of individuals. For example, in an international study of patients treated for cervical cancer, radiation doses were estimated for patients with various types of second cancer, as well as for matched control patients [B1]. An alternative approach is to collect detailed information for cancer cases plus a random sample of the original cohort. The case-cohort study design [P1], which was utilized in an early analysis of cervical cancer patients [H1], is useful when studying the occurrence of several different types of cancer.

10. Cohort-based studies, particularly those performed prospectively, tend to be less susceptible to biases than case-

control studies, which depend on the retrospective collection of data [B18]. Case-control studies can be informative about risks, but particular attention needs to be paid to the potential for biases associated with the fact that the studies are retrospective. Because randomized controlled trials employ an experimental method, they are less susceptible to bias and have fewer methodological limitations than either cohort or case-control studies. However, only a few randomized trials of the effects of radiotherapy in treating cancer have provided information on radiation risks (e.g. [F3]). At the other extreme, results from correlation studies (studies based on data aggregated over, for example, geographical regions) are often unreliable. As will be described later, such studies, which are sometimes referred to as "ecological studies", have high potential for bias, owing to the lack of data on individual exposures and confounders. Therefore cohort-based and case-control studies that contain data at the individual level form the main bases for estimating radiation risks in humans.

11.   To be able to draw substantive inferences from epidemiological studies, it is important to ensure that the potential for bias or confounding is as low as possible and that the statistical precision of the results is reasonably high. In low-dose studies, methodological issues become particularly important, because even a small degree of bias or confounding can distort study results substantially. In spite of the difficulties that can arise in designing and performing epidemiological studies, epidemiology does have the advantage over molecular, cellular and laboratory animal studies of providing direct information on health risks in human populations.

A. BIAS AND CONFOUNDING

12.   Bias can be defined as any process at any stage of inference that tends to produce results or conclusions that differ systematically from the truth [S10]. Although it is possible to address issues such as lack of statistical power or random errors in dose estimates through statistical approaches, described later, bias in an epidemiological study can render its findings meaningless. Bias can arise in a number of ways. One potential source of bias is the failure to obtain follow-up data for all but a very small proportion of the people in a cohort study. Those lost to follow-up are often more likely to have migrated or died than other members of the cohort. If they cannot be identified, they will continue to contribute person-years (PY) to the study beyond the period during which any cancer that had developed (incident or fatal, depending on the type of study) would have been recorded. Thus they will appear, incorrectly, to be immortal. Even if those lost to follow-up can be identified, specifying the date on which they should be withdrawn from the study is not always straightforward. For example, in commenting on a study of second cancers after treatment for Hodgkin's disease in childhood [B16], Donaldson and Hancock [D25] pointed out that in this and other hospital-based studies, patients who develop a second cancer would be more likely to return to the hospital or clinic than patients free of the disease. If the end of follow-up is taken as the date last seen at the hospital, then many of the disease-free patients may be withdrawn from the study at an early time even though, had they later developed the disease, the follow-up would have been longer. Thus, hospital-based studies are susceptible to the possibility of differential follow-up, which may lead to an overestimation of disease rates.

13.   It is also important that the completeness of the follow-up data be uniform and not vary according to the level of exposure. This is a particular concern for diseases that are not immediately apparent, such as thyroid tumours without apparent symptoms. Increased levels of screening in a radiation-exposed population may show a raised disease incidence relative to an unscreened group. Ideally, comparisons would be made between groups with a similar level of screening, as, for example, in a study of irradiation for lymphoid hyperplasia [P8] in which both the exposed and comparison groups were screened. If, however, the level of screening was correlated with dose, examination of any dose-response relationship would be biased.

14.   The issue of differential disease ascertainment can also be important in some occupational studies. If occupational groups have better medical care than the general population, the cause of death for certain diseases (e.g. multiple myeloma and brain cancer) may be determined with greater accuracy in these groups. This could lead to spurious findings if comparison is made with the general population. For example, an apparent excess of brain tumours among a group of workers with potential chemical exposure may have been due to more detailed screening for the disease [G21]. However, this type of problem may be alleviated if disease rates within occupational groups can be compared. As an example in the context of radiation, Ivanov et al. [I13] reported a statistically significant elevated risk of leukaemia incidence among Chernobyl recovery operation workers when compared with risks for the general population. However, the workers received frequent medical examinations, and so the accuracy and completeness of the leukaemia diagnoses are likely to differ from those for the general population [B27]. Indications that differences in the ascertainment of leukaemia may have affected these findings came from a case-control study nested within the cohort of recovery operation workers [I14]. In contrast to the difference in leukaemia rates between these workers and the general population, no correlation between leukaemia risk and either radiation dose or other aspects of their work around Chernobyl was found within the cohort. It is likely that bias arose in the cohort analysis, in part because of the over-ascertainment and misdiagnosis of some leukaemias among the recovery operation workers and under-reporting of leukaemia diagnoses in the general population used for comparison [B27].

15.   The problem of differential disease ascertainment is not restricted to occupational studies. An example is given in Section IV.B.2 of how the recording of cancer on death certificates for the Japanese atomic bomb survivors may have been affected by the knowledge that the person was a survivor [P9]. Even though this type of bias might be small in absolute terms, it could have a particular impact when the risks of

cancer mortality at low doses are being estimated [P9]. The data on cancer incidence for the survivors of the atomic bombings, by contrast, are less susceptible to this type of bias because of the more objective means of ascertaining cancer.

16. Another issue of importance when comparing occupational groups with the general population is the healthy worker effect, whereby individuals selected for employment tend to have better health than the population as a whole [F4]. The healthy worker effect may be intensified because the workers who continue to be employed are healthy individuals and they receive better medical care. As an example of this effect, Carpenter et al. [C19] reported that mortality rates for all cancers combined were significantly lower than national rates among both radiation workers and non-radiation workers in three nuclear industry workforces in the United Kingdom. To overcome the healthy worker problem in studying occupational radiation cohorts, it is preferable to compare radiation workers receiving different levels of dose or dose rates rather than to compare radiation workers with the general population.

17. In case-control studies, it is important that the cases and controls should be chosen from the same well defined population and that the ascertainment of both sets should be complete. In particular, when it is necessary to approach potential study subjects or their next-of-kin for interviews, the refusal rate should be low for both cases and controls if selection bias is to be minimized. It should be noted that in cohort and case-control studies where exposures, both to radiation and other agents, are ascertained retrospectively, it is sometimes necessary to rely on the study subjects themselves or surrogates for such information. This might lead to bias, if the ability to assess exposures accurately depends on whether the disease in question arose or not. For example, in a proportional mortality study of naval shipyard workers in the United States, an increased risk of cancer and leukaemia relative to other causes of death was reported among nuclear workers [N6]. This was based on radiation exposure histories ascertained by newspaper reporters from the next-of-kin of deceased workers. However, the findings were not borne out in a subsequent cohort study in which radiation exposures were determined using employment records [R12]. The epidemiological biases associated with the initial study were discussed in detail by Greenberg et al. [G11]. In particular, the relatives of workers who died from cancer were more likely to have been located and interviewed than the relatives of other deceased workers. This, in combination with the lower all-cause mortality among nuclear workers relative to the comparison group, contributed to the spurious findings. More generally, the use of historical records, where available, is to be preferred to avoid differential ascertainment of exposures.

18. A particular problem when considering a large number of hypotheses in a study is that of multiple comparisons. A statistically significant finding is often referred to as one that would arise only once in 20 times by chance alone, i.e. 5% of the time. Therefore, if 20 non-overlapping cancer categories are examined in an epidemiological study, one of them would be expected to show a statistically significant result at the 5%

level even if the underlying risk was not elevated. This finding could represent either an excess or a deficit if a two-tailed test (i.e. a statistical test that looks in both directions) has been applied. Consequently, it is important to examine the consistency of findings for specific cancer sites across studies, as well as the consistency with other evidence, e.g. from experimental data. Problems of multiple comparisons can arise in studying not only multiple endpoints but also in testing a large number of hypotheses. For example, Jablon et al. [J1] studied cancer around a large number of nuclear facilities throughout the United States. They found that the facility-specific relative risks for childhood leukaemia formed a symmetric distribution, with roughly as many values below 1 as above it. Thus, unless there is prior reason to focus on specific facilities, those results that achieve the nominal levels of statistical significance need to be viewed in the light of the distribution for facilities overall. An extra problem that requires scrutiny is the possibility of selective reporting of results, i.e. the greater tendency for positive findings to be reported than negative findings. It is possible that some reports of highly specific positive findings, based on either small studies or sub-analyses of larger studies, reflect such a publication bias. For example, Carter et al. [C20] published the results of a study that did not show an association between Down's syndrome and maternal radiation only after a positive report appeared in the literature.

19. It is also necessary to address the potential for confounding, which can lead to bias. A confounding factor is correlated with both the disease under study and the exposure of primary interest. While many factors other than ionizing radiation affect cancer rates, in most epidemiological studies of radiation-exposed groups there is no reason to think such factors will be strongly correlated with radiation dose, although weak associations might arise by chance. For example, in studies of the survivors of the atomic bombings and many medically irradiated groups, it is unlikely that there would be a strong association between, say, levels of smoking and the dose received. One possible confounder in occupational studies is time since start of radiation work. This tends to be correlated with cumulative radiation dose and with time-related factors associated with the selection of people into radiation work. However, since the time variation in risks associated with such selection factors tends to be greatest soon after starting work [F4], analyses that omit the first few years of follow-up (when radiation effects would be unlikely to be manifested in any case) may permit resolution of this point. In studies of medical exposures, confounding may arise if the clinical indications that lead to the exposures are related to a subsequent diagnosis of cancer; this is sometimes referred to as "confounding by indication". For example, in a study of patients administered ^{131}I for diagnostic purposes, a slightly elevated risk of thyroid cancer was found [H4]. However, this risk was not related to dose and was concentrated among patients referred because of a suspected thyroid tumour, indicating that the elevated risk was probably due to the underlying condition. Similarly, in a another study, an increased risk of leukaemia and non-Hodgkin's lymphoma that arose shortly after diagnostic x-ray exposures appeared to be due to pre-symptomatic conditions of the diseases that led to the exposures [B24].

20. It is desirable to check for confounding by factors that have a sizeable influence on cancer rates if the level of radiation risk is predicted to be low or if the range of doses is narrow. For example, in case-control studies of indoor radon and lung cancer, it is very important to take account of individual smoking habits. On the other hand, if the level of radiation risk is predicted to be high, instances where data are available on potential confounders may permit not only adjustment for such factors but also examination of how such factors may modify the radiation-induced risk. For example, data on smoking habits among radon-exposed miners can allow examination of the joint influence of radon and smoking on lung cancer risks. Risk modification is discussed later in this Annex and is also covered in Annex H, *"Combined effects of radiation and other agents"*.

21. In contrast to cohort, case-cohort, and case-control studies, which utilize data on specific individuals, correlation studies are based on data averaged over groups. A particular form of this study is the geographical correlation study, in which disease rates in geographical areas are compared with average levels of exposures, e.g. to natural or environmental radiation. An example of such a study, which concerns lung cancer and indoor radon in areas of the United States [C18], is discussed in Section III.E. Since studies of this type do not involve data on individual exposures or confounders, they are susceptible to biases that do not arise in studies for which such data are available [G2]. These biases can be large, although their magnitude is dependent on the particular situation. In addition, migration can be a large problem in geographical correlation studies, because people exposed in one region can die or develop the disease of interest in another region. This suggests that estimates of radiation risks should be based on cohort, case-control, or case-cohort studies. However, correlation studies sometimes can be useful for generating hypotheses or as a means of surveillance for large effects, such as in the study of childhood leukaemia and lymphomas in Europe following the Chernobyl accident [P12], although the potential biases specific to this form of investigation should be borne in mind.

B. STATISTICAL POWER

22. A very important facet of any epidemiological study is its statistical power, i.e. the probability that it will detect a given level of elevated risk with a specific degree of confidence. The power of a cohort study will depend on the size of the cohort, the length of follow-up, the baseline rates for the disease under investigation, and the distribution of doses within the cohort, as well as the predicted level of elevated risk. Similarly, statistical power in a case-control study depends on the number of cases, the number of controls per case, the frequency and level of exposure, and the predicted exposure effect. Statistical power is generally evaluated before a study is conducted. Afterwards it is more correct to refer to statistical precision, which is reflected in the width of the confidence intervals for risk estimates.

23. The following example illustrates how the above factors can influence statistical power. Suppose cancer rates are ascertained in a cohort consisting of two groups, one of which was unexposed (the control group) and the other of which consists of persons who received a single common dose, D (the exposed group). The groups are assumed to have the same distributions for age, gender, and period of follow-up. (For simplicity, the following calculations do not take explicit account of these factors.) Statistical power can be evaluated by simulating the number of cancers in the two groups under a model such that the ratio of the cancer rate in the exposed group to that in the control group (i.e. the relative risk) is $1 + aD$, where a is the excess relative risk (ERR) per unit dose. Given the total number of cancers in the two groups, the statistical power depends only on the product of a and D and on the ratio of the number of cancers expected in the two groups if there were no elevated risk. In particular, power is calculated here by evaluating the proportion of simulations for which the number of cancers in the exposed group is greater than the value which, if there were no increased risk, would be exceeded only 5% of the time. This represents a one-sided test at the 5% level.

24. An approximate form of the power calculation is as follows. Let N denote the total number of cancers in the exposed and unexposed groups, let p denote the proportion of the total number of cancers expected to arise in the exposed group if there were no elevated risk, and let O denote the observed number of cancers in the exposed group. It can be shown that conditional on the value of N, O has expected value $E = Nq$ and variance $V = Nq(1 - q)$, where $q = p(1 + aD)/(1 + paD)$. Furthermore, provided that Nq and $N(1 - q)$ are reasonably large (at least 20 or so), O is approximately normally distributed. Consequently the statistical power (i.e. the probability that O will exceed the value that would be exceeded only 5% of the time if there were no increased risk) can be approximated using tables for the normal distribution. In particular, if there were no elevated risk (i.e. $a = 0$), then $q = p$, and so O would be approximately distributed normally with mean $E_0 = Np$ and variance $V_0 = Np(1 - p)$. Therefore a one-sided test at the 5% level would signal an elevated risk if $T = (O - E_0)/V_0^{1/2}$ exceeds 1.645, where the probability that a normally distributed variable with mean zero and variance 1 would exceed 1.645 is 0.05. More generally, let $C(x)$ denote the probability that a normally distributed variable with mean zero and variance 1 would exceed x. Then the probability that T exceeds 1.645 would be $C(1.645) = 0.05$ if there were no increased risk. More generally this probability, which equates to the power, can be calculated as $C[(E - E_0 + 1.645V_0^{1/2})/V^{1/2}]$.

25. It should be noted that the power is not zero when there is no increased risk, since there is still a chance that a large number of cases might arise in the exposed group, which would lead to a statistically significant (but spurious) finding. Under the above test, the probability of such a finding is set to 0.05. Also, since this example involves an internal comparison group, it does not rely on the validity of, say, published national or regional baseline cancer rates.

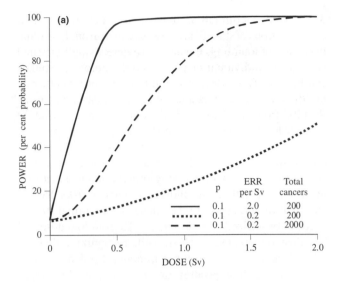

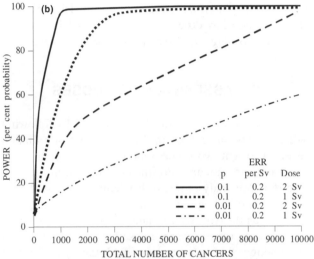

Figure I. Statistical power to detect an increased risk of cancer in an epidemiological study (a) in relation to dose with a baseline cancer incidence of 0.1; (b) in relation to the number of cancers observed with an excess relative risk (ERR) of 0.2 Sv⁻¹.

p denotes the proportion of total cancers expected in the exposed group if there were no raised risk.

26. The upper panel of Figure I shows how the power varies with a and D for various values of the total number of cases in the situation where, in the absence of an elevated risk, the expected number of cancers in the exposed group is 10% of that in the total cohort (i.e. p=0.1). Here the power is expressed as a percentage probability. Usually an analysis with about 80% power would be considered to be quite sensitive in detecting an underlying effect. The first point that should be noted from Figure Ia is the effect of the total number of cancers, N. This number is influenced not only by the size of the combined cohort but also by the baseline cancer rates and the length of follow-up. Thus a study based on a very large cohort may not be particularly informative if a rare cancer is under investigation and the follow-up is short. Conversely, a study based on a fairly small cohort may be quite informative if a common cancer is being considered and the follow-up is long. For the example illustrated in the upper panel of Figure I, if the ERR per Sv, a, is 0.2 and the exposed group

received 1 Sv, then the power to detect an elevated risk is 81% if the total number of cancers in the two groups is 2,000 but only 25% if the total number of cancers is 200.

27. The second point to note from the upper panel of Figure I is the effect of the level of elevated risk. If the overall number of cancers is 200 and the exposed group received a dose of 1 Sv, then the probability of detecting an enhanced risk at the 5% level is 25% if a (the ERR per Sv) is 0.2. In contrast, if a = 2, the corresponding probability is nearly 100%. The same probabilities would arise if, say, the dose D is doubled and a is halved. This is because the ERR can be represented by the product of a and D in this example. However, the calculation is more complex under alternative scenarios in which cohort members receive a range of different doses.

28. The two panels of Figure I are similar, except that in the lower panel the ERR per Sv, a, is fixed at 0.2 and the ratio of expected numbers of cancers in the two groups is allowed to vary. It can be seen that for a given total number of cancers and at a given dose, the power decreases with decreasing values for the proportion, p, of cancers expected in the exposed group in the absence of an elevated risk. However, for given values of a and D, the power tends to be similar if the proportion p and the total number of cancers vary in such a way that the expected number of cancers in the exposed group is roughly constant. For example, based either on p=0.1 and a total of 1,000 cancers or on p=0.01 and a total of 10,000 cancers, the predicted number of cancers in the exposed group is about 120 at a dose of 1 Sv (an excess of roughly 20). The lower panel of Figure I shows that the power is similar in the two instances (58% and 60%, respectively). An exception to this arises if p is very high, owing to the difficulty of establishing baseline cancer rates for a relatively small control group.

29. The above example is intended to show how certain factors can influence statistical power. As indicated earlier, the calculations are often more complex, as when the people in the exposed group receive a range of doses rather than the same dose. Indeed, errors in the assessment of individual doses also affect statistical power, as mentioned in the following Section and as Lubin et al. [L10] illustrated for studies of indoor radon. It should be emphasized that summary measures of the doses received by a population, such as collective dose, are not, by themselves, suitable for determining statistical power. For example, if the same dose is received by all the members of a cohort, then the usual form of analysis that looks for a trend or difference in risk according to the level of dose would not be possible. Indeed, it is essential when calculating statistical power to take account of the distribution of dose within the study population.

30. The above considerations indicate that studies such as the Life Span Study of survivors of the atomic bombings [P4, P9, S3, T1], which are based on large cohorts with doses ranging up to several gray and for which the follow-up has extended over several decades, are particularly informative about radiation-induced cancer risks. The same

holds for medically irradiated cohorts that received a wide range of doses and have a long follow-up, such as in studies of women treated for cervical cancer [B1] or given multiple chest fluoroscopies [B3, M1]. While studies of low-dose chronic exposure are of direct relevance to most occupational, environmental, and diagnostic medical exposures, their power is inherently low, owing to the low predicted level of elevated risk [L3]. In such situations, combining studies with similar designs can be very helpful in attempting to increase power. However, the possible influence of residual bias and confounding needs to be borne in mind, since the gain in precision will not lead to a gain in accuracy if bias still exists. Sometimes a meta-analysis is performed based on published findings from several studies. However, as indicated below, it is preferable, where feasible, to combine the original data and to analyse them using a common format. This approach has been used, for example, to analyse data for about 95,000 radiation workers from Canada, the United Kingdom, and the United States [C11, I2]. It has also been used for studies with greater power and large numbers of excess cancers, such as studies of lung cancer in radon-exposed miners [L4], thyroid cancer following childhood exposure [R4], and breast cancer in medically exposed cohorts [L5], to enhance analyses of effect modification as well as to increase precision.

31. In addition to increasing statistical precision, pooled or meta-analyses may be able to resolve apparently conflicting results from different studies [D1]. By aligning the studies in a parallel fashion and analysing them using a common approach, it may be possible to explain such differences on the basis of, for example, different categorizations of the exposure data. One of the main difficulties that can arise in a meta-analysis is a lack of comparability of the studies under consideration, for example because of differences in the form of the data collected on exposures and potential confounders. Summing many studies with potentially biased results may provide a precise but incorrect estimate of risk; consequently, meta-analyses can produce results that are seriously misleading [B28, B29]. Parallel analyses which address the comparability of data and the potential for bias in the various studies under consideration are therefore important in determining whether it is sensible to perform a pooled analysis. Since such an analysis is easier to perform if the individual studies are of a similar design, a prospective approach whereby studies are constructed around a common protocol is more advantageous than a retrospective pooling exercise. The former approach is being taken for a very large international collaborative study of radiation workers that is being coordinated by the International Agency for Research on Cancer (IARC) [C8]. Another potential problem with retrospective pooling is publication bias, i.e. selective reporting of results depending on whether the outcome was judged to be positive or negative. This bias, however, tends to arise for small or ad hoc studies, which would carry less weight in a meta-analysis if a number of large studies with clear, pre-defined objectives are included.

32. In view of limitations that can arise not only through considerations of statistical power but also through residual bias and confounding, the ability to detect small elevated risks using individual or pooled epidemiological studies can be low. This affects the ability to discern whether or not there is a dose threshold for radiation carcinogenesis. Results from epidemiological studies can be used to indicate levels of dose at which elevated risks are apparent, as well as whether the data are consistent with various dose-response trends [N3]. The inability to detect increases at very low radiation doses using epidemiological methods need not imply that the underlying cancer risks are not elevated; rather, supporting evidence from animal studies needs to be utilized in addressing risks from low-dose and low-dose-rate exposures [N3], while recognizing that not all molecular changes result in tumours. Epidemiological studies of such exposures do, however, enable upper bounds to be placed on radiation-induced risks. Risks at low doses and low dose rates are discussed in detail in Annex G, *"Biological effects at low radiation doses"*.

C. ASSESSMENT OF DOSES

33. A key aspect in estimating cancer risks following radiation exposure relates to the assessment of radiation doses. A recent workshop report reviewed sources of uncertainty in radiation dosimetry and their impact on dose-response analyses [N15]. Epidemiological studies of radiation-exposed groups can differ, depending, for example, on the type of information available on radiation exposure; the time between a dose having been received and making the measurement; and the specificity of assessments of doses to particular organs and particular individuals. Depending on the method of dose assessment, doses estimates could be subject to systematic or random errors or both, which could then affect the dose-response analyses. These issues are now considered in more detail.

34. The assessment of doses received by individuals in epidemiological studies may take several forms. In studies of radiation workers, for example, it is possible to utilize measurements made using personal dosimeters (e.g. [C11, G4]). For doses received from some types of medical exposures, it may be possible to reconstruct organ doses based on patient records, perhaps in combination with computer models, as for example, in an international study of patients treated for cervical cancer [B1]. In other instances, information on the past location of individuals has to be utilized together with measurement data, as for the Japanese atomic bomb survivors and, for example, people exposed to radon in dwellings. In the case of the Japanese survivors, there is still uncertainty about neutron doses at Hiroshima and the associated impact on cancer risk estimates, particularly at low doses [K20]. Further-more, as indicated later, studies of indoor radon are generally hampered by the need to assume that a contemporary measurement of radon concentration can be used to estimate concentrations during the preceding 20 or 30 years.

35. It was emphasized in the UNSCEAR 1994 Report [U2] that the data available for assessing doses were generally not collected with epidemiology in mind. For example, radiation monitoring of workers has often been undertaken to comply with management policies. Consequently, a detailed examination of dosimetry practices, including sources and magnitude of errors, is important in considering whether sufficiently accurate and precise estimates of dose can be obtained for use in an epidemiological study. A recent example is the examination of dosimetry records and practices in Canada, the United Kingdom, and the United States, carried out as part of a study of workers in these three countries [C15]. This addressed issues such as the practices on who should be monitored (e.g. all personnel at a facility or only those workers who were likely to receive doses); how missing dosimeter results should be treated (e.g. by recording zero, the threshold value for the dosimeter, a percentage of the statutory dose limit, or a best estimate of the likely dose); and the recording of a dose near or below the dosimeter threshold (e.g. as zero or by entering a "recording threshold"). Also of relevance is whether data are available on neutron doses and internal exposures. Gilbert and Fix [G4] urged the use of sensitivity analyses to examine the effect of potential sources of bias in dose estimates in epidemiological studies of radiation workers.

36. To examine the risks of specific types of cancer in relation to radiation, it is desirable to use the radiation dose to the organ under study. In some instances, such as external whole-body exposures, it may be possible to use a single value for the dose to an individual and to use this value in analysing the risk for each organ. This approach is commonly used in studies of radiation workers (e.g. [M46]). However, even in the case of external whole-body exposures, attenuation of the radiation may lead to some variation in the absorbed doses to different organs. For example, the DS86 dosimetry system for the Japanese atomic bomb survivors incorporated organ-specific transmission factors to calculate organ absorbed doses [R24]. These factors reflect the circumstances of individual exposures, including posture and orientation of the survivors relative to the explosion hypocentre; average values for the organ gamma-dose transmission factor range from 0.72 for the pancreas to 0.85 for the female breast [S51]. Also, as part of an international study of radiation workers [C15], calculations were made of the ratio of organ to "deep dose" (i.e. dose to 1 cm below the skin [I19]), both for the lung and the red bone marrow, and for various photon energies and rotational exposure geometries. This yielded ratios of approximately 1 for the lung and 0.7–0.8 for the red bone marrow for photon energies between 100 keV and 1 MeV, with the consequence that estimates of the leukaemia risk per unit dose were multiplied by 1.2 whereas no adjustment was made for other cancer types [C15].

37. In situations where the exposure involves radiation over a limited range of energies, it is possible to convert organ absorbed doses (in gray) to organ equivalent doses (in sievert) using the radiation weighting factors cited by ICRP [I1]. For most low-LET radiations, the absorbed and equivalent doses would be numerically equal. In contrast, ICRP recommends, for example, applying a factor 20 to convert organ absorbed doses from high-LET alpha radiation to the corresponding organ equivalent dose. In these situations, it may be more direct to relate organ-specific risks to organ absorbed doses than to include the radiation weighting factor by using equivalent dose. However, if the exposure is totally or virtually all due to low-LET radiation, then the use of absorbed dose or equivalent dose would give the same values for risk per unit dose. Alternatively, if the exposure arises solely from, say, internal alpha irradiation, then estimates of the risk per unit organ absorbed dose can be related by a simple factor to the risk per unit organ equivalent dose. However, if the exposure involves radiations of widely differing energies, including both high- and low-LET radiation, such as arose for workers at the Mayak plant in Russia [K32], then it is desirable to examine organ-specific risks in relation to absorbed doses split by radiation energy. If this information is not available, an alternative may be to use a total equivalent dose, based on applying weighting factors to the component absorbed doses and summing these values. However, it should be recognized that the choice of weighting factors would influence the analysis of risk in relation to dose.

38. An additional difficulty that can arise in studies involving internal high-LET exposure concerns the estimation of organ absorbed (or equivalent) doses. For example, plutonium uptake among potentially exposed workers can be assessed using urine measurements of plutonium excretion, together with information on factors tied to each individual's occupational history [O1, K32]. These assessments are dependent on aspects of the monitoring procedures, such as the level of detection and the sampling periods. To arrive at organ-specific absorbed doses, it is then necessary to use a dosimetric model for the distribution of activity between organs (e.g. [I4]). These calculations depend in turn on factors such as the solubility of plutonium in the workplace at a given time [O1] as well as on physiological factors. It should therefore be recognized that estimates of individual organ-specific doses from internal radiation are subject to uncertainty. However, this may be less of a problem if, as was the case in a study of plutonium workers in the United Kingdom, estimates of organ doses from internal radiation are generally lower than those from external radiation, even after applying a weighting factor to the absorbed doses [O1]. It should also be noted that some epidemiological studies of internal exposures present their results in terms not of organ doses but of some measure of intake (e.g. the amount of thorotrast administered to the patients [V8, V3]) or, say, the plutonium body burden (e.g. [K32]).

39. The use of recent measurements in estimating doses received many years ago, as for example in assessments of indoor radon exposures, carries particular difficulties. Changes in the intervening period (to, say, the structure of the dwelling in the case of radon) may well influence exposure levels. Again, it is important to understand which factors may have a substantial impact on exposure levels and the magnitude of these impacts. For example, investigations have been made of factors affecting temporal concentrations of indoor radon [B7]. Supplementary information may sometimes be available through assess-

ments of contemporary exposures. For example, in the case of radon, there has been interest in whether CR-39 surface measurements using a piece of glass possessed by a person over many years [M7] or *in vivo* measurements of ^{210}Pb in the skull [L25] can assist in assessing cumulative radon exposure. The former approach was used recently in an epidemiological study of indoor radon, alongside traditional track-etch measurements [A24], although further validation of the glass-based approach would be desirable in view of the effects of factors such as smoking [W19]. Furthermore, in contrast to measurements of radon in dwellings, radon exposures to persons can be influenced by occupancy patterns, particle size distributions, and breathing rates, although their effect tends not to be as great as those of factors affecting radon concentrations in houses. In general, it is essential to evaluate in detail the feasibility of estimating exposures accurately enough and precisely enough for the purposes of epidemiology.

40. In addition to the above methods of dosimetry, other biological and physical methods are now being incorporated into epidemiological studies. Such methods include classical cytogenetics for translocations, used for example in a study of women with benign and malignant gynaecological disease [K14]; the glycophorin A mutational assay of red blood cells and the fluorescent *in situ* hybridization (FISH) technique for chromosome stable translocation analysis, used by Bigbee et al. [B19] and Lloyd et al. [L26], respectively, in investigations of Chernobyl recovery operation workers; and electron spin resonance (ESR), also known as electron paramagnetic resonance (EPR), of tooth enamel, used for example in atomic bomb survivors in Japan [I8] and workers at the Mayak facility in Russia [R28]. Several factors can affect the utility of these methods in epidemiology. First, it is generally difficult to evaluate individual doses of less than 100–200 mGy using these methods, although in the case of FISH, for example, it is possible to assess average doses to populations at around these levels. For example, in spite of evaluating more than a quarter of a million metaphases, Littlefield et al. [L31] were unable to detect any increase in chromosome aberrations in lymphocyte cultures from Estonian men who took part in the clean-up of the Chernobyl nuclear power site, compared with men who did not participate in this work. Secondly, it can be difficult and/or expensive to collect, store, and analyse material for thousands or tens of thousand of people. This suggests that collection for only a subgroup of a cohort (e.g. for cancer cases and matched controls) may be a more efficient approach, although the possible effect of cancer treatment on such material needs to be considered. Thirdly, some biological measures can be affected by factors other than radiation. For example, Moore and Tucker [M49] reported that adjusting for age and smoking improved estimates of doses for Chernobyl recovery operation workers based on chromosome translocation frequencies. Fourthly, the effect of radiation on some biological measures, such as dicentric aberrations, is relatively short-lived, so the collection of related materials is unlikely to be useful in studying exposures received many years previously [L26].

41. Provided that assessment of doses is performed "blind" to whether or not the study subjects develop particular diseases, there will not be bias owing to differential misclassification of exposures, as, for example, would arise from selective recall by the subjects of past exposures. However, non-differential misclassification can still lead to bias in estimating dose-response relationships. For example, random errors in individual dose estimates tend to bias the dose response towards the null [A1]. Statistical methods have been developed to allow for such random errors in analyses, based on estimates of the magnitude of the errors, and have been applied to several radiation-exposed groups, such as the survivors of the atomic bombings [P2]. However, such errors can have a profound effect on statistical power, particularly when the predicted elevated level of risk is low [L1].

42. In some studies it is not possible to estimate doses on an individual basis, so average doses for a cohort must suffice. For example, in the study in the United Kingdom of ankylosing spondylitis patients treated with x rays, average doses were estimated for a number of organs, but only for the red bone marrow were doses estimated for a sample of individuals [L2, W1]. However, such studies can still provide information on, for example, the temporal pattern of radiation-induced risks in instances where these risks are large.

D. MORTALITY AND INCIDENCE DATA

43. It is often easier to obtain data on cancer mortality than on cancer incidence, since death certification tends to be more complete than cancer registration. For example, essentially complete follow-up for mortality of the survivors of the atomic bombings can be attained via the compulsory system of family registration *(koseki)* in Japan. Cancer incidence data for these survivors, however, are generally limited to cases arising within the areas covered by the Hiroshima and Nagasaki tumour registries [M2]. It was therefore necessary to allow for migration from these areas when analysing incidence data for the survivors of the atomic bombings [S4]. Elsewhere, complete follow-up for cancer incidence is achievable in several countries; the Nordic countries in particular have long-running cancer registries. Some countries, however, either do not have cancer registries or have strict confidentiality laws that prevent the linkage of names to diagnoses; others, e.g. the United States, have high-quality cancer registries, but only in certain regions [P5].

44. Although mortality data are often more complete, it is well known that the cause of death is recorded incorrectly or with low specificity on a non-trivial proportion of death certificates [H5]. As well as on occasion recording the wrong type of cancer, owing to metastases, there is a general tendency to under-report cancers. This affects estimates of both site-specific and total cancer risks. For example, based on linkage of death certificates to autopsy data for the Life Span Study of survivors of the atomic bombings, Ron et al. [R2] found that 24% of cancers diagnosed at autopsy were missed

on death certificates. Most of these deaths had been assigned to non-neoplastic diseases of the same organ system. Taking account also of non-cancer deaths mistakenly recorded as cancer, Hoel et al. [H5] concluded that total cancer mortality within the Life Span Study had been consistently under-estimated by about 18%. Sposto et al. [S5] showed that adjustment for errors in death certification would increase estimates of radiation-induced cancer deaths in the Life Span Study relative to published values by about 10%. In contrast to mortality, there tend to be fewer diagnostic errors in the registration of incident cancers, and histological subtypes of some cancers can be studied using incidence data. However, the proportion of histologically verified cancers varies among registries, and consideration of completeness as well as accuracy is important in judging the value of incidence data.

45. A particular advantage of incidence data over mortal-ity data is the information they provide for cancers that are often non-fatal. Of special interest within the field of radiation carcinogenesis are cancers of the thyroid, skin, and breast. For the first two of these cancers, elevated risks have been demonstrated only in cancer incidence data for the survivors of the atomic bombings and not in mortality data [R1]. While elevated risks of breast cancer mortality are apparent in this cohort, the larger number of incident breast cancer cases both in this group and in other cohorts permits a much more detailed evaluation of risks for this cancer site, particularly because survival rates may have been increasing over time. Another advantage of incidence data is that latency periods may be determined more accurately, given that the time between exposure and death could be affected by aspects of the cancer treatment.

46. It is clear that high-quality data on cancer incidence should be utilized when these are available. However, careful examination of the completeness and accuracy of data on cancer registrations is important, since mortality data are more reliable than incidence data in some countries. Furthermore, data on total mortality are important as an indicator of the overall health of populations, although incidence data can be of value for site-specific examinations. It is therefore worth considering mortality data, not only to compare levels of incidence and mortality but also as an adjunct to incidence data.

E. FACTORS THAT MODIFY RISK

47. Analyses for several cancer sites have shown that the level of the radiation-induced risk is dependent not solely on the magnitude of the radiation dose but can be modified by factors such as age at exposure and time since exposure. For example, data on the survivors of the atomic bombings [P4, P9] and on some other irradiated groups [U2] show that the ERR per unit dose for leukaemia began to decrease approximately 10–15 years after exposure and that the ERR is greater for people exposed in childhood than in adulthood. The Japanese data, in particular, also show that for all solid cancers combined, the ERR decreases with increasing age at exposure and, among those exposed early in life, tends also to

decrease with increasing time since exposure [P9, T1]. Based on an earlier version of the mortality data, Kellerer and Barclay [K21] suggested that these age and time trends could be described by a model under which the ERR depends simply on attained age. However, Little et al. [L32] showed that the Kellerer-Barclay model is not sufficient to explain the age and time trends in solid cancer risks based on the most recent Japanese incidence data, in contrast to the earlier mortality data; in particular, it is necessary to take account of both age at exposure and time since exposure when modelling the ERR, rather than just the sum of these quantities.

48. As will be discussed later in this Annex, age and temporal factors can also have a large impact on risks for specific types of solid cancer. In the case of radon-exposed miners, Lubin et al. [L4] showed that the ERR of lung cancer decreases with increasing time since exposure and attained age and is also influenced by exposure rate. For thyroid cancer, there is clear difference between the effects of irradiation in childhood and adulthood [R4]. In some instances, it may be possible to associate the effect of age with a specific biological factor; for example, there does not appear to be an elevated risk of breast cancer following post-menopausal irradiation (e.g. [B3, T1]), showing that hormones can modify the radiation risk. Apart from age, other factors that may affect radiation-induced risks are gender and baseline cancer rates, which are considered in more detail later. Indeed, factors that affect cancer rates generally, such as smoking, diet, and chemicals, may also modify the carcinogenic effect of radiation and so have to be borne in mind when, for example, evaluating probability of causation [N1, I12]. Particular examples are smoking in the case of lung cancer and chemotherapy for patients who are also treated with radiation. The combined effects of radiation and other agents are considered in more detail in Annex H, "Combined effects of radiation and other agents".

49. In common with endogenous factors such as gender and age at exposure, the ability to detect a modifying effect of exogenous factors is commonly related to the strength of the separate carcinogenic effects of these factors. For example, studies on smoking among radon-exposed miners [L4] and on chemotherapy for patients treated with radiation, e.g. for leukaemia [C9], are reasonably informative about possible interactions, owing to the high risks associated with both radiation and the other factors on their own. In contrast, epidemiological investigations of the joint effect of radiation and another factor are unlikely to be informative when the effect of either or both is weak, e.g. for a low radiation dose or a weak chemical carcinogen. An exception would be where one agent is a promoter that has an effect only in the presence of a carcinogen; however, such situations are rarely identified in epidemiological studies.

50. In addition to exogenous factors, hereditary factors may affect both baseline and radiation-induced risks. For example, retinoblastoma, a rare cancer of the eye, is frequently caused by inherited mutations of the RB1 tumour-suppressor gene. Radiation treatment for the disease appears to enhance the inborn susceptibility to development of a second cancer,

particularly osteosarcoma and soft-tissue sarcoma [E1, W11], although this effect is seen only at high therapeutic doses (above 5–10 Gy). The extent to which radiation may modify cancer risks associated with other genetic disorders such as ataxia-telangiectasia and Li-Fraumeni syndrome remains to be determined [L28]. The potential for genetic predispositions to influence radiation-induced risks is addressed further in Annex F, "*DNA repair and mutagenesis*", as well as in recent publications by ICRP [I20] and NRPB [N13].

51. Examination of potential modifying factors may be hampered by the relatively small numbers of excess cancers observed for particular sites. First, a lack of statistical power may prevent some modifying effects from being discerned. Secondly, if separate analyses are performed for a large number of cancer sites, some trends with factors such as age at exposure or time since exposure might appear simply through chance variations. To address these difficulties, Pierce and Preston [P6] recommended joint analysis of site-specific cancer risks. In this approach, a general model is fitted simultaneously to data for each of several cancer sites or groupings of sites. This can be achieved by incorporating cancer type as another factor in the usual cross-tabulation of data for analysis. Some of the parameters in this model may be the same for all cancer types; other parameters may be type-specific. Using this approach, significance tests can be performed to examine the compatibility of parameters in the risk model across cancer types. Furthermore, Pierce and Preston [P6] suggested that such comprehensive models may provide a clearer understanding of modifying factors such as gender, age at exposure, and time since exposure.

52. Pierce and Preston [P6] applied this approach to the atomic bomb survivor mortality data that had previously been analysed by the BEIR V Committee [C1]. BEIR V divided solid cancers into four categories (breast cancer, digestive cancers, respiratory cancers, and other cancers) and analysed them separately. The models derived by BEIR V for these categories had different modifying effects of gender, age at exposure, and time since exposure. For example, the ERR for respiratory, but not digestive and other cancers, decreased with increasing time since exposure. Also, the ERR was higher for females than for males in the case of respiratory cancers but was the same for both genders for digestive and other cancers. However, re-analysing these data using a joint analysis approach, Pierce and Preston [P6] showed that the data were consistent with a common model for the ERR for each cancer group except breast cancer. In the model of Preston and Pierce, the difference in relative risk between genders reflected the corresponding difference in baseline rates, the ERR decreased with increasing age at exposure at a common rate for each cancer grouping, and the ERR did not depend on time since exposure.

53. This joint analysis therefore suggested that some of the differences between the risk models developed by BEIR V might be artefacts arising from overinterpretation of the data for separate cancer groupings. On the other hand, there are prior reasons for considering certain cancer sites. For example, leukaemia and other haematopoietic cancers are normally considered separately from solid cancers (as well as from each other) owing to differences in aetiology, in the level of radiation-induced risk, and in the latency period. Also, gender-specific cancers such as breast cancer should be considered separately from non-gender-specific solid cancers, owing to the differences in factors affecting baseline rates as well as (possibly) differences in the radiation-induced risks. It is therefore intended that any modelling of radiation risks conducted in this Annex should be based on either specific cancer sites for which a large amount of data is available (e.g. breast cancer and lung cancer) or groupings such as digestive cancers. However, attention needs to be given to possible differences among cancer sites within such categories (e.g. stomach, colon, and oesophagus in the case of digestive cancers), which may preclude the modelling of combined data. In this regard, reviews of the information available for some individual sites are important, just as they may be for certain cancer subtypes. For example, observations from studies of cancer incidence among the survivors of the atomic bombings, namely that radiation-induced skin cancers are limited primarily to basal-cell carcinomas [R15, T1] and that chronic lymphatic leukaemia and virus-related adult T-cell leukaemia [P4] do not appear to be radiation-inducible, may have significant implications for biomedical research as well as radiological protection. Furthermore, it would still be possible to derive estimates of measures of risk for an individual cancer site by applying the risk model derived for a wider grouping of cancers to baseline rates for the cancer site of interest. One measure of risk that may be calculated is the risk of exposure-induced death (REID), i.e. the probability that an individual will die from a cancer that arose from an exposure [U2]. The approach just outlined has been used, for example, in applying mainly BEIR V-type models to obtain values of REID for specific cancer sites in the population of the United Kingdom [N2].

54. Another point concerns the data available from various studies. Whereas a complete cross-tabulation by factors such as gender, age at exposure, and time since exposure is available for specific cancer sites in the case of the survivors of the atomic bombings, for most studies only summary values in publications are available. Furthermore, these values are sometimes not given separately for different levels of factors such as age at exposure or time since exposure, particularly in small studies. The comparison of risks across studies can therefore be difficult if the levels of these factors differ between studies. The UNSCEAR 1994 Report presented estimates of the ERR and the excess absolute risk (EAR), i.e. the absolute difference in cancer rates, derived from various studies of external low-LET exposures, generally without adjustment for modifying factors (Table 8 of Annex A [U2]). While such a presentation is useful for comparing the general level of risks seen in various studies, it would be helpful, where possible, to consider results specific to particular ranges for age at exposure (e.g. childhood and adulthood) or to each gender, if these factors are likely to be important in modifying risks. Where such factors are important, results that do not allow for them should be interpreted with caution.

II. EVALUATION OF CANCER RISK

A. MEASURES OF RADIATION RISK, INCLUDING LIFETIME RISKS

55. Analyses of epidemiological data on radiation-exposed groups often yield estimates of ERR or EAR. These terms represent the increased cancer rates relative to an unexposed group, measured on proportional and absolute scales, respectively. For example, an ERR of 1 corresponds to a doubling of the cancer rate, while an EAR may be expressed as, for example, the extra annual number of cancers per 10,000 persons. If these values have been derived from a linear dose-response analysis, they may additionally be expressed as amounts per unit dose, e.g. ERR per Sv; otherwise, they may be quoted for a specific dose, e.g. 1 Sv. As was pointed out earlier, the level of radiation-induced cancer risks, either on a relative or an absolute scale, may vary according to various factors. Therefore, one possibility in presenting epidemiological results is to give values for ERR and/or EAR specific to particular values for these factors, for example, specific to gender and age at exposure or time since exposure, when sufficient data are available.

56. Alternatively, it has become increasingly popular in recent years to present models, based on relative or absolute scales, that describe such modifying effects. Particular examples are the models developed by BEIR V [C1] and the models for cancer incidence among the survivors of the atomic bombings [P4, T1]. These models are generally empirical, in that they attempt mainly to provide a good fit to the relevant data. To some extent they can be related to possible biological mechanisms, in that a roughly time-constant relative risk would be predicted if radiation acted at an early stage in a multi-stage process, whereas the EAR would more nearly be constant over time if radiation acted at a late stage [L7]. However, more recent research has focused on the explicit fitting of mechanistic models for carcinogenesis to data on radiation-exposed groups. For example, Little et al. [L8] analysed data on leukaemia among the survivors of the atomic bombings and cervical cancer patients using the Armitage-Doll multi-stage model [A2] and the Moolgavkar-Venzon-Knudson (MVK) two-mutation model [M11, M12]. This analysis suggested that neither model provided an adequate fit to these data, which led to the development of a generalized MVK model involving more than two mutations [L9]. Another type of mechanistic model has been proposed by Pierce and Mendelsohn [P34], in which it is assumed that cancer is caused by mutations that accumulate in a stem cell throughout life and that radiation can cause virtually any of these mutations. This model, in which the relative risk depends mainly on attained age rather than on age at exposure or time since exposure, yields age-specific risks similar to those in the Japanese atomic bomb survivors for all solid cancers combined [P34].

57. As was indicated earlier, the presentation of ERRs or EARs specific to particular levels of factors such as age at exposure and time since exposure can facilitate comparison across studies. A disadvantage of this approach, however, is that the sampling errors in these values may be high if the data are split finely. On the other hand, comparing models fitted to data from different studies may not be straight-forward if the investigators concerned have used different types of models. For example, the respiratory cancer model derived by the BEIR V Committee [C1] for low-LET radiation and the lung cancer models developed by BEIR IV [C2] and BEIR VI [C21] for high-LET radon exposure incorporate different time-since-exposure patterns.

58. If it is desired to make comparisons across studies, one possibility is to incorporate the estimated ERRs or EARs (either specific to certain levels of factors or modelled) into a life-table calculation to produce estimates of the REID, the excess lifetime risk (ELR), or the loss of life expectancy (LLE). These terms, together with a description of their advantages and disadvantages, are given in Annex A of the UNSCEAR 1994 Report [U2]. Some caveats should be attached to this approach, however. First, for the purpose of these calculations it may be necessary to extrapolate beyond the scope of the data, for example, from a limited follow-up period to the end of life, to form a lifetime risk estimate. It is important to be aware of the potential impact of such extrapolations on the comparison of results from different studies. However, lifetime risk estimates such as REID and ELR are of interest in their own right, although additional calculations specific to the types of follow-up periods arising in the studies in question may be desirable. Secondly, it is important to use the same type of life-table calculation for each study, e.g. the calculations must be based on the same baseline cancer rates and survival probabilities. If this is not done, study-to-study differences in values such as REID might arise artefactually as a result of differences in life-tables between countries rather than as a result of variation in radiogenic risks. The aim in these calculations would not, in the first instance, be to derive values of REID etc. that are of general applicability but to provide a basis for inter-study comparison. Thirdly, single values of REID etc. may not encapsulate fully the findings of each study. As a consequence, graphical displays of trends in risk in different studies could usefully complement summary risk estimates.

59. In addition to making comparisons across studies, it would be desirable to compare the main risk estimates calculated in this Annex with those calculated in previous UNSCEAR Reports. This topic is considered in Chapter IV.

B. TRANSFER OF RISKS

60. As indicated earlier, an important factor in the quantification of radiation risks is how to transfer site-specific risks across populations with different baseline rates; in other words, how to take a risk coefficient

estimated for one population and apply it to another population with different characteristics. To give some idea of the likely impact of the method of transfer employed, it is useful to consider variations in baseline rates between different populations. Table 1 builds on the corresponding table in Annex A of the UNSCEAR 1994 Report [U2], showing some of the highest and lowest cancer rates in various populations. Since cancer rates can vary over time, this table is restricted to information over a particular time period (late 1980s and early 1990s) [P5]. Although some of the variation is likely to reflect small numbers and low levels of cancer registration in some areas, broad patterns are discernible. For example, baseline rates for breast and lung cancer are generally higher in North America and western Europe than in Asia, whereas Japan has one of the highest stomach cancer rates in the world [P5]. Even within broad regions, baseline rates may differ in specific areas (e.g. [C22]). For breast, lung, and stomach cancer, Land and Sinclair [L12] showed that, depending on whether the ERR or the EAR (i.e. the multiplicative and additive transfer models, respectively) is assumed to be constant across populations, the values of REID predicted for the United Kingdom and the United States using data on the Japanese atomic bomb survivors can differ by a factor of at least 2. In contrast, differences in the total radiation-induced cancer risk tend to be smaller, reflecting the fact that there is less variation across populations in the baseline rates for all cancers combined. ICRP [I1] compared the risks estimated for five different populations using both of the above approaches in arriving at its most recent risk estimates.

61.   To some extent it is possible to investigate methods for transferring risks across populations by studying the modifying effect of factors known to account for at least some of the differences in baseline rates. Particular examples are smoking in relation to lung cancer (e.g. [C21]) and, for persons living near the Techa River in Russia, the effect of ethnicity on cancer rates [K5]. However, in many instances either little is known about the specific factors responsible for differences in baseline rates or there are few data from analytical (i.e. cohort, case-control, or case-cohort) epidemiological studies on the joint effect of radiation and such factors. As a consequence, it is necessary in most cases to directly compare measures of radiation risk, such as ERR and EAR, obtained from studies conducted in different countries on groups known to have different baseline rates. In doing so, care must be taken to ensure that the data being studied are compatible, so as to avoid confounding due, for example, to temporal changes in baseline rates. It should also be recognized that neither the multiplicative nor the additive transfer model is likely to be "correct", for individual cancer types or for groups of cancers, and that the true modifying effect is probably much more complicated. However, the paucity of relevant data imply that only a descriptive approach comparing fairly simple measures of risk is warranted in this Annex, although the presentation of data in parallel across studies can provide some idea of the influence of baseline rates.

C. TYPES OF EXPOSURE

62.   In the UNSCEAR 1994 Report [U2], epidemiological studies of radiation carcinogenesis were considered for the following types of exposure:

(a)   external low-LET irradiation, subdivided into high-dose-rate and low-dose-rate exposures;
(b)   internal low-LET irradiation; and
(c)   internal high-LET irradiation, subdivided into radon and other exposures.

63.   There are several reasons for considering these studies separately. First, the experimental studies reviewed in Annex F of the UNSCEAR 1993 Report [U3] indicated that the cancer risk per unit dose for external low-LET exposures at high dose rates (taken as >0.1 mGy min^{-1}) tends to be higher than that at low dose rates. Secondly, in addition to being protracted and specific to certain organs, it is important to note that internal exposures generally give rise to heterogeneous irradiation within organs, in contrast to most external instantaneous whole-body exposures. Thirdly, experimental studies as well as some epidemiological results indicate that, relative to low-LET radiation, the relative biological effectiveness (RBE) of high-LET radiation is a complex quantity that depends on radiation type and energy, on the dose and dose rate, and on the endpoint under study [N2].

64.   The procedure adopted in the UNSCEAR 1994 Report [U2], which considered the above types of studies separately, will therefore be used in this Annex as well. However, some studies involve more than one type of exposure, e.g. external and internal exposures to workers at the Mayak plant and to the population around the Techa River in the southern Urals; these studies will therefore be considered under the type of exposure that is of greatest relevance to the cancer in question. In addition, there is value in comparing some of the results from studies based on different types of exposure; for example, in the case of external low-LET radiation, the results of high-dose-rate studies to which a dose and dose-rate effectiveness factor (DDREF) has been applied might be compared with the results of low-dose-rate studies. However, the distinction between high-dose-rate and low-dose-rate studies is not always clear. For example, exposures to diagnostic x rays are often fractionated but are delivered at a high dose rate. As a consequence, a comparison of findings from the instant-aneous exposure of the Japanese atomic bomb survivors and findings from the fractionated exposures of tuberculosis patients who received multiple fluoroscopies [H7, H20, L39] may be more informative about the effects of fractionation than of dose rate. The main difficulty with this type of comparison, as will be shown, is the relatively low statistical power of studies of fractionated or chronic exposures for most individual cancer sites. While other types of comparison can be made concerning, for example, low- and high-LET studies, the complicating factors described above make this exercise difficult. In general, studies of specific types of exposure, such as exposure to radon, are best suited to estimating the associated risks.

D. RELEVANT STUDIES

65. In this Annex, information is examined from well conducted cohort, case-control, and case-cohort studies of radiation-exposed groups that include some assessment of the magnitude of radiation exposures. In describing and comparing these studies, attention is paid to, *inter alia*, the following:

(a) the potential for bias or for confounding by unmeasured factors;

(b) statistical power;

(c) the quality of estimates of radiation doses;

(d) the availability and quality of data on potential confounders and modifiers of radiation risk; and

(e) the availability and quality of data on cancer incidence and on cancer subtypes.

66. Relevant studies of the effects of exposures to low-LET radiation were listed in Table 2 of Annex A of the UNSCEAR 1994 Report [U2], while the strengths and limitations of these studies were summarized in Table 3 of the same Annex. Tables 2 and 3 of the current Annex expand on the tables in the earlier report by including more recent low-LET studies that incorporate estimates of the magnitude of radiation exposure. Chapter III of this Annex focuses on the more informative of these studies, based on the criteria cited in the preceding paragraph. Of these studies, the extended follow-up of mortality among the Japanese atomic bomb survivors [P9] is particularly important, since one of the main uncertainties in the assessment of radiation-induced cancer risks relates to the pattern of risk with time since exposure. Compared to the follow-up to 1985 [S3], this analysis contains 10,500 additional survivors with recently estimated DS86 doses, plus a further five years of follow-up. About 25% of the excess deaths from cancers other than leukaemia during 1950–1990 in this cohort arose in the last five years, between 1986 and 1990; for those exposed as children, this percentage rises to about 50%.

67. Tables 2 and 3 also cover studies of patients with therapeutic or diagnostic exposures, some of which are extensions of studies considered in the UNSCEAR 1994 Report [U2]. These tables are restricted to studies of postnatal and prenatal exposures. However, pertinent results from investigations of preconception irradiation are mentioned in this Annex. It should also be noted that combined analyses of some studies covered in Tables 2 and 3 of Annex A of the UNSCEAR 1994 Report [U2] were published subsequently; these included, in particular, analyses of nuclear workers [C11, I2] and of the effects of external irradiation of the thyroid [R4]. The results from these analyses are described in Chapter III of this Annex.

68. To complement Tables 2 and 3, which are specific to low-LET radiation, Table 4 lists studies of the effects of exposure to high-LET radiation that attempted to quantify levels of exposure, and Table 5 summarizes the strengths and limitations of the studies. Most of these studies were considered in the UNSCEAR 1994 Report [U2], although not in the same format as for the low-LET studies. It is important to note that, in some cases (e.g. for some of the early uranium miners and for exposures to residential radon many years ago), the exposure assessment was performed in the absence of measurements at the time of exposure. However, this caveat also applies to some studies of low-LET exposures.

E. SITE-SPECIFIC RISKS

69. One objective of this Annex is to derive and compare site-specific risk estimates from information provided in the various epidemiological studies. Relative and absolute risk estimates are presented and discussed in Chapter III. There are inherent differences in the exposure conditions, the study populations, and the evaluation procedures. Where the risk estimates available from various studies are in different formats with respect to the classification by factors such as age at exposure or where only fitted models have been presented, some life-table calculations have been performed to derive summary values that can be compared across studies. These calculations have been performed for three types of cancer, namely stomach, colon, and lung cancer. However, potential difficulties in the interpretation of such values need to be borne in mind. One of these concerns the consistency across studies of trends in radiation risks according to dose or modifying factors, which is examined in the relevant sections of Chapter III. Also, the comparison of summary values from studies in different countries with different baseline rates for certain cancer sites may be used in attempting to assess the appropriate means for transferring risks from one population to another.

70. As mentioned earlier, the main aim of these life-table calculations is to permit comparison across studies. To calculate values of measures such as REID that are of general applicability, it is preferable to use models derived for cancer sites for which large amounts of data are available or, possibly, for certain groupings of cancers, although the validity of this approach requires careful assessment. Clearly, the data on the survivors of the atomic bombings, both for mortality and cancer incidence, play a pivotal role in such an exercise. This topic is considered further in Chapter IV.

71. In assessing uncertainties, attention will be paid not only to sampling errors but also to factors such as dose and dose-rate effects, as well as variation with age, gender, and time. The extent to which such factors can explain differences between studies will be examined by, for example, presenting summary risk values based on a lifetime projection and on a period covered by the most recent follow-up. While it is unlikely that all of the uncertainties can be quantified, it is intended that the largest sources of uncertainty for each cancer site can be identified.

III. SITE-SPECIFIC CANCERS

72.   Site-specific cancer risks following radiation exposure are examined in this Chapter. The organs, tissues, or types of cancer considered are those 15 cancer sites for which adequate epidemiological data are available. Each site is discussed in a separate Section, and the summary data and inferred risks are presented in the Tables listed below.

Site of cancer type	ICD number (9th revision)	Table(s)
Oesophagus	150	6
Stomach	151	7, 22
Colon	153	8, 22
Liver	155	9
Lung	162	10, 22–25
Bone and connective tissue	170–171	11
Skin	172–173	12, 26
Female breast	174	13
Prostate	185	14
Urinary bladder	188	15
Brain and central nervous system	191–192	16, 27
Thyroid	193	17, 28, 29
Non-Hodgkin's lymphoma	200, 202	18
Hodgkin's disease	201	19
Multiple myeloma	203	20
Leukaemia	204–208	21, 30

73.   A short description is given of the general epidemiological findings for each cancer site considered, including rates in different countries, trends over time, and factors other than radiation that are known to influence rates. Information on risks in relation to both low-LET and high-LET exposures is then considered in some detail, and conclusions are drawn.

74.   The results included in Tables 6–21 are grouped according to the type of exposure (external or internal) and the radiation quality (low-LET or high-LET). Studies that provide very small numbers of cases or that do not quote sufficient detail have not been included in these Tables. Since the conditions of exposure, the characteristics of the study populations, and the extent and quality of the dosimetry, follow-up, etc. differ widely, the risk estimates are not strictly comparable. They do, however, illustrate the range and significance of estimates obtained and give some indication of the influence of the study-specific factors involved. Where possible, the estimates of the excess relative risk and the excess absolute risk in Tables 6–21 have been taken from the original publications. However, for the Life Span Study and for studies for which estimates were not cited in the associated publications, the methods described in Section I.C of Annex A of the UNSCEAR 1994 Report [U2] have been employed. In particular, if O denotes the observed number of deaths or cancer cases in the exposed population, E denotes the corresponding expected number, D the average dose and PY the number of person-years of follow-up, then the excess relative risk at 1 Sv is estimated by $(O - E)/(E \times D)$, and the excess absolute risk per unit dose and per unit time at risk is estimated by $(O - E)/(PY \times D)$. Instances where this approach has been implemented are indicated by a footnote in

Tables 6–21. It should be noted that the results based on this methodology might differ from those based on a dose-response analysis, if data subdivided into intervals of dose were available for the exposed population.

75.   Lifetime risk estimates for those studies for which estimates of ERR are available are given in Table 22. The values in this table, which are restricted to stomach, colon, and lung cancer, arise from applying the ERR estimates to baseline mortality rates for Japan, as was done in the UNSCEAR 1994 Report, and extrapolating over time both with the ERR remaining constant and with the ERR declining to zero at age 90 years, again in line with the UNSCEAR 1994 Report [U2]. As mentioned earlier, the aim of these calculations is to permit comparison across studies.

A. OESOPHAGEAL CANCER

76.   Cancer of the oesophagus is the ninth most common cancer in the world and is characterized by remarkable variations from country to country and among ethnic groups in individual countries [M40]. Oesophageal cancer rates are generally low in many countries. Extremely high rates are observed in China and among Chinese immigrants and in central Asia; intermediately high rates are seen in black populations in Africa and the United States and in some Caribbean and South American areas [M43]. Oesophageal cancer is almost always fatal, so mortality very closely approximates incidence. Heavy consumption of alcohol and tobacco has long been known to increase the risk of oesophageal cancer, and this contributes to the geographic distribution. Secular trends of oesophageal cancer vary among different populations. There has been a marked decrease in China as the lifestyle changes, a steady increase among blacks in the United States, a possible decline in central Asia, and a slow decline in Finland, India, and Latin America [D28].

77.   Few epidemiological studies have evaluated the role of radiation in the aetiology of cancer of the oesophagus. The limited data for external and internal low-LET exposures are presented in Table 6.

1. External low-LET exposures

78.   Overall, the Life Span Study data do not provide convincing evidence of a link between oesophageal cancer and radiation, although a significant excess in oesophageal cancer mortality occurred in the early years of follow-up, i.e. from 5 to 12 years after exposure. The Life Span Study mortality data also show a higher, although not significant, ERR for this cancer in females than males. Higher relative risks in females have been observed for most other solid cancers [P9]. Cancer incidence data from the Life Span Study, which began 12 years after exposure, do not show a significant excess risk of oesophageal cancer [T1].

79. The ankylosing spondylitis study is the only one to report a significant risk of radiation-associated oesophageal cancer. In contrast to the atomic bomb data, there was no significant variation in risk since first treatment [W1]. Data from other medically exposed populations considered here show no excess oesophageal cancer risk (Table 6).

2. Internal low-LET exposures

80. Very little epidemiological information is available for oesophageal cancer associated with internal low-LET exposures. The data that are available from patients treated with ^{131}I for adult hyperthyroidism [R14] show no increased risk of this cancer, but the doses received by the oesophagus were considered to be small.

3. Internal high-LET exposures

81. Data on oesophageal cancer following high-LET exposures are available from several worker studies, most of which involve small numbers of oesophageal cancers. The most informative studies are those of nuclear workers in the United Kingdom. In a study of the three nuclear industry workforces in the United Kingdom, 23 deaths from oeso-phageal cancer were observed among plutonium workers when 21.3 had been expected [C33]. An analysis of workers who were monitored for exposures to uranium, polonium, actinium, and other radionuclides (apart from tritium), showed 9 deaths from oesophageal cancer compared with 16.1 expected [C33]. Doses to the oesophagus were not available but are considered to be small.

4. Summary

82. Cancer of the oesophagus has been associated with radiation exposure in some studies. Much of the information is for external low-LET exposures, with few data available for internal high-LET exposures. The results from the Life Span Study of survivors of the atomic bombings indicate an excess risk only in the early period following exposure. The ankylosing spondylitis data show a continuing risk, while other medical studies have not demonstrated excess cases of oesophageal cancer. Very few epidemiological data are available on radiation risks for this type of cancer, which is infrequent in many countries. Since oesophageal cancers are extremely common in some parts of the world and for some ethnic groups (e.g. in China and for Chinese populations), more studies are needed to understand the magnitude and nature of the risk, especially the temporal pattern.

B. STOMACH CANCER

83. Incidence rates for stomach cancer vary considerably throughout the world [P5], with particularly high rates in Japan (Table 1). Many countries, including Japan, have seen decreases in incidence and mortality rates during the past few decades [C14]. These changes are likely in large part to reflect changes in diet, in particular, increases in the consumption of fresh vegetables and fruits and de-creases in salt intake, which case-control studies have shown to be linked to reduced stomach cancer risks [K12]. Infection with *Helicobacter pylori* [S43], which in developing countries can often reoccur rapidly following antimicrobial therapy [R36] and which can lead to gastritis, has been associated with elevated stomach cancer risks in descriptive and cohort studies [C14, K12]. In addition, smoking has been linked to modest excesses of stomach cancer in some cohort studies [D11, H28].

84. Several epidemiological studies have shown enhanced stomach cancer risks following exposure to radiation. Studies of external low-LET, internal low-LET, and high-LET radiation are considered separately in this Section.

1. External low-LET exposures

85. Included in Table 7 are the cohort and case-control studies of low-LET exposure for which radiation doses have been assessed. Among these studies, the Life Span Study of the survivors of the atomic bombings in Japan [T1, P9] has the largest number of observed stomach cancers. This is primarily a reflection of the high baseline rates in Japan, since it has been estimated that fewer than 10% of the cancers among exposed survivors are attributable to radiation [T1, P9]. Indeed, compared with national or regional rates, the estimated excess number of cases in the international cervical cancer study [B1] is larger than in the Life Span Study [T1], owing mainly to the higher mean dose in the former study. However, the large numbers of cancers in the Life Span Study make it possible to examine factors that may modify radiation-induced risks. In particular, based on the cancer incidence data, Thompson et al. [T1] showed that the dose response was consistent with linearity and that the ERR per Sv was higher for females than for males, decreased with increasing age at exposure, and did not vary significantly with time since exposure. The mortality findings up to 1990 [P9], summarized in Table 7, accord with the incidence results up to 1987 [T1].

86. Only a few of the other studies listed in Table 7 have sufficient statistical precision to permit meaningful comparison with the Life Span Study. The case-control study of patients treated for cervical cancer [B1] showed a trend in stomach cancer risk with dose that was of borderline significance. It is notable that the ERR per Sv estimated from this study appears to be consistent with that for female survivors of the atomic bombings irradiated in adulthood and that the estimate from the cervical cancer study of the EAR per Sv is lower than that from the Life Span Study, although the confidence intervals are wide. This might suggest that in transferring radiation-induced stomach cancer risks from Japan (which has high baseline rates) to countries in North America and Europe (which contributed to the cervical cancer study and which have lower baseline rates), it would be better to use a multiplicative than an additive model, that is, to transfer the ERR per Sv rather than the EAR per Sv. This is reinforced by Part A of Table 22, which shows that estimates of lifetime risk based on applying estimates of the ERR per Sv to Japanese baseline mortality rates are similar across the Life Span Study, the cervical cancer study, and other studies. Caution is called for, in that a range of other transfer methods would be consistent with these data. However, it is noteworthy

that the estimates of EAR per Sv from the Life Span Study are higher than those from several of the other external low-LET studies in Table 7, whereas the estimates of ERR per Sv are less variable between the studies with the greatest statistical precision.

87.   Another important study that can be compared with the Life Span Study of survivors of the atomic bombings is that of patients in the United Kingdom irradiated for ankylosing spondylitis [W1]. Overall there was no excess of stomach cancer in the latter study, although there was some suggestion of an elevated risk 5–24 years after exposure. While there was no evidence of an increasing trend in risk with the number of treatment courses, data on individual stomach doses were not available [W1], compli-cating the comparison of risk estimates. Another study, that of peptic ulcer patients [G6], showed similar values for males and females of the ERR per Sv, in contrast to the Life Span Study, although the number of cancers in the former study was much smaller and the mean dose (about 15 Sv) was much larger.

88.   Studies of occupational exposure to external low-LET radiation may be of value in examining risks associated with protracted or low-dose-rate exposure. In a combined analysis of radiation workers in Canada, the United Kingdom, and the United States, Cardis et al. [C11] found no statistically significant trend in stomach cancer risk with dose. Although the number of stomach cancers in this study was quite high relative to other the studies considered in Table 7, the generally low radiation doses received by these workers meant that the study had low statistical precision to estimate risks for this type of cancer. Similarly, a study of nuclear workers in Japan [E3], where (as just noted) baseline rates of stomach cancer are higher than in other countries, lacked precision because of the small doses. In contrast, a case-control study of stomach cancer among workers at the Mayak plant in Russia included some individuals with doses in excess of 3 Gy [Z1]. Although the number of cases in this dose category was modest (see Table 7), doses of this magnitude were associated with a statistically significant elevated risk. Comparison of these results for protracted exposure with the estimated ERR per Sv from the studies of acute exposure included in this table is made difficult by the lack of details on, for example, the mean doses in the categories considered in the Mayak study [Z1]. In addition, there was no significant dose response over the full range of external doses in this study, whereas there was weak evidence of an elevated risk associated with the level of plutonium body burden and with occupational chemical exposure [Z1]. Stomach cancer risks among these workers were also reported to be positively associated with gastritis and smoking, in line with other studies referenced earlier. In particular, there was some suggestion that external doses above 3 Gy interacted submultiplicatively with gastritis and multiplicatively with smoking in the incidence of stomach cancer, although as already indicated, the numbers in this dose category are not large. Additional details of the study design, for example, the means by which the study subjects were identified and information on factors such as smoking was collected, would have to be known to evaluate these findings.

89.   Low-dose, protracted exposure from background radiation has been studied in the Yangjiang area of China [T25, T26]. While this did not show an association with stomach cancer risk (see Table 7), the precision of the study was not great, in common with the low-dose occupational studies mentioned above [C11, E3].

2.   Internal low-LET exposures

90.   In a study of about 10,000 Swedish patients treated with ^{131}I for hyperthyroidism, raised incidence [H23] and mortality [H24] from stomach cancer relative to national rates were reported (see Table 7). Furthermore, there were indications of an increasing trend in risk with increasing administered activity of ^{131}I, although this trend was not statistically significant. Some caution should be attached to the interpretation of these findings. The authors examined a range of different cancer sites, so it is quite possible that one of them would show a positive finding by chance. However, it is notable that the mean dose to the stomach in this study, namely 0.25 Gy, was higher than that to other organs apart from the thyroid and was similar to the mean stomach dose among exposed atomic bomb survivors (see Table 7). Some other studies of hyperthyroid patients treated with ^{131}I [F8, G10, H25] have not reported raised rates of stomach cancer, although in some instances their statistical precision was low. Statistical precision was also a problem for studies in Sweden [H26], Italy [D15], Switzerland [G13], and the United Kingdom [E2] of thyroid cancer patients treated with ^{131}I, owing to the small number of subsequent stomach cancers; furthermore, risks in these studies were not analysed according to the level of exposure. In contrast, a large study of hyperthyroidism patients in the United States [R14] has reported rates of stomach cancer mortality that are generally consistent with national rates and that do not appear to show a relation with the level of ^{131}I administered, although there was some suggestion of an elevated risk associated with anti-thyroid drugs.

91.   The relevant part of Table 7 also shows that the estimate of the ERR at 1 Gy from the Swedish hyper-thyroidism study [H23, H24] is consistent with that from studies of external low-LET exposure. However, given the limited number of cases, the study is likely to be consistent with a range of other values. It is therefore difficult, based on this study, to reach a conclusion about how stomach cancer risks from acute, external, low-LET exposure compare with those from protracted internal low-LET exposure.

3.   Internal high-LET exposures

92.   The studies of patients with exposures to radium and thorotrast listed in Table 7 do not tend to indicate elevated risks of stomach cancer relative to unexposed patients. This probably reflects both the modest numbers of cases and, more particularly, the likely low doses to the stomach compared with some other organs. It should also be pointed out that these studies have not analysed risk in relation to individual exposures.

93. Stomach cancer was one of the cancers studied in a collaborative analysis of data from 11 cohorts of underground miners exposed to radon [D8]. Stomach cancer mortality among this group of over 64,000 men was significantly higher than national or local rates (relative risk = 1.33; 95% CI: 1.16–1.52, based on 217 deaths). However, there was no trend in stomach cancer mortality with the cumulative radon exposure received by these miners [D8]. Furthermore, excesses of stomach cancer have been reported in some other groups of miners, such as gold miners [K13] and coal miners [S25]. This, together with the low doses to the stomach from radon exposure, suggests that exposures in mining environments to agents other than radon or other factors such as smoking habits are responsible for these excesses. Among female radium dial workers in the United States, there was a statistically significant increase, relative to regional rates (SMR=3.89), in mortality from stomach cancer among those who started work in 1930 or later, although this was based on only seven deaths [S16]. The absence of an elevated risk among those who started work before 1930 and whose exposures from radium tended to be higher, suggests that the finding for the later workers is not due to ingested radium [S16].

4. Summary

94. Much of the information on stomach cancer risks following radiation exposure comes from the Life Span Study of survivors of the atomic bombings. This reflects not only the large cohort, long follow-up, and wide range of doses but also the high baseline rates for the disease in Japan. The Life Span Study indicates that the dose response is consistent with linearity and that the ERR per Sv decreases with increasing age at exposure, does not appear to vary with time since exposure, and may be higher for females than for males (although one study of medical irradiation may not agree with the latter finding). Some, but not all, studies of external low-LET medical irradiation also show an association between radiation exposure and stomach cancer risk. In particular, the findings from the Life Span Study and the study of cervical cancer patients suggest that it might be more appropriate to transfer relative risks, rather than absolute risks, from Japan to other countries. Studies of low-dose, occupational, low-LET exposure lack precision; a study of protracted, high dose, occupational exposure did indicate an elevated risk, although it is difficult to use it to quantify a dose or dose-rate-effectiveness factor. Studies of internal low-LET and high-LET exposures generally provide little information on stomach cancer risks.

C. COLON CANCER

95. Incidence rates for colon cancer vary considerably around the world [P5, P17] (see Table 1). The highest rates are mainly in North America and western Europe, although some countries with previously low colon cancer rates, such as Japan, now have rates just as high [P17]. Descriptive studies indicate that these patterns are largely associated with diet. Cohort and case-control studies tend to confirm this, with meat consumption being related to an increased risk and vegetable consumption to a decreased risk [P17]. Studies of this type have also shown colon cancer risk to be related inversely to the degree of physical activity [P17]. In addition to lifestyle factors, several rare, genetically determined conditions affect risks [U15]. In particular, familial clustering of colon cancer is thought to be due to an autosomal recessive gene [M22].

96. Colon cancer risks have been examined in various epidemiological studies of radiation-exposed groups. The findings of these studies, classified according to radiation type, are summarized in Table 8. Although it sometimes can be difficult to distinguish rectal cancer from colon cancer, the role of ionizing radiation appears to differ substantially in the aetiology of the two cancers, with cancer of the rectum rarely showing a link with radiation [T1, P9].

1. External low-LET exposures

97. The Life Span Study shows a clear association between external dose and colon risk to the survivors, based both on incidence [T1] and mortality [P9] data. Detailed analysis of the incidence data shows that the dose response is consistent with linearity [T1]. However, it is noticeable from Table 8 that studies with mean colon doses of several Sv or more, namely those of patients treated for cervical cancer [B1] or peptic ulcer [G6], show little or no evidence of an elevated risk. This suggests a possible cell-killing effect at very high doses. However, an excess of stomach cancer was seen among the peptic ulcer patients [G6], whose mean stomach dose exceeded the dose to the colon, although this might be explained by differences from organ to organ in the degree of cell killing. There is no clear pattern in the ERR per Sv by gender or age at exposure among the atomic bomb survivors, which may reflect statistical imprecision. In contrast, while the incidence data suggest that the ERR per Sv may be decreasing with increasing time since exposure [T1], the corresponding values for mortality in Table 8 would suggest, if anything, a trend in the opposite direction, although the confidence intervals are wide. However, it is clear from Table 8 that the EAR per Sv for mortality is increasing with increasing time since exposure.

98. The comparison of risk estimates across studies, in considering how colon cancer risks should be transferred across populations, is complicated by the changing baseline rates in Japan referred to earlier. Furthermore, the confidence intervals for values of the ERR and EAR per Sv estimated from the studies listed in Table 8 are wide and are consistent with various transfer methods. This is confirmed by Part B of Table 22, which shows lifetime risk estimates (based on an implicit multiplicative transfer across populations) that are fairly similar in the Life Span Study and the study of women in the United States treated for benign gynaecological disease [I16] but smaller in other studies of populations in North America and western Europe. It is therefore not possible to come to a firm conclusion on how to transfer colon cancer risks across

populations. It is also not possible to make a meaningful comparison of risks from high- and low-dose-rate exposures, owing to the large imprecision in estimates derived from studying nuclear workers [C11, E3].

2. Internal low-LET exposures

99.   Few data are available on colon cancer following internal exposure to low-LET radiation. Among Swedish patients treated with ^{131}I for hyperthyroidism, the standardized incidence ratio for colon and rectal cancer combined was 1.17 (95% CI: 0.97–1.39) after 10 or more years of follow-up [H23]. However, results for colon cancer alone were not presented or analysed in relation to the level of iodine administered. It should be noted that the mean dose to the colon and rectum from this treatment was estimated to be 0.05 Gy [H23], suggesting that any analysis of risk in relation to the level of exposure would have had low statistical precision. Studies of patients treated with ^{131}I for hyperthyroidism in the United States [G10, H25, R14] and for thyroid cancer in Sweden [H26], together with a study of diagnostic exposures in Sweden [H27], did not report findings specifically for colon cancer. However, a large study of hyperthyroidism patients in the United States provided little indication of an elevated risk of colorectal cancer mortality [R14].

3. Internal high-LET exposures

100.   Numbers of colon cancers reported from studies of thorotrast patients are included in Table 8. Here, as in the combined analysis of underground miners exposed to radon [D8] and in studies of radium patients [N4] and radium dial workers [S16], the very low doses to the colon associated with these exposures preclude meaningful inferences.

4. Summary

101.   Data on the Japanese atomic bomb survivors are consistent with a linear dose response. The effect of gender, age at exposure, and time since exposure on the ERR per Sv is not clear, although the EAR per Sv does increase with increasing time since exposure in the Life Span Study. Changes over time in baseline rates in Japan make it difficult to decide how to transfer risks across populations. Also, the lack of precision in low-dose studies of external low-LET radiation and of internal low-LET and high-LET radiation do not allow conclusions to be drawn.

D. LIVER CANCER

102.   The liver is one of the most frequent sites for metastatic cancer. Since a large proportion (as high as 40%–50%) of liver cancers reported on death certificates are tumours originating in other organs, mortality data are usually a poor measure of the magnitude of primary liver cancer. It is therefore difficult to obtain reliable estimates of the magnitude of liver cancer in many countries and populations. Cancer incidence data, which provide more reliable diagnostic

information, are available for various parts of the world, but their quality also varies. Liver cancer is one of the eight most common cancers in the world, accounting for 5.6% of the new cancers in males and 2.7% in females, but there is a wide geographic variation [M40]. Liver cancer is a common disease in many parts of Asia and Africa but is infrequent in the United States and Europe [P29]. The incidence of liver cancer has been increasing in Japan and the Nordic countries [S45], although some of the increasing trends may be explained by changes in disease classification and coding practices.

103.   The great majority of primary liver cancers in adults are hepatocellular carcinomas. It has been estimated that about 80% of hepatocellular carcinomas are aetiologically associated with chronic infection with hepatitis B virus [L43]. Infection with hepatitis C virus also plays an important role in some countries, notably in Japan. Alcohol consumption and liver cirrhosis have been shown to increase the risk of hepatocellular carcinoma, but their precise roles have yet to be clarified. In general, hepatocellular carcinoma occurs much more frequently in men than in women (male:female ratio of 4–5:1). Other types of liver cancer include cholangiocarcinoma and angiosarcoma, which are rare in adults. The male preponderance is less pronounced for cholangiocarcinoma (male:female ratio of 1–2:1) than for hepatocellular carcinoma. Liver cancer has been associated with infestations with liver flukes in certain areas as well as with exposure to thorotrast [P29, T1].

1. External low-LET exposures

104.   Epidemiological data on liver cancer associated with external exposures to low-LET radiation exposure are limited. Far more information is available on internal high-LET exposure, especially thorotrast (see below). The available data are presented in Table 9. None of the medically and occupationally exposed populations included in this review suggest an association between radiation exposure and liver cancer. Where an increased standardized mortality ratio (SMR) for liver cancer is found, further analyses do not support a dose-response relationship. Furthermore, because a large number of metastatic tumours may be misclassified as liver cancers on death certificates, some of the observed excess liver cancer may be attributable to the inclusion of tumours originating in other organs. The most convincing evidence for excess liver cancer associated with low-LET exposure comes from the Life Span Study. In the latest Life Span Study report [P9], there are 432 deaths from primary liver cancer (939 including those specified as primary and those specified as secondary), the third leading cause following stomach and lung cancers. A significant dose response is found for liver cancer, with an ERR per Sv of 0.52 for males and 0.11 for females, both exposed at age 30 years.

105.   Cancer-incidence-based data obtained from the systematic collection of information reported by hospitals to tumour registries have better diagnostic accuracy. The analysis of the Life Span Study cancer incidence data showed for the first time a significantly increased risk of liver cancer associated with radiation exposure from the

atomic bombings. A subsequent study involved 518 cases of liver cancer, mostly hepatocellular carcinoma, verified by a detailed pathology review of each case [C37]. The dose response was linear and an ERR was estimated to be 0.81 per Sv (liver dose). Males and females had a similar relative risk so that, given a three-fold higher background incidence for males, the radiation-induced excess incidence was substantially higher for males. The excess risk peaked for those exposed in the early 20s with essentially no excess risk in those exposed before age 10 or after 45 years.

2. Internal low-LET exposures

106. Epidemiological data are even more sparse on liver cancer and internal low-LET exposures. In the United States thyrotoxicosis study in which about 21,000 hyperthyroid patients treated with ^{131}I were followed up to 45 years, 39 liver cancer deaths were observed with an SMR of 0.87 [R14]. The doses received by the liver were not estimated but are presumably very low.

3. Internal high-LET exposures

107. Thorium-232 is a primordial, long-lived, alpha-emitting radionuclide. Colloidal (^{232}Th) thorium dioxide (thorotrast) was used widely as an intravascular contrast agent for cerebral and limb angiography in Europe, the United States and Japan from 1928 to 1955. Intravascularly injected thorotrast aggregates tend to be incorporated into the tissues of the reticuloendotheial system, mainly the liver, bone marrow, and lymph nodes. Deposition results in continuous alpha-particle irradiation throughout life at low dose rate. The radiation dosimetry is complex because of the non-uniform distribution of thorium dioxide in the liver, bone marrow and lymph nodes and the possible effects of the colloidal material on cancer risk [C2]. It has been estimated that the typical annual dose from alpha radiation following an injection of 25 ml of thorotrast is 0.25 Gy to the liver [K28, M41], but a re-evaluation of liver organ mass has indicated that the annual dose is 0.40 Gy [K41]. A revised whole-body organ partition of ^{232}Th has shown a small reduction in the relative partition to the liver, but the estimated liver dose remains essentially the same [I25]. Patients who were administered thorotrast from the late 1920s through to 1955 have been followed in Germany, Portugal, Denmark, Sweden, Japan and the United States. The total number of people being followed is approximately 5,500, and over 90% of them have died.

108. In Germany, about 5,000 patients treated with thorotrast for cerebral angiography (about 70%) or arteriography of the limbs (about 30%) between 1937 and 1947 at different hospitals were identified [V3, V7, V8]. As controls, a similar number of age- and gender-matched non-thorotrast treated patients were identified among patients at the same hospitals. When the follow-up study was started in 1968, a large number of the patients had already died. The causes of death among those patients were identified from hospital examinations or death certificates. There were 2,326 thorotrast patients (1,718 males and 608 females) and 1,890 controls (1,407 males and 483 females) who survived three years or more after treatment, could be traced. The patients (899 thorotrast patients and 662 controls) who were still alive at that time have since been followed through clinical examination every two years. The latest follow-up data show 48 thorotrast patients and 239 controls who are still alive [V8]. In the deceased patients, the most common neoplastic disease is liver cancer (454 thorotrast patients compared with 3 controls) [V8]. Previous data showed that cholangiocarcinoma and haemangiocarcinoma, which are norm-ally rare types of liver tumour, accounted for about 54% and hepatocellular carcinoma for only 17%; histological types were unknown for the remaining 29%.

109. In the German study, the cumulative rate of liver cancer was correlated with the mean dose of administered thorotrast, although no formal dose-response analyses were performed. No age-at-exposure effect was observed, as the cumulative rate of liver cancer was similar for the three cohorts having different ages at injection (1–14, 15–29, 30–44, and 45–59 years) [V8]. Recent data suggest an increase in liver cancer among those who received less than one ampoule of thorotrast (less than 6 or 6–12 ml thorotrast) [V8]. Although there was no gender difference with regard to age at injection and mean volume of injected thorotrast and exposure time, the cumulative rate of liver cancer was significantly higher in males than in females. As a measure of the total risk, the cumulative rate was calculated using the sum limit method, i.e. by taking the cumulative number of liver cancers after injection (excluding those dying within the first 15 years of exposure as they are not considered to be due to thorotrast) as the numerator and the cumulative dose of all patients up to 10 or 15 years (wasted dose or time) before clinical manifestation of liver cancer as the denominator. The cumulative risk of liver cancer was estimated to be 607 10^{-4} Gy^{-1} (with 10 years wasted dose) and 774 10^{-4} Gy^{-1} (with 15 years wasted dose) [V8].

110. The continuing follow-up of the Danish thorotrast study, although based on a smaller number of patients than the German study, has provided further detailed epidemiological information [A5, A18, A19]. The thorotrast group consisted of 999 neurological patients treated with thorotrast for cerebral angiography between 1935 and 1947. The group has been followed through linkage to the national death register and the cancer registry in Denmark. Previous analyses of cancer incidence data from this cohort study had been based on SIRs compared to the national cancer data. To avoid possible confounding due to the neurological conditions for which the patients were treated, a control group (1,480 persons) was identified from patients who had been examined during 1946–1963 with cerebral arteriography using contrast agents other than thorotrast [A5].

111. The latest analyses of the Danish thorotrast study data are based on 751 deaths in the thorotrast group and 797 deaths in the control group up to January 1992 [A5]. At the end of follow-up, 40 thorotrast patients and 422 controls were still alive. Since the thorotrast and control groups differed with respect to calendar period and were

not matched for gender, age at arteriography, or neurological condition, multiplicative regression models were fitted, allowing the SMR to vary with gender, age at arteriography, and calendar period. For the thorotrast patients, the models included the amount of thorotrast injected (as a measure of dose rate) and the amount injected multiplied by the time since the injection (as a measure of the cumulative alpha-radiation dose). When evaluated in multiple regression analyses, the effect of injected volume was significant for cancer (relative risk per 10 ml = 11.1; 95% CI: 3.5–34.0) and benign liver conditions (relative risk per 10 ml = 1.2; 95% CI: 1.1–1.3). Analyses of specific cancer types were based on cancer incidence cases (315 cases in the thorotrast group and 201 cases in the control group). Primary liver cancer was the most frequent type of cancer among the thorotrast-exposed patients. There were 84 cases reported as primary liver cancer, 16 reported as liver cancer not specified as primary. A significant effect of injected volume of thorotrast was seen for liver cancer (relative risk per 10 ml = 194; 95% CI: 31–1,220) and as a consequence for all cancers combined (relative risk per 10 ml = 14.7; 95% CI: 5.2–41.5). No effect of the surrogate measure of cumulative dose was seen.

112. The earlier analyses of the Danish thorotrast cancer incidence data showed a positive trend in SIR for liver tumours with young age at injection. However, the cumulative frequency of liver cancer relative to the estimated cumulative radiation dose to the liver showed no significant difference between those injected at different ages (0–25, 26–45, 46–59, and older than 59 years). The female:male ratio for liver cancer was 1.6, but the cumulative frequency of liver cancer relative to the estimated cumulative radiation dose to the liver did not differ for males and females. This is in contrast to the German study, which suggested a larger absolute risk for males than females [V8]. In a separate study in Denmark [A18], cases of primary liver cancer were reclassified by a pathology review. As with the German thorotrast series, cholangiocarcinoma (34%) and haemangiosarcoma (28%) were relatively common, while hepatocellular carcinoma accounted for 38% of cases. However, no significant differences were found between three histological types with respect to such factors as age at injection of thorotrast, mean amount injected, mean time from injection to diagnosis, or mean estimated cumulative alpha-radiation dose. The incidence of all histological subgroups was described most simply as a function of the estimated cumulative dose up to 15 years previously.

113. In Japan, two cohorts of thorotrast patients have been followed. An early study initiated in 1963 involves 262 war veterans who received intravascular injection of thorotrast (with a mean of 17 ml per injection) for diagnosis of injuries during 1937–1945 and a control group of 1,630 war-wounded veterans [M42]. As of 1998, 244 (93%) thorotrast patients had died, of whom 79 died from liver cancer [M47]. The second study began in 1979 after a nationwide survey of thorotrast patients with diagnostic x rays, and this cohort includes 150 thorotrast patients

[K33]. As of 1998, 132 (82%) patients had died, of whom 64 died from liver cancer. Analyses of combined data from these two cohorts show the rate ratio for all causes, compared to controls, to start to increase after a latency period of 20 years after the thorotrast injection [M14]. The rate ratio is highest for liver cancer (35.9) [M14]. Using previous data, the risk of liver cancer was estimated to be 330 10^{-4} Gy^{-1} with a linear dose-response model [U2]. A study of an autopsied series of 106 thorotrast-related liver malignancies showed that 44 (42%) were cholangiocarcinoma, 42 (40%) were angiosarcoma, 17 (16%) were hepatocellular carcinoma, and three were double cancers [K29].

114. The Portuguese thorotrast study was set up in 1961. It involved about 2,500 patients injected with an average of 26 ml of thorotrast between 1929 and 1955 and 2,000 controls [D27, H46]. They were followed for 30 years. Of 1,244 traced thorotrast patients, 955 had died, 137 of them from malignant tumours, including 87 primary liver cancer. The BEIR IV Committee estimated the risk for liver cancer to be 275 10^{-4} Gy^{-1} [C2]. The follow-up of this cohort was interrupted in 1976, but has recently been reactivated. The results of the follow-up extended through 1996 have been made available [D31]. A total of 1,931 patients who received thorotrast systemically and 2,258 unexposed subjects were initially identified from medical records. Follow-up was possible for 1,131 (59%) of the thorotrast patients and 1,032 (46%) of the unexposed patients. By the end of 1996, 92% of the thorotrast patients and 5% of the unexposed patients had died. The relative risk was significantly elevated for liver cancer (70.8) and for leukaemia (15.2), which accounted for most of the excess mortality from malignancies.

115. Liver cancer mortality has been studied among about 11,000 workers exposed to both internally deposited plutonium and to external gamma radiation at the Mayak nuclear plant in the Russian Federation [G23]. Within this cohort, liver cancer risks were elevated among workers with plutonium body burdens estimated to exceed 7.4 kBq, compared to workers with burdens below 1.48 kBq (relative risk 17; 95% CI: 8.0–36), based on 16 deaths in the former group. In addition, trend analyses using plutonium body burden as a continuous variable indicated an increasing risk with increasing burden (p<0.001). However, because of limitations in the current plutonium dosimetry, it was not possible to quantify liver cancer risks form plutonium in terms of organ dose, nor to make a reliable evaluation of the risk from external radiation in this cohort [G23].

4. Summary

116. While an association of liver cancer with radiation exposure has not been demonstrated in medical and worker studies involving external or internal low-LET exposures, the mortality data from the Life Span Study of survivors of the atomic bombings indicate a significant dose response. This relationship is strengthened by the analysis of incidence data based on histologically and clinically

verified primary liver cancer cases. Studies of thorotrast-exposed patients consistently show increased risks of liver cancer from alpha-radiation exposure.

117. While the types of liver cancer associated with thorotrast exposure are typically cholangiocarcinoma, followed by angiosarcoma and hepatocellular carcinoma, the excess risk associated with low-LET exposure in the Japanese atomic bomb survivors is primarily hepatocellular carcinoma. Liver cancer rates are high in Japan, especially in males, and the high rates have been attributed to infection with hepatitis viral infection, particularly hepatitis C virus. In transferring liver cancer risks from one population to another, differences in background liver cancer rates, as affected by the prevalence of hepatitis viral infection, should be considered.

E. LUNG CANCER

118. Although lung cancer was once a rare disease, it is now one of the leading causes of cancer mortality in industrialized countries and is rising in incidence in many developing countries [G1]. Table 1 illustrates the wide variation in rates between different populations. The geographical and temporal differences in incidence and mortality largely reflect cigarette smoking, which has been shown by epidemiological and toxicological evidence to be the main cause of the disease [U17]. Assessments made in the early 1980s indicate that occupational exposures to agents such as arsenic, asbestos, chromium, and nickel may account for 5%–15% of lung cancers in the general population of industrialized countries such as the United States [D6, S6], while outdoor air pollution arising from fuel combustion and industrial sources is thought to be responsible for only a few percent of cases in most areas [D6].

119. In addition to the above factors, ionizing radiation has been shown in numerous epidemiological studies to be a lung carcinogen [U2]. Increased risks have been shown not only with respect to exposure to low-LET radiation but also from exposure to radon and its progeny. Such increases have also been reported in animal studies [C4, U16]. Results from epidemiological studies of low-LET and high-LET exposures are presented in Table 10.

1. External low-LET exposures

120. The results from the latest mortality follow-up of the Japanese atomic bomb survivors [P9] bear out many of the results of the previous mortality and incidence studies. In particular, the dose response is consistent with linearity, and the ERR per Sv is higher for females than for males. However, compared with the previous follow-up, there is more indication now of similarities in the EAR per unit dose for males and females (see Table 10). Taking into account the wide confidence intervals, there is little to suggest that the ERR varies in a consistent fashion with either age at exposure or attained age, either in the

incidence [T1] or the mortality data [P9]. In contrast, Pierce et al. [P9] showed that the EAR per Sv for mortality increases sharply with increasing attained age, reflecting the pattern in baseline rates, whereas (after adjusting for attained age) age at exposure does not appear to influence the EAR per Sv.

121. It should be noted that these analyses do not take account of smoking habits. As indicated above, much of the variation in baseline rates between populations reflects differences in smoking habits, so examination of the joint effect of radiation and smoking is highly pertinent to the issue of how to transfer risks across populations. The UNSCEAR 1994 Report [U2] gave some details of a 1986 study of radiation and smoking among a subgroup of the atomic bomb survivors [K8]. The findings from this study may need to be qualified, since they are based on the use of the previous dosimetry system for the survivors and on data on cancer incidence only up to the end of 1980. Further-more, as in an earlier analysis based on mortality up to 1978 [P13], neither an additive nor a multiplicative model for the joint effect of smoking and radiation was totally inconsistent with the data. However, the suggestions from this analysis of an additive rather than multiplicative effect of low-LET radiation and smoking on lung cancer risk might explain the higher ERR per Sv for females than for males. It is possible that smoking could explain some of the other findings described earlier, such as the lack of trend in the ERR with age at exposure.

122. Further information on the joint effect of radiation and smoking comes from a case-control study of lung cancer incidence among patients treated for Hodgkin's disease in the Netherlands [V2]. In contrast to the Life Span Study findings, there was a statistically significant supramultiplicative effect of radiotherapy dose to the affected lung area and the cumulative amount smoked after diagnosis of Hodgkin's disease. Indeed, a trend in lung cancer risk with radiation dose was evident only among those who had smoked more than a small amount in the period following the original diagnosis. Some caution should be attached to these results. There were only 30 persons in total with lung cancer, of whom 8 were either non-smokers or light smokers. Furthermore, other measures of smoking, such as the number of years smoked before diagnosis of Hodgkin's disease or lifetime consumption, did not show the above supramultiplicative effect. Therefore the possibility of a chance finding cannot be excluded. An alternative interpretation is that smoking may have a strong promoting effect on the induction of lung cancer following an earlier radiation exposure. However, it should be recognized that many of those who smoked after the diagnosis of Hodgkin's disease had also smoked before that time, which makes examination of the interactions even more complicated. A larger, international study of lung cancer incidence following Hodgkin's disease [K9] also collected information on smoking, although this was limited to never/ever smoked and may have been reported more fully for cases than controls. In contrast to some other studies (e.g. [V2]), this international study showed an elevated risk associated with chemotherapy. Risks by type of therapy were reported to be

similar for smokers and for all subjects, although a formal statistical analysis of the joint effect of radiation and smoking was not undertaken. Although, if anything, there appeared to be more evidence of a radiation-induced risk among the patients who did not receive chemotherapy (relative risk = 1.6; 95% CI: 0.66–4.12, for lung doses above 2.5 Gy relative to less than 1 Gy), neither among those patients nor among patients who received chemotherapy did the trend with radiation dose approach statistical significance.

123. The only other study in Table 10 that shows an excess of lung cancer associated with low-LET radiation and that has sufficient numbers to permit examination of modifying factors is that of patients in the ankylosing spondylitis study in the United Kingdom [W1]. It should be borne in mind that, in contrast to other studies, in the ankylosing spondylitis study it was not possible to estimate individual doses to organs other than the bone marrow. This makes it difficult to address the transfer of risks between populations, although it might be worth noting from the above table that the ERR per unit dose estimated for the ankylosing spondylitis study is lower than that from the Life Span Study of the atomic bomb survivors. Indeed, the indications from Part C of Table 22 of higher lifetime risk estimates based on ERR values from the Life Span Study data compared with those from other data sets may suggest that the variation in radiation risks across populations is closer to additive than multiplicative. The latest mortality follow-up of the ankylosing spondylitis study continues to show, in contrast to the Life Span Study, a strong decrease in the relative risk more than 25 years following first treatment. The interpretation of this result is complicated by the absence of smoking data for the ankylosing spondylitis study. However, Weiss et al. [W1] pointed out that relative to national lung cancer rates, the risk among unirradiated patients showed little trend with time since diagnosis of spondylitis. While this suggests that the temporal trend in risk among irradiated patients may not be explained solely by changes over time in smoking habits, the number of lung cancer deaths among unirradiated patients was relatively small.

124. Of particular interest among the low-LET studies is the discrepancy between the lung cancer risks observed among the survivors of the atomic bombings in Japan and the findings from studies of patients who received multiple fluoroscopies in the course of treatment for tuberculosis. Studies of the latter type in both Canada [H7] and the United States (Massachusetts) [D4] found no evidence of a positive association between dose and risk of lung cancer. The Canadian result is particularly important, since it is based on a large cohort of exposed persons (25,000 with lung doses in excess of 10 mSv), while the mean age at exposure, follow-up time, and total number of lung cancer deaths are similar to the corresponding values for the atomic bomb survivors. Table 23 gives details by dose range of lung cancer mortality in both the Canadian fluoroscopy study and the latest follow-up of the atomic bomb survivors. This table clearly shows the lack of evidence for a dose response for lung cancer in the former study, which contrasts with the corresponding results for

breast cancer among female members of the cohort [H20] (see Section III.H.1). Furthermore, the large number of deaths means that the discrepancy with the atomic bomb survivor results cannot be explained by a lack of statistical precision.

125. Howe [H7] addressed a number of possible reasons for the difference between the Canadian and Japanese results. He pointed out that the effect of non-differential measurement errors on estimates of risk per unit dose in the Canadian study was likely to be similar in magnitude to that in the Japanese study for solid tumours, i.e. 4%–11% [P2]. Most of the measurement error was associated with estimating the dose per fluoroscopy, which, since it was not performed individual-by-individual, should not bias risk estimates [A1]. In contrast to breast doses [H20], lung doses were similar for anterior-posterior and posterior-anterior orientations and, consequently, were similar in Nova Scotia (where the former orientation predominated) and in the rest of Canada (where the latter orientation predominated). It is difficult to evaluate the potential for systematic errors in dose estimates, but it seems highly unlikely that such errors could explain the discrepancy with the atomic bomb survivor findings. Howe also addressed the effect of possible misclassification of some lung cancer deaths as deaths from tuberculosis. Had the lack of an association between lung cancer and dose been due to differential misclassification concentrated at higher doses, this would have led to an increasing trend with dose in deaths classified as tuberculosis. However, no such trend was apparent, even among those patients at a minimal or moderate stage of tuberculosis, for whom the potential to detect any such effect is likely to have been greatest [H7]. Finally, although individual data on smoking habits were not available for all members of the Canadian cohort, information for over 13,000 of these patients indicated that heavy smokers had not tended to have received lower doses.

126. Several other possible explanations can be considered for the difference between the results of the Canadian and Massachusetts fluoroscopy studies and the Life Span Study. First, the fluoroscopy studies were performed on groups in North America, in contrast to the atomic bomb survivor study in Japan. In particular, baseline rates for lung cancer in North America are higher than the corresponding values in Japan [P5]. However, elevated risks of lung cancer in other groups exposed to low-LET radiation in North America or western Europe are indicated in Table 10, most notably the ankylosing spondylitis study [W1], demonstrating that genetic factors or differences between countries in smoking habits cannot by themselves explain the difference in risks. Secondly, Howe [H7] drew attention to the differences in the fractionation of dose and in dose rate between the atomic bomb survivors and the fluoroscopy patients. Whereas people in the former group received a single dose averaging several hundred mGy in about one second, the latter group received fractionated doses, with an average dose rate of 0.6 mGy s^{-1} to the lungs. In this regard, Elkind [E5] has suggested that complete repair may occur between fractions of sub-effective lung cancer initiation.

It should be noted that, even if fractionation or low dose rate does considerably reduce the risk of lung cancer from low-LET radiation, this need not imply that similar effects would be seen for other cancers or for high-LET exposures, as will be discussed later. Thirdly, the effect of radiation on inducing cancer in the lung may differ between patients with tuberculosis and healthy persons. However, the lack of an association with radiation dose in the Canadian study was observed for those with tuberculosis in its early stages as well as for those with a more advanced stage of the disease [H7]. On the other hand, even within categories of tuberculosis, the severity of the disease was related to the degree of lung collapse, and hence to both the number of fluoroscopies and the degree of surgery [B15]. The latter would have involved the removal of lung tissue and may have affected the lung cancer risk. Consequently, there remains the possibility that the severity of the tuberculosis may have had some confounding effect.

127. Inferences from the other high-dose-rate, low-LET studies listed in Table 10 are limited by the smaller number of lung cancers and the general lack of data on smoking habits. Furthermore, the comparison of risks at high and low dose rates, even in large studies of radiation workers [C11, E3], is made difficult both by the low statistical precision associated with low doses received and by the lack of data on smoking. However, early workers at the Mayak plant in Russia tended to receive higher cumulative doses than many other groups of radiation workers, so data on them may be more informative. For a group of 1,841 men who started working at the nuclear reactors at Mayak between 1948 and 1958 and who had a mean external whole-body gamma dose of 1.02 Gy (low-LET), there was no indication of an increasing trend in lung cancer risk with gamma dose (see Table 10) [K34]. It should be noted that, in contrast to other groups of Mayak workers, described in Section III.E.3 below, these reactor workers did not have potential for plutonium exposure [K34]. A study of natural radiation in the Yangjiang area of China did not indicate an elevated risk associated with low-dose, protracted exposure [T25, T26] (see Table 10). Although the precision of this study was limited, information on smoking habits collected in an associated survey [Z2] suggested that smoking was not associated with dose and therefore might not be a confounder.

2. Internal low-LET exposures

128. Several studies of patients given ^{131}I have examined the risks of lung and other respiratory cancers. Most of these studies were reviewed in the UNSCEAR 1994 Report [U2]. Among Swedish patients treated for hyperthyroidism, Hall et al. [H24] reported increased mortality relative to national rates more than 10 years after treatment (based on 63 deaths, SMR = 1.80; 95% CI: 1.39–2.31). However, there appeared to be no clear trend in the risk of respiratory cancers with the level of ^{131}I administered. It should be noted that the mean lung dose in this study was only 70 mGy. Studies of hyperthyroid patients in the United States [G10, H25] and of thyroid cancer patients in Sweden

[H26] treated with ^{131}I did not show raised rates of respiratory cancer, although both studies were based on smaller numbers than the study of Hall et al. [H24]. A larger study of hyperthyroid patients in the United States [R14] provided slight evidence of a trend in lung cancer mortality with increasing administered ^{131}I activity, but this was weaker after allowing for a 10-year latency. A study of Swedish patients with diagnostic exposures to ^{131}I [H27] had more respiratory cancers but lower doses than in the Swedish hyperthyroidism study [H24]; the former study again showed no elevated risk. Bearing in mind not only the low risks predicted in these studies but also the general absence of individual lung dose estimates and smoking histories, it is not possible to compare the risks of protracted internal low-LET exposure with the risks of acute external exposure.

129. Kossenko et al. [K5] drew attention to differences between the Techa River cohort and the Japanese atomic bomb survivors with respect to the proportion of cancers of the lung. In particular, lung cancer accounted for 27% of all cancers among men in the former cohort, compared with 10% in the latter. Conversely, among women the corresponding percentages were 4% and 10%, respectively. While differences in the type of exposure and in ethnic background might be responsible for some of these variations, smoking habits are likely to be of importance. However, the available data did not allow investigating this issue.

130. Wing et al. [W14] reanalysed data on cancer incidence near the Three Mile Island nuclear plant in the United States, originally analysed by Hatch et al. [H37]. These data involve scaled estimates of doses associated with the 1979 accident. Wing et al. [W14] suggested that their results, in contrast to those of Hatch et al. [H14], indicate an increasing trend in lung cancer with the radiation dose estimates; they speculated that this may be due to inhaled radionuclides that might be correlated with external doses. However, Hatch et al. [H38] pointed out that their original analysis did indicate an association for lung cancer, and that many of the differences claimed by Wing et al. [W14] were matters of interpretation rather than new findings. In view of the very low doses received (generally less than 1 mSv), the lack of individual doses, the short follow-up (to the end of 1985), the lack of individual smoking data, and the possibility of chance findings when many different cancer types are studied, these data are not informative on radiation and lung cancer.

3. Internal high-LET exposures

131. Results from various studies of radon exposures are included in Table 10. Particularly informative are the studies of radon-exposed miners, in view of the large numbers of excess lung cancers observed. The joint analysis of 11 miner cohorts by Lubin et al. [L4] permitted detailed examination of factors that may modify the risk of radon-induced cancer. This analysis and the component studies were considered in detail in the UNSCEAR 1994 Report [U2]. In summary, the ERR per working level month (WLM) was found to decrease with attained age,

time since exposure, and time after cessation of exposure to radon, but not with age at first exposure. The joint effect of radon and smoking on lung cancer risk was greater than additive, although it is difficult to quantify this further; in particular, only a small proportion of miners never smoked. Similarly, the modifying effect of exposure to other agents encountered in mines is not clear, although the ERR per WLM was lower after adjusting for arsenic exposure [L4].

132. The exposure-response relationship in the various studies of radon-exposed miners is consistent with linearity. However, at relatively high cumulative exposures, the slope of the exposure-response relationship is steeper at lower than at higher exposure rates [L4, L6]. It should be emphasized that this inverse exposure-rate effect does not imply that low exposures carry a greater risk than higher exposures; rather it suggests that for a given total exposure, the risk is higher if the exposure is received over a longer rather than a shorter period of time. Table 24, based on the analysis of Lubin et al. [L6], shows that this inverse exposure-rate effect (as measured by the modification factor γ) is seen, to varying degrees, in all of the studies except the French cohort; workers in the latter study [T8] often worked for many years at low exposure rates. However, a reanalysis of the Beaverlodge data based on revised exposure estimates [H18] provided no evidence of an inverse exposure-rate effect, in contrast to previous analyses. It should be noted that the highest exposure rates, which generally gave rise to the highest cumulative exposures, occurred in the earliest years of mining, when the fewest measurements were made. Furthermore, concentrations of radon rather than radon progeny were measured in the earliest years in many of the studies, requiring assumptions to be made in calculating working levels (WL). Errors in estimating WL were therefore likely to be greatest in the early years of mining and would have tended to lessen the observed effects of high exposure rates, inducing an apparent inverse exposure-rate effect. However, adjustments by Lubin et al. [L4, L6] by calendar year of first exposure, calendar year of exposure, attained age, and years since the last exposure yielded patterns similar to those in Table 24. It, therefore, seems unlikely that WL measurement errors can explain the entire inverse exposure-rate effect. It is also evident from Table 24 that there is wide variation between studies in the estimate of ERR per WLM at an exposure rate of 1 WL, i.e. β. This variation reflects uncertainty in extrapolating to low exposure rates. Another possible explanation for what appears to be an inverse exposure-rate effect actually may be the effect of cell killing at high doses.

133. The BEIR VI Committee [C21] reexamined the data on the radon-exposed miners of Lubin et al. [L4], adding new data from China, the Czech Republic, France, and the United States (Colorado Plateau). Table 25 describes the mathematical format of the models derived by this Committee. In contrast to the model derived by the BEIR IV Committee [C2], the BEIR VI models include an extra time-since-exposure category, so as to distinguish between exposures received 15–24 years earlier and those received 25 or more years earlier. Furthermore, these models allow for effects of either duration of exposure or average radon concentration, again in contrast to the BEIR IV model. Separate models were derived [C21]: "exposure-age-duration" and "exposure-age-concentration", with no preference being given by the BEIR VI Committee to either. Under these models, the ERR associated with a given cumulative exposure increases as the exposure duration increases or the average concentration decreases.

134. Animal studies using very high exposure rates have shown that a longer duration of radon exposure at a lower rate induces more lung cancers than a shorter duration exposure at a higher rate [C3, C4]. As for possible mechanisms, Moolgavkar et al. [M5, M6] suggested, based on the two-stage initiation-progression model for carcinogenesis, that extended duration allows time for the proliferation of initiated cells and thus for higher disease occurrence rates. Furthermore, by incorporating cell killing into such a model, Luebeck et al. [L23] hypothesized that the inverse exposure-rate effect may be reduced in the absence of ore dust, in view of effects on net cell proliferation. Using a different approach, Brenner [B5] postulated that the inverse exposure-rate effect comes from cell cycling, whereby cells in a particular period of their cycle are more sensitive to radiation than at other times. For the same total dose, a greater proportion of cells is predicted to be exposed during the sensitive period if the dose is protracted rather than acute. Multiple traversals of a cell by alpha particles are necessary for such an inverse exposure-rate effect, although it should be recognized that not all traversals will lead to transformation. At sufficiently low exposure rates, there would probably be at most one traversal of any cell. Consequently the inverse exposure-rate effect would be predicted to disappear, owing to the absence of multiple traversals and their associated interaction. Little [L33] outlined a biological justification for using data from epidemiological studies of miners exposed to high radon levels to estimate risks at low exposure rates. This was based on research by Hei et al. [H34], which showed that traversal by a single alpha particle has a low probability of being lethal to a cell, and that many cells survive traversal by one to four alpha particles to express a dose-dependent increase in the frequency of mutations. In a recent cell transformation study, Miller et al. [M53] found that the oncogenic potential of a single alpha particle, with an energy similar to that of radon decay progeny, was significantly less than that from a Poisson distributed mean of one alpha particle. This finding suggests a non-linear response at low doses of high-LET; however, these results need to be replicated by others.

135. Epidemiological backing for the absence of an inverse exposure-rate effect at low exposure rates comes in particular from the study of miners in western Bohemia, which showed that below 10 WL the ERR per WLM did not appear to depend on duration of exposure [T3]. Furthermore, in their joint analysis of the miner studies, Lubin et al. [L6] concluded that the inverse exposure-rate effect diminishes, and possibly disappears, when the duration of exposure becomes very long. Animal data have also been used to address this issue. In a

study of about 3,000 rats, Morlier et al. [M3] found that lung cancer incidence among rats that received a total of 25 WLM appeared to be lower when the exposure was protracted over 18 months rather than 4–6 months; although the corresponding number of cases was small, the study excluded an inverse exposure-rate effect at this level of exposure. Similar conclusions were drawn from an analysis of another data set, based on more than 4,000 rats with a wide range of exposures and exposure rates; namely, risk per unit exposure decreased with increasing duration of exposure for exposure rates below 10 WL [H41].

136. Results from a meta-analysis of eight case-control studies of residential radon and lung cancer published up to the mid-1990s are summarized in Table 10 [L21], together with the results from some more recent large studies. Lubin et al. [L10] pointed out that the results of these studies appear to be consistent with a wide range of underlying risks. The variability in the findings is likely to reflect, at least in part, the impact of errors in assessing radon exposures. In particular, Lubin et al. [L1, L10] showed that errors due, for example, to the use of recent measurements to characterize past levels and gaps in measurements in previous homes can substantially reduce the statistical power of such studies. As was indicated in Section I.C, it is possible to adjust estimates of the exposure-response relationship to allow for the bias towards the null that tends to arise from random errors in exposure assessment. For example, the central estimate of the ERR per 100 Bq m^{-3} in a study in the United Kingdom increased from 0.08 (95% CI: -0.03–0.20) to 0.12 (95% CI: -0.05–0.33) after adjusting for uncertainties in the assessment of radon exposure, although the width of the associated confidence interval also increased [D30]. Another example of the possible effect of errors in exposure assessment occurs in a study in western Germany [W17]. Here the evidence for an association between radon and lung cancer incidence was stronger in a subanalysis of radon-prone areas than in the analysis of the entire study region (see Table 10); the authors suggested that the latter findings may have been biased by the inclusion of many dwellings with low, but imprecisely estimated, radon levels [W17]. In addition, a recent study in Missouri (United States) showed stronger evidence of an association between radon and lung cancer based on CR-39 surface (i.e. glass-based) measurements rather than on the more traditional track-etch measurements, which has been suggested to reflect the effect of the more precise assessments of cumulative exposure achieved using the surface technique [A24]. However, as pointed out in Section I.C, further validation of glass-based techniques would be desirable [W19]. In addition to the weak indications from some of these recent studies, a meta-analysis of eight earlier case-control studies yielded some direct support for an elevated risk from residential radon exposure [L21]. Based on over 4,000 lung cancer cases, the trend in risk in the meta-analysis was significantly greater than zero (p=0.03) and was consistent with the results from the miner studies, as illustrated in Figure II. In particular, the relative risk estimated at 150 Bq m^{-3} was 1.14 (95% CI: 1.0–1.3). It should be noted that a log-linear model was fitted to the case-control and miner data. Importantly, no single study dominated the overall results,

although there were significant differences in the exposure-response trends among the studies considered [L21].

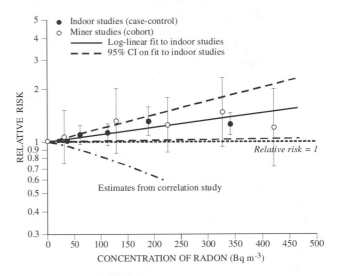

Figure II. Risk estimates of lung cancer from exposure to radon (based on [L21]).
Shown are the summary relative risks from meta-analysis of eight indoor radon studies and from the pooled analysis of underground miner studies, restricted to exposures under 50 WLM [L22] and the estimated linear relative risk from the correlation study of Cohen [C18].

137. Several analyses of lung cancer in the United States in relation to average levels of indoor radon have been published by Cohen (e.g. [C5, C6, C18]). These analyses show decreasing trends in area-specific lung cancer rates with increasing area-averaged radon levels. The findings contrast with those of cohort studies of radon-exposed miners and of case-control studies of indoor radon [L21]. In both the cohort and case-control studies, radon exposures have been estimated for individual study subjects. Furthermore, in the residential studies and some of the miner studies, individual smoking data have been collected. In contrast, the data on radon and the many potential confounders considered in Cohen's correlation studies are averages over geographical areas. Results from such studies are vulnerable to biases not present in results based on individual-level data, such as from cohort or case-control studies. Radon studies are particularly vulnerable to biases associated with the use of geographical area-averaged radon levels because of extreme variation in radon levels within areas. Greenland and Robins [G2] pointed out that a lack of confounding in grouped data need not imply the absence of confounding in data for individuals, and vice versa. This is particularly important in the case of indoor radon, because smoking habits have a much greater impact on lung cancer risk [B35]. Whereas individual smoking habits form the main potential confounder in an individual-level study, the corresponding potential confounder in a geographical correlation study consists of the distribution of all smoking histories across all individuals within each area. Consequently, particularly if the effects of variables such as smoking are non-linear or non-additive at the individual level, the corresponding data available at the area level are unlikely to be sufficiently detailed to adjust for confounding. Furthermore,

the data available in correlation studies do not take account of residential histories. For example, a person who had just moved into an area from another area with different radon levels would be categorized solely by the current area of residence rather than by a time-weighted average exposure. Additionally, residential radon levels often vary widely, even within small geographical areas. On the other hand, Cohen [C7] has drawn attention to the lower statistical uncertainty associated with his studies, relative to case-control studies. However, greater statistical precision needs to be weighed against the potential for substantial bias.

138. The interpretation of geographical correlation studies of radon and lung cancer has continued to be the subject of debate. In examining this issue, it should first be considered whether it is possible mathematically that spurious results could arise from a correlation study. This is possible; there have been numerous mathematical proofs that results from such studies can differ systematically from those based on data for individuals (e.g. [G2, L35]). Secondly, it should be considered whether it is plausible that the results reported by Cohen could be explained simply by the methodological aspects described above. In the absence of data for individuals throughout the regions studied by Cohen, it is difficult to be certain on this point. Lubin [L35] presented examples showing that results of the type described by Cohen can arise even with weak correlations between radon and smoking, but Cohen [C25] stated that correlations far beyond the limits of plausibility cannot explain an appreciable part of the discrepancy with extrapolations from miner data. Smith et al. [S2] reported that a negative correlation seen in the state of Iowa, United States, disappeared when mortality data were replaced by incidence data, although Cohen [C26] was dubious about the value of these data. It should be emphasized that epidemiological studies of all types have their strengths and weaknesses and that none is perfect. As pointed out above, individual residential case-control studies often lack statistical precision, in part because of uncertainties in exposure assessment. However, greater precision should be obtained from planned combined analyses of these studies, which in contrast to the Lubin and Boice's [L21] meta-analysis based on published summary data, will incorporate subject-specific data. For the time being, considering the methodological aspects of the various studies, the data on miners appear to provide the soundest basis for estimating radon-induced risks. Furthermore, it should be noted that risk models based on the full range of miner exposures yield results that are similar to those based on miner exposures of less than 50 WLM [L22].

139. In connection with the development by ICRP of a model for internal doses to the respiratory tract [I4], there has been some interest in comparing risk estimates for lung cancer from studies of low-LET and radon-exposed groups. However, Howe [H7] drew attention to the difficulty of arriving at a single value for the low-LET dose to the lung that could lead to the same lung cancer risk as 1 WLM exposure to radon. In particular, the comparison of data on the Japanese atomic bomb survivors and on fluoroscopy patients referred to earlier suggests a strong fractionation/dose-rate effect for low-LET

radiation, while the data on radon-exposed miners indicate a higher risk per unit exposure at low than at high exposure rates. Furthermore, when attention is confined to low-dose protracted exposures, the derivation of a conversion factor between low-LET and radon exposures is complicated by the paucity of data that are directly relevant.

140. Studies of groups with internal exposures from thorotrast and ^{224}Ra generally provide little evidence of elevated risks of lung cancer; see Table 10. In the case of thorotrast, irradiation of the lung arises principally from exhalation of ^{220}Rn (thoron), one of the daughter nuclides of ^{232}Th [H21]. However, the distribution of dose within the lung is different from that in underground miners exposed to radon. The incidence of lung cancer among neurological patients in Denmark given thorotrast was elevated relative to national rates but not relative to a control group of patients not given thorotrast, after adjusting for gender, age at angiography, and calendar period [A5]. In an analysis of the combined series of Japanese patients, Mori et al. [M14] indicated an elevated risk of lung cancer relative to a control group, although this was based on only 11 deaths. Among female radium dial painters in the United States, there was some suggestion of an increasing trend in lung cancer mortality with increasing intake of ^{226}Ra/^{228}Ra, although there were only 6 deaths in that analysis [S16]. In general, the statistical precision of these studies was limited by the relatively low numbers of lung cancers; furthermore, information on individual smoking habits was not always available.

141. Information from studies of workers with high-LET exposures from plutonium, uranium, and polonium was reviewed in the UNSCEAR 1994 Report [U2]. Since then, more information has been published for workers at the Mayak nuclear plant in the Russian Federation, many of whom were exposed to both plutonium and external low-LET radiation. Koshurnikova et al. [K10, K11] showed that relative to a control group of workers, lung cancer mortality was raised significantly among workers at the radiochemical processing plants and at the plant for plutonium production but not among workers at the nuclear reactors at Mayak, who were exposed predominantly to external gamma radiation (see Table 10). The elevated risk appeared to be concentrated among workers with plutonium body burdens. A subsequent, more detailed analysis of lung cancer deaths among 1,479 men who started work at Mayak during 1948–1958 showed a clear trend in lung cancer risk with estimated alpha dose to the lung, consistent with linearity [K34]. In addition, a separate analysis of data for Mayak workers is consistent with a linear dose response from less than 1 Sv to more than 100 Sv, although it was based on a weighted sum of high- and low-LET doses to the lung rather than the high-LET dose alone [K37]. In contrast to these findings, a case-control study of Mayak radiochemical plant workers [T2] appears to indicate a non-linear dose response. The methodology for this study is summarized in Tables 4 and 5. In particular, in addition to individual measurements of plutonium body burden and gamma dose, information on

smoking habits and other potential confounders was utilized in this analysis [T2]. There was a clear excess of lung cancer among workers with a ^{239}Pu body burden in excess of 5.6 kBq (see Table 10). This association was apparent for adenocarcinoma, squamous-cell carcinoma, and small-cell carcinoma. Further analysis found little evidence of an elevated risk for plutonium body burdens below about 3.7 kBq (corresponding to a lung dose of 0.8 Gy), in qualitative agreement with the form of the dose response reported in animal studies by Sanders et al. [S38]. There was some suggestion of an elevated risk for gamma doses in excess of 2 Gy (low-LET) relative to lower doses, although this finding was not statistically significant [T2, T14]. The wide range of internal doses encountered in the Mayak studies, from less than 0.5 to over 120 Sv [K34], together with the individual data on possible confounders in the case-control study [T2], contribute considerably to the potential ability of the studies to provide information on the carcinogenic effects of plutonium in the lung. The reasons for the differences in the dose-response relationship between the cohort and case-control studies are not clear. One possibility is that the cohort findings have been confounded by smoking. Another possibility relates to the fact that the average lung doses to female workers in the case-control study were higher than those to males, whereas virtually all of the male cases and only one of the female cases were smokers [T2]. Based on this, Khokhryakov et al. [K37] have suggested that curvilinearity in the dose response in the case-control study may be an artefact associated with combining two subgroups with different characteristics, whereas the cohort findings are based exclusively [K34] or largely [K37] on data for males. Further investigation may shed more light on the reasons for the apparent differences in the findings.

142. Other studies of plutonium-exposed workers, such as at the Sellafield plant in the United Kingdom [O1] and at the Los Alamos National Laboratory in the United States [W8], did not show statistically significant elevated risks of lung cancer relative to other workers at the same plants (see Table 10). The internal exposures in these studies were generally much lower than those to Mayak workers; as well as which it was not possible to control for smoking.

4. Summary

143. Results from the Japanese atomic bomb survivors and from several groups of patients with acute high-dose exposures show elevated risks of lung cancer associated with external low-LET radiation. The risk for the atomic bomb survivors is consistent with a linear trend. These data also show similar values for the ERR per Sv by age at exposure and for the EAR per Sv by gender, although without taking account of smoking habits. Indeed the large influence of smoking on lung cancer risks is likely to be of great importance in determining how radiation-induced risks differ from one population to another. There is some suggestion that the joint effect of low-LET radiation and smoking is closer to an additive than a multiplicative relationship, although the data are sparse and not entirely consistent. Studies of

tuberculosis patients who received multiple chest fluoroscopies have not demonstrated increased risks of lung cancer, in spite of the large number of patients with moderate or high lung doses. The fractionation of these exposures, compared with the acute doses received by the atomic bomb survivors, may explain the difference in findings. However, the severity of tuberculosis may have confounded the results for some of the patients with this disease.

144. In contrast to internal low-LET irradiation, there is a substantial amount of information on lung cancer in relation to internal high-LET exposure. Most of this information comes from studies of radon-exposed miners. In particular, the risk appears to increase linearly with cumulative radon exposure, measured in WLM, but the ERR per WLM decreases with increasing attained age and time since exposure. Furthermore, at high cumulative exposures, the ERR per WLM appears to increase with decreasing exposure rate, but both epidemiological and experimental evidence indicate that this phenomenon does not arise at low exposures. Findings from case-control studies of domestic radon exposure have been variable but are consistent with predictions from the miner studies. Among studies of other types of high-LET exposure, the most informative are those of workers at the Mayak plant in the Russian Federation, which show an elevated risk for high lung doses from plutonium; further investigation of the shape of the dose-response relationship would help to understand apparent differences in findings for different groups of workers.

F. MALIGNANT TUMOURS OF THE BONE AND CONNECTIVE TISSUE

145. Malignant tumours of the bone account for about 0.5% of malignant neoplasms in humans [M39], while soft-tissue sarcomas, which include connective tissue malignancies, account for about 1% of all malignancies [Z3]. Among bone sarcomas, dissimilarities in cell type between oesteosarcoma and Ewing's sarcoma indicate that these tumours have different origins. The role of genetic susceptibility has been identified through molecular and cytogenetic studies of the gene loci for these types of sarcomas, as well as by the linkages of osteosarcoma with hereditary retinoblastoma and the Li-Fraumeni syndrome [M39]. Li-Fraumeni syndrome has also been investigated together with connective tissue malignancies [Z3]. As will be described below, a variety of studies with external low-LET and internal high-LET studies exposures have established that bone sarcomas can be induced by radiation. Human and animal studies have suggested a possible association between exposure to chromium and the risk of bone and soft-tissue malignancies [M39].

1. External low-LET exposures

146. The results from studies of bone and connective tissue malignancies following external low-LET exposures are given in Table 11. Among the Japanese atomic bomb survivors overall, the estimated trend in risk per unit dose

is positive but is not statistically significantly greater than zero [P9, R1, T1]. However, there is an indication that the risk is higher for exposure in childhood than in adulthood, although this finding is based on small numbers. Statistically more powerful information comes from studies of patients treated for cancer in childhood. Three studies with reasonably large numbers of cases [H44, T17, W11] have reported a statistically significant increasing trend in risk with dose, based on mean doses between 10 and 30 Gy; another such study reported similar results, although with fewer details [D33]. While the high doses contributed to the detection of an elevated risk, these studies are less informative about risks at doses of a few gray or less, although no excess is apparent at these levels. Compared with some other cancer types, the estimated ERR per Sv of around 0.1–0.2 for bone malignancies and/or soft-tissue sarcomas is not large. A notable finding from the study of Wong et al. [W11] of retinoblastoma patients was that the risk of bone and soft-tissue sarcomas was concentrated among those with hereditary retinoblastoma. Tucker et al. [T17] reported a similar result, and found that the relationship between relative risk and dose was similar for retinoblastoma and other patients; the retinoblastoma patients had a higher absolute excess risk by virtue of their higher baseline risk. These results suggest that genetic predisposition may modify the radiation-associated risk at high doses.

147. Few studies of adult exposure are informative, owing in part to the rarity of malignant tumours of the bone or connective tissue. However, the study of cervical cancer patients involved mean doses comparable to those in the above childhood cancer studies [B1]; in that instance, no significant increasing trend in risk with dose was found. Among ankylosing spondylitis patients in the United Kingdom, the total number of deaths was significantly greater than expected from national rates, but the data were not analysed in relation to estimates of dose [W1]. In a group of over 120,000 women in Sweden treated for breast cancer, the incidence of soft-tissue sarcomas was about double that expected from national rates [K35]. In a case-control study based on this Swedish cohort, which analysed information on the energy imparted from radiotherapy (i.e. the product of the mass of the patient and the dose absorbed) because organ dose estimates were not available, angiosarcoma was not found to be related radiotherapy energy, whereas the risk of other types of soft-tissue sarcomas was found to increase with increasing energy [K35]. A review of medical records at a cancer centre in the United States indicated that fewer than 3% of cases of bone and soft-tissue sarcoma had previously received radiotherapy [B40]. In a study of over 50,000 men in the United States who had received radiotherapy for prostate cancer, the proportion who subsequently developed sarcomas was also low, although there was an elevated risk for sarcomas at sites within the treatment field, in contrast to more distant sites that received lower doses [B42]. An analysis of 53 cases of soft-tissue sarcomas that were identified following radiotherapy showed no definite relation with age at exposure, although there was some suggestion of a shorter latency for therapy involving higher doses [L48].

2. Internal low-LET exposures

148. Studies of groups with medical exposures to radioactive iodine are uninformative about the risks of bone malignancies, owing to the low doses to bone surfaces from this type of exposure and to the rarity of the disease. Even in a large study of patients treated for hyperthyroidism in the United States, deaths from bone malignancies were not listed separately [R14]. More information may be obtained from studies of bone-seeking radionuclides. Residents of the area around the Techa River in the Southern Urals received internal exposures, mainly from ^{90}Sr, which has been shown to induce osteosarcomas in rats [N18, S55], as well as external exposures. In the period 1950–1989, 12 deaths from bone malignancies were observed in a cohort of 26,485 residents in the Techa River region [K5]. This represents about 1% of all cancer deaths in this cohort [K5], compared with a corresponding value of 0.4% in the Life Span Study [P9]. Risk estimation using the Techa River data is made difficult by the absence of information on vital status for over a third of the cohort and by uncertainties in the estimates of individual doses. However, a major dose reconstruction project is in progress that aims to provide more reliable individual dose estimates for cohort members [D37]. Direct measurements of ^{90}Sr have already been made for about half of the population exposed in the Techa River region, either using a whole-body counter or by *in vivo* measurements of teeth. These measurements have shown a clear correlation with year of birth [D37]. Total doses to soft tissue, from external and internal exposures, are likely to be less than 0.1 Sv for most Techa River residents, although a small proportion is estimated to have received doses in excess of 1 Sv [K5]. With further improvements in the quality of the dosimetry and the follow-up, this cohort has the potential to provide quantitative estimates of risks from chronic exposures.

3. Internal high-LET exposures

149. Most of the information on bone tumour risks and internal high-LET irradiation comes from studies of intakes of radium. Data from medical intakes of ^{224}Ra and occupational intakes of predominantly ^{226}Ra are considered in turn.

150. In the early 1950s, Spiess initiated a follow-up study in Germany of 899 patients with ankylosing spondylitis, tuberculosis, or a few other diseases who had received multiple injections of ^{224}Ra [S14]. Up to the end of 1998, a total of 56 malignant bone tumours had occurred in 55 of these patients [N14], whereas less than one tumour would have been expected. Most of the tumours occurred within 25 years of the first ^{224}Ra injection [N14]. Among those cases for which histopathology information was available, about half of the cancers were osteosarcomas. However, there was a relatively high proportion of fibrous-histiocytic sarcomas, compared with spontaneous bone tumours [G22]. In particular, the ratio of osteosarcomas to fibrous-histiocytic sarcomas in this study, 1.8, is similar to that in other groups where radiation-related excesses have been seen, such as the radium dial painters [G22].

151. Bone sarcoma risk among these ^{224}Ra patients was recently analysed [N14] taking into account revised dosimetric calculations [H8]. In particular, these calculations indicate that doses to the bone surface for those exposed at young ages are smaller than had previously been estimated. As a consequence, the new risk analysis indicates that absolute risks of bone tumours decrease with increasing age at exposure [N14]. Nekolla et al. modelled the absolute excess risk in terms of attained age, age at first injection, duration of treatment, and mean absorbed dose to the bone surface; no effect of gender was seen [N14]. A linear dose-response model provided a good fit to the data, although models involving a quadratic component in dose could not be excluded. Also, while the risk for a given cumulative dose was higher if the dose was protracted rather than acute, this difference was estimated to be small for cumulative doses below about 10 Gy. In addition, the excess absolute risk decreased from about 12 years following exposure onwards. Based on this model, the lifetime risk of bone sarcoma incidence for an acute exposure up to several gray (high-LET) of a population aged 0–75 years was estimated to be 1.8 (0.6–2.4) 10^{-3} Gy^{-1}. This value is similar to estimates made previously by, for example, the BEIR IV Committee [C2]. However, as indicated earlier, the new calculations indicate that risks are higher for exposure at younger ages. In particular, the lifetime risk for the incidence of bone sarcomas was estimated to be 4 10^{-3} Gy^{-1} for an acute exposure up to several gray (high-LET) at age 15 years, compared with 0.8 10^{-3} Gy^{-1} for an acute exposure at age 45 years [N14]. It should also be noted that, while these absolute risk coefficients are small, the corresponding estimates of the ERR per Sv (based on a radiation weighting factor of 20), between about 0.45 and 0.04, depending on age at exposure, are consistent with those seen for many other solid tumours [N14].

152. Nekolla et al. [N14] drew attention to uncertainties in the extrapolation of their findings to low doses. In particular, they compared their findings with those of Wick et al. [W20], who studied a more recent group of about 1,500 patients in Germany treated for ankylosing spondylitis with lower activities of ^{224}Ra than patients in the Spiess study. The model of Nekolla et al. [N14] predicts 7.8 excess bone sarcomas in the study of Wick et al. up to 1995, whereas only four malignant tumours of the skeleton, none of them osteosarcomas, have been observed, compared with 1.3 expected spontaneous cases [W20]. Since the mean dose to the bone marrow in [W20] is lower by a factor of about five than that in [N14], the results of this comparison suggest that the linear extrapolation in [N14] may overestimate risks at low doses.

153. Studies of over 4,000 radium dial painters, radium chemists, and patients given ^{226}Ra or ^{228}Ra therapeutically in the United States were reviewed in the UNSCEAR 1994 Report [U2] and by the BEIR IV Committee [C2]; Fry [F9] recently published a detailed history of the radium dial painter studies. Some of these individuals had been internally contaminated with pure ^{226}Ra, which has a half-life of 1,620 years, whereas others received a mixture of ^{226}Ra and ^{228}Ra, which has a half-life of 5.75 years. The BEIR IV Committee reported 87 bone sarcomas in 85 of

4,775 persons whose vital status had been ascertained on at least one occasion [C2]. Among those 2,403 individuals for whom there was an estimate of skeletal dose, 66 sarcomas in 64 persons were reported, whereas fewer than 2 cases of sarcomas would have been expected from national rates [C2]. The elevated risk in dial workers was particularly evident among women who entered the industry before 1930 and whose exposures were higher than those for later workers; among those early workers there were 46 bone sarcoma deaths up to the end of 1990 [C27, R35].

154. Various attempts have been made to model the risks of bone sarcoma in the United States series. Based on 1,468 female radium dial workers who entered the dial industry before 1950 and who were followed to the end of 1979, Rowland et al. modelled the annual rate of bone sarcoma as $(\alpha + \beta A^2)\exp(-\gamma A)$, where A is the activity of radium that entered the blood during the exposure period [R33]. In a later analysis with follow-up to the end of 1990, Rowland suggested that the exponent of A was nearer to 3 than to 2 [R35]. Marshall et al. developed a two-target model, proposing that two successive initiating events are required for osteosarcoma induction and also allowing for the effects of cell killing [M51, M52]. Using information on time to death and average skeletal dose, Raabe et al. drew attention to the effects of dose rate in both human and animal data on bone sarcomas following intakes of ^{226}Ra, in particular to the finding that risks may not be elevated at low dose rates [R34]. More recently, Carnes et al. analysed data on 820 women who started radium dial work before 1930 and who were followed for mortality through to 1990 [C27]. In contrast to some other analyses, the models of Carnes et al. took account of time distributions for both risk and exposure and examined ^{226}Ra and ^{228}Ra separately [C27]. Their preferred model for the excess absolute risk of bone sarcoma consisted of the sum of a quadratic term in the accumulating skeletal dose from ^{226}Ra and a linear term in the accumulating skeletal dose from ^{228}Ra. In addition, the excess relative risk was higher for exposure at ages associated with active bone growth than at older ages, when the skeleton was fully developed, although the excess absolute risk did not appear to vary by age at exposure [C27]. However, all of these analyses should be interpreted with caution: the intake of radium was estimated many years after the event and may be inaccurate; the distribution of radium in the bone is probably non-uniform and hot spots capable of extensive cell killing may have occurred; the continuous receipt of dose makes it difficult to separate out the fraction of dose associated with cancer induction; the contributions from alpha emitters and other radiations accompanying radium decay cannot be separated; and the fraction of the total dose to the endosteal cells cannot be specified precisely [B47].

155. In a group of about 1,200 women in the United Kingdom who worked with paint containing radium from 1939 to 1961, one fatal bone sarcoma occurred up to the end of 1985, compared with 0.17 expected [B14]. The difference between these findings and those from the United States series can be explained by the much lower radium exposures received by the United Kingdom

workers, although it should be noted that both groups would have also received external exposures from the proximal containers of radioactive paint. The results from the United Kingdom study, the models fitted to the United States data that are at least quadratic in dose at low doses, and the findings from animal studies have prompted the suggestion that there is a "practical threshold" of about 10 Gy for the induction of bone sarcomas. However, the UNSCEAR 1994 Report drew attention to a few cases of bone sarcomas and head sinus carcinomas that had arisen at lower doses, down to about 1 Gy, in the United States series [U2]. Furthermore, the bone sarcoma case observed in the United Kingdom study was in a worker with an estimated skeletal dose of 0.85 Gy. It would appear, therefore, that any practical threshold, if it exists, is unlikely to be greater than about 1 Gy [U2].

156. Some studies of thorotrast patients, such as the one in Portugal [D31], have indicated elevated risks of bone sarcomas (see Table 11). However, the numbers of cases in these studies were smaller than among the ^{224}Ra patients and the United States radium dial workers. Based on thorotrast studies, the BEIR IV Committee assessed the lifetime risk of bone cancer to be 1×10^{-2} Gy^{-1} (high-LET) [C2]. This value is somewhat higher than that derived by Nekolla et al. from the ^{224}Ra patients [N14]. However, the estimate based on the thorotrast studies is likely to be more uncertain, because these studies had fewer cases than the studies of the ^{224}Ra patients and because dose estimation may have been more difficult in the thorotrast studies.

157. Studies of workers from the United Kingdom and the United States monitored for exposure to plutonium have reported few if any cases of bone malignancies (e.g. [W8, O1]). In contrast, bone tumour deaths were significantly elevated among plutonium-monitored workers at the Mayak plant in the Russian Federation [K42]. Bone tumour mortality increased with increasing levels of plutonium body burden (p<0.001); however, additional plutonium dosimetry is needed before reliable risk estimates can be calculated.

4. Summary

158. Studies of patients treated for childhood cancer demonstrate an increasing risk of bone sarcomas with dose, over a range of several tens of gray (low-LET). These studies are not informative about risks at doses below a few gray, but a study of retinoblastoma patients in particular indicates that genetic predisposition may affect risks associated with high dose therapeutic radiation exposure. Other studies of external low-LET exposure are less informative, although there is some suggestion that the relative risk is lower for exposure in adulthood than in childhood. Studies of the population living near the Techa River in the Russian Federation may in the future provide more information on bone cancer risks following internal low-LET exposures.

159. There is extremely strong evidence that large intakes of radium have induced bone sarcomas in a group of patients in Germany and in radium dial workers in the

United States. Because of the long half-lives of ^{226}Ra and ^{228}Ra (the source of the high-LET exposures in the United States study) relative to the half-life of ^{224}Ra (the source of exposure in the German study), it is easier to model risks using the latter study. Analysis of the ^{224}Ra data indicates that the excess absolute risk decreases with increasing time since exposure (beyond about 12 years) and age at exposure, and that the effect on risks of exposure rate is small at doses below around 10 Gy. The ^{224}Ra data are consistent with a linear dose response over a range up to more than 100 Gy, although there is uncertainty in extrapolating the findings down to doses of a few gray. The United States study on ^{226}Ra and ^{228}Ra offers little evidence of an elevated risk at these lower doses, although it is difficult to evaluate the dose associated with any "practical threshold" in risk.

G. SKIN CANCER

160. Non-melanoma skin cancers are extremely common in light-skinned populations but relatively rare in populations with highly pigmented skin [A9, S26]. Malignant melanoma incidence is also strongly correlated with skin pigmentation, but it is about 10 times less common than non-melanoma skin cancer. Annual incidence rates for melanoma vary from about 0.5 per 100,000 persons in Asia to over 20 per 100,000 in Australia, whereas rates for non-melanoma skin cancers range from almost 5 per 100,000 in Africa to about 200 per 100,000 in Australia [P5]. Non-melanoma skin cancer incidence rises rapidly with age, with such cancers being common among the elderly. Over the past decades, there has been a dramatic increase in the incidence of both non-melanoma and melanoma skin cancer [A10, M24]. Much of the increase in incidence appears to be due to sun exposure. Total accumulated exposure appears to be the main risk factor for non-melanoma skin cancer [S26], but for melanoma this relationship is not a simple one and may be related to intermittent sun exposure of untanned skin [N8]. Survival for melanoma depends on stage. Non-melanoma skin cancer is a treatable malignancy with a very high cure rate.

161. From a histological standpoint, the two most common types of non-melanoma skin cancer are basal-cell and squamous-cell carcinomas. They are substantially different with respect to demographic patterns, survival rates, clinical features, and aetiological factors. The incidence of both types is higher among males than females [S26].

1. External low-LET exposures

162. Since publication of the UNSCEAR 1994 Report [U2], additional information from the Life Span Study of atomic bomb survivors has become available [R15, Y3]. Data from this and other studies are summarized in Table 12 and Table 26.

163. An association between external ionizing radiation and non-melanoma skin cancer risk has been demonstrated in the Life Span Study of atomic bomb survivors [L29,

R15, T1, Y3], the New York and Israeli tinea capitis studies [R16, S27], the Rochester thymus study [H31], patients irradiated for enlarged tonsils [S28], patients irradiated for various benign head and neck conditions [V4], and the American and British radiologists [M25, S41]. No such relationship has been observed for melanoma, but the number of cases in each study was extremely small. Most of the significantly increased risks observed for non-melanoma skin cancer occurred among people irradiated as children (Table 12).

164. In the latest data from the Life Span Study of atomic bomb survivors, a strong dose-response relationship was demonstrated for basal-cell carcinoma (ERR at 1 Sv = 1.9; 90% CI: 0.83–3.3) (Table 26), but not for squamous-cell carcinoma or melanoma [R15]. There was non-linearity in the basal-cell carcinoma dose response. A dose-response curve having two slopes (with the change in slopes at 1 Sv) marginally improved the fit (p=0.09); a linear model with a threshold at 1 Sv did not fit the data. In earlier evaluations of skin cancer in the Life Span Study, non-linearity was found for all non-melanoma skin cancers combined [L29, T1].

165. For basal-cell carcinoma in the Life Span Study, the ERR decreased substantially with increasing age at exposure, but gender, time since exposure, and attained age had little influence on the risk [R15]. Skin tumour prevalence was assessed among a subgroup of atomic bomb survivors who were clinically examined. A dose-response relationship was found for basal-cell carcinoma and precancerous lesions [Y3]. Consistent with the results from the larger study of all Life Span Study members, age at exposure was predictive of developing a skin neoplasm but gender was not.

166. The substantial increase in skin cancer incidence rates and reporting, as well as the wide variation in incidence depending on region and ethnicity, suggests that relative risks are more suitable than absolute risks for describing radiation-induced skin cancer risks. Analyses of skin cancer conducted by the National Radiological Protection Board in the United Kingdom indicate that a generalized relative risk model describes the data more parsimoniously (i.e. with fewer model parameters) than an absolute risk model [N10]. As seen in Table 12, the ERR at 1 Sv for persons exposed medically ranges from no risk for cervical cancer patients [B1] to 1 for infants treated for enlarged thymus gland [H31, S30]. In the two studies of patients receiving scalp irradiation for tinea capitis, the ERRs were about 0.5 [R16, S27, S30]. For children between the ages of 1 and 15 years, a significant decrease in the ERR with increasing age at exposure was demonstrated in the Israeli tinea capitis study [R16].

167. Several recent studies of medical exposures add to what is known about ionizing radiation and the risk of skin cancers of different histological types. Associations between basal- and squamous-cell skin carcinoma and a history of therapeutic x-irradiation were reported from a case-control study of male skin cancer patients conducted in Alberta Province, Canada [G14]. Most of the exposure was from treatment for benign skin disorders. This is one of the few studies reporting an excess risk for squamous-cell carcinoma. Recall bias or misclassification of the skin disease being treated might account for this finding. The development of a new basal-cell or squamous-cell carcinoma subsequent to therapeutic radiation was evaluated in a study of 1,690 patients diagnosed with an earlier non-melanoma skin cancer in New Hampshire, United States [K16]. A history of radiotherapy was associated with basal-cell carcinoma (relative risk = 2.3; 95% CI: 1.7–3.1) but not squamous-cell carcinoma (relative risk = 1.0; 95% CI: 0.5–1.9). The risk of a second non-melanoma skin cancer was higher among persons exposed early in life.

168. In a follow-up study of bone marrow transplantation patients, an eightfold risk of melanoma was observed among patients treated with high-dose, total-body irradiation [C16]. This finding was based, however, on only nine melanomas. Among Swedish patients treated with ionizing radiation for skin haemangioma, the observed number of melanomas was close to what had been expected [L16], but no data on non-melanoma were available since follow-up was based on the Swedish Cancer Registry, which does not register basal cell carcinomas.

169. Several studies of radiation-exposed medical and nuclear workers have been conducted, but most do not have individual doses. These studies mainly evaluated mortality, so they are not very informative for assessing skin cancer effects. A significantly increased risk of skin cancer mortality was reported for radiologists in the United States [M23] and in the United Kingdom [S41]. The risks were larger for radiologists practicing in the early years, when exposure is thought to have been highest, than for those practicing later. Among radiological technologists in the United States, skin cancer mortality was significantly lower than expected compared with national rates (SMR = 0.62; 95% CI: 0.44–0.84) [D23]. Skin cancer incidence was elevated (SIR = 2.8, p<0.05) among Chinese diagnostic x-ray workers [W10], particularly those who had been employed for 15 years or more. Among 4,151 medical workers in Denmark, whose mean cumulative dose was very small (18.4 mSv whole-body dose equivalent), skin cancer risk was not significantly elevated [A15]. The difference in these findings is probably due to the longer duration of employment among the Chinese workers (69% of the Chinese workers had been employed for five or more years compared with slightly more than 15% of the Danish workers) or their exposure to higher doses during the early years. Although the mean radiation dose is not known for the Chinese workers, it is assumed to be relatively high, since improvements in radiation safety practices were introduced only in the mid-to-late 1960s.

170. The results for nuclear workers are similarly inconsistent. An increased incidence of melanoma was associated with working with radiation sources at the

United States Lawrence Livermore National Laboratory in some studies [A14, S40] but not in others [M34], and no association was observed at the sister laboratory, Los Alamos National Laboratory, or at most other nuclear facilities [W13].

171. Using data from the American Cancer Society database, the frequency of various occupational exposures was evaluated in 2,780 cases of malignant melanoma and approximately three times that number of matched controls. A history of occupational exposure to x rays was significantly more frequent among the cases than the controls [P26]. This study did not, however, distinguish between medical and nuclear workers, and it was not possible to control for confounding due to social class.

172. In a summary of the literature through the late 1980s, Shore [S30] suggested that there is an interaction between ultraviolet and ionizing radiation. One reason for this hypothesis was the fact that black patients treated in New York City with scalp irradiation for ringworm did not develop skin cancer on the scalp or face, while white patients demonstrated a significantly increased risk for developing basal-cell carcinoma [S27]. A recent National Radiological Protection Board publication reported that radiation-associated non-melanoma skin cancer generally develops on areas of the skin exposed to ultraviolet radiation [N10]. It was estimated that for the population of the United Kingdom, the lifetime risk for non-melanoma skin cancer is $2.3 \ 10^{-2} \ Sv^{-1}$. The report concluded that ultraviolet-shielded and heavily pigmented skin would have lower risks than ultraviolet-exposed or lightly pigmented skin. The latest Life Span Study data for basal-cell carcinoma do not support this hypothesis [R15]. First, the ERR for the atomic bomb survivors, who have moderately pigmented skin and very low natural rates of non-melanoma skin cancer, was extremely high; second, the ERR was not larger for ultraviolet-exposed parts of the body than for parts of the body that are generally ultraviolet-shielded [R15]. Yamada et al. [Y3] reported a high risk for the development of skin neoplasia among people with occupational exposure to ultraviolet rays, but they did not report which parts of the body had higher risks. In the New Hampshire study, there did not appear to be a higher risk for ultraviolet-exposed parts of the body compared with ultraviolet-shielded parts [K16]. Thus, the question of a possible interaction between ionizing radiation and ultraviolet radiation remains unresolved. Possibly, ultraviolet radiation exposure plays a less important role in inducing skin cancer in individuals whose skin has a relatively high melanin content, but more data are needed to fully understand this complicated relationship.

2. Internal low-LET exposures

173. Studies of patients receiving ^{131}I diagnostic examinations [H27] or ^{131}I treatment for hyperthyroidism [G18, H23, H25] or thyroid cancer [D15, E2, G13, H26] do not indicate any significantly increased or decreased risks of skin cancer associated with this exposure. Although the amount of ^{131}I administered varies from about 2 to 500 MBq for hyperthyroid treatment to 5.5 GBq for thyroid cancer treatment, the dose to the skin would be relatively small.

3. Internal high-LET exposures

174. A large, significantly elevated risk of non-melanoma skin cancer was observed among uranium miners in Czechoslovakia [S29]. In contrast, neither mortality from melanoma nor non-melanoma skin cancer was significantly elevated or related to cumulative exposure in an international pooled analysis of 11 studies of underground miners [D8]. Although the radon levels in the air were high and the study population large (64,209 miners), the latter study is hampered by the fact that mortality does not reflect the true risks of skin cancer.

175. The major studies of patients treated with internal high-LET exposures were summarized at two international meetings [D31, N4, V1, V8, W20]. These results, as well as results from the Danish thorotrast study [A5], do not suggest that skin cancer is related to exposure from ^{224}Ra, ^{226}Ra, ^{228}Ra, or thorotrast.

4. Summary

176. Ionizing radiation can induce non-melanoma skin cancer, but the relationship is almost entirely due to a strong association with basal-cell carcinoma. To date, there has been little indication of an association between ionizing radiation and malignant melanoma or squamous-cell carcinoma, but the data are sparse. When radiation exposure occurs during childhood, the ERR for basal-cell carcinoma is considerably larger than when the exposure occurs during adulthood. A very strong trend for a decreasing risk of basal-cell carcinoma with increasing age at exposure was observed in the Life Span Study. Data on the dose-response relation for basal-cell carcinoma suggest non-linearity, but more data are needed to better characterize the shape of the dose response, to further evaluate the role of ionizing radiation in the development of squamous-cell carcinoma and melanoma, and to clarify the role of ultraviolet radiation in relation to ionizing radiation.

H. FEMALE BREAST CANCER

177. Breast cancer is the most commonly diagnosed cancer and cause of cancer mortality among women in many countries in North America and western Europe; incidence rates are lower by a factor of 5 or more in Asian countries (see Table 1) [P5]. Breast cancer incidence rates have increased since 1960 at all ages in many countries throughout the world [U14]. In some countries this increase may be explained in part by changes in screening practices. However, particularly outside western Europe and North America, the bulk of the increase is likely to be due to risk factors for the disease. Known risk factors include age, family history of breast cancer, early menarche, late age at first birth, nulliparity, late age at menopause, height, postmenopausal weight, and a history

of benign breast disease [K3]. A recent analysis of more than 50 studies indicated that there is a small increased risk of breast cancer while women are taking combined oral contraceptives, although this does not appear to persist more than 10 years after stopping use [C12]. The potential role of other possible risk factors, such as birth weight [M19], which may be a marker of intrauterine factors, and some components of diet [H19], is still unclear.

178. Ionizing radiation is well documented as a cause of breast cancer in women [U2]. Mammary tumours have also been induced in several studies of mice exposed to radiation (e.g. [S11]). Table 13 presents results from epidemiological studies that have incorporated some assessment of the level of low-LET or high-LET doses.

1. External low-LET exposures

179. Most of the external low-LET studies listed in Table 13 were reviewed in the UNSCEAR 1994 Report [U2]. New findings include those from the extended follow-up for mortality of the Japanese atomic bomb survivors [P9]. However, as a consequence of the high cure rate for this type of cancer, the results for cancer incidence in this cohort [T1] are probably of greater importance, despite the slightly shorter follow-up period for incidence than for mortality. New results have also been reported from a number of studies, including the extended follow-up of Swedish patients irradiated for skin haemangioma in infancy [L46]; this study also incorporated individual estimates of organ doses [L14] and patients from both Stockholm [L17] and Gothenburg [L15].

180. Much of the information that has accumulated since the UNSCEAR 1994 Report relates to exposure in childhood. For example, Bhatia et al. [B16] reported a very high standardized incidence ratio in an international study of breast cancer among patients treated for Hodgkin's disease in childhood, as shown in Table 13. Similar results were reported in studies in the Nordic countries [S23], in France and the United Kingdom [D33], and in the United States [T9]. Furthermore, Bhatia et al. reported evidence of a dose-response trend with relative risks of 5.9 (95% CI: 1.2–30.3) at 20–40 Gy and 23.7 (95% CI: 3.7–152) at more than 40 Gy, relative to those with doses to the mantle region of radiotherapy of less than 20 Gy [B16]. While the study of Hodgkin's disease patients by Hancock et al. [H2] gave a lower ratio of observed to expected breast cancer cases, fewer than 10% of these patients were less than 15 years old when originally diagnosed, and there was no elevated risk among women treated at ages above 30 years. However, as mentioned in Section I.A, there is the possibility in the hospital-based study of Bhatia et al. [B16] that patients with a second cancer were more likely to return to hospital than those who were disease-free [D25]. There is some suggestion of an elevated breast cancer risk following scattered radiation received from radiotherapy for retinoblastoma during infancy [W11], while in a study of patients who underwent bone marrow transplantation (primarily given during childhood to treat leukaemia and lymphoma) inferences are hampered by the limited period

of follow-up (mean of 4.5 years) [C16]. It should be noted that the number of cases in these studies is fairly small, and that the possible role of both chemotherapy and genetic susceptibility in the development of the tumours is unclear. However, from a clinical viewpoint these findings are extremely important, because Bhatia et al. estimate that around 35% (95% CI: 17–53) of the female patients in their study will have developed breast cancer by the age of 40 years [B16]. Although other studies, such as those of the survivors of the atomic bombings in Japan [P9, T1] and of patients who received thymic irradiation [H10], have reported lower risks than that of Bhatia et al. [B16], both the former studies and the latter indicate that the relative risks for females exposed to radiation in childhood are higher than for those exposed in adulthood. In particular, studies of women irradiated after age 40 years [B3, B10, H20, P9, S20, T1] generally show low values for the ERR per Sv. For exposure within childhood, there has been some variation in the findings; for example, the estimate for the ERR per Sv in Swedish skin haemangioma study [L46] is lower than in some other studies (see Table 13), possibly owing to the high proportion of children in the Swedish study who were irradiated in infancy [L46] or to the lower dose rate in this study. A recent follow-up of scoliosis patients in the United States irradiated during childhood and adolescence indicated a relatively high value for the ERR per Sv (see Table 13), although potential confounding associated with the severity of disease and hence reproductive history may explain part of this increase [D34].

181. Several of the studies of medical exposures have a longer follow-up than the Life Span Study. The latest results from an extended follow-up of the Canadian fluoroscopy study [H20] suggest that, after allowing for age at exposure, the ERR per Sv may be lower between 40 and 57 years following exposure compared with the earlier period; however, this difference is not statistically significant. In the Massachusetts fluoroscopy study [B3] the ERR appears to be constant up to 50 years or more after exposure, again after adjusting for age at exposure. A reanalysis of data on women in Sweden irradiated for benign breast disease found no persistent heterogeneity in the ERR over the period up to more than 40 years after exposure [M20]. In contrast to the original analysis [M8], this analysis involved more detailed modelling of internal baseline rates and of age and calendar period effects [M20]. The study of Swedish skin haemangioma patients [L46] also showed that risks were still elevated more than 60 years after exposure. Thus, these studies indicate that, in common with the Life Span Study [T1], the ERR per unit dose is approximately constant up to at least 40 years following exposure, and indeed may be constant at follow-up times of 50–60 years.

182. Howe and McLaughlin [H20] reported results from an extended follow-up of breast cancer mortality among tuberculosis patients in Canada who received multiple chest fluoroscopies. In common with other studies (e.g. [B3, S15, T1]), this study showed a linear dose-response

relationship, although there was some indication of non-linearity in an earlier analysis of this cohort [M1]. As before, the slope of the dose trend was greater for patients in Nova Scotia than that for patients in other parts of Canada. The reason for this difference is not clear. One factor that may be pertinent is the higher doses for the exposures in Nova Scotia. However, Howe and McLaughlin noted that on both a relative and an absolute scale, the risk among Nova Scotia patients appeared to be higher than that among the survivors of the atomic bombings in Japan. Furthermore, the risk per unit dose among the non-Nova Scotia patients is similar to that among the patients in the Massachusetts study [B3]. The quality of the dosimetry for the various sanatoria may also be relevant, although Howe and McLaughlin emphasized that identical protocols were used to estimate doses. It should also be noted that the Nova Scotia findings are driven by data at doses in excess of 10 Gy, so the non-Nova Scotia findings may be more representative of risks at lower doses.

183. As indicated earlier, comparison of the risks seen in studies of the Japanese atomic bomb survivors and of populations elsewhere who received medical exposures may be of value in deciding how to transfer risks across populations. One complication, however, is the different degree of fractionation and radiation quality in the two studies. A parallel analysis of earlier data on breast cancer among the atomic bomb survivors and patients in several of the North American studies indicated that the ERR per unit dose is higher in the latter group, whereas absolute risks are more similar [L5]. Similar results were found by Little and Boice [L39], who analysed more recent incidence data for the Japanese atomic bomb survivors and the Massachusetts cohorts. Little and Boice concluded that these data provide little evidence for a reduction in breast cancer risk after fractionated irradiation [L39]. However, Brenner [B33] has interpreted these findings as being consistent with a lower risk for fractionated compared with acute exposure, based on differences, by a factor of about 2, between the number of *in vitro* cell transformations observed for the relatively soft x rays received in fluoroscopy and other medical exposures and the number observed for the higher energy gamma rays received by the atomic bomb survivors. On the other hand, there is little evidence from animal studies to indicate a difference between x rays and gamma rays in inducing breast cancer [U3]. Also, Elkind [E5] has interpreted the results of Little and Boice [L39] as indicating that breast cancer target cells may be deficient in repair, in line with a radiobiological model that he has proposed [E6]. It should also be emphasized that the comparison of the Japanese and North American cohorts is also influenced by the method of transferring risks across populations. Since the disparity in the ERR per unit dose between the Japanese and Massachusetts cohorts [L39] would be greater rather than smaller if the possible effects of photon energy suggested by Brenner [B33] were allowed for, it would appear to be more appropriate to transfer age-specific absolute (rather than relative) risk coefficients for breast cancer from Japan to North American and possibly other populations.

184. It has been claimed by Gofman [G8] that about 75% of current breast cancer cases in the United States are due to ionizing radiation exposure, mostly from diagnostic medical procedures. This claim is based not on new epidemiological findings but on his estimation of medical doses and breast cancer risk factors. There are a number of flaws and questionable assumptions in his calculations. For example, the risk estimates are based on old mortality data for all cancers among the Japanese atomic bomb survivors, using the previous T65D dosimetry and follow-up to the end of 1982, rather than on recent incidence or mortality data for breast cancer specifically, using the DS86 dosimetry system. The extrapolation to low doses was based on an analysis that failed to take account of competing causes of death in the calculation of cancer rates and that did not adjust for age and gender [M17]; also, a factor introduced into the calculations to allow for a multiplicative transfer of risks from Japan to a United States population was too high and, in the light of the above findings, probably not necessary. Furthermore, while Gofman multiplied the risks from gamma-ray exposure of the atomic bomb survivors by two, in order to arrive at a risk estimate for x-ray exposure, it was noted above that relative risks are lower among women in the United States with x-ray exposures [B3, S15, L39] than among atomic bomb survivors exposed predominately to gamma rays, whereas absolute risks are similar. Given all these considerations, it is likely that Gofman's breast cancer risk estimate is too high by a factor of between 7 and 60 approximately [M18]. Furthermore, doses from past medical practices in the United States are also likely to have been overestimated. Calculations made by Evans et al. [E4] based on scientifically sounder approaches to the estimation of doses and radiation risk factors indicate that the proportion of breast cancers in the United States attributable to diagnostic radiography is closer to 1% than to the much higher values suggested by Gofman [G8].

185. Most of the studies of occupational exposure to low-LET radiation have not been informative about the risks of female breast cancer, owing to the small proportion of women in these studies. The largest amount of information concerns radiation workers in the medical field. Based on a survey of about 79,000 female radiological technologists who had worked in the United States since 1926, Boice et al. [B6] conducted a nested case-control study for 528 women with breast cancer. The study demonstrated associations with known risk factors, such as early age at menarche and family history of breast cancer but did not find correlations with number of years worked or with jobs involving radiotherapy, radioisotopes, or fluoroscopic equipment. However, dosimetry records were available for only 35% of the study subjects, mainly those who had worked in more recent years. Owing to the low level of doses received by these workers (generally below 0.1 Gy), the statistical power to detect an elevated risk was weak. As mentioned, dose data were lacking for earlier workers, whose cumulative doses may have been up to about 1 Gy. A subsequent mortality analysis based on a larger version of the cohort of radiological technologists showed a relative

risk of 1.5 (p<0.05) compared with national rates for women certified before 1940, whereas no enhanced risk was evident for more recent workers [D23]. This might reflect the higher doses received by early workers compared with later workers. However, the early workers were also more likely to be nulliparous than later workers, which may indicate a confounding effect. An elevated risk of breast cancer has also been reported among radiological technologists and radiologists in China; the doses are not known, although measured decreased blood counts suggest that they were generally high [W10].

2. Internal low-LET exposures

186. Several studies of patients given ^{131}I have examined breast cancer risks. Most of these studies were reviewed in the UNSCEAR 1994 Report [U2]. While a study in Massachusetts in the United States showed a higher risk of breast cancer among women treated for hyperthyroidism with ^{131}I compared with patients treated by other methods, there was no consistent trend in risk with the amount of ^{131}I administered [G10]. Similar conclusions were reached in a larger study including this and other hyperthyroid patients in the United States [R14]. In addition, a study of patients treated for hyperthyroidism in Sweden [H23, H24] did not show an elevated breast cancer risk overall, nor did it indicate a trend in risk according to the level of activity administered. It should be noted that the mean dose to the breast in the Swedish study was estimated to be 0.06 Gy [H23], indicating that such studies are unlikely to have sufficient statistical precision to detect an elevated risk. This problem also applies to studies of patients given diagnostic exposures to ^{131}I, where the number of cases was larger but the doses substantially smaller [H27], and of patients treated with ^{131}I for thyroid cancer, where the doses were higher but the number of breast cancers was lower [H26]. In neither of the last two studies were breast cancer rates raised significantly relative to national rates.

187. Among people residing on the banks of the Techa River who received both internal and external low-LET exposures as a consequence of radionuclide releases from the Mayak facility in the southern Urals, the proportion of female cancer deaths from breast cancer (4%) is similar to that among the Japanese atomic bomb survivors [K5]. However, without information on the breast doses in the Techa River cohort, it is difficult to make inferences.

3. Internal high-LET exposures

188. Continued follow-up of the early cohort of [224]Ra patients in Germany [N4] has indicated an excess of female breast cancers compared with the general population, as shown in Table 13. Calculations [H8] have yielded estimated breast doses from [224]Ra of several milligray to about 0.45 Gy [N19], with an average of about 0.1 Gy (high-LET). Analyses of these data indicated that the best fit was with a model in which the relative risk varied linearly with dose and decreased with increasing age at exposure [N4]. In particular, the estimate of the ERR

per Sv was 2.9 among females treated at ages less than 21 years, compared with an ERR per Sv of 0.9 for the full cohort, although these estimates are based on small numbers of cases. To identify potential confounders, a control group was constructed based on 182 patients who had not been treated with [224]Ra. In this group, 7 female breast cancer cases were observed, compared with 3.8 expected. Although the numbers were small, there was a suggestion that some of the cases in the control group may have been associated with repeated fluoroscopic x-ray examinations in the course of pneumothorax therapy [N4]. In contrast, the patients in the [224]Ra cohort had not in general received pneumothorax therapy, so this may not explain the excess seen in this group. Another possible reason for the excess is that patterns of reproductive risk factors may differ between these patients and the general population. In view of the results for the control group of patients, it seems unlikely that this could explain all of the excess, although the severity of the original disease may have affected whether or not radium was used, as well as the patient's subsequent reproductive history (and hence the risk of breast cancer). It may be that a combination of factors has led to the observed increase.

189. The study of neurological patients in Denmark [A5] gave some suggestion of an elevated breast cancer risk among women exposed to thorotrast for cerebral angiography relative to unexposed women, although this increase was not statistically significant (relative risk = 2.1; 95% CI: 0.8–5.7). Autopsy findings suggest that the dose to the breast from thorotrast is likely to be lower than that to many other organs [M21]. There is also some indication of an excess of breast cancer among female dial painters in the United Kingdom who had used a paint containing radium [B13, B14]. While there was no significant excess of breast cancer relative to local rates among radium dial workers in the United States, the cohort included not only dial painters but also women who carried out other tasks in this workplace [S16]. In contrast, a study restricted to the dial painters in the United States provided some suggestion of a raised breast cancer rate [R11]. However, as described in the UNSCEAR 1994 Report [U2], any effect of radiation is more likely to be due to external irradiation of the breast from paint in containers than to exposures arising from intakes of [226]Ra. In addition, reproductive risk factors may be of relevance to the breast cancer findings in these studies.

190. In view of the uncertainties in quantifying breast doses and cancer risks in studies of women exposed to high-LET radiation, it is not possible to directly compare the risks of female breast cancer associated with low-LET and high-LET radiation.

4. Summary

191. Extensive information from the Japanese atomic bomb survivors and several medically exposed groups demonstrates elevated risks of female breast cancer following external low-LET irradiation. The trend in risk with dose is consistent with linearity, and the ERR per Sv is particularly high for exposure

at young ages. In contrast, there is little evidence of increased risks for exposure at ages of more than 40 years. While the ERR per Sv seems to be fairly constant with time since exposure, the EAR per Sv appears to be more stable across populations with differing baseline rates. Examination of data for the atomic bomb survivors and some of the medical studies tend to suggest that dose fractionation has little influence on the risk per unit dose, although different interpretations have been placed on these analyses.

192. Data from studies of low-dose chronic external low-LET irradiation and of internal low-LET and high-LET exposures are limited. The interpretation of some reports of increased risks is complicated by the potential for confounding as a consequence of reproductive factors or other exposures.

I. PROSTATE CANCER

193. Worldwide, prostate cancer is one of the most common malignancies in men, but with wide variations in rates between countries [P5]. Specifically, incidence rates are highest in North America and some European countries and lowest in China and Japan. However, there is less international variation in prostate cancer mortality than in incidence [R32]. Studies of migrants suggest that the variations between countries cannot be explained solely on the basis of genetic predisposition [R32]. Both incidence and mortality rates have increased over the past few decades in many countries, although a substantial proportion of these increases may reflect improved detection of the disease [W6].

194. Prostate cancer is rare before 40 years of age, following which incidence rates double for each subsequent year of life, such that the age-specific curve has a steeper slope than for any other cancer [R32]. Survival rates are related strongly to the stage of the disease at diagnosis. The aetiology of prostate cancer is largely unknown. However, there is some evidence of effects associated with hormonal factors (e.g. levels of testosterone), family history of the disease, and dietary factors (e.g. possibly, fat intake [R32]).

1. External low-LET exposures

195. As indicated in Table 14, there is little evidence of an association between radiation and prostate cancer in the Life Span Study of the Japanese atomic survivors [T1]. In other studies, the point estimate of the ERR per Sv from the study of ankylosing spondylitis patients in the United Kingdom coincides with that for the atomic bomb survivors, but with a tighter confidence interval that excludes values below zero [W1]. However, the latter finding should be viewed cautiously, in that it is based on a combination of the number of x-ray treatments and mean organ dose rather than on individually-based estimates of doses, as in the Life Span Study. Among patients in the United States treated for peptic ulcers, raised mortality from prostate cancer relative to the general population was observed for both those who received radiotherapy and those who did not; rates in the two groups did not differ

significantly [G6]. An international study of patients treated for testicular cancer, many of whom received mean doses of several tens of gray, indicated an elevated risk of prostate cancer (SIR = 1.26, 95% CI: 1.07–1.46). However, this increase was apparent even in the first few years after treatment, and, in the absence of individual dose data, it might be surmised that this result was due to heightened medical surveillance of genitourinary conditions [T21]. Studies of medical exposures in childhood have thus far yielded little information on prostate cancer risks, mainly because a very long follow-up is required to obtain sufficient cases (given that this disease occurs predominantly in older persons).

196. Large studies of radiation workers generally do not show elevated risks of prostate cancer in relation to external low-LET radiation (e.g. [C11, M46]). Instances of worker studies in which increases have been reported may reflect chance variations (e.g. [A15]) or possibly other types of exposure (e.g. [B45, F6, R26], described in more detail below).

2. Internal low-LET exposures

197. In a large study of hyperthyroidism patients in the United States [R14], mortality from prostate cancer among patients treated with ^{131}I was significantly lower than would have been expected from national rates (SMR = 0.68). Furthermore, there was no indication of a trend in risk with the level of ^{131}I administered, although it should be noted that doses to the prostate are likely to have been low. Studies in Sweden of patients with medical exposures to ^{131}I have tended not to present results for prostate cancer specifically [H23, H24, H26, H27]. However, the findings given in these Swedish studies for all male genital cancers combined, most of which are likely to have been prostate cancers, showed overall incidence and mortality to be consistent with national rates. Furthermore, among the group of Swedish patients treated for hyperthyroidism, there did not appear to be a clear trend in mortality from all male genital cancers combined related to the amount of ^{131}I administered [H24]; however, in common with the corresponding study in the United States [R14], the prostate doses are unlikely to have been high.

198. A cohort study of employees of the United Kingdom Atomic Energy Authority showed that while prostate mortality among all radiation workers was consistent with national rates, mortality was raised among those workers who had experienced higher external doses and who had been monitored for internal radiation exposure [B45, F6]. Based on this cohort, a case-control study was conducted that looked at individual assessments of exposure to radionuclides and other substances in the workplace, as well as socio-demographic factors, for 136 workers with prostate cancer and 404 matched controls [R26]. Analyses were conducted for various radionuclides; however, the results were often correlated, because there was simultaneous exposure to some radionuclides in certain working environments. Rooney et al. [R26] reported significantly

elevated relative risks associated either with documented exposure to ^{51}Cr, ^{59}Fe, ^{60}Co, ^{65}Zn, or ^{3}H or with working in environments potentially contaminated by at least one of these radionuclides. The latter finding in particular was based largely on men who worked on heavy water reactors. Exposure to other radionuclides or to chemicals was not associated with an elevated risk. While it was difficult to distinguish the findings for the above five radionuclides, particular attention was paid to ^{65}Zn, because zinc is concentrated in the prostate gland and Auger electrons emitted from ^{65}Zn may give rise to high doses at short range. However, studies of biokinetics and dosimetry [A7, B46] indicate that even with pessimistic assumptions about the uptake of zinc in the prostate and the relative biological effectiveness of Auger electrons, the dose to the prostate from occupational exposures is likely to be 0.1–0.2 Sv at most and, taking account of the findings from the Japanese atomic bomb survivors [T1], would not be sufficient to explain the findings of Rooney et al. [R26].

3. Internal high-LET exposures

199. Few studies have reported results for prostate cancer in relation to internal high-LET exposures. As shown in Table 14, there is little indication of elevated risks among patients with intakes of ^{224}Ra [N4] or thorotrast [V8], although the numbers of cases are not very large. Furthermore, information has rarely been presented about level of exposure. An exception concerns a study of plutonium workers in the United Kingdom, in which there was no increase in risk with the sum of the cumulative organ-specific dose from plutonium and the external dose [O1]. However, in common with many other studies of workers high-LET doses to the prostate are likely to have been low.

4. Summary

200. Data for the Japanese atomic bomb survivors and from most other studies provide little evidence of an elevated risk of prostate cancer following radiation exposure. Elevated risks have occasionally been reported, but it is not clear whether these represent chance findings or facets of particular types of exposure in the workplace, either from radiation and other factors. It should be noted that the statistical precision of some of the medical and occupational studies is limited by small numbers of cases and/or low doses. Also, because prostate cancer is predominantly a disease of the elderly, follow-up studies of exposure in childhood have not been informative to date.

J. CANCER OF THE URINARY BLADDER

201. Bladder cancer accounts for less than 5% of cancer incidence and less than 2% of cancer mortality in industrialized countries. There is wide international variation in bladder cancer incidence, with high rates in Europe and North America and low rates in Latin America and Asia. Incidence increases steeply with age and is more common among men than women. In some countries the gender ratio can reach 5:1

[H47, P5]. The incidence increased from the 1960s to the 1980s, but recently the rates have begun to stabilize. Mortality has been decreasing in both men and women and at all ages. The temporal trends are influenced by changes in detection and improvements in survival.

202. Cigarette smoking is a leading cause of bladder cancer. In Western countries, approximately 50% of the cancer in men and 30% in women have been attributed to smoking. Occupational exposures, particularly to aromatic amines, are also well known bladder cancer risk factors. Urinary tract infections are also associated with an increased risk of bladder cancer, especially among women. Use of phenacetin-containing analgesics and cyclophosphamide, as well as exposure to S. haematobium infection, are also suspected bladder cancer risk factors [H47, M45, S48].

1. External low-LET exposures

203. Estimates of risk for bladder cancer from several studies are given in Table 15. Statistically significant excess risks have been derived for incidence [T1] and mortality data [P9, R1] from the Life Span Study, the cervical cancer case-control study [B1], the anklylosing spondylitis study [W1], the metropathia haemorrhagica study [D7], and the benign gynaecological disease study [I16]. Although the doses are considerably higher in the last two studies (~6 Gy), the risk estimates are about the same as the risk estimate in the ankylosing spondylitis study [W1]. In the Life Span Study, the effects of age and gender on the risks are unclear. In particular, the incidence data exhibit a statistically significant gender difference, with the ERR for females exceeding that for males by a factor of about 5 but the average EAR showing no significant difference [T1]; in the mortality data, the point estimates of the ERRs and EARs for males are higher than those for females, although the differences are not statistically significant [P9]. Neither the mortality data [S3, P9] nor the incidence data [T1] in the Life Span Study exhibit statistically significant variation with age at exposure for either the ERR or the EAR. There is, however, a suggestion of some variation with age in the cervical cancer case-control study [B1].

204. Although individual organ doses frequently are not available, several, but not all, studies of second cancers have reported an association between bladder cancer risk and high therapeutic radiation doses. A non-significant increased risk of bladder cancer was associated with radiotherapy in a large cohort of non-Hodgkin's lymphoma patients [T19] and in a European nested case-control study of 63 women with bladder cancer who had previously been treated for ovarian cancer and 188 ovarian cancer patients who did not develop bladder cancer [K30]. Compared with surgically treated patients, the relative risks were 1.9 (95% CI: 0.77–4.9), 3.2 (95% CI: 0.97–10), and 5.2 (95% CI: 1.6–16) for radiotherapy only, chemotherapy only, and radiotherapy and chemotherapy combined, respectively. Of 32,251 ovarian cancer patients, 20 of the 65 women who developed bladder cancer were treated solely with

radiation, resulting in a significantly increased risk (O/E = 2.1; 95% CI: 1.6–2.6) [T20]. The risks increased with time since exposure, until they were six times greater at 15 or more years. These results are very consistent with those for cervical cancer patients who were treated with similar radiation doses [B1]. Risk was not significantly elevated among ovarian cancer patients treated with chemotherapy only [K30].

205. Among men treated for testicular or prostate cancer, enhanced risks of bladder cancer have been observed. Among testicular patients with seminoma treated with radiotherapy (mean dose = ~22 Gy), a two- to threefold greater risk was found five or more years after treatment. More than 20 years after treatment, the risk rose to 3.2 [T21]. Among non-seminoma patients receiving radio-therapy (mean dose = 45 Gy), the risks were elevated but not statistically significant. Among men treated with high-dose radiotherapy for prostate cancer, a statistically signi-ficant 40% increased risk was noted five or more years after therapy [N11]. No excess risk was found among patients treated surgically. In a reanalysis and update of these data, Brenner et al. [B42] reported a 15% (95% CI: 1.02–1.31) elevated risk of bladder cancer among over 50,000 men treated with high-dose radiotherapy compared with over 70,000 patients who underwent surgery. Risks were much higher, however, for long-term survivors, with radiotherapy patients surviving 10 or more years having a risk of 1.77 (95% CI: 1.14–2.63).

2. Internal low-LET exposures

206. High doses of ^{131}I are often used to treat thyroid cancer. The bladder is one of the organs that concentrate iodine [U2]. The ^{131}I dose to the bladder from treatment for thyroid cancer is about 2 Gy. An excess risk of bladder cancer has been reported in one small study of thyroid cancer patients [E2] but not in two others [D18, H26]. Patients treated with ^{131}I for hyperthyroidism receive 100–200 mGy to the bladder. No significantly increased risks were noted in two studies with a combined study population of about 30,000 patients [H23, H24, R14]. In a recent study of hyperthyroid patients treated with ^{131}I in the United Kingdom, there was a significantly lower risk of bladder cancer than in the general population, but bladder cancer incidence increased (p=0.005) with increasing levels of administered activity [F8].

3. Internal high-LET exposures

207. The recent follow-up of a cohort of German patients treated with ^{224}Ra has demonstrated an excess relative risk of bladder cancer compared with the general population (ERR per Sv = 0.4) [N4]. The relative risk was higher for patients who were older at diagnosis. No excess of bladder cancer has been reported in another cohort of patients treated with ^{224}Ra [W20] or among patients receiving thorotrast as a contrast medium for arteriography [A5, D31, M14, V8].

4. Summary

208. Statistically significant excess risks of cancer of the urinary bladder are seen in several populations exposed to low-LET radiation. The Life Span Study risk estimates are somewhat greater than those seen for cancer patients; however, since the cancer patient studies involve extremely high doses, the differences may reflect cell killing. In addition, second cancer register-based cohort studies often obtain information on initial treatment only. Subsequent treatments can lead to exposure misclassification, which in turn can lead to underestimation of exposure effects. Potential interactions between smoking and radiation remain to be studied.

K. BRAIN AND CENTRAL NERVOUS SYSTEM TUMOURS

209. Depending on tumour location, benign and malignant tumours of the central nervous system (CNS) can have similar symptoms and outcomes. As a result, the two types of tumours are not always easily distinguished, and many tumour registries routinely include both histological types in their CNS incidence rates. [I11, P18]. Annual incidence rates for CNS cancers range from about 1.0 to about 10 per 100,000 persons, but since the quality of medical care varies from country to country and reporting of benign tumours is inconsistent among registries, international comparisons of CNS tumours can be misleading [P5]. The fact that the lower incidence rates are reported primarily from cancer registries with uncertain completeness of ascertainment suggests that country-to-country variation is probably considerably less than current reporting indicates. Over the last few decades, brain tumour incidence and mortality have increased, especially among the elderly, but whether this is a real increase or a result of better diagnosis and reporting is controversial [I11, P18]. With the excep-tion of meningiomas, CNS tumours occur more frequently among men than women [P5]. This Section will consider both benign and malignant CNS tumours occurring within the cranium (brain, cranial nerves, cranial meninges), spinal cord, spinal meninges, and peripheral nervous system because of the potential problem of misclassifica-tion by tumour behaviour. In addition, since the com-parison rates used in some studies are derived from tumour registries that combine all CNS tumours in one category, results are reported for all CNS tumours and not for malignant tumours only.

210. While the aetiology of CNS tumours remains elusive, therapeutic irradiation of the head and neck during childhood is an established risk factor, and social class, trauma, diet, and some chemicals have been identified as potential risk factors [B43, D35, I11, P18]. Primary malignancies of the central nervous system are among the most lethal of all cancers. In the United States, five-year survival for malignant CNS tumours is approximately 30% and shows little relation with stage at diagnosis [K17]. Survival for benign meningiomas has improved

considerably over the last few decades, but depending on tumour size and location, the quality of life can be severely impaired [L30].

1. External low-LET exposures

211. As summarized in Table 16, the epidemiological literature provides evidence for an association between ionizing radiation and tumours of the CNS. Since publication of the UNSCEAR 1994 Report [U2], additional information on the incidence and mortality of CNS tumours in the Life Span Study of atomic bomb survivors has become available [P9, P19]. As in earlier reports, the most recent mortality data from the atomic bomb survivors provide no evidence of a radiation effect for brain tumours but do show a non-significant excess risk for tumours of the CNS outside the brain [P9]. New incidence data that assess histologic types separately demonstrate a strong dose response for neurilemmomas (ERR at 1 Sv = 4.0) and a moderate dose response for meningiomas (Table 27) [P19]. The excess risk for neurilemmomas was observed for persons of all ages at the time of the bombings. Other studies of atomic bomb survivors in Hiroshima and Nagasaki show an association between meningioma incidence and radiation exposure [S33, S39, S42].

212. A significant relationship between radiation dose and CNS tumour risk was demonstrated in the Israeli tinea capitis study [R17]. An average dose of 1.5 Gy from childhood radiotherapy to the scalp was associated with an increased incidence of CNS tumours in the head and neck (relative risk = 8.4). The relative risks ranged from 2.6 for gliomas to 9.5 for meningiomas to 33 for neurilemmomas. Large relative and absolute risks for CNS tumours were also observed in the New York tinea capitis study [A15, S31]. Similarly, an association between radiotherapy and benign CNS tumours was reported following childhood irradiation for inflamed tonsils and other benign head and neck conditions [S28, S46] and irradiation in infancy for an enlarged thymus gland [H31]. Following low doses of radiation from ^{226}Ra treatment for haemangioma during infancy in Stockholm, intracranial tumours were not elevated [L16]. In contrast, the incidence of gliomas and meningiomas was significantly greater in 1,805 infants treated with similar doses of ^{226}Ra for haemangioma in Gothenburg, Sweden, but no clear dose response was observed [K22, L15]. In a recent pooled analysis of the two studies, 86 patients with intracranial tumours were observed among exposed and unexposed patients compared with 61 expected (SIR = 1.42; 95% CI: 1.13–1.75) [K23]. A linear dose-response relationship fit the data best (ERR at 1 Gy = 2.7), and within the narrow age-at-exposure range (0–81 months) the risk increased with decreasing age at exposure. In a small cohort of children treated with nasopharyngeal radium implants to prevent deafness, three adult brain cancers occurred [S32]. Although the incidence was raised, chance could be one explanation for the increase [S47]. CNS mortality was not elevated in a larger study of children treated with smaller doses [V5].

213. A higher-than-expected number of second primary CNS tumours among survivors of childhood cancers has been noted in several studies. Neglia et al. [N9] demonstrated that radiotherapy during childhood was a significant factor in the excess of CNS tumours occurring among acute lymphoblastic leukaemia patients. A cohort of 4,400 childhood cancer survivors in France and the United Kingdom has been followed to evaluate the risk of developing second cancers [D19, L32, L36, L37]. Based on 12 cases with malignant brain tumours and an equal number of cases with benign brain tumours, each matched to 15 controls, a significant dose response was demonstrated for both types of tumours. The risk was higher for benign tumours (ERR = 3.15; 95% CI: 0.37–n.a.) than for malignant tumours (ERR = 0.12; 95% CI: n.a.–0.55), and no modifying effect of age at exposure was found. This pattern of a higher risk for benign tumours has been seen in other studies [P19, R17]. Eng et al. [E1] reported that bilateral retinoblastoma patients treated with radiation had a large excess of mortality from benign and malignant neoplasms of the brain and meninges. More recently, an increased risk of CNS tumour incidence was found among these patients [W11]. In a small study with limited statistical power, no excess risk was observed among retinoblastoma patients [M26]. Young children who received cranial irradiation as a conditioning regimen before bone marrow transplantation were found to have a significantly elevated relative risk of developing brain or other CNS cancers; however, it was likely that earlier cranial radiotherapy to treat acute lymphocytic leukaemia prior to bone marrow transplantation (and associated total-body irradiation) played an important role in the development of these neural malignancies [C16].

214. Data on adult exposures are considerably more limited. Following high-dose (~40 Gy) fractionated radiotherapy, an excess risk of CNS tumours was observed among pituitary adenoma patients [B22, T11]. In several case-control studies of patients with CNS tumours of various histological types, a history of diagnostic x-ray examinations [H32] or x-ray treatments to the head was more often reported for cases than for controls [B23, P20, P21]. In contrast, a mean brain dose of about 0.6 Gy was not associated with an increase in CNS tumour incidence or mortality in two small cohorts of infertile women irradiated to the pituitary gland and ovaries [R18, R30], and ankylosing spondylitics did not have an excess of mortality from spinal cord tumours after being exposed to high radiation doses to their spinal cords [W1].

215. Dental diagnostic x-ray exposures have been assessed in several studies conducted by Preston-Martin et al. in relation to various types of CNS tumours [P20, P21, P22, P23, P24]. They found associations between meningiomas and frequent annual full-mouth x-ray examinations and x-ray examinations performed many years ago, when radiation doses were relatively high. Risks were higher when exposure occurred during childhood. In other studies, however, brain tumour cases did not have a history of dental x-ray exposure significantly more often than controls [K18, M27, R19].

216. Radiation workers in general receive low, fractionated doses with relatively little exposure to the brain. To date, most occupational studies have been negative with respect to this site of cancer [C11, M46, W10]. Brain cancer incidence and mortality rates were elevated among airline pilots in a few studies [B48], but no dose-response relation was observed and confounding due to non-ionizing radiation and socio-economic status has been postulated [G15].

217. The issue of whether CNS tumours are related to fetal exposure to radiation remains controversial. Most recently, Doll and Wakeford [D17] carefully reviewed the literature and concluded that *in utero* exposure to a mean dose of approximately 10 mGy increases the risk of childhood cancer. This conclusion was largely based on the Oxford Survey of Childhood Cancers. In the Oxford Survey, mortality from childhood CNS tumours was associated with fetal irradiation (relative risk = 1.4; 95% CI: 1.2–1.7) [B2]. Miller and Boice [M31] expressed concern about the Oxford Survey results, noting that all childhood cancers were increased about 40%, whereas such commonality is not seen in either animal or human studies. Among atomic bomb survivors exposed *in utero,* an association between dose and cancer mortality has not been found, but the *in utero* survivor cohort is small, and the negative result is compatible with a wide range of risks [D14].

2. Internal low-LET exposures

218. Little is known about brain and CNS tumours following internal exposure to low-LET radiation. A small increased risk of CNS tumours was observed among 35,000 Swedish patients receiving diagnostic ^{131}I examinations (SIR = 1.19; 95% CI: 1.00–1.41) [H27]. Since the dose to the brain was <10 mGy, the observed excess is not likely to be due to the radiation exposure. Significant excess risks were not demonstrated among patients receiving ^{131}I therapy for hyperthyroidism [H23, H24, R14] or thyroid cancer [D15, E2, G13, H26]; however, among ten-year survivors, brain tumour incidence was significantly elevated in the Swedish hyperthyroid patients [H23].

3. Internal high-LET exposures

219. Danish patients exposed to thorotrast had a significantly elevated incidence of brain tumours, but the fact that these tumours developed very soon after the thorotrast examination suggests that they are related to the under-lying disease or better ascertainment rather than to the thorotrast itself [A5]. Thorotrast was given in conjunction with cerebral angiography because of a suspected brain disorder. Often this disorder was later found to be a brain tumour, especially among epileptic patients. Brain malignancies and other CNS tumours have not been linked to exposure to radium [S34] or to radon among miners [D8].

4. Summary

220. Ionizing radiation can induce tumours of the CNS, although the relationship is not as strong as for many other tumours, and most of the observed radiation-associated tumours are benign. Indeed, neurilemmomas, which are highly curable, are the only tumours that consistently exhibit high risks. Overall, exposure during childhood appears to be more effective in tumour induction than adult exposure, but the data on adult exposure are fairly sparse, and the most recent study of atomic bomb survivors demonstrated an excess relative risk for neurilemmomas following exposure at all ages. Little is known about other factors that modify risk. The association between benign tumours, particularly meningiomas and neurilemmomas, and radiation appears to be substantially stronger than with malignant tumours. Malignant brain tumours are seen only after radiotherapy. Additional data are needed to better characterize the dose response for CNS tumours of various histological types.

L. THYROID CANCER

221. Thyroid cancer is one of the less common forms of cancer [P5]. Unlike most cancers, its incidence is relatively high before age 40 years, increases comparatively slowly with age, and is about three times higher in women than men. This female predominance is also observed for benign thyroid tumours. The degree of malignancy varies widely with histological type, ranging from the rapidly fatal anaplastic type to the relatively benign papillary type [F2, R13]. Data from most countries suggest that mortality is falling while incidence is increasing [F1]. Ionizing radiation is a well documented cause of thyroid cancer. The relative risk of thyroid cancer is also substantially increased among persons with a history of benign nodules and goitre. There is some evidence that elevated levels of thyroid-stimulating hormone, multiparity, miscarriage, artificial menopause, iodine intake, and diet also may be risk factors for thyroid cancer[F2, R13].

222. Shore [S8] reviewed the epidemiological studies of radiation and thyroid cancer conducted through the early 1990s. Since then, more information has become available from continued follow-up of some cohorts and from a pooled analysis of seven studies of external radiation [R4]. Additional data on the occurrence of thyroid cancers among children living in radiation-contaminated areas in Belarus [D13], the Russian Federation [I23], and Ukraine [T23] as a result of the Chernobyl nuclear power plant accident have recently been published. New data on Chernobyl recovery operation workers ("liquidators") have also been published in the last few years [K15]. These results are discussed below and in more detail in Annex J, *"Exposures and effects of the Chernobyl accident"*.

1. External low-LET exposures

223. The results for thyroid cancer incidence that were presented in Table 8 of Annex A in the UNSCEAR 1994 Report [U2] are updated here in Table 17. This Table contains findings from a pooled analysis of studies of external irradiation of the thyroid [R4]. This analysis,

which included seven studies and was based on almost 120,000 people with about 700 thyroid cancers and 3 million person-years of follow-up, allowed a more detailed evaluation of the dose-response relationship and of modifying factors than had previously been possible. Nearly 500 thyroid cancers occurred in the half of the study population exposed during childhood or adolescence.

224. In the analysis of the five cohort studies of persons irradiated before age 15 years, 436 thyroid cancers were diagnosed among the exposed population. The pooled ERR per Gy was 7.7 (95% CI: 2.1–28.7). No single study was found to have an undue influence on the overall estimates of risk. The ERR per Gy for females was nearly twice that for males, but the results were not consistent [R4]. Since thyroid cancer naturally occurs two to three times more frequently among females than males, the absolute radiation-induced risk was correspondingly higher among women. Even within the narrow range of ages at exposure, there was strong evidence of a decrease in the ERR with increasing age at exposure, which suggests that the thyroid is particularly sensitive to tumour induction at the time of rapid cell proliferation. The ERR per Gy was highest 15–29 years following childhood exposure, but it remained high for more than 40 years after exposure [R4]. While the latter finding was also reported from an extended follow-up of the Stockholm skin haemangioma cohort [L13], few other studies have more than 40 years of follow-up. In contrast to the well described carcinogenic effects of childhood exposure, there is little evidence of an excess of thyroid cancer associated with external exposure after age 20 years. Among atomic bomb survivors exposed after age 40 years, the ERR was negative [R4, S8, T1].

225. Each of the studies in the pooled analysis was consistent with a linear dose-response relationship, although the range of doses varied considerably among studies [R4]. In the childhood cancer study [T5], which was the only study with doses over 6 Gy, there was some indication that the effects of cell killing flattened the dose response at high doses. Exposures were received in fractions, from all in one day to several years apart in three of the studies included in the pooled analysis. There was very weak evidence that for the same total dose, exposures received in two or more fractions were less carcinogenic than acute exposures by an estimated factor of 1.5, with wide confidence limits [R4]. Although no formal assessment of risk by histology type was conducted, the risk for papillary carcinomas appeared to be higher than for follicular cancer in the individual studies. To date, no clear association between ionizing radiation and either medullary cancer or anaplastic carcinoma has been observed, although there have been reports of anaplastic carcinoma occurring after medical irradiation.

226. An elevated risk of thyroid cancer was reported for patients treated with high-dose radiotherapy for Hodgkin's disease [D33, D36, H9, T5] and for childhood cancers [H30, T5]. New studies emphasize that Hodgkin's disease survivors have a high risk of thyroid cancer if they received radiotherapy as children [B16, S23]. Recently, a large increased risk of thyroid cancer was reported among bone marrow transplantation patients treated with high-dose, total-body irradiation, especially during childhood (4 cases observed compared with 0.02 expected); however, radiotherapy received before bone marrow transplantation might have played a role in the development of these malignancies [C16].

227. Information on fractionated and low-dose-rate exposures mostly comes from studies of high-background areas, diagnostic radiation procedures, and occupational exposures. Studies of residents living in areas of high natural background radiation were conducted in China [T25, T26, W9] and India [P3]. They did not show an association between the prevalence of thyroid nodules and lifetime exposure to elevated background radiation. However, since the doses received in childhood generally were only a few tens of milligray, the statistical power to detect a radiation effect was low. Diagnostic x rays, even those resulting in higher thyroid gland doses or those occurring during childhood, were not linked to thyroid cancer in a study in Sweden [I9]. This study is unique because the ascertainment of diagnostic x-ray procedures was based not on personal recall but on a search of hospital radiation records.

228. While early mortality studies of radiation workers provided no evidence for an elevated risk of thyroid cancer [M23], there have been reports of an increased risk of thyroid cancer among x-ray technologists. Among 27,000 x-ray workers and a similar number of non-radiation medical workers in China, 8 thyroid cancers were found compared with 4.5 expected [W10]. The relative risk was larger for personnel working at relatively early ages and during the period when exposures were greatest. In the United States, a twofold greater risk of thyroid cancer incidence was reported in preliminary results from a survey of over 100,000 predominately female x-ray technologists [B20]. These preliminary results were based on self-reported diagnoses on questionnaires and might have included benign nodules or adenomas. In a recent mortality study of the x-ray technologists, no excess of thyroid cancer deaths was noted [D23]. Consistent with the incidence results are findings from a Swedish record-linkage study in which x-ray technicians had double the risk of thyroid cancer compared with the general population of Sweden [C17] and from a small Italian study in which male hospital radiation workers had a higher prevalence of thyroid nodules than comparable non-exposed workers [A8]. Based on only nine thyroid cancer deaths, a significantly elevated mortality, but no dose response, was observed in mostly male nuclear workers in the United Kingdom [L20]; the evidence for an excess diminished with longer follow-up [M46]. No association was reported for nuclear workers in the United States [G12] or in the combined international analysis of nuclear workers from Canada, the United Kingdom, and the United States [C11]. Since adult, acute radiation exposures have not been linked to thyroid cancer, the reports of excesses are surprising. Each of these studies, however, has methodological weaknesses for studying

thyroid cancer that might have influenced the findings. For example, except for the nuclear worker studies, individual doses were not available; multiple comparisons were tested in most studies; and the number of cases was generally small, which produces unstable risk estimates. Furthermore, the well known association between radiation and thyroid cancer may have led to more complete case ascertainment for radiation workers.

229. As a consequence of the Chernobyl accident, large numbers of men from all over the former Soviet Union were brought in to participate in recovery operations at the reactor and in the surrounding areas. Altogether approximately 600,000 workers were involved, about 240,000 of them during 1986 and 1987. Most of the exposure of the workers came from external gamma and beta irradiation. Internal exposure from radionuclides was minor after the first few weeks [U4]. Several investigations of recovery workers from the Baltic countries have been conducted. In a systematic clinical evaluation, including palpation and ultrasound, of the nearly 2,000 Chernobyl recovery operation workers from Estonia, no excess of thyroid nodularity or cancer was detected [I10]. Doses were estimated for each worker based on medical records, responses to a questionnaire, and biodosimetry. Film badges suggested that workers had been exposed to a mean dose from external sources of approximately 100 mGy, but biodosimetry indicated that the doses might have been considerably lower [L31]. Thyroid cancer incidence and mortality were evaluated in a cohort of nearly 5,000 Estonian workers [R20]. No thyroid cancers were observed, whereas 0.21 would have been expected based on age, gender, and calendar-specific cancer rates in Estonia. In a cohort of Lithuanian Chernobyl workers, the three observed thyroid cancers did not significantly differ from the expected number based on Lithuanian cancer rates [K39]. Given the low dose and late age at exposure, these negative findings are consistent with data from the Life Span Study of atomic bomb survivors [T1].

230. In a much larger study of 168,000 Russian recovery operation workers, Ivanov et al. [I13, I18] reported an increased risk of thyroid cancer compared with the population of Russia. Comparing cancer incidence in these workers to that in a general population is questionable, because the recovery operation workers had a higher level of medical surveillance, especially of their thyroid glands [B4]. However, Ivanov et al. [I17] noted that they adjusted for a screening effect. Further data regarding these findings are needed.

2. Internal low-LET exposures

231. Studies of medical exposures to ^{131}I were reviewed extensively in the UNSCEAR 1994 Report [U2]. Since then, further information has become available from three large follow-up studies of ^{131}I-exposed patients. In addition to an extended period of follow-up (as much as 40 years following exposure), the Swedish study of over 34,000 patients administered ^{131}I for diagnostic purposes now incorporates individual estimates of thyroid doses [H4]. Dose quantification was based on the amount of ^{131}I administered and the 24-hour thyroid uptake. Information on the size of the thyroid gland was available for nearly half of the patients, and adjustments to dose estimates on the basis of these data did not affect the results. Basic details of the study cohort are given in Table 2, Table 3, and Table 17, while Table 28 presents thyroid cancer incidence in relation to dose. Although overall incidence, after excluding the first five years following exposure, was greater than that in the general population, there was no indication of a dose-response trend. Furthermore, analyses based on the reason for the initial referral showed that incidence was higher than expected only among those referred for suspicion of a thyroid tumour. Among those referred for other reasons, thyroid cancer incidence was lower than expected compared with national rates. Among the 34,000 patients evaluated by Hall et al. [H4], 7% were under 20 years of age at the time of exposure and less than 1% were under 10 years of age. Among the 2,408 adolescents and young adults (average thyroid dose of 1.5 Gy), 3 thyroid malignancies were observed compared with 1.8 expected based on national rates (SIR = 1.69; 95% CI: 0.35-4.9). These data do not allow inferences about childhood exposures. No excess of thyroid nodules was detected when 1,005 women who had been examined years before with [131]I (mean thyroid dose of 0.54 Gy) and 248 non-exposed women were screened for thyroid disorders [H36]; however, among the exposed women the prevalence of thyroid nodules was correlated with dose.

232. Studies of patients treated with ^{131}I for hyperthyroidism have dealt almost entirely with adults. Although individual thyroid doses have not been calculated, the intention is to deliver 60-100 Gy to the thyroid [B21]. At doses of this magnitude, the ERR per Gy for children receiving external radiation begins to level off, probably due to cell killing [R4]. Among 10,000 Swedish patients, 18 thyroid cancers were observed, yielding a standardized incidence ratio of 1.29 (95% CI: 0.76-2.03) [H23]. Among 23,000 patients evaluated in a new follow-up of the thyrotoxicosis study in the United States, an increased risk of thyroid cancer mortality was observed [R14]. The excess risk was primarily due to a large risk during the first five years following treatment and was higher among toxic nodular goitre patients than Graves' disease patients. Franklyn et al. [F8] reported an elevated incidence of thyroid cancer and thyroid cancer deaths in a follow-up of 7,417 hyperthyroid patients treated with ^{131}I in England. Compared with the population of England and Wales, both the SIR (3.25; 95% CI: 1.7-6.2) and the SMR (2.78; 95% CI: 1.2-6.7) were elevated, but no dose response was demonstrated. These findings suggest that some of the excess may be due to the underlying thyroid disease.

233. While the data from the medical radioiodine studies are informative, the uncertainties associated with estimating thyroid doses from ^{131}I, especially in persons with thyroid abnormalities, reduce the precision of the risk estimates. The non-uniformity of the dose distribution in the thyroid gland results in some areas of tissue receiving

such high doses that cell killing could occur and other areas receiving extremely low doses [N7]. Thus, the tumorigenic effects of the exposure might be lower than would be expected based on the average dose. Nevertheless, ^{131}I dose estimation in medical studies is far better than for the studies of environmental ^{131}I exposure.

234. Four years after the 1986 accident at the Chernobyl nuclear plant, a substantial increase in childhood thyroid cancer was observed in contaminated regions of the former Soviet Union [S49]. For a more detailed discussion of thyroid cancer risk following the Chernobyl accident, see Annex J, "*Exposures and effects of the Chernobyl accident*". In Belarus, and particularly in the Gomel region to the north of Chernobyl, the number of childhood thyroid cancers diagnosed between 1990 and 1992 was much higher than in 1986-1989 [K6]. The diagnoses of most of the thyroid cancers were confirmed by an international pathology review [W5]. An unusually high frequency of thyroid cancer continues to occur in Belarus [B49, D13] and in heavily contaminated areas in Ukraine [L19, T23] and the Russian Federation [T10, I23, I24] among persons who were less than 15 years of age at the time of the accident. Childhood thyroid cancer rates in these areas in 1991-1994 were higher by a factor of almost 10 than in the preceding five years (Table 29). The number of cases identified among persons born less than 17 years before the accident reached about 1,800 in 1998 (Annex J, "*Exposures and effects of the Chernobyl accident*"). Risk appears to increase with decreasing age at exposure [A3, K31, P31, W16]. Recent data from Belarus suggest that while increases in thyroid cancer incidence are still occurring among individuals who were less than 5 years of age at the time of the accident, rates for older children might be stabilizing [K31]. In the Ukraine, rates are still rising for persons less than 14 years of age, but a similar leveling off of the risk among those 14-18 years old at the time of the accident was observed [T23]. Age-at-exposure effects warrant further investigation.

235. Following early reports of an increased frequency of thyroid cancer, questions were raised about the effects of screening the exposed population [B8, R3, S22]. While the screening programmes being conducted in the contaminated areas are responsible for some increases in thyroid cancer ascertainment, the majority of tumours reviewed by an international panel were not microcarcinomas. In fact, many showed direct invasion of extrathyroidal tissues and lymph node spread [W5].

236. A study of 107 thyroid cancer cases and 214 matched controls was conducted in Belarus [A26]. Taking into account the reason for diagnosis, a strong dose response was demonstrated. Although the estimated doses in the study have considerable uncertainty, the results indicate that the excess of thyroid cancers is related to the radiation exposure.

237. A strong correlation between estimated exposure from ^{131}I and thyroid cancer rates has been reported in several studies [J4, J5, L19, L51]. In a well designed correlation study, Jacob et al. [J5] compared average thyroid doses from ^{131}I exposure in many regions in Belarus and the Russian Federation with 1991-1995 incidence rates for the 1971-1986 birth cohort. A linear dose-response relationship was found (EAR per 10^4 PY Gy = 2.3; 95% CI: 1.4-3.8; ERR per Gy = 23; 95% CI: 8.6-82). Likhtarev et al. [L51] also conducted a correlation study using recent data (1990-1997) from the Ukraine. They reported an EAR per 10^4 PY Gy of 1.6 (95% CI: 0.7-3.4) and an ERR per Gy of 38 (95% CI: 16-97) for the 1971-1986 birth cohorts. While these studies provide reasonable risk estimates, they are based on geographical correlations and are subject to the limitations inherent in such evaluations.

238. No radiation-associated thyroid malignancies have been observed less than five years after external exposure [R4]. Despite some early occurrence of childhood thyroid cancers after the Chernobyl accident, most cases were diagnosed after 1991 (Table 29). A minimal latency period for radiation-induced thyroid tumours of four years might have resulted from the ability to detect an effect because of the millions of children exposed to radiation from the Chernobyl accident or because of the advancement of time to diagnosis due to screening.

239. A high frequency of *RET/PTC* oncogene rearrangements is found in the thyroid cancers occurring in the Chernobyl area. Some studies have reported specific types of *RET/PTC* in Chernobyl cases [B25, K24] compared with tumours associated with external radiation [B26]; however, findings have not been consistent [W4]. Both *RET/PTC1* and *PTC3* rearrangements have been reported in Chernobyl-related patients, and recent research suggests that age at exposure, time since exposure, and morphology may be important in determining the type of PTC rearrangement [P10, S13, T34].

240. Taking all of the data together, screening and other selection effects may explain some of the increase in thyroid tumours seen among the children living around Chernobyl, but radiation exposure from the reactor accident clearly plays a major role. The associated mechanism is not yet well understood, and the magnitude of the risk from ^{131}I *per se* remains uncertain. The geographical distribution of these tumours coincides more closely with the areas of ^{131}I contamination than with the areas of ^{137}Cs contamination, but there is also a correlation with the distribution of shorter-lived radioisotopes (e.g. ^{132}I, ^{133}I, and ^{135}I) [A3]. Other factors that might influence radiation risks have been identified. Many of the regions around Chernobyl are iodine-deficient [G20, P37], and iodide dietary supplementation had been terminated before the accident [W5]. Although large amounts of stable iodine were distributed to the population living near the plant as prophylaxis shortly after the accident, the distribution was incomplete and is thought not to have been very effective [M32]. Genetic susceptibility to radiation-associated thyroid cancer also has been suggested as a potential modifier of risk [C36]. Finally, other potential environmental contaminants need to be investigated.

241. The health effects of exposure to ^{131}I fallout from atmospheric nuclear tests conducted at the Nevada test site in the 1950s have been studied for the last four decades. In the most recent follow-up, 2,500 children were examined and individual doses to the thyroid reconstructed. Nineteen neoplasms, of which eight were malignant, were diagnosed. The ERR per Gy was about 7 (p=0.02). When the analysis was restricted to malignancies, the ERR per Gy was 7.9 but was not statistically significant [K36]. The ^{131}I doses from weapons testing at the Nevada Test Site were assessed by the United States National Cancer Institute [N12]. Iodine-131 is the radionuclide of main concern because it is the principal radionuclide in fallout and is ingested by drinking contaminated milk. Approximately 5.6 EBq of ^{131}I were released into the atmosphere, resulting in radioiodine deposition throughout the United States. Iodine-131 thyroid doses were estimated for each county in the continental United States by age group, gender, and level of milk consumption.

242. The average thyroid dose to the approximately 160 million people living in the United States at the time of testing was 20 mGy. The estimated dose varied substantially depending on geographic location, age at the time of exposure, and quantity, source, and type of milk intake. Doses were highest east of the test site in Nevada and Utah and in some counties in Idaho, Montana, New Mexico, Colorado, and Missouri and were lowest on the West Coast, on the border with Mexico, and in parts of Texas and Florida. Owing to geographic differences, doses ranged from 0.01 to 160 mGy. The average dose to young children was approximately 10 times higher than the estimated adult dose, because the thyroid gland of small children concentrates more iodine and because children drink much more milk than adults. While the uncertainty associated with estimating the average thyroid dose to the population of the United States is about a factor of 2, the uncertainty in dose estimates for individuals is about a factor of 3.

243. Gilbert et al. [G19] related age-, calendar year-, gender-, and county-specific thyroid cancer mortality and incidence rates in the United States to ^{131}I dose estimates, taking geographic location, age at exposure, and birth cohort into account. Neither cumulative dose nor dose received between 1 and 15 years of age was associated with thyroid cancer incidence or mortality, but an association was suggested for dose received before 1 year of age (ERR at 1 Gy = 10.6; 95% CI: 1.1–29 and ERR at 1 Gy = 2.4; 95% CI: 0.5–5.6 for mortality and incidence data, respectively).

244. From 1949 to 1962, the former Soviet Union conducted 133 atmospheric nuclear tests at the Semipalatinsk test site in Kazakhstan [B44, R31]. Local fallout was particularly high from tests carried out in 1949, 1953, and 1962. Approximately 10,000 persons living near the test site and 40,000 living in the Altai region in the Russian Federation were exposed to over 250 mSv effective dose. Effects on the health of populations living near Semipalatinsk in Kazakhstan and in the Altai region are currently being studied. An excess of benign and malignant thyroid tumours has been reported for the Kazakhstan population [B44, R31]. It is expected that new data from the ongoing studies in both Kazakhstan and the Russian Federation will become available soon.

245. Between 1944 and 1957, the Hanford Nuclear Site in Washington State, United States, released 20–25 PBq of ^{131}I into the atmosphere. In January 1999, the results of the Hanford Thyroid Disease Study were released to the public [D29]. In total, 5,199 people born between 1940 and 1946 in seven counties in eastern Washington State were identified for study. Ninety-four percent were located, 4,350 (84%) were alive, and 3,441 (66%) agreed to participate in the study. Study participants provided information on place of residence, consumption of milk and other relevant foods, occupational history, selected lifestyle factors, and medical history. Thyroid doses were estimated for the 3,193 study participants who had lived near Hanford at the time of atmospheric releases based on individual characteristics, e.g. level and type of milk consumption and dosimetry information from the Hanford Environmental Dose Reconstruction project. The other 248 participants had moved from the Hanford area and were considered to have received no exposure. The mean and median doses were 186 mGy and 100 mGy, respectively. The distribution of dose was skewed (range 0 to 2,840 mGy), with a high percentage of participants having low doses and only a small percentage having high doses. Each participant was evaluated clinically by two study physicians. The examination included ultrasound, thyroid palpation, and blood tests. Eleven categories of thyroid disease, ultrasound-detected abnormalities, and hyperparathyroidism were evaluated in terms of estimated ^{131}I radiation dose to the thyroid.

246. A total of 19 participants were diagnosed with thyroid cancer and 249 with benign thyroid nodules. No evidence of a dose-response relationship was found for malignant or benign nodules or any of the other outcomes studied. The final report is yet to be published, and there has been criticism of the large degree of uncertainty in the dose estimates. Nevertheless, the results do not provide evidence that ^{131}I doses on the order of 100 mGy increase the risk of developing thyroid neoplasia.

247. Although some animal studies have suggested that ^{131}I may be less carcinogenic than external radiation [N5], a large study of rats found similar carcinogenic effects for ^{131}I and external radiation [L11]. The strain of rat used has a high rate of developing follicular thyroid carcinomas, yet Royal [R22] noted that the study is particularly relevant, since it was well designed and the rats were the equivalent of young adolescents at the time of exposure and were exposed to low as well as moderate and high radiation doses. In summary, the very limited human data on childhood exposure to ^{131}I and adult exposure to external radiation are insufficient for concluding that there are significant differences between these types of radiation with regard to thyroid cancer induction.

3. Internal high-LET exposures

248. Radium is primarily a bone seeker, and the development of thyroid cancer has not been associated with exposure in most studies, but a statistically significant elevated risk, based on a small number of cases, was observed among radium dial painters in the United States who worked before 1940 [P27] and among patients treated with ^{224}Ra in Germany [N4].

249. Neither thyroid cancer mortality nor incidence was elevated among Danish thorotrast patients [A5]. Radon exposure in mines did not increase the risk of thyroid cancer mortality in a pooled analysis of 11 studies of underground miners [D8].

4. Summary

250. The thyroid gland is highly susceptible to the carcinogenic effects of external radiation during childhood. Age at exposure is an important modifier of risk, and a very strong tendency for risk to decrease with increasing age at exposure is observed in most studies. Although thyroid cancer occurs naturally more frequently among women, the ERR does not appear to differ significantly for men and women. Among people exposed during childhood, the ERR of thyroid cancer is highest 15–29 years after exposure, but elevated risks persist even 40 years after exposure. The carcinogenic effects of ^{131}I are less well understood. Most epidemiological studies have shown little risk following a wide range of exposure levels, but almost all of them looked at adult exposures. Recent results from Chernobyl indicate that radioactive iodine exposure during childhood is linked to thyroid cancer development, but the level of risk is not yet well quantified.

M. NON-HODGKIN'S LYMPHOMA

251. Non-Hodgkin's lymphoma (NHL) is a collection of distinct disease entities that are malignant expansions of lymphocytes. The lymphomas that make up this grouping can generally be separated into those with B-cell or T-cell lineage. The precise definition of NHL has varied over time; a recent classification that is widely used is the Revised European American Lymphomas classification [H42].

252. Rates of NHL have increased in many countries over the past few decades, particularly at older ages [H39]. In part this is likely to be due to changes in the definition of NHL and to improved ascertainment, although these factors are unlikely to explain all of the increases [H39]. Epidemiological studies have shown associations with chronic immunosuppression, for example, among transplant recipients and other patients who received immunopressive therapy [H43, K26]. Associations with certain viruses, such as Epstein-Barr [M37] and HIV [S44], have also been identified. Some studies suggest elevated risks for those employed in agriculture,

particularly those working with pesticides (e.g. [C31]), although other studies have not shown such a link (e.g. [W15]).

1. External low-LET exposures

253. Information on incidence and mortality from NHL following external exposure to low-LET radiation is presented in Table 18. As can be seen from this Table, the results are mixed, with many of the studies listed having failed to show a statistically significant association with radiation exposure. The Life Span Study of survivors of the atomic bombings falls into this category, although Preston et al. [P4] reported some evidence of an increasing dose response for males (p=0.04) but not for females, among whom, if anything, the trend is negative. The latter findings might appear to contradict those for the cervical cancer patients, where there is borderline evidence of a positive dose response; however, among exposed patients, there was little indication of an increasing trend with increasing dose [B1]. Furthermore, studies of women treated for benign gynaecological disorders [D7, I6] have not suggested associations with radiation. Comparison of the Life Span Study findings for males with those findings for the ankylosing spondylitis patients might be informative, given that most of these patients were male. Weiss et al. [W1] reported that NHL mortality among spondylitis patients was raised significantly compared with national rates (relative risk = 1.73; 95% CI: 1.23–2.36), and that this elevated risk appeared to disappear more than 25 years after exposure; however, no dose-response analysis was performed. In another study of a mostly male population, Cardis et al. [C11] did not find an association between NHL and external radiation among nuclear industry workers, although the precision of the study was limited by the generally low doses. The same limitation affected a study of diagnostic x-ray procedures [B39], which also did not show an association when based on a two-year lag; however, this study used numbers of x-ray procedures rather than doses.

254. The Life Span Study also provided no evidence that any elevated risk would be greater for exposure in childhood than in adult life [P4]. There are few other data on childhood exposure. The study of Swedish children treated for benign lesions in the locomotor system [D12] showed rates of NHL incidence and mortality similar to national values, although no dose-response analyses were reported.

2. Internal low-LET exposures

255. There are few data that allow examining the risks of NHL specifically in relation to internal low-LET radiation. The data that are available are for groups with medical exposures to ^{131}I (see Table 18). Among over 35,000 patients with diagnostic exposures, Holm et al. [H27] reported an SIR of 1.21. This value was not significantly different from 1 (at the 5% level), although the SIR of 1.24 for all lymphomas was significantly raised. However, while total cancer risk was

analysed in relation to level of the activity of iodine administered, no results were reported for NHL. Furthermore, doses to this cohort were generally very small (mean bone marrow dose = 0.19 mGy). Doses were higher in a study of Swedish patients treated for hyperthyroidism [H23]. In this instance, the observed number of cases was less than expected from national rates, significantly so after omitting the first 10 years of follow-up (SIR = 0.40, although based on only seven cases). Again, however, NHL incidence was not analysed in relation to level of exposure. Ron et al. [R14] studied NHL mortality among hyperthyroidism patients in relation to estimated bone marrow dose from ^{131}I therapy; most of the patients were from the United States but for this analysis some patients from the United Kingdom were included. There was no evidence of a trend in risk with dose, although the generally low doses limited the precision of the analysis [R14].

3. Internal high-LET exposures

256. There is limited information on NHL risks among groups exposed internally to high-LET radiation. Relevant findings are summarized in Table 18. Among German patients who received thorotrast, van Kaick et al. [V8] reported 15 cases among 2,326 patients, which represented a relative risk of about 2.5 compared to a group of unexposed patients. However, there was no analysis in relation to the level of exposure. Among thorotrast patients in Denmark [A5] and ankylosing spondylitis patients in Germany treated with ^{224}Ra [W3], the numbers of cases were too small to permit detailed inferences. Larger numbers arose in the combined analysis of radon-exposed miners [D8]; here the total number of deaths observed was, if anything, less than that expected from national and regional rates (SMR = 0.80, 95% CI: 0.56–1.10), but no analysis was conducted according to the level of exposure.

4. Summary

257. Results from studies of NHL risk among groups exposed to external low-LET radiation are mixed. The Japanese atomic bomb survivors as a whole do not show an association, although there is some evidence of an increasing trend in incidence with dose among males (but not females). Findings from other studies are variable, with no clear consistency. Overall, there is little evidence of an association between NHL and external low-LET radiation.

258. There is limited information on NHL risk in relation to internal low- or high-LET radiation. The general absence of analyses in relation to level of exposure and the limited statistical precision of one such analysis that was conducted hinders interpretation of the data that are available.

N. HODGKIN'S DISEASE

259. Hodgkin's disease is distinguished from other lymphomas mainly by the presence of giant Reed-Sternberg cells [B34]. While changes over time in the classi-

fication of Hodgkin's disease are likely to have had some effect on analyses of trends in rates, there are indications from various countries of a slight decrease in incidence rates [H39]. More pronounced decreases have been seen in mortality rates during recent decades, reflecting improved treatment [H39]. Internationally, incidence rates tend to be much higher in North America and Europe than in Asia [P5] (see also Table 1). Clustering of cases of Hodgkin's disease has been reported in some studies (e.g. [A17]), and a viral origin has been suggested by associations with certain childhood environments, such as small family size and uncrowded conditions, that could reduce or delay infections [G18]; Epstein-Barr virus has been cited as possibly being relevant [M36].

1. External low-LET exposures

260. The studies of external low-LET radiation included in Table 19 have not always reported estimates of trend based on dose-response analyses but have, at least in some instances, indicated whether there were any statistically significant trends with dose. For the Japanese atomic bomb survivors, Preston et al. [P4] found no evidence of a dose response, although the confidence intervals were fairly wide owing to the small number of cases (see Table 19). Studies of patients treated for benign gynaecological disease [I6] and of nuclear workers [C11] also showed no trend with dose, although based on small numbers of deaths in the former instance and low doses in the latter. For the other studies of external low-LET exposure listed in Table 19, the observed number was sometimes greater than the number expected, although not to a statistically significant extent.

2. Internal low-LET exposures

261. There are few data that allow examining the risks of Hodgkin's disease specifically in relation to internal low-LET radiation. The data that are available concern groups with medical exposures to ^{131}I (see Table 19). Among over 35,000 patients with diagnostic exposures, Holm et al. [H27] reported an SIR of 1.35. This value was not significantly different from 1 (at the 5% level), although the SIR for all lymphomas, 1.24, was significantly raised. However, while total cancer risk was analysed in relation to the activity of iodine administered, no results were reported for Hodgkin's disease. Furthermore, doses to this cohort were generally very small (mean bone marrow dose = 0.19 mGy). Doses were higher in a study of Swedish patients treated for hyperthyroidism [H23]. However, the small number of cases observed, while consistent with national rates, limited inferences. Furthermore, the incidence of Hodgkin's disease was not analysed in relation to level of exposure [H23]. Ron et al. [R14] studied Hodgkin's disease mortality among hyperthyroidism patients, mostly from the United States, in relation to estimated bone marrow dose from ^{131}I therapy. There was no evidence of a trend in risk with dose, although the small number of deaths and the generally low doses limited the precision of the analysis [R14].

3. Internal high-LET exposures

262. Relevant findings are summarized in Table 19. Studies of German [V8] and Danish [A5] thorotrast patients, while not indicating elevated risks, are based on very small numbers of cases. A combined analysis of radon-exposed miners [D8] reported an SMR for Hodgkin's disease of 0.93 (95% CI: 0.54–1.48), but the 17 deaths were not analysed in relation to level of exposure.

4. Summary

263. While dose-response analyses have not always been performed in the relevant studies and the numbers of cases have sometimes been fairly small, the available data do not indicate an association between Hodgkin's disease and radiation, either for external or internal exposures.

O. MULTIPLE MYELOMA

264. This group of conditions consists of plasma cell malignancies, which include Waldenstrom's macro-globulinaemia as well as multiple myeloma [H48]. It is more common among men than women and is rare, particularly at young ages [C23]. Mortality rates have been increasing during the past few decades in various countries, but this increase has largely been confined to older ages and may be due in large part to earlier incompleteness in ascertainment [C23]. Some case-control studies have indicated associations between myeloma and employment in agriculture or in the food industry [B30, B31, C24].

1. External low-LET exposures

265. Table 20 contains information on multiple myeloma following exposure to external low-LET radiation. Of particular note is the discrepancy between the findings for mortality and incidence among the Japanese atomic bomb survivors. The most recent mortality follow-up [P9], in common with an earlier analysis of mortality in this population [S3], showed a statistically significant associa-tion between myeloma risk and dose. However, data on myeloma incidence yield a much lower estimate for the trend in risk with dose; furthermore, it is consistent with there being no effect of dose [P4]. The authors of the incidence report noted that the mortality findings appeared to be heavily dependent on the inclusion of questionable diagnoses and on both second primaries and cases above 4 Gy that were excluded from the incidence analysis [P4]. In view of the care taken to review the myeloma diagnoses in the incidence analysis, it seems reasonable to place greater weight on these findings.

266. Results from the other studies of external low-LET exposure cited in Table 20 are mixed. Some, e.g. the international study of cervical cancer patients [B1], provide no evidence of an elevated risk. On the other hand, Darby et al. [D7] reported a significant elevated risk of myeloma

mortality among metropathia patients in the United Kingdom, although there was less evidence of an association from a similar study in the United States [I6]. The number of myeloma deaths among ankylosing spondylitis patients in the United Kingdom was significantly greater than that expected from national rates but was not analysed in relation to dose [W1]. An international study of cancer mortality among nuclear workers found a significant association with dose [C11], although this finding was influenced strongly by just a few cases with doses above 0.4 Sv. In a study of diagnostic x rays, Boice et al. [B39] found that the risk of myeloma incidence was similar among those who had and those who had not received x rays under two health plans; however, there was some evidence of an increasing trend in risk with an increasing number of x-ray procedures, although actual dose estimates were not available.

267. It is noticeable that of the studies of external low-LET exposures listed above and in Table 20, those that suggest an elevated risk of myeloma tend to be studies of mortality, in contrast to the few studies of incidence. Indeed, in common with the atomic bomb survivors study, the Swedish study of treatment for benign lesions of the locomotor system indicated an elevated risk of mortality relative to national rates, but not of incidence [D12]. It is unclear whether these findings might be due to differential recording of myeloma on death certificates, based on knowledge of prior radiation exposure. However, in view of the greater accuracy in diagnoses of incident cases of myeloma, inferences from incidence data are likely to be more sound.

2. Internal low-LET exposures

268. There are few data on multiple myeloma risks in relation to internal low-LET radiation. In studies of Swedish patients with exposure to ^{131}I for diagnostic purposes [H27] and as treatment for hyperthyroidism [H23], the observed numbers of incident cases were close to those expected from national rates. However, the risk of myeloma was not analysed in relation to level of exposure. Indeed, the bone marrow doses were generally low in the two studies (means of 0.19 mGy and 60 mGy, respectively). Ron et al. [R14] studied myeloma mortality among hyperthyroidism patients, mostly from the United States, in relation to estimated bone marrow dose from ^{131}I therapy. Although the estimated trend was greater than zero, it was not significantly different from zero (p=0.3); the small number of cases and the generally low doses limited the precision of the analysis [R14].

3. Internal high-LET exposures

269. Relevant findings are shown in Table 20. There is some evidence of an excess of myeloma incidence among Danish thorotrast patients, relative both to national rates and to an unexposed control group [A5], although based on only four cases. Among German thorotrast patients, van Kaick et al. [V8] reported ten cases of plasmacytoma among 2,326 patients, which represented a relative risk of about 4.1 compared to a group of unexposed patients.

Among patients in Germany treated with ^{224}Ra, two plasmacytomas were cited in the Spiess study [S14], and one medullary plasmacytoma was reported in the study of Wick et al. [W20]. In the combined analysis of radon-exposed miners [D8], there was an indication of elevated mortality from myeloma relative to national and regional rates, although the difference was not significant (SMR = 1.30; 95% CI: 0.85–1.90); however, the risk of myeloma was not analysed in relation to level of exposure. For a subgroup of these miners, namely uranium miners in western Bohemia, Tomášek et al. [T16] reported a statistically significant positive trend in myeloma risk with increasing cumulative radon exposure, but based on only three deaths. Similarly, while there was a statistically significant excess of multiple myeloma deaths among radium dial workers in the United States (SMR = 2.79; 95% CI: 1.02–6.08), this was based on only six deaths, and the risk did not appear to be related to internal radium body burden [S16].

4. Summary

270. Several mortality studies have indicated an increasing trend in the risk of multiple myeloma with increasing dose from external low-LET radiation. However, such associations are not generally apparent in studies of myeloma incidence, even for groups (such as the atomic bomb survivors) where the corresponding mortality data point towards an elevated risk. This suggests that the classification of myeloma on death certificates may have been conducted differentially, according to whether there was a past radiation exposure, although it is difficult to be certain. Given the generally better quality of diagnoses recorded in incidence data, the findings from the atomic bomb survivors, in particular, would suggest that there is little evidence of an association with low-LET radiation.

271. There is limited information on internal low- and high-LET exposures. Some studies have suggested an elevated risk, but based on small numbers of cases.

P. LEUKAEMIA

272. Although one of the rarer cancers, leukaemia is of particular interest because there is substantial information, both epidemiological and experimental, on the effects of ionizing radiation. In terms of its general epidemiology, it can be seen from Table 1 that the variation in rates between different populations is not as great as for most solid tumours. In considering trends and aetiological factors, it is important to take account of the various subtypes of leukaemia and their different age-specific rates. Modern classifications of leukaemia and other lymphatic and haematopoietic malignancies (e.g. [B32]) are based on cytogenetic and molecular principles that do not always coincide with the International Classification of Diseases. Three main subtypes will be considered here: acute lymphatic leukaemia (ALL), which is a leukaemia of precursor cells of either B-cell or T-cell origin; acute myeloid leukaemia (AML), whose lineage and subtype are

generally defined according to the FAB system [B32]; and chronic myeloid leukaemia (CML), whose predominant haematological feature is an elevated white cell count in the peripheral blood and which is characterized cytogenetically by the Philadelphia chromosome [L52]. Reference will also be made to chronic lymphatic leukaemia (CLL), which has a B-cell or a T-cell lineage [L52].

273. Most leukaemia cases in childhood are ALL, whereas CML and CLL make up a high percentage of cases in adulthood. In the case of childhood ALL, the most striking and consistent trend in different countries since 1950 has been the decline in mortality [K1], reflecting the introduction of effective chemotherapy and cranial radiotherapy. Childhood ALL incidence, in contrast, has been fairly constant or has perhaps shown a small increase over the same period [D2]. Apart from ionizing radiation, risk factors for childhood ALL include alkylating chemotherapeutic agents and genetic factors such as Down's syndrome. Greaves [G5] suggested that the increase in rates during this century would be consistent with many acute lymphatic leukaemias in children being due to delayed exposure to childhood infections. Kinlen suggested, however, that a specific infective agent (or agents) underlies childhood leukaemias, as is true for several animal leukaemias [K1].

274. For adult leukaemia, rates at ages 75–84 years have increased in several countries since 1950 [K1]. These trends are consistent with improvements in cancer registration and in the detail of death certification. Ionizing radiation, benzene, and cytotoxic agents are known causes of leukaemias in adults; there is also some evidence that cigarette smoking is a risk factor, particularly for myeloid leukaemia [K1].

275. Information on the induction of leukaemia by the irradiation of laboratory animals was reviewed in the UNSCEAR 1977 and 1986 Reports [U5, U7]. A variety of lymphatic and myeloid leukaemias have been induced in different animals, although with differing dose-response relationships. However, studies of myeloid leukaemia in mice are consistent in showing a lower risk for a given total dose when exposure to low-LET radiation is protracted rather than acute [U3].

1. External low-LET exposures

276. Risk estimates for leukaemia are presented in Table 21. For the Life Span Study of atomic bomb survivors, only the leukaemia incidence results are shown, because larger numbers are involved relative to the corresponding mortality data [P9]and because the diagnoses of the incident cases have been reviewed [P4]. In the review in the UNSCEAR 1994 Report [U2], it was concluded that the incidence of acute leukaemias or of chronic myelogenous leukaemia exhibits strong associations with exposure to external low-LET radiation. In contrast, several large studies of groups with medical exposures (e.g. [B12, C9, C10, W2]) show no association between radiation and CLL. Although the Life Span Study of atomic bomb survivors also fails to show an association with CLL, the medical studies provide much stronger evidence,

owing to the low baseline rates in Japan. Furthermore, for leukaemia other than CLL, the temporal pattern of radiation-induced risks differs between exposures in childhood and adulthood, although in both instances the minimal latency period is less than for most solid cancers. Further data on the modifying effects of age and time have become available from the extended mortality follow-up of the atomic bomb survivors [P9]. These data show that, whereas both the ERR and EAR decrease soon after exposure in childhood, the decline in risk tends to be less pronounced for exposures in adulthood. Additional information on temporal trends comes from studies of medical exposures in adulthood, such as therapy for cervical cancer [B12], cancer of the uterine corpus [C10] (in this cohort, a considerable number of women were exposed at ages over 65 years), benign gynaecological disease [I6], and ankylosing spondylitis [W2]. The first and last of these studies found the ERR to decrease substantially about 10 years after exposure. However, as in the study of patients with benign gynaecological disease, most of the evidence for this decrease related to CML, whereas the ERR for acute leukaemia (principally AML) was more stable with time since exposure. These results are generally in accord with findings for CML and AML incidence among the atomic bomb survivors [P4], as confirmed by a parallel analysis [L47] of these data in combination with data from the cervical cancer [B12] and ankylosing spondylitis [W2] studies. The combined analysis showed some evidence overall of a decrease in the ERR for AML with increasing time since exposure, but to a lesser extent than for CML [L47]. In connection with this, it can be noted that acute leukaemias formed the majority of the non-CLL leukaemias in the study of uterine corpus patients, for whom there was no clear trend in ERR with time since exposure [C10].

277. Interpretation of the dose-response relationships in studies of groups exposed in adulthood to at least several gray is complicated by the effect of cell killing at high doses. The degree of partial-body irradiation, fractionation, and dose rate may also be relevant, while there is some suggestion (although

based on small numbers) that, for example, the joint effect on leukaemia risk of total-body irradiation and chemotherapy may be more than additive [C9]. Table 30 presents results from modelling of the dose response for leukaemia (other than CLL) in four large, well conducted studies with individual dosimetry. These studies are based on patients treated for cervical cancer, uterine corpus cancer, and ankylosing spondylitis, plus the Japanese atomic bomb survivors. The latter study accords with a linear-quadratic dose response over the range 0–3 Gy, such that the risk per unit dose at low doses is lower than at higher doses. Most of the evidence for this non-linearity arises for AML [P4]. However, a parallel analysis of the atomic bomb data and those from the cervical cancer and ankylosing spondylitis studies [L47] showed that the data for CML and ALL were also consistent with a curvilinear dose response over doses less than 1 Gy (see Figure III). At doses above 3–4 Gy, the risk per unit dose subsequently decreases. This effect is seen at lower doses in the three studies of medical irradiation listed in Table 30. However, while it appears to be particularly strong among the ankylosing spondylitis patients (whose exposures were from x rays given in fractions) and those uterine cancer patients who received brachytherapy (radium implants) alone, it was weaker for the cervical cancer patients, most of whom received a mixture of brachytherapy and external radiation. In addition, Table 30 shows that the estimated ERR at 1 Gy is reasonably similar in the studies of the atomic bomb survivors, the ankylosing spondylitis patients, and the uterine corpus cancer patients given brachytherapy only, but higher than the ERR at 1 Gy for the cervical cancer patients or for the uterine cancer patients treated with external radiation. The risk estimates included in Table 21 from three other large studies of medical exposures in adulthood, namely of breast cancer patients [C9], patients treated for benign lesions in the locomotor system (e.g. arthrosis and spondylosis) [D12], and patients treated for benign gynaecological disease [I6], are also variable, although they are lower than the risk estimates for the atomic bomb survivors.

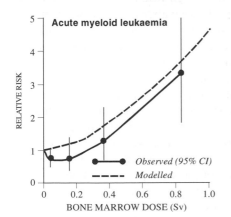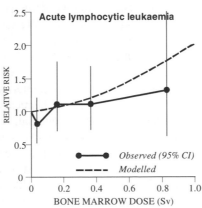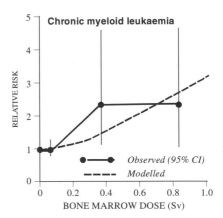

Figure III. Observed and modelled relative risk of acute myeloid, acute lymphocytic and chronic myeloid leukaemia in a combined analysis of data for the Japanese atomic bomb survivors, women treated for cervical cancer, and patients treated for ankylosing spondylitis [L47].
The values are specific to an attained age of 50 years, after exposure at 25 years, and depict the dose-response at doses less than 1 Sv.

278. Reconciling these results is not straightforward. The differing results on the effect of external irradiation and lower-dose-rate brachytherapy make it difficult to explain the findings solely on the basis of dose rate. The possible effect of errors in assessing bone marrow doses should also be borne in mind. However, one potential explanation relates to the degree of partial-body irradiation. Most of the marrow doses for the cervical cancer and uterine corpus cancer patients were to the pelvis, sacrum, and lower lumbar vertebra only. However, a subgroup of the externally irradiated women in the uterine corpus study who received substantial doses to the bone marrow in both the central trunk of the body and the pelvic marrow (as did the ankylosing spondylitis patients) had a greater risk (relative risk = 5.5; 95% CI: 2.0–15.1) than women with more non-uniform exposures (relative risk = 1.90; 95% CI: 1.1–3.2) (p-value for difference = 0.04) [C10]. Furthermore, the estimated ERR per Gy from the study of Swedish patients treated for benign lesions in the locomotor system [D12], in which exposures of the bone marrow were highly non-uniform, appears from Table 21 to be lower than that from the studies of more uniform exposure, although no confidence interval for the former estimate was given. Another important factor concerns differences between leukaemia subtypes. In a parallel analysis of the atomic bomb survivor, cervical cancer, and ankylosing spondylitis studies, Little et al. [L47] showed that there were statistically significant study-to-study differences in the model fitted to data for all leukaemia other than CLL, but that the models fitted to AML, ALL, and CML separately were consistent across the studies. Therefore differences between studies in range of ages at exposure and length of follow-up may explain at least some of the variation in the observed risks.

279. Occupational studies have the potential to provide information on how dose rate influences the risk of leukaemia. In spite of leukaemia being one of the less common cancers, the high ERR per unit dose and the often shorter induction time relative to many other cancers means that the comparison of leukaemia risks among radiation-exposed workers with the risks in groups such as the atomic bomb survivors may well be informative. The UNSCEAR 1994 Report [U2] drew attention to the reports of an association between leukaemia and radiation exposure among workers at the Mayak facility in the Russian Federation, some of whom received substantial bone marrow doses from external gamma irradiation several decades ago [K7]. It can be seen from Table 21 that most of the evidence for an elevated leukaemia risk relates to workers at the radiochemical plant [K10]. Koshurnikova et al. [K11] quoted a preliminary lifetime radiation risk coefficient for men who started work at this plant before 1954 that was similar to that given by ICRP [I1] for workers, although no confidence interval was given for the former value. In interpreting these findings, it should be borne in mind that 10% of the cohort had been lost to follow-up as of the end of 1994, although the cause of death is known for 97% of deaths [K32]. Also, bone marrow doses from plutonium have yet to be calculated for these workers, although they are likely to be lower than those from external gamma radiation [K32].

280. In contrast to the radiation doses received by early Mayak workers, occupational exposures received in various countries in recent years have tended to be low. As a consequence, studies of small groups of such workers have tended to produce varying results, reflecting their low statistical power to detect small increases in risk. To obtain greater statistical precision, it is therefore necessary to assemble as large a cohort with as long a follow-up as possible. In Japan a cohort of nearly 115,000 nuclear industry workers was identified [E3], but it could be followed-up for at most five years, limiting the inferences that could be drawn about leukaemia risks. More powerful information was derived from an international combined study of nuclear industry workers in Canada, the United Kingdom, and the United States [I2, C11]. This study was based on a cohort of over 95,000 workers with individual dosimetry for external radiation and over 2 million person-years of follow-up. As indicated in Table 21, the total number of leukaemias is larger than in many of the other studies listed but the mean dose is lower. Analysis of mortality from leukaemia (other than CLL) showed a statistically significant increasing trend in risk with dose. The central estimate of risk per unit dose corresponded to 0.59 times the value estimated from the atomic bomb survivors based on a linear dose-response model and 1.59 times the value based on a linear-quadratic model fitted to the bomb survivor data; the corresponding 90% confidence interval ranged from about zero up to four times the value from the linear-quadratic bomb survivor model. The evidence for a trend with dose was particularly strong for CML, as has also been reported in a large study of workers in the United Kingdom [M46, L20], some of whom were included in the international study.

281. Several points should be noted when interpreting the results of the international worker study [C11, I2]. First, the statistical significance of the trend in the worker data is based largely on a few cases with cumulative doses above 400 mSv [S24]. Dose-response analyses restricted to lower doses do not show a significant trend, although the estimated trend from these analyses is compatible with that from the full analysis [C11, C13]. However, the small total number, nine, of estimated excess leukaemias should be noted. Secondly, much of the evidence for a trend is based on workers at a reprocessing plant at which there could have been internal exposures not only to radionuclides but also to chemicals. However, excluding workers judged to have potentially received more than 10% of their dose from exposure to neutrons or from intakes of radionuclides did not affect the trend with dose. Indeed, while inspection of the point estimates of the ERR per Sv from the various facilities included in the study might suggest variability [S24], the findings are statistically consistent [C13]. This again reflects the limited statistical precision of the individual studies. Thirdly, dosimetry is an important consideration in a study that draws on data from different countries and in which dosimetry practices have varied over time. A dosimetry committee assembled for the purposes of this study judged that the dose estimates were generally compatible, although bone marrow doses may

have been overestimated by about 20%, implying a slight underestimation of the risk per unit dose [C11]. In addition, as pointed out previously, random errors in ascertaining doses are more likely to bias risk estimates towards the null than away from it. To conclude, this international study of radiation workers is valuable in addressing the risks associated with low-dose and low-dose-rate exposures, and additional investigations of this type should help to reduce uncertainties further.

282. Workers who took part in the recovery operations following the Chernobyl accident often received doses of 0.1–0.2 Gy, i.e. greater than those currently being received by many nuclear industry workers but lower than those of the early Mayak workers. The study of recovery operation workers from Estonia lacked statistical precision, owing to the small cohort and limited follow-up; indeed no leukaemia cases were identified, although one had been expected from population rates [R20]. Further information has been reported from studies of much larger numbers of workers from the Russian Federation. As previously noted in Section I.A, the interpretation of these findings depends on the type of comparison. Ivanov et al. [I13] cited an excess of leukaemia among these workers relative to rates for the general Russian population; however, in a case-control analysis based on comparisons among recovery operation workers, no significant correlations with dose or other aspects of their work were found [I14] (see Table 21). This difference is likely to be due to differences in methodology; in particular, to a probable bias in the cohort study [B27].

283. The study of natural background radiation in the Yangjiang area of China did not show a statistically significant association with leukaemia over all ages [T25, T26] (see Table 21). A subgroup analysis based on an earlier follow-up suggested an excess of leukaemias in the first year of life, but based on very small numbers (three observed compared with 0.4 expected from population rates) and on mortality rather than incidence data [A11]. A small study in Italy found no positive association between adult myeloid leukaemia and levels of background gamma radiation measured in homes, in contrast to earlier suggestions of such an association based on geological inferences of the natural radiation dose levels [F7].

284. Information on the incidence of leukaemia among people living near the Techa River was considered in the UNSCEAR 1994 Report [U2]. Both external and internal exposures were received by these individuals. As indicated previously, these investigations are potentially important sources of risk estimates, particularly for leukaemia. Emigration from this area, the possibly confounding effects of toxic chemicals around the Techa River, and the reconstruction of individual doses are issues pertinent to realizing this potential. Kossenko et al. [K5] noted that the fraction of the total number of deaths due to leukaemia in the Techa River cohort is slightly higher than the corresponding fraction in the Life Span Study. While this result may reflect differences between the cohorts in the

level and type of exposure, the inclusion in the former cohort of leukaemias identified from a wide range of sources may also have influenced the finding [K5]. Studies relating to contamination as a result of the Chernobyl accident are addressed in the Section on internal exposures, although again, both internal and external exposures were received. In contrast, doses to persons exposed to nuclear weapons test fallout in southwestern Utah in the United States were mainly from external radiation. The study of this group, which was discussed in the UNSCEAR 1994 Report [U2], found an association between bone marrow doses from fallout and leukaemia mortality [S17]. This association was restricted primarily to acute leukaemia before the age of 20 years following the period of highest exposure, although the indication of a similar level of risk for CLL in adults may suggest caution in interpreting these findings.

285. Findings from another study at low doses were reported by Boice et al. [B39], who undertook a case-control study of diagnostic x-ray procedures. Relative to persons for whom no such procedures were recorded within two health plans, the relative risk of leukaemia (other than CLL) associated with diagnostic x rays was 1.42 (95% CI: 0.9–2.2), based on a two-year lag. There was no significant trend in risk with the number of procedures, although individual estimates of organ doses were not available [B39].

286. Information on the risks of leukaemia and other cancers from irradiation *in utero* was summarized in the UNSCEAR 1994 Report [U2]. The topic is considered in more detail in Annex G, *"Biological effects at low radiation doses"*. Briefly, various case-control studies of childhood cancer, including leukaemia, have shown elevated relative risks associated with obstetric x-ray examinations of pregnant women of the order of 1.4–1.5 [D17]. Although the relative risk from the Oxford Survey of Childhood Cancers in the United Kingdom, in particular, has high statistical precision, concerns have been raised, most recently by Boice and Miller [B41], about the possibility of bias and confounding. Several of these points, for example, the apparent disparity between the findings of case-control and cohort studies and the similarity of the relative risks for leukaemia and other cancers, have been considered by Doll and Wakeford [D17]. With respect to these specific points, Doll and Wakeford cited problems with some of the cohort obstetric x-ray studies, and noted that the cells that give rise to most childhood cancers other than leukaemia persist and are capable of dividing for only a short time, if at all, after birth [D17]. The doses received in the studies of obstetric x rays are somewhat uncertain, but the mean values are likely to have been 10–20 mGy. It is notable that studies of childhood leukaemia following another type of obstetric examination, namely ultrasound, have not shown elevated risks; for example, a recent national case-control study in Sweden using prospectively assembled data on prenatal exposure to ultrasound reported relative risks close to 1 [N16]. The other main source of information on leukaemia following *in utero* irradiation comes from atomic bomb survivors exposed *in utero*. Delongchamp et al. [D14] have reported some evidence of elevated leukaemia mortality in this group relative to controls in the period from October 1950 to May

1992, although based on only two deaths. No additional leukaemia cases were reported in an earlier analysis of cancer incidence [Y1]. In contrast to survivors exposed in childhood, there was no increasing trend in leukaemia risk with dose among the *in utero*-exposed survivors, owing to the absence of deaths at high doses [D14]. Indeed there were no leukaemia deaths at ages less than 15 years among those exposed *in utero*. However, the low statistical precision associated with the small numbers in this group should be noted. Overall, the available evidence points to an elevated leukaemia risk from *in utero* irradiation, although there is uncertainty over its magnitude.

287. Some studies of childhood leukaemia in relation to paternal preconception irradiation were also mentioned in the UNSCEAR 1994 Report [U2]. Although a case-control study in West Cumbria in the United Kingdom [G7] suggested an association between paternal preconception irradiation and leukaemia in the offspring of workers at the Sellafield plant, this finding was specific to workers in the village of Seascale near Sellafield and was not seen among the offspring of other Sellafield workers with similar preconception doses [H6]. Furthermore, the paternal preconception irradiation result was not replicated in subsequent studies of the children of radiation workers in Scotland [K2] or Canada [M16], and no leukaemia excess has been observed among offspring of the atomic bomb survivors [Y2]. A large study found an elevated risk of leukaemia in the children of nearly 120,000 male radiation workers in the United Kingdom compared with other children (relative risk = 1.83; 95% CI: 1.11–3.04); however, no association was found between leukaemia risk and levels of paternal preconception irradiation [D24]. In a study based on a cohort of nearly 40,000 children of male nuclear industry employees in the United Kingdom, which included workers in the just-mentioned study [D24], the incidence of cancer was found to be similar to national rates [R29]. In this instance, the only suggestion of an elevated risk was based on three cases with total preconception doses of at least 100 mSv [R29], two of which had already been reported in the study in West Cumbria [G7]. In reviews of this topic, Little et al. [L18] and Doll et al. [D10] concluded that the inconsistency not only with the other epidemiological data but also with experimental data makes it highly unlikely that the association observed at Seascale represents a causal relationship.

2. Internal low-LET exposures

288. A study of leukaemia incidence among nearly 47,000 patients in Sweden given ^{131}I for thyroid cancer, hyperthyroidism, or diagnostic purposes [H12], mostly in adulthood, was considered in detail in the UNSCEAR 1994 Report [U2] (see Table 21). Although there was no evidence of an association between bone marrow dose and leukaemia in this study, this may reflect a lack of statistical power associated with the generally low doses (mean 14 mGy). Ron et al. [R14] studied leukaemia mortality among hyperthyroidism patients, mostly from the United States, in relation to estimated bone marrow dose from ^{131}I therapy. There was no evidence of a trend in risk with dose, either for leukaemia excluding CLL (see Table 21) or CLL alone, although the generally low doses (mean of 42 mGy) limited the precision of this analysis [R14]. Statistical precision was even more of a concern in a study of thyroid cancer patients in France [D18], for which, even though the mean bone marrow dose was similar in magnitude (34 mGy), the cohort of 1,771 patients was much smaller than in the aforementioned studies. Although no leukaemias were observed in the French study, the number expected from national rates was only 1.28 [D18].

289. The European Childhood Leukaemia-Lymphoma Incidence Study (ECLIS), set up to monitor trends in rates following the Chernobyl accident, has examined data up to the end of 1991 from 36 cancer registries in 23 countries, including Belarus and parts of the Russian Federation [P12]. This is a geographical correlation study, in which doses and risks have been assessed for geographical areas rather than on an individual basis. As pointed out earlier, this approach may give rise to methodological problems and is not suitable for deriving risk estimates, although it does permit a general description of disease rates. The latest report from ECLIS found an overall increase in age-standardized rates of childhood leukaemias during 1980–1986, which continued at about the same rate during 1987–1991 [P12]. No correlation was found with the geographical distribution of effective dose due to fallout from the accident, based on values published in the UNSCEAR 1988 Report [U4]. In view of the very low bone marrow doses received in most of the areas studied (generally less than 1 mSv), this finding is not surprising. Indeed, to have any hope of detecting very small elevated risks, large studies such as this are required. In contrast, smaller studies often give variable results. For example, Petridou et al. [P15] reported an elevated risk of infant leukaemia in Greece among those *in utero* at the time of or soon after the Chernobyl accident. However, not only was this finding based on a subgroup analysis involving only 12 cases diagnosed in the first year of life, but it is inconsistent with the results of obstetric x-ray studies [U2]. Other small studies, such as those in Finland [A6], Sweden [H22], and Romania [D26], have not shown an association between childhood leukaemia and Chernobyl fallout. In Germany, Michaelis et al. [M30] reported an increased risk of infant leukaemia among those *in utero* at or soon after the time of the accident relative to those born at other times. However, this increase was, if anything, highest in those regions with the lowest levels of contamination, and the authors concluded that *in utero* exposure was not a cause of the elevated risk [M30, S53]. A study in Belarus [I22] has shown that the relative risk for infant leukaemia, while it is greater than 1, is not elevated to a statistically significant extent and is lower than the corresponding values from the studies in Germany [M30] and Greece [P15]. The issue of infant leukaemia following the Chernobyl accident is being examined further using the much larger ECLIS database [P25].

290. While much of the dose to those in western Europe from the Chernobyl accident arose from external exposures, internal exposures may have been more important closer to Chernobyl. Ivanov et al. [I5] reported similar rates of acute

leukaemia among children in areas of Belarus with varying levels of radionuclide contamination. Furthermore, in an analysis of aggregated data from contaminated areas of Belarus, the Russian Federation, and Ukraine, Prisyazhniuk et al. [P16] showed that while age-adjusted leukaemia rates rose from 1980 to 1994, this trend appeared to be similar for the periods before and after the Chernobyl accident; also, rates were similar in areas with different levels of contamination.

291. There has been interest in recent years in reports of cancer clusters in the vicinity of nuclear installations. Many of these reports were considered in the UNSCEAR 1994 report [U2] (see also [M34]). In the United Kingdom, excesses of childhood leukaemia have been reported around some nuclear sites, in particular, the Sellafield [C28] and Dounreay [C29] reprocessing plants. However, environmental assessments suggest that these findings are unlikely to be attributable to radioactive release from the sites. Indeed, while exposures associated with these sites often comprise a mixture of external and internal low-LET and internal high-LET exposures, they have been assessed to be generally less in total than exposures from natural radiation [C28, C29]. Elsewhere, studies in, for example, the United States [J1], Canada [M35], France [H40], western Germany [K25], and Japan [I15], have tended not to show excesses of cancer around nuclear installations, specifically of childhood cancer and/or leukaemia in some instances. Some exceptions have been reported; for example, Wing et al [W14] cited an excess of leukaemia around the Three Mile Island nuclear power plant in the United States. However, as indicated earlier in relation to lung cancer, Hatch et al. [H37, H38] interpreted their original analysis of these data as not providing convincing evidence of an association with the very low doses resulting from radiation emissions from the plant.

292. It should be borne in mind that inferences from studies around nuclear installations are limited by their geographical nature, the very small doses involved, and, as around some of the United Kingdom sites, for example, the relatively small numbers of cases. There are also difficulties in interpretation with the differing analyses performed; for example, with respect to age (0-4, 0-14, or 0-24 years), diagnostic category (leukaemia, leukaemia and NHL, all cancers), time period, and proximity to the installation. When many different analyses are performed, it would not be surprising to obtain a statistically significant finding, i.e. one that would arise 1 in 20 times by chance alone. The unavailability of data can also present a problem; for example, the ascertainment of childhood leukaemias may be incomplete owing to a lack of national incidence data [H40], or small-area data may not be available, e.g. as in parts of the United States studied by Jablon et al. [J1]. Some of these problems can be addressed through case-control or cohort studies, which collect data at the individual level. As mentioned earlier, the case-control approach has been valuable in addressing the issue of paternal preconception irradiation [D24]. However, difficulties can still arise in this type of study. For example, Pobel and Viel [P7] suggested an association between childhood leukaemia and the use of beaches around the La Hague reprocessing plant in France. However, this result was dependent on a small number of cases, relied on the recall of habits stretching back several decades, and involved multiple comparisons [C30, L38]. Furthermore, no such association was found around the Sellafield plant in the United Kingdom [G7].

3. Internal high-LET exposures

293. It has been suggested that uptake of radon by fat cells in the bone marrow might lead to irradiation of the haematopoietic stem cells [R10], and there have been some indications from geographical correlation studies, based on large-area data, of an association between radon exposure in dwellings and leukaemia [H14]. However, this suggestion has not been replicated in geographical studies using small-area data and more refined analyses [M13, R6]. More weight might be given to a large case-control study in the United States by Lubin et al. [L34] that collected data on an individual rather than a geographical basis (see Table 4 for details). No evidence was found of an association between acute lymphoblastic leukaemia in childhood and individual assessments of indoor radon exposure. In particular, for time-weighted average radon concentrations in excess of 148 Bq m^{-3}, the relative risk compared with concentrations of less than 37 Bq m^{-3} was 1.02 (95% CI: 0.5-2.0) based on matched case-control pairs [L34]. A study of childhood acute myeloid leukaemia in the United States [S52] and smaller studies of childhood cancers in Germany [K38] and of acute myeloid leukaemia in Italy [F7], all of which involved measurements of radon in homes, also did not show associations with leukaemia risks overall.

294. To test for any association between radon and the risk of cancers other than lung cancer in a study with individual dosimetry, Darby et al. [D8] performed a collaborative analysis of data from 11 cohorts of underground miners. Further details of the component cohorts are given in Table 4, and Table 21 contains some results for leukaemia. The combined cohort was very large (over 64,000 men) with over 1 million person-years of follow-up. There was an excess in mortality from leukaemia of all types relative to national or regional rates within 10 years of first employment (SMR = 1.93; 95% CI: 1.19-2.95, based on 21 deaths). However, restricting the analysis to the period when the 8th and 9th revisions of the International Classifications of Diseases were in operation, so that leukaemia subtypes could be distinguished, there was no evidence of an elevated level of leukaemia other than CLL (SMR = 1.28; 95% CI: 0.51-2.64, based on seven deaths). The leukaemia subtype with the highest SMR was acute myeloid leukaemia (SMR = 2.42; 95% CI: 0.51-2.64), although based on only three deaths. Perhaps of greater interest than SMRs were analyses in relation to cumulative radon exposure which, although based on small numbers, showed no trend in the risk of all leukaemia, leukaemia excluding CLL, or AML. More than 10 years after first employment, there was no indication of an elevated SMR, either for all leukaemias or specific subtypes. The possibly elevated SMR in the earlier period may well be due to chance, since agents encountered in mines, such as diesel

fumes and arsenic, are thought not to be leukaemogens, and the levels of gamma radiation in mines, although not always known, are likely to be too low to explain this result [D8]. The study therefore provides evidence that high concentrations of radon in air do not cause a material risk of leukaemia mortality.

295.   An elevated level of leukaemia, in particular, myeloid leukaemia, was reported in studies of thorotrast patients in Germany [V8], Denmark [A4, A5], Portugal [D31], and Japan [M14]. Results from these studies are summarized in Table 21. Andersson et al. [A4] showed that the risk of AML and myelodysplastic syndrome increased in relation to cumulative dose, having taken account not only of the amount injected but also the time since injection. There was also a suggestion of a cell-killing effect at high doses, although this was not statistically significant. Based on a mean bone marrow dose of 1.3 Gy (high-LET), Andersson et al. derived a risk estimate for these diseases of 1.7 10^{-2} Gy^{-1}. While this suggests that the RBE of alpha radiation relative to low-LET radiation may be lower than the value of 20 recommended by ICRP [I1], it should be noted that the latter value was chosen to apply at low doses, rather than that at the high doses in this study. Furthermore, there are uncertainties in the risk estimate derived by Andersson et al. [A4] owing to the relatively small number of cases and imprecision in the estimation of individual doses. Hunacek and Kathren [H33] compared published risk coefficients with values determined from dose rates based on post-mortem radiochemical analysis of tissues from a thorotrast patient. Using results from thorotrast studies in Germany, Japan, and Portugal (but not Denmark), they obtained a leukaemia risk coefficient of 3.2 10^{-2} Gy^{-1} [H33], which is somewhat higher than that calculated by Andersson et al. [A4]. However, the former value is likely to be incorrect, owing to an error in calculating bone marrow dose rates based on data for the total skeleton. Furthermore, new dose calculations indicate that the bone marrow dose had been previously underestimated and that the risk per unit dose had been overestimated [I25].

296.   There is some evidence for an excess of leukaemia among patients injected with ^{224}Ra [N4, W20], with one of these studies [W20] indicating an excess more than 30 years after the first injection of ^{224}Ra. However, inferences are restricted by the generally small number of cases and the absence of dose-response analyses. Among radium dial workers in the United States, the number of leukaemias observed was close to that expected in the general population [S7]. Although it has been suggested that the cases among pre-1930 dial painters arose early [S54], the small numbers in both this analysis and an analysis by bone marrow dose [S7] limit the interpretability of these data.

297.   No leukaemias have been observed in the offspring of Danish thorotrast patients [A13]. Although this study was based on a small cohort, its statistical power was enhanced by the high doses to the testes from alpha radiation (mean dose = 0.94 Sv).

4. Summary

298.   There is a substantial amount of information on the risks of leukaemia from radiation exposure. This reflects the high relative increase in risk compared with other cancer types and the temporal pattern in risk, with many of the excess leukaemias occurring within about the first two decades following exposure, particularly among those irradiated at young ages. There are some differences between the Life Span Study of atomic bomb survivors and some large studies of medically exposed groups in estimates of both the magnitude of the radiation risk and the shape of the dose response for external low-LET exposure. These findings may reflect differences between studies in the uniformity of exposure to the bone marrow and in the degree of fractionation and protraction of exposure, as well as differences in the pattern of risk between leukaemia subtypes. There is clear evidence of non-linearity in the dose response for leukaemia, which has a slope that decreases at lower doses.

299.   A large international study of radiation workers suggested an elevated leukaemia risk, although the results were compatible with a range of values. Case-control studies of prenatal x rays indicate an increased risk of leukaemia in childhood due to *in utero* irradiation, although the absence of a dose-related increase in the sparse corresponding data for atomic bomb survivors adds uncertainty to the magnitude of the risk. Epidemiological evidence does not suggest that irradiation prior to conception gives rise to a material risk of childhood leukaemia.

300.   The data available on internal exposures to low-LET radiation do not indicate elevated risks of leukaemia; this may well reflect the low statistical precision associated with generally small doses. There is no convincing evidence of an increased risk of leukaemia due to environmental exposures associated with the Chernobyl accident, although investigations are continuing. Excesses of childhood leukaemia have been reported around some nuclear installations in the United Kingdom, but generally not in other countries; these excesses are based on small numbers of cases and have not been explained on the basis of radioactive releases from the installations. Dose-related increases in leukaemia risk have been seen among patients with large exposures to high-LET radiation arising from injections of thorotrast, a diagnostic x-ray contrast medium. There is less evidence for elevated risks among patients injected with ^{224}Ra and little or no evidence for increased risks among radium dial workers or from studies with individual assessments of radon exposure, either in mines or in homes.

IV. LIFETIME RISK FOR TOTAL CANCER

301. In Chapter III the focus was on risks for specific cancer sites. The aim in this Chapter is to develop risk estimates for total cancer, in line with previous assessments of the Committee, most recently in the UNSCEAR 1994 Report [U2]. Many of the issues associated with producing such estimates were discussed in Chapters I and II. However, some of them are summarized here, together with points that are germane to total cancer risks.

302. The estimation of total cancer risks is in some ways easier than the estimation of risks for specific cancer sites. The most notable difference is the larger number of cancers available from epidemiological studies of all cancers. This means that the statistical precision of estimates based on such data should be greater than the precision for specific cancers. On the other hand, heterogeneity in risks between cancer types may counterbalance this. Indeed, "cancer" is a multitude of different diseases with different aetiologies.

303. As an example of an analysis based on a collection of cancer types, Figure IV shows estimates of the ERR per Sv for various types of solid cancer in survivors of the atomic bombings based on the most recent mortality data [P9]. These values have been adjusted for age at exposure and gender. Pierce et al. [P9] noted that the variation in the ERR per Sv between cancer sites is not statistically significant. However, they cautioned that this should not be taken as substantial evidence that the ERR per Sv is the same for all sites, given the differences in aetiology for different cancer types. Furthermore, the ERR is only one scale of representation, and the EAR per Sv should also be considered. However, because the baseline rates vary for different types of solid cancer, the EAR per Sv is likely to vary much more widely between cancer types than the ERR per Sv.

304. The increased statistical precision associated with an analysis of all solid cancers assists in the development of risk models. In particular, it may be possible to detect variations in risk with factors such as age, time, and gender that are not apparent in data for specific cancer sites. For example, Figure V summarizes models fitted to data on mortality from all solid cancers for the Japanese atomic bomb survivors [P9]. This indicates variations in the ERR per Sv and EAR per Sv with gender, age at exposure, and attained age that may not be evident in analyses for specific cancers. However, it should be recognized that such models might be affected by differences between cancer types in the pattern of risk. On the other hand, as previously mentioned in Section I.E, analyses conducted separately for various cancer sites may yield differences in trends in risk with, for example, age and/or time, simply as a consequence of chance variations. One possibility, suggested by Pierce and Preston [P6], is to analyse data for various cancer sites, or groupings thereof, in parallel. This may allow the development of models for which the level of the relative risk, for example, differs between cancer types but under which the variation with factors such as age and time is the same across cancer types.

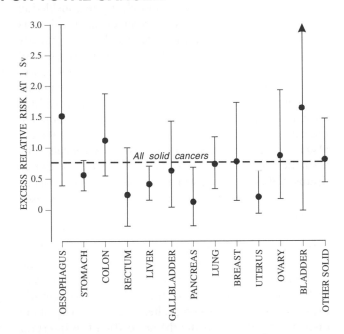

Figure IV.   Excess relative risk (and 90% CI) for mortality from specific solid cancers and all solid cancers together (horizontal line) in survivors of the atomic bombings, standardized for females exposed at age 30 years [P9].

305. Related to the above considerations is the issue of whether one or more data sets should be used to estimate total cancer risks. Artefactual differences might arise if different data sets are used for different cancer types. However, this should be balanced against the quantity of data for a particular cancer type that is available from a given study. Indeed, some studies, such as case-control studies, have focused on only one or a few cancer sites and therefore cannot be used by themselves to estimate total cancer risks.

A. EXPRESSIONS OF LIFETIME RISK

306. There is some confusion in discussions and presentations of lifetime risks associated with radiation exposure. To simplify matters, the following discussion is restricted to mortality, and it is assumed that there are two causes of death: "cancer" and "non-cancer". However, the discussion can be generalized to deal with incident cases and multiple causes, any number of which may be affected by exposure.

307. The obvious definition of a lifetime risk is simply the difference between the proportion of people dying of cancer in an exposed population and the corresponding proportion in a similar population with no radiation exposure. This difference is called the excess lifetime risk (ELR). Formal mathematical expressions for the ELR and related quantities are given by Thomas et al. [T18].

308. While the ELR is of some value, it provides an incomplete summary of the effect of exposure on a population. This can be seen most clearly by considering death from any cause as the outcome of interest. In this case the ELR must be zero, since all people will eventually die of something, even if radiation changes the risk of death. However, the ELR is also misleading for cause-specific mortality. If exposure has the same relative impact on death rates for all causes, then cause-specific ELR estimates will be zero. In contrast, suppose that radiation increases the risk of death for cancer by some fraction but also increases the risk of death from non-cancer causes by a smaller amount. In this instance, the ELR for cancer deaths will be positive, while that for non-cancer deaths will be negative, even though radiation exposure increased non-cancer death rates.

309. One way to address problems with the ELR is to consider how exposed and unexposed populations differ with respect to the expected age at death for all causes or for specific causes of death. However, average life expectancies (or more comprehensive summaries of the distribution of ages at death) are difficult to interpret without a clear understanding of the general pattern of death rates in a population. In particular, what might be considered fairly large increases in death rates are associated with rather small changes in life expectancy. For example, based on death rates in the United States for 1985, a 50% increase in all-cause mortality for 20-years old would reduce their life expectancies by about three years. As another illustration of the problem with changes in life expectancy, consider a situation in which an exposure reduces life expectancy from 75 to 25 years for 1% of the population, for example, as a consequence of leukaemia following an exposure of 1 Gy. In this instance, the average life expectancy for the population would be reduced by 6 months. In general, changes in life expectancy are not a particularly useful summary of the exposure effects. To be useful, loss of life expectancy (LLE) should be related to some measure of the number of people whose life expectancy was affected by the exposure.

310. A useful alternative to the ELR can be developed by considering the (cause-specific) death rate defined by the difference in death rates for exposed and unexposed populations as an additional cause of death that has been introduced into a population. Technically this difference is not a rate function, since it would assume a negative value if exposure had a protective effect. However, by treating the difference as a rate, one can compute the fraction of deaths attributable to this "new" cause of death or the probability that an individual will die from a cancer associated with the exposure. This quantity has been described in [U2, T18] as the risk of exposure-induced death (REID). In contrast to the ELR, the REID is positive if exposure increases death rates and negative if exposure decreases death rates. Furthermore, cause-specific values of the REID are zero for any cause for which the rates are not affected by exposure.

311. The values called excess deaths in recent analyses of the atomic bomb survivor data (e.g. [P9]) are closely related to the REID. In particular, the Life Span Study excess deaths are the sum of REID estimates over the follow-up period, having allowed for gender, age at exposure, and dose, with population background rates determined by the experience of the cohort. The REID estimates presented later in this Chapter are computed using background rates from populations other than the Life Span Study, and they estimate the number of excess deaths for lifetime follow-up after exposure.

312. As in the UNSCEAR 1994 Report [U2] and recent Life Span Study reports, the quantity LLE divided by the REID, which can be thought of as the change in life expectancy per attributable case, provides a helpful summary of the impact of exposure on life expectancy. In the example given earlier, in which 1% of the population is affected, the change in life expectancy per attributable case is 50 years (i.e. 0.5 years/0.01), which is in line with expectations.

313. Mortality computations are, in principle, relatively straightforward; further details were given in the UNSCEAR 1994 Report [U2]. Gender-specific, age-dependent baseline total- and cause-specific death rates for the populations of interest are used to define the baseline survival probability function. For a given age at exposure, gender-specific excess rates for causes affected by radiation are added to the appropriate cause-specific baseline rates to give the cause-specific and total rates for the (hypothetical) exposed population. Conditional on age at exposure, these adjusted total rates define the age-specific survival probability in the exposed population. The risk measures of interest, namely, the ELR, REID and LLE, can be computed from these conditional survival probabilities and cause-specific disease rates.

314. For lifetime incidence computations, the gender- and age-specific survival probabilities for the unexposed populations are replaced by cancer-free survival probabilities. These functions are computed from gender- and age-specific rate functions defined as the sum of the total non-cancer death rate and the total cancer incidence rate. The total non-cancer death rate is defined as the difference between the total death rate and the total cancer death rate.

B.   METHODS AND ASSUMPTIONS OF CALCULATIONS

315. The results presented here are derived from cause-specific attributable risks and the loss of life expectancy per attributable case in five populations: China, Japan, Puerto Rico, the United States, and the United Kingdom. Lifetime mortality risks are computed for the following cancers: oesophagus, stomach, colon, lung, liver, female breast, bladder, other solid cancers, and leukaemia, as well as all other (non-cancer) causes. For incidence, radiation effects on the risk of thyroid cancer are also considered. In the computations presented here, it is assumed that all organs receive the same dose. If exposure is limited to a single organ,

risks for that organ would be only slightly larger than the organ-specific risks discussed below. Even in a whole-body exposure it will be the case that different organs receive different doses; however, the differences in dose-specific risks between those from the joint computa-tion and those from a computation based on the actual doses to each organ will not be large. For example, in a situation in which the breast receives a dose of 1 Gy and the stomach a dose of 0.8 Gy, estimates of the breast cancer risk following a whole-body 1 Gy exposure and of the stomach cancer risk following a 0.8 Gy whole-body exposure will be good approximations to the actual organ-specific risks.

316. Risk estimates for mortality are also given for Chinese and Puerto Rican populations. These estimates make use of life-table and death-rate information given by Land and Sinclair [L12]. In these instances, the computations were carried out in terms of three "causes": non-cancer deaths, non-leukaemia cancer deaths, and leukaemia deaths.

317. Primary results are given for uniform whole-body exposures of 0.1 and 1 Gy for men and women exposed at 10, 30, or 50 years of age. These results depend on the following factors, each of which are discussed briefly below:

(a)   the exposed population for which risk estimates are developed, and the models used to describe the excess risks in this population;

(b)   the models used to describe risks at low doses;

(c)   the method used to extend the excess risk models beyond the period of observation for the population from which these models were developed;

(d)   the cause-specific mortality (or incidence) rates and the age structure of the populations to which the rates are applied;

(e)   the methods used to transport excess risks based on models for one population to another population; and

(f)   the method used to allow for fractionation or dose-rate effects.

1. Risk models

318. As in the UNSCEAR 1994 Report [U2], the risk estimates derived in this Section are based on recent data on the experience of the atomic bomb survivors. The data from Life Span Study Report 12 [P9], which covers the period from 1950 through 1990, were used for the estimation of cause-specific mortality risks for solid cancers. Solid cancer incidence risk estimates are based on linking the Life Span Study survivor cohort and the Hiroshima and Nagasaki tumour registry data [M2] for 1958 through 1987 [T1]. The cause-specific solid cancer mortality and incidence rate models used here were developed specifically for these computations. The method used to estimate risks at low doses is discussed in detail below.

319. Radiation effects are often described by models for cause-specific death rates or hazard functions. The hazard at age a is defined formally in terms of the ratio of the probability of dying from the cause in a short interval (a,

a+l) to the length of the interval (l), given that one is alive at a. The hazard function in the absence of radiation exposure will be called the baseline hazard. It is reasonable to allow the baseline hazard, denoted as $h_0(a,s,p)$, to depend on gender (s) and calendar time period (p) in addition to age. One way to describe the effect of a radiation exposure is to consider the difference between the hazard function in the exposed population and the baseline hazard for this population. This difference is the excess absolute risk (EAR). The ratio of the EAR to the baseline hazard is the excess relative risk (ERR).

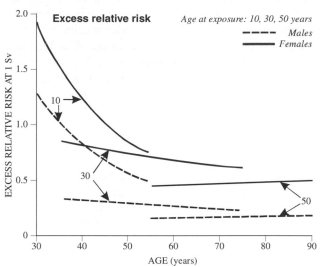

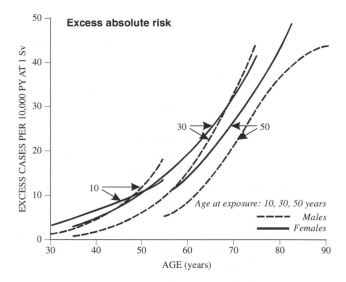

Figure V. ERR and EAR for solid cancer mortality among survivors of the atomic bombings in Japan [P9]. The lines show the patterns of risk in the data.

320. The leukaemia EAR model developed by Preston et al. [P4] was used to describe the effect of radiation on leukaemia risks in both the mortality and incidence computations. To allow for excess leukaemia risks during the first few years after exposure (about which the Life Span Study data provide no direct evidence), it is assumed that excess rates for the first five years are half of those seen five years after exposure.

321. Two types of ERR models were developed for solid cancers. These models are similar to those considered by Pierce et al. [P9] (see Figure V). In the first model, the

ERR depends on gender and age at exposure. For this age-at-exposure model, the cause-specific hazard rate has the form

$$h_d^{LSS}(a,s,e,\cdot) = h_0^{LSS}(\cdot)[1 + \beta\,\theta_s\,d\,\exp(\gamma e)]$$

322. In the model presented here, the baseline hazard, i.e. the hazard in the absence of radiation exposure, denoted here by $h_0(a,s,p)$, depends on age (a), gender (s), and calendar period (p). The dose-response slope is allowed to depend on gender and, for non-gender-specific cancers, is described in terms of the product of the slope for males (β_M) times a gender ratio parameter ($\theta_{F:M}$). Unless there was evidence of a significant lack of fit, the gender ratio and age-at-exposure (e) effects were assumed to be equal to those for all solid cancers as a whole. Lack of fit was defined as a deviance change [M38] of more than 4 for a single parameter or more than 6 for both parameters.

323. Under the second solid cancer risk model, the ERR depends on gender and attained age (i.e. age at death or cancer incidence, denoted by a) but not on age at exposure. For this attained-age model, the cause-specific hazard rate has the form

$$h_d^{LSS}(a,s,e,\cdot) = h_0^{LSS}(\cdot)[1 + \beta\,\theta_s\,d\,a^k]$$

in which the temporal variation in the ERR is modelled as a power function of attained age. As with the age-at-exposure model, the gender ratio and attained-age effects were taken to be equal to those for all solid cancers as a group unless there was evidence of a significant lack of fit.

324. The site-specific solid cancer ERR parameters for the mortality and incidence models are summarized in Table 31. For each site of interest and each model, the Table presents the gender-specific ERR per Gy estimates, together with the gender ratio (female:male) and the age-at-exposure and attained-age effects. The age-at-exposure effect is given as the percentage change in risk associated with a 10-year increase in age at exposure. The attained-age effect is the power of age. For the age-at-exposure model, the gender-specific estimates of ERR per Gy are for a person exposed at age 30 years. For the attained-age model, they are the ERR per Gy estimates for a person at attained age 50 years.

2. Low-dose response

325. The issue of cancer risks at low doses is discussed in detail in Annex G, *"Biological effects at low radiation doses"*. Among the points covered there are the minimum doses at which statistically significant elevated risks have been detected in epidemiological studies. As mentioned earlier in this Annex, the minimum doses for detectable effects depend on the statistical precision of the relevant study and can also be influenced by any potential bias in the study. While statistically powerful studies can allow effects to be detected at lower doses than small studies,

there will be some small doses at which it will not be possible to detect an elevated risk. It is difficult to specify values at which no study will be able to identify an effect, given that, for example, further follow-up of groups such as the Japanese atomic bomb survivors will continue to increase in statistical power and so aid the future investigation of low-dose risks. However, the results cited in Annex G, *"Biological effects at low radiation doses"*, give some idea about minimum doses at which elevated risks can be seen at present.

326. Pierce et al. [P9] reported a statistically significant increasing trend in mortality risks in the 0–50 mSv range for all solid cancers combined among the Japanese atomic bomb survivors, based on follow-up to 1990. However, they also noted that the interpretation of this finding is not straightforward, since it reflects an increasing risk per unit dose in the low dose range not seen for cancer incidence in the survivors [T1]. Observed cancer death rates are increased by about 5% for survivors with doses in the 20–50 mSv range, which is larger than the roughly 2% increase predicted at these doses by linear models fitted to the full dose range. Pierce et al. [P9] suggested that this difference might be due to differential misclassification of cause of death, i.e. a slight bias towards recording cancer rather than other causes on the death certificate for atomic bomb survivors who are known to have been relatively close to the hypocentre. This illustrates how potential biases, while small in absolute terms, can affect the interpretation of low-dose risks. Dose-response relationships for the atomic bomb survivors are discussed further below.

327. Several authors [C35, L40] have raised questions on the statistical support for the low-dose findings in [P9]. However, as indicated by Pierce et al. [P28, P36], the statistical result at low doses is quite robust, although as noted earlier, the relatively small effects in this dose range mean that small biases could distort the inferences about the low-dose response function.

328. Annex G, *"Biological effects at low radiation doses"*, refers to some other studies that provide information on minimum doses for detectable effects. It should be noted that it may be easier to detect elevated risks for particular types of cancer or in specific age-at-exposure groups for which, owing to low background rates, small absolute increases in rates may lead to large relative risks. For example, in a combined analysis of data from seven studies of thyroid cancer after external radiation exposure, Ron et al. [R4] found that a linear dose response provided a good fit to the data on childhood exposure, not only at high doses but also down to 0.1 Gy (low-LET). Annex J, *"Exposures and effects of the Chernobyl accident"*, also reviews studies of childhood cancer following irradiation *in utero* (see also Sections III.K and III.P of this Annex). The Oxford Study of Childhood Cancers shows elevated risks of childhood cancer following prenatal x-ray exposures with a mean dose of 10–20 mGy [D17]; however, concerns have been raised [M31] about the interpretation of this result and the consistency with the findings for the *in-utero*-exposed atomic bomb survivors [D14].

329. Analyses of data across a range of doses usually provide a statistically more powerful approach to considering risks at low doses than focussing on results for specific dose categories. Indeed, the latter approach may yield chance findings owing to multiple significance testing. The mortality risks for all solid cancers combined and for leukaemia among the Japanese atomic bomb survivors, over a wide range of doses, were illustrated in Figure XVIII of Annex G, "*Biological effects at low radiation doses*". Dose-response analyses of data on cancer incidence [P4, T1] and mortality [P9] for the atomic bomb survivors have recently been conducted by Little and Muirhead [L41, L42] and by Hoel and Li [H45] (see also Annex G). It should be noted that in contrast to Hoel and Li [H45], Little and Muirhead [L41, L42] took account of random errors in dose estimates. These analyses showed that for solid cancers, either individually or combined, the atomic bomb survivor data are consistent with a linear dose response and that incorporating a threshold into the dose-response model does not significantly improve the fit [L41, L42]. The only exception may be non-melanoma skin cancer incidence, for which there is some evidence of a threshold at about 1 Sv [L41]. A further analysis of the atomic bomb data by Little and Muirhead [L50] also took account of possible systematic errors in neutron dose estimates for survivors in Hiroshima. This analysis showed little evidence of upward curvature in the dose response for the incidence of all solid tumours combined over the range 0–4 Gy (low-LET); there was more suggestion of upward curvature over the range 0–2 Gy (low-LET), although this was not significant at the 5% level [L50]. For leukaemia, as has been noted previously [P4, P9], a linear-quadratic dose-response model (such that the risk per unit dose is smaller at low than at high doses) provides a significantly better fit than a linear model. However, there is some evidence from the leukaemia incidence data that incorporating a threshold (estimated to be 0.12 Sv; 95% CI: 0.01–0.28) provides a better fit than the linear-quadratic model alone (two-sided p=0.04) [L41]. On the other hand, there is less evidence for such a threshold based on the corresponding mortality data (two-sided p=0.16) [L42]. Since the estimates of relative risk at low dose are similar for the leukaemia incidence and mortality data, Little and Muirhead [L42] suggested that the difference in findings may be due to the finer division of dose groups in the publicly available mortality data than in the corresponding incidence data.

330. In view of the above, the calculations given below are based on linear dose-response models for solid cancers and on a linear-quadratic model for leukaemia. The form of the risk models was described in the preceding Section.

3. Projection methods

331. Generally speaking, the age-at-exposure and attained-age models describe the Life Span Study data equally well. However, as will be seen below, these models lead to different projections of risk beyond the current follow-up period for survivors exposed as children. For people exposed to the atomic bombings after age 50, little projection is needed since their follow-up is close to complete. In Figure V, it can be seen that for this group the ERR basically is constant over

time. For most sites, the age-at-exposure model assumes that ERRs for those exposed as children will remain at their current relatively high values throughout life; in contrast, the attained-age model assumes that ERRs will decline as the survivors get older. Thus, the two models correspond to different methods for projecting risks beyond the current follow-up.

332. In the UNSCEAR 1994 Report [U2], only age-at-exposure models were used, and various ad hoc (i.e. non-model-based) projection methods were used. One of those methods (constant ERR) is equivalent to the use of the age-at-exposure model, while the second method (constant ERR over the current follow-up, with risks declining in the future) is similar to the use of the attained-age model.

4. Populations and mortality rates

333. In Table 32, mortality and incidence estimates are given for five populations: China, Puerto Rico, Japan, the United Kingdom, and the United States. Cause-specific mortality rates for Japan, the United States, and the United Kingdom are based on 1985 national statistics in the three countries. Mortality rates for China and Puerto Rico are taken from Land and Sinclair [L12]. Data on cancer incidence rates were obtained from the current (7th) edition of Cancer Incidence in Five Continents [P5]. For the United States, data for the combined SEER registries were used. Japanese rates were computed as the unweighted average of rates for the Hiroshima, Nagasaki, and Osaka tumour registries. Rates from the Shanghai Cancer Registry were used for China.

334. Population age distributions were used to compute the population risks shown in Tables 33–37. The 1985 age distributions were used for Japan, the United States, and the United Kingdom. Estimates for China and Puerto Rico were based on the summary life-tables given by Land and Sinclair [L12].

5. Transport of risks between populations

335. For each risk model, two methods were used to transport site-specific solid cancer risks estimated for a Japanese population to populations of China, Puerto Rico, the United States or the United Kingdom. These methods will be called relative risk transport and absolute risk transport.

336. For the relative risk transport, the cause-specific hazard rate in the target population, T, was computed as the product of the baseline hazard in the target population and the (age-at-exposure or attained-age) ERR for the Japanese population, J:

$$h_{di}^{T}(a,s,e) = h_{0i}^{T}(a,s) [1 + ERR^{J}(d,s,a,e)]$$

337. For the absolute risk transport, the cause-specific hazard rate in the target population was computed as the

sum of the target-population baseline hazard and the EAR for the Japanese population:

$$h_{di}^T (a,s,e) = h_{0i}^T (a,s) + EAR^J (d,s,a,e)$$

Here the EAR function for the Japanese population was computed as the product of the appropriate ERR function for the model of interest (age-at-exposure or attained-age) and the corresponding Japanese baseline rate, namely

$$EAR^J (d,s,a,e) = h_{0i}^J (a,s) ERR^J (d,s,a,e)$$

For leukaemia, EARs were estimated directly in the survivors, and all transport was done using absolute rates.

6. Fractionation and dose-rate effects

338. Experimental and epidemiological information on cancer risks from fractionated or low-dose-rate exposure, relative to acute or high-dose-rate exposure, was reviewed in the UNSCEAR 1993 Report [U3] and is also considered in Annex G, *"Biological effects at low radiation doses"*. The UNSCEAR 1993 Report indicated that risks associated with low-dose or low-dose-rate exposures may be less than those from acute high doses by a factor of as much as 3. The Committee has not examined all new studies since 1993 to assess potential changes in the range of values. However, some recent information on this topic is provided in this Annex, for example, from the comparison of results from the acutely exposed Japanese atomic bomb survivors and from tuberculosis patients with fractionated x-ray exposures from fluoroscopies. For lung cancer, there is no indication of an elevated risk in the Canadian [H7] and United States (Massachusetts) [D4] fluoroscopy studies, unlike in the atomic bomb survivors [P9, T1]. However, the severity of tuberculosis may have affected the findings for lung cancer in these patients. For breast cancer, it has been suggested, based on comparison of the fluoroscopy and atomic bomb survivor findings, that fractionation may not affect risks [L39], although a different interpretation has been put on this finding [B33].

339. Further information on low-dose-rate occupational and environmental exposures has also become available in recent years and is summarized both in this Annex and in Annex G, *"Biological effects at low radiation doses"*. While it has been possible in some instances to find some evidence of an elevated risk (e.g. for leukaemia among nuclear industry workers [C11]), such studies do not currently have sufficient statistical power to allow those risks to be estimated with great precision. Furthermore, risk estimation based, for example, on groups in the former Soviet Union is sometimes complicated by exposures to both low- and high-LET radiation. Further investigation, including longer follow-up and more detailed analyses, may improve the estimation of risks from fractionated and low-dose-rate exposure. For the time being, however, the values for a reduction factor of less than 3 that were suggested in the UNSCEAR 1993 Report seem to be reasonable, notwithstanding the possibility of differences between some cancer types.

340. For leukaemia, the linear-quadratic dose-response model implies a reduction factor of 2 when extrapolating from acute high doses to low doses or low dose rates. It would, therefore, not be necessary to apply another reduction factor to the leukaemia results given for a dose of 0.1 Sv if the exposure was fractionated or protracted rather than acute. However, the results at 1 Sv for solid cancers could, tentatively, be reduced by a factor of 2 for fractionated or protracted exposures.

C. LIFETIME RISK ESTIMATES

341. The principal results of the calculation of lifetime risks are given in Table 33 for an acute whole-body dose of 1 Sv or 0.1 Sv. This Table presents solid cancer results for the two projection models (age-at-exposure and attained-age dependence of the ERR) and two risk transport models (ERR and EAR transport) for the five populations. As noted above, leukaemia risks always were based on an EAR model. The transport method makes little difference because non-CLL leukaemia rates are similar in the different populations; consequently, results are presented only for the EAR transport model.

342. For comparison, the estimates at 1 Sv for a Japanese population that were derived in the UNSCEAR 1994 Report [U2] are included in Table 33. The UNSCEAR 1994 estimates for the REID (10.9% averaged over gender) were based on an age-at-exposure model applied to Japanese rates and are generally comparable to the current estimates for solid cancers (11.2% averaged over gender). The 1994 leukaemia estimate of 1.1% averaged over gender is slightly higher than the current estimate of 0.9%. This difference arises because of slight differences in the leukaemia risk model and because, for the current computations, leukaemia was included as another "site" in a joint analysis of the impact of a whole-body exposure, while in 1994 leukaemia was considered separately from other causes (i.e. as if only the bone marrow had been exposed). The difference reflects the impact of increased hazards for the competing risks of radiation-associated solid cancers.

343. Although the solid cancer REID estimates are based on a linear dose-response model, the REID estimate for a dose of 0.1 Sv is slightly more than 10% of the estimate for a dose of 1 Sv. For example, considering solid cancer mortality in United States males using an attained age model and relative risk transport, the REID estimates for 1 and 0.1 Sv are 6.2% and 0.7%, respectively. This non-linearity reflects the effect of competing risks at lower doses vs. at higher doses. However, for these models, the REID estimates for solid cancers at lower doses are approximately linear in dose.

344. The use of the attained-age model leads to smaller lifetime risks for solid cancers than the corresponding age-at-exposure model. The reason for this can be seen in Table 34, which is based on a Japanese population. The persistence of high relative risks under the age-at-exposure model leads to large lifetime risks for those exposed as

children. For solid cancer mortality following exposure at age 10 years, the values of REID are 14% for men and 20% for women, while the corresponding gender-specific incidence risks are 31% and 37%, respectively. The attained-age model, which describes the current Life Span Study data as well as the age-at-exposure model, allows the relative risks for those exposed as children to decrease as they reach the ages of high cancer mortality or incidence. As a result, the estimated gender-specific solid cancer mortality and incidence risks following exposure at age 10 years are about half the values predicted by the age-at-exposure model. The population average lifetime risk estimates for the attained-age model are about 70% of those for the age-at-exposure model.

345. Some other measures of radiation detriment, based on mortality in a male Japanese population, are given in Table 35. As expected, the excess lifetime risk ELR is similar to the REID (i.e. the percentage of radiation-associated deaths) for leukaemia, but the former is less than the latter for all solid cancers. Furthermore, the excess lifetime risk is negative for non-cancer causes, since the sum of this measure over all causes must equal zero. The loss of life expectancy per attributable solid cancer death is similar under the attained-age and age-at-exposure projection models.

346. As indicated in Table 33, values of REID for solid cancer mortality in men are generally comparable for the Japanese, United Kingdom, and United States populations: about 9% with the age-at-exposure model and 6% with the attained-age model following a dose of 1 Sv. However, lifetime attributable risks for men in the Chinese and Puerto Rican populations are about 30% lower than those in Japan, the United Kingdom, and the United States. There is greater variability in female rates, but the same general pattern is seen in the magnitude of the risk estimates, with values of REID for Japan, the United Kingdom, and the United States being considerably greater than those for China and Puerto Rico. These differences reflect almost entirely differences in the lifetime probability of cancer mortality in these populations, as presented in Table 32, which in turn reflect the population-to-population variability in baseline rates.

347. Estimates of REID for women are consistently greater than those for men, largely reflecting gender differences in life expectancy and the contribution of breast cancer. REID estimates for cancer mortality in women exhibit greater sensitivity to both the risk projection model and the transport method than do those for men. This difference is primarily due to variations in breast cancer mortality between these populations.

348. Estimates of REID for cancer incidence are slightly lower for Japanese men (19% using the age-at-exposure model and 13% for the attained-age model) than for United States men (15% and 11% for the age-at-exposure and attained-age models, respectively), while estimates for men in the United Kingdom are somewhat higher (26% and 22%, respectively). These differences generally reflect differences in the baseline rates. Since lifetime baseline cancer incidence risks for China and Puerto Rico are more similar to those in Japan, the United Kingdom and the United States than are the corresponding mortality risks, the differences in incidence estimates of REID between these two countries and Japan, the United Kingdom, and the United States are not as marked as they are for mortality. REID estimates associated with relative risk transport tend to be larger for Western women than for Japanese women. This difference is due almost entirely to the higher breast cancer incidence and mortality in the United States and United Kingdom populations than in the Japanese population.

349. Tables 36 and 37 give detailed breakdowns by cancer type of estimates of REID risks for mortality and incidence, respectively, based on one of the above models, namely the attained-age projection model. When relative risk and absolute risk transport are compared for the populations of the United States and the United Kingdom, the main effect of using the latter rather than the former transport method is to reduce the REID estimates for women. This is due principally to reductions in the excess associated with breast and lung cancer, whose background rates are lower for Japanese women than for Western women. With this reduction, the differences in REID for the populations considered are less marked than under the relative risk transport.

350. Since the UNSCEAR 1994 Report [U2], further work has been undertaken to assess uncertainties in cancer risk estimates. In particular, NCRP report 126 [N17] assessed the uncertainty in the total fatal cancer risk for the United States population from external low-LET irradiation at low doses and low dose rates; it took account of the following factors:

(a) statistical uncertainties in the estimation of a risk factor, based on data for the Japanese atomic bomb survivors;

(b) possible bias due to over- or under-reporting of cancer deaths in the atomic bomb survivors;

(c) the effect of both random and systematic errors in dose estimates for the atomic bomb survivors;

(d) uncertainty in the method of transferring risks from Japan to the United States;

(e) uncertainty associated with the projection of risks over time, from the period of follow-up to a complete lifetime;

(f) uncertainty in the DDREF; and

(g) a subjective assessment of any remaining unspecified uncertainties.

351. Uncertainties associated with each of these factors were propagated using a Monte Carlo approach [N17]. For a United States populations of all ages and both genders, the mean value for the total cancer risk at low doses and low dose rates was estimated as 4.0×10^{-2} per Sv, with a 90% confidence interval of $1.2-8.8 \times 10^{-2}$ per Sv. The shape of the total uncertainty distribution was skewed towards higher values, as a consequence of which the median value (3.4×10^{-2} per Sv) was smaller than the mean. A sensitivity

analysis demonstrated that the main contributors to the total uncertainty were the DDREF (about 38% of the total), unspecified uncertainties (about 29%), and the transfer to the United States population (about 19%).

352. In a separate exercise supported by the United States Nuclear Regulatory Commission and the European Commission, uncertainties in cancer risk estimates were elicited from a series of experts [L27]. Using a formal analysis, the uncertainties provided by these experts were combined to obtain an overall distribution of uncertainty that took account of differences between the various subjective assessments. Table 38 shows the estimates of REID elicited

for an acute dose of 1 Gy (low-LET) to a hypothetical European Union/United States population of all ages and both genders, together with the associated 90% confidence interval. For all cancers combined, the limits of the confidence interval range about a factor of three higher and lower around the median of 10.2%. This represents a slightly wider interval than that arising from the NCRP analysis [N17]. For specific cancer types, the uncertainty intervals in the European Union/United States analysis are wider, in relative terms, than the interval for all cancers combined, sometimes ranging from several order of magnitudes lower than the median value up to about an order of magnitude higher [L27]. However, these ranges encompassed previous estimates of risk.

CONCLUSIONS

353. Since the Committee's assessment of the risks of radiation-induced cancer in the UNSCEAR 1994 Report [U2], more information has become available from epidemiological studies of radiation-exposed groups. Some of this information relates to populations exposed to acute doses of external low-LET radiation. For example, mortality data have been updated to the end of 1990 for 86,572 survivors of the atomic bombings at Hiroshima and Nagasaki. As of December 1990, 56% of the survivors were still alive, and it was estimated that 421 excess cancers deaths had occurred; 334 from solid cancer and 87 from leukaemia. Both this study and further follow-up of patients who received medical radiation exposure have provided additional data on cancer risks at long times following irradiation, particularly for those exposed at young ages. However, there are still uncertainties in the projection of risks from the current follow-up periods until the end of life, given that most of the people who were irradiated at young ages are still alive.

354. The increased statistical precision associated with the longer follow-up and the resulting larger number of cancers in the above studies has also assisted in the examination of dose-response relationships, particularly at lower doses. For example, the most recent data for the Japanese atomic bomb survivors are largely consistent with linear or linear-quadratic dose trends over a wide range of doses. However, analyses restricted solely to low doses are complicated by the limitations of statistical precision, the potential for misleading findings owing to any small, undetected biases, and the effects of performing multiple tests of statistical significance when attempting to establish a minimum dose at which elevated risks can be detected. Longer follow-up of large groups such as the atomic bomb survivors will provide more information at low doses. However, epidemiology alone will not be able to resolve the issue of whether there are dose thresholds in risk. In particular, the inability to detect increases at very low doses using epidemiological methods does not mean that the underlying cancer risks are not elevated.

355. New findings have also been published from analyses of fractionated or chronic low-dose exposure to low-LET radiation, although the statistical precision of these studies is low in comparison with high-dose-rate results from the atomic bomb survivors. Analyses of data for nuclear workers indicate that the risk of leukaemia increases with increasing dose, whereas no dose response has been established for solid cancers. A comparison of the atomic bomb survivors with patients who received fractionated x-ray exposures in the course of treatment for tuberculosis suggests that dose fractionation may not reduce the risk of breast cancer, although this interpretation has been questioned in view of the potential effects of radiation quality. It is difficult to arrive at a definitive conclusion on the effects of dose rate on cancer risks, since the relevant epidemiological data are sparse and the effects may differ among cancer types. For example, no elevated risk of lung cancer was observed in tuberculosis patients who received fractionated exposures, whereas a statistically significant elevated risk was found in the atomic bomb survivors; however, the severity of tuberculosis may have affected the results for these patients.

356. Information on the effects of internal exposure, from both low- and high-LET radiation, has increased since the time of the UNSCEAR 1994 Report [U2]. In particular, the early reports of an elevation in thyroid cancer incidence in parts of the former Soviet Union contaminated as a result of the Chernobyl accident have been confirmed and suggest a link with radioactive iodine exposure during childhood. Nevertheless, risk estimation associated with these findings is still complicated by difficulties in dose estimation and in quantifying the effect of screening for the disease. This topic is considered in further detail in Annex J, *"Exposures and effects of the Chernobyl accident"*. Other studies in the former Soviet Union have provided further information relevant to internal exposures; for example, on lung, bone and liver cancers among workers at the Mayak plant and, to a lesser extent, on cancers among the population living near the Techa River, in both instances

in the southern Urals. However, the different sources of radiation exposure (both external and internal) and, in the case of the Techa River studies, the potential effects of migration, affect the quantification of risks. Results from several case-control studies of lung cancer and indoor radon have been published in recent years that, in combination, are consistent with extrapolations from data on radon-exposed miners, although the statistical uncertainties in the findings from the indoor studies are still too large to determine a reliable risk estimate.

357. Particular attention has been paid in this Annex to risks for specific cancer sites. Again, the information that has become available in recent years has helped in the examination of risks. However, there are still problems in characterizing risks for some cancer sites, owing to the low statistical precision associated with relatively small numbers of estimated excess cases. This can limit, for example, the ability to estimate trends in risk in relation to factors such as age at exposure, time since exposure, and gender. Furthermore, data are sometimes lacking or have not been published in a format that is detailed enough to allow an assessment of how risks vary among populations. An exception is breast cancer, where a comparison of data on the Japanese atomic bomb survivors and women with medical exposures in North America points to an absolute transfer of risks between populations. For some other sites, such as the stomach, there are indications that a multiplicative transfer between populations would be appropriate, although the evidence is generally not strong. There are some cancer sites for which there is little evidence for an association with radiation (e.g. non-Hodgkin's lymphoma, Hodgkin's disease, and multiple myeloma). While the risk evaluations for lymphomas are affected by the small numbers of cases in several studies, these results should be contrasted with the clear relation found in many populations between radiation and the risk of leukaemia (excluding CLL), which is also a rare disease.

358. The results presented in Tables 33-37 illustrate the sensitivity of lifetime risk estimates to variations in background rates. These findings suggest that this variability can lead to differences that are comparable to the variations associated with the transport method or method of risk projection. Issues of uncertainty in lifetime risk estimates are discussed in more detail in NCRP report 126 [N17]. The variability in these projections highlights the difficulty of choosing a single value to represent the lifetime risk of radiation-induced cancer. Furthermore, uncertainties in estimates of risk for specific types of cancer are generally greater than for all cancers combined.

359. Despite these difficulties, risk estimates are of considerable value for use in characterizing the impact of a radiation exposure on a population. Using the same approach taken in the UNSCEAR 1994 Report [U2], namely an age-at-exposure model applied to a Japanese population of all ages, the lifetime risk of exposure-induced death from all solid cancers combined following an acute dose of 1 Sv is estimated to be about 9% for men, 13% for women and 11% averaged over genders. The calculations in this Annex show that these values can vary among different populations and with different risk models. Overall, however, the risk estimates are consistent with the value of 10.9% for an acute dose of 1 Sv cited in the UNSCEAR1994 Report [U2]. The uncertainties in the above estimates may be of the order of a factor of 2, higher or lower. The estimates could be reduced by 50% for chronic exposures, again with an uncertainty factor of 2, higher or lower. Using the attained-age model, the estimated lifetime risks of exposure-induced death are about 70% of those based on the age-at-exposure model. Total solid cancer incidence risks can be taken as being roughly twice those for mortality. Lifetime solid cancer risk estimates for those exposed as children might be twice the estimates for a population exposed at all ages. However, continued follow-up of existing irradiated cohorts will be important in determining lifetime risks. The experience of the Japanese atomic bomb survivors is consistent with a linear dose-response for the risk of all solid cancers combined; therefore, as a first approximation, linear extrapolation of the estimates at 1 Sv acute dose can be used for estimating solid cancer risks at lower doses. For specific types of solid cancer, the risks estimated in this Annex are broadly similar to those presented in the UNSCEAR 1994 Report [U2].

360. The computations in this Annex suggest that lifetime risks for leukaemia are relatively invariant to the population used, both because an absolute risk transport model was used and because baseline rates of leukaemia, other than CLL, are less variable among populations than are baseline rates of solid cancers. For either gender, the lifetime risk of exposure-induced leukaemia mortality can be taken as 1% following an acute dose of 1 Sv. This is similar to the value of 1.1% at 1 Sv cited in the UNSCEAR 1994 Report [U2]. Based on a linear-quadratic dose-response model, decreasing the dose tenfold, from 1 Sv to 0.1 Sv, would be expected to reduce the lifetime risk estimate by a fraction of 20. Thus, the lifetime risk of exposure-induced death for leukaemia can be estimated as 0.05%, for either gender, following an acute dose of 0.1 Sv. No further reduction for chronic exposures is necessary. The uncertainty in the leukaemia risk estimate may be on the order of a factor of 2, higher or lower.

Table 1
Examples of high and low cancer rates in various populations [a] [b]
[P5]

Site of cancer	Sex	High cancer incidence		Low cancer incidence	
		Population	Rate	Population	Rate
Nasopharynx	Males	Hong Kong Singapore (Chinese) United States, San Francisco (Chinese)	24.3 18.5 11.6	Canada, Nova Scotia United States, New Mexico (Non-Hispanic white) Ireland, Southern	0.1 0.2 0.2
	Females	Hong Kong Singapore (Chinese) Canada, Northwest Territories	9.5 7.3 5.1	United Kingdom, south-western Finland Norway	0.1 0.1 0.1
Oesophagus	Males	United States, Connecticut (black) Hong Kong France, Haut Rhin	20.1 14.2 14.2	Israel (non-Jews) Italy, Sicily (Ragusa Province) Thailand, Chiang Mai	0.5 1.0 2.3
	Females	India, Bombay China, Tianjin United Kingdom, Scotland, West	8.3 6.2 5.2	United States, New Mexico (American Indian) Spain, Tarragona Israel (non-Jews)	0.2 0.2 0.2
Stomach	Males	Japan, Yamagata China, Shanghai Italy, Romagna	95.5 46.5 39.3	United States, Atlanta (white) Israel (non-Jews) Thailand, Chiang Mai	5.2 6.8 7.5
	Females	Japan, Yamagata Italy, Romagna China, Shanghai	40.1 22.8 21.0	United States, Iowa Israel (non-Jews) Canada, Saskatchewan	2.2 3.2 3.7
Colon	Males	United States, Detroit (black) United States, Hawaii (Japanese) Japan, Hiroshima	35.0 34.4 31.6	India, Madras Thailand, Chiang Mai Peru, Trujillo	1.8 4.2 4.4
	Females	New Zealand (non-Maori) Canada, Newfoundland United States, Detroit (black)	29.6 28.1 27.9	India, Madras Thailand, Chiang Mai Singapore (Indian)	1.3 3.7 4.7
Liver	Males	Japan, Osaka China, Shanghai United States, Los Angeles (Korean)	46.7 28.2 23.9	Canada, Prince Edward Island Netherlands, Eindhoven United Kingdom, south-western	0.7 1.3 1.6
	Females	Japan, Osaka China, Shanghai Thailand, Chiang Mai	11.5 9.8 9.7	Australia, Tasmania Canada, Prince Edward Island India, Madras	0.3 0.3 0.5
Lung and bronchus	Males	United States, New Orleans (black) New Zealand (Maori) Canada, Northwest Territories	110.8 99.7 90.3	United States, New Mexico (American Indian) Peru, Trujillo India, Madras	10.3 11.9 12.6
	Females	New Zealand (Maori) Canada, Northwest Territories United States, San Francisco (black)	72.9 65.6 44.3	India, Madras Spain, Zaragoza Malta	2.4 2.7 3.4
Melanoma of skin	Males	Australia, New South Wales New Zealand (non-Maori) United States, Hawaii (white)	33.1 25.0 19.5	Japan, Osaka China, Shanghai India, Bombay	0.2 0.3 0.4
	Females	New Zealand (non-Maori) Australia, New South Wales Austria, Tyrol	29.8 25.7 15.6	Japan, Osaka China, Shanghai India, Bombay	0.2 0.3 0.3
Breast	Females	United States, Los Angeles (Non-Hisp white) United States, Hawaii (white) Israel (Jews born in Israel)	103.7 96.5 90.5	Thailand, Chiang Mai Israel (non-Jews) United States, Los Angeles (Korean)	14.6 21.3 21.4
Cervix	Females	Peru, Trujillo India, Madras Colombia, Cali	53.5 38.9 34.4	Israel (non-Jews) China, Shanghai Finland	3.0 3.3 3.6

Table 1 (continued)

Site of cancer	Sex	High cancer incidence		Low cancer incidence	
		Population	Rate	Population	Rate
Prostate	Males	United States, Atlanta (black)	142.3	China, Tianjin	1.9
		United States, Hawaii (white)	108.2	India, Madras	3.6
		Canada, British Columbia	84.9	Thailand, Chiang Mai	4.1
Bladder	Males	Italy, Trieste	38.7	United States, New Mexico	2.6
		Spain, Mallorca	36.4	(American Indian)	
		Switzerland, Geneva	32.5	United States, Hawaii (Hawaiian)	3.9
				Canada, British Columbia	11.3
	Females	Italy, Trieste	9.4	United States, New Mexico	0.6
		Denmark	7.7	(American Indian)	
		United Kingdom, Scotland, West	7.5	France, Isere	2.6
				United States, Hawaii (Filipino)	2.7
Brain, central nervous system	Males	Italy, Trieste	9.5	Singapore (Malay)	1.6
		Iceland	9.4	Japan, Yamagata	1.8
		United States, Hawaii (white)	8.7	Thailand, Chiang Mai	2.0
	Females	Italy, Trieste	8.7	United States, Los Angeles (Chinese)	1.1
		Poland, Warsaw city	5.9	India, Madras	1.1
		United States, Atlanta (white)	5.8	Japan, Yamagata	1.7
Thyroid	Males	Iceland	6.1	United Kingdom, Wessex	0.7
		United States, Hawaii (Filipino)	5.1	Estonia	0.7
		United States, Los Angeles (Filipino)	4.0	Denmark	0.8
	Females	United States, Hawaii (Filipino)	25.5	India, Madras	1.6
		United States, Los Angeles (Filipino)	11.2	United Kingdom, Yorkshire	1.7
		Italy, Ferrara	11.1	Netherlands, Eindhoven	1.9
Non-Hodgkin's lymphoma	Males	United States, San Francisco (non-Hisp white)	25.0	India, Madras	3.7
		Italy, Romagna	15.5	Thailand, Chiang Mai	3.8
		United States, Hawaii (white)	15.1	Singapore (Indian)	3.9
	Females	Italy, Ferrara	11.5	India, Madras	2.0
		Israel (Jews born in Israel)	11.1	China, Shanghai	2.5
		United States, San Francisco (Hispanic white)	11.0	Estonia	2.5
Hodgkin's disease	Males	United States, San Francisco (non-Hisp white)	4.3	China, Tianjin	0.3
		Italy, Veneto	4.0	Japan, Miyagi	0.4
		Israel (Jews born in Israel)	3.2	Singapore (Chinese)	0.5
	Females	United States, Connecticut (white)	3.6	Japan, Osaka	0.2
		Italy, Veneto	3.5	China, Shanghai	0.3
		Israel (Jews born in America or Europe)	3.1	Hong Kong	0.3
Multiple myeloma	Males	United States, Los Angeles (black)	9.5	Thailand, Chiang Mai	0.4
		New Zealand (Maori)	5.7	China, Tianjin	0.4
		Australian Capital Territory	5.4	United States, Los Angeles (Japanese)	0.5
	Females	United States, Detroit (black)	6.4	Thailand, Chiang Mai	0.3
		New Zealand (Maori)	5.8	China, Tianjin	0.3
		United States, Hawaii (Hawaiian)	4.2	India, Madras	0.4
Leukaemia	Males	Italy, Trieste	15.0	India, Madras	3.0
		Australia, South Australia	13.3	Singapore (Indian)	3.0
		United States, Detroit (white)	12.7	Japan, Yamagata	4.4
	Females	Italy, Trieste	9.0	United States, Central Louisiana	1.6
		Australia, South Australia	8.9	(black)	
		United States, San Francisco (Filipino)	8.4	India, Madras	2.0
				Japan, Miyagi	3.3

a Numbers given are age-standardized (world) annual incidence per 100,000 population.

b Registries for which IARC [P5] indicated problems in ascertainment have not been included in this Table. However, some of the differences in rates may be due in part to variations in the level of ascertainment and to random variation.

Table 2
Cohort and case-control epidemiological studies of the effects of exposures to low-LET radiation

Study	Type of study	Population studied — Characteristics	Population studied — National origin	Follow-up (years)	Total person years ^{a}	Type of exposure	Type of dosimetry	Cancers studied ^{b}
EXTERNAL HIGH-DOSE-RATE EXPOSURES								
Exposure to atomic bombings								
Life Span Study [P9]	Mortality	50 113 exposed persons ^{c} 36 459 unexposed persons 55.5% females Age: 0->90 (28.4) ^{d}	Japan	5–45	2 812 863 (32.5)	Gamma and neutron radiation from nuclear explosions	Individual estimates derived from detailed shielding histories	Leukaemia*, tongue, pharynx, oesophagus*, stomach*, colon*, rectum, liver*, gallbladder, pancreas, nose, larynx, lung*, bone, skin, female breast*, cervix uteri and uterus, ovary*, prostate, bladder, kidney, brain, other central nervous system, lymphoma, myeloma*
Life Span Study [P4, T1]	Incidence	37 270 exposed persons ^{e} 42 702 unexposed persons 55.5% females Age: 0->90 (26.8)	Japan	13–42 ^{f}	1 950 567 ^{g} (24.4)	Gamma and neutron radiation from nuclear explosions	Individual estimates derived from detailed shielding histories	Leukaemia*, non-Hodgkin's lymphoma*, myeloma, oral cavity, salivary gland*, oesophagus, stomach*, colon*, rectum, liver*, gallbladder, pancreas, lung*, female breast*, non-melanoma skin*, uterus, ovary*, prostate, bladder*, central nervous system, thyroid*
Survivors of atomic bombings (in utero) [D14, Y1]	Mortality/ Incidence	1 078 exposed persons ^{h} 2 211 unexposed persons 50.7% females Exposure: in utero	Japan	5–47	n.a. ^{i}	Maternal exposure to gamma and neutron radiation at high dose rate	Mother's estimated uterus dose	Leukaemia, all solid cancers
Treatment of malignant disease								
Cervical cancer cohort [B11]	Incidence	82 616 exposed women 99 424 unexposed women Age: <30->70 (26.8)	Canada, Denmark, Finland, Norway, Slovenia, Sweden, United Kingdom, United States	1->30	1 278 950 (7.0)	Radiotherapy, including external beam and intra-cavity application and experimental reconstruction	Data on typical range of estimates for specific organs and phantom measurements	Oral cavity, salivary gland, oesophagus*, stomach, small intestine*, colon, rectum*, liver, gallbladder, pancreas*, lung*, breast, uterus, other genital*, kidney, bladder, melanoma, other skin, brain, thyroid, bone, connective tissue, leukaemia (non-CLL)*, myeloma, lymphoma
Lung cancer following breast cancer [I7]	Case-control 61 cases 120 controls from a cohort of 27 106 women	38 exposed women 143 unexposed women Age: 35–72 (50)	United States	10–46 (18 years per case)	n.a.	Radiotherapy	Individual doses from therapy records and experimental measurements	Lung cancer

Table 2 (continued)

Study	Type of study	Population studied		Follow-up (years)	Total person years [a]	Type of exposure	Type of dosimetry	Cancers studied [b]
		Characteristics	National origin					
Cervical cancer case-control [B1, B12, B50]	Case-control 4 188 cases 6 880 controls	10 286 exposed women 782 unexposed women Age: <30–>70 (26.8)	Austria, Canada, Czech Republic, Denmark, Finland, France, Germany, Iceland, Italy, Norway, Slovenia, Sweden, United Kingdom, United States	0–>30 (7.0 years per case)	n.a.	Radiotherapy, including external beam and intra-cavity application and experimental reconstruction	Individual doses from therapy records	Stomach*, pancreas, small intestine, colon, rectum*, breast, uterine corpus*, vagina*, ovary, vulva, bladder*, bone, connective tissue, leukaemia (non-CLL)*, myeloma, lymphoma, thyroid
Contralateral breast cancer, United States [B10]	Case-control 655 cases 1 189 controls from a cohort of 41 109 women	449 exposed women 1 395 unexposed women Age: <45–>60 (51)	United States	7–55 (~13 years per case)	n.a.	Radiotherapy	Individual doses from therapy records and experimental measurements	Contralateral breast among women less than 45 years old at exposure*, contralateral breast in older women
Contralateral breast cancer, Denmark [S20]	Case-control 529 cases 529 controls from a cohort of 56 540 women	157 exposed women 901 unexposed women Age: <45–>60 (51)	Denmark	12–47 (~16 years per case)	n.a.	Radiotherapy	Individual doses from therapy records and experimental measurements	Contralateral breast
Soft tissue sarcoma following breast cancer [K35]	Case-control 107 cases 321 controls from a cohort of 122 991 women	310 exposed women 86 unexposed women 32 women with unknown exposure status Age: 29–86 (59)	Sweden	1–35 (10 years per case)	n.a.	Radiotherapy	Total absorbed energy from radiotherapy, and location of sarcoma in relation to the treatment region	Soft tissue sarcoma
Leukaemia following breast cancer [C9]	Case-control 90 cases 264 controls from a cohort of 82 700 women	110 exposed women 244 unexposed women Age: <50–>70 (61)	United States	<12 (~5 years per case)	n.a.	Adjuvant radiotherapy	Individual doses from therapy records and experimental measurements	Acute non-lymphocytic leukaemia and myelodysplastic syndrome*, chronic myelogenous leukaemia, acute lymphocytic leukaemia
Leukaemia following cancer of the uterine corpus [C10]	Case-control 218 cases 775 controls from a cohort of 110 000 women	612 exposed women 351 unexposed women 30 women with unknown exposure status Age: <55–>75 (62)	Canada, Denmark, Finland, Norway, United States	1–50	n.a.	Radiotherapy	Individual doses from therapy records and experimental measurements	Leukaemia*
Lung cancer following Hodgkin's disease (international) [K9]	Case-control 98 cases 259 controls	303 exposed persons 54 unexposed persons 15% female	Canada, Denmark, Finland, France, Norway, Slovenia, United Kingdom	1–>10	n.a.	Radiotherapy	Individual doses from therapy records and experimental measurements	Lung cancer

Table 2 (continued)

| Study | Type of study | Population studied | | Follow-up (years) | Total person years [a] | Type of exposure | Type of dosimetry | Cancers studied [b] |
		Characteristics	National origin					
Lung cancer following Hodgkin's disease (Netherlands) [V2]	Case-control 30 cases 82 controls from a cohort of 1 939 patients	101 exposed persons 11 unexposed persons 4% female Age: <45->55 (49.4)	Netherlands	1–23	n.a.	Radiotherapy	Individual doses from therapy records and experimental measurements	Lung cancer*
Breast cancer following Hodgkin's disease [H2]	Incidence/ Mortality	855 exposed women 30 unexposed women Age: 4–81 (28)	United States	0–29	8 832 (10)	Radiotherapy	Individual doses from therapy records	Breast cancer*
Leukaemia following Hodgkin's disease (international) [K40]	Case-control 163 cases 455 controls from a cohort of 29 552 patients	36% exposed 35% females Age: (40)	Canada, Denmark, Finland, France, Germany, Italy, Netherlands, Norway, Slovenia, United Kingdom	1–>10	n.a.	Radiotherapy	Individual doses from therapy records and experimental measurements	Leukaemia (non-CLL)
Leukaemia following non-Hodgkin's lymphoma (international) [T6]	Case-control 35 cases 140 controls from a cohort of 11 386 women	123 exposed persons 52 unexposed persons Age: <50–70	Canada, Netherlands, Sweden, United States	2–25 (7.6 years per case)	n.a.	Radiotherapy	Individual doses from therapy records and experimental measurements	Leukaemia
Leukaemia following non-Hodgkin's lymphoma (United States) [T15]	Incidence	61 exposed persons 50% females Age: 18–70 (49.5)	United States	2–22	590 (9.7)	Total body irradiation	Individual doses from therapy records and experimental measurements	Acute non-lymphocytic leukaemia*, all solid cancers
Childhood cancers (international) [T5, T7, T17]	Case-control 23 thyroid cancers / 89 controls 25 leukaemia / 90 controls 64 bone cancers/ 209 controls from a cohort of 9 170 members	112 exposed persons 388 unexposed persons 45% females Age: 0–18 (7)	Canada, France, Netherlands, Italy, United Kingdom, United States	5–48	50 609 (5.5)	Adjuvant radiotherapy	Individual doses from therapy records and experimental measurements	Thyroid*, leukaemia, bone sarcoma*

Table 2 (continued)

Study	Type of study	Population studied		Follow-up (years)	Total person years [a]	Type of exposure	Type of dosimetry	Cancers studied [b]
		Characteristics	National origin					
Childhood cancers (France/U.K.) [D19, D33]	Incidence	3 109 exposed persons 1 291 unexposed persons 45% females Age: 0–16 (7)	France, United Kingdom	3–48	66 000 (15)	External radiotherapy	Individual doses from therapy records and experimental measurements	All solid cancers combined*, breast*, bone*, soft tissue sarcoma*, thyroid*, brain*
Bone cancer after childhood cancer (United Kingdom) [H44]	Case-control, 59 cases 220 controls, largely within a 13 175-member cohort	208 exposed persons 71 unexposed persons Age: 0–14	United Kingdom	3–>20	n.a.	External radiotherapy	Individual doses from therapy records and experimental measurements	Bone cancer
Leukaemia after childhood cancer (UK) [H11]	Case-control 26 cases 96 controls	88 exposed persons 34 unexposed persons Age: 0–14	United Kingdom	1–43	n.a.	External radiotherapy	Individual doses from therapy records and experimental measurements	Leukaemia
Retinoblastoma [W11]	Incidence	962 exposed persons 642 unexposed persons 47% females Age: 0–17	United States	1–>60	n.a. (Median 20)	External radiotherapy	Individual doses from therapy records and experimental measurements	Soft tissue sarcoma*, bone and soft tissue sarcoma*, all other cancers
Thyroid cancer following childhood cancer [D20]	Incidence	2 827 exposed persons	France, United Kingdom	3–29	n.a.	External radiotherapy	Individual doses from therapy records and experimental measurements	Thyroid cancer*
Childhood Hodgkin's disease [B16]	Incidence	1 380 persons 8% unexposed 35% female Age: 1–16 (median 11)	Canada, France, Italy, United Kingdom, United States	0–37 (median 11.4)	15 660 (11.3)	Radiotherapy	Individual doses from therapy records and experimental measurements	Leukaemia*, non-Hodgkin's lymphoma*, breast*, thyroid*, other solid cancers*

Treatment of benign disease

Study	Type of study	Population studied		Follow-up (years)	Total person years [a]	Type of exposure	Type of dosimetry	Cancers studied [b]
		Characteristics	National origin					
Childhood skin haemangioma: Stockholm [K23, L13, L16, L17, L24, L46]	Incidence/ Mortality	14 351 exposed persons [j] 67% females Age: 0–1.5 (0.5)	Sweden	1–67	406 355 (39)	Radiotherapy	Individual organ doses from therapy records and phantom measurements	Thyroid*, breast*, leukaemia, all other sites
Childhood skin haemangioma: Gothenburg [K22, K23, L15, L46]	Incidence	11 914 exposed persons 88% aged <1 year	Sweden	0–69	370 517 (31.1)	Radiotherapy	Individual organ doses from therapy records and phantom measurements	Thyroid*, other endocrine glands*, central nervous system*, all other sites

Table 2 (continued)

Study	Type of study	Population studied		Follow-up (years)	Total person years [a]	Type of exposure	Type of dosimetry	Cancers studied [b]
		Characteristics	National origin					
Benign lesions in locomotor system [D12, J2]	Incidence/Mortality	20 024 exposed persons 49% females Age: <20->70 (53)	Sweden	Up to 38	Incidence: 493 400 (24.6) Mortality: 392 900 (19.6)	X-ray therapy	Individual red bone marrow doses from therapy records and phantoms	Leukaemia*, non-Hodgkin's lymphoma, Hodgkin's disease, multiple myeloma
Ankylosing spondylitis [W1, W2] [k]	Mortality	13 914 exposed persons 16.5% females Age: <20->60	United Kingdom	1-57	245 413 (17.6)	X-ray therapy	Individual doses for leukaemia cases and a 1 in 15 sample of the population	Leukaemia*, other neoplasms* (except colon)
Israel tinea capitis [R5, R9, R16, R17]	Incidence/Mortality	10 834 exposed persons 16 226 unexposed persons 50% females Age: <1-15 (7.1)	Israel	26-38	686 210 (25.3)	X-ray induced epilation	Individual doses from phantom measurements based on institution and age	Incidence: thyroid*, skin*, brain*, salivary gland*, breast Mortality: head and neck*, leukaemia*
New York tinea capitis [S27, S30]	Incidence	2 226 exposed persons 1 387 unexposed persons 16.1% females Age: <1-19 (7.7)	United States	20-39	98 881 (25.4)	X-ray induced epilation	Representative doses based on standard treatment	Thyroid*, skin*, brain, leukaemia, salivary gland
New York acute post-partum mastitis [S15, S30]	Incidence	571 exposed women 993 unexposed women Age: 14->40 (27.8)	United States	20-35	38 784 (25.1)	X-ray therapy	Individual doses from therapy records	Breast*
Rochester thymic irradiation [H10, H31, S30]	Incidence	2 652 exposed persons 4 823 unexposed persons 42% females Age: 0-1	United States	23->50	220 777 (29.5)	X-ray therapy	Individual doses from therapy records	Thyroid*, breast*, skin
Tonsil irradiation [S21, S28, S30]	Incidence	2 634 exposed persons [l] 40.7% females Age: 0-15 (4.3)	United States	0-50	88 101 (33)	X-ray therapy	Individual doses from therapy records and phantom measurements	Skin*, thyroid*, benign parathyroid*, salivary gland*, neural tumours*
Tonsil, thymus or acne irradiation [D5]	Incidence	416 exposed persons Age:(7.1)	United States	n.a.	11 000 (26.4)	Radiotherapy	Individual doses from therapy records	Thyroid*
Swedish benign breast disease [M8, M20, M28]	Incidence	1 216 exposed women 1 874 unexposed women Age: 10->85	Sweden	5-60	56 900 (18)	X-ray therapy	Individual doses from therapy records and phantom measurements	Breast*, all other sites

Table 2 (continued)

| Study | Type of study | Population studied | | Follow-up (years) | Total person years [a] | Type of exposure | Type of dosimetry | Cancers studied [b] |
		Characteristics	National origin					
Metropathia haemorrhagica [D7]	Mortality	2 067 exposed women Age: 35–60	United Kingdom	5–>30	53 144	X-ray therapy	Individual doses from therapy records and phantom measurements	Pelvic sites*, leukaemia* multiple myeloma*, lymphoma, all other sites [m]
Benign gynaecological disease [I6, I16]	Mortality	4 153 exposed women Age: 13–88 (46.6)	United States	0–60	109 910 (26.5)	Intrauterine ²²⁶Ra	Individual doses from therapy records and phantom measurements	Leukaemia*, other haematolymphopoietic cancers, uterus*, bladder*, rectum*, other genital*, colon, bone (in pelvis), liver and gallbladder, stomach, kidney, pancreas*
Lymphoid hyperplasia screening [P8]	Incidence/ Prevalence	1 195 exposed persons 1 063 unexposed persons 40% females Age: 0–17 (6.9)	United States	12–44	66 000 (29)	X-ray therapy	Individual doses from therapy records and phantom measurements	Thyroid nodular disease*
Peptic ulcer [G6]	Mortality	1 831 exposed persons 1 778 unexposed persons 21.2% females Age: <35–>55 (49)	United States	20–51	77 757 (21.5)	X-ray therapy	Individual doses from therapy records and experimental measurements	Stomach*, colon, pancreas*, lung*, leukaemia*, female breast, oesophagus, rectum, liver, larynx, bone and connective tissue, bladder, prostate, kidney, brain, thyroid, non-Hodgkin's lymphoma, Hodgkin's disease, myeloma

Diagnostic examinations

| Study | Type of study | Population studied | | Follow-up (years) | Total person years [a] | Type of exposure | Type of dosimetry | Cancers studied [b] |
		Characteristics	National origin					
Massachusetts TB fluoroscopy [B3, S30]	Incidence	2 367 exposed women 2 427 unexposed women Age: 12–50 (26)	United States	0–>50	54 609 (11.4)	Multiple x-ray chest fluoroscopies	Individual exposures from medical records and doses from phantom measurements and computer simulations	Breast*, skin
Massachusetts TB fluoroscopy [D4]	Mortality	6 285 exposed persons 7 100 unexposed persons 49% females Age: 12–50 (26)	United States	0–>50	331 206 (24.7)	Multiple x-ray chest fluoroscopies	Individual exposures from medical records and doses from phantom measurements and computer simulations	Breast*, oesophagus*, lung, leukaemia
Canadian TB fluoroscopy [H7, H20]	Mortality	25 007 exposed persons 39 165 unexposed persons 50% females Age: <20–>35 (28)	Canada	0–57	1 608 491 (25.1)	Multiple x-ray chest fluoroscopies	Individual exposures from medical records and doses from phantom measurements	Lung, breast*
Diagnostic x rays (US health plans) [B39]	Case-control 565 leukaemia 318 NHL 208 multiple myeloma 1 390 controls	2 203 exposed persons 278 unexposed persons 39% females Age: 15–>50	United States	n.a.	n.a.	Diagnostic x rays	Average dose based on number and type of procedures and estimated doses from published literature	Leukaemia, non-Hodgkin's lymphoma, multiple myeloma

Table 2 (continued)

Study	Type of study	Population studied — Characteristics	Population studied — National origin	Follow-up (years)	Total person years [a]	Type of exposure	Type of dosimetry	Cancers studied [b]
Medical and dental x rays (Los Angeles) [P35]	Case-control 408 cases 408 controls	62% females	United States	2–64	n.a.	Medical and dental diagnostic x rays	Average dose based on number and type of procedures and estimated doses from published literature	Parotid gland*
Diagnostic x rays (Los Angeles) [P10]	Case-control 130 cases 130 controls	39% females	United States	3–20	n.a.	Diagnostic x rays	Average dose based on number and type of procedures and estimated doses from published literature	Chronic myeloid leukaemia*
Diagnostic x rays (Sweden) [I9]	Case-control 484 cases 484 controls	736 exposed persons 232 unexposed persons 77% females Age: <20–>60	Sweden	5–>50	n.a.	Diagnostic x rays	Average dose based on number and type of procedures and estimated doses from published literature	Thyroid
Scoliosis [D34]	Mortality	4 822 exposed women 644 unexposed women Age: <3–≥10 (10.6)	United States	3–>60	218 976 (40.1)	Diagnostic x-rays	Average dose based on number of treatments and estimated doses from published literature	Breast*

EXTERNAL LOW-DOSE OR LOW-DOSE-RATE EXPOSURES

Prenatal exposure

Study	Type of study	Population studied — Characteristics	Population studied — National origin	Follow-up (years)	Total person years [a]	Type of exposure	Type of dosimetry	Cancers studied [b]
Oxford Survey of Childhood Cancers [S1, B2, M29]	Case-control 14 491 cases 14 491 controls	3 797 exposed persons 25 185 unexposed persons 56 % females Exposure: in utero	United Kingdom	16 (max.)	n.a.	Maternal x rays during pregnancy	Number of exposures with a model for dose per exposure	Leukaemia*, all solid tumours*
NE USA childhood cancers [M9]	Case-control 1 342 cases 14 292 controls	1 506 exposed persons 14 130 unexposed persons 49.2 % females Exposure: in utero	United States	20 (max.)	n.a.	Maternal x rays during pregnancy	Number of exposures	Leukaemia*, solid tumours

Occupational exposure

Study	Type of study	Population studied — Characteristics	Population studied — National origin	Follow-up (years)	Total person years [a]	Type of exposure	Type of dosimetry	Cancers studied [b]
Nuclear workers in Japan [E3]	Mortality	114 900 men	Japan	Up to 5	533 168 (4.6)	Exposures in nuclear power plants, fuel processing, and research facilities	Recorded exposures to external radiation	Leukaemia, all other cancers

Table 2 (continued)

Study	Type of study	Population studied — Characteristics	Population studied — National origin	Follow-up (years)	Total person years [a]	Type of exposure	Type of dosimetry	Cancers studied [b]
Nuclear workers in Canada, United Kingdom and United States [C11] [n]	Mortality	95 673 workers 15% females	Canada United Kingdom United States	Up to 43	2 124 526 (22.2)	Exposures in nuclear power plants, fuel processing, and research facilities	Recorded exposures to external radiation	Leukaemia, all other cancers
National Registry for Radiation Workers, U.K. [M46] [o]	Mortality	124 743 monitored workers 9% females	United Kingdom	Up to 47	2 063 300 (16.5)	Exposures in nuclear power plants, fuel cycle, defence, and weapons production	Recorded exposures to external radiation	Leukaemia, all other cancers
Sellafield [C19, D21] [p]	Mortality / Incidence	10 028 monitored workers 3 711 other workers 19% females	United Kingdom	Up to 40	260 000 [q] (26)	Fuel processing and reactor operation	Recorded exposures to external radiation	Leukaemia, all other cancers
UK Atomic Energy Authority [C19, F6] [p]	Mortality / Incidence	21 344 monitored workers 18 071 other workers 8% females	United Kingdom	Up to 42	534 000 [q] (25)	Nuclear and reactor research and fuel processing	Recorded exposures to external radiation	Leukaemia, all other cancers
UK Atomic Weapons Establishment [C19, B24] [p]	Mortality	9 389 monitored workers 12 463 other workers 9% females	United Kingdom	Up to 37	216 000 [q] (23)	Weapons research	Recorded exposures to external radiation	Leukaemia, all other cancers
Chapelcross [B38, M50]	Mortality/ Incidence	2 209 monitored workers 419 other workers 14% females	United Kingdom	Up to 41	63 967 (24.3)	Reactor operation	Recorded exposures to external radiation	Buccal cavity and pharynx, prostate, all cancers combined
National Dose Registry of Canada [A21] [r]	Mortality	206 620 monitored workers 49% females	Canada	Up to 47	2 861 093 (13.8)	Dental, medical, industrial and nuclear power	Recorded exposures to external radiation	Leukaemia, all other cancers
Atomic Energy of Canada Ltd. [C11, G16] [s]	Mortality	11 355 monitored workers 24% females	Canada	Up to 30	198 210 (17.5)	Nuclear and reactor research and related technologies	Recorded exposures to external radiation	Leukaemia, all other cancers
Hanford [G12, G17] [t]	Mortality	32 643 monitored workers 24% females	United States	Up to 43	633 511 (19.4)	Exposures in nuclear fuel cycle and research	Recorded exposures to external radiation	Leukaemia, all other cancers
Oak Ridge: X-10 and Y-12 plants [F5]	Mortality	28 347 men	United States (white)	Up to 40	n.a.	Exposures in nuclear fuel cycle and research	Recorded exposures to external radiation	Leukaemia, all other cancers

Table 2 (continued)

Study	Type of study	Population studied — Characteristics	Population studied — National origin	Follow-up (years)	Total person years[a]	Type of exposure	Type of dosimetry	Cancers studied[b]
Rocky Flats [G12, W12]	Mortality	5 952 men	United States (white)	Up to 32	81 237 (13.6)	Exposures in nuclear fuel cycle and research	Recorded exposures to external radiation	Leukaemia, all other cancers
Portsmouth Naval Shipyard [R12]	Mortality	Males 8 960 monitored workers 15 585 other workers	United States (white)	Up to 26	n.a.	Work on overhauling and building nuclear submarines	Recorded exposures to external radiation	Leukaemia, all lymphatic and haematopoietic neoplasms, all cancers combined
Rocketdyne / Atomics International [R27]	Mortality	4 563 monitored workers 6% females	United States	Up to 45	118 749 (26)	Exposures at a nuclear research and production facility	Recorded exposures to external radiation	Leukaemia, all other cancers
Mound facility [W21]	Mortality	Males 3 229 monitored workers 953 other workers	United States (white)	Up to 33	78 600 (18.8)	Exposures at a nuclear research and production facility	Recorded exposures to external radiation	Leukaemia, all other cancers
Chernobyl clean-up workers: Russian Fed. (cohort) [I13, I21]^u	Incidence	114 504 male workers Age: <2– ≥61	Russian Federation	0–9	797 781 (7.0)	Emergency and recovery work in the vicinity of Chernobyl	Assessed external radiation dose	Digestive*, respiratory, thyroid, all solid tumours combined, leukaemia*
Chernobyl clean-up workers: Russian Fed. (Leukaemia case-control) [I14]	Case-control 34 cases 136 controls from a cohort of 155 680 men	Males 87.3% with doses above 0.05 Sv Age: <20–>55	Russian Federation	2–7	n.a.	Emergency and recovery work in the vicinity of Chernobyl	Assessed external radiation dose	Leukaemia
Chernobyl clean-up workers: Estonia [I10, R20, T13]	Mortality Incidence	4 742 men Age: <30–>60	Estonia	0–7	30 643 (6.5)	Emergency and recovery work in the vicinity of Chernobyl	Recorded radiation doses	Thyroid, all other sites
Mayak workers [K10, K11, K32, K34]	Mortality	15 601 persons monitored for external radiation 3 229 other workers 25% females^v	Russian Federation	0–46	211 427 (31.8)^w	Exposures in nuclear fuel cycle and research	Recorded exposures to external radiation	Lung, leukaemia*
Mayak workers: stomach cancer study [Z1]	Case-control 157 cases 346 controls	40 persons with external doses above 3 Gy 463 with lower doses 10% females	Russian Federation	Up to 37	n.a.	Exposures in nuclear fuel cycle and research	Recorded exposures to external radiation and measurements of plutonium	Stomach*

Table 2 (continued)

Study	Type of study	Population studied — Characteristics	Population studied — National origin	Follow-up (years)	Total person years [a]	Type of exposure	Type of dosimetry	Cancers studied [b]
Japanese radiological technologists [A25]	Mortality	9 179 radiological technologists (2 300 with recorded doses)	Japan	Up to 28	n.a.	Radiology	Recorded exposures to external radiation	All cancers combined*, oesophagus, stomach, colorectal, lung
Danish radio-therapy staff [A15]	Incidence	4 151 persons Age: <20–≥50	Denmark	Up to 32	49 553 (11.9)	Work in radio-therapy departments	Recorded exposures to external radiation	Leukaemia, prostate*, all other cancers
Natural radiation								
Yangjiang [A11, S35, T12, T25, T26, Z2 [x]]	Mortality	89 694 persons in high-background area 35 385 persons in control area 50% females All ages	China	Up to 17	1 698 350 (13.6)	Continuous background radiation	Individual estimates, both direct (TLD measurements) and indirect (environmental measurements and occupancy patterns)	Leukaemia, all other sites
Central Italy [F7]	Case-control 44 cases 211 controls	Males Age at diagnosis: 35–80 (68) 76% with gamma dose rate above 300 nGy h⁻¹	Italy	10	n.a.	Gamma radiation Radon	Measurements in last dwelling occupied and characteristics of dwellings	Acute myeloid leukaemia
INTERNAL LOW-DOSE-RATE EXPOSURES								
Medical exposures								
Diagnostic ¹³¹I [H4, H12, H27] [y]	Incidence	34 104 exposed persons 80% females Age: 1–75 (43)	Sweden	5–39	653 093 (19.1)	Diagnostic ¹³¹I	Individual values of activity administered; organ dose estimates for thyroid	Thyroid, leukaemia, all other sites
Swedish ¹³¹I hyperthyroidism [H23, H24] [z]	Incidence/ Mortality	10 522 exposed persons 82% females Age: 13–70	Sweden	1–26	139 018 (13.6)	Treatment of hyperthyroidism	Average administered activity (multiple treatments)	Stomach*, kidney*, brain*, all other sites [aa]
United States thyrotoxicosis patients [D22, R14, S36] [bb]	Incidence/ Mortality	23 020 exposed persons 12 573 unexposed persons 79% females Age: <10–80	United States	0–45	738 831 (20.8)	Treatment of hyperthyroidism	Individual values of activity administered; organ-dose estimates	Buccal cavity, oesophagus, stomach, colorectal, liver, pancreas, larynx, lung*, breast*, uterus, ovary, prostate, bladder, kidney*, brain and other central nervous system tumours, thyroid*, lymphoma, myeloma, leukaemia

Table 2 (continued)

Study	Type of study	Population studied		Follow-up (years)	Total person years [a]	Type of exposure	Type of dosimetry	Cancers studied [b]
		Characteristics	National origin					
Iodine-131 hyperthyroidism, United Kingdom [F8]	Incidence/ Mortality	7 417 exposed persons 83% females Age: ≤49 – ≥70 (57)	United Kingdom	1 – ≥20	72 073 (9.7)	Treatment of hyper-thyroidism	Individual values of activity administered	Thyroid*, bladder, uterine, small bowel*, all other sites
Swedish ^{131}I thyroid cancer [H26]	Incidence	834 exposed persons 1 121 unexposed persons 75% females Age: 5 – 75 (48)	Sweden	2 – 34	25 830 (13.2)	Treatment of thyroid cancer	Individual values of activity administered	Leukaemia, salivary glands*, kidney*, all other sites
French therapeutic [131]I [D18]	Incidence	846 persons with therapeutic exposures 501 persons with diagnostic exposures 274 unexposed persons 79 % females Age: 5 – 89 (40)	France	2 – 37	14 615 (10)	Diagnostic and therapeutic ^{131}I exposures for thyroid cancer patients	Individual values of activity administered and organ dose estimates	Colon, leukaemia, all other sites

Environmental exposures

Techa River population [K5, K27]	Mortality	26 485 exposed persons 58% females Age: 0 – 96 (29)	Russian Federation (ethnic Russians and Tartars/ Bashkirs)	Up to 39	n.a.	Internal and external exposures to radioactive waste discharged by nuclear weapons production plant	Dose reconstruction based on environmental measurements of gamma dose rate and whole-body counting	Leukaemia*, lymphoma, stomach, liver, lung, breast, bone, all other sites
Chernobyl-related exposure in Belarus [A26]	Case-control 107 cases 214 controls	52% females Age: 0 – 16	Belarus	Up to 6	n.a.	Internal exposure to radioactive iodine in areas contaminated by the Chernobyl accident	[131]I dose estimated from ground deposition of ^{137}Cs and ^{131}I, contemporary thyroid radiation measurements, and from questionnaires and interviews	Thyroid*
Marshall Islands fallout [H35, R21]	Prevalence	2 273 exposed persons 55% females Age: 5 – >60	Marshall Islands	29 – 31	n.a.	Short-lived radionuclides from nuclear explosion	Estimated average dose; distance was also used as a surrogate	Thyroid
Utah ^{131}I fallout: thyroid disease [K36]	Prevalence	2 473 persons	United States	12 – 17 and 32 – 33 [cc]	n.a.	Fallout from nuclear weapons tests	Based on residence histories and fallout deposition records	Thyroid

Table 2 (continued)

Study	Type of study	Population studied		Follow-up (years)	Total person years [a]	Type of exposure	Type of dosimetry	Cancers studied [b]
		Characteristics	National origin					
Utah ^{131}I fallout [S37]	Case-control	92 persons with bone-marrow doses of 6 mGy or more; 6 415 persons with lower doses	United States	Up to 30	n.a.	Fallout from nuclear weapons tests	Based on residence histories and fallout deposition records	Leukaemia
Occupational exposures								
UK Atomic Energy Authority: prostate cancer study [R26]	Case-control 136 cases 404 controls	Males; Age at diagnosis: <65->75; 14% of subjects with documented internal exposure	United Kingdom	n.a.	n.a.	Exposures in nuclear fuel cycle and research	Urine measurements and whole-body monitoring	Prostate*

a   Mean per person in parentheses.
b   An asterisk denotes sites for which statistically significant excesses are reported in the exposed group (cohort studies) or for which a higher proportion of the cases were exposed to radiation (case-control studies).
c   Exposed to more than 0.005 Sv weighted colon dose.
d   Age at exposure, mean in parentheses.
e   Exposed to more than 0.01 Sv weighted colon dose.
f   5–42 years for leukaemia and lymphomas [P4].
g   Based on the follow-up for solid cancer [T1].
h   Figures quoted are for the mortality study [D14]. Exposure denotes doses above 0.01 Sv.
i   Not available.
j   Figures quoted in [L16].
k   Figures quoted are for the leukaemia study [W2].
l   Figures quoted in [S21].
m   Significance tests based on 5-year survivors (2 years for leukaemia).
n   Includes workers in studies [B24, C19, D21, F6, G12, G16, G17, W12, W23].
o   Includes workers in studies [B24, B38, C19, D21, F6].
p   Figures quoted are from [C19].
q   Values for monitored workers only.
r   Includes workers in study [G16].
s   Figures quoted are from [C11].
t   Figures quoted are from [G12].
u   Figures quoted are from [I21].
v   Figures quoted are from [K32].
w   Figures are for males employed before 1959 [K34].
x   Figures quoted are from [T25, T26].
y   Figures quoted are for the thyroid cancer study [H4].
z   Figures quoted are for the incidence study [H23].
aa   Significance tests based on 10-year survivors.
bb   Figures quoted are from [R14].
cc   Periods of thyroid examinations, relative to the peak fallout in 1953 [K36].

Table 3
Strengths and limitations of major cohort and case-control epidemiological studies of carcinogenic effects of exposures to low-LET radiation

Study	Strengths	Limitations
EXTERNAL HIGH-DOSE-RATE EXPOSURES		
Exposures to atomic bombings		
Life Span Study [P4, P9, T1]	Large population of all ages and both sexes not selected because of disease or occupation Wide range of doses Comprehensive individual dosimetry Survivors followed prospectively for up to 45 years Complete mortality ascertainment Cancer incidence ascertainment	Acute, high-dose-rate exposure that provides no direct information on effects of gradual low-dose-rate exposures Restriction to 5-year survivors for mortality (13 years for incidence) Possible contribution of neutrons somewhat uncertain Possible effects of thermal or mechanical injury and conditions following the bombings uncertain
Survivors of atomic bombings (in utero) [D14, Y1]	Not selected for exposure Reasonably accurate estimate of dose Mortality follow-up relatively complete Follow-up into adulthood	Small numbers of exposed individuals and cases Incidence determination may not be complete Mechanical and thermal effects may have influenced results
Treatment of malignant disease		
Cervical cancer cohort [B11, B12, B50]	Large-scale incidence study based on tumour registry records Long-term follow-up Relatively complete ascertainment of cancers Non-exposed comparison patients	Very large doses to some organs result in cell killing and tissue damage Potential misclassification of metastatic disease for some organs Potential misclassification of exposure No individual dosimetry Characteristics of patients with cervical cancer differ from general population
Cervical cancer case-control [B1]	Comprehensive individual dosimetry for many organs Dose-response analyses Other strengths as above [B11]	As above [B11], except problems with individual dosimetry and comparison with general population now removed Small number of non-exposed cases Partial-body and partial-organ dosimetry complex
Lung cancer following breast cancer [I7]	Individual estimates of radiation dose to different segments of the lungs Large number of non-irradiated patients Most patients did not receive chemotherapy Substantial proportion of patients with over 20 years of follow-up	Small number of lung cancers Lack of data on individual smoking habits Potential inaccuracies in partial-body dosimetry
Contralateral breast cancer [B10, S20]	Large numbers of incident cases within population-based tumour registries Individual radiation dosimetry Wide range of doses	Limited number of young women Possibility of over matching, resulting in some concordance of exposure between cases and controls Possible misclassification of metastases or recurrence
Soft-tissue sarcoma following breast cancer [K35]	Incident cases identified from a population-based tumour registry	Analyses based on estimates of energy imparted from radiotherapy (i.e. product of the mass of the patient and the absorbed dose), rather than organ dose
Leukaemia following breast cancer [C9]	Comprehensive individual dosimetry for bone-marrow compartments Comprehensive ascertainment of treatment information to separate chemotherapy risk Dose-response analyses	Very large high-dose partial-body exposure to chest wall, probably resulting in cell-killing
Leukaemia following cancer or the uterine corpus [C10]	Large number of incident cases with population-based cancer registries Comprehensive individual dosimetry for bone-marrow compartments Attempt to adjust for chemotherapy Large non-irradiated comparison group Dose-response analyses covering doses below 1.5 Gy as well as above 10 Gy	Effects of cell-killing at high doses Potential inaccuracies in partial-body dosimetry
Lung cancer following Hodgkin's disease (international) [K9]	Individual estimates of radiation dose to the affected lung Some data on individual smoking habits Detailed information on chemotherapy Relatively large number of cases	Smoking data limited, and reported more fully for cases than for controls Follow-up period generally less than 10 years

Table 3 (continued)

Study	Strengths	Limitations
Lung cancer following Hodgkin's disease (Netherlands) [V2]	Individual estimates of radiation dose to the area of the lung where the tumour developed Individual data on smoking habits Extensive data on doses from chemotherapy	Small number of cases Limited follow-up (median 10 years) Few females
Breast cancer following Hodgkin's disease [H2]	Individual assessment of doses Analysis by age at exposure	Small number of cases Limited follow-up Mostly very high doses (>40 Gy)
Leukaemia following Hodgkin's disease (international) [K40]	Individual radiation dosimetry Detailed information on chemotherapy	Follow-up period generally less than 10 years
Leukaemia following non-Hodgkin's lymphoma (international) [T6]	Comprehensive individual dosimetry for bone marrow compartments Detailed information on chemotherapy	Small number of cases No dose-response analysis, other than separation into two groups
Leukaemia following non-Hodgkin's lymphoma (United States) [T15]	Individual dosimetry for bone marrow Detailed information on chemotherapy	Very small cohort; few cases No comparison group of unexposed patients
Childhood cancers (international) [T5, T7, T17]	Comprehensive individual dosimetry to estimate organ doses Attempt to adjust for drug exposure Dose-response analyses	Only high-dose exposures Potential for some overmatching since hospital-based Complete dosimetry not always available
Childhood cancers (France/United Kingdom) [D19, D33]	Incidence follow-up Doses from radiotherapy and chemotherapy estimated	Individual dose estimates generally not used in analyses Lack of external comparison group Small numbers for specific types of cancers
Bone cancer and leukaemia after childhood cancer (United Kingdom) [H44, H11]	Incidence follow-up Individual dosimetry Information available on chemotherapy	Most of the findings concern doses of 5–10 Gy or more
Retinoblastoma [W11]	Long-term incidence follow-up Individual dose estimates for bone and soft sarcoma sites Wide range of doses	Little information on chemotherapy Most of the findings concern doses of 5 Gy or more
Thyroid cancer following childhood cancers [D20]	Incidence follow-up Individual organ dose estimates Wide range of thyroid doses	Lack of external comparison group
Childhood Hodgkin's disease [B16]	Cohort of persons exposed at young ages to high radiation doses Individual dosimetry Information available on chemotherapy doses	Small numbers of cases No formal modelling of dose-response or of chemotherapy effects
Treatment of benign disease		
Childhood skin haemangioma [K23, L13, L15, L16, L17, L24, L46]	Long-term and complete follow-up Comprehensive individual dosimetry for many organs Incidence ascertained Protracted exposure to radium plaques	Relatively small numbers of specific cancers
Benign lesions in locomotor system [D12, J2]	Long-term and complete follow-up Individual dose estimates Incidence and mortality ascertained	Uncertainties in computing individual doses to sites, based upon a sample of records
Ankylosing spondylitis [W1, W2]	Large number of exposed patients Long-term and complete mortality follow-up Detailed dosimetry for leukaemia cases and sample of cohort Small non-exposed group evaluated for general reassurance that leukaemia risk was unrelated to underlying disease	Comparisons with general population Underlying disease related to colon cancer and possibly other conditions Individual dose estimates available only for leukaemia cases and a 1 in 15 sample of the population
Israel tinea capitis [R5, R9, R16, R17]	Large number of exposed patients Two control groups Ascertainment of cancer from hospital records and tumour registry Individual dosimetry for many organs	Dosimetry for some sites (e.g. thyroid) uncertain, owing to possible patient movement or uncertainty in tumour location Limited dose range

Table 3 (continued)

Study	Strengths	Limitations
New York tinea capitis [S27, S30]	Relatively good dose ascertainment for skin and other cancers	Small number of cancers No recent follow-up information Few females
New York post-partum mastitis [S15, S30]	Individual estimates of breast dose from medical records Breast cancer incidence ascertained Dose-response analyses	All exposed women were parous, but comparison women were not (380 non-exposed and sisters of both exposed and non-exposed) Inflamed and lactating breast might modify radiation effect
Rochester thymic irradiation [H10]	Individual dosimetry for thyroid and some other sites Sibling control group Long follow-up Fractionation effects could be evaluated Dose-response analyses	Radiation treatment fields for newborns varied, and dosimetry uncertain for some sites Adjustment in analysis for sibship size uncertain Questionnaire follow-up may have resulted in under-ascertainment of cases
Tonsil irradiation [S21, S28, S30]	Individual dosimetry for thyroid and some other sites Long follow-up Large numbers of cases for certain sites Dose-responses analyses	Effect of screening on ascertainment of thyroid cancer and nodules No unexposed control group
Tonsil, thymus or acne irradiation [D5]	Long period between exposure and examination Prospective as well as retrospective follow-up	Possible screening effect Small cohort No unexposed control group
Swedish benign breast disease [M8, M20, M28]	Incidence study with long-term follow-up Individual dosimetry for many organs Fractionated exposure Unexposed control group	Lack of data on potential confounders Small numbers for most cancer types, other than breast
Benign gynaecological disease [D7, I6, I16]	Large number of exposed women Non-exposed women with benign gynaecological disease Very long mortality follow-up Individual dosimetry Protracted exposures to radium implants (10-24 hours) Dose-response analyses	Uncertainty in proportion of active bone-marrow exposed Small numbers of specific types of cancer Misclassification on certain cancers on death certificates (e.g. pancreas)
Lymphoid hyperplasia screening [P8]	Individual dosimetry Comparison of questionnaire and clinical examination results Comparison group treated by surgery for the same condition	Apparent bias in questionnaire data, owing to self-selection of subjects Clinical examinations provide data on prevalence rather than incidence Study of thyroid nodules; cancer cases not confirmed
Peptic ulcer [G6]	Individual dosimetry Non-exposed patients with peptic ulcer Exceptionally long follow-up (50 years) Some risk factor information available in records	Standardized radiotherapy precluded dose-response analyses Non-homogeneous dose distribution within organs, such that simple averaging may be misleading Metastatic spread on stomach cancer probably misclassified as liver and pancreatic cancer on death certificates Possible selection of somewhat unfit patients for radiotherapy rather than surgery
Diagnostic examinations		
Massachusetts TB fluoroscopy [B3, D4, S30]	Incidence study with long-term follow-up (50 years) Individual dosimetry based on patient records and measurements Non-exposed TB patients Fractionated exposures occurred over many years Dose-response analyses	Uncertainty in dose estimates related to fluoroscopic exposure time and patient orientation Questionnaire response probably under-ascertained cancers Debilitating effect of TB may have modified radiation effect for some sites, e.g. lung
Diagnostic x rays (US health plans) [B39]	Information on diagnostic x rays abstracted from medical records Surveillance bias unlikely, since cases and controls were at equal risk for having x-ray procedures recorded and malignancy diagnosed	Potential for ascertainment bias, e.g. through early diagnosis of a malignancy Analyses based on number of x-ray procedures rather than actual doses
Canadian TB fluoroscopy [H7, H20]	Large number of patients Non-exposed TB comparison group Individual dosimetry for lung and female breast Fractionated exposures occurred over many years Dose-response analyses	Mortality limits comparisons with breast cancer incidence series, e.g. time response Uncertainties in dosimetry limit precise quantification of risk Different dose responses for female breast cancer between one sanatorium and the rest of Canada may indicate errors in dosimetry, differential ascertainment, or differences in biological response

Table 3 (continued)

Study	Strengths	Limitations
Diagnostic medical and dental x rays (Los Angeles) [P10, P35]	Dosimetry attempted based on number and type of examinations	No available records of x rays Potential for recall bias in dose assessment Doses likely to have been underestimated
Diagnostic x rays (Sweden) [I9]	Information on diagnostic x rays over many years abstracted from medical records	Analyses based on number and type of x-ray procedures rather than actual doses
Scoliosis [D34]	Adolescence possibly a vulnerable age for exposure Dosimetry undertaken based on number of films and breast exposure Dose-response analysis	Comparison with general population potentially misleading, since scoliosis associated with several breast cancer risk factors (e.g. nulliparity) Dose estimates may be subject to bias as well as random error
EXTERNAL LOW-DOSE OR LOW-DOSE-RATE EXPOSURES		
Prenatal exposures		
Oxford Survey of Childhood Cancers [S1, B2, M29]	Very large numbers Comprehensive evaluation of potential confounding Early concerns over response bias and selection bias resolved	Uncertainty in fetal dose from obstetric x-ray examinations Similar relative risks for leukaemia and other cancers may point to possible residual confounding
North-eastern United States childhood cancers [M9]	Large numbers Reliance on obstetric records	Uncertainty in fetal dose
Occupational exposures		
Nuclear workers	Often large numbers Personal dosimetry Low-dose fractionated exposures Could provide useful information in future	Low doses make clear demonstration of radiation effect difficult Possibly confounding influence of chemical and other toxic exposures in workplace Healthy worker effect Mortality follow-up Lifestyle factors (e.g. smoking histories) generally not available
Chernobyl clean-up workers	Often large numbers Low-dose fractionated exposures Could provide useful information in future	Difficulties in assessing individual exposures Possible differences in cancer ascertainment relative to the general population Short period of follow-up so far
Mayak workers [K10, K11, K32, K34, Z1]	Wide range of exposures Individual measurements of external gamma dose and plutonium body burden Individual information on potential confounders in stomach cancer study	Possible uncertainties in assessment of exposures Further details of ascertainment of stomach cancer cases and controls desirable
Medical workers	Often large numbers Low-dose fractionated exposures over long periods	General lack of information on individual doses precludes usefulness to date
Natural radiation		
Yangjiang [T12, A11, Z2, S35, T25, T26]	Large cohorts in high background and control areas Stable population Extensive dosimetry for region Assessment of potential confounders	Mortality follow-up Small numbers for some cancer types Low doses
Central Italy [F7]	Individual measurements of domestic gamma radiation and radon	Small number of cases Mortality data only Measurements only in last home Low doses
INTERNAL LOW-DOSE-RATE EXPOSURES		
Medical exposures		
Swedish ^{131}I thyroid cancer [H23, H24]	Large numbers Nearly complete incidence ascertainment Administered activities of ^{131}I known	Comparison with general population Dose-response not based on organ doses High-dose cell-killing probably reduced possible thyroid effect Patients selected for treatment

Table 3 (continued)

Study	Strengths	Limitations
Diagnostic ^{131}I [H4, H12, H27]	Large numbers Unbiased and nearly complete ascertainment of cancers through linkage with cancer registry Administered activities of ^{131}I known for each patient Organ doses to the thyroid computed with some precision Dose-response analyses for thyroid cancer and leukaemia, based on wide range of doses Low-dose-rate exposure	Comparison with general population only, except for thyroid cancer and leukaemia Reason for some examinations related to high detection of thyroid cancers, i.e. suspicion of thyroid tumour was often correct Doses to organs other than thyroid very low Population under surveillance
United States thyrotoxicosis patients [D22, R14, S36]	Large numbers of patients treated with ^{131}I Large non-exposed comparison groups Comprehensive follow-up effort Administered activities of ^{131}I known	Individual doses computed only for certain organs Mortality follow-up Few patients irradiated at young ages Possibility of selection bias by treatment
Swedish ^{131}I thyroid cancer [H26]	Incidence follow-up Administered activities of ^{131}I known Unexposed group	Individual doses not computed Small numbers for specific cancer types Few patients irradiated at young ages Possibility of selection bias by treatment
French therapeutic ^{131}I [D18]	Incidence follow-up Administered activities of ^{131}I known Exclusion of patients who received external radiotherapy Unexposed group	Individual doses not computed Small numbers for specific cancer types Few patients irradiated at young ages Possibility of selection bias by treatment
Environmental exposures		
Techa River population [K5, K27]	Large numbers with relatively long follow-up Wide range of estimated doses Unselected population; attempted use of local population rates for comparison Possible to examine ethnic differences in cancer risk Potential for future	Dosimetry difficult and not individual Mixture of internal and external exposures complicates dosimetry Follow-up and cancer ascertainment uncertain Contribution of chemical exposures not evaluated
Chernobyl-related exposure [A26]	Large numbers exposed Wide range of thyroid doses within the states of the former Soviet Union	Mixture of radioiodines and availability of data make dose estimation difficult, particularly for individuals Possible differences in cancer ascertainment relative to the general population Fairly short period of follow-up so far
Marshall Islands fallout [H35, R21]	Population unselected for exposure Comprehensive long-term medical follow-up Individual dosimetry attempted	Mixture of radioiodines and gamma radiation preclude accurate dose estimation Surgery and hormonal therapy probably influenced subsequent occurrence of thyroid neoplasms Small numbers
Utah ^{131}I fallout: thyroid disease [K36]	Comprehensive dosimetry attempted Protracted exposures at low rate	Possible recall bias in consumption data used for risk estimation Possible under-ascertainment of disease in low-dose subjects Small number of thyroid cancers
Utah ^{131}I fallout [S37]	Comprehensive dosimetry attempted Large number of leukaemia deaths Protracted exposures at low rate	Uncertainty in estimating bone marrow doses Estimated cumulative doses lower than from natural background radiation
Occupational exposures		
UK Atomic Energy Authority: Prostate cancer study [R26]	Information abstracted for study subjects on socio-demographic factors, exposures to radionuclides, external doses and other substances in the workplace Cases and controls selected from an existing cohort	Exposure to some radionuclides tended to be simultaneous, making it difficult to study them individually

Table 4
Cohort and case-control epidemiological studies of the effects of exposures to high-LET radiation

Study	Type of study	Population studied — Characteristics	Population studied — National origin	Follow-up (years)	Total person-years [a]	Type of exposure	Type of dosimetry	Cancers studied [b]
Treatment of benign disease								
^{224}Ra TB and ankylosing spondylitis patients [H8, N4, N14, S14]	Incidence	899 exposed persons 31% females 24% aged [c] ≤20 years	Germany	0–54	23 400 (28.8) [d]	Injection with ^{224}Ra	Internal dosimetric calculations based on amount injected	Bone*, breast*, connective tissue*, liver*, kidney*, thyroid*, ovary, leukaemia, pancreas, uterus, prostate, bladder*, stomach, colon, lung
^{224}Ra ankylosing spondylitis patients [W3, W20]	Incidence	1 577 exposed persons 1 462 unexposed persons	Germany	0–51	63 500 (20.8)	Injection with ^{224}Ra	Information on amount injected	Bone and connective tissue, leukaemia* non-Hodgkin's lymphoma, Hodgkin's disease, stomach, liver, lung, urinary system, female breast
Diagnostic examinations								
German thorotrast patients [V1, V8]	Mortality	2 326 exposed persons 1 890 unexposed persons 26% females	Germany	3–>50	n.a. [e]	Injection with thorotrast	Hospital records of amounts injected; CT measurements of some patients; x-ray films	Liver*, extrahepatic bile ducts*, gallbladder, myeloid leukaemia*, pancreas*, myelodysplastic syndrome*, non-Hodgkin's lymphoma*, plasmacytoma, larynx, bone sarcoma, lung, mesothelioma*, Hodgkin's disease, lymphatic leukaemia, kidney, bladder, prostate, adrenal, brain, GI tract
Danish thorotrast patients [A4, A5] [f]	Incidence/ Mortality	800 exposed persons 1 236 unexposed persons 49% females 17% aged <20	Denmark	1–>50	49 883 (24.5)	Injection with thorotrast	Records of amount injected; dosimetric factors for dose to liver and bone marrow	Liver*, leukaemia*, brain and nervous system*, lung, breast, ovary, all other sites
Swedish thorotrast patients [M48]	Mortality	693 exposed persons 43% females Ages: 2–66 (35)	Sweden	6–60	10 077 (14.5)	Injection with thorotrast	Records of amount injected	All malignant neoplasms*, all benign neoplasms*
Portuguese thorotrast patients [D3, D31] [g]	Mortality	1 131 exposed persons 1 032 unexposed persons 39% females Age: 5–>65 (34 for exposed, 40 for unexposed)	Portugal	0–68	36 321 (16.8)	Injection with thorotrast	Hospital records of amount injected; dosimetric factors for alpha dose to liver	Liver*, gallbladder, lung, bone*, nervous system, leukaemia (excluding chronic lymphatic)*, unspecific sites, all neoplasms*

Table 4 (continued)

Study	Type of study	Population studied — Characteristics	Population studied — National origin	Follow-up (years)	Total person-years [a]	Type of exposure	Type of dosimetry	Cancers studied [b]
Early Japanese thorotrast patients [M14, M47]	Mortality	262 exposed persons 1 630 unexposed persons Age: 20-39	Japan	18-68	n.a.	Injection with thorotrast	Amount injected	Liver*, lung, bone sarcoma, leukaemia*
Later Japanese thorotrast patients [K33, M14]	Mortality	150 exposed persons Age: 15-39	Japan	34-65	n.a.	Injection with thorotrast	Amount injected	Liver*, lung, leukaemia*
Occupational exposure: radium								
United States radium luminizers [C27, S12, S16, S54] [h]	Incidence/ Mortality	2 543 females	United States	0-71.5	119 020 (46.8)	Ingestion of ^{226}Ra and ^{228}Ra	Body burdens of about 1 500 women assessed by measurement of gamma rays and /or exhaled radon; used for calculation of systemic intake and skeletal dose	Bone sarcoma*, paranasal sinuses and mastoid air cells*, stomach, colon, rectum, liver, lung, breast*, pancreas, brain and other central nervous system tumours, leukaemia, multiple myeloma
United Kingdom radium luminizers [B13, B14]	Mortality	1 203 females	United Kingdom	47 (max.)	44 883	Work with radium	Some measurements of body burdens. Assessments of external doses	Breast, leukaemia, osteosarcoma, all cancers combined
Occupational exposure: plutonium								
Mayak plutonium workers [G23, K10, K11, K32, K34, K42] [i]	Mortality	4 186 persons with measured plutonium body burdens (1 479 males employed before 1959) 14 644 other workers (5 174 males employed before 1959) 25% females [j]	Russian Federation	Up to 47	211 427 (31.8) [k]	Exposures to plutonium in nuclear fuel cycle and research	Measurement of plutonium in urine. Recorded exposures to external radiation	Lung*, liver*, bone*
Mayak radiochemical plant workers: lung cancer study [T2]	Case-control 162 cases 338 controls	283 persons with Pu body burden > 0.75 kBq; 217 persons with lower body burdens 11% females	Russian Federation	42 (max.)	n.a.	Exposures to plutonium in nuclear fuel cycle and research	Measurement of plutonium in urine. Recorded exposures to external radiation	Lung cancer*
Sellafield plutonium workers [O1]	Incidence/ Mortality	5 203 plutonium workers 4 609 of whom had plutonium dose assessed 5 179 other radiation workers 4 003 non-radiation workers 19% females	United Kingdom	Up to 46 for mortality; up to 40 for incidence	415 432 (29)	Exposures to plutonium in nuclear fuel cycle and research	Measurement of plutonium in urine. Recorded exposures to external radiation	Stomach, colon, pancreas, lung, pleura, breast, prostate, bladder, brain and other central nervous system tumours, ill-defined and secondary, non-Hodgkin's lymphoma, leukaemia

Table 4 (continued)

Study	Type of study	Population studied		Follow-up (years)	Total person-years [a]	Type of exposure	Type of dosimetry	Cancers studied [b]
		Characteristics	National origin					
Rocky Flats workers [W12]	Mortality	5 413 males with external and/or plutonium exposures	United States	Up to 28	52 772 (9.7)	Exposures to plutonium in nuclear fuel cycle and research	Measurement of plutonium in urine Recorded exposures to external radiation	Oesophagus, stomach, colon, rectum, liver, pancreas, larynx, lung, skin, prostate, bladder, kidney, brain and other central nervous system tumours, non-Hodgkin's lymphoma, multiple myeloma, myeloid leukaemia
Los Alamos workers [W8]	Mortality	3 775 males with Pu body burden of 74 Bq or more 11 952 males with lower body burdens	United States	Up to 47	456 637 (29)	Exposures to plutonium in nuclear fuel cycle and research	Measurement of plutonium in urine Recorded exposures to external radiation	Oral, stomach, colon, rectum, pancreas, lung, bone, prostate, bladder, kidney, brain and other central nervous system tumours, all lymphatic/haematopoietic cancers

Occupational exposure: others (excluding radon in mines)

Study	Type of study	Population studied		Follow-up (years)	Total person-years [a]	Type of exposure	Type of dosimetry	Cancers studied [b]
		Characteristics	National origin					
Three United Kingdom industry workforces [C33]	Mortality	17 605 workers monitored for radionuclide exposure 23 156 other radiation workers 8% females	United Kingdom	Up to 43	1 020 000 (25)	Exposures in nuclear fuel cycle and research	Data on monitoring for plutonium, tritium, and other radionuclides	Lung, pleura, skin, uterus, prostate, multiple myeloma, leukaemia, other cancers
Oak Ridge, Y-12 workers [C32]	Mortality	Males 3 490 workers with internal exposure monitoring data 3 291 other workers Age at entry: 16–64	United States (white)	Up to 33	133 535 (19.7)	Exposures in nuclear fuel cycle and research	Urine measurements and whole-body monitoring of internally deposited uranium	Lung, brain and other central nervous system
Mound workers [W22]	Mortality	4 402 males	United States (white)	Up to 40	104 326 (23.7)	Exposures in nuclear fuel cycle and research	Measurement of polonium in urine	Oral, oesophagus, stomach, colon, rectum, liver, pancreas, lung, bone, skin, prostate, bladder, kidney, brain and other central nervous system tumours, thyroid, non-Hodgkin's lymphoma, Hodgkin's disease, leukaemia
Florida phosphate workers [C34]	Mortality	23 012 males Age at entry: (median 25)	United States	Up to 44	545 867 (23.7)	Exposures in mining and chemical processing of phosphate ores	Assessments of cumulative exposures to alpha and gamma radiation, based on job histories	Lung

Table 4 (continued)

Study	Type of study	Population studied — Characteristics	Population studied — National origin	Follow-up (years)	Total person-years [a]	Type of exposure	Type of dosimetry	Cancers studied [b]
Fernald workers [D32, R25]	Mortality	4 014 males Age at entry: (30.4)	United States (white)	Up to 49	124 177 (30.9)	Exposures in nuclear fuel cycle and research	Measurement of uranium, thorium and radium compounds in urine, plus environmental area sampling Recorded exposures to external radiation	Lung, respiratory tract, upper gastrointestinal tract, lower gastrointestinal tract, bladder and kidney, haematopoietic and lymphopoietic neoplasms
Chinese iron and steel workers [L49]	Mortality	Males 5 985 exposed 2 849 unexposed	China	Up to 17	111 286 (12.6)	Exposure to thorium-containing dust in an iron and steel company	Assessment of lung doses from inhalation	Lung, leukaemia*
Occupational exposure: radon in mines								
Chinese tin miners [D8, L4, X1] [l]	Mortality	13 649 exposed males 3 494 unexposed males Age: <10–>30	China	0–>25	175 342 (10.2)	Radon in tin mines	Measurements of WL from 1972; reconstruction of earlier values	Lung cancer*, all other cancers
West Bohemia uranium miners [D8, L4, T3, T22] [l]	Mortality	4 284 exposed males Age: <25–>45	Czech Republic	0–>25	107 868 (25.2)	Radon in uranium mines	Measurements of radon progeny after 1967; radon measured since 1949, with equilibrium values based on ventilation	Lung cancer*, all other cancers
Colorado Plateau uranium miners [D8, L4, H17] [l]	Mortality	3 347 exposed males Age: <25–>45	United States	0–>25	82 435 (24.6)	Radon in uranium mines	Measurements of radon progeny during 1951–1968; few measurements earlier	Lung cancer*, all other cancers
Ontario uranium miners [D8, L4, K4] [l]	Mortality	21 346 exposed males Age: <25–>45	Canada	0–>15	380 718 (17.8)	Radon in uranium and gold mines	Measurements of WL in uranium mines from 1957; no measurements in gold mines before 1961, a few during the 1960s and 1970s and extensive measurements thereafter	Lung cancer*, all other cancers
Newfoundland fluorspar miners [D8, L4, M15] [l]	Mortality	1 751 exposed males 337 unexposed males Age: <20–>30	Canada	0–>15	48 742 (23.3)	Radon in fluorspar mines	WL estimated from work environment, architecture and ventilation before 1960. Systematic measurement programme stated in 1968	Lung cancer*, all other cancers

Table 4 (continued)

Study	Type of study	Population studied — Characteristics	Population studied — National origin	Follow-up (years)	Total person-years [a]	Type of exposure	Type of dosimetry	Cancers studied [b]
Swedish iron miners [D8, L4, R8] [l]	Mortality	1 294 exposed males Age: <25->30	Sweden	0->25	33 293 (25.7)	Radon in iron mine	Radon measurements made during 1968–1975, with estimation of WL using characteristics of natural and mechanical ventilation	Lung cancer*, all other cancers
New Mexico uranium miners [D8, L4, S19] [l]	Mortality	3 457 exposed males 12 unexposed males Age: <30->40	United States	0->15	58 949 (17.0)	Radon in uranium mines	Extensive measurements; for individual miners since 1969	Lung cancer*, all other cancers
Beaverlodge uranium miners [D8, L4, H15, H18] [l]	Mortality	6 895 exposed males 1 591 unexposed males Age: <30->45	Canada	0->20	118 385 (14.0)	Radon in uranium mine	Measurements of radon progeny after 1967; radon measurements earlier, mainly for control purposes	Lung cancer*, all other cancers
Port Radium uranium miners [D8, L4, H16] [l]	Mortality	1 420 exposed males 683 unexposed males Age: <30->45	Canada	0->25	52 676 (25.2)	Radon in uranium mine	Measurements for 1945–1958; no exposures estimated before 1940	Lung cancer*, all other cancers
Radium Hill uranium miners [L4, W7] [l]	Mortality	1 457 exposed males 1 059 unexposed males Age: <35->40	Australia	0->25	51 850 (21.9)	Radon in uranium mine	Measurements of radon during 1954–1961, but not earlier	Lung cancer*
French uranium miners [D8, L4, T8] [l]	Mortality	1 769 exposed males 16 unexposed males Age: <30->40	France	0->25	44 043 (24.7)	Radon in uranium mines	Systematic monitoring after 1957; only a few measurements during 1947–1955	Lung cancer*, all other cancers
Cornish tin miners [D8, H13]	Mortality	2 535 males	United Kingdom	0->25	66 900 (26.4)	Radon in tin mines	Monitoring of radon concentrations since 1967; no measurements in previous decades	Lung cancer*, all other cancers

Residential radon studies

Study	Type of study	Population studied — Characteristics	Population studied — National origin	Follow-up (years)	Total person-years [a]	Type of exposure	Type of dosimetry	Cancers studied [b]
United States acute lymphoblastic leukaemia study [L34]	Case-control 505 cases 443 controls	48% females Age at diagnosis: 0–14 10% with time-weighted average radon concentrations above 148 Bq m^{-3}	United States	n.a.	n.a.	Radon in homes	Track-etch detector measurements in homes occupied by subjects	Acute lymphoblastic leukaemia
United States acute myeloid leukaemia study [S52]	Case-control 173 cases 254 controls	51% females Age at diagnosis: 0–17 Mean time-weighted average radon concentration 53 Bq m^{-3} (14% above 100 Bq m^{-3})	United States	n.a.	n.a.	Radon in homes	Track-etch detector measurements in homes occupied by subject at time of diagnosis	Acute myeloid leukaemia

Table 4 (continued)

Study	Type of study	Population studied Characteristics	Population studied National origin	Follow-up (years)	Total person-years [a]	Type of exposure	Type of dosimetry	Cancers studied [b]
West German childhood cancer [K38]	Case control 82 leukaemia cases 82 solid tumour cases 209 controls	Age at diagnosis: 0–14 Mean time-weighted average radon concentration 27 Bq m^{-3}	Germany	n.a.	n.a.	Radon in homes	Track-etch detector measurements in homes occupied by subjects for at least one year	Leukaemia, solid tumours
Central Italy [F7]	Case-control 44 cases 211 controls	Males Age at diagnosis: 35–80 (68) 75% with radon concentration above 100 Bq m^{-3}	Italy	10	n.a.	Radon and gamma radiation in homes	Measurements in last dwelling occupied and characteristics of dwellings	Acute myeloid leukaemia
New Jersey [S50]	Case-control 433 cases 422 controls	Females Age at diagnosis: all (65% aged 58 or more) Mean time-weighted-average radon concentration 26 Bq m^{-3} (20% above 37 Bq m^{-3})	United States	n.a.	n.a.	Radon in homes	Track-etch detector measurements in homes occupied by subjects	Lung cancer
Shenyang [B37]	Case-control 308 cases 362 controls	Females Age at diagnosis: 30–69 Mean time-weighted-average radon concentration 118 Bq m^{-3}	China	n.a.	n.a.	Radon in homes	Track-etch detector measurements in homes occupied by subjects	Lung cancer
Stockholm [P32]	Case-control 210 cases 400 controls	Females Age at diagnosis: all Mean time-weighted-average radon concentration 128 Bq m^{-3}	Sweden	n.a.	n.a.	Radon in homes	Track-etch detector measurements in homes occupied by subjects	Lung cancer
Swedish nationwide [P33]	Case-control 1 281 cases 2 576 controls	47% females Age at diagnosis: 35–74 Mean time-weighted-average radon concentration 107 Bq m^{-3}	Sweden	n.a.	n.a.	Radon in homes	Track-etch detector measurements in homes occupied by subjects	Lung cancer
Winnipeg [L44]	Case-control 738 cases 738 controls	34% females Age at diagnosis: 35–80 Mean time-weighted-average radon concentration 107 Bq m^{-3}	Canada	n.a.	n.a.	Radon in homes	Track-etch detector measurements in homes occupied by subjects	Lung cancer

Table 4 (continued)

Study	Type of study	Population studied — Characteristics	Population studied — National origin	Follow-up (years)	Total person-years [a]	Type of exposure	Type of dosimetry	Cancers studied [b]
Missouri-I [A22]	Case-control 538 cases 1 183 controls	Females Age at diagnosis: 30 – 84 Mean time-weighted-average radon concentration 67 Bq m^{-3}	United States	n.a.	n.a.	Radon in homes	Track-etch detector measurements in homes occupied by subjects	Lung cancer
Finland-I [R23]	Case-control 238 cases 434 controls	Males Age at diagnosis: all Mean time-weighted-average radon concentration 220 Bq m^{-3}	Finland	n.a.	n.a.	Radon in homes	Track-etch detector measurements in homes occupied by subjects	Lung cancer
Finland-II [A23]	Case-control 1 055 cases 1 544 controls (517 matched pairs)	97% males Age at diagnosis: all Mean time-weighted-average radon concentration 96 Bq m^{-3}	Finland	n.a.	n.a.	Radon in homes	Track-etch detector measurements in homes occupied by subjects	Lung cancer
West Germany [W17]	Case-control 1 449 cases 2 297 controls	18% females Age at diagnosis: <75 Mean radon concentration 49 Bq m^{-3} (cases) 50 Bq m^{-3} (controls); corresponding values of 67 Bq m^{-3} (cases) 60 Bq m^{-3} (controls) in radon-prone areas	Germany	n.a.	n.a.	Radon in homes	Track-etch detector measurements in homes occupied by subjects	Lung cancer
East Germany [W18]	Case-control 1 053 cases 1 667 controls	12% females Age at diagnosis: <75 Mean radon concentration 87 Bq m^{-3} (cases) 90 Bq m^{-3} (controls) in living room; corresponding values of 66 Bq m^{-3} (cases) 63 Bq m^{-3} (controls) in bedrooms in radon-prone areas	Germany	n.a.	n.a.	Radon in homes	Track-etch detector measurements in homes occupied by subjects	Lung cancer
Southwest England [D30]	Case-control 982 cases 3 185 controls	33% females Age at diagnosis: <75 Mean time-weighted-average radon concentration 58 Bq m^{-3} (cases) 56 Bq m^{-3} (controls)	United Kingdom	n.a.	n.a.	Radon in homes	Track-etch detector measurements in homes occupied by subjects	Lung cancer

Table 4 (continued)

Study	Type of study	Population studied		Follow-up (years)	Total person-years [a]	Type of exposure	Type of dosimetry	Cancers studied [b]
		Characteristics	National origin					
Missouri-II [A24]	Case-control 512 cases 553 controls	Females Age at diagnosis: <75 Mean time-weighted average radon concentration 64.6 Bq m^{-3} based on CR-39 surface measurements (58.5 Bq m^{-3} based on track-etch measurements)	United States	n.a.	n.a.	Radon in homes	Track-etch and CR-39 detector measurements in homes occupied by subjects	Lung cancer

a   Mean per person in parentheses.

b   An asterisk denotes sites for which statistically significant excesses are reported in the exposed group (cohort studies) or for which a higher proportion of the cases were exposed to radiation (case-control studies).

c   Age at first exposure, mean in parentheses.

d   Figures quoted are for 812 persons with complete information [N14].

e   Not available.

f   Figures quoted are from [A5], for persons eligible for the cancer incidence analysis.

g   Figures quoted are for persons followed up to the end of 1996 [D31].

h   Figures quoted are from [S12].

i   Figures quoted are from [K10]. Preliminary results from a follow-up to the end of 1993 are given in [K11], but not to the same detail.

j   Figures based on [K32, K34].

k   Figures for males employed before 1959 [K34].

l   Figures are from [L45].

Table 5
Strengths and limitations of major cohort and case-control epidemiological studies of carcinogenic effects of exposures to high-LET radiation

Study	Strengths	Limitations
Treatment for benign disease		
[224]Ra patients	Large number of excess bone cancers Long-term follow-up Substantial proportion of patients treated in childhood or adolescence	Uncertainties in organ doses for individual patients Other aspects of treatment may be relevant (e.g. x rays) Comparison group constructed only recently for the Spiess study [S14]
Diagnostic examinations		
Thorotrast patients	Large number of excess cancers Long-term follow-up	Uncertainties in organ doses for individual patients Chemical attributes of thorotrast might influence risks
Occupational exposures		
Radium luminizers	Protracted exposures from [226]Ra Large numbers of excess cancers in United States study	Potential inaccuracies in estimating radium intakes Distribution of radium in bone may be non-uniform External irradiation may be relevant for breast cancers
Mayak workers	Wide range of exposures Individual measurements of plutonium body burden and external gamma dose Information on smoking and other potential confounders in the lung cancer case-control study	Possible uncertainties in assessment of exposures Further details of the ascertainment of subjects in the lung cancer case-control study [T2] would be desirable
United Kingdom and United States nuclear workers	Individual measurements of plutonium body burden or other internally deposited radionuclides, and external gamma dose	General lack of information on smoking and other potential non-radiation confounders Possible uncertainties in assessment of internal exposures
Florida phosphate workers [C34]	Relatively large number of person-years Assessment of exposures to other agents (e.g. silica and acid mists)	Not possible to obtain direct quantitative estimates of exposure levels Absence of data on smoking habits for lung cancer analysis
Chinese iron and steel workers [L49]	Assessments made of lung doses from inhalation of thorium Information available on smoking habits	Lung doses generally low Small number of deaths for specific cancer types
Radon-exposed underground miners	Large numbers Protracted exposures over several years Wide range of cumulative exposures Exposure-response analyses	Uncertainties in assessment of early exposures Possible modifying effect of other types of exposure (e.g. arsenic) Smoking histories limited or not available
Environmental exposures		
Residential radon	Large numbers in most studies Protracted exposures over many years Individual data on radon and smoking	Uncertainties in assessing exposures (measurement error, mobility between dwellings, structural changes to dwellings) Radon concentrations low for many subjects

Table 6
Risk estimates for cancer incidence and mortality from studies of radiation exposure: oesophageal cancer
The number of observed and expected cases as well as the mean dose and person-years for cohort studies are computed throughout this Table for exposed persons only. In the Life Span Study the exposed group included survivors with organ doses of 0.01 Sv or more for incidence and 0.005 Sv or more (weighted colon dose) for mortality

Study		Observed cases	Expected cases	Mean dose (Sv)	Person-years	Average excess relative risk [a] at 1 Sv	Average excess absolute risk [a] $(10^4\ PYSv)^{-1}$
EXTERNAL LOW-LET EXPOSURES							
Incidence							
Life Span Study [T1]							
Sex	Male	68	66.2	0.23	297 452	0.12	0.26
	Female	16	11.2	0.22	491 130	1.95	0.44
Age at exposure	<20 years	8	8.2	0.23	297 452	-0.11	-0.03
	>20 years	76	69.2	0.22	491 130	0.45	0.63
All		84	77.4	0.23	788 582	0.37 (-0.45- 1.31) [b]	0.36 (-0.44- 1.28) [b]
Cervical cancer cohort [B11] [c]		12	11.0	0.35	178 243	0.26 (95% CI: -1.1- 1.3) [b]	0.16 (95% CI: -0.6- 1.3) [b]
Mortality							
Life Span Study [P9]							
Age at exposure							
Males	<20 years	13	15.9	0.21	376 371	-0.87 (-2.44 -1.40)	-0.37 (-1.04 -0.59)
	20-39 years	30	31.2	0.25	117 959	-0.15 (-1.22 -1.20)	-0.40 (-3.22 -3.18)
	>40 years	61	55.0	0.23	132 009	0.48 (-0.50 -1.64)	1.99 (-2.10 -6.82)
Females	<20 years	0	0.0	0.20	416 447	–	–
	20-39 years	14	10.2	0.19	358 988	1.94 (-0.88 -5.96)	0.55 (-0.25 -1.69)
	>40 years	19	12.9	0.17	201 931	2.78 (-0.20 -6.81)	1.77 (-0.13 -4.34)
Time since exposure							
Both sexes	5-10 years	13	12.9	0.22	261 996	0.05 (-1.87 -2.81)	0.02 (-0.92 -1.38)
	11-25 years	52	40.6	0.20	658 705	1.41 (0.02 -3.08)	0.87 (0.01 -1.90)
	26-40 years	51	49.2	0.20	533 369	0.18 (-0.97 -1.57)	0.17 (-0.89 -1.45)
	41-45 years	21	16.2	0.19	144 940	1.55 (-0.67 -4.52)	1.73 (-0.75 -5.04)
All		137	125.2	0.21	1 603 705	0.76 (0.02 -1.59) [b]	0.56 (0.02 -1.16) [b]
Ankylosing spondylitis [W1] [d]		74	38	5.55	287 095	0.17 (95% CI: 0.09- 0.25) [e]	0.23 (95% CI: 0.1- 0.3) [b]
Metropathia haemorrhagica [D7]		9	9.27	0.05	47 144	-0.58 (<-0.2- 13.9) [b]	-0.94 (-7.0 - 22.5) [b]
Massachusetts TB fluoroscopy [D4]		14	6.7	0.80	169 425	n.a. [f]	n.a.
Nuclear workers in Canada, United Kingdom, United States [C11]		104	n.a.	0.04	2 124 526	>0 [g]	n.a.
Nuclear workers in Japan [E3]		25	37.1	0.014	533 168	>0 [g]	n.a.
INTERNAL LOW-LET EXPOSURES							
Mortality							
United States thyrotoxicosis [R14]		25	25	n.a.	385 468	n.a. [h]	n.a.

a   90% CI in parentheses derived from published data for Life Span Study and using exact Poisson methods for the other studies.
b   Estimates based on method described in the introduction to Chapter III.
c   The values given are for 10-year survivors.
d   The values given exclude the period within five years of first treatment.
e   Dose-response analysis based on the number of treatment courses given.
f   Not available.
g   Based on a 10-year lag. Trend not statistically significant..
h   No apparent trend with administered level of ^{131}I, although a significance test was not performed.

Table 7
Risk estimates for cancer incidence and mortality from studies of radiation exposure: stomach cancer
The number of observed and expected cases as well as the mean dose and person-years for cohort studies are computed throughout this Table for exposed persons only. In the Life Span Study the exposed group included survivors with organ doses of 0.01 Sv or more for incidence and 0.005 Sv or more (weighted colon dose) for mortality

Study		Observed cases	Expected cases	Mean dose (Sv)	Person-years	Average excess relative risk [a] at 1 Sv	Average excess absolute risk [a] $(10^4 \ PYSv)^{-1}$
	**EXTERNAL LOW-LET EXPOSURES**						
	**Incidence**						
Life Span Study [T1]							
Sex	Male	679	660.4	0.24	298 700	0.12	2.61
	Female	628	561.3	0.23	493 900	0.52	5.86
Age at exposure	<20 years	167	142.0	0.24	365 200	0.74	2.87
	>20 years	1 140	1 079.7	0.23	427 300	0.24	6.15
All		1 307	1 221.7	0.23	792 500	0.30 (0.2–0.5) [b]	4.68 (2.5–7.4) [b]
Cervical cancer case-control [B1] [c]		348	167.3	2	n.a.	0.54 (0.05–1.5)	1.23
Mayak workers [Z1]		20 [d]	n.a.	>3	n.a.	1.1 (95% CI: 0.01–3.4) [e]	n.a.
Swedish benign breast disease [M28]		14	15.6	0.66	26 493	1.3 (95% CI: 0–4.4)	n.a.
Stockholm skin haemangioma [L16]		5	~6	0.09	406 565	<0	<0
	**Mortality**						
Life Span Study [P9]							
Age at exposure							
Males	<20 years	78	75.2	0.21	369 372	0.18 (−0.7–1.2)	0.37 (−1.4–2.5)
	20-39 years	193	188.2	0.21	116 442	0.12 (−0.4–0.7)	1.96 (−7.1–12.0)
	>40 years	536	527.0	0.21	129 183	0.08 (−0.3–0.4)	3.30 (−10.2–17.8)
Females	<20 years	63	51.6	0.21	414 045	1.05 (−0.1–2.4)	1.31 (−0.1–3.0)
	20-39 years	257	233.6	0.21	357 293	0.48 (−0.0–1.0)	3.12 (−0.3–6.8)
	>40 years	390	369.0	0.21	201 031	0.27 (−0.1–0.7)	4.97 (−2.6–13.0)
Time since exposure							
Both sexes	5-10 years	153	151.6	0.21	186 468	0.04 (−0.6–0.7)	0.35 (−4.7–6.0)
	11-25 years	610	581.0	0.21	725 251	0.24 (−0.1– [f])	1.90 (−0.7– [e])
	26-40 years	606	573.8	0.21	530 897	0.27 (−0.1–0.6)	2.89 (−0.6–6.6)
	41-45 years	148	137.7	0.21	144 740	0.36 (−0.3–1.1)	3.38 (−3.0–10.5)
All		1 517	1 444.1	0.21	1 587 355	0.24 (0.03–0.5) [b]	2.19 (0.30–4.1) [b]
Ankylosing spondylitis [W1] [g]		127	128	3.21	287 095	−0.004 (95% CI: −0.05-0.05) [h]	−0.02 (95% CI: −0.2-0.2)
Yangjiang background radiation [T25, T26]		70	77.8	n.a. [i]	1 246 340	−0.27(95% CI: −1.37-2.69) [j]	n.a.
Peptic ulcer [G6]		40	14.4 [k]	14.8	35 815	0.15	0.25
Metropathia haemorrhagica [D7] [l]		33	26.8	0.23	47 144	1.01 (<−0.2–2.8) [b]	5.72 (<−2.4–16) [b]
Benign gynaecological disease [I16] [m]		23	21.8	0.2	71 958	0.27 (−4.25–4.80) [n]	0.83 (<0–72.7) [b]
Nuclear workers in Canada, United Kingdom, United States [C11]		275	n.a.	0.04	2 124 526	<0 [o]	n.a.
Nuclear workers in Japan [E3]		149	177.2	0.014	533 168	<0 [o]	n.a.

Table 7 (continued)

Study	Observed cases	Expected cases	Mean dose	Person-years	Average relative risk [p]	
INTERNAL LOW-LET EXPOSURES						
Incidence						
Swedish hyperthyroid patients [H23]	58 [q]	43.6	0.25 Gy	n.a.	2.32 [r]	
Mortality						
United States thyrotoxicosis patients [R14]	82	78.0	0.178	385 468	>0 [s]	
INTERNAL HIGH-LET EXPOSURES						
Incidence						
^{224}Ra ankylosing spondylitis patients [W20]	18	12.2	n.a.	32 800	1.56 [t, u]	
^{224}Ra ankylosing spondylitis patients [N4]	13	~11	n.a.	25 000	~1.2	
Danish thorotrast patients [A5]	7	6.9	n.a.	19 365	1.82 (0.61–5.66) [t]	
Mortality						
German thorotrast patients [V3, V8]	30 [v]	n.a.	20.6 ml [w]	n.a.	0.6 [t]	

a   90% CI in parentheses derived from published data for Life Span Study and using exact Poisson methods for the other studies.
b   Estimates based on method described in the introduction to Chapter III.
c   Based on 5-year survivors. The observed and expected numbers are for both exposed and unexposed persons. The excess absolute risk estimate was computed using background incidence rates estimated using the cervical cancer cohort study [B11].
d   Workers with external gamma dose in excess of 3 Gy.
e   ERR among those with external gamma doses in excess of 3 Gy relative to those with lower doses.
f   Calculation of upper confidence limit did not converge.
g   The values given exclude the period within five years of first treatment.
h   Dose-response analysis based on the number of treatment courses given.
i   Mean annual effective dose = 6.4 mSv.
j   Based on a 10-year latent period.
k   Based on unirradiated patients.
l   The values given exclude the period within five years of irradiation.
m   The observed and expected number of cases are for 10-year survivors. The estimated number of expected cases incorporated an adjustment based on the Poisson regression model given in [I16].
n   Wald-type CI.
o   Based on a 10-year lag. Trend not statistically significant.
p   95% CI in parentheses.
q   Restricted to the period 10 or more years after treatment.
r   Relative risk at 1 Gy.
s   No apparent trend with administered activity of ^{131}I , although a significance test was not performed.
t   Relative to unexposed controls.
u   In the control group, 16 stomach cancers were diagnosed, compared with 16.9 expected.
v   Number quoted in an earlier follow-up [V3].
w   Amount of thorotrast administered.

Table 8
Risk estimates for cancer incidence and mortality from studies of radiation exposure: colon cancer
The number of observed and expected cases as well as the mean dose and person-years for cohort studies are computed throughout this Table for exposed persons only. In the Life Span Study the exposed group included survivors with organ doses of 0.01 Sv or more for incidence and 0.005 Sv or more (weighted colon dose) for mortality

Study		Observed cases	Expected cases	Mean dose (Sv)	Person-years	Average excess relative risk[a] at 1 Sv	Average excess absolute risk[a] $(10^4 PYSv)^{-1}$
EXTERNAL LOW-LET EXPOSURES							
Incidence							
Life Span Study [T1]							
Sex	Male	109	90.7	0.23	297 500	0.87	2.66
	Female	114	103.0	0.22	491 100	0.48	1.01
Age at exposure	<20 years	32	28.0	0.23	363 300	0.62	0.48
	>20 years	191	165.7	0.22	425 300	0.70	2.71
All		223	193.7	0.23	788 600	0.67 (0.1-1.3)[b]	1.65 (0.7-3.0)[b]
Cervical cancer case-control [B1][c]		409	409	24	n.a.	0.00 (-0.01-0.02)	0.01 (-0.09-0.18)
Stockholm skin haemangioma [L16]		12	~11	0.07	406 565	0.37[d]	0.11
Mortality							
Life Span Study [P9]							
Age at exposure							
Males	<20 years	18	13.8	0.20	369 372	1.52 (-0.8-4.7)	0.57 (-0.3-1.7)
	20-39 years	25	22.7	0.20	116 442	0.51 (-1.2-2.7)	0.99 (-2.3-5.2)
	>40 years	45	42.7	0.20	129 183	0.27 (-1.0-1.8)	0.88 (-3.2-5.8)
Females	<20 years	9	5.6	0.20	414 045	2.96 (-0.8-8.9)	0.40 (-0.1-1.2)
	20-39 years	49	40.4	0.20	357 283	1.06 (-0.3-2.7)	1.20 (-0.3-3.0)
	>40 years	52	48.0	0.20	201 031	0.42 (-0.8-1.8)	1.00 (-1.8-4.4)
Time since exposure							
Both sexes	5-10 years	9	9.4	0.20	186 468	-0.22 (-2.5-3.3)	-0.11 (-1.3-1.7)
	11-25 years	41	37.7	0.20	725 251	0.44 (-0.9-2.1)	0.23 (-0.5-1.1)
	26-40 years	97	85.3	0.20	530 897	0.69 (-0.2-1.7)	1.10 (-0.4-2.8)
	41-45 years	51	41.9	0.20	144 740	1.08 (-0.2-2.7)	3.14 (-0.7-7.8)
All		198	173.2	0.20	1 587 355	0.71 (0.06-1.4)[b]	0.78 (0.07-1.6)[b]
Benign gynaecological disease [I16][e]		75	46.6	1.3	71 958	0.51 (-0.8-5.61)	3.2 (-0.9-7.1)[b]
Metropathia haemorrhagica [D7][f]		47	33	3.2	47 144	0.13 (95% CI: 0.01-0.26)	0.92 (95% CI: 0.1-1.8)[b]
Peptic ulcer [G6]		31	24.0[g]	6	35 815	0.05 (95% CI: -0.05-0.22)[b]	0.33[b]
Nuclear workers in Canada, United Kingdom, United States [C11]		343	n.a.	0.04	2 124 526	<0[d,h]	n.a.
Nuclear workers in Japan [E3]		51	42.6	0.014	533 168	<0[d,h]	n.a.

Study		Observed cases	Expected cases	Mean dose	Person-years	Average relative risk[i]	
INTERNAL LOW-LET EXPOSURES							
Mortality							
United States thyrotoxicosis patients [R14][j]		282	255	0.108[k]	385 468	n.a.[l]	

Table 8 (continued)

Study	Observed cases	Expected cases	Mean dose	Person-years	Average relative risk [i]	
INTERNAL HIGH-LET EXPOSURES						
Incidence						
Danish thorotrast patients [A5]	9	7.1	n.a.	19 365	1.28 (0.54–2.84) [m]	
Mortality						
German thorotrast patients [V3, V8]	10 [n]	n.a.	20.6 ml [o]	n.a.	~0.5 [m]	

a   90% CI in parentheses derived from published data for Life Span Study and using exact Poisson methods for the other studies.
b   Estimates based on method described in the introduction to Chapter III.
c   Based on 10-year survivors. The observed and expected numbers cover both exposed and unexposed persons. The excess absolute risk estimate was computed using background incidence rates, estimated using the cervical cancer cohort study [B11].
d   Not statistically significantly different from zero.
e   The observed and expected number of cases are for 10-year survivors. The estimated number of expected cases incorporated an adjustment based on the Poisson regression model given in [I16].
f   The values given exclude the period within five years of irradiation.
g   Based on unirradiated patients.
h   Based on a 10-year lag.
i   95% CI in parentheses.
j   Data for colorectal cancer [R14].
k   Value for small intestine [R14].
l   No apparent trend with administered activity of ^{131}I, although a significance test was not performed.
m   Relative to unexposed controls.
n   Number quoted in earlier follow-up [V3].
o   Amount of thorotrast administered.

Table 9
Risk estimates for cancer incidence and mortality from studies of radiation exposure: liver cancer
The number of observed and expected cases as well as the mean dose and person-years for cohort studies are computed throughout this Table for exposed persons only. In the Life Span Study the exposed group included survivors with organ doses of 0.01 Sv or more for incidence and 0.005 Sv or more (weighted colon dose) for mortality

Study		Observed cases	Expected cases	Mean dose (Sv)	Person-years	Average excess relative risk [a] at 1 Sv	Average excess absolute risk [a] $(10^6 \, PYSv)^{-1}$
EXTERNAL LOW-LET EXPOSURES							
Incidence							
Life Span Study [T1] [b]							
Sex	Male	174	150.1	0.24	299 646	0.66	3.32
	Female	110	104.4	0.23	496 606	0.23	0.49
Age at exposure	<20 years	63	48.3	0.24	367 003	1.27	1.67
	>20 years	221	206.2	0.23	429 249	0.31	1.50
All		284	254.5	0.24	796 252	0.48 (0.04–0.96) [c]	1.55 (0.13–3.08) [c]
Cervical cancer cohort [B11] [d]		8	8.8	1.50	178 243	−0.06 (−0.37–0.4) [c]	−0.03 (−0.16–0.2) [c]
Swedish benign breast disease [M28]		12	11.3	0.66	26 493	0.09 (95% CI: <0–1.4)	n.a.
Mortality							
Life Span Study [P9] [e]							
Age at exposure							
Males	<20 years	67	60.2	0.20	371 456	0.57 (−0.50–1.82)	0.92 (−0.81–2.95)
	20-39 years	73	66.5	0.24	116 815	0.41 (−0.44–1.41)	2.34 (−2.52–8.01)
	>40 years	108	99.5	0.21	129 974	0.40 (−0.38–1.28)	3.06 (−2.90–9.82)
Females	<20 years	17	16.2	0.20	416 768	0.23 (−1.62–2.78)	0.09 (−0.63–1.08)
	20-39 years	65	58.1	0.20	359 129	0.60 (−0.50–1.89)	0.96 (−0.81–3.05)
	>40 years	102	97.0	0.17	202 013	0.29 (−0.65–1.37)	1.40 (−3.13–6.56)
Time since exposure							
Both sexes	5-10 years	42	38.9	0.22	186 468	0.36 (−0.82–1.80)	0.75 (−1.71–3.75)
	11-25 years	112	104.1	0.22	725 251	0.34 (−0.39–1.18)	0.49 (−0.56–1.69)
	26-40 years	178	162.8	0.22	530 897	0.42 (−0.17–1.08)	1.30 (−0.53–3.32)
	41-45 years	100	90.5	0.22	144 740	0.48 (−0.32–1.38)	2.97 (−2.01–8.65)
All		432	397.6	0.22	1 596 155	0.42 (0.04–0.83) [c]	1.08 (0.10–2.15) [c]
Ankylosing spondylitis [W1] [f]		11	13.6	2.13	287 095	−0.09 (−0.24–0.2) [c]	−0.04 (−0.11–0.1) [c]
Peptic ulcer [G6]		9	11.4 [g]	4.61	35 815	−0.05 (95% CI: −0.15–0.24) [c]	−0.15 [c]
Benign gynaecological disease [I16] [h]		9 [i]	16.6	0.21	71 958	−2.18 (−3.26--0.3) [c]	−5.03 (−7.52--0.7) [c]
Yangjiang background radiation [T25, T26]		171	213.8	n.a. [j]	1 246 340	−0.99 (95% CI: −1.60–0.10) [k]	n.a.
Nuclear workers in Canada United Kingdom, United States [C11]		33	n.a.	0.04	2 124 526	~0	n.a.
Nuclear workers in Japan [E3]		111	128.9	0.014	533 168	>0 [l]	n.a.

Study	Observed cases	Expected cases	Mean dose	Person-years	Average relative risk [m]	
INTERNAL LOW-LET EXPOSURES						
Mortality						
United States thyrotoxicosis patients [R14]	39	44.8	n.a.	385 468	n.a.	

Table 9 (continued)

Study	Observed cases	Expected cases	Mean dose	Person-years	Average relative risk [m]	
INTERNAL HIGH-LET EXPOSURES						
Incidence						
Danish thorotrast patients [A5]	84	0.7	3.9–6.1 Gy	n.a.	194.2 [n] (31.0–1 216)	
Mortality						
German thorotrast patients [V1, V8]	454	3.6	4.9 Gy	n.a.	25 Gy^{-1}	
Portuguese thorotrast patients [D3]	104	6.6	26 ml thorotrast	16 963	5.7 [n]	
Combined Japanese thorotrast patients [M14]	143	4	n.a.	10 685	n.a.	

a   90% in parentheses derived from published data for Life Span Study and using exact Poisson methods for the other studies.
b   Based on histologically verified cases.
c   Estimates based on method described in the introduction to Chapter III.
d   Based on 10-year survivors.
e   Includes deaths coded as primary liver cancer and liver cancer not specified as secondary.
f   The values given exclude the period within five years of first treatment.
g   Based on unirradiated patients.
h   The estimated number of expected cases incorporated an adjustment based on the Poisson regression model given in [I16].
i   Including gallbladder.
j   Mean annual effective dose = 6.4 mSv.
k   Based on a 10-year latent period.
l   Based on a 10-year lag. Trend not statistically significant.
m   95% CI in parentheses.
n   Per 10 ml injected dose.

Table 10
Risk estimates for cancer incidence and mortality from studies of radiation exposure: lung cancer
The number of observed and expected cases as well as the mean dose and person-years for cohort studies are computed throughout this Table for exposed persons only. In the Life Span Study the exposed group included survivors with organ doses of 0.01 Sv or more for incidence and 0.005 Sv or more (weighted colon dose) for mortality

Study		Observed cases	Expected cases	Mean dose (Sv)	Person-years	Average excess relative risk [a] at 1 Sv	Average excess absolute risk [a] $(10^4 PYSv)^{-1}$
EXTERNAL LOW-LET EXPOSURES							
Incidence							
Life Span Study [T1]							
Sex	Male	245	224.7	0.25	302 000	0.36	2.67
	Female	211	140.1	0.24	500 700	2.08	5.81
Age at exposure	<20 years	30	26.2	0.25	370 000	0.57	0.41
	>20 years	426	338.5	0.24	432 700	1.06	8.27
Time since exposure	5-19 years	85	67.8	0.24	288 566	1.04	2.45
	20-29 years	146	116.3	0.24	317 535	1.05	3.85
	30-42 years	225	186.4	0.24	314 545	0.85	5.05
All		456	364.7	0.25	802 700	1.00 (0.6-1.4) [b]	4.55 (2.4-6.0) [b]
Hodgkin's disease (international) [K9]		79	n.a.	2.2	n.a.	n.a. [c]	n.a.
Hodgkin's disease (Netherlands) [V2]		29	n.a.	7	n.a.	~1 (95% CI: <0- ~10)	n.a.
Breast cancer [I7]		17	n.a.	9.8 [d]	n.a.	0.08 (95% CI: -0.77-0.22) [e]	0.9
Swedish benign breast disease [M28]		10	11.2	0.75	26 493	0.38 (95% CI: <0-0.6)	n.a.
Stockholm skin haemangioma [L16]		11	~9	0.12	406 565	1.4	0.33
Mortality							
Life Span Study [P9]							
Age at exposure							
Males	<20 years	30	28.4	0.23	369 372	0.24 (-1.0-1.9)	0.18 (-0.8-1.4)
	20-39 years	97	90.8	0.23	116 442	0.30 (-0.5-1.2)	2.32 (-3.5-9.0)
	>40 years	182	164.8	0.23	129 183	0.45 (-0.1-1.1)	5.80 (-1.5-1.1)
Females	<20 years	18	16.6	0.23	414 045	0.37 (-1.3-2.6)	0.15 (-0.5-1.1)
	20-39 years	125	115.3	0.23	357 283	0.37 (-0.3-1.1)	1.18 (-1.0-3.6)
	>40 years	132	115.4	0.23	201 031	0.63 (-0.1-1.4)	3.60 (-0.4-8.0)
Time since exposure							
Both sexes	5-10 years	10	8.3	0.23	186 468	0.87 (-1.5-4.5)	0.39 (-0.7-2.0)
	11-25 years	158	143.7	0.23	725 251	0.43 (-0.2-1.1)	0.86 (-0.3-2.2)
	26-40 years	297	268.4	0.23	530 897	0.46 (0.01-0.9)	2.34 (0.07-4.8)
	41-45 years	119	107.4	0.23	144 740	0.47 (-0.2-1.3)	3.50 (-1.7-9.4)
All		584	526.1	0.23	1 587 355	0.48 (0.16-0.8) [b]	1.59 (0.53-2.7) [b]
Ankylosing spondylitis [W1] [f]		563	469	2.54	287 095	0.05 (95% CI: (0.002-0.09) [g]	0.9 (95% CI: 0.0-1.4) [b]
Canadian TB fluoroscopy [H7] [h]		455	473.7	1.02	672 071	0.00 (95% CI: -0.06-0.07)	0.0 (95% CI: -0.4-0.4)
Peptic ulcer [G6]		99	58.2 [i]	1.79	35 815	0.39 (95% CI: 0.11-0.78) [b]	6.36 [b]
Massachusetts TB fluoroscopy [D4]		69	81.8	0.84	169 425	-0.19 (<-0.2-0.04) [b]	-0.90 (<-1.8-0.2) [b]
Yangjiang background radiation [T25, T26]		62	76.5	n.a. [j]	1 246 340	-0.68 (95% CI: -1.58-1.67) [k]	n.a.
Nuclear workers in Canada, United Kingdom, United States [C11]		1 238	n.a.	0.04	2 124 526	<0 [l]	n.a.

Table 10 (continued)

Study	Observed cases	Expected cases	Mean dose (Sv)	Person-years	Average excess relative risk [m] at 1 Sv	Average excess absolute risk [a] $(10^4\ PYSv)^{-1}$
Nuclear workers in Japan [E3]	117	124.9	0.014	533 168	<0 [l]	n.a.
Mayak reactor workers (cohort study) [n] [K34]	47	56.23	1.02	67 097	−0.161 [l]	−11.7 [l]

Study	Observed cases	Expected cases	Mean WLM	Person-years	Average ERR [o] at 100 WLM	
INTERNAL HIGH-LET EXPOSURES (Occupational radon)						
Mortality						
Chinese tin miners [L4, X1] [p]	936	649	277.4	135 357	0.16 (0.1–0.2)	
West Bohemia uranium miners [L4, T22] [q]	702	137.7	219	106 983	0.64 (0.4–1.1)	
Colorado Plateau uranium miners [H17, L4] [o]	327	74	807.2	75 032	0.42 (0.3–0.7)	
Ontario uranium miners [K4, L4] [o]	282	221	30.8	319 701	0.89 (0.5–1.5)	
Newfoundland fluorspar miners [L4, M15] [r]	138	32.1	382.8	48 189	0.70 (0.44–1.14)	
Swedish iron miners [L4, R8] [o]	79	44.7	80.6	32 452	0.95 (0.1–4.1)	
New Mexico uranium miners [L4, S19] [o]	68	23.5	110.3	46 797	1.72 (0.6–6.7)	
Beaverlodge uranium miners [H15, H18, L4] [o]	56	15.4	81.3 [s]	68 040	3.25 (1.0–9.6) [t]	
Port Radium uranium miners [H16, L4] [o]	39	26.7	242.8	31 454	0.19 (0.1–0.6)	
Radium Hill uranium miners [L4, W7] [o]	32	23.1	7.6	25 549	5.06 (1.0–12.2)	
French uranium miners [L4, T8] [o]	45	36.1	68.7	39 487	0.36 (0.0–1.3)	
Cornish tin miners [D8, H13]	82	n.a.	65	66 900	0.045 [u]	

Study	Observed cases	Expected cases	Mean concentration $(Bq\ m^{-3})$	Person-years	Average ERR [k] at 100 $Bq\ m^{-3}$	
INTERNAL HIGH-LET EXPOSURES (Residental radon)						
Incidence						
Meta-analysis of eight case-control studies [L21]	4 263	n.a.	n.a.	n.a.	0.09 (0.0–0.2)	
West Germany [W17] Entire study region	1 449	n.a.	49 [v]	n.a.	−0.02 (−0.18-0.17)	
Radon-prone areas	365	n.a.	67 [o]	n.a.	0.13 (−0.12-0.46)	
East Germany [W18]	1 053	n.a.	87 [w]	n.a.	0.04 (−0.04-0.12)	
Southwest England [D30]	982	n.a.	58 [o]	n.a.	0.08 (−0.03-0.20)	
Missouri-II [A24] Based on track-etch measurements	247	n.a.	58.5	n.a.	0.06 (−0.1-0.6)	
Based on CR-39 surface measurements	372	n.a.	64.6	n.a.	0.65 (0.1-2.0)	

Table 10 (continued)

Study	Observed cases	Expected cases	Mean dose	Person-years	Average relative risk [n]	
INTERNAL HIGH-LET EXPOSURES (other than radon)						
Incidence						
Mayak radiochemical plant workers (case-control study) [T2]	60 [x]	n.a.	n.a.	n.a.	3.1 (1.8–5.1) [y]	
^{224}Ra ankylosing spondylitis patients [W20]	25	35.7	n.a.	32 800	1.20 [z]	
^{224}Ra ankylosing spondylitis patients [N4]	20	30	n.a.	25 500	0.67	
Danish thorotrast patients [A5]	21	10.9 [aa]	0.18 Gy [ab]	19 365	0.7 (0.3–1.7) [ac]	
Mortality						
Mayak workers (cohort study) [j]	105	42.18	6.56 Sv [ad]	31 693	0.321 Sv^{-1} (0.20–0.47)	
Sellafield plutonium workers [O1]	133	145.8	0.19 Sv	415 432	1.12 [ae]	
Japanese thorotrast patients, combined data [M14]	11	n.a.	17 ml [af]	10 685	2.0 (1.0–3.9)	
German thorotrast patients [V1]	53	n.a.	20.6 ml [ag]	n.a.	0.75 [ah]	
Portuguese thorotrast patients [D31]	10	n.a.	26.3 ml [af]	16 963	4.68 (0.24–92.1) [ai]	
Los Alamos workers [aj] [W8]	8	n.a.	n.a.	n.a.	1.78 (0.79–3.99) [ak]	

a 90% CI in parentheses derived from published data for Life Span Study and using exact Poisson methods for the other studies.
b Estimates based on method described in the introduction to Chapter III.
c Relative risks quoted in Section III.E.
d Average dose to both lungs for irradiated controls.
e Wald-type CI; likelihood-based lower confidence bound could not be identified.
f The values given exclude the period within five years of first treatment.
g Dose-response analysis based on the number of treatment courses given.
h The values given exclude the period within ten years of exposure and ages at risk less than 20 years.
i Based on unirradiated patients.
j Mean annual effective dose = 6.4 mSv.
k Based on a 10-year latent period.
l Trend not statistically significant.
m 90% CI in parentheses derived from published data for Life Span Study and using exact Poisson methods for the other studies.
n Results presented here for males only.
o 95% CI in parentheses.
p The values cited are from [L4], unless indicated otherwise, and except for the expected number of cases which has been calculated as O/(1+100 αD), where O is the observed cases, α is the ERR at 100 WLM and D is the mean WLM.
q Values cited are based on data from [T22].
r Values cited are from [M15], and include non-exposed miners.
s Revised value for persons in nested case–control study [H18].
t Values based on case-control analysis with revised exposure estimates [H18].
u Coefficient based on time-weighted cumulative exposure.
v Value for cases.
w Value for cases, based on measurements in living room [W18].
x Workers with plutonium body burden above 5.55 kBq.
y Comparison group consists of workers with plutonium body burden below 5.55 kBq.
z Relative to unexposed controls, among whom 29 cases were observed, compared with 49.6 expected [W20].
aa Based on national rates [A5].
ab As given in [A12].
ac Relative to unexposed controls, with adjustment for sex, age at angiography, and calendar period.
ad Alpha dose to lung, based on a radiation weighting factor of 20 [K34].
ae Relative to other radiation workers at Sellafield; difference is not statistically significant [O1].
af Mean amount of thorotrast administered in the first series of Japanese patients [M47].
ag Amount of thorotrast administered.
ah Relative to unexposed controls.
ai Based on three deaths in the control group, and excluding the first five years after administration of thorotrast [D31].
aj Workers with plutonium body burden of 74 Bq or more.
ak Comparison group consists of workers with plutonium body burden below 74 Bq.

Table 11

Risk estimates for cancer incidence and mortality from studies of radiation exposure: malignancies of the bone and connective tissue

The number of observed and expected cases as well as the mean dose and person-years for cohort studies are computed throughout this Table for exposed persons only. In the Life Span Study the exposed group included survivors with organ doses of 0.01 Sv or more for incidence and 0.005 Sv or more (weighted colon dose) for mortality

Study	Observed cases	Expected cases	Mean dose (Sv)	Person-years	Average excess relative risk [a] at 1 Sv	Average excess absolute risk [a] $(10^4\ PYSv)^{-1}$
EXTERNAL LOW-LET EXPOSURES						
Incidence						
Life Span Study [T1]						
Sex Male	9	6.4	0.23	297 500	1.78	0.38
Female	7	5.7	0.22	491 100	0.99	0.12
Age at exposure <20 years	4	1.1	0.23	363 300	11.0	0.34
>20 years	12	11.0	0.22	425 300	0.42	0.11
All	16	12.1	0.23	788 600	1.42 (<-0.2-4.5) [b]	0.22 (<-0.1-0.7) [b]
Retinoblastoma patients [W11] (bone and soft tissue sarcoma) [c]	81	16.9	0.0 [d]	n.a.	0.19 (95% CI: 0.14-0.32)	n.a.
Childhood radiotherapy, international [T17]	54	20.0	27.0	n.a.	0.06 (0.01-0.2) [b]	n.a.
Childhood cancer, United Kingdom (bone) [e] [H44]	49	18.8	10 [d]	n.a.	0.16 (95% CI: 0.07-0.37)	n.a.
Cervical cancer case-control [B1] (connective tissue) [f]	46	70.8	7.0	n.a.	-0.05 (-0.11-0.13)	-0.01 (-0.03-0.03)
Cervical cancer case-control [B1] (bone) [f]	15	10.4	22	n.a.	0.02 (-0.03-0.21) [b]	n.a.
Mortality						
Life Span Study [R1]						
Sex Male	14	10.8	0.23	471 800	1.26	0.29
Female	10	8.5	0.23	731 300	0.81	0.09
Age at exposure <20 years	3	1.9	0.23	574 500	2.58	0.08
>20 years	21	17.4	0.22	628 600	0.92	0.26
All	24	19.3	0.23	1 203 100	1.07 (<-0.2-3.3) [b]	0.17 (<-0.1-0.5) [b]
Ankylosing spondylitis [W1] [g] (bone and connective and soft tissue)	19	6.3	4.54	287 095	0.44 [b]	0.097 [b]
Nuclear workers in Canada, United Kingdom, United States [C11] (bone)	11	n.a.	0.04	2 124 526	<0 [h]	n.a.
Nuclear workers in Canada, United Kingdom, United States [C11] (connective tissue)	19	n.a.	0.04	2 124 526	>0 [h]	n.a.

Study	Observed cases	Expected cases	Mean dose	Person-years	Average relative risk [i]	Average excess absolute risk [i] $(10^4\ PYSv)^{-1}$
INTERNAL HIGH-LET EXPOSURES						
Incidence						
^{224}Ra TB and ankylosing spondylitis patients (bone) [N14]	55	0.2	30.6 Gy	25 500	n.a.	n.a.
^{224}Ra ankylosing spondylitis patients (bone and connective tissue) [W20]	4	1.3	~6 Gy	32 800	4.3 [j]	n.a.

Table 11 (continued)

Study	Observed cases	Expected cases	Mean dose	Person-years	Average relative risk [i]	Average excess absolute risk $(10^4\ PYSv)^{-1}$
German thorotrast patients (bone sarcoma [V8]	4	n.a.	20.6 ml [k]	n.a.	~3.3 [l]	n.a.
Mortality						
United States radium luminizers [m] (bone) [C27, R35, S12, S16, S54, S56]	46	<1	8.6 Gy	35 819	n.a.	~13
Portuguese thorotrast patients (bone) [D31]	16	n.a.	26.3 ml [k]	16 963	7.08 (1.65–30.3) [n]	n.a.

a   90% CI in parentheses derived from published data for Life Span Study and using exact Poisson methods for the other studies.
b   Estimates based on method described in the introduction to Chapter III.
c   Results are for patients with bone or soft tissue sarcoma for whom dosimetry information was available.
d   Mean dose for controls of bone cancer cases.
e   Results are based on a case-control analysis of bone cancer.
f   Based on one-year survivors. The observed and expected numbers cover both exposed and unexposed persons. The excess absolute risk for connective tissue was computed using baseline incidence data derived from the cohort study [B11].
g   The values given exclude the period within five years of first treatment.
h   Based on a 10-year lag. Trend not statistically significantly different from zero.
i   95% CI in parentheses.
j   Relative to unexposed controls, among whom one case was observed compared with 1.4 expected [W20].
k   Amount of thorotrast administered.
l   Crude relative risk, based on one case in the control group. This relative risk is not significantly different from 1 (p>0.05) [V8].
m   Based on pre-1930 workers with an average skeletal dose greater than zero [C27].
n   Based on five deaths in the control group, and excluding the first five years after administration of thorotrast [D31].

Table 12
Risk estimates for cancer incidence and mortality from studies of radiation exposure: skin cancer
The number of observed and expected cases as well as the mean dose and person-years for cohort studies are computed throughout this Table for exposed persons only. In the Life Span Study the exposed group included survivors with organ doses of 0.01 Sv or more for incidence and 0.005 Sv or more (weighted colon dose) for mortality

Study		Observed cases	Expected cases	Mean dose (Sv)	Person-years	Average excess relative risk [a] at 1 Sv	Average excess absolute risk [a] $(10^4 PYSv)^{-1}$
EXTERNAL LOW-LET EXPOSURES							
Incidence							
Life Span Study [T1]							
Sex	Male	41	31.4	0.33	324 100	0.92	0.89
	Female	57	44.4	0.32	538 900	0.88	0.72
Age at exposure	<20 years	21	7.7	0.32	399 300	5.37	1.04
	>20 years	77	68.2	0.33	463 700	0.39	0.58
All		98	75.9	0.33	863 000	0.88 (0.4–1.9) [b]	0.78 (0.4–1.4) [b]
Childhood exposure							
Israel tinea capitis [R16]		42	10.0	6.8	265 070	0.47 (0.3–0.7) [b]	0.18 (0.1–0.25) [b]
New York tinea capitis (whites) [c] [S27, S30]		83	24.0	5.0	52 000 [d]	0.49 (0.37–0.63) [b]	2.5 (1.9–3.2) [b]
Rochester thymic irradiation [c] [H31, S30]		14	4.2	2.3	87 000 [d]	1.05 (0.50–1.9) [b]	0.50 (0.3–0.9) [b]
Tonsil irradiation [c] [S28, S30]		63	45.0	3.8	96 000 [d]	0.11 (0.03–0.19) [b]	0.50 (0.2–1.0) [b]
Adult exposure							
Cervical cancer cohort [B1]		88	100	10	342 786	−0.01 (−0.02–0.01) [b]	−0.02 (−0.06–0.03) [b]
Massachusetts TB fluoroscopy [c] [D16, S30]		80	75.3	9.6	122 000 [d]	0.01 (0–0.03) [b]	0.04 (0–0.2) [b]
New York mastitis [c] [S30]		14	10.7	2.6	14 000 [d]	0.12 (0–0.8) [b]	0.90 (0–2.8) [b]

a 90% CI in parentheses derived from published data for Life Span Study and using exact Poisson methods for the other studies.
b Estimates based on method described in the introduction to Chapter III.
c From data presented by Shore [S30].
d Person-years estimated from data presented by Shore [S30].

Table 13
Risk estimates for cancer incidence and mortality from studies of radiation exposure: female breast cancer
The number of observed and expected cases as well as the mean dose and person-years for cohort studies are computed throughout this Table for exposed persons only. In the Life Span Study the exposed group included survivors with organ doses of 0.01 Sv or more for incidence and 0.005 Sv or more (weighted colon dose) for mortality. For case-control studies, the observed number of cases covers both exposed and unexposed persons.

Study		Observed cases	Expected cases	Mean dose (Sv)	Person-years	Average excess relative risk [a] at 1 Sv	Average excess absolute risk [a] (10⁴ PYSv)⁻¹

Study		Observed cases	Expected cases	Mean dose (Sv)	Person-years	Average excess relative risk [a] at 1 Sv	Average excess absolute risk [a] $(10^4 \, PYSv)^{-1}$
EXTERNAL LOW-LET EXPOSURES							
Incidence							
Life Span Study [T1]							
Age at exposure	<20 years	122	62.8	0.28	202 600	3.32 (2.3–4.4)	10.3 (7.2–14)
	>20 years	173	137.1	0.27	308 000	0.98 (0.4–1.6)	4.36 (1.8–7.2)
Time since exposure	5–19 years	49	36.9	0.28	161 400	1.19	2.72
	20–29 years	87	63.5	0.27	175 800	1.34	4.86
	30–42 years	159	99.5	0.27	173 400	2.21	12.68
All		295	199.9	0.27	510 600	1.74 (1.1–2.2) [b]	6.80 (4.9–8.7) [b]
Massachusetts TB fluoroscopy [B3]		142	107.6	0.79	54 600	0.40 (0.2–0.7) [b]	7.98 (3.6–13) [b]
New York acute post-partum mastitis [S15]		54	20.8	3.7	9 800	0.43 (0.3–0.6) [b]	9.14 (6.0–13) [b]
Swedish benign breast disease [M8, M20]		115	28.8	8.46	37 400	0.35 (0.3–0.4) [b]	2.72 (2.2–3.3) [b]
Cervical cancer case control [c] [B50]		953 [d]	1083.0	0.31	n.a.	−0.2 (<−0.2–0.3)	<−0.3 (<−0.3–0.2)
Without ovaries		91 [e]	82.6	0.31	n.a.	0.33 (<−0.2–5.8)	n.a.
Contralateral breast							
Denmark [S20]		529	508.7	2.51	n.a.	0.02 (<−0.1–0.2) [b]	n.a.
United States [B10]		655	550.4	2.82	n.a.	0.07 (<−0.1–0.2) [b]	n.a.
Rochester thymic irradiation [f] [H10]		22	7.8	0.76	38 200	2.39 (1.2–4.0) [b]	4.89 (2.4–8.1) [b]
Childhood skin haemangioma [f] [L46]		245	204	0.33	600 000	0.35 (95% CI: 0.18–0.59)	1.44 (95% CI: 0.78–2.28)
Hodgkin's disease (Stanford) [H2]		25	6.1	~44.0	100 057	0.07 (0.04–0.11) [b]	0.04 (0.03–0.07) [b]
Childhood Hodgkin's disease [f] [B16]		17	0.2	20	n.a.	n.a. [g]	n.a.
Mortality							
Life Span Study [P9]							
Age at exposure	<20 years	52	29.1	0.25	414 045	3.16 (1.61–5.0)	2.22 (1.13–3.5)
	20–39 years	57	50.0	0.25	357 283	0.56 (−0.4–1.7)	0.78 (−0.5–2.4)
	>40 years	33	30.2	0.25	201 031	0.37 (−0.8–1.8)	0.55 (−1.2–2.8)
Time since exposure	5–10 years	16	22.3	0.25	108 719	−1.12 (−2.2–0.4)	−2.30 (−4.5–0.8)
	11–25 years	47	40.9	0.25	442 174	0.60 (−0.4–1.19)	0.55 (−0.4–1.7)
	26–40 years	54	36.5	0.25	330 501	1.19 (0.66–3.4)	2.11 (0.72–3.8)
	41–45 years	25	13.5	0.25	90 964	3.43 (1.16–6.4)	5.07 (1.72–9.4)
All		142	107.6	0.25	972 358	1.28 (0.57–2.1) [b]	1.42 (0.63–2.3) [b]
Scoliosis patients [f] [D34]		70	35.7	0.11	184 508	5.4 (95% CI: 1.2–14.1)	12.9 (95% CI: 4.0–21.0)
Ankylosing spondylitis [W1] [h]		42	39.3	0.59	n.a.	0.08 (95% CI: −0.30–0.65) [i]	n.a.
Canadian TB fluoroscopy [H20]		349	237	0.89	411 706	0.90 (95% CI: 0.55–1.39) [j]	3.16 (95% CI: 1.97–4.78) [k]
Nuclear workers in Canada, United Kingdom, United States [C11]		84	n.a.	0.04	n.a.	>0 [l]	n.a.

Table 13 (continued)

Study	Observed cases	Expected cases	Mean dose	Person-years	Average ERR at 1 Sv	
INTERNAL HIGH-LET EXPOSURES						
Incidence						
^{224}Ra TB and ankylosing spondylitis patients [N4]	28	8	~0.1 Gy ^{m}	n.a.	0.9	

a   90% CI in parentheses derived from published data for Life Span Study and using exact Poisson methods for the other studies.

b   Estimates based on method described in the introduction to Chapter III.

c   Excess absolute risk among cervical cancer patients is computed using baseline incidence data derived from the cohort study [B11].

d   Based on 5-year survivors.

e   Based on 10-year survivors.

f   Population exposed as children.

g   Relative risks by dose group quoted in Section III.H.1.

h   The values given exclude the period within five years of first treatment.

i   Dose-response analysis based on the number of treatment courses given.

j   Including a factor to allow for differences between Nova Scotia and other Canadian provinces. Values apply to exposure at age 15 years.

k   Including a factor to allow for differences between Nova Scotia and other Canadian provinces. Values apply 20 years following exposure at age 15 years.

l   Based on a 10-year lag. Trend not statistically significant.

m   High-LET breast dose from radium-224.

Table 14
Risk estimates for cancer incidence and mortality from studies of radiation exposure: prostate cancer
The number of observed and expected cases as well as the mean dose and person-years for cohort studies are computed throughout this Table for exposed persons only. In the Life Span Study the exposed group included survivors with organ doses of 0.01 Sv or more for incidence and 0.005 Sv or more (weighted colon dose) for mortality

Study	Observed cases	Expected cases	Mean dose (Sv)	Person-years	Average excess relative risk [a] at 1 Sv	Average excess absolute risk [a] $(10^4 PYSv)^{-1}$
EXTERNAL LOW-LET EXPOSURES						
Incidence						
Life Span Study [T1]	95	92.01	0.21	297 500	0.14 (−0.6–1.0) [b]	0.44 (−1.8–3.0) [b]
Mortality						
Ankylosing spondylitis [W1] [c]	88	64.7	2.18	n.a.	0.14 (95% CI: 0.02–0.28) [d]	n.a.
Peptic ulcer [G6]	26	18.7 [e]	0.08	n.a.	4.9 (95% CI: −2.5–15.0) [b]	n.a.
Nuclear workers in Canada, United Kingdom, United States [C11]	256	n.a.	0.04	n.a.	<0 [f]	n.a.

Study	Observed cases	Expected cases	Mean dose	Person-years	Average relative risk [g]	
INTERNAL LOW-LET EXPOSURES						
Incidence						
UK Atomic Energy Authority workers: case-control study [R26]	28 [h]	n.a.	n.a.	n.a.	2.36 (1.26–4.43)	
Mortality						
United States thyrotoxicosis patients [R14]	36	52.7	<0.1	n.a.	n.a. [i]	
INTERNAL HIGH-LET EXPOSURES						
Incidence						
^{224}Ra TB and ankylosing spondylitis patients [N4]	16	~12	n.a.	n.a.	~1.3	
Mortality						
German thorotrast patients [V8]	21	n.a.	20.6 ml [j]	n.a.	~0.9 [k]	

a   90% CI in parentheses derived from published data for Life Span Study and using exact Poisson methods for the other studies.
b   Estimates based on method described in the introduction to Chapter III.
c   The values given exclude the period within five years of first treatment.
d   Dose-response analysis based on the number of treatment courses given.
e   Based on unirradiated patients.
f   Based on a 10-year lag. One-sided p-value for increasing trend equals 0.953, based on a normal approximation.
g   95% CI in parentheses.
h   Men who worked in environments potentially contaminated with ^{51}Cr, ^{59}Fe, ^{60}Co, ^{65}Zn or ^{3}H.
i   No apparent trend with administered activity of ^{131}I, although a significance test was not performed.
j   Amount of thorotrast administered.
k   Relative to unexposed controls.

Table 15
Risk estimates for cancer incidence and mortality from studies of radiation exposure: cancer of the urinary bladder
The number of observed and expected cases as well as the mean dose and person-years for cohort studies are computed throughout this Table for exposed persons only. In the Life Span Study the exposed group included survivors with organ doses of 0.01 Sv or more for incidence and 0.005 Sv or more (weighted colon dose) for mortality

Study		Observed cases	Expected cases	Mean dose (Sv)	Person-years	Average excess relative risk [a] at 1 Sv	Average excess absolute risk [a] $(10^4 PYSv)^{-1}$
EXTERNAL LOW-LET EXPOSURES							
Incidence							
Life Span Study [T1]							
Sex	Male	76	70.3	0.23	297 500	0.35	0.84
	Female	39	27.9	0.22	491 200	1.80	1.02
Age at exposure	<20 years	12	10.3	0.23	363 300	0.71	0.20
	>20 years	103	87.8	0.22	425 300	0.79	1.62
All		115	98.1	0.23	788 600	0.76 (0.3–2.1) [b]	0.95 (0.3–2.1) [b]
Cervical cancer case-control [B1] [c]		273	65.8	45	n.a	0.07 (0.02–0.17)	0.12 (0.04–0.3)
Mortality							
Life Span Study [P9]							
Age at exposure							
Males	<20 years	6	3.4	0.20	371 260	3.83 (−1.19–12.50)	0.35 (−0.11–1.15)
	20–39 years	5	4.1	0.23	116 726	0.90 (−2.26–6.63)	0.32 (−0.80–2.35)
	>40 years	39	35.4	0.21	129 809	0.48 (−0.83–2.11)	1.32 (−2.27–5.75)
Females	<20 years	2	1.7	0.20	416 447	1.05 (−3.90–13.98)	0.04 (−0.15–0.55)
	20–39 years	7	8.1	0.19	358 988	−0.69 (−3.07–3.27)	−0.15 (−0.69–0.73)
	>40 years	23	19.5	0.17	201 931	1.04 (−1.14–3.91)	1.00 (−1.10–3.78)
Time since exposure							
Both sexes	5–10 years	4	5.0	0.20	258 146	−1.02 (−3.67–4.14)	−0.20 (−0.71–0.81)
	11–25 years	29	28.3	0.20	658 705	0.12 (−1.34–2.00)	0.05 (−0.58–0.86)
	26–40 years	35	26.1	0.20	533 369	1.75 (−0.04–3.98)	0.85 (−0.02–1.94)
	41–45 years	14	16.6	0.19	144 940	−0.82 (−2.55–1.65)	−0.94 (−2.92–1.89)
All		82	72.2	0.20	1 595 161	0.58 (−0.40–1.72) [b]	0.27 (−0.19–0.79) [b]
Benign gynaecological disease [I16] [d]		19	9.0	6.00	71 958	0.20 (0.08–0.35)	0.24 (0.1–0.4) [b]
Metropathia haemorrhagica [D7] [e]		20	6.7	5.20	47 144	0.40 (95% CI: 0.15–0.66)	0.55 (95% CI: 0.2–0.9) [b]
Ankylosing spondylitis [W1] [f]		71	46.1	2.18	287 095	0.24 (95% CI: 0.09–0.41) [g]	0.39 (95% CI: 0.19–0.54) [b]
Nuclear workers in Canada, United Kingdom, United States [C11]		104	n.a.	0.04	2 142 526	>0 [h]	n.a.

a   90% CI in parentheses derived from published data for Life Span Study and using exact Poisson methods for the other studies.
b   Estimates based on method described in the introduction to Chapter III.
c   Based on 10-year survivors. The observed and expected numbers cover both exposed and unexposed persons. The excess absolute risk estimate was computed using background incidence rates estimated using the cervical cancer cohort study [B11].
d   The observed and expected number of cases are for 10-year survivors. The estimated number of expected cases incorporated an adjustment based upon the Poisson regression model given in [I16].
e   The values given exclude the period within five years of irradiation.
f   The values given exclude the period within five years of first treatment.
g   Dose-response analysis based on the number of treatment courses given.
h   Based on a 10-year lag. Trend not statistically significant.

Table 16
Risk estimates for cancer incidence and mortality from studies of radiation exposure: brain and central nervous system tumours
The number of observed and expected cases as well as the mean dose and person-years for cohort studies are computed throughout this Table for exposed persons only. In the Life Span Study the exposed group included survivors with organ doses of 0.01 Sv or more for incidence.

Study		Observed cases	Expected cases	Mean dose (Sv)	Person-years	Average excess relative risk [a] at 1 Sv	Average excess absolute risk [a] $(10^4 PYSv)^{-1}$
				EXTERNAL LOW-LET EXPOSURES			
				Incidence			
Life Span Study [T1]							
Sex	Male	20	21.7	0.27	307 100	−0.30	−0.21
	Female	51	45.3	0.26	509 300	0.48	0.43
Age at exposure	<20 years	20	15.7	0.26	376 100	1.05	0.44
	>20 years	51	51.4	0.26	440 200	−0.03	−0.03
	All	71	67.1	0.26	816 300	0.22 (<0–1.3) [b]	0.18 (<0–0.8) [b]
Israel tinea capitis [R17]		60	8.4	1.5	283 930	4.08 (3.1–5.2) [b]	1.2 (0.9–1.5) [b]
New York tinea capitis [A16]		8	1.4	1.4	48 115	3.4 (1.3–6.7) [b]	0.98 (0.4–1.9) [b]
Swedish pooled skin haemangioma [K23]		83	58.0	0.07	913 402	2.7 (95% CI: 1.0–5.6)	2.1 (95% CI: 0.3–4.4)
				Mortality			
Pituitary adenoma (UK) [B22]		5	0.5	45	3 760	0.20 (0.07–0.45) [b]	0.27 (0.09–0.59) [b]
Nuclear workers in Canada, United Kingdom, United States [C11]		122	n.a.	0.04	2 142 526	<0 [c]	n.a.

a 90% CI in parentheses derived from published data for Life Span Study and using exact Poisson methods for the other studies.
b Estimates based on method described in the introduction to Chapter III.
c Based on a 10-year lag. Trend not statistically significant.

Table 17
Risk estimates for cancer incidence and mortality from studies of radiation exposure: thyroid cancer
The number of observed and expected cases as well as the mean dose and person-years for cohort studies are computed throughout this Table for exposed persons only. In the Life Span Study the exposed group included survivors with organ doses of 0.01 Sv or more for incidence

Study		Observed cases	Expected cases	Mean dose (Sv)	Person-years	Average excess relative risk [a] at 1 Sv	Average excess absolute risk [a] $(10^4 PYSv)^{-1}$
EXTERNAL LOW-LET EXPOSURES							
Incidence							
Life Span Study [T1]							
Sex	Male	22	14.9	0.27	307 167	1.80	0.87
	Female	110	79.4	0.26	510 388	1.49	2.32
Age at exposure	0–9 years	24	7.6	0.21	185 507	10.25	4.21
	10–19 years	35	14.6	0.31	190 087	4.50	3.46
	20–29 years	18	17.5	0.28	132 738	0.10	0.13
	>30 years	55	54.5	0.25	309 224	0.04	0.06
All		132	94.3	0.26	817 600	1.5 (0.5–2.1) [b]	1.8 (0.8–2.5) [b]
Tuberculosis, adenitis screening [H3, S8]							
Age at exposure	<20 years	6	0.0	8.20	950	36.5 (16–72) [b]	7.7 (3.3–15) [b]
	>20 years	2	0.2	8.20	3 100	1.2 (0.1–3.7) [b]	0.7 (0.1–2.4) [b]
Cohort studies of children							
Life Span Study [T1]							
Age at exposure	0–19 years	59	22.2	0.26	375 600	6.3 (5.1–10.1) [b]	3.8 (2.7–5.4) [b]
Israeli tinea capitis [R9] [c]		43	10.7	0.1	274 180	34 (23–47) [b]	13 (9.0–18) [b]
New York tinea capitis [S8]		2	1.4 [d]	0.1	79 500	7.7(<0–60) [b]	1.3 (<0–10.3) [b]
Rochester thymic irradiation [e] [S18]		37	2.7	1.4	82 204	9.5 (6.9–12.7) [b]	3.0 (2.2–4.0) [b]
Childhood cancer [f] [T5]		23	0.4	12.5	50 609	4.5 (3.1–6.4) [b]	0.4 (0.2–0.5) [b]
Stockholm skin haemangioma [L13]		17	7.5	0.26	406 355	4.9 (95% CI: 1.3–10.2)	0.9 ((95% CI: 0.2–1.9)
Gothenburg skin haemangioma [L15]		15	8	0.12	370 517	7.5 (95% CI: 0.4–18.1)	1.6 (95% CI: 0.09–3.9)
Screening studies of children							
Lymphoid hyperplasia screening [e, g] [P8, S8]		13	5.4 [b]	0.24	34 700	5.9 (1.8–11.8) [b]	9.1 (2.7–18.3) [b]
Thymus adenitis screening [M4, S8]		16	1.1 [b]	2.9	44 310	4.5 (2.7–7.0) [b]	1.2 (0.7–1.8) [b]
Michael Reese, tonsils [h] [S21]		309	110.4	0.6	88 101	3.0 (2.6–3.5) [b]	37.6 (32–43) [b]
Tonsils/thymus/acne screening [D5, S8]		11	0.2 [b]	4.5	6 800	12.0 (6.6–20) [b]	3.5 (2.0–5.9) [b]
Pooled analysis of five studies of children							
Life Span Study Israeli tinea capitis Rochester thymic irradiation Lymphoid hyperplasia screening Michael Reese tonsil [R4]		436	n.a.	n.a.	n.a.	7.7 (95% CI: 2.1–28.7)	4.4 (95% CI: 1.9–10.1)
Studies of adults							
Cervical cancer case-control [d] [B1]		43	18.8	0.11	n.a.	12.3 (<0–76) [b]	6.9 (<0–39.2) [b]
Cervical cancer cohort [d, i] [B11]		16	12.5	0.11	342 786	2.5 (<0–6.8) [b]	0.9 (<0–2.5) [b]
Stanford thyroid [H9]		6	0.4	45	17 700	0.3 (0.1–0.7) [b]	0.07 (0.03–0.1) [b]

Table 17 (continued)

Study	Observed cases	Expected cases	Mean dose	Person-years	Average relative risk	
INTERNAL LOW-LET EXPOSURES						
Incidence						
Diagnostic [131]I [H4]	67	49.7	1.1	653 093	n.a. [j]	

a   90% CI in parentheses derived from published data for Life Span Study and using exact Poisson methods for the other studies.

b   Estimates based on method described in the introduction to Chapter III.

c   Doses to the thyroid in this study may be much more uncertain than doses to organs directly in the x-ray beam.

d   Expected number of cases computed using excess relative risk estimates given in [S8].

e   Known dose. PY and expected number of cases estimated from data given in [S8].

f   Based on cohort members with 15 or more years of follow-up and population-expected rates.

g   This was a study of nodular disease, and cancer cases were not confirmed.

h   Study includes no unexposed controls; estimates of the number of expected cases were computed using the fitted excess relative risk reported in [S21]. Results are based on the new dosimetry described in [S21]. The large excess absolute risk in this study illustrates the impact of screening on thyroid cancer risk estimates. As described in [S21], a special thyroid screening programme in this cohort was initiated in 1974. This screening led to a large increase in the number of incident cases detected among both cases and controls. The paper describes an analysis in which allowance was made for the effect of screening. The screening-adjusted excess absolute risk was estimated as $1.7 \, (10^{-4} \, \text{PYGy})^{-1}$.

i   Excludes cases diagnosed during first 10 years of follow-up.

j   Trend not statistically significant (see Table 28).

Table 18
Risk estimates for cancer incidence and mortality from studies of radiation exposure: non-Hodgkin's lymphoma
The number of observed and expected cases as well as the mean dose and person-years for cohort studies are computed throughout this Table for exposed persons only. In the Life Span Study the exposed group included survivors with organ doses of 0.01 Sv or more for incidence.

Study		Observed cases	Expected cases	Mean dose (Sv) [a]	Person-years	Average excess relative risk [b] at 1 Sv	Average excess absolute risk [b] (10⁴ PYSv)⁻¹
EXTERNAL LOW-LET EXPOSURES							
Incidence							
Life Span Study [P4]							
Sex	Male	41	33.2	0.26	412 400	0.91	0.73
	Female	35	38.3	0.25	664 500	−0.34	−0.20
Age at exposure	<20 years	17	15.8	0.26	478 100	0.30	0.10
	>20 years	59	55.7	0.25	598 800	0.24	0.22
All		76	71.5	0.25	1 076 900	0.25 (<0.2–1.1) [c]	0.17(<−0.3–0.8) [c]
Cervical cancer case-control [d] [B1]		94	37.5	7.10	n.a.	0.21 (−0.03–0.93) [c]	n.a.
Benign lesions in the locomotor system [D12]		81	80.3	0.39	392 900	0.02 [c]	0.05 [c]
Mortality							
Benign lesions in the locomotor system [D12]		50	56.9	0.39	439 400	−0.31 [c]	−0.40 [c]
Ankylosing spondylitis [W1] [e]		37	21.3	4.38	287 095	0.17 [c]	0.77 [c]
Benign gynaecological disease [I6]		40	42.5	1.19	246 821	−0.05 (<−0.2–0.2) [c]	−0.08 (<−0.3–0.3) [c]
Massachusetts TB fluoroscopy [D4]		13 [f]	13.1	0.09	157 578	−0.05 (<−0.2–6.5) [c]	−0.04 (<−0.2–5.4) [c]
Peptic ulcer [G6]		12	6.4 [g]	1.55	35 815	0.57 (95% CI: −0.19–2.6) [c]	1.01 [c]
Nuclear workers in Canada, United Kingdom and United States [C11]		135	n.a.	0.04	2 142 526	<0 [h]	n.a.

Study	Observed cases	Expected cases	Mean dose (Sv)	Person-years	Average excess relative risk at 1 Sv	
INTERNAL LOW-LET EXPOSURES						
Incidence						
Diagnostic ¹³¹I [H27]	95	78.5	0.00019 [i]	527 056	n.a.	
Swedish ¹³¹I hyperthyroid [H23]	22	32.4	0.06	139 018	n.a.	
Mortality						
United States thyrotoxicosis [j] [R14]	74	n.a.	0.042	735 255	0.6 [h]	

Study	Observed cases	Expected cases	Mean dose	Person-years	Average relative risk [k]	
INTERNAL HIGH-LET EXPOSURES						
Incidence						
Danish Thorotrast patients [A5]	2	1.6	n.a.	19 365	1.47 (0.19–8.87) [l]	

Table 18 (continued)

Study	Observed cases	Expected cases	Mean dose (Sv)	Person-years	Average excess relative risk at 1 Sv	
^{224}Ra ankylosing spondylitis patients [W3]	2	0.9-1.8	n.a.	n.a.	~2 [m]	
Mortality						
German Thorotrast patients [V8]	15	n.a.	20.6 ml [n]	n.a.	~2.5 [o]	

a   Mean dose to red bone marrow.
b   90% CI in parentheses derived from published data for Life Span Study and using exact Poisson methods for the other studies.
c   Estimates based on method described in the introduction to Chapter III.
d   Based on 5-year survivors. The observed and expected numbers cover both exposed and unexposed persons.
e   The values given exclude the period within five years of first treatment. Mean dose to bone marrow taken from [W2].
f   Includes deaths from multiple myeloma.
g   Based on unirradiated patients.
h   Not statistically significantly different from zero.
i   Mean dose to bone marrow given in [H12].
j   Some patients from the United Kingdom were included in this analysis [R14].
k   95% CI in parentheses.
l   Risk relative to an unexposed control group, in which three cases were observed compared with 3.5 expected.
m   Risk relative to an unexposed control group, in which one case was observed compared with 1.0-2.3 expected.
n   Amount of thorotrast administered.
o   Crude relative risk, based on five cases in an unexposed control group.

Table 19
Risk estimates for cancer incidence and mortality from studies of radiation exposure: Hodgkin's disease
The number of observed and expected cases as well as the mean dose and person-years for cohort studies are computed throughout this Table for exposed persons only. In the Life Span Study the exposed group included survivors with organ doses of 0.01 Sv or more for incidence.

Study	Observed cases	Expected cases	Mean dose (Sv) [a]	Person-years	Average excess relative risk [b] at 1 Sv	Average excess absolute risk [b] $(10^4 PYSv)^{-1}$
EXTERNAL LOW-LET EXPOSURES						
Incidence						
Life Span Study [P4]	10	9.02	0.23	1 076 500	0.43 (−1.6−3.5) [c]	0.04 (−0.1-0.3) [c]
Cervical cancer case-control [B1] [d]	14	n.a.	7.10	n.a.	n.a. [e]	n.a.
Benign lesions in the locomotor system [D12]	17	22.3	0.39	392 900	−0.61 [c]	−0.35 [c]
Mortality						
Benign lesions in the locomotor system [D12]	21	15.4	0.39	439 400	0.93 [c]	0.33 [c]
Ankylosing spondylitis [W1] [f]	13	7.9	4.38	287 095	0.15 [c]	0.04 [c]
Benign gyenaecological disease [I6]	10	6.6	1.19	246 821	0.43 [c]	0.12 [c]
Nuclear workers in Canada, United Kingdom, and United States [C11]	43	n.a.	0.04	2 142 526	>0 [g]	n.a.

Study	Observed cases	Expected cases	Mean dose (Sv)	Person-years	Average excess relative risk at 1 Sv	
INTERNAL LOW-LET EXPOSURES						
Incidence						
Diagnostic ^{131}I [H27]	27	20.0	0.00019 [h]	527 056	n.a.	
Swedish ^{131}I hyperthyroid [H23]	6	7.2	0.06	139 018	n.a.	
Mortality						
United States thyrotoxicosis [i] [R14]	12	n.a.	0.042	735 255	−1 [g]	

Study	Observed cases	Expected cases	Mean dose	Person-years	Average relative risk [j]	
INTERNAL HIGH-LET EXPOSURES						
Incidence						
Danish thorotrast patients [A5]	1	0.65	n.a.	19 365	1.6 (0.06−40.4) [k]	
^{224}Ra ankylosing spondylitis patients [W3]	1	0.8−1.1	n.a.	n.a.	n.a. [l]	
Mortality						
German thorotrast patients [V8]	2	n.a.	20.6 ml [m]	n.a.	~0.8 [n]	

a Mean dose to red bone marrow.
b 90% CI in parentheses derived from published data for Life Span Study and using exact Poisson methods for the other studies.
c Estimates based on method described in the introduction to Chapter III.
d Based on one-year survivors. The observed number of cases covers both exposed and unexposed persons.
e Unmatched relative risk of 0.63 (90% CI: 0.2-2.6), compared to those with <2 Sv.
f The values given exclude the period within five years of first treatment. Mean dose to bone marrow taken from [W2].
g Trend not statistically significant.
h Mean dose to bone marrow given in [H12].
i Some patients from the United Kingdom were included in this analysis [R14].
j 95% CI in parentheses.
k Risk relative to an unexposed control group, in which one case was observed compared with 1.04 expected.
l In an unexposed control group, no cases were observed compared with 0.8−1.1 expected.
m Amount of thorotrast administered.
n Crude relative risk, based on two cases in an unexposed control group.

Table 20
Risk estimates for cancer incidence and mortality from studies of radiation exposure: multiple myeloma
The number of observed and expected cases as well as the mean dose and person-years for cohort studies are computed throughout this Table for exposed persons only. In the Life Span Study the exposed group included survivors with organ doses of 0.01 Sv or more for incidence and 0.005 Sv or more (weighted colon dose) for mortality

Study	Observed cases	Expected cases	Mean dose (Sv) [a]	Person-years	Average excess relative risk [b] at 1 Sv	Average excess absolute risk [b] $(10^4 PYSv)^{-1}$
EXTERNAL LOW-LET EXPOSURES						
Incidence						
Life Span Study [P4]						
Sex Male	12	9.2	0.26	412 400	0.17	0.26
Female	18	19.3	0.25	664 500	-0.28	-0.08
Age at exposure <20 years	4	3.1	0.26	478 100	1.07	0.07
>20 years	26	25.4	0.25	598 800	0.09	0.04
All	30	28.6	0.25	1 076 900	0.20 (<-0.2-1.7) [c]	0.05 (<-0.05-0.4) [c]
Cervical cancer case-control [d] [B1]	56	n.a.	7.10	n.a.	-0.10 (<0-0.23) [c]	n.a.
Benign lesions in the locomotor system [D12]	65	67.5	0.39	392 900	-0.09 [c]	-0.16 [c]
Mortality						
Life Span Study [P9]						
Sex Male	16	14	0.23	614 997	1.13 (<0-6.41)	0.15 (<0-0.51)
Female	35	31	0.23	972 359	1.16 (0.01-3.9)	0.19 (0.001-0.5)
All	51	45	0.23	1 587 355	1.15 (0.12-3.27) [c]	0.17 (0.02-0.4) [c]
Benign lesions in the locomotor system [D12]	80	63.8	0.39	439 400	0.65 [c]	0.95 [c]
Ankylosing spondylitis [W1] [e]	22	13.6	4.38	287 095	n.a.	n.a.
Benign gynaecological disease [I6]	14	12.4	1.19	246 821	0.11 (<-0.2-0.6) [c]	0.05 (<-0.1-0.3) [c]
Peptic ulcer [G6]	3	2.2 [f]	1.55	35 815	0.23 (95% CI: -0.6-10)	0.13
Metropathia haemorrhagica [D7] [g]	9	3.5	1.30	47 144	1.23 (0.3-2.7) [c]	0.90 (0.2-2.0) [c]
Nuclear workers in Canada, United Kingdom, and United States [C11]	44	n.a.	0.04	2 142 526	4.2 (0.3-14.4)	n.a.

Study	Observed cases	Expected cases	Mean dose (Sv)	Person-years	Average excess relative risk at 1 Sv	
INTERNAL LOW-LET EXPOSURES						
Incidence						
Diagnostic ^{131}I [H27]	50	45.9	0.00019 [h]	527 056	n.a.	
Swedish ^{131}I Hyperthyroid [H23]	21	20.0	0.06	139 018	n.a.	
Mortality						
United States thyrotoxicosis [i] [R14]	28	n.a.	0.042	735 255	11 [j]	

Table 20 (continued)

Study	Observed cases	Expected cases	Mean dose	Person-years	Average relative risk [k]	
INTERNAL HIGH-LET EXPOSURES						
Incidence						
Danish thorotrast patients [A5]	4	0.95	n.a.	19 365	4.34 (0.85–31.3) [l]	
Mortality						
German thorotrast patients [V8]	10 [m]	n.a.	20.6 ml [n]	n.a.	~4.1 [o]	

a   Mean dose to red bone marrow.
b   90% CI in parentheses derived from published data for Life Span Study and using exact Poisson methods for the other studies.
c   Estimates based on method described in the introduction to Chapter III.
d   Based on one-year survivors. The observed number of cases covers both exposed and unexposed persons.
e   The values given exclude the period within five years of first treatment. Mean dose to bone marrow taken from [W2].
f   Based on unirradiated patients.
g   The values given exclude the period within five years of irradiation.
h   Mean dose to bone marrow given in [H12].
i   Some patients from the United Kingdom were included in this analysis [R14].
j   Not statistically significantly different from zero (p=0.3).
k   95% CI in parentheses.
l   Risk relative to an unexposed control group, in which two cases were observed compared with 2.1 expected.
m   Diagnosis of plasmacytoma.
n   Mean amount of thorotrast administered, based on hospital records.
o   Crude relative risk, based on two cases in an unexposed control group (p>0.05).

Table 21
Risk estimates for cancer incidence and mortality from studies of radiation exposure: leukaemia
The number of observed and expected cases as well as the mean dose and person-years for cohort studies are computed throughout this Table for exposed persons only. In the Life Span Study the exposed group included survivors with organ doses of 0.01 Sv or more for incidence.

Study	Observed cases	Expected cases	Mean dose (Sv)	Person-years	Average excess relative risk [a] at 1 Sv	Average excess absolute risk [a] $(10^4 PYSv)^{-1}$
EXTERNAL LOW-LET EXPOSURES						
Incidence						
Life Span Study [P4]						
Sex Male	71	35.3	0.26	412 300	3.91	3.35
Female	70	32.1	0.25	664 500	4.75	2.29
Age at exposure <20 years	46	17.9	0.26	478 100	6.11	2.28
>20 years	95	49.5	0.25	598 700	3.70	3.06
Time since exposure 5-10 years	29	5.1	0.25	160 900	18.69	5.87
11-20 years	45	40.3	0.25	367 200	0.46	0.50
21-30 years	34	18.5	0.25	277 900	3.32	2.21
31-42 years	33	28.1	0.25	270 800	0.70	0.72
All	141	67.4	0.25	1 076 800	4.37 (3.2-5.6) [b]	2.73 (2.0-3.5) [b]
Cervical cancer case-control [c, e] [B12]	141	n.a.	7.2	n.a.	0.74 (0.1-3.8)	0.50 (0.1-2.6)
Cancer of the uterine corpus [d, e] [C10]	118	n.a.	5.4	n.a.	0.10 (95% CI: <0.0-0.23)	n.a.
Benign lesions in the locomotor system [D12]	116	98.5	0.39	392 900	0.46 [b]	1.14 [b]
Hodgkin's disease [e, f] [K40]	60	n.a.	n.a.	n.a.	0.24 (95% CI: 0.04-0.43)	n.a.
Breast cancer therapy [g] [C9]	38	n.a.	7.5	n.a.	0.19 (0.00-0.6)	0.89 (0.00-3.0)
Techa River population [K27]	37	19.3	0.5	388 880	1.84 (0.9-3.1) [b]	0.91 (0.4-15) [b]
UK childhood cancers [f, h] [H11]	21	n.a.	n.a.	n.a.	0.241 (95% CI: 0.01-1.28)	n.a.
International childhood cancer [h, i] [T7]	25	n.a.	10	n.a.	0.0 (0.0-0.004)	n.a.
Chernobyl recovery operation workers in Russian Federation [j] [I14]	24	n.a.	0.115	n.a.	1.67 (-5.90-9.23)	n.a.
Mortality						
Benign lesions in the locomotor system [D12]	115	95.5	0.39	439 400	0.52 [b]	1.14 [b]
Ankylosing spondylitis [e, k] [W2]	53	17.0	4.38	245 413	6.00 [l]	n.a.
Benign gynaecological disease [e] [I6]	47	27.6	1.19	246 821	2.97 (2.2-4.0)	1.25 (0.9-1.7)
Massachusetts TB fluoroscopy [e] [D4]	17	18	0.09	157 578	<-0.2 (<-0.2-4.5) [b]	<-0.2 (<-0.2-5.1) [b]
Israeli tinea capitis [h, m] [R5]	14	6	0.3	279 901	4.44 (1.7-8.7) [b]	0.95 (0.4-1.9) [b]
Stockholm skin haemangioma [h] [L24]	14	~11	0.2	373 542	1.6 (95% CI: (-0.6-5.5) [n]	n.a.
Metropathia haemorrhagica [o] [D7]	12	5.6	1.3	53 144	0.74 (95% CI: -0.11-1.59)	0.85 [b]
Peptic ulcer [e] [G6]	8	2.9 [p]	1.55	35 815	1.13 (95% CI: -0.2-6.5)	0.92 [b]
Nuclear workers [e] in Canada, United Kingdom, United States [C11]	119	n.a.	0.04	2 142 526	2.18 (0.13-5.7) [q]	n.a.

Table 21 (continued)

Study	Observed cases	Expected cases	Mean dose (Sv)	Person-years	Average excess relative risk [a] at 1 Sv	Average excess absolute risk [a] $(10^4 PYSv)^{-1}$
Nuclear workers in Japan [r] [E3]	23	25.5	0.014	533 168	>0 [s]	n.a.
Yangjiang background radiation [T25, T26]	33	29.7	n.a. [t]	1 246 340	1.61 (95% CI: <0-28.4) [u]	n.a.
Mayak workers (cohort study) [K10] Radiochemical plant Plutonium production Reactors	27 11 6	10.8 5.19 6.74	1.71 0.72 0.87	162 556 67 086 87 307	1.65 [v] n.a. n.a.	0.89 [r] n.a. n.a.

Study	Observed cases	Expected cases	Mean dose (Sv)	Person-years	Average excess relative risk at 1 Sv	
INTERNAL LOW-LET EXPOSURES						
Incidence						
Diagnostic and therapeutic ^{131}I [H12]	130	119	0.014	943 944	n.a. [s]	
Mortality						
United States thyrotoxicosis [e] [w] [R14]	82	n.a.	0.042	735 255	~1 [s]	

Study	Observed cases	Expected cases	Mean dose	Person-years	Average relative risk [x]	
INTERNAL HIGH-LET EXPOSURES						
Incidence						
Danish thorotrast patients [A5]	20 [ab]	1.3	n.a.	19 365	12.7 (2.4-138.4) [y]	
^{224}Ra ankylosing spondylitis patients [W20]	13	4.2	n.a.	32 800	2.4 [z]	
Mortality						
Radon-exposed miners[D8]	69	59.5	155 WLM [aa]	1 085 000	n.a. [s]	
German thorotrast patients [V8]	42 [ab]	n.a.	20.6 ml [ac]	n.a.	~4.9 [ad]	
Portuguese thorotrast patients [D31]	11 [ab]	n.a.	26.3 ml [ac]	16 963	15.2 (1.28-181.7) [ae]	
Japanese thorotrast patients (combined data) [M14]	10	n.a.	17 ml [af]	10 685	12.5 (4.5-34.7)	

a   90% CI in parentheses derived from published data for Life Span Study and using exact Poisson methods for the other studies.
b   Estimates based on method described in the introduction to Chapter III.
c   The observed number of cases covers both exposed and unexposed persons. The excess relative risk was estimated using a linear-exponential dose-response model, and the associated CI was estimated from the confidence region curves in [B9]; the excess absolute risk estimate uses incidence estimates from the cohort study [B11].
d   Risk estimate based on a linear dose-response model fitted to data for all radiation types [C10].
e   Excludes cases of chronic lymphatic leukaemia.
f   Risk estimate based on analysis in [L52].
g   The excess absolute risk for this study is computed based on annual incidence estimates and average follow-up times reported in [C9].
h   Population exposed as children.
i   The observed number of cases covers both exposed and unexposed persons. Risk estimates based on an unmatched analysis of data given in [T5].
j   Excludes cases of chronic lymphatic leukaemia. Results are not restricted according to the date of starting work.
k   The values given exclude the one-year period following the treatment.
l   Risk estimate based on a linear exponential dose-response model averaged over the period 1-25 years after exposure [W2].
m   A re-estimate of the dose to bone marrow in this study indicates a mean dose of 0.60 rather than 0.30 Sv. Consequently the excess relative risk becomes 2.22 Sv^{-1} [R7].
n   Based on those with doses above 0.1 Sv.
o   The values given exclude the period within two years of irradiation.

Table 21 (continued)

p	Based on unirradiated patients.
q	Doses lagged by two years.
r	No cases of chronic lymphatic leukaemia (CLL) in cohort. Expected number based on rates for leukaemia excluding CLL.
s	Trend not statistically significant.
t	Mean annual effective dose = 6.4 mSv.
u	Based on a two-year latent period.
v	Based on male workers followed to the end of 1993, as given in [K11].
w	Some patients from the United Kingdom were included in this analysis [R14].
x	95% CI in parentheses.
y	Relative to unexposed controls, adjusted for gender, age at administration and calendar period [A5].
z	In the control group, seven leukaemias were observed, compared with 5.4 expected [W20].
aa	Mean cumulative radon exposure.
ab	Excludes cases of chronic lymphatic leukaemia.
ac	Mean amount of thorotrast administered, based on hospital records.
ad	Crude relative risk, based on seven cases in the control group.
ae	Based on two deaths in the control group, and excluding the first five years after administration of thorotrast [D31].
af	Mean amount of thorotrast administered in the first series of Japanese patients [M47].

Table 22
Estimates of the projected lifetime risk of cancer mortality following an organ dose of 1 Sv, based on studies of radiation exposure

PART A: STOMACH

Study	Gender	Risk of exposure-induced death (REID) (%) [a] for a projection method with a 10-year latent period and a relative risk for exposure at ages			
		<20 years		≥20 years	
		Assumed constant from 10 years after exposure	Declining to zero risk at age 90 years [b]	Assumed constant from 10 years after exposure	Declining to zero risk at age 90 years [b]
EXTERNAL LOW-LET EXPOSURES					
Values based on incidence studies					
Life Span Study [T1]	Both	0.38	0.17	0.10	0.08
Cervical cancer case-control [B1]	Females	–	–	0.18 (0.03-0.49)	0.14 (0.03-0.39)
Mayak workers [Z1]	Males	–	–	0.15 (95% CI: 0-0.5) [c]	0.12 (95% CI: 0-0.4) [c]
Swedish benign breast disease [M28]	Females	–	–	0.43 (0-1.4)	0.34 (0-1.1)
Values based on mortality studies					
Life Span Study [P9]	Males Females Both	0.11 (<0-0.76) 0.40 (<0-0.93) 0.26	0.05 (<0-0.37) 0.17 (<0-0.40) 0.11	0.06 0.13 0.09	0.05 0.09 0.07
Ankylosing spondylitis [W1]	Males	–	–	<0 (95% CI: <0-0.03)	<0 (<0-0.02)
Peptic ulcer [G6]	Both	–	–	0.07	0.05
Metropathia haemorrhagica [D7]	Females	–	–	0.33 (<0-0.92)	0.26 (<0-0.73)
Benign gynaecological disease [I16]	Females	–	–	0.09 (<0-1.57)	0.07 (<0-1.2)

a   Estimated percentage of population that would die of radiation-induced cancer. Computed using relative risks estimated from the relevant studies (split by gender and age at exposure where possible), and applied to Japanese death rates for 1985 [J3]. The calculations have been performed for the gender and age-specific groupings that predominate in the relevant study. 90% CI in parentheses unless otherwise stated.

b   Constant relative risk for first 45 years after exposure. Relative risk then decreases linearly with increasing attained age to zero at age 90 years.

c   Based on the excess relative risk among those with external gamma doses in excess of 3 Gy relative to those with lower doses, divided by an (arbitrary) value of 4 in order to estimate risks at 1 Gy.

Table 22 (continued)

PART B: COLON

Study	Gender	Risk of exposure-induced death (REID) (%) [a] for a projection method with a 10-year latent period and a relative risk for exposure at ages			
		<20 years		≥20 years	
		Assumed constant from 10 years after exposure	Declining to zero risk at age 90 years [b]	Assumed constant from 10 years after exposure	Declining to zero risk at age 90 years [b]
EXTERNAL LOW-LET EXPOSURES					
Values based on incidence studies					
Life Span Study [T1]	Both	0.55	0.25	0.51	0.42
Cervical cancer case-control [B1]	Females	–	–	0.00 (–0.01–0.02)	0.00 (–0.01–0.02)
Stockholm skin haemangioma [L16]	Both	0.33 [c]	0.15	–	–
Values based on mortality studies					
Life Span Study [P9]	Males Females Both	1.5 (<0–4.6) 2.2 (<0–6.7) 1.8	0.73 (<0–2.3) 0.95 (<0–2.9) 0.84	0.35 0.48 0.42	0.28 0.34 0.31
Benign gynaecological disease [I16]	Females	–	–	0.31 (<0–3.5)	0.25 (<0–2.7)
Metropathia haemorrhagica [D7]	Females	–	–	0.08 (95% CI: 0.01–0.21)	0.06 (95% CI: 0.00–0.17)
Peptic ulcer [G6]	Both	–	–	0.04 (95% CI: –0.04–0.18)	0.03 (95% CI: –0.03–0.13)

a   Estimated percentage of population that would die of radiation-induced cancer. Computed using relative risks estimated from the relevant studies (split by gender and age at exposure where possible) and applied to Japanese death rates for 1985 [J3]. The calculations have been performed for the gender and age-specific groupings that predominate in the relevant study. 90% CI in parentheses unless otherwise stated.

b   Constant relative risk for first 45 years after exposure. Relative risk then decreases linearly with increasing attained age to zero at age 90 years.

c   Not statistically significant.

Table 22 (continued)

PART C: LUNG

Study	Gender	Risk of exposure-induced death (REID) (%) [a] for a projection method with a 10-year latent period and a relative risk for exposure at ages			
		<20 years		≥20 years	
		Assumed constant from 10 years after exposure	Declining to zero risk at age 90 years [b]	Assumed constant from 10 years after exposure	Declining to zero risk at age 90 years [b]
EXTERNAL LOW-LET EXPOSURES					
Values based on incidence studies					
Life Span Study [T1]	Both	2.1	1.0	3.3	2.9
Hodgkin's disease (Netherlands) [V2]	Both	–	–	~3 (<0 – ~30)	~3 (<0 – ~30)
Breast cancer [I7]	Females	–	–	0.19 (95% CI: <0–0.52)	0.16 (95% CI: <0–0.45)
Swedish benign breast disease [M28]	Females	–	–	0.43 (95% CI: 0–1.4)	0.34 (95% CI: 0–1.1)
Stockholm skin haemangioma [L16]	Both	5.2	2.5	–	–
Values based on mortality studies					
Life Span Study [P9]	Males Females Both	1.1 (<0–8.7) 1.1 (<0–7.5) 1.1	0.52 (<0–4.1) 0.48 (<0–3.4) 0.50	1.5 1.2 1.3	1.3 1.0 1.2
Ankylosing spondylitis [W1]	Males	–	–	0.20 (95% CI: 0.01–0.36)	0.18 (95% CI: 0.01–0.32)
Canadian TB fluoroscopy [H7]	Both	0.00 (95% CI: <0–0.26)	0.00 (95% CI: <0–0.12)	0.00 (95% CI: <0–0.22)	0.00 (95% CI: <0–0.19)
Peptic ulcer [G6]	Both	–	–	1.2 (95% CI: 0.34–2.4)	1.1 (95% CI: 0.31–2.2)
Massachusetts TB fluoroscopy [D4]	Both	<0 (<0–0.15)	<0 (<0–0.07)	<0 (<0–0.13)	<0 (<0–0.11)

a   Estimated percentage of population that would die of radiation-induced cancer. Computed using relative risks estimated from the relevant studies (split by gender and age at exposure where possible), and applied to Japanese death rates for 1985 [J3]. The calculations have been performed for the gender and age-specific groupings that predominate in the relevant study. 90% CI in parentheses unless otherwise stated.

b   Constant relative risk for first 45 years after exposure. Relative risk then decreases linearly with increasing attained age to zero at age 90 years.

Table 23
Lung cancer mortality in the Canadian fluoroscopy study and in the study of survivors of the atomic bombings

Lung dose (Sv)	Canadian fluoroscopy study (1950–1987) [H7]			Study of survivors of atomic bombings (1950–1990) [P9, P11]		
	Observed deaths	Relative risk [a] [b]	95% CI	Observed deaths	Relative risk [a] [b]	95% CI
0 [c]	723	1.00		349	1.00	
>0–0.49	180	0.87	0.74–1.03	477	1.16	1.02–1.34
0.50–0.99	92	0.82	0.66–1.02	43	1.35	0.97–1.83
1.00–1.99	114	0.94	0.77–1.15	39	2.05	1.40–2.96
2.00–2.99	41	1.09	0.80–1.50	11	2.80	1.41–5.06
≥3.00	28	1.04	0.72–1.53	5	1.65	0.61–3.70

a   Adjusted for age at risk, calendar year at risk and sex.
b   Excludes person-years for age at risk <20 years and deaths and person-years at risk within 10 years of exposure.
c   Defined as less than 0.01 Sv for the fluoroscopy study and less than 0.005 Sv for the study on survivors of atomic bombings.

Table 24
Lung cancer cases and parameters for risk estimates in studies of radon-exposed underground miners [a]
[L6, L45]

Study cohort	Cases [b]	Average cumulative exposure (WLM)	Excess relative risk per 100 WLM ($\beta \times 100$)	Modification factor (γ)	Test of significance (p) [c]
China tin miners	980	277.4	0.59	−0.79	<0.001
Western Bohemia uranium miners	661	198.7	5.84	−0.78	<0.001
Colorado Plateau uranium miners	294	595.7	14.5	−0.79	<0.001
Ontario uranium miners	291	30.8	2.40	−0.55	0.002
Newfoundland fluorspar	118	367.3	5.14	−0.53	<0.001
Sweden iron miners	79	80.6	1.55	−1.02	0.03
New Mexico uranium miners	69	110.3	6.56	−0.30	0.17
Beaverlodge uranium miners	65	17.2 [d]	7.42	−0.67	0.001
Port Radium uranium miners	57	242.8	1.15	−0.42	0.24
Radium Hill uranium miners	54	7.6	5.68	−0.63	0.30
France uranium miners	45	68.7	1.92	0.57	0.57

a   Background lung cancer rates are adjusted for attained age (all studies), other mine exposures [China, France, Ontario, United States (Colorado, New Mexico)], and indicator of radon progeny exposure (Beaverlodge) and ethnicity (New Mexico). United States (Colorado) data are restricted to exposures under 3,200 WLM. The relative risk is modelled by the form RR = 1 + β × WLM × (WL)$^\gamma$.
b   Total number of cases is 2,701 and omits 12 cases that were included in both United States studies (New Mexico and Colorado).
c   P-value for test of significance of continuous variation of ERR/WLM by WL.
d   Howe and Stager [H18] quote a revised mean of 81.3 WLM for exposed miners, compared with an earlier mean of 50.6 WLM for miners with non-zero exposure.

Table 25
Parameter values used by BEIR Committees in risk models for lung cancer following radon exposure
[C2, C21]

Parameter	Parameter value		BEIR IV model
	BEIR VI preferred models [a]		
	Exposure-age-duration	Exposure-age-concentration	
Time since exposure, θ (years)			
5–14	1	1	1
15–24	0.72	0.78	0.5
≥25	0.44	0.51	0.5
Attained age, φ_{age} (years)			
<55	1	1	1
55–64	0.52	0.57	0.83
65–74	0.28	0.29	0.33
≥75	0.13	0.09	0.33
Duration of exposure, γ_z (years)			
<5	1	1	1
5–14	2.78	1	1
15–24	4.42	1	1
25–34	6.62	1	1
≥35	10.20	1	1
Exposure rate (WL)			
<0.5	1	1	1
0.5–1.0	1	0.49	1
1.0–2.99	1	0.37	1
3.0–4.99	1	0.32	1
5.0–14.99	1	0.17	1
≥15.0	1	0.11	1

a   ERR = β ($w_{5-14} + \theta_{15-24} w_{15-24} + \theta_{25+} w_{25+}$)$\varphi_{age} \gamma_z$, i.e. a product of terms representing: (a) exposure in three time periods, i.e. 5-14, 15–24 and 25+ years previously *(Note: BEIR IV used 5–14 and 15+)*; (b) attained age (φ_{age}); (c) duration of exposure or average concentration (γ_z) *(Note: not included in BEIR IV model)*.

Table 26
Basal-cell skin cancer incidence in the Life Span Study
[R15]

Variable	Observed cases [a]	Average excess relative risk at 1 Sv	90% CI
All [b]	80	1.9	0.83–3.3
Gender			
Male	32	2.7	0.5–9.1
Female	48	1.6	0.5–4.1
	(Heterogeneity [c] p > 0.5)		
Age at exposure			
<10 years	3	21	4.1–73
10–19 years	8	6.7	2.1–17
20–30 years	28	1.7	0.5–3.8
>40 years	41	0.7	−0.05–2.2
	(Heterogeneity [c] p = 0.03)		

a   Includes exposed and non-exposed cases.
b   Estimates are for a person exposed to the atomic bombings at age 30 years. The estimates depend on age at exposure with larger risks for those exposed earlier and smaller risks for thoses exposed later in life. The risks change by about 11% for a one-year change in age at exposure.
c   Test of the hypotheses that effects differ across categories.

Table 27
Numbers and rates of tumours of the brain and central nervous system in the Life Span Study of atomic bomb survivors (1958–1994)
[P19]

Histology	Brain dose [a] (Gy)	Number of cases	Incidence rate per 10,000 person years
Glioma, astrocytoma	<0.0005	19	0.24
	0.0005–0.099	12	0.16
	0.1–0.99	7	0.20
	>1	3	0.46
Meningioma	<0.0005	33	0.42
	0.0005–0.099	28	0.38
	0.1–0.99	19	0.53
	>1	5	0.76
Neurilemmoma	<0.0005	18	0.23
	0.0005–0.099	11	0.15
	0.1–0.99	17	0.48
	>1	9	1.37
Not specified and other	<0.0005	15	0.19
	0.0005–0.099	18	0.24
	0.1–0.99	9	0.25
	>1	3	0.46

a   Total person years at <0.0005 Sv: 791,456; at 0.0005–0.099 Sv:7425,831; at 0.1–0.99 Sv: 355,877; and at >1 Sv: 65,844.

Table 28
Thyroid cancer risk in patients receiving diagnostic administration of ^{131}I [a]
[H4]

Dose [b] (Gy)	Observed number of cases	Standardized incidence ratio (SIR)	95% CI
Referred for suspicion of a thyroid tumour			
≤0.25	6	3.57	1.31–7.77
0.26–0.50	12	4.30	2.22–7.51
0.51–1.00	4	1.39	0.38–3.56
>1.00	20	2.72	1.66–4.20
All	42	2.86	2.06–3.86
Referred for other reasons			
≤0.25	5	0.55	0.18–1.29
0.26–0.50	4	0.68	0.18–1.73
0.51–1.00	5	0.47	0.20–1.46
>1.00	11	1.04	0.52–1.86
All	25	0.75	0.48–1.10
All patients			
≤0.25	11	1.03	0.51–1.83
0.26–0.50	16	1.84	1.05–2.98
0.51–1.00	9	0.46	0.38–1.57
>1.00	31	1.60	1.09–2.27
All	67	1.35	1.05–1.71

a   The first five years after exposure were excluded.
b   Estimated without considering thyroid weight.

Table 29
Childhood thyroid cancer in Belarus, Russian Federation and Ukraine before and after the Chernobyl accident [a] [S9]

Country / region	Number of cases			Incidence rate (10^{-6})			Number of children with thyroid cancer born since 1986	Cases found by annual medical examination since 1986 (%) [b]	Range of estimated thyroid doses (Gy)	Cases of papillary cancer (%)	Cases confirmed by international review
	1981–1985	1986–1990	1991–1994	1981–1985	1986–1990	1991–1994					
Belarus [c]	3	47	286	0.3	4.0	30.6	7	62	n.a.	96	91
Gomel	1	21	143	0.5	10.5	96.4	5	n.a.	0.15–5.7	n.a.	n.a.
Russian Federation	n.a.	n.a.	n.a.	n.a.	n.a.	n.a.	n.a.	n.a.	n.a.	n.a.	n.a.
Bryansk and Kaluga regions	0	3	20	0	1.2	10.0	0	n.a.	0.06–1.8	n.a.	n.a.
Ukraine	25	60	149	0.5	1.1	3.4	2	n.a.	n.a.	95	n.a.
Five most northerly regions [d]	1	21	97	0.1	2.0	11.5	n.a.	40	0.05–2.0	n.a.	n.a.

a   Aged under 15 years at diagnosis; rates are expressed as annual averages per million children under 15 in the regions and periods specified.
b   Annual medical examination includes palpation and ultrasound scanning of neck and, in some cases, thyroid hormone tests.
c   Data made available by Drs. Demidchik, Astakhova, Okeanov, and Kenigsberg.
d   Kiev, Chernikov, Cherkassy, Rovno, and Zhitomir.

Table 30
Fitted risks of leukaemia (other than chronic lymphatic leukaemia) in four studies of low-LET irradiation of adults

Risk parameter	Ankylosing spondylitis study [a]		Uterine corpus cancer study [b]		Cervical cancer study [c]	Life Span Study [d]	
	Time since exposure = 10 years	Time since exposure = 25 years	Brachytherapy irradiation	Any external irradiation		Time since exposure = 1–25 years	Time since exposure >25 years
Linear component of excess relative risk Gy^{-1}	12.37 (2.25–52.07) [e]	5.18 (0.81–23.63)	4.69 (1.10–13.4)	0.05 (<0–0.55)	0.88 (−0.50–2.23)	-	-
Reduction in excess relative risk at 1 Gy due to cell sterilization	47% (17–79%)	47% (17–79%)	59% (30–75%)	−4% (−40–23%)	8% (1–14%)	-	-
Predicted excess relative risk at a uniform dose of 1 Gy	6.00 [f]	1.88 [f]	1.91	0.05	0.74	5.13	1.56

a   Estimates derived from the compartmental linear-exponential model allowing for the effect of time since first treatment, assuming that each bone marrow compartment received the same dose and using national rates as baseline risk [W2].
b   Estimates derived from the linear-exponential model [C10].
c   Estimates based on case-control analysis [B12].
d   Estimates for survivors of the atomic bombings in Japan aged over 20 years at exposure [P11].
e   95% CI.
f   Predicted relative risks in the two periods are the average relative risks at 1–25 years and 25–40 years after first treatment, respectively.

Table 31
Models for risks of solid cancer mortality and incidence used in the lifetime risk computations based on the Life Span Study

Cancer type	Age-at-exposure model				Attained-age model			
	Excess relative risk per Sv [a]		Sex ratio (female/male)	Change in risk per 10-year increase in age at exposure (%)	Excess relative risk per Sv [b]		Sex ratio (female/male)	Power of age
	Male	Female			Male	Female		
Cancer mortality risks								
All solid cancer	0.38	0.77	2.1	-32	0.38	0.88	2.3	-1.5
Oesophagus	0.91	1.88	2.1	-32	1.04	2.37	2.3	-1.5
Stomach	0.26	0.54	2.1	-32	0.27	0.63	2.3	-1.5
Colon	0.46	0.95	2.1	-32	0.68	1.56	2.3	-1.5
Liver	0.61	1.66	1.0	-13	0.29	0.29	1.0	1.4
Lung	0.30	0.99	3.3	26	0.68	1.55	2.3	-1.5
Breast	0.00	1.34	-	-32	0.00	2.35	-	-1.5
Bladder	0.46	0.94	2.1	33	0.97	2.21	2.3	-1.5
Other cancer	0.38	0.77	2.1	-32	0.32	0.74	2.3	-1.5
Cancer incidence risks								
All solid cancer	0.38	0.79	2.1	-33	0.58	1.10	1.9	-2.1
Oesophagus	0.41	0.84	2.1	0	0.78	1.48	1.9	-2.1
Stomach	0.29	0.60	2.1	0	0.39	0.73	1.9	-2.1
Colon	0.46	0.95	2.1	0	0.83	1.56	1.9	-2.1
Liver	0.58	0.58	1.0	-7	0.66	0.66	1.0	-0.6
Lung	0.50	2.18	4.3	7	0.51	2.19	4.3	0.2
Breast	0.00	1.55	-	0	0.00	2.22	-	-2.1
Bladder	1.18	0.98	0.8	-61	1.53	2.90	1.9	-2.1
Thyroid	0.89	1.84	2.1	0	1.14	2.15	1.9	-2.1
Other cancer	0.47	0.28	0.6	-50	0.65	0.38	0.6	-3.2

a For age at exposure 30 years.
b For attained age 50 years (ages at exposure <40 years).

Table 32
Estimated lifetime probabilities of solid cancer and leukaemia in unexposed populations

Cancer type	Lifetime probability (%)									
	China		Japan		Puerto Rico		United Kingdom		United States	
	Male	Female	Male	Female	Male	Female	Male	Female	Male	Female
Incidence										
Solid cancer	24.3	16.2	37.2	15.3	26.2	19.9	39.6	33.6	33.9	30.4
Leukaemia	0.3	0.3	0.4	0.3	0.6	0.5	0.7	0.5	0.6	0.5
Mortality										
Solid cancer	12.8	9.5	23.3	25.2	13.9	11.1	24.0	20.1	21.6	17.9
Leukaemia	0.1	0.1	0.4	0.3	0.1	0.1	0.6	0.5	0.6	0.5

Table 33
Estimates of lifetime risk of exposure-induced death (REID) or exposure-induced cancer incidence following an acute whole-body exposure to a population of all ages

Projection model[a]	Risk transport model	China			Japan			Puerto Rico			United Kingdom			United States		
		Male	Female	Both	Male	Female	Both	Male	Female	Both	Male	Female	Both	Male	Female	Both
							REID (%)									
							Dose of 1 Sv									
							Solid cancer mortality									
Age-at-exposure	RR[b]	8.2	11.7	9.9	9.5	12.9	11.2	7.3	11.9	9.6	9.5	19.3	14.4	8.5	16.4	12.5
	AR[c]	8.6	10.5	9.5	9.5	12.9	11.2	9.9	12.8	11.3	10.8	14.4	12.6	8.2	11.5	9.9
Attained-age	RR	4.9	7.1	6.0	6.2	8.5	7.4	4.4	7.9	6.1	6.6	13.5	10.1	6.2	12.4	9.3
	AR	5.3	6.8	6.0	6.2	8.5	7.4	6.1	8.2	7.2	6.7	9.1	7.9	5.4	7.6	6.5
UNSCEAR 1994 [U2]	RR				10.4	11.4	10.9									
							Leukaemia mortality									
Age- and time-varying	AR	0.94	0.56	0.75	1.04	0.79	0.92	0.91	0.52	0.72	1.02	0.87	0.95	1.13	1.25	1.19
UNSCEAR 1994 [U2]	AR						1.1									
							Solid cancer incidence									
Age-at-exposure	RR	14.8	17.6	16.2	18.6	21.0	19.8	17.5	19.8	18.6	22.1	29.9	26.0	18.7	27.4	23.0
	AR	18.7	19.7	19.2	18.6	21.0	19.8	20.4	21.7	21.1	22.1	23.2	22.7	14.1	17.0	15.5
Attained-age	RR	10.2	13.8	12.0	13.3	16.2	14.7	8.6	13.8	11.2	12.9	22.6	17.8	12.3	23.1	17.7
	AR	13.0	14.7	13.8	13.3	16.2	14.7	14.0	16.2	15.1	14.6	17.0	15.8	10.7	13.5	12.1
							Leukaemia incidence									
Age- and time-varying	AR	1.27	0.84	1.06	1.00	0.73	0.87	1.29	1.12	1.21	1.19	0.96	1.08	1.24	0.94	1.09
							Dose of 0.1 Sv[d]									
							Solid cancer mortality									
Age-at-exposure	RR	0.9	1.3	1.1	1.0	1.4	1.2	0.8	1.3	1.1	1.1	2.3	1.7	0.9	1.9	1.4
	AR	0.9	1.2	1.1	1.0	1.4	1.2	1.1	1.4	1.3	1.2	1.6	1.4	0.9	1.3	1.1
Attained-age	RR	0.5	0.7	0.6	0.7	0.9	0.8	0.5	0.8	0.6	0.7	1.5	1.1	0.7	1.4	1.0
	AR	0.6	0.7	0.6	0.7	0.9	0.8	0.6	0.9	0.8	0.7	1.0	0.8	0.6	0.8	0.7

Table 33 (continued)

Projection model [a]	Risk transport model	REID (%)														
		China			Japan			Puerto Rico			United Kingdom			United States		
		Male	Female	Both	Male	Female	Both	Male	Female	Both	Male	Female	Both	Male	Female	Both
Leukaemia mortality																
Age- and time-varying	AR	0.04	0.02	0.03	0.05	0.03	0.04	0.04	0.02	0.03	0.04	0.04	0.04	0.06	0.06	0.06
Solid cancer incidence																
Age-at-exposure	RR	1.8	2.1	1.9	2.4	2.6	2.5	2.3	2.5	2.4	3.0	4.1	3.5	2.6	3.8	3.2
	AR	2.4	2.4	2.4	2.4	2.6	2.5	2.7	2.7	2.7	2.9	2.9	2.9	1.7	2.1	1.9
Attained-age	RR	1.1	1.5	1.3	1.5	1.8	1.6	0.9	1.5	1.2	1.4	2.6	2.0	1.4	2.8	2.1
	AR	1.4	1.6	1.5	1.5	1.8	1.6	1.6	1.8	1.7	1.6	1.9	1.8	1.2	1.5	1.3
Leukaemia incidence																
Age- and time-varying	AR	0.05	0.03	0.04	0.04	0.03	0.04	0.06	0.05	0.06	0.05	0.04	0.05	0.06	0.04	0.05

a   The projection models are described in Section IV.B.1 of this Annex, with the exception of those cited from the UNSCEAR 1994 Report [U2].
b   Relative risk transportation.
c   Absolute risk transportation.
d   The estimates presented for solid cancers at 0.1 Sv are based on a linear-quadratic dose response.
     The estimates presented for solid cancers at 0.1 Sv do not involve a reduction factor for low doses or low dose rates. In contrast, the leukaemia estimates at 0.1 Sv are based on a linear-quadratic dose response.

Table 34
Estimates of REID for an acute whole-body dose of 1 Sv to a Japanese population

Projection model	Age at exposure (years)	REID (%)					
		Solid cancer mortality		Solid cancer incidence		Leukaemia incidence	
		Male	Female	Male	Female	Male	Female
Age-at-exposure model	10	13.9	19.6	31.0	36.5	1.9	1.0
	30	8.6	11.9	15.4	18.8	0.8	0.9
	50	6.2	8.8	9.1	10.7	0.6	0.6
	All	9.5	12.9	18.6	21.0	1.0	0.7
Attained-age model	10	6.7	9.7	14.9	20.1	1.9	1.0
	30	6.7	9.5	13.3	18.1	0.8	0.9
	50	6.3	8.2	11.4	13.0	0.6	0.6
	All	6.2	8.5	13.3	16.2	1.0	0.7

Table 35
Estimates of measures of radiation detriment associated with an acute whole-body dose of 1 Sv to a male Japanese population

Age at exposure (years)	Cause of death	Unexposed	Exposed							
			Age-at-exposure model				Attained-age model			
		Lifetime risk	Lifetime risk	Radiation-associated deaths, REID (%)	Excess lifetime risk, ELR (%)	Loss of life expectancy, LLE (years)	Lifetime risk	Radiation-associated deaths, REID (%)	Excess lifetime risk, ELR (%)	Loss of life expectancy, LLE (years)
10	Solid cancer	23.6	34.6	13.9	11.0	12.7	28.6	6.7	5.0	15.0
	Leukaemia	0.5	2.4	2.0	1.9	53.1	2.4	2.0	1.9	53.0
	Other causes	75.9	63.0	0.0	−12.9	0.0	69.0	0.0	−6.9	0.0
30	Solid cancer	23.8	30.7	8.6	6.9	12.0	29.1	6.7	5.3	13.9
	Leukaemia	0.5	1.3	0.9	0.9	29.0	1.3	0.9	0.9	28.9
	Other causes	75.7	68.0	0.0	−7.7	0.0	69.6	0.0	−6.1	0.0
50	Solid cancer	23.9	28.9	6.2	5.0	10.3	28.9	6.3	5.0	11.4
	Leukaemia	0.4	1.0	0.6	0.6	13.9	1.0	0.6	0.6	14.0
	Other causes	75.7	70.1	0.0	−5.6	0.0	70.1	0.0	−5.7	0.0
All ages	Solid cancer	23.3	30.9	9.5	7.6	11.1	28.2	6.2	4.9	12.8
	Leukaemia	0.4	1.4	1.0	1.0	30.6	1.5	1.0	1.0	30.6
	Other causes	76.3	67.7	0.0	−8.6	0.0	70.4	0.0	−5.9	0.0

Table 36
Estimates of REID for cancer mortality under the attained-age model, based on acute whole-body exposure at age 30 years [a]

Cancer type	REID (%)								
	China		Japan	Puerto Rico		United Kingdom		United States	
	RR [b]	AR [c]		RR	AR	RR	AR	RR	AR
Dose of 1 Sv (males)									
Oesophagus	2.6	0.6	0.7	1.2	0.7	0.5	0.8	0.3	0.7
Stomach	0.7	0.9	1.0	0.5	1.0	0.3	1.1	0.1	0.9
Colon	0.2	0.3	0.4	0.3	0.4	0.6	0.4	0.9	0.4
Liver	0.6	1.5	1.2	0.9	1.7	0.1	1.3	0.2	1.0
Lung	0.5	1.0	1.8	0.2	1.1	3.6	2.0	3.1	1.6
Breast	0.0	0.0	0.0	0.0	0.0	0.0	0.0	0.0	0.0
Bladder	0.1	0.2	0.2	0.2	0.2	0.5	0.3	0.4	0.2
Other solid cancer	0.7	1.1	1.3	1.4	1.2	1.4	1.4	1.8	1.1
All solid cancers	5.3	5.6	6.7	4.8	6.4	7.1	7.3	6.8	5.9
Leukaemia	0.5	0.5	0.9	0.5	0.5	0.7	0.7	1.0	1.0
Total	5.9	6.1	7.6	5.3	6.9	7.8	8.0	7.8	6.9
Dose of 0.1 Sv (males) [d]									
Oesophagus	0.27	0.07	0.08	0.13	0.07	0.06	0.08	0.03	0.07
Stomach	0.07	0.09	0.11	0.05	0.10	0.03	0.12	0.01	0.10
Colon	0.02	0.04	0.04	0.03	0.04	0.06	0.05	0.10	0.04
Liver	0.06	0.16	0.12	0.09	0.18	0.01	0.14	0.02	0.11
Lung	0.05	0.10	0.19	0.02	0.12	0.38	0.21	0.33	0.17
Breast	0.00	0.00	0.00	0.00	0.00	0.00	0.00	0.00	0.00
Bladder	0.01	0.02	0.02	0.03	0.02	0.06	0.03	0.04	0.02
Other solid cancer	0.07	0.11	0.14	0.15	0.13	0.15	0.15	0.19	0.12
All solid cancers	0.55	0.59	0.70	0.50	0.66	0.75	0.77	0.71	0.62
Leukaemia	0.02	0.02	0.04	0.02	0.02	0.03	0.03	0.05	0.05
Total	0.57	0.61	0.74	0.5	0.7	0.78	0.80	0.76	0.67
Dose of 1 Sv (females)									
Oesophagus	3.2	0.4	0.6	1.3	0.5	0.7	0.6	0.2	0.5
Stomach	0.9	1.0	1.4	0.6	1.3	0.4	1.5	0.2	1.2
Colon	0.3	0.5	0.7	0.8	0.6	1.5	0.7	1.9	0.6
Liver	0.6	1.8	0.6	1.0	2.4	0.1	0.6	0.2	0.5
Lung	0.3	0.4	2.5	0.1	0.5	3.5	2.6	3.2	2.3
Breast	0.6	1.2	1.3	2.2	1.3	5.8	1.4	5.2	1.3
Bladder	0.1	0.2	0.2	0.3	0.2	0.5	0.3	0.3	0.2
Other solid cancer	1.8	1.8	2.3	2.8	2.3	2.8	2.5	3.2	2.2
All solid cancers	7.7	7.2	9.5	9.0	9.1	15.2	10.1	14.4	8.8
Leukaemia	0.5	0.5	1.0	0.5	0.5	1.0	1.1	1.5	1.6
Total	8.1	7.6	10.4	9.6	9.7	16.2	11.2	15.9	10.3
Dose of 0.1 Sv (females) [d]									
Oesophagus	0.34	0.04	0.06	0.14	0.06	0.08	0.06	0.03	0.05
Stomach	0.09	0.11	0.14	0.07	0.14	0.04	0.16	0.02	0.13
Colon	0.03	0.05	0.07	0.09	0.07	0.17	0.08	0.22	0.06
Liver	0.06	0.19	0.06	0.10	0.25	0.01	0.06	0.02	0.06
Lung	0.03	0.04	0.26	0.01	0.06	0.38	0.28	0.34	0.24
Breast	0.06	0.12	0.14	0.23	0.14	0.63	0.14	0.56	0.13
Bladder	0.01	0.02	0.02	0.03	0.02	0.05	0.03	0.04	0.02
Other solid cancer	0.19	0.19	0.25	0.29	0.24	0.32	0.27	0.36	0.23
All solid cancers	0.81	0.76	1.00	0.96	0.98	1.68	1.08	1.58	0.93
Leukaemia	0.02	0.02	0.04	0.02	0.02	0.04	0.04	0.06	0.06
Total	0.83	0.78	1.04	1.0	1.0	1.72	1.12	1.64	0.99

a Owing to rounding errors, the sum of the individual values in each column sometimes differs from the total, which has been calculated to greater accuracy. Also, in a few instances, the mortality estimates in this Table are greater than the corresponding incidence values in Table 37 owing to the use of baseline rates that differ by the region studied within a country or that differ by time period (see Section IV.B.4).

b Relative risk transportation.

c Absolute risk transportation.

d The estimates presented for solid cancers at 0.1 Sv do not involve a reduction factor for low doses or low dose rates. In contrast, the leukaemia estimates at 0.1 Sv are based on a linear-quadratic dose response.

Table 37
Estimates of REID for cancer incidence under the attained-age model, based on acute whole-body exposure at age 30 years [a]

Cancer type	REID (%)								
	China		Japan	Puerto Rico		United Kingdom		United States	
	RR [b]	AR [c]		RR	AR	RR	AR	RR	AR
Dose of 1 Sv (males)									
Oesophagus	0.6	0.5	0.5	0.5	0.5	0.5	0.6	0.2	0.4
Stomach	1.1	1.7	1.9	0.4	1.9	0.5	2.1	0.2	1.5
Colon	0.6	1.3	1.4	0.8	1.4	1.2	1.6	1.1	1.2
Liver	0.7	2.4	2.6	0.3	2.5	0.2	2.8	0.1	2.1
Lung	3.2	2.4	2.8	1.3	2.9	5.0	3.4	2.9	2.0
Breast	0.0	0.0	0.0	0.0	0.0	0.0	0.0	0.0	0.0
Thyroid	0.2	0.3	0.3	0.1	0.3	0.4	0.9	0.4	0.6
Bladder	0.5	0.7	0.8	0.7	0.8	0.1	0.3	0.4	0.3
Other solid cancer	2.6	2.7	2.9	4.0	2.9	5.1	3.2	6.8	2.5
All solid cancers	9.4	12.0	13.3	8.0	13.1	12.9	14.9	12.1	10.6
Leukaemia	0.8	0.8	0.8	0.9	0.9	0.9	0.9	0.7	0.7
Total	10.2	12.8	14.1	8.9	14.0	13.8	15.8	19.7	13.8
Dose of 0.1 Sv (males) [d]									
Oesophagus	0.06	0.05	0.05	0.05	0.05	0.05	0.06	0.02	0.04
Stomach	0.12	0.19	0.21	0.04	0.20	0.05	0.23	0.02	0.17
Colon	0.07	0.14	0.16	0.09	0.15	0.13	0.17	0.13	0.13
Liver	0.08	0.26	0.28	0.03	0.27	0.02	0.31	0.02	0.23
Lung	0.35	0.27	0.32	0.14	0.33	0.55	0.39	0.32	0.23
Breast	0.00	0.00	0.00	0.00	0.00	0.00	0.00	0.00	0.00
Thyroid	0.02	0.03	0.04	0.02	0.03	0.04	0.10	0.04	0.07
Bladder	0.05	0.08	0.09	0.07	0.09	0.01	0.04	0.04	0.03
Other solid cancer	0.27	0.29	0.32	0.43	0.31	0.55	0.35	0.74	0.26
All solid cancers	1.01	1.30	1.46	0.85	1.44	1.40	1.65	1.33	1.16
Leukaemia	0.04	0.04	0.04	0.04	0.04	0.02	0.02	0.05	0.05
Total	1.05	1.34	1.50	0.89	1.48	1.42	1.67	1.38	1.21
Dose of 1 Sv (females)									
Oesophagus	0.4	0.1	0.2	0.3	0.2	0.4	0.2	0.1	0.1
Stomach	1.0	1.6	1.9	0.4	1.9	0.4	2.0	0.1	1.6
Colon	1.1	1.6	2.0	1.5	2.0	2.0	2.1	1.9	1.7
Liver	0.6	0.6	0.8	0.1	0.8	0.1	0.9	0.1	0.7
Lung	4.7	3.4	4.6	2.4	4.5	7.4	5.1	7.5	3.5
Breast	4.6	4.9	5.3	8.4	5.3	12.3	5.4	13.6	4.9
Thyroid	0.6	1.2	1.3	0.5	1.3	0.5	0.4	0.5	0.3
Bladder	0.2	0.3	0.4	0.5	0.4	0.3	1.3	1.0	1.2
Other solid cancer	1.1	1.4	1.6	1.4	1.6	2.4	1.6	2.2	1.4
All solid cancers	14.4	15.1	18.1	15.5	17.9	25.7	19.0	27.0	15.4
Leukaemia	0.8	0.8	0.9	1.2	1.2	1.1	1.2	1.0	1.1
Total	15.2	15.9	19.0	16.7	19.0	26.8	20.1	30.2	17.9
Dose of 0.1 Sv (females) [d]									
Oesophagus	0.05	0.02	0.02	0.03	0.02	0.05	0.02	0.01	0.02
Stomach	0.11	0.17	0.21	0.04	0.21	0.05	0.22	0.02	0.17
Colon	0.12	0.18	0.22	0.17	0.22	0.24	0.24	0.24	0.19
Liver	0.07	0.07	0.10	0.02	0.09	0.01	0.10	0.01	0.41
Lung	0.54	0.39	0.54	0.28	0.53	0.91	0.61	0.95	0.07
Breast	0.49	0.52	0.56	0.90	0.56	1.39	0.57	1.57	0.52
Thyroid	0.06	0.12	0.14	0.06	0.14	0.06	0.05	0.06	0.04
Bladder	0.03	0.04	0.05	0.05	0.05	0.03	0.14	0.11	0.13
Other solid cancer	0.12	0.15	0.17	0.15	0.17	0.27	0.18	0.26	0.15
All solid cancers	1.58	1.67	2.02	1.70	1.99	3.01	2.13	3.23	1.70
Leukaemia	0.03	0.03	0.04	0.05	0.05	0.02	0.02	0.05	0.05
Total	1.61	1.70	2.06	1.75	2.04	3.03	2.15	3.28	1.75

a   Owing to rounding errors, the sum of the individual values in each column sometimes differs from the total, which has been calculated to greater accuracy.
b   Relative risk transportation.
c   Absolute risk transportation.
d   The estimates presented for solid cancers at 0.1 Sv do not involve a reduction factor for low doses or low dose rates. In contrast, the leukaemia estimates at 0.1 Sv are based on a linear-quadratic dose response.

Table 38
Comparison of elicited high dose and high-dose-rate lifetime low-LET fatal cancer risks for a general population (European Union / United States) with those derived from other sources
(Risks expressed per 100 at 1 Sv)
(Based on [L27])

Cancer type	Elicited risk [a]	BEIR V [b]	ICRP 60 [c]	UNSCEAR 1994 [d]	UNSCEAR 2000 (Age-at-exposure) [e]	UNSCEAR 2000 (Attained age) [f]
Bone	0.035 (<10^{-3}– 0.88)					
Colon	0.98 (0.011–3.35)		3.24	0.6		0.6
Breast [g]	0.78 (0.11–3.78)	0.35	0.97	1.0		0.6
Leukaemia	0.91 (0.026–2.33)	0.95	0.95	1.1		1.0
Liver	0.86 (<10^{-3}–2.02)			1.2		0.9
Lung	2.76 (0.59–8.77)	1.7	2.92	2.5		2.1
Pancreas	0.17 (<10^{-3}–1.26)					
Skin	0.039 (<10^{-3}–0.37)		0.03			
Stomach	0.30 (<10^{-3}–4.01)		0.51	1.4		1.2
Thyroid	0.059 (<10^{-3}–0.71)					
All other cancers	2.60 (<10^{-3}– 10.8)					
All cancers	10.2 (3.47–28.5)	7.9	12.05	12	12	9

a REID for a joint European Union / United States population (90% CI in parentheses). Elicitation of risks involved questioning a range of experts.
b ELR for a United States population [C1].
c REID averaged over United Kingdom and United States populations, using a relative risk projection model (data extracted from [I1]).
d REID for a Japanese population, using an age-at-exposure model [U2].
e REID for a Japanese population of both genders and all ages, using an age-at-exposure model (derived from Table 34 of this Annex).
f REID for a Japanese population of both genders and all ages, using an attained-age model (derived from Table 36 of this Annex).
g Averaged over genders.

References

A1   Armstrong, B.G. The effects of measurement errors on relative risk regressions. Am. J. Epidemiol. 132: 1176-1184 (1990).

A2   Armitage, P. and R. Doll. The age distribution of cancer and a multistage theory of carcinogenesis. Br. J. Cancer 8: 1-12 (1954).

A3   Averkin, J.I., T. Abelin and J.P. Bleuer. Thyroid cancer in children in Belarus: ascertainment bias? Lancet 346: 1223-1224 (1995).

A4   Andersson, M., B. Carstensen and J. Visfeldt. Leukemia and other related hematological disorders among Danish patients exposed to Thorotrast. Radiat. Res. 134: 224-233 (1993).

A5   Andersson, M., B. Carstensen and H.H. Storm. Mortality and cancer incidence after cerebral arteriography with or without Thorotrast. Radiat. Res. 142: 305-320 (1995).

A6   Auvinen, A., M. Hakama, H. Arvela et al. Fallout from Chernobyl and incidence of childhood leukaemia in Finland, 1976-92. Br. Med. J. 309: 151-154 (1994).

A7   Atkinson, W.D., R.K. Bull, M. Marshall et al. The association between prostate cancer and exposure to ^{65}Zn in UKAEA employees. J. Radiol. Prot. 14: 109-114 (1994).

A8   Antonelli, A., F. Bianchi, G. Silvano et al. Is occupationally induced exposure to radiation a risk factor for thyroid nodule formation? Arch. Environ. Health 51: 177-190 (1996).

A9   Armstrong, B.K. and D.R. English. Cutaneous malignant melanoma. p. 1282-1312 in: Cancer Epidemiology and Prevention (D. Schottenfeld and J.F. Fraumeni Jr., eds.). Oxford University Press, Oxford, 1996.

A10   Armstrong, B.K. and A. Kricker. Cutaneous melanoma. Cancer Surv. 19/20: 219-240 (1994).

A11   Akiba, S., Q. Sun, Z. Tao et al. Infant leukemia mortality among the residents in high-background-radiation areas in Guangdong, China. p. 255-262 in: High Levels of Natural Radiation 96: Radiation Dose and Health Effects (L. Wei et al. eds.). Elsevier, Amsterdam, 1997.

A12   Andersson, M., H. Wallin, M. Jönsson et al. Lung carcinoma and malignant mesothelioma in patients exposed to Thorotrast; Incidence, mortality and p53 status. Int. J. Cancer 63: 330-336 (1995).

A13   Andersson, M., K. Juel, Y. Ishikawa et al. Effects of preconceptional irradiation on mortality and cancer incidence in the offspring of patients given injections of Thorotrast. J. Natl. Cancer Inst. 86: 1866-1870 (1994).

A14   Austin, D.F. and P. Reynolds. Investigation of an excess of melanoma among employees of the Lawrence Livermore National Laboratory. Am. J. Epidemiol. 145: 524-531 (1997).

A15   Andersson, M., G. Engholm, K. Ennow et al. Cancer risk among staff at two radiotherapy departments in Denmark. Br. J. Radiol. 64: 455-460 (1991).

A16   Albert, R.E., R.E. Shore, N. Harley et al. Follow-up studies of patients treated by x-ray epilation for tinea capitis. p. 1-25 in: Radiation Carcinogenesis and DNA Alterations (F.J. Burns, A.C. Upton and G. Silini, eds.). Plenum Press, New York and London, 1986.

A17   Alexander, F.E., T.J. Ricketts, P.A. McKinney et al. Community lifestyle characteristics and incidence of Hodgkin's disease in young people. Int. J. Cancer 48: 10-14 (1991).

A18   Andersson, M., M. Vyberg, J. Visfeldt et al. Primary liver tumors among Danish patients exposed to Thorotrast. Radiat. Res. 137: 262-273 (1994).

A19   Andersson, M. Long-term effects of internally deposited α-particle emitting radionuclides. Dan. Med. Bull. 44: 169-190 (1997).

A20   Andersson, M. and H.H. Storm. Cancer incidence among Danish Thorotrast-exposed patients. J. Natl. Cancer Inst. 84: 1318-1325 (1992).

A21   Ashmore, J.P., D. Krewski, J.M. Zielinski et al. First analysis of mortality and occupational radiation exposure based on the National Dose Registry of Canada. Am. J. Epidemiol. 148: 564-574 (1998).

A22   Alavanja, M.C.R., R.C. Brownson, J.H. Lubin et al. Residential radon exposure and lung cancer among nonsmoking women. J. Natl. Cancer Inst. 86: 1829-1837 (1994).

A23   Auvinen, A., I. Mäkeläinen, M. Hakama et al. Indoor radon exposure and risk of lung cancer: a nested case-control study in Finland. J. Natl. Cancer Inst. 88: 966-972 (1996).

A24   Alavanja, M.C.R., J.H. Lubin, J.A. Mahaffey et al. Residential radon exposure and risk of lung cancer in Missouri. Am. J. Public Health 89: 1042-1048 (1999).

A25   Aoyama, T. Radiation risk of Japanese and Chinese low dose-repeatedly irradiated population. J. Univ. Occup. Environ. Heallth 11 (Suppl.): 432-442 (1989).

A26   Astakhova, L.N., L.R. Anspaugh, G.W. Beebe et al. Chernobyl-related thyroid cancer in children in Belarus: a case-control study. Radiat. Res. 150: 349-356 (1998).

B1   Boice, J.D. Jr., G. Engholm, R.A. Kleinerman et al. Radiation dose and second cancer risk in patients treated for cancer of the cervix. Radiat. Res. 116: 3-55 (1988).

B2   Bithell, J.F. and A.M. Stewart. Prenatal irradiation and childhood malignancy: a review of British data from the Oxford survey. Br. J. Cancer 31: 271-287 (1975).

B3   Boice, J.D. Jr., D. Preston, F.G. Davis et al. Frequent chest x-ray fluoroscopy and breast cancer incidence among tuberculosis patients in Massachusetts. Radiat. Res. 125: 214-222 (1991).

B4   Boice, J.D. Jr. and L.E. Holm. Radiation risk estimates for leukemia and thyroid cancer among Russian emergency workers at Chernobyl. Radiat. Environ. Biophys. 36: 213-214 (1997).

B5   Brenner, D.J. The significance of dose rate in assessing the hazards of domestic radon exposure. Health Phys. 67: 76-79 (1994).

B6   Boice, J.D. Jr., J.S. Mandel and M.M. Doody. Breast cancer among radiologic technologists. J. Am. Med. Assoc. 274: 394-401 (1995).

B7   Bäverstam, U. and G.A. Swedjemark. Where are the errors when we estimate radon exposure in retrospect? Radiat. Prot. Dosim. 36: 107-112 (1991).

B8   Beral, V. and G. Reeves. Childhood thyroid cancer in Belarus. Nature 359: 680-681 (1992).

B9   Blettner, M. and J.D. Boice Jr. Radiation dose and leukaemia risk: general relative risk techniques for dose-response models in a matched case-control study. Stat. Med. 10: 1511-1526 (1991).

B10   Boice, J.D. Jr., E.B. Harvey, M. Blettner et al. Cancer in the contralateral breast after radiotherapy for breast cancer. N. Engl. J. Med. 326: 781-785 (1992).

B11 Boice, J.D. Jr., N.E. Day, A. Andersen et al. Second cancers following radiation treatment for cervical cancer. An international collaboration among cancer registries. J. Natl. Cancer Inst. 74: 955-975 (1985).

B12 Boice, J.D. Jr., M. Blettner, R.A. Kleinerman et al. Radiation dose and leukemia risk in patients treated for cancer of the cervix. J. Natl. Cancer Inst. 79: 1295-1311 (1987).

B13 Baverstock, K.F., D. Papworth and J. Vennart. Risks of radiation at low dose rates. Lancet I: 430-433 (1981).

B14 Baverstock, K.F. and D.G. Papworth. The UK radium luminiser survey. p. 72-76 in: Risks from Radium and Thorotrast (D.M. Taylor et al., eds.). BIR Report 21 (1989).

B15 Boice, J.D. Jr. Communication to the UNSCEAR Secretariat (1996).

B16 Bhatia, S., L.L. Robison, O. Oberlin et al. Breast cancer and other second neoplasms after childhood Hodgkin's disease. N. Engl. J. Med. 334: 745-751 (1996).

B17 Bond, V.P., C.B. Meinhold and H.H. Rossi. Low dose RBE and Q for X-ray compared to gamma ray radiations. Health Phys. 34: 433-438 (1978).

B18 Boice, J.D. Jr. Radiation epidemiology: past and present. p. 7-28 in: Implications of New Data on Radiation Cancer Risk (J.D. Boice Jr., ed.). NCRP Proceedings No. 18 (1997).

B19 Bigbee, W.L., R.H. Jensen, T. Veidebaum et al. Glycophorin A biodosimetry in Chernobyl cleanup workers from the Baltic countries. Br. Med. J. 312: 1078-1079 (1996).

B20 Boice, J.D. Jr., J.S. Mandel, M.M. Doody et al. A health survey of radiologic technologists. Cancer 69: 586-598 (1992).

B21 Becker, D.V. and J.R. Hurley. Radioiodine treatment of hyperthyroidism. p. 943-958 in: Diagnostic Nuclear Medicine (A. Gottschalk, P.B. Hoffer and E.J. Potchen, eds.). Williams & Wilkins, Baltimore, 1996.

B22 Brada, M., D. Ford, S. Ashley et al. Risk of second brain tumour after conservative surgery and radiotherapy for pituitary adenoma. Br. Med. J. 304: 1343-1346 (1992).

B23 Burch, J.D., K.J.P. Craib, B.C.K. Choi et al. An exploratory case-control study of brain tumors in adults. J. Natl. Cancer Inst. 78: 601-609 (1987).

B24 Beral, V., P. Fraser, L. Carpenter et al. Mortality of employees of the Atomic Weapons Establishment, 1951-82. Br. Med. J. 297: 757-770 (1988).

B25 Bongarzone, I., M.G. Butti, L. Fugazzola et al. Comparison of the breakpoint regions of ELE1 and RET genes involved in the generation of RET/PTC3 oncogene in sporadic and in radiation-associated papillary thyroid carcinomas. Genomics 42: 252-259 (1997).

B26 Bounacer, A., R. Wicker, B. Caillou et al. High prevalence of activating *ret* proto-oncogene rearrangements in thyroid tumors from patients who had received external radiation. Oncogene 15: 1263-1273 (1997).

B27 Boice, J.D. Jr. Leukaemia, Chernobyl and epidemiology. J. Radiol. Prot. 17: 129-133 (1997).

B28 Bailar, J.C. III. The practice of meta-analysis. J. Clin. Epidemiol. 48: 149-157 (1995).

B29 Bailar, J.C. III. The promise and problems of meta-analysis. N. Engl. J. Med. 337: 559-561 (1997).

B30 Burmeister, L.F., G.D. Everett, S.F. van Lier et al. Selected cancer mortality and farm practices in Iowa. Am. J. Epidemiol. 118: 72-77 (1983).

B31 Boffetta, P., S.D. Stellman and L. Garfinkel. A case-control study of multiple myeloma nested in the American Cancer Society prospective study. Int. J. Cancer 43: 554-559 (1989).

B32 Bennett, J.M., D. Catovsky, M.T. Daniel et al. The Franco-American-British (FAB) co-operative group. Proposals for the classification of the myelodysplastic syndromes. Br. J. Haematol. 51: 189-199 (1982).

B33 Brenner, D. Commentary: Does fractionation decrease the risk of breast cancer induced by low-LET radiation? Radiat. Res. 151: 225-229 (1999).

B34 Banks, P.M. The pathology of Hodgkin's disease. Semin. Oncol. 17: 683-695 (1990).

B35 Boice, J.D. Jr. and J.H. Lubin. Lung cancer risks: comparing radiation with tobacco. Radiat. Res. 146: 356-357 (1996).

B36 Boice, J.D. Jr., H.H. Storm, R.E. Curtis et al. (eds.). Multiple primary cancers in Connecticut and Denmark. Natl. Cancer Inst. Monogr. 68: (1985).

B37 Blot, W.J., Z.-Y. Xu, J.D. Boice Jr. et al. Indoor radon and lung cancer in China. J. Natl. Cancer Inst. 82: 1025-1030 (1990).

B38 Binks, K., D.I. Thomas and D. McElvenny. Mortality of workers at the Chapelcross plant of British Nuclear Fuels. p. 49-52 in: Radiation Protection – Theory and Practice (E.P. Goldfinch, ed.). Institute of Physics, Bristol, 1989.

B39 Boice, J.D. Jr., M.M. Morin, A.G. Glass et al. Diagnostic x-ray procedures and risk of leukemia, lymphoma and multiple myeloma. J. Am. Med. Assoc. 265: 1290-1294 (1991).

B40 Brady, M.S., J.J. Gaynor and M.F. Brennan. Radiation-associated sarcoma of bone and soft tissue. Arch. Surg. 127: 1379-1385 (1992).

B41 Boice, J.D. Jr. and R.W. Miller. Childhood and adult cancer after intrauterine exposure to ionizing radiation. Teratology 59: 227-233 (1999).

B42 Brenner, D.J., R.E. Curtis, E.J. Hall et al. Second malignancies in prostate carcinoma patients after radiotherapy compared with surgery. Cancer 88: 398-406 (2000).

B43 Berleur, M-P. and S. Cordier. The role of chemical, physical, or viral exposures and health factors in neuro-carcinogenesis: implications for epidemiologic studies of brain tumors. Cancer Causes and Control 6: 240-256 (1995).

B44 Burkart, W., A.M. Kellerer, S. Bauer et al. Health effects. p. 179-228 in: Nuclear Test Explosion: Environmental and Human Impacts (F. Warner and R.J.C. Kirchmann, eds.). John Wiley & Sons Ltd, 2000.

B45 Beral, V., H. Inskip, P. Fraser et al. Mortality of employees of the United Kingdom Atomic Energy Authority, 1946-79. Br. Med. J. 291: 440-447 (1985).

B46 Bingham, D., J.D. Harrison and A.W. Phipps. Biokinetics and dosimetry of chromium, cobalt, hydrogen, iron and zinc radionuclides in male reproductive tissues of the rat. Int. J. Radiat. Biol. 72: 235-248 (1997).

B47 Boice, J.D. Jr., C.E. Land and D.L. Preston. Ionizing radiation. p. 319-354 in: Cancer Epidemiology and Prevention (D. Schottenfeld and J.F. Fraumeni Jr., eds.). Oxford University Press, Oxford, 1996.

B48 Blettner, M., B. Grosche and H. Zeeb. Occupational cancer risk in pilots and flight attendants: current epidemiological knowledge. Radiat. Environ. Biophys. 37: 75-80 (1998).

B49 Buglova, E.E., J.E. Kenigsberg and N.V. Sergeeva. Cancer risk estimation in Belarussian children due to thyroid irradiation as a consequence of the Chernobyl nuclear accident. Health Phys. 71: 45-49 (1996).

B50 Boice, J.D. Jr., M. Blettner, R.A. Kleinerman et al. Radiation dose and breast cancer risk in patients treated for cancer of the cervix. Int. J. Cancer 44: 7-16 (1989).

C1 Committee on the Biological Effects of Ionizing Radiations (BEIR V). Health Effects of Exposure to Low Levels of Ionizing Radiation. National Academy of Sciences, National Research Council. National Academy Press, Washington, 1990.

C2 Committee on the Biological Effects of Ionizing Radiations (BEIR IV). Health Risks of Radon and Other Internally Deposited Alpha-Emitters. National Academy of Sciences, National Research Council. National Academy Press, Washington, 1988.

C3 Chameaud, J., R. Masse and J. Lafuma. Influence of radon daughter exposure at low doses on occurrence of lung cancer in rats. Radiat. Prot. Dosim. 7: 385-388 (1984).

C4 Cross, F.T. A review of experimental animal radon health effects data. p. 476-481 in: Radiation Research: A Twentieth-Century Perspective, Vol. II (J.D. Chapman et al., eds.). Academic Press, San Diego, 1992.

C5 Cohen, B.L. A test of the linear no-threshold theory of radiation carcinogenesis. Environ. Res. 53: 193-220 (1990).

C6 Cohen, B.L. and G.A. Colditz. Tests of the linear no-threshold theory for lung cancer induced by exposure to radon. Environ. Res. 64: 65-89 (1994).

C7 Cohen, B.L. Invited commentary: in defense of ecologic studies for testing a linear no-threshold theory. Am. J. Epidemiol. 139: 765-768 (1994).

C8 Cardis, E., J. Esteve and B.K. Armstrong. Meeting recommends international study of nuclear industry workers. Health Phys. 63: 465-466 (1992).

C9 Curtis, R.E., J.D. Boice Jr., M. Stovall et al. Risk of leukemia after chemotherapy and radiation treatment for breast cancer. N. Engl. J. Med. 326: 1745-1751 (1992).

C10 Curtis, R.E., J.D. Boice Jr., M. Stovall et al. Relationship of leukemia risk to radiation dose after cancer of the uterine corpus. J. Natl. Cancer Inst. 86: 1315-1324 (1994).

C11 Cardis, E., E.S. Gilbert, L. Carpenter et al. Effects of low doses and low dose rates of external ionizing radiation: cancer mortality among nuclear industry workers in three countries. Radiat. Res. 142: 117-132 (1995).

C12 Collaborative Group on Hormonal Factors in Breast Cancer. Breast cancer and hormonal contraceptives: collaborative reanalysis of individual data on 53,297 women with breast cancer and 100,239 women without breast cancer from 54 epidemiological studies. Lancet 347: 1713-1727 (1996).

C13 Cardis, E., E.S. Gilbert, L. Carpenter et al. Response to the letters of Drs. Schillaci and Uma Devi. Radiat. Res. 145: 648-649 (1996).

C14 Correa, P. and V.W. Chen. Gastric cancer. p. 55-76 in: Cancer Surveys. Trends in Cancer Incidence and Mortality, Volumes 19/20 (R. Doll et al., eds.). Cold Spring Harbor Laboratory Press, Imperial Cancer Research Fund, 1994.

C15 Cardis, E., E.S. Gilbert, L. Carpenter et al. Combined analyses of cancer mortality among nuclear industry workers in Canada, the United Kingdom and the United States of America. IARC Technical Report 25 (1995).

C16 Curtis, R.E., P.A. Rowlings, H.J. Deeg et al. Solid cancers after bone marrow transplantation. N. Engl. J. Med. 336: 897-904 (1997).

C17 Carstensen, J.M., G. Wingren, T. Hatschek, et al. Occupational risks of thyroid cancer: data from the Swedish Cancer Environment Register, 1961-1979. Am. J. Ind. Med. 18: 535-540 (1990).

C18 Cohen, B.L. Test of the linear no-threshold theory of radiation carcinogenesis for inhaled radon decay products. Health Phys. 68: 157-174 (1995).

C19 Carpenter, L., C. Higgins, A. Douglas et al. Combined analysis of mortality in three United Kingdom nuclear industry workforces, 1946-1988. Radiat. Res. 138: 224-238 (1994).

C20 Carter, C.O., K.A. Evans and A.M. Stewart. Maternal radiation and Down's syndrome (Mongolism). Lancet ii: 1042 (1961).

C21 Committee on the Biological Effects of Ionizing Radiation (BEIR VI). The Health Effects of Exposure to Indoor Radon. National Academy of Sciences, National Research Council. National Academy Press, Washington, 1999.

C22 Centro de Investigacion en Cancer "Maes Heller". Registro de Cancer Lima Metropolitana, 1990-1991. Instituto de Enfermedades Neoplasicas, Lima, 1995.

C23 Cuzick, J. Multiple myeloma. p. 455-474 in: Cancer Surveys. Trends in Cancer Incidence and Mortality, Volumes 19/20 (R. Doll et al., eds.). Cold Spring Harbor Laboratory Press, Imperial Cancer Research Fund, 1994.

C24 Cuzick, J. and B. de Stavola. Multiple myeloma: a case-control study. Br. J. Cancer 57: 516-520 (1988).

C25 Cohen, B.L. Response to Lubin's proposed explanations of our discrepancy. Health Phys. 75: 18-22 (1998).

C26 Cohen, B.L. Response to criticisms of Smith et al. Health Phys. 75: 23-28 (1998).

C27 Carnes, B.A., P.G. Groer and T.J. Kotek. Radium dial workers: issues concerning dose response and modeling. Radiat. Res. 147: 707-714 (1997).

C28 Committee on Medical Aspects of Radiation in the Environment (COMARE). Fourth Report: The incidence of cancer and leukaemia in young people in the vicinity of the Sellafield site, West Cumbria: Further studies and an update of the situation since the publication of the report of the Black Advisory Group in 1984. Department of Health, Wetherby (1996).

C29 Committee on Medical Aspects of Radiation in the Environment (COMARE). Second Report: Investigation of the possible increased incidence of leukaemia in young people near the Dounreay nuclear establishment, Caithness, Scotland. HMSO, London (1988).

C30 Clavel, J. and D. Hémon. Leukaemia near La Hague nuclear plant: bias could have been introduced into study. Br. Med. J. 314: 1553 (1997).

C31 Cantor, K.P., A. Blair, G. Everett et al. Pesticides and other agricultural risk factors for non-Hodgkin's lymphoma among men in Iowa and Minnesota. Cancer Res. 52: 2447-2455 (1992).

C32 Checkoway, H., N. Pearce, D.J. Crawford-Brown et al. Radiation doses and cause-specific mortality among workers at a nuclear materials fabrication plant. Am. J. Epidemiol. 127: 255-266 (1988).

C33 Carpenter, L.M., C.D. Higgins, A.J. Douglas et al. Cancer mortality in relation to monitoring for radionuclide exposure in three UK nuclear industry workforces. Br. J. Cancer 78: 1224-1232 (1998).

C34 Checkoway, H., N.J. Heyer and P.A. Demers. An updated mortality follow-up study of Florida phosphate industry workers. Am. J. Ind. Med. 30: 452-460 (1996).

C35 Cohen, B.L. The cancer risk from low-level radiation. Radiat. Res. 149: 525-526 (1998).

C36 Cardis, E. and A.E. Okeanov. What is feasible and desirable in the epidemiologic follow-up of Chernobyl. p. 835-850 in: The Radiological Consequences of the Chernobyl Accident. Proceedings of the First International Conference, Minsk, Belarus, March 1996 (A. Karaoglou, G. Desmet, G.N. Kelly et al., eds). EUR 16544 (1996).

C37 Cologne, J.B., S. Tokuoka, G.W. Beebe et al. Effects of radiation on incidence of primary liver cancer among atomic bomb survivors. Radiat. Res. 152: 364-373 (1999).

D1 Dickersin, K. and J.A. Berlin. Meta-analysis: state-of-the-science. Epidemiol. Rev. 14: 154-176 (1992).

D2 Draper, G.J., M.E. Kroll and C.A. Stiller. Childhood cancer. p. 493-517 in: Cancer Surveys. Trends in Cancer Incidence and Mortality, Volumes 19/20 (R. Doll et al., eds.). Cold Spring Harbor Laboratory Press, Imperial Cancer Research Fund, 1994.

D3 dos Santos Silva, I., F. Malveiro, R. Portugal et al. Mortality from primary liver cancers in the Portuguese Thorotrast cohort study. p. 229-233 in: Health Effects of Internally Deposited Radionuclides: Emphasis on Radiation and Thorium (G. van Kaick et al., eds.). World Scientific, Singapore, 1995.

D4 Davis, F.G., J.D. Boice Jr., Z. Hrubec et al. Cancer mortality in a radiation-exposed cohort of Massachusetts tuberculosis patients. Cancer Res. 49: 6130-6136 (1989).

D5 DeGroot, L., M. Reilly, K. Pinnameneni et al. Retrospective and prospective study of radiation-induced thyroid disease. Am. J. Med. 74: 852-862 (1983).

D6 Doll, R. and R. Peto. The causes of cancer: Quantitative estimates of avoidable risks of cancer in the United States today. J. Natl. Cancer Inst. 66: 1191-1308 (1981).

D7 Darby, S.C., G. Reeves, T. Key et al. Mortality in a cohort of women given X-ray therapy for metropathia haemorrhagica. Int. J. Cancer 56: 793-801 (1994).

D8 Darby, S.C., E. Whitley, G.R. Howe et al. Radon and cancers other than lung cancer in underground miners: a collaborative analysis of 11 studies. J. Natl. Cancer Inst. 87: 378-384 (1995).

D9 Darby, S.C., G.M. Kendall, T.P. Fell et al. A summary of mortality and incidence of cancer in men from the United Kingdom who participated in the United Kingdom's atmospheric nuclear weapon tests and experimental programmes. Br. Med. J. 296: 332-338 (1988).

D10 Doll, R., H.J. Evans and S.C. Darby. Paternal exposure not to blame. Nature 367: 678-680 (1994).

D11 Doll, R., R. Peto, K. Wheatley et al. Mortality in relation to smoking: 40 years' observation on male British doctors. Br. Med. J. 309: 901-911 (1994).

D12 Damber, L., L.-G. Larsson, L. Johansson et al. A cohort study with regard to the risk of haematological malignancies in patients treated with X-rays for benign lesions in the locomotor system. I. Epidemiological analyses. Acta Oncol. 34: 713-719 (1995).

D13 Demidchik, E.P., I.M. Drobyshevskaya, L.N. Cherstvoy et al. Thyroid cancer in children in Belarus. p. 677-682 in: The Radiological Consequences of the Chernobyl Accident (A. Karaoglou et al., eds.). EUR 16544 EN (1996).

D14 Delongchamp, R.R., K. Mabuchi, Y. Yoshimoto et al. Mortality among atomic bomb survivors exposed *in utero* or as young children, October 1950-May 1992. Radiat. Res. 147: 385-395 (1997).

D15 Dottorini, M.E., G. Lomuscio, L. Mazzucchelli et al. Assessment of female fertility and carcinogenesis after iodine-131 therapy for differentiated thyroid carcinoma. J. Nucl. Med. 36: 21-27 (1995).

D16 Davis, F.G., J.D. Boice Jr., J.L. Kelsey et al. Cancer mortality after multiple fluoroscopic examinations of the chest. J. Natl. Cancer Inst. 78: 645-652 (1987).

D17 Doll, R. and R. Wakeford. Risk of childhood cancer from fetal irradiation. Br. J. Radiol. 70: 130-139 (1997).

D18 de Vathaire, F., M. Schumberger, M.J. Delisle et al. Leukaemias and cancers following iodine I-131 administra-tion for thyroid cancer. Br. J. Cancer 75: 734-739 (1997).

D19 de Vathaire, F., A. Shamsadin, E. Grimaud et al. Solid malignant neoplasms after childhood cancer irradiation: decrease of the relative risk with time after irradiation. C.R. Acad. Sci. Paris, Sci. Vie 318: 483-490 (1995).

D20 de Vathaire, F., E. Grimaud, I. Diallo et al. Thyroid tumours following fractionated irradiation in childhood. p. 121-124 in: Low Doses of Ionizing Radiation: Biological Effects and Regulatory Control. IAEA-TECDOC-976, Vienna (1997).

D21 Douglas, A.J., R.Z. Omar and P.G. Smith. Cancer mortality and morbidity among workers at the Sellafield plant of British Nuclear Fuels. Br. J. Cancer 70: 1232-1243 (1994).

D22 Dobyns, B.M., G.E. Sheline, J.B. Workman et al. Malignant and benign neoplasms of the thyroid in patients treated for hyperthyroidism: a report of the Cooperative Thyrotoxicosis Therapy Follow-up Study. J. Clin. Endocrinol. Metab. 38: 976-998 (1974).

D23 Doody, M.M., J.S. Mandel, J.H. Lubin et al. Mortality among United States radiologic technologists, 1926-90. Cancer Causes and Control 9: 67-75 (1998).

D24 Draper, G.J., M.P. Little, T. Sorahan et al. Cancer in the offspring of radiation workers: a record linkage study. Br. Med. J. 315: 1181-1188 (1997).

D25 Donaldson, S.S. and S.L. Hancock. Second cancers after Hodgkin's disease in childhood (editorial). N. Engl. J. Med. 334: 792-794 (1996).

D26 Davidescu, D., O. Iacob and C. Diaconescu. Reassessment of childhood leukaemia incidence after Chernobyl accident. Jurnal de Medicina Preventiva 6: 13-18 (1998).

D27 de Motta, L.C., J. da Silva Horta and M.H. Tavares. Prospective epidemiological study of Thorotrast exposed patients in Portugal. Environ. Res. 18: 62-172 (1979).

D28 Day, N.E. and C. Varghese. Oesophageal cancer. in: Trends in Cancer Incidence and Mortality (R. Doll et al., eds.). Cold Spring Harbor Laboratory Press, Imperial Cancer Research Fund, 1994.

D29 Davis, S., K.J. Kopecky, T.E. Hamilton et al. Hanford thyroid disease study. Draft final report (1999).

D30 Darby, S., E. Whitley, P. Silcocks et al. Risk of lung cancer associated with residential radon exposure in south-west England: a case-control study. Br. J. Cancer 78: 394-408 (1998).

D31 dos Santos Silva, I., M. Jones, F. Malveiro et al. Mortality in the Portuguese Thorotrast study. Radiat. Res. 152: S88-S92 (1999).

D32 Dupree, E.A., J.P. Watkins, J.N. Ingle et al. Uranium dust exposure and lung cancer risk in four uranium processing operations. Epidemiology 6: 370-375 (1995).

D33 de Vathaire, F., M. Hawkins, S. Campbell et al. Second malignant neoplasms after a first cancer in childhood: temporal pattern of risk according to type of treatment. Br. J. Cancer 79: 1884-1893 (1999).

D34 Doody, M.M., J.E. Lonstein, M. Stovall et al. Breast cancer mortality following diagnostic x-rays: findings from the U.S. scoliosis cohort study. Spine (2000, in press).

D35 Davis, F.G. and S. Preston-Martin. Epidemiology. p. 5-45 in: Russell and Rubenstein's Pathology of Tumors of the Nervous System (D. Bigner et al. eds.), 1998.

D36 de Vathaire, F., C. Hardiman, A. Shamsaldin et al. Thyroid carcinomas after irradiation for a first cancer during childhood. Arch. Intern. Med. 159:2713-2719 (1999).

D37 Degteva, M.O., V.P. Kozheurov, D.S. Burmistrov et al. An approach to dose reconstruction for the Urals population. Health Phys. 71: 71-76 (1996).

E1 Eng, C., F.P. Li, D.H. Abramson et al. Mortality from second tumors among long-term survivors of retinoblastoma. J. Natl. Cancer Inst. 85: 1121-1128 (1993).

E2 Edmonds, C.J. and T. Smith. The long-term hazards of the treatment of thyroid cancer with radioiodine. Br. J. Radiol. 59: 45-51 (1986).

E3 Epidemiological Study Group of Nuclear Workers (Japan). First analysis of mortality of nuclear industry workers in Japan, 1986-1992. J. Health Phys. 32: 173-184 (1997).

E4 Evans, J.S, J.E. Wennberg and B.J. McNeil. The influence of diagnostic radiography on the incidence of breast cancer and leukemia. N. Engl. J. Med. 315: 810-815 (1986).

E5 Elkind, M.M. Does repair of radiation damage play a role in breast cancer? Radiat. Res. 152: 567 (1999).

E6 Elkind, M.M. Enhanced risks of cancer from protracted exposures to X- or γ-rays: a radiobiological model of radiation-induced breast cancer. Br. J. Cancer 73: 133-138 (1996).

F1 Franceschi, S. and C. La Vecchia. Thyroid cancer. p. 393-422 in: Cancer Surveys. Trends in Cancer Incidence and Mortality, Volumes 19/20 (R. Doll et al., eds.). Cold Spring Harbor Laboratory Press, Imperial Cancer Research Fund, 1994.

F2 Franceschi, S., P. Boyle, C. La Vecchia et al. The epidemiology of thyroid carcinoma. Crit. Rev. Oncol. 4: 25-52 (1993).

F3 Fisher, B., H. Rockette, E.R. Fisher et al. Leukemia in breast cancer patients following adjuvant chemotherapy or post-operative radiation: the NSABP experience. J. Clin. Oncol. 3: 1640-1658 (1985).

F4 Fox, A.J. and P.F. Collier. Low mortality rates in industrial cohort studies due to selection for work and survival in the industry. Br. J. Prev. Soc. Med. 30: 225-230 (1976).

F5 Frome, E.L., D.L. Cragle, J.P. Watkins et al. A mortality study of employees of the nuclear industry in Oak Ridge, Tennessee. Radiat. Res. 148: 64-80 (1997).

F6 Fraser, P., L. Carpenter, N. Maconochie, et al. Cancer mortality and morbidity in employees of the United Kingdom Atomic Energy Authority, 1946-86. Br. J. Cancer 67: 615-624 (1993).

F7 Forastiere, F., A. Sperati, G. Cherubini et al. Adult myeloid leukaemia, geology, and domestic exposure to radon and γ radiation: a case control study in central Italy. Occup. Environ. Med. 55: 106-110 (1998).

F8 Franklyn, J.A., P. Maisonneuve, M. Sheppard et al. Cancer incidence and mortality after radioiodine treatment for hyperthyroidism: a population-based cohort study. Lancet 353: 2111-2115 (1999).

F9 Fry, S.A. Studies of U.S. radium dial workers: an epidemiological classic. Radiat. Res. 150: S21-S29 (1998).

G1 Gilliland, F.D. and J.M. Samet. Lung cancer. p. 175-195 in: Cancer Surveys. Trends in Cancer Incidence and Mortality, Volumes 19/20 (R. Doll et al., eds.). Cold Spring Harbor Laboratory Press, Imperial Cancer Research Fund, 1994.

G2 Greenland, S. and J. Robins. Invited commentary: ecologic studies - biases, misconceptions, and counter-examples. Am. J. Epidemiol. 139: 747-760 (1994).

G3 Greenland, S. and J. Robins. Accepting the limits of ecologic studies: Drs. Greenland and Robins reply to Drs. Piantadosi and Cohen. Am. J. Epidemiol. 139: 769-771 (1994).

G4 Gilbert, E.S. and J.J. Fix. Accounting for bias in dose estimates in analyses of data from nuclear worker mortality studies. Health Phys. 68: 650-660 (1995).

G5 Greaves, M.F. Speculations on the cause of childhood acute lymphoblastic leukaemia. Leukaemia 2: 120-125 (1988).

G6 Griem, M.L., R.A. Kleinerman, J.D. Boice Jr. et al. Cancer following radiotherapy for peptic ulcer. J. Natl. Cancer Inst. 86: 842-849 (1994).

G7 Gardner, M.J., M.P. Snee, A.J. Hall et al. Results of case-control study of leukaemia and lymphoma among young people near Sellafield nuclear plant in West Cumbria. Br. Med. J. 300: 423-429 (1990).

G8 Gofman, J.W. Preventing Breast Cancer: The Story of a Major, Preventable Cause of this Disease. CNR Books, San Francisco, 1995. (Second edition, 1996)

G9 Gofman, J.W. Warning from the A-bomb study about low and slow radiation exposures. Health Phys. 56: 117-118 (1989).

G10 Goldman, M.B., F. Maloof, R.R. Monson et al. Radioactive iodine therapy and breast cancer. A follow-up study of hyperthyroid women. Am. J. Epidemiol. 127: 969-980 (1988).

G11 Greenberg, E.R., B. Rosner, C. Heenekens et al. An investigation of bias in a study of nuclear shipyard workers. Am. J. Epidemiol. 121: 301-308 (1985).

G12 Gilbert, E.S., D.L. Cragle and L.D. Wiggs. Updated analyses of combined mortality data for workers at the Hanford site, Oak Ridge National Laboratory, and Rocky Flats Weapons Plant. Radiat. Res. 136: 408-421 (1993).

G13 Glanzman, C. Subsequent malignancies in patients treated with 131-iodine for thyroid cancer. Strahlenther. Onkol. 168: 337-343 (1992).

G14 Gallagher, R.P., C.D. Bajdik, S. Fincham et al. Chemical exposures, medical history, and risk of squamous and basal cell carcinoma of the skin. Cancer Epidemiol. Biomarkers Prev. 5: 419-424 (1996).

G15 Grayson, J.K. Radiation exposure, socioeconomic status, and brain tumor risk in the US air force: A nested case-control study. Am. J. Epidemiol. 143: 480-486 (1996).

G16 Gribben, M.A., J.L. Weeks and G.R. Howe. Cancer mortality (1956-1985) among male employees of Atomic Energy of Canada Limited with respect to occupational exposure to external low-linear-energy-transfer ionizing radiation. Radiat. Res. 133: 373-380 (1993).

G17 Gilbert, E.S, E. Omohundro, J.A. Buchanan et al. Mortality of workers at the Hanford site: 1945-1986. Health Phys. 64: 577-590 (1993).

G18 Gutensohn, N. and P. Cole. Childhood social environment and Hodgkin's disease. N. Engl. J. Med. 304: 135-140 (1981).

G19 Gilbert, E.S., R. Tarone, A. Bouville et al. Thyroid cancer rates and ^{131}I doses from Nevada atmospheric nuclear bomb tests. J. Natl. Cancer Inst. 90: 1654-1660 (1998).

G20 Gembicki, M., A.N. Stozharov, A.N. Arinchin et al. Iodine deficiency in Belarusian children as a possible factor stimulating the irradiation of the thyroid gland during the Chernobyl catastrophe. Environ. Health Perspect. 105: 1487-1490 (1997).

G21 Greenwald, P., B.R. Friedlander, C.E. Lawrence et al. Diagnostic sensitivity bias – an epidemiologic explanation for an apparent brain tumor excess. J. Occup. Med. 23: 690-694 (1981).

G22 Gössner, W. Pathology of radium-induced bone tumors: new aspects of histopathology and histogenesis. Radiat. Res. 152: S12-S15 (1999).

G23 Gilbert, E.S., N.A. Koshurnikova, M. Sokolnikov et al. Liver cancers in Mayak workers. Radiat. Res. (2000, in press).

H1 Hutchison, G.B. Leukemia in patients with cancer of the cervix uteri treated with radiation. A report covering the first 5 years of an international study. J. Natl. Cancer Inst. 40: 951-982 (1968).

H2 Hancock, S.L., M.A. Tucker and R.T. Hoppe. Breast cancer after treatment of Hodgkin's disease. J. Natl. Cancer Inst. 85: 25-31 (1993).

H3 Hanford, J.M., E. Quimby and V. Frantz. Cancer arising many years after radiation therapy. J. Am. Med. Assoc. 181: 132-138 (1962).

H4 Hall, P., A. Mattsson and J.D. Boice Jr. Thyroid cancer after diagnostic administration of iodine-131. Radiat. Res. 145: 86-92 (1996).

H5 Hoel, D.G., E. Ron, R. Carter et al. Influence of death certificate errors on cancer mortality trends. J. Natl. Cancer Inst. 85: 1063-1068 (1993).

H6 Health and Safety Executive. HSE Investigation of Leukaemia and Other Cancers in the Children of Male Workers at Sellafield. United Kingdom Health and Safety Executive, Sudbury, 1993.

H7 Howe, G.R. Lung cancer mortality between 1950 and 1987 after exposure to fractionated moderate-dose-rate ionizing radiation in the Canadian fluoroscopy cohort study and a comparison with lung cancer mortality in the atomic bomb survivors study. Radiat. Res. 142: 295-304 (1995).

H8 Henrichs, K., L. Bogner, E. Nekolla et al. Extended dosimetry for studies with Ra-224 patients. p. 33-38 in: Health Effects of Internally Deposited Radionuclides: Emphasis on Radium and Thorium (G. van Kaick et al., eds.). World Scientific, Singapore, 1995.

H9 Hancock, S.L., R.S. Cox and R. McDougall. Thyroid diseases after treatment of Hodgkin's disease. N. Engl. J. Med. 325: 599-605 (1991).

H10 Hildreth, N.G., R.E. Shore and P.M. Dvoretsky. The risk of breast cancer after irradiation of the thymus in infancy. N. Engl. J. Med. 321: 1281-1284 (1989).

H11 Hawkins, M.M., L.M. Kinnier Wilson, M.A. Stovall et al. Epipodophyllotoxins, alkylating agents, and radiation and risk of secondary leukaemia after childhood cancer. Br. Med. J. 304: 951-958 (1992).

H12 Hall, P., J.D. Boice Jr., G. Berg et al. Leukaemia incidence after iodine-131 exposure. Lancet 340: 1-4 (1992).

H13 Hodgson, J.T. and R.D. Jones. Mortality of a cohort of tin miners 1941-86. Br. J. Ind. Med. 47: 665-676 (1990).

H14 Henshaw, D.L., J.P. Eatough and R.B. Richardson. Radon as a causative factor in induction of myeloid leukaemia and other cancers. Lancet 335: 1008-1012 (1990).

H15 Howe, G.R., R.C. Nair, H.B. Newcombe et al. Lung cancer mortality (1950-80) in relation to radon daughter exposure in a cohort of workers at the Eldorado Beaverlodge uranium mine. J. Natl. Cancer Inst. 77: 357-362 (1986).

H16 Howe, G.R., R.C. Nair, H.B. Newcombe et al. Lung cancer mortality (1950-80) in relation to radon daughter exposure in a cohort of workers at the Eldorado Port radium uranium mine: possible modification of risk by exposure rate. J. Natl. Cancer Inst. 79: 1255-1260 (1987).

H17 Hornung, R.W. and T.J. Meinhardt. Quantitative risk assessment of lung cancer in U.S. uranium miners. Health Phys. 52: 417-430 (1987).

H18 Howe, G.R. and R.H. Stager. Risk of lung cancer mortality after exposure to radon decay products in the Beaverlodge cohort based on revised exposure estimates. Radiat. Res. 146: 37-42 (1996).

H19 Hunter, D.J. and W.C. Willett. Nutrition and breast cancer. Cancer Causes and Control 7: 56-68 (1996).

H20 Howe, G.R. and J. McLaughlin. Breast cancer mortality between 1950 and 1987 after exposure to fractionated moderate-dose-rate ionizing radiation in the Canadian fluoroscopy cohort study and a comparison with breast cancer mortality in the atomic bomb survivors study. Radiat. Res. 145: 694-707 (1996).

H21 Hofmann, W., J.R. Johnson and N. Freedman. Lung dosimetry of Thorotrast patients. Health Phys. 59: 777-790 (1990).

H22 Hjalmars, U., M. Kulldorff and G. Gustafsson. Risk of acute childhood leukaemia in Sweden after the Chernobyl reactor accident. Br. Med. J. 309: 154-157 (1994).

H23 Holm, L.-E., P. Hall, K. Wiklund et al. Cancer risk after iodine-131 therapy for hyperthyroidism. J. Natl. Cancer Inst. 83: 1072-1077 (1991).

H24 Hall, P., G. Berg, G. Bjelkengren et al. Cancer mortality after iodine-131 therapy for hyperthyroidism. Int. J. Cancer 50: 886-890 (1992).

H25 Hoffman, D.A. Late effects of I-131 therapy in the United States. p. 273-280 in: Radiation Carcinogenesis: Epidemiology and Biological Significance (J.D. Boice Jr. and J.F. Fraumeni Jr., eds.). Raven Press, New York, 1984.

H26 Hall, P., L.-E. Holm, G. Lundell et al. Cancer risks in thyroid cancer patients. Br. J. Cancer 64: 159-163 (1991).

H27 Holm, L.-E., K.E. Wiklund, G.E. Lundell et al. Cancer risk in population examined with diagnostic doses of ^{131}I. J. Natl. Cancer Inst. 81: 302-306 (1989).

H28 Hirayama, T. Life-style and Mortality: A Large-scale Census-based Cohort Study in Japan. Karger, Basel, 1990.

H29 Hazen, R.W., J.W. Pifer, E.T. Toyooka et al. Neoplasms following irradiation of the head. Cancer Res. 26: 305-311 (1966).

H30 Hawkins, M.M. and J.E. Kingston. Malignant thyroid tumours following childhood cancer. Lancet 2: 804 (1988).

H31 Hildreth, N., R. Shore, L. Hempelmann et al. Risk of extra-thyroid tumors following radiation treatment in infancy for thymic enlargement. Radiat. Res. 102: 378-391 (1985).

H32 Howe, G.R., J.D. Burch, A.M. Chiarelli et al. An exploratory case-control study of brain tumors in children. Cancer Res. 49: 4349-4352 (1989).

H33 Hunacek, M.M. and R.L. Kathren. Alpha radiation risk coefficients for liver cancer, bone sarcomas and leukemia. Health Phys. 68: 41-49 (1995).

H34 Hei, T.K, L-J. Wu, S-X. Liu et al. Mutagenic effects of a single and an exact number of α particles in mammalian cells. Proc. Natl. Acad. Sci. U.S.A. 94: 3765-3770 (1997).

H35 Hamilton, T.E., G. Van Belle and J.P. LoGerfo. Thyroid neoplasia in Marshall Islanders exposed to nuclear fallout. J. Am. Med. Assoc. 258: 629-636 (1987).

H36 Hall, P., C.J. Fürst, A. Mattsson et al. Thyroid nodularity after diagnostic administration of iodine-131. Radiat. Res. 146: 673-682 (1996).

H37 Hatch, M.C., J. Beyea, J.W. Nieves et al. Cancer near the Three Mile Island nuclear plant: radiation emissions. Am. J. Epidemiol. 132: 397-412 (1990).

H38 Hatch, M.C., M. Susser and J. Beyea. Comments on "A reevaluation of cancer incidence near the Three Mile Island nuclear plant". Environ. Health Perspect. 105: 12 (1997).

H39 Hartge, P., S.S. Devesa and J.F. Fraumeni. Hodgkin's and Non-Hodgkin's lymphomas. p. 423-453 in: Cancer Surveys. Trends in Cancer Incidence and Mortality, Volumes 19/20 (R. Doll et al., eds.). Cold Spring Harbor Laboratory Press, Imperial Cancer Research Fund, 1994.

H40 Hattchoucel, J.-M., A. Laplanche and C. Hill. Leukaemia mortality around French nuclear sites. Br. J. Cancer 71: 651-653 (1995).

H41 Heidenreich, W.F., P. Jacob, H.G. Paretzke et al. Two-step model for the risk of fatal and incidental lung tumors in rats exposed to radon. Radiat. Res. 151: 209-217 (1999).

H42 Harris, N.H., E. Jaffe and E. Steinleten. A revised European-American classification of lymphoid neoplasms: a proposal from IL. International Lymphoma Study Group. Blood: 1361-1397 (1994).

H43 Hoover, R. and J.F. Fraumeni Jr. Risk of cancer in renal transplant recipients. Lancet ii: 55-57 (1973).

H44 Hawkins, M.M., L.M. Kinnier Wilson, H.S. Burton et al. Radiotherapy, alkylating agents, and risk of bone cancer after childhood cancer. J. Natl. Cancer Inst. 88: 270-278 (1996).

H45 Hoel, D.G. and P. Li. Threshold models in radiation carcinogenesis. Health Phys. 75: 241-250 (1998).

H46 Horta, J., M.E. da Silva, L. da Silva Horta et al. Malignancies in Portuguese thorotrast patients. Health Phys. 35: 137-151 (1978).

H47 Hankey, B.F., D.T. Silverman and R. Kaplan. Urinary bladder. p. xxvi.1-xxvi.17 in: SEER Cancer Statistics Review, 1973-1990 (B.A. Miller, L.A. Gloeckler, B.F. Hankey et al., eds.). NIH Publication No. 93-2789 (1993).

H48 Herrinton, L.J., N.S. Weiss and A.F. Olshan. Multiple myeloma. p. 946-970 in: Cancer Epidemiology and Prevention (D. Schottenfeld and J.F. Fraumeni Jr., eds.). Oxford University Press, Oxford, 1996.

I1 International Commission on Radiological Protection. 1990 Recommendations of the International Commission on Radiological Protection. ICRP Publication 60. Annals of the ICRP 21 (1-3). Pergamon Press, Oxford, 1991.

I2 IARC Study Group on Cancer Risk Among Nuclear Industry Workers. Direct estimates of cancer mortality due to low doses of ionising radiation: an international study. Lancet 344: 1039-1043 (1994).

I3 International Chernobyl Project. Assessment of radiological consequences and evaluation of protective measures. Technical Report. IAEA, Vienna (1991)

I4 International Commission on Radiological Protection. Human Respiratory Tract Model for Radiological Protection. ICRP Publication 66. Annals of the ICRP 24(1-3). Pergamon Press. Oxford, 1994.

I5 Ivanov, E.P., G. Tolochko, V.S. Lazarev et al. Child leukaemia after Chernobyl. Nature 365: 702 (1993).

I6 Inskip, P.D., R.A. Kleinerman, M. Stovall et al. Leukemia, lymphoma and multiple myeloma following pelvic radiotherapy for benign disease. Radiat. Res. 135: 108-125 (1993).

I7 Inskip, P.D., M. Stovall and J.T. Flannery. Lung cancer risk and radiation dose among women treated for breast cancer. J. Natl. Cancer Inst. 86: 983-988 (1994).

I8 Ikeya, M., J. Miyajima and S. Okajima. ESR dosimetry for atomic bomb survivors using shell buttons and tooth enamel. Jpn. J. Appl. Phys. 23: 697-699 (1984).

I9 Inskip, P.D., A. Ekbom, M.R. Galanti et al. Medical diagnostic x rays and thyroid cancer. J. Natl. Cancer Inst. 87: 1613-21 (1995).

I10 Inskip, P.D., M.F. Hartshorne, M. Tekkel et al. Thyroid nodularity and cancer among Chernobyl cleanup workers from Estonia. Radiat. Res. 147: 225-235 (1997).

I11 Inskip, P.D., M.S. Linet and E.F. Heineman. Etiology of brain tumors in adults. Epidemiol. Rev. 17: 382-414 (1996).

I12 International Atomic Energy Agency. Methods for estimating the probability of cancer from occupational radiation exposure. IAEA-TECDOC-870 (1996).

I13 Ivanov, V.K., A.F. Tsyb, A.I. Gorsky et al. Leukaemia and thyroid cancer in emergency workers of the Chernobyl accident: estimation of radiation risks (1986-1995). Radiat. Environ. Biophys. 36: 9-16 (1997).

I14 Ivanov, V.K., A.F. Tsyb, A.P. Konogorov et al. Case-control analysis of leukaemia among Chernobyl accident emergency workers residing in the Russian Federation, 1986-1993. J. Radiol. Prot. 17: 137-157 (1997).

I15 Iwasaki, T., K. Nishizawa and M. Murata. Leukaemia and lymphoma mortality in the vicinity of nuclear power stations in Japan, 1973-1987. J. Radiol. Prot. 15: 271-288 (1995).

I16 Inskip, P.D., R.R. Monson, J.K. Wagoner et al. Cancer mortality following radium treatment for uterine bleeding. Radiat. Res. 123: 331-344 (1990).

I17 Ivanov, V.K. Response to the letter to the editor by J.D. Boice Jr. and L.E. Holm. Radiat. Environ. Biophys. 36: 305-306 (1997).

I18 Ivanov, V.K., A.F. Tsyb, A.I. Gorsky et al. Thyroid cancer among liquidators of the Chernobyl accident. Br. J. Radiol. 70: 937-941 (1997).

I19 International Commission on Radiation Units and Measurements. Determination of dose equivalents from external radiation sources, Part 2. ICRU Report 43 (1988).

I20 International Commission on Radiological Protection. Genetic Susceptibility to Cancer. ICRP Publication 79. Annals of the ICRP 28 (1-2). Elsevier Science, Amsterdam, 1998.

I21 Ivanov, V.K., E.M. Rastopchin, A.I. Gorsky et al. Cancer incidence among liquidators of the Chernobyl accident: solid tumors, 1986-1995. Health Phys. 74: 309-315 (1998).

I22 Ivanov, E., G.V. Tolochko, L.P. Shuvaeva et al. Infant leukaemia in Belarus after the Chernobyl accident. Radiat. Environ. Biophys. 37: 53-55 (1998).

I23 Ivanov, V.K., A.I. Gorsky, A.F. Tysb et al. Dynamics of thyroid cancer incidence in Russia following the Chernobyl accident. J. Radiol. Prot. 19: 305-318 (1999).

I24 Ivanov, V.K., A.I. Gorski, V.A. Pitkevitch et al. Risk of radiogenic thyroid cancer in Russia following the Chernobyl accident. p. 89-96 in: Radiation and Thyroid Cancer. Proceedings of an International Seminar held in St. John's College, Cambridge, UK, 20-23 July 1998 (G. Thomas, A. Karaoglou and E.D. Williams, eds.). World Scientific, Singapore, 1999.

I25 Ishikawa, Y., J.A. Humphreys, C.G. Collier et al. Revised organ partition of thorium-232 in thorotrast patients. Radiat. Res. 152: S102-S106 (1999).

J1 Jablon, S., Z. Hrubec, J.D. Boice Jr. et al. Cancer in populations living near nuclear facilities. NIH Publication No. 90-874 (1990).

J2 Johansson, L., L.-G. Larsson and L. Damber. A cohort study with regard to the risk of haematological malignancies in patients treated with X-rays for benign lesions in the locomotor system. II. Estimation of absorbed dose in the red bone marrow. Acta Oncol. 34: 721-726 (1995).

J3 Japan, Ministry of Health and Welfare, Statistics and Information Department, Minister's Secretariat. Vital statistics, 1985, Japan. Volume 1, Tokyo, 1985.

J4 Jacob, J., G. Goulko, W.F. Heidenreich et al. Thyroid cancer risk to children calculated. Nature 392: 31-31 (1998).

J5   Jacob, P., Y. Kenigsberg, I. Zvonova et al. Childhood exposure due to the Chernobyl accident and thyroid cancer risk in contaminated areas of Belarus and Russia. Br. J. Cancer 80: 1461-1469 (1999).

K1   Kinlen, L.J. Leukaemia. p. 475-491 in: Cancer Surveys. Trends in Cancer Incidence and Mortality, Volumes 19/20 (R. Doll et al., eds.). Cold Spring Harbor Laboratory Press, Imperial Cancer Research Fund, 1994.

K2   Kinlen, L.J., K. Clarke and A. Balkwill. Paternal pre-conceptional radiation exposure in the nuclear industry and leukaemia and non-Hodgkin's lymphoma in young people in Scotland. Br. Med. J. 306: 1153-1158 (1993).

K3   Kelsey, J.L. and M.D. Gammon. Epidemiology of breast cancer. Epidemiol. Rev. 12: 228-240 (1990).

K4   Kusiak, R.A., J. Springer, A.C. Ritchie et al. Carcinoma of the lung in Ontario gold miners: possible aetiological factors. Br. J. Ind. Med. 48: 808-817 (1991).

K5   Kossenko, M.M., M.O. Degteva, O.V. Vyushkova et al. Issues in the comparison of risk estimates for the population in the Techa River region and atomic bomb survivors. Radiat. Res. 148: 54-63 (1997).

K6   Kazakov, V.S., E.P. Demidchik and L.N. Astakhova. Thyroid cancer after Chernobyl. Nature 359: 21 (1992).

K7   Koshurnikova, N.A., L.A. Buldakov, G.D. Bysogolov et al. Mortality from malignancies of the hematopoietic and lymphatic tissues among personnel of the first nuclear plant in the USSR. Sci. Total Environ. 142: 19-23 (1994).

K8   Kopecky, K.J., E. Nakashima, T. Yamamoto et al. Lung cancer, radiation and smoking among A-bomb survivors, Hiroshima and Nagasaki. RERF TR/13-86 (1986).

K9   Kaldor, J.M., N.E. Day, J. Bell et al. Lung cancer following Hodgkin's disease: a case-control study. Int. J. Cancer 52: 677-681 (1992).

K10   Koshurnikova, N.A., G.D. Bysogolov, M.G. Bolotnikova et al. Mortality among personnel who worked at the Mayak complex in the first years of its operation. Health Phys. 71: 90-93 (1996).

K11   Koshurnikova, N.A., N.S. Shilnikova, P.V. Okatenko et al. The risk of cancer among nuclear workers at the 'Mayak' production association: preliminary results of an epidemiological study. p. 113-122 in: Implications of New Data on Radiation Cancer Risk (J.D. Boice Jr., ed.). NCRP Proceedings No. 18 (1997).

K12   Kono, S. and T. Hirohata. Nutrition and stomach cancer. Cancer Causes and Control 7: 41-55 (1996).

K13   Kusiak, R.A., A.C. Ritchie, J. Springer et al. Mortality from stomach cancer in Ontario miners. Br. J. Ind. Med. 50: 117-126 (1993).

K14   Kleinerman, R.A., L.G. Littlefield, R.E. Tarone et al. Chromosome aberrations in lymphocytes from women irradiated for benign and malignant gynaecological disease. Radiat. Res. 139: 40-46 (1994).

K15   Karaoglou, A., G. Desmet, G.N. Kelly et al. (eds.). The Radiological Consequences of the Chernobyl Accident. EUR 16544 EN (1996).

K16   Karagas, M.R., J.A. McDonald, E.R. Greenberg et al. Risk of basal cell and squamous cell skin cancers after ionizing radiation therapy. J. Natl. Cancer Inst. 88: 1848-1853 (1996).

K17   Kosary, C.L., L.A. Gloeckler, B.A. Miller et al. (eds.). SEER Cancer Statistics Review, 1973-1992: Tables and Graphs. NIH Publication No. 96-2789 (1995).

K18   Kuijten, R.R., G.R. Bunin, C.C. Nass et al. Gestational and familial risk factors for childhood astrocytoma: results of a case-control study. Cancer Res. 50: 2608-2612 (1990).

K19   Kendall, G.M., C.R. Muirhead, B.H. MacGibbon et al. Mortality and occupational exposure to radiation: first analysis of the National Registry for Radiation Workers. Br. Med. J. 304: 220-225 (1992).

K20   Kellerer, A.M. and E. Nekolla. Neutron versus gamma-ray risk estimates. Radiat. Environ. Biophys. 36: 73-83 (1997).

K21   Kellerer, A.M. and D. Barclay. Age dependencies in the modelling of radiation carcinogenesis. Radiat. Prot. Dosim. 41: 273-281 (1992).

K22   Karlsson, P., E. Holmberg, L.M. Lundberg et al. Intracranial tumors after radium treatment for skin hemangioma during infancy - a cohort and case-control study. Radiat. Res. 148: 161-167 (1997).

K23   Karlsson, P., E. Holmberg, M. Lundell et al. Intracranial tumors after exposure to ionizing radiation during infancy. A pooled analysis of two Swedish cohorts of 28,008 infants with skin hemangioma. Radiat. Res. 150: 357-364 (1998).

K24   Klugbauer, S., E.P. Demidchik, E. Lengfelder et al. Detection of a novel type of RET rearrangement (PTC5) in thyroid carcinomas after Chernobyl and analysis of the involved RET-fused gene RFG5. Cancer Res. 58: 198-203 (1998).

K25   Kaletsch, U., R. Meinert, A. Miesner et al. Epidemiological studies of the incidence of leukaemia in children in Germany. Report BMU-1997-489. German Environment Ministry (1997).

K26   Kinlen, L.J. Incidence of cancer in rheumatoid arthritis and other disorders after immunosuppressive therapy. Am. J. Med. 78: 44-49 (1985).

K27   Kossenko, M.M. and M.O. Degteva. Cancer mortality and radiation risk evaluation for the Techa river population. Sci. Total Environ. 142: 73-89 (1994).

K28   Kaul, A. and W. Noffz. Tissue dose in thorotrast patients. Health Phys. 35: 113-121 (1978).

K29   Koiro, M. and Y. Ito. Pathomorphological study on 106 autopsy cases of Thorotrast-related hepatic malignancies with comparison to non-Thorotrast-related cases. p. 125-128 in: Risks from Radium and Thorotrast (D.M. Taylor et al., eds.). BIR Report 21 (1989).

K30   Kaldor, J.M., N.E. Day, B. Kittelmann et al. Bladder tumours following chemotherapy and radiotherapy for ovarian cancer: a case-control study. Int. J. Cancer 63: 1-6 (1995).

K31   Kofler, A., T. Abelin, I. Prudyvas et al. Factors related to latency period in post-Chernobyl carcinogenesis. p. 123-130 in: Radiation and Thyroid Cancer. Proceedings of an International Seminar held in St. John's College, Cambridge, UK, 20-23 July 1998 (G. Thomas, A. Karaoglou and E.D. Williams, eds.). World Scientific, Singapore, 1999.

K32   Koshurnikova, N.A., N.S. Shilnikova, P.V. Okatenko et al. Characteristics of the cohort of workers at the Mayak nuclear company. Radiat. Res. 152: 352-363 (1999).

K33   Kido, C., F. Sasaki, Y. Hirota et al. Cancer mortality of Thorotrast patients in Japan: the second series updated 1998. Radiat. Res. 152: S81-S83 (1999).

K34   Koshurnikova, N.A., M.G. Bolotnikova, L.A. Ilyin et al. Lung cancer risk due to exposure to incorporated plutonium. Radiat. Res. 149: 366-371 (1998).

K35   Karlsson, P., E. Holmberg, A. Samuelsson et al. Soft tissue sarcoma after treatment for breast cancer: a Swedish population-based study. Eur. J. Cancer 34: 2068-2075 (1998).

K36   Kerber, R.A., J.E. Till, S.L. Simon et al. A cohort study of thyroid disease in relation to fallout from nuclear weapons testing. J. Am. Med. Assoc. 270: 2076-2082 (1993).

K37 Khokhryakov, V.F., A.M. Kellerer, M. Kreisheimer et al. Lung cancer in nuclear workers of Mayak: a comparison of numerical procedures. Radiat. Environ. Biophys. 37: 11-17 (1998).

K38 Kaletsch, U., P. Kaatsch, R. Meinert et al. Childhood cancer and residential radon exposure – results of a population-based case-control study in Lower Saxony (Germany). Radiat. Environ. Biophys. 38: 211-215 (1999).

K39 Kesminiene, A.Z., J. Kurtinaitis and Z. Vilkeliene. Thyroid nodularity among Chernobyl cleanup workers from Lithuania. Acta Med. Lituanica 2: 51-54 (1997).

K40 Kaldor, J.M., N.E. Day, E.A. Clarke et al. Leukemia following Hodgkin's disease. N. Engl. J. Med. 322: 7-13 (1990).

K41 Kaul, A. Biokinetic models and data. p. 53-67 in: Health Effects of Internally Deposited Radionuclides: Emphasis on Radium and Thorium (G. van Kaick et al., eds.). World Scientific, Singapore, 1995.

K42 Koshurnikova, N.A., E.S. Gilbert, M. Sokolinkov et al. Bone cancers in Mayak workers. Radiat. Res. (2000, in press).

L1 Lubin, J.H., J.M. Samet and C. Weinberg. Design issues in epidemiologic studies of indoor exposure to Rn and risk of lung cancer. Health Phys. 59: 807-817 (1990).

L2 Lewis, C.A., P.G. Smith, I. Stratton et al. Estimated radiation doses to different organs among patients treated for ankylosing spondylitis with a single course of x-rays. Br. J. Radiol. 61: 212-220 (1988).

L3 Land, C.E. Estimating cancer risks from low doses of ionizing radiation. Science 209: 1197-1203 (1980).

L4 Lubin, J.H., J.D. Boice Jr., C. Edling et al. Lung cancer risk in radon-exposed miners and estimation of risk from indoor exposure. J. Natl. Cancer Inst. 87: 817-827 (1995).

L5 Land, C.E., J.D. Boice Jr., R.E. Shore et al. Breast cancer risk from low-dose exposure to ionizing radiation: results from parallel analysis of three exposed populations of women. J. Natl. Cancer. Inst. 65: 353-376 (1980).

L6 Lubin, J.H., J.D. Boice Jr., C. Edling et al. Radon-exposed underground miners and inverse dose-rate (protraction enhancement) effects. Health Phys. 69: 494-500 (1995).

L7 Land, C.E. Temporal distributions of risk for radiation-induced cancers. J. Chronic Dis. 40: 45S-57S (1987).

L8 Little, M.P., C.R. Muirhead, J.D. Boice Jr. et al. Using multistage models to describe radiation-induced leukaemia. J. Radiol. Prot. 15: 315-334 (1995).

L9 Little, M.P. Are two mutations sufficient to cause cancer? Some generalizations of the two-mutation model of carcinogenesis of Moolgavkar, Venzon and Knudson, and of the multistage model of Armitage and Doll. Biometrics 51: 1278-1291 (1995).

L10 Lubin, J.H., J.D. Boice Jr. and J.M. Samet. Errors in exposure assessment, statistical power and the interpretation of residential radon studies. Radiat. Res. 144: 329-341 (1995).

L11 Lee, W., R.P. Chiacchierini, B. Shleien et al. Thyroid tumors following ^{131}I or localised X irradiation to the thyroid and pituitary glands in rats. Radiat. Res. 92: 307-319 (1982).

L12 Land, C.E. and W.K. Sinclair. The relative contributions of different organ sites to the total cancer mortality associated with low-dose radiation exposure. p. 31-57 in: Risks Associated with Ionising Radiation. Annals of the ICRP 22(1). Pergamon Press, Oxford, 1991.

L13 Lundell, M., T. Hakulinen and L-E. Holm. Thyroid cancer after radiotherapy for skin hemangioma in infancy. Radiat. Res. 140: 334-339 (1994).

L14 Lundell, M. Estimates of absorbed dose in different organs in children treated with radium for skin hemangiomas. Radiat. Res. 140: 327-333 (1994).

L15 Lindberg, S., P. Karlsson, B. Arvidsson et al. Cancer incidence after radiotherapy for skin haemangioma during infancy. Acta Oncol. 34: 735-740 (1995).

L16 Lundell, M. and L-E. Holm. Risk of solid tumors after irradiation in infancy. Acta Oncol. 34: 727-734 (1995).

L17 Lundell, M., A. Mattsson, T. Hakulinen et al. Breast cancer after radiotherapy for skin hemangioma in infancy. Radiat. Res. 145: 225-230 (1996).

L18 Little, M.P., M.W. Charles and R. Wakeford. A review of the risks of leukemia in relation to parental pre-conception exposure to radiation. Health Phys. 68: 299-310 (1995).

L19 Likhtarev, I.A., B.G. Sobolev, I.A. Kairo et al. Thyroid cancer in the Ukraine. Nature 375: 365 (1995).

L20 Little, M.P., G.M. Kendall, C.R. Muirhead et al. Further analysis, incorporating assessment of the robustness of risks of cancer mortality in the National Registry for Radiation Workers. J. Radiol. Prot. 13: 95-108 (1993).

L21 Lubin, J.H. and J.D. Boice Jr. Lung cancer risk from residential radon: meta-analysis of eight epidemiologic studies. J. Natl. Cancer Inst. 89: 49-57 (1997).

L22 Lubin, J.H., L. Tomasek, C. Edling et al. Estimating lung cancer mortality from residential radon using data for low exposures of miners. Radiat. Res. 147: 126-134 (1997).

L23 Luebeck, E.G., S.B. Curtis, F.T. Cross et al. Two-stage model of radon-induced malignant lung tumors in rats: effects of cell killing. Radiat. Res. 145: 163-175 (1996).

L24 Lundell, M. and L.-E. Holm. Mortality from leukemia after irradiation in infancy for skin hemangioma. Radiat. Res. 145: 595-601 (1996).

L25 Laurer, G.R., N. Cohen and A. Stark. In-vivo measurements of ^{210}Pb to determine cumulative exposure to radon daughters: a pilot study. Report to U.S. Department of Energy, Office of Science and Technical Information. DE-FG02-90ER60944 (1991).

L26 Lloyd, D.C., A.A. Edwards, A.V. Sevan'kaev et al. Retrospective dosimetry by chromosomal analysis. p. 965-973 in: The Radiological Consequences of the Chernobyl Accident (A. Karaoglou et al., eds.). EUR 16544 (1996).

L27 Little, M.P., C.R. Muirhead, L.H.J. Goossens et al. Probabilistic accident consequence uncertainty analysis. Late health effects uncertainty assessment. NUREG/CR-6555 (1997).

L28 Little, J.B. Inherited susceptibility and radiation exposure. p. 207-218 in: Implications of New Data on Radiation Cancer Risk (J.D. Boice Jr., ed.). NCRP Proceedings No. 18 (1997).

L29 Little, M.P. and M.W. Charles. The risk of non-melanoma skin cancer incidence in the Japanese atomic bomb survivors. Int. J. Radiat. Biol. 71: 589-602 (1997).

L30 Longstreth, W.T., L.K. Dennis, V.M. McGuire et al. Epidemiology of intracranial meningiomas. Cancer 72: 639-648 (1993).

L31 Littlefield, L.G., A.F. McFee, S.I. Salomaa et al. Do recorded doses overestimate true doses received by Chernobyl cleanup workers? Results of cytogenetic analysis of Estonian workers by fluorescence in situ hybridization. Radiat. Res. 150: 237-249 (1998).

L32 Little, M.P., F. de Vathaire, M.W. Charles et al. Variations with time and age in the relative risks of solid cancer incidence after radiation exposure. J. Radiol. Prot. 17: 159-177 (1997).

L33 Little, J.B. What are the risks of low-level exposure to α radiation from radon. Proc. Natl. Acad. Sci. U.S.A. 94: 5996-5997 (1997).

L34 Lubin, J.H., M.S. Linet, J.D. Boice Jr. et al. Case-control study of childhood acute lymphoblastic leukemia and residential radon exposure. J. Natl. Cancer Inst. 90: 294-300 (1998).

L35 Lubin, J.H. On the discrepancy between epidemiologic studies in individuals of lung cancer and residential radon and Cohen's ecologic regression. Health Phys. 4-10 (1998).

L36 Little, M.P., F. de Vathaire, M.W. Charles et al. Variations with time and age in the risks of solid cancer incidence after radiation exposure in childhood. Stat. Med. 17: 1341-1355 (1998).

L37 Little, M.P., F. de Vathaire, A. Shamsaldin et al. Risks of brain tumour following treatment for cancer in childhood: modification by genetic factors, radiotherapy and chemotherapy. Int. J. Cancer 78: 269-275 (1998).

L38 Law, G. and E. Roman. Leukaemia near La Hague nuclear plant: study design is questionable. Br. Med. J. 314: 1553 (1997).

L39 Little, M.P. and J.D. Boice Jr. Comparison of breast cancer incidence in the Massachusetts tuberculosis fluroscopy cohort and in the Japanese atomic bomb survivors. Radiat. Res. 151: 218-224 (1999).

L40 Little, M.P. Comments on the article "Studies of the mortality of atomic bomb survivors. Report 12, Part 1. Cancer: 1950-1990" by D.A. Pierce, Y. Shimizu, D.L. Preston et al. Radiat. Res. 148: 399-400 (1997).

L41 Little, M.P. and C.R. Muirhead. Evidence for curvi-linearity in the cancer incidence dose-response in Japanese atomic bomb survivors. Int. J. Radiat. Biol. 70: 83-94 (1996).

L42 Little, M.P. and C.R. Muirhead. Curvature in the cancer mortality dose response in Japanese atomic bomb survivors: absence of evidence of threshold. Int. J. Radiat. Biol. 74: 471-480 (1998).

L43 London, W.T. and K.A. McGlynn. Liver cancer. p. 772-793 in: Cancer Epidemiology and Prevention, Second edition (D. Schottenfeld and J.F. Fraumaeni Jr., eds.). Oxford University Press, Oxford, 1996.

L44 Létourneau, E.G., D. Krewski, N.W. Choi et al. Case-control study of residential radon and lung cancer in Winnipeg, Manitoba, Canada. Am. J. Epidemiol. 140: 310-322 (1994).

L45 Lubin, J.H., J.D. Boice Jr., C. Edling et al. Radon and lung cancer risk: a joint analysis of 11 underground miners studies. NIH Publication No. 94-3644 (1994).

L46 Lundell, M., A. Mattson, P. Karlsson et al. Breast cancer risk after radiotherapy in infancy: a pooled analysis of two Swedish cohorts of 17,202 infants. Radiat. Res. 151: 626-632 (1999).

L47 Little, M.P., H.A. Weiss, J.D. Boice Jr. et al. Risks of leukemia in Japanese atomic bomb survivors, in women treated for cervical cancer, and in patients treated for ankylosing spondylitis. Radiat. Res. 152: 280-292 (1999).

L48 Laskin, W.B., T.A. Silverman and F.M. Enzinger. Postradiation soft tissue sarcomas: an analysis of 53 cases. Cancer 62: 2330-2340 (1988).

L49 Lili, W., L. Lin, S. Quanfu et al. A cohort study of cancer mortality on workers exposed to thorium-containing dust in Baotou Iron and Steel Company. Chin. J. Radiol. Med. Prot. 14: 93-96 (1994).

L50 Little, M.P. and C.R. Muirhead. Derivation of low dose extrapolation factors from analysis of curvature in the cancer incidence dose response in the Japanese atomic bomb survivors. Int. J. Radiat. Biol. 76: 939-953 (2000).

L51 Likhtarev, I.A., I.A. Kayro, V.M. Shpak et al. Radiation-induced and background thyroid cancer of Ukrainian children (dosimetric approach). Int. J. Radiat. Med. 3-4: 51-66 (1999).

L52 Little, M.P., C.R. Muirhead, R.G.E. Haylock et al. Relative risks of radiation-associated cancer: comparison of second cancer in therapeutically irradiated populations with the Japanese atomic bomb survivors. Radiat. Environ. Biophys. 38: 267-283 (1999).

M1 Miller, A.B., G.R. Howe, G.J. Sherman et al. Mortality from breast cancer after irradiation during fluoroscopic examinations in patients being treated for tuberculosis. N. Engl. J. Med. 321: 1285-1289 (1989).

M2 Mabuchi, K., M. Soda, E. Ron et al. Cancer incidence in atomic bomb survivors. Part I: Use of the tumor registries in Hiroshima and Nagasaki for incidence studies. Radiat. Res. 137: S1-S16 (1994).

M3 Morlier, J.P., M. Morin, G. Monchaux et al. Lung cancer incidence after exposure of rats to low doses of radon: influence of dose rate. Radiat. Prot. Dosim. 56: 93-97 (1994).

M4 Maxon, H.R., E.L. Saenger, S.R. Thomas et al. Clinically important radiation-associated thyroid disease. A controlled study. J. Am. Med. Assoc. 244: 1802-1805 (1980).

M5 Moolgavkar, S.H., F.T. Cross, G. Luebeck et al. A two-mutation model for radon-induced lung tumors in rats. Radiat. Res. 131: 28-37 (1990).

M6 Moolgavkar, S.H., E.G. Luebeck, D. Krewski et al. Radon, cigarette smoking and lung cancer: a re-analysis of the Colorado plateau uranium miner's data. Epidemiology 4: 204-217 (1993).

M7 Mahaffey, J.A., M.A. Parkhurst, A.C. James et al. Estimating past exposure to indoor radon from household glass. Health Phys. 64: 381-391 (1993).

M8 Mattsson, A., B. Ruden, P. Hall et al. Radiation-induced breast cancer and long term follow-up of radiotherapy for benign breast disease. J. Natl. Cancer Inst. 85: 1679-1685 (1993).

M9 Monson, R.R. and B. MacMahon. Prenatal x-ray exposure and cancer in children. p. 97-105 in: Radiation Carcinogenesis: Epidemiology and Biological Significance (J.D. Boice Jr. and J.F. Fraumeni Jr., eds.). Raven Press, New York, 1984.

M10 MacMahon, B. and D. Tricopoulous. Epidemiology, Principles and Methods (Second edition). Little, Brown and Company, Boston, 1996.

M11 Moolgavkar, S.H. and D.J. Venzon. Two-event models for carcinogenesis: incidence curves for childhood and adult tumors. Math. Biosci. 47: 55-77 (1979).

M12 Moolgavkar, S.H. and A.G. Knudson. Mutation and cancer: a model for human carcinogenesis. J. Natl. Cancer Inst. 66: 1037-1052 (1981).

M13 Muirhead, C.R., B.K. Butland, B.M.R. Green et al. Childhood leukaemia and natural radiation. Lancet 337: 503-504 (1991).

M14 Mori, T., C. Kido, K. Fukutomi et al. Summary of entire Japanese Thorotrast follow-up study: updated 1998. Radiat. Res. 152: S84-S87 (1999).

M15 Morrison, H.I., P.J. Villeneuve, J.H. Lubin et al. Radon-progeny and lung cancer risk in a cohort of Newfoundland fluorspar miners. Radiat. Res. 150: 58-65 (1998).

M16 McLaughlin, J.R., W.D. King, T.W. Anderson et al. Paternal radiation exposure and leukaemia in offspring. The Ontario case-control study. Br. Med. J. 307: 959-966 (1993).

M17 Muirhead, C.R. and B.K. Butland. Dose-response analyses for the Japanese A-bomb survivors. Health Phys. 57: 1035-1036 (1989).

M18 Muirhead, C. and C. Sharp. Breast cancer risks. Radiol. Prot. Bull. 168: 11-13 (1995).

M19 Michels, K.B., D. Trichopoulos, J.M. Robins et al. Birthweight as a risk factor for breast cancer. Lancet 348: 1542-1546 (1996).

M20 Mattsson, A., B.I. Ruden, J. Palmgren et al. Dose- and time-response for breast cancer risk after radiation therapy for benign breast disease. Br. J. Cancer 72: 1054-1061 (1995).

M21 Mays, C.W., R.L. Aamodt, K.G.W. Inn et al. External gamma-ray counting of selected tissues from a Thorotrast patient. Health Phys. 63: 33-40 (1992).

M22 McMichael, A.J. and G.G. Giles. Colorectal cancer. p. 77-98 in: Cancer Surveys. Trends in Cancer Incidence and Mortality, Volumes 19/20 (R. Doll et al., eds.). Cold Spring Harbor Laboratory Press, Imperial Cancer Research Fund, 1994.

M23 Matanoski, G.M., P. Sartwell, E. Elliott et al. Cancer risks in radiologists and radiation workers. p. 83-96 in: Radiation Carcinogenesis: Epidemiology and Biological Significance (J.D. Boice Jr. and J.F. Fraumeni Jr., eds.). Raven Press, New York, 1984.

M24 Miller, D.L. and M.A. Weinstock. Non melanoma skin cancer in the United States: Incidence. J. Am. Acad. Dermatol. 30: 774-778 (1994).

M25 Matanoski, G., P. Selser, P. Sartwell et al. The current mortality rates of radiologists and other physician specialists: specific causes of death. Am. J. Epidemiol. 101: 199-210 (1975).

M26 Moll, A.C., S.M. Imhof, L.M. Bouter et al. Second primary tumors in patients with hereditary retinoblastoma: a register-based follow-up study, 1945-1994. Int. J. Cancer 67: 515-519 (1996).

M27 McCredie, M., P. Maisonneuve and P. Boyle. Perinatal and early postnatal risk factors for malignant brain tumours in New South Wales children. Int. J. Cancer 56: 11-15 (1994).

M28 Mattsson, A., P. Hall, B.I. Ruden et al. Incidence of primary malignancies other than breast cancer among women treated with radiation therapy for benign breast disease. Radiat. Res. 148: 152-160 (1997).

M29 Muirhead, C.R. and G.W. Kneale. Prenatal irradiation and childhood cancer. J. Radiol. Prot. 9: 209-212 (1989).

M30 Michaelis, J., U. Kaletsch, W. Burkart et al. Infant leukaemia after the Chernobyl accident. Nature 387: 246 (1997).

M31 Miller, R.W. and J.D. Boice Jr. Cancer after intrauterine exposure to the atomic bomb. Radiat. Res. 147: 396-397 (1997).

M32 Mettler, F.A., H.D. Royal, J.R. Hurley et al. Administration of stable iodine to the population around the Chernobyl nuclear power plant. J. Radiol. Prot. 12: 159-165 (1992).

M33 Musa, B.S., I.K. Pople and B.H. Cummins. Intracranial meningiomas following irradiation - a growing problem? Br. J. Neurosurg. 9: 629-637 (1995).

M34 Muirhead, C.R. Childhood cancer and nuclear installations: a review. Nucl. Energy 37: 371-379 (1998).

M35 McLaughlin, J.R., E.A. Clarke, E.D. Nishri et al. Childhood leukemia in the vicinity of Canadian nuclear facilities. Cancer Causes and Control 4: 51-58 (1993).

M36 Mueller, N.E. Hodgkin's disease. p. 893-919 in: Cancer Epidemiology and Prevention (D. Schottenfeld and J.F. Fraumeni Jr., eds.). Oxford University Press, Oxford, 1996.

M37 Mueller, N.E., A. Mohar and A. Evans. Viruses other than HIV and non-Hodgkin's lymphoma. Cancer Res. 52: 5479s-5481s (1992).

M38 McCullagh, P. and J.A. Nelder. Generalized Linear Models (2nd edition). Chapman and Hall, London, 1989.

M39 Miller, R.W., J.D. Boice Jr and R.E. Curtis. Bone cancer. p. 971-983 in: Cancer Epidemiology and Prevention (D. Schottenfeld and J.F. Fraumeni Jr., eds.). Oxford University Press, Oxford, 1996.

M40 Muir, C.S. and J. Nectoux. International patterns of cancer. p. 141-167 in: Cancer Epidemiology and Prevention (D. Schottenfeld and J.F. Fraumeni Jr., eds.). Oxford University Press, Oxford, 1996.

M41 Mays, C.W., D. Mays and R.A. Guilmette (eds.). Total-body evaluation of a Thorotrast patient. Workshop held on July 1990 at the National Cancer Institute, Bethesda. Health Phys. 63: 1-100 (1992).

M42 Mori, T. and Y. Kato. Epidemiological, pathological and dosimetric status of Japanese Thorotrast patients. J. Radiat. Res. 2 (Suppl.): 34-45 (1991).

M43 Muñoz, N. and N.E. Day. Esophageal cancer. p. 681-706 in: Cancer Epidemiology and Prevention, Second edition (D. Schottenfeld and J.F. Fraumeni Jr., eds.). Oxford University Press, Oxford, 1996.

M44 Moore, D.H.II, H.W. Patterson, F. Hatch et al. Case-control study of malignant melanoma among employees of the Lawrence Livermore National Laboratory. Am. J. Ind. Med. 32: 377-391 (1997).

M45 McCredie, M. Bladder and kidney cancers. p. 343-368 in: Cancer Surveys. Trends in Cancer Incidence and Mortality, Volumes 19/20 (R. Doll et al., eds.). Cold Spring Harbor Laboratory Press, Imperial Cancer Research Fund, 1994.

M46 Muirhead, C.R., A.A. Goodill, R.G.E. Haylock et al. Occupational radiation exposure and mortality: second analysis of the National Registry for Radiation Workers. J. Radiol. Prot. 19: 3-26 (1999).

M47 Mori, T., K. Fukutomi, Y. Kato et al. 1998 results of the first series of follow-up studies on Japanese Thorotrast patients and their relationships to an autopsy series. Radiat. Res. 152: S72-S80 (1999).

M48 Martling, U., A. Mattson, L.B. Travis et al. Mortality after long-term exposure to radioactive Thorotrast: a forty-year follow-up survey in Sweden. Radiat. Res. 151: 293-299 (1999).

M49 Moore, D.H. II and J.D. Tucker. Biological dosimetry of Chernobyl cleanup workers: inclusion of data on age and smoking provides improved radiation dose estimates. Radiat. Res. 152: 655-664 (1999).

M50 McGeoghegan, D. and K. Binks. The mortality and cancer morbidity experience of employees at the Chapelcross plant of British Nuclear Fuels Ltd, 1955-1995. p. 261-264 in: Southport '99: Proceedings of the 6th SRP International Symposium (M. Thorne, ed.). Society for Radiological Protection, London, 1999.

M51 Marshall, J.H. and P.G. Groer. A theory of the induction of bone cancer by alpha radiation. Radiat. Res. 71: 149-192 (1977).

M52 Marshall, J.H., P.G. Groer and R.A. Schlenker. Dose to endosteal cells and relative distribution factors for radium-224 and plutonium-239 compared to radium-226. Health Phys. 35: 91-101 (1978).

M53 Miller, R.C., G. Randers-Pehrson, C.R. Geard et al. The oncogenic transforming potential of the passage of single α particles through mammalian cell nuclei. Proc. Natl. Acad. Sci. U.S.A. 96: 19-22 (1999).

N1   National Institutes of Health. Report of the National Institutes of Health Ad Hoc Working Group to develop radioepidemiological tables. NIH Publication No. 85-2748 (1985).

N2   National Radiological Protection Board. Board statement on diagnostic medical exposures to ionising radiation during pregnancy and estimates of late radiation risks to the UK population. Doc. NRPB 4(4) (1993).

N3   National Radiological Protection Board. Risk of radiation-induced cancer at low doses and low dose rates for radiation protection purposes. Doc. NRPB 6(1) (1995).

N4   Nekolla, E.A., A.M. Kellerer, M. Kuse-Isingschulte et al. Malignancies in patients treated with high doses of radium-224. Radiat. Res. 152: S3-S7 (1999).

N5   National Council on Radiation Protection and Measurements. Induction of thyroid cancer by ionizing radiation. NCRP Report No. 80 (1985).

N6   Najarian, T. and T. Colton. Mortality from leukemia and cancer in shipyard nuclear workers. Lancet I: 1018-1020 (1978).

N7   National Council on Radiation Protection and Measurements. General concepts for the dosimetry of internally deposited radionuclides. NCRP Report No. 84 (1985).

N8   Nelemans, P.J., F.H.J. Rampen, D.J. Ruiter et al. An addition to the controversy on sunlight exposure and melanoma risk: a meta-analytical approach. J. Clin. Epidemiol. 48: 1331-1342 (1995).

N9   Neglia, J.P., A.T. Meadows, L.L. Robison et al. Second neoplasms after acute lymphoblastic leukemia in childhood. N. Engl. J. Med. 325: 1330-1336 (1991).

N10 National Radiological Protection Board. Assessment of skin doses. Doc. NRPB 8(3) (1997).

N11 Neugut, A.I., H. Ahsan, E. Robinson et al. Bladder carcinoma and other second malignancies after radiotherapy for prostate cancer. Cancer 79: 1600-1604 (1997).

N12 National Cancer Institute. Estimated exposures and thyroid doses received by the American people from I-131 in fallout following Nevada atmospheric nuclear bomb tests. National Institutes of Health, Bethesda (1997).

N13 National Radiological Protection Board. Genetic heterogeneity in the population and its implications for radiation risk. Doc. NRPB 10(3) (1999).

N14 Nekolla, E.A., M. Kreisheimer, A.M. Kellerer et al. Induction of malignant bone tumors in radium-224 patients: risk estimates based on the improved dosimetry. Radiat. Res. 153: 93-103 (2000).

N15 National Cancer Institute. Uncertainties in Radiation Dosimetry and Their Impact on Dose-response Analyses (E. Ron and F.O. Hoffman, eds.). National Institutes of Health, Bethesda, 1999.

N16 Naumberg, E., R. Bellocco, S. Cnattingius et al. Prenatal ultrasound examinations and risk of childhood leukaemia: case-control study. Br. Med. J. 320: 282-283 (2000).

N17 National Council on Radiation Protection and Measurements. Uncertainties in fatal cancer risk estimates used in radiation protection. NCRP Report No. 126 (1997).

N18 National Council on Radiation Protection and Measurements. Some aspects of strontium radiobiology. NCRP Report No. 110 (1991).

N19 Nekolla, E., D. Chmelevsky, A.M. Kellerer et al. Malignancies in patients treated with Ra-224. p. 243-248 in: Health Effects of Internally Deposited Radionuclides: Emphasis on Radium and Thorium (G. van Kaick et al., eds.). World Scientific, Singapore, 1995.

O1   Omar, R.Z., J.A. Barber and P.G. Smith. Cancer mortality and morbidity among plutonium workers at the Sellafield plant of British Nuclear Fuels. Br. J. Cancer 79: 1288-1301 (1999).

P1   Prentice, R.L. A case-cohort design for epidemiological cohort studies and disease prevention trials. Biometrika 73: 1-12 (1986).

P2   Pierce, D.A., D.O. Stram, M. Vaeth et al. The errors-in-variables problem: considerations provided by radiation dose-response analyses of the A-bomb survivor data. J. Am. Stat. Assoc. 87: 351-359 (1992).

P3   Pillai, N.K., M. Thangavelu and V. Ramalingaswami. Nodular lesions of the thyroid in an area of high background radiation in coastal Kerala, India. Indian J. Med. Res. 64: 537-544 (1976).

P4   Preston, D.L., S. Kusumi, M. Tomonaga et al. Cancer incidence in atomic bomb survivors. Part III: Leukemia, lymphoma and multiple myeloma, 1950-1987. Radiat. Res. 137: S68-S97 (1994).

P5   Parkin, D.M., S.L. Whelan, J. Ferlay et al. Cancer Incidence in Five Continents. Vol. VII. IARC Scientific Publications No. 143 (1997).

P6   Pierce, D.A. and D.L. Preston. Joint analysis of site-specific cancer risks for the A-bomb survivors. Radiat. Res. 134: 134-142 (1993).

P7   Pobel, D. and J-F. Viel. Case-control study of leukaemia among young people near La Hague nuclear reprocessing plant: the environmental hypothesis revisited. Br. Med. J. 314: 101-106 (1997).

P8   Pottern, L.M., M.M. Kaplan, P.R. Larsen et al. Thyroid nodularity after childhood irradiation for lymphoid hyperplasia: a comparison of questionnaire and clinical findings. J. Clin. Epidemiol. 43: 449-460 (1990).

P9   Pierce, D.A., Y. Shimizu, D.L. Preston et al. Studies of the mortality of A-bomb survivors. Report 12, Part 1. Cancer: 1950-1990. Radiat. Res. 146: 1-27 (1996).

P10 Preston-Martin, S., D.C. Thomas, M.C. Yu et al. Diagnostic radiography as a risk factor for chronic myeloid and monocytic leukaemia (CML). Br. J. Cancer 59: 639-644 (1989).

P11 Preston, D.L. Communication to the UNSCEAR Secretariat (1996).

P12 Parkin, D.M., D. Clayton, R.J. Black et al. Childhood leukaemia in Europe after Chernobyl: five year follow-up. Br. J. Cancer 73: 1006-1012 (1996).

P13 Prentice, R.L., Y. Yoshimoto and M.W. Mason. Relationship of cigarette smoking and radiation exposure to cancer mortality in Hiroshima and Nagasaki. J. Natl. Cancer Inst. 70: 611-622 (1983).

P14 Preston, D.L. Communication to the UNSCEAR Secretariat (1995).

P15 Petridou, E., D. Trichopoulos, N. Dessypris et al. Infant leukaemia after in utero exposure to radiation from Chernobyl. Nature 382: 352-353 (1996).

P16 Prisyazhniuk, A., Z. Fedorenko, A. Okaenov et al. Epidemiology of cancer in populations living in contaminated territories of Ukraine, Belarus, Russia after the Chernobyl accident. p. 909-921 in: The Radiological Consequences of the Chernobyl Accident (A. Karaoglou et al., eds.). EUR 16544 EN (1996).

P17 Potter, J.D. Nutrition and colon cancer. Cancer Causes and Control 7: 127-146 (1996).

P18 Preston-Martin, S. and W. Mack. Nervous system. p. 1231-1281 in: Cancer Epidemiology and Prevention (D. Schottenfeld and J.F. Fraumeni Jr., eds.). Oxford University Press, Oxford, 1996.

P19 Preston, D.L., E. Ron, K. Mabuchi et al. Brain and other tumors of the central nervous system among atomic bomb survivors (2000, in preparation).

P20 Preston-Martin, S., D.C. Thomas, W.E. Wright et al. Noise trauma in the aetiology of acoustic neuromas in men in Los Angeles County, 1978-1985. Br. J. Cancer 59: 783-786 (1989).

P21 Preston-Martin, S., M.C. Yu, B.E. Henderson et al. Risk factors for meningiomas in men in Los Angeles County. J. Natl. Cancer Inst. 70: 863-866 (1983).

P22 Preston-Martin, S., M.C. Yu, B. Benton et al. N-nitroso compounds and childhood brain tumors: a case-control study. Cancer Res. 42: 5240-5245 (1982).

P23 Preston-Martin, S, W. Mack and B.E. Henderson. Risk factors for gliomas and meningiomas in males in Los Angeles County. Cancer Res. 49: 6137-6143 (1989).

P24 Preston-Martin, S., A. Paganini-Hill and B.E. Henderson. Case-control study of intracranial meningiomas in women in Los Angeles County. J. Natl. Cancer Inst. 75: 67-75 (1980).

P25 Parkin, D.M. Communication to the UNSCEAR Secretariat (1998).

P26 Pion, I.A., D.S. Rigel, L. Garfinkel et al. Occupation and the risk of malignant melanoma. Cancer 75 (Suppl. 2): 637-644 (1995).

P27 Polednak, A.P. Thyroid tumors and thyroid function in women exposed to internal and external radiation. J. Environ. Pathol. Toxicol. Oncol. 7: 53-64 (1986).

P28 Pierce, D.A., Y. Shimizu, D.L. Preston et al. Response to the letter of M.P. Little. Radiat. Res. 148: 400-401 (1997).

P29 Parkin, D.M., P. Pisani and J. Ferlay. Estimates of the worldwide incidence of 18 major cancers in 1985. Int. J. Cancer 54: 594-606 (1993).

P30 Parkin, D.M., H. Ohshima, P. Srivatanakul et al. Cholangiocarcinoma: epidemiology, mechanisms of carcinogenesis and prevention. Cancer Epidemiol. Biomarkers Prev. 6: 537-544 (1993).

P31 Pacini, F., T. Vorontsova, E.P. Demidchik et al. Post Chernobyl thyroid carcinoma in Belarus children and adolescents: comparison with naturally occurring thyroid carcinoma in Italy and France. J. Clin. Endocrinol. Metab. 82(11): 3563-3569 (1997).

P32 Pershagen, G., Z.-H. Liang, Z. Hrubec et al. Residential radon exposure and lung cancer in Swedish women. Health Phys. 63: 179-186 (1992).

P33 Pershagen, G., G. Åkerbolm, O. Axelson et al. Residential radon exposure and lung cancer in Sweden. N. Engl. J. Med. 330: 159-164 (1994).

P34 Pierce, D.A. and M.L. Mendelsohn. A model for radiation-related cancer suggested by atomic bomb survivor data. Radiat. Res. 152: 642-654 (1999).

P35 Preston-Martin, S., D.C. Thomas, S.C. White et al. Prior exposure to medical and dental X-rays related to tumors of the parotid gland. J. Natl. Cancer Inst. 80: 943-949 (1988).

P36 Pierce, D.A., Y. Shimizu, D.L. Preston, et al. Response to the letter of Bernard L. Cohen. Radiat. Res. 149: 526-528 (1998).

P37 Parshkov, E.M. Pathogenesis of radiation-induced thyroid cancer in children affected as a result of the Chernobyl accident. Int. J. Radiat. Med. 3-4: 67-75 (1999).

R1 Ron, E., D.L. Preston, K. Mabuchi et al. Cancer incidence in atomic bomb survivors. Part IV: Comparison of cancer incidence and mortality. Radiat. Res. 137: S98-S112 (1994).

R2 Ron, E., R. Carter, S. Jablon et al. Agreement between death certificate and autopsy diagnosis among atomic bomb survivors. Epidemiology 5: 48-56 (1994).

R3 Ron, E., J. Lubin and A.B. Schneider. Thyroid cancer incidence. Nature 360: 113 (1992).

R4 Ron, E., J.H. Lubin, R.E. Shore et al. Thyroid cancer after exposure to external radiation: a pooled analysis of seven studies. Radiat. Res. 141: 259-277 (1995).

R5 Ron, E., B. Modan and J.D. Boice Jr. Mortality after radiotherapy for ringworm of the scalp. Am. J. Epidemiol. 127: 713-725 (1988).

R6 Richardson, S., C. Monfort, M. Green et al. Spatial variation of natural radiation and childhood leukaemia incidence in Great Britain. Stat. Med. 14: 2487-2501 (1995).

R7 Ron, E. Communication to the UNSCEAR Secretariat (1994).

R8 Radford, E.P. and K.G. St. Clair Renard. Lung cancer in Swedish iron miners exposed to low doses of radon daughters. N. Engl. J. Med. 310: 1485-1494 (1984).

R9 Ron, E., B. Modan, D. Preston et al. Thyroid neoplasia following low-dose radiation in childhood. Radiat. Res. 120: 516-531 (1989).

R10 Richardson, R.B., J.P. Eatough and D.L. Henshaw. The contribution of radon exposure to the radiation dose received by red bone marrow. Int. J. Radiat. Biol. 57: 597 (1989).

R11 Rowland, R.E., H.F. Lucas and R.A. Schlenker. External radiation doses received by female radium dial painters. p. 67-72 in: Risks from Radium and Thorotrast (D.M. Taylor et al., eds.). BIR Report 21 (1989).

R12 Rinsky, R.A., R.D. Zumwalde, R.J. Waxweiler et al. Cancer mortality at a naval nuclear shipyard. Lancet I: 231-235 (1981).

R13 Ron, E. The epidemiology of thyroid cancer. p. 1000-1021 in: Cancer Epidemiology and Prevention (D. Schottenfeld and J.F. Fraumeni Jr., eds.). Oxford University Press, Oxford, 1996.

R14 Ron, E., M.M. Doody, D.V. Becker et al. Cancer mortality following treatment for adult hyperthyroidism. J. Am. Med. Assoc. 280: 347-355 (1998).

R15 Ron, E., D.L. Preston, K. Mabuchi et al. Skin tumor risk among atomic-bomb survivors in Japan. Cancer Causes and Control 9(4): 393-401 (1998).

R16 Ron, E., B. Modan, D. Preston et al. Radiation-induced skin carcinomas of the head and neck. Radiat. Res. 125: 318-325 (1991).

R17 Ron, E., B. Modan, J.D. Boice Jr. et al. Tumors of the brain and nervous system after radiotherapy in childhood. N. Engl. J. Med. 319: 1033-1039 (1988).

R18 Ron, E., J.D. Boice Jr., S. Hamburger et al. Mortality following radiation treatment for infertility of hormonal origin or amenorrhea. Int. J. Epidemiol. 23: 1165-1173 (1994).

R19 Ryan, P., M.W. Lee, B. North et al. Amalgam fillings, diagnostic dental x-rays and tumours of the brain and meninges. Eur. J. Cancer 28B: 91-95 (1992).

R20 Rahu, M., M. Tekkel, T. Veidebaum et al. The Estonian study of Chernobyl cleanup workers: II. Incidence of cancer and mortality. Radiat. Res. 147: 653-657 (1997).

R21 Robbins, J. and W. Adams. Radiation effects in the Marshall Islands. p. 11-24 in: Radiation and the Thyroid (S. Nagataki, ed.). Excepta Medica, Tokyo, 1989.

R22 Royal, H.D. Relative biological effectiveness of external radiation vs. I-131: a review of animal data. p. 201-207 in: Radiation and Thyroid Cancer. Proceedings of an International Seminar held in St. John's College, Cambridge, UK, 20-23 July 1998 (G. Thomas, A. Karaoglou and E.D. Williams, eds.). World Scientific, Singapore, 1999.

R23 Ruosteenoja, E., I. Mäkeläinen, T. Rytömaa et al. Radon and lung cancer in Finland. Health Phys. 71: 185-189 (1996).

R24 Roesch, W.C. (ed.). Final U.S.-Japan Joint Workshop for Reassessment of Atomic Bomb Radiation Dosimetry in Hiroshima and Nagasaki. Radiation Effects Research Foundation, Hiroshima, 1987.

R25 Ritz, B. Radiation exposure and cancer mortality in uranium processing workers. Epidemiology 10: 531-538 (1999).

R26 Rooney, C., V. Beral, N. Maconochie et al. Case-control study of prostatic cancer in employees of the United Kingdom Atomic Energy Authority. Br. Med. J. 307: 1391-1397 (1993).

R27 Ritz, B., H. Morgenstern, J. Froines et al. Effects of exposure to external ionizing radiation on cancer mortality in nuclear workers monitored for radiation at Rocketdyne/Atomics International. Am. J. Ind. Med. 35: 21-31 (1999).

R28 Romanyukha, A.A., E.A. Ignatiev, E.K. Vasilenko et al. EPR dose reconstruction for Russian nuclear workers. Health Phys. 78: 15-20 (2000).

R29 Roman, E., P. Doyle, N. Maconochie et al. Cancer in children of nuclear industry employees: report on children aged under 25 years from nuclear industry family study. Br. Med. J. 318: 1443-1450 (1999).

R30 Ron, E., A. Auvinen, E. Alfandary et al. Cancer incidence among women treated with radiotherapy for infertility or menstrual disorders (Israel). Int. J. Cancer 89: 795-798 (1999).

R31 Rosenson, R., B. Gusev, M. Hoshi et al. A brief summary of radiation studies on residents in the Semipalatinsk area 1957-1993. p.127-146 in: Nagasaki Symposium, Radiation and Human Health: Proposal from Nagasaki. Elsevier, Amsterdam, 1996.

R32 Ross, R.K. and D. Schottenfeld. Prostate cancer. p. 1180-1206 in: Cancer Epidemiology and Prevention (D. Schottenfeld and J.F. Fraumeni Jr., eds.). Oxford University Press, Oxford, 1996.

R33 Rowland, R.E., A.F. Stehney and H.F. Lucas. Dose-response relationships for radium-induced bone sarcomas. Health Phys. 44 (Suppl. 1): 15-31 (1983).

R34 Raabe, O., S.A. Book and N.J. Parks. Bone cancer from radium: canine dose response explains data for mice and humans. Science 208: 61-64 (1980).

R35 Rowland, R.E. Dose-response relationships for female radium dial workers: a new look. p. 135-143 in: Health Effects of Internally Deposited Radionuclides: Emphasis on Radium and Thorium (G. van Kaick et al., eds.). World Scientific, Singapore, 1995.

R36 Ramirez-Ramos, A., R.H. Gilman, R. Leon-Barua et al. Rapid reoccurrence of *Helicobacter pylori* infection in Peruvian patients after successful eradication. Clin. Infect. Dis. 25: 1027-1031 (1997).

S1 Stewart, A.M., K.W. Webb and D. Hewitt. A survey of childhood malignancies. Br. Med. J. 1: 1495-1508 (1958).

S2 Smith, B.J., W.R. Field and F.C. Lynch. Residential ^{222}Rn exposure and lung cancer: testing the linear no-threshold theory with ecologic data. Health Phys. 75: 11-17 (1998).

S3 Shimizu, Y., H. Kato and W.J. Schull. Life Span Study Report 11. Part 2. Cancer mortality in the years 1950-85 based on the recently revised doses (DS86). Radiat. Res. 121: 120-141 (1990).

S4 Sposto, R. and D.L. Preston. Correcting for catchment area non-residency in studies based on tumor-registry data. RERF CR/1-92 (1992).

S5 Sposto, R., D.L. Preston, Y. Shimizu et al. The effect of diagnostic misclassification on noncancer and cancer mortality dose response in the RERF Life Span Study. Biometrics 48: 605-617 (1992).

S6 Samet, J.M. and M.L. Lerchen. Proportion of lung cancer caused by occupation: a critical review. p. 55-67 in: Occupational Lung Disease (J.B.L. Gee, W.K.C. Morgan and S.M. Brooks, eds.). Raven Press, New York, 1984.

S7 Spiers, F.W., H.F. Lucas, J. Rundo et al. Leukemia incidence in the U.S. dial workers. Health Phys. 44: 65-72 (1983).

S8 Shore, R.E. Issues and epidemiological evidence regarding radiation-induced thyroid cancer. Radiat. Res. 131: 98-111 (1992).

S9 Stsjazhko, V.A., A.F. Tsyb, N.D. Tronko et al. Childhood thyroid cancer since accident at Chernobyl. Br. Med. J. 310: 801 (1995).

S10 Sackett, D.L. Bias in analytic research. J. Chronic Dis. 32: 51-63 (1979).

S11 Storer, J.B., T.J. Mitchell and R.J.M. Fry. Extrapolation of the relative risks of radiogenic neoplasms across mouse strains to man. Radiat. Res. 114: 331-353 (1988).

S12 Stehney, A.F. Survival times of pre-1950 US women radium dial workers. p. 149-155 in: Health Effects of Internally Deposited Radionuclides: Emphasis on Radium and Thorium (G. van Kaick et al., eds.). World Scientific, Singapore, 1995.

S13 Smida, J., K. Salassidis, L. Hieber et al. Distinct frequency of *ret* rearrangements in papillary thyroid carcinomas of children and adults from Belarus. Int. J. Cancer 80: 32-38 (1999).

S14 Spiess, H. The Ra-224 study: past, present and future. p. 157-163 in: Health Effects of Internally Deposited Radionuclides: Emphasis on Radium and Thorium (G. van Kaick et al., eds.). World Scientific, Singapore, 1995.

S15 Shore, R.E., N. Hildreth, E. Woodard et al. Breast cancer among women given X-ray therapy for acute postpartum mastitis. J. Natl. Cancer Inst. 77: 689-696 (1986).

S16 Stebbings, J.H., H.F. Lucas and A.F. Stehney. Mortality from cancers of major sites in female radium dial workers. Am. J. Ind. Med. 5: 435-459 (1984).

S17 Stevens, W., D.C. Thomas, J.L. Till et al. Leukemia in Utah and radioactive fallout from the Nevada test site. J. Am. Med. Assoc. 264: 485-591 (1990).

S18 Shore, R.E., N. Hildreth, P. Dvoretsky et al. Thyroid cancer among persons given x-ray treatment in infancy for an enlarged thymus gland. Am. J. Epidemiol. 137: 1068-1080 (1993).

S19 Samet, J.M., D.R. Pathak, M.V. Morgan et al. Lung cancer and exposure to radon progeny in a cohort of New Mexico underground uranium miners. Health Phys. 61: 745-752 (1991).

S20 Storm, H.H., M. Andersson, J.D. Boice Jr. et al. Adjuvant radiotherapy and risk of contralateral breast cancer. J. Natl. Cancer Inst. 84: 1245-1250 (1992).

S21 Schneider, A.B., E. Ron, J. Lubin et al. Dose-response relationships for radiation-induced thyroid cancer and thyroid nodules: evidence for the prolonged effects of radiation on the thyroid. J. Clin. Endocrinol. Metab. 77: 362-369 (1993).

S22 Shigematsu, I. and J.W. Thiessen. Childhood thyroid cancer in Belarus. Nature 359: 681 (1992).

S23 Sankila, R., S. Garwicz, J.H. Olsen et al. Risk of subsequent malignant neoplasms among 1,641 Hodgkin's disease patients diagnosed in childhood and adolescence: a population-based cohort study in the five Nordic countries. J. Clin. Oncol. 14: 1442-1446 (1996).

S24 Schillaci, M.E. Comments on 'Effects of low doses and low dose rates of external ionizing radiation: cancer mortality among nuclear industry workers in three countries' by E. Cardis et al. [Radiat. Res. 142: 117-132 (1995)]. Radiat. Res. 145: 647-648 (1996).

S25 Stocks, P. On the death rates from cancer of the stomach and respiratory diseases in 1949-53 among coal miners and other male residents in counties of England and Wales. Br. J. Cancer 16: 592-598 (1962).

S26 Scotto, J., T.R. Fears, K.H. Kraemer et al. Nonmelanoma skin cancer. p. 1313-1330 in: Cancer Epidemiology and Prevention (D. Schottenfeld and J.F. Fraumeni Jr., eds.). Oxford University Press, Oxford, 1996.

S27 Shore, R.E., R.E. Albert, M. Reed et al. Skin cancer incidence among children irradiated for ringworm of the scalp. Radiat. Res. 100: 192-204 (1984).

S28 Schneider, A., E. Shore-Freedman, U. Ryo et al. Radiation-induced tumors of the head and neck following childhood irradiation. Medicine 64: 1-15 (1985).

S29 Sevcova, M., J. Sevc and J. Thomas. Alpha irradiation of the skin and the possibility of late effects. Health Phys. 35: 803-806 (1978).

S30 Shore, R.E. Overview of radiation-induced skin cancer in humans. Int. J. Radiat. Biol. 57: 809-827 (1990).

S31 Shore, R.E., R.E. Albert and B.S. Pasternack. Follow-up study of patients treated by x-ray epilation for tinea capitis. Arch. Environ. Health 31: 17-24 (1976).

S32 Sandler, D.P., G.W. Comstock and G.M. Matanoski. Neoplasms following childhood radium irradiation of the nasopharynx. J. Natl. Cancer Inst. 68: 3-8 (1982).

S33 Shibata, S., N. Sadamori, M. Mine et al. Intracranial meningiomas among Nagasaki atomic bomb survivors. Lancet 344: 1770 (1994).

S34 Stebbings, J.H. and W. Semkiw. Central nervous system tumours and related intracranial pathologies in radium dial workers. p. 63-67 in: Risks from Radium and Thorotrast (D.M. Taylor et al., eds.). BIR Report 21(1989).

S35 Sun, Q., S. Akiba, J. Zou et al. Databases and statistical methods of cohort studies (1979-90) in Yangjiang. p. 241-248 in: High Levels of Natural Radiation 96: Radiation Dose and Health Effects (L. Wei et al., eds.). Elsevier, Amsterdam, 1997.

S36 Saenger, E.L, G.E. Thoma and E.A. Tompkins. Incidence of leukemia following treatment of hyperthyroidism: Preliminary report of the Cooperative Therapy Follow-up Study. J. Am. Med. Assoc. 205: 147-154 (1968).

S37 Stevens, W., D.C. Thomas, J.L. Lyon et al. Leukemia in Utah and radioactive fallout from the Nevada test site. J. Am. Med. Assoc. 264: 585-591 (1990).

S38 Sanders, C.L., K.G. McDonald and J.A. Mahaffey. Lung tumor response to inhaled plutonium and its implications for radiation protection. Health Phys. 65: 455-462 (1988).

S39 Sadamori, N., S. Shibata, M. Mine et al. Incidence of intracranial meningiomas in Nagasaki atomic bomb survivors. Int. J. Cancer 67: 318-322 (1996).

S40 Schwartzbaum, J.A., R.W. Setzer and L.L. Kupper. Exposure to ionizing radiation and risk of cutaneous malignant melanoma. Search for error and bias. Ann. Epidemiol. 4: 487-496 (1994).

S41 Smith, P.G. and R. Doll. Mortality from cancer and all causes among British radiologists. Br. J. Radiol. 54: 187-194 (1981).

S42 Shintani, T., N. Hayakawa and N. Kamada. High incidence of meningioma in survivors of Hiroshima. Lancet 349: 1369 (1997).

S43 Sack, R.B., K. Gyr and R. Leon-Barua. Second international workshop on Helicobater pylori infections in the developing world. Introduction. Clin. Infect. Dis. 25: 971-972 (1997).

S44 Serraino, D., G. Salamina, S. Franceschi et al. The epidemiology of AIDS-associated non-Hodgkin's lymphoma in the World Health Organization European region. Br. J. Cancer 66: 912-916 (1992).

S45 Stuver, S.O. and D. Trichopoulos. Liver cancer. in: Cancer Surveys. Trends in Cancer Incidence and Mortality, Volumes 19/20 (R. Doll et al., eds.). Cold Spring Harbor Laboratory Press, Imperial Cancer Research Fund, 1994.

S46 Sznajder, L., C. Abrahams, D.M. Parry et al. Multiple schwannomas and meningiomas associated with irradiation in childhood. Arch. Intern. Med. 156: 1873-1878 (1996).

S47 Sandler, D.P. Nasopharyngeal radium irradiation: the Washington County, Maryland study. Otolaryngol. Head Neck Surg. 115: 409-414 (1996).

S48 Silverman, D.T., A.S. Morrison and S.S. Devesa. Bladder cancer. p. 1156-1179 in: Cancer Epidemiology and Prevention (D. Schottenfeld and J.F. Fraumeni Jr., eds.). Oxford University Press, Oxford, 1996.

S49 Sobolev, B., I. Likhtarev, I. Kairo et al. Radiation risk assessment of the thyroid cancer in Ukrainian children exposed due to Chernobyl. p. 741-748 in: The Radiological Consequences of the Chernobyl Accident (A. Karaoglou et al., eds.). EUR 16544 (1996).

S50 Schoenberg, J.B., J.B. Klotz, H.B. Wilcox et al. Case-control study of residential radon and lung cancer among New Jersey women. Cancer Res. 50: 6520-6524 (1990).

S51 Shimizu, Y., H. Kato, W.J. Schull et al. Studies of the mortality of A-bomb survivors. 9. Mortality, 1950-1985: Part 1. Comparison of risk coefficients for site-specific cancer mortality based on the DS86 and T65DR shielded kerma and organ doses. Radiat. Res. 118: 502-524 (1989).

S52 Steinbuch, M., C.R. Weinberg, J.D. Buckley et al. Indoor residential radon exposure and risk of childhood acute myeloid leukaemia. Br. J. Cancer 81: 900-906 (1999).

S53 Steiner, M., W. Burkart, B. Grosche et al. Trends in infant leukaemia in West Germany in relation to in utero exposure due to the Chernobyl accident. Radiat. Environ. Biophys. 37: 87-93 (1998).

S54 Stebbings, J.H. Radium and leukemia: is current dogma valid? Health Phys. 74: 486-488 (1998).

S55 Shvedov, V.L. and A.V. Akleyev. Experimental and clinical basis of the risk of 90Sr-induced osteosarcoma for residents of the Southern Urals region. Issues Radiation Safety 8: 31-44 (1998).

T1 Thompson, D.E., K. Mabuchi, E. Ron et al. Cancer incidence in atomic bomb survivors. Part II. Solid tumors, 1958-1987. Radiat. Res. 137: S17-S67 (1994).

T2 Tokarskaya, Z.B., N.D. Okladnikova, Z.D. Belyaeva et al. The influence of radiation and nonradiation factors on the lung cancer incidence among the workers of the nuclear enterprise Mayak. Health Phys. 69: 356-366 (1995).

T3 Tomášek, L., S.C. Darby, T. Fearn et al. Patterns of lung cancer mortality among uranium miners in West Bohemia with varying rates of exposure to radon gas. Radiat. Res. 137: 251-261 (1994).

T4 Taylor, A.M.R., J.A. Metcalfe, J. Thick et al. Leukaemia and lymphoma in ataxia telangiectasia. Blood 87: 423-438 (1996).

T5 Tucker, M.A., P.H. Morris Jones, J.D. Boice Jr. et al. Therapeutic radiation at a young age is linked to secondary thyroid cancer. Cancer Res. 51: 2885-2888 (1991).

T6 Travis, L.B., R.E. Curtis, M. Stovall et al. Risk of leukemia following treatment for non-Hodgkin's lymphoma. J. Natl. Cancer Inst. 86: 1450-1457 (1994).

T7 Tucker, M.A., A.T. Meadows, J.D. Boice Jr. et al. Leukemia after therapy with alkylating agents for childhood cancer. J. Natl. Cancer Inst. 78: 459-464 (1987).

T8 Tirmarche, M., A. Raphalan, F. Allin et al. Mortality of a cohort of French uranium miners exposed to relatively low radon concentrations. Br. J. Cancer 67: 1090-1097 (1993).

T9 Travis, L.B., R.E. Curtis and J.D. Boice Jr. Late effects of treatment for childhood Hodgkin's disease. N. Engl. J. Med. 335: 352-353 (1996).

T10 Tsyb, A.F., E.M. Parshkov, V.V. Shakhtarin et al. Thyroid cancer in children and adolescents of Bryansk and Kaluga regions. p. 691-698 in: The Radiological Consequences of the Chernobyl Accident (A. Karaoglou et al., eds.). EUR 16544 EN (1996).

T11 Tsang, R.W., N.J. Laperriere, W.J. Simpson et al. Glioma arising after radiation therapy for pituitary adenoma. A report of four patients and estimation of risk. Cancer 72: 2227-2233 (1993).

T12 Tao, Z.-F., H. Kato, Y.-R. Zha et al. Study on cancer mortality among the residents in high background radiation area of Yangjiang, China. p.249-254 in: High Levels of Natural Radiation 96: Radiation Dose and Health Effects (L. Wei et al., eds.). Elsevier, Amsterdam, 1997.

T13 Tekkel, M., M. Rahu, T. Veidebaum et al. The Estonian study of Chernobyl cleanup workers: I. Design and questionnaire data. Radiat. Res. 147: 641-652 (1997).

T14 Tokarskaya, Z.B., N.D. Okladnikova, Z.D. Belyaeva et al. Multifactorial analysis of lung cancer dose-response relationships for workers at the Mayak nuclear enterprise. Health Phys. 73: 899-905 (1997).

T15 Travis, L.B., J. Weeks, R.E. Curtis et al. Leukemia following low-dose total body irradiation and chemotherapy for non-Hodgkin's lymphoma. J. Clin. Oncol. 14: 565-571 (1996).

T16 Tomášek, L., S.C. Darby, A.J. Swerdlow et al. Radon exposure and cancers other than lung cancer among uranium miners in West Bohemia. Lancet 341: 919-923 (1993).

T17 Tucker, M.A., G.J. D'Angio, J.D. Boice Jr. et al. Bone sarcomas linked to radiotherapy and chemotherapy in children. N. Engl. J. Med. 317: 588-593 (1987).

T18 Thomas, D.C., S. Darby, F. Fagnani et al. Definition and estimation of lifetime detriment from radiation exposures; principles and methods. Health Phys. 63: 259-272 (1992).

T19 Travis, L.B., R.E. Curtis, B. Glimelius et al. Bladder and kidney cancer following cyclophosphamide therapy for non-Hodgkin's lymphoma. J. Natl. Cancer Inst. 87: 524-530 (1995).

T20 Travis, L.B., R.E. Curtis, J.D. Boice Jr. et al. Second malignant neoplasms among long-term survivors of ovarian cancer. Cancer Res. 56: 1564-1570 (1996).

T21 Travis, L.B., R.E. Curtis, H. Storm et al. Risk of second malignant neoplasms among long-term survivors of testicular cancer. J. Natl. Cancer Inst. 89: 1429-1439 (1997).

T22 Tomášek, L. and V. Plaček. Radon exposure and lung cancer risk: Czech cohort study. Radiat. Res. 152: S59-S63 (1999).

T23 Tronko, M.D., T. Bogdanova, I.V. Komissarenko et al. Thyroid carcinoma in children and adolescents in Ukraine after the Chernobyl nuclear accident. Cancer 86: 149-156 (1999).

T24 Thomas, G.A., H. Burnell, E.D. Williams et al. Association between morphological subtype of post Chernobyl papillary carcinoma and rearrangement of the ret oncogene. p. 255-261 in: Radiation and Thyroid Cancer. Proceedings of an International Seminar held in St. John's College, Cambridge, UK, 20-23 July 1998 (G. Thomas, A. Karaoglou and E.D. Williams, eds.). World Scientific, Singapore, 1999.

T25 Tao, Z., S. Akiba, Y. Zha et al. Analysis of data (1987-1995) from investigation of cancer mortality in high background radiation area of Yangjiang, China. Chin. J. Radiol. Med. Prot. 19: 75-82 (1999).

T26 Tao, Z., Y. Zha, Q. Sun et al. Cancer mortality in high background radiation area of Yangjiang, China, 1979-1995. Natl. Med. J. China 79: 487-492 (1999).

U2 United Nations. Sources and Effects of Ionizing Radiation. United Nations Scientific Committee on the Effects of Atomic Radiation, 1994 Report to the General Assembly, with scientific annexes. United Nations sales publication E.94.IX.11. United Nations, New York, 1994.

U3 United Nations. Sources and Effects of Ionizing Radiation. United Nations Scientific Committee on the Effects of Atomic Radiation, 1993 Report to the General Assembly, with scientific annexes. United Nations sales publication E.94.IX.2. United Nations, New York, 1993.

U4 United Nations. Sources, Effects and Risks of Ionizing Radiation. United Nations Scientific Committee on the Effects of Atomic Radiation, 1988 Report to the General Assembly, with annexes. United Nations sales publication E.88.IX.7. United Nations, New York, 1988.

U5 United Nations. Genetic and Somatic Effects of Ionizing Radiation. United Nations Scientific Committee on the Effects of Atomic Radiation, 1986 Report to the General Assembly, with annexes. United Nations sales publication E.86.IX.9. United Nations, New York, 1986.

U7 United Nations. Sources and Effects of Ionizing Radiation. United Nations Scientific Committee on the Effects of Atomic Radiation, 1977 Report to the General Assembly, with annexes. United Nations sales publication E.77.IX.1. United Nations, New York, 1977.

U14 Ursin, G., L. Bernstein and M.C. Pike. Breast cancer. p. 241-264 in: Cancer Surveys. Trends in Cancer Incidence and Mortality, Volumes 19/20 (R. Doll et al., eds.). Cold Spring Harbor Laboratory Press, Imperial Cancer Research Fund, 1994.

U15 Utsunomiya, J. and H.T. Lynch. Hereditary Colorectal Cancer. Springer-Verlag, New York, 1990.

U16 Ullrich, R.L. and J.B. Storer. Influence of gamma-irradiation on the development of neoplastic disease in mice. II. Solid tumors. Radiat. Res. 80: 317-324 (1979).

U17 United States Department of Health, Education and Welfare. Smoking and health: a report of the Advisory Committee to the Surgeon General. United States Government Printing Office, Washington D.C. (1964).

V1 van Kaick, G., H. Welsch, H. Luehrs et al. Epidemiological results and dosimetric calculations - an update of the German Thorotrast study. p. 171-175 in: Health Effects of Internally Deposited Radionuclides: Emphasis on Radium and Thorium (G. van Kaick et al., eds.). World Scientific, Singapore, 1995.

V2 van Leeuwen, F.E., W.J. Klokman, M. Stovall et al. Roles of radiotherapy and smoking in lung cancer following Hodgkin's disease. J. Natl. Cancer Inst. 87: 1530-1537 (1995).

V3 van Kaick, G., H. Welsh, H. Luehrs et al. The German Thorotrast study - report on 20 years follow-up. p. 98-104 in: Risks from Radium and Thorotrast (D.M. Taylor et al., eds.). BIR Report 21 (1989).

V4 van Vloten, W., J. Hermans and W. van Daal. Radiation-induced skin cancer and radiodermatitis of the head and neck. Cancer 59: 411-414 (1986).

V5 Verduijn, P.G., R.B. Hayes, C. Looman et al. Mortality after nasopharyngeal radium irradiation for eustachian tube dysfunction. Ann. Otol. Rhinol. Laryngol. 98: 839-844 (1989).

V6 van Kaick, G., H. Muth, A. Kaul et al. Recent results of the German Thorotrast study. p. 253-262 in: Radiation Carcinogenesis: Epidemiology and Biological Significance (J.D. Boice and J.F. Fraumeni eds.). Raven Press, New York, 1984.

V7 van Kaick, G., H. Wesch, H. Lührs et al. Neoplastic diseases induced by chronic alpha-irradiation: Epidemiological, biophysical and clinical results of the German Thorotrast Study. J. Radiat. Res. 32 (Suppl. 2): 20-33 (1991).

V8 van Kaick, G., A. Dalheimer, S. Hornik et al. The German Thorotrast study: recent results and assessment of risks. Radiat. Res. 152: S64-S71 (1999).

W1 Weiss, H.A., S.C. Darby and R. Doll. Cancer mortality following x-ray treatment for ankylosing spondylitis. Int. J. Cancer 59: 327-338 (1994).

W2 Weiss, H.A., S.C. Darby, T. Fearn et al. Leukemia mortality after x-ray treatment for ankylosing spondylitis. Radiat. Res. 142: 1-11 (1995).

W3 Wick, R.R., D. Chmelevsky and W. Gössner. Current status of the follow-up of radium-224 treated ankylosing spondylitis patients. p. 165-169 in: Health Effects of Internally Deposited Radionuclides: Emphasis on Radium and Thorium (G. van Kaick et al., eds.). World Scientific, Singapore, 1995.

W4 Williams, G.H., S. Rooney, G.A. Thomas et al. RET activation in adult and childhood papillary thyroid carcinoma using a reverse transcriptase-n-polymerase chain reaction approach on archival-nested material. Br. J. Cancer 74: 585-589 (1996).

W5 Williams, D., A. Pinchera, A. Karaoglou et al. Thyroid cancer in children living near Chernobyl. Expert panel report on the consequences of the Chernobyl accident. EUR 15248 (1993).

W6 Whittemore, A.S. Prostate cancer. p. 309-322 in: Cancer Surveys. Trends in Cancer Incidence and Mortality, Volumes 19/20 (R. Doll et al., eds.). Cold Spring Harbor Laboratory Press, Imperial Cancer Research Fund, 1994.

W7 Woodward, A., D. Roder, A.J. McMichael et al. Radon exposures at the Radium Hill uranium mine and lung cancer rates among former workers, 1952-1987. Cancer Causes and Control 2: 213-220 (1991).

W8 Wiggs, L.D., E.R. Johnson, C.A. Cox-DeVore et al. Mortality through 1990 among white male workers at the Los Alamos National Laboratory: considering exposure to plutonium and external ionizing radiation. Health Phys. 67: 577-588 (1994).

W9 Wang, Z., J.D. Boice Jr., L. Wei et al. Thyroid nodularity and chromosome aberrations among women in areas in China. J. Natl. Cancer Inst. 82: 478-485 (1990).

W10 Wang, J.-X., P.D. Inskip, J.D. Boice Jr. et al. Cancer incidence among medical diagnostic x-ray workers in China, 1950 to 1985. Int. J. Cancer 45: 889-895 (1990).

W11 Wong, F.L., J.D. Boice Jr., D.H. Abramson et al. Cancer incidence after retinoblastoma. J. Am. Med. Assoc. 278: 1262-1267 (1997).

W12 Wilkinson, G.S., G.L. Tietjen, L.D. Wiggs et al. Mortality among plutonium and other radiation-exposed workers at a plutonium weapons facility. Am. J. Epidemiol. 125: 231-250 (1987).

W13 Wilkinson, G.S. Invited commentary: Are low radiation doses or occupational exposures really risk factors for malignant melanoma? Am. J. Epidemiol. 145: 532-535 (1997).

W14 Wing, S., D. Richardson, D. Armstrong et al. A re-evaluation of cancer incidence near the Three Mile Island nuclear plant: the collison of evidence and assumptions. Environ. Health Perspect. 105: 52-57 (1997).

W15 Wiklund, K., J. Dich and L.-E. Holm. Risk of malignant lymphoma in Swedish pesticide appliers. Br. J. Cancer 56: 505-508 (1987).

W16 Williams, E.D., D. Becker, E.P. Dimidchik et al. Effects on the thyroid in populations exposed to radiation as a result of the Chernobyl accident. p. 207-238 in: One Decade after Chernobyl. Summing up the Consequences of the Accident. Proceedings of an International Conference, Vienna, 1996. STI/PUB/1001. IAEA, Vienna, 1996.

W17 Wichmann, H.E., L. Kreienbrock, M. Kreuzer et al. Lungenkrebsrisiko durch Radon in der Bundesrepublik Deutschland (West). Ecomed, Landsberg, 1998.

W18 Wichmann, H.E., M. Gerken, J. Wellman et al. Lungenkrebsrisiko durch Radon in der Bundesrepublik Deutschland (Ost) – Thüringen und Sachsen. Ecomed, Landsberg, 1999.

W19 Weinberg, C.R. Potential for bias in epidemiologic studies that rely on glass-based retrospective assessment of radon. Environ. Health Perspect. 103: 1042-1046 (1995).

W20 Wick, R.R., E.A. Nekolla, W. Gössner et al. Late effects in anklylosing spondylitis patients treated with ^{224}Ra. Radiat. Res. 152: S8-S11 (1999).

W21 Wiggs, L.D., C.A. Cox-DeVore, G.S. Wilkinson et al. Mortality among workers exposed to external ionizing radiation at a nuclear facility in Ohio. J. Occup. Med. 33: 632-637 (1991).

W22 Wiggs, L.D., C.A. Cox-DeVore and G.L. Voelz. Mortality among a cohort of workers monitored for ^{210}Po exposure. Health Phys. 61: 71-76 (1991).

W23 Wing, S., C.M. Shy, J.L. Wood et al. Mortality among workers at Oak Ridge National Laboratory: evidence of radiation effects in follow-up through 1984. J. Am. Med. Assoc. 265: 1397-1402 (1991).

X1 Xuan, X.Z., J.H. Lubin, J.Y. Li et al. A cohort study in southern China of tin miners exposed to radon and radon decay products. Health Phys. 64: 120-131 (1993).

Y1 Yoshimoto, Y., H. Kato and W.J. Schull. Risk of cancer among children exposed in utero to A-bomb radiation, 1950-84. Lancet 2: 665-669 (1988).

Y2 Yoshimoto, Y., J.V. Neel, W.J. Schull et al. Malignant tumors during the first two decades of life in the offspring of atomic bomb survivors. Am. J. Hum. Genet. 46: 1041-1052 (1990).

Y3 Yamada, M., K. Kodama, S. Fujita et al. Prevalence of skin neoplasms among the atomic bomb survivors. Radiat. Res. 146: 223-226 (1996).

Z1 Zhuntova, G.V., Z.B. Tokarskaya, N.D. Okladnikova et al. The importance of radiation and non-radiation-factors for the stomach cancer incidence in workers of the atomic plant Mayak. p. 324-327 in: IRPA9, 1996 International Congress on Radiation Protection. Proceedings, Volume 2. IRPA, Vienna, 1996.

Z2 Zha, Y.-R., J.-M. Zou, Z.-X. Lin et al. Confounding factors in radiation epidemiology and their comparability between the high background radiation areas and control areas in Guangdong, China. p. 263-269 in: High Levels of Natural Radiation 96: Radiation Dose and Health Effects (L. Wei et al., eds.). Elsevier, Amsterdam, 1997.

Z3 Zahm, S.H., M.A. Tucker and J.F. Fraumeni Jr. Soft tissue sarcomas. p. 984-999 in: Cancer Epidemiology and Prevention (D. Schottenfeld and J.F. Fraumeni Jr., eds.). Oxford University Press, Oxford, 1996.

ANNEX J

Exposures and effects of the Chernobyl accident

CONTENTS

INTRODUCTION

1. The accident of 26 April 1986 at the Chernobyl nuclear power plant, located in Ukraine about 20 km south of the border with Belarus, was the most severe ever to have occurred in the nuclear industry. The Committee considered the initial radiological consequences of that accident in the UNSCEAR 1988 Report [U4]. The short-term effects and treatment of radiation injuries of workers and firefighters who were present at the site at the time of the accident were reviewed in the Appendix to Annex G, "Early effects in man of high doses of radiation", and the average individual and collective doses to the population of the northern hemisphere were evaluated in Annex D, "Exposures from the Chernobyl accident", of the 1988 Report [U4]. The objective of this Annex is (a) to review in greater detail the exposures of those most closely involved in the accident and the residents of the local areas most affected by the residual contamination and (b) to consider the health consequences that are or could be associated with these radiation exposures.

2. The impact of the accident on the workers and local residents has indeed been both serious and enormous. The accident caused the deaths within a few days or weeks of 30 power plant employees and firemen (including 28 deaths that were due to radiation exposure), brought about the evacuation of about 116,000 people from areas surrounding the reactor during 1986, and the relocation, after 1986, of about 220,000 people from what were at that time three constituent republics of the Soviet Union: Belorussia, the Russian Soviet Federated Socialist Republic (RSFSR) and the Ukraine [K23, R11, V2, V3] (these republics will hereinafter be called by their present-day country names: Belarus, the Russian Federation and Ukraine). Vast territories of those three republics were contaminated, and trace deposition of released radionuclides was measurable in all countries of the northern hemisphere. Stratospheric interhemispheric transfer may also have led to some environmental contamination in the southern hemisphere [D11]. In addition, about 240,000 workers ("liquidators") were called upon in 1986 and 1987 to take part in major mitigation activities at the reactor and within the 30-km zone surrounding the reactor; residual mitigation activities continued until 1990. All together, about 600,000 persons received the special status of "liquidator".

3. The radiation exposures resulting from the Chernobyl accident were due initially to ^{131}I and short-lived radio-nuclides and subsequently to radiocaesiums (^{134}Cs and ^{137}Cs) from both external exposure and the consumption of foods contaminated with these radionuclides. It was estimated in the UNSCEAR 1988 Report [U4] that, outside the regions of Belarus, the Russian Federation and Ukraine that were most affected by the accident, thyroid doses averaged over large portions of European countries were at most 25 mGy for one-year old infants. It was recognized, however, that the dose distribution was very heterogeneous,

especially in countries close to the reactor site. For example, in Poland, although the countrywide population-weighted average thyroid dose was estimated to be approximately 8 mGy, the mean thyroid doses for the populations of particular districts were in the range from 0.2 to 64 mGy, and individual values for about 5% of the children were about 200 mGy [K32, K33]. It was also estimated in the UNSCEAR 1988 Report [U4] that effective doses averaged over large portions of European countries were 1 mSv or less in the first year after the accident and approximately two to five times the first-year dose over a lifetime.

4. The doses to population groups in Belarus, the Russian Federation and Ukraine living nearest the accident site and to the workers involved in mitigating the accident are, however, of particular interest, because these people have had the highest exposures and have been monitored for health effects that might be related to the radiation exposures. Research on possible health effects is focussed on, but not limited to, the investigation of leukaemia among workers involved in the accident and of thyroid cancer among children. Other health effects that are considered are non-cancer somatic disorders (e.g. thyroid abnormalities and immunological effects), reproductive effects and psychological effects. Epidemiological studies have been undertaken among the populations of Belarus, the Russian Federation and Ukraine that were most affected by the accident to investigate whether dose-effect relationships can be obtained, notably with respect to the induction of thyroid cancer resulting from internal irradiation by ^{131}I and other radioiodines in young children and to the induction of leukaemia among workers resulting from external irradiation at low dose rates. The dose estimates that are currently available are of a preliminary nature and must be refined by means of difficult and time-consuming dose reconstruction efforts. The accumulation of health statistics will also require some years of effort.

5. Because of the questions that have arisen about the local exposures and effects of the Chernobyl accident, the Committee feels that a review of information at this stage, almost 15 years after the accident, is warranted. Of course, even longer-term studies will be needed to determine the full consequences of the accident. It is the intention to evaluate in this Annex the data thus far collected on the local doses and effects in relation to and as a contribution to the broader knowledge of radiation effects in humans. Within the last few years, several international conferences were held to review the aftermath of the accident, and extensive use can be made of the proceedings of these conferences [E3, I15, T22, W7]; also, use was made of books, e.g. [I5, K19, M14, M15], and of special issues of scientific journals, e.g. [K42], devoted to the Chernobyl accident.

6. The populations considered in this Annex are (a) the workers involved in the mitigation of the accident, either during the accident itself (including firemen and power plant personnel who received doses leading to deterministic effects) or after the accident (recovery operation workers); (b) members of the general public who were evacuated to avert excessive radiation exposures; and (c) inhabitants of contaminated areas who were not evacuated.

The contaminated areas, which are defined in this Annex as being those where the average ^{137}Cs ground deposition density exceeded 37 kBq m^{-2} (1 Ci km^{-2}), are found mainly in Belarus, in the Russian Federation and in Ukraine. Information on the contamination levels and radiation doses in other countries will be presented only if it is related to epidemiological studies conducted in those countries.

I. PHYSICAL CONSEQUENCES OF THE ACCIDENT

7. The accident at the Chernobyl nuclear power station occurred during a low-power engineering test of the Unit 4 reactor. Safety systems had been switched off, and improper, unstable operation of the reactor allowed an uncontrollable power surge to occur, resulting in successive steam explosions that severely damaged the reactor building and completely destroyed the reactor. An account of the accident and of the quantities of radionuclides released, to the extent that they could be known at the time, were presented by Soviet experts at the Post-Accident Review Meeting at Vienna in August 1986 [I2]. The information that has become available since 1986 will be summarized in this Chapter.

8. The radionuclide releases from the damaged reactor occurred mainly over a 10-day period, but with varying release rates. An initial high release rate on the first day was caused by mechanical discharge as a result of the explosions in the reactor. There followed a five-day period of declining releases associated with the hot air and fumes from the burning graphite core material. In the next few days, the release rate of radionuclides increased until day 10, when the releases dropped abruptly, thus ending the period of intense release. The radionuclides released in the accident deposited with greatest density in the regions surrounding the reactor in the European part of the former Soviet Union.

A. THE ACCIDENT

9. The Chernobyl reactor is of the type RBMK, which is an abbreviation of Russian terms meaning reactor of high output, multichannel type. It is a pressurized water reactor using light water as a coolant and graphite as a moderator. Detailed information about what is currently known about the accident and the accident sequence has been reported, notably in 1992 by the International Atomic Energy Agency (IAEA) [I7], in 1994 in a report of the Massachusetts Institute of Technology [S1], in 1995 by the Ukrainian Academy of Sciences [P4], and in 1991–1996 by the Kurchatov Institute [B24, C5, K20, K21, S22, V4]. A simplified description of the events leading to the accident and of the measures taken to control its consequences is provided in the following paragraphs. As is the case in an accident with unexpected and unknown events and outcomes, many questions remain to be satisfactorily resolved.

10. The events leading to the accident at the Chernobyl Unit 4 reactor at about 1.24 a.m. on 26 April 1986 resulted from efforts to conduct a test on an electric control system, which allows power to be provided in the event of a station blackout [I2]. Actions taken during this exercise resulted in a significant variation in the temperature and flow rate of the inlet water to the reactor core (beginning at about 1.03 a.m.). The unstable state of the reactor before the accident is due both to basic engineering deficiencies (large positive coefficient of reactivity under certain conditions) and to faulty actions of the operators (e.g., switching off the emergency safety systems of the reactor) [G26]. The relatively fast temperature changes resulting from the operators' actions weakened the lower transition joints that link the zirconium fuel channels in the core to the steel pipes that carry the inlet cooling water [P4]. Other actions resulted in a rapid increase in the power level of the reactor [I7], which caused fuel fragmentation and the rapid transfer of heat from these fuel fragments to the coolant (between 1.23:43 and 1.23:49 a.m.). This generated a shock wave in the cooling water, which led to the failure of most of the lower transition joints. As a result of the failure of these transition joints, the pressurized cooling water in the primary system was released, and it immediately flashed into steam.

11. The steam explosion occurred at 1.23:49. It is surmised that the reactor core might have been lifted up by the explosion [P4], during which time all water left the reactor core. This resulted in an extremely rapid increase in reactivity, which led to vaporization of part of the fuel at the centre of some fuel assemblies and which was terminated by a large explosion attributable to rapid expansion of the fuel vapour disassembling the core. This explosion, which occurred at about 1.24 a.m., blew the core apart and destroyed most of the building. Fuel, core components, and structural items were blown from the reactor hall onto the roof of adjacent buildings and the ground around the reactor building. A major release of radioactive materials into the environment also occurred as a result of this explosion.

12. The core debris dispersed by the explosion started multiple (more than 30) fires on the roofs of the reactor building and the machine hall, which were covered with highly flammable tar. Some of those fires spread to the machine hall and, through cable tubes, to the vicinity of the Unit 3 reactor. A first group of 14 firemen arrived on the scene of the accident at 1.28 a.m. Reinforcements were brought in until about 4 a.m., when 250 firemen were available and 69 firemen participated in fire control activities. These activities were carried out at up to 70 m above the ground under harsh conditions of high radiation levels and dense smoke. By 2.10 a.m., the largest fires on the roof of the machine hall had been put out, while by 2.30 a.m. the largest fires on the roof of the reactor hall were under control. By about 4.50 a.m., most of the fires had been extinguished. These actions caused the deaths of five firefighters.

13. It is unclear whether fires were originating from the reactor cavity during the first 20 h after the explosion. However, there was considerable steam and water because of the actions of both the firefighters and the reactor plant personnel. Approximately 20 h after the explosion, at 9.41 p.m., a large fire started as the material in the reactor became hot enough to ignite combustible gases released from the disrupted core, e.g. hydrogen from zirconium-water reactions and carbon monoxide from the reaction of hot graphite with steam. The fire made noise when it started (some witnesses called it an explosion) and burned with a large flame that initially reached at least 50 m above the top of the destroyed reactor hall [P4].

14. The first measures taken to control the fire and the radionuclide releases consisted of dumping neutron-absorbing compounds and fire-control materials into the crater formed by the destruction of the reactor. The total amount of materials dumped on the reactor was approximately 5,000 t, including about 40 t of boron compounds, 2,400 t of lead, 1,800 t of sand and clay, and 600 t of dolomite, as well as sodium phosphate and polymer liquids [B4]. About 150 t of materials were dumped on 27 April, followed by 300 t on 28 April, 750 t on 29 April, 1,500 t on 30 April, 1,900 t on 1 May, and 400 t on 2 May. About 1,800 helicopter flights were carried out to dump materials onto the reactor. During the first flights, the helicopters remained stationary over the reactor while dumping the materials. However, as the dose rates received by the helicopter pilots during this procedure were judged to be too high, it was decided that the materials should be dumped while the helicopters travelled over the reactor. This procedure, which had a poor accuracy, caused additional destruction of the standing structures and spread the contamination. In fact, much of the material delivered by the helicopters was dumped on the roof of the reactor hall, where a glowing fire was observed, because the reactor core was partially obstructed by the upper biological shield, broken piping, and other debris, and rising smoke made it difficult to see and identify the core location (see Figure I). The material dumping campaign was stopped on day 7 (2 May) through day 10 (5 May) after the accident because of fears that the building support structures could be compromised. If that happened, it would allow the core to be less restrained from

possible meltdown, and steam explosions would occur if the core were to interact with the pressure suppression pool beneath the reactor. The increasing release rates on days 7 through 10 were associated with the rising temperature of the fuel in the core. Cooling of the reactor structure with liquid nitrogen using pipelines originating from Unit 3 was initiated only at late stages after the accident. The abrupt ending of the releases was said to occur upon extinguishing the fire and through transformation of the fission products into more chemically stable compounds [I2].

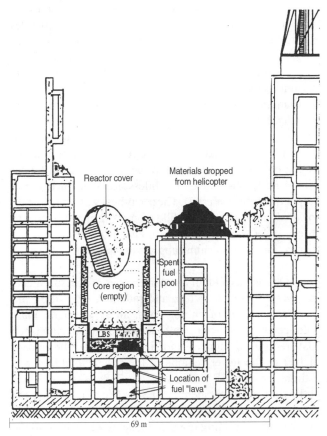

Figure I. Cross-section view of damaged Unit 4 Chernobyl reactor building.

15. The further sequence of events is still somewhat speculative, but the following description conforms with the observations of residual damage to the reactor [S1, S18]. It is suggested that the melted core materials (also called fuel-containing masses, corium, or lava) settled to the bottom of the core shaft, with the fuel forming a metallic layer below the graphite. The graphite layer had a filtering effect on the release of volatile compounds. This is evidenced by a con-centration of caesium in the corium of 35% [S1], somewhat higher than would otherwise have been expected in the highly oxidizing conditions that prevailed in the presence of burning graphite. The very high temperatures in the core shaft would have suppressed plate-out of radionuclides and maintained high release rates of penetrating gases and aerosols. After about 6.5 days, the upper graphite layer would have burned off. This is evidenced by the absence of carbon or carbon-containing compounds in the corium. At this stage, without the filtering effect of an upper graphite layer, the release of volatile fission products from the fuel may have increased,

although non-volatile fission products and actinides would have been inhibited because of reduced particulate emission.

16.   On day 8 after the accident, it would appear that the corium melted through the lower biological shield (LBS) and flowed onto the floor of the sub-reactor region (see Figure I). This rapid redistribution of the corium and increase in surface area as it spread horizontally would have enhanced the radionuclide releases. The corium produced steam on contact with the water remaining in the pressure suppression pool, causing an increase in aerosols. This may account for the peak releases of radionuclides seen at the last stage of the active period.

17.   Approximately nine days after the accident, the corium began to lose its ability to interact with the surrounding materials. It solidified relatively rapidly, causing little damage to metallic piping in the lower regions of the reactor building. The chemistry of the corium was altered by the large mass of the lower biological shield taken up into the molten corium (about 400 of the 1,200-t shield of stainless steel construction and serpentine filler material). The decay heat was significantly lowered, and the radionuclide releases dropped by two to three orders of magnitude. Visual evidence of the disposition of the corium supports this sequence of events.

18.   On the basis of an extensive series of measurements in 1987–1990 of heat flux and radiation intensities and from an analysis of photographs, an approximate mass balance of the reactor fuel distribution was established (data reported by Borovoi and Sich [B16, S1]). The amount of fuel in the lower regions of the reactor building was estimated to be 135 ± 27 t, which is 71% of the core load at the time of the accident (190.3 t). The remainder of the fuel was accounted for as follows: fuel in the upper levels of the reactor building (38 ± 5 t); fuel released beyond the reactor building (6.7 ± 1 t); and unaccounted for fuel (10.7 t), possibly largely on the roof of the reactor hall under the pile of materials dumped by the helicopters.

19.   Different estimates of the reactor fuel distribution have been proposed by others. Purvis [P4] indicated that the amount of fuel in the lava, plus fragments of the reactor core under the level of the bottom of the reactor, is between 27 and 100 t and that the total amount of the fuel in the reactor hall area is between 77 and 140 t. Kisselev et al. [K12, K15] reported that only 24 ± 4 t were identified by visual means in the lower region of the reactor. It may be that most of the fuel is on the roof of the reactor hall and is covered by the material that was dropped on it from helicopters. Only the removal of this layer of material will allow making a better determination of the reactor fuel distribution.

B.   RELEASE OF RADIONUCLIDES

20.   Two basic methods were used to estimate the release of radionuclides in the accident. The first method consists in evaluating separately the inventory of radionuclides in

the reactor core at the time of the accident and the fraction of the inventory of each radionuclide that was released into the atmosphere; the products of those two quantities are the amounts released. The second method consists in measuring the radionuclide deposition density on the ground all around the reactor; if it is assumed that all of the released amounts deposited within the area where the measurements were made, the amounts deposited are equal to the amounts released. In both methods, air samples taken over the reactor or at various distances from the reactor were analysed for radionuclide content to determine or to confirm the radionuclide distribution in the materials released. The analysis of air samples and of fallout also led to information on the physical and chemical properties of the radioactive materials that were released into the atmosphere. It is worth noting, however, that the doses were estimated on the basis of environmental and human measurements and that the knowledge of the quantities released was not needed for that purpose.

1.   Estimation of radionuclide amounts released

21.   From the radiological point of view, ^{131}I and ^{137}Cs are the most important radionuclides to consider, because they are responsible for most of the radiation exposure received by the general population.

22.   Several estimates have been made of the radionuclide core inventory at the time of the accident. Some of these estimates are based on the burn-up of individual fuel assemblies that has been made available [B1, S1]. The average burn-up of 10.9 GW d t^{-1} [B1], published in 1989, is similar to the originally reported value of 10.3 GW d t^{-1} [I2], but with non-linear accumulation of actinides, more detailed values of burn-up allow more precise estimation of the core inventories. In the case of ^{132}Te and of the short-lived radioiodines, Khrouch et al. [K16] took into account the variations in the power level of the reactor during the 24 hours before the accident, as described in [S20]. An extended list of radionuclides present in the core at the time of the accident is presented in Table 1. The values used by the Committee in this Annex are those presented in the last column on the right. For comparison purposes, the initial estimates of the core inventory as presented in 1986 [I2], which were used by the Committee in the UNSCEAR 1988 Report [U4], are also presented in Table 1; these 1986 estimates, however, have been decay-corrected to 6 May 1986, that is, 10 days after the beginning of the accident. The large differences observed between initial and recent estimates for short-lived radionuclides (radioactive half-lives of less than 10 days) are mainly due to radioactive decay between the actual day of release and 6 May, while minor differences may have been caused by the use of different computer codes to calculate the build-up of activity in the reactor core. For ^{137}Cs, the 1986 and current estimates of core inventory at the time of the accident are 290 and 260 PBq, respectively. For ^{131}I, the corresponding values are 1,300 and 3,200 PBq, respectively.

23. There are several estimates of radionuclides released in the accident based on recent evaluations. Three such listings, including two taken from the IAEA international conference that took place at Vienna in 1996 [D8], are given in Table 2 and compared to the original estimates of 1986 [I2]. The estimates of Buzulukov and Dobrynin [B4], as well as those of Kruger et al. [K37], are based on analyses of core inventories [B1, B3]. There is general agreement on the releases of most radionuclides, and in particular those of ^{137}Cs and ^{131}I, presented in the 1996 evaluations. The values used by the Committee in this Annex are those presented in the last column on the right. The release of ^{137}Cs is estimated to be 85 PBq, about 30% of the core inventory and that of ^{131}I is estimated to be 1,760 PBq, about 50% of the core inventory.

24. In the UNSCEAR 1988 Report [U4], estimates were made of the release of ^{137}Cs and ^{131}I in the accident. From average deposition densities of ^{137}Cs and the areas of land and ocean regions, the total ^{137}Cs deposit in the northern hemisphere was estimated to be 70 PBq, which is in fairly good agreement with the current estimate.

25. The release of ^{131}I was estimated in the UNSCEAR 1988 Report to be 330 PBq on the basis of the reported ^{131}I inventory of 1,300 PBq [I2] and of a release fraction of 25% [U4]. This, however, was the inventory of ^{131}I at the end of the release period (6 May 1986). It would have been higher at the beginning of the accident. The ^{131}I inventory is now estimated to be 3,200 PBq, as shown in Table 1, and because the fractional release of ^{131}I is likely to have been about 50%, the ^{131}I release given in the UNSCEAR 1988 Report is lower than the current estimate by a factor of about 5.

26. The results presented in Table 2 are incomplete with respect to the releases of ^{132}Te and of the short-lived radioiodines (^{132}I to ^{135}I). In this Annex, the releases of those radionuclides have been scaled to the releases of ^{131}I, using the radionuclide inventories presented in Table 1 and taking into account the radioactive half-lives of the radionuclides. The following procedure was used: (a) the release rates at the time of the steam explosion were estimated from the radionuclide inventories presented in Table 1, assuming no fractionation for the short-lived radioiodines (^{133}I, ^{134}I and ^{135}I) with respect to ^{131}I, a value of 0.85 for the ratio of the release rates of ^{132}Te and of ^{131}I, and radioactive equilibrium between ^{132}I and ^{132}Te in the materials released. The activity ratios to ^{131}I in the initial release rates are therefore estimated to have been 1.5 for ^{133}I, 0.64 for ^{134}I, 0.9 for ^{135}I, and 0.85 for ^{132}Te and ^{132}I; (b) the variation with time of the release rate of ^{131}I over the first 10 days following the steam explosion was assessed using published data [A4, I6]. The estimated daily releases of ^{131}I are presented in Table 3; and (c) the variation with time of the release rates of the short-lived radioiodines and of ^{132}Te has been assumed to be the same as that of ^{131}I, but a correction was made to take into account the differences in radioactive half-lives. The variation with time of the daily releases of ^{131}I, ^{133}I and ^{132}Te, which are adopted in

this Annex, are illustrated in Figure II; for comparison purposes, the estimated daily releases of ^{137}Cs are also shown in Figure II.

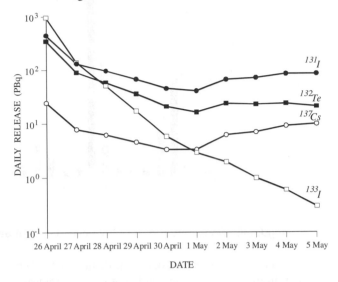

Figure II. Daily release of iodine-131, iodine-133, tellurium-132 and caesium-137 from the Chernobyl reactor.

27. The overall releases of short-lived radioiodines and of ^{132}Te are presented in Table 4; they are found to be substantially lower than those of ^{131}I. This is due to the fact that most of the short-lived radioiodines decayed in the reactor instead of being released.

28. Additional, qualitative information on the pattern of release of radionuclides from the reactor is given in Figure III. The concentrations of radionuclides in air were determined in air samples collected by helicopter above the damaged reactor [B4]. Although the releases were considerably reduced on 5 and 6 May (days 9 and 10 after the accident), continuing low-level releases occurred in the following week and for up to 40 days after the accident. Particularly on 15 and 16 May, higher concentrations were observed, attributable to continuing outbreaks of fires or to hot areas of the reactor [I6]. These later releases can be correlated with increased concentrations of radionuclides in air measured at Kiev and Vilnius [I6, I35, U16].

2. Physical and chemical properties of the radioactive materials released

29. There were only a few measurements of the aero-dynamic size of the radioactive particles released during the first days of the accident. A crude analysis of air samples, taken at 400–600 m above the ground in the vicinity of the Chernobyl power plant on 27 April 1986, indicated that large radioactive particles, varying in size from several to tens of micrometers, were found, together with an abundance of smaller particles [I6]. In a carefully designed experiment, aerosol samples taken on 14 and 16 May 1986 with a device installed on an aircraft that flew above the damaged reactor were analysed by spectrometry [B6, G14]. The activity distribution of the particle sizes was found to be well represented as the superposition of two log-normal functions:

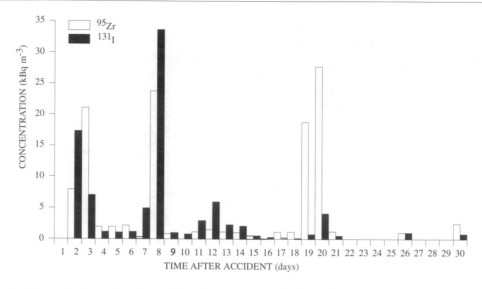

Figure III. Concentration of radionuclides in air measured above the damaged Chernobyl reactor [B6].

one with an activity median aerodynamic diameter (AMAD) ranging from 0.3 to 1.5 μm and a geometric standard deviation (GSD) of 1.6–1.8, and another with an AMAD of more than 10 μm. The larger particles contained about 80%–90% of the activity of non-volatile radionuclides such as ^{95}Zr, ^{95}Nb, ^{140}La, ^{141}Ce, ^{144}Ce and transuranium radionuclides embedded in the uranium matrix of the fuel [K35]. The geometric sizes of the fuel particles collected in Hungary, Finland and Bulgaria ranged from 0.5 to 10 μm, with an average of 5 μm [B35, L40, V9]. Taking the density of fuel particles to be 9 g cm^{-3}, their aerodynamic diameter therefore ranged from 1.5 to 30 μm, with an average value of 15 μm. Similar average values were obtained for fuel particles collected in May 1986 in southern Germany [R20] and for those collected in the 30-km zone in September 1986 [G27].

30. It was observed that Chernobyl fallout consisted of hot particles in addition to more homogeneously distributed radioactive material [D6, D7, K34, S26, S27, S28]. These hot particles can be classified into two broad categories: (a) fuel fragments with a mixture of fission products bound to a matrix of uranium oxide, similar to the composition of the fuel in the core, but sometimes strongly depleted in caesium, iodine and ruthenium, and (b) particles consisting of one dominant element (ruthenium or barium) but sometimes having traces of other elements [D6, J3, J4, K35, K36, S27]. These monoelemental particles may have originated from embedments of these elements produced in the fuel during reactor operation and released during the fragmentation of the fuel [D7]. Typical activities per hot particle are 0.1–1 kBq for fuel fragments and 0.5–10 kBq for ruthenium particles [D6]; a typical effective diameter is about 10 μm, to be compared with sizes of 0.4–0.7 μm for the particles associated with the activities of ^{131}I and ^{137}Cs [D6, D7]. Hot particles deposited in the pulmonary region will have a long retention time, leading to considerable local doses [B33, L23]. In the immediate vicinity of a 1 kBq ruthenium particle, the dose rate is about 1,000 Gy h^{-1}, which causes cell killing; however, sublethal doses are received by cells within a few millimetres of the hot particle. Although it was demonstrated in the 1970s that radiation doses from alpha-emitting hot particles are not more

radiotoxic than the same activity uniformly distributed in the whole lung [B28, L33, L34, L35, R15], it is not clear whether the same conclusion can be reached for beta-emitting hot particles [B33, S27].

C. GROUND CONTAMINATION

1. Areas of the former Soviet Union

31. Radioactive contamination of the ground was found to some extent in practically every country of the northern hemisphere [U4]. In this Annex, contaminated areas are defined as areas where the average ^{137}Cs deposition densities exceeded 37 kBq m^{-2} (1 Ci km^{-2}). Caesium-137 was chosen as a reference radionuclide for the ground contamination resulting from the Chernobyl accident for several reasons: its substantial contribution to the lifetime effective dose, its long radioactive half-life, and its ease of measurement. As shown in Table 5, the contaminated areas were found mainly in Belarus, in the Russian Federation and in Ukraine [I24].

32. The radionuclides released in the accident deposited over most of the European territory of the former Soviet Union. A map of this territory is presented in Figure IV. The main city gives its name to each region. The regions (*oblasts*) are subdivided into districts (*raions*).

33. The characteristics of the basic plume developments were illustrated in the UNSCEAR 1988 Report [U4]. Further details have been presented by Borzilov and Klepikova [B7] and are illustrated in Figure V. The important releases lasted 10 days; during that time, the wind changed direction often, so that all areas surrounding the reactor site received some fallout at one time or another.

34. The initial plumes of materials released from the Chernobyl reactor moved towards the west. On 27 April, the winds shifted towards the northwest, then on 28 April towards the east. Two extensive areas, Gomel-Mogilev-Bryansk and Orel-Tula-Kaluga, became contaminated as a result of

Figure IV. Administrative regions surrounding the Chernobyl reactor.

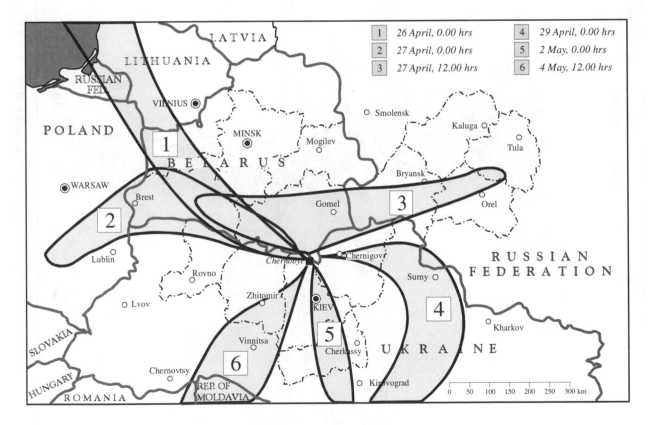

Figure V. Plume formation by meteorological conditions for instantaneous releases on dates and times (GMT) indicated [B7].

deposition of radioactive materials from the plume that passed over at that time (Figure V, trace 3). The contamination of Ukrainian territory south of Chernobyl occurred after 28 April (Figure V, traces 4, 5 and 6). Rainfall occurred in an inhomogeneous pattern, causing uneven contamination areas. The general pattern of ^{137}Cs deposition based on calculations from meteorological conditions has been shown to match the measured contamination pattern rather well [B7].

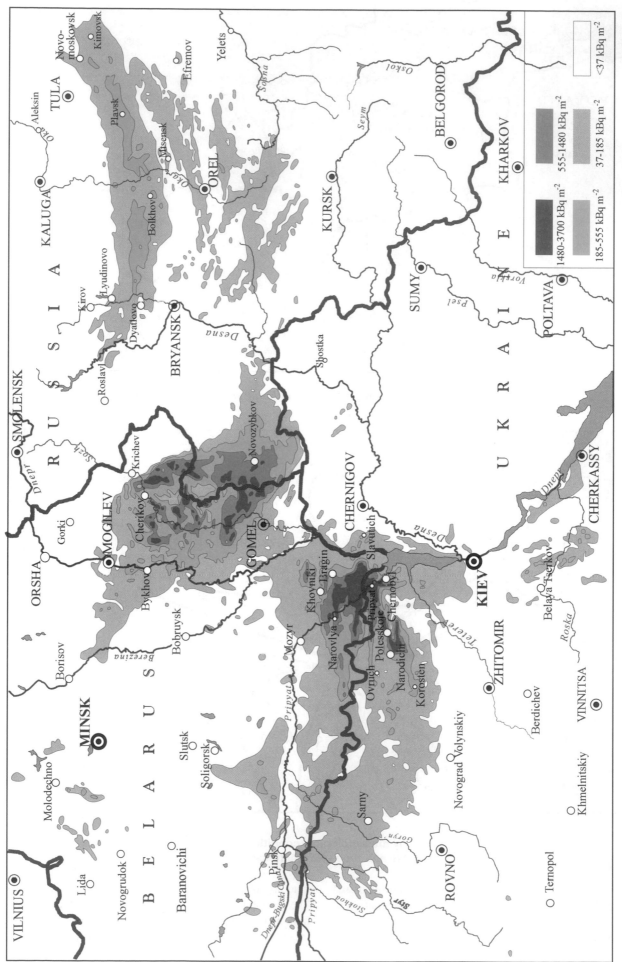

Figure VI. Surface ground deposition of caesium-137 released in the Chernobyl accident [1, 13].

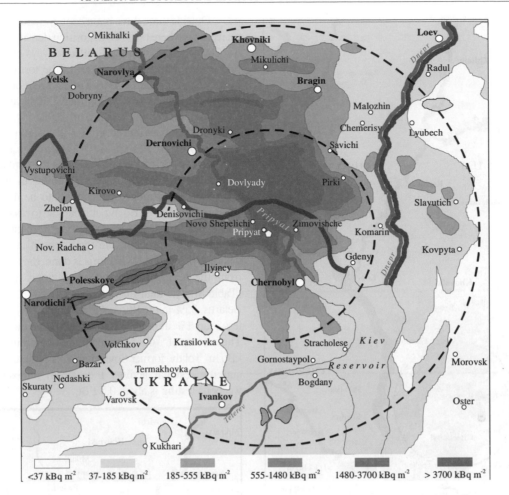

Figure VII. Surface ground deposition of caesium-137 in the immediate vicinity of the Chernobyl reactor [I1, I24].
The distances of 30 km and 60 km from the nuclear power plant are indicated.

35. The detailed contamination patterns have been established from extensive monitoring of the affected territory. The contamination of soil with ^{137}Cs in the most affected areas of Belarus, the Russian Federation and Ukraine is shown in Figure VI, and the ^{137}Cs contamination of soil in the immediate area surrounding the reactor is shown in Figure VII. The deposition of ^{90}Sr and of nuclear fuel particles, usually represented as the deposition of their marker, ^{95}Zr or ^{144}Ce, were relatively localized. The contamination maps for these radionuclides are illustrated in Figures VIII and IX. An important deposition map to be established is that of ^{131}I. Estimated ^{131}I deposition in Belarus and the western part of the Russian Federation is shown in Figure X. Because there were not enough measurements at the time of deposition, the ^{131}I deposition pattern can be only approximated from limited data and relationships inferred from ^{137}Cs deposition. Because the ^{131}I to ^{137}Cs ratio was observed to vary from 5 to 60, the ^{131}I deposition densities estimated for areas without ^{131}I measurements are not very reliable. Measurements of the current concentrations of ^{129}I in soil could provide valuable information on the ^{131}I deposition pattern [S45].

36. The principal physico-chemical form of the deposited radionuclides are: (a) dispersed fuel particles, (b) condensation-generated particles, and (c) mixed-type particles, including the adsorption-generated ones [I22]. The radionuclide

distribution in the nearby contaminated zone (<100 km), also called the near zone, differs from that in the far zone (from 100 km to approximately 2,000 km). Deposition in the near zone reflected the radionuclide composition of the fuel. Larger particles, which were primarily fuel particles, and the refractory elements (Zr, Mo, Ce and Np) were to a large extent deposited in the near zone. Intermediate elements (Ru, Ba, Sr) and fuel elements (Pu, U) were also deposited largely in the near zone. The volatile elements (I, Te and Cs) in the form of condensation-generated particles, were more widely dispersed into the far zone [I6]. Of course, this characterization oversimplifies the actual dispersion pattern.

37. Areas of high contamination from ^{137}Cs occurred throughout the far zone, depending primarily on rainfall at the time the plume passed over. The composition of the deposited radionuclides in these highly contaminated areas was relatively similar. Some ratios of radionuclides in different districts of the near and far zones are given in Table 6.

38. The three main areas of contamination have been designated the Central, Gomel-Mogilev-Bryansk and Kaluga-Tula-Orel areas. The Central area is in the near zone, predominantly to the west and northwest of the reactor. Caesium-137 was deposited during the active period of release, and the deposition density of ^{137}Cs was greater than 37 kBq m^{-2} (1 Ci km^{-2}) in large areas of the

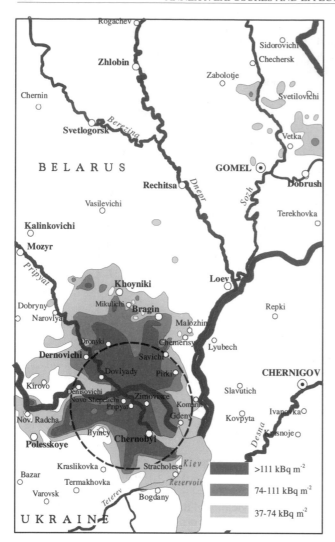

Figure VIII. Surface ground deposition of strontium-90 released in the Chernobyl accident [I1].

Kiev, Zhitomir, Chernigov, Rovno and Lutsk regions of Ukraine and in the southern parts of the Gomel and Brest regions of Belarus. The ^{137}Cs deposition was highest within the 30-km-radius area surrounding the reactor, known as the 30-km zone. Deposition densities exceeded 1,500 kBq m^{-2} (40 Ci km^{-2}) in this zone and also in some areas of the near zone to the west and northwest of the reactor, in the Gomel, Kiev and Zhitomir regions (Figure VII).

39.　The Gomel-Mogilev-Bryansk contamination area is centred 200 km to the north-northeast of the reactor at the boundary of the Gomel and Mogilev regions of Belarus and of the Bryansk region of the Russian Federation. In some areas contamination was comparable to that in the Central area; deposition densities even reached 5 MBq m^{-2} in some villages of the Mogilev and Bryansk regions.

40.　The Kaluga-Tula-Orel area is located 500 km to the northeast of the reactor. Contamination there came from the same radioactive cloud that caused contamination in the Gomel-Mogilev-Bryansk area as a result of rainfall on 28–29 April. The ^{137}Cs deposition density was, however, lower in this area, generally less than 500 kBq m^{-2}.

41.　Outside these three main contaminated areas there were many areas where the ^{137}Cs deposition density was in the range 37–200 kBq m^{-2}. Rather detailed surveys of the contamination of the entire European part of the former Soviet Union have been completed [I3, I6, I24]. A map of measured ^{137}Cs deposition is presented in Figure VI. The areas affected by ^{137}Cs contamination are listed in Table 7. As can be seen, 146,100 km^2 experienced a ^{137}Cs deposition density greater than 37 kBq m^{-2} (1 Ci km^{-2}). The total quantity of ^{137}Cs deposited as a result of the accident in the contaminated areas of the former Soviet Union, including in areas of lesser deposition, is estimated in Table 8 to be 43 PBq. A ^{137}Cs background of 2–4 kBq m^{-2} attributable to residual levels from atmospheric nuclear weapons testing from earlier years must be subtracted to obtain the total deposit attributable to the Chernobyl accident. When this is done, the total ^{137}Cs deposit from the accident is found to be approximately 40 PBq (Table 8). The total may be apportioned as follows: 40% in Belarus, 35% in the Russian Federation, 24% in Ukraine, and less than 1% in other republics of the former Soviet Union. The amount of ^{137}Cs deposited in the contaminated areas (>37 kBq m^{-2}) of the former Soviet Union is estimated to be 29 PBq, and the residual activity there from atmospheric nuclear weapons testing is about 0.5 PBq.

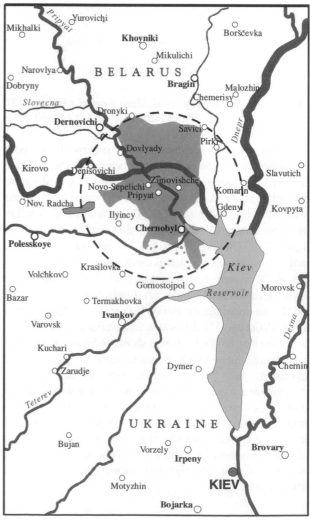

Figure IX. Surface ground deposition of plutonium-239 and plutonium-240 released in the Chernobyl accident at levels exceeding 3.7 kBq m^{-2} [I1].

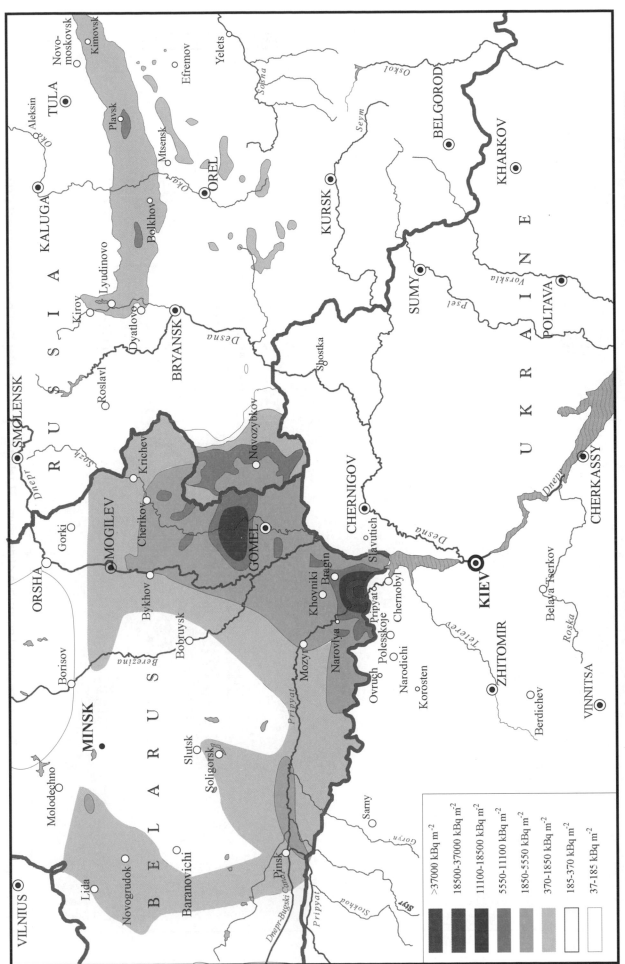

Figure X. Estimated surface ground deposition in Belarus and western Russia of iodine-131 released in the Chernobyl accident [B25, P19].

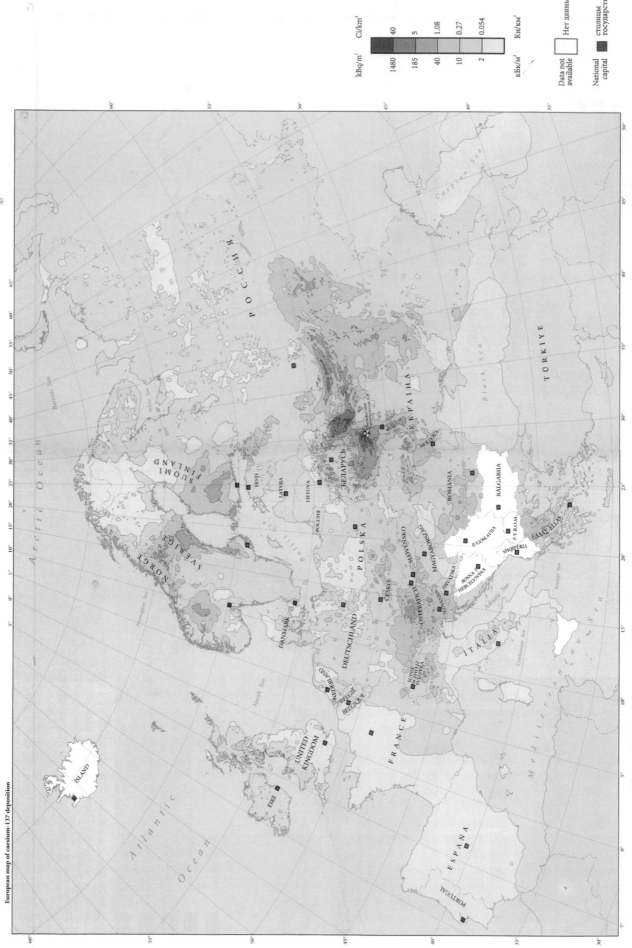

Figure XI. Surface ground deposition of caesium-137 released in Europe after the Chernobyl accident [D13].

42. During the first weeks after the accident, most of the activity deposited on the ground consisted of short-lived radionuclides, of which ^{131}I was the most important radiologically. Maps of ^{131}I deposition have been prepared for Belarus and part of the Russian Federation (Figure X). As indicated in paragraph 35, these maps are based on the limited number of measurements of ^{131}I deposition density available in the former Soviet Union, and they use ^{137}Cs measurements as a guide in areas where ^{131}I was not measured. These maps must be regarded with caution, as the ratio of the ^{131}I to ^{137}Cs deposition densities was found to vary in a relatively large range, at least in Belarus.

43. Deposition of ^{90}Sr was mostly limited to the near zone of the accident. Areas with ^{90}Sr deposition density exceeding 100 kBq m^{-2} were almost entirely within the 30-km zone, and areas exceeding 37 kBq m^{-2} were almost all within the near zone (<100 km). Only a few separate sites with ^{90}Sr deposition density in the range 37–100 kBq m^{-2} were found in the Gomel-Mogilev-Bryansk area, i.e. in the far zone (Figure VIII) [A9, B25, H13].

44. Information on the deposition of plutonium isotopes is not as extensive because of difficulties in detecting these radionuclides. The only area with plutonium levels exceeding 4 kBq m^{-2} was located within the 30-km zone (Figure IX). In the Gomel-Mogilev-Bryansk area, the ^{239,240}Pu deposition density ranged from 0.07 to 0.7 kBq m^{-2}, and in the Kaluga-Tula-Orel area, from 0.07 to 0.3 kBq m^{-2} [A9]. At Korosten, located in Ukraine about 115 km southwest of the Chernobyl power plant, where the ^{137}Cs deposition density was about 300 kBq m^{-2}, the ^{239,240}Pu deposition density due to the Chernobyl accident derived from data in [H8] is found to be only about 0.06 kBq m^{-2}, which is 4–8 times lower than the ^{239,240}Pu deposition density from global fallout.

2. Remainder of northern and southern hemisphere

45. As shown in Table 5, there are also other areas, in Europe, where the ^{137}Cs deposition density exceeded 37 kBq m^{-2}, notably, the three Scandinavian countries (Finland, Norway and Sweden), Austria and Bulgaria. In those countries, the ^{137}Cs deposition density did not exceed 185 kBq m^{-2} except in localized areas (for example, a 2–4 km^2 area in Sweden within the commune of Gävle [E6] and mountainous areas in the Austrian Province of Salzburg [L24]). The pattern of ^{137}Cs deposition density in the whole of Europe is shown in Figure XI [D13, I24].

46. Small amounts of radiocaesium and of radioiodine penetrated the lower stratosphere of the northern hemisphere during the first few days after the accident [J6, K43]. Subsequently, transfer of radiocaesium to the lower atmospheric layers of the southern hemisphere may have occurred as a result of interhemispheric air movements from the northern to the southern stratosphere, followed by subsidence in the troposphere [D11]. However, radioactive contamination was not detected in the southern hemisphere

by the surveillance networks of environmental radiation. Interhemispheric transfer also occurred to a small extent through human activities, such as shipping of foods or materials to the southern hemisphere. Therefore, only very low levels of radioactive materials originating from the Chernobyl accident have been present in the biosphere of the southern hemisphere, and the resulting doses have been negligible.

D. ENVIRONMENTAL BEHAVIOUR OF DEPOSITED RADIONUCLIDES

47. The environmental behaviour of deposited radionuclides depends on the physical and chemical characteristics of the radionuclide considered, on the type of fallout (i.e. dry or wet), and on the characteristics of the environment. Special attention will be devoted to ^{131}I, ^{137}Cs and ^{90}Sr and their pathways of exposure to humans. Deposition can occur on the ground or on water surfaces. The terrestrial environment will be considered first.

1. Terrestrial environment

48. For short-lived radionuclides such as ^{131}I, the main pathway of exposure of humans is the transfer of the amounts deposited on leafy vegetables that are consumed within a few days, or on pasture grass that is grazed by cows or goats, giving rise to the contamination of milk. The amounts deposited on vegetation are retained with a half-time of about two weeks before removal to the ground surface and to the soil. Long-term transfer of ^{131}I from deposition on soil to dietary products that are consumed several weeks after the deposition has occurred need not be considered, because ^{131}I has a physical half-life of only 8 days.

49. Radionuclides deposited on soil migrate downwards and are partially absorbed by plant roots, leading in turn to upward migration into the vegetation. These processes should be considered for long-lived radionuclides, such as ^{137}Cs and ^{90}Sr. The rate and direction of the radionuclide migration into the soil-plant pathway are determined by a number of natural phenomena, including relief features, the type of plant, the structure and makeup of the soil, hydrological conditions and weather patterns, particularly at the time that deposition occurred. The vertical migration of ^{137}Cs and ^{90}Sr in soil of different types of natural meadows has been rather slow, and the greater fraction of radionuclides is still contained in its upper layer (0–10 cm). On average, in the case of mineral soils, up to 90% of ^{137}Cs and ^{90}Sr are found in the 0–5 cm layer; in the case of peaty soils, for which radionuclide migration is faster, only 40% to 70% of ^{137}Cs and ^{90}Sr are found in that layer [I22]. The effective half-time of clearance from the root layer in meadows (0–10 cm) in mineral soils has been estimated to range from 10 to 25 years for ^{137}Cs and to be 1.2–3 times faster for ^{90}Sr than for ^{137}Cs; therefore, the effective clearance half-time for ^{90}Sr is estimated to be 7 to 12 years [A11, A14].

50. For a given initial contamination of soil, the transfer from soil to plant varies with time as the radionuclide is removed from the root layer and as its availability in exchangeable form changes. The ^{137}Cs content in plants was maximum in 1986, when the contamination was due to direct deposition on aerial surfaces. In 1987, ^{137}Cs in plants was 3–6 times lower than in 1986, as the contamination of the plants was then mainly due to root uptake. Since 1987, the transfer coefficients from deposition to plant have continued to decrease, although the rate of decrease has slowed: from 1987 to 1995, the transfer coefficients of ^{137}Cs decreased by 1.5 to 7 times, on average [I22]. Compared with ^{137}Cs from global fallout, ^{137}Cs from the Chernobyl accident in the far zone was found to be more mobile during the first four years after the accident, as the water-soluble fractions of Chernobyl and fallout ^{137}Cs were about 70% and 8%, respectively [H15]. Later on, ageing processes led to similar mobility values for ^{137}Cs from the Chernobyl accident and from global fallout.

51. The variability of the transfer coefficient from deposition to pasture grass for ^{137}Cs is indicated in Table 9 for natural meadows in the Polissya area of Ukraine [S40]. The type of soil and the water content both have an influence on the transfer coefficient, the values of which were found to range from 0.6 to 190 Bq kg^{-1} (dry grass) per kBq m^{-2} (deposition on the ground) in 1988–1989 [S40]. The variability as a function of time after the accident in the Russian Federation has been studied and reported on by Shutov et al. [S41].

52. Contrary to ^{137}Cs, it seems that the exchangeability of ^{90}Sr does not keep decreasing with time after the accident and may even be increasing [B36, S41]. In the Russian Federation, no statistically significant change was found in the ^{90}Sr transfer coefficient from deposition to grass during the first 4 to 5 years following the accident [S41]. This is attributable to two competing processes: (a) ^{90}Sr conversion from a poorly soluble form, which characterized the fuel particles, to a soluble form, which is easily assimilated by plant roots, and (b) the vertical migration of ^{90}Sr into deeper layers of soil, hindering its assimilation by vegetation [S41].

53. The contamination of milk, meat and potatoes usually accounts for the bulk of the dietary intake of ^{137}Cs. However, for the residents of rural regions, mushrooms and berries from forests occupy an important place. The decrease with time of the ^{137}Cs concentrations in those foodstuffs has been extremely slow, with variations from one year to another depending on weather conditions [I22].

2. Aquatic environment

54. Deposition of radioactive materials also occurred on water surfaces. Deposition on the surfaces of seas and oceans resulted in low levels of dose because the radioactive materials were rapidly diluted into very large volumes of water.

55. In rivers and small lakes, the radioactive contamination resulted mainly from erosion of the surface layers of soil in the watershed, followed by runoff in the water bodies. In the 30-km zone, where relatively high levels of ground deposition of ^{90}Sr and ^{137}Cs occurred, the largest surface water contaminant was found to be ^{90}Sr, as ^{137}Cs was strongly adsorbed by clay minerals [A15, M19]. Much of the ^{90}Sr in water was found in dissolved form; low levels of plutonium isotopes and of ^{241}Am were also measured in the rivers of the 30-km zone [A15, M19].

56. The contribution of aquatic pathways to the dietary intake of ^{137}Cs and ^{90}Sr is usually quite small. However, the ^{137}Cs concentration in the muscle of predator fish, like perch or pike, may be quite high in lakes with long water retention times, as found in Scandinavia and in Russia [H16, K47, R21, T23]. For example, concentration of ^{137}Cs in the water of lakes Kozhany and Svyatoe located in severely contaminated part of the Bryansk region of Russia was still high in 1996 because of special hydrological conditions: 10–20 Bq l^{-1} of ^{137}Cs and 0.6–1.5 Bq l^{-1} of ^{90}Sr [K47]. Concentration of ^{137}Cs in the muscles of crucian (Carassius auratus gibeio) sampled in the lake Kozhany was in the range of 5–15 kBq kg^{-1} and in pike (Esox lucius) in the range 20–90 kBq kg^{-1} [K47, T23]. Activity of ^{137}Cs in inhabitants of the village Kozhany located along the coast of lake Kozhany measured by whole-body counters in summer 1996 was 7.4 ± 1.2 kBq in 38 adults who did not consume lake fish (according to interviews performed before the measurements) but was 49 ± 8 kBq in 30 people who often consumed lake fish. Taking into account seasonal changes in the ^{137}Cs whole-body activity, the average annual internal doses were estimated to be 0.3 mSv and 1.8 mSv in these two groups, respectively. Also, the relative importance of the aquatic pathways, in comparison to terrestrial pathways, may be high in areas downstream of the reactor site where ground deposition was small.

E. SUMMARY

57. The accident at the Chernobyl nuclear power station occurred during a low-power engineering test of the Unit 4 reactor. Improper, unstable operation of the reactor allowed an uncontrollable power surge to occur, resulting in successive steam explosions that severely damaged the reactor building and completely destroyed the reactor.

58. The radionuclide releases from the damaged reactor occurred mainly over a 10-day period, but with varying release rates. From the radiological point of view, ^{131}I and ^{137}Cs are the most important radionuclides to consider, because they are responsible for most of the radiation exposure received by the general population. The releases of ^{131}I and ^{137}Cs are estimated to have been 1,760 and 85 PBq, respectively (1 PBq = 10^{15} Bq). It is worth noting, however, that the doses were estimated on the basis of environmental and thyroid or body measurements and that knowledge of the quantities released was not needed for that purpose.

59. The three main areas of contamination, defined as those with ^{137}Cs deposition density greater than 37 kBq m^{-2} (1 Ci km^{-2}), are in Belarus, the Russian Federation and Ukraine; they have been designated the Central, Gomel-Mogilev-Bryansk and Kaluga-Tula-Orel areas. The Central area is within about 100 km of the reactor, predominantly to the west and northwest. The Gomel-Mogilev-Bryansk contamination area is centred 200 km to the north-northeast of the reactor at the boundary of the Gomel and Mogilev regions of Belarus and of the Bryansk region of the Russian Federation. The Kaluga-Tula-Orel area is located in the Russian Federation, about 500 km to the northeast of the reactor. All together, as shown in Table 7 and in Figure XI, territories with an area of approximately 150,000 km^2 were contaminated in the former Soviet Union.

60. Outside the former Soviet Union, there were many areas in northern and eastern Europe with ^{137}Cs deposition density in the range 37–200 kBq m^{-2}. These regions represent an area of 45,000 km^2, or about one third of the contaminated areas found in the former Soviet Union.

61. The environmental behaviour of deposited radionuclides depends on the physical and chemical characteristics of the radionuclide considered, on the type of fallout (i.e. dry or wet), and on the characteristics of the environment. For short-lived radionuclides such as ^{131}I, the main pathway of exposure to humans is the transfer of amounts deposited on leafy vegetables that are consumed by humans within a few days, or on pasture grass that is grazed by cows or goats, giving rise to the contamination of milk. The amounts deposited on vegetation are retained with a half-time of about two weeks before removal to the ground surface and to the soil. For long-lived radionuclides such as ^{137}Cs, the long-term transfer processes from soil to foods consumed several weeks or more after deposition need to be considered.

II. RADIATION DOSES TO EXPOSED POPULATION GROUPS

62. It is convenient to classify into three categories the populations who were exposed to radiation following the Chernobyl accident: (a) the workers involved in the accident, either during the emergency period or during the clean-up phase; (b) inhabitants of evacuated areas; and (c) inhabitants of contaminated areas who were not evacuated. The available information on the doses received by the three categories of exposed populations will be presented and discussed in turn. Doses from external irradiation and from internal irradiation will be presented separately. The external exposures due to gamma radiation were relatively uniform over all organs and tissues of the body, as their main contributors were ^{132}Te-^{132}I, ^{131}I and ^{140}Ba-^{140}La for evacuees, ^{134}Cs and ^{137}Cs for inhabitants of contaminated areas who were not evacuated, and radionuclides emitting photons of moderately high energy for workers. These external doses from gamma radiation have been expressed in terms of effective dose. With regard to internal irradiation, absorbed doses in the thyroid have been estimated for exposures to radioiodines and effective doses have been estimated for exposures to radiocaesiums.

63. Doses have in almost all cases been estimated by means of physical dosimetry techniques. Biological indicators of dose has been mainly used, within days or weeks after the accident, to estimate doses received by the emergency workers, who received high doses from external irradiation and for whom dosemeters were either not operational nor available. Unlike physical dosimetry, biological dosimetric methods are generally not applicable to doses below 0.1 Gy and reflect inter-individual variations in radiation sensitivity. Soon after the accident, biological dosimetry is usually based on the measurement of the frequency of unstable chromosome aberrations (dicentric and centric rings). By comparing the rate of dicentric chromosomes and centric rings with a standard dose-effect curve obtained in an experiment *in vitro*, it is possible to determine a radiation dose. This method has been recommended for practical use in documents of WHO and IAEA. However, the use of dicentric as well as other aberrations of the unstable type for the purposes of biological dosimetry is not always possible, since the frequency of cells containing such aberrations declines in time after exposure.

64. For retrospective dosimetry long after the exposure, biological dosimetry can be a complement to physical dosimetry, but only techniques where radiation damage to the biological indicator is stable and persistent and not subject to biochemical, physiological or immunological turnover, repair or depletion are useful. In that respect, the analysis of stable aberrations (translocations), the frequency of which remains constant for a long time after exposure to radiation, is promising. The probability of occurrence of stable (translocations) and unstable (dicentrics) aberrations after exposure is the same. However, translocations are not subjected to selection during cell proliferation, in contrast to dicentrics. Fluorescence *in situ* hybridization (FISH) or Fast-FISH in conjunction with chromosome painting may be useful in retrospective dosimetry for several decades after exposure.

65. Other biological (or biophysical) techniques for measuring doses are electron spin resonance (ESR) or optically stimulated luminescence (OSL). These techniques are used in retrospective dosimetry to measure the radiation damage accumulated in biological tissue such as bone, teeth, fingernails and hair. Also, the gene mutation glycophorin A that is associated with blood cells may be used. Currently, the detection limits for FISH, ESR and OSL are about 0.1 Gy [P28]. At low dose levels, however, the estimation of the dose due to the radiation accident is

highly unreliable because of the uncertainty in the background dose resulting from other radiation exposures (medical irradiation, natural background, etc.) or, in the case of FISH, from other factors such as smoking.

A. WORKERS INVOLVED IN THE ACCIDENT

66.   The workers involved in various ways in the accident can be divided into two groups: (a) those involved in emergency measures during the first day of the accident (26 April 1986), who will be referred to as emergency workers in this Annex, and (b) those active in 1986–1990 at the power station or in the zone surrounding it for the decontamination work, sarcophagus construction and other clean-up operations. This second group of workers is referred to as recovery operation workers in this Annex, although the term liquidator gained common usage in the former Soviet Union.

1.   Emergency workers

67.   The emergency workers are the people who dealt with the consequences of the accident on the very first day

(26 April 1986), i.e. the staff of the plant, the firemen involved with the initial emergency, the guards and the staff of the local medical facility. Most of them were at the reactor site at the time of the accident or arrived at the plant during the few first hours. In the Russian literature, two other categories of people are referred to: (a) the "accident witnesses", who were present at the plant at the time of the accident and who may or may not have been involved in emergency operations (so that part of them are also classified as "emergency workers") and (b) the "accident victims", who were sent to the local medical facility and then transferred to special hospitals in Moscow and Kiev. All accident victims were emergency workers and/or accident witnesses. The numbers of accident witnesses and emergency workers are listed in Table 10. According to Table 10, on the morning of 26 April, about 600 emergency workers were on the site of the Chernobyl power plant.

68.   The power plant personnel wore only film badges that could not register doses in excess of 20 mSv. All of these badges were overexposed. The firemen had no dosimeters and no dosimetric control. Dose rates on the roof and in the rooms of the reactor block reached hundreds of gray per hour. Measured exposure rates in the vicinity of the reactor at the time of the accident are shown in Figure XII.

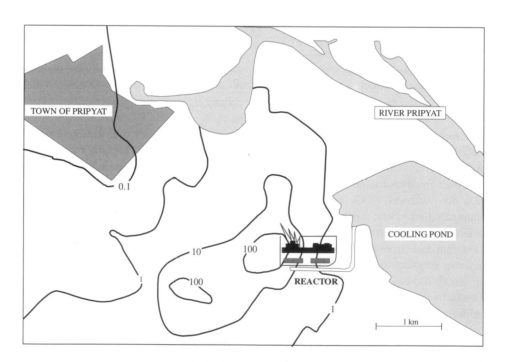

Figure XII. Measured exposure rates in air on 26 April 1986 in the local area of the Chernobyl reactor.
Units of isolines are R h⁻¹.

69.   The highest doses were received by the firemen and the personnel of the power station on the night of the accident. Some symptoms of acute radiation sickness were observed in 237 workers. Following clinical tests, an initial diagnosis of acute radiation sickness was made in 145 of these persons. On further analysis of the clinical data, acute radiation sickness was confirmed later (in 1992) in 134 individuals. The health effects that were observed among the emergency workers are discussed in Chapters III and IV.

70.   The most important exposures were due to external irradiation (relatively uniform whole-body gamma irradiation and beta irradiation of extensive body surfaces), as the intake of radionuclides through inhalation was relatively small (except in two cases) [U4]. Because all of the dosimeters worn by the workers were overexposed, they could not be used to estimate the gamma doses received via external irradiation. However, relevant information was obtained by means of biological dosimetry for the treated

persons. The estimated ranges of doses for the 134 emergency workers with confirmed acute radiation sickness are given in Table 11. Forty-one of these patients received whole-body doses from external irradiation of less than 2.1 Gy. Ninety-three patients received higher doses and had more severe acute radiation sickness: 50 persons with doses between 2.2 and 4.1 Gy, 22 between 4.2 and 6.4 Gy, and 21 between 6.5 and 16 Gy [I5]. As shown in Table 12, the relative errors were 10%–20% for doses greater than 6 Gy; they increased as the dose level decreased, to about 100% for whole-body doses of about 1 Gy, and were even greater for doses of less than 0.5 Gy. The skin doses from beta exposures evaluated for eight patients with acute radiation sickness ranged from 10 to 30 times the dose from whole-body gamma radiation [B10].

71. Internal doses were determined from thyroid and whole-body measurements performed on the persons under treatment, as well as from urine analysis and from post-mortem analysis of organs and tissues. For most of the patients, more than 20 radionuclides were detectable in the whole-body gamma measurements; however, apart from the radioiodines and radiocaesiums, the contribution to the internal doses from the other radionuclides was negligible [U4]. Internal doses evaluated for 23 persons who died of acute radiation sickness are shown in Table 13. The lung and thyroid doses, calculated to the time of death, are estimated to have ranged from 0.00026 to 0.04 Gy and from 0.021 to 4.1 Gy, respectively. Some of the low thyroid doses may be due to the fact that stable iodine pills were distributed among the reactor staff less than half an hour after the beginning of the accident. It is also speculated that the internal doses received by the emergency workers who were outdoors were much lower than those received by the emergency workers who stayed indoors. For comparison purposes, the estimated external doses are also presented in Table 13. The external doses, which range from 2.9 to 11.1 Gy, are, in general, much greater than the internal doses.

72. Internal dose reconstruction was also carried out for 375 surviving emergency workers who were examined in Moscow; the results are presented in Table 14. The average doses were estimated to vary from 36 mGy to bone marrow to 280 mGy to bone surfaces, the maximum doses being about 10 times greater than the average doses. Also, thyroid doses were estimated for the 208 emergency workers admitted to Hospital 6 in Moscow within 3–4 weeks after the accident (Table 15); most of the thyroid doses were less than 1 Gy, but three exceeded 20 Gy. It is interesting to note that the measurements of ^{131}I and ^{133}I among the five emergency workers with the highest thyroid doses showed that ^{133}I contributed less than 20% to the thyroid dose. The specific values of the contributions from ^{133}I were 18% (with 74% from ^{131}I), 11% (81% from ^{131}I), 6% (86% from ^{131}I), 10% (82% from ^{131}I) and 14% (78% from ^{131}I) for the five workers [G12]. The thyroid doses due to internal exposures are estimated to be in the range from several percent to several hundred percent of the external whole-body doses. The median value of the ratio of the thyroid to the whole-body dose was estimated to be 0.3 [K19]. Finally, information is

available for the relative intakes of 16 radionuclides of 116 patients, determined from measurements in urine and in autopsy materials [D12]; according to these measurements, the average intake of ^{132}Te was found to be about 10% that of ^{131}I [D12].

2. Recovery operation workers

73. About 600,000 persons (civilian and military) have received special certificates confirming their status as liquidators, according to laws promulgated in Belarus, the Russian Federation and Ukraine. Of those, about 240,000 were military servicemen [C7]. The principal tasks carried out by the recovery operation workers (liquidators) included decontamination of the reactor block, reactor site, and roads (1986–1990) and construction of the sarcophagus (May-November 1986), a settlement for reactor personnel (May-October 1986), the town of Slavutich (1986–1988, 1990), waste repositories (1986–1988), and dams and water filtration systems (July-September 1986, 1987) [K19]. During the entire period, radiation monitoring and security operations were also carried out.

74. Of particular interest are the 226,000 recovery operation workers who were employed in the 30-km zone in 1986–1987, as it is in this period that the highest doses were received; information concerning these workers is provided in Table 16. About half of these persons were civilian and half were military servicemen brought in for the special and short-term work. The workers were all adults, mostly males aged 20–45 years. The construction workers were those participating in building the sarcophagus around the damaged reactor. Other workers included those involved in transport and security, scientists and medical staff. The distributions of the external doses for the categories of workers listed in Table 16, as well as for the emergency workers and accident witnesses, are shown in Table 17.

75. The remainder of the recovery operation workers (about 400,000), who generally received lower doses, includes those who worked inside the 30-km zone in 1988–1990 (a small number of workers are still involved), those who decontaminated areas outside the 30-km zone, and other categories of people.

76. In 1986 a state registry of persons exposed to radiation was established at Obninsk. This included not only recovery operation workers but evacuees and residents of contaminated areas as well. The registry existed until the end of 1991. Starting in 1992, national registries of Belarus, the Russian Federation and Ukraine replaced the all-union registry. The number of recovery operation workers in the national registries of Belarus, the Russian Federation and Ukraine is listed in Table 18. Some 381,000 workers from these countries were involved in the years 1986–1989. To this must be added the 17,705 recovery operation workers recorded in the registries of the Baltic countries, including 7,152 from Lithuania, 5,709 from Latvia and 4,844 from Estonia [K13]. More detailed information on the registries is provided in Chapter IV. The total number of recovery operation workers

recorded in the registries appears to be about 400,000. This number is likely to increase in the future, as some organizations may not have provided all their information to the central registries; in addition, individuals may on their own initiative ask to be registered in order to benefit from certain privileges. However, the number of recovery operation workers recorded in the national registries is well below the figure of about 600,000, which corresponds to the number of people who have received special certificates confirming their status as liquidators.

(a) External effective doses from gamma radiation

77.   The doses to the recovery operation workers who participated in mitigation activities within two months after the accident are not known with much certainty. Attempts to establish a dosimetric service were inadequate until the middle of June. TLDs and condenser-type dosimeters that had been secured by 28 April were insufficient in number and, in the case of the latter type, largely non-functioning, and records were lost when the dosimetric service was transferred from temporary to more permanent quarters. In June, TLD dosimeters were available in large numbers, and a databank of recorded values could be established. From July 1986 onwards, individual dose monitoring was performed for all non-military workers, using either TLDs or film dosimeters.

78.   The dose limits for external irradiation varied with time and with the category of personnel. According to national regulations established before the accident [M1], for civilian workers, during 1986, the dose limit, 0.05 Sv, could be exceeded by a factor of up to 2 for a single intervention and by a factor of 5 for multiple interventions on condition of agreement by the personnel. The maximum dose allowed during the year 1986 was, therefore, 0.25 Sv. In 1987, the annual dose limits for civilian personnel were lowered to 0.05 or to 0.1 Sv, according to the type of work performed on the site. However, a dose of up to 0.25 Sv could be allowed by the Ministry of Health for a limited number of workers for the implementation of extremely important interventions. In 1988, the annual dose limit was set at 0.05 Sv for all civilian workers, except those involved in the decontamination of the engine hall inside the sarcophagus; for them, the annual dose limit was set at 0.1 Sv. From 1989 onwards, the annual dose limit was set at 0.05 Sv for all civilian workers, without exception [M1, M12]. For military workers, a dose limit of 0.5 Sv, corresponding to radiation exposures during wartime, was applied until 21 May 1986, when the Ministry of Defence lowered the dose limit to 0.25 Sv [C7]. From 1987 onwards, the dose limits were the same for military and civilian personnel.

79.   Estimates of effective doses from external gamma irradiation were generally obtained in one of three ways: (a) individual dosimetry for all civilian workers and a small part of the military personnel after June 1986; (in 1987, they were identified as those working in locations where the exposure rate was greater than 1 mR h^{-1}); (b) group dosimetry (an individual dosimeter was assigned to one member of a group of recovery operation workers assigned to perform a particular task, and all members of the group were assumed to receive the same dose; in some cases, no member of the group wore an individual dosimeter and the dose was assigned on the basis of previous experience); or (c) time-and-motion studies (measurements of gamma-radiation levels were made at various points of the reactor site, and an individual's dose was estimated as a function of the points where he or she worked and the time spent in these places). Methods (b) and (c) were used for the civilian workers before June 1986, when the number of individual dosimeters was insufficient, and for the majority of the military personnel at any time. For example, effective doses from external irradiation have been reconstructed by physical means for the staff of the reactor, as well as for the workers who had been detailed to assist them, exposed from 26 April to 5 May 1986 [K19]. Personnel location record cards filled in by workers were analysed by experts who had reliable information on the radiation conditions and who had personally participated in ensuring the radiation safety of all operations following the accident. Using this method, two values were determined: the maximum possible dose and the expected dose. The maximum possible effective doses ranged from less than 0.1 Sv to a few sievert and were estimated to be about twice the expected doses. It seems that in most cases the maximum possible effective doses are those that were officially recorded.

80.   The main sources of uncertainty associated with the different methods of dose estimation were as follows: (a) individual dosimetry: incorrect use of the dosimeters (inadvertent or deliberate actions leading to either overexposure or underexposure of the dosimeters); (b) group dosimetry: very high gradient of exposure rate at the working places at the reactor site; and (c) time-and-motion studies: deficiencies in data on itineraries and time spent at the various working places, combined with uncertainties in the exposure rates. Uncertainties associated with the different methods of dose estimation are assessed to be up to 50% for method (a) (if the dosimeter was correctly used), up to a factor of 3 for method (b), and up to a factor of 5 for method (c) [P15].

81.   The registry data show that the annual averages of the officially recorded doses decreased from year to year, being about 170 mSv in 1986, 130 mSv in 1987, 30 mSv in 1988, and 15 mSv in 1989 [I34, S14, T9]. It is, however, difficult to assess the validity of the results that have been reported for a variety of reasons, including (a) the fact that different dosimeters were used by different organizations without any intercalibration; (b) the high number of recorded doses very close to the dose limit; and (c) the high number of rounded values such as 0.1, 0.2, or 0.5 Sv [K19]. However, the doses do not seem to have been systematically overestimated, because biological dosimetry performed on limited numbers of workers produced results that are also very uncertain but compatible nonetheless with the physical dose estimates [L18]. It seems reasonable to assume that the average effective dose from external gamma irradiation to recovery operation workers in the years 1986–1987 was about 100 mSv, with individual

effective doses ranging from less than 10 mSv to more than 500 mSv. Using the numbers presented in Table 18, the collective effective dose is estimated to be about 40,000 man Sv.

82. A particular group of workers who may have been exposed to substantial doses from external irradiation is made up of the 1,125 helicopter pilots who were involved in mitigation activities at the power plant in the first three months after the accident [U15]. The doses to pilots were estimated using either personal dosimeters or, less reliably, calculations in which the damaged reactor was treated as a collimated point source of radiation [U15]. The doses obtained by calculation were checked against the results derived from the personal dosimeters for about 200 pilots. That comparison showed a discrepancy of (a) less 0.05 Sv for about 10% of pilots, (b) from 0.05 to 0.1 Sv for about 33%, and (c) more than 0.1 Sv for about 57% [U15]. The simplification used to describe the origin of the radiation emitted from the damaged reactor is the main source of uncertainty in the assessment of the doses received by the helicopter pilots. The average dose estimates are 0.26 Sv for the pilots who took part in the mitigation activities from the end of April to the beginning of May, and 0.14 Sv for the pilots who were exposed after the beginning of May.

83. Another group of workers that may have been exposed to substantial doses from external irradiation is the 672 workers from the Kurchatov Institute, a group that includes those who were assigned special tasks inside the damaged unit 4 before and after the construction of the sarcophagus [S36]. Recorded and calculated doses available for 501 workers show that more than 20% of them received doses between 0.05 and 0.25 Sv, and that about 5% of them received doses between 0.25 and 1.5 Sv [S36]. A number of nuclear research specialists worked in high-radiation areas of the sarcophagus, without formal recording of doses, on their own personal initiative, and were exposed to annual levels greater than the dose limit of 0.05 Sv applicable since 1988. Doses for this group of 29 persons have been estimated using electron spin resonance analysis of tooth enamel as well as stable and unstable chromosome aberration techniques [S42, S47]. It was found that 14 of those 29 persons received doses lower than 0.25 Sv, 5 had doses between 0.25 and 0.5 Sv, 6 between 0.5 and 1 Sv, and 4 greater than 1 Sv [S42]. Additional analyses by means of the FISH technique for three of those nuclear research specialists resulted in doses of 0.9, 2.0 and 2.7 Sv [S48].

84. *Biological dosimetry.* Chromosome aberration levels among Chernobyl recovery operation workers were analysed in a number of additional studies. In a pilot study of a random sample of 60 workers from the Russian Federation, stratified on the level of recorded dose (31 with doses <100 mGy, 18 with 100–200 mGy, and 13 with >200 mGy), no association was found between the percentage of the genome with stable translocations measured by fluorescent *in situ* hybridization (FISH) and

individual recorded physical dose estimates [C1, L18]. A good correlation was found, however, for group (rather than individual) doses. Blood samples of 52 Chernobyl recovery operation workers were analysed by FISH [S32] and simultaneously by conventional chromosome analysis. Based on FISH measurements, individual biodosimetry estimates between 0.32 and 1.0 Gy were estimated for 18 cases. Pooled data for the total group of 52 workers provided an average estimate of 0.23 Gy. For a group of 34 workers with documented doses, the mean dose estimate of 0.25 Gy compared well with the mean documented dose of 0.26 Gy, although there was no correlation between individual translocation frequencies and documented doses. Comparison between the conventional scoring and FISH analyses showed no significant difference. In a study of Estonian workers, Littlefield et al. [L41] did not detect an increase of stable translocation frequencies with reported doses and questioned whether the reported doses could have been overestimated. In conclusion, FISH does not currently appear to be a sufficiently sensitive and specific technique to allow the estimation of individual doses in the low dose range received by the majority of recovery operation workers.

85. Lazutka and Dedonyte [L30], using standard cytogenetic methods, reported no significant overall increase in chromosome aberrations over controls in 183 recovery operation workers from Lithuania with a mean dose estimate of 140 mGy, although ~20% had elevated frequencies of dicentric and ring chromosomes, possibly related to radiation exposure. Lazutka et al. [L31] also evaluated the impact of a number of possible confounders such as age, alcohol use, smoking, recent febrile illness, and diagnostic x-ray exposures on the frequency of chromosome aberrations. When transformed data were analysed by analysis of variance, alcohol abuse made a significant contribution to total aberrations, chromatid breaks, and chromatid exchanges. Smoking was associated with frequency of chromatid exchanges, and age was significantly associated with rates of chromatid exchanges and chromosome exchanges [L31]. In another study [S37], the frequency of chromosomal aberrations was evaluated in more than 500 recovery operation workers. Blood samples were taken from several days to three months after exposure to radiation. The mean frequencies of aberrations for different groups of workers were associated with doses varying from 0.14 to 0.41 Gy, with a good correlation between the doses determined by biological and physical methods [S37].

86. Glycophorin A assay (GPA) was used as a possible biological dosimeter on 782 subjects from Estonia, Latvia and Lithuania with recorded physical dose estimates [B17]. Although a slight increase in the frequency of erythrocytes with loss of the GPA allele was seen among these subjects compared to control subjects from the same countries, this difference was not significant. The pooled results indicate that the average exposures of these workers were unlikely to greatly exceed 100–200 mGy, the approximate minimum radiation dose detectable by this assay.

(b) External skin doses from beta radiation

87.   In addition to effective doses from external gamma irradiation, recovery operation workers received skin doses from external beta irradiation as well as thyroid and effective doses from internal irradiation. The dose to unprotected skin from beta exposures is estimated to have been several times greater than the gamma dose. Ratios of dose rates of total exposures (beta + gamma) to gamma exposures, measured at the level of the face, ranged from 2.5 to 11 (average, around 5) for general decontamination work and from 7 to 50 (average, 28) for decontamination of the central hall of the Unit 3 reactor [O3].

(c) Internal doses

88.   Because of the abundance of ^{131}I and of shorter-lived radioiodines in the environment of the reactor during the accident, the recovery operation workers who were on the site during the first few weeks after the accident may have received substantial thyroid doses from internal irradiation. Information on the thyroid doses is very limited and imprecise. From 30 April through 7 May 1986, *in vivo* thyroid measurements were carried out on more than 600 recovery operation workers. These *in vivo* measurements, which are measurements of the radiation emitted by the thyroid using detectors held or placed against the neck, were used to derive the ^{131}I thyroidal contents at the time of measurement. The thyroid doses were derived from the measured ^{131}I thyroidal contents, using assumptions on the dynamics of intake of ^{131}I and short-lived radio-iodines and on the possible influence of stable iodine prophylaxis. Preliminary thyroid dose estimates (assuming a single intake at the date of the accident and no stable iodine prophylaxis) showed the following distribution [K30]: 64% of workers were exposed to less than 0.15 Gy, 32.9% to 0.15–0.75 Gy, 2.6% to 0.75–1.5 Gy, and the remaining 0.5% to 1.5–3.0 Gy. The average thyroid dose estimate for those workers is about 0.21 Gy. The thyroid doses from internal irradiation are estimated to range from several percent to several hundred percent of the effective doses from external irradiation. The median value of the ratio of the internal thyroid dose to the external effective dose was estimated to be 0.3 Gy per Sv [K19].

89.   It is important to note that information on the influence of stable iodine prophylaxis is limited, as iodine prophylaxis among the recovery operation workers was not mandatory nor was it proposed to everybody. The decision to take stable iodine for prophylactic reasons was made by the individual worker or by the supervisor. The results of interviews of 176 workers (including emergency workers and recovery operation workers who arrived at the plant at the eraly stage of the accident) concerning the time when they took stable iodine for prophylaxis is presented in Table 19. According to this sample of workers, only about 20% took stable iodine before being exposed to radioiodine, while another 10% refused to take stable iodine.

90.   The internal doses resulting from intakes of radionuclides such as ^{90}Sr, ^{134}Cs, ^{137}Cs, ^{239,240}Pu, and others have been assessed for about 300 recovery operation workers who were monitored from April 1986 to April 1987 [K2, K8, P13, S11]. The majority of them were staff of the power plant who took part in the recovery work starting on days 3 and 4 after the accident. The dose assessment was based on the analysis of whole-body measurements and of radionuclide concentrations in excreta. The average value of the effective dose committed by the radionuclide intakes was estimated on the basis of ICRP Publication 30 [I17] to be 85 mSv. The part of the effective dose received between June and September 1986 was estimated to have been about 30 mSv. Internal doses from intakes in later years are expected to be much lower: routine monitoring of the ^{134}Cs + ^{137}Cs body burdens indicated average annual doses from ^{134}Cs + ^{137}Cs of about 0.1–0.2 mSv in 1987 and 1988 [V6].

B. EVACUATED PERSONS

91.   The evacuation of the nearby residents was carried out at different times after the accident on the basis of the radiation situation and of the distance of the populated areas from the damaged reactor. The initial evacuations were from the town of Pripyat, located just 3 km from the damaged reactor, then from the 10-km zone and from the 30-km zone around the reactor (located mostly in Ukraine but also in Belarus). In addition, a number of villages in Belarus, the Russian Federation and Ukraine beyond the 30-km-radius circle centred on the reactor were also evacuated in 1986. The term "exclusion zone" is used in this Annex to refer to the whole area evacuated in 1986, which includes the 30-km zone.

92.   In Ukraine, the residents of Pripyat (49,360 persons) and of the nearest railway station, Yanov (254 persons), 3 km from the reactor, were the first to be evacuated. On the evening of 26 April 1986, the radiation exposures in Pripyat were not considered too alarming. Exposure-rate readings were in the range 1–10 mR h^{-1} [I1], but with the seriousness of the accident becoming evident, the decision to evacuate the residents of the town was taken at 22:00. During the night, arrangements were made for nearly 1,200 buses that would be needed to transport the residents. Around noon on 27 April the evacuation order was broadcast to the people, and the evacuation began at 14:00 and finished at 17:00. The over 40,000 evacuees were taken in by families who lived in settlements in the surrounding districts, especially Polesskoe district of Ukraine. Most people stayed with these families until August 1986. After that they were resettled to apartments in Kiev [I1].

93.   Also in Ukraine, the evacuation of the residents from the southern part of the 10-km zone (10,090 persons) was carried out from 30 April through 3 May. The other Ukrainian residents (28,133) inside the 30-km zone,

including Chernobyl town, were evacuated from 3 May through 7 May. On the basis of exposure-rate criteria (5–20 mR h⁻¹ on 10 May 1986), 2,858 persons who resided outside the 30-km zone in the Kiev and in Zhitomir regions were evacuated from 14 May to 16 August. The last Ukrainian settlement that was evacuated was Bober, with 711 inhabitants, in September 1986. Thus, 91,406 residents from 75 settlements were evacuated in Ukraine in 1986 [S20, U14].

94. The evacuation in Belarus was conducted in three phases. During the first phase (2–7 May), 11,358 residents of 51 villages were evacuated from the 30-km zone. In a second phase (3–10 June), 6,017 residents of 28 villages beyond the 30-km zone were evacuated. In the third phase (August and September 1986), 7,350 residents of 29 villages, also beyond the 30-km zone, were evacuated. In villages evacuated during the second and third phases, the exposure rate was from 5 to 20 mR h⁻¹, corresponding to a projected annual effective dose (26 April 1986 to 25 April 1987) of more than 100 mSv. The total number of Belarusian residents who were evacuated in 1986 was 24,725 from 108 rural settlements. In the Russian Federation, only 186 residents from four settlements in the Krasnaya Gora district of Bryansk region were evacuated, mainly to other settlements of that district. In summary, by the autumn of 1986, about 116,000 residents from 187 settlements had been evacuated (Table 20). By the same time, about 60,000 cattle and other agricultural animals had been relocated from the evacuated zone.

95. The figure of 116,000, adopted in this Annex as the number of evacuees in 1986, is somewhat lower than the figure of 135,000 that was cited by the Committee in the UNSCEAR 1988 Report [U4] and by IAEA in 1996 [I15]. It is believed that the figure of 135,000 was a rough preliminary estimate that was not substantiated.

96. The extent of the exclusion zone was based on two principles: geographical and radiological (dose criteria). A detailed study of the radiation situation carried out in the exclusion zone led to the resettlement of 279 residents of two Ukrainian villages (Cheremoshnya and Nivetskoe) in June 1986. In addition, it was recommended that the residents of 27 other villages might move back after the sarcophagus was constructed (15 settlements in Ukraine and 12 settlements in Belarus). In accordance with these recommendations, 1,612 residents of 12 villages in Belarus had been resettled by December 1986. However, the Ukrainian authorities considered that resettling the residents inside the exclusion zone was economically and socially undesirable. Nevertheless, some people, mainly elderly, resettled by themselves to 15 settlements inside the exclusion zone. The population of those 15 settlements was estimated to be about 900 by spring 1987; about 1,200 by September 1988; and about 1,000 in 1990. In 1996–1997, the number is estimated to be 600–800. The decrease with time is due to migration rather than death.

1. Doses from external exposure

97. The effective doses from external exposure for the persons evacuated from the Ukrainian part of the 30-km zone were estimated from (a) measurements of exposure rates performed every hour at about 30 sites in Pripyat and daily at about 80 sites in the 30-km zone and (b) responses to questionnaires from about 35,000 evacuees from Pripyat and about 100 settlements; the questionnaires asked for information on their locations, types of houses, and activities at the time of the accident and during a few days thereafter [L9, M2, R10]. Individual effective doses were reconstructed in this way for about 30,000 evacuees from the city of Pripyat and settlements in the 30-km zone. The average effective dose from external irradiation for this cohort was estimated to be 17 mSv, with individual values varying from 0.1 to 380 mSv [L9]. This value is concordant with the absorbed dose of 20 mGy estimated for the evacuees of Pripyat using Electron Spin Resonance (ESR) measurements of sugar and exposure rate calculations [N1]. The collective effective dose for the approximately 90,000 evacuees from the Ukrainian part of the 30-km zone was assessed to be 1,500 man Sv [R12].

98. The effective doses and skin doses from external irradiation received by the evacuees from Belarusian territory were estimated on the basis of (a) 3,300 measurements of exposure rates performed in the settlements that were evacuated; (b) 220 spectrometric measurements, carried out mainly in May and June 1986, of the gamma radiation emitted by radionuclides deposited on the ground; (c) measurements of the ¹³⁷Cs ground deposition density for each settlement from the Belarusian data bank [D4]; and (d) responses of about 17,000 evacuees from the territory inside the 30-km zone and from adjoining areas. It was assessed that the doses to evacuees from external irradiation were mainly due to radionuclides deposited on the ground, because external irradiation during the passage of the radioactive cloud played a minor role. The method developed to assess the doses included the reconstruction of the radionuclide composition of the deposition in each of the 108 evacuated settlements in Belarusian territory and the estimation of the contribution to the dose from each radionuclide [S29]. It was assumed that 60%–80% of the effective doses was contributed by the short-lived radionuclides ¹³¹I, ¹³²Te+¹³²I and ¹⁴⁰Ba+¹⁴⁰La, while the contribution from the long-lived radionuclide ¹³⁷Cs was estimated to be only 3%–5%. The distribution of individual doses received by the residents of a given settlement was found to be appropriately described by a log-normal function with a geometric standard deviation of about 1.5. Overall, it is estimated that about 30% of the people were exposed to effective doses lower than 10 mSv, about 86% were exposed to doses lower than 50 mSv, and only about 4% were exposed to doses greater than 100 mSv, with the average dose estimated to be 31 mSv. The highest average effective doses, about 300 mSv, were estimated to be received by the population of two villages located inside the 30-km zone in Khoyniki district: Chamkov and Masany. The uncertainty in the average dose for a settlement is estimated to be characterized with a geometric standard deviation of about

1.3. The main source of uncertainty in the estimation of the average effective doses from external irradiation for the Belarusian evacuees is the assessment of the activity ratios of ^{132}Te and ^{131}I to ^{137}Cs in the deposition. The collective effective dose from external irradiation for the 24,725 evacuees from Belarus is assessed to be 770 man Sv.

99. The average skin doses from beta and gamma radiation are estimated to be 3–4 times greater than the effective doses and to range up to 1,560 mGy. The uncertainty of the average skin doses in a given settlement is estimated to be characterized by a geometric standard deviation of about 1.6.

2. Doses from internal exposure

100. The thyroid doses received from intake of ^{131}I by the evacuees from Pripyat were derived from (a) 4,969 measurements of radioiodine content of their thyroid glands made, on average, 23 days after the accident and (b) responses to questionnaires by 10,073 evacuees on their locations and consumption of stable iodine [G8]. Average individual and collective thyroid doses to the evacuees from Pripyat are shown in Table 21. The thyroid doses from ^{131}I, which were for the most part due to inhalation, were highest for 0–3-year-old children (about 1.4 Gy) and averaged about 0.2 Gy. The main factor influencing the individual dose was found to be the distance of the residence from the reactor [G8].

101. Thyroid doses from intake of ^{131}I to other evacuees from the 30-km zone were also estimated on the basis of measurements of thyroid contents in 10,676 persons [L12, R10]. When dose estimates obtained for the evacuees from Pripyat are compared with those for the evacuees from other settlements of the 30-km zone (Table 21), the doses to the latter are seen to be somewhat higher than those to the evacuees from Pripyat, especially for adults. This may be because Pripyat was evacuated before the rest of the 30-km zone, giving the population of the 30-km zone more time to consume foodstuffs contaminated with ^{131}I. Using for the settlements of the 30-km zone the same age structure as that for Pripyat in Table 21, the collective thyroid dose from ^{131}I intake for the entire population of evacuees from Ukraine is tentatively estimated to be about 30,000 man Gy. Evaluation of thyroid doses to the evacuated population of Belarus is presented in Table 22. The collective thyroid dose estimate for this population is 25,000 man Gy.

102. Inhalation of short-lived radioiodines and of ^{132}Te contributed somewhat to the thyroid dose received by evacuees. According to Goulko et al. [G5], the most important of these short-lived radionuclides is ^{133}I, amounting to about 30% of the contribution of ^{131}I to the thyroid doses. This maximal value was obtained by taking into account an inhalation for one hour occurring one hour after the accident. Khrouch et al. [K16] estimated that the contribution of all the short-lived radioiodines and of ^{132}Te could have represented about 50% of the dose from ^{131}I if

the intake occurred by inhalation during the first day after the accident and about 10% if the intake occurred by both inhalation and the ingestion of contaminated foodstuffs.

103. Internal effective doses from ^{137}Cs were estimated for the Belarusian evacuees on the basis of 770 measurements of gamma-emitting radionuclides in foodstuffs and of 600 whole-body measurements of ^{137}Cs content, in addition to the environmental measurements already mentioned in Section II.B.1 [S29]. The main contribution to dose was from inhalation (about 75% of total internal dose) and radiocaesium intake in milk. The average internal exposure from radiocaesium in milk for the evacuated population is estimated to be 1.4 mSv. The main sources of uncertainty in the assessment of the internal doses from ^{137}Cs are considered to be the dates when the cows were first put on pasture in each settlement and the actual countermeasures that were applied in the settlement. The collective effective dose for the 24,725 Belarusian evacuees from internal exposure was assessed to be 150 man Sv [S29].

3. Residual and averted collective doses

104. Estimates of collective doses for the populations that were evacuated in 1986 from the contaminated areas of Belarus, the Russian Federation and Ukraine are summarized in Table 23. The collective effective and thyroid doses are estimated to be about 3,800 man Sv and 55,000 man Gy, respectively. Most of the collective doses were received by the populations of Belarus and Ukraine.

105. The evacuation of the residents of Pripyat (28 April) and of the rural settlements inside the 30-km zone (beginning of May) prevented the potential occurrence of deterministic effects and resulted in collective doses substantially lower than would have been experienced if there had been no evacuation. A comparison of the external effective doses for the Belarusians, calculated with and without evacuation from the 30-km zone, is presented in Table 24 [S24]. Because of the evacuation, the number of inhabitants with doses greater than 0.4 Sv was reduced from about 1,200 to 28 persons. The collective effective dose from external exposure averted in 1986 for the approximately 25,000 evacuated Belarusian inhabitants was estimated to be 2,260 man Sv (or approximately 75% of the dose that would have been received without evacuation). A similar assessment of averted collective dose for the evacuated Ukrainian inhabitants led to a value of about 6,000 man Sv. Therefore, the averted collective dose from external exposure for the 116,000 persons evacuated in 1986 is estimated to be 8,260 man Sv.

106. The thyroid collective dose was also reduced to some extent. Iodine prophylaxis was mostly effective in Pripyat, where about 73% of the population received iodine tablets on April 26 and 27, i.e. during the very first days after the accident. It is estimated that a single intake reduced the expected thyroid dose by a factor of 1.6–1.7 and that intakes during two consecutive days reduced it by a factor of 2.3 [R10]. In the rural areas close to the nuclear power

plant, about two thirds of the children used iodine tablets for prophylactic reasons. However, they did not start taking the tablets before 30 April, and about 75% of the children who took iodine tablets began to take them on 2–4 May. Thus, because there was a one-week delay in the use of iodine tablets and because only part of the population was covered, the averted collective thyroid dose from ingestion of contaminated milk was about 30% of the expected collective thyroid dose from that pathway, while the thyroid doses from inhalation remained unchanged. An upper estimate of the averted collective thyroid dose for the 116,000 evacuees is about 15,000 man Gy [A10].

C INHABITANTS OF CONTAMINATED AREAS OF THE FORMER SOVIET UNION

107. Areas contaminated by the Chernobyl accident have been defined with reference to the background level of ^{137}Cs deposition caused by atmospheric weapons tests, which when corrected for radioactive decay to 1986, is about 2–4 kBq m^{-2} (0.05–0.1 Ci km^{-2}). Considering variations about this level, it is usual to specify the level of 37 kBq m^{-2} (1 Ci km^{-2}) as the area affected by the Chernobyl accident. Approximately 3% of the European part of the former USSR was contaminated with ^{137}Cs deposition densities greater than 37 kBq m^{-2} [I3].

108. Many people continued to live in the contaminated territories surrounding the Chernobyl reactor, although efforts were made to limit their doses. Areas of ^{137}Cs deposition density greater than 555 kBq m^{-2} (15 Ci km^{-2}) were designated as areas of strict control. Within these areas, radiation monitoring and preventive measures were taken that have been generally successful in maintaining annual effective doses within 5 mSv. Initially, the areas of strict control included 786 settlements and a population of 273,000 in an area of 10,300 km^2 [I3, I4]. The sizes and populations of the areas of strict control within Belarus, the Russian Federation and Ukraine are given in Table 25. Those population numbers applied to the first few years following the accident. Because of extensive migration out of the most contaminated areas and into less contaminated areas, the current population in the areas of strict control is much lower in Belarus and Ukraine and somewhat lower in the Russian Federation. In 1995, the number of people living in the areas of strict control was about 150,000 [K23, R11]. The distribution of the population residing in contaminated areas in 1995 according to ^{137}Cs deposition density interval is provided in Table 26. The total population is about 5 million and is distributed almost equally among the three countries.

109. In the UNSCEAR 1988 Report [U4], the Committee evaluated separately the doses received during the first year after the accident and the doses received later on. The most important pathways of exposure of humans were found to be the ingestion of milk and other foodstuffs contaminated with ^{131}I, ^{134}Cs and ^{137}Cs and external exposure from radioactive deposits of short-lived radionuclides (^{132}Te, ^{131}I, ^{140}Ba, ^{103}Ru, ^{144}Ce, etc.) and long-lived radionuclides (essentially, ^{134}Cs and ^{137}Cs).

110. In the first few months, because of the significant release of the short-lived ^{131}I, the thyroid was the most exposed organ. The main route of exposure for thyroid dose was the pasture-cow-milk pathway, with a secondary component from inhalation. Hundreds of thousands of measurements of radioiodine contents in the thyroids of people were conducted in Belarus, the Russian Federation and Ukraine to assess the importance of the thyroid doses.

111. During the first year after the accident, doses from external irradiation in areas close to the reactor arose primarily from the ground deposition of radionuclides with half-lives of one year or less. In more distant areas, the radiocaesiums became the greatest contributors to the dose from external irradiation only one month after the accident.

112. Over the following years, the doses received by the populations from the contaminated areas have come essentially from external exposure due to ^{134}Cs and ^{137}Cs deposited on the ground and internal exposure due to contamination of foodstuffs by ^{134}Cs and ^{137}Cs. Other, usually minor, contributions to the long-term radiation exposures include the consumption of foodstuffs contaminated with ^{90}Sr and the inhalation of aerosols containing ^{239}Pu, ^{240}Pu and ^{241}Am. The internal exposures to ^{134}Cs and ^{137}Cs result in relatively uniform doses over all organs and tissues of the body. A very large number of measurements of exposure rates, as well as of radiocaesium in soil and in foodstuffs, have been made in Belarus, the Russian Federation and Ukraine to assess the effective doses and have been used to prepare compilations of annual effective doses received by the most exposed residents in the contaminated settlements. These compilations, which were prepared for regulatory purposes, tend to overestimate the average doses that were received during the years 1986–1990.

113. Since 1991, methods for average dose estimation have been introduced to account for observed changes in radiation levels, as evidenced by experimental dose determinations with TLD measurements and ^{134}Cs/^{137}Cs whole-body counting. These methods were introduced in order to make reasonable decisions regarding the radiation protection of the population, and also to obtain dose estimates for use in epidemiological studies, where accurate individual dose estimates are needed, or in risk assessment studies, where collective doses over limited areas are necessary, and they were an improvement in the general state of knowledge in the field of dose reconstruction. These methods are based on as many measurements as possible, either in the area under consideration or for the individual of interest.

114. The experience thus far acquired and the data accumulated are allowing more realistic dose assessment procedures to be formulated. For example, the external

dose estimates may be related to the contributions from each radionuclide present at the time of deposition, the reduction with time due to radioactive decay and penetration of radionuclides into soil, and shielding and occupancy for various types of buildings and population groups (urban, rural, agricultural workers, schoolchildren, etc.) [G1]. Data from whole-body counting of ^{134,137}Cs have allowed a better estimation of ^{137}Cs retention times in relation to sex for adults and in relation to age, body mass and height for children [L1]. A careful analysis of the thyroid activity measurements, along with the consideration of ^{137}Cs deposition densities and of relevant environmental parameters, has improved the reliability of estimated thyroid doses, although much work remains to be done [G7].

115. When the above methods of dose estimation are used, they may yield several estimates of dose, not necessarily comparable, for example maximal projected doses, average projected doses and actual doses. In local areas there could also be wide deviations from the average settlement dose owing to particular control measures or individual behaviour. Estimates of effective doses per unit deposition density from external and internal exposure have been derived for various districts and times following the accident. These effective dose estimates, as well as the thyroid doses from intake of radioiodines, are discussed below.

1. Doses from external exposure

116. Effective doses have been estimated in Belarus, the Russian Federation and Ukraine on the basis of (a) the large number of measurements of exposure rates and of radionuclide concentrations in soil carried out in the contaminated areas and (b) population surveys on indoor and outdoor occupancy as a function of age, season, occupation and type of dwelling. The methodology that was applied [B14] has some similarities to that used by the Committee in the UNSCEAR 1988 Report [U4]. The effective dose for a representative person of age k is calculated as

$$E_k = D_a F_k \sum_i L_{i,k} B_{i,k}$$

where D_a is the absorbed dose in air over the time period of interest at a reference location at a height of 1 m above flat, undisturbed ground; F_k is the conversion factor from absorbed dose in air to effective dose for a person of age k; $L_{i,k}$ is the location factor, which is the ratio of the absorbed doses in air at location i and at the reference location for a person of age k; and $B_{i,k}$ is the occupancy factor, that is, the fraction of time spent at location i. The location i can be indoors (place of work, place of residence, etc.) or outdoors (street, forest, backyard, etc.).

117. The absorbed dose in air at the reference location, D_a, was usually inferred from the measured or assumed radionuclide distribution in deposition. The conversion

factor from absorbed dose in air to effective dose, F_k, was determined using anthropomorphic phantoms simulating individuals from one year of age to adult, containing TLDs in many organs, exposed to radiocaesium outdoors and indoors [E7, G1, G19]. The values of F_k were found to be 0.7–0.8 Sv Gy^{-1} for adults, 0.8–0.9 Sv Gy^{-1} for 7–17-year-old schoolchildren, and about 0.9–1.0 Sv Gy^{-1} for 0–7-year-old pre-schoolchildren [G1].

118. The term $\sum L_{i,k} B_{i,k}$, called the occupancy/shielding or reduction factor, was derived from population surveys. Values obtained for the reduction factor for rural and urban populations in the Russian Federation [B14] are presented in Table 27, along with the values used by the Committee in the UNSCEAR 1988 Report [U4]. There is good agreement between the two sets of values used for representative groups. Detailed information on the location and occupancy factors derived from surveys among the populations of Belarus, the Russian Federation and Ukraine is available [E7]; for example, values of occupancy factors in the summertime for rural populations of the three countries are presented in Table 28.

119. It is clear from Tables 27 and 28 that there are substantial differences in the reduction factor depending on the type of dwelling and occupation. The values used for the representative group are meant to reflect the age and socioprofessional composition of the population living in a typical dwelling. Estimates of external effective doses for specific groups can be obtained by multiplying the dose for the representative group by a modifying factor, as given in Table 29 [B14]. The values of the modifying factor were validated with data from individual dosimetry (TLD measurements) [E8].

120. Values of the overall coefficient used to calculate the average external effective doses, D_b, on the basis of the absorbed dose in air, D_a, are shown in Table 30. These overall coefficients have different values for urban and rural populations, but for both populations, the values are averaged over age, occupation, and type of dwelling.

(a) Doses from external irradiation received during the first year after the accident

121. For times of less than one year after the accident, the reference absorbed dose rate in air was calculated assuming that the radioactive deposit was a plane source below a soil slab with a mass per unit area of 0.5 g cm^{-2} [E7]. During the first few months after the accident, the dose rate in air varied according to the radionuclide composition of the activity deposited, which, as shown in Table 6, varied according to direction and distance from the reactor. As an example, Figure XIII illustrates the variations in the contributions to the absorbed dose rate in air of various radionuclides from a contaminated area of the Russian Federation [G1]. In that case, the radiocaesiums became the greatest contributors to the dose rate in air only one month after the accident, because the short-lived radio-

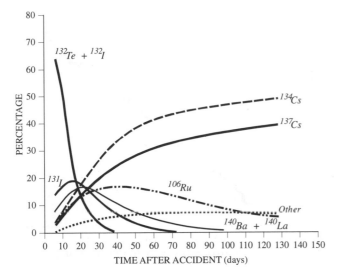

Figure XIII. Contributions of radionuclides to the absorbed dose rate in air in a contaminated area of the Russian Federation during the first several months after the Chernobyl accident [G1].

nuclides and the refractory elements were less important than in areas closer to the reactor. As shown in Figure XIV, the short-lived radioisotopes of refractory elements, such as ^{95}Zr, ^{106}Ru, and ^{141}Ce, played an important role in the doses from external irradiation received during the first year after the accident in areas close to the reactor site [M3]. Following decay of the short-lived emitters, the annual doses per unit ^{137}Cs deposition were similar in all areas, although a slight decrease was observed with increasing distance from the reactor [J1]. Table 31 presents published estimates of normalized effective doses from external irradiation for various periods after the accident and for rural and urban areas in the three countries that were most affected by the accident. The effective doses from external irradiation are estimated to be higher in rural areas than in urban areas by a factor of about 1.5. During the first year after the accident, average values of the normalized effective dose are estimated to have ranged

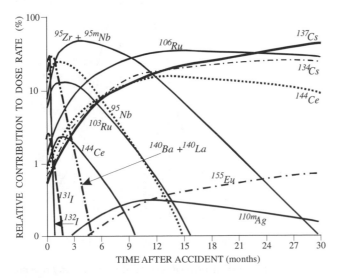

Figure XIV. Contributions of radionuclides to the absorbed dose rate in air in areas close to the Chernobyl reactor site [M3].

from 11 μSv per kBq m^{-2} of ^{137}Cs for urban areas of the Russian Federation to 24 μSv per kBq m^{-2} of ^{137}Cs for rural areas of Ukraine.

122. In summary, during the first year after the accident, the average values of the normalized effective dose are estimated to have been 15-24 μSv per kBq m^{-2} of ^{137}Cs for rural areas and 11-17 μSv per kBq m^{-2} of ^{137}Cs for urban areas, the values for Belarus and Ukraine being higher than those for the Russian Federation because of their closer proximity to the reactor. These values are in agreement with the value of 10 μSv per kBq m^{-2} of ^{137}Cs used by the Committee in the UNSCEAR 1988 Report [U4] for the normalized effective dose equivalent, because most of the data used to derive the 1988 value came from countries further away from the reactor than Belarus, the Russian Federation and Ukraine.

(b) Doses from external irradiation received after the first year following the accident

123. At times greater than one year after the accident, the absorbed dose rate in air came essentially from the gamma radiation from ^{134}Cs and ^{137}Cs. The models used in the three countries (Belarus, the Russian Federation and Ukraine) to derive the variation with time of the normalized absorbed dose rate in air at a height of 1 m above undisturbed ground in the settlements of the contaminated areas are somewhat different. In Belarus, a Monte Carlo method was used; the vertical profile of ^{134}Cs and ^{137}Cs in soil was simulated by a set of infinite isotropic thin sources placed at different depths of soil and an exponential decrease with depth, with an initial relaxation length of 0.5 g cm^{-2} and a linear increase of that value with time after the accident [K38]. In the Russian Federation, the vertical migration of ^{137}Cs to deeper layers of soil was taken into account using a time-varying function r(t), which represents the ratio of the absorbed dose rates in air at a height of 1 m above ground at times t after deposition and at the time of deposition (t = 0), the latter being calculated over flat, undisturbed ground. The variation with time of r(t) may be described as

$$r(t) = a_1 e^{-\ln 2t/T_1} + a_2 e^{-\ln 2t/T_2}$$

with T_1 = 1.5 a, T_2 = 20 a, and a_1 and a_2 equal to 0.4 and 0.42, respectively [B26, M17].

124. In Ukraine, the variation of the normalized absorbed dose rate in air was determined both on the basis of routine measurements of exposure rate at eight reference sites and modelling of the vertical migration of ^{137}Cs. The second approach used the time-varying function given in the above equation but with different parameter values: T_1 = 0.5 a, T_2 = 10 a and a_1 and a_2 equal to 0.18 and 0.65, respectively [M16]. The difference in the estimates obtained for r(t) in the Russian Federation [M17] and in Ukraine [M16] is difficult to explain; it may be partly due to the fact that the measurements were made in different conditions according

to the type of fallout (wet or dry), the distance from the reactor and the type of soil. The values obtained in the three countries for the normalized absorbed dose rate in air are given in Table 32 for each year between 1987 and 1995. The variation with time is fairly similar in the three countries.

125. Average external normalized effective doses for the populations living in contaminated areas are derived from the reference values of normalized absorbed dose rates in air presented in Table 32 and the overall coefficients from dose in air to effective dose presented in Table 30. Results for several time periods are shown in Table 31. Values for rural areas of Belarus for the 1996–2056 time period are estimated in this Annex to be the same as those for the Russian Federation and Ukraine in rural areas; on the basis of data in Table 30, values for urban areas of Belarus are taken to be the same as in rural areas of that country. The selected values of the average normalized external effective doses for urban and rural populations are shown in Table 33. On average, the external doses received during the first 10 years after the accident represent 60% of the lifetime doses (Table 33). The normalized lifetime doses are estimated to range from 42 to 88 μSv per kBq m^{-2} of ^{137}Cs. These values are somewhat lower than the value of 86 μSv per kBq m^{-2} of ^{137}Cs used by the Committee in the UNSCEAR 1988 Report for the normalized effective dose equivalent. This may be due to the fact that in the UNSCEAR 1988 Report, the Committee used the conservative assumption that the vertical profile of ^{137}Cs in soil would be permanently fixed one year after the time of deposition.

126. Average effective doses from external irradiation received during the first 10 years after the accident are estimated to range from 5 mSv in the urban areas of the Russian Federation to 11 mSv in the rural areas of Ukraine. The distributions of the collective effective doses from external irradiation according to region of the country, dose interval, and ^{137}Cs deposition density are presented in Tables 34–36 for Belarus, the Russian Federation and Ukraine. These distributions have been estimated from the databases of radionuclide depositions that are available for each settlement of the contaminated areas of Belarus, the Russian Federation and of Ukraine [B37, L44, M17, S46].

127. The variability of individual external doses can be estimated from the analysis of TLD measurements. Figure XV illustrates the relative distribution of external doses in 1991 and 1992 for 906 inhabitants of 20 Belarusian villages in which the ^{137}Cs deposition density ranged from 175 to 945 kBq m^{-2} [G9, G10]. The individual doses were normalized to the median dose in each settlement. It was found that a log-normal distribution with a geometric standard deviation of 1.54 provides a good approximation of the normalized individual doses from external irradiation. The calculated doses that are recorded in the dose catalogues at that time were in good agreement with the measured median doses, the maximum discrepancy being ±30%.

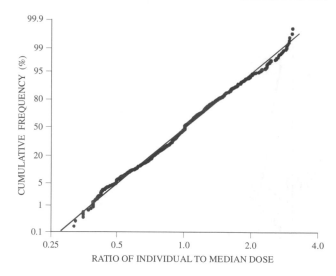

Figure XV. Distribution of ratios of measured external individual doses to median settlement dose for 906 inhabitants of 20 rural settlements of Gomel region in 1991-1992 *(geometric standard deviation: 1.54).*

128. Figure XVI illustrates the distribution of external doses obtained in 1987 in a smaller survey involving the inhabitants of the village of Stary Vyshkov in the Russian Federation [S25]. In that particular case, it was found that a normal distribution with a coefficient of variation of about 1.4 could be used. Individuals who received doses in the upper or lower tenths of the distribution were examined further. The two characteristics found to be important were occupation and the construction of the building in which the individuals spent a large proportion of time. None of the individuals living or working in stone or brick buildings received external doses in the upper tenth percentile of the dose distribution [S25]. In addition, the external dose received as a function of age was also studied for the inhabitants of that Russian village. The results, shown in Figure XVII, indicate a great variability in external dose,

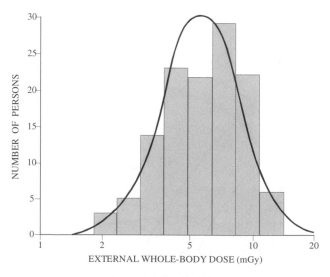

Figure XVI. Distribution of external whole-body doses among 124 residents of Stary Vyshkov, Russian Federation, in 1987 [S25]. *The fitted normal curve is superimposed.*

with an overall trend that suggests an increase in external dose with increasing age [S25]. This may reflect differences in occupational activity, since young people would be expected to spend a large proportion of time indoors at school and, consequently, to receive low external doses, while old people generally spend much time outdoors or inside lightly shielded buildings [S25].

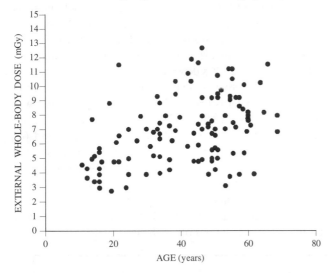

Figure XVII. Variation with age of external whole-body doses among residents of Stary Vyshkov, Russian Federation, in 1987 [S25].

129. The effect of decontamination procedures on external dose was also studied by the analysis of daily external doses calculated from TLD measurements made before and after decontamination of the Belarusian village of Kirov [S25]. Decontamination procedures included replacing road surfaces, replacing roofs on buildings, and soil removal. The results, presented in Table 37, suggest that the decontamination measures were most effective for schoolchildren and field workers (with dose reductions of 35% and 25%, respectively) but had a limited effect on other members of the population [S25]. Similar estimates have been obtained with regard to the decontamination of Russian settlements in 1989 [B38]. The average external dose ratio measured after and before decontamination was found to range from 0.70 to 0.85 for different settlements [B38].

130. The averted collective dose attributable to decontamination procedures was estimated to be about 1,500 man Sv for the first four years after the accident, taking into account the fact that decontamination was only conducted in areas with a ^{137}Cs deposition density greater than 555 kBq m^{-2} and assuming that the doses were reduced by about 20% as a result of the decontamination procedures [A10, I30].

2. Doses from internal exposure

131. The doses from internal exposure came essentially from the intake of ^{131}I and other short-lived radioiodines during the first days or weeks following the accident, and subsequently, from the intake of ^{134}Cs and ^{137}Cs. Other long-lived radionuclides, notably ^{90}Sr and ^{239,240}Pu, have so far contributed relatively little to the internal doses, but

they may play a more important role in the future. Following the Chernobyl accident, about 350,000 measurements of ^{131}I in the thyroids of people [G7, L10, S17] and about 1 million measurements of ^{134,137}Cs whole-body contents [B14, D3, L21] were conducted in the three republics by means of gamma radiation detectors placed outside the body. In addition, thousands of analyses of ^{90}Sr and hundreds of analyses of ^{239}Pu were performed on autopsy samples of tissues.

132. The assessment of the internal doses from radioiodines and radiocaesiums is based on the results of the measurements of external gamma radiation performed on the residents of the contaminated areas. Usually, individuals were measured only once, so that only the dose rate at the time of measurement can be readily derived from the measurement. To calculate the dose, the variation with time of the dose rate needs to be assessed. This is done by calculation, taking into account the relative rate of intake of the radionuclides considered, both before and after the measurement, and the metabolism of these radionuclides in the body, which in the case of thyroid doses from radioiodines may have been modified by the intake of stable iodine for prophylactic purposes. The age-dependent values recommended by the ICRP [I36] for the thyroid mass and the biological half-life of ^{131}I in the thyroid were generally used in thyroid dose assessments based on measurements, although there is evidence of mild to moderate iodine deficiency in some of the contaminated areas [A16].

(a) Thyroid doses from radioiodines and tellurium-132

133. The same methodology as described in the preceding paragraph was used in the three countries to reconstruct the thyroid doses of the persons with thyroid measurements [W7]. There were, however, practical differences related to the quantity and quality of the thyroid measurements and the assumptions used to derive the temporal variation of the radioiodine intake. For the individuals who were not measured but who lived in areas where many persons had been measured, the thyroid doses usually are reconstructed on the basis of the statistical distribution of the thyroid doses estimated for the people with measurements, together with the knowledge of the dietary habits of the individuals for whom the doses are reconstructed. Finally, the thyroid doses for people who lived in areas with very few or no direct thyroid measurements within a few weeks after the accident are being reconstructed by means of relationships using available data on ^{131}I or ^{137}Cs deposition, exposure rates, ^{137}Cs whole-body burdens, or concentrations of ^{131}I in milk. The largest contribution to the thyroid dose was from the consumption of fresh cows' milk contaminated with ^{131}I. Short-lived radioiodines (^{132}I and ^{133}I) in general played a minor role for the populations that were not evacuated within a few days after the accident; the contribution of the short-lived radioiodines and of ^{132}Te is estimated to have been up to 20% of the ^{131}I thyroid dose if the radionuclide intake occurred only through inhalation and of the order of 1% if the consumed foodstuffs (milk in particular) were contaminated [K16]. Although many initial estimates of

thyroid doses are available, they need to be refined using all the relevant and scientifically reviewed information that is available [L42, L43]. In order to obtain better information on the pattern of deposition density of ^{131}I, measurements of the ^{129}I concentrations in soil are envisaged [P25, S45].

134. The influence of having taken stable iodine for prophylactic purposes has usually not been taken into account in the determination of thyroid doses. Based on a survey conducted in 1990 of 1,107 persons living in contaminated areas, the number of persons who indicated that they actually took potassium iodide (KI) for prophylactic purposes is about one quarter of the population [M5]. Forty-five percent of those who took KI indicated that they took it only once, 35% more than once, and 19% could not remember details of their KI prophylaxis [M5]. The exact day that administration of KI was begun was poorly recalled by the subjects, making it difficult to use these data for dose reconstruction purposes. In another survey, conducted in the three most contaminated districts of the Gomel region of Belarus, it was found that 68% of the children who consumed fresh cow's milk took KI

pills between 2 and 4 May 1986 [S43]. However, in a survey performed in 1992 on about 10,000 individuals from 17 Ukrainian districts of the Chernigov region, only about 1% of the respondents reported that they took stable iodine between 1 May and 20 May 1986 [L25].

135. For several reasons, thyroid dose estimates were made independently of ^{137}Cs measurements and not only in areas where the ^{137}Cs deposition density exceeded 37 kBq m^{-2}: (a) the thyroid measurements were carried out within a few weeks after the accident, that is, in large part before an accurate and detailed pattern of ^{137}Cs deposition density was available; (b) the ^{131}I to ^{137}Cs activity ratio in fallout was markedly variable, especially in Belarus (Figure XVIII); (c) the milk that was consumed within a few weeks after the accident was not necessarily of local origin, at least in urban areas; and (d) there is a large variability of the individual thyroid doses according to age and dietary habits. The thyroid dose estimates reported in the scientific literature are for populations with thyroid measurements, for populations that resided at the time of the accident in ill-defined "contaminated areas", and for the entire populations of the three republics.

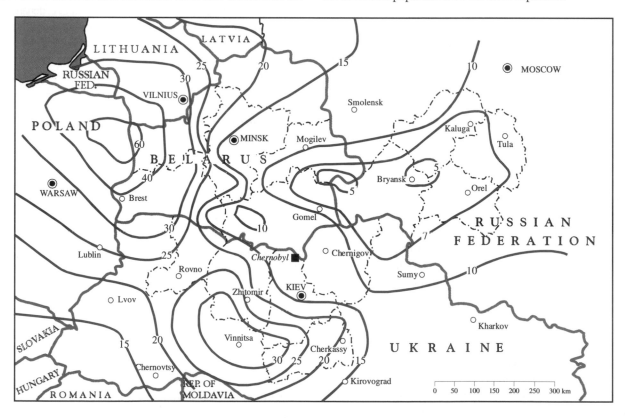

Figure XVIII. Estimated pattern of iodine-131/caesium-137 activity ratio over the European territory of the former USSR resulting from the Chernobyl accident [S30]. *(Values decay corrected to 1 May 1986).*

136. The ratio of ^{131}I to ^{137}Cs deposited by dry processes (i.e. in the absence of precipitation) in Poland was assessed from measurements of air concentrations [K39]. From 28 April to 1 May, the measured time-integrated concentrations of ^{131}I and ^{137}Cs were 187 and 18.2 Bq d m^{-3}, respectively. This ratio of about 10 for the time-integrated concentrations of ^{131}I and ^{137}Cs is consistent with that found in a previous estimation [Z5]. The measured physico-chemical forms of ^{131}I were 62% aerosol-bound, 34%

elemental, and 4% organic. Assuming that (a) all of the ^{137}Cs is aerosol-bound, (b) the deposition velocity of elemental iodine is five times greater than that of the aerosol-bound fraction, and (c) the deposition velocity of organic iodine is negligible, the ratio of ^{131}I to ^{137}Cs deposition can be estimated to be 23 [K39]. The measurements of ^{131}I and ^{137}Cs in soil sampled in a few locations in central and southern Poland yield a ratio of 20 (95% CI: 20–40). These values are in agreement with both the ratio derived from air concentrations and the values shown in Figure XVIII. Measurements of deposition are lacking for the northeastern part of Poland, but the measured concentrations in milk indicate that the ratio of ^{131}I to ^{137}Cs deposition was greater there than in central and southern Poland, again in agreement with Figure XVIII. There are also other reports on the composition of ^{131}I species in the air in different countries. Some of them were presented in the UNSCEAR 1988 Report [U4]. The results indicate that the distribution of the physico-chemical forms changed with time, distance, and weather conditions. Because the transfer of radioiodine from the air to vegetation is highly influenced by its chemical forms, it is important to consider the distribution of iodine species in the assessment.

137. *Belarus.* The main contaminated areas of Belarus are located in the Gomel and Mogilev regions. Within a few weeks after the accident, direct thyroid measurements (i.e. measurements of gamma radiation emitted by the thyroid using detectors placed outside the body) were made on approximately 130,000 persons, including 39,500 children, living in the most contaminated areas of Gomel and Mogilev regions, as well as in the city of Minsk [G6]. The content, at the time of measurement, of ^{131}I in the thyroid of these 130,000 persons was derived from the direct thyroid measurements. The thyroid dose estimation was then performed for the measured individuals, supplementing the results of the direct thyroid measurements with standard radio-ecological and metabolic models, for ^{131}I intake with inhalation and with ingestion of fresh milk following a single deposition of fallout on pasture grass [G6, S44]. Unfortunately, most of the thyroid measurements are of poor quality, as they were made by inexperienced people with uncollimated detectors. The uncertainties in the thyroid dose estimates obtained in this manner in Belarus are reported to be characterized by a geometric standard deviation of up to 1.7 [G7]. A detailed breakdown of the thyroid dose distribution for approximately 32,000 children with thyroid measurements is presented in Table 38. In each age category, the thyroid dose estimates are found to lie in a very wide range (from <0.02 Gy to >2 Gy). As shown in Figure XIX, doses to adults also show a large variability, even if the samples are taken from a single village or town [G6].

138. Limited information is available on *in utero* thyroid doses. In a study of 250 children born during the period from May 1986 to February 1987 from mothers who lived at the time of the accident in areas with ^{137}Cs deposition densities greater than 600 kBq m^{-2} in Gomel region (222 mothers), in Mogilev region (14 mothers) and in Pripyat town (14 mothers who were evacuated to Belarus), thyroid doses were estimated to range up to 4.3 Gy, with 135 children exposed to less than 0.3 Gy, 95 children between 0.3 and 1.0 Gy, and 20 children with doses greater than 1.0 Gy [I37]. Uncertainties in the estimated doses were characterized by a geometric standard deviation of 1.7 to 1.8.

139. Average and collective thyroid doses for the rural and urban populations of the contaminated areas of the Gomel and Mogilev regions were derived from an analysis of dose estimates obtained from direct thyroid measurements [I28].

The results, presented in Table 39 for children 0–7 years old and for the total population, show that the thyroid doses are about two times greater in rural areas than in urban areas and also two times greater in Gomel region than in Mogilev region.

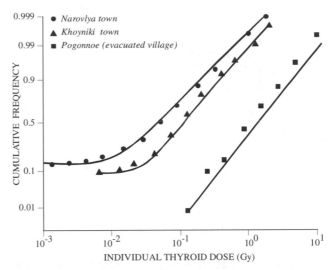

Figure XIX. Cumulative distribution of individual thyroid doses for adults of selected towns and villages of Belarus [G6].

140. Because very few or no thyroid measurements were available for many villages and towns, whereas ^{137}Cs deposition densities were measured in practically all inhabited areas of Belarus, a model was developed to establish a relationship between the ^{137}Cs deposition densities, F(^{137}Cs), and the mean thyroid doses to adults, D_{ad}, in areas where abundant thyroid measurements had been performed. This model enabled the estimation of thyroid dose to be made for the populations of any area in Belarus. In Figure XX, the values of D_{ad} are plotted against those of F(^{137}Cs) for 53 villages of the Khoyniki district. A proportional relationship between the ^{137}Cs deposition density and the mean thyroid dose to adults seems to be inadequate; however, there is a weak tendency shown by the solid line, although characterized by large uncertainties. Similar relationships were observed for all areas of Belarus where abundant thyroid measurements had been performed. That there is no proportional relationship between D_{ad} and F(^{137}Cs) in Belarus is likely to be partly due to the fact that the fraction of ^{131}I intercepted by pasture grass differs according to whether deposition occurs with or without rainfall and varies also as a function of rainfall intensity. The fraction of ^{131}I intercepted by pasture grass is greater when the deposition occurs in the absence of rainfall (usually associated with low levels of deposition) than when deposition occurs with rainfall (generally associated with high levels of deposition). Because of this, the thyroid dose per unit ^{137}Cs deposition density is found to decrease as the ^{137}Cs deposition density increases. A confounding factor is that the ratio of ^{131}I to ^{137}Cs in deposition also varied according to whether deposition occurred in the presence or absence of rainfall. However, similar relationships are observed when the thyroid dose is plotted against either ^{131}I or ^{137}Cs deposition density, suggesting that the variation in the interception coefficient is the dominant factor.

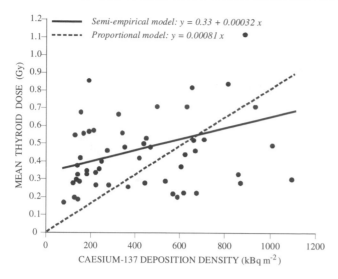

Figure XX. Thyroid dose to adults in relation to caesium-137 deposition density in the district of Khoyniki (Gomel region, Belarus) [G17].

141. Using these relationships for areas with no or few direct thyroid measurements, estimates of collective thyroid dose have been calculated for the entire population of each region of Belarus and for the entire population of the country [G7]. The collective thyroid dose to the entire population of Belarus is roughly estimated to be about 500,000 man Gy (Table 40).

142. *Russian Federation.* The main areas of contamination in the Russian Federation are located 150–250 km to the northeast of Chernobyl in the Bryansk region and at a 500 km distance in the Kaluga-Tula-Orel regions. The ratio of ^{131}I to ^{137}Cs varied little in this area, which indicated that the contamination originated from a single plume. The plume arrived 1–2 days after release from the

reactor. During this time period, most of the short-lived iodine isotopes had decayed. Rainfall in the area decreased the concentrations in air and reduced the inhalation intake. Therefore, the dose to thyroid was due primarily to ^{131}I intake with milk and leafy vegetables, and the pattern of doses was similar throughout the region.

143. About 45,000 direct thyroid measurements were made in May-July 1986 in the Bryansk, Kaluga, Tula and Orel regions [B39, Z1]. These measurements showed a maximum on 16 and 17 May of up to 300 kBq in the thyroid of some individuals in the villages of Barsuki and Nikolayevka in the Krasnogorsk district of the Bryansk region. The ^{137}Cs deposition at these locations was 2.6–3 MBq m^{-2} [Z1]. In other areas, the content of ^{131}I in the thyroid was considerably less, owing to lower contamination and also to earlier implementation of protective measures, including the ban on consumption of local milk and leafy vegetables and the administration of stable iodine. Activities of ^{131}I in the thyroid were calculated from the results of direct thyroid measurements and were corrected for the contribution of the gamma radiation due to radiocaesium incorporated in the entire body.

144. In the absence of protective measures, the temporal variation of the ^{131}I intake, taking into account inhalation and the ingestion of contaminated milk, is calculated from standard radio-ecological models shown in Figure XXI (left panel). However, for the purposes of dose reconstruction, a simplified representation has been adopted (Figure XXI, right panel) [B14]. The thyroid mass was determined from autopsies in the Novozybkov district hospital in the Bryansk region. The average value for adults was 26.7 g, suggesting a mildly endemic goiter area. The Tula and Orel regions are not in endemic areas, and as direct measurements were unavailable, the standard thyroid mass for adults of 20 g was used in dose calculations [Z1].

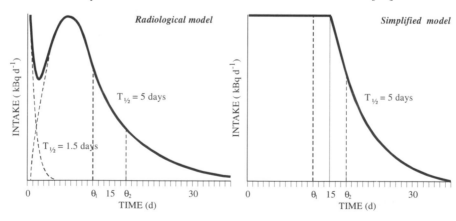

Figure XXI. Models of iodine-131 intake to inhabitants of contaminated areas in Russia [B14].

145. Thyroid doses were estimated in this manner for six age groups: <1, 1–2, 3–5, 7–11, 12–17 and >18 years. Within each age group the distribution of dose was asymmetrical, approximately log-normal. The maximum individual doses often exceeded the mean dose by a factor of 3–5. The variations between age groups were different for towns and villages, reflecting not only the age-related iodine metabolism but also differences in social and nutritional habits. As

presented in Table 41, the average thyroid doses for children less than one year old were greater than those for adults by factors of 13 in towns and 5 in villages [B14, Z1].

146. Where measurements were insufficient or lacking, correlations were used to estimate the thyroid doses. The uniformity of contamination allowed correlation analyses to be used to relate the thyroid doses to the deposition of ^{137}Cs, the

air kerma rate on 10–12 May 1986, the concentrations of ^{131}I in milk, and the body content of ^{137}Cs in adults measured within a few months after the accident. The analysis of the results of the direct thyroid measurements for inhabitants of the Kaluga region showed that the thyroid doses of people who did not consume local milk was about 15% of the thyroid doses received by the people who consumed local milk. The type and number of data available in the Russian Federation are presented in Table 42.

147. The analysis of the direct thyroid measurements and of data from personal interviews for 600 inhabitants of the Bryansk region showed a significant correlation with milk consumption. From 80% to 90% of ^{131}I intake appeared to be derived from this source and only 10% to 20% from vegetables and inhalation. Thus, estimates of doses to individuals could be derived by normalizing 80% of the average dose for the settlement by the actual volumes of milk consumed (litres per day times days) relative to the average consumed volume.

148. Estimates of thyroid doses in contaminated areas of the Russian Federation are presented in Table 43. In areas where there were no limitations on ^{131}I intake (e.g. Plavsk district of Tula region), the thyroid dose for children less than 3 years old reached 0.35 to 0.7 Gy, on average, with individual doses up to 4 Gy. In the most contaminated areas of the Orel region, the thyroid doses were approximately 0.3 Gy, on average, for young children. The highest doses were received by inhabitants of the most contaminated areas of the Bryansk region even though local milk consumption was banned in those areas in early May 1986. In some villages average doses in children exceeded 1 Gy and individual doses exceeded 10 Gy. Average thyroid doses of rural inhabitants were higher than those received by urban populations in areas with similar radioactive contamination. The age distribution of the collective dose for the population of the Bryansk region is presented in Table 44. About 40% of the collective thyroid dose in rural areas and 60% of the collective thyroid dose in urban areas were received by children under 15 years of age.

149. *Ukraine.* Over 150,000 direct thyroid measurements of the radioiodine content of the thyroid gland were made in May-June 1986 in the areas of Ukraine closest to Chernobyl. Most of the thyroid measurements were made in eight districts surrounding Chernobyl and the town of Pripyat: Polesskoe, Ivanov, Chernobyl and Pripyat in the Kiev region; Kozeletsk, Repkine and Chernigov in the Chernigov region; and Narodichi and Ovruch in the Zhitomir region [L12]. Between 30% and 90% of the children and 1% and 10% of the adults from these areas were measured [L2, L11, L12, R4]. When the quality of these measurements was reviewed, over 80% were found to be of acceptably high quality [L13, L14].

150. Preliminary estimates of the thyroid doses received by the persons with direct thyroid measurements were made using standard methods [A3, I13]. Except for the city of Kiev, where the ^{131}I concentrations in air, water and milk were monitored extensively [L15], very few measurements of ^{131}I in the environment are available from Ukraine. Two models have been used to describe the variation of the temporal intake

of ^{131}I. According to the most conservative model, a single intake of ^{131}I was assumed to have occurred on the first day of the accident, and a single exponential is used to describe the iodine retention in the thyroid gland. The more realistic model for calculating thyroid doses assumes that ^{131}I intake occurred during the entire period of stay in the contaminated areas [L12]. The intake function was determined assuming a single initial contamination event by ^{131}I in an area. A two-exponential function is used to represent the dynamics of intake by milk consumption. The effective decay constants from grass and milk are 0.15 and 0.63 d^{-1}, respectively [A3]. The period of intake was taken to be the period until relocation or, if there was no relocation or the information is lacking, the whole period until ^{131}I decay.

151. Results of the thyroid dose evaluations for children and adults of the Ukraine indicate that the highest absorbed doses (1.5–2.7 Gy) were received by children of the Narodichi and Ovruch districts of Zhitomir region and of Pripyat and the Polesskoe districts of Kiev region. Doses to children 7–15 years old were, in general, 2.5 times lower than doses to the 0–7-year-old group. The adult doses were lower by a factor of 2–8 [L2, L12, R4]. According to the conservative, single-exponential model, there were 38,000 children (>40%) with doses lower than 0.3 Gy and 79,500 (nearly 90% of children) with doses lower than 2 Gy. Use of the more realistic model generally shifts the distribution to lower doses. In this case, 63% of children had doses below 0.3 Gy [L12]. The distribution of thyroid doses in a settlement usually was found to be log-normal.

152. The estimation of doses to individuals living in the city of Kiev was performed using direct thyroid measurements for approximately 5,000 residents and measured ^{131}I concentrations in air, water and milk during May-June 1986 [L16, L17]. The individual thyroid doses were found to vary by an enormously wide range of up to four orders of magnitude [L15]. The average thyroid doses to individuals of five age groups were as follows: 0.10 Gy (birth years 1983–1986), 0.06 Gy (1979–1982), 0.2 Gy (1975–1978 and 1971–1974), and 0.04 Gy (those born before 1974) [L15].

153. To estimate the thyroid doses received by the persons without direct thyroid measurements living in areas other than the city of Kiev, two procedures were used, depending on the abundance of the thyroid measurements in the area considered. In the three regions where most of the thyroid measurements were performed (Chernigov, Kiev and Zhitomir), the following empirical relationship between the measured thyroid doses, D, and the ^{137}Cs deposition density, as well as the location relative to the Chernobyl reactor, was determined to be D(n) = K a^t [L10, L25], where t = e^{-bn}, D is the thyroid dose (Gy), n is the age of the individual (years), and K is a scaling parameter (Gy). The parameters a (dimensionless) and b (a^{-1}) describe the age dependence of the thyroid dose in the locality considered.

154. The parameter values for K, a, and b were established for the towns and villages of each district of the three regions. As examples, Table 45 presents the values obtained for three districts of the Chernigov region as well as the mean thyroid

doses for infants and adults [L25], while the measured individual doses are compared in Figure XXII with the calculated mean thyroid doses for various age groups in Rudka village of the Chernigov district [L25]. As is the case in Belarus and the Russian Federation, the mean thyroid doses in villages are about twice those in urban areas (Table 45). However, the variability of the individual doses, within a given village and a given age group, is very great (Figure XXII).

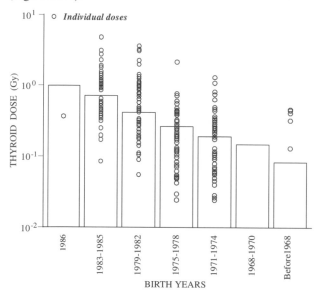

Figure XXII. Comparison of individual doses estimated from thyroid measurements and of calculated mean doses (histogram) for the Ukrainian village of Rudka [L25].

155. Another method of thyroid dose assessment was used for the regions of Cherkasy and Vinnytsia, where very few thyroid measurements were carried out. Those regions were subdivided into sectors and segments with relatively uniform ^{131}I intake functions [L10]. The relationships established for areas with thyroid measurements were then extrapolated to other areas.

156. The age-dependent thyroid dose distribution obtained for the population of the five regions (Cherkasy, Cherni-gov, Kiev, Vinnytsia and Zhitomir) is given in Table 46. Most of the thyroid doses are estimated to be less than 0.3 Gy. Doses exceeding 2 Gy are found only among children less than 4 years old [L10]. An estimate of the collective thyroid dose to residents of Ukraine is presented in Table 47.

(b) Effective doses from caesium-134 and caesium-137

157. Internal effective doses from [134]Cs and ^{137}Cs have been estimated by two methods: (a) estimation of dietary intake from measured concentrations in foods and standard consumption assumptions and (b) whole-body counting. The foodstuffs that contribute the most to the effective dose are milk, meat, potatoes and mushrooms [F4]. From 1986 to 1990, the internal doses were calculated from the measured ^{137}Cs concentrations in milk and potatoes, assuming that the

intake of radiocaesium by ingestion is adequately represented by consumption rates of 0.8 l d^{-1} of milk and 0.9 kg d^{-1} of potatoes. The concentrations used to calculate the doses were those corresponding to the 90th percentiles of the distributions [A12]. The relationship between the 90th percentiles to the average ^{137}Cs concentrations in milk and potatoes was 1.7±0.1 [B9]. Beginning in 1991, the average ^{137}Cs concentrations in milk and potatoes, rather than the 90th percentiles, were used to estimate the effective doses from internal irradiation.

158. Calculated internal doses assuming consumption only of locally produced foods (no imported, uncontaminated foods) have been recognized to overestimate actual doses. The dose estimates calculated in this manner are used only for decision-making purposes. The ratios of calculated doses using the 90th percentiles of the distributions to the doses determined by whole-body counting range from 2.5 to 25 for most settlements in the zone of strict control (where the ^{137}Cs deposition density is greater than 555 kBq m^{-2}), with a median value of 7 [B9]. In areas with lower ^{137}Cs deposition density, where most of the consumed foodstuffs are of local origin, this ratio is estimated to be in the range 1.5–15, with a median of 4. For this reason, the internal effective dose estimates provided currently in the official dose catalogues in Belarus and the Russian Federation are based on whole-body measurements. However, the dose estimates presented in the Ukrainian dose catalogues are still based on the assessment of the dietary intakes.

159. The transfer of radiocaesium from soil to milk depends substantially on the type of soil. For example, in Ukrainian territory, as a result of radio-ecological monitoring carried out in 1991, four zones with typical values of soil-milk transfer coefficient for radiocaesium ranging from less than 1 to greater than 10 Bq l^{-1} per kBq m^{-2} were delineated [K23]. The corresponding normalized effective doses from internal irradiation are shown in Table 48 for the first 10 years after the accident; the estimated normalized effective doses for the zone with highest values of the transfer coefficient are about 20 times greater than those obtained in the zone with the lowest values of the transfer coefficient. The territories with peat-swampy soil that are characterized with the highest values of soil-milk transfer coefficient are mainly located in the northern part of the Rovno and Zhitomir regions (Ukraine) and in the eastern part of the Brest region and the southwestern part of the Gomel region (Belarus).

160. In the Russian Federation, normalized doses were estimated for the sodic-podzol sand soil found in some areas of the Bryansk and Kaluga regions and for the chernozem soil found in the Tula and Orel regions [B25, R9]. Here again, the values of the transfer coefficients and of the normalized effective doses vary by factors of 10–20 (Table 49).

161. Estimated internal effective doses, normalized to unit deposition density of [137]Cs (1 kBq m^{-2}), are given in Table 50 for various periods after the accident and for areas with different degrees of contamination in the Russian Federation. More detailed information for the population of Belarus is presented in Table 51. It is

recognized that the internal effective doses normalized per unit deposition density of ^{137}Cs have to be treated with caution, because of the large differences in the transfer of ^{137}Cs from soil to milk and because of the influence of protective measures. Estimates of projected doses from internal exposures are highly uncertain, as they depend on local soil conditions for the transfer to foodstuffs, on the composition of the diet, and on the extent to which local foods are supplemented by imported foods [Z4]. In particular, it is to be noted that the importance of forest products (mushrooms, berries, wild game) increases with time, since the ^{137}Cs concentration in these products generally have longer ecological half-times than food products from agricultural systems (milk, vegetables, meat from domestic animals) [K7, S16]. Average internal effective doses received during the first 10 years after the accident are estimated to range from 4 mSv in the rural areas of Tula region in the Russian Federation to 13 mSv in the rural areas of Bryansk region in the Russian Federation. On average, the internal doses received during the first 10 years after the accident represent 90% of the lifetime doses (Table 50). The distributions of the collective effective doses from internal irradiation according to the region of the country, dose interval, and ^{137}Cs deposition density are presented in Tables 34–36 for Belarus, the Russian Federation and Ukraine.

162. Several measures were taken to reduce the internal exposure to the residents of the contaminated areas. During the first year after the accident, the most important measures were taken in the territories of strict control (territories with a ^{137}Cs deposition density exceeding 555 kBq m^{-2}). In Belarus, the resulting factor of decrease in the internal dose was estimated to be 3.2–3.4 [S24], while in the Bryansk region it was about 3.6. In the following years, the corresponding factors of decrease were maintained at approximately the same levels: 4.1 [I30] and 3.7 [I22], respectively. Therefore, the averted collective dose for the strict control zone (inhabited by 273,000 residents) can be assessed to be 6,000 man Sv. For the territories with a ^{137}Cs deposition density from 185 to 555 kBq m^{-2} (inhabited by about 1,300,000 people), the factor of decrease in the internal dose was estimated to be approximately 2 [S24], resulting in an averted dose of 7,000 man Sv. Thus, the averted collective dose for about 1,500,000 residents of the areas contaminated with a ^{137}Cs level exceeding 185 kBq m^{-2} was assessed at nearly 13,000 man Sv. It should be noted that self-imposed measures also led to a decrease in the internal doses [L21].

163. Within the framework of the Chernobyl Sasakawa Health and Medical Cooperation Project, about 120,000 children, aged 0–10 years at the time of the accident, were examined in two Belarusian centres, two Ukrainian centres and one Russian centre [N2, S2, S7]. The examinations were essentially of a medical nature but included a measurement of the whole-body concentration of ^{137}Cs. Average values were similar from centre to centre and from year to year, with an overall average of about 50 Bq kg^{-1} of ^{137}Cs; the corresponding internal effective dose rate from ^{137}Cs is about 0.1 mSv a^{-1}.

164. The Ministry of the Environment, Protection of Nature, and Reactor Safety of Germany also organized a campaign of whole-body counting in Belarus, the Russian Federation and Ukraine. About 300,000 persons were monitored from 1991 to 1993 for their ^{137}Cs whole-body content [H1, H4, H5]. For 90% of the persons monitored, the internal effective dose rates from ^{137}Cs were found to be less than 0.3 mSv a^{-1}. The analysis of the results for Kirov, in Belarus, shows that the population monitored could be classified in one of five groups, according to the nature of their diet and the origin of the consumed milk, with increasing ^{137}Cs content from one group to the next; the dietary characteristics of those groups are defined as (a) no milk, no forest products; (b) local milk, no forest products; (c) non-local milk, forest products; (d) local milk, forest products but no wild game; and (e) local milk and forest products including wild game [S25].

(c) Internal doses from strontium-90

165. Because of the relatively small release of ^{90}Sr and because a large fraction of the ^{90}Sr activity released was deposited within the 30-km zone, internal doses from ^{90}Sr are relatively small. It is estimated that the ^{90}Sr contribution to the effective dose from internal exposure does not exceed 5%–10%, according to intake calculations based on measurements of ^{90}Sr concentrations in foodstuffs, as well as measurements of ^{90}Sr in human bones [B15].

(d) Lung doses from transuranics

166. The fuel particles that deposited on the ground contained alpha-emitting transuranics, such as ^{238}Pu, ^{239}Pu, ^{240}Pu and ^{241}Am, as well as beta-emitting transuranics, such as ^{241}Pu. Lung doses from transuranics may be caused by inhalation of radioactive particles during the passage of the cloud and following resuspension of deposited materials. A pathway of concern is the potential hazard from the resuspension from soil to air of radioactive aerosols containing ^{239}Pu. In an assessment of the equivalent doses to lungs for agricultural workers, it was concluded that even at sites inside the 30-km zone, lifetime committed equivalent doses to the lungs per year of work will not exceed 0.2 mSv for most individuals [J1]. In the future, the relative importance of ^{241}Am among the alpha-emitting transuranics is going to increase as a result of the decay of its shorter-lived parent, ^{241}Pu.

D. INHABITANTS OF DISTANT COUNTRIES

167. Information on doses received by populations other than those of Belarus, the Russian Federation and Ukraine is not as complete. In the UNSCEAR 1988 Report [U4], the Committee estimated first-year thyroid and effective doses for most of the countries of the northern hemisphere and lifetime effective doses for three latitude bands (northern, temperate and southern) of the northern hemisphere. These data have been used in this Annex to estimate crude average thyroid and bone-marrow doses received by the populations considered in

epidemiological studies of thyroid cancer and of leukaemia. These populations usually resided in areas where the deposition densities of ^{137}Cs resulting from the Chernobyl accident were the highest (except for contaminated areas of Belarus, the Russian Federation and Ukraine).

1. Thyroid doses

168. Populations of Croatia, Greece, Hungary, Poland and Turkey have been considered in epidemiological studies of thyroid cancer [S12]. The average thyroid doses received by those populations have been estimated, using the approximation that the population-weighted average thyroid dose is three times that to adults. Results are presented in Table 52. The estimates of average thyroid dose range from 1.5 to 15 mGy.

2. Bone-marrow doses

169. Populations of Bulgaria, Finland, Germany, Greece, Hungary, Romania, Sweden and Turkey have been considered in epidemiological studies of leukaemia [S12]. The average bone-marrow doses received by those populations have been estimated from data in the UNSCEAR 1988 Report [U4], using the approximation that the bone-marrow dose is numerically equivalent to the effective dose equivalent from all radionuclides for external irradiation and from radiocaesium for internal irradiation. The values of effective dose equivalents per unit ^{137}Cs deposition density that are calculated in the UNSCEAR 1988 Report for the populations of three latitude bands have been assigned in this Annex to the populations of the countries considered, as appropriate. In each country, the ^{137}Cs deposition density corresponding to the region of highest fallout has been selected, when that information is provided in the UNSCEAR 1988 Report. The resulting estimates of average bone-marrow dose for the populations considered range from about 1 to 4 mGy (Table 52).

E. COLLECTIVE DOSES

170. In this Section, the collective doses received by the populations of the contaminated areas of Belarus, the Russian Federation and Ukraine are summarized.

1. Collective doses from external exposure

171. The collective effective doses from external exposure received by the inhabitants of the contaminated areas during the first 10 years after the accident have been estimated using the average ^{137}Cs deposition densities in each district and estimated average annual effective doses from external exposure in each district. Detailed estimates of the collective effective doses and of their distributions are presented in Tables 34-36 for Belarus, the Russian Federation and Ukraine. The totals for each country are presented in Table 53, while the distribution of the collective effective doses for the three republics is summarized in Table 54. The total collective effective dose

received during the first 10 years after the accident by the approximately 5.2 million people living in the contaminated areas of Belarus, the Russian Federation and Ukraine is estimated to be 24,200 man Sv. Assuming that this collective dose represents 60% of the lifetime collective dose, on the basis of the data presented in Table 33, the lifetime collective dose from external irradiation received by the inhabitants of the contaminated areas of the three republics would be 40,300 man Sv.

2. Collective doses from internal exposure

172. *Collective effective doses.* The collective effective doses from internal exposure received by the inhabitants of the contaminated areas during the first 10 years after the accident have also been estimated using the average ^{137}Cs deposition densities in each district and estimated average annual effective doses from internal exposure in each district. They are found to be about 5,500 man Sv for Belarus [D3], 5,000 man Sv for the Russian Federation and 7,900 man Sv for Ukraine [L7]. Detailed estimates of the collective effective doses and of their distributions are presented in Tables 34-36 for Belarus, the Russian Federation and Ukraine. Whether those estimates were obtained using similar methodologies has not been thoroughly clarified. From the data presented in Table 50, the doses from internal exposure received during the first 10 years after the accident represent about 90% of the lifetime doses. The collective effective doses from internal exposure received by the population of the contaminated areas can thus be estimated to be about 18,400 man Sv for the first 10 years after the accident and about 20,400 man Sv over lifetime; this corresponds to an average lifetime effective dose of 3.9 mSv.

173. *Collective thyroid doses.* Estimated collective thyroid doses, as reported for populations of Belarus, the Russian Federation and Ukraine, are presented in Table 40. Collective thyroid doses are estimated to be about 550,000, 250,000 and 740,000 man Gy for the entire populations of Belarus, the Russian Federation and Ukraine, respectively.

3. Total collective doses

174. Estimated collective effective doses received during the 1986-1995 time period by the inhabitants of the contaminated areas of Belarus, the Russian Federation and Ukraine are presented in Table 53. The collective effective doses that were delivered during the first 10 years after the accident are estimated to be about 24,200 man Sv from external exposure and 18,400 man Sv from internal exposure, for a total of 42,600 man Sv and an average effective dose of 8.2 mSv. Assuming that the doses delivered during the first 10 years represent 60% of the lifetime dose for external exposure and 90% of the lifetime dose for internal exposure, the estimated lifetime effective doses for the populations of the three countries living in contaminated areas are about 40,300 man Sv from external exposure and 20,400 man Sv from internal exposure, for a total of about 60,700 man Sv. This total corresponds to an average lifetime effective dose of 12 mSv.

175. These figures do not include the collective thyroid doses, which were delivered in their totality during 1986, and which are estimated to be 1,500,000 man Gy in total for the three countries. Taking the population size of the three republics to be 215 million, the average thyroid dose is found to be 7 mGy. Much larger thyroid doses, however, were received by a small fraction of the population. For example, the distribution of the thyroid doses received in Ukraine is such that very low thyroid doses were received in a large part of the country, while average thyroid doses greater than 500 mGy were received in twelve districts.

F. SUMMARY

176. Doses have been estimated for: (a) the workers involved in the mitigation of the accident, either during the accident itself (emergency workers) or after the accident (recovery operation workers) and (b) members of the general public who either were evacuated to avert excessive radiation exposures or who still reside in contaminated areas, which are found mainly in Belarus, in the Russian Federation and in Ukraine. A large number of radiation measurements (film badges, TLDs, whole-body counts, thyroid counts, etc.) were made to evaluate the radiation exposures of the population groups that are considered.

177. The highest doses were received by the approximately 600 emergency workers who were on the site of the Chernobyl power plant during the night of the accident. The most important exposures were due to external irradiation, as the intake of radionuclides through inhalation was relatively small in most cases. Acute radiation sickness was confirmed for 134 of those emergency workers. Forty-one of these patients received whole-body doses from external irradiation of less than 2.1 Gy. Ninety-three patients received higher doses and had more severe acute radiation sickness: 50 persons with doses between 2.2 and 4.1 Gy, 22 between 4.2 and 6.4 Gy, and 21 between 6.5 and 16 Gy. The skin doses from beta exposures evaluated for eight patients with acute radiation sickness ranged from 10 to 30 times the whole-body doses from external irradiation.

178. About 600,000 persons (civilian and military) have received special certificates confirming their status as liquidators (recovery operation workers), according to laws promulgated in Belarus, the Russian Federation and Ukraine. Of those, about 240,000 were military servicemen. The principal tasks carried out by the recovery operation workers included decontamination of the reactor block, reactor site, and roads, as well as construction of the sarcophagus, of a town for reactor personnel, and of waste repositories. These tasks were completed by 1990.

179. A registry of recovery operation workers was established in 1986. This registry includes estimates of doses from external irradiation, which was the predominant pathway of exposure for the recovery operation workers. The registry data show that the average recorded doses decreased from year to year, being about 0.17 Sv in 1986, 0.13 Sv in

1987, 0.03 Sv in 1988, and 0.015 Sv in 1989. It is, however, difficult to assess the validity of the results that have been reported for a variety of reasons, including (a) the fact that different dosimeters were used by different organizations without any intercalibration; (b) the high number of recorded doses very close to the dose limit; and (c) the high number of rounded values such as 0.1, 0.2, or 0.5 Sv. Nevertheless, it seems reasonable to assume that the average effective dose from external gamma irradiation to recovery operation workers in the years 1986–1987 was about 0.1 Sv.

180. The doses received by the members of the general public resulted from the radionuclide releases from the damaged reactor, which led to the ground contamination of large areas. The radionuclide releases occurred mainly over a 10-day period, with varying release rates. From the radiological point of view, the releases of ^{131}I and ^{137}Cs, estimated to have been 1,760 and 85 PBq, respectively, are the most important to consider. Iodine-131 was the main contributor to the thyroid doses, received mainly via internal irradiation within a few weeks after the accident, while ^{137}Cs was, and is, the main contributor to the doses to organs and tissues other than the thyroid, from either internal or external irradiation, which will continue to be received, at low dose rates, during several decades.

181. The three main areas of contamination, defined as those with ^{137}Cs deposition density greater than 37 kBq m^{-2} (1 Ci km^{-2}), are in Belarus, the Russian Federation and Ukraine; they have been designated the Central, Gomel-Mogilev-Bryansk and Kaluga-Tula-Orel areas. The Central area is within about 100 km of the reactor, predominantly to the west and northwest. The Gomel-Mogilev-Bryansk contamination area is centred 200 km to the north-northeast of the reactor at the boundary of the Gomel and Mogilev regions of Belarus and of the Bryansk region of the Russian Federation. The Kaluga-Tula-Orel area is located in the Russian Federation, about 500 km to the northeast of the reactor. All together, territories from the former Soviet Union with an area of about 150,000 km^2 were contaminated. About five million people reside in those territories.

182. Within a few weeks after the accident, approximately 116,000 persons were evacuated from the most contaminated areas of Ukraine and of Belarus. The thyroid doses received by the evacuees varied according to their age, place of residence and date of evacuation. For example, for the residents of Pripyat, who were evacuated essentially within 48 hours after the accident, the population-weighted average thyroid dose is estimated to be 0.17 Gy, and to range from 0.07 Gy for adults to 2 Gy for infants. For the entire population of evacuees, the population-weighted average thyroid dose is estimated to be 0.47 Gy. Doses to organs and tissues other than the thyroid were, on average, much smaller.

183. Thyroid doses have also been estimated for residents of the contaminated areas who were not evacuated. In each of the three republics, thyroid doses exceeding 1 Gy were estimated for the most exposed infants. For residents of a

given locality, thyroid doses to adults were smaller than those to infants by a factor of about 10. The average thyroid dose received by the population of the three republics is estimated to be 7 mGy.

184. Following the first few weeks after the accident when ^{131}I was the main contributor to the radiation exposures, doses were delivered at much lower dose rates by radionuclides with much longer half-lives. Since 1987, the doses received by the populations of the contaminated areas have resulted essentially from external exposure from [134]Cs and ^{137}Cs deposited on the ground and internal exposure due to contamination of foodstuffs by [134]Cs and [137]Cs. Other, usually minor, contribu-

tions to the long-term radiation exposures include the consumption of foodstuffs contaminated with ^{90}Sr and the inhalation of aerosols containing isotopes of plutonium. Both external irradiation and internal irradiation due to [134]Cs and ^{137}Cs result in relatively uniform doses in all organs and tissues of the body. The average effective doses from [134]Cs and ^{137}Cs that were received during the first 10 years after the accident by the residents of contaminated areas are estimated to be about 10 mSv. The median effective dose was about 4 mSv and only about 10,000 people are estimated to have received effective doses greater than 100 mSv. The lifetime effective doses are expected to be about 40% greater than the doses received during the first 10 years following the accident.

III. EARLY HEALTH EFFECTS IN REACTOR AND EMERGENCY WORKERS

185. The first information on the early manifestations and outcomes of acute radiation sickness in persons who were exposed to ionizing radiation in the early phase of the Chernobyl accident was provided to the international community in Vienna in August 1986 [I2]. A detailed and comprehensive review of these effects was included in the UNSCEAR 1988 Report (Appendix to Annex G, "Early effects in man of high doses of radiation") [U4]. This Chapter describes the health effects observed in this group in the years since the accident. Dose estimations for those working at the Chernobyl nuclear power plant on 26 April, 1986, are given in Section II.A.1.

186. Among the staff members of the reactor and emergency workers at the site at the time of the accident, a total of 237 were initially examined for signs and symptoms of acute radiation sickness, defined here as having at least minimal bone-marrow suppression as indicated by depletion of blood lymphocytes. This diagnosis was later confirmed in 134 patients, the others being designated as unconfirmed. A computerized questionnaire for patients with acute radiation sickness was developed in 1990 [F11] and later extended, incorporating other nuclear accidents [F3]. The fate of the whole group of 237 patients has been monitored up to the present, although not always systematically, with accurate data available for most patients for the acute phase and incomplete information available for the follow-up period of 1986-1996.

187. The definition of acute radiation sickness is well established and based on clinical observations and the degree of pancytopenia [B22, K26]. A reliable assessment of the severity of acute radiation sickness from mild (Grade I) to severe (Grade IV) is possible at three days following exposure. To predict the likelihood of bone marrow recovery, damage to the stem cell pool must be determined [F12]. Seven to ten days after exposure, patients in whom prolonged myelosuppression was diagnosed were selected for bone marrow transplantation.

188. Among 37 patients considered for transplantation treatment, all had severe radiation damage to the skin and,

in 15 cases, gastrointestinal tract symptoms. Cutaneous lesions and/or oropharyngeal mucositis were the primary causes of death in the majority of these patients who later died as an immediate consequence of the accident. As might be expected, there was a clear relationship between the extent of local skin radiation injury, the grade of acute radiation sickness and mortality. Patients not selected for bone marrow transplantation received supportive therapy such as transfusions and antibiotics.

189. A total of 13 patients with estimated whole-body doses of 5.6 to 13 Gy received bone marrow transplants at the Institute of Biophysics of the Ministry of Health and Clinical Hospital, Moscow [B40]. Two transplant recipients, who received estimated radiation doses of 5.6 and 8.7 Gy, were alive more than three years after the accident. The others died of various causes, including burns (n = 5), interstitial pneumonitis (n = 3), graft-vs-host disease (n = 2), and acute renal failure and respiratory distress syndrome (n = 1).

190. Stable chromosome aberrations in circulating stem cells, indicating residual damage in the stem and progenitor cells, were used for retrospective dosimetry [K45]. Unstable chromosome aberrations seemed to be a less reliable proxy for average whole-body dose unless evaluated shortly after exposure [T15].

191. The distribution of patients with acute radiation sickness by severity of disease and range of absorbed dose from whole-body gamma radiation is given in Table 11. Among the 134 cases, 28 died within the first four months of the accident. The causes of death are listed in the UNSCEAR 1988 Report, Appendix to Annex G, "Early effects in man of high doses of radiation" [U4]. In the early period (14-23 days after exposure), 15 patients died of skin or intestinal complications and 2 patients died of pneumonitis. In the period 24-48 days after exposure, six deaths from skin or lung injury and two from secondary infections following bone-marrow transplantation

occurred. Between 86 and 96 days following the accident, two patients died from secondary infections and one patient from renal failure [U4]. Underlying bone marrow failure was the main contributor to all of these deaths.

192. There have been eleven deaths between 1987 and 1998 among confirmed acute radiation sickness survivors who received doses of 1.3–5.2 Gy. The causes of death are presented in Table 55. There were three cases of coronary heart disease, two cases of myelodysplastic syndrome, two cases of liver cirrhosis, and one death each of lung gangrene, lung tuberculosis and fat embolism. One patient who had been classified with Grade II acute radiation sickness died in 1998 from acute myeloid leukaemia.

193. At exposures below 6 Gy, the bone-marrow depletion was not the direct cause of death when prompt and adequate treatment of the complications could be provided. The therapy decreased the incidence and severity of infectious complications and haemorrhagic manifestations [G21, S33, W2].

194. Inflammation of the oropharynx (mucositis) was apparent even at relatively low doses (1–2 Gy) of gamma radiation 4–6 days following exposure. An unknown influence of beta radiation could explain these findings. The incidence of mucositis increased and reached 100% in patients receiving gamma doses of 6–13 Gy. The pathogenesis of the oropharyngeal syndrome is complex, as it is determined by the initial radiation damage to the skin and mucosa and further complicated by infections of viral, bacterial and fungal species. Recovery of the mucosal epithelium was observed even in severe acute radiation sickness survivors, which is typical of beta radiation effects [G22, P10]. Acute gastrointestinal symptoms were observed in 15 Chernobyl accident victims and were the most severe symptoms in 11 patients who received doses higher than 10 Gy.

195. Radiation skin burns were observed in 56 patients, including 2 patients with combined radiation-thermal burns. Alopecia, onycholysis, mucositis, conjunctivitis and acute radiation ulcers were seen. There was a clear correlation between extent of skin injury and severity of acute radiation sickness. Skin damage varied from patient to patient in terms of occurrence, severity, course and extent. The clinical course of skin damage was shown to be dependent on the skin exposure conditions, the beta/gamma ratio, and the radionuclide contamination on skin and clothing and in the environment [B29]. Absorbed doses in skin exceeded bone marrow doses by a factor of 10–30 in some victims, corresponding to doses up to 400–500 Gy. From detailed analysis of clinical morphology

data, it can be stated that severe skin injury by beta radiation of moderate energy (1–3 MeV) could be a major cause of death if the damaged area exceeds 50% of the body surface. Relatively smaller areas of injury (10%–15% of body surface) from high-energy beta exposure (^{134}Cs, ^{137}Cs, ^{106}Ru, ^{90}Y, ^{90}Sr) with early development of necrosis-ulceration require surgery and can cause long-term disability [B29, G21, N3]. Surgical treatment was provided to fifteen acute radiation sickness survivors with extensive cutaneous radiation injuries, including ulcerations and fibrosis, at University of Ulm between 1990 and 1996. Follow-up of these survivors has not shown a single case of skin cancer.

196. Cataracts, scarring and ulceration are the most important causes of persistent disability in acute radiation sickness survivors. The consequence of severe skin ulceration is cutaneous fibrosis, which has been successfully treated with low-dose interferon [P26]. The recovery of physical ability is related to the severity of the initial symptoms of acute radiation sickness. To limit occupational radiation exposures of the acute radiation sickness survivors, legal measures adopted in the Russian Federation and other countries of the former Soviet Union have restricted their activities or caused them to change their occupations.

197. Sexual function and fertility among acute radiation sickness survivors was investigated up to 1996 [G2]. In the majority of cases, functional sexual disturbances predominated, while fourteen normal children were born to acute radiation sickness survivor families within the first five years after the accident (in one family, the first newborn died from sepsis, but a second, healthy child was born subsequently).

198. Patients with acute radiation sickness Grades III and IV were severely immunosuppressed. Whereas haemopoietic recovery occurs within a matter of weeks or, at most, months, full reconstitution of functional immunity may take at least half a year, and normalization may not occur for years after exposure. This does not necessarily mean that after the acute phase, i.e. the first three months, recovering patients display major immunodeficiency, and it is not surprising that studies of immune status did reveal pattern of changes in the blood cell concentrations without clinical manifestations of immunodeficiency [N3]. Nineteen parameters of the immune system were investigated five and six years after the accident in acute radiation sickness survivors (1–9 Gy) and in persons without acute radiation sickness (0.1–0.5 Gy) as well. For higher doses of radiation, T-cell immunity may show protracted abnormalities; however, these abnormalities are not necessarily associated with clinically manifest immunodeficiency.

IV. REGISTRATION AND HEALTH MONITORING PROGRAMMES

199. In order to mitigate the consequences of the Chernobyl accident, registration followed by continuous monitoring of the exposed populations was one of the priorities set by the Ministry of Health of the former Soviet Union. In the summer of 1986, Chernobyl registries were established for continuous monitoring of the health status of the exposed populations [M7, T12, W1]. Specialized population-based disease registers were created to monitor haematological tumours [I10, W1]. These activities have continued in Belarus, the Russian Federation and Ukraine since the dissolution of the USSR in 1991, with financial and technical support from many countries of the world.

200. Following the first reports of an increased incidence of thyroid cancer in children exposed to the radioactive fallout during the early 1990s [B18, K11, P8], thyroid cancer registries were developed in Belarus, the Russian Federation and Ukraine [D2, R2, T4]. More recently, specialized childhood cancer registries began to be developed in these countries [V7].

201. Considering the potential long-term health consequences of the Chernobyl accident, the existing general population cancer surveillance systems in Belarus, the Russian Federation and Ukraine have attracted particular attention. International researchers started to assess the functioning of these systems as well as their quality and completeness in comparison with the cancer registries in many Western countries and to evaluate their potential for research purposes [S6, W10]. To provide a basis for interpreting the health effects reported from epidemiological studies, this Chapter reviews the available information on registers and their follow-up.

A. REGISTRATION AND MONITORING OF EXPOSED POPULATIONS

1. The Chernobyl registries

202. In May 1986, the Ministry of Health of the USSR convened a conference of Soviet experts on the treatment and follow-up of radiation-exposed individuals [M7]. This conference recommended the establishment of a special registry to assist in the delivery of primary health care, treatment, and follow-up and to provide a basis for the long-term monitoring of the Chernobyl-exposed populations. Governmental orders issued by the Ministry of Health in 1987 provided the basis for creating the All-Union Distributed Clinico-Dosimetric Registry and for appointing the Medical Radiological Research Centre at Obninsk as the institution responsible for the development and maintenance of this registry [T12]. Compulsory registration and continuous monitoring of the health status was introduced for four population groups (primary registration groups): group 1, persons engaged in the

recovery operations following the accident (liquidators); group 2, persons evacuated from the most contaminated areas (^{137}Cs deposition >1,480 kBq m^{-2}); group 3, residents of highly contaminated areas (^{137}Cs deposition >555 kBq m^{-2}); and group 4, children born after the accident to those registered in groups 1–3.

203. For the purpose of data collection, registries were established at the national, regional and district levels, as well as in certain ministries that provide health care for their employees independent of the general health care network. Four special data collection forms were introduced [M7, W1]: registration, clinical examination, dosimetry and correction form.

204. Persons were registered in the All-Union Registry upon presentation of their official documents of work or residence in the Chernobyl zone, mainly during the compulsory annual medical examination in the outpatient department of the district hospital responsible for their place of residence (see Section IV.A.2). In return, each person obtained a special registration document enabling him/her to obtain special Chernobyl-related social benefits. At the time of the dissolution of the USSR, the All-Union Registry had accumulated data on 659,292 persons, 43% of whom were recovery operation workers (group 1), 11% of whom were persons evacuated from the most contaminated areas (group 2), and 45% of whom were persons living in contaminated areas (group 3); the remaining 1% were children of groups 1–3 (group 4) [W7].

205. Since 1992, the national Chernobyl registries have continued to operate, but independently, with only basic data items in common. Although the registries continue to employ the general registration techniques and categories developed during the Soviet era, they have evolved separately in terms of the population groups and/or data items that are covered, data quality, dose-reconstruction methodology and follow-up mechanisms [W12]. This must be kept in mind when interpreting and comparing results from the three countries most heavily exposed as a consequence of the accident.

206. Successive publications using Chernobyl registry data sources show ever-increasing numbers of persons registered. Whereas at the beginning, the Chernobyl registry of Belarus contained information on 193,000 persons, 21,100 of whom were reported to have worked as recovery operation workers [W7], at the beginning of 1995 this number had risen to 63,000 [O2]. Similarly, the number of persons registered in the Russian National Medical and Dosimetric Registry as belonging to one of the four primary registration groups has increased steadily over the years (Figure XXIII). The legal statute regulating medical criteria for disability and invalidity has changed over the years, and as a result the number of individuals included in the registries has increased [O4, W7].

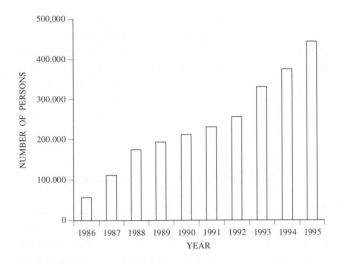

Figure XXIII. Number of persons registered in the Russian National Medical and Dosimetric Registry as exposed to ionizing radiation as a consequence of the Chernobyl accident [I14].

207. All persons included in the Chernobyl registries continue to receive active follow-up. This follow-up was centralized for a number of years in the former Soviet Union. Subsequently it has been performed in the three successor countries. Originally, the Chernobyl registration process was grafted onto an existing infrastructure, but the follow-up of the exposed populations is more specific and concerns all age ranges, including retired persons, who would not normally have received medical follow-up, since this is confined to the working population.

208. Compulsory annual medical examinations are conducted by a general practitioner in the outpatient department of the district hospital at the official place of residence. The information ascertained during these examinations is systematically reported to the Chernobyl registry by means of the specially devised clinical examination form [T12, W1]. In case a more severe health condition is suspected, the patient is referred to specialized health-care institutions for diagnosis and treatment.

2. Specialized registries

209. In addition to the Chernobyl registries, the Russian Federation keeps specialized registries of Chernobyl-exposed populations. Whereas the primary information on persons included in these registries is systematically reported to the Russian National Medical and Dosimetric Registry, the specialized registries generally contain more detailed information on exposure and follow-up.

210. The follow-up mechanisms for the specialized registries follow the same principles as those for the Chernobyl registries. However, medical services and annual medical examinations are provided at special medical facilities of the respective Ministry. The provision and quality of services at these special facilities are generally considered to be better than those offered by the general health-care network [W10]. Follow-up in the special medical facilities ceases with

termination of employment (other than retirement), after which the person returns to the general health-care regime.

211. The *Registry of Professional Radiation Workers* is maintained by the Institute of Biophysics in Moscow. It contains information on 22,150 professional radiation workers who participated in recovery work. Approximately 18,600 of them currently live in the Russian Federation, and 13,340 worked in the Chernobyl area in 1986–1987. The radiation doses for these workers were monitored with personal dosimeters. The Registry currently contains doses for approximately 50% of the workers; doses for the remainder are being collected and entered into the database.

212. Medical examinations of persons included in the *Registry of Professional Radiation Workers* are carried out in about 70 medical facilities of the Ministry of Health. The oncology service of the Institute of Biophysics verifies the diagnosis of all cases of cancer and extracts information from case histories. Before 1993, no information was collected on the date of diagnosis or cause of death. Of the 22,150 workers registered at the beginning, only 18,430 are currently being followed [T13].

213. The *Registry of Military Liquidators,* operated by the Military Academy at St. Petersburg, contains information on persons from all over the former USSR who were drafted by the Ministry of Defence to help in the recovery work after the accident. Their doses were mainly determined through time and motion studies. Information on places and dates of work in the Chernobyl area has been recorded in the registers. The medical follow-up of the approximately 15,000 servicemen included in the registry is carried out at local polyclinics and specialized dispensaries of the Ministry of Defence. Another group of approximately 40,000 persons who were sent to work in the Chernobyl area by the Ministry of Defence are followed in the general civilian network of health care.

214. Among the recovery operation workers, there are about 1,250 helicopter pilots and crew. All individuals with doses of more than 250 mGy were hospitalized for careful medical examination as soon as their work was finished. No sign of radiation sickness was found. A list of persons is maintained by the Russian Aviation Medicine Institute [U15]. The Aviation Medicine Institute carries out periodic examinations of helicopter pilots; results have been published on groups of 80 to 200 of them, including military pilots and crew members who flew over Unit 4 of the Chernobyl reactor in April and May 1986 [U15].

B. REGISTRATION OF MORTALITY AND DISEASE IN THE GENERAL POPULATION

1. Mortality

215. Mortality statistics are one of the main measures of health outcome used in epidemiological studies. In the countries of the former USSR, and throughout the world, registration of death is the responsibility of vital statistics

departments. For each death, a medical death certificate (medical document) is completed by the attending physician or by a medically trained person. The death is then registered at the district vital registration department by completing a death registration act (legal document), which provides authorization for burial. The medical death certificate in use in the countries of the former USSR has followed international recommendations since the 1980s [W10]. The quality of the death certificates depends on the competency of the person completing them, and the validity of the death certificates has probably changed over time.

216. A copy of the death registration act is forwarded to the regional vital registration department, which is in charge of the coding and the preparation of annual regional mortality statistics. The cause of death is coded using a slightly modified version of the Ninth Revision of the International Classification of Diseases B-List [S6]. Regional mortality statistics are submitted to the national statistical authority for the compilation of annual national mortality statistics.

217. As part of a general policy of censorship of demographic and health statistics, the use of mortality statistics for research purposes was severely restricted during the Soviet era [R1]. The main use of the mortality statistics was health planning. Since the dissolution of the Soviet Union into independent countries, however, such data have been readily available.

2. Cancer incidence

218. Cancer diagnosis and treatment were the responsibility of oncological dispensaries throughout the Soviet Union. Cancer patients were followed by the district hospital at the official place of residence. Certain malignant neoplasms, such as haematological neoplasms, childhood cancers, and certain rare tumours, e.g. brain and eye, are diagnosed and treated outside the network of oncological dispensaries [W10].

219. Based on the existing oncological infrastructure, cancer registration was made compulsory in 1953 throughout the USSR [N6]. The cancer registration process, which still applies today in the successor countries, involves the passive reporting of newly diagnosed cancer cases and information on their follow-up to regional cancer registries at the regional oncological dispensaries in the patient's place of residence [W10].

220. The patient cards maintained at the regional cancer registries are continuously updated with the information received from the cancer registration documents. After a cancer patient dies, the card is removed from the active cancer registry for storage in archives. Information on cancer deaths is gathered at regular intervals from the vital registration departments, and trace-back procedures are initiated for persons not registered during their lifetimes.

221. Annual cancer statistics are compiled at both the regional and national levels. Two basic reports are compiled manually each year. The cancer incidence report provides the number of cases according to sex, age group and cancer site, and the cancer patient report provides basic information on prevalence as well as on the diagnosis, treatment and survival of the patient. Death certificates and autopsy reports as information sources have been used since 1961, but the registration of leukaemia became compulsory only in 1965 [R1]. Thus, cancer statistics that meet the Western definition of cancer registries have existed only since 1966 in the countries of the former Soviet Union [W10].

3. Specialized cancer registries

222. Haematological cancer registries were set up in Belarus, the Russian Federation and Ukraine shortly after the accident, but thyroid cancer registries only started to appear during the early 1990s, after first reports of an increase in thyroid cancer following the Chernobyl accident. Specialized childhood cancer registries were set up only recently because the quality of the general registration was questionable [V7].

223. These specialized cancer registries are for cancers that are mainly diagnosed outside the network of regional oncological dispensaries. The registration of these cancers in the national cancer registries is likely to be less complete than for cancers directly diagnosed and registered in the oncological dispensaries. An important difference between the national cancer registries and the specialized cancer registries is that case ascertainment is passive in the former and active in the latter. Active reporting systems generally achieve a higher degree of completeness, as cancer registry personnel directly and systematically abstract the information from the source. Furthermore, specialized cancer registries generally record more detailed information on the clinical features of the tumour than do general cancer registries.

(a) Haematological cancer registries

224. In Belarus, the *National Register of Blood Diseases* is operated by the Institute of Haematology and Blood Transfusion in Minsk. Although activities only started in 1987, data since 1979 have been collected retrospectively. The registry covers the entire population of Belarus and receives details of cases of leukaemia, lymphoma and related blood diseases from haematological departments and oncological dispensaries and from autopsies. Information in this registry has, however, only recently been computerized.

225. The *Registry of Leukaemia and Lymphomas* in the Bryansk region (Russian Federation) records all relevant cases in all ages in this region, including a retrospective assessment since 1979. Before 1986, however, the registry is reported to be incomplete [W1].

226. Little is known about the *Ukrainian Registry of Haemoblastosis* (term used in the Russian language) that covers leukaemia, lymphomas and related non-malignant diseases, such as polycythaemia vera, aplastic anaemia and sideroblastic anaemia. The registry is operated by the Ukrainian Centre for Radiation Medicine and is designed to record all cases of haemoblastosis in the Ukrainian resident population [W1].

(b) Thyroid cancer registries

227. In Belarus, the *Thyroid Surgery Registry* is operated by the Scientific and Practical Centre for Thyroid Tumors at the Institute of Medical Radiology and Endocrinology, Minsk [D2, P2]. This registry includes only patients with thyroid disorders treated at the Centre. A comparison between the registry data and data obtained from pathologists indicated that it is incomplete for thyroid cancers diagnosed during adolescence and adulthood, as these are mostly diagnosed and treated at the regional level [C6]. The number of thyroid cancer cases reported from Belarus in different scientific publications varies according to the age range considered and depends on whether case series were abstracted from the *Thyroid Surgery Registry* only [P2], whether they were complemented by incidence data [B11], or whether only cases known to the national cancer registry were used [K40].

228. In the Russian Federation, the *Thyroid Cancer Registry* has been created for the Bryansk and Tula regions; it includes data on all cases of thyroid cancer since 1981 [R2]. In Ukraine, a clinical morphological register of thyroid cancers has been created at the Institute of Endocrinology and Metabolism in Kiev [T4]. This registry includes data on all thyroid cancer cases in which the patients were under the age of 18 years at the time of the Chernobyl accident, as well as retrospective data since 1986. Information is gathered from two sources: (a) data on all cases treated at the surgery department of the Institute of Endocrinology and (b) data on thyroid cancer cases from the national cancer registry, diagnosed throughout Ukraine. In 1986–1994, one third of the total of 531 cases recorded was identified from surgical records. Data quality and the amount of detail available in this registry is likely to vary, depending on the data sources. Also, thyroid cancer cases occurring in the Chernobyl-contaminated areas are more likely to undergo surgery in Kiev than cases occurring in other areas.

(c) Childhood cancer registries

229. Publications on childhood thyroid cancer or haematological malignancies generally are based on data from the corresponding specialized cancer registries [D2, I10, I31, T4] or, in the case of childhood thyroid cancers, from population screening programmes [I26, T5]. A new specialized registry, the *Belarusian Childhood Cancer Registry*, was created at the Institute of Oncology and Medical Radiology in 1996. Data are reported from centres diagnosing and treating children. Similarly, childhood cancer registries are also being created for the Bryansk and Tula regions of the Russian Federation [R2].

(d) Registers of hereditary disorders and congenital malformation

230. A national monitoring system for hereditary diseases exists in Belarus at the Institute for Hereditary and Congenital Diseases [L8]. Reporting is compulsory since 1979 for anencephaly, spina bifida, cleft lip and palate, polydactyly, limb reduction defects, oesophageal and anorectal atresia and Down's syndrome. In addition, reporting of multiple congenital malformations became compulsory in 1983. Morphogenesis defects in embryos and early fetuses have been recorded since 1980.

231. The comparability of registries of congenital malformations with similar registries from countries outside the former Soviet Union is generally considered to be poor owing to the great variation in definitions of the included parameters. Differences are found in defining the geographical location (e.g. place of residence of the mother or place of birth of the child), in the definition and coding of the diagnosis, and in coverage (live births, stillbirths, induced abortions, fetal deaths, spontaneous abortions) [L8, L20]. There has been no independent assessment of the quality and completeness of this registry, in particular for the period before the Chernobyl accident.

C. QUALITY AND COMPLETENESS OF REGISTRATION

1. Registers of exposed populations

232. The ever-increasing number of persons registered in the Chernobyl registries raises the crucial question of whether and when the registration of the exposed population groups can ever be reasonably complete. The legal statute regulating medical criteria for disability and invalidity in citizens of the Russian Federation was promulgated in 1991. According to this law, the causal association between the Chernobyl accident and disease, death, or invalidity can be established independently of the absorbed dose received or the health status of the person before the accident. As a result of this law, the number of individuals included in the register has increased over the past years [O4, W7]. Some registration increases occur as the people exposed as a result of the accident grow older, become ill, and seek registration to obtain social benefits. Any difference between the health status of exposed persons registered in recent years in the Chernobyl registries and that of exposed persons registered in earlier years could introduce bias into epidemiological studies using these data. However, preliminary analysis in Belarus and the Russian Federation appears to indicate that the health status distribution of the newly registered workers is similar to that of the workers registered previously [P18].

233. To obtain a more reliable and complete enumeration of a cohort of Chernobyl recovery workers for epidemiological purposes, Estonian researchers used four independent data sources:

(a) records of military personnel, both regular personnel and reservists, who were sent to the Chernobyl area;
(b) The Estonian Chernobyl Radiation Registry;
(c) members of the Estonian Chernobyl Committee, a non-governmental organization that aims to obtain compensation and health-care benefits for the recovery workers;

(d) files from the Ministry of Social Welfare of Estonia, which had conducted an independent registration of recovery operation workers for social benefits.

By using multiple data sources, the information on 83% of the 4,833 Estonian men who worked in the Chernobyl area could be ascertained from two or more sources [T6].

234. It remains unclear to what extent a verified diagnosis in the more specialized institutions is actually reported to the Chernobyl registries. One study indicated that almost half of the leukaemia diagnoses in the Russian Federation Chernobyl registry were not confirmed [O1]. By contrast, in Belarus, where a national registry of haematological malignancies was created following the Chernobyl accident, this information has been periodically reported to the Chernobyl registry in recent years.

235. Caution should be observed when comparing the mortality experience in the Chernobyl registry populations with national background mortality rates. The Chernobyl registries obtain mortality information from next-of-kin whenever the person under observation does not present himself/herself for the annual medical examination and/or from official death certificates, which are screened for information. The Chernobyl registry does not use the coding rules implemented by the official demographic authorities but often uses the existing medical information to code the appropriate cause of death in the Chernobyl registry. As a result, the mortality information available in the Chernobyl registries is often more specific and has a higher degree of completeness than official demographic sources.

236. Understanding the registration processes, coding and quality changes over time is essential for researchers to appropriately conduct and interpret epidemiological studies using Chernobyl related data. More research is needed to document the methods of data collection and the quality of data collected in the framework of the Chernobyl registries.

2. Registers of the general population

237. *Mortality.* Very little information exists on the quality of mortality statistics from countries of the former Soviet Union. During the late 1970s, a study was conducted to compare the quality of cause-of-death coding in seven countries, including the USSR [P22]. The results showed that for a standard set of 1,246 causes of death related to cancer there were no differences in the quality of coding in the USSR and in the other participating countries. However, for this particular study, an expert pathologist performed the coding in the USSR, whereas in the rest of the countries a vital statistics official did the coding. A study is currently under way comparing the coding for the same 1,246 cause-of-death series as performed by the actual coders of a number of regional vital registration departments in Belarus, the Russian Federation and Ukraine.

238. Mortality rates in the three countries declined in 1984–1987, increased in 1987–1994 [L38], and were more stable in recent years [L37]. International comparisons of mortality statistics show higher rates for deaths from cardiovascular disease and violence in the Russian Federation. [M8]. Although differences in data quality cannot be excluded entirely, recent evidence suggests that increased alcohol consumption may account for a substantial portion of the difference in mortality patterns between the countries of the former Soviet Union and Western countries [L37, L38, W11].

239. Mortality data are used in epidemiological studies as endpoints for individual follow-up of exposed populations or in aggregated form as mortality rates in geographically defined populations. Although changes in mortality patterns in the countries of the former Soviet Union are not related to radiation exposure, they must be taken into account when interpreting epidemiological studies using these data. In these countries, individual mortality data are available only on paper at the district level. The absence of centralized, computerized mortality registries poses considerable difficulty, particularly in tracing subjects who have moved from one area to another [O1].

240. *Cancer incidence.* A survey of cancer registration techniques in the countries of the former Soviet Union [W10] showed a number of important differences compared to Western countries. First, cancer registries in Western countries were created for research purposes, whereas in the Soviet Union they were created for health-planning purposes. The system in operation in the countries of the former Soviet Union functions as a public health surveillance system with fast reporting of statistics (three months after the end of a reporting year). The impact of the fast reporting on the quality of the information is unknown. Second, many data items are coded independently of the internationally recommended and accepted classification schemes. This is particularly relevant for one of the most important data items in cancer registration: tumour pathology. This may have important implications for the quality of the data recorded. Third, standard concepts for verifying the quality and completeness of registries, widely used for Western cancer registries, are virtually unknown in the countries of the former Soviet Union. Finally, personal identification is based on names. Differences in spelling, the increasing use of national languages rather than the Russian language, and the increasing mobility of the population raise questions about the correctness of the identifications. Whereas probabilistic record-linkage procedures are used for this purpose in Western cancer registries, these procedures remain unknown in the countries of the former Soviet Union.

241. Of the countries most contaminated by the Chernobyl accident, only Belarus had a centralized cancer registry in operation before the accident. In the 1990s, the Russian Federation and Ukraine began developing computerized cancer registration software and regional registration networks, and Belarus worked to modernize the system in operation there. This work of gradually adjusting the cancer registration techniques in these countries to satisfy international standards is being conducted within a framework of international collaboration to ensure adequate training [S6].

242. A computerized central cancer registry was established in Belarus as early as 1973, and computerized cancer incidence data have been available since 1978 [O4]. However, owing to the lack of resources, the data were computerized anonymously during the early years of operation. In 1985 personal-computer-based cancer registration was developed, including appropriate information on the individuals. Since 1991 a computerized network has provided the basis for cancer registration in Belarus.

243. Efforts to computerize cancer registration in Ukraine started at the beginning of the 1990s [W10]. By the end of 1997, computerized cancer registration had been implemented in many regions. Approximately 82% of the population of Ukraine was covered at that time, and nationwide coverage is expected by the year 2000. In recent years, the Russian Federation has also started to develop a computerized cancer registration system [W10]. Full coverage of the population of the Russian Federation with computerized cancer registration technology has to be seen as a long-term goal. In any case, priority will be given to computerizing the Chernobyl-contaminated regions and a number of control regions.

244. Probably the main source of international cancer incidence statistics is the series Cancer Incidence in Five Continents, published by the International Agency for Research on Cancer (IARC) at five-year intervals since the late 1960s. Although cancer registries all over the world are invited to contribute data, only data sets satisfying the defined quality standards are accepted for publication. No data from the USSR were included in any of the early volumes, although a supplement to Volume III, "Cancer incidence in the USSR", was published jointly by IARC and the USSR Ministry of Health in 1983 [N6]. The two most recent volumes, presenting data for 1983–1987 and 1988–1992, include data from Belarus, Estonia and Latvia, and data from St. Petersburg and Kyrgyzstan were published in the earlier of the two volumes. However, the editors warned that all data from the countries of the former Soviet Union (except data from Estonia) may under-ascertain the number of cases, may lack validity, or may be inaccurate with respect to the denominators of the rates [P17, P23].

D. INTERNATIONAL COLLABORATIVE SCREENING PROJECTS

245. For more than three years after the Chernobyl accident, efforts to mitigate its consequences were considered by the Soviet Union to be exclusively an internal matter. International collaborations started to develop in 1990 and have since played a substantial role in the assessment of the health consequences of the Chernobyl accident.

246. Soviet health authorities initiated screening activities in June 1986 throughout the areas most affected by the accident. Initially, these activities were locally organized; large-scale screening was not started until the early 1990s. A description of the organization, completeness, and results of these efforts has not yet been published.

1. The International Chernobyl Project

247. In 1990–1991, IAEA organized the International Chernobyl Project [I1] at the request of the Government of the USSR. International experts were asked to assess the concept the USSR had developed that would enable the population to live safely in areas affected by radioactive contamination following the Chernobyl accident; they were also asked to evaluate the effectiveness of the steps taken in these areas to safeguard the health of the population. The International Chernobyl Project investigated the general health of the population of seven rural settlements in Belarus, the Russian Federation and Ukraine and of six control settlements using an age-matched study design [M10]. Psychological health, cardiovascular, thyroid and haematological disorders, cancer, radiation-induced cataracts and fetal anomalies were also investigated, and cytogenetic studies were carried out.

2. The IPHECA project

248. In 1992–1995, WHO conducted an International Programme on the Health Effects of the Chernobyl Accident (IPHECA). A number of pilot projects surveying registration activities, radiation- dose reconstruction, haematological disorders, thyroid disorders, brain damage, and oral health were carried out. The IPHECA Dosimetry Project reconstructed radiation doses from measurement surveys of the populations of contaminated and evacuated areas [W1, W13].

3. The Chernobyl Sasakawa Health and Medical Cooperation Project

249. Between 1991 and 1996, the Sasakawa Memorial Health Foundation sponsored the largest international programme of screening of children following the Chernobyl accident. The project aimed at assessing the anxiety and health effects in the people affected by the Chernobyl accident through large-scale population screening. Children were selected as the target population for the screening. Regional diagnostic centres were set up in Gomel and Mogilev in Belarus, in Klincy (Bryansk region) in the Russian Federation, and in Kiev and Korosten (Zhitomir region) in Ukraine to screen children by means of mobile examination units [S38].

250. Standard examination protocols and questionnaires focussing on thyroid disorders, haematological disturbances and radiation dose were used. The examination included collection of disease history and anthropometric data, dosimetric measurements using whole-body counters, ultrasonography of the thyroid, general blood count, determination of thyroid hormones in the serum, determination of iodine and creatinine in the urine, and examination by a paediatrician. When abnormalities were found, the child was referred to the regional diagnostic

centre for comprehensive examination and appropriate treatment. During the course of the project (May 1991-April 1996), approximately 120,000 children were examined. The results are given in Chapter V [H2, I16, I21, I26, S7, Y1].

251. International collaborative efforts are continuing, and several studies of the effects of the accident are underway. An example is a project to establish a tissue bank of thyroid carcinomas from all of the three most affected countries, which has as participating organizations the European Union, National Cancer Institute of the United States, the Sasakawa Memorial Health Foundation, the World Health Organization, and the health ministries of Belarus, the Russian Federation and Ukraine..

E. SUMMARY

252. Following the Chernobyl accident, compulsory registration and continuous health monitoring of recovery operation workers and residents of the most contaminated areas, including their offspring, were initiated throughout the Soviet Union. Until the end of 1991, the All-Union Distributed Clinico-Dosimetric Registry recorded information on 659,292 persons. After the dissolution of the Soviet Union into independent states, national Chernobyl registries have continued to operate, but independently. Changes in national registration criteria, compensation laws, dose-reconstruction methods, and follow-up mechanisms increasingly limit the comparability of data from the different national sources. More detailed registries of exposed populations exist in the Russian Federation (Registry of Professional Radiation Workers, Registry of Military Workers and the cohort of Helicopter Pilots and Crew). The quality and completeness of these registries remain largely unknown, however.

253. The number of people registered in the national Chernobyl registries continues to increase, even in recent years, which raises questions about the completeness and accuracy of registration. Information on mortality and cancer incidence is collected from many different sources and is coded independently of international guidelines. Evidence from recent cohort studies suggests that the Chernobyl health outcome data cannot be successfully compared with health data obtained from official statistical sources.

254. Systematic linkage of the Chernobyl registry population data with existing mortality and/or cancer incidence registries and the subsequent comparison of the health outcome experience in the cohort with the corresponding national reference statistics could be a valuable tool for epidemiological research. Internal comparisons, e.g. using a low-dose comparison group, are likely to provide information on risks associated with ionizing radiation in the future. However, complete information on, e.g. previous exposure to ionizing radiation in an occupational setting, will most probably only be available for small sub-cohorts.

255. Health outcome registries are an important source of information for assessing the consequences of the Chernobyl accident. Their primary advantage is that the information was collected in a systematic way before and after the accident and that the criteria for data collection are the same in all countries of the former Soviet Union. However, most of these registries, whether related to mortality, cancer incidence, or special diseases, continue to be largely operated manually, which seriously limits their use for epidemiological research purposes. The Chernobyl accident led to major international efforts to computerize cancer incidence and special disease registries and to improve their registration methods so as to comply with international standards. However, mortality registration systems have received little attention so far. Information on the quality and completeness of these systems remains scarce.

256. Compulsory cancer registration was introduced throughout the former Soviet Union in 1953. The system relies on passive reporting of information on all newly diagnosed cancer cases to the regional cancer registry for the patient's place of residence. Since the early 1990s, there have been efforts to computerize the existing systems and to gradually improve their quality to satisfy international standards. Belarus has been covered with a network of computerized cancer registries since 1991. Computerization is well advanced in Ukraine, and full population coverage is expected soon. In the Russian Federation, efforts to develop computerized cancer registration started only recently and will be concentrated in contaminated areas and control areas.

257. Specialized population-based registries for haematological malignancies and thyroid cancer were set up in the wake of the Chernobyl accident and in response to the unknown quality and lack of detail for these sites in the general registries. Childhood cancer registries were recently developed for the same reasons. Quality assessments of these registries are underway. Other registries for hereditary disorders and malformations exist, but their quality and completeness have not so far been independently assessed.

258. Shortly after the Chernobyl accident, efforts were devoted mainly to developing adequate registration systems for future follow-up of those population groups most affected by the radionuclide deposition. More recently, international collaborations have helped to modernize the existing disease registration infrastructure. However, information on the quality and completeness of all these registries is still very scarce. The usefulness of the vast amount of data collected will become clearer in the coming decades as the long-term consequences of the accident are studied. In particular, matching the health outcomes with the dosimetric data described in Chapter II, will be of great importance.

V. LATE HEALTH EFFECTS OF THE CHERNOBYL ACCIDENT

259. The studies of late health consequences of the Chernobyl accident have focussed on, but not been restricted to, thyroid cancer in children and leukaemia and other cancer in recovery operation workers and residents of contaminated areas. Many studies have been descriptive in nature, but until individual dosimetry is completed, proper controls established, and methodological requirements satisfied, the results will be difficult to interpret. Quantitative estimates and projections will certainly be very unreliable without individual and reliable dose estimates.

260. The late health effects of the Chernobyl accident are described in this Chapter. These effects include malignancies, especially thyroid cancer and leukemia, non-malignant somatic disorders, pregnancy outcome and psychological effects. The focus will be on health effects in the most contaminated areas, but possible effects in other parts of the world will also be considered.

A. CANCER

1. Thyroid cancer

(a) Epidemiological aspects

261. Thyroid carcinomas are heterogeneous in terms of histology, clinical presentation, treatment response and prognosis. Although rare, they are nevertheless one of the most common cancers in children and adolescents. Thyroid cancer is known to be more aggressive in children than in adults, but paradoxically, the prognosis is supposed to be better in children [V8]. Several risk factors have been suggested for thyroid cancer, but only ionizing radiation has been found to have a causative effect, although a history of benign nodules, miscarriages, iodine deficiency or excess, and an elevated level of thyroid-stimulating hormones have been discussed as causative factors [F9, R18]. Risk factors for thyroid cancer are discussed in Annex I, "*Epidemiological evaluation of radiation-induced cancer*".

262. The childhood thyroid gland is, besides red bone marrow, premenopausal female breast, and lung, one of the most radiosensitive organs in the body [U2]. Age at exposure is the strongest modifier of risk; a decreasing risk with increasing age has been found in several studies [R7, T20]. Among survivors of the atomic bombings, the most pronounced risk of thyroid cancer was found among those exposed before the age of 10 years, and the highest risk was seen 15–29 years after exposure and was still increased 40 years after exposure [T20]. The carcinogenic effect of ^{131}I is less understood, and the effects of radioiodine in children have never been studied to any extent, since medical examinations or treatments rarely include children [H6]. Similar to the studies of atomic

bomb survivors, there is little evidence of an increasing risk for exposures occurring after age 20 years [H6, T20].

263. As seen in Tables 56–58 an increasing number of thyroid cancers among children and adolescents living in areas most contaminated by the accident have been diagnosed in the last 12 years. Among those less than 18 years of age at exposure, 1,791 thyroid cancers were diagnosed during 1990–1998 (complete information is not available for the Russian Federation). The increase in all three countries for 1990–1998 was approximately fourfold, with the highest increase seen in the Russian Federation. The increase in absolute numbers seems to have leveled off, particularly for the older age-at-exposure cohorts. It should be emphasized that the source population in Belarus is approximately 1.3 million, in the Russian Federation 300,000 (only one region) and in Ukraine 9 million children.

264. As previously discussed in this Annex, there are considerable uncertainties in the estimates of individual thyroid doses to the population in the contaminated areas. More than 80% of the dose from internal exposure was estimated to be from [131]I [Z1]; external exposure contributed only a small proportion of the thyroid dose [M4] (see Section II.C.2.a).

265. Two recent studies [F13, R17] found an elevated risk of thyroid cancer mortality following adult ^{131}I treatment for hyperthyroidism, which is in contrast to previous studies of hyperthyroid patients [H14] or patients examined with [131]I [H6]. The reason for referral, i.e. the underlying thyroid disorder, could have influenced the risk, since the highest risk was seen less than five years after exposure. The thyroid dose (60–100 Gy) received by most hyperthyroid patients had previously been considered as having a cell-killing rather than a carcinogenic effect.

266. In a recent paper by Gilbert et al. [G23], thyroid cancer rates in the United States were related to ^{131}I doses from the Nevada atmospheric nuclear weapons tests. The analysis involved 4,602 thyroid cancer deaths and 12,657 incident cases of thyroid cancer. An elevated risk was found for those exposed before the age of one year and born during the 1950s, but no association with radiation exposure was seen in older children in the age range 1–15 years. It could be that migration complicated the dose assessment and case identification. The authors concluded that the increase was most likely due to the ^{131}I exposure, but the geographical correlation approach, i.e. lack of individual doses and information on residency, precluded them from making quantitative estimates of risk related to exposure. Further, ^{131}I exposures from nuclear weapon tests in other countries were not taken into account, nor were other releases such as those that occurred from the Hanford site [R22].

267. Between 1944 and 1957, the Hanford site in the United States released large quantities of ^{131}I into the atmosphere during fuel processing. In the Hanford Thyroid Disease Study, thyroid doses were estimated for 3,193 individuals, and the mean thyroid dose was 186 mGy [R22]. Initial results of the study indicated that a diagnosis of thyroid cancer could be made for 19 participants, but no dose-response relationship was apparent. The final report is yet to be published.

268. Prisyazhiuk [P8] described in 1991 three cases of childhood thyroid cancer in Ukraine that were diagnosed in 1990, in contrast to no diagnosed cases during the preceding eight years. A screening effect was discussed, and it was postulated that "these thyroid cancers might represent the beginning of an epidemic". In the following year, Kazakov et al. [K11] reported 131 cases of childhood thyroid cancer in Belarus. The geographical distribution suggested a relationship with the ionizing radiation caused by radionuclides released from the Chernobyl accident, but unexplained differences existed. Most cases had been confirmed by a panel of international pathologists [B18]. Increased risks of childhood thyroid cancer were later reported in Ukraine [L6, T2] and, more recently, in the part of the Russian Federation most contaminated by the Chernobyl accident [S19, T1].

269. The large number of cases appearing within five years of the accident was surprising, since it had been believed that thyroid cancer needed an induction and latency period of at least 10 years after exposure to ionizing radiation [U2]. The findings were challenged, the major concerns being the influence of an increased awareness and of thyroid cancer screening [B23, R8, S5].

270. The numbers of thyroid cancers in children born before 26 April 1986 and who were less than 15 years of age at that time by country/region and year of diagnosis are given in Table 56 and illustrated in Figure XXIV. The corresponding incidence rates are given in Table 57 and Figure XXV Childhood thyroid cancers in heavily contaminated areas show 5-10-fold increases in incidence in 1991-1994 compared to the preceding five-year period. The number of childhood thyroid cancers occurring from 1990 to 1998 in a wider age range of children (0-17 years old at time of the accident) is presented in Table 58. It should be noted that many of the cases were diagnosed in adolescents and young adults and that only few of them have undergone histopathological review. The numbers given in Tables 56-58 are not entirely consistent, since various sources of information have been used (see Section IV.B.3.b).

271. Kofler et al. [K41] recently described thyroid cancer in children 0-14 years old at the time of the accident in Belarus. Through the Belarus cancer registry, 805 thyroid cancers were found by the end of 1997. The distribution with time and by age at exposure is presented in Figure XXVI. It can be seen that children 0-4 years old at the time of the accident still have an increase in absolute numbers of thyroid cancers, while the number of thyroid cancers diagnosed among those who were 5-9 years old

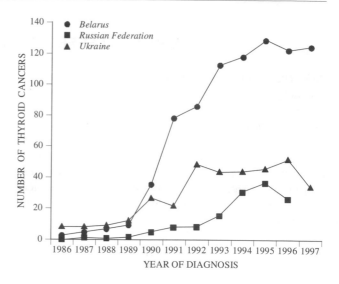

Figure XXIV. Number of thyroid cancers in children exposed before the age of 14 years as a result of the Chernobyl accident [I23, K41, T2, T16].

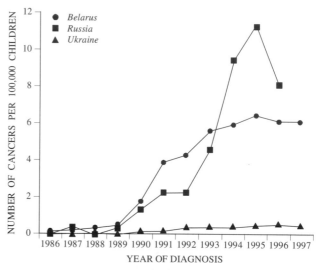

Figure XXV. Thyroid cancer incidence rate in children exposed before the age of 14 years as a result of the Chernobyl accident [I23, K41, T2, T16].

seems to decrease after 1995. In those 10-14 years of age at exposure, the number of thyroid cancers seems to be stable for the period 1991-1997.

272. In a recent report on the childhood thyroid cancer cases in the Russian Federation, Ivanov et al. [I27] described findings similar to those revealed above. In total, 3,082 thyroid cancer cases in persons less than 60 years of age at diagnosis were recorded between 1982 and 1996 in the four most heavily contaminated regions of the Russian Federation (Bryansk, Kaluga, Orel and Tula). Among those 0-17 years of age at the time of the accident, 178 cases were found. A significantly lower incidence of thyroid cancer in women in these four areas compared to women in the Russian Federation as a whole was found for the period 1982-1986 (Figure XXVII). In the next five years, described as an induction period, the risk was, on

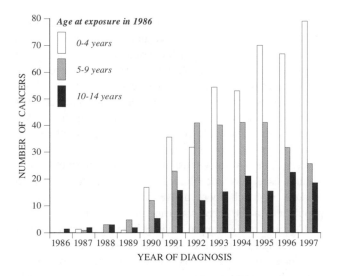

Figure XXVI. Number of diagnosed thyroid cancer cases in Belarus as a result of the Chernobyl accident [K41].

average, 1.6 times the Russian Federation baseline risk, probably reflecting increased surveillance and screening. Separate figures for those less than 14 years of age were not given, but the elevated risk seen for the period 1992-1996 was said to reflect the high rates found in those less than 14 years old at the time of the Chernobyl accident. A nearly 14-fold increased risk was found among girls 0-4 years of age at exposure compared to adult rates.

273. Jacob et al. [J2] reported a correlation between collective dose and incidence of thyroid cancer in 5,821 settlements in Belarus, the Russian Federation and Ukraine in 1991-1995. Using the southern half of Ukraine as a reference, the excess thyroid cancer risk was found to be linear in the dose interval 0.07-1.2 Gy. The study was extended in an attempt to take thyroid surveillance, background incidence, age, gender and appropriate methods into consideration [J5]. Thyroid doses in two cities and 2,122 settlements in Belarus and one city and 607 settlements in the Russian Federation were reconstructed. Thyroid cancers diagnosed during 1991-1995 in individuals between the age of 0-18 years at the time of the accident were included in the study. Information on residency at the time of the accident was collected for all cases, and 243 thyroid cancers were found, giving an excess absolute risk of 2.1 (95% CI: 1.0-4.5) per 10^4 person-year Gy (Table 59).

274. In a pooled analysis of children exposed to external photon radiation [R7], the excess absolute risk of thyroid cancer was 4.4 cases per 10^4 person-year Gy (95% CI: 1.9-10.1). The excess relative risk found by Jacob et al. [J5] was 23 Gy^{-1} (95% CI: 8.6-82), which is non-significantly higher than the excess relative risk of 7.7 Gy^{-1} (95% CI: 2.1-28.7) found by Ron et al. [R7]. It could be that indolent thyroid cancers detected only at screening were included. The possible shortening of the latency period by screening, which served to "harvest" thyroid cancers, could partly explain the differences. Another explanation for the differences could be the dosimetry,

since collective thyroid doses were used, and individual dose measurements in Belarus were available for only a minority of the study population. Dose calculations were based on environmental measurements and radio-ecological models.

275. A recent case-control study by Astakhova et al. [A6] included the initial 131 Belarus thyroid cancers presented by Kazakov et al. in 1992 [K11] but excluded 24 cases for different reasons. Eleven cases did not have pathological confirmation, 8 cases were not in Belarus at time of the accident, 2 cases had no information on pathology, 2 cases were diagnosed before 1987 and 1 patient was deceased. For the remaining 107 thyroid cancers included in the study, the male:female ratio was 1:1.1, and 105 cancers (98%) were of papillary origin. Two sets of controls were chosen, both matched on age, sex, rural/urban residency, taking reason for diagnosis and area of residency into consideration. A strong relationship between estimated thyroid dose and thyroid cancer was found, even when reason for diagnosis, gender, age, year of diagnosis, and ^{131}I level in soil were taken into consideration [A6]. The mean doses were different for the cases and for the controls selected to represent the general population of children exposed to fallout from the accident (Table 60), and the odds ratio was 3.1 (95% CI: 1.7-5.8) when comparing the lowest and the two highest dose groups. The highest odds ratio, based on 19 cases and controls, was seen for those diagnosed incidentally (Table 60) when using the other set of controls, i.e. those having the same opportunity for diagnosis as the cases. The odds ratio for those thyroid cancers found at routine screening (OR = 2.1, 95% CI: 1.0-4.3) indicated that screening was conducted in high fallout areas, since no large difference in dose was noted.

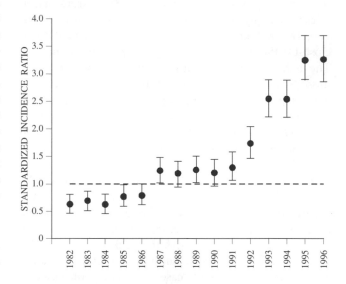

Figure XXVII. Standardized incidence ratios (SIR) of thyroid cancers among women less than 60 years of age at diagnosis in the four most contaminated areas (Bryansk, Kaluga, Orel, Tula) of Russia to Russia as a whole [I27].

276. The Astakhova case-control study [A6] is one of the more reliable published to date. Individual thyroid doses received by the children were inferred from established

relationships between adult thyroid doses and ^{137}Cs deposition on a village basis and then modified according to age-adjusted intake rates and dose coefficients. Location of the children before and after the accident was established on the basis of interviews. Although the questionnaire was designed to acquire individual information on milk-consumption rates and the administration of thyroid-blocking agents, such information was incomplete in many cases and could not be used. Age-dependent default values of milk-consumption rates were therefore used for all children in order not to introduce bias into the study; the use of blocking agents was assumed either not to have occurred at all or to have occurred too late to have been effective. Uncertainty in the individual estimates of thyroid dose is difficult to quantify, but is estimated to be at least a factor of three.

277. In a study of Ukrainian thyroid cancer patients less than 15 years old at diagnosis, registered at the Institute of Endocrinology and Metabolism, Kiev, the thyroid cancer rate for 1986–1997 exceeded the pre-accident level by a factor of ten [T18]. A total of 343 thyroid cancers occurred in patients born between 1971 and 1986, and the thyroid cancer rate for this age cohort was 0.45 per 100,000 compared with 0.04–0.06 per 100,000 before the accident. For the slightly older group of patients 15–18 years old at diagnosis in 1986–1997, 219 cases of thyroid cancers were found, and the average incidence was three times higher than that in the group diagnosed before the accident.

278. Descriptions of the dosimetric methods or dose-response models are unavailable, but it was stated that 22% of the patients aged 0–14 years at the time of the accident received a thyroid dose of >0.3 Gy [T18]. Thyroid cancer rates were analysed for different areas, and a rate of 27 per 100,000 was found in children evacuated from the villages closest to the accident (including Pripyat and Chernobyl). The authors concluded that the highest risk was found in those less than 5 years of age at the time of the accident, but it is questionable whether the methods used allow such a statement. Only cases below the age of 19 years at diagnosis are registered, thus excluding those older than 7 years of age in 1986 to be registered in 1997.

279. In a recent paper, Shirahige et al. [S34] compared 26 children diagnosed with thyroid cancer in Belarus within the framework of the Sasakawa project with 37 children diagnosed with thyroid cancer in Japan between 1962 and 1993. A peculiar finding was the peak incidence at 10 years of age at time of diagnosis and a drop thereafter among the Belarus cancer cases compared with a steady increase between the ages of 8 and 14 among the Japanese cases. This could reflect a difference in age distribution, since those exposed very early in life seemed to have an increased risk many years after exposure [G23, K41]. The differences could also indicate a different growth pattern caused by an alternative process of carcinogenesis, differences in screening routines, or the manner in which the children were selected for screening, since some registries only follow children until the age of 15.

280. In a study of male recovery operation workers from Belarus, the Russian Federation and Ukraine who worked within the 30-km zone, an increased incidence of thyroid cancer was noted, based on a total number of 28 cases [C2]. Significant thyroid doses may have been received from short-lived iodines during the first days after the accident, but information on the period workers spent within the 30-km zone was not taken into consideration. Histopathology and mode of confirmation were not available for the 28 thyroid cancers. These results must therefore be interpreted with caution, especially since the follow-up of recovery operation workers is much more active than that of the general population in the three countries (see Section IV.A.1 and IV.C.2). In a study of approximately 34,000 patients receiving ^{131}I for diagnostic purposes, 94% being older than 20 years at exposure, no increased risk of thyroid cancer related to radioiodine exposure could be found [H6]. As previously discussed, the intensity of screening may greatly influence the observed incidence of thyroid cancer in adults [S4].

281. Ivanov et al. [I8, I14, I33] extensively studied the late effects in Russian recovery operation workers. The Russian National Medical and Dosimetric Registry (the former Chernobyl registry) was used, and a significantly increased risk of thyroid cancer was found when the number of observed cases was compared to the number expected from national incidence figures. With the exception of the recent papers by Ron [R17] and Franklyn [F13], this is the first time an increased risk of thyroid cancer has been reported in adults after exposure to ionizing radiation. The study has been criticized for not using individual doses and internal comparisons [B30, B31]. The increased medical surveillance and active follow-up of the emergency workers most likely influenced the results, particularly when observed numbers are contrasted to national background rates.

282. In a descriptive study of cancer incidence in the six most contaminated regions of the Russian Federation [R2], the thyroid cancer incidence increased over time in adults, and the increase was larger than that observed in the whole of the Russian Federation. The highest values were found in the Bryansk region in 1994; the thyroid cancer incidence for women there was 11 per 100,000 compared with 4.0 per 100,000 for the Russian Federation as a whole. The corresponding figures for males were 1.7 and 1.1 per 100,000, respectively. It was concluded that no correlation was found between adult thyroid cancer and the levels of radioactive contamination [R2]. A pronounced difference was observed in children based on 14 cases; the incidence was 2.5 in the Bryansk region compared with 0.2 per 100,000 for the whole of the Russian Federation. Before the accident, the incidence of thyroid cancer in children in Bryansk and in the Russian Federation was the same. The registration and follow-up of the Russian population as a whole are probably not comparable to that in the highly contaminated regions surrounding Chernobyl (see Section IV.A.1 and IV.C.2), and this influences the results.

283. Thyroid examinations, including ultrasound and fine needle aspiration, were conducted in 1,984 Estonian recovery

operation workers nine years after the accident [I19]. The average age on arrival at Chernobyl was 32 years and at thyroid examination, 40 years. The mean documented dose from external irradiation of the thyroid was 108 mGy, but a poor correlation was found with biological indicators of exposures such as loss of expression of the glycophorin A gene in erythrocytes. Doses from incorporated iodine were not taken into consideration. Two cases of papillary carcinoma were identified and referred for treatment. Both men with thyroid cancer had worked at Chernobyl in May 1986, when the potential of exposure to radioactive iodine was highest.

284. In a study of 3,208 Lithuanian recovery operation workers in 1991-1995, three thyroid cancers (two papillary carcinomas and one mixed papillary-follicular carcinoma) were detected [K9]. There was no significant difference compared with the Lithuanian male population and no association with level of radiation dose or duration of stay in the area of Chernobyl.

285. Three years after the Chernobyl accident, Jewish residents of the former USSR began to emigrate in large numbers to Israel. Between 1989 and 1996, about 140,000 persons from contaminated regions of Belarus, the Russian Federation and Ukraine moved to Israel. The thyroid status of 300 immigrant children brought to the clinic voluntarily by parents was evaluated [Q1], and enlarged thyroid glands were found in about 40% of subjects. One 12-year-old girl from Gomel was found to have a malignant papillary carcinoma of the thyroid.

286. Several studies on the late health effects of the Chernobyl accident were carried out in Europe and have been critically reviewed and summarized [S12]. No increase in thyroid cancer among children was observed, although no study focussed specifically on childhood thyroid cancer, since the disease is so rare and a small increase could have gone undetected in these studies.

287. Screening programmes have increased the ascertainment of occult thyroid tumours through the use of ultrasound examination [B23, S5], a possibility discussed in one of the original reports [B18]. Thyroid screening was locally organized in the most contaminated areas after the accident, but large-scale screening with ultrasound examination, supported by the Sasakawa and IPHECA programmes, did not start until 1991 and 1992. Soviet health authorities initiated a national screening programme shortly after the accident, and each country later continued the thyroid screening [S19]. It is anticipated that 40%-70% of the diagnosed childhood thyroid cancer cases have been found through these programmes. In the case-control study of 107 childhood thyroid cancers [A6], 63 cases were found through endocrinological screening (Table 60). In a survey of 50 thyroid childhood cancers in Belarus [A5, C1], 12 cases were detected by targeted screening and another 23 cases were found incidentally in other examinations (Table 61). A study by Ron et al. [R8] supports these findings of a screening effect. Increased screening of a cohort given external photon radiation for benign head and neck diseases in the United States resulted in a roughly sevenfold increase in thyroid cancers.

288. Although rarely fatal, the aggressiveness of the thyroid cancers found in the Chernobyl area, which is frequently present with periglandular growth and distant metastases [E1, K11, W8], argues against the findings being entirely a result of screening. Although thyroid tumours in adults are usually tumours of relatively low malignancy, they tend to be more aggressive in children [S3], so it could be argued that the growth pattern would have led to the diagnosis of a thyroid cancer sooner or later.

289. Although ultrasound screening is not sufficiently widespread to explain the majority of the thyroid cancer cases observed until now, it is clear that increased awareness and medical attention to thyroid disorders during routine medical examinations in the contaminated territories influence the findings. Enhanced examinations in schools may have advanced the time at which some tumours were recognized [A5]. However, the continuing increase in the number of cases [C4, I27, K41] and the observation that the increase appears to be confined to children who were born before the accident support the conclusion that ascertainment bias could not fully explain the increased rates. Rates among those conceived after the accident appear to be similar to pre-accident rates in the affected countries.

290. Factors other than screening and lack of individual dosimetry may modify the risk of radiation-induced thyroid cancer, one of them being iodine saturation. The iodine deficiency in some of the affected areas will affect not only the level of dose received by the thyroid gland at the moment of exposure but also, if continued, thyroid function in the years after exposure [Y1]. The risk of thyroid cancer may be enhanced when the excretion of radioactive iodine is limited owing to reduced thyroid hormone synthesis and blocked thyroid hormone secretion [R6]. Iodide dietary supplementation had been terminated in the former USSR approximately 10 years before the accident. The relationship between iodine deficiency and goiter is discussed in paragraph 345.

291. Another risk-modifying influence might be a genetic predisposition to radiation-induced thyroid cancer, perhaps related to ethnicity. A recent survey indicated that cases occurred in siblings in at least three families in Belarus [A5, C1], a finding unlikely to be explained by chance alone. In a study of children exposed in the Michael Reese hospital in the United States, the risk of radiation-induced thyroid cancer appeared to vary by ethnic origin of the children [S4]. The genetic susceptibility to thyroid cancer and its familial aggregation were studied in relatives of 177 patients with thyroid cancer [V8]. No significantly increased rate of thyroid cancer was seen compared with controls. In a recent study, 7 of 119 patients with papillary thyroid microcarcinoma had a family history of thyroid carcinoma, and they experienced less favourable tumour behaviour than patients without a family history of thyroid carcinoma [L39]. The possible existence of a genetic predisposition to radiation-induced thyroid cancer might be important.

(b) Clinical and biological aspects

292. A large proportion of the childhood thyroid cancers in Belarus and Ukraine were reported to be locally aggressive; extrathyroidal growth was seen in 48%–61% of the cases, lymph node metastases in 59%–74% and distant metastases (mainly lung) in 7%–24% [F5, P2, P6, T4, T18]. Comparisons with characteristics of tumours from other countries (France, Italy, Japan, the United Kingdom and the United States) indicate a higher percentage of extrathyroidal extension for tumours from Belarus and the Russian Federation but similar percentages of cases with metastases [N5, P2, P6, V8, Z3].

293. In a recent pooled analysis of 540 thyroid cancers diagnosed before the age of 20 years (mean age at diagnosis, 14 years) that included nine Western centres, the average male:female ratio was 1:3.2 and the mean follow-up was 20 years [F7]. Eighty-six percent were papillary thyroid carcinomas, 79% showed evidence of lymph node metastases, 20%–60% had extracapsular invasion and 23% were diagnosed with distant metastases. In nearly all cases the presenting sign was a neck mass. Thirteen of the patients died as a consequence of the disease.

294. Two of the above-mentioned centres participated in a study in which 369 Italian and French thyroid cancers were compared to 472 Belarusian cancers [P2]. The Belarusian cases were diagnosed between May 1986 and end of 1995 and included approximately 98% of the cases in the country diagnosed in that period. The Belarusian patients were younger at diagnosis, and 95% of the cancers were of papillary origin compared to 85% in Italy/France. Extrathyroidal extension and lymph node metastases were more frequent in Belarus, 49% and 65%, compared to Italy/France, where the corresponding figures were 25% and 54%, respectively. Thyroid lymphocytic infiltration and circulating antithyroidperoxidase were more frequent in the Belarusian patients, possibly indicating a higher rate of autoimmune thyroid disorders. The male:female ratio was 1:2.5 in Italy/France, in contrast to 1:1.6 in Belarus. The low male:female ratio could be a screening effect, since no difference in sex ratio of occult thyroid cancer was found at autopsy in young individuals [F10]. The age distribution could probably also explain some of the difference in stage, the histopathological distribution and gender ratio.

295. A number of thyroid cancer cases in Belarus were treated with radioiodine at the university clinics of Essen and Würzburg [R23]. All 145 patients had undergone operations at the Centre for Thyroid Tumors in Minsk; lymph node metastases were found in 140 patients and distant metastases in 74 of them. The mean age at diagnosis was 12 years. Among 125 children subsequently followed, 90 were classified as in complete remission and the others had partial remissions.

296. In a study of 577 Ukrainian thyroid cancer cases diagnosed in patients less than 19 years of age [T18],

histopathology was evaluated in 296 cases (123 were analysed by non-Ukrainian pathologists, who confirmed the initial diagnosis in all cases). Ninety-three percent were papillary carcinomas, and 65% were found to be of the more aggressive solid/follicular type. In 55% of cases, lymph node metastases were found, and in 17% lung metastases were found either at initial diagnosis or in later follow-up. Difference in TNM classification [H17] over time did not show a significant trend towards more advanced stages (Table 62), as could have been anticipated if radiation-associated cancers are indeed more aggressive. Cancers diagnosed in 1996 and 1997 were more likely to be locally aggressive, stage T4, but they revealed the same pattern of lymph node metastases and distant spread. The male:female ratio was found to be influenced by age at the time of diagnosis (Table 63): the ratio was 1.1:1 for those less than 5 years of age at time of diagnosis and 1:2.7 for those 15–18 years. However, age at time of the accident did not seem to influence the male:female ratio (Table 63). A possible sex difference in the susceptibility of the thyroid tissue to ionizing radiation did not seem to influence the gender ratio, since age at diagnosis and not age at exposure influenced the distribution.

297. In a study in the United States of 4,296 patients previously irradiated for benign disorders, 41 childhood (mean age at diagnosis, 16 years) and 77 adult (mean age at diagnosis, 27 years) thyroid cancers were found [S4]. The childhood cancers more often presented themselves with lymph node metastases and vessel invasion but were significantly smaller in adults and found incidentally when benign nodules were operated. Of the childhood cancers, 95% were papillary carcinomas compared with 84% of the adult cancers. Thirty-nine percent of the childhood cancers relapsed compared with 16% of the adult cancers. After a mean follow-up of 19 years, there was only one death due to thyroid cancer, and this was in the adult group.

298. The histopathology of over 400 post-Chernobyl thyroid cancers diagnosed in children under the age of 15 years was reviewed by pathologists from the United Kingdom and from the countries where the children were diagnosed [E2]. Virtually all cases were papillary carcinomas, in contrast to thyroid cancers in British children of similar age who were not exposed to ionizing radiation, where 68% of the carcinomas were of papillary origin [H3]. The solid variant of papillary carcinoma, indicating a low level of differentiation, was particularly prevalent in those cases believed to have resulted from radiation exposure following the Chernobyl accident [E2]. This finding is not in agreement with a study of 19 cases in the Gomel area carried out by the Chernobyl Sasakawa Health and Medical Cooperation Project, where no specific morphological evidence of radiation-induced thyroid cancer was observed [I26].

299. Shirahige et al. [S34] examined 26 Belarusian and 37 Japanese children diagnosed with thyroid cancer and found the mean tumour diameter to be smaller in Belarus (1.4 cm) than in Japan (4.1 cm). The solid growth pattern was seen in 62% of the Belarusian papillary carcinomas

compared to 18% of the Japanese carcinomas. All cancers from Belarus showed a papillary growth pattern, compared to 92% of the cancers found in Japan.

300. Recent advances in the field of molecular biology have improved the understanding of the mechanisms underlying the thyroid carcinomas. Cellular signalling has become a major research area, and signalling via protein tyrosine kinases has been identified as one of the most important events in cellular regulation. Protein tyrosine kinases are thus important in the development of cancer, and the *RET* proto-oncogene is one of the genes coding for a receptor tyrosine kinase. Rearrangements of the tyrosine kinase domain of the *RET* proto-oncogene have been found in some thyroid cancers and at a higher rate among those supposed to be associated with ionizing radiation [F2, I21, K14]. Other mutations are those seen in the *RAS* gene, which probably represent an early event in the carcinogenic process; the mutations of the *RAS* gene therefore possess an early gate-keeper function, as reflected by the fact that they are found in similar frequencies in both thyroid carcinomas and adenomas. Point mutations in the *TP53* gene are rare in differentiated thyroid carcinomas but are found in anaplastic thyroid cancer [F8].

301. A number of studies have been carried out to determine whether any specific molecular biological alterations characterize the childhood papillary cancers in the Chernobyl area. A high frequency of *RET/PTC3*-type rearrangement in post-Chernobyl papillary carcinoma was suggested in two early studies [F2, K14]. However, a study including more cases showed that a *RET/PTC1* rearrangement was more frequent than *PTC3* [E2] and that there was no significant increase in the proportion of papillary carcinomas showing RET rearrangement, when compared with a non-irradiated population of a similar age [W3]. In a French study, a higher frequency of *RET/PTC* rearrangements (84%) was found among 19 thyroid cancer patients who had received previous external radiotherapy for a benign or malignant condition than among 20 "sporadic" thyroid cancers (15%) [B12]. The most frequently observed chimeric gene was *RET/PTC1*, and for the first time *RET/PTC* rearrangements were found in follicular adenomas.

302. Recent findings indicate a strong dependence of age at exposure and latency period in the distribution of *RET/PTC* rearrangements [S35]. When comparing 51 Belarusian childhood papillary thyroid carcinomas (mean age at exposure: three years) with 16 Belarusian and 16 German adult thyroid cancers patients, only *RET/PTC1* mutations were found in adults (Table 64), while similar frequencies of *RET/PTC1* and *RET/PTC3* were found in children. The authors suggested that thyroid cancers expressing *RET/PTC3* may be a feature of cancers detected soon after exposure, while *RET/PTC1* may be a marker of later-occurring, radiation-associated papillary thyroid carcinoma in both children and adults. It also seems like *RET/PTC3* rearrangements are more common in younger individuals. These findings are supported by a study comparing *RET/PTC* rearrangements in Belarusian

radiation-associated thyroid cancers diagnosed in 1991–1992 and in 1996 [P27]. A switch occurred from *RET/PTC3* rearrangements in patients diagnosed in the early 1990s to *RET/PTC1* alterations in patients diagnosed later on.

303. The morphological subtype may also influence the pattern of *RET* rearrangements in papillary thyroid carcinomas, as shown by Thomas et al. [T19]. Among 116 Chernobyl-related childhood papillary thyroid cancers, *RET/PTC1* and *RET/PTC3* mutations were found in 9% and 19%, respectively, of the solid follicular subtype. The corresponding figures for the non-solid follicular subtype (classic papillary carcinoma and diffuse sclerosing carcinoma) were 46% and 0%, respectively. It could be that the solid, less differentiated, follicular subtype associated with *RET/PTC3* rearrangements have a shorter induction period.

304. Despite the large amount of information accumulated on *RET* activations in radiation-associated thyroid cancers, little is known about the clinical significance of the deletions. No significant differences were found, e.g. in tumour size, multicentricity, extrathyroidal growth, vascular invasion, lymphocytic invasion, or lymph node invasion, when eight *RET* positive thyroid cancers were compared with 25 carcinomas not displaying the activation [S39]. A lower proliferation rate was, however, seen in the *RET*-activated tumours.

305. Karyotype abnormalities were studied in 56 childhood thyroid tumours from Belarus, and clonal structural aberrations were seen in 13 cases [Z2]. In particular, aberrations in 1q, 7q, 9q and 10q were found. It is interesting to note that the 10q chromosomal band harbours the *RET* proto-oncogene.

306. Two new rearrangements have been described in three post-Chernobyl thyroid carcinomas. One is the *RET/PTC4*, which involves a different breakpoint in the *RET* gene [F6] and *PTC5*, which involves fusion of *RET* with another ubiquitously expressed gene, *rgf5* [K31]. However, *PTC4* and *PTC5* appear to be present in only a small number of post-Chernobyl papillary cancers. Just recently, two novel types *RET* rearrangements have been described, *RET/PTC6* and *PTC7* in childhood papillary thyroid carcinomas [K27].

307. No association has previously been shown between mutations of other types of genes, e.g. *RAS*, *bcl-2*, and *TP53*, and thyroid carcinogenesis [E2, S9]. However, in a study of 22 papillary thyroid cancers associated with ionizing radiation, four mutations of *TP53* were found compared to none in the 18 thyroid cancers not known to have been exposed to ionizing radiation [F8]. In three of the four mutation carriers, invasion beyond the thyroid capsule was found compared with 2 out of 17 in the rest of the radiation-associated thyroid cancer patients. In a study of thyroid adenomas and of well and poorly differentiated thyroid cancers, the rates of *TP53* mutations were 0%, 11% and 63%, respectively [P1]. However, no correlation with age, sex, stage, or survival was seen.

(c) Summary

308. There can be no doubt about the relationship between the radioactive materials released from the Chernobyl accident and the unusually high number of thyroid cancers observed in the contaminated areas during the past 14 years. While several uncertainties must be taken into consideration, the main ones being the baseline rates used in the calculations, the influence of screening, and the short follow-up, the number of cases is still higher than anticipated based on previous data. This is probably partly a result of age at exposure, iodine deficiency, genetic predisposition, and uncertainty that surrounds the role of ^{131}I compared with that of short-lived radioiodines. The exposure to short-lived radioiodines is entirely dependent on the distance from the release and the mode of exposure, i.e. inhalation or ingestion. It was only in the Gomel region, the area closest to the Chernobyl reactor, that Astakhova et al. [A6] found a significantly increased risk of thyroid cancer. It has been suggested that the geographical distribution of thyroid cancer cases correlates better to the distribution of shorter-lived radioisotopes (e.g. ^{132}I, ^{133}I and ^{135}I) than to that of ^{131}I [A7].

309. The identification of a genomic fingerprint that shows the interaction of a specific target cell with a defined carcinogen is a highly desirable tool in molecular epidemiology. However, a specific molecular lesion is almost always missing, probably because of the large number of factors acting on tumour induction and progression. Signalling via protein tyrosine kinases has been identified as one of the most important events in cellular regulation, and rearrangements of the tyrosine kinase domain of the *RET* proto-oncogene have been found in thyroid cancers thought to be associated with ionizing radiation [F2, I21, K14]. However, the biological and clinical significance of *RET* activation remains controversial, and further studies of the molecular biology of radiation-induced thyroid cancers are needed before the carcinogenic pathway can be fully understood.

2. Leukaemia

310. As discussed in Annex I, *"Epidemiological evaluation of radiation-induced cancer"*, the risk of leukaemia has been found to be elevated after irradiation for benign and malignant conditions, after occupational exposure (radiologists), as well as among the survivors of the atomic bombings. Leukaemia, although a rare disease, is the most frequently reported malignancy following radiation exposure. However, not all subtypes of leukaemia are known to be associated with ionizing radiation, e.g. chronic lymphatic leukaemia and adult T-cell leukaemia. The naturally occurring subtypes of leukaemia have an age dependency, with acute lymphatic leukaemia most common in childhood and acute myeloid leukaemia predominating in adulthood.

311. The incidence of leukaemia was increasing in countries of the former USSR even before the accident. Prisyazhiuk et al. [P7] found an increase starting already in 1981, which was most pronounced in the elderly. The

increase may simply be the result of improved registration and diagnosis, but this cannot yet be exactly quantified. The underlying trend must, however, be taken into account when interpreting the results of studies focussing on the period after the accident. Most existing studies have not addressed this problem and suffer from a number of other limitations and methodological weaknesses, making it premature to attempt a quantitative risk assessment based on the results.

312. A number of publications have presented details of the medical and dosimetric follow-up of the large number of workers who took part in recovery operations following the Chernobyl accident [K6, O2, T3, T7, T8, T9, T10]. Cardis et al. [C2] analysed cancer incidence in 1993 and 1994 among male recovery operation workers who worked within the 30-km zone of the reactor during 1986 and 1987. The observed numbers of cancers were obtained from the national cancer registry in Belarus and from the Chernobyl registries in the Russian Federation and Ukraine. In total, 46 leukaemia cases were reported in the three countries in the two-year period (Table 65), and non-significant increases were observed in Belarus and in the Russian Federation. In Ukraine, a significant increase was reported (28 cases observed, 8 cases expected) [C2]. It is most likely that the increase reflects the effect of increased surveillance of the recovery operation workers and under-registration of cases in the general population, since no systematic centralized cancer registration existed in the three countries at the time of the accident (see Section IV.A.1 and IV.C.2). It could also be that different registries define haematological malignancies differently, including, for example, myelodysplastic syndrome, which could result in a leukaemia.

313. Ivanov et al. [I8, I9, I14] studied the late effects in 142,000 Russian recovery operation workers. The Russian National Medical and Dosimetric Registry (formerly the Chernobyl registry) was used, and a significantly increased risk of leukaemia was found when the observed cases were compared with those expected from national incidence rates. The studies have been criticized for not using individual doses and internal comparisons and for including chronic lymphatic leukaemia, a malignancy not linked to radiation exposure [B30, B31]. The increased medical surveillance and active follow-up of the emergency workers, coupled with under-reporting in the general population, most likely influenced the results.

314. In contrast to their findings for the above-mentioned cohort of recovery operation workers, the same investigators [I29] did not find an increased risk of leukaemia related to ionizing radiation in a case-control setting. From 1986 to 1993, 48 cases of leukaemia were identified through the Russian National Registry and 34 of these patients (10 cases were diagnosed as chronic lymphatic leukaemia) were selected for the case-control study. The same registry was used when four controls were chosen for each case, matched on age and region of residence at diagnosis. For cases occurring among those

who were working in 1986 and 1987, controls had to have worked during the same period, which is a questionable approach, since dose is highly dependent on the period of work in the Chernobyl area. The mean dose for the cases was 115 mGy compared with 142 mGy for controls. No association was found between leukaemia risk and radiation.

315. These studies [I8, I9, I14, I29] suggest that, at least in the case of the Russian Federation, cancer incidence ascertainment in the exposed populations differs from that in the general population. Future epidemiological investigations might be more informative if they are based on appropriate Chernobyl registry-internal comparison populations, although care must be taken if recent additions to the register have been made because of disease diagnosis and compensation (see Section IV.C.1).

316. In discussing the discrepancy between the findings of the case-control and cohort studies, Boice and Holm [B31] claimed that the increased incidence in the cohort analyses reflected a difference in case ascertainment between recovery operation workers and the general population and not an effect of radiation exposure. Boice [B30] further argued that the results of the case-control study and of a study of Estonian recovery operation workers [R13] indicate that leukaemia risk among recovery operation workers is not consistent with predictions from atomic bomb survivors. He postulated that this may be due to an overestimation of official doses received by recovery operation workers and/or to the effect of protracted exposure. In response, Ivanov [I18] questioned the interpretation of the case-control analyses because of the considerable uncertainty surrounding the accuracy and quality of the official estimates of radiation doses available in the Chernobyl registry of the Russian Federation. Cardis et al. [C2] estimated that 150 cases of leukaemia should occur within 10 years of the accident among 100,000 recovery operation workers exposed to an average dose of 100 mSv. Such numbers have not been apparent in any studies or reports.

317. Shantyr et al. [S31] examined 8,745 Russian recovery operation workers involved in operations from 1986 to 1990. Dosimetry records were available for 75% of the workers, and the doses generally fell in the range 0–250 mSv. Although cancer incidence increased, particularly 4–10 years after the accident, no evidence of a systematic dose-response relationship was found, and it was suggested that the aging of the cohort influenced the findings. Tukov and Dzagoeva [T11] observed no increased risk of haematological diseases, including acute forms of leukaemia, in a careful study of Russian recovery operation workers and workers from the nuclear industry.

318. Osechinsky et al. [O6] studied the standardized incidence rates of leukaemia and lymphoma in the general population of the Bryansk region of the Russian Federation for the period 1979–1993 on the basis of an *ad hoc* registry of haematological diseases established after the Chernobyl accident. The results were not adjusted for age, and the rates in the six most contaminated districts (more than 37 kBq m^{-2} of ^{137}Cs deposition density) did not exceed the rates in the rest of the region or in Bryansk city, where the highest rates were observed. Comparisons of crude incidence rates before and after the accident (1979–1985 and 1986–1993) showed a significant increase in the incidence of all leukaemia and non-Hodgkin's lymphoma, but this was mainly due to increases in the older age groups in rural areas. The incidence of childhood leukaemia and non-Hodgkin's lymphoma was not significantly different in the six most contaminated areas from the incidence in the rest of the region.

319. The health status of 174,812 Ukrainian recovery operation workers (96% males) was examined by Buzunov et al. [B21]. Information on diagnosis was obtained from the State register of Ukraine and on leukaemia from an *ad hoc* registry for haematological disorders. The majority (77%) of the recovery operation workers were exposed in 1986–1987, and information on radiation exposure was available for approximately 50% of the workers. A total of 86 cases of leukaemia were reported in the period 1987–1992, and the highest number of cases was found among those employed in April-June 1986. The average rate of leukaemia among male recovery operation workers was 13.4 per 100,000 among those employed in 1986 and 7.0 per 100,000 among those employed in 1987. No apparent trend over time was seen among those employed in 1986. Eighteen cases of acute leukaemia among recovery operation workers exposed to 120–680 mGy were recorded as occurring 2.5–3 years after exposure in 1986–1987. No difference in histopathology or response to treatment was found compared with cases that occurred before the accident [B20].

320. Leukaemia and lymphoma incidence among adults and children in the regions of Kiev and Zhitomir, Ukraine, during 1980–1996 was examined by Bebeshko et al. [B32]. Total incidence in adults increased from 5.1 per 100,000 during 1980–1985 to 11 per 100,000 during 1992–1996, but there were no excess cases in contaminated areas of the regions. Likewise, no excess cases among children who resided in contaminated districts were found.

321. The incidence of leukaemia and lymphoma in the three most contaminated regions of Ukraine increased from 1980 to 1994 [P12]. This result should be viewed cautiously, since the findings were based on only a few cases, and lymphomas have not previously been known to be induced by radiation. Increased awareness and better health-care facilities and diagnosis most likely influenced the findings.

322. Childhood leukaemia in Belarus during 1982–1994 was investigated with regard to area of residency [I20]. Approximately 75% of the leukaemia cases were of the acute lymphatic subtype. No evidence of an increasing number of childhood leukaemia cases over time was noted. When the two most heavily contaminated areas, Gomel and Mogilev, were compared with the rest of the country, no difference was seen.

323. The incidence of leukaemia and lymphoma in the general population of Belarus was studied for 1979–1985 and 1986–1992 [I11]. Among children, no difference was observed either over time or in relation to ^{137}Cs ground contamination. In adults, significant increases were noted in the post-accident period for most subtypes of leukaemia and for lymphoma, but no relationship with the level of radioactive contamination was found.

324. An analysis of the mortality and cancer incidence experience of Estonian Chernobyl recovery operation workers took a different approach. First, a cohort consisting of 4,833 recovery operation workers was constructed using multiple data sources not based solely on Chernobyl registry data [R13, T6]. Second, mortality data were ascertained from vital registration sources, and death certificates in the cohort were coded following the same coding rules used by the Statistical Office of Estonia and compared to the official national mortality statistics [T6]. Furthermore, cancer incidence data in the cohort, along with the corresponding national cancer incidence rates, were obtained from the Estonian cancer registry. During 1986–1993, 144 deaths were identified compared with 148 expected [R13, T6]. A non-significant excess of non-Hodgkin's lymphoma was observed, based on three cases, while no case of leukaemia was found.

325. Fujimura et al. [F1] reported the results of haematological screening organized in the framework of the Sasakawa project. By the end of 1994, 86,798 children who were less than 10 years of age at the time of the accident were examined, and four cases of haematological malignancies were found. No correlation was observed between the prevalence of any haematological disorder and either the level of environmental contamination or ^{137}Cs measured by whole-body counting.

326. The European Childhood Leukaemia–Lymphoma Incidence Study (ECLIS), coordinated by IARC, was set up to monitor trends in childhood leukaemia and lymphoma [P5, P11]. Incidence rates from European cancer registries were related to the calculated radiation dose in the large geographical regions for which environmental dose estimates were provided in the UNSCEAR 1988 Report [U4]. Thirty-six cancer registries in 23 countries are collaborating in ECLIS by supplying an annual listing of cases in children less than 15 years of age. Data for 1980–1991 indicated a slight increase in the incidence of childhood leukaemia in Europe. This increase was, however, not related to the estimated radiation dose from the accident [P11, P21]. No indication of an increased incidence among those exposed *in utero* was noticed.

327. A study of infant leukaemia incidence in Greece after *in utero* exposure to radiation from the Chernobyl accident was based on *ad hoc* registration of childhood leukaemia cases diagnosed throughout Greece since 1980 by a national network of pediatric oncologists [P3]. Based on 12 cases, a statistically significant 2.6-fold increase in the incidence of infant leukaemia (from 0 to 11 months after

birth) was observed among the 163,337 live births exposed *in utero* (i.e. born between 1 January 1986 and 31 December 1987) compared with 31 cases among non-exposed children. Those born to mothers residing in areas of high radioactive fallout were at significantly higher risk of developing leukaemia. The reported association, which is not consistent with risk estimates from other studies of prenatal exposures, is based on the selective grouping of data. It is unclear how the authors chose the group <1 year to represent "infant leukaemia", as there is little *a priori* aetiologic reason for limiting to this age group. No significant difference in the incidence of leukaemia among children aged 12–47 months born to presumably exposed mothers was found.

328. Michaelis et al. [M6] conducted a study of childhood leukaemia using the population-based cancer registry in Germany. Cohorts were defined as exposed and non-exposed, based on dates of birth using the same criteria as Petridou et al. [P3] in the Greek study. The cohorts were subdivided into three categories based on level of ^{137}Cs ground deposition (<6, 6–10 and >10 kBq m^{-2}). These categories corresponded to the estimated *in utero* doses of 0.55 mSv for the lowest exposure category and 0.75 mSv for the highest exposure category. Overall, a significantly elevated risk was seen (1.48, 95% CI: 1.02–2.15) for the "exposed cohort" compared with the "non-exposed", based on 35 cases observed in a cohort of 900,000 births. However, the incidence was higher for those born in April to December 1987 than for those born between July 1986 and March 1987, although *in utero* exposure levels in the latter group would have been much higher than in the former group. The authors concluded that the observed increase was not related to radiation exposure from the Chernobyl accident.

329. A cluster effect described several decades ago could explain the Greek findings. A total of 13,351 cases of childhood leukaemia diagnosed between 1980–1989 in 17 countries was included in a study aimed at relating childhood leukaemia to epidemic patterns of common infectious agents [A2]. A general elevation of the incidence was found in densely (but not the most densely) populated areas, and weak, but significant, evidence of clustering was found. When seasonal variation in the onset of childhood leukaemia was studied in the Manchester tumour registry catchment area, the onset of acute lymphatic leukaemia (n = 1,070) demonstrated a significant seasonal variation, with the highest peak found in November and December [W6]. Both studies provide supportive evidence for an infectious aetiology for childhood leukaemia.

330. In a Swedish study of cancer incidence among children [T21] in areas supposed to have been contaminated as a consequence of the Chernobyl accident, 151 cases of acute lymphatic leukaemia were found during 1978–1992 in those 0–19 years at diagnosis. The areas were divided into three exposure categories, and the lowest risk was found in the supposedly highest exposure group. A non-significant decreasing trend with calendar year was also noted. A Finnish study covering nearly the same

period analysed leukaemia risks among those 0-14 years of age for the whole country [A13]. The estimated population-weighted mean effective dose was 0.4 mSv. No increased incidence of childhood leukaemia could be seen for 1976-1992, and no risk could be related to exposure to ionizing radiation. These results are consistent with the magnitude of effects expected.

331. *Summary*. Although leukaemia has been found to be one of the early carcinogenic effects of ionizing radiation with a latency period of not more than 2-3 years [U4], no increased risk of leukaemia related to ionizing radiation has been found among recovery operation workers or in residents of contaminated areas. Numerous reports have compared incidence and mortality data from the registers described in Chapter IV with national rates not taking the differences in reporting into consideration. A case-control study would diminish this bias, and a recent paper by Ivanov et al. [I29] failed to show an increased risk of leukaemia related to ionizing radiation in 48 cases of leukaemia in recovery operation workers identified through the Russian National Registry.

3. Other solid tumours

332. Given the doses received by the recovery workers, described in Chapter II, and the previous data on radiation-associated cancer in exposed populations, reviewed in Annex I, *"Epidemiological evaluation of radiation-induced cancer"*, an increased number of solid tumours could be anticipated in the years to come. The induction and latency period of 10 years [U4] and the protracted nature of the exposure probably explain why no radiation-associated cancers have been noticed so far.

333. The numbers of observed and expected cases of cancer in 1993-1994 among residents of the territories with ^{137}Cs contamination in excess of 185 kBq m^{-2} included in the Chernobyl registries of Belarus, the Russian Federation and Ukraine are presented in Table 65 [C2]. The observed numbers of cancer cases were obtained from the national cancer registry in Belarus and from the Chernobyl registries in the Russian Federation and Ukraine. Age- and sex-standardized expected numbers were based on rates for the general national population. Fewer solid tumours than anticipated were seen in workers in Belarus, while the workers in the Russian Federation and Ukraine revealed higher risks. The different registries used in the three countries could probably explain these differences. No increased risk was seen among those residing in contaminated areas [C2].

334. The crude incidence of malignant diseases per 100,000 persons among Russian recovery operation workers, excluding leukaemia, was estimated by Tukov and Shafransky [T13]. It rose from 152 in 1989, 193 in 1991 and 177 in 1993, to 390 in 1995. This increase was interpreted as an effect of age, but no attempt was made to adjust for age. The cause of death has changed over time. Accidents and trauma were the main cause of death in 1989-1990, while cardiovascular diseases

were responsible for 43% of all deaths in 1996, followed by cancers (20%) and accident and trauma (15%) [T14]. The increase found in the report is similar to and consistent with that reported for the population of the Russian Federation as a whole [L26].

335. In a recent paper covering 114,504 of the approximately 250,000 Russian recovery operation workers, 983 cases of solid tumours were found during the years 1986-1996 [I32]. The observed number of cases was compared with the Russian national rates, and the overall standardized incidence ratio was 1.23 (95% CI: 1.15-1.31), with a significant excess relative risk per gray of 1.13. The only individually elevated site was the digestive tract (n = 301), and the corresponding figures were 1.11 (95% CI: 1.01-1.24) and 2.41, respectively. No increased risk of respiratory tract tumours was noticed. Increased ascertainment could influence the data but probably not explain the dose-response relationship. The excess relative risk is higher than what has been reported for survivors of the atomic bombings [T20], which might indicate uncertainty in the individual dose estimates.

336. When the same investigators presented data on individuals living in contaminated areas, somewhat contradictory results were seen [I12]. Cancer risks in three of the most contaminated districts of the Kaluga region were compared with the region as a whole. The population of the three regions contained approximately 40,000 individuals, and incidence rates before and after the accident were compared. The increase over time was similar for almost all sites regardless of exposure status. No increased risk of gastrointestinal cancer was seen for men, but an increased risk for respiratory tract cancers was suggested for women, based on 31 cases. However, the overall cancer risk among women in the contaminated areas was only one third of the incidence for the region in 1981-1985, a sign of previous under-reporting.

337. A descriptive study of cancer incidence in 1981-1994 was performed for the six most contaminated regions of the Russian Federation (Bryansk, Kaluga, Orel, Tula, Ryazan and Kursk) [R2]. Information on cancer incidence was gathered from the local oncological dispensaries and compared with Russian national statistics. It is unclear whether the analyses were adjusted for age, and the absolute number of cases was not reported. An increased incidence of all cancers was observed over the study period, both in the contaminated regions and for the Russian Federation as a whole. However, from 1987 onwards, the increase was more pronounced in the six study regions than in the rest of the country. The incidence rate of solid cancer among men in 1994 in Bryansk and Ryazan was 305 per 100,000 compared with 272 per 100,000 for the Russian Federation as a whole. The corresponding figures for women were 180 per 100,000 and 169 per 100,000, respectively.

338. An increase in dysplasia and urinary bladder cancer was seen in 45 Ukrainian males living in contaminated

areas and compared to 10 males living in uncontaminated areas of the country [R16]. Forty-two of the exposed individuals had signs of irradiation cystitis. It was reported that the incidence of bladder cancer in the Ukrainian population gradually increased, from 26 to 36 per 100,000, between 1986 and 1996. Among other histopathological features, increased levels of p53 were noted in the nucleus of the urothelium in the exposed individuals, indicating either an early transformation event or an enhancement of repair activities. Further analyses of the exposed patients, including urine sediments collected 4–27 months after the first biopsy, indicated a novel type of *p53* mutation not seen in the first analyses and showed that the mutation carriers could be identified [Y2]. The authors concluded that screening would be required.

339. The health status of the 45,674 recovery operation workers from Belarus registered in the Chernobyl registry was studied by Okeanov et al. [O2]. Eighty-five percent of them were men, and 31,201 (90%) had worked in the 30-km zone. For 1993 and 1994, the overall cancer incidence was lower than anticipated for both male (standardized incidence ratio (SIR) = 77; 95% CI: 65–90) and female (SIR = 90; 95% CI: 59–131) workers compared with the general population. Among men, a significant excess of urinary bladder cancer was seen (SIR = 219; 95% CI: 123–361), and non-significant increases were seen for cancers of the stomach, colon, and thyroid and for leukaemia. The numbers of cases on which these comparisons are based were small, particularly for thyroid cancer (n = 4). Recovery operation workers who worked in the 30-km zone more than 30 days had a slightly higher incidence of all cancers than other recovery operation workers.

340. Breast cancer incidence data in different time periods for the Mogilev region of Belarus were recently presented [O5]. A steadily increasing incidence was noted for the whole follow-up period, 1978–1996, and when the period 1989–1992 was compared with 1993–1996, a difference was found, but only for those 45–49 years at diagnosis, i.e. 35–42 years at the time of the accident. The findings could be due to increased awareness, documentation, or accessibility to screening, since the rates are lower in all age categories compared with the ages found in Western data. It is peculiar that younger age groups were not affected, and continued follow-up is warranted. However, as in all studies of radiation-associated cancer, individual dosimetry is essential, and individual doses were not used in this study.

341. Several studies of the effects of the Chernobyl accident outside the former Soviet Union have been carried out [E5, P5, P11]. The evaluations have mainly been done on a local or national level. Most studies have focussed on various possible health consequences of the accident, ranging from changes in birth rates to adult cancer. Studies related to cancer have been critically reviewed and summarized [S12]. Overall, no increase in cancer incidence or mortality that could be attributed to the accident has been observed in countries of Europe outside the former USSR.

342. *Summary*. The occurrence of solid tumours other than thyroid cancers in workers or in residents of contaminated areas have not so far been observed. The weaknesses in the scientific studies, the uncertainties in the dose estimates, the latency period of around 10 years and the protracted nature of the exposures probably explain why no radiation-associated cancers have been noticed so far. Some increase in incidence of solid tumours might have been anticipated in the more highly exposed recovery operation workers.

B. OTHER SOMATIC DISORDERS

1. Thyroid abnormalities

343. The first report of non-malignant thyroid disorders in the Chernobyl area was published in 1992 [M10]. The prevalence of thyroid nodules among individuals in seven contaminated villages in Belarus, the Russian Federation and Ukraine was compared with the prevalence in six uncontaminated villages, and 1,060 individuals, in total, were examined. Ultrasound examinations revealed an overall rate of discrete nodules of 15% in adults and 0.5% in children, with a higher prevalence in women. No difference related to exposure status was found, but it was suggested that it might be helpful to screen selected groups such as recovery operation workers and individuals living in contaminated areas.

344. The Chernobyl Sasakawa Health and Medical Cooperation Project started in May 1991 as a five-year programme, and through April 1996, approximately 160,000 children had been examined [S2, S7, Y1]. The examined children were all born in Belarus, the Russian Federation and Ukraine between 26 April 1976 and 26 April 1986. The thyroid examinations included ultrasound, serum free thyroxine, thyroid-stimulating hormone (TSH), antithyroperoxidase, antithyroglobulin and urine iodine concentration. A total of 45,905 thyroid abnormalities were found in 119,178 examined children (Table 66) [Y1]. Ninety-one percent of the abnormalities were diagnosed as goiter, and 62 thyroid cancers were found. The incidence rates in Gomel, the area of Belarus with the highest contamination, had the lowest incidence of goiter but the highest incidence of abnormal echogenity, cystic lesions, nodular lesions, and cancer, the latter two known to be related to radiation (Table 67). No association between thyroid antibodies, hypo- or hyperthyroidism, and ^{137}Cs activity in the body or soil contamination was seen in 114,870 children examined [Y1].

345. The contaminated areas around Chernobyl have been recognized as iodine-deficient areas, but the influence on the goiter prevalence has not been clear. In an extended study of the 119,178 children included in the Chernobyl Sasakawa Health and Medical Cooperation Project [A8], urinary iodine excretion levels were measured in 5,710 selected cases. The study did not reveal any correlation between goiter and whole-body ^{137}Cs content or ^{137}Cs

contamination level at the place of residence either at the time of examination or the time of the accident. However, a significant negative correlation was indicated between prevalence of goiter and urinary iodine excretion levels. The highest prevalence of goiter (54%) was found in Kiev, where the incidence of childhood cancer was relatively low. However, the Kiev area was also identified as an endemic iodine-deficient zone. The opposite was found in Gomel, i.e. no profound iodine deficiency and a lower rate of thyroid nodules (18%) in an area with a higher rate of childhood thyroid cancer.

346. For inhabitants of the Bryansk region in the Russian Federation who were born before the accident and examined with ultrasound, the overall prevalence of thyroid abnormalities did not differ when contaminated and uncontaminated areas were compared [K10]. A difference was revealed when age was taken into consideration. For those 0–9 years of age at the time of the accident, the prevalence of thyroid abnormalities was 8.1% in the exposed cohort compared with 1.6% in the non-exposed. The corresponding figures for individuals 10–27 years in 1986 were 18.8% and 17.7%, respectively. Approximately half of the pathological findings identified through ultrasound were also noticed at palpation.

347. The prevalence of thyroid antibodies (antithyroglobulin and antithyroperoxidase) in children and adolescents in Belarus was measured in 287 individuals residing in contaminated areas (average ^{137}Cs contamination, 200 kBq m^{-2}) and compared to the findings in 208 individuals living in uncontaminated areas (average ^{137}Cs contamination, <3.7 kBq m^{-2}) [P20]. All individuals were younger than 12 years at time of the accident. Significantly elevated concentrations of thyroid antibodies were found among the exposed individuals with most pronounced concentrations in girls of puberty age. No indication of thyroid dysfunction was found, but the future development of clinically relevant thyroid disorders was thought to be a possibility.

348. Blood samples from 12,803 children living in the Kaluga region were studied for antibodies to thyroid antigen with a modifying reaction of passive haemoagglutination [S15]. In the sixth year after the accident, the reaction showed positive results in only a small percentage of samples (1.2%–4.8%). However, in children from contaminated areas, the percentage of positive results was consistently higher than in children from uncontaminated areas.

349. Similar results were found in a Russian study in which 89 exposed and 116 non-exposed children were examined [K28]. There was no apparent alteration in thyroid function, but a higher rate of thyroid antibodies was found in the exposed group. This group also had a lower percentage of individuals with iodine deficiency, as defined by urinary iodine excretion (76%), than groups living in uncontaminated areas (92%), but at the same time a fivefold greater rate of thyroid enlargement was identified by ultrasound. No sex difference was seen for goiters in the exposed group compared with a 1:2 male to female ratio in the non-exposed group.

350. Fifty-three Ukrainian children (0–7 years of age at the time of the accident) living in contaminated areas were compared with 45 children living in supposedly uncontaminated regions [V1]. The level of antithyroglobulin, thyroid-stimulating hormone and abnormal findings at ultrasound were higher in the exposed individuals, and it was concluded that there was a dose-response relationship. In contrast, no difference in thyroid function was noticed when 888 Belarusian schoolchildren living in contaminated areas were compared with 521 age-matched, non-exposed controls [S10]. Both groups lived in iodine-deficient areas, and the prevalence of diffuse goiter was significantly higher in the exposed group. Thyroid antibodies were not measured.

351. When 143 children 5 to 15 years old at examination living in a contaminated area were compared with 40 age- and sex-matched controls living in clean areas of the Tula region of the Russian Federation, a higher prevalence of thyroid autoimmunity was found in the exposed group [V10]. The difference was only noticed in those less than 5 years of age at the time of the accident, and no difference in thyroid function was noticed.

352. A total of 700,000 persons from the former Soviet Union have immigrated to Israel, approximately 140,000 of whom come from territories affected by the Chernobyl accident [Q1, Q2]. The thyroid status of 300 immigrant children voluntarily brought for thyroid examination by their parents was evaluated. Enlarged thyroid glands were found in about 40% of subjects, irrespective of whether they came from the contaminated or uncontaminated areas, i.e. ^{137}Cs greater or less than 37 kBq m^{-2} [Q1, Q2]. Thyroid-stimulating hormone levels, although within normal limits, were significantly higher (p < 0.02) for girls from the more contaminated regions.

353. Thyroid screening was performed in 1,984 Estonian recovery operation workers through palpation of the neck by a thyroid specialist and high-resolution ultrasonography by a radiologist [I19]. Fine-needle biopsy was carried out for palpable nodules and for nodules larger than 1 cm found by ultrasound; enlarged nodules were observed in 201 individuals. The prevalence of nodules increased with age at examination but was not related to recorded dose, date of first duty at Chernobyl, duration of service at Chernobyl, or the activities carried out by the recovery operation workers. Two cases of papillary carcinoma and three benign follicular neoplasms were identified and referred for treatment [I19].

354. Thyroid examinations by ultrasound were performed on 3,208 Lithuanian recovery operation workers in 1991–1995, and thyroid nodularity (nodules > 5 mm) was detected in 117 individuals [K9]. There was, however, no significant difference in the prevalence of thyroid nodularity compared with the Lithuanian male population as a whole and no association with level of radiation dose or duration of stay in the Chernobyl area.

355. The development of hypothyroidism following high-level external or internal exposures to ionizing radiation is well known. A change in hypothyroid rates in newborns has been shown in some areas of the United States and was supposed to be related to fallout from the Chernobyl accident [M9]. These findings were challenged [W4] on the grounds that the received doses were far too low to induce hypothyroidism. Doses received in the northwestern part of the United States were approximately 1/10,000 of that received in the Chernobyl area, and extensive examinations of children in Belarus, the Russian Federation and Ukraine have not shown a relationship between dose and either hyper- or hypothyroidism [Y1].

356. *Summary.* Other than the occurrence of thyroid nodules in workers and in children, which is unrelated to radiation exposure, there has been no evidence of thyroid abnormalities in affected populations following the Chernobyl accident. Even the large screening programme conducted by the Chernobyl Sasakawa Health and Medical Cooperation Project in 1991–1996, involving 160,000 children, less than 10 years of age at time of the accident, there was no increased risk of hypothyroidism, hyper-thyroidism or goiter that could be related to ionizing radiation. Neither was an increase in thyroid antibodies noticed, which is in contradiction with some other minor studies.

2. Somatic disorders other than thyroid

357. The first study of health effects other than cancer and thyroid disorders on a representative sample of the popula-tions from contaminated and control districts was carried out in the framework of the International Chernobyl Project [I1]. The conclusion of this project was that, although there were significant health disorders in the populations of both contaminated and control settlements, no health disorder could be attributed to radiation exposure.

358. As discussed previously, between 1991 and 1996, the Sasakawa Memorial Health Foundation of Japan funded the largest international screening programme of children in five medical centres in Belarus, the Russian Federation and Ukraine [Y1]. In all, haematological investigations were carried out for 118,773 children. White blood cells, red blood cells, haemoglobin concentration, haematocrit, mean corpus-cular volume and concentration, and platelets were measured. The prevalence of anaemia was higher in girls than in boys and ranged from 0.2% to 0.5 %. An extended examination of 322 children suggested that iron deficiency was the cause of one third of these cases. The prevalence of leukopenia was somewhat lower in girls than boys (overall range, 0.2%–1.1%), while no sex difference was seen for leukocytosis (range, 2.8%–4.9%). No differences between sexes or centres were seen for trombocytopenia (range, 0.06%–0.12%), trombocytosis (range, 1.0%–1.3%), or eosinophilia (range, 12.2%–18.9%). The prevalence of eosinophilia changed dramatically during the five years of follow-up, from 25% in

1991 to 11% in 1996. There are probably several reasons for this decline, among them better socioeconomic conditions, a greater awareness of health, and improved medical conditions. The frequency of haematological disorders showed no difference by level of ^{137}Cs contamination at the place of residency at the time of the accident, current residency, or ^{137}Cs concentration in the body [Y1].

359. A number of studies have addressed the general morbidity of populations living in contaminated areas [I8, O2, W1]. When individuals in the contaminated areas were compared with the general population in these countries, increased morbidity due to diseases of the endocrine, haematopoietic, circulatory and digestive systems was found. A higher rate of mental disorders and disability has also been noted. It is difficult to interpret these results, since the observations may be at least partly explained by the active follow-up of the exposed populations and by the fact that age and sex are not taken into account in these studies. On the other hand, they may reflect a real increase in morbidity following the Chernobyl accident, which would mainly be an effect of psycho-social trauma, since existing epidemiological studies of radiation-exposed populations are not consistent with these findings. Stress and economic difficulties following the accident are most likely influencing the results.

360. The demographic situation in Belarus has changed since the accident. People have moved to the cities to a larger extent, and the population is, on average, older as a result of the low birth rate. Mortality due to accidents and cardiovascular diseases has increased, particularly among evacuated populations and people living in zones recommended for relocation [W1]. Mortality rates in the Russian Federation regions of Kaluga and Bryansk are close to those in the rest of the country and are relatively stable over time; however, infant mortality is steadily decreasing. Except for an increase in accidental deaths in the contaminated areas, no significant difference in cause of death was found [W1]. Population growth in Ukraine, as in other parts of the former Soviet Union, has become negative. General mortality in contaminated areas is higher (14–18 per 1,000) than in the whole of Ukraine (11–12 per 1,000). The pattern in causes of death in Ukraine is stable, with some decrease in cardiovascular mortality [W1].

361. Since 1990, 4,506 children (3,121 from Ukraine, 1,018 from the Russian Federation and 367 from Belarus) from the Chernobyl area have received medical care at the Centre of Hygiene and Radiation Protection, in Cuba [G18]. Measured body burdens of ^{137}Cs were in the range 1.5–565 Bq kg^{-1} (90% of children had levels below 20 Bq kg^{-1}). Doses from external irradiation were estimated to range from 0.04 to 30 mSv (90% < 2 mSv), 2 to 5.4 mSv from internal irradiation and thyroid doses from 0 to 2 Gy (44% < 40 mGy). Assessment of overall health condition, including haematological and endocrinological indicators, did not differ when the children were divided into five groups on the basis of ^{137}Cs contamination (<37, 37–185, >185 kBq m^{-2}, evacuated, unknown).

362. The incidence of non-malignant disorders in children was evaluated using the Belarus Chernobyl registry [L19]. The children were divided into three groups: evacuated from the 30-km zone, residing (or previously residing) in areas with contamination >555 kBq m^{-2}, and born to exposed parents. Increased rates of gastritis, anaemia and chronic tonsillitis were found among all exposure categories compared to Belarus as a whole, and the highest rates were for children in the Gomel region. The authors concluded that the increases were most probably due to psycho-social factors, lifestyle, diet and increased medical surveillance and suggested that further analyses would be needed to establish aetiological factors.

363. When hormonal levels, biologically active metabolites and immunoglobulins in 132 Russian recovery operation workers were stratified by absorbed doses, no differences related to ionizing radiation were seen except for so-called biomarkers of oxidative stress, e.g. conjugated dienes [S8]. These biomarkers are, however, not specific for radiation damage and can be seen in several pathological conditions.

364. In an Estonian cohort of 4,833 recovery operation workers, 144 deaths were identified in the period 1986–1993, compared with 148 expected [R13, T6]. A relatively high number of deaths were due to accidents, violence and poisoning. In nearly 20%, the cause of death was suicide, and the relative risk of 1.52 (n = 28) was statistically significant [R13, T6].

365. A Lithuanian cohort of 5,446 recovery operation workers was followed regularly at the Chernobyl Medical Centre during the years 1987–1995, and 251 deaths were observed [K3]. The major causes of death were injuries and accidents, and the overall mortality rate of the recovery operation workers was not higher than that of the total population.

3. Immunological effects

366. Acute as well as fractionated exposures to low doses of ionizing radiation have been reported to alter several immunological parameters in experimental animals. It is, however, not clear what effects are found in humans. Many papers have been published in the last decade on the immunological effects of exposure to radiation from the Chernobyl accident. Since it is unclear, however, if possible confounding factors have been taken into account, including, in particular, infections and diet, it is difficult to interpret the results.

367. The immunological status of 1,593 recovery operation workers was studied by Kosianov and Morozov [K24]. A moderate decrease in the number of leukocytes was observed, as well as a decrease in T-lymphocytes and periodic decreases in the number of B-lymphocytes and in immunoglobulin level. These disturbances lasted for 4–6 months in individuals with a dose <2.5 mGy and about a year in those with doses from 2.5 to 7 mGy.

368. The immune status of 85 recovery operation workers who were professional radiation workers from the Mayak plant was studied carefully between 9 and 156 days after they finished work in the Chernobyl area [T17]. The radiation doses were between 1 and 330 mGy. Only some decreases in T-lymphocytes and increases in null-lymphocytes showed causal relations to the radiation dose.

369. A three-year study of 90 recovery operation workers living in the town of Chelyabinsk [A1] showed that the average numbers of leukocytes, neutrophyls and lymphocytes in the whole period were the same as in the control group, consisting of the general population of Chelyabinsk with the same age and sex distribution. During the first and second years, a moderate increase of IgM level in blood was found, while a slight decrease was seen in the first month; complete recovery was seen in the third year [A1].

370. A five-year study of the immunological status of 62 helicopter pilots exposed to radiation doses from 180 to 260 mGy did not reveal significant quantitative changes in functional characteristics of T- and B-lymphocytes [U15]. Among persons with the highest doses and with some chronic diseases, however, an increase in the functional activity of B-lymphocytes and other non-specific changes in immune status were observed.

371. A careful immunological study of 500 healthy children evacuated from Pripyat with doses from 0.05 to 0.12 Gy did not show significant differences in comparison with Kiev-resident children of the same age [B8]. This study considered the T-lymphocyte subpopulation, natural killer activity, levels of immune complexes, and of interleukin-1 and -2. Some changes in immunoglobulin-A with hypoglobulineaemia and other functional changes were found in children who suffered from respiratory allergy and chronic infections.

372. No differences in absolute and relative levels of T-lymphocytes were found in more than 1,000 examined children living in contaminated areas of the Gomel and Mogilev regions [G20]. A slight increase in serum Ig-G level and B-lymphocytes was observed in children 3–7 years old at the time of examination. The study was conducted in the second year after the accident.

373. Immune status was studied in 84 children 7–14 years old (at the time of the accident) and living in contaminated areas of Belarus and in a control group of 60 children (with the same age and sex distribution) living in uncontaminated areas [K25]. The study was conducted four and half years after the accident. A direct association was observed between the T-lymphocyte levels in children from contaminated areas and the average reconstructed dose from radioiodine to the thyroid in the settlements where the children resided.

374. While evaluating the significance of these various findings, it must be borne in mind that the doses received by the subjects were unlikely to directly affect constituents of the immune system. The long period over which disturbances of

immune function were observed are not consistent with the understanding of recovery of immune functions following acute exposure of experimental animals. It is quite likely, therefore, that psychological stress mediated by neuro-endocrine factors, cytokines, respiratory allergies, chronic infections, and autoimmunity related imbalances could have caused the fluctuations in some immunological parameters in different groups of subjects.

375. *Summary*. With the exception of the increased risk of thyroid cancer in those exposed at young ages, no somatic disorder or immunological defect could be associated with ionizing radiation caused by the Chernobyl accident.

C. PREGNANCY OUTCOME

376. In a group of Belarusian children born to exposed mothers with *in utero* doses ranging from 8 to 21 mSv, no relationship between birth defects and residency in contaminated areas was seen [L5]. The observations that the defects were largely of multifactorial origin and varied according to the residency of the mother appeared to reflect the influence of complex and multiple non-radiation factors. No consistent relationship was seen between the detected rate of chromosome and chromatid aberrations in children and the level of radioactive ground contamination.

377. Later studies of birth defects and malformations in Belarus yielded conflicting results [L8]. The studies conducted on all legal medical abortions from 1982 to 1994 revealed increased rates of polydactyly, limb reduction, and multiple malformations in highly contaminated areas (>555 kBq m^{-2}) when pre- and post-accidental rates were compared [L8]. In the less contaminated areas (<37 kBq m^{-2}), increased rates of anenchephaly, spina bifida, cleft lip/palate, poly-dactyly, limb reduction and multiple malformations were noted. The city of Minsk was used as a control, and spina bifida, polydactyly, multiple malformations, and Down's syndrome were found to have increased. No changes in birth defects over time could be related to exposure to ionizing radiation.

378. One explanation to the findings of Lazjuk [L8] could be that classification of birth disorders has not been consistent over time, probably reflecting the lack of clarity of diagnostic criteria and the significant improvement in diagnostic procedures. Only a few reliable clinical studies have been undertaken in representative groups and regions [K5], and these studies suggest that the observed shifts in the health status of children are unlikely to have been caused by radiation exposure only.

379. Somewhat conflicting results have also been reported when reproductive outcomes in contaminated areas of the Russian Federation were examined [B19, L27, L28, L29].

The outcomes before and after the accident were compared in regions of different contamination levels. The results are summarized in Table 68. Birth rates decreased in all three regions and were related to severity of contamination, while spontaneous abortions increased in two of the three regions. Congenital malformations, stillbirths, premature births and perinatal mortality were studied, but no consistency or apparent relationship to ionizing radiation was noticed.

380. The frequency of unfavourable pregnancy outcomes for 1986–1992 was studied through interviews of 2,233 randomly selected women from 226 contaminated settlements of Belarus and the Russian Federation [G11]. In the contaminated areas of Gomel and Mogilev (Belarus) and of Bryansk (the Russian Federation), a decrease in the birth rate in both urban and rural populations was reported. This corresponds to an increase in the number of medical abortions in both populations.

381. Studies of chromosomal aberrations in distant populations have been critically reviewed [L32, V5]. Increased numbers of cases of Down's syndrome were reported in West Berlin in January 1987 [S23], in the region of Lothian, in Scotland [R14] and in the most contaminated areas of Sweden [E4]. All studies were based on a small number of cases and were later challenged [B13]. The doses in Berlin and Scotland reached 10% of the natural background irradiation, and it is not likely that this contribution was enough to cause the non-disjunction in oocytes during meiosis that is needed for the specific aneuploidy of Down's syndrome. The findings have not been confirmed in larger and more representative series in Europe [D9, L32]. In particular, no peak in Down's syndrome among children exposed at time of conception was observed in equally contaminated zones of Europe (e.g. Finland) or even in Belarus [B34, V5]. In a careful study of birth defects in Belarus, no increased rate of Down's syndrome was found when pre- and post-accident figures were compared in contaminated areas [L8].

382. According to a recent paper [K4], perinatal mortality in Germany showed a statistically significant increase in 1987, and it was concluded that this was an effect of the Chernobyl accident fallout. The findings were later questioned, since whole-body doses from incorporated caesium were found to be 0.05 mSv [R19]. No effect of the Chernobyl accident could be found when temporal patterns of perinatal mortality in Bavaria were correlated to different fallout levels and subsequent exposures [G24].

383. *Summary*. Several studies on adverse pregnancy outcomes related to the Chernobyl accident have been performed in the areas closest to the accident and in more distant regions. So far, no increase in birth defects, congenital malformations, stillbirths, or premature births could be linked to radiation exposures caused by the accident.

D. PSYCHOLOGICAL AND OTHER ACCIDENT-RELATED EFFECTS

384. Many aspects of the Chernobyl accident have been suggested to cause psychological disorders, stress and anxiety in the population. The accident caused long-term changes in the lives of people living in the contaminated districts, since measures intended to limit radiation doses included resettlement, changes in food supplies and restrictions on the activities of individuals and families. These changes were accompanied by important economic, social and political changes in the affected countries, brought about by the disintegration of the former Soviet Union. These psychological reactions are not caused by ionizing radiation but are probably wholly related to the social factors surrounding the accident.

385. The decisions of individuals and families to relocate were often highly complex and difficult. The people felt insecure, and their lack of trust in the scientific, medical and political authorities made them think they had lost control [H9]. Experts who tried to explain the risks and mollify people were perceived as denying the risk, thus reinforcing mistrust and anxiety.

386. The environmental contamination created widespread anxiety that should be referred to not as radiophobia, as it initially was, but as a real, invisible threat, difficult to measure and localize. The key to how people perceive risk is the degree of control they exert over it. Once measures are taken to improve the quality of life for those still living in contaminated areas, the climate of social trust improves, probably because of the better cooperation between inhabitants and local authorities [H9].

387. Psychological effects related to the Chernobyl accident have been studied extensively [I1, L3, L4]. Symptoms such as headache, depression, sleep disturbance, inability to concentrate, and emotional imbalance have been reported and seem to be related to the difficult conditions and stressful events that followed the accident.

388. The psychological development of 138 Belarusian children who were exposed to radiation from the Chernobyl accident *in utero* was compared with that of 122 age-matched children from uncontaminated areas [K46]. The children were followed for 6–12 years and the study included neurological, psychiatric and intellectual assessments of children and parents. The exposed group was found to have a slightly lower intellectual capability and more emotional disorders. A correlation was found between anxiety among parents and emotional stress in children. It was concluded that unfavourable psychosocial factors, such as broken social contacts, adaptation difficulties, and relocation, explained the differences between the exposed and non-exposed groups. No differences could be related to ionizing radiation.

389. Many individuals affected by the Chernobyl accident are convinced that radiation is the most likely cause of their poor health [H7]. This belief may cause or amplify psycho-somatic distress in these individuals. When studying the impact of the accident in exposed areas of Belarus, Havenaar et al. [H11, H12] found that depression, general anxiety and adjustment disorders were more prevalent among those evacuated and in mothers with children under 18 years of age. It was concluded that the Chernobyl accident had had a significant long-term impact on psychological well-being, health-related quality of life, and illness in the exposed populations [H10]. However, none of the findings could be directly attributed to ionizing radiation.

390. Post-traumatic stress is an established psychiatric diagnostic category involving severe nightmares and obsessive reliving of the traumatic event. Although it is widely perceived by victims of disasters, such stress is supposed to occur only in those persons who were directly and immediately involved. The uncertainty, threat and social disruption felt by the wider public has been termed chronic environmental stress disorder by Lee [L3], who compared the consequences of the Chernobyl accident with the consequences of other destructive events and accidents.

391. Among recovery operation workers, those without occupational radiation experience suffered a higher rate of neurotic disturbances than the general population [R5, S13]. Clinically expressed disturbances with significant psychosomatic symptoms were predominant in this group, but the increased medical attention, which leads to the diagnosis of chronic somatic diseases and subclinical changes that persistently attract the attention of the patient, complicates the situation. The possibility of rehabilitation decreased correspondingly, while unsatisfactory and unclear legislation exacerbated the conflicts and tended to prolong the psychoneurotic reactions of the patients [G4, S13]. The health status of recovery operation workers who were nuclear industry professionals did not seem to be different from that of the rest of the cohort [N3].

392. Social and economic suffering among individuals living in contaminated areas has exacerbated the reactions to stressful factors. Although the incidence of psycho-somatic symptoms in the population of highly con-taminated areas is higher than that in populations of less contaminated areas, no direct correlation with radiation dose levels has been observed. The self-appraisal of this group is low, as is their general physical health, as observed in systematic screening programmes, including the International Chernobyl Project [I1]. This makes the individuals functionally unable to solve complicated social and economic problems and aggravates their psychological maladaptation. The tendency to attribute all problems to the accident leads to escapism, "learned helplessness", unwillingness to cooperate, overdependence, and a belief that the welfare system and government authorities should solve all problems. It also contributes to alcohol and drug abuse. There is evidence of an increased incidence of accidents (trauma, traffic incidents, suicides, alcohol intoxication and sudden death with unidentified cause) in this population, as well as in recovery operation workers, compared with the populations of unaffected regions.

393. A follow-up study of the psychological status of 708 emigrants to Israel from the former Soviet Union was carried out over a two-year period [C3]. A total of 374 adults who had lived in contaminated areas and for whom body-burden measurements had been carried out and 334 non-exposed emigrants matched by age, sex and year of emigration were compared. The subjects from exposed areas were categorized into two exposure groups: high and low (^{137}Cs greater or less than 37 kBq m^{-2}) on the basis of the map of ground caesium contamination [I1]. The prevalence of post-traumatic stress disorders, depression, anxiety and psychosomatic effects, such as high blood pressure and chronic illness, were measured. Interviews were carried out during the initial contact and approximately one year later with 520 of the original respondents. The results obtained in the first interview showed that psychological symptoms were much more prevalent in the exposed groups than in the non-exposed group; in the second interview, a decline in the prevalence of disorders was noted. The proportion of those who reported three or more chronic health problems was 48% among the high-exposure group, 49% in the low-exposure group, and 31% in the non-exposed group (p < 0.0003). Based on these results, the authors concluded that the Chernobyl accident had had a strong impact on both the mental and physical health of the immigrants from contaminated areas of the former Soviet Union.

394. *Summary*. The Chernobyl accident caused long-term changes in the lives of people living in the contaminated areas, since measures intended to limit radiation dose included resettlement, changes in food supplies, and restrictions on the activities of individuals and families. These changes were accompanied by important economic, social, and political changes in the affected countries, brought about by the disintegration of the former Soviet Union. The anxiety and emotional stress among parents most likely influenced the children, and unfavourable psychosocial factors probably explain the differences between the exposed and non-exposed groups.

E. SUMMARY

395. A majority of the studies completed to date on the health effects of the Chernobyl accident are of the geographic correlation type that compare average population exposure with the average rate of health effects or cancer incidence in time periods before and after the accident. As long as individual dosimetry is not performed no reliable quantitative estimates can be made. The reconstruction of valid individual doses will have to be a key element in future research on health effects related to the Chernobyl accident.

396. The number of thyroid cancers in individuals exposed in childhood, particularly in the severely contaminated areas of the three affected countries, is considerably greater than expected based on previous knowledge. The high incidence and the short induction period have not been experienced in other exposed populations, and factors other than ionizing radiation are almost certainly influencing the risk. Some such factors include age at exposure, iodine intake and metabolic status, endemic goitre, screening, short-lived isotopes other than ^{131}I, higher doses than estimated, and, possibly, genetic predisposition. Approximately 1,800 thyroid cancer cases have been reported in Belarus, the Russian Federation and Ukraine in children and adolescents for the period 1990–1998. Age seems to be an important modifier of risk. The influence of screening is difficult to estimate. Approximately 40%–70% of the cases were found through screening programmes, and it is unclear how many of these cancers would otherwise have gone undetected. Taking the advanced stage of the tumours at time of diagnosis into consideration, it is likely that most of the tumours would have been detected sooner or later.

397. The present results from several studies indicate that the majority of the post-Chernobyl childhood thyroid carcinomas show the intrachromosomal rearrangements characterized as *RET/PTC1* and *3*. There are, however, several questions left unanswered, e.g. the influence of age at exposure and time since exposure on the rate of chromosome rearrangements.

398. The risk of leukaemia has been shown in epidemiological studies to be clearly increased by radiation exposure. However, no increased risk of leukaemia linked to ionizing radiation has so far been confirmed in children, in recovery operation workers, or in the general population of the former Soviet Union or other areas with measurable amounts of contamination from the Chernobyl accident.

399. Increases in a number of non-specific detrimental health effects other than cancer in recovery operation workers and in residents of contaminated areas have been reported. It is difficult to interpret these findings without referring to a known baseline or background incidence. Because health data obtained from official statistical sources, such as mortality or cancer incidence statistics, are often passively recorded and are not always complete, it is not appropriate to compare them with data for the exposed populations, who undergo much more intensive and active health follow-up than the general population.

400. Some investigators have interpreted a temporary loss of ability to work among individuals living in cotaminated areas as an increase in general morbidity. High levels of chronic diseases of the digestive, neurological, skeletal, muscular and circulatory systems have been reported. However, most investigators relate these observations to changes in the age structure, the worsening quality of life, and post-accident countermeasures such as relocation.

401. Many papers have been published in the last decade on the immunological effects of exposure to radiation from the Chernobyl accident. Since it is unclear, however, if possible confounding factors have been taken into account, including, in particular, infections and diet, it is difficult to interpret these results.

CONCLUSIONS

402. The accident of 26 April 1986 at the Chernobyl nuclear power plant, located in Ukraine about 20 km south of the border with Belarus, was the most serious ever to have occurred in the nuclear industry. It caused the deaths, within a few days or weeks, of 30 power plant employees and firemen (including 28 with acute radiation syndrome) and brought about the evacuation, in 1986, of about 116,000 people from areas surrounding the reactor and the relocation, after 1986, of about 220,000 people from Belarus, the Russian Federation and Ukraine. Vast territories of those three countries (at that time republics of the Soviet Union) were contaminated, and trace deposition of released radionuclides was measurable in all countries of the northern hemisphere. In this Annex, the radiation exposures of the population groups most closely involved in the accident have been reviewed in detail and the health consequences that are or could be associated with these radiation exposures have been considered.

403. The populations considered in this Annex are (a) the workers involved in the mitigation of the accident, either during the accident itself (emergency workers) or after the accident (recovery operation workers) and (b) members of the general public who either were evacuated to avert excessive radiation exposures or who still reside in contaminated areas. The contaminated areas, which are defined in this Annex as being those where the average ^{137}Cs ground deposition density exceeded 37 kBq m^{-2} (1 Ci km^{-2}), are found mainly in Belarus, in the Russian Federation and in Ukraine. A large number of radiation measurements (film badges, TLDs, whole-body counts, thyroid counts, etc.) were made to evaluate the exposures of the population groups that are considered.

404. The approximately 600 emergency workers who were on the site of the Chernobyl power plant during the night of the accident received the highest doses. The most important exposures were due to external irradiation (relatively uniform whole-body gamma irradiation and beta irradiation of extensive body surfaces), as the intake of radionuclides through inhalation was relatively small (except in two cases). Acute radiation sickness was confirmed in 134 of those emergency workers. Forty-one of these patients received whole-body doses from external irradiation of less than 2.1 Gy. Ninety-three patients received higher doses and had more severe acute radiation sickness: 50 persons with doses between 2.2 and 4.1 Gy, 22 between 4.2 and 6.4 Gy, and 21 between 6.5 and 16 Gy. The skin doses from beta exposures, evaluated for eight patients with acute radiation sickness, were in the range of 400–500 Gy.

405. About 600,000 persons (civilian and military) have received special certificates confirming their status as liquidators (recovery operation workers), according to laws promulgated in Belarus, the Russian Federation and

Ukraine. Of those, about 240,000 were military servicemen. The principal tasks carried out by the recovery operation workers included decontamination of the reactor block, reactor site and roads, as well as construction of the sarcophagus and of a town for reactor personnel. These tasks were completed by 1990.

406. A registry of recovery operation workers was established in 1986. This registry includes estimates of effective doses from external irradiation, which was the predominant pathway of exposure for the recovery operation workers. The registry data show that the average recorded doses decreased from year to year, being about 170 mSv in 1986, 130 mSv in 1987, 30 mSv in 1988, and 15 mSv in 1989. It is, however, difficult to assess the validity of the results that have been reported because (a) different dosimeters were used by different organizations without any intercalibration; (b) a large number of recorded doses were very close to the dose limit; and (c) there were a large number of rounded values such as 0.1, 0.2, or 0.5 Sv. Nevertheless, it seems reasonable to assume that the average effective dose from external gamma irradiation to recovery operation workers in the years 1986–1987 was about 100 mSv.

407. Doses received by the general public came from the radionuclide releases from the damaged reactor, which led to the ground contamination of large areas. The radionuclide releases occurred mainly over a 10-day period, with varying release rates. From the radiological point of view, the releases of ^{131}I and ^{137}Cs, estimated to have been 1,760 and 85 PBq, respectively, are the most important. Iodine-131 was the main contributor to the thyroid doses, received mainly via internal irradiation within a few weeks after the accident, while ^{137}Cs was, and is, the main contributor to the doses to organs and tissues other than the thyroid, from either internal or external irradiation, which will continue to be received, at low dose rates, during several decades.

408. The three main contaminated areas, defined as those with ^{137}Cs deposition density greater than 37 kBq m^{-2} (1 Ci km^{-2}), are in Belarus, the Russian Federation and Ukraine; they have been designated the Central, Gomel-Mogilev-Bryansk and Kaluga-Tula-Orel areas. The Central area is within about 100 km of the reactor, predominantly to the west and northwest. The Gomel-Mogilev-Bryansk contaminated area is centred 200 km north-northeast of the reactor at the boundary of the Gomel and Mogilev regions of Belarus and of the Bryansk region of the Russian Federation. The Kaluga-Tula-Orel area is in the Russian Federation, about 500 km to the northeast of the reactor. All together, territories from the former Soviet Union with an area of about 150,000 km^2 were contaminated with ^{137}Cs deposition density greater than 37 kBq m^{-2}. About five million people reside in those territories.

409. Within a few weeks after the accident, more than 100,000 persons were evacuated from the most contaminated areas of Ukraine and of Belarus. The thyroid doses received by the evacuees varied according to their age, place of residence, dietary habits and date of evacuation. For example, for the residents of Pripyat, who were evacuated essentially within 48 hours after the accident, the population-weighted average thyroid dose is estimated to be 0.17 Gy and to range from 0.07 Gy for adults to 2 Gy for infants. For the entire population of evacuees, the population-weighted average thyroid dose is estimated to be 0.47 Gy. Doses to organs and tissues other than the thyroid were, on average, much smaller.

410. Thyroid doses also have been estimated for the residents of the contaminated areas who were not evacuated. In each of the three republics, thyroid doses are estimated to have exceeded 1 Gy for the most exposed infants. For residents of a given locality, thyroid doses to adults were smaller than those to infants by a factor of about 10. The average thyroid dose was approximately 0.2 Gy; the variability of the thyroid dose was two orders of magnitude, both above and below the average.

411. Following the first few weeks after the accident, when ^{131}I was the main contributor to the radiation exposures, doses were delivered at much lower dose rates by radionuclides with much longer half-lives. Since 1987, the doses received by the populations of the contaminated areas came essentially from external exposure from ^{134}Cs and ^{137}Cs deposited on the ground and internal exposure due to the contamination of foodstuffs by ^{134}Cs and ^{137}Cs. Other, usually minor, contributions to the long-term radiation exposures include the consumption of foodstuffs contaminated with ^{90}Sr and the inhalation of aerosols containing plutonium isotopes. Both external irradiation and internal irradiation due to ^{134}Cs and ^{137}Cs result in relatively uniform doses in all organs and tissues of the body. The average effective doses from ^{134}Cs and ^{137}Cs that were received during the first 10 years after the accident by the residents of contaminated areas are estimated to be about 10 mSv.

412. The papers available for review by the Committee to date regarding the evaluation of health effects of the Chernobyl accident have in many instances suffered from methodological weaknesses that make them difficult to interpret. The weaknesses include inadequate diagnoses and classification of diseases, selection of inadequate control or reference groups (in particular, control groups with a different level of disease ascertainment than the exposed groups), inadequate estimation of radiation doses or lack of individual data and failure to take screening and increased medical surveillance into consideration. The interpretation of the studies is complicated, and particular attention must be paid to the design and performance of epidemiological studies. These issues are discussed in more detail in Annex I, *"Epidemiological evaluation of radiation-induced cancer"*.

413. Apart from the substantial increase in thyroid cancer after childhood exposure observed in Belarus, in the Russian Federation and in Ukraine, there is no evidence of a major public health impact related to ionizing radiation 14 years after the Chernobyl accident. No increases in overall cancer incidence or mortality that could be associated with radiation exposure have been observed. For some cancers no increase would have been anticipated as yet, given the latency period of around 10 years for solid tumours. The risk of leukaemia, one of the most sensitive indicators of radiation exposure, has not been found to be elevated even in the accident recovery operation workers or in children. There is no scientific proof of an increase in other non-malignant disorders related to ionizing radiation.

414. The large number of thyroid cancers in individuals exposed in childhood, particularly in the severely contaminated areas of the three affected countries, and the short induction period are considerably different from previous experience in other accidents or exposure situations. Other factors, e.g. iodine deficiency and screening, are almost certainly influencing the risk. Few studies have addressed these problems, but those that have still find a significant influence of radiation after taking confounding influences into consideration. The most recent findings indicate that the thyroid cancer risk for those older than 10 years at the time of the accident is leveling off, the risk seems to decrease since 1995 for those 5–9 years old at the time of the accident, while the increase continues for those younger than 5 years in 1986.

415. There is a tendency to attribute increases in cancer rates (other than thyroid) over time to the Chernobyl accident, but it should be noted that increases were also observed before the accident in the affected areas. Moreover, a general increase in mortality has been reported in recent years in most areas of the former USSR, and this must also be taken into account in interpreting the results of the Chernobyl-related studies. Because of these and other uncertainties, there is a need for well designed, sound analytical studies, especially of recovery operation workers from Belarus, the Russian Federation, Ukraine and the Baltic countries, in which particular attention is given to individual dose reconstruction and the effect of screening and other possible confounding factors.

416. Increases of a number of non-specific detrimental health effects other than cancer in accident recovery workers have been reported, e.g. increased suicide rates and deaths due to violent causes. It is difficult to interpret these findings without reference to a known baseline or background incidence. The exposed populations undergo much more intensive and active health follow-up than the general population. As a result, using the general population as a comparison group, as has been done so far in most studies, is inadequate.

417. Adding iodine to the diet of populations living in iodine-deficient areas and screening the high-risk groups could limit the radiological consequences. Most data suggest that the youngest age group, i.e. those who were less than five years old at the time of the accident, continues to have an increased risk of developing thyroid cancer and should be closely monitored. In spite of the fact that many thyroid cancers in childhood are presented at a

more advanced stage in terms of local aggressiveness and distant metastases than in adulthood, they have a good prognosis. Continued follow-up is necessary to allow planning of public health actions, to gain a better understanding of influencing factors, to predict the outcomes of any future accidents, and to ensure adequate radiation protection measures.

418. Present knowledge of the late effects of protracted exposure to ionizing radiation is limited, since the dose-response assessments rely heavily on high-dose exposure studies and animal experiments. The Chernobyl accident could, however, shed light on the late effects of protracted exposure, but given the low doses received by the majority of exposed individuals, albeit with uncertainties in the dose estimates, any increase in cancer incidence or mortality will most certainly be difficult to detect in epidemiological studies. The main goal is to differentiate the effects of the ionizing radiation and effects that arise from many other causes in exposed populations.

419. Apart from the radiation-associated thyroid cancers among those exposed in childhood, the only group that received doses high enough to possibly incur statistically detectable increased risks is the recovery operation workers. Studies of these populations have the potential to contribute to the scientific knowledge of the late effects of ionizing radiation. Many of these individuals receive annual medical examinations, providing a sound basis for future studies of the cohort. It is, however, notable that no increased risk of leukaemia, an entity known to appear within 2–3 years after

exposure, has been identified more than 10 years after the accident.

420. The future challenge is to provide reliable individual dose estimates for the subjects enrolled in epidemiological studies and to evaluate the effects of doses accumulated over protracted time (days to weeks for thyroid exposures of children, minutes to months for bone-marrow exposures of emergency and recovery operation workers, and months to years for whole-body exposures of those living in contaminated areas). In doing this, many difficulties must be taken into consideration, such as (a) the role played by different radionuclides, especially the short-lived radioiodines; (b) the accuracy of direct thyroid measurements; (c) the relationship between ground contamination and thyroid doses; and (d) the reliability of the recorded or reconstructed doses for the emergency and recovery operation workers.

421. Finally, it should be emphasized that although those exposed as children and the emergency and recovery operation workers are at increased risk of radiation-induced effects, the vast majority of the population need not live in fear of serious health consequences from the Chernobyl accident. For the most part, they were exposed to radiation levels comparable to or a few times higher than the natural background levels, and future exposures are diminishing as the deposited radionuclides decay. Lives have been disrupted by the Chernobyl accident, but from the radiological point of view and based on the assessments of this Annex, generally positive prospects for the future health of most individuals should prevail.

Table 1
Radionuclide inventory in Unit 4 reactor core at time of the accident on 26 April 1986

Radionuclide	Half-life	Activity (PBq)			
		1986 estimates [a] [I2]	Estimates by [B1, I2]	Estimates by [S1]	Estimates by [B2, B3, B4] [b]
^{3}H	12.3 a			1.4 [d]	
^{14}C	5 730 a			0.1 [d]	
^{85}Kr	10.72 a	33	33	28	
^{89}Sr	50.5 d	2 000	2 330	3 960	
^{90}Sr	29.12 a	200	200	230	220
^{95}Zr	64.0 d	4 400	4 810	5 850	
^{95}Nb	35 d			5 660	
^{99}Mo	2.75 d	4 800	5 550	6 110	
^{103}Ru	39.3 d	4 100	4 810	3 770	
^{106}Ru	368 d	2 100	2 070	860	850
^{110m}Ag	250 d			1.3	
^{125}Sb	2.77 a			15	
^{129m}Te	33.6 d			1 040	
^{132}Te	3.26 d	320	2 700	4 480	4 200 [f]
^{129}I	15 700 000 a			0.000081 [d]	
^{131}I	8.04 d	1 300	3 180	3 080	3 200 [f]
^{132}I	2.3 h			4 480	4 200 [f]
^{133}I	20.8 h			6 700	4 800 [f]
^{134}I	52.6 min				2 050 [f]
^{135}I	6.61 h				2 900 [f]
^{133}Xe	5.25 d	1 700	6 290	6 510	
^{134}Cs	2.06 a	190	190	170	150
^{136}Cs	13.1 d			110 [e]	
^{137}Cs	30.0 a	290	280	260	260
^{138}Cs	32.2 min			6 550	
^{140}Ba	12.7 d	2 900	4 810	6 070	
^{140}La	40.3 h			6 070	
^{141}Ce	32.5 d	4 400	5 550	5 550	
^{144}Ce	284 d	3 200	3 260	3 920	3 920
^{147}Nd	11.0 d			2 160	
^{154}Eu	8.6 a			14	
^{235}U	704 000 000 a			0.000096 [d]	
^{236}U	23 400 000 a			0.0085 [d]	
^{238}U	4 470 000 000 a			0.0023 [d]	
^{237}Np	2 140 000 a			0.00026	
^{239}Np	2.36 d	140	49 600 [c]	58,100	58 100
^{236}Pu	2.86 a			0.0001	
^{238}Pu	87.74 a	1	1.0	1.3	0.93
^{239}Pu	24065 a	0.8	0.85	0.95	0.96
^{240}Pu	6537 a	1	1.2	1.5	1.5
^{241}Pu	14.4 a	170	170	180	190
^{242}Pu	376 000 a		0.0025	0.0029	0.0021
^{241}Am	432 a			0.17	0.14
^{243}Am	7 380 a			0.0097	0.0056
^{242}Cm	163 d	26	15 [c]	43	31
^{244}Cm	18.1 a			0.43	0.18

a   Decay-corrected to 6 May 1986.
b   Values used in this Annex.
c   Corrected to account for burnup of individual fuel assemblies.
d   Reference [K1].
e   Corrected value.
f   Reference [K16].

Table 2
Estimates of the principal radionuclides released in the accident

Radionuclide	Activities released (PBq)			
	1986 estimates [a] [I2]	1996 estimates [b] [B4, B27, D8]	1996 estimates [K37]	1996 estimates [c] [d] [D5, D8, N4]
Noble gases				
^{85}Kr	33	33	33	
^{133}Xe	1 700	6 500	6 500	6 500
Volatile elements				
^{129m}Te		240		~1 150
^{132}Te	48	1 000		
^{131}I	260	1 200–1 700	1 800	~1 760
^{133}I		2 500		
^{134}Cs	19	4 4–48	50	~54
^{136}Cs		36		
^{137}Cs	38	74–85	86	~85
Intermediate				
^{89}Sr	80	81	80	~115
^{90}Sr	8	8	8	~10
^{103}Ru	120	170	120	>168
^{106}Ru	63	30	25	>73
^{140}Ba	170	170	160	240
Refractory (including fuel particles)				
^{95}Zr	130	170	140	196
^{99}Mo	96	210		>168
^{141}Ce	88	200	120	196
^{144}Ce	96	140	90	~116
^{239}Np	4.2	1 700		945
^{238}Pu	0.03	0.03	0.033	0.035
^{239}Pu	0.024	0.03	0.0334	0.03
^{240}Pu	0.03	0.044	0.053	0.042
^{241}Pu	5.1	5.9	6.3	~6
^{242}Pu	0.00007	0.00009		
^{242}Cm	0.78	0.93	1.1	~0.9
Total (excluding noble gases)	1 000–2 000	8 000 [e]-	–	5 300

a Decay-corrected to 6 May 1986.
b Estimate of release, decay-corrected to 26 April 1986.
c Estimate of total release during the course of the accident.
d Values used in this Annex.
e Decay correction to beginning of accident allows more short-lived radionuclides to be included, giving a higher estimate of total release, which, however, is a probable overestimate since many of these radionuclides would have decayed inside the damaged core before any release to the atmosphere could occur.

Table 3
Estimated daily releases of iodine-131 during the accident

Day of release	Percentage (based on [A4, I6])	Daily releases (PBq)
26 April	40.0	704
27 April	11.6	204
28 April	8.5	150
29 April	5.8	102
30 April	3.9	69
1 May	3.5	62
2 May	5.8	102
3 May	6.1	107
4 May	7.4	130
5 May	7.4	130
Total	100	1 760

Table 4
Estimated amounts of radioiodines [a] and tellurium-132
[K16]

Radionuclide	Half-life	Amount in the reactor core at the time of the accident (PBq)	Activity released [b] (PBq)
^{132}Te	78.2 h	4 200	1 040
^{132}I [c]	2.3 h	4 200	1 040
^{133}I	20.8 h	4 800	910
^{134}I	52.6 min	2 050	25
^{135}I	6.61 h	2 900	250

a   The activity of iodine-131 in the reactor core at the time of the accident is taken to be 3,200 PBq. The release of iodine-131 is assumed to be 1,760 PBq.
b   With decay correction.
c   Iodine-132 is assumed to be in radioactive equilibrium with tellurium-132.

Table 5
Contaminated areas in European countries following the accident
[I24]

Country	Area in deposition density ranges (km²) [a]			
	37–185 kBq m⁻²	185–555 kBq m⁻²	555–1 480 kBq m⁻²	>1 480 kBq m⁻²
Russian Federation	49 800	5 700	2 100	300
Belarus	29 900	10 200	4 200	2 200
Ukraine	37 200	3 200	900	600
Sweden	12 000	–	–	–
Finland	11 500	–	–	–
Austria	8 600	–	–	–
Norway	5 200	–	–	–
Bulgaria	4 800	–	–	–
Switzerland	1 300	–	–	–
Greece	1 200	–	–	–
Slovenia	300	–	–	–
Italy	300	–	–	–
Republic of Moldova	60	–	–	–

a   The ^{137}Cs levels include a small contribution (2-4 kBq m⁻²) from fallout from the atmospheric weapons tests carried out mainly in 1961 and 1962.

Table 6
Composition of radionuclide deposition in the near and far zones around the reactor
[B5, I6]

Radionuclide	Ratio to ^{137}Cs release [B4]	Ratio to ^{137}Cs deposition [a]					
		Near zone (<100 km)			Far zone		
		North	South	West	Northeast [b]	South [c]	Southeast [d]
^{89}Sr	1.0	0.7	12	4	0.14	0.3	1.0
^{90}Sr	0.1	0.13	1.5	0.5	0.014	0.03	0.1
^{91}Y		2.7	8	5	0.06	0.17	0.6
^{95}Zr	2.0	3	10	5	0.06	0.3	1.0
^{99}Mo		3	25	8	(0.11)	(0.5)	(1.5)
^{103}Ru	2.0	2.7	12	4	1.9	2.7	6
^{106}Ru	0.4	1.0	5	1.5	0.7	1.0	2.3
^{110m}Ag		0.01	0.01	0.005	0.008	(0.01)	(0.01)
^{125}Sb		0.02	0.1	0.05	0.05	0.1	0.1
^{131}I	20	17	30	15	10	(1)	(3)
^{132}Te	5	17	13	18	(13)	(1)	(3)
^{134}Cs	0.5	0.5	0.5	0.5	0.5	0.5	0.5
^{137}Cs	1.0	1.0	1.0	1.0	1.0	1.0	1.0
^{140}Ba	2.0	3	20	7	0.7	(0.5)	(1.5)
^{141}Ce	2.3	4	10	5	0.11	0.5	1.8
^{144}Ce	1.6	2.3	6	3	0.07	0.3	1.2
^{239}Np	20	7	140	25	(0.6)	(3)	(10)

a   Decay-corrected to 26 April 1986; values from indirect data in parentheses.
b   Areas of higher ^{137}Cs deposition in Belarus and Russian Federation.
c   District south of Kiev.
d   Northern Caucasus.

Table 7
Areal extent of ^{137}Cs contamination from the accident in the European part of the former USSR
[G16, I3]

Country / Region	Area in deposition density ranges (km²)				
	37–185 kBq m⁻²	185–555 kBq m⁻²	555–1 480 kBq m⁻²	>1 480 kBq m⁻²	Total
Belarus					
Gomel	16 900	6 700	2 800	1 625	
Mogilev	5 500	2 900	1 400	525	
Brest	3 800	500			
Grodno	1 700	12			
Minsk	2 000	48			
Vitebsk	35				
Total	29 900	10 200	4 200	2 200	46 500
Russian Federation					
Bryansk	6 750	2 630	2 130	310	
Kaluga	3 500	1 420			
Tula	10 320	1 270			
Orel	8 840	130			
Other	20 350				
Total	49 760	5 450	2 130	310	57 650
Ukraine	37 200	3 200	900	600	41 900
Other republics	60				60
Total area	116 920	18 850	7 230	3 110	146 110

Table 8
Estimated ^{137}Cs deposit from the accident

^{137}Cs deposition density (kBq m^{-2})		Area [13] (km^2)	^{137}Cs deposit (PBq)		
Range	Mean ^{a}		Total	From fallout ^{b}	From Chernobyl accident
7.4–19	12	654 200	7.8	1.3–2.6	5.2–6.5
19–37	26	211 850	5.6	0.4–0.8	4.8–5.2
37–185	83	116 920	9.7	0.2–0.4	9.3–9.5
185–555	320	18 850	6.0	0.05–0.1	5.9
555–1 480	910	7 230	6.6	0.02–0.04	6.6
>1 480	2 200	3 110	6.8	0.008–0.02	6.8
Total			42.5	2.0–4.0	38.5–40.5

a   Assumed to be geometric mean of range.
b   The estimated residual range in 1986 of ^{137}Cs deposition density from atmospheric nuclear weapons fallout is 2–4 kBq m^{-2}.

Table 9
Measured transfer coefficients for ^{137}Cs in natural meadows of the Polissya region in Ukraine, 1988–1989
[S40]

Type of soil	Type of meadow	Transfer coefficient ^{a}
Black soil loam	Floodplain humid	0.6
Loamy sand	Dry valley normal	2–3
	Floodplain humid	8–11
Soddic-podzolic loam	Dry valley normal	1–4
	Dry valley normal	5–9
Soddic-podzolic sand	Dry valley, water-saturated	13–22
	Floodplain humid	25–39
Peaty-gley	Peaty drained	30–45
	Peaty flooded	58–82
	Peaty callows	135–190

a   Bq kg^{-1} of dry grass per kBq m^{-2} deposited on the ground.

Table 10
Staff on site and emergency workers in initial hours of the accident
[K23]

Professional group	Accident witnesses	Emergency workers (at 8 a.m. on 26 April 1986)
Staff of the power plant (Units 1, 2, 3 and 4)	176	374 ^{c}
Construction workers at Units 5 and 6	268	–
Firemen	14 ^{a}, 10 ^{b}	69
Guards	23	113
Staff of the local medical facility	–	10

a   Arrived on the site of the accident at 1.27 a.m.
b   Arrived on the site of the accident at 1.35 a.m.
c   Excluding the accident victims, the numbers of whom are given in Table 11.

Table 11
Emergency workers with acute radiation sickness following the accident
[I5]

Degree of acute radiation sickness		Range of dose (Gy)	Number of patients treated [a]		Number of deaths [b]	Number of survivors
			Moscow	Kiev		
Mild	(I)	0.8–2.1	23	18	0 (0%)	41
Moderate	(II)	2.2–4.1	44	6	1 (2%)	49
Severe	(III)	4.2–6.4	21	1	7 (32%)	15
Very severe	(IV)	6.5–16	20	1	20 (95%)	1
Total		0.8–16	108	26	28	106

a   Acute radiation sickness was not confirmed in a further 103 treated workers.
b   Percentage of treated patients in parentheses.

Table 12
Error range of estimated external doses evaluated by cytogenetic analysis to patients admitted to Hospital 6 in Moscow
[P14]

Dose range (Gy)	Number of persons sampled	Number of counted cells per sample	Statistical error range (%)
10.1–13.7	7	19–100	11–18
6.1–9.5	12	19–101	11–16
4.0–5.8	16	65–630	8.6–36
2.1–3.8	33	30–300	22–56
1.0–1.9	19	30–300	33–100
0.5–0.9	17	65–900	–100; +100
0.1–0.4	25	50–350	–100; +300

Table 13
Estimated internal and external doses to victims of the accident

Personal code	Internal absorbed dose until time of death [a] (Gy) [K17, K18]		External dose [b] (Gy) [G25]
	Thyroid	Lungs	
25	0.021	0.00026	8.2
18	0.024	0.0028	6.4
22	0.054	0.00047	4.3
5	0.062	0.00057	6.2
9	0.071	0.00077	5.6
21	0.077	0.00068	6.4
8	0.13	0.0015	3.8
2	0.13	0.0022	2.9
19	0.21	0.0035	4.5
23	0.31	0.0023	7.5
1	0.34	0.0087	11.1
15	0.32	0.0027	6.4
16	0.47	0.0041	4.2
3	0.54	0.0068	7.2
17	0.60	0.12	5.5
4	0.64	0.034	6.5
7	0.78	0.0047	10.2
10	0.89	0.0094	8.6
11	0.74	0.029	9.1
14	0.95	0.02	7.2
20	1.9	0.019	5.6
24	2.2	0.021	3.5
13	4.1	0.04	4.2

a   The relative errors in the organ doses are estimated to be less than 30%.
b   Evaluated by chromosome analysis of peripheral blood lymphocytes.

Table 14
Estimates of internal doses received by surviving emergency workers [a]
[K17, K18]

Tissue	Absorbed dose (Gy)	
	Average [b]	Maximum
Bone surface	0.28	3.6
Lungs	0.25	2.4
Wall of lower large intestine	0.22	2.9
Thyroid gland	0.096	1.8
Wall of upper large intestine	0.090	1.2
Liver	0.056	0.73
Red bone marrow	0.036	0.46

a   Doses estimated from measurements with whole-body counters; the relative error does not exceed 45%.
b   The doses were averaged over 375 individuals.

Table 15
Distribution of thyroid doses in workers treated at Hospital 6 in Moscow during April and May 1986
[G12]

Thyroid dose range [a] (Gy)	Number of workers
0-1.2	173
1.2-3.7	18
3.7-6.1	4
6.1-8.6	4
8.6-11	2
11-13	2
13-16	0
16-18	2
18-21	0
21-23	1
>23	2

a   Doses assessed by repeated *in vivo* thyroid counting, except for the two workers with estimated doses greater than 23 Gy; for those two workers, the doses were assessed by repeated gamma spectrometry of bioassay samples (blood and urine). The relative error in the estimated individual thyroid doses does not exceed 30%.

Table 16
Estimated effective doses from external irradiation received by recovery operation workers in the 30-km zone during 1986-1987
[I25]

Group	Number of workers		Average dose (mSv)		Collective dose (man Sv)	
	1986	1987	1986	1987	1986	1987
Staff of nuclear power plant	2 358	4 498	87	15	210	70
Construction workers	21 500	5 376	82	25	1 760	130
Transport, security workers	31 021	32 518	6.5	27	200	870
Military servicemen	61 762	63 751	110	63	6 800	4 000
Workers from other power plants		3 458		9.3		30
Annual total or average	116 641	109 601	77	47	8 970	5 100
Total or average	226 242		62		14 070	

Table 17
Distribution of the external doses received by emergency and recovery operation workers
[K44]

Group	Number of persons	Percentage in the dose interval (mSv)						
		0-10	10-50	50-100	100-200	200-250	250-500	>500
Emergency workers and accident witnesses	820 [a]	–	–	2	4	–	7	87
Staff of nuclear power plant 1986	2 358	13	45	24	14	2	2	–
Staff of nuclear power plant 1987	4 498	66	42	1	1	–	–	–
Construction workers 1986	21 500	23	24	11	18	11	13	–
Construction workers 1987	5 376	47	23	24	4	1	1	–
Military servicemen 1986	61 762	13	22	16	23	19	19	–
Military servicemen 1987	63 751	15	15	49	15	6	6	–
Workers from other power plants 1987	3 458	78	21	1	–	–	–	–

a   Number of persons included in the registry of the Institute of Biophysics in Moscow.

Table 18
Distribution of doses to recovery operation workers [a] as recorded in national registries
[C2, M13]

Area and period	Number of recovery operation workers	Percentage for whom dose is known	Effective dose (mSv)			
			Mean	Median	75th percentile	95th percentile
Belarus						
1986-1987	31 000	28	39	20	67	111
1986-1989	63 000	14	43	24	67	119
Russian Federation						
1986	69 000	51	169	194	220	250
1987	53 000	71	92	92	100	208
1988	20 500	83	34	26	45	94
1989	6 000	73	32	30	48	52
1986-1989	148 000	63	107	92	180	240
Ukraine						
1986	98 000	41	185	190	237	326
1987	43 000	72	112	105	142	236
1988	18 000	79	47	33	50	134
1989	11 000	86	35	28	42	107
1986-1989	170 000	56	126	112	192	293
Total						
1986	187 000	45	170			
1987	107 000	65	130			
1988	45 500	80	30			
1989	42 500	80	15			
1986-1989	381 000	52	113			

a   Including those who worked outside the 30-km zone, but excluding those for whom the year of service is not recorded.

Table 19
Stable iodine prophylaxis among a sample of workers involved in early phases of the accident
[K44]

Time of iodine prophylaxis [a] (h)	Number of workers	Percentage of sample
Before arrival at the plant	19	11
Upon arrival at the plant	22	12.5
0-1	22	12.5
1-3	16	9
3-10	27	15
10-30	22	12.5
30-100	23	13
100-300	3	2
Did not accept prophylaxis	22	12.5
Total	176	100

a   Counted from the time of arrival of the worker at the plant.

Table 20
Population groups evacuated in 1986 from contaminated areas

Country	Area	Date	Number of evacuees
Belarus [M4, S24]	51 villages within the 30-km zone 28 villages outside the 30-km zone 29 villages outside the 30-km zone Total of 108 villages	2–7 May 3–10 June August/September	11 358 6 017 7 350 24 725
Russian Federation [S20]	4 villages of Krasnaya Gora district, Bryansk region	August	186
Ukraine [U14]	Pripyat town Railway station Yanov Burakovka village 15 villages within the 10-km zone Chernobyl town 43 villages within the 30-km zone 8 villages outside the 30-km zone 4 villages outside the 30-km zone Bober village Total of 75 settlements	27 April 27 April 30 April 3 May 5 May 3–7 May 14–31 May 10 June–16 August September	49 360 254 226 9 864 13 591 14 542 2 424 434 711 91 406
Former USSR	Total of 187 settlements		116 317

Table 21
Estimates of thyroid doses from intake of ^{131}I received by the Ukrainian evacuees of towns and villages within the 30-km zone
[G8, R12]

Age at time of accident (years)	Pripyat town [G8]			Chernobyl town [a]			Evacuated villages [a]			Total collective dose (man Gy)
	Number of persons	Arithmetic mean dose (Gy)	Collective dose (man Gy)	Number of persons	Arithmetic mean dose (Gy)	Collective dose (man Gy)	Number of persons	Arithmetic mean dose (Gy)	Collective dose (man Gy)	
<1	340	2.18	741	219	1.5	329	369	3.9	1 439	2 509
1–3	2 030	1.28	2 698	653	1	653	1 115	3.6	4 014	7 265
4–7	2 710	0.54	1 463	894	0.48	429	1 428	1.7	2 428	4 320
8–11	2 710	0.23	623	841	0.15	126	1 360	0.62	843	1 592
12–15	2 710	0.12	325	846	0.11	93	1 448	0.46	666	1 084
16–18	2 120	0.066	140	650	0.09	59	941	0.39	367	566
>18	36 740	0.066	2 425	9 488	0.16	1 518	21 794	0.40	8 718	12 661
Total	49 360		8 315	13 591		3 206	28 455		18 475	29 996

a   Assumes same age distribution of population as Pripyat.

Table 22
Estimates of thyroid doses from intake of ^{131}I received by the evacuees of Belarusian villages
[G15]

Age at time of accident [a] (years)	Number of measured persons	Arithmetic mean thyroid dose (Gy)	Median thyroid dose (Gy)	Estimated number of residents [b]	Collective thyroid dose (man Gy)
<1	145	4.3	2.3	586	2 519
1-3	290	3.7	1.7	966	3 573
4-7	432	2.1	1.2	1 199	2 517
8-11	460	1.4	0.86	1 105	1 548
12-15	595	1.1	0.61	1 392	1 531
16-17	221	1.0	0.59	704	704
>17	7 332	0.68	0.38	18 773	12 766
Total	9 475			24 725	25 158

a Derived from information on year of birth; e.g. age <1 includes children born in 1986 and 1985.
b Based on the age distribution available for 17,513 evacuees.

Table 23
Summary of estimated collective effective and thyroid doses to populations of areas evacuated in 1986

Country	Number of persons evacuated	Collective dose (man Sv)		
		Thyroid [a]	External effective dose	Internal effective dose
Belarus	24 725	25 000	770	150
Russian Federation	186	<1 000	< 10	< 10
Ukraine	91 406	30 000	1 500	1 300
Total	116 317	55 000	2 300	1 500

a Units: man Gy.

Table 24
Distribution of the estimated first-year doses from external irradiation to inhabitants of Belarus evacuated from the exclusion zone
[S24]

Dose interval (mSv)	Number of persons in the dose interval	
	In the absence of evacuation (calculated)	In the evacuated population (actual)
0-10	1 956	7 357
10-20	4 710	7 652
20-30	3 726	3 480
30-40	2 552	1 764
40-50	1 795	1 094
50-60	1 226	761
60-70	961	556
70-80	726	416
80-90	565	314
90-100	453	239
100-150	1 513	605
150-200	1 015	195
200-250	814	109
250-300	646	67
300-350	494	42
350-400	371	26
>400	1 204	28

Table 25
Inhabitants in 1986 and 1987 of the areas of strict control
[I3, I4]

Country	Region	Population	Total population
Belarus	Gomel Mogilev	85 700 23 300	109 000
Russian Federation	Bryansk	111 800	111 800
Ukraine	Kiev Zhitomir	20 800 31 200	52 000
Total			272 800

Table 26
Distribution of the inhabitants in 1995 of areas contaminated by the Chernobyl accident
[K23, R11, V3]

^{137}Cs deposition density $(kBq\ m^{-2})$	Population [a]			
	Belarus	Russian Federation	Ukraine	Total
37–185	1 543 514	1 654 175	1 188 800	4 386 389
185–555	239 505	233 626	106 700	579 831
555–1 480	97 593	95 474	300	193 367
Total	1 880 612	1 983 275	1 295 600	5 159 487

a   For social and economic reasons, some of the populations living in areas contaminated below 37 kBq m⁻² are also included.

Table 27
Values of occupancy/shielding factors used in evaluation of external exposure 0–1 year and >1 year after the accident
[B14]

Population group	Occupancy/shielding factor			
	Rural areas		Urban areas	
	0–1 year after accident	>1 year after accident	0–1 year after accident	>1 year after accident
Wooden houses				
Indoor workers	0.32	0.26	0.23	0.29
Outdoor workers	0.41	0.36	0.29	0.25
Schoolchildren	0.39	0.34		
Brick houses				
Indoor workers	0.24	0.22	0.15	0.13
Outdoor workers	0.34	0.31	0.23	0.20
Schoolchildren	0.31	0.29		
Both types of houses				
Representative value	0.36	0.31	0.22	0.20
UNSCEAR assessment [U4]	0.36 [a]	0.36	0.18 [a]	0.18

a   Estimated for time period from one month to one year after the accident.

Table 28
Values of occupancy factor in the summer for rural populations of Belarus, Russian Federation, and Ukraine
[E7]

Location	Occupancy factor								
	Indoors			Outdoors in same village			Outdoors elsewhere		
	Belarus	Russia	Ukraine	Belarus	Russia	Ukraine	Belarus	Russia	Ukraine
Indoor workers	0.77	0.65	0.56	0.19	0.32	0.40	0.04	0.03	0.04
Outdoor workers	0.40	0.50	0.46	0.25	0.27	0.29	0.35	0.23	0.25
Retired people	0.44	0.56	0.54	0.42	0.40	0.41	0.14	0.04	0.05
Schoolchildren	0.44	0.57	0.75	0.45	0.39	0.21	0.11	0.04	0.04
Preschool children		0.64	0.81		0.36	0.19		0	0

Table 29
Ratio of external effective dose of specific population groups to that of the representative group
[B14]

Population living in	Dose ratio to representative group			
	Indoor workers	Outdoor workers	Herders, foresters	Schoolchildren
Wooden dwellings	0.8	1.2	1.7	0.8
One- or two-storey brick houses	0.7	1.0	1.5	0.9
Multi-storey buildings	0.6	0.8	1.3	0.7

Table 30
Estimated values of the overall coefficient used to calculate external effective dose on the basis of the absorbed dose in air

Country	Type of settlement	Dose coefficient (mSv mGy^{-1})	
		1986 [a]	1987–1995
Former USSR [M11]	Rural	0.28	0.28
	Small town	0.19	0.19
	Large town	–	0.14
Belarus [K38]	Rural	0.19	0.12
	Urban	0.12	0.12
Russian Federation [R9]	Rural	0.24	0.19
	Small town	0.15	0.12
	Large town	0.13	0.10
Ukraine [M16]	Rural	0.23	0.23
	Small town	0.16	0.16
	Large town	0.10	0.10

a   Covers the time span from 26 April 1986 to 25 April 1987.

Table 31
Estimates of external effective doses per unit deposition density of ^{137}Cs for the residents of contaminated areas

Country	Type of settlement	Normalized effective dose (μSv per $kBq\ m^{-2}$)				
		1986 [a]	1987–1995	1986–1995	1996–2056	1986–2056
Former USSR [b] [A12, G3, M11]	Rural	13–28	34	47–62	48	95–108
	Urban	7–15	23	30–38	33	63–71
Belarus [K38]	Rural	19	36	55		
	Urban	12	36	48		
Russian Federation [M17]	Rural	15	22	37	28	65
	Urban	11	14	25	17	42
Ukraine [M16]	Rural	24	36	60	28	88
	Urban	17	25	42	19	61

a   Covers the time span from 26 April 1986 to 25 April 1987.
b   The estimates were obtained mainly in the empirical manner by relating the results derived from monitoring the residents of settlements in Belarus, Russian Federation, and Ukraine by means of TLD devices to the ^{137}Cs deposition density in those settlements (37–2,200 $kBq\ m^{-2}$).

Table 32
Reference values of absorbed dose rate in air per unit deposition density of ^{137}Cs

Country	Normalized absorbed dose rate in air [a] ($nGy\ h^{-1}$ per $kBq\ m^{-2}$)								
	1987	1988	1989	1990	1991	1992	1993	1994	1995
Belarus [K38] [b]	4.5	3.1	2.3	1.9	1.6	1.4	1.3	1.2	1.0
Russian Federation [B26, M17]	3.4	2.4	1.9	1.5	1.3	1.1	1.0	0.93	0.85
Ukraine [L36]	2.1	1.8	1.8	1.7	1.7	1.6	1.6	1.5	1.5
[M16]	4.5	3.0	2.3	1.9	1.6	1.4	1.3	1.2	1.1
Former USSR [A12, G3] [b]	3.0	2.4	1.9	1.6	1.3	1.1	1.0	0.89	0.79

a   Estimated for May of the corresponding year.
b   In publications [A12, G3, K38], the normalized absorbed dose rate in air is expressed in $\mu R\ h^{-1}$ per $Ci\ km^{-2}$. The conversion from exposure to absorbed dose in air was made using the relationship 8.7 nGy per μR.

Table 33
Selected values of external effective dose normalized to the deposition density of ^{137}Cs for the residents of contaminated areas

Country	Type of settlement	Normalized effective dose (μSv per $kBq\ m^{-2}$)				
		1986	1987–1995	1986–1995	1996–2056	1986–2056
Belarus	Rural	19	22	41	28	69
	Urban	12	14	26	17	43
Russian Federation	Rural	15	22	37	28	65
	Urban	11	14	25	17	42
Ukraine	Rural	24	36	60	28	88
	Urban	17	25	42	19	61

Table 34
Estimated collective effective doses to the populations of contaminated areas 1986–1995, (excluding thyroid dose)

Region	^{137}Cs deposition density $(kBq\ m^{-2})$	Population	Collective effective dose (man Sv)		
			External	Internal	Total
Belarus [S46]					
Brest	37–185	151 312	404	311	715
	185–555	16 183	159	118	277
	Total	167 495	563	429	992
Gomel	37–185	1 246 613	2 792	1 676	4 468
	185–555	140 732	1 556	1 237	2 793
	≥ 555	77 571	2 315	822	3 137
	Total	1 464 916	6 663	3 735	10 398
Grodno	37–185	28 060	71	57	128
	185–555	282	2.6	2.2	4.8
	Total	28 342	74	59	133
Minsk	37–185	23 809	68	61	129
	185–555	880	7	6	13
	Total	24 689	75	67	142
Mogilev	37–185	93 658	347	302	649
	185–555	81 428	796	584	1 380
	≥ 555	20 022	1 118	328	1 446
	Total	195 108	2 261	1 214	3 475
Vitebsk	37–185	62	0.12	0.11	0.23
Total	≥ 37	1 880 612	9 636	5 504	15 140
Russian Federation [B37, M17]					
Belgorod	37–185	73 350	114	48	162
Bryansk	37–185	205 625	472	1 414	1 886
	185–555	150 002	1 440	1 023	2 463
	≥ 555	95 462	2 611	799	3 410
	Total	451 089	4 523	3 236	7 759
Kaluga	37–185	91 801	201	74	275
	185–555	12 654	108	45	153
	Total	104 455	309	119	428
Kursk	37–185	133 720	278	118	396
Leningrad	37–185	8 434	18	43	61
Lipetsk	37–185	50 732	58	47	105
Mordovia	37–185	10 909	20	20	40
Orel	37–185	151 008	328	183	511
	185–555	14 435	64	34	98
	Total	165 443	392	217	609
Penza	37–185	9 910	16	16	32
Pyazan	37–185	199 687	275	313	588
Tambov	37–185	16 832	23	23	46
Tula	37–185	667 168	1 916	649	2 565
	185–555	56 535	453	115	568
	≥ 555	12	0.25	0.03	0.28
	Total	723 715	2 369	764	3 133
Ulyanovsk	37–185	2 805	4	4	8
Voronezh	37–185	32 194	55	23	78
Total	≥ 37	1 983 275	8 454	4 991	13 445

Table 34 (continued)

Region	^{137}Cs deposition density (kBq m^{-2})	Number of population		Collective effective dose (man Sv)					
				External		Internal		Total	
		Rural	Urban	Rural	Urban	Rural	Urban	Rural	Urban
Ukraine									
Vinnyts'ka	37–185	66 000	17 100	261	34.5	73.6	36.0	335	70.5
Volyns'ka	37–185	15 400		49.7		214.3		264	
Zhytomyrs'ka	37–185	207 200	18 300	1 012	81.2	1 785	38.7	2 797	120
	185–555	21 300	66 300	332	769	264	140	596	909
	≥555	300		10.8		13.7		24.6	
Ivano-Frankivs'ka	37–185	8 500		45.1		15.4		60.5	
Kyivs'ka	37–185	243 500	142 400	941	483	570	288	1 510	771
	185–555	11 500		176		56.4		233	
Rivnens'ka	37–185	221 700	49 600	903	121	3 710	188	4 613	309
	185–555	1 400		18.1		30.8		48.9	
Sums'ka	37–185	4 500		22.7		21.2		44.0	
Ternopils'ka	37–185	7 400		29.6		11.7		41.2	
Khmel'nyts'ka	37–185	2 000		10.0		3.0		13.0	
	185–555	100		0.6		0.2		0.8	
Cherkas'ka	37–185	61 700	69 600	252	255	70.7	147	323	401
	185–555	2 900		43.2		10.7		53.9	
Chernivets'ka	37–185	18 500		87.2		31.3		119	
	185–555	3 000		35.5		10.2		45.6	
Chernihivs'ka	37–185	25 400	9 900	101	24.9	105	20.8	206	45.7
	185–555	300		4.7		3.3		7.9	
Total		922 600	373 200	4 336	1 768	7 000	858	11 340	2 626

Table 35
Distribution of estimated total effective doses received by the populations of contaminated areas,
1986–1995 (excluding thyroid dose)

Dose interval (mSv)	Number of persons in dose interval	Percentage of persons in dose interval	Cumulative percentage of persons in dose interval
Belarus [S46]			
<1	133 053	7.07	7.07
1–2	444 709	23.65	30.72
2–3	362 510	19.28	50
3–4	221 068	11.75	61.75
4–5	135 203	7.19	68.94
5–10	276 605	14.71	83.65
10–20	163 015	8.67	92.32
20–30	63 997	3.40	95.72
30–40	32 271	1.71	97.43
40–50	17 521	0.93	98.36
50–100	25 065	1.33	99.69
100–200	5 105	0.27	99.96
>200	790	0.04	100
Total	1 880 912	100	
Russian Federation [B37, M17, S46]			
<1	155 301	7.83	7.83
1–2	445 326	22.45	30.28
2–3	383 334	19.32	49.60
3–4	258 933	13.06	62.66
4–5	165 537	8.35	71.01
5–10	317 251	16	87.01
10–20	156 925	7.91	94.92
20–30	50 010	2.52	97.44
30–40	21 818	1.10	98.54
40–50	11 048	0.55	99.09
50–100	14 580	0.74	99.83
100–200	2 979	0.15	99.98
>200	333	0.02	100
Total	1 983 375	100	

Dose interval (mSv)	Number of persons in dose interval		Percentage of persons in dose interval		Cumulative percentage of persons in dose interval
	Rural	Urban	Rural	Urban	
Ukraine [L44]					
1–2					0
2–3	26 100		2.0		2.0
3–4	57 100	38 800	4.4	3.0	9.4
4–5	97 200	111 700	7.5	8.6	25.6
5–10	294 700	145 700	22.7	11.2	59.5
10–20	290 500	77 000	22.4	6.0	87.9
20–30	99 100		7.7		95.6
30–40	31 400		2.4		98.0
40–50	18 200		1.4		99.4
50–100	7 700		0.59		99.97
100–200	400		0.03		100
Total	922 400	373 200	71.2	28.8	100

Table 36
Summary of estimated collective effective doses to the populations of contaminated areas, 1986–1995 (excluding thyroid dose)

^{137}Cs deposition density $(kBq\ m^{-2})$	Population	Collective effective dose (man Sv)		
		External	Internal	Total
Belarus [S46]				
37–185	1 543 514	3 682	2 409	6 091
185–555	239 505	2 521	1 945	4 466
≥555	97 593	3 433	1 150	4 583
Total	1 880 612	9 636	5 504	15 140
Russian Federation [B37, M17, S46]				
37–185	1 654 175	3 778	3 009	6 787
185–555	233 626	2 065	1 183	3 248
≥555	95 474	2 611	799	3 410
Total	1 983 275	8 454	4 991	13 445

^{137}Cs deposition density $(kBq\ m^{-2})$	Population		Collective effective dose (man Sv)					
			External		Internal		Total	
	Rural	Urban	Rural	Urban	Rural	Urban	Rural	Urban
Ukraine [L44]								
37–185	881 800	306 800	3 715	999	6 610	717	10 330	1 717
185–555	40 400	66 300	610	769	375	140	986	909
>555	300		11.0		13.8		24.8	
Total	922 500	373 100	4 336	1 768	7 000	857	11 340	2 626

Table 37
Distribution of external dose rates before and after decontamination measures in Kirov, Belarus, in 1989 [S25]

Population group	Measured mean dose rate ($\mu Gy\ d^{-1}$)		External dose ratio after/before
	Before decontamination	After decontamination	
Cattle breeders	12.3	11.9	0.97
Field workers	17.5	13.2	0.75
Office workers	12.1	11.8	0.98
Housewives and retired persons	12.9	12.8	0.99
Schoolchildren	15.0	9.7	0.65
Tractor drivers	12.8	12.7	0.99
Average	13.7	12.1	0.89

Table 38
Distribution of estimated individual doses in the thyroid of children in contaminated districts of Belarus
[G13]

Absorbed dose in thyroid (Gy)	Number of children in age range [a]						
	<1 year	1–3 years	4–7 years	8–11 years	12–15 years	16–18 years	All children
Gomel district							
<0.05	134 (6.7)	198 (6.1)	452 (7.4)	518 (8.4)	540 (8.8)	596 (16)	2 438 (8.9)
0.05–0.1	58 (2.9)	107 (3.3)	362 (5.9)	399 (6.5)	485 (7.9)	354 (9.4)	1 765 (6.4)
0.1–0.3	224 (11)	449 (14)	1 089 (18)	1 385 (22)	1 613 (26)	1 086 (29)	5 846 (21)
0.3–1	587 (30)	963 (29)	2 023 (33)	2 365 (38)	2 364 (38)	1 119 (30)	9 421 (34)
1–2	318 (16)	590 (18)	1 075 (18)	868 (14)	695 (11)	383 (10)	3 929 (14)
>2	3 667 (34)	965 (29)	1 095 (18)	643 (10)	464 (7.5)	230 (6.1)	4 064 (15)
Total	1 988 (100)	3 272 (100)	6 096 (100)	6 178 (100)	6 161 (100)	3 768 (100)	27 463 (100)
Mogilev district							
<0.05	33 (13)	43 (9.1)	210 (19)	273 (28)	326 (29)	227 (37)	1 112 (24)
0.05–0.1	31 (12)	93 (20)	215 (19)	157 (16)	207 (19)	103 (17)	806 (18)
0.1–0.3	65 (26)	170 (36)	351 (31)	324 (33)	372 (33)	169 (28)	1 451 (32)
0.3–1	74 (29)	127 (27)	275 (25)	190 (20)	195 (17)	99 (16)	960 (21)
1–2	36 (14)	28 (5.9)	55 (4.9)	24 (2.5)	15 (1.3)	14 (2.3)	172 (3.8)
>2	14 (5.5)	14 (3.0)	16 (1.4)	1 (0.1)	1 (0.09)	1 (0.2)	47 (1.0)
Total	253 (100)	475 (100)	1 122 (100)	969 (100)	1 116 (100)	613 (100)	4 548 (100)

a   Percent of total in parentheses.

Table 39
Thyroid doses to 0–7-year-old children and to the total population in contaminated areas of Belarus
[I28]

Region	Number of persons		Average absorbed dose (Gy)		Collective dose (man Gy)	
	Children	Total	Children	Total	Children	Total
Gomel						
Rural	23 900	238 600	1.1	0.4	25 000	98 000
Urban	8 600	85 600	0.4	0.2	3 800	15 000
Mogilev						
Rural	9 300	93 700	0.4	0.2	4 100	17 000
Urban	4 900	48 700	0.2	0.08	1 100	4 000
Total	46 700	466 600	0.7	0.3	34 000	134 000

Table 40
Estimates of collective thyroid doses to populations of Belarus, the Russian Federation, and Ukraine

Country/region		Population	Collective thyroid dose (man Gy)
Belarus [D1, G7]			
Brest		1 400 000	101 000
Gomel		1 700 000	301 000
Grodno		1 200 000	49 000
Minsk		3 200 000	68 000
Mogilev		1 300 000	32 000
Vitebsk		1 400 000	2 000
Entire country		10 000 000	553 000
Russian Federation [Z1]			
Bryansk		1 500 000	55 000
Orel		900 000	15 000
Tula		1 900 000	50 000
Bryansk, Kaluga, Kursk, Leningrad, Orel, Ryaza, and Tula [a]	[S21]	3 700 000	234 000
Entire country	[B14]	150 000 000	200 000–300 000
Ukraine			
Kiev	[L15]	3	110 000
Eight districts [b] plus Pripyat	[L12]	0.5	190 000
Entire country	[L6]	55	740 000 [c]

a   Only the territories with ^{137}Cs deposition densities greater than 37 kBq m^{-2} were considered.
b   Situated around the Chernobyl nuclear power plant.
c   Derived from an estimated collective thyroid dose of 400,000 man Gy for children aged 0–18 years in the entire Ukraine.

Table 41
Ratios of the average thyroid doses in children and teenagers to those in adults for the populations of contaminated areas in the Russian Federation
[Z6, Z7]

Age group (years)	Urban population	Rural population
<1	15±5	7±2
1–2	10±3	6±2
3–7	5±2	3±1
8–12	2.0±0.5	1.0±0.5
13–17	1.5±0.5	2.0±0.5
>17	1	1

Table 42
Input data for internal thyroid dose reconstruction in the Russian Federation
[B39]

Region	Number of measurements					Personal interviews
	^{131}I in thyroid	^{131}I in milk	^{137}Cs in soil	^{137}Cs in food products	^{137}Cs in human body	
Bryansk	12 700	2 100	2 081	217 000	300 000	17 000
Tula	644	2 157	2 308	2 000	17 000	1 800
Orel	3 600	872	1 577	17 000	10 000	–
Kaluga	28 000	256	578	18 000	28 000	6 000
Total	45 000	5 385	6 544	250 000	360 000	25 000

Table 43
Collective and average thyroid doses in population of the Russian Federation
[R3, Z1]

Region	Population (millions)			Collective thyroid dose (man Gy)			Average thyroid dose (mGy)		
	Urban	Rural	Total	Urban	Rural	Total	Urban	Rural	Total
Bryansk	1.0	0.5	1.5	33 000	27 000	60 000	33	54	40
Kaluga	0.8	0.3	1.1	4 000	3 000	7 000	5	10	6
Orel	0.6	0.3	0.9	8 000	5 000	13 000	13	17	15
Tula	1.5	0.4	1.9	14 000	6 000	20 000	9	15	11
Other regions [a]	–	–	16	–	–	110 000 ±30 000	–	–	7
Total	–	–	21	–	–	110 000 ±30 000	–	–	10

a   Other eleven contaminated regions: Belgorod, Voronezh, Kursk, Leningrad (without city of St. Petersburg), Lipetsk, Mordovia, Penza, Ryazanj, Smolensk, Tambov, Uljanov.

Table 44
Age distribution of the collective thyroid doses to inhabitants of the Bryansk region in the Russian Federation
[Z6]

Age (years)	Population		Collective thyroid dose (man Gy)	
	Urban	Rural	Urban	Rural
0–2	47 000	21 000	9 000	4 000
3–5	47 000	19 000	5 000	3 000
6–9	59 000	23 000	3 000	2 000
10–15	85 000	36 000	2 000	2 000
16–19	59 000	17 000	1 000	1 000
20	694 000	368 000	13 000	15 000
Total	992 000	483 000	33 000	27 000

Table 45
Calculated mean thyroid doses for infants and adults in three districts of the Chernigov region, Ukraine
[L25]

District	Type of settlement	Parameter values			Mean thyroid doses (Gy)	
		K (Gy)	a	b (a^{-1})	Infants (1 a)	Adults (20 a)
Repkine	Towns	0.031	7.4	0.049	0.21	0.065
	Villages	0.082	5.0	0.094	0.35	0.11
Chernigov	City	0.030	10	0.064	0.26	0.057
	Villages	0.17	3.0	0.079	0.47	0.22
Kozelets	Towns	0.012	18	0.15	0.14	0.014
	Villages	0.047	7.4	0.062	0.30	0.085

Table 46
Estimated age-dependent thyroid dose distribution in the population of the Ukrainian regions of Cherkassy, Chernigov, Kiev, Vinnitsa, and Zhitomir [a]
[L10]

Age at the time of the accident (years)	Percentage of age groups within dose intervals					
	0–0.049 Gy	*0.05–0.099 Gy*	*0.1–0.29 Gy*	*0.3–0.99 Gy*	*1–1.99 Gy*	*>2 Gy*
<1	2.8	10.4	55.5	23.3	7.4	0.5
1–3	3.2	10.3	58.6	24.9	2.1	0.9
4–7	10.8	27.2	47.7	13.4	0.9	0.0
8–11	19.3	39.6	34.0	6.1	0.9	0.0
12–15	35.3	35.2	24.4	4.3	0.7	0.0
16–18	45.7	26.8	23.5	4.1	0.0	0.0
>18	54.6	21.0	21.6	2.8	0.0	0.0

a   Excludes the 30-km zone and city of Kiev.

Table 47
Distribution of absorbed dose in the thyroid of the population of Ukraine from the Chernobyl accident
[L12]

Absorbed dose (Gy)	Number of persons	Collective thyroid dose (man Gy)
<0.01	7 325 000	36 625
0.01–0.05	3 400 000	102 000
0.05–0.1	1 312 000	98 400
0.1–0.3	228 000	45 600
0.3–0.5	131 000	52 400
0.5–1.0	26 000	19 500
1.0–1.5	28 000	35 000
Total	12 450 000	390 000 [a]

a   Estimate is less by a factor of 2 for more realistic intake model.

Table 48
Normalized effective doses from radiocaesium via internal exposure to rural inhabitants of Ukraine in different soil zones
[K23]

Soil zone	Transfer coefficient from soil to milk in 1991 (Bq l^{-1} per kBq m^{-2})	Normalized doses (μSv per kBq m^{-2})		
		1986	*1987–1995*	*1986–1995*
I	<1	9	26	35
II	1–5	42	144	186
III	5–10	95	320	415
IV	>10	176	591	767

Table 49
Values of the transfer coefficient and of the normalized effective dose from internal exposure for sodic-podzol sand and chernozem soils in the Russian Federation
[B25, R9]

Type of soil	Transfer coefficient (Bq kg^{-1} per kBq m^{-2})				Normalized effective dose (μSv per kBq m^{-2})		
	Milk		Potatoes				
	1987	1993	1987	1993	1986	1987–1994	1995–2056
Sodic-podzol sand	5	0.2	0.16	0.04	90	78	16
Chernozem	0.07	0.01	0.03	0.004	28	2	1

Table 50
Rounded estimates of internal effective doses per unit of ^{137}Cs deposition density for inhabitants of contaminated areas of the Russian Federation

Region	Normalized effective dose (μSv per kBq m^{-2})			
	1986	1987–1995	1996–2056	1986–2056
Bryansk [a]	36 (10)	48 (13)	9 (9)	93 (32)
Tula	15	6	1.8	23
Orel	15	8	2.4	25

a   The values within parentheses correspond to areas with a ^{137}Cs deposition density greater than 555 kBq m^{-2}.

Table 51
Average effective doses from internal exposure per unit deposition density of ^{137}Cs for population subgroups in Belarus
[D3, K22]

Location	Age group	Normalized effective dose (μSv per kBq m^{-2})			
		1986	1987–1990	1991–1995	1986–1995
Rural areas	0–6 years	3–17	6–32	3–7	12–56
	7–17 years	5–19	8–45	2–6	15–70
	>18 years	5–24	12–62	3–10	20–96
Towns	0–6 years	2	5	2	9
	7–17 years	2	4	2	8
	>18 years	3	7	4	14

Table 52
Populations in Europe examined in epidemiological studies
[S12]

Country	Study region	Age group	Average absorbed dose (mGy) [a]
Thyroid studies			
Croatia	Whole country	All ages	15 [b]
Greece	Whole country	20-60 years	5
Hungary	Whole country	All ages	3 [b]
Poland	Krakow, Nowy Sacz	All ages	4 [b]
Turkey	Five most affected areas on Black Sea coast and Edirne province	All ages	1.5 [b]
Leukaemia studies			
Bulgaria	Whole country	Adults	2
		Children 0-14 years	2
Finland	Whole country	Children 0-14 years	2
Germany	Bavaria	Children 0-14 years	4
Greece	Whole country	Children 0-14 years	1
Hungary	Six counties	All ages	0.7
Romania	Whole country	Children 0-14 years [c]	3
Sweden	Whole country	Children 0-14 years	4
Turkey	Five most affected areas on Black Sea coast and Edirne province	All ages	0.7

a   To thyroid in thyroid studies and to bone marrow in leukaemia studies; assumes bone marrow dose is numerically equal to effective dose and dose in children is the same as in adults.

b   Assumes population-weighted thyroid dose is three times that to adults.

c   Age at death.

Table 53
Summary of estimated collective effective doses to populations of areas contaminated by the Chernobyl accident (1986-1995)

Country	Population	Collective effective dose (man Sv)			Average effective dose (mSv)		
		External exposure	Internal exposure	Total	External exposure	Internal exposure	Total
Belarus	1 880 000	9 600	5 500	15 100	5.1	2.9	8.0
Russian Federation	1 980 000	8 500	5 000	13 500	4.3	2.5	6.8
Ukraine	1 300 000	6 100	7 900	14 000	4.7	6.1	10.8
Total	5 160 000	24 200	18 400	42 600	4.7	3.5	8.2

Table 54
Distribution of the estimated total individual effective doses received by the populations of contaminated areas, 1986–1995 (excluding thyroid dose)

Dose interval (mSv)	Number of persons in dose interval in			
	Belarus	Russian Federation	Ukraine	Total
<1	133 053	155 301	0	288 804
1–2	444 709	445 326	0	890 035
2–3	362 510	383 334	26 100	771 944
3–4	221 068	258 933	95 900	575 901
4–5	135 203	165 537	208 900	509 640
5–10	276 605	317 251	440 400	1 034 056
10–20	163 015	156 925	367 500	687 440
20–30	63 997	50 010	99 100	213 107
30–40	32 271	21 818	31 400	85 489
40–50	17 521	11 048	18 200	46 769
50–100	25 065	14 580	7 700	47 345
100–200	5 105	2 979	400	8 484
>200	790	333	0	1 123
Total	1 880 912	1 983 375	1 295 600	5 159 887

Table 55
Deaths of survivors of acute radiation sickness during 1986–1998
[W5]

Year of death	Grade of acute radiation sickness	Disease recorded and/or cause of death
1987	II	Lung gangrene
1990	II	Coronary heart disease
1992	III	Coronary heart disease
1993	I	Coronary heart disease
	III	Myelodysplastic syndrome
1995	I	Lung tuberculosis
	II	Liver cirrhosis
	1	Fat embolism
	III	Myelodysplastic syndrome
1998	II	Liver cirrhosis
	II	Acute myeloid leukaemia

Table 56
Thyroid cancer cases in children under 15 years old at diagnosis

Region	Number of cases												
	1986	1987	1988	1989	1990	1991	1992	1993	1994	1995	1996	1997	1998
Belarus [P9]													
Brest	–	–	1	–	8	7	9	28	19	20	16	17	16
Vitebsk	–	–	–	–	1	3	3	–	2	–	–	2	–
Gomel	1	2	2	2	15	38	28	34	38	41	37	33	14
Grodno	2	1	1	2	1	4	6	5	4	5	4	5	5
Minsk city	–	–	1	–	4	5	7	9	3	9	1	4	7
Minsk	–	1	1	1	–	4	8	5	7	2	6	8	3
Mogilev	–	–	–	–	2	1	1	6	4	5	3	4	3
Total	3	4	6	5	31	62	62	87	77	82	67	73	48
Russian Federation [I23]													
Bryansk region													
Novozybkovsky	–	–	–	–	–	–	–	–	4	–	–	–	
Klintsovsky	–	–	–	–	–	1	1	1	–	5	2	3	
Klimovsky	–	–	–	–	–	–	1	–	–	1	–	–	
Sturodubsky	–	–	–	–	–	–	1	–	–	–	–	–	
Unechsky	–	–	–	–	–	–	–	–	–	–	–	–	
Komarichsky	–	–	–	–	–	–	–	–	–	–	–	–	
Dyatkovsky	–	1	–	–	–	–	–	–	1	–	–	2	
Surazhsky	–	–	–	–	–	–	–	–	–	–	–	–	
Vygonechsky	–	–	–	–	–	–	–	–	1	–	–	–	
Suzemsky	–	–	–	–	1	–	–	–	–	1	–	–	
Total	–	1	–	–	1	1	3	1	6	7	2	5	
Ukraine [T2, T16]													
Zhitomir	–	–	–	–	2	1	3	3	5	7	6	5	7
Kiev	2	–	–	2	4	4	10	10	3	9	18	8	7
Kiev city	1	–	1	1	2	3	9	5	5	9	4	4	6
Rovno	–	–	–	–	–	1	5	2	–	–	3	2	2
Cherkassy	–	–	1	–	–	2	4	4	2	2	2	2	1
Chernigov	–	–	–	1	3	2	3	4	11	4	4	6	3
Other regions	5	7	6	7	15	9	15	16	18	16	19	9	18
Total	8	7	8	11	26	22	49	44	44	47	56	36	44

Table 57
Thyroid cancer incidence rates in children under 15 years old at diagnosis

Region	Number of cases per 100 000 children												
	1986	1987	1988	1989	1990	1991	1992	1993	1994	1995	1996	1997	1998
Belarus [P9]													
Brest	–	–	0.4	–	3.3	2.9	3.7	11.6	8.1	8.8	7.3	8.2	8.0
Vitebsk	–	–	–	–	0.5	1.4	1.4	–	1.0	–	–	1.2	–
Gomel	0.4	0.7	0.7	0.7	5.6	15.0	11.2	13.9	15.9	17.9	16.9	16.0	7.1
Grodno	1.1	0.6	0.5	1.1	0.6	2.2	3.2	2.7	2.2	2.8	2.4	3.1	3.3
Minsk city	–	–	0.4	–	1.5	1.9	2.7	3.5	1.2	3.9	0.5	2.0	3.7
Minsk	–	0.4	0.4	0.4	–	1.7	3.2	2.0	3.0	0.9	2.8	3.9	1.6
Mogilev	–	–	–	–	1.0	0.5	0.5	3.2	2.2	2.9	1.8	2.6	2.0
Total	0.2	0.3	0.4	0.3	1.9	3.9	3.9	5.5	5.1	5.6	4.8	5.6	3.9
Russian Federation [I23]													
Bryansk region													
Novozybkovsky									26.6				
Klintsovsky	–	–	–	–	–	4.3	4.3	4.3	–	21.7	8.6	12.9	
Klimovsky	–	–	–	–	–	–	12.1	–	–	12.1	–	–	
Sturodubsky	–	–	–	–	–	–	9.9	–	–	–	–	–	
Unechsky	–	–	–	–	–	–	–	–	–	–	–	–	
Komarichsky	–	–	–	–	–	–	–	–	–	–	–	–	
Dyatkovsky	–	9.1	–	–	–	–	–	–	9.1	–	–	18.2	
Surazhsky	–	–	–	–	–	–	–	–	–	–	–	–	
Vygonechsky	–	–	–	–	–	–	–	–	20.2	–	–	–	
Suzemsky	–	–	–	–	23.2	–	–	–	–	23.2	–	–	
Krasnogorsky	–	–	–	–	–	–	–	–	–	18.8	–	–	
Navlinsky	–	–	–	–	–	–	–	–	–	–	–	14.3	
Karachevsky	–	–	–	–	–	–	–	–	10.5	–	–	10.5	
Bryansk city	–	–	–	–	–	–	–	–	10.0	–	–	–	
Total	–	0.3	–	–	0.3	0.3	0.9	0.3	2.8	2.5	0.6	2.2	
Ukraine [T2, T16]													
Zhitomir	–	–	–	–	0.6	0.3	0.9	1.0	1.6	2.3	2.0	1.7	2.5
Kiev	0.5	–	–	0.5	1.0	1.0	2.5	2.5	0.8	2.4	4.9	2.2	2.0
Kiev City	0.2	–	0.2	0.2	0.4	0.6	1.7	1.0	1.0	1.8	0.8	0.9	1.4
Rovno	–	–	–	–	–	0.3	1.7	0.7	–	–	1.1	0.7	0.7
Cherkassy	–	–	0.3	–	–	0.7	1.3	1.3	0.7	0.7	0.7	0.7	0.4
Chernigov	–	–	–	0.4	1.1	0.8	1.2	1.6	4.4	1.7	1.7	2.6	1.4
Other regions	0.1	0.1	0.1	0.1	0.2	0.1	0.2	0.2	0.2	0.2	0.2	0.1	0.2
Total	0.2	0.1	0.1	0.1	0.2	0.2	0.5	0.4	0.4	0.5	0.6	0.4	0.5

Table 58
Thyroid cancer cases diagnosed among children 0–17 years old at the time of the Chernobyl accident

Region	Sex	Number of cases in year of diagnosis									
		1990	1991	1992	1993	1994	1995	1996	1997	1998	Total
Belarus [P9]											
Brest	M	3	6	5	11	11	9	10	8	14	77
	F	4	5	5	26	18	27	23	18	20	146
Gomel	M	9	13	19	15	16	25	16	24	19	156
	F	7	35	15	31	41	37	41	42	34	283
Grodno	M	1	3	6	–	5	4	3	3	1	26
	F	–	4	6	8	5	4	7	5	8	47
Minsk city	M	1	5	5	3	2	4	5	5	9	39
	F	4	5	5	14	14	15	10	11	22	100
Minsk	M	1	2	4	2	5	3	5	9	2	33
	F	–	3	8	11	5	5	6	5	8	51
Mogilev	M	3	1	2	2	1	3	1	1	3	17
	F	–	–	–	5	8	7	4	8	12	44
Vitebsk	M	–	2	5	–	1	–	1	2	2	13
	F	2	2	5	2	5	4	4	5	6	35
Total		35	86	90	130	137	147	136	146	160	1 067
Russian Federation [I23]											
Bryansk	M	–	1	2	4	5	6	3	6		27
	F	–	2	4	5	17	11	7	15		61
Kaluga	M	–	–	1	–	2	1	2	1		7
	F	1	1	–	–	2	–	1	2		7
Orel	M	–	1	–	–	3	1	1	–		6
	F	2	–	3	6	10	9	14	2		46
Tula	M	–	3	2	1	–	5	1	–		12
	F	2	2	–	3	8	8	7	9		39
Total		5	10	12	19	47	41	36	35		205
Ukraine [T16]											
Zhitomir	M	1	2	–	1	4	4	3	4	5	24
	F	1	1	4	2	7	9	5	5	9	43
Kiev	M	2	3	5	6	2	10	13	5	2	48
	F	7	6	11	7	9	14	26	17	10	107
Kiev city	M	3	1	5	3	4	3	3	5	7	34
	F	4	5	12	13	10	25	14	16	15	114
Rovno	M	–	1	3	2	3	3	4	4	1	22
	F	–	1	2	1	1	1	2	2	5	16
Cherkassy	M	–	–	2	1	1	1	1	–	2	8
	F	1	3	3	6	5	1	4	5	2	30
Chernigov	M	2	1	2	2	7	1	1	3	3	22
	F	1	3	4	5	6	7	6	7	12	51
Total		22	27	55	49	59	79	82	73	73	519

Table 59
Thyroid cancers and risk for children 0–18 years old at the time of the Chernobyl accident for the years 1991–1995 in three cities and 2,729 settlements in Belarus and the Russian Federation
[J5]

Thyroid dose	Person years at risk	Observed number of cases	Expected number of cases [a]	Excess absolute risk [b] $(10^4 \, PY \, Gy)^{-1}$
0–0.1 (0.05)	1 756 000	38	16	2.6 (0.5–6.7)
0.1–0.5 (0.21)	1 398 000	65	13	1.9 (0.8–4.1)
0.5–1.0 (0.68)	386 000	52	3.6	2.0 (0.9–4.2)
1.0–2.0 (1.4)	158 000	50	1.5	2.3 (1.1–4.9)
>2.0 (3.0)	56 000	38	0.5	2.4 (1.1–5.1)

a   Calculated by multiplying the age-specific incidence observed in Belarus in 1983–1987 by three.
b   95% confidence intervals in parentheses.

Table 60
Odds ratio for thyroid cancer cases in Belarusian children compared with age-, location-, and exposure-matched controls
[A6]

Pathway to diagnosis	Number of cases or controls at estimated thyroid dose from ^{131}I				
	<0.30 Gy	0.30–0.99 Gy	>0.99 Gy	Total	Odds ratio [a]
Routine endocrinological screening					
Cases	32	16	15	63	2.1
Control	43	16	4	63	(1.0–4.3)
Incidental findings					
Cases	13	4	2	19	8.3
Control	18	1	0	19	(1.1–58)
Enlarged or nodular thyroid					
Cases	19	6	0	25	3.6
Control	23	2	0	25	(0.7–18)
Cases	64	26	17	107	3.1
Controls	84	19	4	107	(1.7–5.8)
Total	152	41	21	214	

a   Comparing lowest and two highest groups. 95% confidence interval in parentheses.

Table 61
Circumstances of diagnosis of thyroid problems
[A5]

Probable circumstance of diagnosis	Ultrasound diagnosis			
	Unknown	No	Yes	Total
Formal screening programme	2 (29%)	7 (20%)	3 (38%)	12
Incidental to other examination	2 (29%)	18 (51%)	3 (38%)	23
Consulted since unwell	1 (14%)	8 (23%)	1 (13%)	10
Consulted even though well	2 (29%)	2 (6%)	1 (13%)	5
Total	7	35	8	50

Table 62
Pathological types of thyroid cancers in Ukrainian children less than 15 years old at diagnosis
[T18]

Pathological classification [a]	Number of cancers in relation to year of diagnosis [b]			
	1986–1990	*1991–1995*	*1996–1997*	*Total*
T1	–	4 (2)	1 (1)	5 (1)
T2	9 (18)	49 (20)	18 (16)	76 (19)
T3	6 (12)	27 (11)	6 (5)	39 (10)
T4	14 (29)	69 (28)	38 (35)	121 (30)
N1	9 (18)	45 (18)	22 (20)	76 (19)
N2	9 (18)	45 (18)	22 (20)	76 (19)
M1	2 (4)	6 (2)	3 (3)	11 (3)
Total	49 (100)	245 (100)	110 (100)	404 (100)

a   Staging system of the World Health Organization [H17].
b   Percentage of total in parentheses.

Table 63
Age and sex distribution of Ukrainian children and adolescents who underwent surgery for thyroid carcinoma during 1986–1997
[T18]

Age at time of surgery (years)	Number		Ratio female : male
	Females	*Males*	
0–4	5	5	1.0 : 1
5–9	46	42	1.1 : 1
10–14	176	84	2.1 : 1
15–18	159	61	2.7 : 1
0–14	227	131	1.7 : 1

Age at time of exposure (years)	Number		Ratio female : male
	Females	*Males*	
0–4	146	88	1.7 : 1
5–9	149	58	2.6 : 1
10–14	69	30	2.3 : 1
15–18	15	7	2.1 : 1
0–14	364	176	2.1 : 1

Table 64
Frequency of *RET* rearrangements in thyroid papillary carcinomas in relation to age and radiation history
[S35]

Patient group	Number of cases	Mean age at exposure (years)	Mean age at surgery (years)	Number of cases	
				with RET/PTC1 mutations	*with RET/PTC3 mutations*
Children (Belarus)	51	2	12	12 (23.5%)	13 (25.5%)
Adults (Belarus)	16	22	31	11 (69%)	–
Adults (Germany) [a]	16	–	48	3 (19%)	–

a   Negligible radiation exposure from Chernobyl accident.

Table 65
Incidence of leukaemia and all cancer during 1993–1994 among recovery operation workers and residents of contaminated areas
[C2]

Country	Leukaemia cases [a]		All cancer cases [a]		Standardized incidence ratio (SIR)	
	Observed	Expected	Observed	Expected	Leukaemia	All cancer
Recovery operation workers [b]						
Belarus	9	4.5	102	136	200	75
Russian Federation	9	8.4	449	405	108	111
Ukraine	28	8	399	329	339	121
Residents of contaminated areas [c]						
Belarus	281	302	9 682	9 387	93	103
Russian Federation	340	328	17 260	16 800	104	103
Ukraine	592	562	22 063	22 245	105	99

a   ICD9 codes: 204-208 (leukaemia) and 140-208 (all cancer); expected cases are for age- and sex-matched members of the general population.
b   Males who worked in the 30-km zone during 1986 and 1987.
c   Areas with ^{137}Cs deposition density > 185 kBq m^{-2}.

Table 66
Thyroid abnormalities diagnosed by ultrasonography in children 0–10 years old at the time of the accident screened by the Chernobyl Sasakawa project during 1991–1996
[Y1]

Region	Number of children screened	Number of children with diagnosis					
		Goitre	Abnormal echogenity	Cystic lesion	Nodular lesion	Cancer	Anomaly
Mogilev	23 531						
Boys		2 231	91	19	5	1	16
Girls		2 391	188	25	19	1	19
Gomel	19 273						
Boys		1 355	332	59	130	12	67
Girls		2 053	604	63	212	25	73
Bryansk	19 918						
Boys		3 666	172	56	46	3	7
Girls		4 480	251	51	53	5	8
Kiev	27 498						
Boys		6 634	246	18	13	2	4
Girls		8 194	588	40	33	4	7
Zhitomir	28 958						
Boys		4 473	38	34	23	4	17
Girls		6 453	87	137	43	5	19
Total	119 178	41 930	2 597	502	577	62	237

Table 67
Incidence of thyroid abnormalities diagnosed by ultrasonography in children screened by the Sasakawa project during 1991–1996
[Y1]

Region	Incidence rate per 1000 children examined					
	Goitre	Abnormal echogenity	Cystic lesion	Nodular lesion	Cancer	Anomaly
Mogilev	219	11.9	1.9	1.0	0.08	1.5
Gomel	177	48.6	6.3	17.7	1.9	7.3
Bryansk	409	21.2	5.4	5.0	0.4	0.8
Kiev	539	30.3	2.1	1.7	0.2	0.4
Zhitomir	377	4.3	5.9	2.3	0.3	1.2
Total	352	21.8	4.2	4.8	0.5	2.0

Table 68
Comparison of reproductive effects in population groups in the Russian Federation during 1980–1993
[B19, L27, L28, L29]

Parameter / effect	Ratio of effect before and after accident [a]							
	Bryansk region			Tula region			Ryazan region	
	<37 kBq m^{-2}	$37-185$ kBq m^{-2}	$185-555$ kBq m^{-2}	<37 kBq m^{-2}	$37-185$ kBq m^{-2}	$185-555$ kBq m^{-2}	<37 kBq m^{-2}	$37-185$ kBq m^{-2}
Birth rate	0.81	0.83	0.75	0.87	0.73	0.69	1.0	0.90
Spontaneous abortions	1.27	1.34	1.34	0.90	1.03	1.18	1.22	0.91
Congenital anomalies	0.66	1.41	1.67	1.32	1.28	0.91	1.43	0.91
Stillbirths	0.66	1.39	1.29	1.50	0.93	1.41	0.90	0.97
Perinatal mortality	1.18	1.13	0.91	0.77	1.57	1.21	1.13	1.00
Premature births	1.07	0.95	1.39	0.88	0.86	0.71	0.83	1.23
Overall diseases in newborns	1.02	1.03	1.42	1.06	1.32	1.29	1.00	1.38
Overall unfavourable pregnancy outcome	1.07	1.16	1.35	0.95	0.97	0.92	1.07	1.00

[a] Number of women examined before and after accident: Byransk region: 3,500-4,100 in each area; Tula region, 2,400 (<37 kBq m^{-2}), 2,100 ($37-185$ kBq m^{-2}) and 810-860 ($185-555$ kBq m^{-2}); Ryazan region, 1,600-1,00 (<37 kBq m^{-2}) and 1,200-1,400 ($37-185$ kBq m^{-2}).

References

A1   Akleev, A.V. and M.M. Kosenko. Quantitative functional and cytogenetic character of lymphocytes and some indecis of immunological status of persons participated in recovery operation works in Chernobyl. J. Haematol. Transfusiol. 36: 24-26 (1991).

A2   Alexander, F. Clustering of childhood acute leukaemia. The EUROCLUS Project. Radiat. Environ. Biophys. 37: 71-74 (1998).

A3*   Arephjeva, Z.S., V.I. Badjin, Y.I. Gavrilin et al. Guidance on Thyroid Dose Estimation during the Intake of the Iodine Radioisotopes into the Human Organism. Energoatomizdat, Moscow, 1988.

A4*   Abagyan, A.A., V.G. Asmolov, A.K. Gusyikova et al. The information on the Chernobyl accident and its consequences, prepared for IAEA. At. Ener. 3(5): (1986).

A5   Astakhova, L.N., E. Cardis, L.V. Shafarenko et al. Additional documentation of thyroid cancer cases (Belarus): Report of a Survey, International Thyroid Project. IARC Internal Report 95/001 (1995).

A6   Astakhova, L.N., L.R. Anspaugh, G.W. Beebe et al. Chernobyl-related thyroid cancer in children of Belarus: a case-control study. Radiat. Res. 150: 349-356 (1998).

A7   Averkin, Y.I., T. Abelin and J. Bleuer. Thyroid cancer in Belarus: ascertainment bias? Lancet 346: 1223-1224 (1995).

A8   Ashikawa, K., Y. Shibata, S. Yamashita et al. Prevalence of goiter and urinary iodine excretion levels in children around Chernobyl. J. Clin. Endocrinol. Metab. 82: 3430-3433 (1997).

A9   All-Russian Medical and Dosimetric State Registry. Contamination of Russian territory with radionuclides ^{137}Cs, ^{90}Sr, ^{239}Pu+^{240}Pu, and ^{131}I. Bulletin of the All-Russian Medical and Dosimetric State Registry, Issue 3, Appendix 1. Moscow (1993).

A10   Alexakhin, R.M., L.A. Buldakov, V.A. Gubanov et al. Major Radiation Accidents: Medical and Agroecological Consequences (L.A. Ilyin and V.A. Gubanov, eds.). IzdAt, Moscow, 2000.

A11   Alexakhin, R.M. and N.A. Korneev. Agricultural radio-ecology. p. 397-407 in: Ecology. Moscow, 1991.

A12   Avetisov, G.M., R.M. Barkhudarov, K.I. Gordeev et al. Methodological basis for the prediction of doses from cesium radionuclides that could be received by the population living permanently in the areas contaminated as a result of the Chernobyl accident. Document of the National Commission on Radiation Protection, Ministry of Health of the USSR, September 20, 1988. p. 54-61 in: Proceedings of Methodological Materials "Methodological Principles and Recommendations on How to Calculate External and Internal Doses to Populations Living in the Contaminated Areas and Exposed as a Result of the Chernobyl Accident". Ministry of Health, Institute of Biophysics, Moscow, 1991.

A13   Auvinen, A., M. Hakama, H. Arvela et al. Fallout from Chernobyl and incidence of childhood leukaemia in Finland, 1976-92. Br. Med. J. 309: 151-154 (1994).

A14   Ageyets, V.Yu., N.N. Shuglya and A.A. Shmigelskiy. Migration of caesium and strontium radionuclides in various types of soil in Belarus. p. 26-36 in: Proceedings of the 4th Conference on Geochemical Pathways of Migration of Artificial Radionuclides in Biosphere. Pushino, 1991.

A15   Amano, H., T. Matsanuga, S. Nagao et al. The transfer capability of long-lived Chernobyl radionuclides from surface soil to river water in dissolved forms. Org. Geochem. 30: 437-442 (1999).

A16   Astakhova, L.N., T.A. Mityukova and V.F. Kobzev. Endemic goiter in Belarus following the accident at the Chernobyl nuclear power plant. p. 67-94 in: Nagasaki Symposium Radiation and Human Health. Elsevier Science B.V., Amsterdam, 1996.

B1   Begichev, S.N., A.A. Borovoi, E.V. Burlakov et al. Radio-active releases due to the Chernobyl accident. Presentation at International Seminar on Fission Product Transport Processes in Reactor Accidents, Dubrovnik, May 1989.

B2   Belyaev, S., A.A. Borovoi, V. Demin et al. The Chernobyl source term. p. 71-91 in: Proceedings of Seminar on Comparative Assessment of the Environmental Impact of Radionuclides Released during Three Major Nuclear Accidents: Kyshtym, Windscale, Chernobyl. EUR 13574 (1991).

B3*   Begichev, S.N., A.A. Borovoi, E.V. Burlakov et al. Fuel of unit 4 reactor at the Chernobyl NPP. USSR Institute of Atomic Energy, Preprint 5268/3 (1990).

B4   Buzulukov, Yu.P. and Yu.L. Dobrynin. Release of radionuclides during the Chernobyl accident. p. 3-21 in: The Chernobyl Papers. Doses to the Soviet Population and Early Health Effects Studies, Volume I (S.E. Merwin and M.I. Balonov, eds.). Research Enterprises Inc., Richland, Washington, 1993.

B5   Balonov, M.I. Overview of doses to the Soviet population from the Chernobyl accident and the protective actions applied. p. 23-45 in: The Chernobyl Papers. Doses to the Soviet Population and Early Health Effects Studies, Volume I (S.E. Merwin and M.I. Balonov, eds.). Research Enterprises Inc., Richland, Washington, 1993.

B6   Borisov, N.B., V.V. Verbov, G.A. Kaurov et al. Contents and concentrations of gaseous and aerosol radionuclides above Unit-4 of the Chernobyl NPP. in: Questions of Ecology and Environmental Monitoring, Volume 1. Moscow, 1992.

B7   Borzilov, V.A. and N.V. Klepikova. Effect of meteorological conditions and release composition on radionuclide deposition after the Chernobyl accident. p. 47-68 in: The Chernobyl Papers. Doses to the Soviet Population and Early Health Effects Studies, Volume I (S.E. Merwin and M.I. Balonov, eds.). Research Enterprises Inc., Richland, Washington, 1993.

B8   Bebeshko, V.G., E.M. Bryzlova, E.P. Vinnizkaya et al. The state of immune system in children evacuated from town Pripyat after Chernobyl accident. in: Proceeding of the Whole-Union Conference on Human Immunology and Radiation, Gomel, 1991.

B9   Barkhudarov, R.M., L.A. Buldakov, K.I. Gordeev et al. Characterization of irradiation levels of the population in the controlled areas within the first four years after the Chernobyl NPP accident. Institute of Biophysics, Moscow (1994).

B10   Barabanova, A. and D. Osanov. The dependence of skin lesions on depth-dose distribution from beta-irradiation of people in the Chernobyl nuclear power plant accident. Int. J. Radiat. Biol. 57(4): 775-782 (1990).

B11 Bleuer, J.P., Y.I. Averkin, A.E. Okeanov et al. The epidemiological situation of thyroid cancer in Belarus. Stem Cells 15 (Suppl. 2): 251-254 (1997).

B12 Bounacer, A., R. Icker, B. Caillou et al. High prevalence of activating *ret* proto-oncogene rearrangements, in thyroid tumors from patients who had received external radiation. Oncogene 15: 1263-1273 (1997).

B13 Burkart, W., B. Grosche and A. Schoetzau. Down syndrome clusters in Germany after the Chernobyl accident. Radiat. Res. 147: 321-328 (1997).

B14 Balonov, M.I. Chernobyl dose for population of areas radiocontaminated after the Chernobyl accident. p. 207-243 in: Environmental Dose Reconstruction and Risk Implications. NCRP Proceedings No. 17 (1996).

B15 Balonov, M., P. Jacob, I. Likhtarev et al. Pathways, levels and trends of population exposure after the Chernobyl accident. p. 235-249 in: The Radiological Consequences of the Chernobyl Accident. Proceedings of the First International Conference, Minsk, Belarus, March 1996 (A. Karaoglou, G. Desmet, G.N. Kelly et al., eds.). EUR 16544 (1996).

B16 Borovoi, A.A. and A.R. Sich. The Chornobyl accident revisited, Part II: The state of the nuclear fuel located within the Chornobyl sarcophagus. Nucl. Saf. 36(1): (1995).

B17 Bigbee, W.L., R.H. Jensen, T. Veidebaum et al. Glyco-phorin A biodosimetry in Chernobyl cleanup workers from the Baltic countries. Br. Med. J. 312: 1078-1079 (1996).

B18 Baverstock, K., B. Egloff, A. Pinchera et al. Thyroid cancer after Chernobyl. Nature 359: 21-22 (1992).

B19 Buldakov, L.A., A.M. Liaginskaya, S.N. Demin et al. Radiation epidemiology study of reproductive health, oncological morbidity and mortality in population exposed to ionizing radiation caused by Chernobyl accident and industrial activities of Urals "Majak" facility. State Research Centre of Russian Federation, Institute of Biophysics, Moscow (1996).

B20 Bebeshko, V.G., V.I. Klimenko and A.A. Chumak. Clinical-immunological characteristics of leukosis in persons exposed to radiation as a result of the Chernobyl accident. Vestn. Akad. Med. Nauk SSSR 8: 28-31 (1991).

B21 Buzunov, V., N. Omelyanetz, N. Strapko et al. Chernobyl NPP accident consequences cleaning up participants in Ukraine - health status epidemiologic study - main results. p. 871-878 in: The Radiological Consequences of the Chernobyl Accident. Proceedings of the First International Conference, Minsk, Belarus, March 1996 (A. Karaoglou, G. Desmet, G.N. Kelly et al., eds.). EUR 16544 (1996).

B22* Baranov, A.E., A.K. Guskova, N.M. Nadejina et al. Chernobyl experience - biological indicators of exposure to ionizing radiation. in: Stem Cells, Volume 13 (Suppl. 1) Alpha Medica Press, 1995.

B23 Beral, V. and G. Reeves. Childhood thyroid cancer in Belarus. Nature 359: 680-681 (1992).

B24 Belyaev, S.T., A.A. Borovoi and Yu.P. Buzulukov. Work at the industrial site of the Chernobyl NPP. Current and future state of the "SARCOPHAGUS". p. 37-58 in: Selected Proceedings of the International Conference on Nuclear Accidents and Future of Energy. The Lessons of Chernobyl, Paris, April 1991.

B25 Belarussian National Report, Ministry of the Emergency Situations and Academy of Science. The consequences of the Chernobyl catastrophe in the Republic of Belarus. Minsk (1996).

B26 Balonov, M.I., G.Ya. Bruk, V.Yu. Golikov et al. Exposure of the population of Russian Federation as a result of the Chernobyl accident. p. 39-71 in: Bulletin of the All-Russian Medical and Dosimetric State Registry, Issue 7. Moscow-Obninsk, 1996.

B27 Borovoy, A. Characteristics of the nuclear fuel of power unit No. 4 of Chernobyl NPP. p. 9-20 in: Radioecological Consequences of the Chernobyl Accident (I.I. Kryshev, ed.). Nuclear Society International, Moscow, 1992.

B28 Bair, W.J. Toxicology of plutonium. Adv. Radiat. Biol. 4: 255-315 (1974).

B29 Barabanova, A. and A.K. Guskova. The diagnosis and treatment of skin injuries and other non-bone-marrow syndromes in Chernobyl victims. in: The Medical Basis for Radiation Accident Preparedness II. Clinical Experience and Follow-up since 1979 (R.C. Ricles and Sh.H. Fry, eds.). Elsevier, New York, 1990.

B30 Boice, J.D. Jr. Leukaemia, Chernobyl and epidemiology (Invited editorial). J. Radiol. Prot. 17(3): 129-133 (1997).

B31 Boice, J.D. Jr. and L.E. Holm. Radiation risk estimates for leukaemia and thyroid cancer among Russian emergency workers at Chernobyl. Letter to the editor. Radiat. Environ. Biophys. 36: 213-214 (1997).

B32 Bebeshko, V.G., E.M. Bruslova, V.I. Klimenko et al. Leukemias and lymphomas in Ukraine population exposed to chronic low dose irradiation. p. 337-338 in: Low Doses of Ionizing Radiation: Biological Effects and Regulatory Control. Contributed papers. International Conference held in Seville, Spain, November 1997. IAEA-TECDOC-976 (1997).

B33 Burkhart, W. Dose and health implications from particulate radioactivity (hot particles) in the environment. p. 121-129 in: Hot Particles from the Chernobyl Fallout. Band 16. Schriftenreihe des Bergbau- und Industriemuseums, Ost-bayern Theuern, 1988.

B34 Boice, J.D. Jr. and M. Linet. Chernobyl, childhood cancer and chromosome 21, probably nothing to worry about. Br. Med. J. 309: 139-140 (1994).

B35 Balashazy, I., I. Feher, G. Szabadyne-Szende et al. Exami-nation of hot particles collected in Budapest following the Chernobyl accident. Radiat. Prot. Dosim. 22: 263-267 (1988).

B36 Bogdevitch, I.M., V.Yu. Ageyets and I.D. Shigelskaya. Accumulation of radionuclides of ^{137}Cs and ^{90}Sr by farm crops depending on soil properties. p. 20-30 in: Belarus-Japan Symposium, Minsk, 3-5 October 1994.

B37 Balonov, M.I., M.N. Savkin, V.A. Pitkevich et al. Mean effective cumulated doses. Bulletin of the National Radiation and Epidemiological Registry, Special issue. Moscow – Obninsk, 1999.

B38 Balonov, M.I., V.Yu. Golikov, V.G. Erkin et al. Theory and practice of a large-scale programme for the decontamination of the settlements affected by the Chernobyl accident. p. 397-415 in: Proceedings of the International Seminar on Intervention Levels and Countermeasures for Nuclear Accidents. EUR-14469 (1991).

B39 Balonov, M., G. Bruk, I. Zvonova et al. Internal dose reconstruction for the Russian population after the Chernobyl accident based on human and environmental measurements. Presented at the Workshop on Environmen-tal Dosimetry, Avignon, France, 22-24 November 1999.

B40 Baranov, A., R.P. Gale, A. Guskova et al. Bone marrow transplantation after the Chernobyl nuclear accident. N. Engl. J. Med. 321(4): 205-212 (1989).

C1 Cardis, E. and A.E. Okeanov. What is feasible and desirable in the epidemiologic follow-up of Chernobyl. p. 835-850 in: The Radiological Consequences of the Chernobyl Accident. Proceedings of the First International Conference, Minsk, Belarus, March 1996 (A. Karaoglou, G. Desmet, G.N. Kelly et al., eds.). EUR 16544 (1996).

C2 Cardis, E., L. Anspaugh, V.K. Ivanov et al. Estimated long term health effects of the Chernobyl accident. p. 241-279 in: One Decade After Chernobyl. Summing up the Consequences of the Accident. Proceedings of an International Conference, Vienna, 1996. STI/PUB/1001. IAEA, Vienna, 1996.

C3 Cwikel, J., A. Abdelgani, J.R. Goldsmith et al. Two-year follow-up study of stress-related disorders among immigrants to Israel from the Chernobyl area. Environ. Health Perspect. 105 (Suppl. 6): 1545-1550 (1997).

C4 Cardis, E., E. Amoros, A. Kesminiene et al. Observed and predicted thyroid cancer incidence following the Chernobyl accident - evidence for factors influencing susceptibility to radiation induced thyroid cancer. in: Radiation and Thyroid Cancer. Proceedings of an Internal Seminar held in St. John's College, Cambridge, UK, 20-23 July 1998 (G. Thomas, A. Karaoglou and E.D. Williams, eds.). World Scientific, Singapore, 1999.

C5* Cherkasov, Yu.M., O.Yu. Novoselsky, N.I. Zhukov et al. Post accident state of Chernobyl-4. p. 11 in: International Seminar "Chernobyl Lessons – Technical Aspects", Desnogorsk, April 1996.

C6 Cardis, E. Communication to the UNSCEAR Secretariat (1998).

C7 Chvyrev, V.G. and V.I. Kolobov. Organization of the radiation-hygiene operations conducted by the military personnel to decontaminate the Chernobyl reactor after the 1986 accident. Military Medical Journal 4: 4-7 (1996).

D1 Drozdovitch, V.V. Estimation of the thyroid doses resulting from the atmospheric releases of ^{131}I during the Chernobyl accident. National Academy of Sciences of Belarus, Report IPE-37. Minsk (1998).

D2 Demidchik, E.P., I.M. Drobyshevskaya, E.D. Cherstvoy et al. Thyroid cancer in children in Belarus. p. 677-682 in: The Radiological Consequences of the Chernobyl Accident. Proceedings of the First International Conference, Minsk, Belarus, March 1996 (A. Karaoglou, G. Desmet, G.N. Kelly et al., eds.). EUR 16544 (1996).

D3 Drozdovitch, V.V. and V.F. Minenko. Assessing internal exposure in Belarus from the ingestion of radiocaesium following the Chernobyl accident. (1997, in preparation).

D4* Drozdovitch, V.V., V.F. Minenko and A.V. Ulanovsky. The results of measurement of the body-burden of radioactive materials among the population of Gomel and Mogilev oblasts. Ministry of Health of Belarus, Minsk (1989).

D5 Devell, L., S. Güntay and D.A. Powers. The Chernobyl reactor accident source term. Development of a consensus view. NEA/CSNI/R(95)24 (1996).

D6 Devell, L. Nuclide composition of Chernobyl hot particles. p. 23-34 in: Hot Particles from the Chernobyl Fallout. Band 16. Schriftenreihe des Bergbau- und Industriemuseums, Ostbayern Theuern, 1988.

D7 Devell, L. Composition and properties of plume and fallout materials from the Chernobyl accident. p. 29-46 in: The Chernobyl Fallout in Sweden - Results from a Research Programme on Environmental Radiology (L. Moberg, ed.). Swedish Radiation Protection Institute, 1991.

D8 Dreicer, M., A. Aarkrog, R. Alexakhin et al. Consequences of the Chernobyl accident for the natural and human environments. p. 319-366 in: One Decade After Chernobyl. Summing up the Consequences of the Accident. Proceedings of an International Conference, Vienna, 1996. STI/PUB/1001. IAEA, Vienna, 1996.

D9 Dolk, H., P. de Wals, Y. Gillerot et al. The prevalence at birth of Down's syndrome in 19 regions of Europe 1980-1986. p. 3-11 in: Key Issues in Mental Retardation Research. Routledge, London, 1990.

D10 Dubrova, Y.E., V.N. Nesterov, N.G. Krouchinsky et al. Human minisatellite mutation rate after the Chernobyl accident. Nature. 380: 683-686 (1996).

D11 Dibb, J.E., P.A. Mayewski, C.S. Buck et al. Beta radiation from snow. Nature 345: 25 only (1990).

D12 Dementyev, S.I., V.V. Kupstov, A.V. Titov et al. Assessment of internal intake of radioactive products in victims of the accident at the Chernobyl NPP using the results of spectrometry analysis of urine samples. p. 448-455 in: The Nearest and Delayed Consequences of the Radiation Accident at the Chernobyl NPP (L.A. Ilyin and L.A. Buldakov, eds.). Institute of Biophysics, Moscow, 1987.

D13 De Cort, M., G. Dubois, Sh.D. Fridman et al. Atlas of caesium deposition on Europe after the Chernobyl accident. EUR 16733 (1998).

E1 European Commission. Thyroid cancer in children living near Chernobyl. Expert panel report on the consequences of the Chernobyl accident (D. Williams, A. Pinchera, A. Karaoglou et al.eds.). EUR 15248 (1993).

E2 European Commission. Molecular, cellular, biological characterization of childhood thyroid cancer. Experimental collaboration project No.8 (E.D. Williams and N.D. Tronko, eds.). EUR 16538 (1996).

E3 European Commission and the Belarus, Russian and Ukrainian Ministries on Chernobyl Affairs, Emergency Situations and Health. The Radiological Consequences of the Chernobyl Accident. Proceedings of the First International Conference, Minsk, Belarus, March 1996 (A. Karaoglou, G. Desmet, G.N. Kelly et al., eds.). EUR 16544 (1996).

E4 Ericson, A. and B. Kallen. Pregnancy outcome in Sweden after the Chernobyl accident. Environ. Res. 67: 149-159 (1994).

E5 European Registration of Congenital Anomalies. EUROCAT Report 4. Surveillance of congenital anomalies 1980-1988. EUROCAT Working Group, Brussels (1991).

E6 Edvarson, K. Fallout over Sweden from the Chernobyl accident. p. 47-65 in: The Chernobyl Fallout in Sweden - Results from a Research Programme on Environmental Radiology (L. Moberg, ed.). Swedish Radiation Protection Institute, 1991.

E7 European Commission. Pathway analysis and dose distributions. Joint study project no. 5 (P. Jacob and I. Likhtarev, eds.). EUR 16541 (1996).

E8 Erkin, V.G. and O.V. Lebedev. Thermoluminiscent dosimeter measurements of external doses to the population of the Bryansk region after the Chernobyl accident. p. 289-311 in: The Chernobyl Papers. Doses to the Soviet Population and Early Health Effects Studies, Volume I (S.E. Merwin and M.I. Balonov, eds.). Research Enterprises Inc., Richland, Washington, 1993.

F1 Fujimura, K., T. Shimomura, S. Kusumi et al. Chernobyl Sasakawa Health and Medical Cooperation Project. Report 3: Haematological findings. Presented in the International Conference on Health Consequences of the Chernobyl and other Radiological Accidents, WHO, Geneva, November 1995.

F2   Fuggazola, L., S. Pilotti, A. Pinchera et al. Oncogenic rearrangements of the RET proto-oncogene in papillary thyroid carcinomas from children exposed to the Chernobyl nuclear accident. Cancer Res. 56: 5617-5620 (1995).

F3   Fischer, B., D.A. Belyi, M. Weiss et al. A multi-centre clinical follow-up database as a systematic approach to the evaluation of mid- and long-term health consequences in Chernobyl acute radiation syndrome patients. p. 625-628 in: The Radiological Consequences of the Chernobyl Accident. Proceedings of the First International Conference, Minsk, Belarus, March 1996 (A. Karaoglou, G. Desmet, G.N. Kelly et al., eds.). EUR 16544 (1996).

F4   Frank, G., P. Jacob, G. Pröhl et al. Optimal management routes for the restoration of territories contaminated during and after the Chernobyl accident. Draft Report of the European Commission (May 1997).

F5   Furmanchuk, A.W., J. Averkin, B. Egloff et al. Pathomorphological findings in thyroid cancers of children from the Republic of Belarus: a study of 86 cases occurring between 1986 ('post-Chernobyl') and 1991. Histopathology 21: 401-408 (1992).

F6   Fugazzola, L., M.A. Pierotti, E. Vigano et al. Molecular and biochemical analysis of RET/PTC4, a novel oncogenic rearrangement between RET and ELE1 genes, in a post-Chernobyl papillary thyroid cancer. Oncogene 13: 1093-1097 (1996).

F7   Feinmesser, R., E. Lubin, K. Segal et al. Carcinoma of the thyroid in children - a review. J. Pediatr. Endocrinol. Metab. 10: 561-568 (1997).

F8   Fogelfeld, L., T.K. Bauer, A.B. Schneider et al. *p53* gene mutations in radiation-induced thyroid cancer. J. Clin. Endocrinol. Metab. 81: 3039-3044 (1996).

F9   Franceschi, S., P. Boyle, P. Maisonneuve et al. The epidemiology of thyroid carcinoma. Crit. Rev. Oncog. 4: 25-53 (1993).

F10   Franssila, K.O. and H.R. Harach. Occult papillary carcinoma of the thyroid in children and young adults. A systemic autopsy study in Finland. Cancer 58: 715-719 (1986).

F11   Fliedner, T.M., H. Kindler, D. Densow et al. The Moscow-Ulm radiation accident clinical history data base. Adv. Biosciences 94: 271-279 (1994).

F12   Fliedner, T.M., B. Tibken, E.P. Hofer et al. Stem cell responses after radiation exposure: a key to the evaluation and prediction of its effects. Health Phys. 70: 787-797 (1996).

F13   Franklyn, J.A., P. Maisonneuve, M. Sheppard et al. Cancer incidence and mortality after radioiodine treatment for hyperthyroidism: a population-based cohort study. Lancet 353: 2111-2115 (1999).

G1   Golikov, V.Yu., M.I. Balonov and A.V. Ponomarev. Estimation of external gamma radiation doses to the population after the Chernobyl accident. p. 247-288 in: The Chernobyl Papers. Doses to the Soviet Population and Early Health Effects Studies, Volume I (S.E. Merwin and M.I. Balonov, eds.). Research Enterprises Inc., Richland, Washington, 1993.

G2   Guskova, A.K. Ten years after the accident at Chernobyl (retrospective assessment of clinical findings and of counter-measures for mitigating consequences). Clin. Med. 3: 5-8 (1996).

G3   Gordeev, K.I., R.M. Barkhudarov, I.K. Dibobes et al. Methodological principles of assessment of external and internal irradiation doses for the population living in the areas contaminated as a result of the Chernobyl accident. Document of the National Commission on Radiation Protection, Ministry of Health of the USSR, June 18, 1986. p. 27-37 in: Proceedings of Methodological Materials "Methodological Principles and Recommendations on How to Calculate External and Internal Doses to Populations Living in the Contaminated Areas and Exposed as a Result of the Chernobyl Accident". Ministry of Health, Institute of Biophysics, Moscow, 1991.

G4   Guskova, A.K. Radiation and the human brain. p. 23 in: International Conference on the Mental Health Consequences of the Chernobyl Disaster: Current State and Future Prospects, May 1995. Ukraine, Kiev, 1995.

G5*   Goulko, G.M., I.A. Kairo, B.G. Sobolev et al. Methods of the thyroid dose calculations for the population of the Ukraine. p. 99-103 in: Proceedings: Actual Questions of the Prognosis, Current and Retrospective Dosimetry after the Chernobyl Accident, Kiev, October 1992. Kiev, 1993.

G6   Gavrilin, Yu., V. Khrouch, S. Shinkarev et al. Chernobyl accident: reconstruction of thyroid dose for inhabitants of the Republic of Belarus. Health Phys. 76: 105-119 (1999).

G7   Gavrilin, Yu., V. Khrouch, S. Shinkarev et al. Estimation of thyroid doses received by the population of Belarus as a result of the Chernobyl accident. p. 1011-1020 in: The Radiological Consequences of the Chernobyl Accident. Proceedings of the First International Conference, Minsk, Belarus, March 1996 (A. Karaoglou, G. Desmet, G.N. Kelly et al., eds.). EUR 16544 (1996).

G8   Goulko, G.M., V.V. Chumak, N.I. Chepurny et al. Estimation of 131-I doses for the evacuees from Pripjat. Radiat. Environ. Biophys. 35: 81-87 (1996).

G9   Grinev, M.P., T.G. Litvinova, A.A. Kriminski et al. The results of the individual dosimetric monitoring in the settlements located in areas with different ^{137}Cs level of deposition density and various distribution over vicinity of a settlement. Technical Report N51-10-16/B-9178. Institute of Biophysics, Moscow (1991).

G10   Grinev, M.P., A.Kh. Mirkhaidarov, A.A. Molin et al. Individual dosimetric monitoring of the residents of some settlements in the Gomel Region. Report on Research Study, NN175/131. Scientific - Production enterprise "ALARA", Moscow (1992).

G11*   Golovko, O.V. and P.V. Izhevski. The study of reproductive behaviour in the Russian and Belarussian population exposed to radiation as a result of the Chernobyl accident. Radiat. Biol. Radioecol. 36: 3-8 (1996).

G12*   Gusev I.A. Internal exposure of thyroid caused by radioiodines in accidental staff of Chernobyl NPP. p. 440-447 in: Proceedings of All-Union Symposium - Results of the Work of Scientific and Practical Medicine Facilities for the Elimination of the Consequences of Accident at the Chernobyl Nuclear Power Plant (L.A. Ilyin and L.A. Buldakov, eds.). Moscow, 1987.

G13   Gavrilin, Yu.I., V.T. Khrusch and S.M. Shinkarev. Communication to the UNSCEAR Secretariat (1997).

G14   Gaziev, Ya.I., L.E. Nazarov, A.V. Lachikhin et al. Investigation of physical and chemical characteristics of the materials in gaseous and aerosol forms released as a result of the Chernobyl accident and assessment of the intensity of their releases. in: Proceedings of the First All-Union Conference "Radiation Aspects of the Chernobyl Accident", Obninsk 1988. Volume 1. Hydrometeoizdat Publishing House, Saint Petersburg, 1993.

G15   Gavrilin, Yu.I. Communication to the UNSCEAR Secretariat. Institute of Biophysics, Moscow (1997).

G16   Germenchuk, M. Communication to the UNSCEAR Secretariat. Belgidromet, Minsk (1998).

G17* Gavrilin, Yu.I., V.T. Khrouch and S.M. Shinkarev. Internal thyroid exposure. Chapter 4.1. p. 91-136 in: Radioiodine - Thyroid (I.I. Dedov and V.I. Dedov, eds.). Moscow Publishing House, 1996.

G18 Garcia, O., R. Cruz, M. Valdes et al. Cuban studies of children from areas affected by the Chernobyl accident. p. 313-314 in: One Decade After Chernobyl: Summing up the Consequences of the Accident. Poster CN-63/207. Book of Extended Synopses. EU/IAEA/WHO, Vienna, 1996.

G19 Grinev, M.P., I.K. Sokolova, A.V. Titov et al. Estimates of external effective doses for personnel working in the 5-km zone of the Chernobyl NPP by anthropomorphic phantom modelling. Technical Report 92-A-4-2. ALARA Limited, Moscow, 1992.

G20 Galizkaya, N.N. et al. Evaluation of the immune system of children in zone of heightening radiation. Zdravookhr. Beloruss. 6: 33-35 (1990).

G21 Gavrilov, O. (ed.). Chernobyl nuclear power plant after 1986 accident. p. 29-102 in: Hematology Review. Russian Academy of Medical Sciences (Section C). Moscow, 1996.

G22* Guskova, A.K., A.E. Baranov, A.V. Barabanova et al. Diagnosis, clinical picture and treatment of the acute radiation disease in Chernobyl accident victims. J. Ther. Arch. I: 95-103 (1989).

G23 Gilbert, E.S., R. Tarone, A. Bouville et al. Thyroid cancer rates and ^{131}I doses from Nevada atmospheric nuclear bomb tests. J. Natl. Cancer Inst. 90: 1654-1660 (1998).

G24 Grosche, B., C. Irl, A. Schoetzau et al. Perinatal mortality in Bavaria, Germany, after the Chernobyl reactor accident. Radiat. Environ. Biophys. 36: 129-136 (1997).

G25 Gusev, I. Communication to the UNSCEAR Secretariat (1998).

G26 Gerasko, V.N., A.A. Kluchnikov, A.A. Korneev et al. "Sarcophagus" Object. History, State and Prospects (A.A. Kluchnikov, ed.). Kiev, 1997.

G27 Garger, E.K., V. Kashpur, H.G. Paretzke et al. Measurement of resuspended aerosol in the Chernobyl area. Part II. Size distribution of radioactive particles. Radiat. Environ. Biophys. 36: 275-283 (1998).

H1 Hill, P. and R. Hille. Meßprogramm der Bundesrepublik Deutschland. Ergebnisse der Ganzkörpermessungen in Rußland, Weißrußland und der Ukraine in der Zeit vom 17. Mai bis 15. September 1993 und vom 8. Oktober bis zum 1. November 1993. Report Jül-3046 (1995).

H2 Hoshi, H., M. Shibata, S. Okajima, et al. [137]Cs concentration among children in areas contaminated with radioactive fallout from the Chernobyl accident: Mogilev and Gomel oblasts, Belarus. Health Phys. 67: 272-275 (1994).

H3 Harach, H.R. and E.D. Williams. Childhood thyroid cancer in England and Wales. Br. J. Cancer 72: 777-783 (1995).

H4 Hill, P. and R. Hille. Meßprogramm der Bundesrepublik Deutschland. Ergebnisse der Ganzkörpermessungen in Rußland in der Zeit vom 17. Juni bis 4. Oktober 1991. Report Jül-2610 (1992).

H5 Hill, P. and R. Hille. Meßprogramm der Bundesrepublik Deutschland. Ergebnisse der Ganzkörpermessungen in Rußland, Weißrußland und der Ukraine in der Zeit vom 13. Mai bis 6. Oktober 1992. Report Jül-3042 (1995).

H6 Hall, P., A. Mattsson and J.D. Boice Jr. Thyroid cancer after diagnostic administration of iodine-131. Radiat. Res. 145: 86-92 (1996).

H7* Havenaar, J.M., G.M. Rumyantsyeva and J. Van den Bout. Problems of mental health in Chernobyl area. J. Soc. Clin. Psychiatr. 3: 11-17 (1993).

H8 Hoshi, M., M. Yamamoto, H. Kawamura et al. Fallout radioactivity in soil and food samples in the Ukraine: measurements of iodine, plutonium, cesium, and strontium isotopes. Health Phys. 67(2): 187-191 (1994).

H9 Heriard Dubreuil, G., J. Lochard, P. Girard et al. Chernobyl post-accident management: The ETHOS project. Health Phys. 77(4): 361-372 (1999).

H10 Havenaar, J.M., G.M. Rumyantzeva, A. Kasyanenko et al. Health effects of the Chernobyl disaster: illness or illness behavior? A comparative general health survey in two former Soviet regions. Environ. Health Perspect. 105 (Suppl.): 1533-1537 (1997).

H11 Havenaar, J.M., G.M. Rumyantzeva, W. van den Brink et al. Long-term mental health effects of the Chernobyl disaster: an epidemiologic survey in two former Soviet regions. Am. J. Psychiatry 154: 1605-1607 (1997).

H12 Havenaar, J.M., W. van den Brink, J. van den Bout et al. Mental health problems in the Gomel region (Belarus): an analysis of risk factors in an area affected by the Chernobyl disaster. Psychol. Med. 26: 845-855 (1996).

H13 Hydrometeorology Committee of Ukraine. Contamination of the Ukrainian territory with radionuclides ^{90}Sr, [239]Pu+[240]Pu. Hydrometeorology Committee of Ukraine, Kiev, 1998.

H14 Hall, P., G. Berg, G. Bjelkengren et al. Cancer mortality after iodine-131 therapy for hyperthyroidism. Int. J. Cancer 50: 1-5 (1992).

H15 Hilton, J., R.S. Cambray and N. Green. Chemical fractionation of radioactive caesium in airborne particles containing bomb fallout, Chernobyl fallout and atmospheric material from the Sellafield site. J. Environ. Radioact. 2: 103-111 (1992).

H16 Hakanson, L. Radioactive caesium in fish in Swedish lakes after Chernobyl – Geographical distributions, trends, models, and remedial measures. p. 239-281 in: The Chernobyl Fallout in Sweden - Results from a Research Programme on Environmental Radiology (L. Moberg, ed.). Swedish Radiation Protection Institute, 1991.

H17 Hedinger, C., E.D. Williams and L.H. Sobin. Histopathological Typing of Thyroid Tumors. 2nd edition, WHO. Springer, Berlin, 1988.

I1 International Advisory Committee. The International Chernobyl Project. Assessment of radiological consequences and evaluation of protective measures. Technical Report. IAEA, Vienna (1991).

I2 International Atomic Energy Agency. Summary report on the post-accident review meeting on the Chernobyl accident. Safety Series No. 75-INSAG-1. IAEA, Vienna (1986).

I3 Izrael, Y., E. Kvasnikova, I. Nazarov et al. Global and regional pollution of the former European USSR with caesium-137. Meteorol. Gidrol. 5: 5-9 (1994).

I4 Ilyin, L., M.I. Balonov, L. Buldakov et al. Radio-contamination patterns and possible health consequences of the accident at the Chernobyl nuclear power station. J. Radiol. Prot. 10: 3-29 (1990).

I5 Ilyin, L.A. Realities and Myths of Chernobyl. ALARA Limited, Moscow, 1994.

I6* Izrael, Yu.A., S.M. Vakulovskii, V.A. Vetrov et al. Chernobyl: Radioactive Contamination of the Environment. Gidrometeoizdat, Leningrad, 1990.

I7 International Atomic Energy Agency. The Chernobyl accident: updating of INSAG-1. Safety Series No. 75-INSAG-7. IAEA, Vienna (1992).

I8   Ivanov, V.K. and A.F. Tsyb. Morbidity, disability and mortality among persons affected by radiation as a result of the Chernobyl accident: radiation risks and prognosis. in: Proceedings of the 4th Symposium on Chernobyl-related Health Effects, Tokyo, 1996.

I9   Ivanov, V.K., A.F. Tsyb, A.I. Gorsky et al. Leukaemia and thyroid cancer in emergency workers of the Chernobyl accident: estimation of radiation risks (1986-1995). Radiat. Environ. Biophys. 36: 9-16 (1997).

I10   Ivanov, E.P., G. Tolochko, V.S. Lazarev et al. Child leukaemia after Chernobyl. Nature 365: 702 only (1995).

I11   Ivanov, E.P., D. Parkin, G. Tolochko et al. Haematological diseases in the Belarus Republic after the Chernobyl accident. Presented at the International Conference on Health Consequences of the Chernobyl and other Radiological Accidents, WHO, Geneva, November 1995.

I12   Ivanov, V.K., A.F. Tsyb, E.V. Nilova et al. Cancer risks in the Kaluga oblast of the Russian Federation 10 years after the Chernobyl accident. Radiat. Environ. Biophys. 36: 161-167 (1997).

I13* Ilyin, L.A., G.V. Arkhangelskay, Yu.O. Konstantinov et al. Radioiodine in the Problem of Radiation Safety. Atomizdat, Moscow, 1972.

I14   Ivanov, V. Health status and follow-up of the liquidators in Russia. p. 861-870 in: The Radiological Consequences of the Chernobyl Accident. Proceedings of the First International Conference, Minsk, Belarus, March 1996 (A. Karaoglou, G. Desmet, G.N. Kelly et al., eds.). EUR 16544 (1996).

I15   International Atomic Energy Agency, European Commission and World Health Organization. One Decade After Chernobyl. Summing up the Consequences of the Accident. Proceedings of an International Conference, Vienna, 1996. STI/PUB/1001. IAEA, Vienna (1996)

I16   Ito, M., S. Yamashita, K. Ashizawa et al. Childhood thyroid diseases around Chernobyl evaluated by ultrasound examination and fine needle aspiration cytology. Thyroid 5(5): 365-368 (1995).

I17   International Commission on Radiological Protection. Limits for intakes of radionuclides by workers. ICRP Publication 30. Pergamon Press, Oxford, 1979.

I18   Ivanov, V.K. Response to the letter to the editor by J.D. Boice and L.E. Holm. Radiat. Environ. Biophys. 36: 305-306 (1998).

I19   Inskip, P.D., M.F. Hartshorne, M. Tekkel et al. Thyroid nodularity and cancer among Chernobyl cleanup workers from Estonia. Radiat. Res. 147: 225-235 (1997).

I20   Ivanov, E.P., G.V. Tolochko, L.P. Shuvaeva et al. Childhood leukemia in Belarus before and after the Chernobyl accident. Radiat. Environ. Biophys. 35: 75-80 (1996).

I21   Ito, M., T. Seyama, K.S. Iwamoto et al. Activated RET oncogene in thyroid cancers of children from areas contaminated by Chernobyl accident. Lancet 344: 259 (1994).

I22   International Atomic Energy Agency. Present and future environmental impact of the Chernobyl accident. IAEA-TECDOC (2000, to be published).

I23   Ivanov, V. Communication to the UNSCEAR Secretariat (1999).

I24   Izrael, Yu.A., M. De Cort, A.R. Jones et al. The atlas of caesium-137 contamination of Europe after the Chernobyl accident. p. 1-10 in: The Radiological Consequences of the Chernobyl Accident. Proceedings of the First International Conference, Minsk, Belarus, March 1996 (A. Karaoglou, G. Desmet, G.N. Kelly et al., eds.). EUR 16544 (1996).

I25* Ilyin, L.A., V.P. Krjuchkov, D.P. Osanov et al. Exposure levels for persons involved in recovery operations following the Chernobyl accident in 1986-87 and dosimetric data verification. J. Radiat. Biol. Radioecol. 35(6): 803-828 (1995).

I26   Ito, M., S. Yamashita, K. Ashizawa et al. Histopathological characteristics of childhood thyroid cancer in Gomel, Belarus. Int. J. Cancer 65: 29-33 (1996).

I27   Ivanov, V.K., A.I. Gorski, V.A. Pitkevitch et al. Risk of radiogenic thyroid cancer in Russia following the Chernobyl accident. p. in: Radiation and Thyroid Cancer. Proceedings of an Internal Seminar held in St. John's College, Cambridge, UK, 20-23 July 1998 (G. Thomas, A. Karaoglou and E.D. Williams, eds.). World Scientific, Singapore, 1999.

I28   Ilyin, L.A. Public dose burdens and health effects due to the Chernobyl accident. Paper Presented at the International Meeting Organized Jointly by Soviet and French Nuclear Societies with the Participation of the European Nuclear Society, Paris, April 1991.

I29   Ivanov, V.K., A.F. Tsyb, A.P. Konogorov et al. Case-control analysis of leukaemia among Chernobyl accident emergency workers residing in the Federation, 1986-1993. J. Radiol. Prot. 17: 137-157 (1997).

I30   Ilyin, L.A. Doses to population and medical consequences as a result of the Chernobyl accident. p. 35-47 in: Information Bulletin N10. Center of Social Information on Atomic Energy, Moscow (1992).

I31   Ivanov, E.P., G.V. Tolochko, L.P. Shuvaeva et al. Childhood leukemia in Gomel, Mogilev, Vitebsk and Grodno oblasts (regions) of Belarus prior and after Chernobyl disaster. Hematology Reviews 9: 169-179 (1995).

I32   Ivanov, V.K., E.M. Rastopchin, A.I. Gorsky et al. Cancer incidence among liquidators of the Chernobyl accident: solid tumors, 1986-1995. Health Phys. 74: 309-315 (1998).

I33   Ivanov, V.K., A.F. Tsyb, A.I. Gorsky et al. Thyroid cancer among "liquidators" of the Chernobyl accident. Br. J. Radiol. 70: 937-941 (1997).

I34   Ivanov, V.K., A.F. Tsyb and S.I. Ivanov. Liquidators of the Chernobyl Catastrophe: Radiation Epidemiological Analysis of Medical Consequences. Galanis, Moscow, 1999.

I35   Institute of Physics, Academy of Sciences of Lithuania. Atmospheric Physics, collected volumes. Volume 14. Vilnus, Mokslas, 1989.

I36   International Commission on Radiological Protection. Age-dependent doses to members of the public from intake of radionuclides: Part I. Ingestion dose coefficients. Annals of the ICRP 23(3/4). ICRP Publication 56. Pergamon Press, Oxford, 1990.

I37   Igumnov, S.A. The prospective investigation of a psychological development of children exposed to ionizing radiation in utero as a result of the Chernobyl accident. Ph.D. dissertation (1999).

J1   Jacob, P., P. Roth, V. Golikov et al. Exposures from external radiation and from inhalation of resuspended material. p. 251-260 in: The Radiological Consequences of the Chernobyl Accident. Proceedings of the First International Conference, Minsk, Belarus, March 1996 (A. Karaoglou, G. Desmet, G.N. Kelly et al., eds.). EUR 16544 (1996).

J2   Jacob, P., G. Goulko, W.F. Heidenreich et al. Thyroid cancer risk to children calculated. Nature 392: 31-32 (1998).

J3 Jaracz, P., E. Piasecki, S. Mirowski et al. Analysis of gamma-radioactivity of "Hot Particles" released after the Chernobyl accident. I. Calculation of fission products in hot particles (A detective approach). J. Radioanal. Nucl. Chem. 141(2): 221-242 (1990).

J4 Jaracz, P., E. Piasecki and S. Mirowski. Analysis of gamma-radioactivity of "Hot Particles" released after the Chernobyl accident. II. An interpretation. J. Radioanal. Nucl. Chem. 141(2): 243-259 (1990).

J5 Jacob, P., Y. Kenigsberg, I. Zvonova et al. Childhood exposure due to the Chernobyl accident and thyroid cancer risk in contaminated areas of Belarus and Russia. Br. J. Cancer 80(9): 1461-1469 (1999).

J6 Jaworowski, Z. and L. Kownacka. Tropospheric and stratospheric distribution of radioactive iodine and cesium after the Chernobyl accident. J. Environ. Radioact. 6: 145-150 (1988).

K1 Kirchner, G. and C.C. Noack. Core history and nuclide inventory of the Chernobyl core at the time of accident. Nucl. Saf. 29: 1-5 (1988).

K2 Kutkov, V.A. and Yu.B. Muravyev. Dosimetry of internal exposure to hot particles from the Chernobyl accident. Med. Radiol. 4: 4-9 (1994).

K3 Kesminiene, A.Z., J. Kurtinaitis, Z. Vilkeliene et al. Thyroid nodularity among Chernobyl cleanup workers from Lithuania. Acta Med. Lituanica 2: 51-54 (1997).

K4 Körblein, A. and H. Küchenhoff. Perinatal mortality in Germany following the Chernobyl accident. Radiat. Environ. Biophys. 36(1): 3-7 (1997).

K5 Kylkova, L.V., E. Ispenkov, I. Gutkovsky et al. Epidemiological monitoring of the state of children's health living in the radioactive contaminated territory of Gomel district. Med. Radiol. 2: 12-15 (1996).

K6 Kvachyeva, Yu.E. and P.V. Vlasov. Pathomorphological investigation of circulatory organs in persons who died of acute leukaemia as a result of the Chernobyl accident. Arkh. Anat. Gistol. Ehmbriol. 2: 60-63 (1989).

K7 Kenigsberg, Ya., M. Belli, F. Tikhomirov et al. Exposures from consumption of forest produce. p. 271-281 in: The Radiological Consequences of the Chernobyl Accident. Proceedings of the First International Conference, Minsk, Belarus, March 1996 (A. Karaoglou, G. Desmet, G.N. Kelly et al., eds.). EUR 16544 (1996).

K8 Kutkov, V.A., I.A. Gusev, S.I. Dyementyev et al. Internal exposures from the Chernobyl accident caused by inhalation of radioactive aerosols. p. 77-78 in: Proceedings of All-Russian Conference on Radioecological, Medical and Socio-economical Consequences of the Chernobyl Accident. Rehabilitation of Territory and Population, May 1995. Golitsino, Moscow, 1995.

K9 Kesminiene, A.Z., J. Kurtinaitis and G. Rimdeika. The study of Chernobyl clean-up workers from Lithuania. Acta Med. Lituanica 2: 55-61 (1997).

K10 Kumpusalo, L., E. Kumpusalo, S. Soimakallio et al. Thyroid ultrasound findings 7 years after the Chernobyl accident. Acta Radiol. 37: 904-909 (1996).

K11 Kazakov, V.S., E.P. Demidchik and L.N. Astakhova. Thyroid cancer after Chernobyl [see comments]. Nature 359: 21 only (1992).

K12 Kisselev, A.N. and K.P. Checherov. Lava-like fuel-containing masses in Unit 4 of Chernobyl NPP (based on research in the period 1986-1993). p. 54-63 in: OECD Documents "Sarcophagus Safety '94". The State of the Chernobyl Nuclear Power Plant Unit 4. Proceedings of an International Symposium, Zeleny Mys, Chernobyl, Ukraine, March 1994. NEA/OECD, France, 1995.

K13 Kesminiene, A. Communication to the UNSCEAR Secretariat (1997).

K14 Klugbauer, S., E. Lengfelder, E.P. Demidchik et al. High prevalence of RET rearrangements in thyroid tumours of children from Belarus after the Chernobyl reactor accident. Oncogene 11: 2459-2467 (1995).

K15 Kisselev, A.N., A.I. Surin and K.P. Checherov. Post-accident research on Unit 4 of the Chernobyl nuclear power plant. At. Energ. 80(4): 240-247 (1996).

K16 Khrouch, V., Yu. Gavrilin, S. Shinkarev et al. Case-control Study of Chernobyl-related Thyroid Cancer Among Children of Belarus: Estimation of Individual Doses. Part I. (2000, submitted for publication).

K17 Kutkov, V.A., I.A. Gusev and S.I. Dementiev. Internal exposure of the staff involved in 1986 in the liquidation of the accident of the Chernobyl nuclear power plant. Presented in the International Conference on Health Consequences of the Chernobyl and other Radiological Accidents, WHO, Geneva, November 1995.

K18 Kutkov, V.A., I.A. Gusev and S.I. Dementiev. Doses of internal irradiation of the persons involved in April-May 1986 in the liquidation of the consequences of the accident on Chernobyl nuclear power plant. Med. Radiol. 3: (1996).

K19* Krjutchkov, V.P., A.V. Nosovsky et al. Retrospective Dosimetry of Persons Involved in Recovery Operations Following the Accident at the Chernobyl NPP. Seda-Style, Kiev, 1996.

K20* Kiselev, A.N. Post-accident balance of nuclear fuel in the fourth block of the Chernobyl NPP. p. 17 in: IAE-5716/3. RRC-Kurchatov's Institute, Moscow, 1994.

K21* Kiselev, A.N., A.I. Surin and K.P. Checherov. The results of additional studies of lava localities in the fourth block of the Chernobyl NPP. p. 59 in: IAE-5783/3. RRC-Kurchatov's Institute, Moscow, 1994.

K22 Kenigsberg, Ya., V. Minenko, E. Buglova et al. Collective doses to the Belorussian population exposed due to the Chernobyl accident and prognosis of stochastic effects. p. 61-70 in: Proceedings of Scientific Publications "Ten Years of Chernobyl. Medical Consequences", Issue 2. Minsk, 1995.

K23 Kholosha, V.I., N.G. Koval'skij and A.A. Babich. Social, economic, institutional and political impacts. Report for Ukraine. p. 429-444 in: One Decade After Chernobyl. Summing up the Consequences of the Accident. Proceedings of an International Conference, Vienna, 1996. STI/PUB/1001. IAEA, Vienna, 1996.

K24 Kosianov, A.D. and V.G. Morozov. Characteristic of immunological state of liquidators of industrial accident with radiation components. p. 120-121 in: Proceedings of the Whole-Union Conference on Human Immunology and Radiation, Gomel, 1991.

K25 Khmara, I.M., L.N. Astakhova, L.L. Leonova et al. Indices of immunity in children suffering with autoimmune thyroiditis. J. Immunol. 2: 56-58 (1993).

K26* Konchalovsky, M.V., A.E. Baranov and V.Yu. Soloviev. Dose response of neutrophils and lymphocytes in whole-body homogeneous human gamma-irradiation (based on Chernobyl accident data). J. Med. Radiol. Radiat. Safety 36(1): 29-31 (1991).

K27 Klugbauer, S. and H. Rabes. The transcription coactivator HTIF1 and a related protein are fused to the RET receptor tyrosine kinase in childhood papillary thyroid carcinomas. Oncogene 18: 4388-4393 (1999).

K28 Kasatkina, E.P., D.E. Shilin, A.L. Rosenbloom et al. Effects of low level radiation from the Chernobyl accident in a population with iodine deficiency. Eur. J. Pediatr. 156: 916-920 (1997).

K29   Kodaira, M., C. Satoh, K. Hiyama et al. Lack of effects of atomic bomb radiation on genetic instability of tandem-repetitive elements in human germ cells. Am. J. Hum. Genet. 57: 1263-1266 (1995).

K30   Khrouch, V.T., Yu.I. Gavrilin, Yu.O. Konstantinov et al. Characteristics of the radionuclides inhalation intake. p. 76 in: Medical Aspects of the Accident at the ChNPP. Proceedings of the International Conference, Kiev, May 1988. Zdorovie Publishing House, 1988.

K31   Klugbauer, S., E.P. Demidchik, E. Lengfelder et al. Detection of a novel type of RET rearrangement (PTC5) in thyroid carcinomas after Chernobyl and analysis of involved RET-fused gene RFG5. Cancer Res. 58: 198-203 (1998).

K32   Krajewski, P. Effect of administering stable iodine to the Warsaw population to reduce thyroid content of iodine-131 after the Chernobyl accident. p. 257-271 in: Recovery Operations in the Event of a Nuclear Accident or Radiological Emergency. Proceedings of a Symposium, Vienna, 1989. STI/PUB/826. IAEA, Vienna, 1990.

K33   Krajewski, P. Assessment of effective dose equivalent in thyroid for Polish population due to iodine-131 intakes after the Chernobyl accident. Estimation of thyroid blocking effect with stable iodine. Pol. J. Endocrinol. 42(2): 189-202 (1991). (In Polish).

K34   Kuriny, V.D., Yu.A. Ivanov, V.A. Kashparov et al. Particle-associated fall-out in the local and intermediate zones. Ann. Nucl. Energy 20(6): 415-420 (1993).

K35   Kutkov, V.A., Z.S. Arefieva, Yu.B. Muravev et al. Unique form of airborne radioactivity: nuclear fuel "hot particles" released during the Chernobyl accident. p. 625-630 in: Environmental Impact of Radioactive Releases. Proceedings of a Symposium, Vienna, 8-12 May 1995. STI/PUB/971. IAEA, Vienna, 1995.

K36   Kutkov, V.A. Application of human respiratory tract models for reconstruction of the size of aerosol particles through the investigation of radionuclides behaviour in the human body. Radiat. Prot. Dosim. 79: 265-268 (1998).

K37   Kruger, F.W., L. Albrecht, E. Spoden et al. Der Ablauf des Reaktorunfalls Tchernobyl 4 und die weitraumige Verfrachtung des freigesetzten Materials: Neuere Erkenntnisse und die Bewertung. p. 7-22 in: Zehn Jahre nach Tschernobyl, eine Bilanz. Gustav Fischer, 1996.

K38   Korneev, S., S. Tretyakevich, V. Minenko et al. External exposure. Retrospective dosimetry and dose reconstruction. Final report of the ECP10. p. 104-110 in: EUR-16540EN (1996).

K39   Krajewski, P. Evaluation and verification of dose assessment model for radioiodine and radiocaesium environmental releases. PhD. Thesis. Warsaw (1999).

K40   Kenigsberg, J., E. Buglova, H.G. Paretzke et al. Perspectives of development of thyroid cancers in Belarus. p. 771-775 in: The Radiological Consequences of the Chernobyl Accident. Proceedings of the First International Conference, Minsk, Belarus, March 1996 (A. Karaoglou, G. Desmet, G.N. Kelly et al., eds.). EUR 16544 (1996).

K41   Kofler, A., T.H. Abelin, I. Prudyves et al. Factors related to latency period in post Chernobyl carcinogenesis. in: Radiation and Thyroid Cancer. Proceedings of an Internal Seminar held in St. John's College, Cambridge, UK, 20-23 July 1998 (G. Thomas, A. Karaoglou and E.D. Williams, eds.). World Scientific, Singapore, 1999.

K42   Kelly, G.N. and V.M. Shershakov (eds.). Environmental contamination, radiation doses and health consequences after the Chernobyl accident. Radiat. Prot. Dosim. 64(1/2): (1996).

K43   Kownacka, L. and Z. Jaworowski. Nuclear weapon and Chernobyl debris in the troposphere and lower stratosphere. Sci. Total Environ. 144: 201-215 (1994).

K44   Kryuchkov, V. Communication to the UNSCEAR Secretariat (1999).

K45   Kreja, L., K.M. Greulich, T.M. Fliedner et al. Stable chromosomal aberrations in haemopoietic stem cells in the blood of radiation accident victims. Int. J. Radiat. Biol. 75(10): 1241-1250 (1999).

K46   Kolominsky, Y., S. Igumonov and V. Drozdovitch. The psychological development of children from Belarus exposed in the prenatal period to radiation from the Chernobyl atomic power plant. J. Child Psychol. Psychiat. 40(2): 299-305 (1999).

K47   Konoplev, A.V., A.A. Bulgakov, V.G. Zhirnov et al. Study of the behavior of Cs-137 and Sr-90 in Svyatoe and Kozhanovskoe lakes in the Bryansk region. Meteorol. Gidrol. 11: (1998).

L1   Lyaginskaya, A.M., L.A. Buldakov, A.T. Ivannikov et al. Modified effect of ferroxin on kinetics of exchange of ^{134}Cs and ^{137}Cs in children. p. 34 in: Abstracts of All Union Conference on Actual Problems of Internal Dosimetry, Gomel, 1989.

L2   Likhtarev, I.A., L. Kovgan, O. Bobilova et al. Main problem in post-Chernobyl dosimetry. p. 27-51 in: Assessment of the Health and Environmental Impact from Radiation Doses due to Released Radionuclides. NIRS-M-102 (1994).

L3   Lee, T.R. Environmental stress reactions following the Chernobyl accident. p. 283-310 in: One Decade After Chernobyl. Summing up the Consequences of the Accident. Proceedings of an International Conference, Vienna, 1996. STI/PUB/1001. IAEA, Vienna, 1996.

L4   Lee, T.R. Social and psychological consequences of the Chernobyl accident: an overview of the first decade. in: Proceedings of Conferences on Health Consequences of the Chernobyl and other Radiological Accidents, Geneva, 1995. WHO, Geneva (1996).

L5   Lazuk, G.I., K.A. Bedelbaeva and J.N. Fomina. The cytogenetic effects of the additional radiation exposure at low doses of ionizing radiation. Zdravookhr. Beloruss. 6: 38-41 (1990).

L6   Likhtarev, I.A., B.G. Sobolev, I.A. Kairo et al. Thyroid cancer in the Ukraine. Nature 375: 365 only (1995).

L7   Likhtarev, I.A. Exposure of different population groups of Ukraine after the Chernobyl accident and main health-risk assessments. Presented at "Ten Years after Chernobyl, a Summation". Seminar of the Federal Office for Radiation Protection (BfS) and the Commission on Radiological Protection (SSK), Munich, March 1996.

L8   Lazjuk, G.I., D.L. Nikolaev and I.V. Novikova. Changes in registered congenital anomalies in the Republic of Belarus after the Chernobyl accident. Stem Cells 15 (Suppl. 2): 255-260 (1997).

L9   Likhtarev, I.A., V.V. Chumak and V.S. Repin. Retrospective reconstruction of individual and collective external gamma doses of population evacuated after the Chernobyl accident. Health Phys. 66: 643-652 (1994).

L10   Likhtarev, I., B. Sobolev, I. Kairo et al. Results of large scale thyroid dose reconstruction in Ukraine. p. 1021-1034 in: The Radiological Consequences of the Chernobyl Accident. Proceedings of the First International Conference, Minsk, Belarus, March 1996 (A. Karaoglou, G. Desmet, G.N. Kelly et al., eds.). EUR 16544 (1996).

L11 Likhtarev, I.A., N.K. Shandala, G.M. Gulko et al. Hygienic assessment of thyroid doses of UKSSR population after the Chernobyl accident. Vestn. Akad. Med. Nauk SSSR 8: 44-47 (1991).

L12 Likhtarev, I.A., N.K. Shandala, G.M. Gulko et al. Ukrainian thyroid doses after the Chernobyl accident. Health Phys. 64(6): 594-599 (1993).

L13* Likhtarev, I.A., G.M. Gulko, B.G. Sobolev et al. The main methods for the thyroid dose reconstruction in the Ukraine. p. 92-98 in: Proceedings: Actual Questions of the Prognosis, Current and Retrospective Dosimetry after the Chernobyl Accident, Kiev, October 1992. Kiev, 1993.

L14 Likhtarev, I.A., G.M. Gulko, B.G. Sobolev et al. Reliability and accuracy of the 131-I thyroid activity measurements performed in the Ukraine after the Chernobyl accident in 1986. GSF-19/93 (1993).

L15 Likhtarev, I.A., G.M. Gulko, I.A. Kairo et al. Thyroid doses resulting from the Ukraine Chernobyl accident. Part I: Dose estimates for the population of Kiev. Health Phys. 66: 137-146 (1994).

L16 Likhtarev, I.A., N.K. Shandala, A.E. Romanenko et al. Radioactive iodine concentrations in elements of the environment and evaluation of exposure doses to the thyroid among inhabitants of Kiev after the Chernobyl accident. p. 351-353 in: Environmental Contamination Following a Major Nuclear Accident, Volume 2. Proceedings of a Symposium, Vienna, October 1989. STI/PUB/825. IAEA, Vienna, 1990.

L17* Likhtarev, I.A., N.K. Shandala, G.M. Gulko et al. Prognosis of possible medical-genetic consequences to residents of Kiev as a result of the Chernobyl accident. p. 134 in: International Conference on Biological and Radioecological Aspects of Consequences of the Chernobyl Accident, Zeleny Mys, September 1990. Moscow, 1990.

L18 Lloyd, D.C., A.A. Edwards, A.V. Sevan'kaev et al. Retrospective dosimetry by chromosomal analysis. p. 965-973 in: The Radiological Consequences of the Chernobyl Accident. Proceedings of the First International Conference, Minsk, Belarus, March 1996 (A. Karaoglou, G. Desmet, G.N. Kelly et al., eds.). EUR 16544 (1996).

L19 Lomat, L., G. Galburt, M.R. Quastel et al. Incidence of childhood disease in Belarus associated with the Chernobyl accident. Environ. Health Perspect. 105 (Suppl. 6): 1529-1532 (1997).

L20 Lechat, M.F. and H. Dolk. Registries of congenital anomalies: EUROCAT. Environ. Health Perspect. 101 (Suppl. 2): 153-157 (1993).

L21 Likhtarev, I.A., L.N. Kovgan, S.E. Vavilov et al. Internal exposure from the ingestion of foods contaminated by Cs-137 after the Chernobyl accident. Report 1. General model: ingestion doses and countermeasure effectiveness for the adults of Rovno Oblast of Ukraine. Health Phys. 70(3): 297-317 (1996).

L22 Likhtarev, I.A. Dosimetry catalogue of the Ukrainian settlements contaminated as a result of the Chernobyl accident. Official document of the Ukrainian Ministry of Public Health, Kiev (1997).

L23 Likhtariov, I.A., V.S. Repin, O.A. Bondarenko et al. Radiological effects after inhalation of highly radioactive fuel particles produced by the Chernobyl accident. Radiat. Prot. Dosim. 59(4): 242-254 (1995).

L24 Lettner, H. Post-Chernobyl distribution of ^{137}Cs concentration in soil and environmental samples in mountainous and plain areas of the Province of Salzburg, Austria. p. 193-203 in: Environmental Contamination Following a Major Nuclear Accident. IAEA, Vienna, 1990.

L25 Likhtarev, I.A., G.M. Gulko, B.G. Sobolev et al. Thyroid dose assessment for the Chernigov region (Ukraine): estimation based on ^{131}I thyroid measurements and extrapolation of the results to districts without monitoring. Radiat. Environ. Biophys. 33: 149-166 (1994).

L26 Linge, I.I., E.M. Melekhov and O.A. Pavlovsky. Systematic and information supply for analysis of medico-demographic situation in regions of Russia contaminated after Chernobyl accident. Preprint N 51-30-94. Nuclear Safety Institute, Moscow (1994).

L27 Ljaginskaja, A.M. and V.A. Osipov. Comparison of estimation of reproductive health of population from contaminated territories of Bryansk and Ryazan regions of the Russian Federation. p. 91 in: Thesis on the Radio-ecological, Medical and Socio-economical Consequences of the Chernobyl Accident. Rehabilitation of Territories and Populations, Moscow, 1995.

L28 Ljaginskaja, A.M., P.V. Izhewskij and O.V. Golovko. The estimate reproductive health status of population exposed in low doses in result of Chernobyl disaster. p. 62 in: IRPA 9. Proceedings of the International Congress on Radiation Protection, Volume 2 (1996).

L29 Ljaginskaja, A.M., O.V. Golovko, V.A. Osipov et al. Criteria for estimation of early deterministic effects in exposed populations. p. 73-74 in: Third Congress of Radiation Research, Moscow, 1997.

L30 Lazutka, J.R. and V. Dedonyte. Increased frequency of sister chromatid exchanges in lymphocytes of Chernobyl clean-up workers. Int. J. Radiat. Biol. 67: 671-676 (1995).

L31 Lazutka, J.R., J. Mierauskiene, G. Slapsyte et al. Cytogenetic study in peripheral blood lymphocytes of Chernobyl clean-up workers. Acta Med. Lituanica 2: 62-67 (1997).

L32 Little, J. The Chernobyl accident congenital anomalies and other reproductive outcomes. Paediatr. Perinat. Epidemiol. 7: 121-151 (1993).

L33 Lafuma, J., J.-C. Nenot, M. Morin et al. Respiratory carcinogenesis in rats after inhalation of radioactive aerosols of actinides and lanthanides in various physico-chemical forms. in: Experimental Lung Cancer (E. Karbe and J.F. Park, eds.). Springer, Berlin, 1974.

L34 Lafuma, J., J.-C. Nenot, M. Morin et al. An experimental study on the toxicity of several radionuclides inhaled at different doses. in: Criteria for Radiation Protection. Third European Congress of the International Radiation Protection Association, Amsterdam, May 1975.

L35 Little, J.B., B.N. Grossman and W.F. O'Toole. Factors influencing the induction of lung cancer in hamsters by intratracheal administration of Po-210. in: Radionuclide Carcinogenesis (C.L. Sanders et al., eds.). AEC Symposium Series 29. CONF-720505 (1973).

L36 Likhtarev, I., L. Kovgan, D. Novak et al. Effective doses due to external irradiation from the Chernobyl accident for different population groups of Ukraine. Health Phys. 70(1): 87-98 (1996).

L37 Leon, D.A. and V.M. Shkolnikov. Social stress and the Russian mortality crises. J. Am. Med. Assoc. 279: 790-791 (1998).

L38 Leon, D.A., L. Chenet, V.M. Shkolnikov et al. Huge variation in Russian mortality rates 1984-94: artefact, alcohol, or what? Lancet 350: 383-388 (1997).

L39 Lupoli, G., G. Vitale, M. Caraglia et al. Familial papillary thyroid microcarcinoma: a new clinical entity. Lancet 353 (9153): 637-639 (1999).

L40 Lehtinen, S., S. Luokkawn, R. Raunemaa et al. Hot particles of Chernobyl in Finland. p. 77-82 in: Hot Particles from the Chernobyl Fallout. Band 16. Schriftenreihe des Bergbau- und Industrie-museums. Ostbayern Theuern, 1988.

L41 Littlefield, L.G., A.F. McFee, S.I. Salomaa et al. Do recorded doses overestimate true doses received by Chernobyl clean up workers? Results of cytogenetic analyses of Estonian workers by fluorescence in situ hybridization. Radiat. Res. 150: 237-249 (1998).

L42 Likhtarev, I.A., I.A. Kayro, V.M. Shpak et al. Radiation-induced and background thyroid cancer of Ukrainian children (dosimetric approach). Int. J. Radiat. Med. 3-4(3-4): 51-66 (1999).

L43 Likhtarev, I.A., I.A. Kayro, V.M. Shpak et al. Thyroid retrospective dosimetry problems in Ukraine: achievements and delusions. p. 71-78 in: Radiation and Thyroid Cancer. Proceedings of an Internal Seminar held in St. John's College, Cambridge, UK, 20-23 July 1998 (G. Thomas, A. Karaoglou, E.D. Williams, eds.). World Scientific, Singapore, 1999.

M1 Ministry of Atomic Energy, Moscow. Standards of Radiation Safety SRS-76 and General Sanitary Guidance of Working with Radioactive Materials and Sources, GSG-72/80. Energoatomizdat, Moscow, 1981.

M2 Meckbach, R. and V.V. Chumak. Reconstruction of the external dose of evacuees from the contaminated areas based on simulation modelling. p. 975-984 in: The Radiological Consequences of the Chernobyl Accident. Proceedings of the First International Conference, Minsk, Belarus, March 1996 (A. Karaoglou, G. Desmet, G.N. Kelly et al., eds.). EUR 16544 (1996).

M3 Ministry of the Chernobyl Affairs of Ukraine. Ten years after the accident at the Chernobyl nuclear power plant. Ukrainian National Report, Kiev (1996).

M4 Minenko, V.F., V.V. Drozdovich and S.S. Tret'yakevich. Methodological approaches to calculation of annual effective dose for the population of Belarus. p. 246-252 in: Bulletin of the All-Russian Medical and Dosimetric State Registry, Issue No. 7. Moscow-Obninsk, 1996.

M5 Mettler, F.A. Jr., H.D. Royal, J.R. Hurley et al. Administration of stable iodine to the population around the Chernobyl nuclear power plant. J. Radiol. Prot. 12(3): 159-165 (1992).

M6 Michaelis, J., U. Kaletsch, W. Burkart et al. Infant leukaemia after the Chernobyl accident. Nature 387: 246 (1997).

M7 Morgenstern, W., V.K. Ivanov, A.I. Michalski et al. Mathematical Modelling with Chernobyl Registry Data. Registry and Concepts. Springer Verlag, Heidelberg, 1995.

M8 Meslé, F., V. Shkolnikov, V. Hertrich et al. Tendences récentes de la mortalité par cause en Russie, 1965-1993. Dossiers and Recherches No. 50. INED, Paris, 1995.

M9 Mangano, J.J. Chernobyl and hypothyroidism. Letter. Lancet. 347: 1482-1483 (1996).

M10 Mettler, F.A. Jr., M.R. Williamson, H.D. Royal et al. Thyroid nodules in the population living around Chernobyl. J. Am. Med. Assoc. 268: 616-619 (1992).

M11 Ministry of Health of the USSR. Estimation of the annual total effective equivalent doses received by the population living in controlled areas in the Russian Federation, Ukraine and Belarus contaminated with radionuclides as a result of the Chernobyl accident. Methodological guidance. N5792-91 (Approved on July 5, 1991). Ministry of Health of the USSR, Moscow (1991).

M12 Ministry of Health of the USSR. Protocol of a meeting of the Ministry of Health of the former USSR (1987).

M13 Ministry of Health of Ukraine. Chernobyl State Registry Data. Kiev, 1999.

M14 Moberg, L. (ed.). The Chernobyl Fallout in Sweden - Results from a Research Programme on Environmental Radiology. Swedish Radiation Protection Institute, 1991.

M15 Merwin, S.E. and M.I. Balonov (eds.). The Chernobyl Papers. Doses to the Soviet Population and Early Health Effects Studies, Volume I. Research Enterprises Inc., Richland, Washington, 1993.

M16 Ministry of Health of Ukraine. Reconstruction and prognosis of doses for inhabitants of the Ukrainian areas contaminated following the accident at the Chernobyl NPP. Methodic directives. Adopted on 10 December 1997. Kiev (1997).

M17 Ministry of Health of Russia. Reconstruction of mean effective cumulative dose in 1986-1995 for inhabitants of settlements contaminated following the accident at the Chernobyl NPP in 1986. Methodic directives. Adopted on 12 November 1996. Moscow (1996).

M18 Mikhalevich, L.S., D.C. Lloyd, A.A. Edwards et al. Dose estimates made by dicentric analysis for some Belarussian children irradiated by the Chernobyl accident. Radiat. Prot. Dosim. 87: 109-114 (2000).

M19 Matsanuga, T., T. Ueno, H. Amano et al. Characteristics of Chernobyl-derived radionuclides in particulate form in surface waters in the exclusion zone around the Chernobyl nuclear power plant. J. Contaminant Hydrology 35: 101-113 (1998).

N1 Nakajima, T. Estimation of absorbed dose to evacuees at Pripyat-City using ESR measurements of sugar and exposure rate calculations. Appl. Radiat. Isot. 45(1): 113-120 (1994).

N2 Nagataki, S. Communication to the UNSCEAR Secretariat (1997).

N3* Nadyezina, N.M., V.G. Lyelyuk, I.A. Galstyan et al. Basic principles of medical rehabilitation of injuries caused by radiation accidents. J. Med. Catastrophe 1-2(9/10): 150-157 (1995).

N4 Nuclear Energy Agency of the Organisation for Economic Cooperation and Development. Chernobyl: Ten Years on Radiological and Health Impact. OECD, Paris, 1995.

N5 Nagataki, S. and K. Ashizawa. Thyroid cancer in children: comparison among cases in Belarus, Ukraine, Japan and other countries. p.169-175 in: Chernobyl: A Decade. Proceedings of the Fifth Chernobyl Sasakawa Medical Cooperation Symposium, Kiev, 14-15 October 1996 (S. Yamashita and Y. Shibata, eds.). Elsevier Science B.V., Amsterdam, 1997.

N6 Napalkov, N.P., G.F. Tserkovny, V.M. Merabishvili et al. Cancer incidence in the USSR. IARC Scientific Publication No. 48 (1983).

O1 Okeanov, A.E., V.C. Ivanov, E. Cardis et al. Study of cancer risk among liquidators. Report of EU Experimental collaboration project No.7: Epidemiological investigations including dose assessment and dose reconstruction. IARC Internal Report 95/002 (1995).

O2 Okeanov, A.E., E. Cardis, S.I. Antipova et al. Health status and follow-up of the liquidators in Belarus. p. 851-859 in: The Radiological Consequences of the Chernobyl Accident. Proceedings of the First International Conference, Minsk, Belarus, March 1996 (A. Karaoglou, G. Desmet, G.N. Kelly et al., eds.). EUR 16544 (1996).

O3   Osanov, D.P., V.P. Krjutchkov and A.I. Shaks. Determination of beta radiation doses received by personnel involved in the mitigation of the Chernobyl accident. p. 313-348 in: The Chernobyl Papers. Doses to the Soviet Population and Early Health Effects Studies, Volume I (S.E. Merwin and M.I. Balonov, eds.). Research Enterprises Inc., Richland, Washington, 1993.

O4   Okeanov, A.E., S.M. Polyakov, A.V. Sobolev et al. Development of the cancer registration system in Belarus. p. 925-927 in: The Radiological Consequences of the Chernobyl Accident. Proceedings of the First International Conference, Minsk, Belarus, March 1996 (A. Karaoglou, G. Desmet, G.N. Kelly et al., eds.). EUR 16544 (1996).

O5   Ostapenko, V.A., E.J. Dainiak, K.L. Hunting et al. Breast cancer rates among women in Belarus prior to and following the Chernobyl catastrophe. Presented at the International Conference on Diagnosis and Treatment of Radiation Injury, Rotterdam, August 1998.

O6   Osechinsky, I.V. and A.R. Martirosov. Haematological diseases in the Belarus Republic after the Chernobyl accident. Presented at the International Conference on Health Consequences of the Chernobyl and other Radiological Accidents, WHO, Geneva, November 1995.

P1   Pollina, L., F. Pacini, G. Fontanini et al. bcl-2, p53 and proliferating cell nuclear antigen expression is related to the degree of differentiation in thyroid carcinomas. Br. J. Cancer 73: 139-143 (1996).

P2   Pacini, F., T. Vorontsova, E.P. Demidchik et al. Post Chernobyl thyroid carcinoma in Belarus children and adolescents: comparison with naturally occurring thyroid carcinoma in Italy and France. J. Clin. Endocrinol. Metab. 82(11): 3563-3569 (1997).

P3   Petridou, E., D. Trichopoulos, N. Dessypris et al. Infant leukemia after in utero exposure to radiation from Chernobyl. Nature 382: 352-353 (1996).

P4   Purvis, E.E. III. The Chernobyl 4 accident sequence: update - April 1995. Report of the Ukrainian Academy of Sciences Intersectorial Scientific and Technical Center (ISTC) "UKRYTIE". Kiev (1995).

P5   Parkin, D.M., D. Clayton, R.J. Black et al. Childhood leukemia in Europe after Chernobyl: 5 year follow-up. Br. J. Cancer 73: 1006-1012 (1996).

P6   Pacini, F., T. Vorontsova, E.P. Demidchik et al. Diagnosis, surgical treatment and follow-up of thyroid cancers p. 755-763 in: The Radiological Consequences of the Chernobyl Accident. Proceedings of the First International Conference, Minsk, Belarus, March 1996 (A. Karaoglou, G. Desmet, G.N. Kelly et al., eds.). EUR 16544 (1996).

P7   Prisyazhiuk, A., Z. Fedorenko, A. Okeanov et al. Epidemiology of cancer in populations living in contaminated territories of Ukraine, Belarus, Russia after the Chernobyl accident. p. 909-921 in: The Radiological Consequences of the Chernobyl Accident. Proceedings of the First International Conference, Minsk, Belarus, March 1996 (A. Karaoglou, G. Desmet, G.N. Kelly et al., eds.). EUR 16544 (1996).

P8   Prisyazhiuk, A., O.A. Pjatak, V.A. Buzanov et al. Cancer in the Ukraine, post-Chernobyl. Lancet 338(8778): 1334-1335 (1991).

P9   Piliptsevich, N.N. Communication to the UNSCEAR Secretariat (1999).

P10   Protasova, T.G. Pathomorphological findings in acute radiation syndrome (Chernobyl accident patients). Report for International Congress of Pathologists, Copenhagen (1995).

P11   Parkin, D.M., E. Cardis, E. Masuyer et al. Childhood leukaemia following the Chernobyl accident: the European Childhood Leukaemia-Lymphoma Incidence Study (ECLIS). Eur. J. Cancer 29A: 87-95 (1993).

P12   Prisyazhiuk, A., V. Gristchenko, V. Zakordonets et al. The time trends of cancer incidence in the most contaminated regions of the Ukraine before and after the Chernobyl accident. Radiat. Environ. Biophys. 34: 3-6 (1995).

P13   Popov, V.I., O.A. Kuchetkov and A.A. Molokanov. Formation of dose from internal exposure in personnel of the Chernobyl nuclear power station and workers in 1986-1987. Med. Radiol. 2: 33-41 (1990).

P14   Piatkin, E.K., V.Ju. Nugis and A.A. Chirkov. Evaluation of absorbed dose using the results of cytogenetic studies of the culture of lymphocytes in Chernobyl accident victims. Med. Radiol. 34(6): 52-57 (1989).

P15   Pitkevich, V.A., V.K. Ivanov, A.F. Tsyb et al. Dosimetric data of the All-Russian Medical and Dosimetric State Registry for emergency workers. p. 3-44 in: Special Issue of Bulletin of the All-Russian Medical and Dosimetric State Registry. Moscow, 1995.

P16   Peter, R.U. Communication to the UNSCEAR Secretariat (1998).

P17   Parkin, D.M., S.L. Whelan, J. Ferlay et al. Cancer incidence in five continents. IARC Scientific Publications No. 143, Vol. VII (1997).

P18   Poliakov, S.M. and V.K. Ivanov. Communication to the UNSCEAR Secretariat (1998).

P19   Pitkevic, V.A., V.V. Duba, V.K. Ivanov et al. Reconstruction of absorbed doses from external exposure of population, living in areas of Russia contaminated as a result of the accident at the Chernobyl nuclear power plant. Epidemiological Registry Pilot Project. WHO/EOS/94.10. WHO, Geneva (1994).

P20   Pacini, F., T. Vorontsova, E. Molinaro et al. Prevalence of thyroid autoantibodies in children and adolescents from Belarus exposed to the Chernobyl radioactive fallout. Lancet 352: 763-766 (1998).

P21   Peterson, L.E., Z.E. Dreyer, S.E. Plon et al. Design and analysis of epidemiological studies of excess cancer among children exposed to Chernobyl radionuclides. Stem Cells 15 (Suppl. 2): 211-230 (1997).

P22   Percy, C. and A. Dolman. Comparison of the coding of death certificates related to cancer in seven countries. Public Health Rep. 93: 335-350 (1978).

P23   Parkin, D.M., C.S. Muir, S.L. Whelan et al. Cancer incidence in five continents. IARC Scientific Publication No. 120, Vol. VI (1992).

P24   Philippot, J.C. Fallout in snow. Nature 348: 21 (1990).

P25   Paul, M. et al. Measurement of ^{129}I concentrations in the environment after the Chernobyl reactor accident. Nucl. Instrum. Methods Phys. Res. B29: 341-345 (1987).

P26   Peter, R., P. Gottlöber, N. Nadeshina et al. Interferon gamma in survivors of the Chernobyl power plant accident: new therapeutic options for radiation-induced fibrosis. Int. J. Radiat. Oncol. Biol. Phys. 45(1): 147-152 (1999).

P27   Pisarchik, A., G. Ermak, E. Demidchik et al. Low prevalence of the ret/PTC3r1 rearrangement in a series of papillary thyroid carcinomas presenting in Belarus ten years post-Chernobyl. Thyroid 11(8): 1003-1008 (1998).

P28   Philipps, A. and D.B. Chambers. Review of current status of biological dosimeters. Report prepared for the Atomic Energy Control Board, Canada, March (2000).

Q1   Quastel, M., J.R. Goldsmith and J. Cwikel. Lessons learned from the study of immigrants to Israel from areas of Russia, Belarus and Ukraine contaminated by the Chernobyl accident. Environ. Health Perspect. 105(6): 1523-1527 (1997).

Q2   Quastel, M.R., J.R. Goldsmith, L. Mirkin et al. Thyroid-stimulating hormone levels in children from Chernobyl. Environ. Health Perspect. 105 (Suppl. 6): 1497-1498 (1997).

R1   Rahu, M. Cancer epidemiology in the former USSR. Epidemiology 3(5): 464-470 (1992).

R2   Remennik, L.V., V.V. Starinsky, V.D. Mokina et al. Malignant neoplasms on the territories of Russia damaged owing to the Chernobyl accident. p. 825-828 in: The Radiological Consequences of the Chernobyl Accident. Proceedings of the First International Conference, Minsk, Belarus, March 1996 (A. Karaoglou, G. Desmet, G.N. Kelly et al., eds.). EUR 16544 (1996).

R3   Ramzaev, P.V., M.I. Balonov, A.I. Kacevich et al. Radiation doses and health consequences of the Chernobyl accident in Russia. p. 3-25 in: Assessment of the Health and Environmental Impact from Radiation Doses due to Released Radionuclides. NIRS-M-102 (1994).

R4   Romanenko, A.E., I.A. Likhtarev, N.K. Shandala et al. Thyroid doses and endocrinological monitoring of the UKSSR population after the Chernobyl accident. Med. Radiol. 2: 41-49 (1991).

R5   Rumyantsyeva, G.M. The Chernobyl accident and psychological stress. p. 59 in: Proceedings of All-Russian Conference on Radioecological, Medical and Socio-economical Consequences of the Chernobyl Accident. Rehabilitation of Territory and Population, May 1995. Golitsino, Moscow, 1995.

R6   Reiners, C. Prophylaxis of radiation-induced thyroid cancers in children after the reactor catastrophe of Chernobyl. Nuklearmedizin 33: 229-234 (1994).

R7   Ron, E., J. Lubin, R.E. Shore et al. Thyroid cancer after exposure to external radiation: a pooled analysis of seven studies. Radiat. Res. 141: 259-277 (1995).

R8   Ron, E., J. Lubin and A. Schneider. Thyroid cancer incidence. Nature 360: 113 only (1992).

R9   Russian State Committee of Sanitary and Epidemiology Supervision. Reconstruction of cumulative effective doses received during the 1986-1995 time period by the population of contaminated settlements following the accident at the Chernobyl nuclear power plant. Methodological guidance. Official issue of the Russian State Committee of Sanitary and Epidemiology Supervision, Moscow (1996).

R10   Repin, V.S. Radiation hygiene conditions and doses for the population of the 30-km zone after the Chernobyl accident. Dissertation, Ukrainian Scientific Center for Radiation Medicine, 1996.

R11   Rolevich, I.V., I.A. Kenik, E.M. Babosov et al. Social, economic, institutional and political impacts. Report for Belarus. p. 411-428 in: One Decade After Chernobyl. Summing up the Consequences of the Accident. Proceedings of an International Conference, Vienna, 1996. STI/PUB/1001. IAEA, Vienna, 1996.

R12   Repin, V.S. Dose Reconstruction and Assessment of the Role of Some Factors in Radiation Exposure to Inhabitants, Evacuated Outside the 30-km Zone After the Chernobyl Accident. Problems of Chernobyl Exclusion Zone. Naukova Dumka Publishing House, Kiev, 1996.

R13   Rahu, M., M. Tekkel, T. Veidebaum et al. The Estonian study of Chernobyl cleanup workers: II. Incidence of cancer mortality. Radiat. Res. 147: 653-657 (1997).

R14   Ramsay, C.N., E.N. Ellis and H. Zealley. Down's syndrome in the Lothian region of Scotland 1978-1989. Biomed. Pharmacother. 45: 267-272 (1991).

R15   Richmond, C.R. The importance of non-uniform dose distribution in an organ. Health Phys. 29: 525-537 (1975).

R16   Romanenko, A., C. Lee, S. Yamamoto et al. Urinary bladder lesions after the Chernobyl accident: immunhisto-chemical assessment of p53, proliferating cell nuclear antigen, cyclin D1 and p21$^{WAF1/Cip1}$. Jpn. J. Cancer Res. 90: 144-153 (1999).

R17   Ron, E., M.M. Doody, D.V. Becker et al. Cancer mortality following treatment for adult hyperthyroidism. J. Am. Med. Assoc. 280: 347-355 (1998).

R18   Ron, E. The epidemiology of thyroid cancer. p. 1000-1021 in: Cancer Epidemiology and Prevention (D. Schottenfeld and J.F. Fraument Jr., eds.). Oxford University Press, Oxford, 1996.

R19   Rossi, H.H. A response to "Perinatal mortality in Germany following the Chernobyl accident". Radiat. Environ. Biophys. 36: 137 (1997).

R20   Rudhard, J., B. Schell and G. Lindner. Size distribution of hot particles in the Chernobyl fallout. p. 6-15 in: Proceedings of Symposium on Radioecology – Chemical Speciation – Hot Particles (CEC). Znojmo, Czech Republic, 1992.

R21   Ryabov, I., N. Belova, L. Pelgunova et al. Radiological phenomena of the Kojanovskoe Lake. p. 213-216 in: The Radiological Consequences of the Chernobyl Accident. Proceedings of the First International Conference, Minsk, Belarus, March 1996 (A. Karaoglou, G. Desmet, G.N. Kelly et al., eds.). EUR 16544 (1996).

R22   Ron, E. Radiation effects on the thyroid: Emphasis on iodine-131. in: Proceedings of the NCRP Thirty-fifth Annual Meeting, 7-8 April 1999.

R23   Reiners, C., J. Biko and E. Demidchik. Results on radioiodine treatment in 145 children from Belarus with thyroid cancer after the Chernobyl accident. in: Radiation and Thyroid Cancer. Proceedings of an Internal Seminar held in St. John's College, Cambridge, UK, 20-23 July 1998 (G. Thomas, A. Karaoglou and E.D. Williams, eds.). World Scientific, Singapore, 1999.

S1   Sich, A.R., A.A. Borovoi and N.C. Rasmussen. The Chernobyl accident revisited: source term analysis and reconstruction of events during the active phase. MITNE-306 (1994).

S2   Sasakawa Memorial Health Foundation. A report on the 1994 Chernobyl Sasakawa Project Workshop, Moscow, 16-17 May 1994. Sasakawa Memorial Health Foundation, Tokyo (1994).

S3   Schlumberger, M., F. de Vathaire, J.P. Travaglia et al. Differentiated thyroid carcinoma in childhood: long term follow-up of 72 patients. J. Clin. Endocrinol. Metab. 65: 1088-1094 (1987).

S4   Schneider, A.B., E. Ron, J. Lubin et al. Dose-response relationships for radiation-induced thyroid cancer and thyroid nodules: evidence for the prolonged effects of radiation on the thyroid. J. Clin. Endocrinol. Metab. 77(2): 362-369 (1993).

S5   Shigematsu, I. and J.W. Thiessen. Childhood thyroid cancer in Belarus. Nature 359: 680-681 (1992).

S6   Storm, H.H., R.A. Winkelmann, A.E. Okeanov et al. Development of infrastructure for epidemiological studies in Belarus, the Russian Federation and Ukraine. p. 879-893 in: The Radiological Consequences of the Chernobyl Accident. Proceedings of the First International Conference, Minsk, Belarus, March 1996 (A. Karaoglou, G. Desmet, G.N. Kelly et al., eds.). EUR 16544 (1996).

S7 Sasakawa Memorial Health Foundation. A report on the 1995 Chernobyl Sasakawa Project Workshop, St. Petersburg, 7-8 July 1995. Sasakawa Memorial Health Foundation, Tokyo (1996).

S8 Souchkevitch, G. and L. Lyasko. Investigation of the impact of radiation dose on hormones, biologically active metabolites and immunoglobulins in Chernobyl accident recovery workers. Stem Cells 15 (Suppl. 2): 151-154 (1997).

S9 Suchy, B., V. Waldmann, S. Klugbauer et al. Absence of RAS and p53 mutations in thyroid carcinomas of children after Chernobyl in contrast to adult thyroid tumours. Br. J. Cancer 77: 952-955 (1998).

S10 Sugenoya, A., K. Asanuma, Y. Hama et al. Thyroid abnormalities among children in the contaminated area related to the Chernobyl accident. Thyroid 5: 29-33 (1995).

S11 Simakov, A.V., G.V. Fomin, A.A. Molokanov et al. Inhalation intake of radioactive aerosols in work of liquidation of consequences of the Chernobyl accident. p. 22-29 in: Proceedings of All-Union Conference of Radiation Aspects of the Chernobyl Accident. Obninsk, 1988.

S12 Sali, D., E. Cardis, L. Sztanyik et al. Cancer consequences of the Chernobyl accident in Europe outside the former USSR: a review. Int. J. Cancer 67: 343-352 (1996).

S13 Simonova, L.I., S.A. Amirazyan, M.U. Tikhomirova et al. Social-psychological factor affecting ability to work in Chernobyl accident clean-up participants. p. 269 in: International Conference on the Mental Health Consequences of the Chernobyl Disaster: Current State and Future Prospects, May 1995. Ukraine, Kiev, 1995.

S14 Sevan'kaev, A.V., D.C. Lloyd, H. Braselmann et al. A survey of chromosomal aberrations in lymphocytes of Chernobyl liquidators. Radiat. Prot. Dosim. 58: 85-91 (1995).

S15 Shinkarina, A.P., V.K. Podgorodnichenko and V.C. Poverenniy. Antibodies to the thyroid antigen in children and teenagers exposed to radiation as a result of Chernobyl accident. J. Radiat. Biol. Radioecol. 34(4-5): 603-609 (1994).

S16 Strand, P., M. Balonov, L. Skuterud et al. Exposures from consumption of agricultural and semi-natural products. p. 261-269 in: The Radiological Consequences of the Chernobyl Accident. Proceedings of the First International Conference, Minsk, Belarus, March 1996 (A. Karaoglou, G. Desmet, G.N. Kelly et al., eds.). EUR 16544 (1996).

S17 Stepanenko, V., Yu. Gavrilin, V. Khrousch et al. The reconstruction of thyroid dose following Chernobyl. p. 937-948 in: The Radiological Consequences of the Chernobyl Accident. Proceedings of the First International Conference, Minsk, Belarus, March 1996 (A. Karaoglou, G. Desmet, G.N. Kelly et al., eds.). EUR 16544 (1996).

S18 Sich, A.R. Chernobyl accident management actions. Nucl. Saf. 35(1): 1-24 (1994).

S19 Stsjazhko, V.A., A.F. Tsyb, N.D. Tronko et al. Childhood thyroid cancer since accident at Chernobyl. Br. Med. J. 310: 801 only (1995).

S20* Sivintsev, Yu.V. and V. Kachalov (eds.). Chernobyl. p. 381 in: Five Hard Years: Book of Proceedings. Publishing House, Moscow, 1992.

S21 Stepanenko, V.F., A.F. Tsyb and E.M. Parshkov. Dosimetric estimation of childhood thyroid cancer cases in Russia after Chernobyl accident. p. 210-211 in: International Conference. One Decade After Chernobyl: Summing up the Consequences of the Accident. Book of Extended Synopses. IAEA, Vienna, 1996.

S22* Surin, A.I. and K.P. Checherov. Heat flux sources at the "Ukrytie". p. 10 in: IAE-5818/3. RRC-Kurchatov's Institute, Moscow (1994).

S23 Sperling, K.S., J. Pelz, R.D. Wegner et al. Significant increase in trisomy 21 in Berlin nine months after the Chernobyl reactor accident: temporal correlation or causal relation? Br. Med. J. 309: 157-161 (1994).

S24 Savkin, M.N., A.V. Titov and A.N. Lebedev. Distribution of individual and collective exposure doses for the population of Belarus in the first year after the Chernobyl accident. p. 87-113 in: Bulletin of the All-Russian Medical and Dosimetric State Registry, Issue 7. Moscow, 1996.

S25 Skryabin, A.M., M.N. Savkin, Y.O. Konstantinov et al. Distribution of doses received in rural areas affected by the Chernobyl accident. NRPB-R277 (1995).

S26 Steinhäusler, F. Hot particles in the Chernobyl fallout. p. 15-21 in: Hot Particles from the Chernobyl Fallout. Band 16. Schriftenreihe des Bergbau- und Industrie-museums, Ostbayern Theuern, 1988.

S27 Steinhäusler, F. Summary of the present understanding of the significance of hot particles in the Chernobyl fallout. p. 143-144 in: Hot Particles from the Chernobyl Fallout. Band 16. Schriftenreihe des Bergbau- und Industrie-museums, Ostbayern Theuern, 1988.

S28 Sandalls, F.J., M.G. Segal and N.V. Victorova. Hot particles from Chernobyl: a review. J. Environ. Radioact. 18: 5-22 (1993).

S29 Savkin, M.N. Retrospective assess of external and internal exposure to population of the Republic of Belarus at the first stage of the Chernobyl accident. Technical Report of the Institute of Biophysics, Moscow (1993).

S30 Stepanenko, V.F., A.F. Tsyb, E.M. Parshkov et al. Retrospective thyroid absorbed dose estimation in Russia following the Chernobyl accident: Progress and application to dosimetric evaluation of children thyroid cancer morbidity. p. 31-84 in: Proceedings of the Second Hiroshima International Symposium "Effects of Low-level Radiation for Residents near Semipalatinsk Nuclear Test Site", Hiroshima, Japan, 23-25 July 1986 (M. Hoshi, J. Takada, R. Kim et al., eds.). Hiroshima University, Japan, 1996.

S31 Shantyr, I.I., N.V. Makarova and E.B. Saigina. Cancer morbidity among the emergency workers of the Chernobyl accident. p. 366-368 in: Low Doses of Ionizing Radiation: Biological Effects and Regulatory Control. International Conference held in Seville, Spain, November 1997. IAEA-TECDOC-976 (1997).

S32 Snigirova, G., H. Brasselmann, K. Salassidis et al. Retrospective biodosimetry of Chernobyl clean-up workers using chromosome painting and conventional chromosome analysis. Int. J. Radiat. Biol. 71(2): 119-127 (1997).

S33* Selidovkin, G.D. Current methods of treatment of patients with acute radiation syndrome in a specialized hospital. J. Med. Catastrophe 1-2: 135-149 (1995).

S34 Shirahige, Y., M. Ito, K. Ashizawa et al. Childhood thyroid cancer: comparison of Japan and Belarus. Endocrin J. 45: 203-209 (1998).

S35 Smida, J., K. Salassidis, L. Hieber et al. Distinct frequency of ret rearrangements in papillary thyroid carcinomas of children and adults from Belarus. Int. J. Cancer 80: 32-38 (1999).

S36 Shikalov, V.F., A.F. Usatiy, L.V. Kozlova et al. Medical and dosimetric database for the liquidators from the Kurchatov Institute. p. 181-194 in: International Congress on the Sarcophagus. Slavutitch, 1997.

S37 Schevchenko, V.A., E.A. Akayeva, I.M. Yeliseyeva et al. Human cytogenetic consequences of the Chernobyl accident. Mutat. Res. 361: 29-34 (1996).

S38 Sasakawa Memorial Health Foundation. A report of the 1st Chernobyl Sasakawa Medical Symposium, June 1992. Sasakawa Memorial Health Foundation, Tokyo (1992).

S39 Soares, P., E. Fonesca, D. Wynford-Thomas et al. Sporadic ret-rearranged papillary carcinoma of the thyroid: A subset of slow growing, less aggressive thyroid neoplasms? J. Pathol. 185: 71-78 (1998).

S40 Science and Technology Center in Ukraine. Comprehensive risk assessment of the consequences of the Chornobyl accident. Ukrainian Radiation Training Center, Project No. 369. Kyiv, 1998.

S41 Shutov, V.N., G.Y. Bruk, M.I. Balonov et al. Cesium and strontium radionuclide migration in the agricultural ecosystem and estimation of internal doses to the population. p. 167-218 in: The Chernobyl Papers. Doses to the Soviet Population and Early Health Effects Studies, Volume I (S.E. Merwin and M.I. Balonov, eds.). Research Enterprises, Inc., Richland, Washington, 1993.

S42 Shikalov, V.F. and A.F. Usatiy. Paper submitted to Medical Radiation and Radiation Protection (1999).

S43 Shinkarev, S. Communication to the UNSCEAR Secretariat (2000).

S44 Shinkarev, S., Y. Gavrilin, V. Khrouch et al. Preliminary estimates of thyroid dose based on direct thyroid measurements conducted in Belarus. in: Proceedings of the 10th International Congress of the International Radiation Protection Association, Hiroshima, Japan, 14-19 May 2000.

S45 Straume, T., A.A. Marchetti, L.R. Anspaugh et al. The feasibility of using ^{129}I to reconstruct ^{131}I deposition from the Chernobyl reactor accident. Health Phys. 71(5): 733-740 (1996).

S46 Savkin, M.N. and A.N. Lebedev. Communication to the UNSCEAR Secretariat (2000).

S47 Shikalov, V.F. Database of medical, biological and dosimetric data of staff members of the Russian Research Center - Kurchatov's Institute who participated in recovery operations of Chernobyl accident. in: Proceedings of the International Conference on Radioactivity of Nuclear Blasts and Accidents, Moscow, Russia, 24-26 April 2000.

S48 Shevchenko, V.A. et al. Reconstruction of radiation doses in population and radiation workers applying cytogenetic techniques. in: Proceedings of the International Conference on Radioactivity of Nuclear Blasts and Accidents, Moscow, Russia, 24-26 April 2000.

T1 Tsyb, A.F., E.M. Parshkov, V.K. Ivanov et al. Disease indices of thyroid and their dose dependence in children and adolescents affected as a result of the Chernobyl accident. p. 9-19 in: Nagasaki Symposium on Chernobyl: Update and Future (S. Nagataki, ed.). Elsevier Science B.V., Amsterdam, 1994.

T2 Tronko, N., Ye. Epstein, V. Oleinik et al. Thyroid gland in children after the Chernobyl accident (yesterday and today). p. 31-46 in: Nagasaki Symposium on Chernobyl: Update and Future (S. Nagataki, ed.). Elsevier Science B.V., Amsterdam, 1994.

T3 Tsyb, A.F., V.K. Ivanov, S.A. Airanetov et al. Epidemiological analysis of data on liquidators of the Chernobyl accident living in Russia. Med. Radiol. 9/10: 44-47 (1992).

T4 Tronko, N.D., T. Bogdanova, I. Komissarenko et al. Thyroid cancer in children and adolescents in Ukraine after

the Chernobyl accident (1986-1995). p. 683-690 in: The Radiological Consequences of the Chernobyl Accident. Proceedings of the First International Conference, Minsk, Belarus, March 1996 (A. Karaoglou, G. Desmet, G.N. Kelly et al., eds.). EUR 16544 (1996).

T5 Tsyb, A.F., E.M. Parshkov, V.V. Shakhtarin et al. Thyroid cancer in children and adolescents of Bryansk and Kaluga regions. p. 691-698: The Radiological Consequences of the Chernobyl Accident. Proceedings of the First International Conference, Minsk, Belarus, March 1996 (A. Karaoglou, G. Desmet, G.N. Kelly et al., eds.). EUR 16544 (1996).

T6 Tekkel, M., M. Rahu, T. Veidbaum et al. The Estonian study of Chernobyl clean-up workers: design and questionnaire data. Radiat. Res. 147: 641-652 (1997).

T7* Tkashin, V.S. Evaluation of the thymus and of the vegeto-nerval system of liquidators of the Chernobyl accident based on questionnaires and computer evaluation of information using the program vegetative balance. p. 44 in: The Chernobyl Catastrophe. Belarus Committee of Children of Chernobyl, Minsk, 1994.

T8* Tsyb, A.F., V.K. Ivanov, A.I. Gorsky et al. Assessment of parameters of morbidity and mortality of participants in liquidation of the consequences of the Chernobyl accident. p. 114-129 in: Medical Aspects of Eliminating the Consequences of the Chernobyl Accident. Central Scientific Research Institute, Moscow, 1993.

T9 Tsyb, A.F., V.K. Ivanov, S.A. Airanetov et al. System of radiation-epidemiological analysis of data of the Russian national medico-dosimetric register of participants in the liquidation of consequences of the Chernobyl accident. Bull. Radiat. Risk 2: 69-109 (1992).

T10 Tsyb, A.F., V.K. Ivanov, S.A. Airanetov et al. Radiation-epidemiological analysis of the national register of persons exposed to radiation as a result of the Chernobyl accident. Vestn. Akad. Med. Nauk SSSR 11: 32-36 (1991).

T11* Tukov, A.R. and L.G. Dzagoeva. Morbidity of atomic industry workers of Russia who participated in the work of liquidating the consequences of the Chernobyl accident. p. 97-99 in: Medical Aspects of Eliminating the Consequences of the Chernobyl Accident. Central Scientific Research Institute, Moscow, 1993.

T12 Tsyb, A.F. et al. Registry material. Radiat. Risk 1: 67-131 (1992).

T13 Tukov, A.R. and I.L. Shafransky. Approaches to morbidity and mortality radiation risk estimation for people of exposed selected cohorts. Med. Radiol. 5: 11-13 (1997).

T14 Tukov, A.R. and A.K. Guskova. Analysis of experience and errors in assessment of health status of persons involved in radiation accident. Med. Radiol. 5: 5-10 (1997).

T15 Tucker, J.D., D.A. Eastmond and L.G. Littlefield. Cytogenetic end-points as biological dosimeters and predictors of risk in epidemiological studies. IARC Scientific Publication 142 (1997).

T16 Tronko, N.D. Communication to the UNSCEAR Secretariat (1999).

T17 Telnov, V.I., I.A. Vologodskaya and N.A. Kabasheva. Influence of the complex of factors on immunological state of recovery operation workers participated in Chernobyl. Med. Radiol. 2: 8-11 (1993).

T18 Tronko, M., T. Bogdanova, I. Komissarenko et al. Thyroid carcinoma in children and adolescents in Ukraine after the Chernobyl nuclear accident: Statistical data and clinicomorphologic characteristics. Cancer 86(1): 149-156 (1999).

T19 Thomas, G.A., H. Bunnell, E.D. Williams et al. Association between morphological subtype of post Chernobyl papillary carcinoma and rearrangement of the ret oncogene. in: Radiation and Thyroid Cancer. Proceedings of an Internal Seminar held in St. John's College, Cambridge, UK, 20-23 July 1998 (G. Thomas, A. Karaoglou and E.D. Williams, eds.). World Scientific, Singapore, 1999.

T20 Thompson, D.E., K. Mabuchi, E. Ron et al. Cancer incidence in atomic bomb survivors. Part II: Solid tumors, 1958-1987. Radiat. Res. 137: S17-S67 (1994).

T21 Tondel, M., G. Carlsson, L. Hardell et al. Incidene of neoplasms in ages 0-19 y in parts of Sweden with high ^{137}Cs fallout after the Chernobyl accident. Health Phys. 71: 947-950 (1996).

T22 Thomas, G., A. Karaoglou and E.D. Williams (eds.). Radiation and Thyroid Cancer. EUR 18552 (1999).

T23 Travnikova, I.G., G.Ya. Bruk and M.I. Balonov. Communication to the UNSCEAR Secretariat (2000).

U2 United Nations. Sources and Effects of Ionizing Radiation. United Nations Scientific Committee on the Effects of Atomic Radiation, 1994 Report to the General Assembly, with scientific annexes. United Nations sales publication E.94.IX.11. United Nations, New York, 1994.

U4 United Nations. Sources, Effects and Risks of Ionizing Radiation. United Nations Scientific Committee on the Effects of Atomic Radiation, 1988 Report to the General Assembly, with annexes. United Nations sales publication E.88.IX.7. United Nations, New York, 1988.

U14 Ukrainian Civil Defense. The Chernobyl Accident. Events. Facts. Numbers. April 1986 through March 1990. Guide of the Ukrainian Civil Defense, Kiev, 1990.

U15 Ushakov, I.B., B.I. Davydov and S.K. Soldatov. A Man in the Sky of Chernobyl. A Pilot and a Radiation Accident. Rostov University Publishing House, Rostov at Don, 1994.

U16 Ukrainian Research Center of Radiation Medicine. Medical consequences of the Chernobyl accident. in: Information Bulletin of the Ukrainian Research Center of Radiation Medicine. URCRM, Kiev, 1991.

V1 Vykhovanets, E.V., V.P. Chernyshov, I.I. Slukvin et al. ^{131}I dose-dependent thyroid autoimmune disorders in children living around Chernobyl. Clin. Immunol. Immunopathol. 84: 251-259 (1997).

V2 Voznyak, V.Ya. Social, economic, institutional and political impacts. Report for the Soviet period. p. 369-378 in: One Decade After Chernobyl. Summing up the Consequences of the Accident. Proceedings of an International Conference, Vienna, 1996. STI/PUB/1001. IAEA, Vienna, 1996.

V3 Voznyak, V.Ya. Social, economic, institutional and political impacts. Report for the Russian Federation. p. 379-410 in: One Decade After Chernobyl. Summing up the Consequences of the Accident. Proceedings of an International Conference, Vienna, 1996. STI/PUB/1001. IAEA, Vienna, 1996.

V4 Velikhov, E.P., N.N. Ponomarev-Stepnoy, V.G. Asmolov et al. Current understanding of occurrence and development of the accident at the Chernobyl NPP. p. 12-36 in: Selected Proceedings of the International Conference on Nuclear Accidents and the Future of Energy. The Lessons of Chernobyl, Paris, April 1991.

V5 Verger, P. Down's syndrome and ionising radiation. Health Phys. 73(6): 882-893 (1997).

V6 Vasilychenko, D.L., Yu.P. Ivanov, S.V. Kazakov et al. Exposure to the staff of the prohibited zone. Bulletin of Ecological Status of the Prohibited Zone. Chernobyl, 1992.

V7 van Hoff, J., Y.I. Averkin, E.I. Hilchenko et al. Epidemiology of childhood cancer in Belarus: review of data 1978-1994, and discussion of the new Belarusian Childhood Cancer Registry. Stem Cells 15 (Suppl. 2): 231-241 (1997).

V8 Viswanathan, K., T.C. Gierlowski and A.B. Schneider. Childhood thyroid cancer. Characteristics and long-term outcome in children irradiated for benign conditions of the head and neck. Arch. Pediatr. Adolesc. Med. 148: 260-265 (1994).

V9 Vapirev, E.J., T. Kamenova, G. Mandjoukov et al. Visualisation, identification and spectrometry of a hot particle. Radiat. Prot. Dosim. 30: 121-124 (1990).

V10 Vermiglio, F., M. Castagnia, E. Vonova et al. Post-Chernobyl increased prevalence of humoral thyroid autoimmunity in children and adolescents from a moderately iodine-deficient area in Russia. Thyroid 9(8): 781-786 (1999).

W1 World Health Organisation. Report of the International Project for the Health Effects of the Chernobyl Accident. WHO, Geneva, 1995.

W2 Wagemaker, G. Heterogeneity of radiation sensitivity of hemopoietic stem cell subsets. Stem Cells 13 (Suppl. 1): 257-60 (1995).

W3 Williams, G.H., S. Rooney, G. Thomas et al. RET activation in adult and childhood papillary carcinoma. Br. J. Cancer 74: 585-589 (1996).

W4 Williams, D. Chernobyl and hypothyroidism. Lancet 348: 476-477 (1996).

W5 Wagemaker, G., A.K. Guskova, V.G. Bebeshko et al. Clinically observed effects in individuals exposed to radiation as a result of the Chernobyl accident. p. 173-196 in: One Decade After Chernobyl. Summing up the Consequences of the Accident. Proceedings of an International Conference, Vienna, 1996. STI/PUB/1001. IAEA, Vienna, 1996.

W6 Westerbeek, R.M.C., V. Blair, O.B. Eden et al. Seasonal variations in the onset of childhood leukemia and lymphoma. Br. J. Cancer 78: 119-124 (1998).

W7 World Health Organization. Health consequences of the Chernobyl accident. Results of the IPHECA pilot projects and related national programmes (G.N. Souchkevitch and A.F. Tsyb, eds.). Scientific Report, WHO/EHG 95-19 (1996).

W8 Williams, E.D., D. Becker, E.P. Dimidchik et al. Effects on the thyroid in populations exposed to radiation as a result of the Chernobyl accident. p. 207-238 in: One Decade After Chernobyl. Summing up the Consequences of the Accident. Proceedings of an International Con-ference, Vienna, 1996. STI/PUB/1001. IAEA, Vienna, 1996.

W9 Wutke, K., C. Streffer, W.U. Muller et al. Micronuclei in lymphocytes of children from the vicinity of Chernobyl before and after 131-I therapy for thyroid cancer. Int. J. Radiat. Biol. 69(2): 259-268 (1996).

W10 Winkelmann, R.A., A. Okeanov, L. Gulak et al. Cancer registration techniques in the NIS of the former Soviet Union. IARC Technical Report No. 35 (1998).

W11 Walberg, P., M. McKee, V. Shkolnikov et al. Economic change, crime, and mortality crises in Russia: regional analysis. Br. Med. J. 317: 312-318 (1998).

W12 Winkelmann, R.A. and C. Dora. WHO summary of studies of the Chernobyl accident. Part 1: Registry information sources for research on health effects of the Chernobyl accident. WHO, European Centre for Environment and Health, Rome (1999).

W13 Winkelmann, R.A. and C. Dora. WHO summary of studies of the Chernobyl accident. Part 2: State of the Chernobyl cancer research in exposed population groups of Belarus, Russian Federation and Ukraine. WHO, European Centre for Environment and Health, Rome (1999).

Y1   Yamashita, S. and Y. Shibata (eds.). Chernobyl: A Decade. Proceedings of the Fifth Chernobyl Sasakawa Medical Cooperation Symposium, Kiev, 14-15 October 1996. Elsevier Science B.V., Amsterdam, 1997.

Y2   Yamamoto, S., A. Romanenko, M. Wei et al. Specific p53 gene mutations in urinary bladder epithelium after the Chernobyl accident. Cancer Res. 59: 3606-3609 (1999).

Z1   Zvonova, I.A. and M.I. Balonov. Radioiodine dosimetry and prediction of consequences of thyroid exposure of the Russian population following the Chernobyl accident. p. 71-125 in: The Chernobyl Papers. Doses to the Soviet Population and Early Health Effects Studies, Volume I (S.E. Merwin and M.I. Balonov, eds.). Research Enterprises Inc., Richland, Washington, 1993.

Z2   Zitzelsberger, H., L. Lehmann, L. Hieber et al. Cytogenetic changes in radiation-induced tumors of the thyroid. Cancer Res. 59(1): 135-140 (1999).

Z3   Zimmerman, D., I. Hay and E. Bergstrahl. Papillary thyroid cancer in children. p. 3-10 in: Treatment of Thyroid Cancer in Childhood (J. Robbins, ed.). Huddersfield, Springfield, 1992.

Z4   Zvonova, I.A., T.V. Jesko, M.I. Balonov et al. ^{134}Cs and ^{137}Cs whole-body measurements and internal dosimetry of the population living in areas contaminated by radioactivity after the Chernobyl accident. Radiat. Prot. Dosim. 62: 213-221 (1995).

Z5   Zarnowiecki, K. Analysis of radioactive contamination and radiological hazard in Poland after the Chernobyl accident. CLOR Report 120/D. Warsaw (1988).

Z6   Zvonova, I.A., M.I. Balonov, A.A. Bratilova et al. Methodology of thyroid dose reconstruction for population of Russia after the Chernobyl accident. in: Proceedings of the 10th International Congress of the International Radiation Protection Association, Hiroshima, Japan, 14-19 May 2000.

Z7   Zvonova, I.A., M.I. Balonov and A.A. Bratilova. Thyroid dose reconstruction for the population of Russia after the Chernobyl accident. Radiat. Prot. Dosim. 79: 175-178 (1998).

(References marked with an asterisk have been published in Russian language).